HANDBOOK OF SYSTEMS ENGINEERING AND MANAGEMENT

HANDBOOK OF SYSTEMS ENGINEERING AND MANAGEMENT

Edited by

Andrew P. Sage
George Mason University

William B. Rouse
Enterprise Support Systems

A WILEY-INTERSCIENCE PUBLICATION

JOHN WILEY & SONS, INC.

New York   •   Chichester   •   Weinheim   •   Brisbane   •   Singapore   •   Toronto

This book is printed on acid-free paper.

For ordering and customer service, call 1-800-CALL WILEY.

Library of Congress Cataloging-in-Publication Data:

Handbook of systems engineering and management/edited by Andrew P.
  Sage, William B. Rouse.
      p. cm.
    Includes index.
    ISBN 0-471-15405-9 (cloth : alk. paper)
    1. Systems engineering—Handbooks, manuals, etc.—2. Industrial
  management—Handbooks, manuals, etc.   I. Sage, Andrew P.
  II. Rouse, William B.
  TA168.H33   1998
  658.5—dc21   98-35045
                                                                   CIP

Printed in the United States of America
10 9 8 7 6 5 4 3 2 1

CONTENTS

CONTRIBUTORS

JAMES E. ARMSTRONG, JR.   Operations Research Center, Department of the Army, United States Military Academy, West Point, New York

A. TERRY BAHILL,   Department of Systems and Industrial Engineering, University of Arizona, Tucson, Arizona

JOSEPH D. BEN-DAK,   Global Technology Group, UNDP, One United Nations Plaza, New York, New York

DEAN M. BEHRENS,   Department of Sociology, University of Toronto, Toronto, Canada

BENJAMIN S. BLANCHARD,   Department of Industrial and Systems Engineering, Virginia Polytechnic Institute and State University, Blacksburg, Virginia

PEGGY BROUSE,   Department of Systems Engineering and Operations Research, George Mason University, Fairfax, Virginia

DENNIS M. BUEDE,   Department of Systems Engineering and Operations Research, George Mason University, Fairfax, Virginia

RUTH BUYS,   Department of Systems Engineering and Operations Research, George Mason University, Fairfax, Virginia

KATHLEEN M. CARLEY,   Department of Social and Decision Sciences, Carnegie Mellon University, Pittsburgh, Pennsylvania

FRANK F. DEAN,   Systems Engineering and Project Management, Sandia National Laboratories, Albuquerque, New Mexico

YACOV Y. HAIMES,   Department of Systems Engineering, University of Virginia, Charlottesville, Virginia

CHARLES S. HARRIS,   Department of Sociology, Marymount University, 2807 North Glebe Road, Arlington, Virginia

BETTY K. HART,   Organization Change, 1410 North Johnson Street, Arlington, Virginia

KEITH W. HIPEL,   Department of Systems Design Engineering, University of Waterloo, Waterloo, Ontario, Canada

ANNE J. JENSEN,   MITRE Corporation, McLean, Virginia

PATRICIA M. JONES,   Department of Mechanical and Industrial Engineering, University of Illinois at Urbana-Champaign, Urbana, Illinois

D. MARC KILGOUR,   Department of Mathematics, Wilfred Laurier University, Waterloo, Ontario, Canada

CRAIG W. KIRKWOOD, Department of Management, Arizona State University, Tempe, Arizona

ANDREW KUSIAK, Department of Industrial Engineering, University of Iowa, Iowa City, Iowa

NICK LARSON, Department of Industrial Engineering, University of Iowa, Iowa City, Iowa

ALEXANDER H. LEVIS, Department of Electrical and Computer Engineering, George Mason University, Fairfax, Virginia

STEPHEN C. LOWELL, Department of Defense Standardization Program Division, 5203 Leesburg Pike, Falls Church, Virginia

JAMES L. MELSA, College of Engineering, Iowa State University, Ames, Iowa

CHRISTINE M. MITCHELL, Center for Human-Machine Systems Research, School of Industrial and Systems Engineering, Georgia Institute of Technology, Atlanta, Georgia

JUDITH M. ORASANU, Aviation Safety Branch, NASA Ames Research Center, Moffett Field, California

JAMES D. PALMER, Professor Emeritus, School of Information Technology and Engineering, George Mason University, Fairfax, Virginia

F. G. PATTERSON, JR., Office of Training and Development, NASA Headquarters, Washington, D.C.

MICHAEL PECHT, Computer Aided Life Cycle Engineering (CALCE) Center, University of Maryland, College Park, Maryland

SIAMAK RAJABI, Department of Systems Design Engineering, University of Waterloo, Waterloo, Ontario, Canada

WILLIAM B. ROUSE, Enterprise Support Systems, 3295 River Exchange Drive, Norcross, Georgia

ANDREW P. SAGE, Department of Systems Engineering and Operations Research School of Information Technology and Engineering, George Mason University, Fairfax, Virginia

MICHAEL G. SHAFTO, Human Automation Interaction Branch, NASA Ames Research Center, Moffett Field, California

AARON J. SHENHAR, Stevens Institute of Technology, Hoboken, New Jersey

THOMAS B. SHERIDAN, Human Machine Systems Laboratory, Massachusetts Institute of Technology, Cambridge, Massachusetts

JOYCE SHIELDS, Hay/McBer North America, Hay Management Consultants, 4301 North Fairfax Drive, Arlington, Virginia

WIL A. THISSEN, Department of Systems Engineering, Policy Analysis, and Management, Delft University of Technology, The Netherlands

JAMES M. TIEN,   Department of Decision Sciences and Engineering Systems, Rensselaer Polytechnic Institute, Troy, New York

C. ELS VAN DAALEN,   Department of Systems Engineering, Policy Analysis, and Management, Delft University of Technology, The Netherlands

ALEXANDER VERBRAECK,   Department of Systems Engineering, Policy Analysis, and Management, Delft University of Technology, The Netherlands

K. PRESTON WHITE, JR.,   Department of Systems Engineering, University of Virginia, Charlottesville, Virginia

PREFACE

The primary purpose of this handbook is to support use of the theory and practice of systems engineering and systems management. There are many ways in which we can describe systems engineering. It can be described according to structure, function, and purpose. It can be characterized in terms of efforts needed at the levels of systems management, life-cycle processes, and methods and tools. We can speak of systems engineering organizations in terms of their organizational management facets, business processes or product lines, or specific products or services. We can explain systems engineering in terms of the knowledge principles, practices, and perspectives necessary for present and future success in systems engineering. We can describe systems engineering in terms of the human dimensions associated with the stakeholders within systems—that is, investors, developers, users, maintainers, and the like. We can describe systems engineering in terms of a large variety of relevant applications. Systems engineering is clearly a multidimensional transdisciplinary endeavor.

This handbook takes the multifaceted view of systems engineering. It describes systems engineering in terms of this relatively large number of dimensions, and especially from process and systems management perspectives. Systems engineering methods and tools are discussed, as are a variety specific products that have been fielded from a systems engineering perspective. However, it is not our intent to produce a catalog of either systems engineering methods and tools or products. Our focus is on a process and management view of systems engineering. We expand this view in some detail within the context of the structure of systems engineering and management.

Within this framework, a large number of necessary roles for systems engineering are described. The 30 chapters in the handbook present definitive discussions of systems engineering from many of this wide array of perspectives. The needs of systems engineering and systems management practitioners in industry and government, as well as students aspiring to careers in systems engineering and management, provide the motivation for the majority of the chapters.

The handbook begins with a comprehensive Introduction to the coverage that follows. This was written by the editors after receiving and editing individual contributions to the handbook. It provides not only an introduction to systems engineering and management but also a brief overview and integration of the 30 chapters that follow in terms of a knowledge map. This framework is intended to be used as a "field guide" that indicates why, when, and how to use the material contained in the 30 chapters.

There are many roles for systems engineers in industry and government. Among the common and not-so-common names used for systems engineering are: Systems Engineers, Systems Architects, Systems Integrators, Systems Management Engineers, Systems Infrastructuralists, Systems Quality Assurance Engineers, Systems Theorists, and Systems Reengineers. There are also a number of closely associated professions, such as Operations Research and Management Science, that have much in common with systems engineers.

The handbook contains chapters describing knowledge principles and practices from each of these perspectives.

The major themes and objectives addressed by the handbook are

1. To develop an appreciation and understanding of the role of systems engineering processes and systems management in producing products and services that meet user needs, and are reliable, trustworthy, and of high quality.
2. To document systematic measurement approaches for generally cross-disciplinary development efforts, and to discuss capability assessment models that allow organizations to first evaluate and then improve their systems engineering maturity or capability.
3. To document the knowledge base of effective systems engineering processes and systems management strategies, and expand the knowledge base needed to support these processes.
4. To advance understanding of the complexity and roles of advanced information technologies and new organizational structures in enhancing productivity and quality of systems for both products and services.
5. To discuss tools, methods, and technologies available for support of complex high-technology systems engineering efforts.
6. To provide perspectives on systems engineering and management for the Twenty-first Century.

The handbook also is intended for systems engineers in industry and government, and to serve as a university reference handbook in systems engineering. In particular, the handbook will be useful for a wide range of professionals involved in systems engineering and management efforts:

- Systems Engineers in Government and Industry
- Software Engineers in Government and Industry
- Human Factors Engineers in Government and Industry
- Systems and Software Development Managers in Government and Industry
- Systems Engineering Graduate Programs
- Software Engineering Graduate Programs
- Computer Science and Engineering Graduate Programs
- Business Administration and Management Programs in Information Technology
- Business Administration and Management Programs in Technology Development and Management
- Organizational and Enterprise Computing Personnel

The handbook book is reasonably self-contained. It is primarily focused on systems engineering and systems management for fielding systems of all types, especially systems that are information-technology and knowledge intensive. Thus, the handbook is not primarily focused on traditional systems analysis, theory, and operations research concerns. Instead, these topics are addressed in the broader context of systems engineering and systems management processes. In this way, we have designed the handbook to be of much value for those concerned with the technical direction, development, and management of complex projects of large scale and scope.

ANDREW P. SAGE
WILLIAM B. ROUSE

Fairfax, Virginia
Norcross, Georgia
March 1999

An Introduction to Systems Engineering and Systems Management

ANDREW P. SAGE AND WILLIAM B. ROUSE

This Introduction provides a perspective on all of systems engineering and, within that, systems management. This is a major challenge for a single chapter, but our overall purpose is to introduce and provide a summary perspective of the 30 much more detailed chapters that follow. Appreciation for the overall process of systems engineering leads naturally to a discussion of the important role for systems management, and the applications of this to realistic contemporary issues of diverse types. Following our introductory comments, we briefly describe the organization of the handbook.

Here, as throughout this handbook, we are concerned with the engineering of systems, or *systems engineering*. We are also and especially concerned with strategic level systems engineering, or *systems management*. In this introduction, we begin our effort by first discussing the need for systems engineering, and then providing several definitions of systems engineering. We next present a structure describing the systems engineering process. The result of this is a *life-cycle model* for systems engineering processes. This model is used to motivate discussion of the functional levels, or considerations, involved in systems engineering efforts:

- Systems engineering *methods and tools,* or technologies
- Systems methodology, or *process,* as a set of phased activities that support efforts to engineer the system, and
- *Systems management*

Figure 1 illustrates the natural hierarchical relationship among these levels. There are many discussions throughout this handbook on systems engineering methods and tools, or technologies. Our primary focus, however, is on systems engineering and management processes, and technical direction of these efforts. These are intended to result in appropriate products or services, or systems that fulfill client needs. It is a bit cumbersome to continually refer to both products and services, and so we will generally use the term *product* to refer both to physical products as well as to services. The products that are engineered result from an appropriate set of systems engineering methods and tools, or technologies, and an appropriate product line or process effort. These are guided by efforts at systems management as suggested in Figure 1, and as we will discuss often in this handbook.

Handbook of Systems Engineering and Management, Edited by A. P. Sage and W. B. Rouse
ISBN 0471-15405-9   ©1999 John Wiley and Sons, Inc.

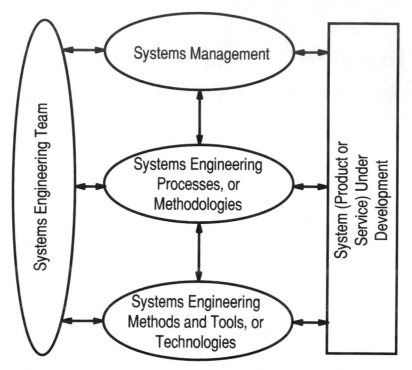

Figure 1   Conceptual illustration of the three levels of systems engineering.

SYSTEMS ENGINEERING

Systems engineering is a management technology. Technology is the organization, application, and delivery of scientific and other forms of knowledge for the betterment of a client group. This is a functional definition of technology as a fundamentally human activity. A technology inherently involves a purposeful human extension of one or more natural processes. For example, the stored program digital computer is a technology in that it enhances the ability of a human to perform computations and, in more advanced forms, to process information.

Management involves the interaction of the organization with the environment. A purpose of management is to enable organizations to better cope with their environments so as to achieve purposeful goals and objectives. Consequently, a management technology involves the interaction of technology, organizations that are collections of humans concerned with both the evolvement and use of technologies, and the environment. Figure 2 illustrates these conceptual interactions.

Information is the glue that enables the interactions shown in this figure. Information is a very important quantity that is assumed to be present in the management technology that is systems engineering. This strongly couples notions of systems engineering with those of technical direction or systems management of technological development, rather than exclusively with one or more of the methods of systems engineering, important as they may be for the ultimate success of a systems engineering effort. It suggests that

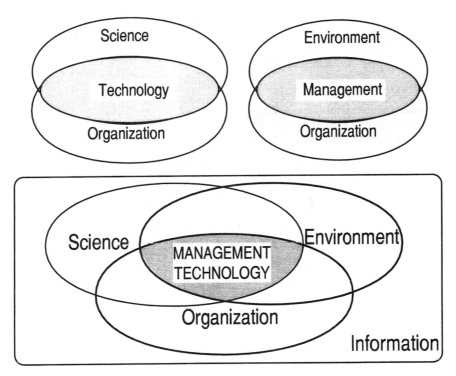

Figure 2   Systems engineering as a management technology.

> Systems engineering is the management technology that controls a total system life-cycle process, which involves and which results in the definition, development, and deployment of a system that is of high quality, is trustworthy, and is cost-effective in meeting user needs.

This process-oriented notion of systems engineering and systems management is emphasized here.

Figure 3 illustrates our view that systems engineering knowledge comprises:

- *Knowledge Principles,* which generally represent formal problem-solving approaches to knowledge, generally employed in new situations and/or unstructured environments;
- *Knowledge Practices,* which represent the accumulated wisdom and experiences that have led to the development of standard operating policies for well-structured problems.
- *Knowledge Perspectives,* which represent the view that is held relative to future directions and realities in the technological area under consideration.

Clearly, one form of knowledge leads to another. Knowledge perspectives may create the incentive for research that leads to the discovery of new knowledge principles. As knowledge principles emerge and are refined, they generally become imbedded in the form of knowledge practices. Knowledge practices are generally the major influences of the sys-

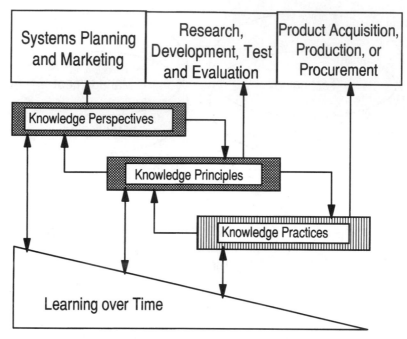

Figure 3 Systems engineering knowledge types.

tems that can be acquired or fielded. These knowledge types interact together as suggested in Figure 3, which illustrates how these knowledge types support one another. In a non-exclusive way, they each support one of the principal life cycles associated with systems engineering. Figure 3 also illustrates a number of feedback loops that are associated with learning to enable continual improvement in performance over time. This supports our view, and a major premise of this handbook, that it is a serious mistake to consider these life cycles in isolation from one another.

The use of the term *knowledge* is very purposeful here. It has long been regarded as essential in systems engineering and management to distinguish between data and information. *Information* is generally defined as data that are of value for decision making. For information to be successfully used in decision making, it is necessary to associate context and environment with it. The resulting information, as enhanced by context and environment, results in knowledge. Appropriate information management and knowledge management are each necessary for high-quality systems engineering and management.

It is on the basis of the appropriate use of the three knowledge types depicted in Figure 3 that we are able to accomplish the technological system planning and development, and the management system planning and development, that lead to a new, innovative product or service. All three types of knowledge are needed. The environment associated with this knowledge needs to be managed, and this is generally what is intended by use of the term *knowledge management*. Also, the learning that results from these efforts is very much needed, both on an individual and an organizational basis.

We will soon discuss these three different primary systems engineering life cycles for technology growth and change:

- System planning and marketing
- Research, development, test, and evaluation (RDT&E)
- System acquisition, production, or procurement

Each of these life cycles is generally needed, and each primarily involves the use of one of the three types of knowledge. We discuss these life cycles briefly here, and will illustrate how and why they make major but nonexclusive use of knowledge principles, practices, and perspectives. There are a number of needed interactions across these life cycles for one particular realization of a system acquisition life cycle. It is important that efforts across these three major systems engineering life cycles be integrated. There are many illustrations of efforts that were dramatically successful efforts in RDT&E, but where the overall results represent failure because of lack of consideration of planning, or of the ultimate manufacturing needs of a product while it is in RDT&E.

It is important when studying any area, new or not, to define it. We have provided one definition of systems engineering thus far. It is primarily a structural definition. A related definition, in terms of purpose, is that

Systems engineering is management technology to assist and support policymaking, planning, decision making, and associated resource allocation or action deployment.

Systems engineers accomplish this by quantitative and qualitative *formulation, analysis,* and *interpretation* of the impacts of action alternatives upon the needs perspectives, the institutional perspectives, and the value perspectives of their clients or customers. Each of these three steps is generally needed in solving systems engineering problems.

- Issue formulation is an effort to identify the needs to be fulfilled and the requirements associated with these in terms of objectives to be satisfied, constraints and alterables that affect issue resolution, and generation of potential alternate courses of action.
- Issue analysis enables us to determine the impacts of the identified alternative courses of action, including possible refinement of these alternatives.
- Issue interpretation enables us to rank the alternatives in terms of need satisfaction and to select one for implementation or additional study.

This particular listing of three systems engineering steps and their descriptions is rather formal. Often, issues are resolved this way. The steps of formulation, analysis, and interpretation may also be accomplished on an "as-if" basis by application of a variety of often-useful heuristic approaches. These may well be quite appropriate in situations where the problem solver is experientially familiar with the task at hand, and the environment into which the task is imbedded, such as to enable development of an appropriate context for issue resolution. This requires information. It also requires knowledge, such as information imbedded into experienced-based context and environment, and this may lead to very useful heuristic approaches to use of this knowledge.

We can apply these systems engineering steps to a variety of situations that should enable us to develop an appreciation for systems engineering design and problem solving.

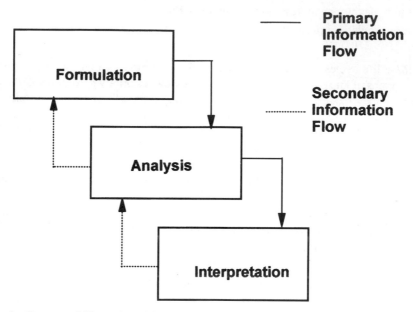

Figure 4   Conceptual illustration of formulation, analysis, and interpretation as the primary systems engineering steps.

Generally, there is iteration among the steps, and they follow more or less in the sequence illustrated in Figure 4.

The key words in this purposeful definition are formulation, analysis, and interpretation. In fact, all of systems engineering can be thought of as consisting of formulation, analysis, and interpretation efforts, together with the systems management and technical direction efforts necessary to bring this about. We may exercise these in a formal sense, or in an *as-if* or experientially based intuitive sense. These are the stepwise or micro level components that make up a part of the structural framework for systems methodology.

In our first definition of systems engineering, we indicated that systems engineers are concerned with the appropriate

- Definition
- Development, and
- Deployment of systems.

These aspects compose a set of phases for a systems engineering life-cycle, as illustrated in Figure 5. There are many ways to characterize the life-cycle phases of systems engineering processes, and a considerable number of them are described in Chapter 1. Each of these life-cycle models, and those that are outgrowths of them, comprise these three phases in one way or another. For pragmatic reasons, a typical life cycle will generally contain more than three phases, as we shall soon indicate.

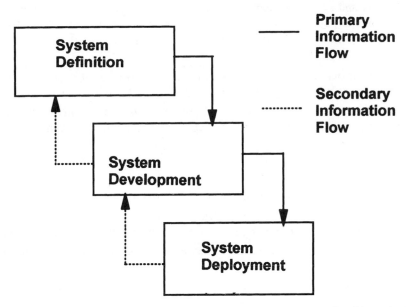

Figure 5 Conceptual illustration of the three primary systems engineering life-cycle phases.

THE IMPORTANCE OF TECHNICAL DIRECTION AND SYSTEMS MANAGEMENT

In order to resolve large scale and complex problems, or to manage large systems of humans and machines, we must be able to deal with important contemporary issues that involve and require

- Many considerations and interrelations
- Many different and perhaps controversial value judgments
- Knowledge from several disciplines
- Knowledge at the levels of principles, practices, and perspectives
- Considerations involving planning or definition, development, and deployment
- Considerations that cut across the three different life cycles associated with systems planning and marketing, RDT&E, and system acquisition or production
- Risks and uncertainties involving future events that are difficult to predict
- Fragmented decision-making structures
- Human and organizational need and value perspectives, as well as technology perspectives, and
- Resolution of issues at the level of institutions and values as well as the level of symptoms.

The professional practice of systems engineering must use a variety of formulation, analysis, and interpretation aids for evolvement of technological systems and management systems. Clients and system developers alike need this support to enable them to cope with multifarious large-scale issues. This support must avoid several potential pitfalls. These include the following 12 deadly systems engineering transgressions.

1. There is an overreliance upon a specific analytical method or tool, or a specific technology, that is advocated by a particular group.
2. There is a consideration of perceived problems and issues only at the level of symptoms, and the development and deployment of "solutions" that only address symptoms.
3. There is a failure to develop and apply appropriate methodologies for issue resolution that will allow:
   (a) Identification of major pertinent issue formulation elements;
   (b) A fully robust analysis of the variety of impacts on stakeholders and the associated interactions among steps of the problem solution procedure; and
   (c) An interpretation of these impacts in terms of institutional and value considerations.
4. There is a failure to involve the client, to the extent necessary, in the development of problem resolution alternatives and systemic aids to problem resolution.
5. There is a failure to consider the effects of cognitive biases that result from poor information processing heuristics.
6. There is a failure to identify a sufficiently robust set of options, or alternative courses of action.
7. There is a failure to make and properly utilize reactive, interactive, and proactive measurements to guide the systems engineering efforts.
8. There is a failure to identify risks associated with the costs and benefits, or effectiveness, of the system to be acquired, produced, or otherwise fielded.
9. There is a failure to properly relate the system that is designed and implemented with the cognitive style and behavioral constraints that impact the users of the system, and an associate failure of not properly designing the system for effective user interaction.
10. There is a failure to consider the implications of strategies adopted in one of the three life cycles (RDT&E, acquisition and production, and planning and marketing) on the other two life cycles.
11. There is a failure to address quality and sustainability issues in a comprehensive manner throughout all phases of the life cycle, especially in terms of reliability, availability, and maintainability.
12. There is a failure to properly integrate a new system together with heritage or legacy systems that already exist and that the new system should support.

All of these may be, and generally are, associated with systems management failures. There needs to be a focus upon quality management to avoid these failures and, as has been noted, quality management refers as much or more to the quality of management as it does to the management of quality.

The need for systematic measurements is a very real one for the appropriate practice of systems management. The use of such terms as reactive measurements, interactive measurements, and proactive measurements may seem unusual. We may, however, approach measurements, and management in general, from at least four perspectives.

- *Inactive.* Denotes an organization that does not use metrics, or that does not measure at all except perhaps in an intuitive and qualitative manner.
- *Reactive.* Denotes an organization that will perform an outcomes assessment, and after it has detected a problem, or failure, will diagnose the cause of the problem and often will get rid of the symptoms that produce the problem.
- *Interactive.* Denotes an organization that will measure an evolving product as it moves through various phases of the life cycle process in order to
  (a) Detect problems as soon as they occur;
  (b) Diagnose their causes; and
  (c) Correct the difficulty through recycling, feedback, and retrofit to and through that portion of the life-cycle process in which the problem occurred.
- *Proactive.* Proactive measurements are those that are designed to predict the potential for errors and synthesis of an appropriate life-cycle process that is sufficiently mature such that the potential for errors is minimized.

Actually, we can also refer to the systems management style of an organization as inactive, reactive, interactive, and/or proactive. All of these perspectives on measurement purpose, and on systems management, are needed. Inactive and reactive measurements are associated with organizations that have a low level of process maturity, a term we will soon define. As one moves to higher and higher levels of process maturity, the lower level forms of measurement purpose become less and less used. In part, this is so because a high level of process maturity results in such appropriate metrics for systems management that final product errors, which can be detected through a reactive measurement approach, tend to occur very infrequently. While reactive measurement approaches are used, they are not at all the dominant focus of measurement. In a very highly mature organization, they might be only needed on the rarest of occasions.

In general, we may also approach systems management issues from an inactive, reactive, interactive, or proactive perspective.

- *Inactive.* Denotes an organization that does not worry about issues and that does not take efforts to resolve them. It is a very hopeful perspective, but generally one that will lead to issues becoming serious problems.
- *Reactive.* Denotes an organization that will examine a potential issue only after it has developed into a real problem. It will perform an outcomes assessment, and after it has detected a problem, or failure, will diagnose the cause of the problem and often will get rid of the symptoms that produce the problem.
- *Interactive.* Denotes an organization that will attempt to examine issues while they are in the process of evolution so as to detect problems at the earliest possible time. Issues that may cause difficulties will not only be detected, but efforts at diagnosis and correction will be implemented as soon as they have been detected. This will detect problems as soon as they occur, diagnose their causes, and correct any difficulty through recycling, feedback, and retrofit to and through that portion of the life-cycle process in which the problem occurred. Thus, the term interactive is, indeed, very appropriate.

> • *Proactive.* Denotes an organization that predicts the potential for debilitating issues and that will synthesize an appropriate life-cycle process that is sufficiently mature such that the potential for issues developing is as small as possible.

It should be noted that there is much focus on process in the two more appropriate—interactive and proactive—of these perspectives on organizational effort.

Thus, management of systems engineering processes, which we call *systems management,* is very necessary for success. There are many evidences of systems engineering failures at the level of systems management. Often, one result of these failures is that the purpose, function, and structure of a new system are not identified sufficiently before the system is defined, developed, and deployed. These failures generally cause costly mistakes that could truly have been avoided. Invariably this occurs because either the formulation, the analysis, or the interpretation efforts (or all of them perhaps) are deficient. A major objective of systems engineering, at the strategic level of systems management, is to take proactive measures to avoid these difficulties.

Now that we have introduced some of the flavor of systems engineering, let us turn to some definitions of the professional area of effort we call systems engineering. There are many of these. As we will see, it is possible to define a term from any of several perspectives, or to combine these perspectives. In a great many cases, misunderstandings occur because terms have not been clearly defined.

ADDITIONAL DEFINITIONS OF SYSTEMS ENGINEERING

Concerns associated with the *definition, development,* and *deployment* of systems, including product systems and service systems, such that they can be used efficiently and effectively, have always been addressed to some extent. Often, this has been on an implicit and "trial-and-error" basis. When tool designers or product developers were also tool or product users, which was more often than not the case for the simple tools, machines, and products of the past, the resulting designs were often good initially, or soon evolved into good designs through this trial-and-error effort. When the scale or scope of products and services is large, then it is generally not possible for the engineers of systems to also be the users of systems, and a more formal approach is needed.

The phased efforts of definition, development, and deployment represent the macrolevel structure of a systems engineering framework, as shown in Figure 5. They each need to be employed for each of the three life-cycles illustrated in Figure 3. And within each life-cycle phase, there are a number of steps, as illustrated in Figure 6. Thus, we see that our relatively simple description of systems engineering is becoming more and more complex. Figure 7 illustrates how these 3 steps, 3 phases, and 3 life cycles make up a more complete methodological, or structural, or process-oriented view of systems engineering. Even in this relatively simple methodological framework, which is simultaneously incomplete but relatively complex, we have a total of 27 cells of activity. In a much more realistic view of the steps and phases, as would need to be the case in actual systems development, we might well have 7 phases and 7 steps of effort. This yields a total of 147 cells of activity (i.e., $7 \times 7 \times 3$).

In the early days, when the designers of a product were also the ultimate end users, things were even simpler than this 27-cell model. Definition, development, and deployment could often be accomplished on a simple trial-and-error basis. When physical tools, ma-

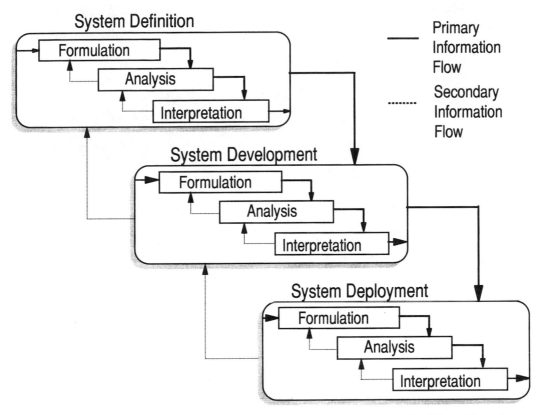

Figure 6   One representation of three systems engineering steps within each of three life-cycle phases.

chines, and systems become so complex that it is no longer possible to design them by a single individual who might even also be the intended user of the tool, and a design team is necessary, then a host of new problems emerge. This is, very much, the condition today. To cope with this, a vast number of tools and methods associated with systems engineering have emerged. Through use of these, it has been possible to develop processes for systems engineering that allow us to decompose the engineering of a large system into smaller component subsystem engineering issues, engineer the subsystems, and then build the complete system as a integrated collection of these subsystems. This emphasizes the importance, both of the processes used for systems engineering and the systems integration efforts that are needed to assure the complete functioning of the resulting system.

Even so, problems remain. There are many instances of failures due to approaches of this sort, generally because they are incomplete. Just simply connecting together individual subsystems often does not result in a system that performs acceptably, either from a technological efficiency perspective, or from a effectiveness perspective. This has led to the realization that *systems integration engineering* and *systems management* throughout an entire system life cycle will be necessary. Thus it is that contemporary efforts in *systems engineering* contain a focus on

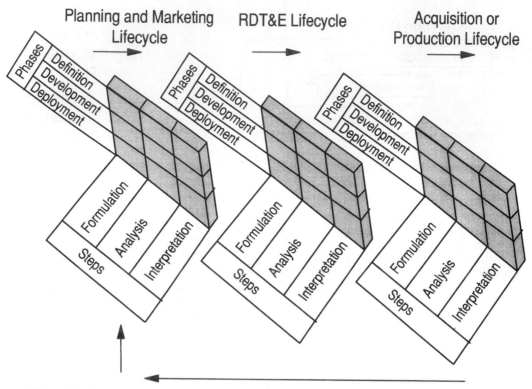

Figure 7   Three systems engineering life cycles and phases and steps within each life cycle.

- Tools and methods, and technologies for the engineering of systems
- Systems methodology for the life-cycle process of definition, development, and deployment that enables appropriate use of these tools, methods, and technologies, and
- Systems management approaches that enables the proper imbedding of systems engineering product and process evolution approaches within organizations and environments.

In this way, systems engineering and management provide very necessary support to the role of conventional and classic engineering endeavors associated with application of physical and material science principles to accomplish the detailed design and implentation of products in various specialty areas for the betterment of humankind.

Figure 1 illustrated this conceptual relationship among the three levels of systems engineering, and also showed each as a necessary facet that the various members of a systems engineering team must apply to define, develop, and deploy an evolving system. Each of these three levels

- Systems engineering methods and tools
- Systems engineering processes
- Systems management

is necessarily associated with appropriate environments in order to assure an appropriate systems engineering process, including the very necessary client interaction during system definition, development, and deployment. The use of appropriate systems methods and tools as well as systems methodology and systems management constructs enables the engineering of systems for more efficient and effective human interaction.

System management and integration issues are of major importance in determining the effectiveness, efficiency, and overall functionality of system designs. To achieve a high measure of functionality, it must be possible for a system, meaning a product or a service, to be efficiently and effectively produced, used, maintained, retrofitted, and modified throughout all phases of a life cycle. This life cycle begins with conceptualization and identification, then proceeds through specification of system requirements and architectures to the ultimate system installation, operational implementation, or deployment, evaluation, and maintenance throughout a productive lifetime. It is important to note that a system, product, or service that is produced by one organization may well be used as a process, or to support a process, by another organization.

In reality, there are many difficulties associated with the production of functional, reliable, and trustworthy systems of large scale and scope. There are many studies that indicate that

- It is very difficult to identify the user requirements for a large system.
- Large systems are expensive.
- System capability is often less than promised and expected.
- System deliveries are often quite late.
- Large-system cost overruns often occur.
- Large-system maintenance is complex and error prone.
- Large-system documentation is inappropriate and inadequate.
- Large systems are often cumbersome to use and system design for human interaction is generally lacking.
- Individual new subsystems often cannot be integrated with legacy or heritage systems.
- Large systems often cannot be transitioned to a new environment or modified to meet the evolving needs of clients.
- Unanticipated risks and hazards often materialize.
- There may be no provision for risk management or crisis management associated with the use of a system.
- Large systems often suffer in terms of their reliability, availability, and maintainability.
- Large-system performance is often of low quality.
- Large systems often do not perform according to specifications.
- It is difficult to identify suitable performance metrics for a large system that enable determination of system cost and effectiveness.
- There is often poor communication among program management, designers, and customers or program sponsors.
- System specifications often do not adequately capture user needs and requirements.
- Large systems may not be sustainable over time.
- Large systems may be rejected by users as not appropriate to their needs.

These potential difficulties, when they are allowed to develop, can create many problems that are difficult to resolve. Among these are inconsistent, incomplete, and otherwise imperfect system requirements or specifications; system requirements that do not provide for change as user needs evolve over time; and poorly defined management structures for product design and delivery. These lead to delivered products that are difficult to use, that do not solve the intended problem or resolve the stated issue, that operate in an unreliable fashion, that are not maintainable or sustainable, and that, as a result, are not used. Sometimes these failures are so great that operational products and systems are never even fully developed, much less operationally deployed, before plans for the product or system are abruptly canceled.

These same studies generally show that the major problems associated with the production of trustworthy systems often have much more do with the *organization and management of complexity* than with direct technological concerns that affect individual subsystems and specific physical science areas. Often the major concern should be more associated with the definition, development, and use of an appropriate process, or product line, for production of a product than it is with actual product implementation itself. Direct attention to the product or service without appropriate attention to the process leads to the fielding of a low-quality and expensive product or service.

A functional definition of systems engineering is also of interest and we provide a simple one here:

> Systems engineering is the art and science of creating a product or service, based on phased efforts that involve definition, design, development, production, and maintenance activities. The resulting product or service is functional, reliable, of high quality, and trustworthy, and has been developed within cost and time constraints.

There are, of course, other definitions. Two closely related and appropriate definitions are provided by two military standards, MIL-STD-499A and MIL-STD-499B:

Systems engineering is the application of scientific and engineering efforts to

- Transform an operational need into a description of system performance parameters and a system configuration through the use of an iterative process of definition, synthesis, analysis, design, test, and evaluation;
- Integrate related technical parameters and ensure compatibility of all physical, functional, and program interfaces in a manner that optimizes the total system definition and design;
- Integrate reliability, maintainability, safety, survivability, human engineering, and other factors into the total engineering effort to meet cost, schedule, supportability, and technical performance objectives.

Systems engineering is an interdisciplinary approach to evolve and verify an integrated and life-cycle-balanced set of system product and process solutions that satisfy the customer's needs. Systems engineering

- Encompasses the scientific and engineering efforts related to the development, manufacturing, verification, deployment, operations, support, and disposal of system products and processes;

- Develops needed user training equipment, procedures, and data;
- Establishes and maintains configuration management of the system;
- Develops work breakdown structures and statements of work, and provides information for management decision making.

These two definitions attempt to combine structural, functional, and purposeful views of systems engineering. While they are not inappropriate, they tend to be more descriptions of the development phase of a systems acquisition life cycle than they are definitions that embrace all three of these phases. Nor do they closely relate to the RDT&E and planning and marketing life cycle. They also tend to assume that the systems engineer is not necessarily, perhaps not at all, the engineer or builder of systems, but more or less their architect only. These definitions carry over to the current versions of these standards as well. We feel there is a need to also stress the integrative role needed across humans, organizations, technologies, and their environments that is essential for effective systems engineering and management. We attempt to stress this knowledge integration role in the 30 chapters of this handbook.

It is generally accepted that we may define things according to either structure, function, or purpose. Often, definitions are incomplete if they do not address structure, function, and purpose. Our discussion of systems engineering, in this Introduction and throughout the 30 chapters of this handbook, include structural, purposeful, and functional definitions of systems engineering. These aspects are based on the three definitions we have provided earlier. Table 1 presents these three definitions. Each of these definitions is important and all three are generally needed, as we have noted. In our three-level hierarchy of systems engineering there is generally a non-mutually exclusive correspondence between function and tools, structure and methodology, and purpose and management, as we note in Figure 8.

TABLE 1 Definitions of Systems Engineering

Structure	Systems engineering is management technology to assist clients through the formulation, analysis, and interpretation of the impacts of proposed policies, controls, or complete systems upon the need perspectives, institutional perspectives, and value perspectives of stakeholders to issues under consideration.
Function	Systems engineering is an appropriate combination of the methods and tools of systems engineering, made possible through use of a suitable methodological process and systems management procedures, in a useful process-oriented setting that is appropriate for the resolution of real-world problems, often of large scale and scope.
Purpose	The purpose of systems engineering is information and knowledge organization and management to assist clients who desire to develop policies for management, direction, control, and regulation activities relative to forecasting, planning, development, production, and operation of total systems to maintain overall quality, integrity, and integration as related to performance, trustworthiness, reliability, availability, and maintainability.

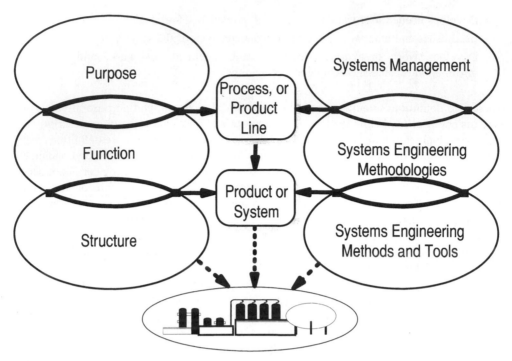

Figure 8   Relations between purpose and systems management, structure and systems methodology, and function and methods, and their product and process results.

This figure, unlike Figure 1, illustrates the reality that there is no truly sharp distinction between these three levels of systems engineering. Structural, functional, and purposeful definitions and discussions will necessarily overlap, also, as these terms are not mutually exclusive. Generally, purpose relates to systems management, structure relates to process, and function relates to product, but the relationships are not at all mutually exclusive. It is of much interest to also note in Figure 8 that a specific process, or product line, results from the interaction of systems management and systems methodology. The product is the result of the use of a number of methods and tools, and systems methodology as it eventuates from use of life-cycle process. Formally, the product itself results from the acquisition process, but a specific acquisition process needs necessarily to be shaped by RDT&E and planning and marketing considerations.

Thus, it is quite correct to view an abstraction of this figure, as shown in Figure 9, in which appropriate systems management results in a process, or product line, and the result of using a systems engineering process is the production of a product. Not explicitly shown here are the interactions across all three levels with the environment. Also not shown in this figure is the fact that there are three life cycles associated with the general systems engineering process: planning and marketing, RDT&E, and acquisition (or manufacturing or production). This does not suggest that the same individuals are associated with all three of these life cycles, nor are they even necessarily associated with all efforts across all phases within a given life cycle. At a sufficiently high level of systems management, there is doubtlessly this overall responsibility. However, and as in the systems engineering

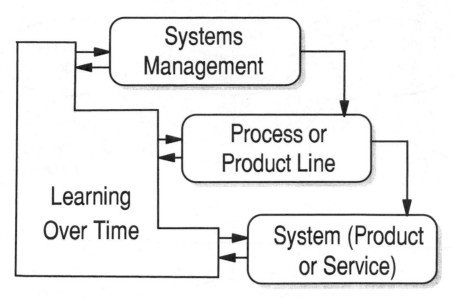

Figure 9 Another view of the hierarchical levels of systems engineering.

military standards just cited, many of the efforts of a given systems engineering team may be associated with the development phase of systems acquisition. Learning over time, such as to enable continuous improvement in quality and effectiveness across all three levels is, however, necessary and shown in this illustration.

We have illustrated three hierarchical levels for systems engineering in Figure 1 and again in Figure 8. We now expand on this to indicate some of the ingredients at each of these levels. The functional definition, or lowest level, of systems engineering says that we will be concerned with the various tools and techniques and methods, and technologies, that enable us to engineer systems that comprise products or services. Often, these will be systems analysis and operations research tools that enable the formal analysis of systems. They can also include specific product-oriented methods, tools, and technologies, such as is illustrated in Figure 10. These could include a variety of computer science and programming tools or methods. They could include modern manufacturing technologies. It should be, strictly speaking, more appropriate to refer to these as product-level methods. Then we could also refer to the methods associated with systems methodology, or process methods, and systems management methods. When the term "method(s)" is used alone and without a modifier, we will generally be referring to product-level methods. The specific nature of the most useful methods and tools will naturally depend greatly on the particular life cycle that is being considered and the particular product, service, or system, that is ultimately to be acquired.

The functional definition of systems engineering also says that we will be concerned with a combination of these tools. In systems engineering, we obtain this combination as the result of using systems methodology to engineer an appropriate process. For our purposes, a methodology is an open set of procedures for problem solving or issue resolution. This brings about such important notions as appropriate development life cycles, operational quality assurance issues, and configuration management procedures which are very

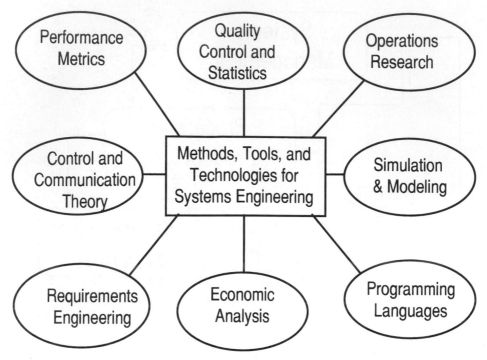

Figure 10 Methods, tools, and technologies for systems engineering products.

important and are discussed in more detail in the various chapters of this handbook. Each of these reflects structural, architectural, or methodological perspective on systems engineering in the sense that the result of functional or process oriented efforts is the evolution of the structure or architecture of the system being engineered. Figure 11 illustrates some of the process-based methods associated with systems methodology. How to best bring these about will vary from product to product and across each of the three life cycles leading to that product or service system. We will have more to say about these in later chapters.

Finally, the functional definition of systems engineering says that we will accomplish this in a useful and appropriate setting. Systems management provides this useful setting, for evolution of an appropriate process for a specific systems engineering effort. We will use the term systems management to refer to the cognitive strategy and organizational tasks necessary to produce a useful process from the variety of methodologies that are available. There are many drivers of systems management strategy, including organizational culture, the internal strengths and weaknesses of the organization, and the external threats and opportunities that it faces. The result of systems management is an appropriate combination of the methods and tools of systems engineering, including their use in a methodological setting, together with appropriate leadership in managing system process and product development, to ultimately field a system that can be used to resolve issues. There are many interesting issues associated with systems management. Figure 12

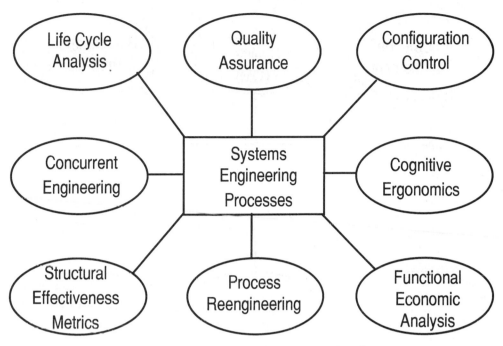

Figure 11   Methods, tools, and technologies for systems engineering process supports.

illustrates some of the many process-related concerns associated with systems management.

We should note that some of these elements, such as economic analysis and assessment, may appear to be at all levels. At the level of the product or systems methods, we use functional economic analysis. This might be associated with cost control of the evolving product. At the level of systems methodology, we would use structural economic analysis, such as is associated with a work breakdown structure (WBS). Structured economic analysis would have to do with life-cycle cost and effectiveness determination. Finally, at the level of systems management, we use a number of strategic cost management approaches that deal with strategic positioning in terms of process costs and effectiveness tradeoffs. These will enable selection of an appropriate development process, including as a special case the decision not to enter a particular market. Thus, economic systems analysis is needed at each of the three systems engineering hierarchical levels. A different form of economic systems analysis and assessment is, however, needed at each of these levels. We need to be concerned with these three levels (management, methodology, method), or the somewhat equivalent levels of systems management, process, and product, across the three life cycles of systems planning and marketing, RDT&E, and system acquisition or production. The particular life cycle in question, as well as the phase of development, will strongly influence the actual approaches that are appropriate for use at each of these levels.

The structural definition of systems engineering tells us that we are concerned with a framework for problem resolution that, from a formal perspective at least, consists of three

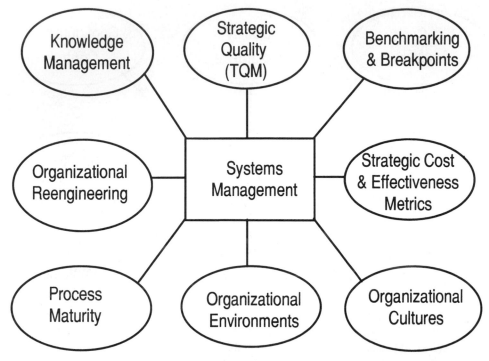

Figure 12   Systems management and associated supports.

fundamental steps:

- Issue formulation
- Issue analysis
- Issue interpretation

These are each conducted at each of the life-cycle phases that have been chosen for definition, development, and deployment. Regardless of the way in which the systems engineering life-cycle process is characterized, and regardless of the type of product or system or service that is being designed, all characterizations of systems engineering life cycles will necessarily involve:

- *Formulation of the problem,* in which the situation or issue at hand is assessed, needs and objectives of a client group are identified, and potentially acceptable alternatives, or options, are identified or generated;
- *Analysis of the alternatives,* in which the impacts of the generated options are identified, assessed, and evaluated; and
- *Interpretation and selection,* in which the options, or alternative courses of action, are compared by means of an evaluation of the impacts of the alternatives and how these are valued by the client group. The needs and objectives of the client group are necessarily used as a basis for evaluation. The most acceptable alternative is selected

for implementation or further study in a subsequent phase of systems engineering effort.

We note these three steps again, because of their great importance in systems engineering. Our model of the steps of the fine structure of the systems process, shown in Figure 4, is based upon this conceptualization. As we shall also indicate later, these three steps can be disaggregated into a number of others.

Each of these steps of systems engineering is accomplished for each of the life-cycle phases. Chapter 24 discusses a number of approaches for issue formulation. Chapters 25, 26, and 27 are generally concerned with issue analysis. Chapters 28 and 29 are concerned with issue interpretation, including decision assessment and planning for action. As we have noted, there are generally three different and major systems engineering life cycles needed to engineer a system. Thus we may imagine a three-dimensional model of systems engineering that is composed of steps associated with each phase of a life cycle, the phases in the life cycle, and the life cycles that make up the coarse structure of systems engineering. Figure 7 illustrated this across three distinct but interrelated life cycles, for the three steps, and three phases that we have described here. This is one of many possible morphological frameworks for systems engineering.

The words morphology and methodology may be unfamiliar. The word *morphology* is adapted from biology and means a study of form or structure. It is an important concept and a needed one in both biology and systems engineering. As noted earlier, a *methodology* is an open set of procedures for problem solving. Consequently, a methodology involves a set of methods, a set of activities, and a set of relations between the methods and the activities. To use a methodology we must have an appropriate set of methods. Generally, these include a variety of qualitative and quantitative approaches from a number of disciplines that are appropriate for the specific product, service, or system to be acquired. Associated with a methodology is a structured framework into which particular methods are associated for resolution of a specific issue. The specific structural framework used for engineering a system is the process or life cycle used to evolve the product under consideration. The RDT&E, acquisition or manufacturing, and planning life cycles are the major ones that result from systems management efforts.

Without question, this is a formal rational model of the way in which these three systems engineering steps—formulation, analysis, and interpretation—are accomplished. Even within this formal framework, there is the need for much iteration from one step back to an earlier step when it is discovered that improvements in the results of an earlier step are needed in order to obtain a quality result at a later step, or phase, of the systems engineering effort. Also, this description does not emphasize the key role of information and knowledge and the determination of requirements for information and knowledge.

Even when these realities are associated with an identified and appropriate morphological framework, they still represent an incomplete view of the way in which people do, could, or should accomplish planning, design, development, or other problem-solving activities. The most that can be argued is that this framework is correct in an as-if manner. One result of a systems management effort is the identification of an appropriate process, or life cycle, to be used to engineer a system. Also involved in systems management are the determination of how best to use this process to engineer a system.

We can also add a third dimension, made up of the three life cycles associated with the systems engineering process, to our framework, as we illustrated in Figure 7, and is as

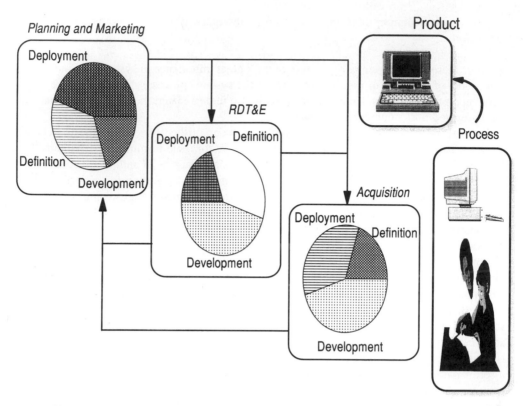

Figure 13   Generic representation of the three basic systems engineering life cycles.

further illustrated conceptually in Figure 13. The use of this process is intended to result in the engineering of an appropriate system. The choice of a process is affected by systems management concerns. Thus, systems management can be regarded as a driver of the process. The product or service system, process, and systems management are supported by a number of methods, tools, technologies, and measurements, as represented by Figure 14. This representation of systems engineering and management shows systems management as exercising strategic management controls over the three evolving systems engineering processes. The instantiation of these processes results in the systems engineering product. In systems planning and marketing, the product of planning and marketing is a determination of whether RDT&E and/or acquisition of a product or system should be undertaken and in what amounts and configurations. It is for this reason that we portray varying sized pie slices for definition, development, and deployment efforts associated with each of these phases in Figure 13.

Figure 15 represents an expansion of the planning and marketing effort and illustrates how the efforts in RDT&E and acquisition must necessarily be based on results from planning and marketing if the overall set of processes is to be effective.

In many cases, the proper application of one or more technologies is necessary to ameliorate an existing problems or fulfill an existing need. Technology is a world need, but it must be human- and organization-centered technology. This requires systems engineering and management efforts that result in definition, development, and deployment of

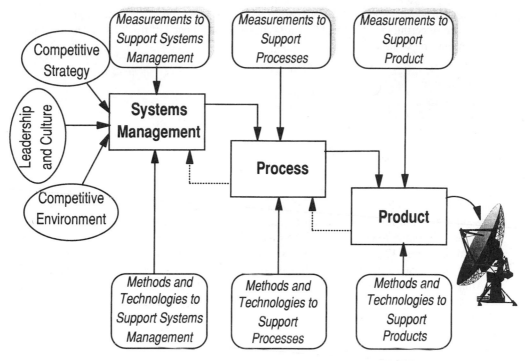

Figure 14   A model of systems engineering and management.

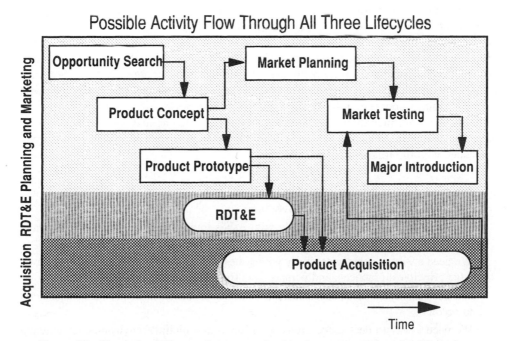

Figure 15   Illustration of interactions across three basic systems engineering life cycles.

a system. However, successful application of technology and successful application of systems engineering to major problem areas must consider each of the three levels at which solution may be sought:

- Symptoms
- Institutions
- Values

Or we may well be confronted by a technological solution looking for a problem. This is especially the case when we approach problems only at the level of symptoms: unemployment, bad housing, inadequate health-care delivery, pollution, hunger, and unsustainable economic development in general. A technological fix is found that addresses symptoms only and the resulting "solution" creates the illusion that the outpouring of huge quantities of moneys to resolve symptoms will actually resolve the fundamental underlying problem. Chapters 17 through 21 concentrate particularly on the need for organizational and human focused solutions to problems, as contrasted with technological fix solutions.

Attacking problems at the level of institutions would allow the design of new institutions and organizations to make full and effective use of new technologies. This is generally a major improvement on what results when we only consider problem abatement at the level of removal of symptoms. With respect to measurements and management, symptomatic approaches are generally reactive in nature. Measurement and management approaches at the level of institutions are generally interactive and process related.

Also of vital importance is the need to deal with problems at the level of values. Systems engineers and systems managers who are serious about resolving major problems must appreciate the significance of human values and be able to identify basic issues in terms of conflicting values. Further, value elements and systems, as well as institutions and purely technological components, must be utilized in determining useful problem solutions. All of this again illustrates the major need for incorporation of systems management approaches into the development of process, and product evolvement efforts. Proactive management and proactive measurements are necessarily value oriented in that they must necessarily be anticipatory. Approaches based on proactivity are generally associated with double learning, a concept described in Chapters 13 and 30.

Some necessary ingredients, which must exist in order to engineer large systems, solve large and complex problems, resolve complicated issues, or manage large systems, are associated with the following needs.

- We need a way to deal successfully with issues involving many considerations and interrelations, including changes over time.
- We need a way to deal successfully with issues in which there are far-reaching and controversial value judgments.
- We need a way to deal successfully with issues, the solutions to which require knowledge principles, practices, and perspectives from several disciplines.
- We need a way to deal successfully with issues in which future events are difficult to predict.
- We need a way to deal successfully with issues in which the environments, external and internal, are difficult to predict.

- We need a way to deal successfully with issues in which structural and human institutional and organizational elements are given full consideration.
- We need to deal with integrated knowledge management across the diversity of stakeholders and perspectives that are associated with major issues of large scale and scope.
- We need to engineer systems that are sustainable.

We believe that systems engineering, through empowered systems management, possesses the necessary characteristics to fulfill these needs. Thus, systems engineering is potentially capable of exposing not only technological perspectives associated with large-scale systems but also needs and value perspectives. Further, it can relate these to knowledge principles, practices, and perspectives such that the result of its application is successful planning and marketing, RDT&E, and acquisition or production of high-quality, trustworthy systems.

Systems engineering efforts are very concerned with technical direction and management of systems definition, development, and deployment. As we have noted, we call this effort *systems management*. By adopting the management technology of systems engineering and applying it, we become very concerned with making sure that correct systems are designed, and not just that system products are correct according to some potentially ill-conceived notions of what the system should do. Appropriate metrics to enable efficient and effective error prevention and detection at the level of systems management, and at the process and product level will enhance the production of systems engineering products that are "correct" in the broadest possible meaning of this term. To ensure that correct systems are produced requires that considerable emphasis be placed on the front end of each of the systems engineering life cycles.

In particular, there needs to be considerable emphasis on the accurate definition of a system, including identification of what it should do and how people should interact with it, before the system is developed and deployed. In turn, this requires emphasis upon conformance to system requirements specifications, and the development of standards to ensure compatibility and integratibility of system products. Such areas as documentation and communication are important in all of this. Thus, we see the need for the technical direction and management technology efforts that compose systems engineering, and the strong role for process- and systems-management-related concerns in this.

LIFE-CYCLE METHODOLOGIES, OR PROCESSES, FOR SYSTEMS ENGINEERING

As we have noted, systems engineering is the creative process through which products, services, or systems that are presumed to be responsive to client needs and requirements are conceptualized or specified or defined, and ultimately developed and deployed. There are at least 15 primary assertions implied by this not uncommon definition of systems engineering, and they apply especially to the development of knowledge-intensive systems and software-intensive systems, as well as to more conventional hardware and physical systems.

- Systems planning and marketing is the first strategic level effort in systems engineering. It results in the determination of whether or not a given organization should undertake the engineering of a given product or service. It also results in at least a preliminary determination of the amount of effort to be devoted to RDT&E and the amount devoted to actual system acquisition or production.

- Creation of an appropriate process or product line for RDT&E and one for acquisition is one result of system planning and marketing. The initial systems planning and marketing efforts determine the extent to which RDT&E is a need; they also determine the acquisition process characteristics that are most appropriate.

- An appropriate planning process leads to efficient and effective RDT&E, and to the actual system acquisition that follows appropriate RDT&E.

- The first phase of any systems engineering life-cycle effort results in the identification or definition of specifications for the product or service that is to result from the process.

- Systems architecting is a very important endeavor that leads to the conceptual structure of the system to be engineered.

- Systems engineering is a creative and process-based effort.

- Systems engineering activities are conceptual in nature at the initial phases of effort, for either of the three generic life cycles, and become operational in later phases.

- A successful systems engineering product or service must be of high quality and responsive to client needs and requirements.

- A successful systems engineering product, or service, generally results only from use of a successful systems engineering process (or set of processes).

- An appropriate systems engineering acquisition or RDT&E process is, generally, the result of successful systems management, and appropriate planning and marketing.

- Appropriate systems engineering efforts need necessarily be associated with systematic measurements to ensure high-quality information and knowledge as a basis for decision making across the three generic systems engineering life cycles.

- Appropriate systems engineering efforts are necessarily attuned to organizational and environmental realities as they affect both the client organization or enterprise, the systems engineering organization, and the product implementation organization(s).

- Systems engineering is inherently associated with knowledge integration and knowledge management across transdisciplinary boundaries.

- Systems engineering efforts must lead to sustainable products and services if they are to have lasting value.

- Systems engineering efforts are of necessity interactive. However, they transcend interactivity to include proactivity.

Good systems engineering practice requires that the systems engineer be responsive to each of these 15 ingredients for quality effort. Clearly, not all members of a systems engineering team are responsible for, and participate in, each and every systems engineering activity. Nevertheless, the activities across each of the phases and each of the life cycles must be integrated if the systems engineering effort is to be of high quality.

It is of value to expand on the simple three-phase and three-step model we have used to characterize a framework for systems engineering. Figure 16 illustrates a typical sequence of phases, seven in this case, that would generally be appropriate for a system

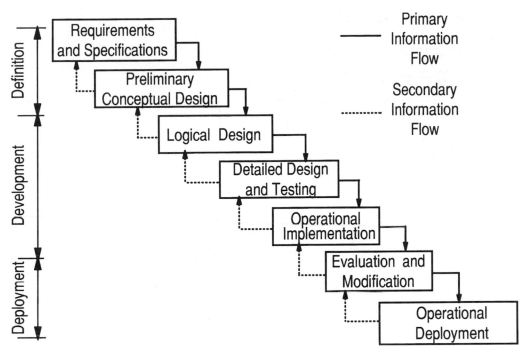

Figure 16   One of several possible life-cycle models for acquisition in systems engineering.

acquisition and procurement life cycle. Figure 17 illustrates a possible manufacturing life cycle. The particular words associated with the phases in Figures 16 and 17 are not the most appropriate for a systems planning and marketing life cycle or an RDT&E life cycle. However, they could easily be modified to indicate identification of societal needs for a product, and the evolution of a product marketing strategy, such that they are, indeed, appropriate for systems planning and marketing. They also could be modified to more appropriately represent an RDT&E effort.

In Figure 16, we have identified a phased system engineering life cycle that consists of seven phases:

- Requirements and specifications identification
- Preliminary conceptual design and system-level functional architecting
- Logical design and physical architecture specification
- Detailed design of the physical architecture and associated production, and testing
- Operational implementation of the resulting product or service system in the field
- Evaluation and modification
- Operational deployment

These life-cycle phases are sequenced in the iterative manner as shown in Figure 16. There are many descriptions of systems engineering life cycles and associated methodologies

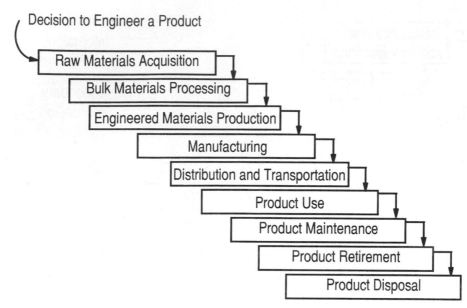

Figure 17   A possible life cycle for manufacturing and related efforts.

and frameworks for systems engineering, and we outline only one of them here. Chapter 1 expands a great deal on these ideas.

In general, a simple conceptual model of the overall process, for a single life cycle, may be structured as in Figure 18, which illustrates an effort to accommodate the three steps we have described and the seven phases illustrated here. Systems methodology and the systems engineering process are described in the middle box in this figure, and the representation used for this is a two-dimensional morphological box. In this particular framework, there are 21 activity cells. There should be one or more methods, tools, or technologies associated with each of these activity cells. Choice of an appropriate mix of methods is an important challenge in systems engineering. One of these morphological box frameworks needs to be associated with each of the three life cycles that are associated with an overall systems engineering effort, and this would lead to 63 specific activity cells (21 activity cells for each of the three life cycles) specific activity cells. And, this is a simplified model of the process.

If we were to restate the steps of the fine structure as seven rather than three, we would obtain a morphological box comprising 49 elements for each life cycle. Figure 19 illustrates a not untypical 49-element morphological box. This is obtained by expanding our three systems engineering steps to a total of seven. These seven steps, but not the seven phases that we associate with them, are essentially those identified by Arthur Hall in his pioneering efforts in systems engineering of 30 years ago. They may be described as follows.

Issue Formulation
- *Problem definition* involves isolating, quantifying, and clarifying the needs that create the issue at hand, and describing that set of environmental factors that

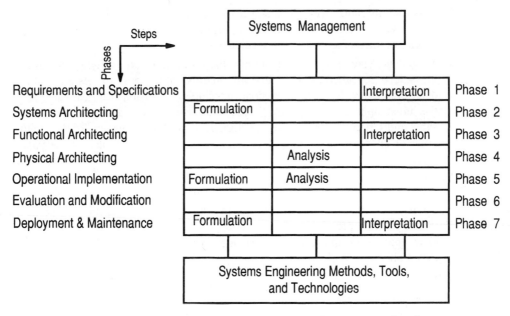

Figure 18   The steps, phases, and activity levels in systems engineering.

	Formulation			Analysis		Interpretation	
	Problem Definition	Value System Design	Systems Synthesis	Systems Analysis	Alternative Refinement	Decision Making	Planning for Action
Requirements and Specifications							
Systems Architecting and Preliminary Conceptual Design							
Logical Design and Functional Architecting							
Detailed Design, Physical Architecting, and Testing							
Operational Implementation							
Evaluation and Modification							
Deployment & Maintenance							

Figure 19   The phases and steps in one 49-element, two-dimensional systems engineering framework.

constrain alterables for the system to be developed. In addition to identifying needs, we need to identify constraints, those facets that cannot be changed, and alterables, those facets that can be changed.

- *Value system design* involves selection of the set of objectives or goals that guide the search for alternatives. Very importantly, value system design enables determination of the multidimensional attributes or decision criteria for selecting the most appropriate system. These are also known as objectives measures and represent, or may be transformed into needed metrics for the evaluation of alternatives courses of action in terms of their impact on the value system

- *Systems synthesis* involves searching for, or hypothesizing, a set of alternative courses of action, or options. Each alternative must be described in sufficient detail to permit analysis of the impacts of implementation, and subsequent evaluation and interpretation with respect to the objectives. We also need to associate appropriate metrics with each identified alternative, and these may be called alternatives measures.

Analysis

- *Systems analysis* involves determining specific impacts or consequences that were specified as relevant by the value system. These impacts may relate to such important concerns as product quality, market, reliability, cost, and effectiveness or benefits. Modeling and simulation approaches are often used in systems analysis.

- *Refinement* of the alternatives refers to adjusting, sometimes by optimizing the system parameters for each alternative in order to meet system objectives, ideally in a "best" fashion, and to satisfy system constraints.

Interpretation

- Decision assessment as a conscious activity will lead to *decision-making*. This involves evaluating the impacts or consequences of the alternatives developed in analysis relative to the value system. This enables interpretation of these evaluations such that all alternatives can be compared relative to these values. One or more alternatives or courses of action can be selected for advancing to the next step.

- *Planning for action,* or to implement the next phase, includes communicating the results of the effort to this point. It includes such pragmatic efforts as scheduling subsequent efforts, allocating resources to accomplish them, and setting up system management teams. If we are conducting a single-phase effort, this step would be the final one. More generally, it leads to a new phase of effort.

The specific methods we need to use in each of these seven steps is clearly dependent upon the phase of activity that is being completed.

Using a seven phase and seven step framework raises the number of activity cells to 49 for a single life cycle, or 147 when we consider that there are generally 3 life cycle phases. A very large number of systems engineering methods, tools, and technologies may be needed to fill in this matrix, especially since more than one of these will almost invariably be associated with many of the entries.

We now describe the seven phases selected here. The (user) *requirements and* (system) *specifications phase* of the systems engineering life cycle has as its goal the identification of client, customer, or stakeholder needs, activities, and objectives for the functionally operational system. This phase should result in the identification and description of preliminary conceptual design considerations for the next phase. It is necessary to translate operational deployment needs into technical requirements and system specifications so that these needs can be addressed by the system development efforts. If this is not accomplished well, the developed system may be such that deployment is not easily, if at all, possible. Thus, the requirements and specifications phase, or systems definition phase, is affected by, and affects, each of the other phases of the systems engineering life cycle. The notion that these phases are distinct and that they are not interrelated is simply incorrect. We could easily disaggregate the system definition phase into two phases:

- Identification of user-level requirements for a system; and
- Translation of these user-level requirements into system-level specifications.

As a result of the requirements specifications phase, there should exist a clear definition of development issues such that it becomes possible to make a decision concerning whether to undertake the *preliminary conceptual design phase,* or *functional architecting phase.* If the requirements specifications effort indicates that client needs can be satisfied in a functionally satisfactorily manner, then documentation is typically prepared concerning system-level specifications for the preliminary conceptual design phase. Initial specifications for the following three phases of effort are typically also prepared, and a concept design team is selected to implement the next phase of the life-cycle effort. This effort is sometimes called functional system level architecting and is described in Chapter 12.

Preliminary conceptual system design typically includes, or results in, an effort to specify the content and associated enterprise level functional architecture for the system product in question. The primary goal of the next phase, which is here called the *logical design and physical architecting phase,* is to develop some sort of prototype that is responsive to the specifications and functional level architecture previously identified in earlier phases of the life cycle concerned with technical specifications and systems architectures. Implementation of the preliminary conceptual design, one that is responsive to user requirements for the system and associated technical system specifications, should be obtained in this phase.

The desired product of the fourth phase of activity is a set of detailed design and implementation architecture specifications that should result in a useful system product. There should exist a high degree of user confidence that a useful product will result from the efforts following detailed design, or the entire effort should be redone or possibly abandoned. Another product of this phase is a refined set of specifications for the evaluation and operational deployment phases of the life cycle. In the fourth phase, these are translated into detailed representations in physical architecture form such that system production, manufacturing, or implementation may occur.

A product, process, or system is produced in the *detailed design and implementation architecture phase* of the life cycle. This is not necessarily the final system, but rather the result of implementation of the functional architecture and physical architecture efforts of the last two phases, perhaps in the form of a production prototype. User guides for the product should be produced such that realistic operational test and evaluation can be conducted during the *operational implementation phase.* In this phase, the system is manufactured, or otherwise assembled, and otherwise made operational.

Evaluation and modification of the implementation architecture and detailed design is achieved in the next phase. This yields the final product or service as a result of this *evaluation and implementation phase* of the systems engineering life cycle. The seventh and last phase noted in Figure 19 is the *deployment and maintenance phase.*

System acquisition is an often-used word to describe the entire systems engineering process that results in an operational systems engineering product. There are many other systems acquisition life cycles. A number of these are described in Chapter 1.

Generally, an acquisition life cycle involves primarily knowledge practices, or standard procedures to produce or manufacture a product based on established practices. An RDT&E life cycle is generally associated with maturing an emerging technology and involves knowledge principles. A planning and marketing life cycle is concerned with product planning and other efforts to determine market potential for a product or service, and generally involves knowledge perspectives. Generally this life cycle is needed to identify emerging technologies chosen to enter an RDT&E life cycle. Figure 15 indicated information flow across these three life cycles. As we have indicated in many of our discussions, there needs to be feedback and iteration across these three life cycles, and within the phases and steps of a given life cycle.

Figure 20 presents a three-gateway model that has been used to describe the necessary conceptual relations among these three life cycles. Here, the systems planning and market

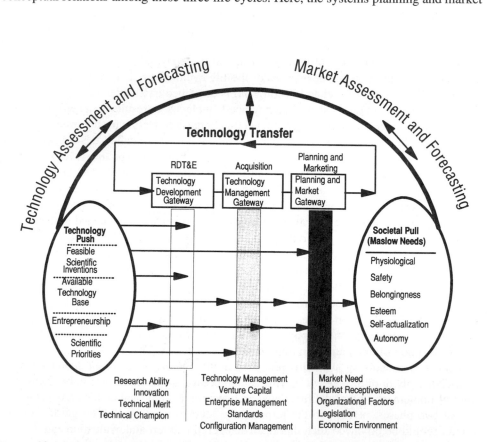

Figure 20 A three-gateway model for innovation and technology transfer through systems engineering.

planning gateway is shown downstream, in the sense of being to the right of the other two gateways. Much of systems planning and marketing should proceed both RDT&E and acquisition, however, if these latter developments are to be effective, efficient, and to supply useful products and services. This is represented in this manner in Figure 15. The positioning of this gateway upstream or downstream in Figure 20 depends on the extent of technology push vs. market pull. Different incentives are associated with technology push than with market or societal pull. These latter are based primarily on Abraham Maslow's Hierarchy of Needs.

Evaluation of systems engineering efforts is very important. Evaluation criteria are initially obtained as a part of the requirements specifications phase and modified during the subsequent phases of a systems engineering effort. Evaluation must be adapted to other systems engineering phases so that it becomes an integral and functional part of the overall systems engineering process and product. Generally, the critical issues for evaluation involve adaptations of the elements present in the requirements and specifications phase of the design process. A set of specific evaluation test requirements and tests are evolved from the objectives and needs determined in requirements specifications. These requirements and tests should be such that each objective measure and critical evaluation issue component can be measured by at least one evaluation test instrument. Evaluation is discussed in some detail in Chapter 22. Generic approaches to identify metrics and measurements appropriate for evaluation are discussed in Chapter 15.

If it is determined, perhaps through an operational evaluation, that the resulting systems process or product cannot meet user needs, the systems engineering life cycle process reverts iteratively to an earlier phase, and the effort continues. An important by-product of system evaluation is determination of ultimate performance limitations for an operationally realizable system. Often, operational evaluation is the only realistic way to establish meaningful information concerning functional effectiveness of the result of a systems engineering effort. Successful evaluation is dependent upon having predetermined explicit evaluation standards.

The *deployment and maintenance phase* of the systems life-cycle effort includes final user acceptance of the system, and operational deployment and associated maintenance. Maintenance and retrofit can be defined either as additional phases in life cycles, or as part of the operational deployment phase. Either is an acceptable way to define the system life cycle, and there are many possible systems engineering life cycles, some of which are discussed in subsequent chapters of this handbook and especially in Chapter 1. Maintenance is related to, but is somewhat different from product-level reengineering.

THE REST OF THE *HANDBOOK OF SYSTEMS ENGINEERING AND MANAGEMENT*

As we have already noted, there are many ingredients associated with the engineering of trustworthy systems. The 30 chapters that follow provide detailed discussions of many of these. In concluding this introductory writing, we summarize these contributions and provide a road map to the *Handbook*.

Chapter 1—*Systems Engineering Life Cycles: Life Cycles for Research, Development, Test,* and *Evaluation; Acquisition; and Planning and Marketing.* This chapter examines basic types of life cycle models and extensions:

- Waterfall, and the use of feedback and prototyping to improve the system definition
- V model;
- Incremental and evolutionary approaches, such as the repeated waterfall with feedback, the spiral model, and rapid development;
- Concurrent engineering;
- Models that are specific to domains or organizations, requiring process measurement for certification and for process improvement.

This chapter includes consideration of relationships between organizational characteristics and life-cycle purpose, and the interaction of process models with other dynamic organizational forces. Of particular interest is the determination of several fundamental relationships along a continuum of four fundamental types of acquisition:

- Commercial-off-the-shelf
- Performance-based contracting
- Cost-plus contracting
- In-house development

There is also discussion of contemporary issues associated with life cycle use; such as those related to reengineering, reuse, and automated software generation. We are moving into more domain-specific ways of conducting the business of systems engineering, including a move to associate the process with the product requirements, such as in the integrated product and process development (IPPD) efforts that are also discussed in Chapter 23. From an organizational point of view this is a welcome development, since the organization is then liberated to improve processes according to their own critical core capabilities.

Chapter 2—*Systems Engineering Management: The Multidisciplinary Discipline*. Systems engineering is a multidisciplinary function that is associated with requirements, specifications, design, development, building, and operation so that all elements are integrated into an optimum overall system. A systems engineer is a person who is capable of integrating knowledge from different disciplines and seeing problems with a holistic view through application of the "systems approach." Since no system of any complexity is created by a single person, systems engineering is strongly linked to management and team efforts. This chapter deals with two strongly linked efforts: systems engineering and systems management. This multidisciplinary field is termed systems engineering management in this chapter. The questions addressed are:

- What is systems engineering management?
- What are the roles and activities associated with systems engineering management?
- How is it applied to the various types of systems?
- What knowledge and skills does it require?
- How can it be developed and taught, and trained for?

The chapter outlines a framework for a description of context, content, roles, and activities in systems engineering management. It uses a two-dimensional taxonomy in which systems are classified according to four levels of technological uncertainty and three levels of system scope, thereby allowing discussion of different systems engineering management roles for different types of systems. The skills and disciplines involved in systems engineering management are identified and the chapter presents an integrative framework that encompasses the theories, tools, and disciplines involved. It also provides guidelines for educational and training programs in systems engineering.

Chapter 3—*Risk Management.* Risk-based decision making and risk-based approaches in decision making are generally premised on the assumption that there are available approaches to deal with uncertainties, and that these are used to formulate policy options and assess their various impacts and ramifications. There are two fundamental reasons for the complexity of this subject. One is that decision-making under uncertainty is pervasive and affects personal, organizational, and governmental decisions. Uncertainties are ubiquitous throughout the decision making process, regardless of whether there are one or more concerned parties, or whether the decision is constrained by economic or environmental considerations, or whether it is driven by social, political, or geographical forces, directed by scientific or technological know-how, or influenced by various power brokers and stakeholders. The second reason risk-based decision making is complex is that it is cross-disciplinary; it involves human, social, economic, political, and institutional concerns as well as technology and management concerns. The subject has been further complicated by the development of diverse approaches of varying reliability.

Consider, for example, the risks associated with natural hazards. Causes for major natural hazards are many and diverse. The risks associated with them affect human lives, the environment, the economy, and the country's social well-being. Many of these risks are manifested in engineering-based physical infrastructures: dams, levees, water distribution systems, wastewater treatment plants, transportation systems, communication systems, and hospitals. When addressing the risks associated with such natural hazards as earthquakes and major floods, we must also account for the impact of these hazards on the integrity, reliability, and performance of engineering-based physical infrastructures. Then we must assess the consequences or impacts on human and nonhuman populations, on the socioeconomic fabric of large and small communities, and on decisions that may be taken.

The premise that risk assessment and management must be an integral part of the overall decision-making process necessitates following a systematic, holistic approach to dealing with risk. Such a holistic approach builds on the principles and philosophy upon which systems engineering and management are grounded. This chapter discusses the principles, philosophies, and practice of risk management and risk-based decision making.

Chapter 4—*Discovering System Requirements.* Cost and schedule overruns are often caused by poor requirements for the product system or service system that is to be developed. These are often produced by people who do not understand the requirements process. Due to this, they often produce such poor user and system requirements and system specifications that these need to be modified at a later phase in the life cycle when the system is under development at great cost to budgets and time schedules. This chapter provides a high-level overview of the system requirements process. It explains the types, sources, and characteristics of good requirements. Unfortunately, system requirements are seldom stated well initially by the customer or system user. This chapter shows ways to help work with customers and users to discover the system requirements. It also explains terminology commonly used in the requirements definition field, such as verification, validation, technical performance measures, and various design reviews. There is a uniform and identifiable process for logically discovering the system requirements regardless of system purpose, size, or complexity. The purpose of this chapter is to reveal this process.

Chapter 5—*Configuration Management.* Configuration management is an accepted part of the systems engineering and management infrastructure that is necessary for efficient and effective support of systems development. Configuration management is the art of identifying, organizing, and controlling modifications to the product or service that is being engineered. The goal is to maximize productivity by minimizing mistakes. This requires identification, organization, and control of modifications of a system while it is being built or engineered.

Configuration management is formally considered to be the discipline used in the development phase of the systems engineering life cycle, generally the acquisition or production life cycle, to classify and control artifacts associated with the evolving system. The primary objective is to ensure that changes to these artifacts are controlled. Uncontrolled changes can result in far-reaching problems that may be compounded throughout the life cycle. For example, uncontrolled changes to user or technical system requirements that are not under configuration control may result in code being written based on old requirements or specifications that have not been properly updated with new approved changes to these. The programmer will code to the incorrect specification, which in turn will lead to problems during systems integration, or elsewhere. In general, lack of attention to configuration management will result in unsatisfied user needs. It may also result in poor performance in satisfying either the functional or nonfunctional requirements, or both.

Changes to all baseline systems documentation as well as system hardware, software, and other artifacts must be controlled in configuration management. During the greater part of any project, the user and developer will rely on the project documents for monitoring progress and ensuring that the system under development complies with user requirements and system specifications. Documentation also serves as the baseline from which to trace requirements throughout the systems engineering life cycle. A standard configuration management process and information for tailoring a project-specific configuration management plan are very important to success. These and other facets of configuration management are discussed in this chapter.

Chapter 6—*Cost Management.* Life-cycle costing includes the consideration of all future costs associated with research and development, construction, production, distribution, system operation, sustaining maintenance and support, retirement, and material disposal and/or recycling. It involves the costs of all applicable technical and management activities throughout the system life cycle.

Many of the day-to-day decisions pertaining to the engineering of new systems or the reengineering of existing systems are based on technical performance-related factors alone. Economic considerations, if addressed at all in many traditional efforts, often dealt primarily or only with initial procurement and acquisition costs. They may ignore costs associated with changes to accommodate new requirements specifications identified after the initial set has been developed, or costs brought about by the need for product maintenance and upgrade. The consequences of this "short-term" approach have been detrimental to overall project success. Many of the engineering and management decisions made in the early phases of a systems engineering life cycle have a large impact on the sustaining operation and maintenance support that the deployed system will require. These "downstream" activities often constitute a large percentage of the total cost of a system.

It is essential that we extend our planning and decision making to address system requirements from a total life-cycle perspective, particularly if we are to properly assess the risks associated with such decisions. The application of life-cycle analysis methods and *total* cost management must be inherent within the various processes discussed throughout this handbook if the objectives of systems engineering and management are to be met. This chapter deals with the need for life-cycle cost management, the life-cycle cost analysis process, some applications, and the benefits derived from such. The aspects of functional economic analysis, development of a work breakdown structure, activity-based costing, and the implementation of cost and operational effectiveness methods in the design and/or reengineering of systems are addressed.

Chapter 7—*Total Quality Management.* It is critical that institutions think about the directions that they are headed and where those directions might lead. Quality improvement and cycle time reduction are no longer fads or slogans, but have become organizational survival issues. This chapter indicates why quality improvements and cycle time reductions can be critical to organizational success. The four stages of the quality movement are described with an emphasis on why "inspection" in all of its forms cannot guarantee outstanding quality. The basic tools and concepts of the Total Quality Management are presented. Quality management is really much more about the quality of the management than it is about the management of quality. The latter generally follows as a consequence of the former. The use of quality systems such as ISO 9000 and processes and criteria associated with the Malcolm Baldrige National Quality Award are discussed.

Chapter 8—*Reliability, Maintainability, and Availability.* The objective of this chapter is to provide an overview of reliability, maintainability and availability (RMA) issues, techniques, and tasks associated with the engineering of systems. Reliability, maintainability, and availability engineering is an important facet of systems engineering and management. It concerns the processes whereby data are gathered and models are formulated to make decisions concerning system failure, operational readiness and success, maintenance and service requirements, and the collection and organization of information from which the effectiveness of a system can be evaluated and improved. RMA engineering thus includes and applies to engineering design, manufacturing, testing, and analysis.

RMA metrics can be considered to be characteristics of a system in the sense that they can be estimated in design, controlled in manufacturing, measured during testing, and sustained in the field. When utilized early in the concept development stage of a system's development, RMA efforts can be used to estimate acquisition feasibility and risk. In the design phases of system development, RMA analysis involves methods to enhance performance over time through the selection of materials, design of structures, choice of design tolerances, manufacturing processes and tolerances, assembly techniques, shipping and handling methods, and maintenance and maintainability guidelines. Such engineering concepts as strength, fatigue, fracture, creep, tolerances, corrosion, and aging play a role in these design analyses. The use of physics-of-failure concepts coupled with the use of factors of safety and worst case studies are often required to understand the potential problems, tradeoffs, and corrective actions. In the manufacturing stage of a system, RMA analysis involves determining quality control procedures, reliability improvements, maintenance modifications, field-testing procedures, and logistics requirements. In the system's operational phases, RMA involves collecting and using test and field information obtained from field service, safety inspection, failure analysis, manufacturing records, suppliers, and customers.

It has been well documented that while the costs incurred during the concept and preliminary conceptual design stages of the systems acquisition life cycle are about 5−8 percent of the total costs, the decisions made in these early phases affect over 80 percent of the total system costs. The earlier a potentially problematic event is addressed, the more prepared the systems engineering team can be to correct or alleviate possible problems. When a system is being designed, it is assumed that the required investment will be justified according to how well the system performs its intended functions over time. This assumption cannot be justified when a system fails to perform upon demand or fails to perform repeatedly. For example, it is not enough to show that a computer can conduct a mathematical operation; it must also be shown that it can do so repeatedly over many years. Furthermore, higher life cycle costs and stricter legal liabilities have also made system RMA considerations of greater importance. As the attitude toward production and engineering of systems has changed, RMA engineering has established itself as one of the key elements that must be considered when designing, manufacturing, and operating a system.

Chapter 9—*Concurrent Engineering.* Product design and implementation, or manufacturing, have been traditionally viewed as separate entities. Recent initiatives, however, have sought to integrate all functions of product development into a single, cooperative process. The new, multifunctional approach to design is commonly referred to as *concurrent engineering.* This chapter provides an introduction to the topic of concurrent engineering and discusses several challenging problems inherent to concurrent-engineering environments.

In general, concurrent engineering requires the integration of people, processes, problem-solving mechanisms, and information. This massive integration task is generally a responsibility of those involved in systems engineering and management. The chapter presents examples of methods, tools, and technologies that target several of the most common problems faced by systems engineers when attempting to build and manage concurrent engineering environments. Topics covered include selection of concurrent engineering teams, negotiation in engineering design, design process modeling, reengineering design processes, approaches to design problem decomposition, information and data modeling, engineering database management systems, enterprise integration for concurrent engineering, product life-cycle risk assessment, and implementation of concurrent engineering systems.

Chapter 10—*The Logical Framework in a Project Management System.* The logical framework is a conceptual tool, or systematic process, that helps translate common sense, or intuition-based experience, or tacit knowledge, into a logical connection that provides more formal or explicit knowledge. It is a process that helps formalize problems in order to make it possible for decision makers to see what needs to be changed in order to make the engineering of a system more efficient and effective and to bring about resolution of identified issues. It clarifies situations. It converts intuition into a rational logical decision framework that can be more easily communicated. It supports identification of the dynamics involved in any form of organization and the transformation of these dynamics into an orderly, logical sequence of steps for issue formalization and solution management. Thus, the logical framework is a management tool intended to provide greater insight into organizational thinking, the allocation of responsibilities, and unambiguous communication.

The logical framework consists of agreements that bind different parties on the basis of the achievement of mutual expectations of these parties. More precisely, the source of inspiration in both the logical framework and contract law is the necessity to ensure that agreements are enforced over time. This task requires a precise assessment of the extent to which the commitments or expectations have been fulfilled. This is the concept of performance measurement, which is one of the components of evaluation. The logical framework is a foundation to allow defining evaluation essentially as two activities:

- Comparing expectations and observations; and
- Explaining the differences between expectations and observations if expectations have not been fulfilled.

In this chapter, the logical framework, or LogFrame, is presented as four related tools that enable:

- Decomposition of a project concept;
- Measurement of performance;
- Evaluation of projects; and
- Developing project management guidelines.

A comparative assessment of this framework is presented in the context of project management in developing countries.

Chapter 11—*Standards in Systems Engineering.* There are many different specifications and standards. In the context of systems engineering, specifications and standards are technical documents that are used to define requirements for the development, manufacture, and integration of products, processes, and services. Specifications describe the technical requirements for products or services, and the verification procedures to determine that the requirements have been met. *Specifications* are sometimes referred to as product standards in that they represent binding agreements concerning product performance characteristics, at least at a structural and functional level.

Standards are process-oriented technical documents. They are often referenced in specifications to establish or limit engineering processes and practices in the preparation, manufacture, and testing of materials, designs, and products. Standards establish uniform methods for testing procedures; define the engineering characteristics of parts and materials; provide for physical and functional interchangeability and compatibility of items; and establish quality, reliability, maintainability, and safety requirements.

This chapter provides an overview of the role of standards in systems engineering. It discusses key standards that can be used to develop and assess the major elements of the technical development process that must be integrated as part of the systems engineering effort. In addition, there is information on tailoring of requirements, proper application of standards, and how to locate and obtain standards. Additional discussions of standards may be found in other chapters of the handbook, especially in Chapter 7.

Chapter 12—*System Architectures.* Systems architecting has been defined as the process of creating complex, unprecedented systems. This description fits well the systems, especially information systems and knowledge-intensive systems, that are being created or planned today for use in industry, government, and academia. The requirements of the marketplace are often ill-defined, and rapidly evolving information technology has increasingly made it possible to offer new services at a global level. At the same time, there is increasing uncertainty as to the way in which the resulting systems will be used, what components will be incorporated, and the interconnections that will be made. Generating systems architecting, as part of the systems engineering process, can be seen as a deliberate approach to cope with the uncertainty that characterizes these complex, unprecedented systems. One thinks of system architectures when the system in question consists of many diverse components. A system architect, while knowledgeable about the individual components or subsystems that will make up the physical architecture of the system to be engineered, needs to have a good understanding of the interrelationships and interfaces among these components. While there are many methods, tools, and technologies to aid the systems architect, and there is a well-defined systems architecture development process, systems architecting requires creativity and vision because of the unprecedented nature of the systems being developed and the initially ill-defined requirements.

Many of the methodologies for systems engineering have been designed to support a traditional system development model. A set of requirements is defined; several options are considered, and through a process of elimination a final design emerges that is well-defined. This approach, based on structured analysis and design, has served well the needs of systems engineers and has produced many of the complex systems in use today. This is effective when the requirements are well defined and remain essentially constant during the system development period. When the time between requirements definition and operational implementation of the system is long, however, two things often happen:

- The requirements change because of changing needs; and
- New technology offers different alternatives and opportunities.

This conventional well-focused approach cannot handle change well. Its strength lies in its efficiency in designing a system that meets the sets of both fixed and unchanging user requirements and technological specifications.

Two basic paradigms are available for designing and evaluating systems architectures for systems, especially information systems: the structured analysis and the object-oriented approaches. Both require multiple models to represent the architecture, and both lead, through different routes, to executable models. The latter are appropriate for analyzing the behavior of the architecture and for evaluating performance. At this time, the goal of a seamless methodology supported by a set of well-integrated tools has not been reached, but promising progress is being made. This chapter describes progress in systems architecture and provides guidelines for use of this important facet of systems engineering and management.

Chapter 13—*Systems Design.* Design is the creative process by which our understanding of logic and science is joined with our understanding of human needs and wants to conceive and refine artifacts that serve specific human purposes. Design is central to the practice of systems engineering and management. Systems engineers understand that design is a creative, iterative, decision-making process.

In this chapter, the steps in the design process are discussed. These include:

- The specification of measurable goals, objectives, and constraints for the design;
- The conceptualization and parameterization of alternative candidate designs that meet or surpass specifications;

- The analysis and ranking of design alternatives; and
- The selection, implementation, and testing of the most preferred alternative.

The characteristic activities, methods, tools, and technologies of the systems engineer can be interpreted in terms of how these activities support the central mission of systems design and how design fits in with and can be integrated into the overall process for the engineering of systems. We describe some of these design tools and refer the reader to state-of-the-art expositions of many others in the surrounding chapters of this handbook.

Researchers from many backgrounds and traditions have attempted to elicit and capture the fundamental nature of design. While there is no unifying or universally accepted theory of design, this chapter briefly reviews some of the major influences on current design research. Of particular interest is the axiomatic theory of design developed from the perspective of manufacturing engineering. This theory ultimately leads to a set of design principles, rules, and practices that are embodied in design for manufacture.

Chapter 14—*Systems Integration*. Systems integration is essential to the deployment and procurement of large, complex engineered systems in that it provides an organized, sensible, accountable, and workable approach to otherwise seemingly incomprehensible programs. Implementation of systems integration applies the methodologies of systems engineering and systems management to large, complex engineered systems that include hardware, software, facilities, personnel, procedures, and training. The purpose of this is the integration of existing and new capabilities to achieve specific goals and objectives.

This chapter defines systems integration, examines its role in large, complex engineered systems, and provides life-cycle perspectives on systems integration. In addition, it indicates the tasks that must be completed, examines personnel requirements, explores a strategy for success, investigates several applications, defines characteristics necessary for configuration management, reviews relevant management characteristics, discusses how disparate subsystems may be brought together and tested, and considers risk management strategies.

Chapter 15—*Systematic Measurements*. One major need in all of systems engineering and systems management, at all levels of the organization, is to obtain the information and the knowledge that is necessary to organize and direct individual programs associated with the creation of products and services. These products and services result from the use of systems engineering process life cycles that are associated with research, development, test, and evaluation; systems acquisition, production, procurement, or manufacturing; and systems planning and marketing. This information can only be obtained through an appropriate program of systematic measurements and the development of appropriate models for use in processing this information.

This chapter discusses frameworks for systematic measurements at the level of process and product, and systems management of process and product. Associated with each of these three levels—systems management, process, and product—and as illustrated in Figure 14, are a variety of methods, tools, and technologies, and associated metrics. There is not an abrupt transition from activity at one level to activity at another, nor should there be. Measurements are needed at each of these three levels. In this chapter, we discuss approaches and frameworks for measurements and models that support organizational success through effective use of innovations brought about in systems engineering and management and the related subjects of information technology and knowledge management.

Chapter 16—*Human Supervisory Control.* This chapter summarizes the concept of supervisory control as an emerging form of control by people within automated systems. The chapter begins by defining what is meant by supervisory control and relating this concept to other popular terms that relate human performance to automation, such as human-centered automation. A brief history of supervisory control is given and its salience is discussed. Task analysis and function allocation, as well as the allocation of functions between human and machine, are critical efforts for the system designer.

There are five dominant aspects of supervisory phase control:

- Planning
- Command and programming
- Monitoring and failure detection
- Intervention
- Learning and system evolution

These are aspects that the human supervisory controller must understand in order to be a good operator. As a consequence, the systems engineer must also understand them in order to engineer a successful system. The discussions also concern several overriding considerations—including special aspects about design of human–machine systems, problems of authority; trust of automation; social implications; confusion about "human centeredness" of automation; and limits to our ability to model supervisory control—that are very important today.

Chapter 17—*Designing for Cognitive Task Performance.* This chapter is about cognitive systems engineering. It addresses the question of why designers should be concerned with the cognitive capabilities of potential users of their products and how designers might take these capabilities into account as part of the total process that leads to the engineering of a product system or a service system.

The major hypothesis is that designers and engineers of systems ignore the human element of the system at their peril. Technology-centered design can lead to at least three consequences:

- Systems may be impossible to use;
- Technology-centered rather than human-centered design can increase the effort or difficulty of operating a system;
- Technology-centered design may increase the likelihood of errors and compromise the effectiveness and safety of the system rather than increase it.

This chapter describes the capabilities and limitations of the human cognitive system and the implications for systems design and systems engineering. Several examples of incidents and accidents in engineered high-risk domains such as aviation, medicine, space, and transportation are used to illustrate what can happen when the human element is insufficiently considered during the design process. Three approaches to appropriate cognitive engineering, or human-centered design, are described. The chapter also discusses why design has not always taken the user's cognitive requirements into account and the fallacious assumptions behind technology-centered reasoning.

Chapter 18—*Organizational and Individual Decision Making.* Organizational decision making is a product of the way that individuals make decisions and the context in which these individual decisions are made. A study of this important subject requires knowledge from behavioral decision theory, social networks, information processing, cognitive psychology, artificial intelligence, and computational organization theory. Taken together, this work suggests that limits to cognition and rationality, and the structure of relations among individuals and organizations, are each vitally important in determining what decisions are made. Advances in this area are facilitated by the use of computational models of organizations. In these models, organizational decisions result from the concurrent actions and interactions among multiple distributed intelligent and adaptive agents.

Organizations do not make decisions; people do. This observation recognizes both structural and operational fact. Organizations are made by and are composed of people. Technology may support them. The associated infrastructure affects what information people have access to and how this information is interpreted as knowledge and, as a consequence, what decisions are made. This chapter discusses a number of important facets concerning organizational decision making and how context and environment greatly influence the rationality of decisions made. It sets forth a framework for the use of information technology and knowledge-management precepts and prescriptions to support enhanced decision making.

Chapter 19—*Human Error and Its Amelioration.* Human error, as a topic in its own right, has come about partly because of its prominence as a cause or contributing factor in such disasters and accidents as Three Mile Island, Chernobyl, and Bhopal. Investigations of human error are inseparable from those of human performance; the same mechanisms and processes that presumably give rise to "correct" human performance give rise to "incorrect" human performance. Thus, conceptual distinctions in generative mechanisms of and remedies for human error are integral parts of human–machine systems engineering.

Many researchers and practitioners have examined human error from a variety of perspectives. Safety engineers and risk assessment experts have focused on quantifying probabilities of different sorts of errors and the consequences associated with those errors. Typically, the analysis of human error is in terms of phenomenological categories such as errors of omission, commission, and substitution. Cognitive and engineering psychologists concerned with cognitive, as contrasted with physiological, performance have focused primarily on the generative cognitive mechanisms for error, such as lapses in memory and attention, that give rise to observable error types and the interactions between system design and human performance as the nexus for error. In looking at group work and interpersonal interaction, human communication researchers have studied misunderstandings in conversational and face-to-face interaction. More recently, sociologists and organizational theorists, as well as a growing number of cognitive engineering researchers, have focused on the historical and cultural contexts in which organizations make decisions and assess risks. They also have examined the ways in which practices, norms, and the like have influenced actions that, in retrospect, may have contributed to disaster. Inquiries into human error may focus on describing different kinds of errors, on underlying causes of error, and on how to prevent or minimize the impact of errors. A comprehensive approach to examining human error spans perspectives, from simple "physical" or physiological categories to cultural and organizational influences. This chapter discusses the many findings in this area and their role in the effective engineering of systems.

Chapter 20—*Culture, Leadership, and Organizational Change.* This chapter outlines a strategy for achieving successful organization-wide change through application of a behavioral model. This model is based on the premise that when efforts to implement major change fail, a common cause is insufficient attention to the people side of the change, or reengineering, effort. Work cultures shape individual and group behaviors; successful change entails moving from traditional work cultures to newer, more responsive forms. Starting from the top, leadership often separates successful from unsuccessful change efforts. A three phase, seven-step process to achieve change is detailed. The process outlines the way leaders communicate, convince, and move their people to new work patterns and practices. The chapter profiles three cases of successful change, illustrating both the model presented and strategy recommended for the engineering of systems.

Chapter 21—*Model-based Design of Human Interaction with Complex Systems.* This chapter provides a methodology for use by systems engineers concerned with human interaction with complex systems. Human system interaction addresses the issues concerned with the ways in which users or operators of a system interact with the system in order to meet established objectives. These issues include a wide variety of systems and of humans who interact with them. The associated design issues may be very context sensitive; or contingency task structure dependent, which is to say task and user and environment dependent.

This chapter sets forth important design issues in a world in which the rapid pace of technological evolution allows and encourages systems and organizations, especially knowledge-capital-intensive organizations, to change rapidly. Design guidance and levels of human–system interaction are discussed. A set of important design issues is then identified. A model-based design methodology is described, and the generic formulation of the methodology is presented. Specific implementation guidelines are provided based on what is called an operator function model. Inexpensive and powerful computer technology creates opportunities for human–system interaction research with much operational significance, and the chapter concludes with some realistic challenges for development of knowledge principles and knowledge practices concerning this.

Chapter 22—*Evaluation of Systems.* Almost any device, product, service, operation, or program can be considered to be a system or subsystem. While the evaluation of electromechanical devices or products important and necessary, the more challenging problem is that of evaluating human-centered and knowledge-intensive systems. The focus in this chapter is on the evaluation of such product systems or service systems.

The chapter discusses the evaluation field, the evaluation framework, the evaluation elements, and an example evaluation modeling effort. Given a growing need for evaluation, it is critical that proper procedures exist for the development of valid evaluations. The systems engineering and evaluation modeling approach presented here provides such procedures. Additionally, an evaluation modeling application is included in this chapter.

Chapter 23—*Systems Reengineering.* Virtually all of the discussions in this handbook have indicated the need for continual revitalization in the way in which we do things, such that performance continually improves over time. This is the case even if the external environment were static and unchanging. When we are in a period of high-velocity environments, however, continual organizational change and associated change in processes and product must be considered as a fundamental rule of the game for progress. There is a variety of change models and change theories. Some seek to change in order to survive; others seek to change in order to establish or to retain competitive advantage. In this chapter, we discuss change in the form of reengineering. There are a variety of names given to change efforts including reengineering, restructuring, downsizing, rightsizing, and redesign.

We can approach a discussion of reengineering from several perspectives. First, we can discuss the structural, functional, and purposeful aspects of reengineering. Alternately, or in addition, we can examine reengineering at the level of systems management, process, or product. We can examine reengineering issues across any, or all, of the three fundamental systems engineering life cycles. Within each of these life cycles, we could consider reengineering at any or all of the three generic phases of definition, development, or deployment. At the level of systems management, we examine the enterprise as a whole and consider all organizational processes within the company for improvement through strategic level change. At the level of process reengineering, as we define it here, only a single process is reengineered, and with no fundamental or radical changes in the structure or purpose of the organization as a whole. Changes, when they occur, may be radical and revolutionary, or incremental and evolutionary, at the level of systems management, process, product, or any combination of these. The scale of improvement efforts may vary from incremental and continuous improvement, such as is generally advocated by quality management efforts to radical change efforts that affect systems management itself.

One fundamental notion of reengineering, however, is the reality that it must be top down directed, and well coordinated and communicated, if it is to achieve the significant and long lasting effects that are possible. Thus, there should be a strong purposeful and systems management orientation to reengineering, even though it may have major implications for such lower level concerns as structural facets of a particular product. Several contemporary approaches to reengineering are examined in this chapter which is particularly concerned with reengineering to enhance the incorporation of knowledge intensive activities and efforts.

Chapter 24—*Issue Formulation.* Problem and issue formulation is the most important part of any systems engineering effort, because the formulation effort shapes the identification of the set of feasible alternatives, one of which will be eventually selected for implementation. No matter how elegantly the alternatives are analyzed and modeled or how thoroughly we plan for implementation, the value of the entire effort depends on having developed good alternatives in the first place. Critical to formulation is a situation assessment that defines the goals the stakeholders want to accomplish, the current state of affairs relative to those goals, and the relevant factors in the environment and their associated time frame that may impact accomplishment of the goals.

This chapter discusses in detail a number of very useful techniques for the three logical steps of issue formulation: problem definition, value system design, and system synthesis. These are the generic efforts in issue formulation, and they are followed by issue analysis and issue interpretation, so as to complete one of the phases of a systems engineering life cycle. This issue formulation is needed a number of times within a given systems engineering process, as were illustrated in Figures 18 and 19.

Chapter 25—*Functional Analysis.* Functional analysis is performed in systems engineering and management as a portion of the logical design process that results in identification of the functional architecture. It may be used for the issue analysis step at each of the phases of a systems engineering effort. Its major use, however, is in the development phase of a systems acquisition life cycle. Typical systems engineering life-cycle processes for acquisition may be described in terms of such initial phases as requirements definition and analysis; functional analysis; physical or resource definition; and operational analysis. This last phase of operational analysis involves the marriage of functions with resources to determine if the requirements are met. Functional analysis addresses the activities that the system, software, or organization must perform to achieve its desired outputs; that is, what transformations are necessary to turn the available inputs into the desired outputs. Additional elements include the flow of data or items between functions; the processing of instructions that are available to guide the transformations, and the control logic that dictates the activation and termination of functions. Various diagrammatic methods for functional analysis have been developed to capture some or all of these concepts.

This chapter examines the elements of functional analysis, functional decomposition, systems engineering requirements statements and functional analysis, and diagrams and software support for functional analysis. One result of an appropriate functional analysis is verification and validation of the fitness of a given functional architecture.

Chapter 26—*Methods for the Modeling and Analysis of Alternatives.* Mathematical modeling may serve many purposes in a systems engineering project. It may be used for analyzing an existing system and for specifying and analyzing design alternatives. In this chapter, a general cycle for developing mathematical models is presented. A number of phases in modeling are identified: defining goals and functions, conceptualization, model construction, validation and model use. The discussion focuses on issues that are common to many types of model. Following this, several different modeling methodologies are described according to the model cycle. Aspects specific to these methodologies are discussed, as are requirements for use of the methodology.

Three of these methodologies are aimed at developing causal system models, and allow investigation of alternative system configurations: physical systems modeling, system dynamics and discrete event simulation modeling. Physical systems modeling is intended to be used in modeling technical or physical systems, whereas system dynamics is a methodology that has been developed primarily for analyzing and modeling business and socioeconomic systems. Discrete-event simulation models are stochastic models in which separate entities are recognized. These three methodologies result in causal models that are able to explain system behavior. Time-series models, on the other hand, are black-box models that are not aimed at explaining, but at forecasting only. Black-box models may be used when it is not necessary to describe the exact internal workings of the system, or when the underlying causal mechanisms are unknown, but sufficient data are available.

The preceding modeling methodologies allow investigation of system behavior over time, but do not include financial aspects. When designing a system, economic aspects should be taken into account. Therefore, cost−benefit analysis is discussed briefly as a complement to the discussion in Chapter 6.

Although the various modeling methodologies are discussed in separate sections, a combination of various approaches is usually necessary for large-scale systems. Some recent advances in modeling are discussed, including group model building, and animation and visualization possibilities.

Chapter 27—*Operations Research and Refinement of Courses of Action.* The objective of this chapter is to explain how the methods and tools of operations research can relate systems engineering approaches to decision making by generating and refining alternative solutions to complex engineering problems. Subsequent to a general discussion of operations research and its utilization in problem solving, the philosophical underpinnings of operations research and systems engineering are compared and evaluated. Their complementary capabilities allow the reductionist-scientific approach of operations research to be coupled with the holistic methodologies of systems engineering to furnish a powerful tool to solve many real-world problems. A range of useful operations research techniques is described next, followed by a discussion of how operations research can be used to generate and screen actions and alternatives. Because of the great importance of multiple-criteria decision making and multiple-participant decision making in operations research, these sets of useful tools are discussed in separate sections. The last section of this chapter is devoted to a brief discussion of heuristic methods in operations research.

Chapter 28—*Decision Analysis.* This chapter reviews formal knowledge-based decision analysis methods and their use in support for decision making in systems engineering and management. Methods are presented that assist in determining evaluation considerations and specifying evaluation measures. Procedures are also presented that help in developing better alternatives. Multiobjective value analysis procedures are reviewed, as well as methods for quantifying uncertainties in decision making. The roles of risk aversion and probabilistic interdependencies are also considered, and methods are presented for taking them into account in decision making.

A formal technoeconomic rationality-based systematic approach to quantitative decision analysis includes the following steps:

- Specifying objectives and scales for measuring achievement with respect to these objectives
- Developing alternatives that potentially might achieve the objectives
- Determining how well each alternative achieves each objective
- Evaluating tradeoffs among the objectives
- Selecting the alternative that, on balance, best achieves the objectives

These activities are considered in detail in this chapter and guidelines for accomplishing decision analysis are presented.

Chapter 29—*Project Planning and Planning for Action.* The core elements of planning are familiar to most systems engineering professionals. The idea of identifying tasks to be accomplished, finding staff to execute those tasks, and the resources to support them are part of every systems engineer's professional experience and set of skills. For small projects, as well as for individual efforts, these plans remain largely ad hoc and unstructured. The execution of plans developed in this manner is entirely an individual effort. Each effort is different based on a uniquely individual set of preferences, constraints, and tolerances. These may be, and usually are, expressed in a unique vocabulary with unique assumptions and knowledge about the implementation alternatives.

However, when a project is undertaken by, or on behalf of, more than a few individuals, when the objectives are complex and there are many variables, planning for action becomes a team activity. It must become a more formal process in these instances. There must be a common vocabulary, common approach for identifying the work to be done and products to be produced, and a common way to communicate the status and results of the effort.

This commonality requires standard, formal methods that can be understood by a wide range of system engineering professionals and shared by colleagues using standard methods of data, information, and knowledge exchange.

Standard, formal planning techniques generally share common goals. They allow the definition and tracking of such items as activities, milestones, and deliverables. If the project is large enough, this may require the ability to divide the project into phases. Activities or tasks must be organized into time frames or schedules for their accomplishment. There must also be a mechanism to support the transition of a project between phases.

There are many methods, tools, and technologies available to systems engineers for planning a project's activities. These include familiar techniques such as work breakdown structures (WBSs), as well as more recent methodologies made possible by technological advances such as network-based planning and project management. Some of the methods focus on tracking the progress of projects. These include techniques such as the critical path method (CPM) and the program evaluation and review technique (PERT). There are also techniques to support the definition of roles and selection of personnel to staff projects and to assist in the organization of teams to complete the work. This chapter discusses each of these efforts. It complements the systems engineering management discussions in Chapter 2 and the configuration management discussions in Chapter 5.

Chapter 30—*Information Technology and Knowledge Management.* In the early days of human civilization, development was made possible through the use of human effort, or labor, primarily. Human ability to use natural resources led to the ability to develop based not only on labor, but also on the availability of land, which was the classic economic term that implied natural, or physical, resources. At that time, most organizations were made up of small proprietorships. The availability of financial capital during the Industrial Revolution led to this being a third fundamental economic resource, and also to the development of large, hierarchical corporations. This period is generally associated with centralization, mass production, and standardization.

Major availability of technologies for information capture, storage, and processing has led to information, as well as the resulting knowledge that comes from information associated with context and environment, as a fourth fundamental technoeconomic resource for development. This is the era of total quality management, mass customization of products and services, reengineering at the level of product and process, and decentralization and horizontalization of organizations, and systems management. While information technology has enabled these changes, much more than just information technology is needed to bring them about satisfactorily.

Major growth in power of computing and communicating, and associated networking is quite fundamental and has changed relationships among people, organizations, and technology. These capabilities allow us to study much more complex issues than was formerly possible. They provide a foundation for dramatic increases in learning and associated increases in both individual and organizational effectiveness. In large part, this is due to the networking capability that enables enhanced coordination and communications among humans in organizations. It is also due to the vastly increased potential availability of knowledge to support individuals and organizations in their efforts.

At first glance, the increased power may appear due to information-technology-based capability increases *only*. However, information technologies need to be appropriately integrated within organizational frameworks and human networks if they are to be broadly useful. This poses a transdisciplinary challenge of unprecedented magnitude if we are to move from high-performance information technologies to high-performance and knowledge-intensive organizations.

In just the past few years, the pace has quickened quite substantially and the need for integration of information technology issues with organizational issues has led to the creation of a field of study, the objectives of which generally include:

- Capturing human information and knowledge needs in the form of system requirements and specifications;
- Developing and deploying systems that satisfy these requirements;
- Supporting the role of cross-functional teams in work;
- Overcoming behavioral and social impediments to the introduction of information technology systems in organizations;
- Enhancing human communication and coordination for effective and efficient workflow through knowledge management.

Because of the importance of both information and knowledge to an organization, two related areas of study have arisen. The first of these is concerned with technologies associated with the effective and efficient acquisition, transmission, and use of information, or information technology. When associated with organizational use, this is sometimes called *organizational intelligence,* or *organizational informatics.* The second area, known as *knowledge management,* refers to an organization's capacity to gather information, generate knowledge, and act effectively and in an innovative manner on the basis of that knowledge. This provides the capacity for success in the rapidly changing or highly competitive environments of knowledge organizations. Developing and leveraging organizational knowledge is a key competency and, as noted, it requires information technology as well as many other supporting capabilities that are associated with knowledge management. Knowledge management refers to management of the environment for effective and efficient transfer of information to knowledge and not to management of knowledge itself.

The human side of knowledge management is very important. Knowledge capital is sometimes used to describe the intellectual asset value of employees and is a real, demonstrable asset. We have used the term *systems ecology* to suggest managing organizational change to create a knowledge organization, and enhance and support the resulting intellectual property for the production of sustainable products and services. Managing information and knowledge effectively to facilitate a smooth transition into the Information Age calls for this systems ecology, a body of methods for systems engineering and management that is based on analogous models of natural ecologies. Such a systems ecology would enable the modeling, simulation, and management of truly large systems of information and knowledge, technology, humans, organizations, and the environments that surround them. This chapter considers several broad trends, which we feel will persist regardless of the specific technologies that enable them. In particular, it focuses on the effects of information technology on knowledge management, and on the organizational implications of these effects. It is safe to predict that these efforts will see much development and evolution in the early part of the 21st Century. It is therefore fitting to conclude our presentation in this Handbook with a discussion of the continued evolution of systems engineering and management and its role in knowledge integration and knowledge management.

KNOWLEDGE MAP OF THE SYSTEMS ENGINEERING AND MANAGEMENT HANDBOOK

This handbook includes an immense amount of information with which to approach the issues, phases, and life cycles discussed in this chapter. The numerous illustrations in this chapter indicate how this information is organized in general. In this section, we briefly describe how this information is organized in particular in this handbook.

Each chapter was reviewed and analyzed to identify the 10–15 concepts that the chapter treats with most depth. These could have been assigned weights, but here only binary interactions are assumed. Only the most important concepts are identified for each chapter, and this is clearly a subjective listing. The result of this effort is a listing of almost 400 concepts covered by this handbook. Similar concepts were then grouped into almost 60 categories. This enabled depicting the coverage of the handbook shown in Table 2. Note that each concept in the first column of Table 2 is expandable to 6 to 10 more detailed concepts. While such expansions are straightforward by computer, they are obviously not possible in this hardcopy book.

Relationships among chapters can be depicted as a knowledge map. If such a map is drawn at the level of 400 concepts, the resulting picture resembles a massive bowl of spaghetti. If we limit the grain size of the concepts, and also impose a threshold on the strength of linkages, a more manageable map result, as represented in Figure 21. It should be noted that a better way to create such maps is by using the intent of the user, in the form of a contextual relationship, which will enable the pruning of irrelevant portions of a larger, more complex map. This results in intent-driven maps that are, for obvious reasons, much more useful for mining knowledge bases such as this handbook.

Figure 21 indicates five clusters of chapters, where we impose the restriction that the clusters not overlap:

- Systems Engineering and Management Perspectives
- Systems Engineering Processes and Systems Management for Definition, Development, and Deployment
- Systems Engineering and Systems Management Methods, Tools, and Technologies
- Humans and Organizations
- Economics

There are a variety of minor linkages among these clusters. However, the intracluster linkages are much more frequent than the intercluster linkages. Also, there are a number of other clusters that could be chosen.

An exception to this linkage observation occurs relative to the Economics cluster. It is the only cluster linked to every other cluster, and has many more external linkages than do the other clusters. This is not surprising in that most every facet of systems engineering and management is linked to economics in one way or another. Interestingly, systems engineering is not generally viewed in most of the chapters in this handbook as a revenue generator—or contributor to sales—even though revenue/sales are the flip side of costs, and hence determine profits. This is probably due to the traditional focus of systems engineering on large, complex systems projects, often for government customers. For such projects, systems engineering tends to be done after the sale. With the military moving toward increased commercial practices and solutions, however, we may expect that systems engineering may have an increasing impact on the "top line" as well as the "bottom line."

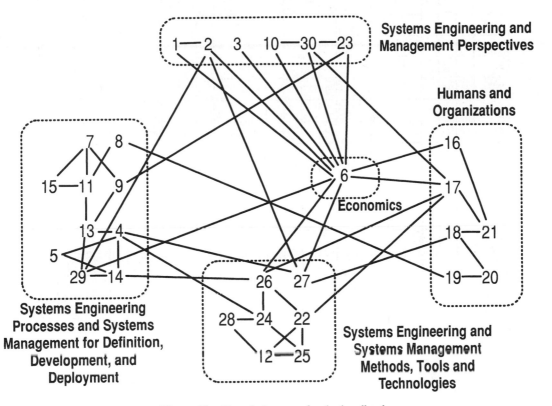

Figure 21 Knowledge map for the handbook.

Knowledge maps such as exemplified in Figure 21 provide a basis for much richer interactions with bodies of knowledge such as are captured in this handbook. As electronic versions of handbooks and encyclopedias become the standard for accessing and utilizing large amounts of knowledge, we expect that knowledge mapping will become a highly valuable means of using intentions to drive access and tailor utilization. The inherent linear constraints of hardcopy book technology will become less and less acceptable.

THE MANY DIMENSIONS OF SYSTEMS ENGINEERING

The knowledge map that we have just obtained and illustrated in Figure 21 is but one of a large number that could be obtained. We could associate other than binary weights with the interaction terms used to obtain Table 2, and then we could show information linkages at various sensitivity thresholds. There are also a large number of different knowledge clusters that could be obtained. For example, we could have clusters for:

- Knowledge principles, practices, and perspectives
- Systems management, processes, and methods and tools
- Functional architecture, physical architecture, implementation architecture

TABLE 2 Systems Engineering Concepts Included in Various *Handbook* Chapters

Concept/Chapter	1	2	3	4	5	6	7	8	9	10	11	12	13	14	15	16	17	18	19	20	21	22	23	24	25	26	27	28	29	30
Acquisition																			X											X
Aiding														X			X				X									
Analysis							X							X		X								X	X	X	X			
Architecture												X															X			
Automation																X	X				X									
Complexity		X																											X	X
Concurrent Engineering	X								X				X										X							
Configuration					X						X																			
Control							X									X					X									
Costs and Economics			X			X											X						X			X			X	X
Culture																			X	X										
Customer Satisfaction															X															
Decisions												X						X										X		
Design					X				X			X	X	X		X							X		X			X		X
Diagrams							X					X												X	X				X	
Ecology																														
Engineering	X	X											X				X		X											X
Evaluation														X		X	X	X								X				
Failures			X					X											X			X	X							
Functions												X				X					X	X	X		X					X
Human												X		X									X	X						
Information																							X							
Issue																								X						
Knowledge							X										X							X						X
Maintainability											X		X																X	
Management		X								X			X	X					X	X			X							X

Manpower, Personnel, and Training																				X										
Manufacturing							X		X			X											X		X					
Measurement										X		X			X							X						X		X
Methods																		X			X			X			X			
Modeling	X		X									X					X	X		X	X	X		X	X	X	X			X
Multiple Criteria			X																								X	X		
Networks																		X						X			X		X	
Objectives										X														X	X			X		
Optimization			X															X									X			
Organizations	X				X													X	X	X				X						X
Planning	X	X				X				X						X											X		X	X
Problem				X									X													X				
Process							X																X							
Prototypes													X				X							X						
Quality							X		X		X		X	X	X									X						
Reengineering	X																						X							X
Requirements				X	X									X										X						
Risk			X											X					X									X	X	
Safety			X																X											
Scenarios																								X	X					X
Simulation													X											X		X	X			
Software	X										X				X															X
Stakeholders																								X	X					
Standards					X		X				X				X															
Systems	X	X			X									X									X	X		X	X			
Teams							X		X																					
Testing					X								X	X			X							X						X
Uncertainty and Imprecision			X																											X
Value							X						X											X				X		X

- Humans, organizations, and technologies
- Purpose, function, and structure

and so forth. Even these clusters could have clusters associated with them. For example, processes could be described by:

- RDT&E, acquisition or production, and planning and marketing

And each of these processes could be described in terms of the phases of

- definition, development, and deployment

Within each of these phases, we have steps of

- formulation, analysis, and interpretation.

An interpretive picture of these dimensions is shown in Figure 22. Even this is not at all complete. When we examine further the human element in systems engineering and look at cognitive functioning, we can define two extremes for reasoning: experience-based intuitive thought and analytical thought. Thus, we see that systems engineering is very multidimensional.

Two precautionary comments are needed at this point. There is no standard systems engineering vocabulary, so many authors use different terms in place of the terms used here. For example, we use the term phase to describe the distinct yet integrated activities associated with a systems engineering life-cycle process. The term phase is not at all common across all systems engineering literature. In a similar way, the term step is not

Three Levels
- ★ Systems Management
- ★ Processes
- ★ Products

Three Lifecycles
- ★ RDT&E
- ★ Acquisition
- ★ Systems Planning

Three Phases within Each Lifecycle
- ★ Definition
- ★ Development
- ★ Deployment

Three Steps Within Each Phase
- ★ Formulation
- ★ Analysis
- ★ Interpretation

Three Systems Architecture Efforts
- ★ Functional Architecure
- ★ Physical Architecture
- ★ Implementation Architecture

Three Definitions and Descritptions
- ★ Purposeful
- ★ Functional
- ★ Structural

Figure 22   Some of the many dimensions of systems engineering.

always used by all authors to distinguish the distinct but integrated activities associated with one of the systems engineering phases. We can have "designs" of various types, as we have discussed here. While there is not a great difference in terminology across the 30 chapters in this handbook, there is necessarily some difference. This is unavoidable at this time. Careful attention to the context in which terms are used will generally suggest their specific meaning and relationship to the terms used here.

PEOPLE, ORGANIZATIONS, TECHNOLOGY, AND ARCHITECTURES

The unprecedented technological advances in the information technologies of computation, communication, software, and networking create numerous opportunities for enhancing our life quality, the quality of such critical societal services such as health and education, and the productivity and effectiveness of organizations. We are witnesses to the emergence of new human activities that demand new processes and management strategies for the engineering of systems. The major need is for appropriate management of people, organizations, and technology as a social system and, without this, errors of the third kind, or wrong problem solutions, associated with technological fixes are almost a foregone conclusion.

Systems engineering and systems management basically are concerned with finding integrated solutions to issues of large scale and scope. Fundamentally, systems engineers are brokers of information and knowledge leading to the definition, development, and deployment of systems of all types. The major objective in systems engineering and man-

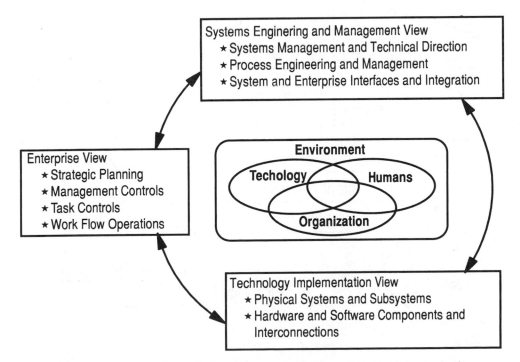

Figure 23 The three views of a large system of technology, humans, and organization.

agement is to provide appropriate products, services, and processes that fulfill client needs. This is accomplished through the engineering of a "system" that meets these needs. Generally, the needs of a client, usually an organization or an enterprise, need to be defined in functional form and are often expressed in terms of a functional architecture. Systems engineers generally construct this functional architecture or functional design to be responsive not only to enterprise needs but also to constraints imposed by regulations and social customs.

The functional architecture is generally transformed into a physical architecture that represents the major systems that will ultimately be engineered. This physical architecture, often in what is called "block diagram" form is a high-level logical picture of the overall product, service, or process, that will be delivered to the customer. This product must be implemented, and this leads to a third architecture form, generally called the implementation architecture or operational architecture. Systems engineers generally work with implementation contractors, from a technical direction, various configuration control and management efforts, to insure successful realization and implementation of an operational system that is responsive to the needs of the enterprise. Just as the enterprise is, broadly speaking, composed of a number of organizational entities, with a variety of perspectives, so also are implementation engineering contractors, especially when one considers the role of subcontractors and outsourced suppliers of systems components.

Figures 23 and 24 attempt to represent some of this complexity. Figure 23 illustrates the three major stakeholders associated with the fielding of a large system:

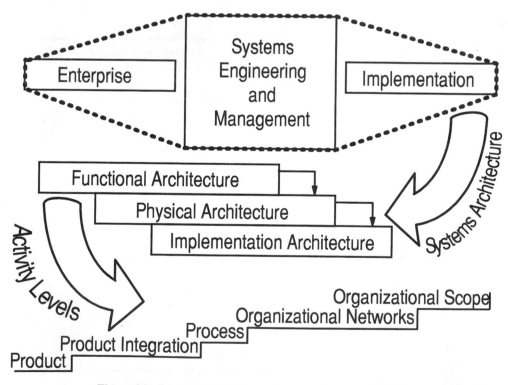

Figure 24   Systems engineering and management perspectives.

- The enterprise for whom the system is to be engineered,
- The systems engineering and management team responsible for overall technical direction, integrity, and integration of the effort, and
- Implementation specialists, who generally represent the plethora of classical engineering disciplines in performing invaluable roles in the actual realization of specific technologies.

It also provides some detail on the perspectives taken by these three major stakeholders.

A major challenge for the systems engineering and management team is to build a system in terms of the functional, physical, and engineering architecture. The composite of this architecture may be called the *systems architecture*. The resulting system, to be implemented, may well be a physical product or service. Generally it is rare that a completely new physical product is produced. Usually, there are a variety of legacy systems or legacy products and the "new" product must be capable of being integrated with these legacy systems. Also, products are generally used to support some organizational process and an important role in systems engineering is the engineering of appropriate processes to effectively accommodate humans, organizations, and technologies. Often today, there is a major need for considering organizational networks and organizational scope issues in the engineering of large systems. Thus, we immediately see that all of the knowledge integration and management issues discussed in the previous section arise. Immediately,

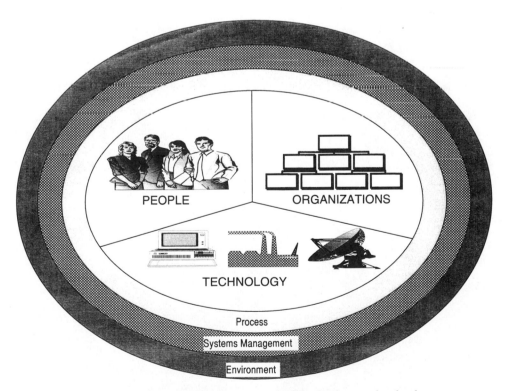

Figure 25   Systems engineering as people, organizations, and technology.

we see that systems engineering and management is an inherently transdisciplinary profession. Figure 24 only begins to represent this complexity. It does show systems engineering and management as an integration effort to support evolution of the three architectural perspectives associated that comprise the systems architecture: functional, physical, and enterprise. It also shows levels at which issue resolution efforts may occur: product, product integration, process and process integration, and organizational networks and scope.

This handbook covers much ground. Numerous academic disciplines and domains of application are discussed and a wide range of concepts, principles, methods, and tools are presented. At this level of detail, systems engineering and management may seem overwhelming. At another level, however, the nature of this area is quite straightforward. Put simply, as shown in Figure 25, systems engineering and management are concerned with people, organizations, and technology as they interact to fulfill purposes and to add value.

1 Systems Engineering Life Cycles: Life Cycles for Research, Development, Test, and Evaluation; Acquisition; and Planning and Marketing

F. G. PATTERSON, JR.

1.1 INTRODUCTION

In this chapter we discuss a number of process models, which we also refer to as *life cycles*. In so doing we discuss what a life cycle is with respect to our concepts of systems engineering, systems thinking, and the systems approach. As engineers, we ask "What is a life cycle good for?" and "How does it work?" As students we are constantly looking for fundamental principles and models with which to structure an understanding of our subject (Bruner, 1960). Thus, a study of life-cycle models is an appropriate way to begin an investigation of systems engineering.

Our understanding of life cycles is based upon our understanding of systems engineering or the systems engineering approach. Johnson et al. define a system as "an organized or complex whole; an assemblage or combination of things or parts forming a complex or unitary whole . . . more precisely, an array of components designed to accomplish a particular objective according to plan" (1967). An unorganized collection of parts has no purpose. As pointed out as early as in the writings of Aristotle (Ogle, 1941), it is exactly the identification of purpose with the organization of components that defines the approach that we refer to as systems engineering.

Ackoff uses the term "systems thinking" to describe a means for understanding an entity in terms of its purpose. He formulates the concept of systems thinking as three steps (Ackoff, 1981):

1. Identify a containing whole (system), of which the thing to be explained is a part.
2. Explain the behavior or properties of the containing whole.
3. Explain the behavior or properties of the thing to be explained in terms of its *role(s)* or *function(s)* within its containing whole.

Strategic planning is purposeful, goal-oriented, goal-setting. Systems engineering is the analysis and synthesis of parts, with an eye toward the goal set by strategic planning. Focusing on the problem is *analysis,* or taking the problem apart to identify and understand the pieces with which to do synthesis.

Handbook of Systems Engineering and Management, Edited by A. P. Sage and W. B. Rouse
ISBN 0471-15405-9 ©1999 John Wiley and Sons, Inc.

When we apply Ackoff's formulation of the systems approach to explain systems engineering life cycles, we first identify the containing whole, and we find that life cycles are attributes of processes, processes are elements of enterprises, and enterprises are actions of organizations. A life cycle is an application of the systems approach for the purpose of understanding and implementing processes. Every process has a life cycle. A *process* is a collection of one or more actions whose purpose is to accomplish a goal. In a systems engineering organization, this goal contributes to the fulfillment of a strategic plan. If we define a process very generally as a strict partial ordering in time of the strategically important activities of the enterprise, then we can regard a set of processes that includes all such activities of the enterprise in some process as a partitioning of the enterprise. To be interesting a process should have at least one input and at least one output, and each activity in a process containing more than one activity should either depend upon or be depended upon by some other activity in the same process.

A process is an element of an enterprise, and an enterprise is itself a process. Every enterprise (process) may be approached as a system (Thome, 1993) and has a life cycle. Using a basic enterprise model due to Sage (1995), a process will be one of three basic types: system acquisition or production; research, development, test, and evaluation (RDT&E); or planning and marketing.

For a number of reasons, many of which relate to the systems engineering organization, it is useful to regard a process as an ordered collection of phases that, when taken together, produce a desired result. We refer to this ordered collection of phases as a *process life cycle*. Moreover, it is possible and desirable to identify life cycles for each of the three basic classes of processes. Thus, we may speak of a system production or system acquisition life cycle, an RDT&E life cycle, and a planning and marketing life cycle.

An enterprise is a collection of one or more processes undertaken by an organization to achieve a goal. An organization may have one or more enterprises. Every enterprise also has an organization, whose purpose is closely tied to the enterprise. Indeed, from the standpoint of purpose, the enterprise and the organization are identical, since each is an element of the other. Thus, to understand the purpose of the systems engineering life cycle, it is reasonable to consider it part of an organization and to examine it in that role.

Life cycles are both descriptive and prescriptive. In our study of systems engineering life cycles, it is important to abstract and examine life-cycle models not only for their simplicity for describing processes, but also for their *prescriptive* applicability to the organization of an enterprise. Bass and Stogdill (1990) attribute to Deming the idea that managers need to know whether the organizational system in which they are involved is stable and predictable. A well-defined systems engineering life cycle model that is imposed on the system from the beginning can give the maturity derived from lessons learned in using the time-tested model, thus adding an important element of stability to the organizational structure. Bass and Stogdill (1990) also cite the need for strong leadership in an organization's early, formative period.[1] This can be related to the idea of the relatively

---

[1]Bass and Stogdill write, "Hersey and Blanchard's life cycle theory of leadership synthesizes Blake and Mouton's managerial grid, Reddin's 3-D effectiveness typology, and Argyris's maturity-immaturity theory. According to this theory, the leader's behavior is related to the maturity of the subordinates. As the subordinates mature, the leader's behavior should be characterized by a decreasing emphasis on task structuring and an increasing emphasis on consideration. As the subordinates continue to mature, there should be an eventual decrease in consideration. Maturity is defined in terms of subordinates' experience, motivation to achieve, and willingness and ability to accept responsibility" (Bass and Stogdill, 1990).

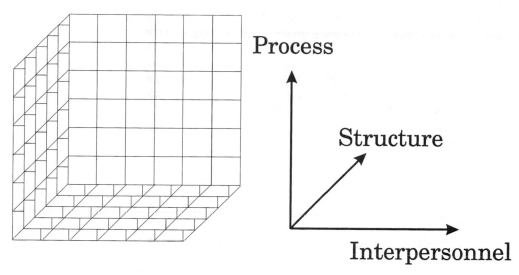

Figure 1.1 An organizational model by Stogdill.

strong need for a predefined systems engineering life cycle in early, immature stages of the organization, where "strong need" equates to strong leadership requirements. Leadership and structure can help an organization to avoid many problems, by eliminating several sources of conflict. For example, Bradford et al. (1980) enumerate three of the most common group problems, each of which may be reduced by strong leadership and well-defined structure:

1. Conflict or fight.
2. Apathy and nonparticipation.
3. Inadequate decision-making.

Burns (1978) associates both power and leadership with purpose:

> Leadership over human beings is exercised when persons with certain motives and purposes mobilize, in competition or conflict with others, institutional, political, psychological, and other resources so as to arouse, engage, and satisfy the motives of followers. This is done in order to realize goals mutually held by *both* leaders and followers. . . .

Stogdill (1966), describing the classic view of the relationship of purpose to organizational structure, writes, "One of the most stable and enduring characteristics of organization, purpose serves as a criterion or anchorage in terms of which a structure of positions is differentiated and a program of operations is designed." Stogdill models the organization as an interbehavioral system in three dimensions:[2] interpersonnel, structure, and operations, shown in Figure 1.1.

---

[2]This is an example of *morphological analysis,* which, according to Hall (1977b), refers to the decomposition of a problem into its orthogonal basic variables, each becoming a dimension of a morphological box.

The *interpersonnel* dimension consists of five variables: interaction, intercommunication, interexpectation, interpersonal affect, and intersocial comparisons.

1. *Interaction* denotes the tendency of initially unstructured groups to progress through systematic stages in the development of positions and roles. This differentiation of structure results from the interaction process itself.

2. *Intercommunication* is a critical factor in formally structured organizations. Stogdill identifies formal structure with formally defined status levels that inhibit the upward flow of certain types of information and the downward flow of other types. Flow of information tends to follow lines of authority vertically. Horizontal flow of information, especially between and among departments, tends to diminish. As a result there are three classes of common problems related to the communications process:

    (a) Planning and policy changes reach subordinates before the information needed to react properly to the changes.

    (b) Problems are not visible to supervisors unless operations are negatively affected.

    (c) Loss of horizontal exchange of information inhibits cooperative efforts.

    Stogdill asserts that continuous attention to the communication process is necessary to prevent problems of these three types. The difficulty is complicated by the reality that the informal relationships among members in an organization are rarely the same as the relationships described by the formal organization chart.

3. *Interexpectation* refers to the tendency of a group, whose structure of positions and roles has been determined through interaction, to develop group norms of behavior for members of the group. When group norms have been established, they define limits of acceptable behavior to which members are expected by other members to conform.

4. *Interpersonal affect* comes from Stogdill, who cites research that indicates that a leadership position within a group is most likely to be bestowed upon the group member who exhibits spontaneity and enables spontaneity and freedom of action in others, and who expresses positive rather than negative attitudes about the group task and group members.

5. *Intersocial comparisons* refers to the tendency of individuals to join or to form groups with values and norms that are similar to their own. According to Stogdill (1986).

> Although individual value systems and reference group identifications are highly subjective in nature, it appears that they are easily perceived by individuals seeking membership and by groups considering individuals for membership. . . . Intersocial comparisons are important factors in the formation and composition of organizations. They can provide the basis for a harmonious or for a disruptive organization. The right of an organization to determine the composition of its membership may be a critical factor in its capacity for survival.

The *structure* dimension consists of six variables: purpose and policies; task structures; positions and roles; communication nets; work group structures and norms; and other subgroup structures and norms.

1. *Purpose and Policies.* The existence of purpose and policies may be inferred from the activities of the group and serve to bring about a desired class of outcomes. We take it as axiomatic that strategic purpose is reflected in organizational structure, designed to implement business processes in such a way as to provide unity of action through community of purpose.

2. *Task Structures.* A variety of generally highly structured tasks must be carried out to realize the purpose of an organization. These tasks are often denoted as functions of organization. There are direct functions and support functions. Responsibility, authority, and status often derive from function.

3. *Positions and Roles.* Positions and roles can be clearly distinguished from each other in an organization. Organizations are composed of structures of positions and structures of roles. As Williamson (1970) observes, "One of the problems of organizational design, therefore, is to provide a structure that elicits the desired role characteristics." Stogdill regards responsibility and authority as dimensions of role performance rather than as attributes of position.[3]

4. *Communication Nets.* Stogdill observed that communication networks with centralized access enabled an organization to carry out its operations better than networks that provided all members with equal access to information. In any organization where complex operations must be coordinated, centralized information access is needed.

5. *Work Group Structures and Norms.* Just as organizations develop structures of positions and roles, work groups develop structures of positions and roles. Stogdill points out that work group members performing the same functions tend to evolve different status rank within the work group. Whether or not members perform the same functions, they tend to assume different roles. Also, each group may develop its own set of norms that may differ from those of the larger organization.

6. *Other Group Structures and Norms.* In addition to work groups, members usually organize various informal groups, each with its own structure and norms. While such groups do not have a formal relationship to the formal organizational structure, each one influences and is influenced by other individuals and groups and, using a systems approach, should be treated as part of the structure of the organization.

The *operations* dimension is the third dimension of Stogdill's model, the one concerned with processes. The operations dimension includes business processes that both describe and determine the enterprise activities of the organization. In introducing this third axis, Stogdill notes that the operations side of the model will determine to a large degree the structural and interpersonnel aspects of his model. Thus, operations may lead, rather than follow, the other dimensions. We may introduce the systems approach that views a process as a synthesis of organizational elements to achieve a purpose. By introducing processes, as represented by their life-cycle models, we can purposefully influence the formation of the organization in all of its dimensions.

Stogdill points out that classic organizational theory is concerned mostly with the subdivision of work and with the differentiation of responsibility and authority. It constitutes an empirical and logical analysis of these aspects of organizations that has been proven to

---

[3]Stogdill differentiates the concept of *role* from the concept of *position.* The alternative, the concept of unifying position with role, is central to a form of organizational maintenance known as *deengineering* (Wheatley, 1992).

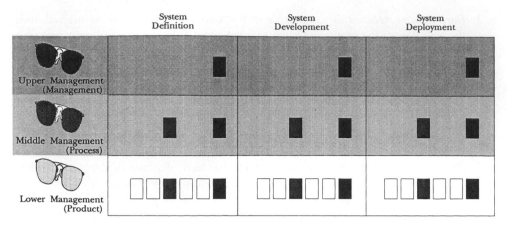

Figure 1.2   Management views of life-cycle products by life-cycle phases.

be highly effective as a template or set of principles for structuring new organizations. With adequate domain knowledge, including knowledge of the nature, technology, size, and complexity of the tasks to be achieved, it is possible to create positions and to specify their structure, function, and status interrelationships; design communication channels; and chart the flow of operations (processes) necessary to perform the tasks. Training of existing personnel or hiring of new personnel to fill the positions is accomplished as a next step.

Life cycles are tools of management that allow the organization of enterprise activities around a purpose. When an organization undertakes an enterprise, members need to structure themselves according to their goals. Organizations organize themselves around processes according to some partition of the enterprise.[4] Their strategic plan is much enhanced by introducing a life-cycle model with proven success for the domain. An ad hoc organization will occur—this is the message of Stogdill—if not templated, especially if the organization is involved in an enterprise with which it is unfamiliar. Introducing the template saves much time and helps to assure success. Galbraith and Nathanson (1978) carry on the tradition of Chandler with their claim that "Effective financial performance is obtained by the achievement of congruence between strategy, structure, processes, rewards, and people." Thus, in terms of Stogdill's model, the development of the operations side of the model can be enhanced by injecting a life cycle (or set of process life cycles) with proven effect in every aspect of the organization.

Now that we have identified the systems engineering organization as the whole that embodies corporate purpose and have identified many of the dynamic forces within the organization, we can continue our application of systems thinking by focusing on overall behavior. Life cycles model organizational behavior. Behavior is characterized by products, which may be organized into phases for manageability. Each phase is usually characterized by one or more major products emerging from the organization during that phase. In Figure 1.2 three classic views of the system life cycle are depicted. At the lowest level

---

[4]This is a basic assumption of business process reengineering, a procedure that involves examining, combining, replacing, or adding enterprise activities as necessary to repartition the organization into processes in a way that is more profitable. Reorganization during reengineering follows repartitioning of enterprise tasks.

of management, each phase of the life cycle terminates with completion of one or more major product, shown as blackened rectangles. Many intermediate products may be identified that are important to product-level management, not because they are marketable products, but because they represent components of the finished product, or checkpoints, that are associated with progress and productivity, sometimes referred to as earned value. The intermediate products are also shown as visible to the lowest level of management. This level of the organization is responsible for ensuring the quality of the product through product inspection, measurement of partial products or at checkpointed intervals, and other means that require all product details to be visible.

Middle management typically has much less responsibility, visibility, and cognizance relative to intermediate products than product-level management. In general, the focus of middle management is on process rather than product. This is depicted in Figure 1.2 by reducing the number of intermediate products shown to middle management to those deemed to be important for assessing process quality, and through this product quality, shown as rectangles shaded in gray. Thus, middle management is concerned with integrating product-level resources into a high-quality process. At the upper management level, the concern is with integrating process to achieve an organizational goal, a strategic purpose. Thus, upper management requires even less visibility into intermediate products, perhaps none at all. Instead, the focus is on the coordination and integration of production and acquisition, RDT&E, and planning and marketing life cycles.

1.2   CLASSIFICATION OF ORGANIZATIONAL PROCESSES

Drucker (1974) references several traditional methods of classifying and grouping activities in organizations. From a systems engineering point of view, most are analytic, in that they tend to direct attention inward to the skills required to accomplish tasks by grouping like skills together. An outward-looking, synthetic approach, that Drucker clearly finds to be more useful, results from grouping activities by their type of contribution to the organization. He defines four major groups:

1. Result-producing activities
2. Support activities
3. Hygiene and housekeeping activities
4. Top-management activity

Of Drucker's four categories, only *result-producing activities* have nontrivial life cycles. The other three types are on-going, organic elements of the organization, whose activities are necessary, but whose efficacy cannot be measured by their contribution to corporate revenue.

Drucker identifies three subclassifications of result-producing activities:

(a) Those that directly generate revenue
(b) Those that do not directly generate revenue but are still fundamentally related to the results of the enterprise
(c) Information activities that feed the corporate appetite for innovation

Each one of Drucker's categories corresponds to a basic systems engineering process type identified by Sage (1992a), each with a life cycle that we will examine in detail. Respectively, we will discuss

(a) The Planning and Marketing life cycle
(b) The Acquisition or Production life cycle
(c) The RDT&E life cycle

With respect to each life cycle type, it is necessary to recognize, analyze, and synthesize a response to the market, which may be external or internal to the organization.

For example, one high-level representation due to Sage (1992a) of a generic systems engineering life cycle is a three-phase model: definition, development, and deployment (DDD). Each phase can be further described using recognize, analyze, and synthesize (RAS) to categorize activities contained in the phase. Activities and products can be organized in a matrix representation as suggested by Figure 1.3.

The DDD model is itself a prime example of RAS. System definition is essentially a recognition (R) activity during which requirements are recognized; development is an analysis (A) activity in which requirements, and alternatives for their realization, are considered, analyzed, and developed; and deployment is a synthesis (S) step, in that the results of the development phase are purposefully integrated into an operational system. RAS is *recursive* in that each phase can be considered to be a stand-alone process, with inputs and outputs. Each phase may then be further analyzed using RAS. For example, the definition phase of a generic acquisition process recognizes stakeholder and environmental needs and other constraints using such tools as requirements elicitation and classification. The analysis subphase analyzes and, as necessary, prototypes requirements to develop a requirements list. The synthesis subphase transforms the finished product of the analysis subphase into a specification. We can recursively apply RAS to "requirements elicitation and classification," since it is a process with an input that needs to be recognized and an output that must be synthesized based upon an analysis step. The DDD model can also be referred to as *formulate, analyze,* and *interpret* (FAI) (Sage, 1992a). DDD, FAI, and RAS are congruent to one another.

The three basic systems engineering life cycles of Sage can be seen as three filters through which a product concept must pass to be deployed successfully by an organization (Fig. 1.4). Each successive filter subsets the universe of possibilities based upon both internal and external factors.

In the RDT&E cycle, the feasibility of ideas is measured by the standards of existing technology, constrained only by the goals of the organization. These goals are normally more liberating than limiting, since the organization that is focused on a goal will in general

	Recognize	Analyze	Synthesize
Definition			
Development			
Deployment			

Figure 1.3 Abstract matrix representation of a generic process.

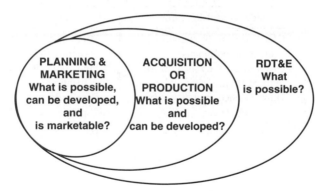

Figure 1.4   Narrowing the universe of possible products.

be better endowed, in terms of its technology and its organization, and thus better positioned than competitors to create successful products. From the standpoint of the acquisition or production life cycle, the view is inward toward the capabilities of the organization. Finally, from the viewpoint of the planning and marketing cycle, the direction of focus is outward to the external market. We may note that these three successive filters have, in terms of their direction and focus, the same characteristics as the three parts of our RAS framework: "recognize" looks at possibilities within a goal orientation; "analyze" looks inward and considers what resources are available; and "synthesize" looks outward, answering patterns of need from the market in terms of feasible alternatives based on organizational capabilities.

We now look at three classes of systems engineering life cycles: the RDT&E life cycle; the acquisition, or production, life cycle; and the planning and marketing life cycle.

1.3   RESEARCH, DEVELOPMENT, TEST, AND EVALUATION LIFE CYCLES

We represent RDT&E as a separate life cycle from System Acquisition and Planning and Marketing, since the staff, activities, goals, and constraints of research projects are sufficiently different to require special treatment. Research by its nature is proactive. Whether it is independent or directed, research is based upon an innovative idea, the worth of which is to be tested by the RDT&E organization and evaluated by the Planning and Marketing organization. The innovation may be the result of independent, or basic, research, whose work may result in either a product or a process innovation. Alternately in the case of directed research, often referred to as *commercial development,* the innovation can be found in terms of a poorly understood requirement for a system under development. In this case, prototype development may help articulate the requirement, which can be returned to the Acquisition organization for development or to the RDT&E organization for further directed research. Poorly understood requirements may also originate in the marketing organization, who, having perceived a market need, subsequently wish to explore possibilities with the research organization.

RDT&E is often viewed as unnecessary overhead by poor management who, when faced with market pressures, reduce RDT&E expenditures in an effort to reduce cost

	Independent Research	Directed Research
Management	Obsolescence by new management methods Obsolescence by new workforce norms Obsolescence by new process requirements	Poor understanding of market requirements Poor understanding of organization Poor understanding of business processes
Process	Obsolescence by new methodologies Obsolescence by new tools Obsolescence by new product requirements	Poor understanding of models Poor understanding of new methods Poor understanding of new tools
Product	Obsolescence by new technology Obsolescence by changing demand Obsolescence by new market requirements	Poor understanding of requirements Poor understanding of design Poor understanding of environment

Figure 1.5 A classification of problems addressed by RDT&E.

through eliminating a probable loss. A well-managed RDT&E program is better regarded as a tool for risk mitigation, the use of which probably results in profit (Cooper, 1992). Risk can be identified for each fundamental element of a system: management, business processes, and products. Research is fundamentally either independent or directed. Within these categories, we can systematically examine areas of risk and corresponding techniques of addressing risk through RDT&E. Figure 1.5 depicts some of the relationships.

RDT&E provides a framework within which to manage research and development. The concept of an RDT&E life cycle can be defined abstractly. For example, we can postulate three major phases: definition, development, and deployment. These phases exist apart from the acquisition life cycle, however, and create inputs to it: proactively in the case of independent research, and interactively in the case of directed research.[5] As suggested by Figure 1.5, independent research and directed research mitigate different classes of risk. While directed research solves short-term problems and yields answers on demand, independent research deals with the long-term problems caused, ironically, by success, and attendant complacence, inertia, and lack of innovation in a competitive market. From the standpoint of life cycles, these two types of research are different, in that independent research can properly be said to have a life cycle of its own with definition, development, and deployment phases; but directed research "borrows" the life cycle of the process that it serves.

Pearce and Robinson (1985) describe four basic decision areas for Research and Development (R&D):

[5]For completeness we can speak of reactive research, such as failure analysis.

1. Basic research vs. commercial development.
   - To what extent should innovation and breakthrough research be emphasized? In relation to the emphasis on product development , refinement, and modification?
   - What new projects are necessary to support growth?
2. Time horizon.
   - Is the emphasis short term or long term?
   - Which orientation best supports the business strategy? Marketing and production strategy?
3. Organizational fit.
   - Should R&D be done in-house or contracted out?
   - Should it be centralized or decentralized?
   - What should be the relationship between the R&D unit(s) and product managers? Marketing managers? Production managers?
4. Basic R&D posture.
   - Should the firm maintain an offensive posture, seeking to lead innovation and development in the industry?
   - Should the firm adapt a defensive posture, responding quickly to competitors' developments?

In terms of Pearce and Robinson's decision areas, both basic (independent) research and commercial development (directed research) are systems engineering processes. Each is a systems process in that it employs systems thinking to synthesize the goals of the organization and the requirements of the marketplace. Basic research is an engineering process in that it solves the general problem of generating a corporate response to the market at the levels of management, process, and product. The inputs to its definition phase may be provided by strategic planning. These inputs may be general, necessitating the search of a broad solution space, or very specific. Often the outputs from the definition phase are poorly understood and require further definition. While it is true that the refinement of requirements is an active task throughout the all-systems engineering life cycles, the RDT&E life cycle is especially tolerant of imprecision and lack of clarity in early stages of research.

Figure 1.6 depicts an RDT&E process model, organized in terms of the familiar three-phase systems engineering life cycle. In the definition phase, Melcher and Kerzner (1988) differentiate basic research as either well defined or non-well-defined. An example of well-defined research is *defensive* research, undertaken to protect the organization's market position from market competition. Non-well-defined research is more likely to result in product diversification.

During the development phase, in which an actual product is designed and built, the constraining influences of organizational goals may be felt and perhaps measured in terms of the capacity and necessity for corporate change that the potential new product represents. In addition to new insights about the product, the prospect of realigning business processes to accommodate new production may provide both positive and negative insights needed by strategic planners to guide future research efforts. Product-level insights cause iteration on the requirements phase until an acceptable product is defined and built. In a very real sense, the development phase might be regarded as the prototyping element of the require-

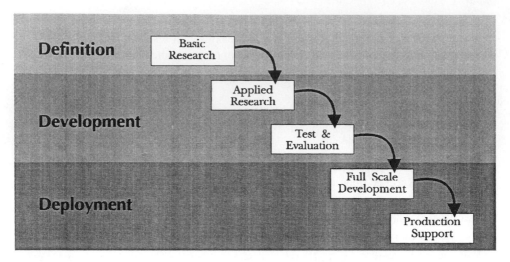

Figure 1.6 A life-cycle model for RDT&E.

ments phase. At the end of the development phase, requirements will be much more stable than they were at the end of the definition phase, but still not immutable. The purpose of the development phase is to provide support for a decision on whether the potential product is feasible and desirable.

Test and evaluation of the product provide the content of the deployment phase of the RDT&E life cycle. The goal of this phase is to deploy a useful model of a potential product for the consideration of management. The model provides much more information than would requirements alone about the impact potential of the product upon the organization in terms of start-up costs, perturbation of existing functions, and applicability of existing assets. Testing in this phase may take many forms, including product testing; product validation; and review by management, marketing, and process elements of the organization. This "internal marketing" is essential to assess the level of organizational buy-in, which is an indirect measure of consistency with strategic goals.

The output of the RDT&E life cycle may be viewed as input to the acquisition life cycle. Successful candidates from the RDT&E life cycle are moved to the requirements phase of the acquisition life cycle. While it is possible that, because a candidate has been chosen for development, requirements are complete, it may not be assumed. It is more often the case that "life-cycle issues," such as maintainability, safety, and reliability, have not been considered adequately and must be added in the definition phase of the acquisition life cycle. RDT&E provides a proof of concept that is generally not of production quality.

The Planning and Marketing cycle influences and thus provides input to the RDT&E process. Marketing may evaluate an existing product or line of products and services according to their market share and their potential for growth. Figure 1.7 catalogs classic market share and market growth rate concepts (Sage, 1995).

An organization should understand its position in the market to properly coordinate its RDT&E efforts. According to the dynamics of the market, management, through the planning and marketing process, may pursue any of three mathematically possible strategies: (1) increase market share, (2) maintain market share, or (3) reduce market share. We examine each in turn.

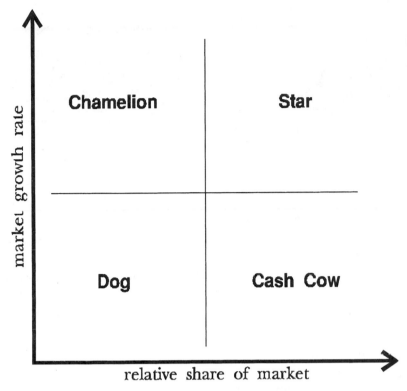

Figure 1.7 Market share vs. market growth rate.

1. Pursuing an increase in market share is a cost-effective strategy in the case of a selected "star" or a carefully selected "chameleon" (Fig. 1.7). The RDT&E process will not be seeking a new product, but rather a way to save costs (especially in the case of a star) or improve quality (especially in the case of a chameleon). Solutions may be found in improvements to either process or product. Cash cows and dogs characteristically have low potential for growth. However, a dramatic product improvement or a new product based upon either may be provided by RDT&E.

2. Maintaining market share is an appropriate strategy in the case of a selected star or a cash cow. In view of the relatively low potential for market share, a relatively defensive RDT&E strategy may be necessary. Pressure from competing products, such as product innovations, pricing pressure, or product availability, may be countered through product innovations from RDT&E that match or exceed competitive innovation or through process improvements that lower cost and allow defensive pricing without significant loss of profitability. RDT&E may also innovate through new applications of the product.

3. Reducing market share is appropriate for dogs, where profitability is modest or negative. A decision to remain in the market could be based upon an expected breakthrough from RDT&E, with respect either to product or process, that would improve quality or lower the cost of production.

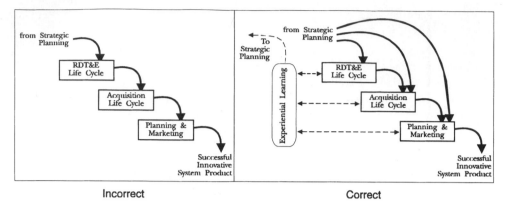

Figure 1.8   The wrong and right ways to picture life cycle interrelationships (Sage, 1995).

The output of the RDT&E life cycle may also be viewed as input to the Planning and Marketing life cycle. Successful candidates from the RDT&E life cycle provide a principal input for the marketing function, in which a marketing plan is developed based both upon market research and upon the characteristics of the product prototype. Unsuccessful candidates from the RDT&E life cycle are also valuable, although their value is not directly measurable. Virtually all RDT&E activities contribute valuable data to the corporate knowledge base. In recognition of this contribution, Sage (1995) depicts both a right and wrong view of the interrelationships among RDT&E, acquisition, and planning and marketing (Fig. 1.8).

1.4   SYSTEM ACQUISITION OR PRODUCTION LIFE CYCLES

A large number of life cycles have been proposed for systems production or acquisition. For example, Cleland and King (1983) give a five-step USAF Systems Development Life Cycle consisting of the following phases:

1. Conceptual phase
2. Validation
3. Full-scale development
4. Production
5. Deployment

King and Cleland (1983) also describe a slightly different five-phase life cycle consisting of the following:

1. The conceptual phase
2. The definition phase
3. The production or acquisition phase

4. The operational phase
5. The divestment phase

In addition, both the U.S. Department of Defense (DoD) and the National Aeronautics and Space Administration (NASA) (Shishko, 1995) have multiphased acquisition life cycles that have evolved over time and incorporated many expensive lessons. A comparison of the DoD and NASA acquisition life cycles to each other and to a generic six-phase commercial process model has been constructed by Grady (1993). There are various other names for the acquisition process model: production cycle, procurement process, project cycle,[6] and implementation process are often used interchangeably. In this discussion, we refer to all of these as *acquisition*. The basic shape of the acquisition life cycle follows the basic definition–development–deployment shape that is the basis of many action models. During the definition phase, we create and generally prefer to distinguish between its two basic products: requirements and specifications. The requirements for a product or service form a statement of need that is expressed by a person or group of persons who will be affected, either positively or negatively, by the product or service. Hall (1977a) refers to requirements as *the problem definition:*

> One may start with problem definition, the logical conclusion of which is the setting of definite objectives. Objectives function as the logical basis for synthesis. They also serve to indicate the types of analyses required of the alternate systems synthesized, and finally they provide the criteria for selecting the optimum system.

Requirements characteristically suffer from the ambiguity and imprecision usually attributed to natural language. Nevertheless, the requirements are the most accurate representation of actual need and should be treated as the standard by which the resulting product or service will be judged or measured. Specifications are engineering products, and mark the end, rather than the beginning, of the requirements life cycle, a subset of the acquisition life cycle. This is a valid life cycle.[7] It sets forth steps in a process of eliciting requirements and then transforming them into a system requirements specification. The transition points between any two successive phases are well-defined and represent a change in the activities of the requirements engineer and usually a change in the product as well. We may identify five phases in this life cycle, represented graphically in Figure 1.9:

1. Elicitation of requirements
2. Classification
3. Analysis
4. Prototyping
5. Requirements documentation and specification

---

[6]Kerzner (1992) notes that "there is no agreement among industries, or even companies within the same industry, about the life-cycle phases of a project. This is understandable because of the complex nature and diversity of projects."

[7]Although a life cycle for requirements is given here with a discrete beginning and end and steps in between, we must keep in mind that requirements issues are persistent and arise throughout the development and maintenance of the product.

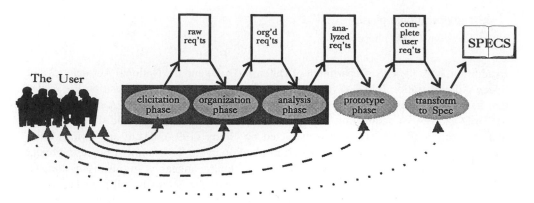

Figure 1.9 A requirements life cycle.

An initial requirements list can be generated using a team approach or, for smaller endeavors, by interviewing stakeholders and discussing their needs. In either case, the starting point of the requirements engineering process is an elicitation process that involves a number of people to ensure consideration of a broad scope of potential ideas and candidate problems (Hill and Warfield, 1977). During requirements elicitation the following questions must be addressed:

- Who are potential end users? What do they do?
- What is currently available? What is new? What is old?
- What are the boundary conditions? (This is determined by the customer and governs the range of elicitation.)
- What work does the developer need to do?

The second step is the organization of the elicited requirements. In this step there is no transformation of the requirements, but simple classification and categorization. For example, requirements may be grouped into functional vs. nonfunctional requirements.

The third step is analysis of the requirements. This represents a transformation: a *transformation* occurs if the requirements engineer makes any change that did not come from the stakeholder. For example, the adding of detail is a transformation. During this step, a number of checks are typically performed, such as to enable determination of the following:

Ambiguity	Performance
Completeness	Redundancy
Conflict or consistency	Reusability
Correctness	Testability
Orthogonality	Tradeoff analyses

It is worth noting that if any of these analyses are done at all, they are usually done manually. NASA (Radley and Wetherholt, 1996) has cataloged a number of significant manual requirements analysis methods that are related to software safety.

The fourth step is the development of a prototype, a synthesis step in which some part of the problem is developed to some level of completion. In this way poorly understood requirements may be tested and perhaps strengthened, corrected, or refined. This activity is often done as a proof of concept and serves to induce feedback from both the stakeholders and engineers.

The fifth step represents the requirements as the finished product of the stakeholder requirements team. The requirements are compiled into a requirements list or into some equivalent document format. These collected requirements are then transformed into a specification.

The transformation from a requirements list to a specification represents a significant risk in the requirements engineering process. In this operation the requirements that were elicited from the user and refined through analysis, synthesis, prototyping, and feedback are transformed into a requirements specification. It is a transition of form (according to some prescribed document standard), substance (through addition of details necessary to support successful implementation), and language (through the use of rigid requirements language and nonnatural language representations, such as state charts and various mathematical notations).

The system specification is the engineering definition of the product to be developed or acquired. In the acquisition phase the specification is partitioned, enlarged, and refined in accordance with the detailed life cycle that is used. Typically, the specification is decomposed in the way that best utilizes the resources of the development organization, unless specific guidance for decomposition is provided in the requirements definition. For systems that are primarily composed of hardware, the benefits in terms of cost and quality of allowing the development organization a high degree of freedom in interpreting the specification are passed along to the customer. Software systems differ in that the product architecture must support the evolutionary nature of software maintenance, the part of the software life cycle that predominates in terms of both duration and cost. One of the principal advantages of in-house software development, as opposed to acquisition of either custom or commercial-off-the-shelf (COTS) software, is that, because the development organization and the maintenance organization are the same, product cost and quality dividends are effectively realized twice, once during development and again during the maintenance phase, over a much longer time period.

There are several additional interesting points to be noticed about the requirements life cycle as depicted in Figure 1.9. First, we refer to the definition phase of the systems engineering life cycle as containing the requirements life cycle. That the requirements life cycle is a "proper life cycle" may be seen by noting that each of the steps is well-defined, each with a product that characterizes the step. Moreover, the steps may be grouped into the familiar definition–development–deployment paradigm. Requirements elicitation defines the requirements. Classification, analysis, and prototyping develop the requirements, which are then deployed as a requirements list and transformed into a requirements specification. None of this is to say, however, that activities that have been ascribed to one phase or step, rather than to another, may not be performed or revisited in a subsequent phase or step. Problem solving is the essential nature of engineering; the more we know about a problem, the more we may define, develop, and deploy. Thus the partitioning of activities into phases, and the resultant linear ordering of activities, is done to facilitate understanding, but certainly not to levy boundary conditions on phased process activities. Second, we may note that the more transformations that are performed on the elicited requirements, the less communication is possible with the stakeholders (represented as "users" in Fig. 1.9). This constraining influence represents risk, in that the stakeholders may be unaware until a failure of possible misapplication of requirements occurs. Figure 1.9 depicts lines of communication with the stakeholders using solid and broken lines, where solid lines represent relatively good communication, and increasingly broken lines represent increasingly poor communication. Third, we may note Figure 1.9 associates the first three steps, frames them together, and shows solid lines of communications to the stakeholders. This suggests that these activities may all be done together as a stakeholder team activity, thus further investing the stakeholders in the process of developing high-quality requirements. Because of the great potential benefits associated with keeping the stakeholder team involved, it is important to discipline the requirements engineer to use the language (terminology, examples, associations among requirements, etc.) of the stakeholder and to refrain from introducing engineering models, templates, paradigms, and tools during stakeholder involvement.

The definition phase of the system acquisition or production life cycle contains, in addition to requirements definition, three other possible elements, each of which creates a link to other processes in the enterprise:

1. Preparation of acquisition documents
2. Linkage with RDT&E life cycle
3. Linkage with Planning and Marketing life cycle

Acquisition documents create a link within the acquisition life cycle between the requirements definition process and the development of the product. Jackson (1995) describes specifications as the interface between the problem space (requirements) and the solution space (implementation). In the case of in-house development, the requirements specification is all that is required. In other cases, however, where two or more different organizations must cooperate, formal structure may include a request for proposal and a proposal containing a statement of work, a systems engineering management plan, a work breakdown structure, and a negotiated list of other deliverables.

The linkage with the RDT&E process may be of two basic types. If the research was directed by the requirements development process either to clarify requirements or to solve a technical problem, then the research results are needed to complete the requirements life cycle. Or, if the research product was the result of independent and innovative work of an on-going nature through which the organization creates new products, then the product definition must be handed off to the requirements team to allow the product to be brought to market.

Finally, requirements from Planning and Marketing may influence product definition, especially to incorporate plans for and the results of market research. Input from the Planning and Marketing process is critical to setting and maintaining cost goals. Michaels and Wood (1989) discuss "design to cost." The basic idea of "design to cost" is that, through the analysis of design alternatives (additional analysis that can add 20 percent to design costs), tradeoffs of functionality for affordability, and allocation of cost goals to the work breakdown structure, and through dedication and discipline on the part of management, cost goals can be built into a design-to-cost plan, and the plan can be carried out successfully. Marketing studies that provide projections of sales, selling price, and sales life cycle,[8] allow such cost goals to be set intelligently. Output from the Acquisition process to the Planning and Marketing process allows early product information to be used to create marketing plans, tools, and schedules.

The second phase of the system acquisition life cycle is the development phase. A great deal of work has been done in this area, and there are many life-cycle models. This phase begins with the system specification and ends with the delivery of the developed system. Basically, a divide-and-conquer approach is almost universally employed. In this, the system architecture represented in the system specification is divided and subdivided into components that are individually specified, designed, built, tested, and integrated to compose the desired system. We can say that systems are designed from the top down, then implemented from the bottom up. Ould (1990) represents this approach as the "V" process model for software engineering, but which clearly applies more generally to systems. It is sometimes argued that new software requirements may be introduced throughout development; hence, the need for a flexible process model. As noted earlier, however, every stage of the development process is itself a process that is amenable to an RAS approach that recognizes and satisfies needs repeatedly. The conclusion that the software engineering process model is a rediscovery and alternate description of the systems engineering process model seems inescapable.

Figure 1.10 depicts the V as a composition of three layers or views of the system in increasing engineering detail. The top view is the *user model,* which associates system requirements with their realization as a delivered system. The user model is the perspective of the customer or stakeholder, who is interested in submitting a list of requirements and receiving a finished product that meets the requirements. The second or middle layer of detail is the *architectural model,* which addresses the decomposition of the system-level specification into system design and subsystem specifications and designs; paired together with built and tested subsystems, and finally the tested system. The architectural model is the perspective of the systems engineer, who is interested in decomposing the whole into manageable parts, respecifying and designing the parts, and integrating the parts to compose the finished system. The third and lowest level is the *implementation model,* which couples component specifications and designs with fully tested components. The implementation model is the perspective of the contractor, who is interested in component-level specifications, designs, and products. The subtlety of these models lies in the role of the systems engineer, whose activities must encompass three fundamental activities: recognizing the product or component as a system, analyzing the system requirements, and synthesizing the system components. These three activities—recognize, analyze, synthesize—are the necessary parts of the systems engineering process, not only in development but also in definition and deployment. For example, the reengineerability of a software system can be shown to be directly related to the ability of the system to be recognized, analyzed, and synthesized; and metrics may be ascribed to each of these dimensions, thus providing indirect measures of reengineerability (Patterson, 1995a).

Ould (1990) also describes a variation of the V process model known as the VP, where P refers to prototyping. When requirements are not well understood for system elements, prototypes may be developed to provide insights and answers to problems and questions.

---

[8]The sales life cycle refers to a set of phases through which the product moves after it has been placed into the market. For example, a four-phase sales cycle contains (1) establishment, (2) growth, (3) maturation, and (4) declining sales phases (King and Cleland, 1983).

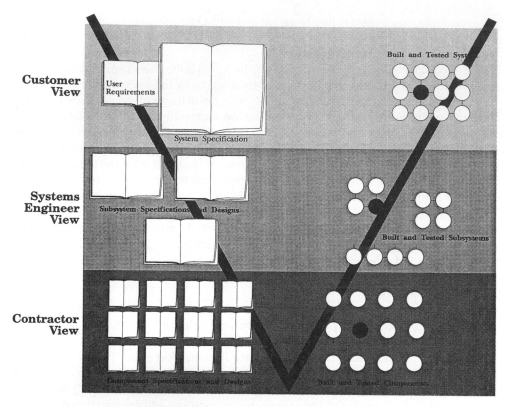

Figure 1.10 Acquisition process life cycle V.

The use of a prototype may be viewed as a form of directed research and development, just as in the definition phase.

Both verification and validation (V&V) can be visualized using the V process model (Whytock, 1993). Verification, a form of document tree analysis that ascertains that every requirement in a higher level specification is decomposed successfully into requirements at a lower level in the tree (and vice versa), follows the shape of the V. It is essentially a vertical concern, as depicted in Figure 1.11. According to ANSI/IEEE Standard 729, software verification is "the process of determining whether or not the products of a given phase of the software development cycle fulfill the requirements established during the previous phase" [American National Standards Institute/Institute of Electrical and Electronics Engineers (ANSI/IEEE), 1983; Boehm, 1990a]. The horizontal dimension asks a different question, namely whether the built and tested product at some level of completion successfully represents the specifications and designs *at the same level of the model:* this is validation. Again according to the ANSI/IEEE standard definition, software validation is "the process of evaluating software at the end of its software development process to ensure compliance with software requirements" (ANSI/IEEE, 1983; Boehm, 1990a). These software specific definitions may be, and often are, generalized to systems. As shown in Figure 1.11, considerations of risk often result in a departure in practice from these definitions. In both the case of verification and the case of validation, the cost of detecting and correcting problems later, rather than earlier, in the life cycle is nonlinear with the

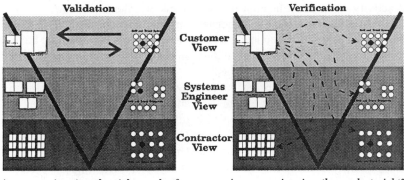

THEORY

PRACTICE

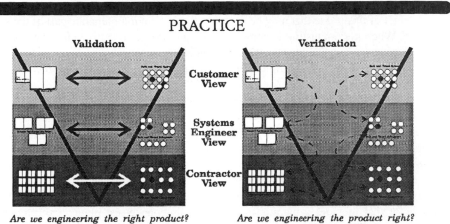

Figure 1.11   Theory and practice of verification and validation in the acquisition life cycle.

distance between the beginning of the project and the point of problem detection. As a result, the added overhead expense of reducing this distance by validating at each stage and by verifying at each engineering step is likely to result in cost savings.

Forsberg and Mooz (1991, 1995) have developed a more comprehensive V chart, including a repeated V for incremental strategies (Shishko, 1995). On the downside of the V, definition and development phase activities, including requirements specification and design, lead to the fabrication activity, located at the bottom of the V. Inspection, verification, integration, and validation activities are found on the upside.

The system development phase may be described as a complete process with its own life cycle. The development phase begins with the principal products of the definition phase, namely, the system specification, the systems engineering management plan, and various contractual vehicles. The system specification is the ultimate source of guidance for the development phase. A system architecture should be derived from the system specification, and it should be consistent with the system engineering management plan. Both the specification and the plan determine the architecture both of the product and of the system engineering organization. Thus, the derivation of architectural detail can be viewed as a management activity. The engineering activity that corresponds to this management activity is the identification of subsystems and the beginning of configuration control. To minimize effort and the associated expense of testing and integration, subsystems should be largely self-contained. Interfaces among subsystems should be few, small, and very well understood. This understanding should be captured in the subsystem specifications and subsystem test requirements.

The next level of detail comprises the specification and design of subsystem components, including hardware, software, firmware, and test procedures. Following design approval comes the fabrication and testing of hardware components, and coding and unit testing of software and firmware components. These activities may include the development of test equipment for hardware testing and scaffolding[9] for software unit testing.

Integration of components into subsystems is the next step, which involves not only integration testing, but also validation of the built and tested subsystems with respect to subsystem specifications. Next, the subsystems are integrated into a system that can be integration tested and validated against the system specification. Integration test procedures are a product of this phase that may be useful in the system maintenance part of the system life cycle. When system testing is complete, full attention can be given to other aspects of preparing to deliver the system, such as installation materials and procedures, training manuals and special training equipment and procedures, maintenance equipment and manuals, and user's guides.

The third part of the systems engineering life cycle is deployment. This part of the life cycle begins with the actual delivery and installation of the system and ends with the decision to decommission the system. As before, this phase can be viewed as a process with its own life cycle. The deployment life cycle begins with delivery and installation, acceptance testing, operational testing and evaluation, and acceptance by the customer. The second phase accommodates changes to the system. The software community refers to change as software maintenance. There may be a formal maintenance agreement with a contractor that includes a maintenance plan, maintenance test plans, and maintenance test procedures. The third and final phase of the deployment life cycle is the decision to decommission the system. This phase is often preceded by a tradeoff study to compare options for continuing to support the business process, such as hardware upgrade or replacement using existing, reengineered, or COTS products, or the acquisition of a new system. This version of the systems engineering life cycle is summarized in Figure 1.12.

The end of the deployment life cycle as depicted in Figure 1.12 is very much linked to the Planning and Marketing process. This part of the life cycle is concerned with how a deployed system is used to support a business process in some organization. In step 6, the concept of a tradeoff study to compare options for continuing to support the business process is a very general notion. In the case of a manufacturing process, market factors are the primary criteria for continuing, renovating, replacing, or retiring the system that supports the process. More subtly, in the case of nonmanufacturing processes, the performance of the organization is the primary consideration. In both cases, change in the business process drives change in the system that supports it, although technological improvements alone may result in changes, especially where improvements in economy, reliability, or safety are possible.

---

[9]Scaffolding refers to software written to simulate the essential aspects of the environment of the software that is to be tested. Normally scaffolding is not a part of delivered software. Incremental and evolutionary development, however, may deploy scaffolding as a part of an interim product. For example, in incremental software reengineering, scaffolding may be delivered as part of early increments to provide an interface between new software and a legacy system that is gradually being replaced.

Definition

1. Elicitation of requirements
2. Classification of requirements
3. Analysis of requirements
4. Prototyping of requirements
5. Requirements documentation and specification
6. Preparation of transition documents: request for proposal and a proposal containing a statement of work; a systems engineering management plan; a work breakdown structure; and a negotiated list of other deliverables

Development

1. Creation of a system architecture and an organizational structure consistent with the system engineering management plan
2. Establishment of configuration control
3. Identification of subsystems
4. Production of subsystem specifications and subsystem test requirements
5. Specification of subsystem components
6. Design of subsystem components
7. Fabrication and testing of hardware components
8. Coding and unit testing of software and firmware components
9. The development of environments and test equipment for hardware testing and scaffolding for software unit testing
10. Integration of components into subsystems
11. Integration testing of each subsystem
12. Validation of the built and tested subsystems
13. Integration of subsystems into a system
14. Integration testing of the system
15. System validation against the system specification
16. Integration test procedures
17. Provide product information to Planning and Marketing
18. Preparation of transition documents and other aspects of preparing to deliver the system, such as installation materials and procedures, training manuals and special training equipment, maintenance equipment and manuals, and user's guides

Deployment

1. Delivery and installation of the system
2. Acceptance testing of the system
3. Operational testing and evaluation of the system
4. System acceptance by the customer
5. Formal maintenance agreement with a contractor that includes a maintenance plan, maintenance test plans, and maintenance test procedures
6. Tradeoff study to compare options for continuing to support the business process
7. Decision to decommission the system
8. Provide current system description to requirements team (linkage to new cycle)

Figure 1.12 Steps in a systems acquisition life cycle.

System *Definition*

1. Perception of need
2. Requirements definition
3. Draft request for proposal (RFP)
4. Comments on the RFP
5. Final RFP and statement of work
6. Proposal development
7. Source selection

System Design and *Development*

8. Development of refined conceptual architectures
9. Partitioning of the system into subsystems
10. Subsystem-level specifications and test requirements development
11. Development of components
12. Integration of subsystems
13. Integration of the overall system
14. Development of user training and aiding supports

System *Deployment,* Operation and Maintenance

15. Operational implementation or fielding of the system
16. Final acceptance testing
17. Operational test and evaluation
18. Final system acceptance
19. Identification of system change requirements
20. Bid on system changes or prenegotiated maintenance support
21. System maintenance change development
22. Maintenance testing by support contractor.

Figure 1.13   The 22-step acquisition cycle in Sage (1992a).

Sage (1992a) describes a 22-phase acquisition life-cycle model. The steps can be organized into three general phases: definition, design and development, and operations and maintenance. The 22 steps are shown in Figure 1.13.

1.5   THE PLANNING AND MARKETING LIFE CYCLE

According to Kast and Rosenzweig (1970), *planning* is the systematic, continuous process of making decisions about the taking of risks, calculated on the basis of forecasts of future internal and external market conditions. Planning is modeled in several dimensions:

1. Repetitiveness, for which the authors visualize a continuum from single-event to continuous
2. Organizational level

3. Scope, ranging from a functionally oriented activity to a total organizational endeavor
4. Distance in time into the future

The model yields an eight-step approach to business planning. A similar list is given by Briner and Hastings (1994), who refer to the steps as preconditions for effective strategic management:

1. Appraising the future political, economic, competitive, or other environment.
2. Visualizing the desired role of the organization in this environment.
3. Perceiving needs and requirements of the clientele.
4. Determining changes in the needs and requirements of other interested groups — stockholders, employees, suppliers, and others.
5. Providing a system of communication and information flow whereby organizational members can participate in the planning process.
6. Developing broad goals and plans that will direct the efforts of the total organization.
7. Translating this broad planning into functional efforts on a more detailed basis — research, design and development, production, distribution, and service.
8. Developing more detailed planning and control of resource utilization within each of these functional areas — always related to the overall planning effort.

We can identify basic phases in the Planning and Marketing process and organize the eight steps into twelve activities within the familiar framework of the generic three-phase life-cycle model:

1. Definition
   - Assessment of the market
   - Perception of demand
   - The strategic response to the market in the light of organizational goals
   - The articulation of market demand into product requirements
2. Development
   - Providing a system of communication
   - Promoting the flow of information
   - Research
   - Design and development
   - Production
3. Deployment
   - Distribution
   - Service
   - Planning in greater detail

Effective planning requires careful external market analysis. This is a principal reason that we treat planning and marketing as part of a single life cycle. While an organization has a uniquely informed perspective on its own capabilities, goals, and interests, a free market does not respect individual organizations. Demand is an unbiased market force. Thus, an organization, especially its leadership,[10] must perform exceptionally to recognize, analyze, and synthesize a unique response to market demand. The first three steps of Kast and Rosenzweig's approach address assessment of the market, perception of demand, the strategic response of the organization to the market in the light of its own goals, and the articulation of market demand into product requirements. Together with other similarly inspired activities, such as product prototyping, user testing, and test marketing, we can refer to these steps collectively as the external market analysis phase.

There is also what can be called the internal market. This is the collection of interests within the organization and its support environment, such as the stockholders, employees (individuals, formal groups, and informal groups), suppliers and other creditors, governmental authorities, and other special interests. The responsiveness of the organization depends upon a collection of internal market requirements that must be satisfied to assure the success of the organization. By providing a system of communication, promoting the flow of information, and encouraging broad intraorganizational participation in goal development and planning, total quality management (TQM) (Deming, 1982; Sage, 1992b, 1995) connects internal and external requirements in a manner designed to promote growth in all areas of the internal marketplace, the goal of which is the creation of high quality products that satisfy requirements in the external marketplace. Steps 4 and 5 of Kast and Rosenzweig's approach address these issues.

After having decided upon the destination, the organization must plan its journey. Goals must be formulated into functional efforts. Referring to steps 6, 7, and 8, these efforts include research, design and development, production, distribution, and service. Again the internal organization must be consulted to develop more detailed planning and control of resource utilization within each functional area. Kast and Rosenzweig believe that providing organizational members with a voice in the planning process, resulting in an investment in the plans, serves to integrate the organization.

Planning and Marketing is tightly coupled with both RDT&E and with the Acquisition process. Blake and Mouton (1961, 1980) define a process for achieving an objective that may be vague initially, and that may change over time with refinement. Given this statement of a goal, planning may proceed, leading to an action step. The action step enables the fact-finding or feedback stage that is critical to a goals-oriented management program, because it provides "steering data." Three qualitative measures may be evaluated as well: (1) progress toward the goal; (2) adequacy of planning; and (3) quality or effectiveness of any given action step. These measures can be quantified using a 9-point scale system or some similar heuristic. Experience and feedback using a prototype in the context of the RDT&E process is extremely useful for evaluating both products and processes for eventual deployment. Mistakes corrected in the RDT&E cycle are much less costly than those detected in the marketplace. In this light, RDT&E can be seen as part of the overall definition phase for the organization's strategic planning process.

---

[10]The leader's role is essentially that of a politician. As Buckley noted, "The successful political leader is one who 'crystallizes' what the people desire, 'illuminates' the rightness of that desire, and coordinates its achievement" (1979).

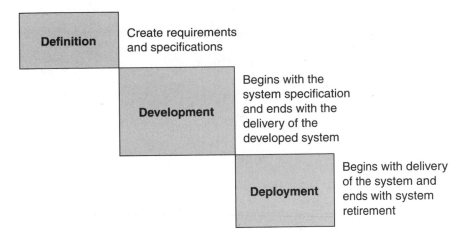

Figure 1.14   Three-level acquisition model.

1.6   SOFTWARE ACQUISITION LIFE CYCLE MODELS

It is our opinion that software engineering is a subdiscipline of systems engineering. Some authors refer to *software systems engineering* (Sage and Palmer, 1990). Like with many issues, however, there are two opposite points of view on this opinion. Undoubtedly software engineering, whose roots are in computer programming, has been slow to emerge as a discipline at all and for many years was regarded more as art than engineering. Even in recent years software development processes have seemed to diverge rather than to converge to a single methodology.[11] Different process models have different life cycles. In this section we present several important representatives.

Many system acquisition or production life cycles have been proposed and used. A three-level model, shown in Figure 1.14, defines the basic shape.

It is widely acknowledged that Royce (1970) presented the first published *waterfall* model, depicted in Figure 1.15. The principal idea is that each phase results in a product that spills over into the next phase. The model is pleasing for several important reasons. First, the stages follow the basic systems engineering life-cycle template. Requirements analysis and specification fit into the definition phase. Design and implementation make up the development phase, while test and maintenance constitute the deployment phase. Second, the stepwise nature of the waterfall suggests that only one phase is active at any one time, reducing the scope of possible interests, and thus simplifying the engineering process. Third, the logical ordering and the temporal ordering of the steps are identical. Fourth, and perhaps most importantly from an engineering management point of view, each step is represented as being largely self-contained, suggesting that different organizations may be designated for each phase without loss of continuity. Since the introduction of the waterfall model by Royce, many other software process models have adopted a similar format. For example, at the Software Engineering Laboratory at NASA, a seven-

---

[11]A *methodology* is defined as a collection of three sets: (1) tools, (2) activities, and (3) relations among the tools and activities (Warfield and Hill, 1972).

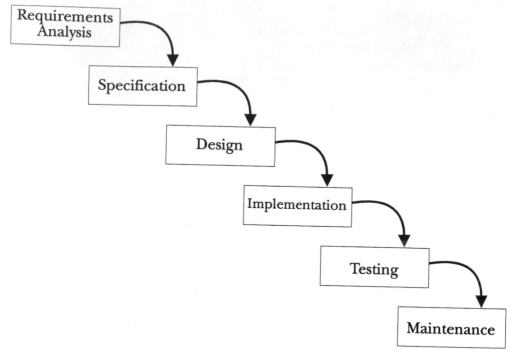

Figure 1.15 Royce's waterfall process model.

phase life cycle model is used as the basis for cost estimation (McGarry et al., 1984). More recently, an essentially identical life cycle was used as the basis of the *NASA Software Configuration Management Guidebook* [National Aeronautics and Space Administration (NASA), 1995]. The seven phases are:

1. Requirements analysis
2. Preliminary design
3. Detailed design
4. Implementation
5. System integration and testing
6. Acceptance testing
7. Maintenance and operation

The purpose of any model is to provide a framework that is better understood than the reality that it models, and whose outcomes are sufficiently close to reality to counter the easy, obvious, and devastating objection: "Does this reflect the way things really are?" Most variations of the waterfall model may be represented as attempts to overcome this objection.

When we apply the binary "reality measure" to the waterfall model of Royce, several incongruities may be noticed. One problem is the duration of the phases: although phases tend to begin in the logically predictable sequence, in reality they never end. The products from each phase, if perfect, would indeed bring an end of the phase.

Yet all products are fatally flawed and must be dealt with throughout the software development life cycle. To see that they are flawed, consider software requirements, which are subject to many metrics. To present our case, we need only consider one: *completeness.* Requirements are complete if they anticipate all needs of the user. This is an impossible goal, one that is addressed by various techniques that are essential for progress, but which constrain the engineering process, such as configuration management, and verification and validation. Therefore, requirements are never complete and are with us throughout the life cycle.

A second problem is the suggestion that only one phase is active at one time. Once again, we need only consider requirements and recognize that this phase is active throughout the development of the product. A third problem is the possible claim that separate organizations may have complete control of the project at different times in the development process without the loss of continuity. It is clear, however, that a requirements team must remain intact throughout the development process to avoid *loss of information,* subsequent retraining, and learning curve inefficiencies.

Boehm (1981) presents a waterfall model, shown in Figure 1.16, that addresses these problems. It is based on nine products, the creation of which Boehm characterizes as sequential, and which achieve necessary software engineering goals:

1. *Concept.* Concept definition, consideration of alternative concepts, and determination of feasibility.
2. *Requirements.* Development and validation of a complete specification of functional, interface, and performance requirements.

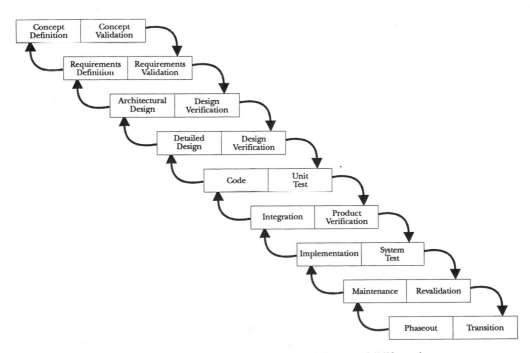

Figure 1.16   Boehm's nine-level version of the waterfall life cycle.

3. *Architectural Design.* Development and verification of a complete specification of the overall system architecture (hardware and software), including control and data structures.

4. *Detailed Design.* Development and verification of a complete specification of the control and data structures, interfaces, algorithms for each program component (defined as a program element of 100 lines of code or less).

5. *Code.* Development and verification of a complete set of program components.

6. *Integration.* Development of a working software system.

7. *Implementation.* Development and deployment of a working hardware and software system.

8. *Maintenance.* Development and deployment of a fully functional update of the hardware and software system.

9. *Phaseout.* The retirement of the system and the functional transition to its replacement.

In this version of the waterfall life-cycle model, as in Royce's version, Boehm includes a return path from each step in the model to the previous step to account for the necessity to change the product of a previous step. This rendition offers an orderly procedure for making changes as far back as necessary, both to meet the standards of verification and validation and to satisfy the underlying concept definition.

Boehm's waterfall life cycle is logically complete, insofar as that, at any given step, it successfully deals with deficiencies of previous steps. Moreover, it does so without violating the twin quality goals and constraints of configuration management and of validation and verification. The inadequacies remaining in the waterfall model are largely temporal and organizational.

A temporal problem is created, as noted in the Royce waterfall model, when a phase is deemed to have ended. The most striking example is the requirements phase. This phase does in fact logically precede the design phase. However, the requirements phase does not end when the design phase begins. It is of the nature of the waterfall model that each step begins with a set of requirements upon which action is required. This set of requirements is the result of actions in a previous step on a previous set of requirements. Likewise, when the current step is complete, the result will be requirements for the following step. Thus, at each successive step in the waterfall, requirements beget requirements (as well as changes to previous sets of requirements), the requirements for the whole governing the derivation of requirements for its parts. Every step is a requirements step, and not one of them ever ends.

One can picture the organizational consequences of living with this problem over the lifetime of a project. At least four scenarios can be generated, as shown in Figure 1.17, depending upon the size of the engineering problem and the size of the organization. For simplicity, we can characterize problems, as well as organizations, as large or small, examining each case in turn.

Case 1. This is the case in which a small organization has a small engineering problem to solve. Provided that the organization has both the engineering knowledge (skill) and the tools required, a small team can perform all life-cycle tasks with great efficiency and with a very high probability of success. There are many success stories about

	Small Engineering Problem	Large Engineering Problem
Small Organization	1. Small team may perform all steps in the life cycle	3. Small team may perform all steps in the life cycle; or other small teams may be employed
Large Organization	2. One or more small teams may perform the steps in the life cycle	4. One large team, many small teams, or a compromise organizational structure may be used to perform the steps in the life cycle

Figure 1.17   Organizational size vs. problem size.

significant systems that have been specified, designed, built, tested, and successfully marketed by entrepreneurs with meager, but adequate, resources. There are doubtlessly countless untold success stories about more typical efforts of small engineering organizations.

Case 2.   In this case, a large organization has a small engineering problem to solve. Management may decide to exploit the small problem size to advantage by using the approach of Case 1. Alternately, several small teams may be engaged, thus introducing additional learning curve costs, necessitating the continuous employment and communication among a larger number of engineers, and lowering the probability of successful completion.

Case 3.   When a small organization undertakes a large engineering task, management must make several decisions based upon available resources. If time is not a critical factor, a small team approach offers the high quality, the low risk, and the economy of Case 1. In effect the large problem has become manageable by subdivision into small problems without introducing the problems of using subcontracts. Unfortunately, time is almost always a critical factor, and management is forced into the role of general contractor, and assuming the high risk of using independent contractor teams, each with its own layer of management, its own development process, and its own strategic goals.

Case 4.   Assuming that the large organization has the required skills, methods, and tools, the high risk of Case 3 can be mitigated somewhat by performing all tasks without resort to external organizations. In this way, a single development process, and a single set of strategic goals, improves communication and cooperation among groups of engineers. However, large organizations are inherently plagued by problems of management complexity and communication complexity (Brooks, 1975). Moreover, because of the need to retain the organizational expertise from previous steps at each successive step in the life cycle, the total organization may be very large, and largely unoccupied, at any given time.

We can see that problem management is a logical issue, and that organizational management is a temporal issue. Partially in recognition of the organizational problem, Boehm (1981, 1990b) devised the spiral model, depicted in Figure 1.18.

We have looked at the structure and the function of the waterfall. An insight into the nature of the possible problems with the waterfall can be gained by reflecting on the purpose. Life-cycle models represent an attempt to manage complexity through division of resources. The ordering of processes and products into phases, the organizational struc-

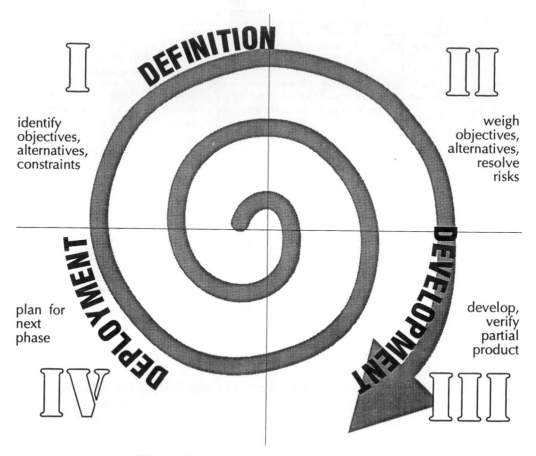

Figure 1.18 Boehm's spiral development model.

turing, are necessary to managing the complexity of the solution. The complexity of the problem, however, cannot be reduced by dealing exclusively with the resources assembled to solve the problem. The problem itself must also be reduced into a tractable form and size, where tractability varies according to whether the domain is well understood. The waterfall model is well able to be applied to manageable problems in familiar domains.

As previously noted, Boehm identified a repetitive aspect to the spiral model. Whytock (1993) describes the spiral as comparable to a repeated waterfall model. As shown in Figure 1.18, the spiral model attempts to abstract a set of activities that every layer of the spiral has in common. Boehm divides the model into four quadrants, shown in Figure 1.18. Quadrant I is the beginning of each level of the spiral and is concerned with identifying needs, possible ways of satisfying the needs, and factors that may constrain the solution. If Quadrant I may be viewed as the early part of the requirements development life cycle, Quadrant II represents the latter part, during which requirements are analyzed, problems are resolved, and tradeoffs are made. These two quadrants represent the parts of the waterfall model concerned with definition of the system to be built. In Quadrant III, a partial product is developed and verified, and this product is "deployed" in Quadrant IV, although such deployment simply represents a stepping stone for beginning a new definition effort in a new layer of the spiral model.

It should be noted that the spiral is a variation of the waterfall model that adds generality by including repetition as its basic feature. By unwinding the spiral we may recover the waterfall model, as well as solutions based upon partial implementation, such as incremental development and evolutionary development.

The waterfall process model can be seen to be most appropriate for the development of products in a familiar domain, where the risk of building poor products is reduced by an experience base that may include reusable specifications and designs. In an unfamiliar domain, or in large or complex projects, an incremental approach reduces risk, since the cost of each increment is relatively small. An increment may even be discarded and redeveloped without catastrophic cost consequences. The great strength of the spiral model is the capability to develop increments, or prototypes, with each full turn of the spiral. The prototype that is specified, planned, built, tested, and evaluated is now a working core version of the final system. Subsequent possible failures in later turns of the spiral will likely not impact the successfulness of previous increments.

Incremental development is a variation of the divide-and-conquer strategy in which software is built in increments of functional capability. In this approach, as defined by Boehm (1981), the first increment will be a basic working system. Each successive increment will add functionality to yield a more capable working system. The several advantages of this approach include ease of testing, usefulness of each increment, and availability during development of user experiences with previous increments. Boehm's modification of the waterfall to allow incremental development is to provide a number of successive copies of the section of the waterfall model, starting with detailed design; followed by code, integration and product verification, implementation and system test; and ending with deployment and revalidation. The spiral model provides a more succinct representation in which each successive increment is assigned to the next higher level of the spiral.

The evolutionary development model is an attempt to achieve incremental development of products whose requirements are not known in advance (U.S. Department of Defense, 1994; Rubey, 1993). Boehm discusses a process that can be used with iterative rapid prototyping, especially automatic program generation, and user feedback to develop a full-scale prototype (Boehm, 1981). This prototype may be refined and delivered as a production system, or it may serve as a de facto specification for new development. More generally, where evolutionary development is possible, it can be represented using a waterfall model in the fashion used by Boehm to represent incremental development. This is the approach taken in MIL-STD-498 (U.S. Department of Defense, 1994). However, representing evolutionary development entails repetition of the requirements and specification development activities; thus, substantially more of the waterfall must be repeated. Again, the spiral model is a more natural representation, since deployment for a given level of the spiral is directly linked to the definition portion of the next higher level, allowing necessary requirements growth in an orderly fashion.

From the standpoint of organizational management, the incremental software development model affords a more economical approach than the "grand design" strategy (U.S. Department of Defense, 1994) by allowing better utilization of fewer people over a comparable period of time (Boehm, 1981). Referring again to the Stogdill model in Figure 1.1, stability of processes is closely related to stability of structure in the organization. The changes in organizational size and structure that are suggested by the use of the grand design model can have a destabilizing influence on the organization. Davis et al. (1988a,b) suggest metrics for use as a possible basis for deciding among alternative software process models.

A similar use of the prototyping concept is *rapid development* (Reilly, 1993). The term *rapid development* refers to a sequence of prototype development efforts, each brought to operational status using a waterfall process, and deployed into actual operational environments. This is not to imply that the entire system is built rapidly, but that operational subsets are engineered in 10- to 18-month cycles. Each deployed system after the first is a replacement for its predecessor. This incremental process ensures that meaningful user feedback and operational testing will take place prior to the final release of the system. Moreover, as new requirements emerge, stimulated by operational experience with previous deliveries, they can be incorporated into subsequent iterations of the waterfall.

A standard released by the United States Department of Defense, MIL-STD-498, *Software Development and Documentation* (1994), was designed to accommodate three basic development strategies, discussed earlier: *grand design, incremental,* and *evolutionary development.* These strategies are also found in related standards DoDI 5000.2 (U.S. Department of Defense, 1991) and DoDI 8120.2 (U.S. Department of Defense, 1993). In this way MIL-STD-498 represents an improvement over its predecessors, such as DoD-STD-2167a (U.S. Department of Defense, 1988a), DoD-STD-7935A (U.S. Department of Defense, 1988b), and DoD-STD-1703 (U.S. Department of Defense, 1987). To achieve this flexibility, MIL-STD-498 abstracts from all three development strategies the concept of a "build," one or more of which is necessary to implement the mature software product. Each successive build incorporates new capabilities as specified and planned into an operational product.

A new joint standard, *Software Life Cycle Processes* (ISO/IEC 12207), was recently approved by the International Organization for Standardization and the International Electrotechnical Commission (1995). This standard contains 17 processes grouped into three sets:

    I. Primary Processes (5):
- (1) Acquisition
- (2) Supply
- (3) Development
- (4) Operation
- (5) Maintenance

   II. Support Processes (8):
- (6) Documentation
- (7) Configuration management
- (8) Quality assurance
- (9) Verification
- (10) Validation
- (11) Joint review
- (12) Audit
- (13) Problem resolution

III. Organizational Processes (4):
  (14) Management
  (15) Infrastructure
  (16) Improvement
  (17) Training

Each process contains activities, and each activity contains tasks. There are 74 activities and 224 tasks. Each process incorporates a TQM "plan–do–check–act" cycle. The processes, activities, and tasks are building blocks that must be assembled according to the project or organization. According to Sorensen (1996), MIL-STD-498 was intended to be replaced by a new standard, IEEE/EIA 12207, which is an adaptation of ISO/IEC 12207 that implements the intent of MIL-STD-498. This did occur in the summer of 1998.

1.7   TRENDS IN SYSTEMS ENGINEERING LIFE CYCLES

The essence of life-cycle management is division of resources (Buck, 1966). Kingsley Davis (1949) recognized division of labor as one of the fundamental characteristics of socialized humans. The division of tasks into subtasks, the documentation of progress through intermediate products, these are important concepts that did not begin in abstraction, but in concrete problems. The essence of engineering is problem solving; and divide-and-conquer strategies, process definition and improvement, process modeling, and life-cycle management are all problem-solving methods that begin by reducing complex challenges into tractable parts and conclude by integrating the results.

The failures of some systems engineering efforts have led some theoreticians to criticize or abandon the traditional ways of dividing resources. The trend is broad and can be seen in many areas. Life cycles are criticized for many reasons. For example, the concept that a requirements phase is self-contained, self-supportive, and separable from other phases has come under sharp attack both from academia and industry. The deeply ingrained tradition in industry that requirements must not dictate any given specific design is also coming into question. The justification for this belief may be readily derived from our basic activity model, shown in Figure 1.19.

Each of the three orthogonal activities is necessary for successful systems engineering. It is not clear, however, that separation in time, separation by assignment to more than one action team, or separation in terms of any other resource based on the orthogonality of a process model (that is to say, a life cycle) is effective in reducing engineering complexity issues. In fact, the opposite seems to be true. In the natural course of raising and resolving problems, it is much more efficient to solve problems when and where they arise. The inefficiencies associated with following each life-cycle step to completion before beginning the next step tend to result in a loss of information, and this tendency is exacerbated in proportion not only to the size of the *engineering* problem but also to the size of the engineering resources (Brooks, 1975). This is a major obstacle in engineering big systems.

A solution based upon partitioning big systems into a number of small systems reduces the inefficiency inherent in life-cycle-based approaches. However, this solution does not

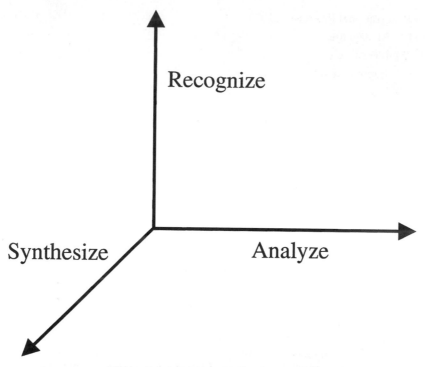

Figure 1.19 Engineering activity model.

answer the emerging generation of systems engineering theorists who maintain that life-cycle activities are dependent upon each other in a nontrivial way that neither a waterfall model, nor even a spiral model in its full generality, can adequately address. This suggests that specialties, such as *requirements engineering,* are undesirable, unless their scope of activity spans the entire life cycle: definition, development, and deployment. An alternate way to view this suggestion is that the number of specialties is effectively reduced to one: *systems engineering.* The systems engineer is evercognizant that a system must be specified, built, tested, and deployed in terms not only of its internal characteristics, but also in terms of its *environment.* Thus, the job of integrating becomes a global concern that subsumes all of the steps in the life cycle.

1.7.1 Process Improvement

The trend away from the use of prescribed standard life cycles, while widespread in the private sector, has only recently begun to affect government acquisitions. The U.S. Department of Defense no longer supports the use of MIL-STD-499a (1974), the successor of MIL-STD-499 (U.S. Department of Defense, 1969) for the acquisition of systems. [MIL-STD-499b (U.S. Department of Defense, 1992), in "draft" form since 1992, has never been approved.] Moreover, the new software development standard, MIL-STD-498, after years of development, has been canceled, a loss that will be unlamented because of the standard's virtual replacement by the U.S. Commercial Standard (Sorensen, 1996) as previously noted. Other military standards are receiving similar treatment. Developers have their own processes that are specific to their own organizations. Also, commercial standards are emerging. In terms of Stogdill's model of organizational dynamics (Fig. 1.1), it is no longer clear that process models should precede the development of structure and inter-personnel attributes, since, without a customer-mandated process model, a conceivably more efficient and efficacious process may evolve naturally. This change in emphasis should not be viewed as freedom from the use of a disciplined approach, but rather freedom to customize the approach to optimize quality attributes.

The wide acceptance of ISO-9000 as an international standard is actually a trend away from standardized process models. The basic philosophy of ISO-9000 has been summa-rized as "Say what you do; then do what you say" (Ince, 1994; Rabbitt and Bergh, 1994). ISO-9000 certification (International Benchmarking Clearinghouse, 1992) is a goal to which many companies aspire in order to gain competitive advantage. Certification dem-onstrates to potential customers the capability of a vendor to control the processes that determine the acceptability of the product or service being marketed.

The Software Engineering Institute has developed a model, known as the Capability Maturity Model (CMM) (Humphrey, 1989; Paulish and Carleton, 1994; Software Engi-neering Institute, 1991), upon which methods and tools for process assessment and im-provement in the software development arena have been based. As the name suggests, the model is based upon the existence of a documented and dependable process that an or-ganization can use with predictable results to develop software products. In effect, the details of the process are of little interest, as long as the process is repeatable.

The CMM was adapted from the five level model of Crosby (1979) to software devel-opment by Humphrey. The five levels of the CMM model are:

1. *The Initial Level.* Ad hoc methods may achieve success through heroic efforts; little quality management, no discernible process; nothing is repeatable except, perhaps, the intensity of heroic efforts; results are unpredictable.
2. *The Repeatable Level.* Successes may be repeated for similar applications; thus, a repeatable process is discovered which is measurable against prior efforts.
3. *The Defined Level.* Claims to have understood, measured, and specified a repeatable process with predictable cost and schedule characteristics.
4. *The Managed Maturity Level.* Comprehensive process measurements enable inter-active risk management.
5. *The Optimization Level.* Continuous process improvement for lasting quality. Ac-cording to Sage (1995), "There is much double loop learning and this further sup-ports this highest level of process maturity. Risk management is highly proactive, and there is interactive and reactive controls and measurements."

The CMM helps software project management to select appropriate strategies for pro-cess improvement by (1) examination and assessment of its level of maturity, according to a set of criteria; (2) diagnosis of problems in the organization's process; and (3) pre-scription of approaches to cure the problem by continual improvement.

Even though managers may be seasoned veterans, fully knowledgeable about the prob-lems and pitfalls in the engineering process, they may disagree with each other on how to cope with problems as they occur. If agreement is difficult to produce in an organization, the resultant lack of focus is taxing on organizational resources and may endanger the product. Thus, the management of an organization must be greater than the sum of its managers by providing strategies for management to follow and tools for management to utilize. Such strategies and tools will be the result of previous organizational successes and incremental improvements over time, and measured by a level of maturity. The CMM, developed at the Software Engineering Institute (SEI) at Carnegie Mellon University, provides a framework that is partitioned by such levels of maturity. Although the CMM was developed to measure the maturity of software development processes, the ideas upon which it is based are quite general, applying well to systems engineering, and extensible, applying to such processes as software acquisition management and even unto the software engineer's own personal software process (Humphrey, 1995).

In analyzing the CMM, it is helpful to look closely at the five levels or stages in quality maturity attributable to Crosby (1979):

1. *Uncertainty*. Confusion, lack of commitment. "Management has no knowledge of quality at the strategic process level and, at best, views operational level quality control inspections of finished products as the only way to achieve quality."
2. *Awakening*. Management wakes up and realizes that quality is missing. "Statistical quality control teams will conduct inspections whenever problems develop."
3. *Enlightenment*. Management decides to utilize a formal quality improvement process. "The cost of quality is first identified at this stage of development which is the beginning of operational level quality assurance."
4. *Wisdom*. Management has a systematized understanding of quality costs. "Quality related issues are generally handled satisfactorily in what is emerging as strategic and process oriented quality assurance and management."
5. *Certainty*. Management knows why it has no problems with quality.

In each of these environments, a particular kind of person is required. There is a shifting in focus from one type of key individual to another as we move from one CMM level to the next. The progression seems to be, roughly as follows:

1. *Heroes*. Necessary for success in a relatively unstructured process, the hero is able to rise above the chaos and complete a product.
2. *Artists*. Building on the brilliance of the heroes, the artists begin to bring order, resulting through repetition in a codifiable process.
3. *Craftsmen*. These are the people who follow the process, learning from experience handed down from previous successes.
4. *Master Craftsmen*. These people are experts in their respective facets of the development process, who understand and appreciate nuances of process and their relationship to quality.
5. *Research Scientists*. Finally, master craftsmen appear, who, through experiential learning and attention to process integration, are able to fine tune the process, improving the overall process by changing steps in the process, while avoiding harmful side effects.

The characteristics of the organizational culture are directly related to organizational learning. The organization appears to depend primarily upon two factors: (1) the people who compose the organization, and (2) the environment internal to the organization. Of course, a great case may be made for including the external environment, since the overall success of the organization (and its probability of survival) are directly related to its correct adaptation to external factors, including both the market and technological factors. Following the CMM model, some of the organizational characteristics may be organized in five stages, as follows:

1. *Heroes and Supporters.* Dependent upon the ability of heroes to rise above the chaos, the organization grows up around the activities of each hero, each of whom may require low-level support services. The hero's processes are largely self-contained, and very loosely coupled with other heroes' processes. While there are very efficient aspects of this kind of organization (viz., the hero's own activities), there is no overall efficiency induced by integration of activities into an overall process. Thus, at this CMM level, there are really two levels of workers: heroes and others.

2. *Artist Colony.* Through mutual respect and attention to the successful practices of the previous generation of heroes, these people work together to recreate the successes of the past, creating new processes along the way. Management begins to be able to track progress.

3. *Professional Cooperative Organization.* Through long service and attention to process, master craftsmen have emerged, creating more hierarchical structure in the organization, as less experienced individuals are able to learn from the more experienced. There now exists the concept of "the way to do the job," a concept that must be measurably adhered to. Management's role is to control adherence to the process by defining metrics and implementing a metrics program.

4. *Society of Professionals.* At this point, the organization is mature enough to be able to receive from its individual members meaningful suggestions on how to improve selected parts of its process and to implement them in the overall process. This is largely a shift in the organization's ability to learn.

5. *Institute of Professionals.* The organization is now so mature that it is able to look continuously for ways to improve processes. Outside influences are no longer repelled, but are welcomed and evaluated.

The NASA Software Engineering Laboratory has also developed a methodology for software process improvement (NASA, 1996). NASA's three-phase approach, unlike the CMM approach, is domain dependent and focuses upon improving software products. Software product metrics and measures are used to assess indirectly and to improve the software development process. The three-step approach is iterative, and consists of the following:

1. *Understanding.* In this step, a baseline is established using product metrics and measures.

2. *Assessing.* Change in the targeted attributes, which is due to a change in the development process, is measured and its impact is assessed with respect to the desired goal and to the baseline established in the first step.

3. *Packaging.* If the change justifies a change in the organization's development process, the process improvements are incorporated.

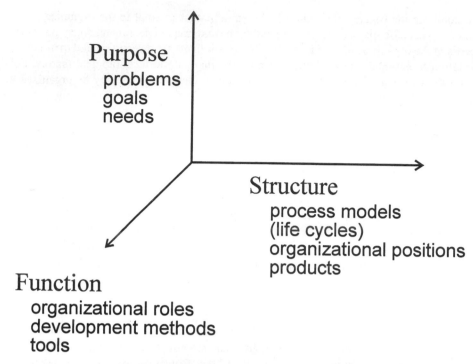

Figure 1.20 Dimensions of systems acquisition.

The trend away from rigid standardization of systems engineering process models can be understood better when viewed in the context of system acquisition. A relatively new trend in acquisition management is the widespread use of performance-based contracting (PBC), a procurement tool based upon the systems engineering approach that emphasizes the *purpose* of the acquisition. Figure 1.20 is based on a systems engineering model of Sage that relates three orthogonal dimensions of systems engineering projects: purpose, structure, and function (Sage, 1992a, 1995). By focusing attention upon purposeful aspects (goals, needs of the consumer, the *why*), as opposed to the structure (the *what*) and function (the *how*) of the solution, the vendor is not constrained to life-cycle models, development tools, methodologies, and products specified by the consumer.

In effect, the consumer and the vendor enter a partnership agreement at a higher level of abstraction than in conventional contracting methods. Figure 1.21 depicts a relationship among four acquisition strategies:

1. In-house development, using contracting strategies such as "level of effort (LOE)";
2. Conventional strategies, based upon "cost plus" contracting;
3. PBC;
4. COTS acquisition.

The key determinant in Figure 1.21 is the level of the agreement between the consumer and the vendor. In the case of in-house development, the needs of the consumer are satisfied through internal means, and any externally procured components, personnel, or other re-

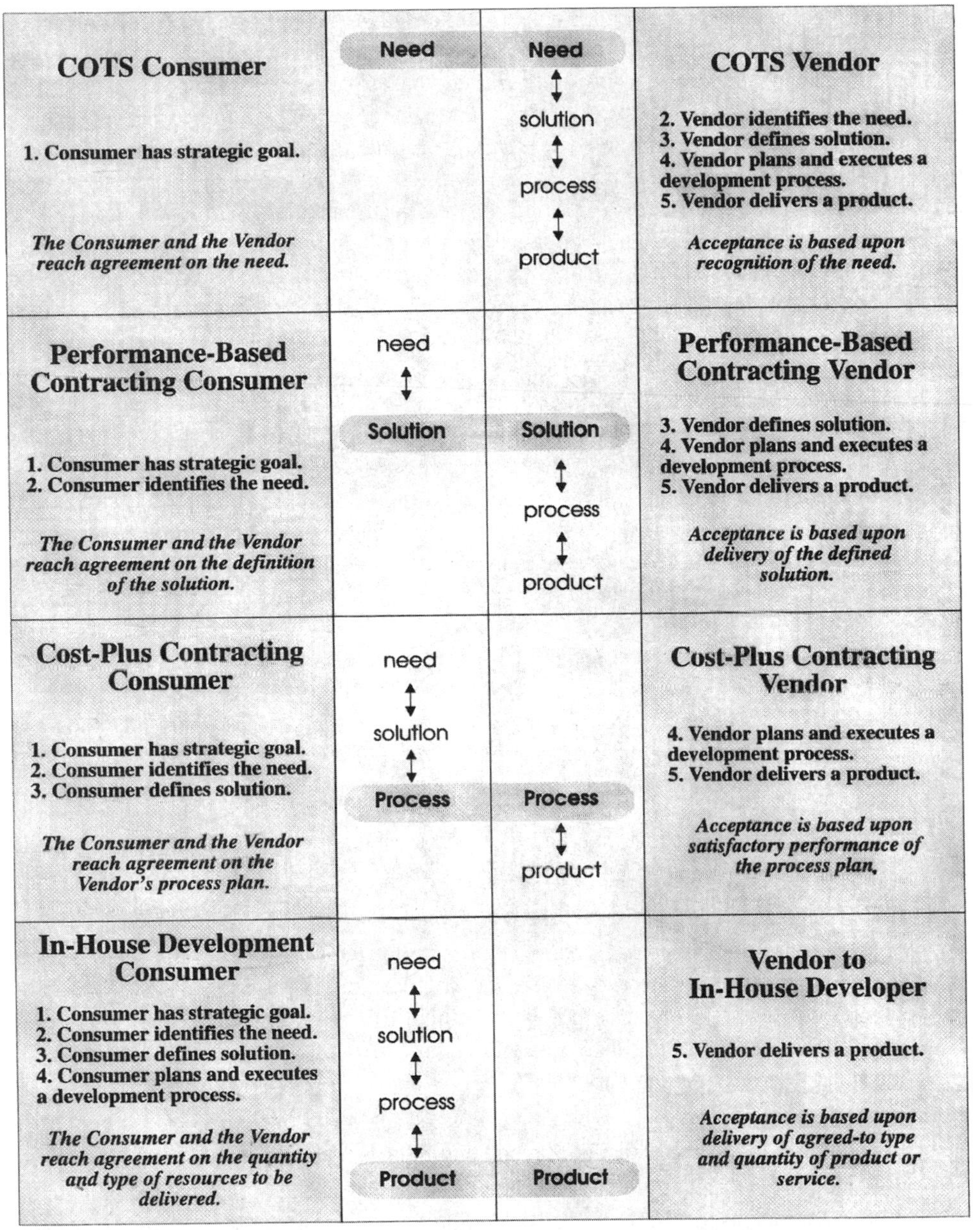

COTS Consumer	Need	Need	COTS Vendor
1. Consumer has strategic goal.		↕ solution ↕ process ↕ product	2. Vendor identifies the need. 3. Vendor defines solution. 4. Vendor plans and executes a development process. 5. Vendor delivers a product.
The Consumer and the Vendor reach agreement on the need.			*Acceptance is based upon recognition of the need.*
Performance-Based Contracting Consumer	need ↕ Solution	Solution ↕ process ↕ product	**Performance-Based Contracting Vendor**
1. Consumer has strategic goal. 2. Consumer identifies the need.			3. Vendor defines solution. 4. Vendor plans and executes a development process. 5. Vendor delivers a product.
The Consumer and the Vendor reach agreement on the definition of the solution.			*Acceptance is based upon delivery of the defined solution.*
Cost-Plus Contracting Consumer	need ↕ solution ↕ Process	Process ↕ product	**Cost-Plus Contracting Vendor**
1. Consumer has strategic goal. 2. Consumer identifies the need. 3. Consumer defines solution.			4. Vendor plans and executes a development process. 5. Vendor delivers a product.
The Consumer and the Vendor reach agreement on the Vendor's process plan.			*Acceptance is based upon satisfactory performance of the process plan.*
In-House Development Consumer	need ↕ solution ↕ process ↕ Product	Product	**Vendor to In-House Developer**
1. Consumer has strategic goal. 2. Consumer identifies the need. 3. Consumer defines solution. 4. Consumer plans and executes a development process.			5. Vendor delivers a product.
The Consumer and the Vendor reach agreement on the quantity and type of resources to be delivered.			*Acceptance is based upon delivery of agreed-to type and quantity of product or service.*

Figure 1.21 A comparison of contracting strategies based upon the level of agreement.

sources are managed by the consumer. The agreement of the consumer with the vendor is at the product level.

In the case of conventional contracting strategies, the consumer and vendor enter into a partnership at the process level of agreement. The consumer, driven by an unsatisfied need, analyzes the problem, synthesizes a solution to the problem, and creates a product specification. The specification is presented to the contractor, who writes a development plan. It is agreement upon the vendor's development plan that is the basis of the transaction between consumer and vendor. The development plan sets forth a process model (development life cycle), methods, tools, and, in some cases, organizational roles and responsibilities that the vendor agrees to follow. The consumer accepts an oversight role and a responsibility to convey to the vendor enough information from the problem space (the *purpose* dimension) to ensure the successful completion of the solution. Whenever a problem is reformulated in terms of a solution, however, information is lost. All the answers to the "why" question have been replaced by a specification. The "lost" information is retained by the *consumer* organization. Unfortunately, the organization that needs this information is not the consumer, but the vendor. While this lossy contracting procedure may work for the procurement of small and well-understood systems, large, innovative efforts magnify the loss and improve the probability of failure.

PBC reduces the loss of information by raising the level of abstraction of the information handed off from the consumer to the vendor. As shown in Figure 1.21, the organizational interface is at the *management* level. Both organizations address the problem space. It is the role of the vendor to define, develop, and deliver a solution, beginning with a vendor-developed specification. It is this specification that is the basis of agreement between the consumer and the vendor. By reversing the roles of the contracting parties, the vendor is responsible for the entire solution to the consumer's problem. The probability of failure has been reduced by eliminating an extremely complex and difficult process, the communication of purpose from one organization to another. Of course, both the consumer and the vendor can still fail, by reaching agreement on a poor solution.

Even this risk can be substantially reduced by the fourth contracting strategy shown in Figure 1.21. COTS procurement raises the level of communication to an even higher level of abstraction, so that a product can be purchased that directly satisfies the purpose for which it is desired.

1.7.2 Concurrent Engineering

Concurrent (or simultaneous) engineering is a technique that addresses the management of total life-cycle time (Carter and Baker, 1992; Rosenau, 1990; Slade, 1993), focusing on the single most critical resource in a very competitive market: time to market (or time to deployment). This is accomplished primarily by shortening the life cycle through the realization of three engineering subgoals:

1. Introduction of customer evaluation and engineering design feedback during product development.
2. A greatly increased rate of focused, detailed technical interchange among organizational elements.
3. Development of the product and creation of an appropriate production process in parallel rather than in sequence.

Concurrent engineering is a metaprocess in which domain experts from all the departments concerned for developing a product at any stage of the life cycle work together as a concurrent engineering (CE) team, integrating all development activities into one organizational unit. The formation of the team does not per se shorten the engineering life cycle; however, through early involvement with the CE team, organizational learning and

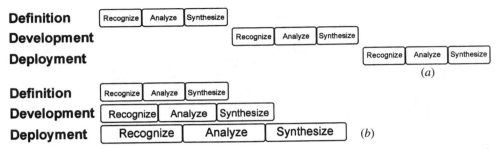

Figure 1.22   (*a*) Waterfall representation of abstract life cycle. (*b*) The compressing effect of concurrent engineering upon the waterfall model.

analysis activities can be removed from the critical path to market. There is an explicit tradeoff of workers for time to market. That is, the CE team involves more personnel for a greater fraction of the life cycle[12] than in the case of the waterfall model. However, the time to market can be greatly reduced. In terms of the abstract life-cycle model, the activities labeled "recognize," "analyze," and "synthesize" can occur concurrently for all organizational elements involved in the development of the product. Of course there will be some activities that have temporal, as well as logical, sequential dependence upon other activities (see Fig. 1.22*a* and *b*). Marketing, drawing upon organizational expertise, including RDT&E products, begins the process through the generation of an idea of a product, based upon market analysis. Marketing will generate targets for the selling price and the production costs of the proposed product to support management in deciding whether to proceed with product development. During development, the CE team work simultaneously with the design team to generate in parallel a design for the manufacturing *process*.

A CE life-cycle model is shown in Figure 1.23. The principal feature of this process model is the concurrent development of the product, the manufacturing process, and the manufacturing system through the continuous participation of the CE team (Yeh, 1992). A notable feature of the life cycle is the absence of a return path from production to design. This deliberate omission is in recognition of the extremely high risk of losing market share because of engineering delays due to design errors.

The organizational response to a change to concurrent engineering from traditional methods is likely to be fraught with difficulty. An organization that has formed around a particular life-cycle model, and that has experienced a measure of success, perhaps over a period of many years, will almost certainly resist change. Effort in several specific areas appears to be basic to any transition:

1. We have noted that life-cycle models are purposeful, in that they reflect, organize, and set into motion the organizational mission. A change of life-cycle model should be accompanied by a clearly stated change of mission that may be digested, assimilated, and rearticulated by all organizational elements: individuals, formal and informal team structures, and social groups.

---

[12]Although the CE team remains together for a greater percentage of the total life cycle, the life cycle is significantly shorter than in traditional models. Consuming a greater portion of a smaller resource may not increase cost and, in some cases, may actually decrease cost.

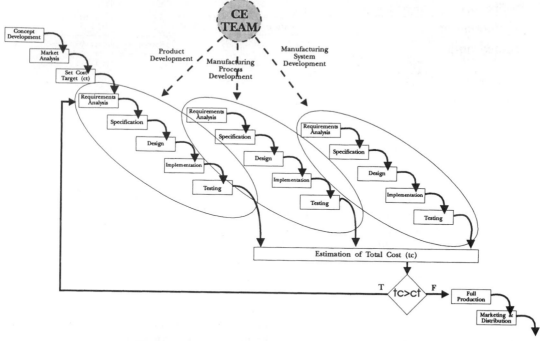

Figure 1.23   A concurrent engineering life-cycle model.

2. Formal team structures should be examined, destroyed and rebuilt, replaced, or supplemented as necessary to conform to the new life-cycle model. Informal team structures and individual expectation, such as those described by Stogdill, may be replicated, and thus preserved, by the formal organizational structure to minimize loss.

3. Particular attention should be paid to intramural communication and cooperation among individuals, teams, and departments in the organization (Carter and Baker, 1992). Communication and cooperation are essential elements of concurrency. Acquisition or improved availability of communications tools, developed through advances in communications technology, may reduce the cost and increase the rate of concurrency.

The NASA *Systems Engineering Handbook* (Shishko, 1995) defines concurrent engineering very generally as "the simultaneous consideration of product and process downstream requirements by multidisciplinary teams." A slightly more conservative view is that of Grady (1993), who characterizes concurrent engineering as the correction of a mistake that has been made by many organizations, which he refers to as "transom engineering," wherein engineers working in isolation throw their respective products over transoms to their successors in a linear process. Grady believes that this is a gross distortion of the correct concept of systems engineering, and that concurrent engineering is merely a return to what systems engineering should be. Additional discussions of Concurrent Engineering are presented in Chapter 9.

1.7.3   Software Reengineering for Reuse

On the basis of two types of reuse identified by Barnes and Bollinger (1991), it is useful to distinguish between two different types of software reengineering (Chikofsky and Cross, 1990) for reuse:

1. Software reengineering for maintenance (adaptive reusability) improves attributes of existing software systems, sometimes referred to as legacy systems, which have been correlated to improvements that improve software maintainability (Basili, 1990; Sneed, 1995).
2. Software reengineering for reuse (compositional reuse) salvages selected portions of legacy systems for rehabilitation to enable off-the-shelf reuse in assembling new applications (Arnold and Frakes, 1992; Cimitile, 1992).

Both types of reengineering for reuse share a common life cycle (Patterson, 1995a), (shown in Fig. 1.24, and discussed below) for reengineering to an object-oriented software architecture.

The life cycle is divided into three successive phases: reengineering concept development, reengineering product development, and deployment. During the concept development phase, reengineering is considered as an alternative to new software development. Considerations of scope and of level of reengineering allow planning and cost estimation prior to the development phase.

Reengineering product development proceeds according to the scope and level of reengineering planned in the previous phase. Reverse engineering of the old software is followed by forward engineering to create a new product. During the reverse engineering stage, products are recreated at the design and specification levels as needed to recapture implementation decisions that may have been lost over the lifetime of the legacy software. During the entire reverse engineering stage, candidate objects are repeatedly created, modified, or deleted as necessary to provide the basis of an object-oriented design for the forward engineering stage.

During the forward engineering stage, the candidate objects from the reverse engineering stage are used to create an object-oriented specification and design. Implementation through coding and unit testing complete the development phase. During the deployment phase, software integration and testing, followed by system integration and testing, allow production and deployment to proceed.

Reengineering Concept Development			Reengineering Product Development						Deployment		
			Reverse Engineering			Forward Engineering					
Feasibility	Scope	Level of Reen- gineering	from Code to Design	from Design to Spec	Identification of Candidate Objects	Re- specification	Design	Coding and Unit Testing	Software Integration and Testing	System Integration and Testing	Production and Deployment

Figure 1.24   A software reengineering life cycle.

Software reengineering is often associated with business process reengineering (Davenport, 1993; Hammer and Champy, 1993; Morris and Brandon, 1993). A recent study (Patterson, 1995b) shows that there is a reciprocal relationship between business process reengineering and software reengineering. The enabling role of information technology makes possible the expansion of the activities of business processes. Moreover, changes in support software may influence changes in the business process. In particular, changes in support software make the software more useful or less useful to a given business process, so that the business process (or, indeed, the software) must adapt. Changes in the potential for software functionality that are due to improvements in the technology may enable changes in business processes, but must not drive them. In general, successful technology follows, rather than leads, humans, and organizations.

Similarly, changes in the business process create changes in the requirements for support software. However, this cause-and-effect relationship between business process reengineering and software reengineering cannot be generalized and is inherently unpredictable. Each case requires independent analysis. The effect of business process reengineering on software can range from common perfective maintenance to reconstructive software reengineering. New software may be required in the event that the domain has changed substantially. Because the software exists to automate business process functions, the purpose of the support software can be identified with the functions making up the process. Therefore, reengineering the process at the function level will in general always require reengineering the software at the purpose level. This is equivalent to changing the software requirements. Software reengineering can be the result of business process reengineering, or it can be the result of a need to improve the cost-to-benefit characteristics of the software. An important example of software reengineering that may have little impact on the business process is the case of reengineering function-oriented software products into object-oriented products, thereby choosing the more reactive paradigm to reduce excessive cost due to poor maintainability.

There are many levels of business process reengineering and of software reengineering, ranging from redocumentation to using business process reengineering as a form of maintenance. In both there is a continuum between routine maintenance (minor engineering) and radical, revolutionary reengineering. At both ends of the spectrum, change should be engineered in a proactive, not a reactive manner. As Sage (1995) notes, reengineering ". . . must be top down directed if it is to achieve the significant and long-lasting effects that are possible. Thus, there should be a strong purposeful and systems management orientation to reengineering." Additional commentaries on systems re-engineering are presented in Chapter 23.

1.7.4 Knowledge-Based Software Engineering

Lowry and Duran (1989), in assessing the adequacy of the waterfall model, cite the lack of adequate support for incremental and evolutionary development, such as many artificial intelligence applications. The spiral model is much more natural, especially as computer-aided software engineering (CASE) tools shorten the production cycle for the development of prototypes. In terms of the spiral model (Fig. 1.18), the amount of time needed to complete one turn has been shortened for many of the turns through the use of CASE tools. A potentially greater savings can be realized by reducing the number of turns as well through the development and use of knowledge-based tools. Lowry (1991, 1992) believes that much of the process of developing software will be mechanized through the application of artificial intelligence technology. Ultimately, the specification-to-design-to-code process will be replaced by domain-specific specification aids that will generate code directly from specifications. This interesting vision is based upon much current reality, not only in the CASE arena, but also in the area of domain-based reuse repositories (Gaffney and Cruickshank, 1992; Kozaczynski, 1990; Moore and Bailin, 1991). In terms of the life cycle implications, the waterfall will become shorter, and the spiral will have fewer turns.

1.8   CONCLUSIONS

The use of the engineering term *life cycle* is relatively new, although the concept is as old as humankind. One of the earliest uses of the term was Juran's reference in 1951 to life-cycle costs, which he referred to as a military term used to describe the operational costs of a system (Juran et al., 1951), or the cost of ownership. The term "life-cycle cost" has come to include the cost of system acquisition (Aslaksen and Belcher, 1992; Blanchard, 1991; Blanchard and Fabrycky, 1998). Today the term "life cycle" is synonymous with "process model." Because of engineering failures, some spectacular, on many systems engineering projects, the engineering community has given a great deal of attention both to the structure and the function of life cycles and, to a lesser degree, their purpose.

In this chapter we have examined basic types of life-cycle models and their variants: the waterfall, and the use of feedback and prototyping to improve the system definition; the V model; the incremental and evolutionary approaches, such as the repeated waterfall with feedback, the spiral model, and rapid development; concurrent engineering; and models that are specific to domains or organizations requiring process measurement for certification and for process improvement. We have identified the organization with the purpose of the life cycle, and noted the interaction of process models with other dynamic organizational forces. We have discussed the life cycle as a tool for the division of resources: for the division and interrelationships of the human resources, through formation of structural and interpersonnel relationships both within and among organizations; for the division and interrelationships of types of business processes; and for division of business process into phases, phases into steps, and for the orderly completion of steps through sequencing of the definition, development, and integration of partial products. We have discerned several fundamental relationships along a continuum of four fundamental types of acquisition: commercial-off-the-shelf, performance-based contracting, cost-plus contracting, and in-house development. Finally, we have recognized several trends in life cycles, such as those related to reengineering, reuse, and automated software generation.

It seems apparent that we are moving to a more domain-specific way of conducting the business of systems engineering. We are seeing confirmation in the software reuse area, also in the ISO-9000 standard and the CMM. We are seeing a move to associate the process with the product requirements from the domain of choice. Ultimately this leads to automation of the kind that Lowry (1991, 1992) is advocating for automated software generation from requirements. But, meanwhile, we are specifying a process, which in the most mature domains is most likely to be well-defined, when we specify our requirements. This is good news, since a mature process moves system creation a step farther along the path from art to engineering. From an organization point of view this is also good, because an organization is liberated in a sense to optimize processes according to its own strengths. This is the same sense in which process standards are sometimes regarded as denigrating to quality.

TABLE 1.1 Five Frequently Cited Reasons for Systems Engineering Failures

Complaint	Possible Cause	Recommended Approach
1. Ineffective management	Poor leadership, which could be caused by many factors related to organizational processes Shortage of domain knowledge which could be caused by either unfamiliarity with the domain or loss of information between organizations	Use in-house development, or development by a single organization Use an organization with a mature process for the domain Inject initial or early use of life cycle with proven success in the domain
2. Uncertain funding	Long development cycles during which no product is deployed or demonstrated	Use a life cycle that fields or demonstrates an early prototype Automate processes
3. Unmanaged complexity	Addressed by division of labor; division of the problem; division of processes into phases Inadequate process controls	Use a life cycle that divides the problem as well as the resources into manageable parts
4. Poor requirements	Addressed by spiral and rapid development models, which enable incremental and evolutionary approaches; domain engineering	Use a life cycle that repeats or extends the requirements cycle, e.g., concurrent engineering; spiral model
5. Poor specification or other products	Loss of information caused by using several independent organizations	Use acquisition strategy that uses a single organization, e.g., in-house development

There are several persistent themes associated with systems engineering failures, especially in large systems, including the five addressed in Table 1.1. Many of these problems have their roots in the life-cycle issues that have been discussed in this chapter. We have the luxury of having observed and having learned from many of the mistakes of the past. We conclude that systems engineering life cycles are purposeful tools that should not be substituted for the systems approach, but rather used as part of an integrated systems approach that addresses not only the what and the how, but also the why.

REFERENCES

Ackoff, R. L. (1981). *Creating the Corporate Future*. New York: Wiley.

American National Standards Institute/Institute of Electrical and Electronics Engineers (ANSI/IEEE). (1983). *IEEE Standard Glossary of Software Engineering Terminology*. New York: IEEE.

Arnold, R. S., and Frakes, W. B. (1992). Software reuse and re-engineering. In *Software Reengineering* (R. S. Arnold, ed.), pp. 476–484. Los Alamitos, CA: IEEE Computer Society Press.

Aslaksen, E., and Belcher, R. (1992). *Systems Engineering*. New York: Prentice-Hall.

Barnes, B. H., and Bollinger, T. B. (1991). Making reuse cost-effective. *IEEE Software,* **8**(1) pp. 13–24.

Basili, V. R. (1990). Viewing maintenance as reuse-oriented software development. *IEEE Software,* pp. 19–25.

Bass, B. M., and Stogdill, R. M. (1990). *Handbook of Leadership: Theory, Research, and Managerial Applications,* 3rd ed. New York: Macmillan (Free Press).

Blake, R. R., and Mouton, J. S. (1961). *Group Dynamics—Key to Decision Making,* Chapter 8, pp. 97–112. Houston, TX: Gulf Publ. Co.

Blake, R. R., and Mouton, J. S. (1980). Power styles within an organization. In *Models for Management: The Structure of Competence* (J. A. Shtogren, ed.), pp. 60–79. The Woodlands, Texas: Teleometrics Int.

Blanchard, B. S. (1991). *System Engineering Management*. New York: Wiley.

Blanchard, B. S., and Fabrycky, W. J. (1998). *Systems Engineering and Analysis,* 3rd ed. Englewood Cliffs, NJ: Prentice-Hall.

Boehm, B. W. (1981). *Prentice-Hall Advances in Computing Science and Technology Series. Software Engineering Economics*. Englewood Cliffs, NJ: Prentice-Hall.

Boehm, B. W. (1990a). Verifying and validating software requirements and design specifications. In *System and Software Requirements Engineering* (M. Dorfman and R. H. Thayer, eds.), pp. 471–484. Los Alamitos, CA: IEEE Computer Society Press; reprinted from *IEEE Software* **1**(1), 75–88 (1984).

Boehm, B. W. (1990b). A spiral model of software development and enhancement. In *System and Software Requirements Engineering* (M. Dorfman and R. H. Thayer, eds.), pp. 513–527. Los Alamitos, CA: IEEE Computer Society Press; reprinted from *Software Engineering Project Management,* pp. 128–142. Los Alamitos, CA: IEEE Computer Society Press, 1987.

Bradford, L. P., Stock, D., and Horwitz, M. (1980). How to diagnose group problems. In *Models for Management: The Structure of Competence* (J. A. Shtogren, ed.), pp. 373–390. The Woodlands, TX: Teleometrics Int.

Briner, W., and Hastings, C. (1994). The role of projects in the strategy process. In *Global Project Management Handbook* (D. I. Cleland and R. Gareis, eds.), Chapter 15, pp. 1–24. New York: McGraw-Hill.

Brooks, F. P., Jr. (1975). *The Mythical Man-Month: Essays on Software Engineering*. Reading, MA: Addison-Wesley (reprinted with corrections, January 1982).

Bruner, J. S. (1960). *The Process of Education*. Cambridge, MA: Harvard University Press.

Buck, V. E. (1966). A model for viewing an organization as a system of constraints. In *Approaches to Organizational Design* (J. D. Thompson, ed.), 1971 paperback ed., pp. 103–172. Pittsburgh, PA: University of Pittsburgh Press.

Buckley, W. F., (1979). Let's define that 'leadership' that Kennedy says we need (quoted in Bass and Stogdill, 1990). *Press-Bulletin,* Binghamton, NY, September 22.

Burns, J. MacG. (1978). *Leadership*. New York: Harper and Row.

Carter, D. E., and Baker, B. S. (1992). *Concurrent Engineering: The Product Development Environment for the 1990s*. Reading, MA: Addison-Wesley.

Chikofsky, E. J., and Cross, J. H., II. (1990). Reverse engineering and design recovery: A taxonomy. *IEEE Software,* **7**(1) pp. 11–12.

Cimitile, A. (1992). Towards reuse reengineering of old software. In *Proceedings IEEE Fourth International Conference on Software Engineering and Knowledge Engineering,* pp. 140–149. Los Alamitos, CA: IEEE Computer Society Press.

Cleland, D. I., and King, W. R. (1983). *Systems Analysis and Project Management,* 3rd ed. New York: McGraw-Hill.

Cooper, R. G. (1992). *Winning at New Products: Accelerating the Process from Idea to Launch,* 2nd ed. Reading, MA: Addison-Wesley.

Crosby, P. B. (1979). *Quality is Free.* New York: McGraw-Hill.

Davenport, T. H. (1993). *Process Innovation: Reengineering Work Through Information Technology.* Boston: Harvard Business School Press.

Davis, A. M., Bersoff, E. H., and Comer, E. R. (1988a). A strategy for comparing alternative software development life cycle models. *IEEE Trans. Software Eng.,* **14**(10), 1453–1461.

Davis, A. M., Bersoff, E. H., and Comer, E. R. (1988b). In *System and Software Requirements Engineering* (R. H. Thayer and M. Dorfman, eds.), pp. 496–503. Los Alamitos, CA: IEEE Computer Society Press.

Davis, K. (1949). *Human Society.* New York: Macmillan.

Deming, W. E. (1982). *Out of the Crisis.* Cambridge, MA: MIT Press.

Drucker, P. F. (1974). *Management: Tasks, Responsibilities, Practices.* New York: Harper & Row.

Forsberg, K., and Mooz, H. (1991). The relationship of system engineering to the project cycle. *Proc. Jt Conf. of Natl. Counc. Syst. Eng. (NCOSE)/Am. Soc. Eng. Manage. (ASEM),* Chattanooga, TN, *1991,* pp. 57–65.

Forsberg, K., and Mooz, H. (1995a). Application of the 'Vee' to incremental and evolutionary development. *Proc. Natl. Counc. Syst. Eng. (NCOSE), 5th Annu. Int. Symp.,* St. Louis, MO, *1995* pp. 801–808.

Forsberg, K., and Mooz, H. (1991). The Relationship of Systems Engineering to the Project Cycle. *Proceedings of the National Council for Systems Engineering (NCOSE)* 1st Annu. Int. Symp., Chattanooga, TN, pp. 57–65.

Gaffney, J. E., Jr., and Cruickshank, R. D. (1992). A general economics model of software reuse. In *14th Proceedings of the International Conference on Software Engineering.* New York: Association for Computing Machinery.

Galbraith, J. R., and Nathanson, D. A. (1978). *Strategy Implementation: The Role of Structure and Process.* St. Paul, MN: West Publ. Co.

Grady, J. O. (1993). *System Requirements Analysis.* New York: McGraw-Hill.

Hall, A. D., III. (1977a). Systems engineering from an engineering viewpoint. In *Systems Engineering: Methodology and Applications* (A. P. Sage, ed.), pp. 13–17. New York: IEEE Press, reprinted from *IEEE Trans. Syst. Sci. Cybern.,* **SSC-1,** 4–8 (1965).

Hall, A. D., III. (1977b). Three-dimensional morphology of systems engineering. In *Systems Engineering: Methodology and Applications* (A. P. Sage, ed.), pp. 18–22. New York: IEEE Press, reprinted from *IEEE Trans. Syst. Sci. Cybern.,* **SSC-5,** 156–160 (1969).

Hammer, M., and Champy, J. (1993). *Reengineering the Corporation: A Manifesto for Business Revolution.* New York: HarperCollins.

Hill, J. D., and Warfield, J. N. (1977). Unified program planning. In *Systems Engineering: Methodology and Applications* (A. P. Sage, ed.), pp. 23–34. New York: IEEE Press, reprinted from *IEEE Trans. Syst., Man, Cybern.,* **SMC-2,** 610–621 (1972).

Humphrey, W. S. (1989). *Managing the Software Process.* Reading, MA: Addison-Wesley (reprinted with corrections August, 1990).

Humphrey, W. S. (1995). *A Discipline for Software Engineering.* Reading, MA: Addison-Wesley.

Ince, D. (1994). *ISO 9001 and Software Quality Assurance.* Maidenhead, Berkshire, England: McGraw-Hill Book Company Europe.

International Benchmarking Clearinghouse. (1992). *Assessing Quality Maturity: Applying Baldrige, Deming, and ISO 9000 Criteria for Internal Assessment.* Houston, TX: American Productivity and Quality Center.

International Organization for Standardization/International Electrotechnical Commission. (1995). *Software Life Cycle Processes* (ISO/IEC 12207). Geneva: IOS/IEC.

Jackson, M. (1995). *Software Requirements and Specifications.* Wokingham, England: Addison-Wesley.

Johnson, R. A., Kast, F. E., and Rosenzweig, J. E. (1967). *The Theory and Management of Systems.* New York: McGraw-Hill.

Juran, J. M., Gryna, F. M., and Bingham, R. S. (eds.). (1951). *Quality Control Handbook.* New York: McGraw-Hill.

Kast, F. E., and Rosenzweig, J. E. (1970). *McGraw-Hill Series in Management. Organization and Management: A Systems Approach.* New York: McGraw-Hill.

Kerzner, H. (1992). *Project Management: A Systems Approach to Planning, Scheduling, and Controlling* 4th ed. New York: Van Nostrand-Reinhold.

King, W. R., and Cleland, D. I. (1983). Life cycle management. In *Project Management Handbook* (D. I. Cleland and W. R. King, eds.), pp. 209–221. New York: Van Nostrand-Reinhold.

Kozaczynski, W. (1990). The 'catch 22' of reengineering. In *12th International Conference on Software Engineering,* p. 119. Los Alamitos, CA: IEEE Computer Society Press.

Lowry, M. R. (1992). Software engineering in the Twenty-First Century. *AI Mag.* **14,** 71–87.

Lowry, M. R., and Duran, R. (1989). A tutorial on knowledge-based software engineering. In *The Handbook of Artificial Intelligence* (A. Barr, P. R. Cohen, and E. A. Feigenbaum, eds.), Vol. 4. Reading, MA: Addison-Wesley.

McGarry, F. E., Page, J., Card, D., Rohleder, M., and Church, V. (1984). *An Approach to Software Cost Estimation.* Natl. Aeron. Space Admin., Software Eng. Lab. Ser. (SEL-83-001). Greenbelt, MD: Goddard Space Flight Center.

Melcher, B. H., and Kerzner, H. (1988). *Strategic Planning: Development and Implementation.* Blue Ridge Summit, PA: TAB Books.

Michaels, J. V., and Wood, W. P. (1989). *New Dimensions in Engineering. Design to Cost.* New York: Wiley.

Moore, J. M., and Bailin, S. C. (1991). Domain analysis: Framework for reuse. In *Domain Analysis and Software System Modeling* (R. Prieto-Diaz and G. Arango, eds.). Los Alamitos, CA: IEEE Computer Society Press.

Morris, D., and Brandon, J. (1993). *Re-engineering Your Business.* New York: McGraw-Hill.

National Aeronautics and Space Administration. (NASA) (1995). *Software Configuration Management Guidebook* (NASA-GB-9503 or NASA-GB-A501). Washington, DC: NASA, Office of Safety and Mission Assurance.

National Aeronautics and Space Administration. (NASA). (1996). *Software Process Improvement Guidebook,* NASA Software Eng Program (NASA-GB-001-95). Greenbelt, MD: Goddard Space Flight Center.

Ogle, W. (1941). De partibus animalium. In *The Basic Works of Aristotle* (R. McKeon, ed.), pp. 641–661. New York: Random House.

Ould, M. A. (1990). *Wiley Series in Software Engineering Practice. Strategies for Software Engineering: The Management of Risk and Quality.* Chichester: Wiley.

Patterson, F. G., Jr. (1995a). Reengineerability: Metrics for software reengineering for reuse. Unpublished Ph.D. Dissertation, George Mason University, Fairfax, VA.

Patterson, F. G., Jr. (1995b). A study of the relationship between business process reengineering and software reengineering. *Inf. Syst. Eng.* **1**(1), 3–22.

Paulish, D. J., and Carleton, A. D. (1994). Case studies of software-process-improvement measurement. *IEEE Comput.* **27,** 50–57.

Pearce, J. A., II, and Robinson, R. B., Jr. (1985). *Strategic Management: Strategy Formulation and Implementation* 2nd ed. Homewood, IL: R. D. Irwin.

Rabbitt, J. T., and Bergh, P. A. (1994). *The ISO 9000 Book: A Global Competitor's Guide to Compliance and Certification,* 2nd ed. New York: American Management Association (AMACOM).

Radley, C. F., and Wetherholt, M. S. (1996). *NASA Guidebook for Safety Critical Software Analysis and Development* (NASA-GB-1740.13-96). Cleveland, OH: National Aeronautics and Space Administration, Lewis Research Center.

Reilly, N. B. (1993). *Successful Systems Engineering for Engineers and Managers.* New York: Van Nostrand-Reinhold.

Rosenau, M. D., Jr. (1990). *Faster New Product Development: Getting the Right Product to Market Quickly.* New York: American Management Association (AMACOM).

Royce, W. W. (1970). Managing the development of large software systems: Concepts and techniques. *Proc. of WESCON,* pp. 1–70.

Rubey, R. J. (1993). *Software Management Guide,* 3rd ed., Vol. 2. Hill Air Force Base, UT: U.S. Air Force Software Technology Support Center.

Sage, A. P. (1992a). *Systems Engineering.* New York: Wiley.

Sage, A. P. (1992b). Systems engineering and information technology: Catalysts for total quality in industry and education. *IEEE Trans. on Syst., Man, Cybernet.* **22**(5), 833–864.

Sage, A. P. (1995). *Systems Management for Information Technology and Software Engineering.* New York: Wiley.

Sage, A. P., and Palmer, J. D. (1990). *Software Systems Engineering.* New York: Wiley.

Shishko, R. (1995) (Ed.). *NASA Systems Engineering Handbook (SP-6105).* Washington, DC: National Aeronautics and Space Administration.

Slade, B. N. (1993). *Compressing the Product Development Cycle.* New York: American Management Association (AMACOM).

Sneed, H. M. (1995). Planning the reengineering of legacy systems. *IEEE Software,* pp. 24–34.

Software Engineering Institute. (1991). *Capability Maturity Model* (CMU/SE-91-TR-24). Pittsburgh, PA: Carnegie-Mellon University.

Sorensen, R. (1996). Adopting MIL-STD-498: The steppingstone to the U. S. commercial standard. *Crosstalk: J. Def. Software Eng.* **9**(3) (Hill AFB, UT: Software Technology Support Center).

Stogdill, R. M. (1966). Dimensions of organization theory. In *Approaches to Organizational Design* (J. D. Thompson, ed.), 1971 paperback ed., pp. 1--56. Pittsburgh, PA: University of Pittsburgh Press.

Thome, B. (1993). Definition and scope of systems engineering. In *Wiley Series in Software Based Systems. Systems Engineering: Principles and Practice of Computer-Based Systems Engineering* (B. Thome, ed.), pp. 1–23. Chichester: Wiley.

U.S. Department of Defense. (1969). *Engineering Management* (MIL-STD-499). Washington, DC: USDoD.

U.S. Department of Defense. (1974). *Engineering Management* (MIL-STD-499A). Washington, DC: USDoD (canceled notice 1, 27 February 1995).

U.S. Department of Defense. (1987). *Software Products Standard* (DoD-STD-1703). Washington, DC: USDoD (canceled December 5, 1994).

U.S. Department of Defense. (1988a). *Defense System Software Development* (DoD-STD-2167A). Washington, DC: USDoD (replaced December 5, 1994 by MIL-STD-498).

U.S. Department of Defense. (1988b). *DoD Automated Information Systems (AIS) Documentation*

Standard (DoD-STD-7935a). Washington, DC: USDoD (replaced December 5, 1994 by MIL-STD-498).

U.S. Department of Defense. (1991). *Defense Acquisition Management Policies and Procedures* (DoDI 5000.2) Washington, DC: USDoD.

U.S. Department of Defense. (1992). *Systems Engineering* (MIL-STD-499B). Washington, DC: USDoD (DRAFT May 6, 92; never approved by OSD; final decision to not approve was in April 1994).

U.S. Department of Defense. (1993). *Automated Information System (AIS) Life-Cycle Management (LCM) Process, Review, and Milestone Approval Procedures* (DoDI 8120.2). Washington, DC: USDoD.

U.S. Department of Defense. (1994). *Software Development and Documentation* (MIL-STD-498). (December 5, 1994 ed.). Washington, DC: USDoD.

Warfield, J. N., and Hill, J. D. (1972). *A Unified Systems Engineering Concept*. Columbus, OH: Battelle Memorial Institute.

Wheatley, M. J. (1992). *Leadership and the New Science*. San Francisco: Berrett-Koehler Publishers.

Whytock, S. (1993). The development life-cycle. In *Wiley Series in Software Based Systems. Systems Engineering: Principles and Practice of Computer-Based Systems Engineering* (B. Thome, ed.), pp. 81–96. Chichester: Wiley.

Williamson, O. E. (1970). *Corporate Control and Business Behavior: An Inquiry into the Effects of Organization Form on Enterprise Behavior*. Englewood Cliffs, NJ: Prentice-Hall.

Yeh, R. T. (1992). Notes on concurrent engineering. *IEEE Trans. Knowl. Data Eng.* **4**(5), 407–414.

2 Systems Engineering Management: The Multidisciplinary Discipline

AARON J. SHENHAR

2.1 INTRODUCTION

The creation of complex human-made systems clearly has its historical roots in early civilization. Major undertakings such as the Egyptian Pyramids or the Chinese Wall have long been proof of humanity's system ingenuity. Today, with the development of management theory, such creations are often linked to modern concepts of systems engineering. As a problem-driven field, the practice of systems engineering is rapidly growing, while more standards are written and additional tools are being developed. As a theoretical discipline, however, systems engineering is quite new and probably not well understood. Systems engineering is considered a branch of technology management and engineering dedicated to controlling the design, creation, and use of complex human-made systems. The systems engineering process involves a logical sequence of technical activities and decisions of identifying an operational need, transforming this need into a description of system performance parameters, creating a preferred system configuration and a final system design, controlling the development and building of the system, and validating its performance against the original requirements and specifications. Within this process, one thing must be clear at the onset: no complex system can be created by a single person, thus systems engineering is strongly linked to management. We therefore need to combine the two fields and talk about *systems engineering management*.

But what exactly is systems engineering management? Is it a discipline by itself? Is it an attitude, or a process? What does it entail and how is it applied? How could it be developed and learned? The development and study of systems engineering is relatively new. Its origins are usually traced to the 1950s and 1960s during the large military and space development programs. Today, however, it is widely applied in various industries as a mean of integrating distinct disciplines and technologies into an overall complicated but optimized outcome. Recently, a new professional organization—the International Council on Systems Engineering (INCOSE)—was established, to foster and exchange knowledge in this evolving and important field.

Although, as a practical tool, systems engineering is widely in use, as a scholarly and formal discipline, it hardly exists. Usually, most systems engineers grow on the job; only a handful of institutions offer educational programs in such interdisciplinary areas; and there is a great variety among the few programs that do exist (Swain, 1991). Furthermore, there is not yet a common definition of the discipline, let alone an agreement on its exact content (Brill, 1994, 1999).

The purpose of this chapter is to present a framework for the description of an inter-

Handbook of Systems Engineering and Management, Edited by A. P. Sage and W. B. Rouse
ISBN 0471-15405-9   ©1999 John Wiley and Sons, Inc.

disciplinary discipline of systems engineering management (SEM), its application for various types of systems, and its possible development in terms of theories, tools, and educational programs. We shall (1) present a comprehensive definition for SEM, (2) describe the roles, activities, and responsibilities associated with SEM, (3) present a framework for the application of SEM for various types of systems, and (4) suggest a framework for the integration of the relevant theories and tools into a possible multidisciplinary education and training program in SEM.

This work is based on our on-going studies in systems engineering and project management and on our experience as educators of systems engineers (Shenhar, 1990, 1994). Our research involved qualitative and quantitative data collection and extensive interviews with hundreds of project managers and systems engineers (Shenhar, 1990, 1993, 1994; Shenhar and Bonen, 1997; Shenhar and Dvir, 1996).

2.2 DEFINING SYSTEMS ENGINEERING MANAGEMENT

2.2.1 Systems Engineering Definitions

Various definitions of systems engineering exist in the literature. Some see it as a process; others as a function or discipline (e.g., Blanchard, 1998; Booton and Ramo, 1984; Chambers, 1985, 1986; Chase, 1985; U.S. Department of Defense, 1992). For a collection of ten variations on the definition of systems engineering, see Brill (1994). We quote two definitions:

Definition 1   (McPherson, 1986). Systems engineering is a hybrid methodology that combines policy analysis, design, and management. It aims to ensure that a complex man-made system, selected from the range of options on offer, is the one most likely to satisfy the owner's objectives in the context of long-term future operational or market environments.

Definition 2   (Sage, 1980a). This definition consists of three parts:

Structure. Systems engineering is management technology to assist clients through the formulation, analysis, and interpretation of the impacts of proposed policies, controls, or complete systems upon the need perspectives, institutional perspectives, and value perspectives of stakeholders to the issue under consideration.

Function. Systems engineering is an appropriate combination of the mathematical theory of systems and the behavioral theory, in a useful setting appropriate for the resolution of real-world problems, which are often of large scale and scope.

Purpose. The purpose of systems engineering is to develop policies for management direction, control, and regulation of activities relative to forecasting, planning, development, production, and operations of total systems to maintain overall integrity and integration as related to performance and reliability.

2.2.2 Systems Engineering Management Definition

Complex systems can only be created through the combined efforts of many people. Therefore, controlling the design and creating such systems will always involve an engineering part as well as a managerial part (Sage, 1981, 1994). Similarly, the systems engineering

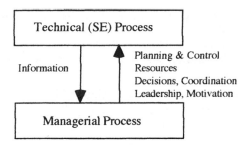

Figure 2.1   Systems engineering management: The two processes.

process can be seen as consisting of a technical (engineering) process and a managerial process (see Fig. 2.1). Furthermore, as Sage proposed (1980b), systems engineering is solving problems, not only for clients but with clients. In view of these assertions and using some components from previous definitions (e.g., U.S. Department of Defense, 1992), we propose a revised definition for SEM:

Definition. Systems engineering management is the application of scientific, engineering and managerial efforts to:

a. Identify an operational need (in a commercial or military area, and in the public or private sector), together with a marketing and technological opportunity that leads to the creation of a system that will address this need;

b. Transform an operational need into a description of system performance parameters and a system configuration through the use of an iterative process of definition, synthesis, analysis, design, test, and evaluation;

c. Integrate related technical parameters and components and ensure compatibility of all physical, functional, and program interfaces in a manner that optimizes the total system definition and design;

d. Integrate reliability, maintainability, safety, survivability, human, and other such factors into the total engineering effort to meet cost, schedule, and technical performance objectives;

e. Work with clients to ensure that the system created is qualified to address required needs and solve clients' problems.

This wider definition suggests that:

1. The systems engineering effort does not start with the system design, nor with the writing of system requirements or specifications. Rather, it starts with the identification of a need, together with an opportunity. To identify an opportunity, the need must be matched with a technical feasibility of a system that will be capable of addressing this need. Very often, needs are identified by potential clients or users, but system engineers in the modern environment are playing an active role in looking for system creation opportunities and identifying market needs. Sometimes, they may even envision a need that was not realized by the customer, but is rather based on a new technical capability (a solution looking for a problem).

2. Systems engineering management directs the design, development, synthesis, and creation of a system, rather than the analysis of it, and the transforming of an operational or customer need into an existing system. It therefore requires the creation of the system architecture and acting as "chief designer" of the system, rather than just being the system analyst.

3. Systems engineering management means taking a "holistic view" (the design of the whole as distinguished from the design of the parts). Such a view is multidisciplinary in nature, rather than disciplinary or interdisciplinary.

4. Systems engineering involves a technical part and a managerial part; that is, it requires making technical decisions and tradeoffs while controlling and managing the efforts of different experts and teams from various disciplines.

5. In addition to its technical and managerial elements, systems engineering also consists of economic and analytical elements, as well as social, environmental and political aspects.

6. Finally, the system engineer must be in close contact with the customer, not only in identifying the need, but also in seeing that the system produced is going to meet these needs.

2.3 ACTIVITIES AND ROLES OF THE SYSTEMS ENGINEERING MANAGER

2.3.1 Activities

The creation of a system is characterized by numerous activities. These activities are performed at various phases of the system's life cycle, as well as within the well-established steps of the system engineering process (Chase, 1985; Blanchard and Fabrycky, 1998; Blanchard, 1991; U.S. Department of Defense, 1992; Mar, 1994; Lake, 1994; Sheard, 1996). Here are the most common ones:

1. The identification of an operational need with an opportunity to create a system to respond to this need.

2. Setting the exact system and functional requirements to ensure the best fit to customer needs.

3. Dividing and allocating the functional requirements into different subfunctions and modes of operation.

4. Choosing the system concept that will best fit these requirements.

5. Designing the system architecture, based on the chosen concept.

6. Dividing the system into separate subsystems and components to ensure overall optimization, least interfaces, and least mutual effects of the various subsystems.

7. Optimizing the specifications of the various subsystems through simulation, analysis, and trade studies.

8. Managing the interaction with various engineering groups that perform the subsystems' design while integrating various people and disciplines.

9. Performing the integration of the various subsystems into a total system.

10. Evaluating the performance and qualifications of the final system through simulation and testing activities.
11. Demonstrating the operating system to customers and convincing them that it responds to their needs.

Brian Mar (1994) has grouped these activities into four basic-steps used repeatedly in the process of systems engineering to describe the development of an answer to any type of problem. These steps are (1) functions—define what functions the solution must perform; (2) requirements—define how well each function must be performed; (3) answers—search for a better answer and manage risk associated with that answer; and (4) *tests*—demonstrate that the answer performs the needed functions. For a more detailed description of the systems engineering process, see Blanchard and Fabricky (1998), U.S. Department of Defense (1992), and Lake (1994).

To guide the work and activities of systems engineering, a set of simple but powerful principles has been identified (Ottenberg, 1993). The first principle is iterative top-down (or hierarchical) design: a complex system is designed by breaking the system into its component subsystems and then repeating the process on each subsystem. The second principle is bottom-up integration: large systems are built by taking the lowest level components and putting them together one level at a time; bottom-up integration is the same as top-down design in reverse. The third principle is that of the system life cycle: it is the understanding and tracking of the progression of a system from inception to design, to construction, implementation, operation, maintenance, and eventually to its shutdown, disassembly, and disposal. The forth and last principle is "user perspective": systems engineering must take into account what the user wants, needs, prefers, is happy with, and is capable of.

2.3.2   The Roles of a Systems Engineering Manager

A systems engineering manager has to perform at least ten different roles during the process of creating the system (Shenhar, 1994; Mar, 1994; Sheard, 1996). These roles can be described as follows:

1. *Need Identification and Customer Linkage.* To identify the need and the system opportunity by matching need and technical feasibility and be the link bond between customer needs and system idea and design during the entire process of system creation.
2. *Requirements Management.* To develop a set of system and functional requirements based on customer needs (Sheard, 1996).
3. *Architecture and System Design.* To be the lead person in envisioning the system's concept, and to create the link between the system's requirements and the system's configuration (Rechtin, 1991).
4. *Integration.* To see the entire picture and how each part is contributing to the performance and feasibility of the system as a whole (Grady, 1994; Hunger, 1995). Also, to coordinate the work of the various disciplines and professions involved and manage the interfaces among them such that the result is an overall optimal system.

5. *Analysis, Development, and Testing.* To collect data from various sources, perform modeling and simulation and analyze them as a basis for decision making to confirm that the system is designed to its requirements; and to test and verify that the system built will meet these requirements as designed.

6. *Process Management.* To plan, document, direct, and follow the systems engineering process.

7. *Technical and Risk Management.* The process of systems engineering involves technical and tradeoff decisions and the resolution of technical problems and conflicts at different interface points. These conflicts are primarily professional rather than personal, and reflect the different views of the distinct disciplines involved in creation of the system. This role also involves risk assessment on various system elements and overall risk management during the system creation.

8. *Leading, Coordinating, and Managing.* In addition to being a technical manager, the systems engineer must be a manager of activities and leader and coordinator of people. The job includes dealing with work plans, schedules, and budgets, but also working with people—organize their work, motivate them, communicate with them, and deal with their needs.

9. *Logistics and Operations Management.* To consider and include maintenance, operation, logistics, and disposal concerns during the requirements, design, and development phases, and to "escort" the users during the operational phase of the system, to "break then in," to answer questions, and solve anomalies.

10. *Information Management.* To see the overall information needs of the system, plan the forms and means in which information will be created, disseminated and stored, and direct the process of information sharing and configuration control.

Thus, the systems engineering manager's role is not limited to dealing with functional requirements and system synthesis only. It should be seen in a wider sense as identifying opportunities, managing the total process of the system's creation, and dealing with the customer. The system engineering manager is therefore an entrepreneur, as well as an architect, designer, planner, analyzer, integrator, manager, tester, decision maker, persuader, and marketer. The degree to which each of these roles is performed during the process of system creation depends on the specific kind of system involved. In the following sections we provide a framework for the classification of different kinds of systems and for the implementation of specific styles of systems engineering management associated with these systems.

2.4 TOWARD A COMPREHENSIVE FRAMEWORK FOR THE IMPLEMENTATION OF SYSTEMS ENGINEERING MANAGEMENT: THE TWO-DIMENSIONAL TAXONOMY

We turn now to the implementation of systems engineering management. One of the fundamental elements in understanding its nature is the need to distinguish between the system type and its strategic, as well as its systems engineering and managerial problems. Obviously, there are great differences among systems and among the processes of their creation. Consider, for example, the case of building a new metropolitan highway, as compared to the development of a new space vehicle, or the case of making a new model

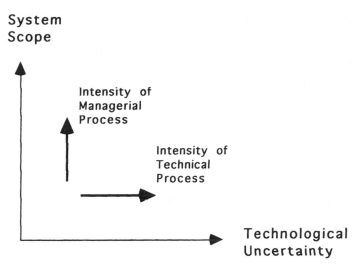

Figure 2.2   The intensity and importance of the technical and managerial processes depend on the levels of technological uncertainty and system scope.

of a small appliance such as a CD player, as compared to the effort of building the English Channel Tunnel. Naturally, these efforts are called projects, and they employ system engineering procedures; however, it is clear that their differences outweigh their resemblance.

Following some previous distinctions among systems and projects (e.g., Ahituv and Neumann, 1984; Blake, 1978; Steele, 1975; Wheelwright and Clark, 1992), our premise is that the systems engineering process is not universal and that management attitudes must be adapted to the specific system type. We will use a two-dimensional framework for the classification of systems and for the selection of a specific systems engineering style (Shenhar and Bonen, 1997). In this framework, systems are classified according to their technological uncertainty at the moment of project initiation and system scope (or complexity), which is their location on a hierarchical ladder of systems and subsystems.

As claimed, the systems engineering task includes both a technical process and a managerial process. The extent and intensity of each process depends on the level of the system's technological uncertainty and system scope (see Fig. 2.2). To identify the exact systems engineering process we must refine the two dimensions, as will be described next.

The innovation literature has often used a traditional distinction between radical and incremental innovation (Abernathy and Utterback, 1978; Dewar and Dutton, 1986; Ettlie et al., 1984; Zaltman et al., 1973). Radical innovation is frequently based on new technology, and it often results in products new to the firm and new ventures unrelated to existing businesses. In contrast, incremental innovation introduces relatively minor changes in existing products, and generally takes the form of product modifications, upgrades, derivatives, and line extensions. In practice, however, this low–high dichotomy is insufficient to capture the full spectrum of modern engineering tasks (Shenhar, 1993; Shenhar and Dvir, 1996), and there is a need to use a more refined classification.

The extent of technological uncertainty is assessed by the organization at the time of the project's initiation (Galbraith, 1977). Our empirical studies showed a clear distinction among four different levels, designated as: A—low-tech; B—medium-tech; C—high-

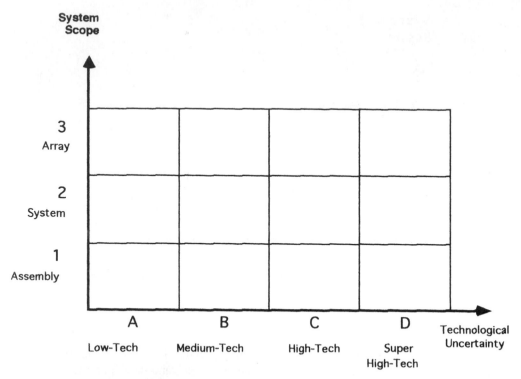

Figure 2.3   A two-dimensional taxonomy of engineering projects and systems.

tech; and D—super high-tech (Fig. 2.3). The second dimension—system scope—is based on the complexity of the system as expressed by the different hierarchies inside a product. Since products are composed of components and systems of subsystems, hierarchies in systems may include many levels (Boulding, 1956; Shenhar, 1991; Van Gigch, 1978). As found, however, systems engineering practices exhibit a predominant split between three clusters of engineering project styles: assembly, system, and array (Fig. 2.3). In the following section we define these project types.

2.4.1   The Technological Uncertainty Dimension

Type A: Low-Tech Projects.   Type A are those projects that rely on existing and well-established technologies to which all industry players have equal access. Although such projects may be very large in scale, no new technology is employed by the firm at any stage. Typical projects in this category are construction, road building, and other utility works that are common in the construction industry.

Type B: Medium-Tech Projects.   Type B rest mainly on existing technologies, yet they incorporate some new technology or a new feature of limited scale. Typical projects of this kind include engineering efforts of incremental innovation, as well as improvements and modifications of existing products and systems. They may involve new models in well-established product lines, such as automobiles or consumer electronics; they could also include defense projects that involve upgrades of existing systems. Such projects are found in almost all industries, including mechanical, electronics, aerospace, and chemical.

Type C: High-Tech Projects.   Type C are defined as projects in which most of the technologies employed are new, but existent, having been developed prior to the project's initiation. Many projects in the high-tech industry that involve a new family of products or a new technology generation produce products of this kind. Similarly, most new defense development efforts would normally be characterized as high-tech system projects, incorporating new technologies that have been recently developed (Fox, 1988). Typical industries are electronics, computers, and aerospace.

Type D: Super High-Tech Projects.   Type D are based primarily on new, not entirely existent technologies. Some of these technologies are emerging, others are unknown even at the time of the project's initiation. The project's execution period is therefore devoted in part to the development of new technologies, testing, and selection among alternatives. This type of development project obviously entails extreme levels of uncertainty and risk; it is relatively rare, and is usually carried out by only a few and probably large organizations or government agencies in leading electronics and aerospace industries. Typical products are the development of a new nonproven concept, or a completely new family of systems. The most familiar example of such a project was the Apollo Moon-landing mission; however, other smaller scale projects can also be found in this category.

2.4.2   The System Scope (Complexity) Dimension

Scope 1: An Assembly.   Scope 1 is a collection of components and modules, combined together into a single unit. A typical assembly may perform a well-defined function within a larger system, hence constituting one of its subsystems; it can also be an independent, self-contained product that performs a single function of a limited scale. A radar receiver, a missile's guidance and control unit, or a computer's hard disk are common examples of assemblies (subsystems) within larger systems; CD players, coffee makers, washing machines, and other household appliances can be considered independent assemblies of the second kind. Such products are either functioning as autonomous, or they require only a limited amount of human interaction.

Scope 2: A System.   Scope 2 is a complex collection of interactive units and subsystems within a single product, jointly performing a wide range of independent functions to meet a specific operational mission or need. A system consists of many subsystems (and assemblies), each performing its own function and serving the system's major mission. Radar, computers, missiles, and, for that matter, entire aircraft are typical examples of systems performing independent tasks. Systems are capable of performing a complex mission, and their use involves considerable man–machine interaction.

Scope 3: An Array.   Scope 3 is a large collection of systems functioning together to achieve a common purpose. A different terminology for an array may be a "super-system," a "system of systems," or to mention a Greek word, a "systema," all being an expression of the array's nature as a conjunction or conglomeration of systems. Usually arrays are dispersed over wide geographical areas and normally include a variety of systems of many kinds. A nation's air defense system, consisting of early-warning radar, command and control centers, combat aircraft, and ground-to-air missiles is a good example of such a supersystem. Similarly, the public transportation network of a large city, or the city's highway system may also be considered typical arrays. Arrays achieve wide-range missions, are based on the simultaneous functioning of many systems, and their operation involves the association and interaction of many people.

2.4.3 A Retrospective Classification of Well-known Programs

Several previous programs can be easily coded according to this framework. For example, IBM's project of building the personal computer in the early 1980s would be characterized in this framework as a high-tech system development project, since it was based on new, but existent technologies. Similarly, development projects of defense systems, such as the Patriot missile, are usually based on recent technologies that were developed prior to the main project effort. Such projects are normally contracted following a well-structured procedure of defense acquisition to contractors that demonstrate a viable capability and sufficient maturity of technologies. Other examples of high-tech projects were the development efforts of the first generation of videocassette recorders introduced by JVC and Sony in the mid-1970s (Rosenbloom and Cusumano, 1987). According to our framework, however, it is an assembly. Lockheed's SR-71 Blackbird Reconnaissance Aircraft initiated in the late 1950s with no clear technology available, and almost no known configuration (Johnson and Smith, 1985), can in retrospect be classified as a super high-tech system. New York City's Transit Authority Capital Program of modernizing the city's transit infrastructure, would be classified as a low-tech array (Manne and Collins, 1990); the English Channel Tunnel is a medium-tech array, since it involves some new challenges and risks of digging such a long undersea tunnel (Lemley, 1992). Finally, the U.S. Strategic Defense Initiative, or as it was often called Star Wars (Lawrence, 1987), is one of a few examples we can find of super high-tech array programs.

2.5 DIFFERENT SYSTEMS ENGINEERING MANAGEMENT ROLES FOR VARIOUS PROJECT TYPES

What are the distinct project characteristics and systems engineering management styles used for different kinds of systems and projects? As found, different concerns characterize the two dimensions. In this section we outline the different systems engineering management roles that were defined previously and how they change with different kinds of projects according to the two-dimensional taxonomy. These roles are summarized in Tables 2.1 and 2.2.

2.5.1 Different SEM Roles Along the Technological Uncertainty Dimension

Since low tech projects (Type A) use no new technology, their main characteristic is that no development work is required and no changes are made in the design—they are simply designed and built, and usually no testing is required. System needs are often set by the customer prior to signing the contract and before the formal initiation of the project execution phase. System functional requirements are usually simple, straightforward, and often static. The architecture and system design efforts are often performed in a standard, well-established field, and with very few or no design iterations. Integration is usually of

TABLE 2.1   SEM Roles and Functions for Different Levels of Technological Uncertainty

	Project Type			
	A	B	C	D
SEM Roles	Low-Tech	Medium-Tech	High-Tech	Super High-Tech
Need identification and customer linkage	Need is determined by customer. Simple customer links	Need determined by market; usually incremental change in previous product. Fast customer feedback is needed	Need either required by customer or identified by contractor. Customer expects extensive gain and may be actively involved	Far-reaching need; enormous opportunity. Customer often not aware of technical feasibility—may be less involved
Requirements management	Straightforward, standard process of requirements creation; no changes	Requirements must reflect the advantage of system; minor changes expected	State-of-the art requirements; may be set and changed in an iterative process	Pushing the state of the art; requirements could stay flexible during development
Architecture and system design	Standard architecture design	Designing an improved system; no breakthroughs in system design	New system design concepts; looking for system advantages	Breakthrough system; unknown technologies and configuration
Integration	Usually no complex integration problems	Some integration problems, usually fixed after a short while	Extensive integration problems; interference and mutual effects	Enormous problems from integrating new and unknown technologies
Analysis, development and testing	Standard analysis; no need for development or prototyping; almost no testing	Some analysis and development; one or two design cycles; prototypes are usually helpful; some testing is needed	Complex analysis and development; three or more design cycles; prototypes and testing are necessary	Extensive analysis and development of new technologies; must build a small-scale prototype to test new ideas in many cycles
Process and life-cycle management	Simple, well-known process	Follow a structured process, based on industry experience	Often a nonstandard process—determined at inception of system design	No previous guidelines; must adapt a unique process
Technical and risk management	No specific technical problems	Some technical decisions and problems; risk management may be helpful	Extensive technical knowledge needed; major technical problems; must use risk management	Enormous technical challenges and risks; many tradeoff decisions; intensive risk control and management
Leading, coordinating, and managing	Standard management in a well-established field; good planning is key	Leading and managing the team for efficient results; planning with some slack	Leading and motivating technical professionals; difficulty to plan with precision	Leading in a dynamic, ambiguous technical environment; leave much room for slack
Logistics and operations management	Standard, well-established logistics problems	Some new logistics, based on system's new features	Mostly new logistics and operational problems	Totally new logistics and operations; must be developed with system
Information management	Standard information sharing; a fixed and low frequency of interaction	Moderate extent of new information to share and store	Extensive new information is created; must be recorded and disseminated promptly	Continuous changes—must track, control, and share information on an ongoing basis

123

no specific concern, nor are the process or the technical management needs. While this kind of project may still have problems, they are mainly encountered at the higher levels of the system scope ladder (at the System and Array levels), as a result of system complexity and not of technological difficulty (this is discussed later).

Being an incremental improvement over previous practices, the need for a medium-tech (Type B) project is usually determined by the market. The opportunity is recognized from the need to improve a previous product or when the time comes to present a new model in a well-established field. Customer feedback is extremely important and should be sought as soon as possible. The requirements are mainly set in advance and should reflect the advantage of the new or improved system. Some changes may be introduced during the product development phase, but product design is frozen relatively early. Architecture and system design should be focused on the improved system and no technical breakthroughs are expected in this kind of projects. There may be some integration problems, but they are usually fixed after a short period. The analysis, development, and testing are moderate, may be completed after only one or two cycles, and some prototyping may be helpful. The systems engineering process must be structured, and usually should be based on the industry practice. Technical and risk management are important to a certain extent, and should emphasize the product advantage and distinction from previous generations. Leadership should be focused on efficient results and getting the work done as soon as possible, and resources may be planned with high accuracy while leaving some minor slack. The logistics and operation management in this kind of project should be focused on previous experience, with emphasis on the new features of the product, and the information management requirements are relatively moderate.

High-tech (Type C) projects usually produce a new generation of a product based on a collection of newly developed technologies. Integrating several new technologies for the first time obviously leads to a high level of uncertainty and risk, and it affects most roles and functions of systems engineering. The need for a Type C project may be determined by the customer or the contractor. Since it is a new type of system, however, the customer expects extensive gain over previous means and is usually actively involved in setting the needs as well as during the system development period. System requirements are focused on meeting the state of the art, and are usually derived iteratively with the customers or potential users, and with many changes before the final decision is made. The system architecture and design are new and bear little resemblance to previous system designs. Designers are looking for full-system advantages based on the new technologies installed. Integration may be of great concern in this kind of project, since the inclusion of new technologies for the first time may cause severe interference and mutual effect problems. Analysis, development, and testing are complex, usually completed only after three or more design cycles, and with several prototypes built along the way. The systems engineering and life-cycle processes in this type of project are often nonstandard—it depends on the specific system—and are determined during the system design phase. The technical knowledge required is extensive; it requires the solving of many technical problems; and it most likely needs a risk management program. The leadership in this type of project must acknowledge the expertise of the professionals involved. It must know how to motivate them and integrate their skills into an overall system effort. The planning of resources becomes difficult and requires more slack than in Type B projects. Similarly, the logistics and operations management are of a new kind and must be adapted to the specific system type. And the information created is extensive and must be managed carefully and recorded and disseminated promptly.

Super high-tech (Type D) projects are build to address a far-reaching need, one that is usually beyond the existing state of the art. Such projects are "pushing the envelope," and therefore require extensive development, both of new, nonexisting technologies, and of the actual system. Typically, they offer an enormous opportunity for all parties, but often the requirements are stated by the customer. In many cases, however, the customer may not be aware of the technical feasibility or potential problems and may end up less involved in the development process compared to lower levels of technological uncertainty. The requirements may need to be changed several times, and they should stay flexible for an extended period until they are finally set. System architecture of this type is usually of a breakthrough nature and it requires enormous creativity and a large problem-solving capability to combine and integrate unknown technologies in a completely new configuration. The analysis, development work, and testing are extensive. A great effort is put into the product design and development, as well as into the development of the new technologies. In most cases, this requires building a small-scale prototype on which to test the new technologies. The development process may therefore require up to five or more cycles of design–build–test. The process in these projects is completely new and must be planned for the particular system with no previous guidelines. The technical challenges and risk in this kind of project are enormous. They require first-line technical expertise, intensive tradeoff decision making, and careful risk control and management. Leading the technical teams is also a unique challenge. The environment is dynamic and ambiguous, and extensive slack must be left for absorbing contingencies and surprises. The logistics and operations management is also unique and must be developed specifically for the new type of system. Finally, the information created in this type of project is extensive, with continuous changes and an ongoing need for information sharing and documentation.

2.5.2   Different SEM Roles Along the System Scope Dimension

The execution of an assembly-type project (Scope 1) is usually performed within one organization, and often under the responsibility of a single functional group. It involves mostly technical people, with limited staff support or no staff at all. The need addressed by this type of system scope is usually a limited one, and the functional requirements are quite simple, as they are for the system design, the integration, the analysis and testing, and the system engineering process. Scope 1 is mostly concerned with design-related issues, such as design for manufacturability and reliability, and much less with total system optimization. The core of the systems engineering activity is the process of concurrent engineering and the involvement of cross-functional teams. However, most technical problems are solved within one discipline or a small number of groups. The logistics and operations management are also simple, and the extent of information management required is quite limited, with a lot of informal interaction within the team involved. In conclusion, the systems engineering style at this level is fairly simple and very informal, and it is based on close interaction among members within a relatively small team.

Organizing a system-type project (Scope 2) usually involves a main contractor who establishes a project management team and who is, in turn, responsible for managing and coordinating many internal and external subcontractors. The management team includes technical personnel, administrative staff, and systems engineering people. The systems engineering roles at this level encompass all the activities that are performed at the lower level and much more, and they primarily involve the entire spectrum of SEM roles. Defining the need is a complex activity that requires many iterations and continuous inputs

TABLE 2.2 SEM Roles and Functions for Different Scope Levels

SEM Roles	System Scope Level		
	1 Assembly	2 System	3 Array
Need identification and customer linkage	Addressing a simple need, either for the consumer or as a subsystem in a larger system	Addressing a complex system need; many inputs from the customer are needed	A large need of many individuals and organizations; customer represented by a small group or coordinating organization
Requirements management	Simple requirements, usually performing one function	Complex, multifunctional requirements; several modes of operation	Full requirements are hard to specify; often added or modified as the array is built gradually
Architecture and system design	Simple configuration and system design	Complex architecture; several iterations are needed	General architecture and interface specified; specifics are left to subunits at the system level
Integration	No particular integration problems	Extensive integration problems; require several iterations and enough time	Fewer integration problems; systems are added gradually
Analysis, development, and testing	Simple analysis and testing of functions of operation	May require complex analysis and testing of many functions and modes of operation	Significant analysis during design; some testing as systems are added gradually
Process and life-cycle management	Usually simple SE process	Must follow a well-planned SE process and life-cycle management; use existing standards	Specific process depends on the unique array built; no standards exist at this level
Technical and risk management	Technical problems usually in just one or a few technical disciplines	Many technical areas involved; may require extensive risk management	Technical problems usually solved at the lower system level; at this level must deal with legal, environmental, and social issues
Leading, coordinating, and managing	Leading and managing a small team, usually in one or a few areas of expertise	Leading many teams in different areas of expertise; extensive planning and resources management	Remote from leading the teams; management focuses on guidelines and policies as well as financial and schedule control
Logistics and operations management	Simple logistics and operational problems	Complex logistics and operations management	Logistics and operational problems are usually left to the system level
Information management	Simple information sharing; a lot of informal interaction within the team	Formalize and bureaucratize the information flow; some informal interaction with subcontractors and customers	Very formal and tight system for information management; no informal interaction at all

from the customer or user. The requirements are complex, multifunctional, and usually involve several modes of operation. System architecture, design, and integration are complex, too, and will also require several iterations. So would system analysis, development, and testing. One of the major difficulties in integrating this system level is the need to achieve an overall optimized and effective system rather than a suboptimization of its different subsystems or components. The systems engineering process for this level must be performed carefully and follow a well-planned life-cycle model. The process may benefit from one of the existing standards that were primarily developed for this level [e.g., U.S. Department of Defense, 1992; Institute of Electrical and Electronics Engineers (IEEE), 1994]. However, complexities at this level involve the consideration of many areas and an extensive effort of risk management. Leadership, too, requires leading teams in different areas and the careful planning of resource allocation and scheduling. The logistics and operational management are fairly complex, and the information management must be carried in a rather formal and bureaucratic way to handle systematically all the facets of information that are created in various parts of the systems. Two major concerns are configuration control and configuration management, to keep track of all the changes and the detailed baseline of design. In summary, the management style of system creation must be very formal and bureaucratic, yet, some informal relationships always develop at this level between the management team and some subcontractors and customers.

Array-type projects (Scope 3) are usually designed and built to address a large need of many individuals and organizations. To manage the effort of building or even improving an array, however, a small representative group is usually established as an umbrella organization to speak for the needs of the users and to coordinate the efforts of the various system project organizations that are involved. The full set of requirements is usually hard to specify in advance, and in many cases it grows iteratively together with the development of the array and in cases where its components are build and added onto the system gradually. The coordinating organization is typically staffed with legal, administrative, and finance experts, who provide the proper control in each discipline. This organization is usually less concerned, however, with technical matters. It provides the general system architecture and guidelines, and leaves the specifics and the technical details to the separate projects that build the system subunits. It is also much less concerned with system integration. After some initial, usually significant analysis, the development and testing is left to the subunits. The process must be developed specifically for each array effort, and usually no existing standards are found for this level. A major factor of risk in array projects involves environmental and government regulations, which become critical due to the dispersed nature of the final product. The leadership in these projects is usually remote from actual working teams, and planning involves a general and central master plan, which is followed by detailed planning at the systems level. Most management activities in these projects include controlling, financial, and contractual issues, which require extensive amounts of documentation, while the logistics and operational problems are left to the system level.

Finally, managing the information of such programs is primarily done in a formal, bureaucratic, and rigid way, with most of the interaction being formal and in written form.

2.5.3 Interaction Between the Two Dimensions: Combining High Uncertainty and High Scope

As can be seen from the previous discussion, when moving from lower to higher levels of technology while keeping the system scope unchanged, managers' and system engineers' concerns shift toward additional technical issues—more design cycles, more testing, and longer periods of ambiguity in requirements and specifications. Similarly, increased scope with constant uncertainty results in additional concerns for formal administrative issues—more planning, tighter control, more subcontracting, and more bureaucracy. In addition, when one moves simultaneously along both dimensions, from low- and medium-tech assemblies toward high-tech and super high-tech systems and arrays, new challenges must be addressed, and there is a strong interaction between the two dimensions.

First, it is at this end of the model that systems engineering must be practiced to its full extent. The development of large, highly uncertain, multidisciplinary systems, which consist of many subsystems and components, involve the incorporation of most modern systems engineering tools and processes to optimally harmonize an ensemble of subsystems and components. Such tools are almost not present in low-tech or medium-tech system projects. Systems engineers in high-end projects are a natural part of the management team, and they are responsible for creating the system's architecture and configuration and incorporating a multitude of technologies that have to cohere in the final design. They are also accountable for the system's reliability, maintainability, supportability, human factors, and economic feasibility of the entire engineering effort.

In lower tech systems, assembling, integrating, and testing all subsystems into one unit is usually not an issue, since all pieces readily fit together. Integration difficulties become prominent, however, in the higher-uncertainty-type system projects. In these projects, the successful production of the separate subunits is one thing, while integrating them into one working piece is another. Problems of interfaces, energy dissipation, electromagnetic interference, and even lack of space, require long and tedious processes of assembly, integration, frequent testing, and necessary tradeoffs in many of these projects.

Finally, there are problems of configuration and risk management. Special software must be used in high-uncertainty, high-scope projects to keep track of all the decisions and changes, and to identify the potential interactions that would occur with each change. Such projects develop a cumulative database for their design baseline, which, in some cases, must be updated on a day-to-day basis. As for risks, although all projects involve a certain level of risk, higher-scope, higher-tech projects are more sensitive to the need for systematic risk analysis and risk mitigation. In such projects, a risk management procedure is usually integrated into the program. It often involves issues of design control, personnel qualifications, procurement management, quality control, and budget and schedule control.

2.5.4 The Danger of Misclassification

Systems engineering and program management must be conducted according to the proper style and be adapted to the system type. Such style should be chosen according to the classification presented earlier. In that sense one may see the preceding discussion as a description of project and system "ideal types." When the proper style is employed, we claim, that the chances for project success are much higher. When a wrong style is utilized, however, or the when the system is misclassified, this may result in substantial difficulties and delays in the process of the system creation. For example, working on a project perceived as a Type C, when it is really a Type D, may lead to problems that are beyond the competence within the framework of the organization. It may require higher technical skills, additional resources, longer periods of development and testing, and a much later design freeze. A retrospective analysis of NASA's Space Shuttle program suggests that this program had all the ingredients of a super-high-tech project, while it was managed as a lower, high-, or even medium-tech program. A D-type philosophy would have vested the program managers and engineers with a different systems engineering attitude toward risks, possible development problems, and probability of failure, and might have help eliminate many of the circumstances that eventually led to the *Challenger* tragedy (Shenhar, 1993). In contrast, when a project is actually a Type C, and it is wrongly perceived as Type D, many overruns and difficulties may be attributed to technology, while they are simply the result of inept administration and weak leadership.

2.6   THE SKILLS, TOOLS, AND DISCIPLINES INVOLVED IN SYSTEMS ENGINEERING MANAGEMENT

How can we make all this work? How should we select, educate, and train system engineering managers? What kind of skill, knowledge, tools, theories, and disciplines do we need to develop in order to assure the required skills and capabilities for a successful implementation of systems engineering management processes? We start with the skills.

2.6.1   The Skills

What skills does one need to be a good systems engineering manager? A systems engineering manager is an engineer and much more, being a person who has the capability of integrating various disciplines into an overall view of a complete system, an engineer who "knows enough about many things," who is capable of utilizing a "holistic view" and making extensive technical, managerial, financial, and many other decisions.

A systems engineering manager must know about many different things, without necessarily being an expert in any field, with one exception—the system itself. But to be the system's expert, a systems engineer must first of all be proficient and have experience in at least one area—technical or other—that is relevant to the system at hand (Booton and Ramo, 1984). By being (initially) good in one area, the engineer can develop this technical expertise and gain the required confidence for further and more complex multidisciplinary decisions. However, a systems engineering manager must be much more than this. Such a person must be able to look beyond this initial area of expertise, to become a "generalist" rather than a specialist, and possess extensive conceptual skills. The systems engineering manager must also be able to understand, apply, and interpret various analytical tools and methods of system analysis. Finally, such person should also possess "the managerial touch," that is, be able to manage people—motivate, lead, and communicate, with them— as well as be a proficient administrator and a good decision maker.

2.6.2   The Tools and Disciplines

The multidisciplinary nature of systems engineering management suggests that it involves the application and combination of various theories, tools, and disciplines. The disciplines, and methods and tools, of systems engineering can be grouped into the following clusters:

1. Mathematical and statistical tools of analysis
2. Technical and engineering disciplines
3. Information technology, computers, and human-machine interfaces

   4. Economic and financial disciplines
   5. Management and business-related disciplines
   6. Behavioral sciences and organization theory
   7. Systems engineering procedures, standards, and checklists

A short description of these tools and disciplines follows:

   1. *Mathematical and Statistical Tools.* These include the major mathematical tools that are used both in analysis and in synthesis of systems. They involve the mathematical theory of systems, differential equations, linear algebra, linear and dynamic programming, complex variables, mathematical transformations, morphological descriptions, flowcharts, and other representation tools. They also include probabilistic tools for decision analysis, risk analysis, and stochastic systems design, as well as statistical tools for simulation and for reliability and quality analysis.
   2. *Technical and Engineering Disciplines.* This is the technical disciplinary portion of the systems engineering field. Every system involves a specific major engineering discipline, and managing the creation of the system requires the understanding and application of this discipline (e.g., mechanics, electronics, computers, aeronautics, or chemical processes).
   3. *Information Technology, Computers, and Human–Machine Interfaces.* Although many systems are purely information systems—namely, their technical portion involves computers, data, and software—almost every kind of complex system today is heavily dependent on modern computing and information technology (Sage, 1987). The relevant disciplines for systems engineering include computer science and computer architecture, information systems, computer networks, data administration, and software engineering. In addition, systems engineers must be familiar with the modern principles of human–machine interfaces and cybernetics.
   4. *Economic and Financial Disciplines.* Obviously, this portion involves the economical and cost aspects of the systems engineering effort. It may include the disciplines of finance, macro- and micro-economics, economic analysis, accounting, costing, and financial control.
   5. *Management and Business-related Disciplines.* This portion includes many additional managerial and business administration elements, such as marketing, strategic and business planning, contract management, operations management, project management, quality management, and logistics.
   6. *Behavioral Sciences and Organization Theory.* The systems engineering manager is required to have an understanding of the theories relevant to human behavior, such as motivation, leadership, and communication. This portion also involves the wide basis of organization theory, like organization structures, and group and team building.
   7. *Systems Engineering Procedures, Standards, and Checklists.* This portion involves the procedures and standards that have been developed over the years and are used during the systems engineering process. It includes a description of the various system life-cycle phases and the role of the system engineering manager during the life cycle of the project. It includes techniques for mission and functional analysis, functional allocation, and the application of trade studies. It also involves reviewing

systems engineering and other related standards (e.g., U.S. Department of Defense, 1992; IEEE, 1994).

Although it is not easy to characterize the disciplines involved in the systems engineering process, it is even less clear how to characterize the need for a holistic view. As already mentioned, this is a central part of the systems engineering activity, but what exactly is it all about? And how can one acquire it? We deal with this question in more detail in the coming paragraphs.

2.7 DEVELOPING EDUCATIONAL AND TRAINING PROGRAMS IN SYSTEMS ENGINEERING MANAGEMENT

Any educational and training program in systems engineering management must provide an interdisciplinary and multidisciplinary framework or curriculum. It must emphasize system sciences, operations research, mathematical methods, system design methodologies, business, management, and many other approaches that can be expected to have important applications in the design, development, and implementation of complex large-scale systems. Such multidisciplinary programs should be offered at the graduate level for candidates who have acquired extensive maturity and experience in their field, as described in the following discussion.

2.7.1 Candidates for Systems Engineering Management Education

Before discussing a suggested framework for a systems engineering management curriculum development, we may ask: Who should be the typical participants in such a program? Can they be freshmen or young engineers who have recently acquired their first engineering degree? The answer is absolutely no. Any program of this type can only enhance, develop, and nurture systems' related skills that already exist, and have been previously demonstrated. A typical candidate for an educational program in systems engineering management should therefore:

1. Be an engineer, a scientist or have at least a Bachelor's degree in an area that is relevant to the system involved;
2. Have several years of proven experience and achievements in one of these disciplinary areas;
3. Have some experience in interdisciplinary or multidisciplinary work, performed in conjunction with other professionals belonging to different areas one's own;
4. Have demonstrated interpersonal skills in teamwork, motivation, communication, and leading other people.

An ideal candidate for a systems engineering management program is a person who possesses a B.Sc. and preferably an M.Sc. degree in engineering, science, or another field that is relevant to the system involved, as well as have previously worked in that person's own discipline as an independent worker for several years and for several additional years as a group leader or project manager on a interdisciplinary but small-scale project. At this stage the candidate is ready to gain most from systems engineering management education.

2.7.2 The Objectives of an Educational Program in Systems Engineering Management

Based on the candidate's initial knowledge and experience, the objectives of the program should be to augment knowledge and develop skills in the following areas:

1. Recognize operational needs, identify market and technological opportunities, forecast the development of operational and technological processes.
2. Formulate new concepts and devise systems' solutions capability of analyzing and designing large-scale systems while integrating various disciplines.
3. Manage projects of design and development of systems while considering the aspects of cost, quality, reliability, manufacturing, marketing, maintenance, service, and an overall view of the system's life cycle.

By considering the previously mentioned areas and disciplines required in the systems engineering effort, a curriculum in systems engineering management should be directed toward developing the following components of knowledge and skills:

1. A holistic thinking capability and conceptual analysis;
2. Methodologies for performing systems engineering in its various stages;
3. Analytical, mathematical, and statistical tools that are used for systems' analysis, systems' design, and systems' problem solving;
4. Basic principles and theory in different technological and engineering disciplines;
5. Skills in computers, networks, and information systems;
6. Economic, financial, and other nontechnical managerial and business disciplinary areas that are relevant to systems engineering;
7. Interpersonal skills;
8. Leadership, organization, and administrative skills.

2.7.3 The Possible Structure for a Graduate Educational Program in Systems Engineering Management

An academic educational program in systems engineering management should be directed toward the joint development of two parts—the technical part and the managerial part—and it should be multidisciplinary in nature. Since most areas of academic endeavor are discipline-oriented, there is a need for cooperation between different schools at the university, usually engineering and management. It could be led by either one, but each school should be responsible for different elements. In view of the components typically offered by such schools, the program may be based on three stages and five sections. (For a suggested list of courses see Shenhar, 1990, 1994). The five suggested sections are as follows (see Fig. 2.4):

1. *Basic Studies.* This preliminary section is designed to bring all participants to a common level and to refresh their knowledge on the basics of mathematics, statistics, computers, and the like. This refresher is usually needed since most students in such a program have been away from formal studies for several years.

School Responsible:	Management School	Engineering School
Stage 1:	Section A Basic Studies	
Stage 2:	Section B Disciplinary Studies	Section C Engineering Systems, Technologies, and Information
Stage 3:	Section D Management Studies	Section E Systems Engineering Concepts and Techniques

Figure 2.4 The logical structure of a systems engineering management educational program.

2. *Disciplinary Studies.* This section should include courses in various disciplines that are adjunct to systems engineering. The section involves tools and techniques of analysis, evaluation and decision making, as well as various additional disciplines, such as economic analysis, accounting, finance, contracting, pricing, and law. The knowledge acquired in this section will enable the systems engineers to apply these tools on their own or it will help them to communicate with other professionals working on the system and to make decisions regarding their work and its effect on the system at large.

3. *Engineering Systems, Technology, and Information.* In this section students of systems engineering gains familiarity with various kinds of systems. Each course in this section will provide the theoretical background for a certain family of systems or technologies. Students will learn the critical factors involved in different systems and gain an understanding of their impact on a large-scale system that includes several subsystems. Systems will be surveyed in broad terms, rather than in depth, and the list of courses will include electives. Obviously, students who were originally familiar with a certain discipline will not participate in courses that involve their original area of focus. Also, while all other sections are common to all student groups, in this section a distinction can be made, for example, between military and commercial systems and between public and private-sector systems, as well as other classes of systems. In addition to specific technologies, students will be exposed in this section to the study of computers, information technology, computer networks, and computer applications.

4. *Management Studies.* This section will include various management courses that are relevant to the systems engineering process. It will include courses in operations management, marketing, human resources management, project management, and

business policy and planning. It will emphasize planning skills, administrative skills, and the related human skills of motivation, negotiation, presentation, and writing.

5. *Systems Engineering Concepts and Techniques.* This is the final and integrative section of the program. It should develop the participants' system thinking and their "holistic view," and furnish the tools, standards, and technique used during the systems engineering process. It will present the framework for distinction among systems and the distinct systems engineering management roles that must be applied for each system type. In particular, this section will emphasize the system's life cycle and the process of system requirements writing, methods of functional analysis and decomposition, and the iterative nature of the systems engineering process for various types of systems. Finally, this section will present famous systems building case studies, which will be followed by a final project to be performed and submitted by the students.

2.8 CONCLUSION

Systems engineering management is a broad multidisciplinary activity, and it should not be exercized in the same form for all kinds of systems. A systems engineering manager must have knowledge in many areas, but above all about systems. Systems engineering requires a proper attitude, concept, and philosophy, and the combination of activities and skills from many fields. Such skills can only be accumulated by doing and by being exposed to various experiences. Formal studies may come later, after a person has gained a certain level of "system maturity." Systems thinking is also a quality gained through experience. No one is born a systems thinker, let alone a systems engineer. To design and manage systems effectively, one must understand the nature of various systems, choose the appropriate concept, and adopt the right attitude. Only then can the engineer choose the specific tools and practices for executing the project. As we have seen, different systems require different concepts and distinctive roles for effective systems engineering and management practices.

The frameworks presented in this chapter are not conclusive. They should provide the conceptual basis for the understanding of systems engineering management and be the basis for further refinement in the world of practice. For example, one may consider additional uncertainties for the analysis of system management, such as market uncertainties. In this case, the uncertainty dimension may represent a different kind of uncertainty and would require different skills. There is a need to develop better and more advanced tools for systems engineering in specific classes of systems, as well as better ways for teaching and training systems engineers. The models presented, however, provide a framework upon which new insights can be added in the future.

REFERENCES

Abernathy, W. J., and Utterback, J. M. (1978). Patterns of industrial innovation. *Technol. Rev.,* June-July; pp. 40–47.

Ahituv, N., and Neumann, S. (1984). A flexible approach to information system development. *MIS Q.,* June; pp. 69–78.

Blake, S. B. (1978). *Managing for Responsive Research and Development*. San Francisco: Freeman.

Blanchard, B. S. (1998). *System Engineering Management*. 2nd ed. New York: Wiley.

Blanchard, B. S., and Fabricky, W. J. (1998). *System Engineering and Analysis,* 3rd ed. Englewood Cliffs, NJ: Prentice-Hall.

Booton, R. C., and Ramo, S. (1984). The development of systems engineering. *IEEE Trans. Aerosp. Electron. Syst.* **AES-20**(4), 306–309.

Boulding, K. (1956). General systems theory: The skeleton of science. *Manage. Sci.,* **2**(3) April; 197–208.

Brill, J. H. (1994). Systems engineering—a semantic jungle, *Syst. Eng., J. N. Counc. Syst. Eng.,* **1**(1); 29–33.

Brill, J. H. (1999). Systems engineering—A retrospective view, *Syst. Eng.,* **1**(4), 258–266.

Chambers, G. J. (1985). What is a systems engineer? *IEEE Trans. Syst., Man. Cybernet.* **15,** 517–521.

Chambers, G. J. (1986). The systems engineering process: A technical bibliography. *IEEE Trans. Syst., Man. Cybernet.* **16**(5), 712–721.

Chase, W. P. (1985). *Management of Systems Engineering,* Malabar, FL: Krieger Publ. Co.

Dewar, R. D., and Dutton, J. E. (1986). The adoption of radical and incremental innovations: An empirical analysis. *Manage. Sci.* **32,** 1422–1433.

Ettlie, J. E., Bridges, W. P., and O'Keffe, R. D. (1984). Organizational strategy and structural differences for radical vs. incremental innovation. *Manage. Sci.* **30,** 682–695.

Fox, R. J. (1988). *The Defense Management Challenge: Weapons Acquisition.* Boston: Harvard Business School Press.

Galbraith, J. R. (1977). *Organization Design.* Reading, MA: Addison-Wesley.

Grady, J. O. (1994). *System Integration.* Boca Raton: CRC Press.

Hunger, J. W. (1995). *Engineering the System Solution.* Upper Saddle River, NJ: Prentice Hall.

Institute of Electrical and Electronics Engineers (IEEE). (1994). *IEEE Standard for Systems Engineering* (IEEE P1220). New York: IEEE Standards Dept.

Johnson, C. L. "Kelly," and Smith, M. (1985). *Kelly: More Than My Share of It All.* Washington, DC: Smithsonian Institution Press.

Lake, J. G. (1994). Axioms for systems engineering. *Syst. Eng., J. Nat. Counc. on Syst. Eng.* **1**(1); 17–28.

Lawrence, R. M. (1987). *Strategic Defense Initiative.* Boulder, CO: Westview Press.

Lemley, J. K. (1992). The channel tunnel: Creating a modern wonder of the world. *PmNetwork,* July, pp. 14–22.

Manne, S. J., and Collins, L. (1990). Reconstructuring an aging infrastructure. *Proj. Manage. J.,* April, pp. 9–24.

Mar, B. W. (1994). Systems engineering basics. *Syst. Eng., J. Nat. Counc. Syst. Eng.* **1**(1), 7–16.

McPherson, P. (1986). Systems engineering: A proposed definition. *Inst. Electr. Eng.* **133,** 130–133.

Ottenberg, M. (1993). Toward a new generation education: An application of systems engineering principles. *Proc. Annu. Symp. Na. Counc. Syst. Eng. (NCOSE),* Washington, DC.

Rechtin, E. (1991). *Systems Architecting.* Englewood Cliffs, NJ: Prentice-Hall.

Rosenbloom, R. S., and Cusumano, M. A. (1987) Technological pioneering and competitive advantage: The birth of the VCR Industry. *Calif. Manage. Rev.* **24**(4), 51–73.

Sage, A. P. (1980a). From philosophical perspectives to practice in the design of program planning linkages for systems engineering education. *IEEE Trans. Syst., Man, Cybernet.* **SMC-10**(11), 693–695.

Sage, A. P. (1980b). Desiderata for systems engineering education. *IEEE Trans. Syst., Man, Cybernet.* **SMC-10**(12), 777–780.

Sage, A. P. (1981). Systems engineering: Fundamental limits and future prospects. *Proc. IEEE* **69**(2), 158–166.

Sage, A. P. (1987). Knowledge transfer: An innovative role for information engineering education. *IEEE Trans. Syst., Man, Cybernet.* **17**, 725–728.

Sage, A. P. (1994). The many faces of systems engineering. *Syst. Eng., J. Na. Counc. Syst. Eng.* **1**(1), 43–60.

Sheard, S. A. (1996). Twelve systems engineering roles. *Proc. Annu. Symp. Int. Counc. Syst. Eng.* pp. 481–488.

Shenhar, A. J. (1990). *A Multidisciplinary Program in Systems Engineering Management.* Tel-Aviv: Tel-Aviv University, Faculty of Management.

Shenhar, A. J. (1991). On system properties and systemhood. *Int. J. Gen. Syst.* **18**(2), 167–174.

Shenhar, A. J. (1993). From low to high-tech project management. *R&D Manage.* **2**(3), 199–214.

Shenhar, A. J. (1994). Systems engineering management: A framework for the development of a multidisciplinary discipline. *IEEE Trans. Syst., Man, Cybernet.* **24**(2), 327–332.

Shenhar, A. J., and Bonen, Z. (1997). The new taxonomy of systems: Toward an adaptive systems engineering framework. *IEEE Trans. Syst., Man, Cybernet.* **27**(2), 137–145.

Shenhar, A. J., and Dvir, D. (1996). Toward a typological theory of project management. Res. Policy **25**, 607–632.

Steele, L. W. (1975). *Innovation in Big Business.* New York: Elsevier.

Swain, R. E. (1991). A comparative evaluation of interdisciplinary systems engineering Master's programs in the United States. *Pap., 1st Ann. Conf. Nat. Counc. Syst. Eng. (NCOSE),* Chattanooga, TN.

U.S. Department of Defense. (1992). *Engineering Management* (MIL-STD-499b). (Draft). Washington, DC: USDoD.

Van Gigch, J. P. (1978). *Applied General Systems Theory,* 2nd ed. New York: Harper & Row.

Wheelwright, S. C., and Clark, K. B. (1992). *Revolutionizing Product Development.* New York: Free Press.

Zaltman, G., Duncan, R. L., and Holbek, J. (1973). *Innovations and Organizations.* New York: Wiley.

3 Risk Management

YACOV Y. HAIMES

3.1 THE PROCESS OF RISK ASSESSMENT AND MANAGEMENT

Risk-based decision making and risk-based approaches in decision making are terms frequently used to indicate that some systematic process that deals with uncertainties is being used to formulate policy options and assess their various distributional impacts and ramifications. Today an ever-increasing number of professionals and managers in industry, government, and academia are devoting a larger portion of their time and resources to the task of improving their understanding and approach to risk-based decision making. In this pursuit they invariably rediscover (often with considerable frustration) the heuristic truism: the more you know about a complex subject, the more you realize how much still remains unknown.

There are two fundamental reasons for the complexity of this subject. One is that decision making under uncertainty literally encompasses every aspect of our lives. It affects us at the personal, corporate, and governmental levels, and during the planning, development, design, operation, and management phases of a project. Uncertainty colors the decision-making process, whether it involves one or more parties or is constrained by economic or environmental considerations, driven by sociopolitical or geographical forces, directed by scientific or technological know-how, or influenced by various power brokers and stakeholders. Uncertainty is inherent when the process attempts to answer the set of questions posed by William W. Lowrance: "Who should decide on the acceptability of what risk, for whom, in what terms, and why?" (1976). The second reason risk-based decision making is complex is that it is cross-disciplinary. The subject has been further complicated by the development of diverse approaches of varying reliability. Some methods, which on occasion produce fallacious results and conclusions, have become entrenched and are hard to eradicate.

Consider, for example, the risks associated with natural hazards. Causes for major natural hazards are many and diverse and the risks associated with them affect human lives, the environment, the economy, and the country's social well-being. Many of these risks are manifested in engineering-based physical infrastructure—dams, levees, water distribution systems, wastewater treatment plants, transportation systems (roads, bridges, freeways, and ports), communication systems, and hospitals, to cite a few. Thus, when addressing the risks associated with natural hazards, such as earthquakes and major floods, one must also account for the impact of these hazards on the integrity, reliability, and performance of engineering-based physical infrastructure. The next step is to assess the consequences—the impact on human and nonhuman populations and on the socioeconomic fabric of large and small communities.

Thus, risk assessment and management must be an integral part of the decision-making

Handbook of Systems Engineering and Management, Edited by A. P. Sage and W. B. Rouse
ISBN 0471-15405-9   ©1999 John Wiley and Sons, Inc.

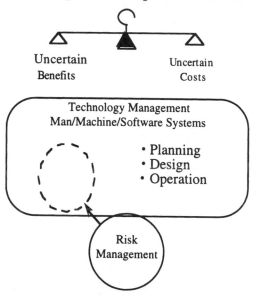

Figure 3.1 Risk management within technology management.

process, rather than a gratuitous add-on technical analysis. Figure 3.1 depicts this concept and indicates the ultimate need to balance all the uncertain benefits and costs.

For the purpose of this chapter, *risk* is defined as a measure of the probability and severity of adverse effects (Lowrance, 1976). Lowrance also makes the following distinction between risk and safety: "Measuring risk is an empirical, quantitative, scientific activity (e.g., measuring the probability and severity of harm). Judging safety is judging the acceptability of risks—a normative, qualitative, political activity." This suggests that those private and public organizations that can successfully address the risk inherent in their business—whether in environmental protection, resource availability, natural forces, the reliability of man/machine systems, or in future use of new technology—will dominate their markets.

The premise that risk assessment and management must be an integral part of the overall decision-making process necessitates following a systematic, holistic approach to dealing with risk. Such a holistic approach builds on the principles and philosophy upon which systems analysis and systems engineering are grounded.

A common question is: "What constitutes a good risk assessment effort?" Numerous studies have attempted to develop criteria for what might be considered a "good" risk analysis, the most prominent of which is the Oak Ridge study (Fischhoff et al., 1980). Risk studies may be judged against the following list of ten criteria: to be good, the study must be comprehensive; adherent to evidence; logically sound; practical and politically acceptable; open to evaluation; based on explicit assumptions and premises; compatible with institutions; conducive to learning; attuned to risk communication; and innovative.

3.2 THE HOLISTIC APPROACH TO RISK ANALYSIS

Good management of technological systems must address the holistic nature of the system in terms of its hierarchical, organizational, and fundamental decision-making structure. Also to be considered are the multiple noncommensurate objectives, subobjectives, and sub-subobjectives, including all types of important and relevant risks, the various time horizons, the multiple decision makers, constituencies, power brokers, stakeholders, and users of the system, and a host of institutional, legal, and other socioeconomic conditions. Thus, risk management raises several fundamental philosophical and methodological questions.

Engineering systems are almost always designed, constructed, and operated under unavoidable conditions of risk and uncertainty, and are often expected to achieve multiple and conflicting objectives. The identification, quantification, evaluation, and trading off of risks, benefits, and costs should constitute an integral and explicit component of the overall managerial decision-making process, and not be a separate, cosmetic afterthought. The body of knowledge in risk assessment and management has gained significant attention during the last two decades; it spans many disciplines and encompasses empirical and quantitative as well as normative, judgmental aspects of decision making. Does this constitute a new discipline that is separate, say, from systems engineering/systems analysis? Or has systems engineering/systems analysis been too narrowly defined? When risk and uncertainty are addressed in a practical decision-making framework, has it been properly perceived that the body of knowledge known as risk assessment and management markedly fills a critical void that supplements and complements the theories and methodologies of systems engineering/systems analysis? Reflecting on these and other similar questions on the nature, role, and place of risk assessment and management in the management of technological systems and in the overall managerial decision-making process should provide a way to bridge the gaps and remove some of the barriers that exist between the various disciplines.

Integrating and incorporating risk assessment and management of technological systems within the broader holistic approach to technology management also requires examining the expected-value concept when it is used as the sole representation of risk. Many agree that commensurating high-frequency/low-damage and low-frequency/catastrophic-damage events markedly distorts their relative importance and consequences as they are viewed, perceived, assessed, evaluated, and traded off by managers, decision makers, and the public. Some are becoming more and more convinced of the grave limitations of the traditional and commonly used expected-value concept, and are complementing and supplementing it with conditional expectation, where decisions about extreme and catastrophic events are not averaged out with more commonly occurring events.

3.2.1 Sources of Failure

To be effective and meaningful, risk management must be an integral part of the overall management of a system. This is particularly important in technological systems, where the failure of the system can be caused by failure of the hardware, the software, the organization, or the humans involved.

The term risk *management* may vary in meaning according to the discipline involved and/or the context. *Risk* is often defined as a measure of the probability and severity of adverse effects. *Risk management* is commonly distinguished from *risk assessment,* even though some may use the term risk management to connote the entire process of risk assessment and management. In risk assessment, the analyst often attempts to answer the following set of three questions: "What can go wrong? What is the likelihood that it would go wrong? What are the consequences?" (Kaplan and Garrick, 1981). Answers to these questions help risk analysts identify, measure, quantify, and evaluate risks and their consequences and impacts. Risk management builds on the risk assessment process by seeking answers to a second set of three questions: "What can be done? What options are available, and what are their associated tradeoffs in terms of all costs, benefits, and risks? What are the impacts of current management decisions on future options?" (Haimes, 1991). Note that the last question is a most critical one for any managerial decision making. This is so because unless the negative and positive impacts of current decisions on future options are assessed and evaluated (to the extent possible), these policy decisions cannot be deemed "optimal" in any sense of the word. Indeed, the assessment and management of risk is essentially a synthesis and amalgamation of the empirical and normative, the quantitative and qualitative, and the objective and subjective effort. Only when these questions are addressed in the broader context of management, where all options and their associated tradeoffs are considered within the hierarchical organizational structure, can total risk management (TRM) be realized. (The term TRM is formally defined later.) Indeed, evaluating the total tradeoffs among all important and relative system objectives in terms of costs, benefits, and risks cannot be done seriously and meaningfully in isolation from the broader resource allocation perspectives of the overall organization.

Good management must thus incorporate and address risk management within a holistic and all-encompassing framework that incorporates and addresses all relevant resource allocations and other related management issues. A total risk management approach that harmonizes risk management with the overall system management must address the first four of the following internal sources of failure (see Figure 3.2):

1. Hardware failure
2. Software failure
3. Organizational failure
4. Human failure
5. External failure

The set of four failure sources 1–4 is intended to be internally comprehensive (i.e., comprehensive within the system's own internal environment). External sources of failure 5 are not discussed here because they are commonly system-dependent. These four elements are not necessarily independent of each other, however. The distinction between software and hardware is not always straightforward, and separating human and organizational failure is often not an easy task. Nevertheless, these four categories of failure sources provide a meaningful foundation upon which to build a total risk management framework. In his premier book on quality control, *Kaizen,* Masaaki (1986) states: "The three building blocks of business are hardware, software, and 'humanware'". He further notes that total quality control "means that quality control effects must involve people, organization, hardware, and software."

Organizational errors are often at the root of failures of critical engineering systems. Yet, when searching for risk management strategies, engineers tend to focus on technical

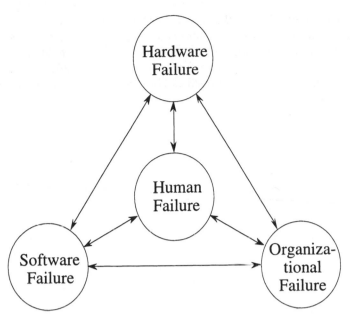

Figure 3.2   System failure.

solutions, in part because of the way risks and failures are analyzed. In her study of offshore drilling rigs, Pate-Cornell (1990) found that over 90 percent of the failures documented were caused by organizational errors. The following is a list of common organizational errors:

- Overlooking and/or ignoring defects
- Tardiness in correcting defects
- Breakdown in communication
- Missing signals or valuable data due to inadequate inspection or maintenance policy
- Unresolved conflict(s) between management and staff
- Covering up mistakes due to competitive pressure
- Lack of incentives to find the problem
- The "kill the messenger" syndrome instead of "reward the messenger"
- Screening information, followed by denial
- Tendency to accept the most favorable hypothesis
- Ignoring long-term effects of decisions
- Loss of institutional memory
- Loss of flexibility and innovation

The importance of considering the four sources of failure is twofold. First, they are comprehensive, involving all aspects of the system's planning, design, construction, operation, and management. Second, they require the total involvement in the risk assessment and management process of everyone concerned—blue- and white-collar workers and managers at all levels of the organizational hierarchy. To ensure total quality control and productivity throughout the organizational structure, Deming (1986) advocates appointing a leader knowledgeable in statistical methodology, who reports directly to the president of the organization. Deming states:

> He will assume leadership in statistical methodology throughout the company. He will have authority from top management to be a participant in any activity that in his judgment is worth his pursuit. He will be a regular participant in any major meeting of the president and his staff.

Indeed, total quality control is very much commensurate and harmonious with total risk management. Total risk management builds on the concept of, and experience gained from, total quality control that is widely used in manufacturing. Here is the Japanese definition of quality control according to the Japan Industrial Standards (Masaaki, 1986):

> Implementing quality control effectively necessitates the cooperation of all people in the company, including top management, managers, supervisors, and workers in all areas of corporate activities such as market research and development, product planning, design, preparations for production, purchasing, vendor management, manufacturing, inspection, sales and after-services, as well as financial control, personnel administration, and training and education. Quality control carried out in this manner is called company-wide quality control or total quality control.

3.2.2 Total Risk Management

Total risk management (TRM) can be defined as a systemic, statistically-based, holistic process that builds on a formal risk assessment and management. It answers the previously introduced two sets of three questions for risk assessment and risk management, and addresses the set of four sources of failures within a hierarchical/multiobjective framework. Figure 3.3 depicts the total risk management paradigm.

The term hierarchical/multiobjective framework can be explained in the context of TRM. Most, if not all, organizations are hierarchical in their structure, and consequently in the decision-making process that they follow. Furthermore, at each level of the organizational hierarchy, multiple conflicting, competing, and noncommensurate objectives drive the decision-making process. Thus, within the organization, there are commonly several sets of objectives, subobjectives, and sub-subobjectives corresponding to the levels of hierarchical structure and to its various units of subsystems (Haimes et al., 1990). At the heart of good management decisions is the "optimal" allocation of the organization's resources among its various hierarchical levels and subsystems. The "optimal" allocation is meant in the Pareto optimal sense, where tradeoffs among all costs, benefits, and risks are evaluated in terms of hierarchical objectives (and subobjectives) and in terms of their temporal impacts on future options. Methodological approaches for such hierarchical frameworks are discussed in Haimes et al. (1990).

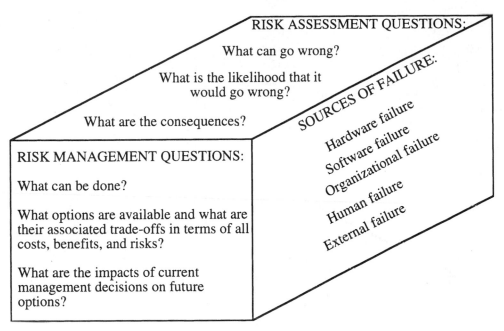

Figure 3.3 Total risk management.

3.2.3 Multiple Objectives

The tradeoffs among multiple noncommensurate and often conflicting and competing objectives are at the heart of risk management. One is invariably faced with deciding the level of safety (the level of risk that is deemed acceptable) and the acceptable cost associated with that safety. The following student dilemma is used to demonstrate the fundamental concepts of Pareto optimality and tradeoffs in a multiobjective framework (Chankong and Haimes, 1983).

A student working part-time to support her college education is faced with the following dilemma that is familiar to all of us:

$$\text{Maximize} \begin{cases} \text{income from part-time work} \\ \text{grade-point average} \\ \text{leisure time} \end{cases}$$

In order to use the two-dimensional plane for graphic purposes, we will restrict our discussion to two objectives: maximize income and maximize grade-point average (GPA). We will assume that a total of 70 hours per week are allocated for studying and working. The remaining 98 hours per week are available for "leisure time," covering all other activities. Figure 3.4 depicts the income generated per week as a function of hours of work, while Figure 3.5 depicts the relationship between studying and GPA. Figure 3.6 is a dual plotting of both functions (income and GPA) vs. work time and study time, respectively.

The concept of optimality in multiple objective situations differs in a fundamental way

$f_1(\bullet)$

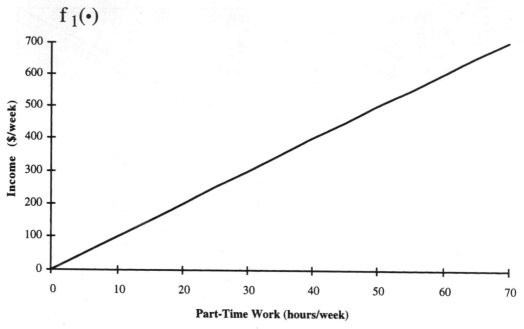

Figure 3.4 Income from part-time work.

$f_2(\bullet)$

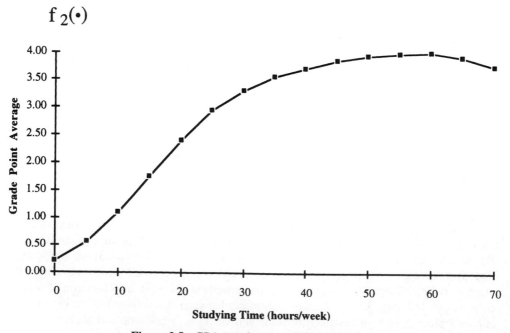

Figure 3.5 GPA as a function of studying time.

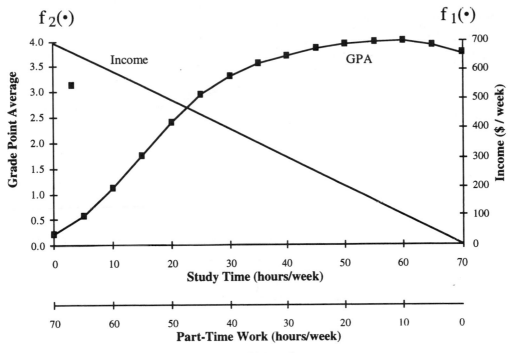

Figure 3.6 GPA vs. income.

from that in single-objective optimization situations. Pareto optimality in a multiobjective framework is where improving one objective solution, policy, or option is obtained only at the expense of degrading another. A Pareto optimal solution is also known as a non-inferior, nondominated, or efficient solution. In Figure 3.6, for example, studying up to 60 hours per week (and correspondingly working 10 hours per week) is Pareto optimal, since in this range income is sacrificed for a higher GPA. On the other hand, studying over 60 hours per week (or working less than 10 hours per week) is a non-Pareto optimal policy, since in this range both income and GPA are diminishing. A non-Pareto optimal solution is also known as an inferior, dominated, or nonefficient solution. Figure 3.7 further distinguishes between Pareto and non-Pareto optimal solutions by plotting income vs. GPA. The line connecting all the square points is called the *Pareto optimal frontier*. Note that any point interior to this frontier is non-Pareto optimal. Consider, for example, policy option A. At this point the student makes $300 per week at a GPA of just above 1.0, whereas at point B she makes $600 per week at the same GPA level. One can easily show that all points (policy options) interior to the Pareto optimal frontier are inferior points.

Consider the risk of groundwater contamination as another example. We can generate the Pareto optimal frontier for this risk-based decision making. Minimizing the cost of contamination prevention and the risk of contamination is similar in many ways to generating the Pareto optimal frontier for the student dilemma problem. Determining the best work-study policy for the student can be compared to determining the level of safety, that

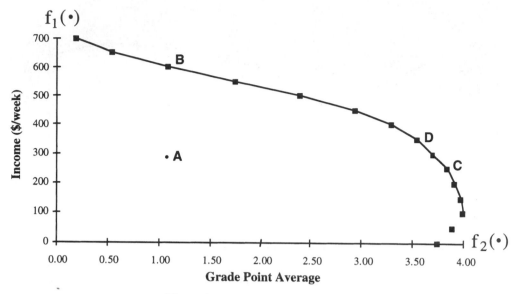

Figure 3.7   Pareto optimal frontier.

is, the level of acceptability of risk of contamination and the cost associated with the prevention of such contamination. To arrive at this level of acceptable risk, we will again refer to the student dilemma problem illustrated in Figure 3.7. At point B the student is making about $600 per week at a GPA of just above 1.0. Note that the slope at this point is about $100 per week for each one GPA point. Thus the student will opt to study more. At point C, the student can achieve a GPA of about 3.6 and a weekly income of about $250. The tradeoff (slope) at this point is very large—by sacrificing about 0.2 of a GPA point the student can increase her income by about $200 per week. Obviously, the student may choose neither policy B nor C; rather, she may settle for something like policy D, with an acceptable level of income and GPA. In a similar way, and short of strict regulatory requirements, a decision maker may determine the amount of resources to allocate for the prevention of groundwater contamination at an acceptable level of risk of contamination.

In summary, the question is: Why should we expect environmental or other technologically-based problems involving risk/cost/benefit tradeoffs to be any easier than solving the student dilemma?

A single decision maker as in the student dilemma problem is not common, especially when dealing with public policy; rather, the existence of multiple decision makers is more prevalent. Indeed, policy options on important and encompassing issues are rarely formulated, traded off, evaluated, and finally decided upon at one single level in the hierarchical decision-making process. Rather, a hierarchy that represents various constituencies, stakeholders, power brokers, advisors, administrators, and a host of shakers and movers constitutes the true players in the complex decision-making process. For more on multiobjective analysis, refer to Haimes and Hall (1974), Chankong and Haimes (1983), and Haimes (1997).

TABLE 3.1 Comparative Costs of Safety and Health Regulations

Regulation	Statute and Year	Initial Annual Risk Estimate[a]	Cost per Life Saved ($ thousand, 1984)
Safety			
Oil and gas well service	Proposed 1983	1.1 in 10^3	100
Underground construction	Proposed 1983	1.6 in 10^3	300
Servicing wheel rims	Final 1984	1.4 in 10^3	500
Crane-suspended personnel platform	Proposed 1984	1.8 in 10^3	900
Concrete and masonry construction	Proposed 1985	1.4 in 10^6	1,400
Hazard communication	Final 1983	4.0 in 10^5	1,800
Grain dust	Proposed 1984	2.1 in 10^4	2,800
Health			
Asbestos	Final 1972	3.9 in 10^4	7,400
Benzene	Final 1987	8.8 in 10^4	17,100
Ethylene oxide	Final 1984	4.4 in 10^6	25,600
Acrylonitrile	Final 1978	9.4 in 10^4	37,600
Coke ovens	Final 1976	1.6 in 10^4	61,800
Asbestos	Final 1986	6.7 in 10^6	89,300
Arsenic	Final 1978	1.8 in 10^3	92,500
Ethylene dibromide	Proposed 1983	2.4 in 10^4	15,600,000
Formaldehyde	Proposed 1985	6.8 in 10^7	72,000,000

[a] Annual deaths per exposed population. [*Source:* Morall (1987)]

3.2.4 The Perception of Risk

The enormous discrepancies and monumental gaps in the dollars spent by various federal agencies in their quest to save human lives can no longer be justified under today's austere budgetary constraints. These expenditures vary within five to six orders of magnitude. For example, according to Morall (1987), the cost per life saved by regulating oil and gas well service was $100,000 (1984 dollars); for formaldehyde it was $72 billion; and for asbestos, $7.4 million (see Table 3.1).

A natural and logical set of questions arises: What are the sources of these gaps and discrepancies? Why do they persist? What can be done to synchronize federal agency policies on the value of human life? A somewhat simplistic, albeit pointed explanation may be found in the lexicon of litigation—intimidation, fear, and public pressure in the media as well as in the electoral and political processes. U.S. companies have ample statistical information on the costs of improved product safety, but are most careful to keep their analyses secretive and confidential. Our litigious society has effectively prevented industry and government from both explicitly developing and publicly sharing such analyses.

What is needed is at least a temporary moratorium on litigation in this area. We should extend immunity and indemnification to all analysts and public officials engaged in quantifying the cost-effectiveness of all expenditures aimed at saving human lives and/or preventing sickness or injury. In sum, we ought to generate a public atmosphere that is conducive to open dialog and reason, and to a holistic process of risk assessment and management.

3.3 RISK OF EXTREME EVENTS

In a society that is slowly adjusting to the risks of everyday life, most analysis and decision theorists are beginning to recognize a simple yet fundamental philosophical truth. In the face of such unforeseen calamities as bridges falling, dams bursting, and airplanes crashing, we are more willing to acknowledge the importance of studying "extreme" events. Modern decision analysts are no longer asking questions about expected risk; instead, they are asking questions about expected maximum risk. These analysts are focusing their efforts on forming a more robust treatment of extreme events, in both a theoretical and a practical sense. Furthermore, managers and decision makers are most concerned with the risk associated with a specific case under consideration, and not with the likelihood of the average adverse outcomes that may result from various risk situations. In this sense, the expected value of risk, which until recently has dominated most risk analysis, is not only inadequate, but can lead to fallacious results and interpretations. Indeed, people in general are not risk-neutral. They are often more concerned with low-probability catastrophic events than with more frequently occurring but less severe accidents. In some cases, a slight increase in the cost of modifying a structure might have a very small effect on the unconditional expected risk (the commonly used business-as-usual measure of risk), but would make a significant difference to the conditional expected catastrophic risk. Consequently, the expected catastrophic risk can be of significant value in many multiobjective risk problems.

Two difficult questions—how safe is safe enough? and, what is an acceptable risk?—underlie the normative, value-judgment perspectives in risk-based decision making. No mathematical, empirical knowledge base today can adequately model the perception of risks in the minds of decision makers. In the study of multiple-criteria decision making (MCDM), we clearly distinguish between the quantitative element in the decision-making process, where efficient (Pareto optimal) solutions and their corresponding tradeoff values are generated, and the normative value-judgment element, where the decision makers make use of these efficient solutions and tradeoff values to determine their preferred (compromise) solution (Chankong and Haimes, 1983). In many ways, risk-based decision making can and should be viewed as a type of stochastic MCDM in which some of the objective functions represent risk functions. This analogy can be most helpful in making use of the extensive knowledge already generated by MCDM, as available in the abundance of publications and conferences on the subject.

It is worth noting that there are two modalities to the considerations of risk-based decision making in a multiobjective framework. One is viewing risk (e.g., the risk of dam failure) as an objective function to be traded off with the cost and benefit functions. The second modality concerns the treatment of damages of different magnitudes and different probabilities of occurrence as noncommensurate objectives, which thus must be augmented by a finite, but small, number of risk functions (e.g., a conditional expected-value function as will be formally introduced in subsequent discussion). Probably the most demonstrable aspect of the importance of considering risk-based decision making within a stochastic MCDM framework is the handling of extreme events.

To dramatize the importance of understanding and adequately quantifying the risk of extreme events, the following statements are adopted from Runyon (1977).

Imagine what life would be like if:

- Our highways were constructed to accommodate the average traffic load of vehicles of average weight.
- Mass transit systems were designed to move only the average number of passengers (i.e., total passengers per day divided by 24 hours) during each hour of the day,
- Bridges, homes, and industrial and commercial buildings were constructed to withstand the average wind or the average earthquake,
- Telephone lines and switchboards were sufficient in number to accommodate only the average number of phone calls per hour,
- Your friendly local electric utility calculated the year-round average electrical demand and constructed facilities to provide only this average demand,
- Emergency services provided only the average number of personnel and facilities required during all hours of the day and all seasons of the year, or
- Our space program provided emergency procedures for only the average type of failure.

Chaos is the word for it. Utter chaos.

Lowrance (1976) makes an important observation on the imperative distinction between the quantification of risk, which is an empirical process, and the determination of safety, which is a normative process. In both of these processes, which are seemingly dichotomous, the influence and imprint of the analyst cannot and should not be overlooked. The essential role of the analyst, sometimes hidden but often explicit, is not unique to risk assessment and management; rather, it is indigenous to the process of modeling and decision making.

The major problem for the decision maker remains one of information overload: for every policy (action or measure) adopted, there will be a vast array of potential damages as well as benefits and costs with their associated probabilities. It is at this stage that most analysts are caught in the pitfalls of the unqualified expected-value analysis. In their quest to protect the decision maker from information overload, analysts precommensurate catastrophic damages that have a low probability of occurrence with minor damages that have a high probability. From the perspective of public policy, it is obvious that a catastrophic dam failure, which might cause flooding of, say, 10^6 acres of land with associated damage to human life and the environment, but which has a very low probability, (say, 10^{-6}) of happening, cannot be viewed by decision makers in the same vein as minor flooding of, say, 10^2 acres of land that has a high probability (say 10^{-2}) of happening. Yet this is exactly what the expected-value function would ultimately generate. Most importantly, the analyst's precommensuration of these low-probability/high-damage events with high-probability/low-damage events into one expectation function (indeed some kind of a utility function) markedly distorts the relative importance of these events and consequences as they are viewed, assessed, and evaluated by decision makers. This is similar to the dilemma that used to face theorists and practitioners in the field of MCDM (Haimes et al., 1990).

3.3.1 The Fallacy of the Expected Value

One of the most dominant steps in the risk assessment process is the quantification of risk, yet the validity of the approach most commonly used to quantify risk—its expected value—has received neither the broad professional scrutiny it deserves nor the hoped-for wider mathematical challenge that it mandates. The conditional expected value of the risk of extreme events (among other conditional expected values of risks) generated by the partitioned multiobjective risk method (PMRM) (Asbeck and Haimes, 1984) is one of the few exceptions.

Let $P_x(x)$ denote the probability density function of the random variable X, where X is, for example, the concentration of the contaminant trichloroethylene (TCE) in a groundwater system, measured in parts per billion (ppb). The expected value of the containment concentration (the risk of the groundwater being contaminated by TCE at an average concentration of TCE), is $E(x)$ ppb. If the probability density function is discretized to n regions over the entire universe of contaminant concentrations, then $E(x)$ equals the sum of the product of p_i and x_i, where p_i is the probability that the ith segment of the probability regime has a TCE concentration of x_i. Integration (instead of summation) can be used for the continuous case. Note, however, that the expected-value operation commensurates contaminations (events) of low concentration and high frequency with contaminations of high concentration and low frequency. For example, events $x_1 = 2$ ppb and $x_2 = 20,000$ ppb that have the probabilities $p_1 = 0.1$ and $p_2 = 0.00001$, respectively, yield the same contribution to the overall expected value: $(0.1)(2) + (0.00001)(20,000) = 0.2 + 0.2$. However, to the decision maker in charge, the relatively low likelihood of a disastrous contamination of the groundwater system with 20,000 ppb of TCE cannot be equivalent to the contamination at a low concentration of 0.2 ppb, even with a very high likelihood of such contamination. Due to the nature of mathematical smoothing, the averaging function of the contaminant concentration in this example does not lend itself to prudent management decisions. This is because the expected value of risk does not accentuate the catastrophic events and their consequences, thus misrepresenting what would be perceived as unacceptable risk.

It is worth noting that the number of "good" decisions managers make during their tenure is not the only basis for rewards, promotion, and advancement; rather, they are likely to be penalized for any disastrous decisions, no matter how few, made during their career. The notion of "not on my watch" clearly emphasizes the point. In this and other senses, the expected value of risk fails to represent a measure that truly communicates the manager's or the decision maker's intentions and perceptions. The conditional expected value of the risk of extreme events generated by the PMRM, when used in conjunction with the (unconditional) expected value, can markedly contribute to the total risk-management approach. In this case, the manager must make tradeoffs not only between the cost of the prevention of contamination by TCE vs. the expected value of such risk of contamination, but also between the cost of the prevention of contamination vs. the conditional expected value of a risk of extreme level of TCE contamination. Such a dual multiobjective analysis provides the manager with more complete, more factual, and less-aggregated information about all viable policy options and their associated tradeoffs (Haimes, 1991).

This act of commensurating the expected value operation is analogous in some sense to the commensuration of all benefits and costs into one monetary unit. Indeed, few today would consider benefit/cost analysis, where all benefits, costs, and risks are aggregated into monetary units, as an adequate and acceptable measure for decision making when it is used as the sole criterion for excellence. Multiple-objective analysis has been demonstrated as a superior approach to benefit/cost analysis (Haimes and Hall, 1974). In many respects, the expected value of risk is similar in its theoretical–mathematical construct to the commensuration of all costs, benefits, and risks into monetary units.

Consider the dam safety problem mentioned earlier, with precipitation being the single random variable evaluated in the damage risk function. Let us discretize the universe of events for this random variable into J segments. Then the damage expectation function, $U(\mathbf{x})$, can be written as (Haimes, 1988):

$$U(\mathbf{x}) = \sum_{j=1}^{J} p_j f_j(\mathbf{x}) \tag{3.1}$$

where $f_j(\mathbf{x})$ is the damage associated with the jth segment given as a function of the decision variables $\mathbf{x}$, p_j is the probability associated with the jth segment, and

$$p_j \geq 0, \qquad \sum_{j=1}^{J} p_j = 1 \tag{3.2}$$

One might argue that if it were practical, decision makers would rather consider the risk-based decision-making problem in the following multiobjective optimization format:

$$\underset{\mathbf{x} \in \mathbf{X}}{\text{minimize}} \{f_1(\mathbf{x}), f_2(\mathbf{x}), \ldots, f_j(\mathbf{x})\} \tag{3.3}$$

where $f_j(\mathbf{x})$ represents a specific range of the damage function that corresponds to a specific range of probabilities of exceedance. Additional objectives representing costs and benefits should be added as appropriate.

Each damage function, $f_j(\mathbf{x})$, associated with the jth segment of the probability axis can be viewed as a noncommensurate objective function. In their totality, these damage functions constitute a set of noncommensurate objective functions. Clearly, at one extreme we can consider an infinite number of such objective functions, and at the other extreme we can consider a single-objective function—namely, the expected-value function. A compromise between the two extremes must be made for tractability and also for the benefit of the decision makers.

In the field of MCDM, the profession has accepted the fact that tradeoffs exist between the consideration of a very large number of objectives and the ability of decision makers to comprehend these objectives. Consequently, the relevant objective functions, which are commonly numerous, are either organized into a hierarchy of objectives and subobjectives (see Peterson, 1974) or some augmentation is followed in order to limit the number of objectives to between five and seven.

To further demonstrate the fallacy of the expected-value approach, consider a design problem where four design options are being considered. Associated with each option are cost, the mean of a failure rate (i.e., the expected value of failures for a normally distributed probability density function of a failure rate), and the standard deviation (see Table 3.2). Figure 3.8 depicts the normally distributed probability density functions (PDF) of failure rates for each of the four options. Clearly, on the basis of the expected value alone, the least-cost design (Option 4) seems to be preferred, at a cost of $40,000. However, con-

TABLE 3.2   Design Options Data and Results

Option Number	Cost ($)	Mean (m) Expected Value	Standard Deviation (s)
1	100,000	5	1
2	80,000	5	2
3	60,000	5	3
4	40,000	5	4

sulting the variances, which provide an indication of extreme failures, reveals that this choice might not be the best after all, and a more in-depth tradeoff analysis is needed. This problem will be revisited when extreme-event analysis is discussed.

3.4   THE PARTITIONED MULTIOBJECTIVE RISK METHOD

The *partitioned multiobjective risk method* (PMRM) is a risk analysis method developed for solving multiobjective problems of a probabilistic nature (Asbeck and Haimes, 1984). Instead of using the traditional expected value of risk, the PMRM generates a number of conditional expected-value functions, termed *risk functions,* which represent the risk given that the damage falls within specific ranges of the probability of exceedance. Before the PMRM was developed, problems with at least one random variable were solved by computing and minimizing the unconditional expectation of the random variable representing damage. In contrast, the PMRM isolates a number of damage ranges (by specifying so-called partitioning probabilities) and generates conditional expectations of damage given

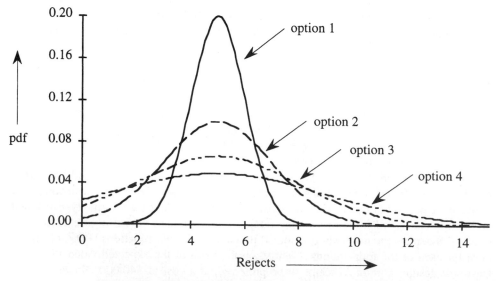

Figure 3.8   PDF of failure-rate distributions for four designs.

that the damage falls within a particular range. In this manner, the PMRM generates a number of risk functions, one for each range, which are then augmented with the original optimization problem as new objective functions.

The conditional expectations of a problem are found by partitioning its probability axis and mapping these partitions onto the damage axis. Consequently, the damage axis is partitioned into corresponding ranges. A conditional expectation is defined as the expected value of a random variable given that this value lies within some prespecified probability range. Clearly, the values of conditional expectations are dependent on where the probability axis is partitioned. The choice of where to partition is made subjectively by the analyst in response to the extreme characteristics of the problem. If, for example, the analyst is concerned about the once-in-a-million-years catastrophe, the partitioning should be such that the expected catastrophic risk is emphasized. Although no general rule exists to guide the partitioning, Asbeck and Haimes (1984) suggest that if three damage ranges are considered for a normal distribution, then the $+1s$ and $+4s$ partitioning values provide an effective rule of thumb. These values correspond to partitioning the probability axis at 0.84 and 0.99968; that is, the low-damage range would contain 84 percent of the damage events, the intermediate range would contain just under 16 percent, and the catastrophic range would contain about 0.032 percent (probability of 0.00032). In the literature, catastrophic events are generally said to be events with a probability of exceedance of 10-5 [see, for instance, the NRC Report on dam safety (National Research Council, 1985)]. This probability corresponds to events exceeding $+4$ Sigma.

A continuous random variable X of damages has a cumulative distribution function (CDF) $P(x)$ and a probability density function (PDF) $p(x)$, which are defined by the relationships

$$P(x) = \text{prob}[X \leq x] \tag{3.4}$$

and
$$p(x) = \frac{dP(x)}{dx} \tag{3.5}$$

The CDF represents the *nonexceedance probability* of x. The *exceedance probability* of x is defined as the probability that X is observed to be greater than x and is equal to 1 minus the CDF evaluated at x.

The expected value, average, or mean value of the random variable X is defined as

$$E[X] = \int_0^\infty xp(x)\,dx \tag{3.6}$$

In the PMRM, the concept of the expected value of damage is extended to generate multiple *conditional expected-value functions*, each associated with a particular range of exceedance probabilities or their corresponding range of damage severities. The resulting conditional expected-value functions, in conjunction with the traditional expected value, provide a family of risk measures associated with a particular policy.

Let $1 - \alpha_1$ and $1 - \alpha_2$, where $0 < \alpha_1 < \alpha_2 < 1$, denote exceedance probabilities that partition the domain of X into the following three ranges. On a plot of exceedance probability, there is a unique damage β_1 on the damage axis that corresponds to the exceedance probability $1 - \alpha_1$ on the probability axis. Similarly, there is a unique damage β_2 that corresponds to the exceedance probability $1 - \alpha_2$. Damages less than β_1 are considered to be of low severity, and damages greater than β_2 are of high severity. Similarly, damages

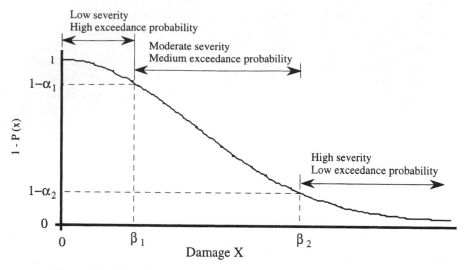

Figure 3.9 Mapping of the probability partitioning onto the damage axis.

of a magnitude between β_1 and β_2 are considered to be of moderate severity. The partitioning of risk into three severity ranges is illustrated in Figure 3.9. If the partitioning probability α_1 is specified, for example, to be 0.05, then β_1 is the 5th percentile. Similarly, if α_2 is 0.95, that is, $1 - \alpha_2$ is equal to 0.05, then β_2 is the 95th percentile.

For each of the three ranges, the conditional expected damage (given that the damage is within that particular range) provides a measure of the risk associated with the range. These measures are obtained through the definition of the *conditional expected value*. Consequently, the new measures of risk are: $f_2(\cdot)$, of high exceedance probability and low severity; $f_3(\cdot)$, of medium exceedance probability and moderate severity; and $f_4(\cdot)$, of low exceedance probability and high severity. The function $f_2(\cdot)$ is the expected value of X, given that x is less than or equal to β_1:

$$f_2(\cdot) = E[X|x \le \beta_1]$$

$$= \frac{\int_0^{\beta_1} xp(x)\, dx}{\int_0^{\beta_1} p(x)\, dx} \tag{3.7}$$

Similarly, for the other two risk functions, $f_3(\cdot)$ and $f_4(\cdot)$,

$$f_3(\cdot) = E[X|\beta_1 \le x \le \beta_2]$$

$$f_3(\cdot) = \frac{\int_{\beta_1}^{\beta_2} xp(x)\, dx}{\int_{\beta_1}^{\beta_2} p(x)\, dx} \tag{3.8}$$

and

$$f_4(\cdot) = E[X|\beta_2 \le x]$$

$$f_4(\cdot) = \frac{\displaystyle\int_{\beta_2}^{\infty} xp(x)\,dx}{\displaystyle\int_{\beta_2}^{\infty} p(x)\,dx} \tag{3.9}$$

Thus, for a particular policy option, there are three measures of risk, $f_2(\cdot)$, $f_3(\cdot)$, and $f_4(\cdot)$, in addition to the traditional expected value denoted by $f_5(\cdot)$. The function $f_1(\cdot)$ is reserved for the cost associated with the management of risk. Note that

$$f_5(\cdot) = \frac{\displaystyle\int_{0}^{\infty} xp(x)\,dx}{\displaystyle\int_{0}^{\infty} p(x)\,dx}$$

$$= \int_{0}^{\infty} xp(x)\,dx \tag{3.10}$$

since the probability of the sample space of X is necessarily equal to 1. In the PMRM, all or some subset of these five measures are balanced in a multiobjective formulation. The details are made more explicit in the next two sub-sections.

3.4.1 General Formulation of the PMRM

Assume that the damage severity associated with the particular policy s_j, $j \in \{1, \ldots, q\}$ can be represented by a continuous random variable X, where $p_X(x;s_j)$ and $P_X(x;s_j)$ denote the PDF and the CDF of damage, respectively. Two partitioning probabilities, α_i, $i = 1, 2$, are preset for the analysis and determine three ranges of damage severity for each policy s_j. A unique damage, β_{ij}, corresponding to the exceedance probability $(1 - \alpha_i)$, can be found due to the monotonicity of $P_X(x;s_j)$. The policies s_j, the partitions α_i, and the bounds β_{ij} of damage ranges are related by the expression

$$P_X(\beta_{ij};s_j) = \alpha_i \qquad i = 1, 2 \tag{3.11}$$

This partitioning scheme is illustrated in Figure 3.10 for two hypothetical policies s_1 and s_2. The ranges of damage severity include high exceedance probability and low damage, $\{X: x \in [\beta_{0j},\beta_{1j}]\}$, the set of possible realizations of X for which it is true that $x \in [\beta_{0j},\beta_{1j}]$; medium exceedance probability and medium damage, $\{X: x \in [\beta_{1j},\beta_{2j}]\}$; and low exceedance probability and high damage (extreme event), $\{X: x \in [\beta_{2j},\beta_{3j}]\}$, where β_{0j} and β_{3j} are the lower and upper bounds of damage X.

The conditional expected-value risk functions f_i, $i = 2, 3, 4$, are given by

$$f_i(s_j) = \ E\ \{X|p_X(x;s_j), x \in [\beta_{i-2,j},\beta_{i-1,j}]\} \qquad i = 2, 3, 4; \qquad j = 1, \ldots, q \tag{3.12}$$

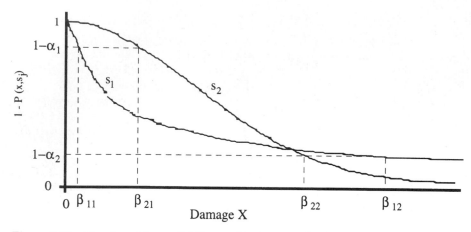

Figure 3.10 Mapping of the probability partitioning onto the damage axis for two policies.

and equivalently,

$$f_i(s_j) = \frac{\displaystyle\int_{\beta_{i-2,j}}^{\beta_{i-1,j}} x\, p_X(x;s_j)\, dx}{\displaystyle\int_{\beta_{i-2,j}}^{\beta_{i-1,j}} p_X(x;s_j)\, dx} \qquad i = 2, 3, 4; \qquad j = 1, \ldots, q \qquad (3.13)$$

The denominator of Equation (3.13) is defined to be q_i, $i = 2,3,4$ as follows:

$$q_2 = \int_0^{\beta_1} p_X(x;s_j)\, dx$$

$$q_3 = \int_{\beta_1}^{\beta_2} p_X(x;s_j)\, dx$$

$$q_4 = \int_{\beta_2}^{\infty} p_X(x;s_j)\, dx$$

If the unconditional expected value of the damage from policy s_j is defined to be $f_5(s_j)$, then the following relationship holds:

$$f_5(s_j) = q_2 f_2(s_j) + q_3 f_3(s_j) + q_4 f_4(s_j) \qquad (3.14)$$

with the $q_i \geq 0$ and $q_2 + q_3 + q_4 = 1$. The q_i are the probabilities that X is realized in each of the three damage ranges and is independent of the policies s_j.

The preceding discussion has described the partitioning of three damage ranges by fixed exceedance probabilities α_i, $i = 1, 2$. Alternatively, the PMRM provides for the partitioning of damage ranges by preset thresholds of damage. For example, the meaning of $f_4(s_j)$ in partitioning by a fixed damage becomes the expected damage resulting from policy j given that the damage exceeds a fixed magnitude. For further details on the partitioning of damage ranges, see Asbeck and Haimes (1984) and Karlsson and Haimes (1988a,b).

The conditional expected-value functions in the PMRM are multiple, noncommensurate measures of risk, each associated with a particular range of damage severity. In contrast, the traditional expected value commensurates risks from all ranges of damage severity and represents only the central tendency of the damage.

Combining any one of the generated conditional expected risk functions or the unconditional expected risk function with the cost objective function f_1 creates a set of multiobjective optimization problems:

$$\min \, [f_1, f_i]', \qquad i = 2, 3, 4, 5$$

This formulation offers more information about the probabilistic behavior of the problem than the single formulation $\min \, [f_1, f_5]'$. The tradeoffs between the cost function f_1 and any risk function f_i, $i \in \{2, 3, 4, 5\}$, allow decision makers to consider the marginal cost of a small reduction in the risk objective, given a particular risk assurance for each of the partitioned risk regions and given the unconditional risk functions f_5. The relationship of the tradeoffs between the cost function and the various risk functions is given by

$$\frac{1}{\lambda_{15}} = \frac{q_2}{\lambda_{12}} + \frac{q_3}{\lambda_{13}} + \frac{q_4}{\lambda_{14}} \tag{3.15}$$

where

$$\lambda_{1i} = -\frac{\partial f_1}{\partial f_i} \tag{3.16}$$

with q_2, q_3, and q_4 as defined earlier. A knowledge of this relationship among the marginal costs provides decision makers with insights that are useful for determining an acceptable level of risk. Any multiobjective optimization method can be applied at this stage—for example, the surrogate worth tradeoff (SWT) method (Haimes and Hall, 1974).

It often has been observed that expected catastrophic risk is very sensitive to the partitioning policy. This sensitivity can be quantified using the statistics of extremes approach suggested by Karlsson and Haimes (1988a,b). In many applications, if given a database representing a random process (e.g., hydrological data related to flooding), it is very difficult to find a specific distribution that represents this database. In some cases, one can exclude some PDFs or guess that some are more representative than others. Quite often, one is given a very limited database that does not contain information about the extreme events. In flood control, for example, records have only been kept for the last 50–100 years, and it is virtually impossible to draw any definite conclusions about floods with return periods exceeding 100 years. In particular, nothing can be said with certainty about the probable maximum flood (PMF), which corresponds to a flood with a return period between 104 and 106 years. Events of a more extreme character are very important because they determine the expected catastrophic risk. The conditional expectations in the PMRM are dependent on the probability partitions and on the choice of the PDF representing the probabilistic behavior of the data (Karlsson and Haimes, 1988a,b).

To illustrate the usefulness of the additional information provided by the PMRM, consider Figure 3.11, where the cost of prevention of groundwater contamination f_1 is plotted against the conditional expected value of contaminant concentration at the low probability

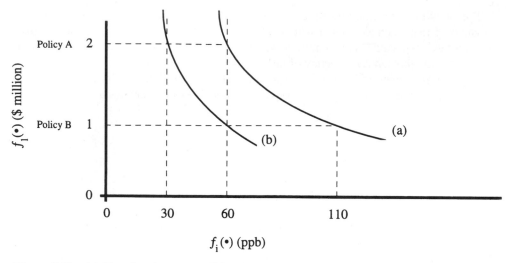

Figure 3.11 (*a*) Cost function vs. conditional expected value of contaminant concentration $f_4(\cdot)$. (*b*) Cost function vs. expected value of contaminant concentration $f_5(\cdot)$.

of exceedance/high-concentration range f_4 (Fig. 3.11*a*), and the unconditional expected value of contaminant concentration f_5 (Fig. 3.11*b*). Note that with Policy A, an investment of $2 million in the prevention of groundwater contamination results in an expected value of contaminant concentration of 30 parts per billion (ppb); however, under the more conservative view (as presented by f_4), the conditional expected value of contaminant concentration (given that the state of nature will be in a low probability of exceedance/high-concentration region) is twice as high (60 ppb). Policy B, $1 million of expenditure, reveals similar results: 60 ppb for the unconditional expectation f_5, but 110 ppb for the conditional expectation f_4. Also note that the slopes of the noninferior frontiers with Policies A and B are not the same. The slope of f_5 between Policies A and B is smaller than that of f_4, indicating that a further investment beyond $1 million would contribute more to a reduction of the extreme-event risk f_4 than it would to the unconditional expectation f_5. The tradeoffs λ_{1i} provide a most valuable piece of information. More specifically, the decision maker is provided with an additional insight into the risk tradeoff problem through f_4 (similarly through f_2 and f_3). The expenditure of $1 million may not necessarily result in a contaminant concentration of 60 ppb; it may instead have a nonnegligible probability resulting in a concentration of 100 ppb. (If, for example, the partitioning were made on the probability axis, and in addition a normal probability distribution were assumed, then this likelihood can be quantified in terms of a specific number of standard deviations.) Furthermore, with an additional expenditure of $1 million (Policy A), even the extreme event of likely concentration—60 ppb—is within the range of acceptable standards. It is worth remembering that the additional conditional risk functions provided by the PMRM do not invalidate the traditional expected-value analysis per se—they improve on it by providing additional insight into the nature of risk assessment and management.

Let us revisit the design problem with its four alternatives. Table 3.3 summarizes the values of the conditional expected value of extreme failure, f_4. Figure 3.12 depicts the cost of each design vs. the unconditional expected value, f_5, and the cost vs. the conditional

TABLE 3.3 Conditional Expected Values of Extreme Failure $f_4(\,\cdot\,)$

Option Number	Cost ($)	Mean (m) Expected Value	Standard Deviation (s)	Conditional Expected Value, $f_4(\,\cdot\,)$
1	100,000	5	1	8.37
2	80,000	5	2	11.73
3	60,000	5	3	15.10
4	40,000	5	4	18.47

expected value, f_4. Clearly, the conditional expected value, f_4, provides much more valued additional information on the associated risk than the unconditional expected value, f_5, where the impact of the variance of each alternative design is captured by f_4.

To further demonstrate the value of the additional information provided by the conditional expected value $f_4(\mathbf{x})$, consider the following results obtained by Petrakian et al. (1989) on the Shoohawk dam study. Two decision variables are considered: (1) raising the dam's height, and (2) increasing the dam's spillway capacity. Although Petrakian et al. considered several policy options or scenarios, only a few are discussed here. Table 3.4 presents the values of $f_1(\mathbf{x})$ (the cost associated with increasing the dam's height and the spillway capacity), and $f_4(\mathbf{x})$ and $f_5(\mathbf{x})$ (the conditional and unconditional expected value of damages, respectively). These values are listed for each of the selected scenarios. Note that the range of the unconditional expected value of the damage, $f_5(\mathbf{x})$, is $161.5-161.7$ million for the various scenarios. On the other hand, the range of the low-frequency/high-damage conditional expected value, $f_4(\mathbf{x})$, varies between $719 million and $1,260 million—a marked difference. Thus, while an investment in the safety of the dam at a cost, $f_1(\mathbf{x})$, ranging from $0 to $46 million, does not appreciably reduce the unconditional expected value of damages, such an investment markedly reduces the conditional expected value of extreme damage from about $1,260 million to $720 million.

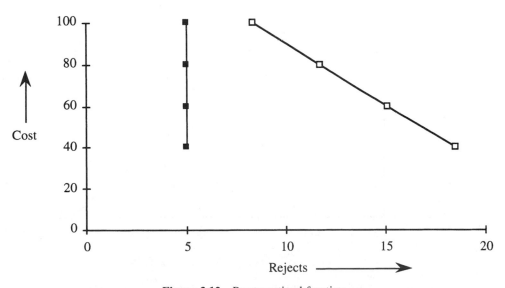

Figure 3.12 Pareto optimal frontier.

TABLE 3.4 Cost of Improving the Dam's Safety and Corresponding Conditional and Unconditional Expected Damages

Scenarios	$f_1(x)$	$f_4(x)$ ($ millions)	$f_4(x)$
1	0	1,260	161.7
2	20	835	161.6
3	26	746	161.6
4	36	719	161.5
5	46	793	160.5

This significant insight into the probable effect of different policy options on the safety of the Shoohawk dam would have been completely lost without considering the conditional expected value derived by the PMRM. Figure 3.13 depicts the plotting of $f_1(x)$ vs. $f_4(x)$ and $f_5(x)$. Note that the unusually high values of $f_1(x)$, on the order of $160 million, are attributed to assumptions concerning antecedent flood conditions (in compliance with the guidelines and recommendations established by the U.S. Army Corps of Engineers).

In sum, new metrics to represent and measure the risk of extreme events are needed to supplement and complement the expected value measure of risk, which represents the central tendency of events. There is much work to be done in this area, including the extension of the PMRM. Research efforts directed at making use of results from the area of statistics of the extremes in representing risk of extreme events, however, have been proven to be very promising and should be continued.

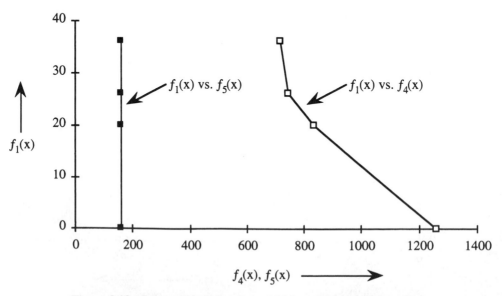

Figure 3.13 Pareto optimal frontiers of $f_1(x)$ vs. $f_4(x)$ and $f_1(x)$ vs. $f_5(x)$.

3.5 THE CHARACTERISTICS OF RISK IN HUMAN-ENGINEERED SYSTEMS

In spite of some commonalties, there are inherent differences between natural systems (e.g., environmental, biological, and ecological systems) and human-made, engineering systems. It is constructive to focus in this chapter on the characteristics of risk associated with human-engineered, or engineering-based, systems.

The following twelve risk characteristics are endemic to most engineering-based systems, including software engineering-based systems:

1. *Organizational failures of engineering-based systems are likely to have dire consequences.* Risk management of technological systems must be an integral part of overall systems management. Organizational failures often constitute a major source of risk of overall system failure.

2. *Risk of extreme events is misrepresented when it is solely measured by the expected value of risk.* The pre-commensuration of rare but catastrophic events of low probability with much less adverse events of high probability in the expected value measure of risk can lead to misrepresentation and mismanagement of catastrophic risk.

3. *Risk of project cost overrun and schedule delay.* Projects involving engineering-based systems have been experiencing major cost overruns and delays in schedule completion, particularly for software-intensive systems. The process of risk assessment and management is also the sine qua non requirement for ensuring against unwarranted delay in a project's completion schedule, cost overrun, and failure to meet performance criteria.

4. *Risk management as a requisite for engineering-based systems integration.* Effective systems integration necessitates that all functions, aspects, and components of the system must be accounted for along with an assessment of most risks associated with the system. Furthermore, for engineering-based systems, systems integration is not only the integration of components, but also an understanding of the functionality that emerges from the integration.

5. *Rehabilitation and maintenance of physical infrastructure.* Maintenance and rehabilitation of water distribution networks have become important issues as the nation addresses the risk of failure of its infrastructure. Recent public criticism has charged that maintenance/rehabilitation funds have not been appropriately spent. Accurate assessment of the risks of failure of this deteriorating physical infrastructure is a prerequisite for the optimal allocation of limited resources.

6. *Multiple failure rates and multiple reliability measures for engineering-based systems.* Engineering-based systems often have any number of paths to failure. Evaluating the interconnected consequences of multiple modes in which potential failures can occur is at the heart of risk assessment and engineering systems management. This includes addressing multiple reliability measures of the systems.

7. *Risk in software engineering development.* The development of software engineering—an intellectual, labor-intensive activity—has been marred by the risks of not meeting performance criteria, cost overruns, and time and delivery delays. An integrated and holistic approach to software risk management is imperative.

8. *Safe failure of engineering-based systems.* Assessing risk management associated with critical engineering-based systems cannot be complete without ensuring that even in the remote likelihood of a system failure, there will be a safe shutdown without catastrophic consequences to people or facilities. Examples of such critical systems include transportation systems, space projects, the nuclear industry, and chemical plants.

9. *Cross-disciplinary nature of engineering-based systems.* All engineering-based systems are built to serve the well-being of people; incorporating knowledge-based expertise from other disciplines is essential if the risk of failure of such systems is to be reduced.

10. *Risk management: a requisite for sustainable development.* Sustainable development is the ultimate manifestation of long-term protection of the ecology and the environment, and of ensuring the harmony of this protection with economic development. This cannot be realized without a systemic process of risk assessment and management.

11. *Evidence-based risk assessment.* Sparse databases and limited information often characterize most large-scale engineering systems, especially during the conception, planning, design, and construction phases. The reliability of specific evidence, including the evidence upon which expert judgment is based, is essential for effective management of the risks assessed for these systems.

12. *Impact analysis.* Good technology management necessarily incorporates good risk-management practices. Determining the impacts of current decisions on future options, however, is the imperative in decision making.

3.6 SELECTED CASES OF RISK-BASED ENGINEERING PROBLEMS

The following are summaries of selected risk analyses of engineering-based systems. Citations of the complete detailed studies are included with each case.

3.6.1 Risk Identification of System Acquisition Using Hierarchical Holographic Modeling

This section documents the applications of hierarchical holographic modeling (HHM) (Haimes, 1981) to risk identification for the acquisition of a large database system. HHM provided a tool to identify analysts' and managers' perceptions of what could go wrong in the acquisition process. In this project, the identified risks were ranked on the basis of their likelihoods and potential consequences, and ultimately generated and evaluated alternatives for risk mitigation. For a more detailed discussion of the ranking process, see Haimes et al. (1995) and Haimes (1998).

A mathematical model may be viewed in the abstract as a one-sided, limited image of the real system that it portrays. With single-model analysis and interpretation, it is quite impossible to clarify and document the multiple components, objectives, and constraints of a system, and also its numerous societal aspects (functional, temporal, geographic, economic, political, legal, environmental, sectoral, institutional, etc.). In the HHM scheme, plural models represent the various aspects of the system, with each model termed a holographic submodel. The HHM approach recognizes that no single vision or perspective of a system is adequate to represent a system and its component parts. Instead, the HHM approach identifies and coordinates multiple, complementary decompositions of a complex system. A decomposition is a hierarchy of the system's components, subcomponents, and sub-subcomponents that captures the structure of a particular view of the system (Haimes and Macko, 1973; Haimes et al., 1990).

HHM enables the enumeration of the risks in a complex system—risks in terms of its technical aspects, organizational and functional decision-making structures, the various time horizons, the multiple decision makers, stakeholders, and users of the system, and the host of institutional, legal, and other socioeconomic conditions that require consideration (Haimes, 1991). The synthesis of the identified risks from different perspectives can then generate a more complete image of the overall system risk. The HHM concept has been applied elsewhere to software project development (Chittister and Haimes, 1993; Schooff, 1996) and to global sustainable development (Haimes, 1992).

3.6.1.1   *Acquisition Problem Description.*   Managers of a large database acquisition project commissioned the University of Virginia's Center for Risk Management of Engineering Systems to provide support to the risk assessment and management effort of a large-scale engineering-based system. The complexity of the project involved advanced hardware and software, translation of a massive database, personnel from many organizational units, transitional program phases spanning more than five years in implementation, and hundreds of millions of dollars in investment. Our results have been modified for presentation here and some details of the effort have been omitted.

The earliest stage of the study required a common identification of the program's risks by the system managers and the analysts. Later, program managers had to agree on priorities to reduce the likelihood of the program's failure to meet its schedule, cost, and performance criteria. A ranking methodology was suggested to improve the allocation of limited resources for risk mitigation. Finally, there was the need to generate and compare alternative policies for risk management. Numerous interviews with program managers and technical experts were conducted at the work site by the Center's team. The many oral discussions and reviews of internal documents were essential to the process of risk identification, prioritization, and mitigation.

Figure 3.14 depicts the multiple views of the risk identification problem for this system using the HHM approach. The methodological framework for identifying most, if not all, sources of risk associated with the system consisted of eight major perspectives. The strategy for risk identification revolved around the multiple decompositions, or visions. After each main-level vision was introduced, a more detailed and comprehensive discussion of the entire risk assessment structure began. In an interview with an expert to identify new sources of risk to the large-scale technological system, an initial subset of two or more of the hierarchy's decompositions was used to formalize and structure the risk identification process. Later, inclusion of additional decompositions provided increased detail and focus to the risk identification process.

For example, one vision or decomposition of the risk for the database system was the *functional* perspective, focusing on the various services that the system will provide. From a functional view, the database system was decomposed into five major subsystems. These functional areas were then evaluated for sources of risk by cross-reference to other decompositions.

The results of the HHM identification process were consolidated in a master list of over 250 sources of risks. Items on the list ranged in nature from technology issues to specifications documents, to schedule inconsistencies, to personnel, and to managerial leadership

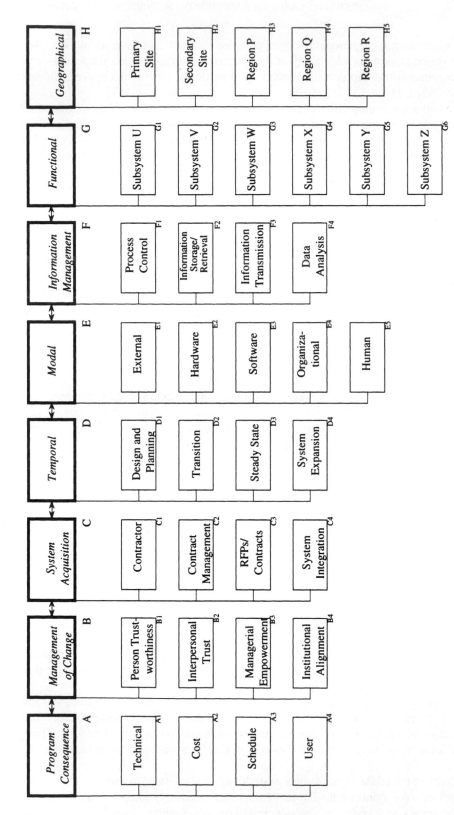

Figure 3.14 Hierarchical holographic model for risk identification.

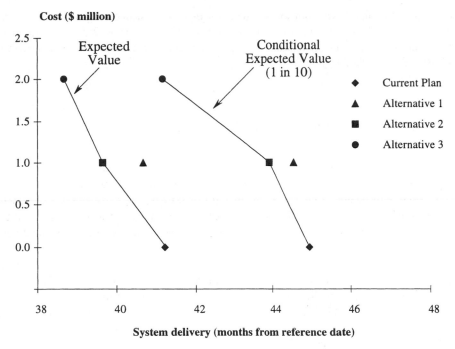

Figure 3.15 Tradeoffs among alternatives and the current plan: implementation costs vs. delay of system delivery for the expected completion date and the one-in-ten worst-case completion.

problems. There was considerable redundancy, which indicated the connectedness of various levels and different perspectives in the system. Thus, the master list gave an unfiltered impression of the perceived importance of a great number of risks to the system.

With program managers, we developed a hierarchy of criteria that was used to prioritize the risk items in terms of their likelihoods of occurrence, their potential consequences to the program, and the efficacy and immediacy of risk-reduction efforts. Risks identified in the master list using HHM were grouped into categories of related items, reducing the over 250 items to approximately 20 broad issues. The overall system's risks were divided into three categories: Program Risk, User Risk, and Risk Mitigation. The consolidated categories of risk issues were ranked using the analytic hierarchy process (Saaty, 1980).

Figure 3.15 illustrates the tradeoff between the implementation costs of the options for risk mitigation and the completion date, both the expected date and the one-in-ten worst case (conditional expected value) of the system delivery. Notice that in studying the tradeoff between a short-term implementation cost and the long-term issue of program delay, Alternative 1 is dominated by Alternative 2. The up-front cost of these two alternatives is the same, while the system delivery is later for the dominated alternative, both on the average and in considering the one-in-ten worst case.

It was not sufficient to consider only the scenario where subsystems X and Y were on the critical schedule path. We considered four scenarios, of which the baseline scenario is Scenario 1. The three additional scenarios were specified by the program managers and accounted for possible delays of subsystems V and W so that they were potentially on the critical schedule path. It was not known which scenario would actually be the case. No probabilities could be assigned to the likelihoods of the four scenarios. Instead, simulations of the system delivery and the analogs to Figure 3.2 were generated to represent each of the additional scenarios.

3.6.2 Evaluation of Automobile Safety Features in a Multiobjective Risk Framework

With over 168 million licensed drivers and 140 million registered automobiles on the roads in the United States today, the investigation of factors affecting automobile safety has taken on great significance (Federal Highway Administration, 1992). In fact, the frequency and severity of motor vehicle accidents are of such consequence that they constitute the sixth leading cause of fatality in the United States and the number one cause of death due to injury.

The statistics in Blincoe and Falgin (1990) point to the necessity of focusing research energies on the development and subsequent adoption of new automotive technologies to improve safety on highways in the United States and elsewhere. Before this can be done, however, the causes of vehicle accidents must be identified. Here we describe a framework for analyzing the interacting factors that contribute to the risk of automobile accidents within a multiobjective framework.

The objective of this study (supported by the National Science Foundation and General Motors) was to examine the functional factors that contribute to automobile accident occurrence and to model the causation structure in order to manage the risks associated with automobile accidents. A fault-tree model provided an intuitive qualitative as well as quantitative framework for decomposing possible pathways to accident occurrence (Kuzminski et al., 1995; Eisele et al., 1996). Fault-tree analysis also provided a statistical representation of how interacting driver, vehicle, and environmental factors contribute to the likelihood of automobile accident occurrence (see Figure 3.16). The application of this model facil-

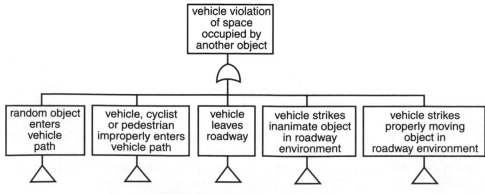

Figure 3.16 Top level of fault tree.

itated pinpointing those factors that most contribute to accident causation and subsequently enabled the identification and comparison of potential crash-avoidance technologies.

The frequency and severity of highway accidents are influenced by a variety of factors, including a vehicle's dynamics and crashworthiness, environmental conditions, and driver factors. All of these interactive causal factors ultimately influence the onset and outcome of an accident. Improving highway safety through the introduction of crash-avoidance measures in vehicle design requires a thorough investigation of causation/consequence relationships, the availability of appropriate multiobjective decision-making methodologies, and optimal resource allocation. Such methodologies can evaluate the tradeoffs between reducing the risk level of vehicle accidents and bearing the associated cost. Indeed, the ultimate quantification of the costs, risks, and benefits associated with the vehicle's design, reliability, and response to environmental and driver factors is a multiobjective optimization problem in nature, and is at the heart of the challenge facing the automotive industry today.

Using fault-tree logic, an extensive network of events leading to the occurrence of an automobile accident was developed. Environmental, driver, and vehicle failures and their interactions were all brought together to describe accident causation (Joshua and Garber, 1992; Kuzminski et al., 1995; Eisele et al., 1996).

Select vehicle design modifications were evaluated in terms of their cost and driver-safety tradeoffs using multiobjective decision analysis. For each prospective combination of redesign options, a framework was used that examined the minimization of four noncommensurate measures of harm and the minimization of vehicle production costs.

A general multiobjective decision-making framework incorporating the preferences of design engineers and other potential decision makers was used to evaluate a selected set of crash-avoidance design modifications. These designs were evaluated for their mitigating influences on the contribution of driver factors to accident causation. Further, the concept of tradeoffs with crashworthiness was considered, where crashworthiness was reported in terms of an expected accident injury severity distribution, an expected distribution of workdays lost, a total body injury distribution, and an expected hospitalization stay distribution. The multiobjective framework used is based on the surrogate worth tradeoff (SWT) method (Haimes and Hall, 1974; Chankong and Haimes, 1983).

3.6.3 Reliability Modeling for Water Distribution Rehabilitation

The deterioration of the United States' water distribution systems has become a focal issue within the nation's overall infrastructure problem. At the same time, achieving the goal of maintaining large-scale water distribution systems at optimum standards is an extremely difficult and complex task. The complexity associated with the selection of the best replacement/repair strategy is due to the dynamic evolution of the failure modes of water mains, budget and other resource constraints, the large scale and intricacy of the problem, and the competing and often conflicting demands of the various constituencies.

The overall goal of this project (supported by the National Science Foundation) was the development of effective operational management methodologies and procedures for large-scale public facilities—facilities that are characterized by their hierarchical structure, their multiple objectives, and their elements of risk and uncertainty (Li and Haimes, 1992a, b; Li et al., 1993). In particular, special attention was paid to optimal maintenance-related decision making for deteriorating water distribution systems.

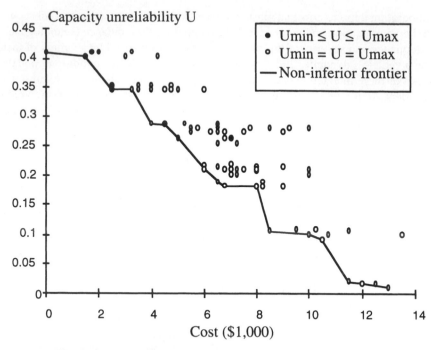

Figure 3.17 Rehabilitation scenarios and Pareto optimal frontier.

A water distribution system often consists of a large number of components. In general, however, resources are not available to completely update all the deteriorating water pipes and other system components. Limited resources have to be optimally distributed among the system's components in order to achieve the highest systems availability.

For evaluating the capacity reliability, the data needed are nodal or system supply, nodal demand, and pipe-carrying capacity. Nodal demand and system supply are assumed to be deterministic, while pipe-carrying capacity is a probabilistic parameter that corresponds to the system uncertainty. In assessing the maximum flow capacity for each arc, expert opinion may be combined with historical data, either about the actual water distribution network in question or about similar ones elsewhere.

Maintenance and rehabilitation of pipes take two primary forms: relining and replacing. Relining pipes restores carrying capacity to near the original level, since the layer with which the pipe's interior is covered is thin, smooth, and durable, and there is little change in diameter. This is a very useful strategy in situations where carrying capacity is the greatest concern. Of course, replacing pipes can return the carrying capacity to the original level or better—at a much greater cost. The set of rehabilitation actions for this study was limited to relining pipes.

The nonseparable resource allocation problem for a deteriorating water distribution system was embedded into a separable, albeit multiobjective, resource allocation problem of a type easy to solve by a decomposition method. For the problem of the availability network, the optimal solution to the original nonseparable resource allocation problem was proven to be in the set of noninferior solutions of the corresponding multiobjective problem (see Figure 3.17). One important product of this three-level decomposition approach is that it gives an importance measure for each system component, which provides the priority rank for the maintenance of various system components.

3.6.4 Channel Reliability of the Upper Mississippi River Navigation System

The Upper Mississippi River (UMR) originates out of Lake Itasca in Minnesota and flows generally southward, fed by several tributaries such as the Minnesota, St. Croix, Wisconsin, Rock, Des Moines, and Illinois rivers. Just above St. Louis, the UMR meets the Missouri and then joins the Ohio River near Cairo, Illinois (Tweet, 1983). The first 500 miles of the river downstream from Lake Itasca to Minneapolis are not navigable. The stretch from Minneapolis to Cairo, a distance of over 800 miles, forms the Upper Mississippi River Navigation System. The distance from St. Louis to Minnesota is navigable by means of a series of low navigation dams and associated locks that form a sequence of pools in the river. The stretch from St. Louis downstream forms the open river and is navigable without requiring any locks or dams. Currently, the navigation channel on the UMR is mandated by the U.S. Congress to be 300 feet wide and 9 feet deep.

Sedimentation in the navigation channel reduces the depth available for navigation. The U.S. Army Corps of Engineers maintains the required navigation standard through the use of structural measures, such as wing dams and closing dams, as well as through the use of maintenance dredging; however, these measures have an associated cost. In addition, there are several environmental concerns associated with the disposal of dredged material. Aside from these concerns, the physical deterioration of the various structures, including wing and closing dams, can also impact the need for dredging of the channel.

An important purpose of the reliability portion of the UMR navigation study (supported by the U.S. Army Corps of Engineers) is to project general funding requirements to maintain the navigation system in the future. The objective was not to decide exactly which projects should be built; this role remains in the domain of professional engineering judgment, personal maintenance experience, and models of the physical processes involved. Nor was the purpose to give accurate forecasts of needed resources in the short term. Rather, the role of the models was to provide foundations upon which to quantify the benefits of increased rehabilitation funding for wing dams and closing dams on a system-wide basis and over a period of many years. Two reliability models for the navigation channel were developed: one associated with the need for dredging of the *pool*, and the other with the dredging of the *reach*. These models are complementary approaches to demonstrate the reduced-dredging benefits associated with rehabilitation of channelization structures (Tulsiani et al., 1995).

3.6.4.1 Dredge-Capacity Reliability Model. The dredge-capacity reliability model generated a probabilistic description of the annual dredge need for a given pool based on an assumed relationship between dredging and underlying features of the hydrograph. Underlying features that were considered included the total annual discharge through the pool, the number of hydrograph dips, and the number of flow peaks, where the hydrograph features were modeled as random variables. The dredge-capacity model also estimated a probability distribution of annual dredge need for the pool that could be expected if *significant* rehabilitation were performed poolwide. The two probability density functions, one of the unrehabilitated and one of the rehabilitated pool, were useful to characterize the variable cost of dredging the pool, or a system of pools that were similarly evaluated.

A function relating the pool-dredging amount to the cost of dredging is required for this purpose.

In principle, one can use the probability function of dredge need for the pool to evaluate the dredge-capacity reliability in the following two-step process: (1) define the annual dredge capacity for the given pool, and (2) calculate the probability that the dredge need exceeds the capacity. In this study, a capacity-based approach was used to allow economists to distinguish and characterize a failure of the system as an exceedance of the normal operating budget for dredging.

An intermediate result from this model gives insight into the need for dredging on the poolwide scale: it relates the annual dredge amount to the average daily discharge for the year (or vs. some other extremal-oriented hydrograph feature). From this curve it is useful to study the impact of poolwide rehabilitation on the relationship between dredging amount and the hydrograph reading in the year preceding dredging.

3.6.4.2 Reach Reliability Model. As a complement to the pool model just described, the evaluation of channel reliability can be extended down to the level of individual reaches to better understand and characterize the impact of channel sedimentation associated with individual structural rehabilitations. Thus, a reliability model was developed that uses data on individual reaches as the statistical basis for an ideal characterization of sedimentation in the channel. This reach model can be used to generate a chart of the UMR on which the estimates of channel reliability are provided for all reaches together with the potential for improving reach reliability by rehabilitating structures. Application of the reach model yields a general picture of the benefits of rehabilitation—based on the identified significance of the parameters affecting sedimentation at the reach level—but the model is not able to recommend projects at specific reaches. The amount of dredging was not considered in this model because it was assumed that setup costs (between dredging events) dominated the cost differences attributable to dredging volume for a particular reach.

The interdredge reliability model describes the probability that in some time interval no dredging is required in a particular reach. This model also estimates the improvement in interdredge reliability to be expected if the reach is rehabilitated. It assumes that a reach can be characterized by a small set of parameters representing the channel morphology. A weighted sum of the parameter values, with weighting coefficients estimated from the real system, gives both the estimate of reliability and the expected improvement in reliability from rehabilitation. This ideal-process model of interdredge reliability generates the frequency of the need to dredge expected from an idealized reach-by-reach model. It is important to distinguish this from the observations of the real system, which are dredging records influenced by dredging policy shifts and other factors not related to the need to dredge.

3.7 SUMMARY

The comprehensiveness of total risk management makes the systemic assessment and management of risk tractable from many perspectives. Available theories and methodologies developed and practiced by various disciplines can be adopted and/or modified as appropriate for total risk management. Fault-tree analysis, for example, which has been developed for the assessment and management of risk associated with hardware, can and is being modified and applied to assess and manage all four sources of failure: hardware, software, organizational, and human. Hierarchical/multiobjective tradeoff analysis can and is being applied to risk associated with public works and the infrastructure. As the importance of risk is better understood and its analysis is incorporated within a broader and more holistic management framework, the field of risk management will closely merge with the field of systems analysis. There are a few off-the-shelf software tools that can be helpful in the risk assessment and management process, such as Risk and Crystal Ball®.

Some key ideas advanced in this chapter are summarized below:

1. Risk assessment and management is a process that must answer the following questions:

    What can go wrong?

    What is the likelihood?

    What are the consequences?

    What can be done?

    What options are available and what are their associated tradeoffs?

    What are the impacts of current decisions on future options?

2. Organizational failures are major sources of risk.
3. The perception of risk and its importance in risk-based decision making should not be overlooked.
4. Risk management should be an integral part of technology management leading to multiple-objective tradeoff analysis.
5. The expected value of risk when used as the sole criterion for risk measurement leads to erroneous results. Also, risk of extreme and catastrophic events should not be made to coexist with high-probability, low-consequence events.

Finally, it is important to keep in mind Heisenberg's uncertainty principle, which states that the position and velocity of a particle in motion cannot simultaneously be measured with high precision, and Einstein's statement: "So far as the theorems of mathematics are about reality, they are not certain; so far that they are certain, they are not about reality." By projecting Heisenberg's principle and Einstein's statement to the field of risk assessment and management, we assert that:

To the extent that risk assessment is precise, it is not real.

To the extent that risk assessment is real, it is not precise.

ACKNOWLEDGMENT

The technical editorial work of Grace Zisk is greatly appreciated. This article is primarily based on excerpts from *Risk Modeling, Assessment, and Management*, by Yacov Y. Haimes (Wiley, 1998).

REFERENCES

Asbeck, E., and Haimes, Y. Y. (1984). The partitioned multiobjective risk method. *Large Scale Syst.* **6,** 13–38.

Blincoe, C., and Falgin, S. (1990). *The Economic Cost of Motor Vehicle Crashes.* (DOT-HS-807-876). U.S. Department of Transportation, Washington, DC: U.S. Government Printing Office.

Chankong, V., and Haimes, Y. Y. (1983). *Multiobjective Decision Making: Theory and Methodology.* New York: North-Holland.

Chittister, C., and Haimes, Y. (1993). Risk associated with software development: a holistic framework for assessment and management. *IEEE Trans. Syst., Man, Cybernet.* **23**(3), 710–723.

Deming, W. E. (1986). *Out of the Crisis.* Cambridge, MA: MIT Press.

Eisele, J. S., Haimes, Y. Y., Schwing, R., Garber, N., Li, D., Lambert, J. H., Kuzminski, P., and Chowdhury, M. (1996). The impact of improved vehicle design on highway safety. *Reliab. Eng. Syst. Saf.* **54**(1), 65–76.

Federal Highway Administration. (1992). *Highway Statistics 1991* (Document FHWA-P1-92-025) Washington, DC: U.S. Government Printing Office.

Fischhoff, B., Lichtenstein, S., Slovic, P., Keeney, R., and Derby, S. (1980). *Approaches to Acceptable Risk: A Critical Guide.* Oak Ridge, TN: Oak Ridge National Laboratory, U.S. Department of Commerce.

Haimes, Y. Y. (1981). Hierarchical holographic modeling. *IEEE Trans. Syst., Man, Cybernet.* **SMC-11**, 606–617.

Haimes, Y. Y. (1988). Alternatives to the precommensuration of costs, benefits, risks, and time. In *The Role of Social and Behavioral Sciences in Water Resources Planning and Management* (D.D. Bauman and Y. Y. Haimes, eds.). New York: ASCE.

Haimes, Y. Y. (1991). Total risk management. *Risk Anal.* **11**(2), 169–171.

Haimes, Y. Y. (1992). Sustainable development: A holistic approach to natural resource management. *IEEE Trans. Syst., Man, Cybernet.* **22**(3), May/June, 413–417.

Haimes, Y. Y., (1998). *Risk: Modeling, Assessment, and Management.* New York: Wiley. **11**(2), 169–171.

Haimes, Y. Y., and Hall, W. A., (1974). Multiobjectives in water resources systems analysis: the surrogate worth trade-off method. *Water Resour. Res.* **10**(4), 615–624.

Haimes, Y. Y., and Macko, D. (1973). Hierarchical structures in water resources systems, *IEEE Trans. Syst., Man, Cybernet.* **SMC-3**(4), 396–402.

Haimes, Y. Y., Tarvainen, K., Shima, T., and Thadathil, J. (1990). *Hierarchical Multiobjective Analysis of Large-Scale Systems.* New York: Hemisphere Publishers.

Haimes, Y. Y., Lambert, J. H., Li, D., Schooff, R., and Tulsiani, V. (1995). *Identification, Ranking, and Mitigation of Risk in a System Acquisition: A Case Study.* Charlottesville: University of Virginia, Center for Risk Management of Engineering Systems.

Joshua, S. C., and Garber, N. (1992). A causal analysis of large vehicle accidents through fault-tree analysis. *Risk Anal.* **23**(2).

Kaplan, S., and Garrick, J. (1981). On the quantitative definition of risk. *Risk Anal.* **1**, No.1.

Karlsson, P. O., and Haimes, Y. Y. (1988a). Risk-based analysis of extreme events. *Water Resour. Res.* **24**, No. 1, 9–20.

Karlsson, P. O., and Haimes, Y. Y. (1988b). Probability distributions and their partitioning. *Water Resour. Res.* **24**, No. 1, 21–29.

Kuzminski, P., Eisele, J. S., Garber, N., Schwing, R., Haimes, Y. Y., Li, D., and Chowdhury, N. (1995). Improvement of highway safety: Identification of causal factors through fault-tree modeling. *Risk Anal.* **15**(3), 293–312.

Li, D., and Haimes, Y. Y. (1992a). Optimal maintenance-related decision making for deteriorating water distribution systems: Semi-Markovian model for water main. *Water Resour. Res.* **28**, 1053–1061.

Li, D., and Haimes, Y. Y. (1992b). Optimal maintenance-related decision making for deteriorating water distribution systems: Multilevel decomposition approach. *Water Resour. Res.* **28,** 1063–1070.

Li., D., Dolezal, T., and Haimes, Y. Y. (1993). Capacity reliability of water distribution networks. *Reliab. Eng. Syst. Saf.* **42,** 29–38.

Lowrance, W. W. (1976). *Of Acceptable Risk*. Los Altos, CA: William Kaufmann.

Masaaki, I. (1986), *Kaizen: The Key to Japan's Competitive Success*. New York: Random House Business Division.

Morall, J. F., III. (1987). *Chem. Eng. News,* September 14.

National Research Council, Committee on Safety Criteria for Dams. (1985). *Safety of Dams—Flood and Earthquake Criteria*. Washington, DC: National Academy Press.

Pate-Cornell, M.E., (1990). Organizational aspects of engineering system safety: The case of offshore platforms. *Science* **250,** 1210–1217.

Peterson, D. F. (1974). *Water Resources Planning, Social Goals, and Indicators: Methodological Development and Empirical Tests* (PRWG 131-1). Logan: Utah State University, Utah Water Research Lab.

Petrakian, R., Haimes, Y. Y., Stakhiv, E. Z., and Moser, D. A. (1989). Risk analysis of dam failure and extreme floods. In *Risk Analysis and Management of Natural and Man-Made Hazards* (Y. Y. Haimes and E. Z. Stahkiv, eds.) New York: ASCE, pp. 81–122.

Runyon, R. P. (1977). *Winning the Statistics*. Reading, MA: Addison-Wesley.

Saaty, T. L. (1980). *The Analytic Hierarchy Process*. New York: McGraw-Hill.

Schooff, R. M. (1996). Hierarchical holographic modeling for software acquisition: Risk assessment and management. Ph.D. Dissertation, University of Virginia, Systems Engineering Department, Charlottesville.

Tulsiani, V., Haimes, Y. Y., Lambert, J. H., Li, D., and Nanda, S. K. (1995). *Channel Reliability of the Navigation System in the Upper Mississippi River: The Pool Model,* Working Paper. Charlottesville: University of Virginia, Center for Risk Management of Engineering Systems.

Tweet, R. D. (1983). *History of Transportation on the Upper Mississippi and Illinois Rivers,* Navigation History NWS-83-6. Fort Belvoir, VA: National Waterways Study, Institute of Water Resources, U.S. Army Engineer Water Resources Support Center.

4 Discovering System Requirements

A. TERRY BAHILL and FRANK F. DEAN

4.1 INTRODUCTION

No two systems are exactly alike in their requirements. However, there is a uniform and identifiable process for logically discovering the system requirements regardless of system purpose, size, or complexity (Grady, 1993). The purpose of this chapter is to reveal this process.

Problem Formulation. What is the problem that this chapter solves? A lot of engineers are confused about system requirements. These engineers cannot answer these questions: Who is responsible for writing requirements? Who uses them? What are they? When should they be written? Where do they come from? Why are they written? How are they written? How are they organized? How are they discovered?

Solution Requirements. What are the requirements for this chapter? It should answer the who, what, when, where, why, and how questions. It should provide examples. It should explain existing nomenclature. It should present a process for discovering requirements.

This chapter presents the philosophy and terminology used by the New Mexico Weapons Systems Engineering Center at Sandia National Laboratories for discovering system requirements. Other organizations may use different procedures and terminology. However, we think a consensus is developing in the Systems Engineering community. It is hoped that this chapter is consistent with that consensus. Like systems engineering in general, the statements in this chapter are not dogmatic. Each statement has been rightfully violated many times (see, for example, Martin, 1995). However, these statements are generalizations of good engineering practices.

This chapter only explains a part of the systems requirements process. Large projects should use a computer tool to help write, decompose, and maintain system requirements. Many such computer-based tools are commercially available. For examples, see any recent Proceedings of the International National Council on Systems Engineering (INCOSE). Each project team will select a specific tool and then provide training for it. Because such training is tool specific, this chapter will not discuss such tools. Another part of the requirements process is modeling the proposed system. Dozens of tools are available (Bahill, et al 1998; Rechtin and Maier, 1997); two recently popular ones are *object-oriented design* and *functional analysis and decomposition* (Bharathan et al., 1995; Chapter 25 of this Handbook). This chapter does not discuss tools for modeling systems, because of the sheer magnitude of the task. Methodologies for constructing systems engineering models are discussed in Chapter 26.

Handbook of Systems Engineering and Management, Edited by A. P. Sage and W. B. Rouse
ISBN 0471-15405-9   ©1999 John Wiley and Sons, Inc.

4.2 STATING THE PROBLEM

Stating the problem is one of the systems engineer's most important tasks. The problem must be stated in a clear, unambiguous manner.

State the problem in terms of what the world would be like if the problem did not exist, and not in terms of preconceived solutions. In 1982 a flood washed out a bridge across the Santa Cruz River, near Tucson, AZ, and made it difficult for the Indians at Mission San Xavier del Bac to get to the Bureau of Indians Affairs Health Center. A common way of stating this problem was "We must rebuild the bridge across the Santa Cruz River." A better way, however, would be to say "The Indians at San Xavier Mission do not have a convenient way to get to their health center."

It is good engineering practice to state the problem in terms of the top-level functions that the system must perform. However, it is better to state the problem in terms of the deficiencies that must be ameliorated. This stimulates consideration of more alternative designs.

Example 1

Top-Level Function. The system shall hold together 2 to 20 pieces of 8½-by-11-inch, 20-pound paper.

Alternatives. Stapler, paper clip, fold the corner, put the pages in a folder.

Example 2

The Deficiency. My reports are typically composed of 2 to 20 pieces of 8½-by-11-inch, 20-pound paper. The pages get out of order and become mixed up with pages of other reports.

Alternatives. Stapler, paper clips, fold the corner, put the pages in folders, number the pages, put them in envelopes, put them in three-ring binders, throw away the reports, convert them to electronic form, have them bound as books, put them on audiotapes, distribute them electronically, put them on floppy disks, put them on microfiche, transform the written reports into videotapes.

Do not believe the first thing your customer says. Verify the problem statement with the customer, and expect to iterate this procedure several times. For an excellent (and enjoyable) reference on stating the problem, see Gause and Weinberg (1990).

4.2.1 Do Not Use the Word "Optimal"

The word "optimal" should not appear in the statement of the problem, because there is no single optimal solution for a complex systems problem. Most system designs have several performance and cost criteria. Systems engineering creates a set of alternative designs that satisfies these performance and cost criteria to varying degrees. Moving from one alternative to another will usually improve at least one criterion and worsen at least one criterion, that is, there will be tradeoffs. None of the feasible alternatives is likely to optimize all the criteria (Szidarovszky et al., 1986). Therefore, we must settle for less than optimality.

It might be possible to optimize some subsystems, but when they are interconnected, the overall system may not be optimal. The best possible system may not be that made up of optimal subsystems. An all-star team may have optimal people at all positions, but is it likely that such an all-star team could beat the world champions? For example, a Pro Bowl football team is not likely to beat the Super Bowl champions.

Humans are not optimal animals. Shrews are smaller; elephants are bigger; cheetahs can run faster; porpoises can swim faster; dolphins have bigger brains; bats have wider bandwidth auditory systems; deer have more sensitive olfaction systems; pronghorn antelope have sharper vision. Humans have not used evolution to optimize these systems, but have remained generalists. The frog's visual system has evolved much farther than man's; frogs have cells in the superior colliculus that are specialized to detect moving flies. Leaf-cutting ants had organized agricultural societies millions of years before humans. Although humans are not optimal in any overall sense, they seem to rule the world.

If the system requirements demanded an optimal system, data could not be provided to prove that any resulting system was indeed optimal. In general, it can be proven that a system is at a local optimum, but it cannot be proven that it is at a global optimum.

If it is required that optimization techniques be used, then they should be applied to subsystems. Total system performance must be analyzed, however, to decide if the cost of optimizing a subsystem is worthwhile. Furthermore, total system performance should be analyzed over the whole range of operating environments and tradeoff functions, because what is optimal in one environment with one tradeoff function will probably not be optimal with others.

Because of the rapid rate at which technology is advancing, flexibility is more important than optimality. A company could buy a device, spend man-years optimizing its inclusion into their system, and then discover that a new device is available that performs better and costs less than the optimized system.

4.2.2 Define the Customer

The term *customer* includes anyone who has a right to impose requirements on the system. This includes end users, operators, bill payers, owners, regulatory agencies, victims, and sponsors. Because systems engineering delivers both a *product* and a *process* for manufacturing it, we must also consider the customer for the process.

We now illustrate some of these customer roles for a commercial airliner, such as the Boeing 777. The users are the passengers who fly on the airplane, and the operators are the crew who fly the plane and the mechanics who maintain it. The bill payers are the airline companies, such as United or TWA. The owners are the stockholders of these companies. The Federal Aviation Administration (FAA) writes the regulations and certifies the airplane. Among others, people who live near the airport are victims of noise and air pollution. If the plane is tremendously successful, AirBus (the manufacturer of a competing airplane) would also be a victim. The sponsor of the systems engineering effort in this example would be the corporate headquarters of Boeing.

The users and operators of the process would be the employees in the manufacturing plant. The bill payer would be Boeing. The owner would be the stockholders of Boeing. Occupational Safety and Health Administration (OSHA) would be among the regulators. Victims would include physically injured workers and, according to Deming, workers who have little control of the output but who are reviewed for performance (Deming, 1982; Latzko and Saunders, 1995).

4.2.3 Identify the Audience

Before writing a document you should identify the audience for the document. For a requirements document, the audience is the client and the designers.

System requirements communicate the customer's needs to the technical community that will design and build the system; therefore, they must be understandable by both. One of the most difficult tasks in creating a system is communicating with all subgroups within both groups [Institute of Electrical and Electronics Engineers (IEEE), 1993].

The client and the designers have different backgrounds and needs. Wymore (1993) suggests two different documents for these two different groups: The Operational Need Document for the client and the System Requirements Document for the design engineers.

> The Operational Need Document is a detailed description of the problem in plain language. It is intended for management, the customer and systems engineering. . . . The Systems Requirement Document is a succinct mathematical description or model of the . . . requirements as described in the Operational Need Document. Its audience is systems engineering. [Chapman et al., 1992]

Sometimes these are referred to as user (or customer) requirements and technical (or system) requirements, or system specifications, respectively.

4.3 WHAT ARE REQUIREMENTS?

Requirements are the necessary attributes defined for a system before and during design. The customer's need is the ultimate system requirement from which all other requirements flow (Grady, 1993). In addition, requirements are statements that identify the essential needs of a system in order for it to have value and utility. Requirements may be derived or based upon interpretation of other stated requirements to assist in providing a common understanding of the desired characteristics of a system. Finally, requirements should state what the system is to do, but they should not specify how the system is to do it. Section 4.3.1 presents an example of a requirement.

4.3.1 Example of a Requirement (Sommerville, 1989)

The Graphic Editor Facility. To assist in positioning items on a diagram, the user can turn on a grid in either centimeters or inches, by an option on a control panel. Initially the grid is off. The grid can be turned on or off at any time during an editing session, and can be toggled between inches and centimeters at any time. The grid option will also be provided on the reduce-to-fit view, but the number of grid lines shown will be reduced to avoid filling the diagram with grid lines.

Good Points About this Requirement. It provides rationale for the items; it explains why there should be a grid. It explains why the number of grid lines should be reduced for the reduce-to-fit view. It provides initialization information; initially the grid is off.

Bad Points. The first sentence has three different components: (1) it states that the system should provide a grid; (2) it gives detailed information about grid units (centimeters and inches); and (3) it tells how the user will activate the grid. This requirement provides initialization information for some but not all similar items; it specifies that initially the grid is off, but it does not specify the units when it is turned on. Section 4.3.2 shows how this requirement might be improved.

4.3.2 Example of an Improved Requirement (Sommerville, 1989)

4.3.2.1 The Grid Facility

1. The graphic-editor grid facility shall produce a pattern of horizontal and vertical lines forming squares of uniform size as a background to the editor window. The grid shall be passive rather than active. This means that alignment is the responsibility of the user and the system shall not automatically align items with grid lines.

    Rationale. A grid helps the user to create a neat diagram with well-spaced entries. Although an active grid might be useful, it is best to let the user decide where the items should be positioned.

2. When used in the "reduce-to-fit" mode, the logical grid line spacing should be increased.

    Rationale. If the logical grid line spacing were not increased, the background would become cluttered with grid lines.

    Specification. Eclipse/Workstation/Defs:Section 2.6.

This requirement definition refers to the requirement specification, which provides details such as units of centimeters and inches and the initialization preferences.

4.4 CHARACTERIZATIONS

There are many orthogonal characterizations of system requirements. Four of these are types, sources, expressions or modalities, and input–output trajectories. A summary of these characterizations follows.

4.4.1 Types

There are two types of system requirements: mandatory and preference.

Mandatory requirements:

1. Specify the necessary and sufficient conditions that a minimal system must have in order to be acceptable and are usually expressed with *shall* and *must;*
2. Are passed or failed (do not use scoring functions); and
3. Must not be susceptible to tradeoffs between requirements.

The following is a typical mandatory requirement: "The system shall not violate federal, state or local laws." After mandatory requirements have been identified, systems engineers propose alternative candidate designs, all of which satisfy the mandatory requirements. Preference requirements are then evaluated to determine the "best" designs.

Preference requirements:

1. State conditions that would make the customer happier and are often expressed with *should* and *want;*

2. Should use scoring functions (Chapman et al., 1992) to produce figures of merit (see Fig. 4.1), and

3. Should be evaluated with a multicriteria decision technique (Szidarovszky et al., 1986), because none of the feasible alternatives is likely to optimize all the criteria, and there will be tradeoffs among these requirements.

Figure 4.2 shows an example of tradeoffs in the investigation of alternative laser printers. Many printers were below and to the left of the circular arc. They were clearly inferior and were dismissed. Three printers lay on the circular arc: they were the best. No printers were above and to the right of the circular arc. The question now becomes, Which of these three printers is the best? With the present data, there is no answer to that question. The customer will have to say which preference requirement (or figure of merit) is more important before an answer can be obtained. Moving from one alternative to another will improve at least one criterion and worsen at least one criterion, that is, there will be tradeoffs. An arc like this (or a surface when there are more than two criteria) is called a Pareto optimal contour.

Sometimes there is a relationship between mandatory and preference requirements in which a mandatory requirement is a lower threshold of a preference requirement. For example, for one computer program 8 Mbytes of random access memory (RAM) are required, but 12 Mbytes are preferred.

A scoring function may be used to give a system a normalized score that reflects how the requirement has been met for each criterion. The value of the figure of merit, using the example of Mbytes of RAM, is put into the scoring function and a normalized score is returned. The use of scoring functions allows different criteria to be compared and traded

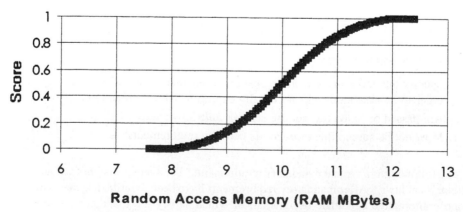

Figure 4.1 A scoring function for the amount of RAM.

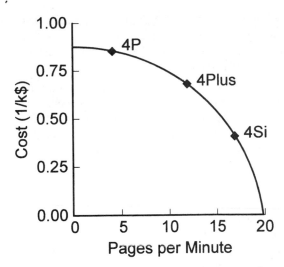

Figure 4.2   A typical tradeoff between preference requirements.

off against each other. In other words, scoring functions allow apples to be compared to oranges and nanoseconds to be compared to billions of dollars.

4.4.2   Sources

In this section we list two dozen sources of requirements. However, Wymore (1993) says that only the first six sources are necessary: input–output, technology, performance, cost, tradeoff, and system test. He says all of the other sources can be put into one of these six. Grady (1993) says we should have only five sources: functional, performance, constraints, verification, and programmatic. He thinks that most of our sources are constraints. The EIA-632 Standard on Systems Engineering says there are only three: functional, performance, and constraints (EIA-632, 1998; Martin, 1996). Project managers say that there are only three: cost, schedule, and performance (Kerzner, 1995). We leave it to the reader to decide whether or not our list of sources can be condensed.

   1. *Input–Output.* Perhaps the most common requirements relate the inputs of the system to its outputs. For example, an input–output requirement for an electronic amplifier could be stated as "The ratio of the output to the input at 10 kHz shall be +20 dB." Wymore (1993) maintains that functional requirements are a subset of input–output requirements. A well-stated input–output requirement describes a function. The preceding input–output requirement describes the function "Amplify the input signal." The functional requirement "The system shall fasten pieces of paper" is covered by the input–output requirement "The system shall accept 2 to 20 pieces of 8½-by-11-inch, 20-pound paper and secure them so that the papers cannot get out of order." One function of an automobile is to accelerate. The input is torque (perhaps developed with an engine), and the output is a change in velocity.

   2. *Technology.* The technology requirement specifies the set of components—hardware, software, and bioware—that is available to build the system. The technology re-

quirement is usually defined in terms of types of components that *cannot be used,* that *must be used,* or both. For example, Admiral Rickover required that submarine nuclear instrumentation be done with magnetic amplifiers. Your Purchasing Department will often be a source of technology constraints.

3. *Performance.* Performance requirements include quantity (how many, how much), quality (how well), coverage (how much area, how far), timeliness (how responsive, how frequent), and readiness [ability, mean time between failures (MTBF)]. System functions often map to performance requirements. For example, "The car shall accelerate from 0 to 60 mph in 7 seconds or less." Performance is an attribute of products and processes. Its requirements are initially defined through requirements analyses and trade studies using customer need, objective, and/or requirements statements (MIL-STD-499B, 1993).

4. *Cost.* There are many types of cost, such as manpower, resources, and monetary cost. An example cost requirement would be that the purchase price cannot be more than $10,000, and the total life-cycle cost cannot exceed $18,000.

5. *Tradeoff.* Tradeoff between performance and cost is defined as the different relative value assigned to each factor. For example, the performance figures of merit may have a weight of 0.6, and the cost figures of merit may be given a weight of 0.4.

6. *System Test.* The purpose of the system test is to verify that the design and the system satisfy the requirements. For example, in an electronic amplifier, a 3-mV, 10-kHz sinusoid will be applied to the input, and the ratio of output to input will be calculated.

7. *Company Policy.* Company policy is another way of stating requirements. For example, Learjet Inc. has stated, "We will make the airframe, but we will buy the jet engines and the electronic control systems."

8. *Business Practices.* Corporate business policies might require work breakdown structures, PERT charts, quality manuals, environmental safety and health plans, or a certain return on investment.

9. *Systems Implementation Engineering.* Systems or software implementation engineering practices might require that every transportable disk (e.g., floppy, zip or Bernoulli) have a Readme file that describes the author, date, contents, software program, and version (e.g., Word 7.0 or Excel 4.0).

10. *Project Management.* Access to source code for all software might be a project management requirement. It takes time and money to install new software. This investment might be squandered if the supplier went bankrupt and the customer could no longer update and maintain the system. Therefore, most customers would like to have the source code. Few software houses are willing to provide source code, however, because it might decrease their profits and complicate customer support. When there is any possibility that the supplier might stop supporting a product, the source code should be provided and placed in escrow. This source code remains untouched as long as the supplier supports the product. But if the supplier ceases to support the product, the customer can get the source code and maintain the product in-house. Therefore, placing the source code in escrow can be a requirement.

11. *Marketing.* The marketing department wants features that will delight the customer. They may be features that customers did not know they wanted. In the 1970s, IBM queried customers to discover their personal computer (PC) needs. No one mentioned portability, so IBM did not make it a requirement. Compaq made a portable PC and then a laptop, thereby dominating those segments of the market. In the 1950s IBM could have

bought the patents for Xerox's photocopy machine, but they did a market research study and concluded that no one would pay thousands of dollars for a machine that would replace carbon paper. They did not realize that they could delight their customers with a machine that provided dozens of copies in just minutes.

12. *Manufacturing Processes*. Sometimes we might require a certain manufacturing process or environment. We might require our semiconductor manufacturer to have a Class 10 clean room. Someone might specify that quality function deployment (QFD) be used to help elicit customer desires (although this would be in bad form, because it states how not what). Recently, minimization of the waste stream has become a common requirement.

13. *Design Engineering*. Design engineers impose requirements on the system. These are the "build to," "code to," and "buy to" requirements for products and "how to execute" requirements for processes.

14. *Reliability*. Reliability could be a performance requirement, or it could be broken out separately.

15. *Safety*. Some requirements may come from safety considerations. These may state how the item should behave under both normal and abnormal conditions.

16. *The Environment*. Concern for the environment will produce requirements, such as forbidding the use of chlorofluorocarbons (CFCs) or tetraethylchloride (TEC).

17. *Ethics*. Ethics could require physicians to obtain informed consent before experimenting on human subjects.

18. *Intangibles*. Sometimes the desires of the customer will be hard to quantify, for example: for intangible items such as aesthetics, national or company prestige (e.g., putting a man on the moon in the Apollo project), or ulterior motives such as trying to get a foot in the door using a new technology (e.g., the stealth airplanes), or starting business in a new country (e.g., China).

19. *Common Sense*. Many requirements will not be stated because they are believed to be common sense. For example, characteristics of the end user are seldom stated. If we are designing a computer terminal, it would not be stated that the end user would be a human with two hands and ten fingers. Common sense also dictates that computers not be damaged if they are stored at temperatures as high as 140°F. Furthermore, we do not write that there can be no exposed high-voltage conductors on a personal computer, but it certainly is a requirement. Many of these requirements can be found in de facto standards.

20. *Laws or Standards*. Requirements could specify compliance with certain laws or standards, such as the National Electrical Code, City/County Building codes, ISO-9000, or the IEEE 1220 Standard for Systems Engineering (1994).

21. *The Customer*. Some requirements are said to have come from the customer, such as statements of fact and assumptions that define the expectations of the system in terms of mission or objectives, environment, constraints, and measures of effectiveness. These requirements are defined from a validated needs statement (customer's mission statement), from acquisition and program decision documentation, and from mission analyses.

22. *Legacy Requirements*. Sometimes the existence of previous systems creates requirements. For example, "Your last system was robust enough to survive a long trip on a dirt road, so we expect your new system to do the same." Many new computer programs must be compatible with COBOL, because so many COBOL business programs are still running. Legacy requirements are often unstated.

23. *Data Collection Activities*. If an existing system is similar to the proposed new

system, then existing data collection activities can be used to help discover system requirements, because each piece of data that is collected should be traceable to a specific system requirement. Often it is difficult to make a measurement to verify a requirement. It might be impossible to meet the stated accuracy. Trying to make a measurement to verify a requirement might reveal more system requirements.

24. *Other Sources.* There are many other sources of system requirements, such as human factors, the environment (e.g., temperature, humidity, shock, vibration), the end user, the operator, potential victims, management, company vision, future expansion, schedule, logistics, politics, the U.S. Congress, public opinion, business partners, past failures, competitive intelligence, liability, religion, culture, government agencies (e.g., DoE, DoD, OSHA, FAA, EPA), industry standards (e.g., ANSI, SAE, IEEE, EIA), availability, maintainability, compatibility, service, maintenance, need to provide training, competitive strategic advantage, time to market, time to fill orders, inventory turns, accident reports, deliverability, reusability, future expansion, politics, society, standards compliance, standards certification (e.g., ISO 9000), effects of aging, the year 2000 problem, user friendly, weather (e.g., must be installed in the summer), need to accommodate system abuse by humans, security as in government classification, security as in data transmission, it must fit into a certain space, secondary customer, retirement, and disposal.

4.4.3 There Are Many Ways to Express Requirements

For some purposes, the best expression of the requirements will be a narrative in which words are organized into sentences and paragraphs. Such documents are often called *operation concepts* or *operational needs*. But all descriptions in English will have ambiguities, both because of the language itself and the context in which the reader interprets the words. Therefore, for some purposes the best description of a system will be a list or string of *shall* and *should* statements. Such a list would be useful for acquisition or acceptance testing. However, it is still very difficult to write with perfect clarity so that all readers have the same understanding of what is written.

Other modalities that can be used instead of written descriptions include:

- Wymorian Notation (Wymore, 1993)
- Finite-State Machines (Katz, 1994; Bahill, et al., 1998)
- Algorithmic State Machine Notation (Katz, 1994)
- Hardware
- Object-oriented Models (Booch, 1994; Rumbaugh et al., 1991; Jacobson et al., 1995; Bahill, et al., 1998)
- Special purpose, requirements management, computer programs

The big advantage of these modalities over the English language is that they can be rigorous and executable by computer. This greatly helps to point out contradictions and omissions. It also allows us to perform a sensitivity analysis of the set of requirements to learn which requirements are the real cost drivers (Karnavas et al., 1993).There are three important factors to consider in expressing requirements.

1. *A Prototype Expresses Requirements.* A publicly assessable prototype can express the system requirements as they are currently understood. This technique is very popular in the software community where a computer can be placed in the building lobby. Of

course, many functions of the final system will not be implemented in the prototype; instead there will be a statement of what the functions are intended to do. A publicly assessable prototype is easy to update, and it helps everyone understand what the requirements are. The purpose of building a prototype is to reduce project risk. Therefore, the first functions that are prototyped should be (but usually are not) the most risky functions (Rechtin and Maier, 1997; Chapman and Bahill, 1996).

2. *Consider Bizarre Alternatives.* During concept exploration, encourage consideration of bizarre alternatives. Studying unusual alternatives leads to a better and deeper understanding of the requirements by both the systems engineer and the design engineer. Likewise, studying models and computer simulations will help you understand the requirements. Concept exploration is one of the most fruitful phases in requirements discovery.

3. *Preparing the Users Manual Flushes Out Requirements.* The users manual should be written by future users early in the system design process (Shand, 1994). This helps get the system requirements stated correctly and increases user "buy in."

4.4.4 Input and Output Trajectories

Input and output trajectories are descriptions of input and output values as functions of time.

4.4.4.1 Behavioral Scenarios Describe the System. A powerful technique for describing the behavior of a system and for discovering requirements is creating typical sequences of events that the proposed system will go through. Such descriptions of behavior as a function of time are called trajectories, behavioral scenarios, use cases, threads, operational scenarios, logistics, or interaction diagrams.

A *behavioral scenario* for an automated teller machine (ATM) is shown in Figure 4.3. Several other examples are given in Appendix A. The basis of these diagrams is to list the system's objects (or components) along the top of the diagram. Then, with time running from top to bottom, list the messages that are exchanged between the objects. Alternatively, the arrows can be labeled with data that are exchanged between the components or the functions that are performed. These ATM examples were derived using object-oriented modeling. This technique relies on collecting a large number of behavioral scenarios. This collection then describes the desired system behavior. Additional scenarios can be incrementally added to the collection. Behavioral scenarios are easy for people to describe and discuss, and it is easy to transform them into a system design.

Incorrect PIN Scenario

1. The Customer inserts a bank card, the Card Input sends the card's information to the Card Transaction Handler, which detects that the card is valid (not invalid; if no message is returned, the card is assumed valid).
2. The Card Transaction Handler instructs the Graphical User Interface (GUI) to display a message requesting the customer's Personal Identification Number (PIN).
3. The GUI requests the PIN and the customer enters his or her PIN, which is then passed to the Card Transaction Handler.
4. The Card Transaction Handler checks if the PIN is correct. In this scenario it is not, and the GUI is instructed to inform the customer that the PIN is invalid.

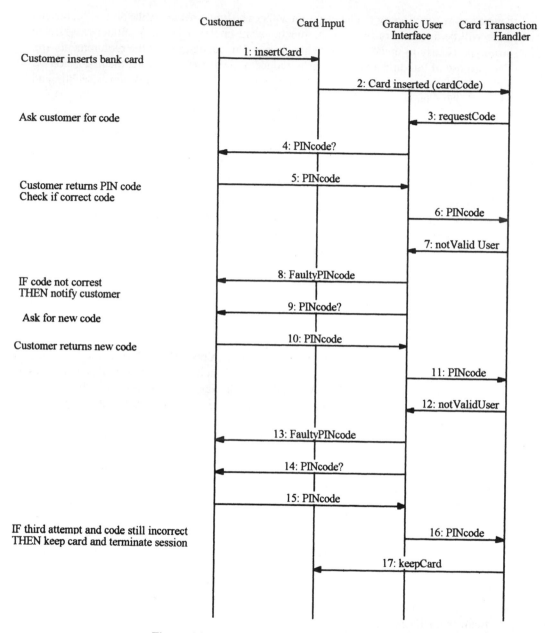

Figure 4.3 Behavioral scenario for an incorrect PIN.

5. The customer is then asked to input his or her PIN number again, and step 4 is repeated.

6. If the customer has not supplied the correct PIN number in three attempts (as is the case in this scenario), the Card Input is instructed to keep the card and the session is terminated.

4.4.4.2   Input-Output Relationships. Wymore (1993) describes the following six techniques for writing input–output relationships. These techniques have different degrees of precision, comprehensibility, and compactness.

1. For each input value, produce an appropriate output value. For example, multiply the input by 3: output(t + 1) = 3 * input(t).
2. For each input string, produce an output value. For example, compute the average of the last three inputs: output(t + 1) = [input(t − 2) + input(t − 1) + input(t)]1/3.
3. For each input string, produce an output string. For example, collect inputs and label them with their time of arrival:

   For an input string of 1, 1, 2, 3, 5, 8, 13, 21, the output string shall be (1,1), (2,1), (3,2), (4,3), (5,5), (6,8), (7,13), (8,21). All strings are finite in length.

4. For each input trajectory, produce an output trajectory. For example, collect inputs and label them with their time of arrival.

   For an input trajectory of 1, 1, 2, 3, 5, 8, 13, 21, . . . the output trajectory would be (1,1), (2,1), (3,2), (4,3), (5,5), (6,8), (7,13), (8,21). . . A trajectory may be infinite in length.

5. For each state and input, produce a next state and next output. For example, design a Boolean system where the output is asserted whenever the input bit stream has an odd number of 1s. An Odd Parity Detector can be described as:

   Z1 = (SZ1, IZ1, OZ1, NZ1, RZ1), where

   SZ1 = {Even, Odd}, /* The 2 states are named Even and Odd. */

   IZ1 = {0, 1}, /* A 0 or a 1 can be received on this input port. */

   OZ1 = {0, 1}, /* The output will be 0 or 1. */

   NZ1 = {((Even, 0), Even), /* If the present state is Even and the input is 0, then the next state will be Even. */ ((Even, 1), Odd), ((Odd, 0), Odd), ((Odd, 1), Even)},

   RZ1 = {(Even, 0), (Odd, 1)} /* If the state is Even the output is 0, if the state is Odd the output is 1. */

6. Input–output relationships may be stated as qualitative descriptions, which includes words, sentences, paragraphs, blueprints, pictures, and schematics.

4.5   TOOLS FOR GATHERING REQUIREMENTS

The following tools are used to help discover and write requirements. See Appendix B for a comparison of these tools.

- Affinity diagrams
- Force-field analysis

- Ishikawa fishbone (cause-and-effect) diagrams
- Pugh charts
- Quality function deployment (QFD)
- Functional decomposition
- Wymorian T3SD
- RDD-100
- CORE
- Slate
- Doors
- RTM

Grady (1995) and the INCOSE Proceedings discuss many more tools that systems engineers can use to gain insight into the system and to derive appropriate requirements.

4.6 THE REQUIREMENTS DEVELOPMENT PROCESS

There are many aspects of requirements development processes. We will now discuss two of them. First you have to find out what the requirements are and then you use the requirements that you have discovered.

4.6.1 The Requirements Discovery Process

4.6.1.1 Define and State the Problem. The first steps in discovering system requirements are identifying the customer, understanding the customers needs, and stating the problem.

Identify Customers and Stakeholders. The first step in developing requirements is to identify the customer. The term *customer* includes anyone who has a right to impose requirements on the system. This includes end users, operators, bill payers, owners, regulatory agencies, victims, and sponsors. All facets of the customer must be kept in mind during system design. For example, in evaluating the cost of a system, the total life-cycle cost and the cost to society should be considered. Frequently, the end user does not fund the cost of development. This often leads to products that are expensive to own, operate, and maintain over the entire life of the product, because the organization funding development saves a few dollars in the development process. It is imperative that the systems engineer understands this conflict and exposes it. The sponsor and user can then help trade off the development costs against the cost to use and maintain. Total life-cycle costs are significantly larger than initial costs. For example, in one of their advertisements, Compaq proclaimed, "80% of the lifetime cost of your company's desktops comes after you purchase them." In terms of the personal computer, if total life-cycle costs were $10,000, purchase cost would have been $2,000 and maintenance and operation $8,000.

Understand the Customer's Needs. The system design should begin with a complete understanding of the customer's needs. The information necessary to begin a design usually comes from preliminary studies and specific customer requests. Frequently the customer is not aware of the details concerning what is needed. Systems engineers must enter the customer's environment, discover the details, and explain them. Flexible designs and rapid prototyping facilitate identification of details that might have been overlooked. Talking to the customer's customer and the supplier's supplier can also be useful. This activity is frequently referred to as *mission analysis*.

It is the systems engineer's responsibility to ensure that all relevant information concerning the customer's needs is collected. The systems engineer must also ensure that the definitions and terms used have the same meaning for everyone involved. Several direct interviews with the customer are necessary to ensure that all of the customer's needs are stated and that they are clear and understandable. The customer might not understand the needs, and so may be responding to someone else's requirements. Often, customers will misstate their needs; for example, a person might walk into a hardware store and say he needs a half-inch drill bit. But what he actually needs is a half-inch *hole* in a metal plate, and a chassis-punch might be more suitable.

State the Problem. What is the problem we are trying to solve? Answering this question is one of the systems engineer's most important and often overlooked tasks. An elegant solution to the wrong problem is less than worthless.

Early in the process, the customer frequently fails to recognize the scope or magnitude of the problem that is to be solved. The problem should not be described in terms of a perceived solution. It is imperative that the systems engineer help the customer develop a problem statement that is completely independent of solutions and specific technologies. Solutions and technologies are, of course, important; however, there is a proper place for them later in the systems engineering process. It is the systems engineer's responsibility to work with the customer, asking the questions necessary to develop a complete "picture" of the problem and its scope. The Air Force customer did not know that they wanted a stealth airplane until after the engineers showed that they could do it.

4.6.1.2 Write System Requirements.
The systems engineer must interact with the customer to write the system requirements. The systems engineer must involve the customer in the process of defining, clarifying, and prioritizing the requirements. It is prudent to involve users, bill payers, regulators, manufacturers, maintainers, and other key players in the process.

Next, systems engineering must discover the functions that the system must perform in order to satisfy its purpose. The system functions form the basis for dividing the system into subsystems. QFD is useful for identifying system functions (Bahill and Chapman, 1993; Bicknell and Bicknell, 1994; Lawton, 1993).

Although it may seem that requirements are transformed into functions in a serial manner, that is not the case. It is actually a parallel and iterative process. First we look at system requirements, then at system functions. Then we reexamine the requirements and then re-examine the functions. Then we reassess the requirements and again the functions, and so on. Identifying the system's functions helps us to discover the system's requirements.

4.6.1.3 Review System Requirements.
The system requirements must be reviewed with the customer many times. At a minimum, requirements should be reviewed at the end of

the requirements modeling phase, after testing the prototypes, before commencement of production, and after testing production units.

The main objectives of these reviews are to find missing requirements, eliminate unneeded requirements, ensure that the requirements have been met, and verify that the system satisfies customer needs. At these reviews, tradeoffs will usually have to be made between performance, schedule and cost. Additional objectives include assessing the maturity of the development effort, recommending whether to proceed to the next phase of the project, and committing additional resources. These reviews should be formal. The results and conclusions of the reviews should be documented. The systems engineer is responsible for initiating and conducting these reviews.

The following definitions based on Sage (1992) and Shishko (1995) might be useful. They are arranged in chronological order. Although these definitions are written with a singular noun, they are often implemented with a collection of reviews. Each system, subsystem, subsubsystem, and so on, will be reviewed and the totality of these constitutes the indicated review.

Mission Concept Review. The mission concept review (MCR) and the mission definition review are the first formal reviews. They examine the mission objectives and the functional and performance requirements. If the organization does not have a vision or mission statement, then you should write one.

System Requirements Review. The system requirements review (SRR) demonstrates that the product development team understands the mission and the system requirements. It confirms that the system requirements are sufficient to meet mission objectives. It ensures that the performance and cost figures of merit are realistic, and that the verification plan is adequate. At the end of the system requirements review the requirements are placed into a formal configuration management system with appropriate approvals required for changes. Changing requirements after this review will impact schedule and cost.

System Definition Review. The system definition review (SDR) examines the proposed system architecture, the proposed system design, and the flow of functions down to the major subsystems. It also ensures that the verification plan is complete.

Preliminary Design Review. The preliminary design review (PDR) demonstrates that the preliminary design meets all the system requirements with acceptable risk. System development and verification tools are identified, and the work breakdown structure is examined. Full-scale engineering design begins after this review.

Critical Design Review. The critical design review (CDR) verifies that the design meets the requirements. The CDR examines the system design in full detail, ensures that technical problems and design anomalies have been resolved, checks the technical performance measures, and ensures that the design maturity justifies the decision to commence manufacturing. Few requirements should be changed after this review.

Production Readiness Review. For some systems there is a long phase when prototypes are built and tested. At the end of this phase, and before production begins, there is a production readiness review (PRR).

System Test. At the end of manufacturing and integration, the system is tested to verify that it satisfies its requirements. Technical performance measures are compared to their goals. The results of these tests are presented at the system acceptance and operational readiness reviews.

Figure 4.4 shows the timing of some of these reviews.

4.6.1.4   Ask Why Each Requirement Is Needed.   At these reviews it is important to ask why each requirement is needed. This can help eliminate unneeded requirements. It can also help reveal the requirements behind the stated requirements. It may be easier to satisfy the requirements behind the requirements, than the stated requirements themselves.

4.6.1.5   Define Performance and Cost Figures of Merit.   Figures of merit are the criteria on which the different designs will be "judged." Each figure of merit must have a fully described unit of measurement. Units of power could be horsepower, for example, and units of cost could be dollars (or inverse dollars if it is desirable to consistently have "more is better" situations). Suppose a figure of merit were acceleration, then the unit of measurement could be seconds taken to accelerate from 0 to 60 mph. The units of measurement can be anything, as long as they measure the appropriate criteria, are fully described, and are used consistently for all designs. The value of a figure of merit describes how effectively a preference requirement has been met. For example, the car went from 0 to 60 in 6.5 seconds. These values are the ones put into the scoring functions, as shown in Figure 4.2, to give the requirements scores, which are in turn used to perform tradeoff studies. Such measurements are made throughout the development of the system.

4.6.1.6   Validate System Requirements.   Validating requirements means ensuring that the set of requirements is consistent, that a real-world solution can be built that satisfies the requirements, and that it can be proven that such a system satisfies its requirements. If systems engineering discovers that the customer has requested a perpetual-motion machine, the project should be stopped. Requirements are often validated by reference to an existing system that meets most of the requirements.

Figure 4.5 summarizes this requirements discovery process of Section 4.6.1.

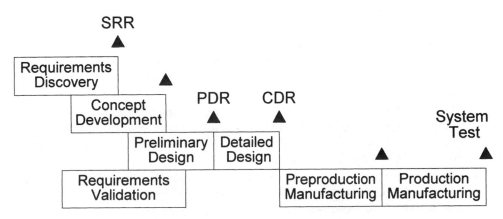

Figure 4.4   Timing of the major reviews.

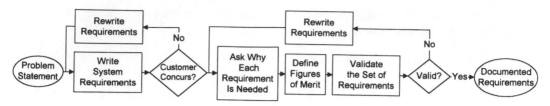

Figure 4.5 The requirements discovery process.

4.6.2 Using System Requirements

Once the system requirements have been discovered, there are several ways that they are used. We verify and validate system requirements, define technical performance measures, and mitigate project development risk as part of this use effort.

4.6.2.1 Verify System Requirements. A critical element of the requirements development process is describing the tests, analysis, or data that will be used to prove compliance of the final system with its requirements. Each test must explicitly link to a specific requirement; this will help expose untestable requirements. Describing the system tests informs the producers how the system will be tested, so that they know how they will be "graded." This process frequently uncovers overlooked requirements. The following definitions are important.

Validating a System. Building the right system; making sure that the system does what it is supposed to do. It determines the correctness of an end product, compliance of the system with the customer's needs, and completeness of the system.

Validating Requirements. Ensuring that the set of requirements is consistent, that a real-world solution can be built that satisfies the requirements, and that it can be proven that such a system satisfies its requirements. For example, if it is discovered that the customer has requested a perpetual-motion machine, the project should be stopped.

Verifying a System. Building the system right; ensuring that the system complies with its requirements. Verifying a system determines the conformance of the system to its design requirements. It also guarantees the consistency of the product at the end of each phase, with itself and with the previous prototypes. In other words, it guarantees the honest and smooth transition from model to prototype to preproduction unit to production unit.

Verifying Requirements. Inspection, analysis, test, simulation, or demonstration that proves whether a requirement has been satisfied. This process is iterative. The requirements should be verified with respect to the model, the prototype, the preproduction unit, and the production unit.

Verification and Validation. Sometimes, the words verification and validation are used in almost the exact opposite fashion (Grady, 1994). For systems engineers, to validate requirements is to prove that it is possible to satisfy them. System verification, on the other hand, is a process of proving that a system meets its requirements. The

NASA Systems Engineering Handbook indicates that verification consists of proving that a system (or a subsystem) complies with its requirements, whereas validation consists of proving that the total system accomplishes its purpose (Shishko, 1995). It is necessary to agree on the definitions of verification and validation, as these terms pertain to your system.

4.6.2.2 *Define Technical Performance Measures.* Technical performance measures (TPMs), or metrics, are used to track the progress of the design and manufacturing process. TPMs are measurements that are made during the design and manufacturing process to evaluate the likelihood of satisfying the system requirements. Not all requirements have TPMs, just the most important ones. In the beginning of the design and manufacturing process, the prototypes will not meet the TPM goals. Therefore the TPM values are only required to be within a tolerance band. It is hoped that as the design and manufacturing process progresses, the TPM values of the prototypes and preproduction units will come closer and closer to the goals.

As an example, let us consider the design and manufacture of solar ovens (Funk and Larson, 1994). In many societies, particularly in Africa, many women spend as much as 50 percent of their time acquiring wood for their cooking fires. To ameliorate this poor use of human resources, people have been designing and building solar ovens. Let us now examine the solar oven design and manufacturing process that we followed in a freshman engineering design class at the University of Arizona.

First, we defined a TPM for our design and manufacturing process. When a loaf of bread is finished baking, its internal temperature should be 95°C (203°F). To reach this internal temperature, commercial bakeries bake the loaf at 230°C (446°F). As initial values for our oven temperature TPM, we chose a lower limit of 100°C, a goal of 230°C, and an upper limit of 270°C. The tolerance band shrinks with time as shown in Figure 4.6.

In the beginning of the design and manufacturing process, our day-by-day measurements of this metric increased because of finding better insulators, finding better glazing materials (e.g., glass and Mylar), sealing the cardboard box better, aiming at the sun better, and so on.

At the time labeled "Design Change-1," there was a jump in performance caused by adding a second layer of glazing to the window in the top of the oven. This was followed by another period of gradual improvement as we learned to stabilize the two pieces of glazing material.

At the time labeled "Design Change-2," there was another jump in performance caused by a design change that incorporated reflectors to reflect more sunlight onto the window in the oven top. This was followed by another period of gradual improvement as we found better shapes and positions for the reflectors.

But in this case it seemed that we might not attain our goal. Therefore we reevaluated the process and the requirements. Bread baking is a complex biochemical process that has been studied extensively. Millions of loaves have been baked each day for the last 4000 years. These experiments have revealed the following consequences of insufficient oven temperature.

1. Enzymes are not deactivated soon enough, and excessive gas expansion causes coarse grain and harsh texture.

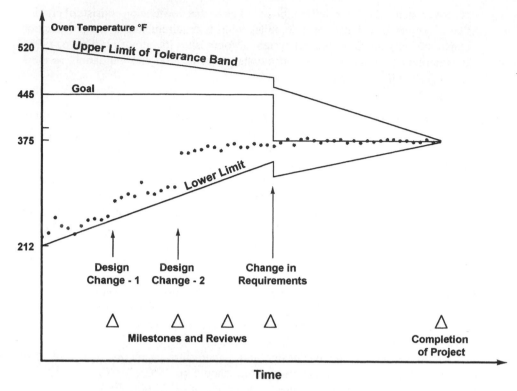

Figure 4.6 A technical performance measure.

2. The crust is too thick, because of drying caused by the longer duration of baking.

3. The bread becomes dry, because prolonged baking causes evaporation of moisture and volatile substances.

4. Low temperatures cannot produce carmelization, and crust color lacks an appealing bloom.

After consulting some bakers, our managers decided that a temperature of 190°C (374°F) would be sufficient to avoid the preceding problems. Therefore, the requirements were changed at the indicated spot and our TPM was then able to meet our goal. Of course, this change in requirements forced a review of *all* other requirements and a change in many other facets of the design. For example, the duration weight tables had to be recomputed.

If sugar, eggs, butter, and milk were added to the dough, we could get away with temperatures as low as 175°C (347°F). But we decided to design our ovens to match the needs of our customers, rather than try to change our customers to match our ovens.

4.6.2.3 Mitigate Risk. Identifying and mitigating project risk is the responsibility of management at all levels in the company. Each item that poses a threat to the cost, schedule, or performance of the project must be identified and tracked. The following information should be recorded for each identified risk: name, description, type, origin, probability, severity, impact, identification number, identification date, work breakdown structure element number, risk mitigation plan, responsible team, needed resolution date, closure criteria, principal engineer, current status, date, signature of team leader. Forms useful in identifying and mitigating risk are given in chapter 17 of Kerzner (1995), section 4.10 of Grady (1995), and Chapter 3 of this Handbook. For the solar oven project we identified the following risks.

1. Insufficient internal oven temperature was a performance risk. Its origin was Design and Manufacturing. It had high probability and high severity. We mitigated it by making it a technical performance measure, as shown in Figure 4.6.
2. High cost of the oven was a cost risk. Its origin was the Design process. Its probability was low, and its severity was medium. We mitigated it by computing the cost for every design.
3. Failure to have an oven ready for testing posed a schedule risk. Its origin was Design and Manufacturing. Its probability was low, but its severity was very high. We mitigated this risk by requiring final designs seven days before the scheduled test date and a preproduction unit three days in advance.

Models, and associated computer simulations, are often used to reduce risk. Low-risk portions of the system should be modeled at a high level of abstraction, whereas high-risk portions should be modeled with fine resolution.

Figure 4.7, based on Grady (1995), shows the whole requirements development process.

4.6.3 Fitting the Requirements Process into the Systems Engineering Process

The requirements discovery process of Figure 4.5 is one subprocess of the systems design process shown in Figure 4.8.

Systems engineering is a fractal process, where in there is a vertical hierarchy. This process is applied at levels of greater and greater detail: it is applied to the system, then to the subsystems, then to the components, and so on. It is applied to the system being designed and also to the enterprise in which the system will operate. It is also applied horizontally. It is applied to alternative-1, then to alternative-2, then to alternative-3, and so on. It is applied to component-1, component-2, component-3, and so on. This process is recursive, iterative, and much of it is done in parallel.

The fact that discover requirements (Figure 4.5) is a subprocess in the system design process (Figure 4.8), and that the system design process is a subprocess in the systems engineering process of Figure 4.9, illustrates the hierarchical nature of system engineering. We now explain the repetitive aspect of the system engineering process. In Figure 4.9 the system design process is applied to preliminary designs, models, prototypes and to the real system. However, this process is not serial. Each of the loops will be executed many times. Execution of the redesign loop in the upper left is very inexpensive and should be exercised often. Execution of the redesign loop in the lower right is expensive, but should be exercised, whereas, execution of the redesign loop from the lower right all the way back to the upper left is very expensive and should seldom be exercised.

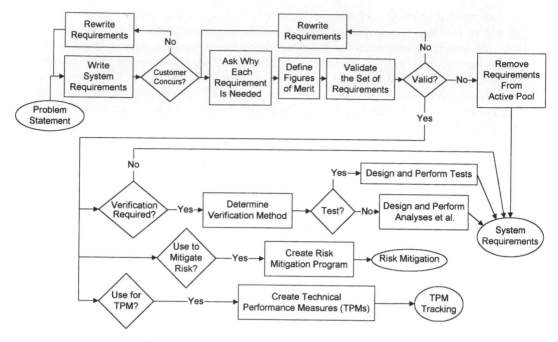

Figure 4.7 The requirements development process.

Figure 4.9, of course, only shows a part of the systems engineering process. It omits logistics, maintenance, risk management, reliability, project management, documentation, and other needed activities.

Although it seems that the systems design process has been applied in a routine and repetitive manner to preliminary designs, models, prototypes, and the real system, the

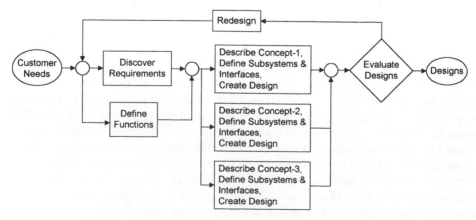

Figure 4.8 The system design process tailored for the preliminary design phase.

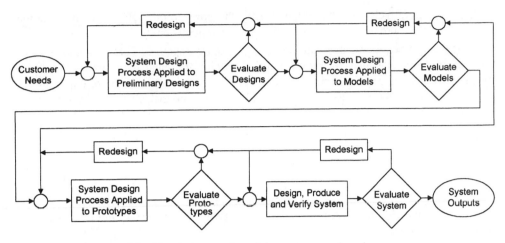

Figure 4.9   The design portion of the systems engineering process.

emphasis should be different in each application. In the preliminary design phase, the emphasis should be on discovering requirements and defining functions, with some effort devoted to alternative concepts. In the modeling phase the emphasis should be on describing, analyzing, and evaluating alternative concepts, with some effort devoted to rewriting requirements and redefining functions. In the prototype phase the emphasis should be on evaluating the prototypes, with some effort devoted to rewriting requirements, redefining functions, and if there are multiple prototypes, describing alternatives and analyzing alternative concepts. And finally, when applied to the real system, the emphasis should be on evaluating the system. When the system design process is applied in the four phases, the inputs and outputs must also be relabeled. For the preliminary design phase the input is the Customer Request and the output is Final Designs. For the modeling phase the input is Final Designs and the output is models and modeling results that are the inputs for the prototype phase, and so on.

4.7   CHARACTERISTICS OF A GOOD REQUIREMENT

As stated in the Introduction, the statements in this section are not dogmatic. Each statement has been rightfully violated many times. Kar and Bailey (1996) provide a complementary discussion to the material in this section.

1. *Describes What, Not How.* There are many characteristics of a good requirement. First and foremost, a good requirement defines what a system is to do and to what extent, but does not specify how the system is to do it. A statement of a requirement should not be a preconceived solution to the problem that is to be solved. To avoid this trap, ask why the requirement is needed, then derive the real requirements. For example, it would be a mistake to require a relational database for the requirements. The following requirements state what is needed, not how to accomplish it: provide the ability to store, provide the ability to sort, provide the ability to add attributes. It should be noted that because QFD

is often used iteratively to define requirements, the *hows* in one QFD chart become the *whats* in the next, possibly making the preceding statements confusing.

2. *Atomic.* A requirement should be "atomic," not compound. That is, it should have a single purpose (one idea per requirement). Furthermore, each requirement should be allocated to a single physical entity. It is acceptable to assign two or more requirements to one physical component. However, it would be a mistake to assign one requirement to two physical components.

3. *Unique.* A requirement should have a unique label, a unique name, and unique contents. Avoid repeating requirements.

4. *Documented and Accessible.* A requirement must be documented (writing, pictures, images, databases, etc.) and the documentation must be accessible. In situations where confidentiality is important, each requirement should clearly indicate classification status. Only individuals with the appropriate clearance and the need to know should have access to classified requirements.

5. *Identifies Its Owner.* A good requirement will identify its owner and custodian, which could be the same person. The requirement's owner must approve of any change in the requirement.

6. *Approved.* After a requirement has been revised, reviewed, and rewritten, it must be approved by its owner. Furthermore, each top-level requirement must be approved by the customer.

7. *Traceable.* A good requirement is traceable; it should be possible to trace each requirement back to its source. A requirement should also identify related requirements (i.e., parents, children, siblings) and requirements that would be impacted by changes to it. A requirements document should have a tree (or graph) structure, and this structure should be evident.

8. *Necessary.* All requirements should be necessary. Systems Engineers should ask: "Is this requirement really necessary? Will the system necessarily be better because of this requirement?" Avoid overspecifying the system, writing pages and pages that no one will probably ever read. There are two common types of overspecification: gold plating and specifying unnecessary things. For example, requiring that the outside of a CPU box be gold-plated is not a good requirement, because something far less expensive would probably be just as effective. Also, requiring that the inside of the CPU box be painted pink is probably an unnecessary request. Overspecification (of both types) is how $700 toilet seat covers and $25,000 coffee pots are created (Hooks, 1994). The documentation should include a complete statement of the rationale behind each requirement.

9. *Complete.* The documentation must be as clear, concise, and complete as possible.

10. *Unambiguous.* Avoid the use of synonyms (e.g., The software requires 8 Mbytes of RAM but 12 Mbytes of memory are recommended) and homonyms (e.g., summaries of disk X rays should be stored on disk). There should be only one interpretation of the meaning of a requirement.

11. *Is Not Always Written.* It must be noted that all systems will undoubtedly have many "common sense" requirements that will not be written. This is acceptable as long as the requirements really are common sense. An exhaustive list of requirements would take years upon years and use reams of paper, and even then you would probably never finish.

12. *Quantitative and Testable.* Quantitative values must be given in requirements. A

requirement states a necessary attribute of a system to be designed. The designer cannot design the system if a magnitude is not given for each attribute. Without quantification, system failure could occur because (1) the system exceeded the minimum necessary cost due to over design, or (2) it failed to account for a needed capability. Quantitative values for attributes are also necessary in order to test the product to verify that it satisfies its requirements (Grady, 1993).

Each requirement must be verifiable by test, demonstration, inspection, logical argument, analysis, modeling, or simulation and therefore must have a well-defined figure of merit. Qualitative words like *low* and *high* should be (at least roughly) defined. What is low cost to a big corporation and what is low cost to a small company may be very different. Only requirements that are clear and concise will be easily testable. Requirements with ambiguous qualifiers will probably have to be refined before testing will be possible. Furthermore, the value given should be fully described as, for example, an expected value, a median, a minimum, a maximum, or the like. A requirement such as "reliability shall be at least 0.999" is a good requirement because it is testable, quantified, and the value is fully described as a minimum. Also the requirement "the car's gas mileage should be about 30 miles per gallon" is a good requirement as it establishes a performance measure and an expected value. Moody et al. (1997) present a few dozen metrics that can be used to evaluate performance requirements.

Note that often the customer will state a requirement that is not quantified. For example: "The system should be aesthetically pleasing." It is then the engineer's task to define a requirement that is quantified, such as, "The test for aesthetics will involve polling two hundred potential users; at least 70% should find the system aesthetically pleasing."

It is also important to make the requirements easily testable. NASA once issued a request for proposals for a radio antenna that could withstand earthquakes and high winds. It was stated that the antenna shall not deflect by more than 0.5 degree in spite of a 0.5 G force, 100 knot steady winds or gusts of up to 150 knots. They expected bids around $15 million. But all of their bids were around $30 million. NASA asked the contractors why the bids were so high, and the contractors said testing the system was going to be very expensive. NASA revised the requirements to "When 'hit with a hammer,' the antenna shall have a resonant frequency less than 0.75 Hz." Then they got bids between $12 and $15 million (Rechtin, 1996).

13. *Identifies Applicable States.* Some requirements only apply when the system is in certain states or modes. If the requirement is only to be met sometimes, the requirement statement should reflect when. There may be two requirements that are not intended to be satisfied simultaneously, but they could be at great expense.

For example: The vehicle shall

1. Be able to tow a 2,000-pound cargo trailer at highway speed (65 mph),

2. Accelerate from 0 to 60 mph in less than 9.5 seconds.

It would be expensive to build a car that satisfied both requirements simultaneously.

Are Your Lights On?

However, as with everything, you can take this principle too far, as illustrated by the following, which is probably a true story. We first saw it in Gause and Weinberg (1990).

Recently the highway department tested a new safety proposal. They asked motorists to turn on their headlights as they drove through a tunnel. However, shortly after exiting the tunnel the motorists encountered a scenic-view overlook. Many of them pulled off the road to look at the reflections of wildflowers in pristine mountain streams and snow-covered mountain peaks 50 miles away. When the motorists returned to their cars, they found that their car batteries were dead, because they had left their headlights on. So the highway department decided to erect signs to get the drivers to turn off their headlights.

First they tried "Turn your lights off." But someone said that not everyone would heed the request to turn their headlights on. And it would be impossible for these drivers to turn their headlights off.

So they tried "If your headlights are on, then turn them off." But someone objected that would be inappropriate if it were night time.

So they tried "If it is daytime and your headlights are on, then turn them off." But someone objected that would be inappropriate if it were overcast and visibility was greatly reduced.

So they tried "If your headlights are on and they are not required for visibility, then turn them off." But someone objected that many new cars are built so that their headlights are on whenever the motor is running.

So they tried "If your headlights are on, and they are not required for visibility, and you can turn them off, then turn them off." But someone objected. . . .

So they decided to stop trying to identify applicable states. They would just alert the drivers and let them make the appropriate actions. Their final sign said, "Are your lights on?"

14. *States Assumptions.* All assumptions should be stated. Unstated bad assumptions are one cause of bad requirements.

15. *Use of Shall, Should, and Will.* A mandatory requirement should be expressed using the word *shall* (e.g., "The system shall conform to all state laws"). A preference requirement can be expressed using *should* or *may* (e.g., "The total cost for the car's accessories should be about 10% of the total cost of the car"). The term *will* can be used to express a declaration of purpose on the part of a contracting agency, to express simple future tense, and for statement of fact (e.g., "The resistors will be supplied by an outside manufacturer") (Grady, 1993).

16. *Avoids Certain Words.* The words *optimize, maximize,* and *minimize* should not be used in stating requirements, because we could never prove that we had achieved them. Consider the following criteria: (1) we should minimize human suffering, and (2) we should maximize the quality and quantity of human life. A starving child should be fed, even if the child continues to live in misery. However, the criterion of minimal suffering could lead to the conclusion that the child should die.

Requirements should not use the word *simultaneous* because it means different things to different people. It might mean within a few fempto seconds to a physicist, on the same clock cycle to a computer engineer, or to a paleontologist studying the extinction of the dinosaurs, within the same millennium.

17. *Might Vary in Level of Detail.* The amount of detail in the requirements depends upon the intended supplier. For in-house work or work to be done by a supplier with well-established systems engineering procedures, the requirements can be written at a high level. However, for outside contractors with unknown systems engineering capabilities, the requirements might be broken down to a very fine level of detail.

Requirements also become more detailed with time. In the beginning the requirements should describe a generic process that many alternative designs could satisfy. As time progresses, the problem will become better understood, the acceptable number of alternatives will decrease and the requirements should become more detailed.

18. *Contains Date of Approval.* The name of the approver and the date of approval should be included in each requirement.

19. *States its Rationale.* Although it is seldom done, it would be nice if each requirement stated why it was written and what it was supposed to ensure.

20. *Respects the Media.* Newspaper journalists quote out of context, and headlines do not reflect the content of their stories. It is important to write each requirement so that it cannot spark undue public criticism of your project.

4.8   RELATED ITEMS

1. *Requirements Versus Constraints.* The terms *requirements* and *constraints* are sometimes used interchangeably. However, a design constraint can be defined as a boundary condition within which the designer must remain while satisfying the performance requirements (Grady, 1993). With this definition, almost all of the requirements mentioned in this document (except for performance and system test) could alternatively be called constraints.

2. *Requirements vs. Goals.* The term goal is often used for a requirement that cannot be tested. Some people say there are requirements and desirements. For example, a requirement may be that "The hole shall be 5 mm in diameter, plus or minus 0.5 mm." According to Taguchi, a goal would say, "The hole shall be 5 mm in diameter and the standard deviation should be as small as feasible." Some people use "goal" as a specific value for a preference requirement.

3. *External vs. Internal.* Some engineers characterize requirements as external and internal. External requirements are driven by customer need; internal requirements are driven by company practices and resources. For example, a company might require certain processes or technologies.

4. *Outcomes, Environments, and Constraints.* Some engineers also characterize requirements as outcomes, environments, and constraints. Outcomes are related to the customer's statement of the problem. Environmental requirements change as the system design progresses. Finally, constraints, such as laws that have to be obeyed or standards that have to be followed, are often left unstated for the sake of brevity.

5. *Requirement Definition vs. Specification.* A requirements definition set, which we usually call the requirements, describes the functions the systems should provide, the constraints under which it must operate, and the rationale for the requirements. It should be written in plain language. It is intended to describe the proposed system to both the customer and the designers. It should be broad so that many alternative designs fit into its feasibility space.

The requirements specification, which we usually call the specification, provides a precise description of the system that is to be built. It should have a formal format and might be written in a specialized language. It is intended to serve as the basis of a contract between Purchasing and Manufacturing. It should narrow the feasibility space to a few points that describe the system to be manufactured.

The set of requirements determines the boundaries of the solution space. The specifications define a few solutions within that space. The requirements say what; the specifications say how.

These definitions came out of the software engineering literature (Sommerville, 1989). However, some in the software community disagree. Because of the variable usage in the literature, if someone uses the term specification, you should ask them what *they* mean by the term.

Why do so many people write the requirements after the system has been built? Perhaps they (1) write the requirements up front, (2) develop the requirements into specifications, and (3) build the system, continually updating the specifications but not the requirements. Consequently, when they deliver the system and the customer asks for the requirements, they must go back and write them.

6. *Performance, Functional, and Design Requirements.* In olden days, there was a progression from performance requirements to functional requirements to design requirements. For example, a teenage boy might express the operational need this way: "Hey, Dad, We need speakers in the car that will make your insides rumble during drum solos." The father would translate this into the performance requirement: "For bass frequencies, we need 110 dB of sound output." Then the systems engineer would convert this into the functional requirement: "Amplify the radio's output to produce 115 watts in the frequency range 20 to 500 Hz." Finally, after a trip to the audio shop, the design engineer would transform this into the design requirement: "Use Zapco Z100S1VX power amplifiers with JL Audio 12W1-8 speakers." But this implies a sequential process, and the requirements process is concurrent and iterative.

7. *Figures of Merit, Technical Performance Measures, and Metrics.* Figures of merit, technical performance measures, and metrics are all used to quantify system parameters. These terms are often used interchangeably, but we think a distinction is useful, Figures of merit are used to quantify requirements. Technical performance measures are used to mitigate risk. Metrics are used to help manage a company's processes.

Performance and cost figures of merit show how well the system satisfies its requirements, for example, "In this test the car accelerated from 0 to 60 in 6.5 seconds." Such measurements are made throughout the evolution of the system: based first on estimates by the design engineers, then on models, simulations, prototypes, and finally on the real system. Figures of merit are used to help select among alternative designs, and they are used to quantify system requirements. During concept selection, figures of merit are traded off, that is, going from one alternative to another increases the value of one figure of merit and decreases the value of another.

Technical performance measures (TPMs) are used to track the progress of design and manufacturing. They are measurements that are made during the design and manufacturing process to evaluate the likelihood of satisfying the system requirements. Not all requirements have TPMs. They are usually associated only with high-risk requirements, because they are expensive to maintain and track. Early prototypes will not meet TPM goals. Therefore the TPM values are only required to be within a tolerance band. It is hoped that as the design and manufacturing process progresses the TPM values will come closer and closer to the goals.

Metrics are often related to the process, not the product (Moody et al., 1997). Therefore, they do not always relate to specific system requirements. Rather, some metrics relate to the company's mission statement and subsequent goals. A useful metric is the percentage of requirements that have changed after the System Requirements Review.

8. *Grouping of Requirements.* Requirements should be organized into categories, subcategories, and so forth. Requirements that are correlated should be grouped together. Suppose a young couple wants to buy a new car. The man says his most important requirement is horse power and the woman says her most important requirement is gas mileage. Although these are conflicting requirements, with a negative correlation, there is no problem. Their decision of what car to buy will probably be based on a tradeoff between these two requirements. Now, however, assume there is another couple where the woman says her only requirement is safety (as measured by safety claims in advertisements), but the man says his most important requirements are lots of horse power, lots of torque, low time to accelerate 0 to 60 mph, low time to accelerate 0 to 100 mph, low time for the standing quarter mile, large engine size (in liters), and many cylinders. Assume the man agrees that the woman's requirement is more important than his. So they give safety the maximum importance value of 10, and they only give his requirements importance values of 3 and 4. What kind of a car do you think they will buy? The man's requirements should have been grouped into one subcategory, and this subcategory should have been traded off with the woman's requirement. In summary, similar, but independent, requirements ought to be grouped together into subcategories. Quality function deployment can help you to group requirements (Bahill and Chapman, 1993).

4.9   A HEURISTIC EXAMPLE OF REQUIREMENTS

4.9.1   An Automated Teller Machine Example

Earlier we discussed several ways to express requirements, such as narratives, shall and should statements, and computer models. Here is another example, one that uses formal logic notation. LaPlue et al. (1995) state that a requirement should contain (1) the description of a system output, (2) the name of the system that accepts this output, (3) conditions under which the requirement must be met, (4) external inputs associated with the requirement, and (5) all conditions that determine if the system output is correct. The authors have organized this into a standard template:

```
The system shall ⟨function⟩
    for use by ⟨users⟩,
    if ⟨conditions⟩,
    using ⟨inputs⟩,
    where ⟨conditions⟩.
Where ⟨function⟩ is usually of the form ⟨verb⟩ ⟨output⟩
```

They offer the following example.

Requirements for an Automated Teller Machine

3.0   Transaction Requirements
3.1   Related to the ATM User
    3.1.1   Produce Receipt
    3.1.2   Dispense Cash

The ATM shall dispense cash
- For use by the ATM user
- If the ATM user requested a withdrawal
- And if the Central Bank verified the account and PIN
- And if the Central Bank validated the withdrawal amount
- And if the ATM cash on hand is greater than or equal to the cash requested
- Using the Withdrawal Validation Message from the Central Bank
  and the Account Verification Message from the Central Bank
  and the Withdrawal Request from the user
- Where the amount of cash produced equals the amount requested
- And where the cash is dispensed within 10 seconds of the receipt of the Withdrawal Validation Message from the Central Bank.

3.1.3   Eject Card

3.1.3.1   Eject bank card at end of session

The ATM shall eject the bank card
- For use by the ATM user
- If the ATM user has inserted a bank card
- And if the ATM user has requested termination of session
- Using the Bank Card and the Terminate Request
- Where the Bank Card is ejected within 1 second of the receipt of the Terminate Request

3.1.3.2   Eject unreadable cards

The ATM shall eject the bank card
- For use by the ATM user
- If the ATM user has inserted a bank card
- And if the bank card does not contain a valid code
- Using the bank card
- Where the code reading and validation is as specified in Bank Card Specifications, Section 4.1.2

3.1.4   Produce Error Messages

3.2   Related to the Central Bank

3.2.1   Verify Account Message

The ATM shall produce the Verify Account Message
- For use by the Central Bank
- If the ATM user has entered a PIN
- And if the bank card contains a readable code
- Using the bank card and user-entered PIN
- Where the content and format is as specified in the Central Bank Interface Specification, Section 4.2.21
- And where the message is issued within 1 second of the final digit of the PIN

This example shows many of the features of good requirements that were mentioned in this chapter. The numbering scheme manifests the tree structure of this set of requirements: parent, child, and sibling relationships are clear. References are made to the specifications. In each requirement the customer is identified: for example, the ATM user, the central bank. Many behavioral scenarios were used to elicit these requirements. Performance figures of merit are given, they are specified as maximum values, units are given, and they are testable: for example, cash must be dispensed within 10 seconds. The requirements state what, not how: for example, The ATM shall dispense cash. The requirements identify applicable states with the conjunctive if clauses. The word choice is correct. It is unfortunate that there is no allowance for the rationale.

ACKNOWLEDGMENTS

We thank Patty Guyer for technical editing, Ron Andreé for technical illustrations, and Bo Bentz for helping us write an earlier version.

APPENDIX 4A   BEHAVIOR SCENARIOS OF ATM TRANSACTIONS

The following six diagrams depict behavioral scenarios for an automated teller machine (ATM). The scenarios were adapted from *The Object Advantage* by Jacobson, Ericsson, and Jacobson (1995). The scenarios were derived using object-oriented modeling.

An ATM is a machine that performs basic banking transactions without the need for a human teller. In each of the following scenarios a bank customer attempts to perform a withdrawal transaction. Each diagram describes a different possible scenario for the customer–ATM interaction. In these diagrams time runs from top to bottom.

4A.1   Scenario 1: Invalid Card

The Customer inserts a bank card, the Card Input sends the card's information to the Card Transaction Handler, which detects that the card is invalid. The Card Transaction Handler

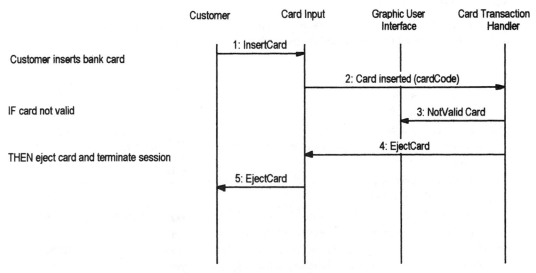

Figure 4A.1   Scenario 1: Invalid card.

instructs the Graphical User Interface (GUI) to display a message to the customer stating that the card is invalid. The Card Transaction Handler then instructs the Card Input to eject the card. The customer then removes the card from the ATM, and the transaction is terminated.

4A.2 Scenario 2: Incorrect PIN

Note: The text set in italics is identical to text in the previous scenario. It is not necessary to reread the italicized text if the scenarios are being read in order.

1. *The Customer inserts a bank card, the Card Input sends the card's information to the Card Transaction Handler, which detects that the card is* valid (not invalid, if no message is returned, the card is assumed valid).
2. The Card Transaction Handler instructs the Graphical User Interface (GUI) to display a message requesting the customer's Personal Identification Number (PIN).
3. The GUI requests the PIN and the customer enters the PIN, which is then passed to the Card Transaction Handler.
4. The Card Transaction Handler checks if the PIN is correct. In this scenario it is not, and the GUI is instructed to inform the customer that the PIN is invalid.
5. The customer is then asked to input his or her PIN number again and step 4 is repeated.
6. If the customer has not supplied the correct PIN number in three attempts (as is the case in this scenario), the Card Input is instructed to keep the card and the session is terminated.

4A.3 Scenario 3: No Cash in ATM

Note: The text set in italics is identical to text in the previous scenario. It is not necessary to reread the italicized text if the scenarios are being read in order.

1. *The Customer inserts a bank card, the Card Input sends the card's information to the Card Transaction Handler, which detects that the card is valid (not invalid, if no message is returned, the card is assumed valid).*
2. *The Card Transaction Handler instructs the Graphical User Interface (GUI) to display a message requesting the customer's Personal Identification Number (PIN).*
3. *The GUI requests the PIN and the customer enters the PIN, which is then passed to the Card Transaction Handler.*
4. *The Card Transaction Handler checks if the PIN is correct.* In this case it is and the GUI is instructed to display the customer's options.
5. The customer requests a withdrawal transaction. This information is returned to the Card Transaction Handler, which in turn calls upon the Withdrawal Handler.

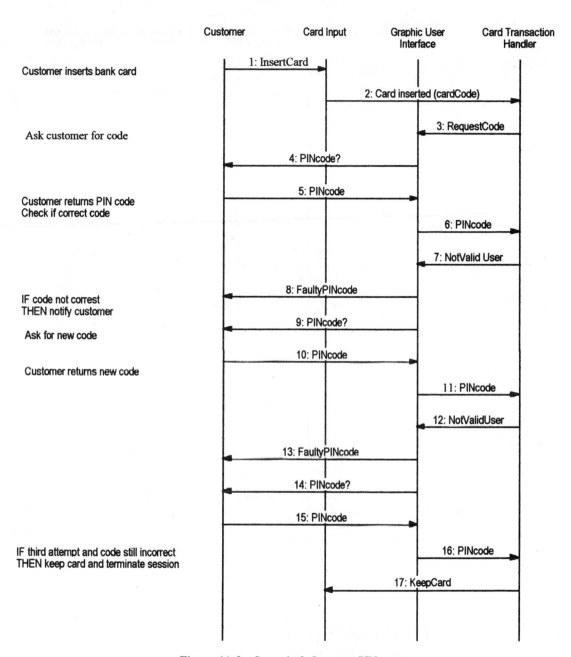

Figure 4A.2 Scenario 2: Incorrect PIN.

6. The Withdrawal Handler checks whether there is cash in the machine by querying the Cash Handler. The Cash Handler replies that the ATM does not have any cash in it.

7. The transaction is terminated and the card is ejected.

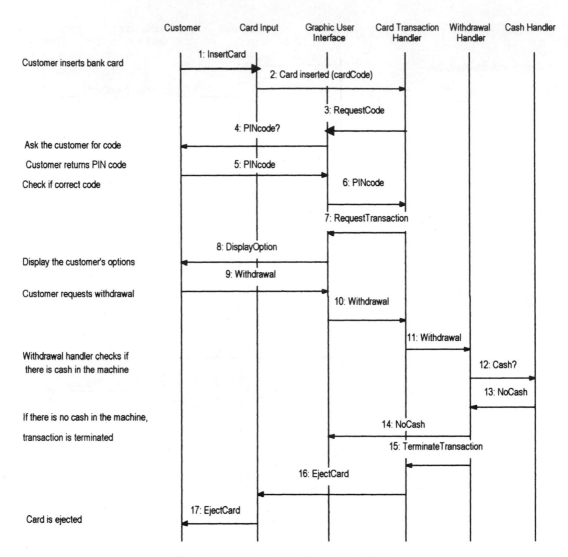

Figure 4A.3   Scenario 3: No cash in ATM.

4A.4   Scenario 4: Not Enough Cash in ATM

Note: The text set in italics is identical to text in the previous scenario. It is not necessary to reread the italicized text if the scenarios are being read in order.

1. *The Customer inserts a bank card, the Card Input sends the card's information to the Card Transaction Handler, which detects that the card is valid (not invalid, if no message is returned, the card is assumed valid).*
2. *The Card Transaction Handler instructs the Graphical User Interface (GUI) to display a message requesting the customer's Personal Identification Number (PIN).*

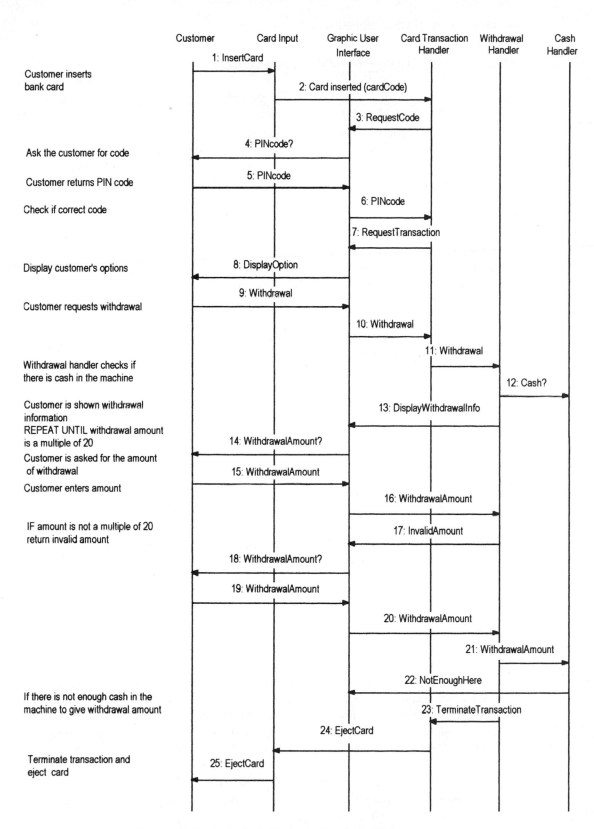

Figure 4A.4 Scenario 4: Not enough cash in ATM.

3. *The GUI requests the PIN and the customer enters the PIN, which is then passed to the Card Transaction Handler.*

4. *The Card Transaction Handler checks if the PIN is correct. In this case it is and the GUI is instructed to display the customer's options.*

5. *The customer requests a withdrawal transaction. This information is returned to the Card Transaction Handler, which in turn calls upon the Withdrawal Handler.*

6. *The Withdrawal Handler checks whether there is cash in the machine by querying the Cash Handler.* There is cash in the machine in this scenario (not No Cash; if no message is returned, it is assumed that there is money in the machine).

7. The Withdrawal Handler instructs the GUI to display the withdrawal information. The GUI requests the amount of withdrawal. The customer enters the withdrawal amount.

8. If the amount is not a multiple of 20 (as is the case), the customer is asked again to enter the amount of withdrawal. This will be repeated until the amount is a multiple of 20 (in this case the customer inputs a valid amount the second time) or the customer terminates the transaction.

9. After a valid amount is entered, the Withdrawal Handler passes that amount to the Cash Handler. The Cash Handler returns that there is not enough cash in the ATM, the GUI informs the customer, the transaction is terminated, and the card ejected.

4A.5 Scenario 5: Bank Denies Withdrawal

Note: The text set in italics is identical to text in the previous scenario. It is not necessary to reread the italicized text if the scenarios are being read in order.

1. *The Customer inserts a bank card, the Card Input sends the card's information to the Card Transaction Handler, which detects that the card is valid (not invalid, if no message is returned, the card is assumed valid).*

2. *The Card Transaction Handler instructs the Graphical User Interface (GUI) to display a message requesting the customer's Personal Identification Number (PIN).*

3. *The GUI requests the PIN and the customer enters the PIN, which is then passed to the Card Transaction Handler.*

4. *The Card Transaction Handler checks if the PIN is correct. In this case it is and the GUI is instructed to display the customer's options.*

5. *The customer requests a withdrawal transaction. This information is returned to the Card Transaction Handler, which in turn calls upon the Withdrawal Handler.*

6. *The Withdrawal Handler checks whether there is cash in the machine by querying the Cash Handler. There is cash in the machine in this scenario (not NoCash; if no message is returned, it is assumed that there is money in the machine).*

7. *The Withdrawal Handler instructs the GUI to display the withdrawal information. The GUI requests the amount of withdrawal. The customer enters the withdrawal amount.*

8. *If the amount is not a multiple of 20, the customer is asked again to enter the amount of withdrawal. This will be repeated until the amount is a multiple of 20*

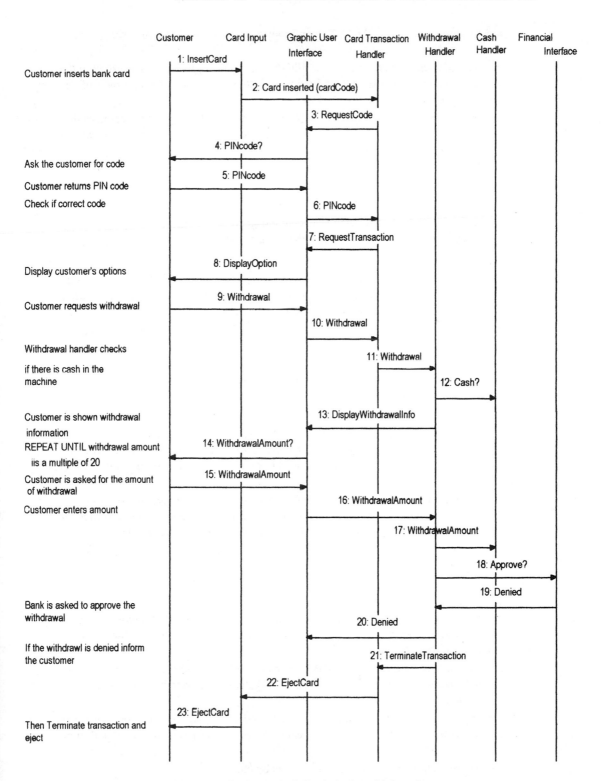

Figure 4A.5   Scenario 5: Bank denies withdrawal.

(in this case, however, the customer input a valid amount the first time) *or the customer terminates the transaction.*

9. *After a valid amount is entered, the Withdrawal Handler passes that amount to the* Cash Handler, then calls on the Financial Systems Interface for approval of the transaction.

10. The bank denies the withdrawal. The GUI is instructed to inform the customer that the transaction was denied. The transaction is then terminated and the card is ejected.

4A.6 Scenario 6: Successful Withdrawal

Note: The text set in italics is identical to text in the previous scenario. It is not necessary to reread the italicized text if the scenarios are being read in order.

1. *The Customer inserts a bank card, the Card Input sends the card's information to the Card Transaction Handler, which detects that the card is valid (not invalid, if no message is returned, the card is assumed valid).*

2. *The Card Transaction Handler instructs the Graphical User Interface (GUI) to display a message requesting the customer's Personal Identification Number (PIN).*

3. *The GUI requests the PIN and the customer enters the PIN, which is then passed to the Card Transaction Handler.*

4. *The Card Transaction Handler checks if the PIN is correct. In this case it is and the GUI is instructed to display the customer's options.*

5. *The customer requests a withdrawal transaction. This information is returned to the Card Transaction Handler, which in turn calls upon the Withdrawal Handler.*

6. *The Withdrawal Handler checks whether there is cash in the machine by querying the Cash Handler. There is cash in the machine in this scenario (not NoCash; if no message is returned, it is assumed that there is money in the machine).*

7. *The Withdrawal Handler instructs the GUI to display the withdrawal information. The GUI requests the amount of withdrawal. The customer enters the withdrawal amount.*

8. *If the amount is not a multiple of 20, the customer is asked again to enter the amount of withdrawal. This will be repeated until the amount is a multiple of 20 (in this case, however, the customer input a valid amount the first time) or the customer terminates the transaction.*

9. *After a valid amount is entered, the Withdrawal Handler passes that amount to the Cash Handler, then calls on the Financial Systems Interface for approval of the transaction.*

10. The bank approves the withdrawal. The Withdrawal Handler instructs the Cash Dispenser to dispense the cash. The Cash Handler is updated and the Receipt Printer is instructed to print a receipt.

11. The transaction is complete and the card is ejected.

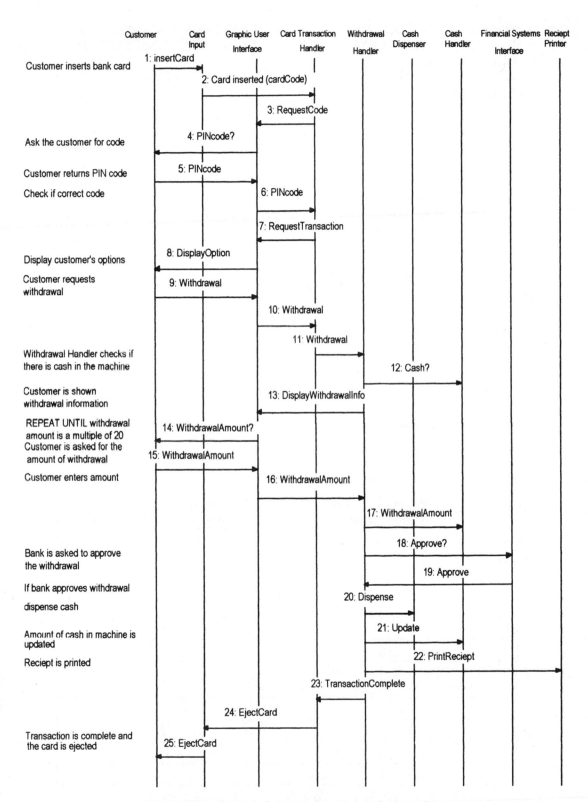

Figure 4A.6   Scenario 6: Successful withdrawal transaction.

APPENDIX 4B A COMPARISON OF SOME SYSTEMS ENGINEERING TOOLS THAT AID IN DEVELOPING AND UNDERSTANDING SYSTEM REQUIREMENTS

4B.1 Introduction

This Appendix compares several of the tools used by systems engineers to analyze or design complex systems. This Appendix provides a brief synopsis of each of the tools, a comparison of their utility in a variety of real-world scenarios that a systems engineer will likely encounter, and an evaluation of each tool against a common set of criteria. None of the tools is a systems engineering panacea; each of the tools likely addresses only a small part of the problem domain. However, knowledgeable application of one or more of these tools may provide the systems engineer with information that is critical to developing solutions that meet requirements.

4B.2 Tools Examined in This Appendix

The following tools are examined:

- Affinity diagrams
- Force-field analysis
- Ishikawa fishbone (cause-and-effect) diagrams
- Pareto diagrams
- Pugh charts
- Quality function deployment (QFD)

TABLE 4B.1 Comparison of Tools

Tool	Failure Analysis	Concept Development	Best Concept	Requirements Analysis	Market Forces	Service Systems
Affinity		•		•		
Force field					•	
Ishikawa	•					
Pareto	•					
Pugh		•	•			
QFD		•		•	•	•
Functional decomposition		•		•		
Wymore		•	•	•		•
RDD-100	•			•		
Slate				•		
CORE				•		

- Functional decomposition
- Wymorian T3SD
- RDD-100
- CORE
- Slate

Before attempting a comparison or use–case analysis, it is useful to provide a brief description of each of the tools.

1. *Affinity diagrams* implement organized brainstorming. A group of people write ideas on small pieces of paper. These ideas are then grouped by natural relationships, and names are derived for these relationships. Using this method, everybody has a chance to give input. Since the ideas are "anonymous," people tend to be more willing to express them, resulting in more complete discussions.

2. *Force-field analysis diagrams* present the major forces that influence a problem or situation under study. In a given domain of interest, forces may be loosely divided into two groups: driving forces, which promote some type of change or departure from the status quo, and opposing forces, which may resist change or promote change in another direction. The force-field diagram presents these forces in opposing columns. Such a presentation provides unique insight into the probability of achieving some desired change. If the driving forces appear overwhelming, the impetus for strategic change may be present. If, however, opposing forces appear more formidable, the change in question is unlikely.

3. *Ishikawa fishbone diagrams* are examples of cause-and-effect diagramming. Cause-and-effect diagramming presents a clearly organized, graphical representation of the possible factors contributing to a given problem. Then, the possible contributing factors to each identified possible problem factor are recursively enumerated until the diagram provides a clear representation of the root causes of any problem within the system. In an Ishikawa fishbone diagram, the cause-and-effect relationship is presented concisely and completely, allowing the user to quickly identify and discard unlikely causes in favor of more likely alternatives.

4. *Pareto diagrams* help to identify which problems are dominant. It is a frequency distribution (or histogram) of data arranged by category. The problems are listed on the x-axis and their corresponding frequency on the y-axis. It operates on the idea that 80 percent of the problems are caused by 20 percent of the factors. The most significant problems, which should be worked on first, are easily seen as the tallest columns.

5. *Pugh charts* provide a qualitative concept comparison. A matrix is formed with the alternatives on the columns and the factors used to judge the alternatives on the rows. A base alternative is then chosen. All other alternatives will be compared to the base alternative. The elements of the matrix are filled with plus signs $(+)$, minus signs $(-)$, and S's. A plus sign is used if a given alternative is judged to be better than the base alternative for a given factor, a minus sign is used if it is worse; and an S is used if it is the same as the base. The number of plus signs, minus signs, and S's are totaled for each column. An alternative is chosen based on the totals.

6. *Quality function deployment (QFD)* is a matrix-based system for the evaluation of customer needs and requirements providing a systematic means of analyzing customer requirements and deploying them into product, service, and business operations. QFD uses

a matrix structure to map the customer "wants" to the "hows," progressively passing through a series of matrices known as Houses of Quality. The relationship between the wants and hows are weighted, and on each successive level the process is further refined.

QFD provides a means of tracking the process of concept refinement from inception to realization. The traditional use has been to improve product acceptability as an element of a larger quality improvement structure. Qualitative and quantitative measures are used for both the mapping of the wants to the hows as well as the weights applied. In addition, the process can be used to identify correlations within a set of either wants or hows, providing a means of establishing sets and subsets of the system requirements.

7. *Functional decomposition* is the process where the top-level function of the system is described first. This top-level function is then broken down into subfunctions. Each of the subfunctions is then decomposed into sub-subfunctions. This process is continued until the functions are small enough that a team of engineers can design a system to implement the function.

8. *Wymorian T3SD design* is a tool based in the theory of discrete systems and the tricotyledon theory of systems design. The process incorporates a seven-document set to record the various steps in the design or analysis process. These documents include the following: a lay description of the problem; a mathematical formulation of the feasibility of system requirements; an evaluation of the system alternatives; a functional decomposition; and a physical synthesis of the systems designs.

The process involves qualitative and quantitative measures for establishing the design or designs that best meet system requirements. The application of mathematically based evaluations regarding characteristics and attributes of the candidate systems provides a means of analytical comparison using a combination of the degrees of fulfillment of each requirement.

9. *RDD-100, Slate, and CORE* are large software packages that were designed to help engineers design complex systems and satisfy U.S. Government systems engineering requirements. RDD-100 has a dynamic modeling facility, while Slate and CORE do not. Otherwise, there is little difference.

4B.2.1 *Failure Analysis: Analysis performed to determine why a system does not meet requirements or in any way fails to operate as planned.* The Ishikawa fishbone diagram is an ideal tool for failure analysis. The diagram displays factors that influence a particular part of the system. By tracing through the diagram from the failed or nonconforming system component, the diagram presents probable and improbable suspect factors.

The Pareto diagram is also a good tool. The groupings of reported errors indicate the sources of the most common errors. Systems engineering can use this information to identify likely causes of each type of error. The Pareto diagram is particularly useful in warning of a process that is deteriorating or that has gone out of control.

Because large systems can be simulated on RDD-100, it is a good tool for failure analysis.

4B.2.2 *Concept Development: Identify potential system concepts that satisfy system requirements.* Affinity diagrams encourage concept brainstorming. They generate concepts, but have no mechanism for rejecting concepts that do not satisfy requirements. However, brainstorming is an effective tool for encouraging innovative design.

Functional decomposition helps to identify the functions the system must perform. This allows models to be developed and alternative concepts to be explored.

As a tool for establishing requirements and evaluating criteria, Wymorian methodology can provide the means for defining and selecting alternative designs from a variety of options. Through an iterative process, requirements are obtained from the client, which, in turn, are used to establish the design requirements upon which the alternatives can be measured through an analytical process. Alternative designs can be evaluated at various stages in the design and prototyping phases of development, allowing those most promising to move forward. Further, the Wymorian process examines life-cycle and retirement issues often overlooked in many design processes.

QFD, as a tool to evaluate issues, is well adaptable for use in product design, primarily in establishing design requirements and in identifying quality information. QFD may provide the best means of interpreting client or public input at the requirements development phase of the design process.

4B.2.3   Best Concept: Identify the best concept alternative from a list of candidate concepts.

Pugh charts generate a qualitative comparison with a null (usually existing) concept and any number of proposed concepts. No formal method for choosing the "best" concept exists; however, Pugh charts provide information that may be valuable in an initial rough concept comparison.

Wymorian T3SD provides a quantitative procedure for identifying, validating, and testing a system component. It is an excellent tool for best concept selection, but is considerably more complex than Pugh chart methods. For large, complex systems, however, the benefits of improved requirements tractability and the integrated select–build–test features of T3SD usually outweigh the initial simplicity of Pugh charts.

4B.2.4   Requirements Analysis: Generate a complete and concise statement of the system requirements.

Requirements analysis, particularly requirements validation, is an area in which QFD has typically been applied in industry. Through the identification of the various goals and objectives of the clients as expressed in the Houses of Quality, QFD provides a means of interrupting and translating the input to a set of requirements. These requirements ultimately translate into performance measures or may map into various manufacturing processes. In either case, the requirements are readily traceable from their origins.

Functional decomposition helps assure that all the requirements have been discovered. Affinity diagrams help gather input from a wide range of customers.

The Wymorian methodology, through the Systems Requirements Validation (Document 4) and Concept Exploration (Document 5), also provides a means of analyzing requirements. In many ways, these capabilities are unique for the systems under consideration. This is particularly true when numerical values can be assigned to the performance of various systems for each of the requirements. Additionally, the scoring functions allow for the measurement and assessment of system sensitivity to parameter changes. The ability to validate requirements and to address sensitivity by analytical measures provides a powerful requirements analysis tool.

4B.2.5   Market Forces: Analyze the potential demand for a system that satisfies a given set of requirements.

A force-field diagram generates a clear picture of the forces opposing and encouraging a particular system. QFD market analysis may help to determine what characteristics are essential to a successful product.

4B.2.6 Service Systems: A system designed to provide some form of nontangible utility to its customers. Wymorian theory, as a result of its general nature, can be applied to various types of service systems, but its application to those systems may be too difficult and provide similar information to less complex and readily usable operations research techniques. As the system grows in complexity, however, these models may be more accurate.

As with any system that affects, and therefore would benefit from the ability to use client input, the application of QFD can be highly beneficial in developing performance criteria and potential sources of difficulty in service-type systems. To a great extent, QFD use can be carried through the requirements development phase of virtually any system, but its usefulness tends to wane as the process moves into a design phase. QFD can take a leadership role after a system is in place by providing a means of accepting client input about an operation after it commences on the performance, so that further improvements can be accommodated in the operation.

REFERENCES

Bahill, A. T., Alford, M., Bharathan, K., Clymer, J., Dean, D. L., Duke, J., Hill, G., LaBudde, E., Taipale, E., and Wymore, A. W. (1998). The design-methods comparison project, *IEEE Trans Sys., Man, Cybernet, Part C: Applications and Reviews,* **28**(1), 80–103.

Bahill, A. T., and Chapman, W. L. (1993). A tutorial on quality function deployment. *Eng. Manage. J.* **5**(3), 24–35.

Bharathan, K., Poe, G. L., and Bahill, A. T. (1995). Object-oriented systems engineering. *Syst. Eng. Global Market Place, Proc. 5ht Annu. Symp. Natl. Counc. Syst. Eng. (NCOSE),* St. Louis, MO, pp. 759–765.

Bicknell, K. D., and Bicknell, B. A. (1994). *The Road Map to Repeatable Success: Using QFD to Implement Changes.* Boca Raton, FL: CRC Press.

Booch, G. (1994). *Object-Oriented Analysis and Design.* Menlo Park, CA: Benjamin/Cummings.

Chapman, W. L., and Bahill, A. T. (1996). Design modeling and production. In *The Engineering Handbook* (R. C. Dorf, ed.), pp. 1732–1737. Boca Raton, FL: CRC Press.

Chapman, W. L., Bahill, A. T., and Wymore, W. (1992). *Engineering Modeling and Design.* Boca Raton, FL: CRC Press.

Deming, W. E. (1986). *Out of the Crisis.* Cambridge, MA: MIT Press.

EIA Standard, Process for Engineering a System, EIA-632, Version 1.0, 1998.

Funk, P. A., and Larson, D. L. (1994). Design features influencing thermal performance of solar box cookers. *Int. Winter Meet., Amer. Soc. Agric. Eng.,* Pap. No. 94-6546.

Gause, D. C., and Weinberg, G. M. (1990). *Are Your Lights On? How to Figure Out What the Problem Really Is.* New York: Dorset House.

Grady, J. O. (1993). *System Requirements Analysis.* New York: McGraw-Hill.

Grady, J. O. (1994). *System Integration.* Boca Raton, FL: CRC Press.

Grady, J.O. (1995). *System Engineering Planning and Enterprise Identity.* Boca Raton, FL: CRC Press.

Hooks, I. (1994). Writing good requirements. *Proc. Natl. Counc. Syst. Eng. (NCOSE),* pp. 197–203.

Institute of Electrical and Electronics Engineers (IEEE). (1993). *Guide For Developing System Requirements Specifications* (IEEE P1233). New York: IEEE Standards Dept.

Institute of Electrical and Electronics Engineers (IEEE). (1994). *Standard for Systems Engineering* (IEEE P1220). New York: IEEE Standards Dept.

International Council on Systems Engineering (INCOSE). (1996). *Proc. 6th Annu. Symp. Int. Counc. Syst. Eng. (INCOSE),* Boston.

Jacobson, I., Ericsson, M., and Jacobson, A. (1995). *The Object Advantage: Business Process Re-engineering with Object Technology.* New York: Addison-Wesley.

Kar, P., and Bailey, M. (1996). Characteristics of good requirements, systems engineering practices and tools. *Proc. 6th Annual Int. Symp. Int. Counc. Syst. Eng. (INCOSE),* Boston, Vol. 2, pp. 284–291.

Karnavas, W. J., Sanchez, P., and Bahill, A. T. (1993). Sensitivity analyses of continuous and discrete systems in the time and frequency domains. *IEEE Trans. Syst., Man, Cybernet.,* **23,** 488–501.

Katz, R. (1994). *Contemporary Logic Design.* Menlo Park, CA: Benjamin Cummings.

Kerzner, H. (1995). *Project Management: A Systems Approach to Planning, Scheduling, and Controlling.* New York: Van Nostrand-Reinhold.

LaPlue, L., Garcia, R. A., and Rhodes, R. (1995). A rigorous method for formal requirements definition. *Syst. Eng. Global Market Place, Proc. 5th Annu. Symp. Natl. Counc. Syst. Eng. (NCOSE),* St. Louis, MO: pp. 401–406.

Latzko, W. J., and Saunders, D. M. (1995). *Four Days with Dr. Deming.* Reading, MA: Addison-Wesley.

Lawton, R. (1993). *Creating a Customer-Centered Culture.* Milwaukee, WI: ASQC Press.

Martin, J. (1995). Requirements methodology: Shattering myths about requirements and the management thereof. *Syst. Eng. Global Market Place, Proc. 5th Annu. Symp. Natl. Counc. Syst. Eng. (NCOSE),* St. Louis, MO, pp. 473–480.

Martin, J. (1996). *Systems Engineering Guideline.* Boca Raton, FL: CRC Press.

Moody, J. A., Chapman, W. L., Van Voorhees, F. D., and Bahill, A. T. (1997). *Metrics and Case Studies for Evaluating Engineering Designs.* Upper Saddle River, NJ: Prentice-Hall.

National Council on Systems Engineering (NCOSE). (1995). *Systems Engineering in the Global Market Place,* Proc. 5th Annu. Symp. Natl. Counc. Syst. Eng., St. Louis, MO. Washington, DC.

Rechtin, E. (1996). Personal communication.

Rechtin, E., and Maier, M. (1997). *The Art of Systems Architecting.* Boca Raton, FL: CRC Press.

Rumbaugh, J., Blaha, M., Premerlani, W., Eddy, F., and Lorenson, W. (1991). *Object Oriented Modeling and Design.* New York: Prentice-Hall.

Sage, A. P. (1992). *Systems Engineering.* New York: Wiley.

Shand, R. M. (1994). User manuals as project management tools. *IEEE Trans. Profess. Commun.,* **37,** 75–80, 123–142.

Shishko, R. (ED.) (1995). *NASA Systems Engineering Handbook* (SP-6105). Washington, DC: National Aeronautics and Space Administration.

Sommerville, I. (1989). *Software Engineering.* Reading, MA: Addison-Wesley.

Szidarovszky, F., Gershon, M., and Duckstein, L. (1986). *Techniques for Multiobjective Decision Making in Systems Management.* Amsterdam: Elsevier.

Wymore, W. (1993). *Model-Based Systems Engineering.* Boca Raton, FL: CRC Press.

5 Configuration Management

PEGGY BROUSE

5.1 INTRODUCTION

Configuration management (CM) is an accepted part of the infrastructure that is necessary in the support of systems development. It has been defined by a number of different sources, both commercial and noncommercial. Babich defines CM in the following way: On any team project, a certain degree of confusion is inevitable. The goal is to minimize this confusion so that more work can get done. The art of coordinating software development to minimize this particular type of confusion is called configuration management. Configuration management is the art of identifying, organizing, and controlling modifications to the software being built by a programming team. The goal is to maximize productivity by minimizing mistakes (Babich, 1986). This definition encompasses the identification, organization, and control of modifications of software in a development effort. Tichy also addresses the management of software components. He states that software configuration management [SCM] is a discipline whose goal is to control changes to large software system families, through the functions of component identification, change tracking, version selection and baselining, software manufacture, and managing simultaneous updates (teamwork) (Tichy, 1988). Tichy expands on Babich in that he explicitly addresses component identification, before the tracking of changes to these components. Neither definition, however, addresses hardware or documentation configuration management.

In IEEE-Std-610, the definition is expanded to include reporting capabilities. It allows for the identification, documentation, control, and verification of configuration items associated with systems development and the subsequent reporting of these activities [Institute of Electrical and Electronics Engineers (IEEE), 1990]. This definition is not complete in that the management of the development process is not addressed. It has been noted by Dart, that CM is not complete if it does not address process management (Software Engineering Institute (SEI), 1992).

In the following sections, CM is considered to be the discipline used in the system development life cycle (SDLC) to classify and control artifacts associated with the system. The primary objective of CM is to ensure that changes to these artifacts, including requirements, design, software, documentation, hardware, or other components, are controlled. Uncontrolled changes can result in far-reaching problems that may be compounded throughout the life cycle. For example, uncontrolled changes to requirements may result in code being written based on old requirement specifications that have not been properly updated with new changes. The programmer will code to the incorrect specification, which in turn will lead to problems during integration and unsatisfied user needs.

Changes to all baselined systems documentation as well as system hardware, software, and other artifacts must be controlled. During the greater part of any project, the user and

Handbook of Systems Engineering and Management, Edited by A. P. Sage and W. B. Rouse
ISBN 0471-15405-9 ©1999 John Wiley and Sons, Inc.

developer will rely on the project documents for monitoring progress and ensuring that the system is complying with system requirements. Documentation also serves as the baseline from which to trace requirements throughout the life cycle. In addition to direct control of the documentation suite, CM is responsible for the release of production software through software libraries. CM is a critical part of the infrastructure in systems engineering (Sage, 1992). This has been validated with the requirement for CM in many of the certified process models. For example, a robust CM process must be in place for a company to be considered a Level 2 company in the SEI Capability Maturity Model (Paulk, 1993). A standard configuration management process and information for tailoring a project-specific configuration management plan are very important to the success of this discipline.

5.2 CONFIGURATION MANAGEMENT PROCEDURES

Within the system development infrastructure, there are a number of procedures that need to be in place from the beginning of the life cycle. The process is explained within the framework of the traditional waterfall life cycle model. To successfully control change to systems, a CM process should be put in place. A standard process should be developed for the organization and used for all system development and maintenance within the organization. In addition, training should be provided for staff to ensure understanding and compliance with the CM process. A suggested process flow is illustrated in Figure 5.1. The process as well as individual steps to be conducted within each phase of the life cycle are defined. Many of the terms used to explain the CM process are generally used in the literature and accepted by CM practitioners and will be used to explain the CM process. To better understand the process, configuration management responsibilities and roles of those involved in systems development will be defined, as well as those tools that may aid in the configuration management process.

1. *Project Initiation Phase.* Before the requirements are gathered for subsequent development, project infrastructure steps must be undertaken. As it relates to CM, there are several activities that occur. As part of the planning process, the project manager should plan for and delegate responsibility for the creation of a configuration management plan. The CM plan is developed specifically for the given project, but may have been tailored from the standard CM plan of the organization. Many organizations, both commercial and noncommercial, require compliance with a specific standard, such as ISO 12207, Information Technology—Software Life Cycle Processes (ISO/IEC, 1995) or MIL-STD-973, Configuration Management (U.S. Department of Defense, 1995). The project-specific CM plan should contain the change control process as well as those procedures needed to support the process. CM is a critical component of system success, so project team members need to be introduced to the CM process that is to be used on the project. Any automated tools that will be used in the CM process should be acquired at this time. The group, usually the Configuration Control Board [CCB], that will be responsible for the determination of allowable changes to the system components should be notified of the project.

2. *Requirements Analysis Phase.* The requirements analysis phase is a critical phase in configuration management. It is in this phase that the requirements are identified to be baselined and the functional baseline is created. The functional baseline is the initially approved documentation describing a system's or item's functional, interoperability, and

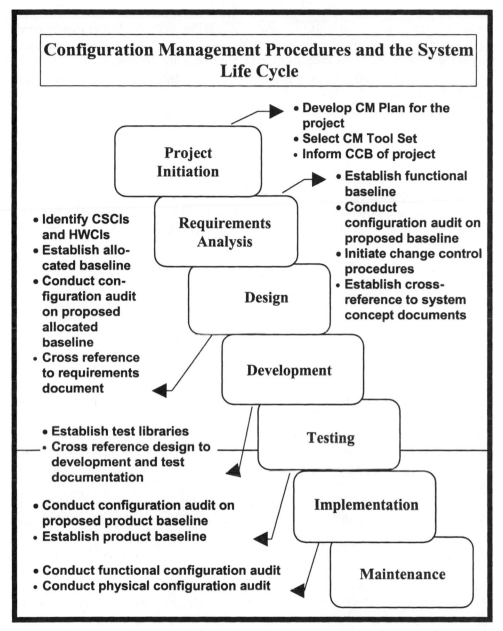

Figure 5.1   Configuration management procedures.

interface characteristics and the verification required to demonstrate the achievement of those specified characteristics (U.S. Department of Defense, 1995). This document contains the user requirements and is used throughout the remainder of the life cycle as the baseline document from which to ensure system compliance with user requirements.

The process of audits is also initiated during this phase with the conduct of the configuration audit on the proposed baseline. An audit is an independent examination of a work product or set of work products to assess compliance with specifications, standards, contractual agreements, or criteria (IEEE, 1990). Traceability is established as requirements are traced back to the original system concept documents, and responsibilities of individual project team members for the integrity of the CM process are assigned.

3. *Design Phase.* This phase builds on the requirements analysis phase. Identified during this phase are:

Computer software configuration items (CSCIs), an aggregation of software that satisfies an end-use function and is designated for separate configuration management by the acquirer (U.S. Department of Defense, 1994).

Computer documentation configuration items (CDCIs), technical data or information, including computer listings, regardless of media, that documents the requirements, design, or details of computer software; explains the capabilities and limitations of the software; or provides operating instructions for using or supporting computer software during the software's operational life cycle (U.S. Department of Defense, 1995).

Hardware configuration items (HWCIs), an aggregation of hardware that satisfies an end-use function and is designated for separate configuration management by the acquirer (U.S. Department of Defense, 1994).

An allocated baseline is created. This baseline is traced back to the requirements of the previous phase. A library of configuration items (CIs) should be established to more easily manage the process. A configuration audit is conducted on the allocated baseline.

4. *Development Phase.* The development phase entails the creation of code to meet the design and the original requirements. The code must be traced back to the design. This is also the phase in which the test libraries are created for subsequent testing of the code.

5. *Testing Phase.* The testing phase validates that the system that has been developed meets the original requirements of the user. A configuration audit is conducted on the product baseline before the baseline is established. It is from this that future releases of the system are baselined.

6. *Implementation Phase.* The implementation phase includes moving the developed system to the user sites. Both the functional and physical configuration audit are performed during this phase.

7. *Maintenance Phase.* During maintenance, the system undergoes a number of changes as dictated by the user or by failures within the existing system. These changes must be controlled just as they were during the original development life cycle. A release schedule must be instituted that will allow for periodic releases of the system to incorporate new changes.

5.3 CONFIGURATION MANAGEMENT RESPONSIBILITIES

In each phase of the life cycle, there are clear responsibilities for all the staff participating on the project, including the project manager, programmers, and organization staff, that are responsible for configuration management throughout the organization. Table 5.1 provides descriptions of these teams members.

TABLE 5.1 Descriptions of Team Members

Team Member(s)	Responsibility
Project Manager (PM)	The project-level person who has authority for the deliverables and schedules associated with the project. The PM is responsible for reviewing and authorizing all changes to the baselines
CM Process Auditor (CMPA)	The organizational-level person who reviews project CM activities and ensures compliance with organization CM policies. The CMPA also tracks engineering change proposals (ECPs)
CM Project Engineer (CMPE)	The project-level person who provides the project's technical liaison for CM issues. The CMPE participates in Configuration Control Board meetings
CM Librarian (CML)	The project-level person who has CM maintenance management responsibility, including periodic releases of documentation and software
Configuration Control Board (CCB)	A team consisting of organization, project, and acquirer personnel who evaluate ECPs for both technical and resource impact on the project. The CCB recommends approval or disapproval of a given ECP

5.4 CONFIGURATION MANAGEMENT ACTIVITIES

Because CM can be very labor-intensive, activities have been developed to help project staff in their CM organization efforts. These activities may encompass every phase of the life cycle. Descriptions of these major CM activities are as follow.

5.4.1 Configuration Identification

Configuration identification includes the selection of those components or configuration items to be managed; the unique identification of each configuration item (CI); the definition of the CI's configuration, including internal and external interfaces; the release of CIs and their associated configuration documentation; and the establishment of configuration baselines for CIs (U.S. Department of Defense, 1995).

Several subactivities must be undertaken to identify the products created for a development project. These include:

Computer Software Configuration Item Selection. This is the partitioning of software into CSCIs. Small programs are generally defined as single CSCIs; larger programs may require partitioning. Judicious choice of CSCIs is encouraged. Too many CSCIs may result in unwise resource utilization, including an increase in coordination activities. Too few CSCIs may result in change identification at too high a level. CSCIs, once identified, should be given unique labels so they can be traced through future phases and releases of the system. CSCIs that are changed should be given a new number that is traceable to the previous CSCI.

Computer Software Documentation Item Selection. This is the selection of the documentation suite associated with the system. Document creation is often dictated by the contract and associated deliverables for the system. As part of the organization's generic process, however, additional documents may be required. Once identified, CDSIs should be given unique labels so they can be traced through future phases and releases of the system.

Computer Hardware Configuration Item Selection. This is the selection of the hardware components associated with the system. This hardware may be both hardware associated with the development process and that will be part of the operational system. Hardware may also be used on more than one system, in which case, it may need to be tracked at an organizational level. Once identified, CHSIs should be given unique labels so they can be traced through future phases and releases of the system.

5.4.2 Configuration Change Control

Configuration control is the systematic proposal, justification, evaluation, coordination, approval or disapproval of proposed changes, and the implementation of all approved changes, in the configuration of a CI after establishment of the configuration baseline (U.S. Department of Defense, 1995).

Configuration control follows a process by which to accomplish these tasks. Figure 5.2 contains a flow that can be used in configuration control. It begins with the awareness of a need for change to a configuration item, and follows the change through the approval and implementation phases. The following paragraphs describe this process in more detail.

5.4.2.1 Change Classification. Changes are categorized as emergency or standard. An emergency change is a nontrivial change that affects the operational system or may jeopardize safety. These changes if not implemented may provide considerable risk to the system. Standard changes are those changes that may easily be considered and implemented in a future release of the configuration.

Emergency changes will be submitted automatically for customer approval before implementation. Standard changes require customer notification and verification of classification. Selected standard changes also may be submitted to the customer for approval before implementation; if, for example, they have a significant impact on budget or schedule. The ECP is usually the standard form used to document the need for change to the configuration. A project may use the standard organizational form or other required contract form. An example form used by the originator of the change is shown in Figure 5.3. The ECP is logged by the CML into the CM database for tracking throughout the life of the system. The ECP is then given to project technical staff, the designated CMPE, for review, impact analysis, and classification. In an emergency situation, both the user and the CCB will be notified of the ECP. The CCB will meet to review the ECP, approve or disapprove the recommendation, and authorize the implementation of the change.

5.4.2.2 ECP Preparation. Identifying information regarding the proposed change will be documented on the ECP form by the originating user or development staff. They will enter project name, CI name, date or release within which change should be accomplished, and their classification of the change. They will also given a short description of the suggested change, their name, and identifying information.

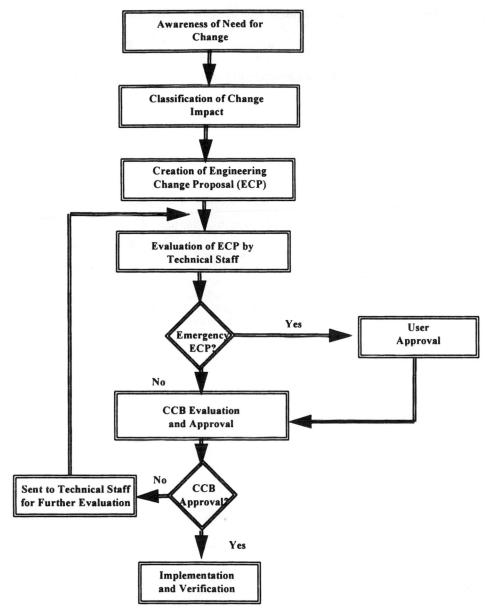

Figure 5.2   Engineering change proposal process.

The ECP form is then given to the CML, who logs the ECP, assigns it a unique number, notifies the organizational level CMPA, and notifies the project-level CMPE.

5.4.2.3   *ECP Evaluation.* If the change is deemed an emergency, the CMPA or CMPE may immediately notify the user and convene an emergency CCB meeting. For standard ECPs, the ECP will be reviewed in a regularly scheduled CCB meeting.

The second page of the ECP form, Figure 5.4, which will already have the unique

Engineering Change Proposal   **Page 1 of 2**

ECP NUMBER: _____
[*will be assigned*]

System Name: _____

Configuration Item: _____

Required Date: _____

Classification: [*Please check one of the boxes*]

Very High Risk   Prevents accomplishment of an essential capability or jeopardizes operator safety

High Risk   Adversely affects performance of an essential capability; no work-around solution is available

Medium Risk   Adversely affects performance of an essential capability; a known work-around solution is available

Nuisance   An inconvenience that does not affect essential capabilities

Other   All other errors

Short Description of Problem [*including documentation/software/hardware to be changed*]:

Originator Name: _____

Originator Organization: _____

Originator Phone: _____

Date Originated: _____

Figure 5.3   ECP form documented by originator.

Engineering Change Proposal	Page 2 of 2

ECP NUMBER: xxxxxxxxxxxxx

CMPE Name: _____

CMPE Organization: _____

CMPE Phone: _____

Analysis of Problem:

Impact of Problem:

Recommended Solution:

CCB ☐ Approval Date:
Approval:

Name of Correction Analyst: _____

Analyst Organization: _____

Analyst Phone: _____

Date Received: _____ Date Completed: _____

Verification _____ Date: _____
Name:

Figure 5.4 ECP form documented by CMPE.

identifier noted, is initiated by the CMPE assigned to analyze and suggest a solution to implement the change. Both pages of the form will be given to the CCB. The CMPE will enter their identifying information, date of problem receipt, recommended solution, and impact (if known).

5.4.2.4 ECP Implementation. After the solution receives the approval of the CMPE and the CCB, an analyst will be assigned to correct the problem. The second page of the form will contain approval by the CCB, information filled in by the analyst who implements the suggested solution, and verification of solution.

5.4.2.5 Change Notification. When the change is approved and changes have been made to the affected CI, the CML records the change and notifies user and developer staff of the change. If the change is an emergency, implementation may occur immediately; otherwise, the change will be assigned to a future release.

5.4.3 Configuration Baseline Control

In is incumbent upon all staff to help maintain control of the configuration baseline. Individual team members will be responsible for their own files, backup, and unit test libraries. At this level, the individual is the only person allowed to make changes to the configuration item. At the point that the item is satisfactory to the team member, they will notify the project manager and the CI will be incorporated into the project-level environment. The CMPE is responsible for control at this point, but may assign others to modify project-level CIs, though the CMPE is ultimately responsible for the integrity of the products. Configuration management tools may be available to prevent simultaneous modifications by more than one staff member. Upon completion of project-level testing, the CIs will be formally baselined and included in the system baseline. Changes may not occur until the formal ECP process just outlined is followed.

5.4.4 Configuration Status Accounting

Configuration status accounting (CSA) is the recording and reporting of information needed to manage configuration items effectively (U.S. Department of Defense, 1995). The accounting function is an important part of CM and provides for traceability and product integrity. The considerations in this function are contained in the following paragraphs.

5.4.4.1 Collection, Recording, and Maintenance of Data. The CML will be responsible for ensuring the collection, recording, and maintenance of CI data. Collection of data will be performed by gathering the documents generated by Configuration Identification, Configuration Control, and Configuration Audits. Configuration Identification will identify all CIs associated with the various system baselines; Configuration Change Control will generate all ECPs, minutes of CCB meetings, and the various forms associated with reporting and implementing the change; and Configuration Audits will generate minutes and lists of action items from the audits. When a CI is ready to be added to the database, the CML will assign an identifier and enter it into the database.

5.4.4.2   Status of Proposed Engineering Changes. Change status reports, which will be generated automatically by the configuration control function, will be used to log the status of all requested ECPs. Requests for deviations and waivers also will be included in this log. Waivers are written authorization to accept an item, which during manufacture, or after having been submitted for government inspection or acceptance, is found to depart from specified requirements, but nevertheless is considered suitable for use "as is" or after repair by an approved method (U.S. Department of Defense, 1995).

5.4.4.3   Change Traceability. A CCB log will record proposed changes, and a ECP log will record internal reports and requests for changes. For approved changes, the ECP log will track their effectiveness and installation status.

5.4.5   Configuration Audits

The developer should perform a formal functional configuration audit (FCA) and a physical configuration audit (PCA) if required by the user.

5.4.5.1   Functional Configuration Audit. The FCA is defined as the formal examination of functional characteristics of a configuration item, prior to acceptance, to verify that the item has achieved the requirements specified in its functional and allocated configuration documentation (U.S. Department of Defense, 1995). The FCA validates that the CI has been developed and is traceable to a requirement specified by the user.

   After the audit is concluded, the developer documents the audit findings and reports to the user.

5.4.5.2   Physical Configuration Audit. The PCA is the formal examination of the "as-built" configuration of a configuration item against its technical documentation to establish or verify the configuration item's product baseline (U.S. Department of Defense, 1995). The PCA is used for the production and acceptance of CIs. In certain systems, the FCA and the PCA may be combined into a single audit.

5.5   CONCLUSION

Configuration management is an important activity that is essential for the success of system development projects. Without configuration management, uncontrolled changes may result in systems that do not meet user needs, need countless changes, and are often not fielded.

APPENDIX: APPLICABLE STANDARDS

MIL-STD-490A *Specification Practices*

This standard establishes the format and contents of specifications for program-peculiar configuration items, processes, and materials.

MIL-STD-498 *Software Development and Documentation*

The purpose of this document is to establish uniform requirements for software development and documentation. The EIA and IEEE have developed a commercial replacement for this standard, the U.S. implementation of ISO/IEC 12207, Information Technology — Software Life Cycle Processes.

MIL-STD-973 *Configuration Management*

This standard defines configuration management requirements that are to be selectively applied, as required, throughout the life cycle of any configuration item (CI).

MIL-STD-2549 *Configuration Management Data Interface Standard*

This standard defines data interface requirements for configuration management. It will replace MIL-STD-973.

ISO 9001 *Quality Systems Model for Quality Assurance in Design, Development, Production, Installation, and Servicing*

This document specifies activities necessary to produce systems that meet user requirements.

ISO 12207 *Information Technology—Software Life-Cycle Processes*

This standard provides a software life-cycle framework for the definition, development, and deployment of software.

REFERENCES

Babich, W. (1986). *Software Configuration Management: Coordination for Team Productivity.* Boston: Addison-Wesley.

Institute of Electrical and Electronics Engineers (IEEE). (1990). *IEEE Standard Glossary of Software Engineering Terminology* (IEEE STD-610). New York: IEEE.

International Organization for Standardization/International Electrotechnical Commission. (1995). *Information Technology—Software Life Cycle Processes*, (ISO/IEC 12207).

Paulk, M., Curtis, B., Chrissis, M. B., and Weber, C. (1993). Capability maturity model, version 1.1. *IEEE Software,* **10**(4) July, pp. 18–27.

Software Engineering Institute. (1992). *The Past, Present, and Future of Configuration Management,* SEI Tech. Rep. (CMU/SEI-92-TR-8). Pittsburgh, PA: Carnegie-Mellon University.

Sage, A. P. (1992). *Systems Engineering,* New York: Wiley

Tichy, W. F. (1988). Tools for software configuration management. *Proc. 2nd Int. Workshop Software Version Config. Control.*

U.S. Department of Defense. (1994). *Software Development and Documentation* (MIL-STD-498). Washington, DC: USDoD.

U.S. Department of Defense. (1995). *Configuration Management* (MIL-STD-973). Washington, DC: USDoD.

BIBLIOGRAPHY

Ayer, S. (1992). *Software Configuration Management: Identification, Accounting, Control, and Management.* New York: McGraw-Hill.

Bazelmans, R. (1985). Evolution of configuration management. *ACM SIGSOFT Software Eng. Notes* **10**(5), 37–46.

Ben-Menachem, M. (1994). *Software Configuration Management Guidebook.* New York: McGraw-Hill.

Berlack, H. R. (1992). *Software Configuration Management.* New York: Wiley.

Brumm, E. (1994). *Managing Records for ISO Compliance.* Milwaukee, WI: ASQC.

Buckley, F. J. (1993). *Implementing Configuration Management.* New York: IEEE Press.

Compton, S. B., and Conner, G. R. (1994). *Configuration Management for Software.* New York: Van Nostrand-Reinhold.

Eggerman, W. V. (1991). *Configuration Management Handbook.* Blue Ridge Summit, PA: TAB Books.

Electronic Industries Association. (1998). *National Consensus Standard for Configuration Management* (ANSI/EIA 649). New York: ANSI.

Freedman, D. P., and Weinberg, G. M. (1990). *Handbook of Walkthroughs, Inspections and Technical Reviews,* 3rd ed. London: Dorset House.

Gilb, T. (1988). *Principles of Software Engineering Management.* Reading, MA: Addison-Wesley.

Humphrey, W. S. (1989). *Managing The Software Process.* Reading, MA: Addison-Wesley.

International Organization for Standardization. (1987). *Quality Systems Model for Quality Assurance in Design/Development, Production, Installation, and Servicing* (ISO 9001). Washington, DC: ISO.

International Organization for Standardization. (1991). *Guidelines for the Application of ISO 9001 to the Development, Supply, and Maintenance of Software* (ISO 9000-3). Washington, DC: ISO.

Jacobsen, I., Griss, M., and Jonsson, P. (1997). *Software Reuse: Architecture, Process, and Organization for Business Success.* Reading, MA: Addison-Wesley.

Kimball, J. (1993). Configuration management and process management. *Proc. 4th Int. Workshop Software Configuration Manage.* Baltimore, MD, *1993,* pp. 138–140.

Pressman, R. S., (1992). *Software Engineering: A Practitioner's Approach.* New York: McGraw-Hill.

Sage, A. (1995). *Systems Management for Information Technology and Software Engineering.* New York: Wiley.

Software Engineering Institute. (1990). *Spectrum of Functionality in Configuration Management Systems,* SEI Tech. Rep. (CMU/SEI-90-TR-11). Pittsburgh, PA: Carnegie-Mellon University.

Software Engineering Institute. (1993). *Key Practices of the Capability Maturity Model, Version 1.1,* SEI Tech. Rep. (CMU/SEI-93-TR-25). Pittsburgh, PA: Carnegie-Mellon University.

Strauss, S. H., and Ebenau R. G. (1993). *Software Inspection Process.* New York: McGraw-Hill.

Tichy, W. (1994). *Configuration Management.* New York: Wiley.

Watts, H. S. (1995). *A Discipline for Software Engineering.* Reading, MA: Addison-Wesley.

Whitgift, D. (1991). *Software Configuration Management: Methods and Tools.* New York: Wiley.

6 Cost Management

BENJAMIN S. BLANCHARD

6.1 INTRODUCTION

Many of our day-to-day decisions, as they pertain to the design and development of new systems and/or the reengineering of existing systems, are based on *technical* performance-related factors alone. *Economic* considerations, if addressed at all, have dealt primarily with initial procurement and acquisition costs only. The consequences of this "short-term" approach have been rather detrimental overall, as experience has indicated that many of the engineering and management decisions made in the early phases of a system life cycle have had a great impact on the sustaining operation and maintenance support of that system later on. Further, these "downstream" activities often constitute a large percentage of the total cost of a system. Thus, it is essential that we extend our planning and decision making to address system requirements from a total life-cycle perspective, particularly if we are to properly assess the risks associated with such decisions. The application of life-cycle analysis methods and *total* cost management must be inherent within the various processes discussed throughout this handbook if the objectives of systems engineering and management are to be met. This chapter deals with the need for life-cycle cost management, the life-cycle cost analysis process, some applications, and the benefits derived from such. The aspects of functional economic analysis, development of a work breakdown structure, activity-based costing, and the implementation of cost and operational effectiveness methods in the design and/or reengineering of systems are addressed.

6.2 LIFE-CYCLE COSTING

Life-cycle costing includes the consideration of *all* future costs associated with research and development (i.e., design), construction, production, distribution, system operation, sustaining maintenance and support, retirement, and material disposal and/or recycling. It involves the costs of all applicable technical and management activities throughout the system life cycle; that is, producer activities, contractor and supplier activities, and consumer or user activities. While individual design and management decisions may be based on some aspect of cost (e.g., purchase price or initial acquisition cost), the consequences must be assessed in terms of *total* cost in order to identify the true risks associated with these decisions. Full-cost visibility is essential in fulfilling the requirements of risk management as described in Chapter 3.

Handbook of Systems Engineering and Management, Edited by A. P. Sage and W. B. Rouse
ISBN 0471-15405-9 ©1999 John Wiley and Sons, Inc.

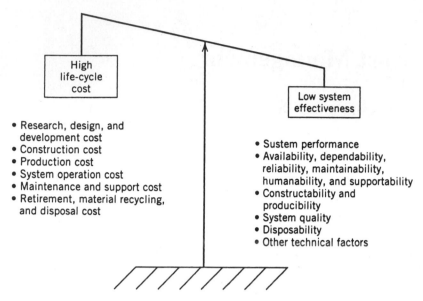

Figure 6.1 The cost-effectiveness balance.

6.2.1 The Need for Life-Cycle Costing

Do you know the actual *true cost* of your system? Can you identify the cost of each *functional* element? Are you aware of the *high-cost contributors?* Can you determine the *causes* for these high-cost areas? Can you truly assess the *risks* associated with the development, construction or production, operation, and/or support of your system? The answer to these and many questions of a related nature is a definite "No!"

To complicate matters, experience in recent years has indicated that the complexity and the costs of systems in general have been increasing. A combination of introducing new technologies in response to a constantly changing set of performance requirements, the increase in external social and political pressures associated with environmental issues, the requirements to reduce the time that it takes to design and construct a new plant or deliver a new product to the consumer, and the need to extend the life cycle of systems already in operation constitutes a major challenge! Further, many of the systems currently in use today are not adequately responding to the needs of the consumer, nor are they cost-effective in terms of their operation and support, resulting in an imbalance, as conveyed in Figure 6.1. This is occurring at a time when available resources are dwindling and international competition is increasing worldwide.

In addressing the issue of cost-effectiveness, one often finds that there is a lack of total cost *visibility,* as illustrated by the "iceberg" in Figure 6.2. For many systems, the costs associated with the design, construction, the initial procurement and installation of capital equipment, production, and so forth, are relatively well known. We deal with, and make decisions based on, these costs every day. However, the costs associated with utilization and the maintenance and support of the system throughout its planned life cycle are somewhat hidden. In essence, we have been successful in addressing the short-term aspects of cost, but have not been very responsive to the long-term effects.

At the same time, it has been found that a large percentage of the total life-cycle cost

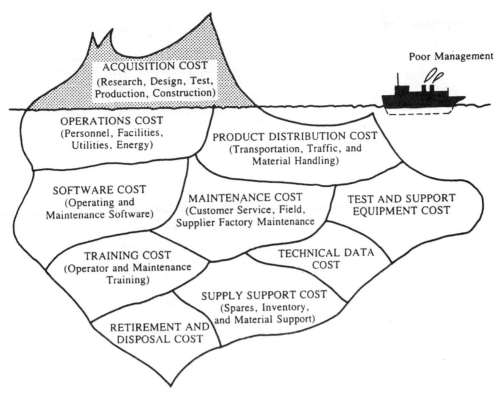

Figure 6.2   Total cost visibility.

for a given system is attributed to operating and maintenance activities (e.g., up to 75 percent for some systems). When looking at "cause-and-effect" relationships, one often finds that a significant portion of this cost stems from the consequences of decisions made during the early phases of advance planning and conceptual design. Decisions pertaining to the design of a process, the selection of materials, the selection of an item of capital equipment, equipment packaging schemes, determination of manual vs. the incorporation of automated applications, the design of maintenance and support equipment, and so on, have a great impact on the "downstream" costs and, thus, life-cycle cost. Additionally, the ultimate maintenance and support infrastructure selected for a system throughout its period of utilization can significantly impact the overall cost-effectiveness of that system. There are many interactions that occur when dealing with systems and their respective elements (Blanchard and Fabrycky, 1998).

Given these relationships, in today's environment, where resources are dwindling and international competition is increasing, there is a need to reevaluate our methods used not only in the design and construction (or production) of *new* systems but in the sustaining operation and maintenance of *existing* systems already in use. Addressing system requirements from a total life-cycle perspective is essential, and the application of life-cycle cost analysis methods can be highly beneficial in facilitating this objective. While these issues have been discussed throughout this handbook, the aspect of "total cost" is highlighted in this chapter.

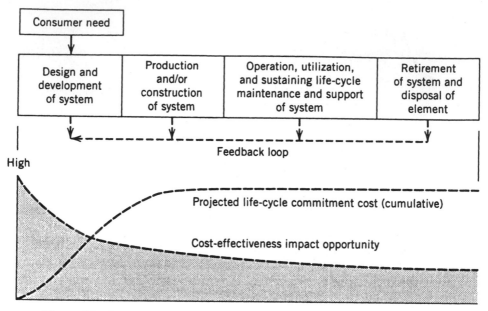

Figure 6.3 Opportunity for impacting cost-effectiveness in the system life cycle.

6.2.2 The Life-cycle Cost Analysis Process

The application of life-cycle costing methods can be effectively implemented in:

1. *The design, development, and construction (or production) of a* new *system.* Given an identified need, there are design and development activities (i.e., definition of system requirements, functional analysis and allocation, tradeoff studies and design optimization, and synthesis), construction and/or production activities, and so forth. As illustrated in Figure 6.3, it is at the early stages in a program where the greatest gains can be realized in terms of the ultimate total life-cycle cost of a system. The objective is to establish, from a top-down perspective, a quantitative "design-to-cost" requirement, and then to design, build, operate, and maintain to meet that requirement. A system-level cost goal can be established, costs can be allocated to activities and/or elements of a system, and life-cycle costs can be controlled from the beginning of a project (Fabrycky and Blanchard, 1991).

2. *The evaluation of an* existing *system capability, with the objective of implementing a "continuous-product/process-improvement" approach to increase the effectiveness while reducing the life-cycle cost of that system.* This involves the initial determination of some quantitative goals based on a defined need (i.e., the establishment of some technical performance measure or "metric" for the purposes of "benchmarking"), describing the system and its processes in functional terms, collecting the appropriate data and identifying the resources being consumed in accomplishing the various functions, identifying the high-cost contributors and determining cause-and-effect relationships, and initiating the necessary recommendations for improvement of the system and its operation. This is an on-going iterative process. In

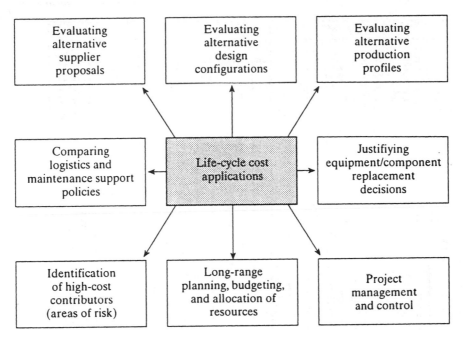

Figure 6.4 Life-cycle cost applications.

other words, life-cycle cost analysis methods can be effectively employed in the reengineering of existing systems through the identification of high-cost (high-risk) areas requiring management attention.

Throughout the process of developing new systems and/or the reengineering of existing systems, there are numerous applications of a more specific nature where life-cycle cost methods can be effectively employed. Some of these applications are identified in Figure 6.4. The objective is, of course, to extend the "thought processes" of the decision maker (whether the design engineer or a manager) to view things from a *total* cost perspective.

When performing a life-cycle cost analysis, there are a series of steps that one might follow. In the development of new systems, the process illustrated in Figure 6.5 can be applied. Life-cycle cost (LCC) and/or design-to-cost (DTC) goals are established, along with other technical performance measures, in conceptual design. The architect and the design engineer must view "cost" as a *design* parameter (and not just initial price). Given a top-level requirement, costs can be allocated to cover the appropriate functional activities and elements of the system. Life-cycle cost analyses can then be accomplished throughout the design process initially for the purposes of prediction and later for the purposes of assessment. The "tracking" of costs must not only include the initial acquisition costs associated with most design and development projects, but the projected costs associated with the downstream activities of system operation and support.

In applying life-cycle cost analysis methods in the evaluation of an existing system, the steps presented below may be applied. First, one must define the new requirements for the system. Why is there a need to upgrade or improve the existing capability? What new "benchmarks" have been established for system operation and support? Given a new set

240

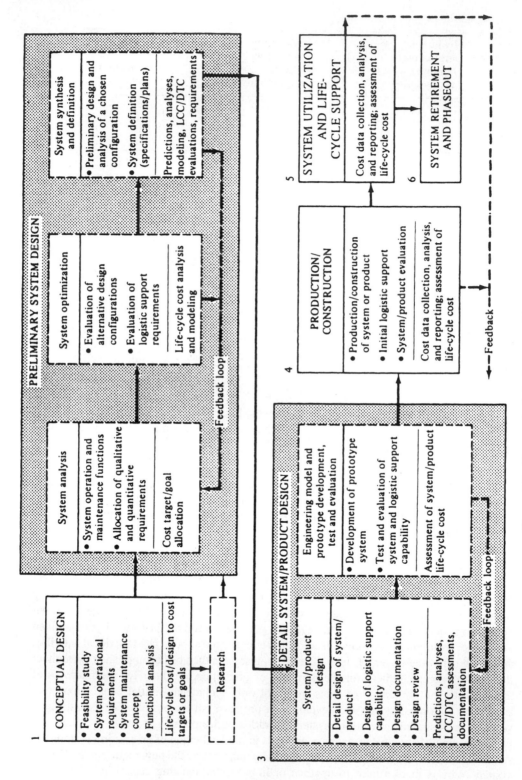

Figure 6.5 Cost emphasis in the system life-cycle process.

of requirements, one must first assess the existing "baseline" (i.e., the current capability and associated costs), and then determine the steps that will be necessary in order to evolve from this initial baseline to the newly established goal(s). In the area of cost, what are your current costs and what are the new cost goals for the future? Assuming that a "cost reduction" goal has been specified, the identification of high-cost contributors and the causes for such is a necessary first step. In any event, following the steps presented below represents an approach for the implementation of a "continuous process improvement" effort.

The Basic Steps in a Life-cycle Cost Analysis

Step 1. Describe the system configuration being evaluated in *functional* terms, and identify the appropriate technical performance measures (TPMs) or applicable "metrics" for the system.

Step 2. Describe the system life cycle and identify the major activities in each phase as applicable (system design and development, construction and/or production, utilization, maintenance and support, retirement, and disposal).

Step 3. Develop a work breakdown structure (WBS), or cost breakdown structure (CBS), covering *all* activities and work packages throughout the life cycle.

Step 4. Estimate the appropriate costs for each category in the WBS (or CBS), using activity-based costing (ABC) methods, or equivalent.

Step 5. Develop a computer-based model to facilitate the life-cycle cost analysis process.

Step 6. Develop a cost profile for the "baseline" system configuration being evaluated.

Step 7. Develop a cost summary, identifying the high-cost contributors (i.e., high-cost "drivers").

Step 8. Determine the "cause-and-effect" relationships, and identify the "causes" for the high-cost areas.

Step 9. Conduct a sensitivity analysis to determine the effects of input factors on the analysis results, and identify the high-risk areas.

Step 10. Construct a Pareto diagram and rank the high-cost areas in terms of relative importance and requiring immediate management attention.

Step 11. Identify feasible alternatives (potential areas for improvement), construct a life-cycle cost profile for each, and construct a break-even analysis showing the point in time when a given alternative assumes a point in preference.

Step 12. Recommend a preferred approach, and develop a plan for system modification and improvement (this may entail a modification of equipment or software, a facility change, and/or a change in some process). This constitutes an ongoing iterative approach for *continuous process improvement.*

Regardless of the application, the system in question must be defined in *functional* terms, the appropriate measures must be established for each functional area, and the resource requirements and associated costs for completion of each function must be identified. The functional analysis, accomplished as a major step in the system engineering process (refer to Chapter 25), provides a necessary foundation. Cost-generating activities are then identified through a WBS, costs are estimated for all WBS elements and work packages, and life-cycle cost analyses are accomplished to either ensure the development of a new system that is cost-effective or to improve an existing system capability.

6.3 FUNCTIONAL ECONOMIC ANALYSIS

Having identified the need for a new and/or reengineered system, it is essential that one proceed with a good comprehensive definition of system requirements. For instance, what functions must the system perform (missions to be completed)? When will the system be required to perform these functions? Where is this to be accomplished and for how long? In what environment(s) must the system operate? What are the system availability requirements? How is the system to be supported? What are the budgetary requirements and/or limitations (if any)?

The answers to these and related questions can be acquired through the definition of system operational requirements and the maintenance concept, leading to the identification and prioritization of TPMs or the "metrics" for the system overall (Blanchard, 1998). Operational requirements include a description of the appropriate mission scenarios and utilization profiles, the application of performance factors and their relationship to the identified scenarios, the anticipated deployment and geographical locations where system elements are to be utilized, relevant effectiveness factors, the projected system life cycle, and environmental factors as they pertain to system utilization. The maintenance concept, on the other hand, represents a "before-the-fact" description of how the system is to be designed for supportability (i.e., establishing the goals or benchmarks for maintenance and support) and generally covers the anticipated levels and responsibilities for the accomplishment of maintenance and support, design criteria for the development and/or assessment of the various elements of support (e.g., spares and related inventories, test equipment, facilities), effectiveness factors as they apply to the overall support infrastructure, and environmental factors relative to maintenance activities.

The objective is to define the requirements for the system in terms of the "whats" during the initial stages of a program, and the aspect of *cost* must be considered from the beginning! Definition of system operational requirements and the maintenance concept must be addressed early, not only to identify design requirements for system elements but for the purposes of establishing a necessary "baseline" upon which to develop cost estimates for activities downstream in the life cycle; that is, the costs represented by the lower part of the iceberg in Figure 6.2.

The next step is the development of a *functional* description of the system, and the translation of system-level requirements from the "whats" into an applicable set of "hows" through the iterative process of functional analysis, requirements allocation, tradeoffs and design optimization, synthesis, and so on. The functional analysis can best be accomplished through the utilization of functional block diagrams, as described in Chapter 25. The top-level system configuration (or system architecture) is first defined; individual functions are successively broken down to lower-levels depicting greater and greater detail; similar functions are combined and integrated, leading to a packaging concept and the identification of major system components; and the requirements (i.e., TPMs) specified for the system are allocated or apportioned down to the subsystem and below as appropriate.

Figure 6.6 illustrates a simplified flow diagram showing the major functions that must be accomplished, commencing with the identification of a need and evolving through system design and development, construction or production, system operation and support,

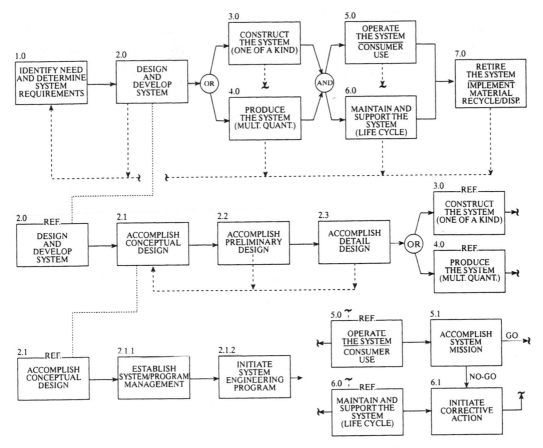

Figure 6.6   Partial functional block diagram (example).

and ultimate retirement. Series and parallel relationships may be established, and each functional block is "action-oriented." Individual functional blocks can be broken down as necessary to gain the desired level of visibility relative to a given system element, component, and/or a process. Referring to the figure, the design function (block 2.0) is broken down into a series of subfunctions (blocks 2.1, 2.2, etc.), and the system operating function (block 5.0) is broken down to accomplishing the mission (block 5.1). This, in turn, is extended into a series of functions shown in Figure 6.7, which represents the operating characteristics of a manufacturing system. Further, a GO/NO-GO situation is illustrated primarily to show the translation from "operating" functions into "maintenance" functions (refer to block 5.1 leading into block 6.1 in a NO-GO situation).

Each individual function (in Figs. 6.6 and 6.7) may then be evaluated in terms of inputs, expected outputs, constraints and/or controls, and the resources (mechanisms) that are required to accomplish the function in question. Figure 6.8 shows these relationships as they may be applied to a typical functional block, with the "resources" reflecting the methods by which a function may be completed (i.e., the "hows"). Resources may take the form of equipment, people, materials, facilities, software, data, or combinations thereof.

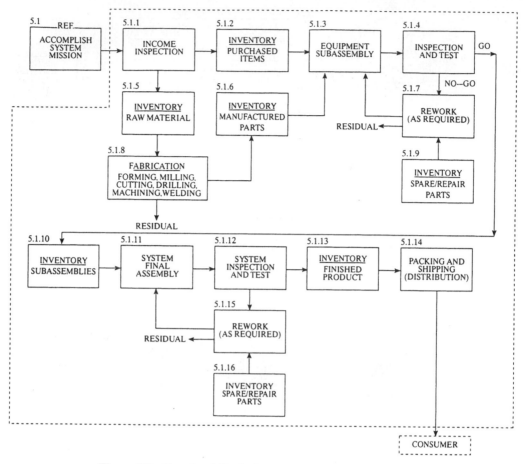

Figure 6.7 Functional block diagram of manufacturing system.

Tradeoff studies are conducted to help ensure that the best approach is selected in the resource-identification or -utilization process.

The functional analysis provides an excellent baseline for the initial top-down allocation of costs in the design of new systems, enabling the early establishment of DTC objectives and the implementation of the process illustrated in Figure 6.3. Costs may be allocated for design functions, construction and/or production functions, operating functions, and maintenance functions; that is, to each block in Figures 6.6 and 6.7. Referring to Figure 6.8, an allocated cost factor then becomes an economic constraint which, in turn, will influence the design (i.e., product output) and the resources selected. This will force the system architect, or designer, to consider economics in the day-to-day decision making.

In the evaluation and reengineering of existing systems, the functional analysis serves as a basis for developing work breakdown structures (or cost breakdown structures), leading to the collection of costs by functional area. Referring to Figure 6.8, the determination of resource requirements for each functional block and the costs of these resources over the life cycle will lead to the identification of high-cost contributors and candidate areas for improvement.

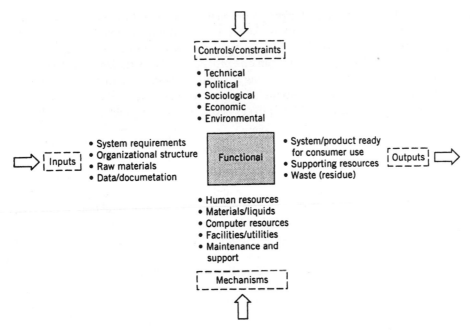

Figure 6.8   The identification of resource requirements by function.

Finally, the functional analysis helps to identify all major activity areas throughout the entire life cycle. While the emphasis on most projects in the past has been relegated primarily to design, development, and production activities (i.e., Fig. 6.6, blocks 1.0, 2.0, 3.0, and 4.0), there is a need to address the downstream activities as well. The program activities that occur during design have a large influence on the costs later on. Further, the system requirements reflected in Figure 6.7, for example, must be addressed in blocks 1.0 and 2.0, and will determine the activities that will be required in 2.0, 3.0, and 4.0. The functional analysis tends to force a life-cycle approach, aids in establishing the proper series-parallel relationships, shows the interrelationships among functions and feedback provisions, and facilitates a *total-cost* approach.

6.4   WORK BREAKDOWN STRUCTURE

The functions described through the block diagrams presented in Figures 6.6 and 6.7 can be broken down into subfunctions, categories of work, work packages, and ultimately the identification of physical elements (as illustrated in Fig. 6.8). From a planning and management perspective, it is necessary to establish a top-down framework that will allow for the initial allocation and subsequent collecting, accumulating, organizing, and computing of costs. Thus, a WBS can be prepared to show, in a hierarchical manner, all of the work elements that are necessary to complete a given program. The WBS may show the subdivision of work, leading from a functional definition to a physical element (Kerzner, 1995).

The WBS is *not* an organizational chart in terms of project personnel assignments and responsibilities, but does represent an organization of work packages prepared for the purposes of program planning, budgeting, contracting, and reporting. In the context of the system life cycle, there may be work packages associated with system design and development activities, construction and/or production activities, system operation and support activities, and so on. These will vary from one program to the next, depending on whether a new system is being designed or an existing capability is being reengineered and modified for improvement. Further, when initially allocating budgets, the assigned cost must consider *all* applicable activities in the life cycle because of the interaction effects that may occur between the various segments of cost.

Thus, in developing a WBS, one should first construct a summary work breakdown structure (SWBS) to include all life-cycle activities as identified through the functional analysis described in Section 6.3. This may cover customer activities, contractor activities, supplier activities, consumer or user activities, and/or combinations thereof. A SWBS will usually include up to three levels, starting at the top with the identification of the system and working down to various categories of activity (e.g., research and development, reengineering, production) and then down to a more specific set of activities (e.g., systems engineering management, reliability engineering, system operation, system modification). Given a top-level structure, which should reflect the functions described in Figures 6.6 and 6.7, then it will be necessary to develop an individual lower-level contract work breakdown structure (CWBS) to cover a specific element of work to be conducted through the establishment of a formal "contract" or some type of "procurement" action. In essence, the SWBS serves as the overall mechanism for identifying work elements and associated costs (with a life-cycle perspective in mind), while a number of individual CWBSs may be prepared and "tailored" to specific company, agency, and/or organizational interests. Figure 6.9 illustrates the relationships between the SWBS and the CWBS.

Figure 6.10 presents an example of a SWBS reflecting the activities depicted in Figure 6.6. The elements of work identifed can be broken down into specific duties, jobs, and tasks involving the consumption of resources as illustrated in Figure 6.8. Initially, costs can be allocated from the system-level and on down when a DTC requirement exists in the development of a new system. Figure 6.11 shows the relationships between the traditional WBS and a partial project organizational structure, where costs may be allocated to an element of a system and then transferred to one or more organizational activities. Additionally, costs may be collected (from the bottom-up) for each activity reflected in the applicable CWBSs, and ultimately accumulated at the SWBS level. This network represents an overall "cost breakdown structure (CBS)," a term used in a number of references dealing with the subject of life-cycle costing (Blanchard and Fabrycky, 1998).

While it is not the intent to cover the concepts of basic project management and the methods used in developing WBSs herein, it is felt that the objectives described in Section 6.2 can only be met *if* the development of the initial SWBS includes all activities in the system life cycle. A common occurrence is to utilize the WBS-approach only when new systems are being acquired, and then to only cover research and development (and possibly construction and production) activities, leading down to the development of one or more CWBSs. Then, when it comes to individual contract negotiations, specific elements of work may be deleted because of the short-term budget limitations. If a given work package is eliminated from the CWBS, then, of course, it should still be covered under the requirements specified through the SWBS. This doesn't always happen, however, as the deleted work requirements are often forgotten until later when "panic" prevails. A good example

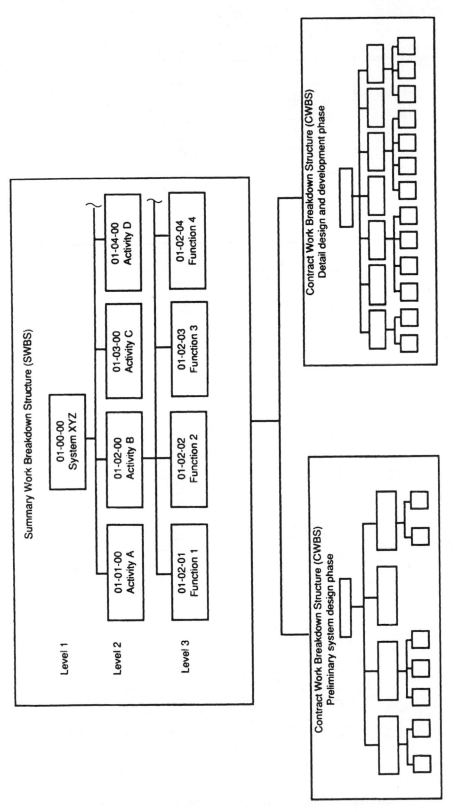

Figure 6.9 Work breakdown structure development (partial).

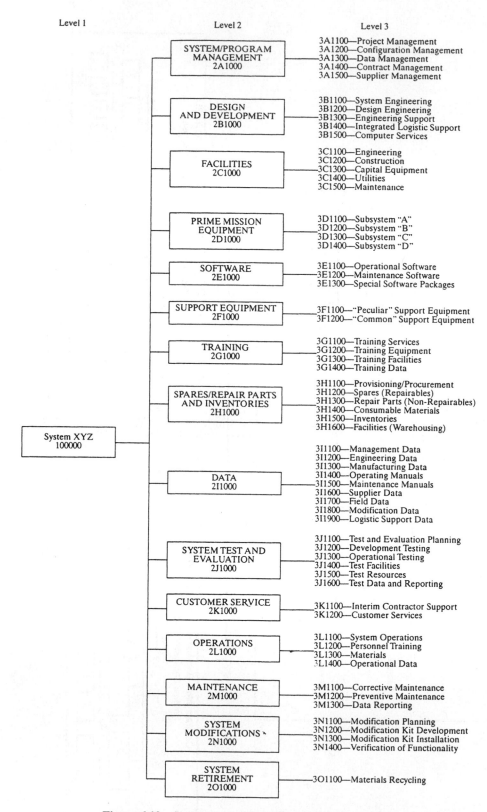

Figure 6.10 Summary work breakdown structure (example).

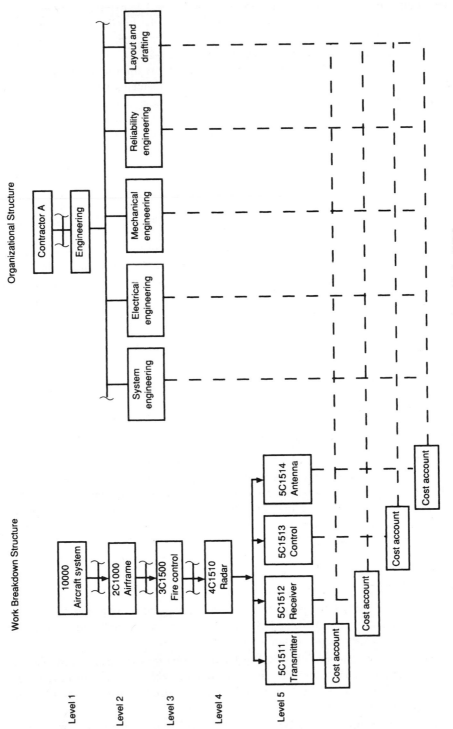

Figure 6.11 Organizational integration with the WBS.

is the elimination of "logistics and the elements of support" in deference to the short-term needs. The applicable contractor will then forget about the implications of maintenance and support in the design process, thus creating a potential high-cost high-risk situation, as reflected by the bottom part of the iceberg in Figure 6.2.

Development of the SWBS constitutes a critical step within the overall spectrum of total cost management. The functions and applicable elements of work are identified, broken down to the appropriate level in order to gain the necessary visibility for proper cost management, cost estimating methods are applied in determining the costs associated with each work package, and the necessary cost monitoring and control provisions are implemented accordingly. The WBS network serves as the framework that enables both the initial allocation and the subsequent collection and accumlation of costs later on.

6.5 ACTIVITY-BASED COSTING

To be effective in total cost management (and in the accomplishment of cost-effectiveness analyses) requires full-cost visibility allowing for the traceability of all costs back to the activities, processes, and/or products that generate these costs. In the traditional accounting structures employed in most organizations, a large percentage of the total cost cannot be traced back to the "causes!" For example, "overhead" or "indirect" costs, which often constitute greater than 50 percent of the total, include a lot of management costs, supporting organization costs, and other costs that are difficult to trace and assign to specific objects. With these costs being allocated across the board, it is impossible to identify the actual "causes" and to pinpoint the *true* high-cost contributors. As a result, the concept of *activity-based costing* has been introduced (Kidd, 1994).

6.5.1 Objectives of Activity-based Costing

Activity-based costing (ABC) is a methodology directed toward the detailing and assignment of costs to the items that cause them to occur. The objective is to enable the "traceability" of *all* applicable costs to the process or product that generates these costs. The ABC approach allows for the initial allocation and later assessment of costs by function, and was developed to deal with the shortcomings of the traditional management accounting structure where large overhead factors are assigned to all elements of the enterprise across the board without concern for whether they directly apply or not. More specifically, the principles of ABC are noted below (Canada et al., 1996).

1. Costs are directly traceable to the applicable cost-generating process, product, and/ or a related object. Cause-and-effect relationships are established between a cost factor and a specific process or activity.

2. There is no distinction between direct and indirect (or overhead) costs. While 80 to 90 percent of all costs are traceable, those nontraceable costs are not allocated across the board, but are allocated directly to the organizational unit(s) involved in the project (Kidd, 1994).

3. Costs can be easy allocated on a *functional* basis; that is, the functions identified in Figures 6.6 and 6.7. It is relatively easy to develop cost-estimating relationships in terms of the cost of activities per some activity measure (i.e., the cost per unit output).

4. The emphasis in ABC is on "resource consumption" (vs. "spending"). Processes and products consume activities, and activities consume resources. With resource consumption being the objective, the ABC approach facilitates the evaluation of day-to-day decisions in terms of their impact on resource consumption downstream.

5. The ABC approach fosters the establishment of "cause-and-effect" relationships and, as such, enables the identification of the "high-cost contributors." Areas of risk can be identified with some specific activity and the decisions that are being made within.

6. The ABC approach tends to eliminate some of the cost doubling (or double counting) that occurs when attempting to differentiate on what should be included as a "direct" cost or as an "indirect" cost. By not having the necessary visibility, there is the potential of including the same costs in both categories.

Implementation of the ABC approach, or something of an equivalent nature, is essential if one is to do a good job of total cost management. Costs are tied to objects and viewed over the long-term, and the use of such facilitates the life-cycle cost analysis process described in Section 6.2. An objective for the future is to convince the accounting organizations in various companies/agencies to supplement their current end-of-the-year financial reporting structure to include the objectives of ABC.

6.5.2 Cost Estimating Methods

The next step is to estimate costs, by category in the WBS (or CBS), for each year in the system life cycle. Such estimates must consider the effects of inflation, learning curves when repetitive processes or activities occur, and any other factors that are likely to cause changes in cost, either upward or downward. Cost estimates are derived from a combination of accounting records (i.e., historical records using the ABC and/or traditional approaches for accounting), project cost projections, supplier proposals, and predictions in one form or another.

Referring to Figure 6.3, the early stage in the system life cycle is the preferred time to commence with the estimation of costs, since it is at this point when the greatest impact on total system life-cycle cost can be realized. However, the availability of good historical cost data at that time is almost nonexistent in most organizations, particularly the type of data that pertain to the downstream activities of operations and support for similar systems in the past. Thus, one must depend heavily on the use of various cost estimating methods in order to accomplish the end objectives.

Referring to Figure 6.12, as the system configuration becomes better defined in a new developmental effort, the use of direct engineering and manufacturing standard factors based on past experience can be applied as is the case for any "cost-to-complete" projection on a typical project today (e.g., cost per labor-hour). On the other hand, in the earlier stages of the life cycle, when the system configuration has not been well defined, the analyst must rely on the use of a combination of analogous and/or parametric methods developed from experience on similar systems in the past. The objective is to collect data on a "known entity," identify the major functions that have been accomplished and the costs associated with these functions, relate the costs in terms of some functional or physical parameter of the system, and then use this relationship in attempting to estimate the costs for a new system. As a goal, one should identify the applicable TPMs for the system in question and

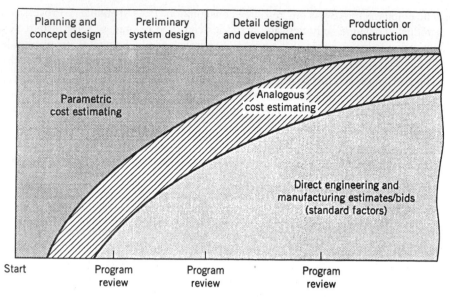

Figure 6.12 Cost estimation by program phase.

estimate the cost per a given level of performance (e.g., cost per unit of product output, cost per mile of range, cost per unit of weight, cost per volume of capacity used, cost per unit of acceleration, cost per functional output). Costs can be related to the appropriate functional blocks identified in Figures 6.6 and 6.7. Figure 6.13 provides a simple illustration of some of the relationships that can be established to facilitate the cost estimation process. However, care must be exercised to ensure that the historical information used in the development of cost estimating relationships (CERs) is relevant to the system configuration being evaluated today. CERs based on the mission and performance characteristics of one system may not be appropriate for another system configuration, even if the configuration is similar in a physical sense. Thus, costs must be related from a *functional* perspective.

6.5.3   Development of a Cost Profile

Once the costs are determined for each WBS category and for each year in the projected system life cycle, the appropriate inflationary factors are applied, and a cost profile is developed as shown in Figure 6.14 (Section 6.2.2, step 6). All recurring and nonrecurring costs, variable and fixed costs, producer and consumer costs, and so on, are included. Thus, a "budgetary" profile is developed and can be utilized for the identification of resources for the future.

In evaluating profiles, it may be feasible to start out with the development of one in terms of *constant* dollars first and then add the inflationary factors to reflect a *budgetary* profile. When comparing alternative profiles, the appropriate economic analysis methods must be applied in converting the various alternative money streams to the *present value,*

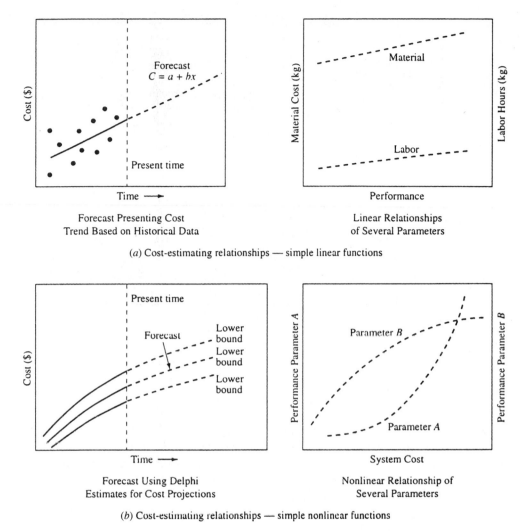

Forecast Presenting Cost
Trend Based on Historical Data

Linear Relationships
of Several Parameters

(*a*) Cost-estimating relationships — simple linear functions

Forecast Using Delphi
Estimates for Cost Projections

Nonlinear Relationship of
Several Parameters

(*b*) Cost-estimating relationships — simple nonlinear functions

Figure 6.13 (*a*) Cost estimating relationship—simple linear functions. (*b*) Cost estimating relationships—simple nonlinear functions.

or to the point when the decision will be made in selecting a preferred approach. It is necessary to evaluate alternative profiles on the basis of some form of *equivalence* (Thuesen and Fabrycky, 1993).

In order to gain some insight relative to the costs for each major category in the WBS, it may be appropriate to view the results presented in a tabular form as illustrated in Table 6.1 (Section 6.2.2, step 7). Referring to the table, the major categories of the WBS are identified, along with the percent contribution of each. Through inspection, one can readily identify the high-cost contributors (e.g., the inspection and test function at 17 percent). These are the areas where a more in-depth analysis is required relative to the "causes" for these high costs (Section 6.2.2, step 8).

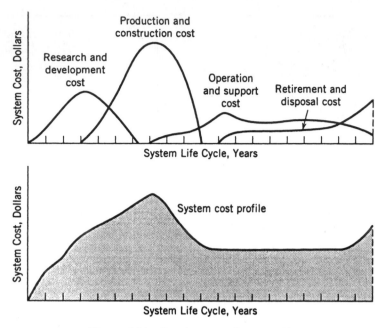

Figure 6.14   Development of cost profile.

6.6   COST AND EFFECTIVENESS ANALYSIS

Referring to Figure 6.1, there are two sides of the spectrum to include, the *cost* side and the *effectiveness* side, and each has an impact on the other. The objective is, of course, to attain the proper "balance" between the two. While the emphasis in this chapter is primarily on the aspect of cost, it should not be assumed that the cost factors are independent with

TABLE 6.1   Life-Cycle Cost Breakdown Summary

Cost Category	Cost × 1,000 ($)	% of Total
1. Architecture and design	2,248	7
2. Construction	12,524	39
a. Facilities	6,744	21
b. Capital equipment	5,780	18
3. Factor operation and maintenance	17,342	54
a. Incoming inspection	963	3
b. Fabrication	3,854	12
c. Subassembly	1,927	6
d. Final assembly	3,533	11
e. Inspection and test	5,459	17
f. Packing and shipping	1,606	5
Grand Total	32,114	100

respect to levels of effectiveness. The assumption has been that a desired level of "effectiveness" has been attained, and that one needs to do something to minimize (or reduce) the cost side of balance. With this in mind, it is appropriate first to complete the life-cycle cost analysis coverage and then address some of the tradeoffs between cost and effectiveness.

As conveyed in Figure 6.4 there are a wide variety of areas where life-cycle analysis methods can be effectively applied, for example, the evaluation of alternative operational and/or maintenance support concepts, alternative design configurations, alternative sources of supply for commercial and standard items, alternative production approaches, and so on. In each situation, there is a *process* that one follows that is relatively common across the board. Figure 6.15, an extension of the list of steps in Section 6.2.2, illustrates this process. Given a definition of the problem and the goals of the analysis, the next step is to describe the baseline configuration and the alternatives that are to be evaluated. The final steps are to select the appropriate tool/model as an aid in completing the analysis, collect the right data, and proceed with the evaluation effort.

Relative to the selection of a model, one must ensure that the tool selected does what is expected, is sensitive to the problem at hand, and allows for the visibility needed in addressing the system as an entity as well as any of its major components on an individual-by-individual basis. The model must enable the comparison of *many* different alternatives and aid in selecting the best among them rapidly and efficiently. The model must be *comprehensive*, allowing for the integration of many different parameters; *flexible* in structure, enabling the analyst to look at the system as a whole or any part of the system; *reliable* in terms of repeatability of results; and be *user-friendly*. So often, one selects a computer model based on the advertised brochure material alone, purchases the necessary equipment and software, uses the model to manipulate data, and believes in the output results without having any idea as to how the model was put together, the internal analytical relationships established, whether it is sensitive to the variation of input parameters in terms of output results, and so on. The results of a survey indicates that there are over 350 computer-based tools available in the commercial marketplace and intended for use in accomplishing different levels of analysis. Each was developed on a relatively "independent" or "isolated" basis in terms of selected platform, the language used, input data needs, and interface requirements. In general, the models do not "talk to each other," are not user friendly, and are too complex for use in early system design and development (Blanchard et al., 1994).

In selecting a model, it is essential that the analyst becomes thoroughly familiar with the tool, know how it was put together, and understand what it can do. For the purposes of accomplishing a life-cycle cost analysis, it may be appropriate to select a group of models, combined as illustrated in Figure 6.16, and integrated in such a manner that will enable the analyst to look at not only the cost for the system overall but at some of the key functional areas representing potential high-cost contributors. The model(s) must be structured around the WBS and in such a way that will allow the analyst to look at the costs associated with each of the major functions in Figure 6.6. Further, it must be *adaptable* for use during the early stages of conceptual design as well as in the detail design and development phase. The subject of models is discussed further in Chapter 26.

With the requirements defined and a model available to facilitate the LCC analysis process, the various alternatives being considered can now be evaluated. For the purposes of illustration, it is assumed that a new ground vehicular system is in the development phase and requires the incorporation of a communications capability. The communication

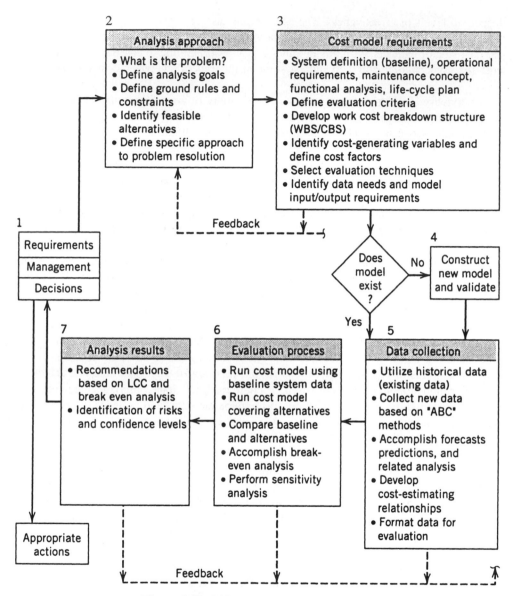

Figure 6.15   Life-cycle cost analysis process.

equipment must meet certain performance and effectiveness requirements, and at a specified unit life-cycle cost. There are two alternative design configurations being considered to fulfill the stated need, and a decision must be made as to whether to select Alternative *A* or Alternative *B* (Blanchard and Fabrycky, 1998).

It is necessary to first establish a system-level baseline description considering operational requirements, the maintenance concept, and a life-cycle plan (refer to block 3 of Figure 6.15). The communication equipment is to be installed in a light vehicle; there are

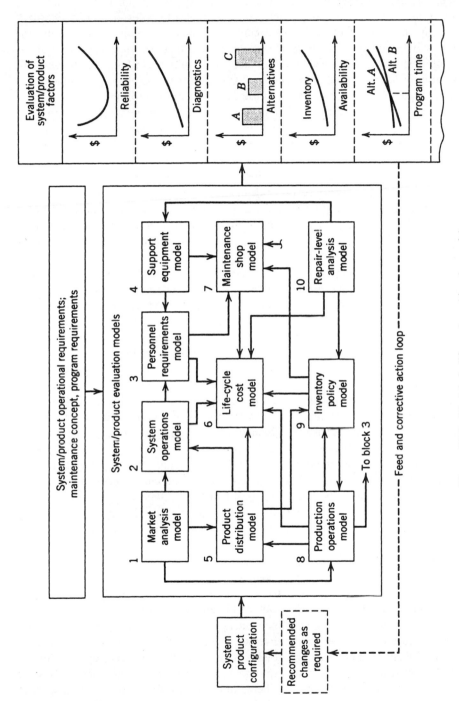

Figure 6.16 Models in life-cycle costing (example).

65 vehicles deployed in three different geographical areas; the equipment must enable the communications with other ground vehicles at a range of 200 miles, overhead aircraft at an altitude up to 10,000 feet, and with a centralized area communication facility; the equipment will be utilized 4 hours per day; the reliability mean time between failures (MTBF) must be at least 450 hours; the M̄ct must be 30 minutes or less; and the unit cost (based on life-cycle cost) should not exceed $20,000. Further, the planned life-cycle is to be 12 years. Figure 6.17 illustrates some of the initial planning information.

Referring to the life-cycle plan in Figure 6.17, the specific functions and supporting activities under research and development, investment (or production), operations and maintenance must be broken down and included in the form of WBS, "tailored" to the communications system. Cost and supporting data are collected by category in the WBS, spread over the planned 12-year period, inflationary factors are included, and individual cost profiles are developed for each of the two configurations being evaluated. Reliability, maintainability, logistic support, and related factors are utilized to aid in the determination of operations and maintenance costs. These factors may be derived through a combination of allocations, predictions, and/or actual assessments from field data (refer to Chapter 8).

Given the *budgetary* cost profiles for each Configuration A and Configuration B, they must then be converted to the *present value* to allow for comparison on an equivalent basis. Table 6.2 presents a breakdown summary of the present-value costs by major WBS category, and identifies the relative percent contribution of each category in terms of the total. A 10% interest rate was used in determining present-value costs (Thuesen and Fabrycky, 1993, or any basic text dealing with economic analysis methods).

While a review of Table 6.2 might lead one to immediately select Configuration A as being preferable, prior to making such a decision the analyst needs to project the two cost streams in terms of the life cycle and determine the point when Configuration A assumes the position of *preference*. Figure 6.18 shows the results of a breakeven analysis, and it appears that A is preferable after approximately 6½ years into the future. The question arises as to whether this "breakeven" point is reasonable when considering the type of system and its mission, the technologies being utilized, the length of the planned life cycle, and the possibilities of obsolescence. For systems where the requirements are changing constantly and obsolescence could become a problem two to three years hence, the selection of Configuration B may be preferable. On the other hand, for larger systems with longer life cycles (e.g., 10 to 15 years and greater), the selection of Configuration A may represent the best choice.

In this case, it is assumed that Configuration A is preferable. When the cost profile for this alternative is converted back to a *budgetary* projection, however, it is realized that a further reduction of cost is necessary. This, in turn, leads the analyst to Table 6.2 and the identification of potential *high-cost* contributors. Given that a large percentage of the total cost of a system is often in the area of maintenance and support (refer to Fig. 6.2), one might investigate the categories of "maintenance personnel" and "spares/repair parts," representing 23.4 and 11.5 percent of the total cost, respectively. The next step is to identify the applicable cause-and-effect relationships and to determine the actual "causes" for such high costs. This may be accomplished by being able to trace the costs back to a specific function, process, product design characteristic, or a combination thereof. The analyst also needs to refer back to the WBS (or CBS) and review how the costs were initially derived and the assumptions that were made at the input stage. In any event, the problem may be

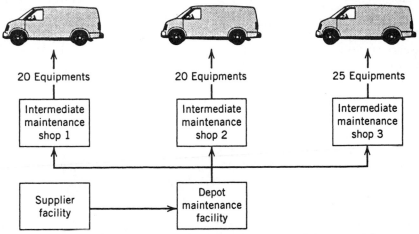

Deployment: Three geographical areas (flat and mountainous terrain)
Utilization: Four (4) hr/day throughout year (average)

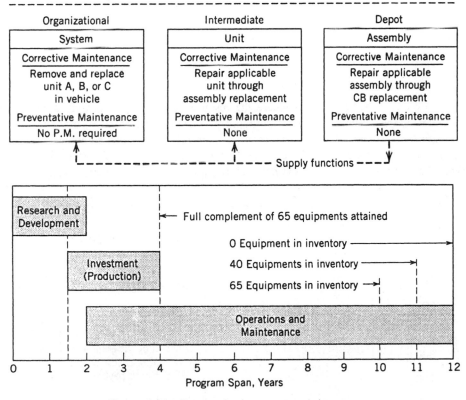

Figure 6.17   Communication system requirements.

TABLE 6.2 Life-Cycle Cost Breakdown (Evaluation of Two Alternative Configurations)

Cost Category	Configuration A		Configuration B	
	Present Cost	% of Total	Present Cost	% of Total
1. Research and development	70,219	7.8	53,246	4.2
a. Management	9,374	1.1	9,252	0.8
b. Engineering design	45,552	5.0	28,731	2.3
c. Test and evaluation	12,176	1.4	12,153	0.9
d. Technical data	3,117	0.3	3,110	0.2
2. Production	407,814	45.3	330,885	26.1
a. Construction	45,553	5.1	43,227	3.4
b. Manufacturing	362,261	40.2	287,658	22.7
3. Operations and maintenance	422,217	46.7	883,629	69.4
a. Operations	37,811	4.2	39,301	3.1
b. Maintenance	382,106	42.5	841,108	66.3
Maintenance personnel	210,659	23.4	407,219	32.2
Spares/repair parts	103,520	11.5	228,926	18.1
Test equipment	47,713	5.3	131,747	10.4
Transportation	14,404	1.6	51,838	4.1
Maintenance training	1,808	0.2	2,125	0.1
Facilities	900	0.1	1,021	Neg
Field data	3,102	0.4	18,232	1.4
4. Phaseout and disposal	2,300	0.2	3,220	0.3
Grand Total	900,250	100	1,267,760	100

traced back to a specific function where the resource consumption is high, a particular component of the system with a low reliability and requiring frequent maintenance, a specific system operating function that requires a lot of highly skilled personnel, or something of an equivalent nature. Various design tools can be effectively utilized to aid in making these "causes" visible, and to help identify areas where improvement can be made,

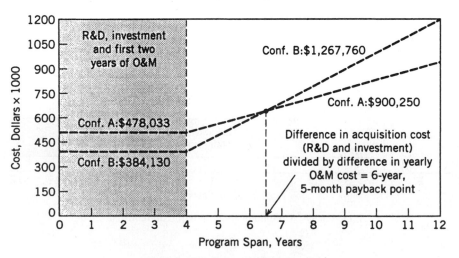

Figure 6.18 Breakeven analysis.

such as, the failure mode, effects, criticality analysis, and the detailed task analysis (Blanchard et al., 1995).

As a final step, the analyst needs to conduct a sensitivity analysis in order to properly assess the risks associated with the selection of Configuration A (refer to Fig. 6.15, block 6). One can challenge the validity of the input data (i.e., the factors used and the assumptions made in the beginning) and determine their impact on the analysis results. This can be accomplished by identifying the critical factors at the input stage (i.e., those parameters that are known to have a large impact on the results), introduce variations over a designated range, and determine the output results. For example, if the initially predicted reliability MTBF value is "suspect," it may be appropriate to apply variations at the input stage and determine the changes in cost at the output. Figure 6.19 illustrates this approach as it applies to the "maintenance personnel" and "spares/repair parts" categories addressed earlier. The objective is to identify those areas where a small variation at the input stage will cause a large delta cost at the output. This, in turn, leads to the identification of potential high-risk areas, a necessary input to the risk management program described in Chapter 3.

In other words, an unreliable and critical component in the system design, a function that to accomplish consumes a lot of resources, or a series of activities in a given process may be responsible for the high costs of system operation and support later on. There may be a number, or combination, of contributing "causes," and it is the objective to identify these causes and establish a priority in terms of those problem areas needing management attention the most. A Pareto diagram, an example of which is included in Figure 6.20, may be used to help prioritize those problem areas where the greatest potential cost savings

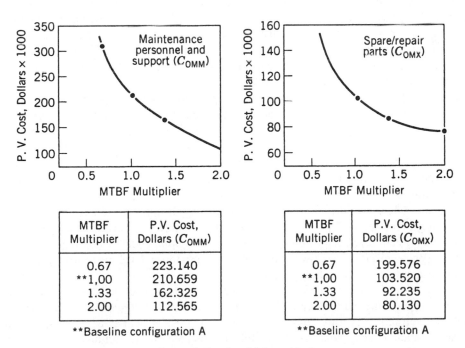

MTBF Multiplier	P.V. Cost, Dollars (C_{OMM})
0.67	223.140
**1.00	210.659
1.33	162.325
2.00	112.565

**Baseline configuration A

MTBF Multiplier	P.V. Cost, Dollars (C_{OMX})
0.67	199.576
**1.00	103.520
1.33	92.235
2.00	80.130

**Baseline configuration A

Figure 6.19 Sensitivity analysis.

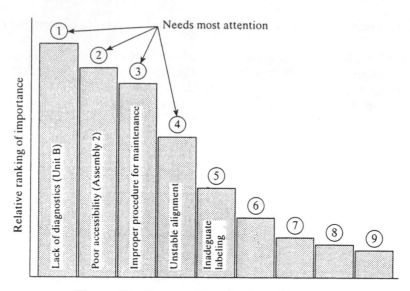

Figure 6.20 Pareto ranking of major problem areas.

can be realized. Each problem, in turn, is subject to further investigation, leading to a possible equipment design change, software modification, and/or the reengineering of a process. Through this iterative approach of identifying high-cost "drivers" (based on total life-cycle cost), relating these to the "causes" through good cost management, and initiating the appropriate corrective action, one can over time realize many benefits. While there is nothing new here relative to the concepts and the techniques presented, we just have not done a very good job in terms of implementation, nor have we addressed the cost issue from a total life-cycle perspective.

Having addressed the cost issue, it is necessary to view the results in the context of the overall cost-effectiveness balance illustrated in Figure 6.21 (an extension of Fig. 6.1). The term, "system effectiveness," refers to the *technical* characteristics of a system as related to the mission or functions that are to be performed. System effectiveness can be expressed in terms of performance, availability, capacity, capability, dependability, reliability, and/ or other comparable measures that may be applicable for the system in question. Operational requirements and the maintenance concept are defined, mission scenarios are described, and TPMs are identified and prioritized through implementation of the system engineering process. The TPMs, which address the technical or performance-related aspects at the system level, can be combined within the system *effectiveness* segment of the balance. Top-level TPMs can then be broken down into lower-level measures through the functional analysis and allocation process.

As one proceeds through the design and evaluation process for both new and/or reengineered systems, there are many tradeoffs that are conducted. Regardless of the type of problem, the ultimate decision-making process must consider both sides of the spectrum; such as, *cost* and *effectiveness*. As an example, the two alternative communication system options discussed earlier are identified in Figure 6.22, both with respect to the specified $20,000 cost goal and the reliability MTBF goal of 450 hours. The shaded area represents

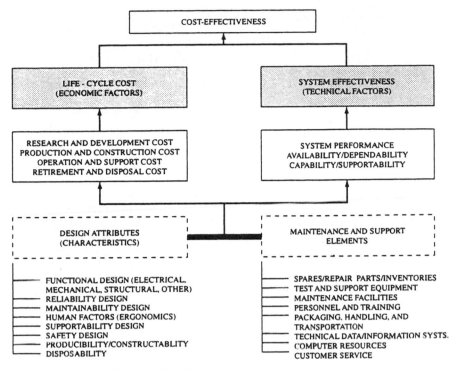

Figure 6.21   The elements of cost-effectiveness.

the allowable design tradeoff "space." The various illustrations in Figure 6.23 present some additional examples of tradeoffs where cost and some measure of effectiveness are considered. In any event, both sides of the balance in Figure 6.21 must be inherent factors when performing cost-effectiveness analyses. The interrelationships between the two are numerous!

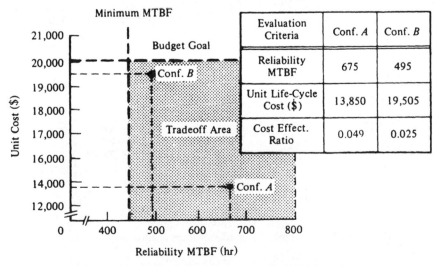

Evaluation Criteria	Conf. *A*	Conf. *B*
Reliability MTBF	675	495
Unit Life-Cycle Cost ($)	13,850	19,505
Cost Effect. Ratio	0.049	0.025

Figure 6.22   Reliability vs. unit life-cycle cost.

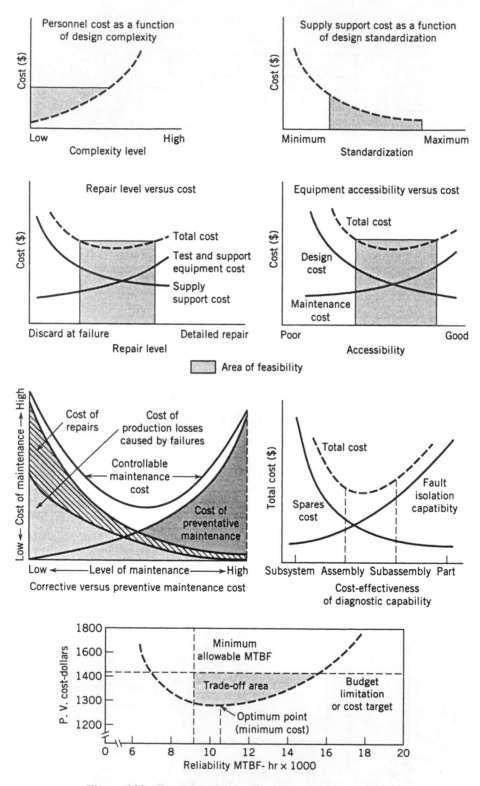

Figure 6.23 Examples of cost-effectiveness analysis applications.

6.7 SYSTEM EVALUATION AND COST CONTROL

Whether one is involved in the development of a new system or the reengineering of an existing capability, cost targets are established in the beginning through a requirements analysis and/or by specifying a benchmark objective. Given the appropriate cost goals, it is necessary to consider these factors along with the other TPMs as *requirements* within the overall management and program planning process described in Chapter 2. Further, cost goals must be included within the systems engineering management plan (refer to Chapter 31), and must be a major factor in the system evaluation process described in Chapter 22.

The aspect of *cost* must be considered as a *design* parameter along with the many other performance-related parameters that are typically addressed and, as such, the day-to-day program management review and control process must incorporate the necessary provisions that will allow for the "tracking" of costs as one evolves through a given project. While this concept is not new, the usual approach for a typical program is to monitor "cost, schedule, and performance" goals, with the aspect of cost being limited only to the short-term "costs to complete" associated with some functions during the early phases of research and development or production, such as, a specific element in the CWBS for a contracted work package. The objective for the future is to "build into" the management process a means for the review and evaluation of system development in terms of the *total* cost objective. This should be inherent within the requirements described in Chapter 29.

In the design and development of new systems, cost may initially be specified as a "DTC" goal (refer to Figure 6.5). Then, as the system definition evolves, life-cycle cost predictions are generated based on the specific configuration being evaluated at the time. The predicted values are compared with the ultimate objective, and the results should be addressed within the formal design-review structure. If "cost" is to be addressed as a *design* parameter, it must be considered and discussed along with the other applicable TPMs in the environment where engineering design decisions are being consummated. Otherwise, there is a tendency to forget about the economic issues in design until it is too late. Figure 6.24 identifies life-cycle cost as being one of the critical TPMs being evaluated as part of a scheduled design review. It can be seen that the projected LCC for the design configuration being evaluated is high, and that there is a risk associated with not meeting the requirement unless corrective action is initiated accordingly.

At the same time, the element of cost must be addressed in the overall management review and program evaluation process. Costs associated with the various segments of the summary work breakdown structure (refer to Fig. 6.10) and allocated to organizational elements (refer to Fig. 6.11) are summarized, and the results are evaluated through the normal program management structure. Once again, the objective is to view the aspect of cost from a life-cycle perspective. Figure 6.25 illustrates the "tracking" of costs from the program management view.

The important issue in total cost management is to first *influence* the design and/or reengineering process to ensure that the system being produced will be cost-effective in terms of its follow-on operation and support, and then to *manage* the acquisition processes to ensure that the initial objectives are met.

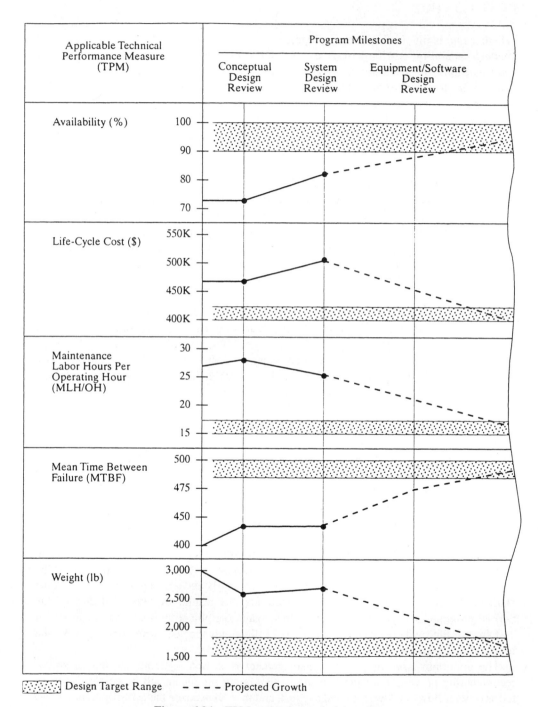

Figure 6.24 TPM evaluation at design review.

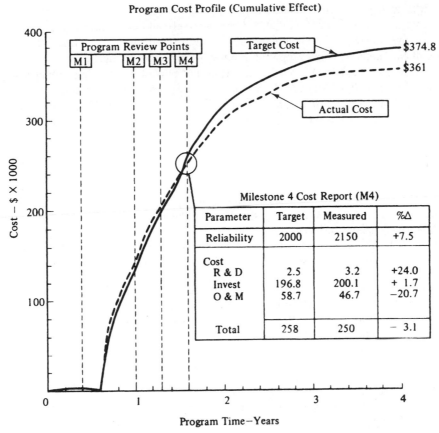

Program Cost Profile (Cumulative Effect)

Figure 6.25   Program cost projections and reviews.

6.8   SUMMARY

The purpose of this chapter is to address the issue of "cost management," not from the usual "short-term" project management perspective, but including the consideration of *total* cost in the day-to-day decision-making processes. If one is to properly assess the "risks" associated with a given decision, then it is essential that the possible consequences be evaluated prior to initiating action(s). In other words, early decisions pertaining to design, the procurement of capital equipment or software, the construction of facilities, and so on, can have a great impact on the overall life-cycle cost of a given system. Good cost management requires that one first *influence* the design and/or reengineering of a system to be cost-effective in the long term, and then to *manage* the acquisition processes in order to meet this objective. This is certainly a goal in implementing the principles of systems engineering as conveyed throughout this handbook.

With this purpose in mind, it is felt that the reader should first address the issue of life-cycle costing and the possible applications for life-cycle cost analyses. Then, it is necessary to view costs from a *functional* perspective, and to apply the principles of activity-based costing in order to gain the necessary visibility for good cost management. Given this, cost-effectiveness analysis methods can be applied both in the design and reengineering of systems to produce a product that can be effectively and efficiently operated and supported by the consumer. The communication system "case study" is included to illustrate the use of the methods being proposed. Finally, it should be recognized that many of the principles and concepts conveyed herein, while not new, are not being properly implemented (if at all). It is believed, however, that these principles and concepts must be implemented in the future *if* we are to survive in this day where resources are dwindling and international competition is increasing.

REFERENCES

Blanchard, B. S. (1998). *System Engineering Management,* 2nd ed. New York: Wiley.

Blanchard, B. S., and Fabrycky, W. J. (1998). *Systems Engineering and Analysis.* 3rd ed. Upper Saddle River, NJ: Prentice-Hall.

Blanchard, B. S., Fabrycky, W. J., and Verma, D. (1994). *Application of the System Engineering Process to Define Requirements for Computer-Based Tools,* Monogr. New Carrollton, MD: Society of Logistics Engineers (SOLE).

Blanchard, B. S., Verma D., and Peterson, E. (1995). *Maintainability: A Key to Effective Serviceability and Maintenance Management.* New York: Wiley.

Canada, J. R., Sullivan, W. G., and White, J. A. (1996). *Capital Investment Analysis for Engineering and Management.* 2nd ed. Upper Saddle River, NJ: Prentice Hall.

Fabrycky, W. J., and Blanchard, B. S. (1991). *Life-Cycle Cost and Economic Analysis.* Upper Saddle River, NJ: Prentice-Hall.

Kerzner, H. (1995). *Project Management: A Systems Approach to Planning, Scheduling, and Controlling,* 5th ed. New York: Van Nostrand-Reinhold.

Kidd, P. T. (1994). *Agile Manufacturing: Forging New Frontiers.* Reading, MA: Addison-Wesley.

Thuesen, G. J., and Fabrycky, W. J. (1993). *Engineering Economy,* 8th ed. Upper Saddle River, NJ: Prentice-Hall.

7 Total Quality Management

JAMES L. MELSA

7.1 INTRODUCTION

If we don't change our direction, we might end up where we are headed.

—Chinese Proverb

It is critical that institutions begin to think about the directions that they are headed and where those directions might lead them if they don't change. Quality improvement and cycle-time reduction are no longer *fads* or *slogans*, but have become the survival issues of the 1990s. This chapter begins with some information on why quality improvements and cycle-time reductions can be critical to business success. The four stages of the quality movement are described with an emphasis on why "inspection," in all of its forms, cannot guarantee outstanding quality. The basic tools and concepts of total quality management (TQM) are presented. Section 7.2 explains how quality programs are no longer isolated aspects of a corporation, but have become embedded into the very fabric of the corporate operation. Stated another way, quality management is about the *quality of the management* not the management of quality. The use of quality systems such as ISO 9000 and the Malcolm Baldrige National Quality Award (MBNQA) are discussed in Appendices 7A and 7B.

The U.S. Government Accounting Office's review (House of Representatives: Management Practices, 1991) of 20 companies that were among the highest-scoring applicants in 1988 and 1989 for the MBNQA indicated that the companies that adopted quality management practices experienced an *overall improvement in corporate performance,* including better employee relations, higher productivity, greater customer satisfaction, increased market share, and improved profitability.

Each of the companies studied developed its practices in a unique environment with its own opportunities and problems. There were, however, common features in their quality management systems that were major contributing factors to improved performance. These features included corporate focus on meeting customer needs, management that led the way, the empowerment of employees to seek continuous process improvement, a flexible and responsive corporate culture, fact-based decision making, and partnerships with suppliers. It is important to note that many different kinds of companies benefited from putting specific TQM practices in place; however, none of these companies reaped those benefits immediately. Allowing *sufficient time for results* to be achieved was as important as initiating a quality management program.

Handbook of Systems Engineering and Management, Edited by A. P. Sage and W. B. Rouse
ISBN 0471-15405-9 ©1999 John Wiley and Sons, Inc.

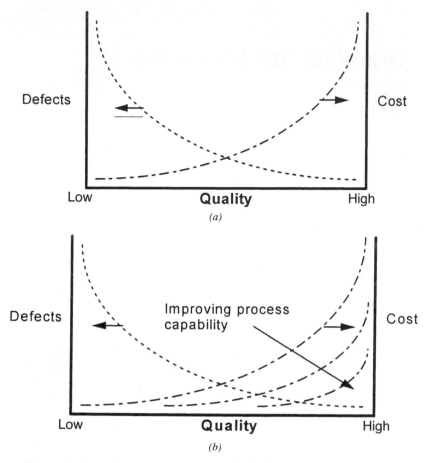

Defects · Cost

Low **Quality** High

(a)

Defects · Improving process capability · Cost

Low **Quality** High

(b)

Figure 7.1   (*a*) Old cost vs. defect tradeoff. (*b*) New cost vs. defect tradeoff.

Quality, unfortunately, is a very misunderstood concept. To many, improved quality means that there must be more inspection. Others believe that the only important quality issues have to do with manufacturing operations. There has been a strongly held belief that quality costs both time and money. Some organizations would argue that if one wants a higher quality product, it will take longer to design and manufacture and it will cost more. Interestingly, *all* of these beliefs have been proven to be wrong. Inspection does not improve quality! And it turns out that the real quality issues pervade the entire organization. Finally, many companies are now demonstrating that high-quality products can be produced more rapidly and at lower cost.

In arguing the tradeoff of quality vs. costs, some people would draw the diagram shown in Figure 7.1*a*. This diagram clearly shows that decreasing defects also increases costs. They would say that it is necessary to select a point in the middle where there is a reasonable balance between these two opposing curves. If one moves too far to the left, a low-cost, low-quality product is produced. Moving to the right side produces a high-priced, high-quality product.

The TQM (new) approach to this tradeoff is shown in Figure 7.1*b*. By improving the process capability, defects will actually decrease while maintaining the same cost or perhaps even decreasing it. This is a very powerful insight and has been one of the driving forces behind the TQM movement.

A study of the Profit Impact of Market Strategies (PIMS) Data Base (Strategic Planning Institute, 1986), which contains financial and strategic information for around 3,000 businesses over a 20-year period, indicates that in the long run, the most important single factor affecting a business unit's performance is the quality of its products and services, relative to those of its competitors. The study goes on to point out that:

- Businesses that offer premium quality products and services usually have large market shares and are early entrants into their markets. The clear conclusion is that quality does not cost time!
- Quality is positively and significantly related to a higher return on investment for almost all kinds of products and market situations. Companies with superior relative quality receive almost three times the return on investment (ROI) as compared with those companies with inferior relative quality (7 vs. 20 percent).
- Product quality is an important determinant of business profitability. High-quality producers can usually charge premium prices for their products. As we will see later, by using the right processes, quality products can, in fact, be made at a lower cost. As a result, it is clear that quality does not cost money!

A 1996 study by the National Institute of Standards and Technology (NIST), the governmental agency responsible for the MBNQA, (see Appendix 7B), shows that quality management can result in impressive financial returns. This is the second study; the first was done in 1995 with similar results (Helton, 1995). NIST "invested" a hypothetical sum of money in Standard & Poor's (S&P) 500 and in each of the publicly traded companies (five whole companies and nine parent companies of subsidiaries) that have won the MBNQA since 1988. The investment was tracked from the first business day in April of the year the Baldrige winner received the award, or the date they began publicly trading, through August 1, 1995. As a group, the 14 companies soundly outperformed the S&P 500 by greater than 4 to 1, achieving a 248.7 percent return on investment compared to 58.5 percent for the S&P 500. The five whole company winners did even better, outperforming the S&P 500 by greater than 5 to 1 (279.8 to 55.7 percent). NIST also conducted a similar investment study for the 41 publicly traded applicants that qualified for Baldrige site visits. That group outperformed the S&P 500 by greater than 2 to 1.

Having decided that quality is good, one is faced with the question of what is good enough. In school, 90 percent was a pretty good grade and generally caused one to receive an A in the exam or the course. Yet people are not happy to receive products that only work 90 percent of the time; for example, they wouldn't be happy to buy a new car that only starts nine times out of ten.

If 90 percent is not good enough, then what about 99 percent. That seems pretty good. However, 99 percent quality still means that one defective product will be shipped for every 99 good ones. However, no one wants the defective part. The following story illustrates this point.

A company was selecting a vendor for a purchase. After the usual negotiations, the Japanese supplier agreed to a 99 percent quality level for the shipment. The shipment was for a 1,000 pieces to be made in three weeks. After exactly three weeks, two boxes were received from the vendor, one large and one small. The supplier was asked why two boxes were sent when there was only one order. They replied, "In the large box, you will find 990 perfect pieces. Since you ordered one defective part for every 99 perfect parts, the 10 defective parts are in a separate box since we were not sure what you planned to do with them."

Customers really do not want their defective parts in a separate box! They do not want them at all. Clearly 99 percent does not seem to be good enough either. Let us look at 99.9 percent! The following shows what doing things right 99.9 percent of the time might lead to in the United States today:

- There would be one hour of unsafe drinking water per month.
- There would be two short or long landings at most major airports each day.
- There would be 16,000 lost pieces of mail per hour.
- There would be 20,000 incorrect drug prescriptions each year.
- There would 22,000 checks deducted from the wrong account each hour.

None of these things sounds acceptable. Even 99.9 percent does not seem to be good enough. Motorola and others have set a goal of six sigma (99.99966 percent) or 3.4 defects per million opportunities for all of their operations. They call this *practically perfect*. It takes incredible commitment and organizational skill to achieve quality at this level, but many of organizations have done it. In fact, some have done even better. The following rough comparisons of 99.9 percent quality with six sigma quality may be instructive to understand the magnitude of this jump in performance.

- In terms of area: if 99.9 percent is equal to the area of a small hardware store, then six sigma is equal to the area of the diamond in a typical engagement ring
- In terms of time: if 99.9 percent is equal to 3.5 months per century, then six sigma is six seconds per century
- In terms of spelling: If 99.9 percent is 1.5 misspelled words per page in a book, then six sigma is one misspelled word in all of the books in a small library.

There are significant costs for not improving. To gauge the cost of problems that are swept downstream to the customer, consider that for every wronged customer who complains, 26 others remain silent. A high percentage of dissatisfied customers will never purchase goods or services from that company again. The average wronged customer will tell 8 to 16 others. It costs about five times as much to attract new customers as it costs to keep old ones.

7.2 HISTORICAL BACKGROUND OF THE QUALITY MOVEMENT

> It is not products but the processes that create products that bring companies long-term success. Good products don't make winners; winners make good products
> —Michael Hammer

It is helpful to look at the evolution of the total-quality movement as a way to understand the degree to which quality has to be inclusive to be effective. In his seminal book, David Garvin (1988) defines the following four stages of quality:

- Inspection
- Quality control
- Quality assurance
- Strategic quality

Garvin points out that it is possible that different activities in an organization are at different stages of quality at the same time.

In the first stage, *inspection*, the focus is on sorting good products from bad or defective products. Once a defective product is identified, a decision is made to either discard it or rework/repair it. This often leads to extensive facilities and resources being committed to the reworking of products or results in a high percentage of scrap from a factory.

The *quality control* stage involves the use of statistical sampling techniques to reduce the amount of inspection. This approach is particularly valuable when it is impossible to inspect each product. Statistical process control is a keystone of this stage; it provides a way to determine if a process is operating properly.

In the *quality assurance* stage, the organization begins to address the quality problem proactively by looking at the root cause of problems. The focus begins to shift from the detection of defects to the prevention of defects. In the hardware design process, for example, this approach leads to the derating or tolerancing of component parts. In this stage, the quality movement begins to involve organizations outside of manufacturing.

The final stage of the quality movement is *strategic quality* or *total quality management*. This stage is characterized by five driving philosophies:

- Quality is defined from the customers' point of view;
- Quality is linked with profitability on both the market and cost side;
- Quality is viewed as a competitive weapon;
- Quality is linked with the strategic planning process;
- Quality requires an organization-wide commitment.

In the strategic quality stage, organizations finally realize that quality must be defined from the customer's point of view. If the customer is not satisfied, then it is not important that internal measurements are suggesting that you are providing a high-quality product. It is critical that the customer's voice be heard regarding what is a quality product. There are no minor defects. If the product does not meet the customer's expectations, then it is defective.

Customers' expectations regarding quality are growing rapidly, and almost every organization is beginning to talk about quality. When customers purchase a television today, they expect it to work. Customers no longer accept having the final assembly done on a car after they have bought it. The quality required by the customer is not a static target either. If your quality is not improving, its deteriorating. An absence of defects is no longer sufficient; customers are now seeking the presence of value.

The strategic quality philosophy also emphasizes that quality is linked with organization profitability on both the market and cost sides. It is accepted that quality products can be priced at a premium. By reducing scrap and all other forms of waste (see Section 7.4.5), many organizations are now finding that quality products can also be made at a reduced cost. It is clear that if one can sell quality products at a premium price and can produce them at the same or lower cost, then profits will be significantly higher.

In each of the three previous stages, quality is viewed as a problem to be solved. In the strategic quality stage, quality becomes a competitive weapon. Quality is linked with the strategic planning process and requires an organization-wide commitment. Quality programs must be included in the complete planning process of the corporation, not added as an afterthought.

Total quality management encompasses both a philosophy and a set of guiding principles that represent the foundation of a continuously improving organization. Total quality management applies both quantitative methods *and* human resources. Total quality management integrates fundamental management techniques, existing improvement efforts, and technical tools. It is important to understand this duality of *philosophy* (soft people issues) and *tools* (quantitative/decision-making methods). Doing one without the other will not be successful. The integration of existing tools and techniques is critical if TQM is to be something other than an overlay program. The next two sections of this chapter elaborate on these two interrelated concepts—tools and philosophy. Related discussions may be found in Chapters 15 and 23.

7.3 TOTAL QUALITY MANAGEMENT TOOLS

If the only tool that you have is a hammer, then every problem will look like a nail.

—Anonymous

There are a wide range of TQM tools (Brassard, 1989), but the size of this chapter does not permit a detailed discussion of them along with appropriate examples. The following is a list of widely used tools. There is no tool that is best for every application; the knowledgeable practitioner is aware of a rich variety of tools and uses the appropriate one(s).

- *Process Maps.* One of the important keys to understanding how to improve a process is to map the process. While there are several different approaches to process mapping, the key is to determine who does what at each step of the process. Often, the simple drawing of a process map is sufficient to solve many quality problems because the map makes it so obvious where defects can be introduced.
- *"Poke-A-Yoke".* The concept of this Japanese management philosophy is to make a process foolproof. The idea is to design the process in such a way that it is self-checking or that process steps are included that cause immediate detection and pos-

sible correction of any defect. Simple examples include color coding and special keying of parts to ensure that they are assembled the correct way.

- *Statistical Tools.* One of Deming's major contributions to the quality movement was the introduction of statistically grounded approaches to the analysis of defects. Without the use of these tools, one can often make incorrect decisions regarding the cause of a problem (see Appendix 7C). This can often lead to exactly the opposite effect of the one being sought. Included in this set of tools are statistical process control (SPC) charts, Pareto charts, and histograms.

- *Force-Field Analysis.* This tool asks one to diagram the forces (policies, culture, and so forth) that are resisting a desired change and the forces that support the change. This assists one in clearly determining the degree of difficulty of making the change and exactly where effort will be needed. The supporting forces are places where assistance can be expected.

- *Root-Cause Analysis (Five Whys).* This tool was popularized by the Japanese. It consists of asking a series of questions (whys) until one uncovers the root cause of a defective product. The objective is to determine why a defective product was produced; this is to be contrasted with the usual approach of just fixing the defective product or replacing it.

- *Fishbone Diagram (Ishakawa Diagram).* This tool is also called a cause-and-effect diagram. It is used in a brainstorming session to examine factors that may influence a given situation or outcome. The causes are often grouped into categories such as people, material, method or process, and equipment. The resulting diagram takes the shape of a fishbone, hence the name.

- *Loss Functions.* In many manufacturing situations, one creates tolerance limits for a product. Products that fall outside of the limits are defective, and those that are inside the limits are deemed good. Several difficulties arise with this approach. First, there is always the temptation to reclassify products that are just outside the limits into the acceptable category, especially if there is a great push for quantity. Second, and perhaps more important, the accumulative effect of several parts that are all on the extreme limits of acceptability may lead to defective performance. The loss function tool is used to recognize that there is a cost associated with any deviation from the ideal value.

- *The Plan-Do-Check-Act (PDCA) Cycle.* This tool is also known as the Shewhart cycle. It was popularized in Japan by Deming; as a result the Japanese refer to it as the Deming cycle. The tool emphasizes a new plan for change. It carries out tests to make the change on a small scale, observes the effects, and finally studies the results to determine what has been learned. The cycle is repeated as needed.

- *Brainstorming.* This process has become a staple of the TQM movement. The concept is to invite participants to suggest "solutions" to a problem without any evaluation of the usefulness or correctness of their ideas. Several approaches are possible, including open suggestions, rotating suggestions, or blind suggestions. There are several computer tools that have been developed to assist in this process. After a fixed period of time, or after all suggestions have been made, there is discussion of the "value" of the suggestions.

- *Affinity Diagram.* The affinity diagram tool is used to organize large amounts of nonquantitative (ideas, opinions, issues, etc.) information into groupings based on natural relationships between the items. It is largely a creative rather than a logical process. In a very loose sense, the affinity diagram does for ideas what statistics do for numbers, namely, extract meaning from raw data. The affinity diagram process is often used with the results of a brainstorming session to organize the resulting ideas.

- *Interrelation Digraph.* This tool takes complex, multivariable problems, or desired outcomes, and explores and displays all of the interrelated factors involved. It graphically shows the logical and often causal relationship between factors. It is often used in conjunction with the results of an affinity diagram exercise to seek causes and effects in order to determine why corrective action needs to be applied.

- *Tree Diagram.* This tool is used to systematically map out in increasing detail the full range of paths and tasks that need to be accomplished to achieve a primary goal and every related subgoal. Graphically, it resembles an organization chart or family tree.

- *Prioritization Matrices.* Prioritization matrices are one of a group of decision-making tools that help to prioritize tasks, issues, or possible actions on the basis of agreed-upon criteria. While these tools cannot make decisions, they can help to ensure that all factors are evaluated and that logical decisions are reached.

- *Activity Network Diagram.* This class of tools includes a wide range of project management tools used to plan the most appropriate schedule for a complex project. Typical examples are Gantt charts and PERT charts. These tools project likely completion time and associated effects and provide a method for judging compliance with a plan. Several excellent computer programs exist for automating the work associated with this class of tools.

7.4 TOTAL QUALITY MANAGEMENT PHILOSOPHIES

If most of us are ashamed of shabby clothes and shoddy furniture, let us be more ashamed of shabby ideas and shoddy philosophies.

—Albert Einstein

As noted earlier, total quality management must become the management method! The TQM philosophy involves nine interrelated concepts:

1. Management must lead the way;
2. Focus attention on meeting customer needs;
3. Ensure simultaneous improvement of quality and reduction in cycle time;
4. Concentrate on prevention rather than inspection;
5. Create an environment of continuous process improvements;
6. Make decisions based on data;
7. Use benchmarking to set stretch goals;
8. Empower employees;
9. Utilize cross-functional teamwork.

The following nine subsections develop each of these items in more detail.

7.4.1 Management Must Lead the Way

We are what we repeatedly do. Excellence, then, is not an act but a habit.

—Aristotle

As noted earlier, quality management is about the quality of the management not the management of quality. Management's role appears at the top of this list because it is the most critical element. Without direct and meaningful involvement of the leadership, some progress can be made on the quality journey, but it will be limited and local.

As one embraces the TQM philosophy, it is necessary to forget old ideas about leadership. The most successful corporations of the 1990s will be something called *learning organizations* (Garvin, 1993). The ability to learn faster than your competitors may be the only sustainable competitive advantage. Leadership in the TQM organization will deal with how to learn quickly and how to encourage others to do so.

It is critical that we remember that leadership is not about what you know, but what you do with what you know. Leaders in TQM organizations must learn that they are part of the problem and that the organization is not likely to change until the leaders do. Too often, the leaders set about changing their organizations, but do not change themselves. Employees watch the leaders' feet not their mouths; the leaders' actions must match their words. Some call this "walking the talk." An organization reflects the leaders' behavior. Everyone is trying to figure out how to do what the leaders want or would do.

Extorting employees to improve performance, without addressing the method to be used, will be fruitless, frustrating, and counterproductive. Some leaders feel that the way to improve quality is to put up banners extolling employees to reduce defects by 3 percent or by declaring a zero defect day. Deming (Walton, 1986) stated that more than 85 percent of the defects that are produced by employees are due to the system in which they operate, not the employee. It is the job of the leader to create the system so that the employees can achieve the desired objectives.

A simple story may illustrate the wisdom of this teaching. A farmer does not beat up on his corn because it did not perform. Now, everyone knows the performance objectives for corn. Surely the corn knows what it is supposed to do: The corn should be knee-high by the Fourth of July and the corn should be as high as an elephant's eye by the end of the season.

The corn does not always achieve these objectives! But the farmers do not attack the corn, because they are systems thinkers and understand that much of the corn's performance is related to system issues such as the way that the corn was planted and cultivated (the methods), the weather during the season (the environment), the soil in which the corn was planted (the culture), the fertilizer used (rewards and recognition), and the selection of seeds (recruiting). None of these aspects of the systems were controlled by the corn plant and yet all aspects affect its performance.

The role of leadership is to set the vision using profound knowledge or systems thinking (Senge, 1990). Leaders must tell employees what is going to be different, for what people, and at what cost. Finally, they must pull the organization into the future by getting it to understand what it can be.

7.4.2 Focus Attention on Meeting Customer Needs

Quality is having customers that do return and products that don't.

—Stanley Marcus

Customers now recognize quality as the presence of value, *not* the absence of defects. Mere adherence to specifications isn't enough. Customer loyalty is created by customers who are excited by the performance of the product, not just satisfied. Customer requirements can be classified into three categories. First, there are *implicit* specifications. These are requirements that are generally not stated by the customer, but the customer will be very unhappy if they are not provided. For example, if a group of people are asked for important specifications on a new car they are going to buy, almost no one will say brakes. Yet, these customers would be very dissatisfied if the car did not have brakes. While the implicit requirements for a car may seem clear, they can become very difficult to determine when one begins to address more complex relationships regarding customer service. This class of requirement is often known as "dissatisfiers," because one can only dissatisfy a customer if one fails to provide for all implicit requirements. The provision of implicit specifications is expected and does not assist with customer satisfaction.

The second class of customer requirements is *explicit* specifications. Here the customer explicitly states what is wanted. To continue the car-buying example, explicit specifications might include a CD player or special color. Once again, there is little opportunity to create customer satisfaction because there is a clear expectation that these requirements will be met. There is also little room for competitive advantage, since all competitors are aware of these needs.

The final class of requirements is known as *exciting* features. These are things that the customer did not expect but is delighted to find included. At one time, features such as automatic transmission, air conditioning, and cup holders were exciting features. The presence of exciting features in both the product and the related service can be a source of significant customer satisfaction. The difficult part about exciting features is that they quickly become explicit and even implicit requirements. Stated in another way, unanticipated quality over time becomes expected quality. This, of course, ensures that there is always a need to develop new exciting features to continue to drive customer satisfaction.

Institutions that have fully embraced the TQM philosophy establish true partnerships with customers. In these partnerships, they work with their customers to discover how they can contribute to the customers goals. It is critical that they work as hard for their customer's success as they work for their own. It is critical that the organization begins to believe that the customer is the *boss* rather than the boss being the boss.

Since the high percentage of customer interactions are at the front line, many organizations have used the process of inverting their organization charts. In this case, they show customers and employees who have direct contact with customers at the top of the chart and the "high-level" management at the bottom, playing supporting roles. In a competitive environment, the difference between winning and losing is service. Service is rendered by those serving the customer. All others must support those who serve the customer. Managers do not run the corporation, the people who serve customers do. Therefore, managers' main functions are to offer support, encouragement, and resources to the people who indirectly or directly serve customers. Organizations will fail in rendering and sustaining superior customer service unless those at the top truly recognize and appreciate those who render service to customers. Furthermore, it is critical that leadership does not tolerate those in the organization who do not share their feelings—not for a millisecond.

Studies by Heskett (1987) and others have revealed a three-step value chain relating employee satisfaction to company profits:

- Employee satisfaction is essential to customer satisfaction;
- Customer satisfaction is essential to customer loyalty;
- Customer loyalty is essential to company profits.

It is important to note the nature of the cause-and-effect relationship implied in this value chain. Customer satisfaction will not occur without employee satisfaction; however, it is possible to have satisfied employees, at least in the short run, that do not satisfy customers. For example, one can pay employees twice the normal salary and ask for very little work. This may create many satisfied employees, but it will surely not satisfy customers.

Employees must believe that they can deliver satisfaction/value to customers; employees must feel that they have significant authority to act on behalf of customers. The lack of this feeling is a main reason for high employee turnover in many service industries. Technology (see below), if used properly, can be important, but the real key is giving the employees the central role in setting priorities to meet customer needs. While many organizations require first-level employees to ask their supervisors if they can say "yes" to a customer's request, some of the leading organizations now require that first-level employees ask permission if they plan to say "no" to a customer request.

The following are reasons that customers stop dealing with an organization:

- 3 percent move away
- 5 percent develop other relationships
- 9 percent leave for competitive reasons
- 14 percent are dissatisfied with product
- 68 percent are treated indifferently by their current supplier

Note that the huge percentage (68 percent) of customers stop dealing with an organization because they feel that the organization does not pay attention to their needs.

Customer satisfaction is essential to customer loyalty, that is, customers will not become loyal if they are not satisfied. Stated in another way, satisfied customers are necessary, but not sufficient for success. It is necessary to move beyond just meeting expectations in order to create loyal, excited customers. Customer loyalty demands that one not only meet the customers' needs, but also anticipate them. This requires that organizations determine how they can address the exciting expectations of customers so that they will say, "I cannot believe that you really did that!"

Organizations can create loyal customers, or "raving fans," (Blanchard and Bowles, 1993) by understanding what drives customers' satisfaction. This requires that one focus on eliminating all dissatisfiers (see below), making sure that all satisfiers are fully addressed, and finding a continuing array of exciters. In customer relations, individuals can make a big difference. It is important to remember to serve hundreds of customers one person at a time. "Raving" fans talk to others and become true partners.

Finally, loyal customers are a key to profits. There is significant leverage in increasing this group by even a small amount. Some data suggest that 20% of an organization's customers generate 240% of the profits. The other 80% create negative 140% of profits. Other studies (Reichheld and Sasser, 1990) on the effects of customer loyalty have shown that even a 5% increase in customer retention can raise profitability by 25% to 85%.

Figure 7.2 shows a typical hierarchy of customer service needs. If the elements at the

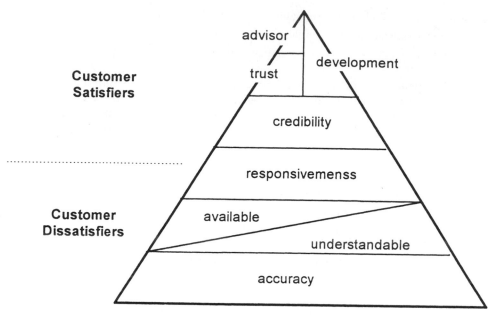

Figure 7.2 Hierarchy of service needs.

lower levels, such as accuracy, are missing, they create customer dissatisfaction. Their presence has little or no effect on customer satisfaction, however, while the presence of those elements on the top levels can have significant effect on customer satisfaction.

Aggregation is generally deceiving; good customer satisfaction is driven by good one-on-one interaction, not by average performance. Global issues also complicate the issue of customer satisfaction. Different cultures often bring different expectations.

Technology may provide a way to provide better service with fewer employees. Many organizations are using knowledge systems to integrate people, process and technology in order to improve customer service. In many cases, the knowledge already exists in the organization, but it is not available to the right people or at the right time or place. Knowledge systems can be used to disseminate specialized knowledge available to everyone involved in a process. It must be remembered that while technology and systems can play valuable support roles to people, the best systems are useless without people. Before a knowledge system is deployed, it is critical that the customer services processes have been analyzed to ensure that the needs of customers are being addressed. Otherwise, one may find that technology has been used to automate a nonresponsive system and customers are being given messages like, "The computer does not allow me to get that information for you." This is not a very satisfactory answer to a customer's inquiry.

7.4.3 Ensure Simultaneous Improvement of Quality and Reduction in Cycle Time

"If you knew time as well as I do," said the Mad Hatter, "you wouldn't talk about wasting it."

—*Through the Looking Glass*, Lewis Carroll

Perhaps the critical element of this component of the TQM philosophy is to recognize that simultaneous improvement in quality and reduction in cycle time can be and has been done. This philosophy has also been called *high-velocity performance*. Many of the most successful companies have found that quality improvement and cycle-time reduction are really two sides of the same coin. They argue that to reduce cycle time, one must do things right the first time—a critical element of quality improvement. In a similar fashion, reducing cycle time can only be achieved if one does not need to have inspection and rework activities. In a simplistic way, if you are going to do something faster, then you have less time to correct mistakes. It has been very clearly established that things that are done right the first time are better than things finally made right by repeated inspection and correction.

While the early successes in cycle-time reduction and simultaneous quality improvement were accomplished in manufacturing using tools such as just-in-time (JIT) (Schonberger, 1986) and lean manufacturing (Womack et al., 1990), these tools achieve their greatest leverage in white collar activities, *not* manufacturing. In particular, significant success has been noted in accelerated product design (Smith and Reinertsen, 1991) and service redesign (Hammer and Champy, 1993).

To make the changes needed, it is important to shift from a *cost-based* to *time-based* mindset. For example, many organizations keep detailed accounting records regarding the cost of a product design, but pay little attention to the time required. Yet, early delivery of a product to the marketplace may add significantly to the overall profitability of the product.

Time is the ultimate nonrenewable resource: there is no such thing as too little time, there is just a question of how well time is used. In the new world order, it is no longer the *haves* and *have-nots*. Today the difference between success and failure on a global scale depends on whether an organization is *fast* or *slow*.

Time-based strategies are becoming the competitive advantage. Many companies previously competed on cost, many now compete on quality, and the next frontier is to compete on time! Companies such as Hewlett-Packard, Motorola, General Electric, Toyota, and Honda already do so! As the world's main system for producing wealth speeds up, companies that wish to sell will have to operate at the pace of those in a position to buy. This means that slow enterprises will have to speed up their responses or find they will lose contracts and investments. Some will drop out of the race entirely.

When an organization starts to reduce cycle time, it uncovers problems such as time needed for machine setup. These are called rocks, in the JIT vernacular, in the sense that lowering the water in a pool uncovers hidden rocks. One approach to such problems is to add buffering or to increase lot size. This amounts to adding water to cover the rocks. The high-velocity performance approach is to use process improvement to eliminate the root cause *(remove the rocks)* of the longer cycle time. As one continues to lower the water, new rocks are uncovered and must be similarly removed.

There are several competitive advantages of high-velocity performance. The important positive relationship between high-velocity performance and quality has already been noted. In order to reduce cycle time, one must do things right the first time.

High-velocity performance is also very important to market responsiveness. The ability to respond to newly determined customer needs greatly enhances customer satisfaction. It allows input from customers in a more timely way, and it significantly improves the stability of product definitions. It is much easier to predict what a customer will want 18 months from now rather than 3 years. And, of course, the first company to market has a clear leadership advantage.

Competitive stature is closely related to ability to learn. Shorter cycle time gives many more opportunities to learn. Every time a project is completed, one learns something about the process used. The more cycles per unit time, the more chances to learn and improve. Hence high-velocity performance leads to greater ability to get better faster; the result is a positive upward spiral of performance.

High-velocity performance also drives improved profitability. Early entrants into a market are often able to set the pricing, which can often be high relative to mature life-cycle pricing. When competition does catch up and begins to drive down prices, the early entrants are further down the manufacturing experience curve and can usually continue to be more profitable than the competitors. As indicated earlier, early entrants to a market usually enjoy larger-than-average market share and longer product life cycles.

7.4.4 Concentrate Prevention Rather than Inspection

An ounce of prevention is worth a pound of cure.

—Benjamin Franklin

As Figure 7.3 illustrates, prevention has a great deal more leverage than inspection. Deming states this concept as one of his famous 14 points: "Cease dependence on mass inspection" (See Appendix 7C). It has been estimated that the cost of fixing a defect, after a design is complete, is 10 to 100 times the cost of fixing the defect if it is found during the design process. If the defect escapes from the organization, is found by the customer, and needs to be fixed, then the cost is 10 to 100 times the costs of fixing the error before the release of the product. In other words, the cost of fixing an error that is found by a customer is two to four orders of magnitude greater than the cost of preventing the defect. None of these costs reflects the loss of customer loyalty (see Section 7.4.2).

Quality must be designed into a product or service rather than inspected into a product or service through a series of testing cycles. Many managers will state that they know that a design review or simulation is a good idea, but they did not have time for it. It is critical to break this "cycle of poverty." A manager who fails to provide resources and time for prevention activities is practicing false economy.

The old quality control paradigm relied on inspection to eliminate defective parts. Ineffective practices were left in place while more and more effort was put into better inspection practices. To improve quality, one would add inspectors and related inspection stations. Often one found one inspector inspecting the results of another inspector. Sometimes the process continued through several iterations. Of course, people were clever (or perhaps not clever) enough to make sure that these inspection points were separated in

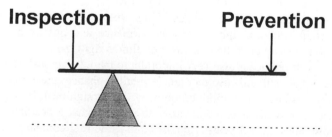

Figure 7.3 The leverage of prevention.

time and/or space so that the lunacy of this concept was not obvious. Even with all of these efforts, defective products were shipped. In fact, one can never get truly good quality via inspection; there will always be test escapes and no testing cycle can be foolproof. If it is not possible for an organization to make a product without defect, then it is highly unlikely that an organization can be expected to inspect with perfect accuracy.

Focus a little bit more on the use of inspection as a quality approach (Womack et al., 1990). The usual mass-production practice is to pass on errors to keep the line running. Every worker can reasonably think that errors will be caught at the end of the line and that any action that stops the line likely to be disciplined. The initial error, whether a bad part or a good part improperly installed, is quickly compounded by assembly workers farther down the line. Once a defective part is embedded in a complex product, an enormous amount of rework may be needed to fix it. Because the problem is not discovered until the very end of the line, a large number of similarly defective products have been built before the problem is found.

Software development offers a similar lesson. Many organizations believed that they could deliver software products with high quality if they increased the testing of the product after it was complete. However, they quickly found that any reasonable amount of testing, and in some cases even unreasonable amounts, did not ensure that the product would be delivered with high quality. Instead, they found that they could only truly improve the quality of the finished product by careful design of the software development process with decreased emphasis on final-stage testing.

7.4.5 Create an Environment of Continuous Process Improvement

Insanity is doing the same things the same way and expecting different results.
—Roger Miliken

Process improvement is concerned with the elimination of waste. Here *waste* is identified as anything that does not add value for the customer. Identification of waste is a critical first step. Many of the things that people think of as essential are, in fact, waste. Waste includes movement, inspection, storage, as well as the more common scrap. It is important to remember that the movement and storage of knowledge may be waste, just like the movement and storage of physical things.

View every activity as a *process* that can be improved. Employees must be educated in the tools of process improvement and expected to use them as a matter of course (see Section 7.3). Leadership is critical here; if the leadership does not provide an example, the impact will be short-lived. There is no finish line in the run for perfection. Fact-based decision making using carefully chosen metrics and benchmark information is critical (see Sections 7.4.6 and 7.4.7). The metrics are used to determine if real improvement is being made, while the benchmark data can drive breakthrough innovations and stretch improvements.

One must not only ask how to do a job better, but also whether the job needs to be done at all (Hammer, 1990). It is never of great value to do something with quality that did not need to be done. Chris Argyris (1991) addresses this issue by what he calls two-loop learning. In Argyris's thinking, thermostats are one-loop learners; they keep the system running as specified. A two-loop learning thermostat would ask if the temperature was set correctly. Many individuals have been trained to be one-loop learners; these people will unfailingly do what they are told to do. They will never question whether what they are doing is adding value for the customer.

Success can often be an enemy! The very things that created success yesterday may prove to be the downfall in the future. So what should we do? Peter Drucker (1993) provided some valuable insight when he wrote that "every organization has to prepare for the abandonment of everything it does." This is a pretty unequivocal statement. Drucker went on to say that every organization has to learn to ask of every process, every program, every procedure, every policy: "If we did not do this already, would we begin it now, knowing what we now know?" If the answer is *no,* the organization has to do something.

7.4.6 Make Decisions Based on Data

How would you run a high jump competition without metrics?

—Anonymous

The key principle of making decisions based on data is, "If it is not measured, it will not improve." However, if attention is not given to the measurements, it will not improve even if it is measured. An organization cannot deal with too many metrics at one time and effectively improve them all. Metrics must be cascaded throughout an organization in order to meaningfully engage all employees in the improvement process.

Measurement must precede improvement. Activity-based improvement programs have failed to achieve intended results because they are not keyed to specific results and often are too large scale and diffused. The results may not be integrally linked to the activity. Activity measurement is confused with results. Activity programs are based on "involvement" rather than results. Results-oriented change programs, keyed by measurements, do produce results and sustain the improvement process. It is critical that management pay attention to the measurements since reinforcement energizes improvement. Empirical results help to sort out effective activity from ineffective activity. Managers must participate in prioritizing targeted goals.

Individuals cannot effectively address more than a few key metrics at one time. In Tom Peters' (1987) words, "More appropriate measurement is achieved with fewer measures." It is critical that these metrics be related to important customer and organizational issues. These metrics might include customer satisfaction, effectiveness of planning, product realization, product deployment, and financial performance. A word of warning may be appropriate. Measurement is always prone to suboptimization; in an optimized organization, no part may be performing optimally. Every part of the organization doing its best may not lead to optimal performance. Individuals inherently see the world through our senses. As a result, they have limited ability to perceive changes that take place over extended periods of time or space.

Once a set of critical metrics is selected for the organization, these metrics must be cascaded to all levels of the organization. Key corporate metrics should be identified and ratified at the leadership level. Stakeholder and owner groups should be identified for each key metric area. Owner groups should define decomposition of each key metric so that each contributing organization understands its contribution to the corporate performance.

7.4.7   Use Benchmarking to Set Stretch Goals

If you know your enemy and know yourself, you need not fear the result of a hundred battles.
—Sun Tzu

Benchmarking enables the best practices from any industry to be creatively incorporated into the processes of the benchmarked function. It can provide stimulation and motivation to the professionals whose creativity is required to perform and implement benchmark findings. Benchmarking breaks down ingrained reluctance to change. Benchmarking may identify a technology breakthrough that would not have been recognized, and thus not applied, in one's own industry for some time to come. Finally, those involved in the benchmarking process often find their professional contacts and interactions from benchmarking are invaluable for future professional growth.

David T. Kearns, former CEO of Xerox, defines benchmarking as ". . . the continuous process of measuring products, services, and practices against the toughest competitors or those companies recognized as industry leaders." This definition points out four important aspects of benchmarking. To be most effective, benchmarking should be a *continuing* process, not a one-time activity. Benchmarking is concerned with *measuring,* that is, gathering specific numerical data; benchmarking needs to focus on *all activities* of an organization, its products, its services, and its processes or practices. In fact, it is the benchmarking of practices that is often the most valuable. Finally, benchmarking should be done with organizations that are *recognized leaders.*

Benchmarking is not a mechanism for making resource reductions. It is possible that reductions may occur because many operations do not emulate best practices. On the other hand, benchmarking may require resource increases. Benchmarking is not a panacea or program. It must be an ongoing management process with a structured methodology. To be successful, benchmarking is almost always a two-way exchange of information. In order for an organization to share information, it will expect to receive similar information. As the degree of trust between the two organizations grows, the level of sharing can become quite high.

Benchmarking is not a cookbook process that requires only looking up ingredients and using them for success. It must become a discovery process coupled with a learning experience. Benchmarking requires observing what the best practices are and projecting what performance should be in the future.

More and more companies are finding that benchmarking is not a fad, but is, in fact, a winning business strategy. It assists managers in identifying practices that can lead to superior performance. Benchmarking is a new way of doing business that forces an external view to ensure objectivity. It forces constant testing of internal actions against external standards of industry practices. It promotes teamwork by directing attention to business practices to remain competitive rather than personal, individual interests. It removes subjectivity from decision making. By creating proof that things can be much better, it makes it possible to set goals that are high enough that they cannot be achieved by doing business the same old way.

There are several different types of benchmarking activities. One can do *internal* benchmarking, in which comparisons are made between the best internal operations. This is often a good way to initiate benchmarking activities since it seems less threatening and is generally less expensive than other types of benchmarking. It can be particularly effective if there are internal organizations that are recognized as performing well above the organizational average.

In *competitive* benchmarking the objective is to study specific competitor-to-competitor comparisons for the product or function of interest. For example, one might compare the functionality of a competitor's product to one's own. This type of benchmarking is perhaps the most difficult because it requires a degree of cooperation and trust between organizations that are direct competitors. Amazingly, many organizations have found that competitors are willing to share information on a quid pro quo basis. They have also found that with careful management and well-understood rules, competitive benchmarking can be very successfully deployed.

For *functional* benchmarking, comparisons are made to similar functions within the same broad industry or to industry leaders. A typical example of this type of benchmarking might be safety or on-time performance of airlines. Here, it is often possible for independent third parties to collect the information and, in some cases, remove potentially sensitive data. This has the potential of making organizations more willing to share information. Unfortunately, when third parties are used, it is often difficult to get much information regarding processes. The consequence is a measurement of what was accomplished without much indication of how it was accomplished.

Finally, *generic* benchmarking is based on comparison of business functions or processes that are the same regardless of industry. Here, for example, a company unloading and loading an airplane at an airport might study an Indianapolis 500 pit crew. While this type of benchmarking requires some difficult and creative thinking, it can often be very productive in finding completely new approaches that can become competitive advantages. Because there is little or no competition between the two organizations, there is usually no problem with open exchanges of information.

Figure 7.4 illustrates a typical benchmarking process. Note that the benchmarking of metrics provides information on how much and where changes are needed. However, the benchmarking of practices provides the process information needed to make changes. For example, it is interesting to know that someone can run a four-minute mile and you can only run a mile in eight minutes. It is much more valuable, however, to understand the training methods (process) used; that information gives you some knowledge of how to change performance.

To begin a benchmarking process, organizations need to identify what is to be benchmarked. The choice of benchmarking targets should be driven by strategic business drivers. There is little value in benchmarking a nonstrategic function. Next, identify comparative companies both within and outside the industry. This selection is also critical and creativity is important. Finally, determine a data-collection method and collect data. Some trial and error may be needed here, especially when an organization is new to the benchmarking process. In general, collecting very targeted information is more useful than a generic "fishing" expedition.

After data have been collected, the next step is to analyze those data. The goal is to determine if a performance "gap" currently exists, either positive or negative. Because performance changes over time, it is important to attempt to develop sources that update the analysis through periodic reviews. It is also important to project future performance levels. Meeting the competition's current performance level a year from now will be of little value if they have made significant performance enhancements.

In some cases, one will find that their organization's performance is the best-in-class. This is not necessarily a time to do nothing. Rather, it may be an opportunity to extend a

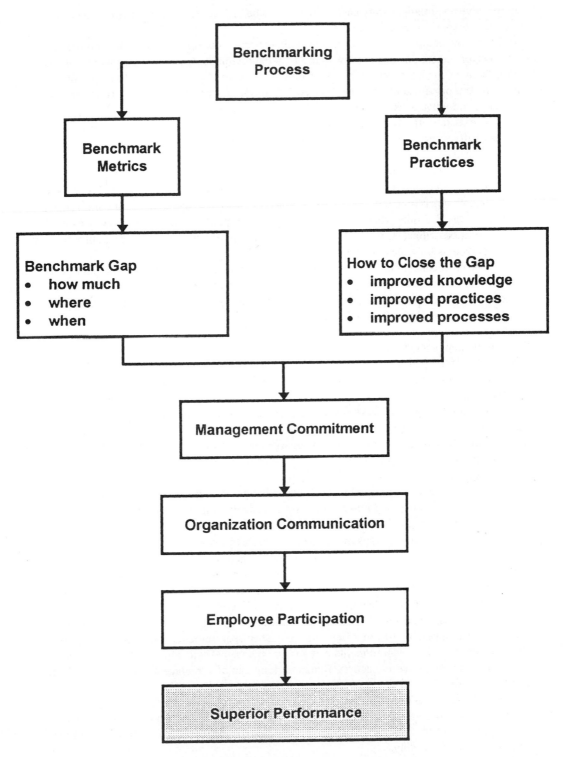

Figure 7.4   Generic benchmarking process.

competitive advantage. In any case, one needs to be vigilant lest the competition meet and exceed one's performance level.

If the conclusion is that the competition is better or that one has not achieved best-in-class performance, one needs to examine why the competition is better and what one can learn from them. Next, one needs to determine how to apply what has been learned to improve the organization. It may also be helpful to determine if the gap is wider than expected and what is likely to happen if no action is taken.

The next steps are to communicate benchmark findings and to gain acceptance of the extent of the gaps. These data are used to establish functional goals for the organization and to weave these goals into the formal planning processes of the organization. The most critical first step is to gain senior leadership's acceptance of the results of the benchmarking analysis and its commitment to develop action plans for improvement. This may well take more than one iteration and, on occasion, a need to go back and review data and related analyses. This is a time when a knowledge of the competitor's practices, as well as their results, will be critical. It is much easier to accept the performance of a competitor once one understands how they achieve those results.

Once senior leadership has committed, it is appropriate to develop an action plan using any variety of problem-solving techniques. It is important that one continue to question whether the plan is a realistic one for closing the gap. The plans should have measurable goals and targets clearly established.

Armed with an action plan, complete with measurable goals and targets, the next step is to implement specific improvement actions and monitor progress. The action plan may need to be developed in more detail so that it can be effectively deployed to all related levels of the organization. It is critical that an employee reward system be linked with the implementation.

Periodically the benchmarks need to be updated in order to ensure that the plan is appropriately closing the competitive gap. If an appropriate foundation has been laid, this update can often be considerably less extensive than the original benchmarking activity. This update will also help to assess whether the projection of future competitive performance is accurate.

The maturity level is achieved when a leadership position is attained and benchmarking practices are fully integrated into the organization's planning and management processes. The maturing level also indicates that the organization understands what is required to remain the leader and has plans in place to accomplish this goal.

There are several reasons for conducting benchmarking activities. First, benchmarks can help to define customer requirements. Without benchmarking, these requirements are often based on history, gut feelings, or internal perceptions. With benchmarking, one can begin to understand market reality, namely, how well others, including competitors, are at addressing customers needs. For example, without benchmarking, it is possible to believe that no one is meeting customer delivery or quality requirements. The existence of data that indicate that these requirements can and are being met will often make it possible to create appropriate corrective action. Benchmarking can also replace a gut feeling with an objective evaluation.

Second, benchmarking can help to establish effective goals and objectives. Without benchmarking, these goals can lack external focus and be reactive, leading to performance that lags the industry. With benchmarking, goals can become credible, unarguable, proactive, and industry leading.

Third, benchmarking can help to develop true measures of productivity. Without benchmarking, it is possible to spend a lot of time pursuing pet projects, and the sources of strengths and weaknesses are not understood. The path of least resistance may be chosen rather than one that leads to true superior performance relative to the competition. With benchmarking, one can focus on solving real problems and understanding outputs. Plans and objectives are based on best industry practices.

Finally, without benchmarking, competitive practices are often internally focused and change is slow and evolutionary. There is often a lack of commitment to improvement. When the realities are fully understood through benchmarking, a concrete understanding of competition develops. One finds new ideas of proven practices and technology. The result is a high commitment to the changes needed to reach world-class or best-in-industry performance.

The knowledge of industry best practices can be a strong driver for change. Without benchmarking, people are often hung up in a *not invented here* mentality. Few solutions are found; performance settles into the average of the industry progress followed by frantic catch-up activity when competitors push ahead. With benchmarking, options can be identified in a proactive fashion. One often finds business practice breakthroughs leading to superior performance.

7.4.8 Empower Employees

Few things help an individual more than to place responsibility upon him, and let him know that you trust him.

—Booker T. Washington

Empowerment is a strategic process of building a *partnership* between people and the organization, and fostering trust, responsibility, authority, and accountability throughout the organization *to best serve the customer*. Note the importance of the concept of partnership in pursuit of corporate objectives. The role of management is to exercise its responsibility by involving employees in the decision-making processes, especially those that impact customers. Effective empowerment creates a culture in which management can focus on strategic issues rather than routine, day-to-day issues.

Empowerment does not eliminate management responsibility, but it does shift its focus. While empowerment will have a direct impact on employees, the greater impact is on management. Many people have worked their way "up" to management so that they could be the boss and tell everyone else what to do. For those people, empowerment may pose a difficult or, in some cases, insurmountable change. Empowerment is not the solution to all business problems and will not by itself give an organization a competitive advantage.

One of the important revelations of the TQM movement is that people want to be responsible for their performance and, if given a chance, they will do good work. Empowerment generally leads to flatter organizations, resulting in closer coupling of problem identifiers with problem solvers. Empowerment implies a commitment to learning and training, and requires open sharing of information. It is important to think in terms of transferring ownership, not delegating work. Delegating implies that the employee does the work and the manager will decide if it is correct. Transferring ownership implies that the employee owns the outcome.

Empowerment embodies the belief that the answer to the latest crisis lies within each individual and therefore everyone must buckle up for the adventure. Perhaps the critical advantage of empowerment is that one cannot dictate resourcefulness. If one wants people to put their hearts and minds into a project, then they had better be empowered.

Empowerment must be built on trust. Trust, in turn, depends on capability and creditability. That is, employees must be adequately trained (capable) to do the job that they have been asked to do; empowerment without process leads to anarchy. Empowerment is not something that a manager can grant from the top (Stayer, 1990). Employees must want to be empowered. Learning to let employees lead is hard work, and it is generally unnatural to management training.

Empowered employees feel that it is acceptable to report problems to their management without the possibility of reprisal. Without this belief, employees may hide issues from management, and these may be exactly the issues that need to be addressed significantly improve customer satisfaction.

Many people equate empowerment with letting employees do whatever they want. Empowerment is not the right to *do as one pleases,* but to be *pleased to do what is right* for the customer. Empowerment must operate within the boundaries established by management regarding policies, key initiatives, and business goals. Contrary to popular belief, "contented cows" do *not* necessarily give more milk. If one wants to increase productivity and quality, however, then it is necessary to have satisfied employees.

One must avoid the "stop sign syndrome" in managing employees. The stop sign syndrome adds a new check point, a new level of signature, or a new level of inspection every time that a bad thing happens. Followed long enough, the result is that nothing will get done! Deming argues that when one follows this approach, one is treating a "special cause" with a general cause solution (see Appendix 7C); this is what he calls tampering.

There is another exciting, and some might think scary, aspect of employee involvement. There comes a point where employee involvement is impossible to turn off. To quote Rodger Bricknell (Stewart, 1991) of GE's Power Systems, "If you teach a bear to dance, you'd better be prepared to keep dancing till the bear wants to stop."

7.4.9 Utilize Cross-functional Teamwork

People acting together as a group can accomplish things which no individual acting alone could ever hope to bring about.

—Franklin D. Roosevelt

The word *team* conveys different things to different people. Some think entirely of sports teams, where coaching, *individual bests,* and practicing hard to win matter the most. Some think about teamwork values like sharing, cooperating, and helping one another, while others think that any group that works together is a team. In the interest of helping to crystallize this thinking, the following is offered as a definition of a team (Katzenback and Smith, 1993).

A team is a small number of people with complementary skills who are committed to a common purpose, performance goals, and approach for which they hold themselves mutually accountable.

This definition contains five key concepts: (1) small number, (2) complementary skills, (3) commitment to a common purpose and performance goals, (4) commitment to a common approach, and (5) mutual accountability. The following paragraphs examine each of these aspects in a little more detail to see how they support the concept of a high-performance team.

Small Number. This is the most controversial part of the preceding definition. Surely some people will say that they have been on a team with 50 members. While the other four aspects of teams are absolute necessities, *small number* is a pragmatic guide. The facts are that most effective teams have less than 10 members. While a larger number of people, say 50, can theoretically become a team, groups of this size often break into subteams of smaller size. Large numbers of people, by virtue of size alone, have trouble interacting constructively as a group, much less agreeing on actionable specifics. Ten, or fewer, people are far more likely than fifty to resolve successfully their individual, functional, and hierarchical differences toward a common plan and hold themselves jointly accountable for the results. Large teams also face logistical issues, like finding enough physical space and time to meet together.

Complementary Skills. Teams must develop the right mix of skills, that is, each of the complementary skills necessary to do the team's job. The team must possess the necessary *technical or functional expertise* to address the problem at hand. Teams must have *problem-solving and decision-making skills,* so that they can identify the problems and opportunities they face, evaluate the options, and make the necessary tradeoffs and decisions. A common understanding and purpose cannot arise without effective communication and constructive conflict, which depend on *interpersonal skills.* These skills include risk taking, helpful feedback, objectivity, active listening, giving the benefit of the doubt, support, and recognizing the interests and achievements of others. While it is important to consider the skill mix in the selection of the team, one should avoid becoming overly concerned about skills in team selection. In many cases, skills can be developed *on the fly* by on-the-job training. Teams are a powerful vehicle for personal learning and development. Performance focus helps teams quickly identify skill gaps and the specific development needs of team members to fill them.

Commitment to a Common Purpose and Performance Goals. A team's purpose and its performance goals go together. The team's short-term performance goals must always relate directly to its overall purpose; otherwise, team members become confused, pull apart, and revert to mediocre performance behaviors. A common, meaningful purpose sets the tone and aspiration; specific performance goals are an integral part of the purpose. Transforming broad directives into specific and measurable performance goals is the surest first step for a team trying to shape a common purpose meaningful to its members. The combination is essential to performance; a team's purpose and specific performance goals have a symbiotic relationship. Each depends on the other to stay relevant and vital.

Commitment to a Common Approach. Teams also need to develop a common approach, that is, how they will work together to accomplish their purpose. Indeed, they should invest just as much time and effort crafting their working approach as shaping their purpose. A team's approach must include both economic and administrative aspects, as well as a social aspect. To meet the economic and administrative challenge, every member of a team must do similar amounts of real work beyond commenting, reviewing, and deciding. Team members must agree on who will do particular jobs, how schedules will be set and adhered to, what skills need to be developed, how continuing membership is to be earned, and how the group will make and modify decisions.

Mutual Accountability. No group ever becomes a team until it can hold *itself* accountable *as a team*. Like common purpose and approach, this is a stiff test! At its core, team accountability is about the sincere promises individuals make to themselves and others, promises that underpin two critical aspects of teams: commitment and trust. By promising to hold themselves accountable to the team's goals, the team members earn the right to express their views about all aspects of the team's effort and to have their views receive a fair and constructive hearing. Most people enter a potential team situation cautiously; ingrained individualism discourages them from putting their fates into the hands of others. Teams do not succeed by ignoring or wishing away such behavior. Mutual promises and accountability cannot be coerced any more than people can be made to trust one another. Accountability arises from and reinforces the time, energy, and action invested in figuring out what the team is trying to accomplish and how best to get it done. When people do real work together toward a common objective, trust and commitment follow. Accountability, then, provides a useful litmus test of the quality of a team's purpose and approach. Groups that lack mutual accountability for performance have not shaped a common purpose and approach that can sustain them as a team.

It is important to note that teams do not just happen. Careful training of team leaders and team members is critical to the success of a team. In fact, if people are put into groups and asked to perform as a team without such training, the result may be a decrease in performance from that which could have been achieved by the members acting individually.

One of the important realizations of the TQM movement is that most important quality problems cross normal functional boundaries. As a result, organizations increasingly find that creating teams with cross-functional membership is critical to solving these problems. The creation of such teams does a great deal to eliminate "finger pointing" and to eliminate many forms of waste associated with handoffs between functional organizations. This is particularly effective in the area of new-product development. In this case, teams made up of design, manufacturing, service, marketing, and often actual customer representatives have significantly decreased cycle time and improved quality.

7.5   SUMMARY

"Come to the edge," the leader said. "We're afraid," they said. "Come to the edge," the leader repeated. "We're afraid," they said. Finally they came to the edge, the leader pushed them over and they flew like eagles.

Often when people talk about how they truly embraced the TQM strategy, they talk about a *compelling event*. This is often driven by a crisis such as the massive loss of business to major competitors or a significant drop in profitability. Companies such as Motorola and Xerox adopted the TQM journey after experiencing sharp competition from several Japanese companies. The big-three automakers began to embrace TQM after they had lost significant market share to Honda and Toyota. Samuel Johnson may have captured the essence of this need best when he said, "Nothing heightens a man's senses as the prospect of being hanged in the morning." Or in the words of George Fisher, ex-CEO of Motorola, "It is easy to get religion when you are being led to the gallows."

Others have embraced the TQM philosophy based on a motivating vision. For example, Disney has used the vision: "The Happiest Place on the Earth," to motivate employees to new heights of customer satisfaction. While a crisis creates stress, a motivating vision creates opportunity or romance. In either case, the most important issue for success is to have *resolve,* as illustrated by the following story.

This story deals with an interview with a survivor of an oil-rig explosion in the North Sea several years ago. Over 100 people were killed in the incident. The survivor described how he had been thrown out of his bed in the middle of the night by the explosion. He immediately ran to the edge of the platform and jumped into the water.

The first thing that one is taught upon arriving on an oil rig in the North Sea is "Don't ever get into the water." This is the North Sea not the Caribbean. One will die within 15 to 20 minutes in the water. In spite of this warning, the survivor jumped.

But there is more to the story! The top of the platform was 150 feet from the water, so he jumped 15 stories in the middle of the night into the water. In addition, there was burning oil on the surface of the water and a lot of debris that would have killed him if he landed on it. He also did not know if there was a boat that would pick him up. In spite of all of this, he jumped.

When the reporter asked him why he had jumped, he answered without hesitation, "I was going to fry if I stayed on the platform. I chose probable death over inevitable death." That is the type of *resolve* that is needed to make a quality program truly effective! One will never accomplish the work needed to travel the quality journey because it feels good.

In summary, there is no magic formula to success in the quality journey. It is hard work, but it is fun! The work needed to implement TQM may, in many cases, seem unnatural in the sense that it is very different than anything that one has been led to believe is true. One will have to begin doing some new things, and, perhaps even more importantly, stop doing some things. It is clear that one must allow sufficient time for results to be achieved. Do not expect instantaneous results, although there is always some "low-hanging fruit" that will give some positive reinforcement. It is important to allow sufficient time for results to be achieved. This is a journey of unending length, but it is critical to start now.

Finally, nothing will change if one does not start. Deming perhaps said it most succinctly, "Can we make the change in our lifetime? We don't have much longer to do it." The 1990's was a time when companies were differentiated not on the basis of "lack of defects" in their products, but on the basis of understanding and servicing customer needs.

APPENDIX 7A   THE ISO 9000 STANDARDS

The ISO 9000 standards (Rabbit and Bergh, 1993) are a set of internationally accepted standards that define a quality management system. The ISO 9000 standards establish a framework for operating a business that ensures goods and services consistently meet the customer's expected level of quality. In simplistic terms, the ISO 9000 standards consist of three elements:

- Say what you are going to do, that is, what process you are going to use, and show that the process meets the standard;
- Do what you said that you were going to do; and finally;
- Be able to prove that you have done it.

Training is a critical element in the preparation for ISO 9001 Registration. This activity can be summed up by the following quote: "If it moves, train it. If it doesn't move, calibrate it. And, if it isn't documented, it doesn't exist or didn't happen."

ISO Registration represents a significant cultural shift for many organizations. Except for manufacturing, many do not often write down the processes they use. Even when they write down procedures, there are many in the organization who feel that they can ignore them because they are special, the procedures apply to everyone else, or this is a special case. Finally, keeping records is not something that many organizations have done completely and routinely. However, this shift of culture is an important step in the maturing of a quality program.

The ISO 9000 standard consists of three quality assurance models and a series of supplemental guidelines. The three models apply to three different types of companies:

- ISO 9001: Companies that design, produce, install, and service products and services;
- ISO 9002: Companies that produce and install products and services;
- ISO 9003: Companies engaged in final assembly and distribution of products and services.

In registering a quality system, an accredited registrar will use the applicable model (standard) in deciding the necessary requirements. It is important to note that ISO 9000 Registration is for a quality system (a set of processes) *not for products*. The appropriate ISO standard sets the minimum criterion needed to run a business. ISO registration is a milepost, not the destination.

From a customer perspective, ISO 9000 ensures that a company has defined procedures and operates according to them. ISO 9000 Registration will give an organization's customers confidence that the products and services that they acquire are likely to meet their quality requirements because they are designed and produced by a system that has been shown to provide high quality. For customers, ISO 9000 Registration provides an independent third-party assessment that ensures that an organization's quality system meets a set of minimum, established standards. In using this standard, the organization and its customers share a common language. In meeting the standard, an organization ensures that its products and services are designed, manufactured, and delivered under controlled processes, and that it, in turn, manages the quality of supplies that will eventually reach its customers.

ISO 9000 Registration is of value to customers because it reduces their need to qualify their vendors via audits. This also benefits the vendor because it saves the time spent on those audits. It should also reduce the customers' need for inspection and testing of incoming products. It should provide a framework for capturing their product requirements in a timely manner, which enables an organization to produce the product right the first time. Finally, ISO Registration will give customers the comfort of knowing that when problems arise, the organization has a means of resolving those problems and preventing them from occurring again.

The ISO 9000 Registration process provides value to the organization. In some cases, it is necessary to counter a competitive threat or is necessary because registration is required by the customer. Going through the ISO 9000 Registration process accelerates quality, cycle-time, and productivity improvements. Critical to making process improvements is an accurate understanding of the current process. ISO 9000 Registration will make this happen. The simple activity of documenting a process often leads to obvious improvements. Besides meeting requirements, the ISO 9000 culture provides a framework for documenting improved ways of accomplishing activities. It is a well-established fact that certified companies are more focused on continuous process improvement.

The General Accounting Office compiled data on the value of ISO 9000 Registration from 20 companies that scored the highest on the Malcolm Baldrige applications. Some findings are:

- A 4.7 percent improvement in on-time deliveries;
- Reduction in errors by 10.3 percent;
- Cost reduction of 9 percent;
- A 11.6 percent decline in customer complaints;
- Market share increase of 13.7 percent;
- Annual average increase in sales per employee of 8.6 percent.

While there is some paperwork associated with ISO 9000 Registration, the companies that have implemented ISO 9000 have found that it has ultimately saved time and often reduced the amount of paperwork. The existence of quality documents also makes it possible for employees to find what they need quickly and easily.

A recently published report by CEEM Information Services of Fairfax, Virginia, found that the average cost of ISO 9000 Registration is $245,000, including the cost of hiring consultants, publishing documents and manuals, hiring an auditor, and paying internal staff assigned to prepare for accreditation. The process, from start to finish, took slightly longer than a year. The average annual savings attributed to accreditation came to $179,000, mostly from improved efficiency in training, control, and documentation. Including the costs of maintaining certification, half of the survey respondents said payback came in about 40 months.

The ISO 9000 Registration process creates a companywide ownership by all employees of the corporation's quality program. It makes the quality program a living, breathing part of the business. Without a program such as ISO 9000 Registration, it is possible for many people in the company to think that quality is someone else's problem.

ISO 9000 Registration improves employee satisfaction by keeping people better informed. Employees know what is expected. It reduces fire drills because jobs are done right the first time. It improves the general level of training and competence of all employees. It helps to knock down functional barriers and unifies teams across departments. The discipline required for ISO Registration leads to more cooperation and respect among departments because they know each other's responsibilities.

ISO 9000 Registration increases public awareness of an organization's quality program. The visibility of the ISO 9000 Registration gives a clear market perception of fitness. In several cases, achievement of ISO 9000 Registration will eliminate or greatly simplify quality audits.

APPENDIX 7B   MALCOLM BALDRIGE AWARD CRITERIA

The Malcolm Baldrige National Quality Award (MBNQA) (NIST, 1996) was created by the U.S. Congress in 1987 in response to foreign competitive challenges. The award was named after the late Secretary of Commerce, Malcolm Baldrige, who was a strong advocate of improving the competitiveness of U.S. companies. Annually, the award can be given to no more than two companies in each of three categories: manufacturing companies, service companies, and small (less that 500 employees) businesses. The award is currently open to for-profit organizations only.

One of the purposes of the award is to increase awareness of companies regarding the role of quality programs in improving their competitiveness. The award provides a recognition system for outstanding companies, and perhaps most importantly, the MBNQA set up a system of information transfer that allows companies to share a wide variety of experiences. All award recipients are required to share information on their successful performance and quality strategies with other U.S. organizations.

The MBNQA is based on a unique private/public partnership. The Foundation for the MBNQA was created to foster the success of the program. The Foundation's main objective is to raise funds to endow the award program. The direct responsibility for the award is assigned to the U.S. Department of Commerce. The National Institute of Standards and Technology (NIST), an agency of the Department of Commerce, manages the award program. The American Society for Quality Control (ASQC) assists in administering the award program under contract to NIST.

The MBNQA Board of Overseers evaluates all aspects of the award program, including the adequacy of the criteria and the procedures for making awards. The Board is appointed by the Secretary of Commerce, and serves as an advisory organization on the award to the Department of Commerce. An important part of the Board's responsibility is to assess how well the award is serving the national interest.

The MBNQA Board of Examiners evaluates award applications, prepares feedback reports, and makes award recommendations to the Director of NIST. The members of the Board of Examiners are volunteers, primarily from the private sector, who are selected by NIST through a competitive application process. All members of the Board of Examiners are required to take an examiner preparation course.

The MBNQA criteria are directed toward results. As a consequence, there is considerable weight given to the results sections of the criteria (see Table 7B.1). The criteria are intended to be nonprescriptive, that is, the criteria speak to "what" is to be achieved, not "how" it is to be achieved. The criteria are based on interrelated process and results learning cycles. The criteria are designed to help a company align its products and services with customer requirements.

The criteria are intended to be part of a diagnostic system that a company can use to determine how well it is doing. Many organizations are using the criteria in a variety of ways for self-assessment. The criteria help them to learn more about quality as a factor in business success. In some cases, the application of the criteria has revealed weaknesses of which the company was not aware; this led to many new ideas. Companies report that the criteria has helped to bring an external perspective to their thinking—thinking that can too often become inwardly focused. Being familiar with the MBNQA "language and culture" has enabled them to enter into helpful information sharing.

In contrast to some quality systems, the MBNQA criteria are aimed at both customer value and organizational effectiveness. On the customer value side, the criteria address product and service improvements and relationship development. On the operations side, they address productivity growth, waste and error reduction, and workforce development and utilization. The MBNQA criteria are based on the following important concepts; the reader should note the similarity of these items to the TQM philosophy discussed in Section 7.4.

- *Customer-driven Quality.* Quality is defined by the customer. All product and service attributes are important; the important issues relate to the total purchase and ownership experiences. Delivered features must include implicit, explicit, and exciting requirements.
- *Long-range View of the Future.* Progress cannot be judged by short-time improvements.
- *Continuous Improvement.* Planning is built on goals and facts. There are clear implementation plans with appropriate measurement, assessment, review, and analysis that lead to refinement of plans.
- *Employee Empowerment.* There is clear evidence that quality and productivity cannot be achieved unless employees are empowered and satisfied.
- *Importance of Benchmarking and Stretch Objectives.* This is a key route to breakthrough thinking. Benchmarking opens the window to new ideas and approaches. It provides a shock value and highlights gaps and improvement potential.
- *Management by Fact—Quantitative.* Organizations need to be "data driven." The data need to include information about customers, market, employees, operations, and benchmarks. The data collection must be followed by appropriate analysis to extract actionable information. There must be fast response to problems uncovered by this analysis.
- *Cycle Time as a Driver of Quality and Productivity.* Quality enables cycle-time reduction. A focus on cycle time will help to force quality improvements. It will also force assessment of the value added by each step in a process. Quality improvements and cycle-time reduction often come from process simplification or process elimination.

The MBNQA has had far-reaching impacts. Over ten thousand presentations have been made by MBNQA winners, and over one million copies of the criteria have been printed by NIST. The award has spawned numerous networks of people to share information and provide assistance to each other. Over 30 states have created state awards based on the MBNQA criteria. There is considerable international interest in Europe, South America, Mexico, the Pacific Rim, and Canada; in several cases awards based on the MBNQA criteria have been developed.

There is significant scholarly research on the effect of quality programs based on the MBNQA criteria by business schools and others. These studies have revealed the critical importance of "hands-on" leadership by senior executives. The MBNQA winners also make training and education investments well above the average in their industries. The winners all have demonstrable improvement in all indices of involvement, with major improvements in quality and productivity. These researchers find huge gaps between the best and the average companies. However, very few of the best companies practice quality management in all operations.

The MBNQA criteria are divided into seven interrelated *categories* as shown in Table

TABLE 7B.1 1996 Malcolm Baldrige National Quality Award Categories and Point Values

Category Number	Category Title	Point Value
1.0	Leadership	90
2.0	Information and Analysis	75
3.0	Strategic Planning	55
4.0	Human Resource Development and Mangement	140
5.0	Process Management	140
6.0	Business Results	250
7.0	Customer Focus and Satisfaction	250
Total		1,000

7B.1. The point value associated with each category is also shown. Note the heavy weighting given to categories six and seven, which are associated with results.

Each of the categories contains two or more *areas to be addressed;* these are specific requests for description of performance. In the 1996 criteria, there were a total of 24 areas to be addressed. Each of the items to be addressed is identified as an approach/deployment item or a results item. The type of questions to be addressed by each type of item is shown in Table 7B.2.

The award process involves a four-stage review. The first stage involves an independent, multiexaminer screen with a written evaluation by each examiner. Each written evaluation contains specific comments on areas of strength and areas for improvement. These results are collected and evaluated by the MBNQA judges. If a sufficient level of performance is evident, then the application goes to the second step; if not, a feedback report is prepared and sent to the applicant.

The second stage is a consensus review. Here the examiners compare results and derive a consensus score. This consensus score is examined by the judges. If a sufficient level of performance is evident, the applicant is selected for a site visit; if not, a feedback is prepared and sent to the applicant.

TABLE 7B.2 Three Types of Items to Address

Type of Item	Typical Questions to be Addressed in the Response
Approach	Is the approach prevention based?
	Are appropriate tools and techniques used?
	How is the approach integrated with other activities?
	Is there a process for improvement, and have cycles of improvement been shown?
Deployment	Have all transactions with customers, suppliers, and the public been included?
	Have all operations and processes been addressed?
	Do the approaches involve all products and services?
Results	What are the current quality levels?
	Has there been a sustained and significant rate of improvement?
	Are appropriate competitive comparisons known, and do the results compare favorably?

TABLE 7B.3   1995 MBNQA Distribution of Written Scores

Range	% Applicants in Range
0–250	6
251–350	15
351–450	11
451–550	32
551–650	30
651–750	6
751–1000	0

The third stage involves a five-day site visit by eight to ten examiners. During the site visit, the examiners validate the information provided in the written application and address open items developed during the earlier stages of the process. A detailed written report of the site visit is prepared. Finally, the judges evaluate all of the preceding material and recommend award winners to the Board of Overseers. Any organization recommended for an award undergoes detailed financial and ethical scrutiny before an award is made.

The range of scores from the 1995 round of applications is shown in Table 7B.3. It should be noted that no applicant has ever scored higher than 750. The small number of scores in the lower ranges are a function of a self-selection process that normally keeps organizations that would score low from applying.

The applicants have made several observations about participation in the award process. First, they note that their rate of improvement has increased. They observe improved morale and cooperation among their employees as well as improved communications. The applicants now view quality more as a strategic issue that must be embedded in all activities. They note that the feedback received from the application process is very valuable to their improvement programs.

APPENDIX 7C   DEMING'S QUALITY PHILOSOPHY

While many individuals throughout the world have shaped the TQM movement, perhaps the most influential is Dr. W. Edwards Deming (Walton, 1986). Trained as a statistician, Deming combined his statistical background with unique insight into individuals and systems to create an extremely powerful philosophy of how organizations can perform better. This Appendix is a brief summary of Deming's teachings; the reader is urged to compare the comments here with the material presented earlier to see the deep influence that Deming has had.

Deming emphasizes the need for leaders of organizations to have *profound knowledge*. He argues that most (85 percent or more) of an organization's defects are caused by problems in the design of the system. These system problems are the responsibility of managers, not workers. He points out many examples of how best-effort management without profound knowledge has led to huge financial losses.

Deming bases much of his teachings on statistical process control (SPC) concepts. These concepts can be used to determine whether a system is in statistical control (stable) or not in statistical control (unstable). There are two types of causes for variations in outcomes of a system. In stable systems, variations are due to *common causes*. Common causes generate outcomes that fall within the SPC control limits. This type of variation can only be reduced by making changes in the process (system). The other types of causes, called *special causes,* generate outcomes that are outside of the control limits.

It is important to note that a stable system can produce defective outcomes (i.e., outcomes that do not meet the needs of the system). If specifications are set tighter than systems control limits, there is a tendency to think of outcomes outside of the specifications as being due to special causes, thereby leading to tampering and the potential for significant deterioration in performance. One of the principal uses of SPC concepts is to help separate variations due to a special cause from variations due to common causes, and to respond appropriately.

Organizations make two types of mistakes in dealing with an outcome that is unacceptable. In the first case, they react to the outcome as if it came from a special cause when, in fact, it came from common causes of variation. This type of mistake leads to tampering; tampering with a system in statistical control will cause its performance to deteriorate and to potentially become erratic. In the second case, organizations react to the unacceptable outcome as if it came from common causes when, in fact, it came from a special cause. This leads to a lack of response to a true problem in the system.

Innovation, whether it be in product design or process improvement, only comes from freedom. Freedom must include freedom from fear. Management needs to indicate that no one will lose their job due to process improvements. Freedom also includes empowerment of workers.

Management must predict the outcomes of stable systems based on known process capabilities. Insanity is doing the same things the same way and expecting different results. Slogans exhorting improvement without specifying a method are dangerous and demoralizing. One should always ask by what method is an improvement to be achieved. Optimal performance can only be achieved by systems thinking (profound knowledge).

Deming summarized this teaching in his famous 14 points (Scherkenbach, 1994).

1. *Create Constancy of Purpose.* Create constancy of purpose toward improvement of products and service with the aim to become competitive, stay in business, and provide jobs.

2. *Adopt the New Philosophy.* We are in a new economic age, created by the Japanese. Western management must awaken to the challenge, must learn their responsibilities, and take on leadership for change.

3. *Cease Dependence on Mass Inspection.* Cease dependence on inspection to achieve quality. Eliminate the need for inspection on a mass basis by building quality into the product in the first place.

4. *Stop Awarding Business on Pricetag.* End the practice of awarding business on the basis of price tag. Instead, minimize total cost. Move toward a single supplier for any one item on a long-term relationship of loyalty and trust.

5. *Improve the System.* Improve constantly and forever the system of production and service, resulting in improving quality and productivity, and thus constantly decreasing costs.

6. *Institute Training on the Job.*

7. *Institute Leadership.* The aim of leadership should be to help people, machines, and gadgets to do a better job. Supervision of management is in need of overhaul, as well as supervision of production workers.

8. *Drive Out Fear.* Drive out fear, so that everyone may work effectively for the company.

9. *Break Down Barriers Between Departments.* People in research, design, sales, and production must work as a team to foresee problems in production and in use that may be encountered with the product or service.

10. *Eliminate Slogans.* Eliminate slogans, exhortation, and targets for the work force that ask for zero defects and new levels of productivity.

11. *Eliminate Work Standards.* Eliminate work standards (quotas) on the factory floor. Substitute leadership for work standards. Eliminate management by objective. Eliminate management by numbers and numerical goals.

12. *Remove Barriers.* Remove barriers that rob the hourly worker of the right to pride of workmanship. The responsibility of supervisors must be changed from stressing sheer numbers to quality. Remove barriers that rob people in management and engineering of their right to pride of workmanship. This means, *inter alia,* abolishment of the annual merit rating and of management by objective.

13. *Institute Education and Retraining.* Institute a vigorous program of education and self-improvement.

14. *Accomplish the Transformation.* Put everybody in the organization to work to accomplish the transformation. The transformation is everybody's job.

Deming complemented his 14 points with his 7 deadly diseases:

1. *Lack of Constancy of Purpose.* A company that lacks constancy of purpose does not think beyond the next quarterly report.

2. *Emphasis on Short-term Profits.* Companies are too driven by quarterly earning reports to see the need for long-term corrective actions.

3. *Evaluation of Performance, Merit Ratings, or Annual Review.* Deming argues that the popular management by objectives (MOB) approach is really management by fear.

4. *Mobility of Top Management.* The critically important profound knowledge can only be developed by leaders with in-depth knowledge of an organization. The interchangeable philosophy of management does not work. Deming noted the contrast with Japanese companies where the top executives work their way to the top over many years.

5. *Running a Company on Visible Figures Alone ("Counting the Money").* While financial performance is important, it cannot be the sole measure used to determine a company's performance.

6. *Excessive Medical Costs.* It is worth noting that Deming pointed out this problem well before it became a national issue.

7. *Excessive Costs of Warranty, Fueled by Lawyers Who Work on Contingency Fees.* Deming pointed out that the United States is the most litigious country in the world; there is no clear correlation to value added.

REFERENCES

Argyris, C. (1991). Double loop learning. *Harv. Bus. Rev.* **68**(5), 99–109.

Blanchard, K., and Bowles, S. (1993). *Raving Fans,* New York: William Morrow.

Brassard, M. (1989). *The Memory Jogger Plus+.* Methuen, MA: GOAL/QPC.

Drucker, P. F. (1993). *Post Capitalist Society.* New York: Harper Business.

Garvin, D. A. (1988). *Managing Quality.* New York: Free Press.

Garvin, D. A. (1993). Building a learning organization. *Harv. Bus. Rev.* **71**(4), 78–91.

Hammer, M. (1990). Reengineering work: Don't automate, obliterate. *Harv. Bus. Rev.* **67**(4), 104–112.

Hammer, M., and Champy, J. (1993). *Reengineering the Corporation.* New York: HarperCollins.

Helton, B. R. (1995). The Baldie play. *Qual. Prog.* **28**(2), 43–45.

Heskett, J. L. (1987). Lessons in the service sector. *Harv. Bus. Rev.* **64**(2), 118–126.

House of Representatives: Management Practices. (1991). *GAO Report to the Honorable Donald Ritter, US Companies Improve Performance Through Quality Efforts* (GAO/NSIAD-91-190). Washington, DC: House of Representatives.

Katzenback, J. R., and Smith, D. K. (1993). *The Wisdom of Teams.* Boston: Harvard Business School Press.

National Institute of Standards and Technology (NIST). (1996). *Malcolm Baldrige National Quality Award, 1996 Award Criteria.* Washington, DC: NIST.

Peters, T. (1987). *Thriving on Chaos.* New York: Knopf.

Rabbit, J. T., and Bergh, P. A. (1993). *The ISO 9000 Book.* White Plains, NY: Quality Resources.

Reichheld, F. F., and Sasser, W. E., Jr. (1990). Zero Defections: Quality comes to services. *Harv. Bus. Rev.* **67**(5), 105.

Scherkenbach, W. W. (1994). *The Deming Route to Quality and Productivity.* Washington, DC: CEEP Press Books.

Schonberger, R. J. (1986). *World Class Manufacturing.* New York: Free Press.

Senge, P. M. (1990). *The Fifth Discipline.* New York: A Currency Book.

Smith, P. G., and Reinertsen, D. G. (1991). *Developing Products in Half the Time.* New York: Van Nostrand-Reinhold.

Stayer, R. C. (1990). How I learned to let my workers lead. *Harv. Bus. Rev.* **67**(6), 66–83.

Stewart, T. A. (1991). GE keeps those ideas coming. *Fortune,* August 12, pp. 41–49.

Strategic Planning Institute. (1986).

The PIMS Letter on Business Strategy, No. 4. Cambridge, MA: SPI.

Walton, M. (1986). *The Deming Management Method.* New York: A Perigee Book.

Womach, J. P., Jones, D. T., and Roos, D. (1990). *The Machine that Changed the World.* New York: Rawson Associates.

8 Reliability, Maintainability, and Availability

MICHAEL PECHT

8.1 INTRODUCTION AND MOTIVATION

Reliability, maintainability, and availability (RMA) engineering is a discipline in which data are gathered and models are formulated to make decisions concerning system failure, operational readiness and success, maintenance and service requirements, and the collection and organization of information from which the effectiveness of a system can be evaluated and improved. RMA engineering thus includes engineering design, manufacturing, testing, and analysis. RMA metrics can be considered to be characteristics of a system in the sense that they can be estimated in design, controlled in manufacturing, measured during testing, and sustained in the field (Pecht, 1995).

When utilized early in the concept development stage of a system's development, RMA serve to determine feasibility and risk. In the design stage of system development, RMA analysis involves methods to enhance performance over time through the selection of materials, design of structures, choice of design tolerances, manufacturing processes and tolerances, assembly techniques, shipping and handling methods, and maintenance and maintainability guidelines. Engineering concepts such as strength, fatigue, fracture, creep, tolerances, corrosion, and aging play a role in these design analyses. The use of physics-of-failure concepts coupled with the use of factors of safety and worst case studies are often required to understand the potential problems, tradeoffs, and corrective actions. In the manufacturing stage of a system, RMA analysis involves determining quality control procedures, reliability improvements, maintenance modifications, field testing procedures, and logistics requirements. In the system's operational phases, RMA involves collecting and using test and field information obtained from field service, safety inspection, failure analysis, manufacturing records, suppliers, and customers.

It has been well documented that while the costs incurred during the concept and early design stages of the system development cycle are about 5–8 percent of the total costs, these stages affect over 80 percent of the total system costs. For purposes of cost-effective and timely system development, RMA engineering has reduced value if the design phase is completed. The earlier a potential event is addressed, the more prepared the engineering team can be to correct or alleviate potential problems.

When a system is being designed, it is assumed that the required investment will be justified according to how well the system performs its intended function over time. This assumption cannot be justified when a system fails to perform upon demand or fails to perform repeatedly. For example, it is not enough to show that a computer can conduct a mathematical operation; it must also be shown that it can do so repeatedly over many

Handbook of Systems Engineering and Management, Edited by A. P. Sage and W. B. Rouse
ISBN 0471-15405-9   ©1999 John Wiley and Sons, Inc.

years. Furthermore, higher life-cycle costs and stricter legal liabilities have also made system RMA considerations of greater importance. As the attitude toward production of engineering systems has changed, RMA engineering has established itself as one of the key elements when designing, manufacturing, and operating a system.

8.2 EVOLUTION OF RMA ENGINEERING

Interest in establishing a quantitative measure of system RMA began in World War II with the development of the German V-1 missile and the reliability design concept that a chain is only as strong as its weakest link. After the war, in the period between 1945 and 1950, the U.S. Armed Forces found that electronic equipment was operative only 30 percent of the time during Navy maneuvers and the Air Force spent about ten times the original cost of a system for repair and maintenance. Moreover, for every vacuum tube in operation, one was on the shelf and seven in transit, and one electronics technician was required for every 250 vacuum tubes. Many of these problems were a result of pre−World War II design concepts and a lack of highly skilled workers on the production lines. However, the need for some form of RMA analysis was clear.

Reliability engineering for electronics started with the establishment of the Ad Hoc Group on Reliability of Electronic Equipment on December 7, 1950. However, the modern field of reliability is often traced to the Advisory Group on the Reliability of Electronic Equipment (AGREE) formed by the U.S. Department of Defense in 1952. At this time, engineers were asked to estimate system RMA in order to meet the government procurement needs. Government procurement agencies sought standardization of RMA requirement specifications in fear that each contractor could develop individual estimates based on in-house data and it would be impossible to evaluate system estimates against requirements from different suppliers or to compare competitive designs for the same system.

Although statisticians had focused on the analysis of equipment lifetime data and parametric families of distribution that could be used as lifetime models, it was not until 1953 when the application of the exponential distribution to life testing and statistical problems was made available in a popular manner, that significant advances became possible. Other distributions, such as the Weibull distribution developed originally to describe the breaking strength of materials for both material strength and life strength, were well known but infrequently used by RMA engineers and logisticians. The fundamental reason for the popularity of the exponential was the simplicity of analysis of RMA functions. By the 1960s, however, it was pointed out that many life-test procedures based on the exponential distribution were not practical and the exponential-based statistical techniques were found to be sensitive to departures from initial assumptions. The application of these techniques to life-test data, when the exponential failure law was not satisfied, resulted in an increase in the probability of accepting systems having poor mean time to failure.

Motivated by nuclear power reactor safety considerations, the 1970s marked the birth of fault-tree analysis. Papers concerning minimum cut-set algorithms, marked the beginning of the efforts to avoid Monte Carlo methods for finding minimum cut-sets. Also, set theoretic, combinatorial algorithms for analyzing very large fault trees were developed. The interested reader is referred to Coudert and Madre (1995) and Dugan et al. (1992) for details on fault-tree analysis.

The Bayesian approach made it possible to obtain an estimate of RMA parameters and to upgrade estimates in a systematic way. Engineers utilized handbooks, expert opinions, and experience to formulate a probability distribution referred to as a *prior distribution*. With Bayesian analysis, the test and field data were used to modify distributions to yield previous or posterior distributions. Given the availability of test results, the posterior distribution represented a new state of knowledge, and its mean or median represented an improved point estimate. For more information on Bayesian statistics see Martz and Waller (1982).

The 1990s marked the widespread development of physics-of-failure, particularly by the avionics and automotive communities. Physics-of-failure uses facts from root-cause failure processes in order to prevent the failures of the products by robust design approaches and better manufacturing practices (Pecht and Dasgupta, 1995). This approach proactively incorporates reliability into design and manufacturing.

8.3 ALLOCATION

RMA allocation is conducted in the system concept phase. Commencing with the assignment of RMA goals to the system, there is subsequent assimilation of these goals to subsystems, but generally not to assemblies and parts. The process can be modeled by the equation:

$$f(X_1, X_2, \ldots, X_n) \geq X_A \tag{8.1}$$

where the X_i represent some RMA metric of the subsystems that compose the system, f is a function that describes how the subsystem metrics interact together to make up system metrics, and X_A is the system RMA goals.

The RMA goals are usually based on the expected system life-cycle application profile. Allocation estimates should be made based on hystorical data of similar systems. The estimates must not be used to define maximum attainable RMA levels, and allocation information should not go to the contractor as an estimate of the "deliverable RMA."

8.4 DESIGN FOR RELIABILITY

Reliability hinges on the recognition that an organized, disciplined, and time-phased approach to design, development, qualification, manufacture, and in-service management of a system is required to achieve mission performance over time, safety, supportability, and cost effectiveness. The foundation of the approach consists of tasks, each having total engineering and management commitment and enforcement. The tasks are:

1. *Define Realistic System Requirements.* Every system must operate through a range of application conditions for a specified length of time. Requirements may be determined by the life-cycle application profile, performance expectations, size, weight, or cost. The manufacturer and the customer must jointly define system requirement in the light of both the customer's needs and the manufacturer's capabilities to meet those needs.

2. *Define the System Usage Environment.* The manufacturer and the customer must jointly specify all relevant life-cycle operating, shipping, and storage conditions, and assess available tradeoffs. The traditional use of standard environmental categories

should be replaced by measured data for temperature, temperature changes, operable duty cycle, humidity, vibration, applied voltage, and any other key electrical, thermal, radiation, and mechanical conditions.

3. *Identify Potential Failure Sites and Failure Mechanisms.* Potential failure modes, sites, and mechanisms, as well as potential architectural and stress interactions, must be identified early in the design process. Ishikawa diagrams and failure modes effects analysis (FMEA) can be used for this purpose. Pareto diagrams can be used to assess the relative importance of each type of failure and to determine a hierarchy of design features. Once expected failures are identified, appropriate measures must be implemented to reduce, eliminate, or accommodate them.

4. *Characterize the Materials and the Manufacturing and Assembly Processes.* All materials must be characterized and their key characteristics controlled. These include types and levels of defects, as well as expected variations in properties and dimensions of materials, and variabilities of the manufacturing and assembly processes.

5. *Design Reliable Systems Within the Capabilities of the Materials and Manufacturing Processes Used.* The design must be evaluated and optimized for manufacturability, quality, reliability, and cost-effectiveness before production begins. It is unrealistic and potentially dangerous to assume that structures are defect-free because materials often have naturally occurring defects, and manufacturing processes can induce additional flaws. These concerns can be addressed concurrently, using experimental step-stress and other accelerated life-testing methods.

6. *Qualify the Manufacturing and Assembly Processes.* All manufacturing and assembly processes must be optimized and capable of producing the system. Key process characteristics must be identified, measured, and optimized.

7. *Control the Manufacturing and Assembly Processes.* The manufacturing process must be monitored and controlled.

8. *Manage the Life Cycle of the System.* Closed-loop procedures must be used to collect data from tests performed in design, manufacturing, accelerated life testing, and field operation, and to continuously assess and improve the quality, reliability, and cost-effectiveness of the system.

8.5 SYSTEM ARCHITECTURE

Although functional and performance characteristics often play the dominant role in developing a system, RMA needs must be considered. Structures, interconnections, and interfaces (both hardware and software) must be properly selected. Furthermore, the interactions of the constituents of a system must be a part of system architecture development. For example, a system may be reliable but may be difficult to use and interpret.

When developing the system architecture, the use of redundant subsystems may be necessary to reduce the risk of a failure. Redundancy may thus be deemed necessary to satisfy the application requirements when the estimates on RMA indicate improbable success or unacceptable risk. The form of redundancy may be hot spares, cold (standby) spares, polled circuits, and fault-tolerant designs, all to enhance mission success. In Section 8.12 we discuss the concepts of redundancy modeling and tradeoffs.

In some cases, such as where safety is an issue, it may be desirable to design-in a means for preventing a system or subsystem from failing or from causing further damage if it fails. Protective architectures can be used to sense failure and protect against possible secondary effects. In some cases, self-healing techniques self-check and self-adjust to permit continued operation after a failure (Kelkar et al., 1996). For example, thermostats may be used to sense critical temperature limiting conditions and shut down the system until the temperature returns to normal. In some systems, self-checking circuitry can also be incorporated to sense abnormal conditions and operate controls to restore normal conditions or activate switching means to compensate for the malfunction.

In some instances, means can be provided for preventing a failed subsystem from completely disabling the system. For example, fuses and circuit breakers are examples used in electronic systems to sense excessive current drain and disconnect power. Fuses within circuits safeguard systems against voltage transients or excessive power dissipation and protect power supplies from shorts. A fuse or circuit breaker can disconnect a failed subsystem from a system in such a way that it is possible to permit partial operation of the system, in preference to total system failure. By the same reasoning, degraded performance of a system after failure of a subsystem is often preferable to complete stoppage. An example is the shutting down of a failed circuit whose function is to provide precise trimming adjustment within a dead band of another control system. Acceptable performance may thus be permitted, perhaps under emergency conditions, with the deadband control product alone.

In the use of protective techniques, the basic procedure is to take some form of action, after an initial failure or malfunction, to prevent additional or secondary failures. By reducing the number of failures, such techniques can be considered as enhancing system reliability, although they also affect availability and product effectiveness. No less a consideration is the impact of maintenance, repair, and system replacement. If a fuse protecting a circuit is replaced, what is the impact when the system is reenergized? What protective architectures are appropriate for postrepair operations? What maintenance guidance must be documented and followed when fail-safe protective architectures have or have not been included?

8.6 STRESS ANALYSIS AND MANAGEMENT

Given the system architecture, the engineering team must assess the influence of the magnitude and duration of the stresses on the reliability of the system. This allows stress and environment management, and derating techniques, to be implemented properly. Temperature, humidity, vibration, and radiation are environmental stress variables affecting reliability. In electrical systems, stresses caused by power, current, and voltage must also be considered. Although lowering the stresses is an alternative of design teams, the cost and complexity of lowering stresses must be balanced against the total system cost and complexity.

A properly designed system should be capable of operating satisfactorily with subsystems that may drift or change with time, temperature, humidity, and altitude, so long as the parameters of the systems and the interconnects are within their rated tolerances. To guard against out-of-tolerance failures, the design team must consider the combined effects of tolerances on systems to be used in manufacture, subsequent changes due to the range of expected environmental conditions, drifts due to aging over the period of time specified in the reliability requirement, and tolerances in systems used in future repair or maintenance functions. Systems should be designed to operate satisfactorily at the extremes of the parameter ranges, and allowable ranges must be included in the procurement specifications.

In general, there are no distinct stress boundaries for voltage, current, temperature, power dissipation, and so on, above which immediate failure can occur and below which the system will operate indefinitely. Instead, the life of some systems increases in a continuous manner as the stress level is decreased. However, there is often a minimum stress below which the system will not function properly or at which the increased complexity required to allow the lower stress level will not offer an advantage in cost-effectiveness, and may lower reliability, especially if there is a need to increase the complexity required to step up performance.

Manufacturers' ratings or users' procurement specifications are generally used to determine the maximum rating value. The published values are often conservative estimates of the actual stress level of the product based on the design, the manufacturing uniformity and repeatability, and the desired margin of safety.

Derating and stress management are methods of increasing a system's life by decreasing applied thermal, electrical, mechanical, and other stresses. The premise is that the greater the stress level, the more accelerated the failure mechanisms; and the lower the stress level, the longer the system will operate successfully.

Other methods of dealing with parameter variations are statistical analysis and worst case analysis. In statistical design analysis, a functional relationship is established between the output characteristics of the system and the parameters of one or more of its subsystems. In worst case analysis, the effect that a subsystem has on a system output is evaluated on the basis of end-of-life performance values or out-of-specification replacement subsystems.

There are various ways in which both the operating and environmental stresses can be controlled to improve system reliability. Methods can be applied to keep harmful stresses (high temperatures, high shock loads, high humidity, and high radiation levels) away from sensitive devices and structures. Methods can also be applied to manage the system environment to obtain specific stress conditions.

Thermal stress management addresses the failure mechanisms that are sensitive to temperature, temperature changes, temperature gradients, and the overall temperature history. Once the heat dissipation per unit heat-transfer area is estimated and a need for thermal management is recognized, the design team identifies appropriate thermal management mechanisms based on cost, size, weight, and reliability criteria. Cooling and heating parameters can then be bounded by examining the limiting temperatures dictated by inherent performance and reliability requirements. Once the design parameters associated with a cooling mechanism have been bounded, the design team can conduct tradeoff studies. Analysis and tradeoffs must be extended to the assembly, box, and system. To control temperature, heat dissipation devices or cooling systems as well as thermal insulation and materials that withstand temperature extremes may be used. A combination of these techniques can be implemented for thermal shock ambient.

In cases of mechanical shock, a reduction of inertia (moments) and shock absorbing structures can contribute to increased reliability. For controlling vibration, damping control of resonance is often necessary.

For ambients composed of water and contaminant spray, hermetic sealing, moisture-resistant materials, and nonmetallic protective covers may be used. In general, dehumidification is not advisable. For systems used in low-pressure environments, pressurization of the equipment may be advised along with the use of liquids that have low volatility.

8.7 QUALIFICATION

The process of qualification establishes the ability of the nominal design and manufacturing specifications of the product to meet the customer's needs. In order to successfully qualify a product or technology for a given requirement, it is important to understand the design, the manufacturing process, the loads throughout the life-cycle, the potential failure mechanisms, and the cost penalties associated with each design and manufacturing decision. The term loads is used in a generic sense and includes mechanical, thermal, electrical, radiation, and chemical stimuli that can affect performance during manufacture, including rework, testing, screening, storage and handling, transportation, operational use, and repair. Design integrity models, based on physics-of-failure mechanisms, provide a vehicle for designing qualification tests and interpreting test results. A qualification test can use accelerated stress levels to precipitate the same failure mechanisms that could compromise the product's long-term reliability in the field. However, evaluating the distribution of failures and life expectancy can therefore be difficult, because of the very long test periods required under actual operating conditions to obtain sufficient failure data. For this reason, overstresses and accelerated wear-out stresses are often employed. Overstress tests can be conducted when qualifying a product for potential overstress service loads and when investigating the impact of a manufacturing process on the overstress strength of a material and/or subassembly. On the other hand, qualification for wear-out failure mechanisms is usually accomplished by accelerated testing that allows for compression of test time. Accelerated testing involves measuring the reliability characteristics of the product quantitatively under more severe stress conditions than the normal operating level, in order to induce wear-out failures within a reduced time period. The advantages of accelerated life tests are both economic savings and quick turnaround during the development of new products or of mature products subjected to manufacturing and workmanship change. The severity of the applied stress in accelerated wear-out testing is usually selected based on achieving a reasonable test time compression without altering the fundamental failure mechanism. The results from the tests are then extrapolated using a quantitative acceleration transform, to give a lower bound estimation of life for the product. In some cases, the stress type used in the test need not necessarily be the same as that causing the failure mechanism in the field, provided that a correlation can be established between the damage caused during the test and under field conditions. Such correlations explicitly utilize quantitative failure mechanism models, based on experimentally characterized properties of the materials and structures being tested.

Qualification should be performed only during initial product development and immediately after any design or manufacturing changes in an existing product. Once the product is qualified, routine lot-to-lot requalification is redundant and an unnecessary cost item. The manufacturing volume, however, may not be high enough for adequate prototyping and qualification during the developmental phase. In such cases, there may be some unavoidable overlap between the qualification and quality conformance tasks. In order to maintain a cost-effective operation, these overlaps should be minimized.

8.8 RELIABILITY TESTING

This section provides brief descriptions of various environmental conditions and the associated tests that can be conducted to ensure that a system is properly designed and assembled. In addition to the tests listed, some form of preconditioning can be used prior to the tests to accelerate the environmental conditions. Reliability qualification and quality assurance testing for electronics can be found in the book titled *Quality Conformance and Qualification of Microelectronic Packages and Interconnects,* by Pecht et al. (1994).

1. *High Temperature.* High-temperature tests are primarily performed to assess failure mechanisms that are thermally activated. In electromechanical and mechanical systems, high temperatures may soften insulation, jam moving parts due to thermal expansion, blister finishes, thermally age coatings, oxidize materials, reduce viscosity of fluids, evaporate lubricants, and cause structural overloads due to physical expansions. In electrical systems, high temperature can cause variations in resistance, inductance capacitance, power factor, and dielectric constant.

2. *Low Temperature.* In mechanical and electromechanical systems, low temperatures can cause plastics and rubber to lose flexibility and become brittle, cause ice to form when moisture is present, increase viscosity of lubricants and gels, crack finishes, and cause structural damage due to physical contraction. In electrical systems, low-temperature tests are performed primarily to accelerate threshold shifts and parametric changes, due to variation in electrical material parameters.

3. *Temperature Cycle.* Temperature cycle testing is most often employed to assess the effects of thermal expansion mismatch among the different elements within a system, materials' overstressing and cracking, crazing, and delamination. Systems or subsystems inserted into a cycling system are held at the cold-dwell temperature long enough to thermally stabilize and creep and stress relax as appropriate to the application. Following this cold dwell, the systems or subsystems are heated to the hot dwell, where they remain for another minimum time period. The dwell at each extreme, plus the two transition times, constitute one cycle.

   In electronic systems, power cycling may also be performed. Tests may be performed at a constant ambient temperature with operating voltage(s) periodically applied and removed. Common failure modes include parametric shifts and catastrophic events.

4. *Thermal Shock.* While thermal shock testing is one form of temperature cycling, thermal shock provides additional stress, in that the system is exposed to a sudden change in temperature due to the rapid transfer time. Failure mechanisms caused by temperature transients and temperature gradients can be detected with this test. Systems or subsystems are typically placed in a fluorocarbon cold bath for some minimum period of time, and are then transferred to an adjacent chamber filled with fluorocarbon at the maximum specified temperature. Common failure modes include parametric shifts and catastrophic events.

5. *Humidity.* Humidity can cause leakage paths between electrical conductors, oxidation, corrosion, swelling in materials such as gaskets, embrittlement and granulation due to excessive loss of humidity. In electronic systems, humidity is often coupled with high temperature and a voltage, as an environmental test designed to measure corrosion/moisture resistance.

6. *Mechanical Shock.* Some systems must be able to withstand a sudden change in mechanical stress typically due to abrupt changes in motion as seen in handling, transportation, or actual use. Mechanical shock can lead to overstressing of mechanical structures causing weakening, collapse, or mechanical malfunction. In some cases, items may be ripped from their mounts.

7. *Variable Frequency Vibration.* Some systems must be able to withstand deterioration due to mechanical resonance. Vibration may lead to the deterioration of mechanical strength due to fatigue or overstress; electrical signals may be erroneously modulated; materials and structure may be cracked, displaced, or shaken loose from mounts; mechanical functions may be impaired; finishes may be scoured by other surfaces; and wear may be increased.

8. *Atmospheric Contaminants.* The atmosphere contains many contaminants such as airborne acids and salts that can lower electrical and insulation resistance, oxidize materials, and accelerate chemical corrosion. Mixed flowing gas tests are often used to assess the reliability of products subjected to these environments.

9. *Electromagnetic Radiation.* Electromagnetic radiation can cause spurious and erroneous signals from electronic components and circuitry. In some cases, it may cause complete disruption of normal electrical equipment such as communication and measuring systems.

10. *Nuclear/Cosmic Radiation.* Nuclear/cosmic radiation can cause heating and thermal aging; alter chemical, physical, and electrical properties of materials; produce gasses and secondary radiation; oxidize and discolor surfaces; and damage electronic components and circuits.

11. *Sand and Dust.* Sand and dust can scratch and abrade finished surfaces; increase friction between surfaces, contaminate lubricants, clog orifices, wear materials, crack and chip materials, and cause abrasion and corona paths.

12. *Low Pressure.* Low pressure can cause overstress of structures such as containers and tanks that can explode or fracture; seals may leak; air bubbles in materials may explode; internal heating may increase due to lack of cooling medium; insulations may suffer arcing breakdown; ozone may be formed; and outgassing is more likely to occur.

8.9 QUALITY CONFORMANCE

Quality conformance ensures that previously qualified parameters are maintained within specified tolerances through the monitoring, verification, and control of critical material variables and process parameters. Quality conformance is not intended to check the ability of the nominal product attributes to meet usage criteria, but to ensure that all variabilities beyond a specified tolerance (established during the qualification program) are identified and controlled and to eliminate any product whose attributes are out of tolerance. Parameter variability may be due to any one or a combination of the following factors:

- Raw material property variability;
- Variability in a manufacturing process parameter due to inaccuracies of process monitoring and control devices;
- Human error and workmanship inadequacies; and
- Unintended stresses (e.g., contaminants, particles, vibration) in the manufacturing environment.

Understanding the potential defects introduced prior to and during the product manufacturing cycle is a fundamental aspect of a quality conformance program. A defect arises whenever a critical design variable (material or structural) is beyond acceptable (previously qualified) tolerances. Defects can lead to performance degradations, outright failure, and other causes of customer dissatisfaction.

Product and materials evaluation testing is an important quality conformance task that addresses defect detection and is often called *screening*. Screening is an audit process to ensure that the product's materials and manufacturing conform to the control limits of the production processes. Screening involves both the early detection of product parameters that are out of tolerance and the precipitation of defects. Defect detection is most effective if conducted at the time the defect is created. Thus, in order to be proactive, screening should be part of the in-line manufacturing processes associated with quality control. Screening at individual process stages ensures that detected defects can be attributed to a specific manufacturing step, thereby facilitating immediate corrective action and minimizing troubleshooting and rework costs. In some cases, a defective part may be eliminated immediately, preventing additional costs from accruing on a poor-quality product.

Generally, there are limitations on when, or at which process, a product can be screened. In other words, the engineer has to decide whether screening should be employed at every process or after a set of processes, based on schedule, cost, and risk constraints. However, because the exact cause of defects becomes increasingly difficult to identify at higher levels of assembly, screening should be conducted at as many stages as feasible within the cost constraints.

Failure analysis of defects detected during the screening process is the key to formulating effective corrective action to remove the root cause of the defect. Such analysis should be comprehensive and should associate the defect with one or more of the input functions, such as the material or manufacturing process, in order to allow suitable corrective action. A typical defect analysis should include

- Details of the defect or failure event;
- The circumstances leading to the defect or failure;
- Process irregularities;
- Mechanisms inducing defects or failures; and
- Recommendations for eliminating the defects.

Defect detection methods can be broadly classified into three groups. In order of preference, these are

- Nondestructive examination screens
- Overstress screens
- Accelerated wear-out screens

Overstress screens use a single application of a controlled stress, designed to overstress only the defective elements of a product. Wear-out screens are designed to precipitate a defect that could lead to field failures from time-dependent stresses.

Overstress and accelerated wear-out screens have the potential to cause damage to good products and therefore must be planned and implemented so that they cause minimal damage to properly manufactured products. This requires careful determination of screening parameters and their magnitude or intensity. The parameters should be selected to stimulate only defects causing premature customer dissatisfaction, while consumption of the useful life of acceptable products is minimized. Some overstress and wear-out screens can cause residual stresses that may have a potentially detrimental effect on product reliability later during field usage. The magnitudes of the stresses must be tailored to the specific defect through the use of screening transforms, rather than by using the same generic stress magnitudes for all screening situations. A screening transform is a quantitative failure model that allows the testing engineer to compute the load required to precipitate all defects above qualified thresholds in a given product.

The fact that screens must be designed and tailored in order to be effective does not necessarily mean that a given screen cannot be effective at more than one site or for more than one type of defect. However, to use the same screen in multiple roles requires that the screen's effectiveness be clearly understood. For example, a visual screen can address structural, material, and cosmetic defects; temperature-cycling stress screens can effectively address defects that cause failures associated with fatigue and rachetting; and steady-state temperature can address some defects that cause failures associated with creep and aging (caused by diffusion, interdiffusion, intermetallic formation, depolymerization, corrosion, dendritic growth, electromigration, etc.). Screens for each potential defect site will vary, depending on the dominant failure mechanism induced by the dominant defect.

8.10 RELIABILITY ASSESSMENT

Failures are governed by fundamental mechanical, electrical, thermal, and chemical processes. For this reason, potential problems in new and existing technologies can be identified and solved before they occur, by understanding the possible failure mechanisms. This approach, called *physics-of-failure,* is illustrated schematically in Figure 8.1. It begins within the first stages of design. A design team defines the product requirements, based on the customer's needs and the supplier's capabilities. These requirements can include the product's functional, physical, testability, maintainability, safety, and serviceability characteristics. At the same time, the service environment is identified, first broadly as aerospace, automotive, business office, storage, or the like, and then more specifically as a series of defined temperature, humidity, vibration, shock, and other conditions. The conditions are either measured or specified by the customer. From this information, the design team, usually with the aid of a computer, can model the thermal, mechanical, electrical, and electrochemical stresses acting on the product.

Next, stress analysis is combined with knowledge about the stress response of the chosen design materials to identify where failure might occur (failure sites), what form it might take (failure modes), and how it might take place (failure mechanisms). Prior analyses of the root causes of actual field failures is of particular use at this stage. Failures are generally caused by one of four types of stresses—mechanical, electrical, thermal, chemical—and generally result either from the application of a single overstress, or by the accumulation of damage over time from lower-level stresses. Once the potential failure mechanisms have been identified, specific failure mechanism models are employed. The validity of these models can be tested by conducting accelerated aging tests. If no models are available, or if the models are found to be inaccurate, then new models are developed using a series of statistically designed experiments, which identify the most important design and environmental factors governing failure and the mathematical relationship linking those factors to the time to failure. Wherever possible, variabilities in the factors are specified using distribution functions. The effect of material properties, the geometry at

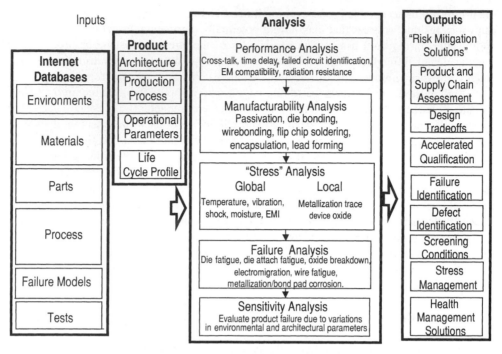

Figure 8.1   The physics-of-failure process.

failure sites, any damage to the part, and any manufacturing flaws and defects are addressed at this point as well.

Once the models are established, a reliability assessment can be conducted on the product. This consists of calculating the time to failure for each potential failure mechanism, and then using the principle that a chain is only as strong as its weakest link, choosing the dominant failure site and mechanism as those resulting in the least time to failure (see Fig. 8.1). The information from this assessment can be used to determine whether a product will survive for its intended application life, or it can be used to redesign a product for increased robustness against the dominant failure mechanisms and sites. It can even be used in the traditional way to assess the fraction of parts that will fail after a given period of time, more accurately, however, than traditional models. The physics-of-failure approach is also used to qualify design and manufacturing processes to ensure that the nominal design and manufacturing specifications meet or exceed reliability targets. It is employed throughout the development cycle to assess stress margins and establish process controls as a means of continuously improving the product's reliability from early prototype testing through final manufacture.

Today, this new type of reliability model is demanded by suppliers, to measure how well they are doing and to determine what kind of reliability assurances they can give to a customer; by customers, to determine that the suppliers know what they are doing and that they are likely to deliver what is desired; and by both groups, to assess and minimize their risks. This knowledge is essential, because the supplier of a product that fails in the field loses the customer's confidence and often its repeat business, while the customer who buys a faulty product endangers its business and possibly the safety of its customers.

8.11   SYSTEM RELIABILITY ASSESSMENT MODELING

System reliability modeling is a means of evaluating possible system design configurations based on subsystem reliabilities obtained from reliability assessments. Reliability block diagrams are drawn to aid in the development of the system reliability models and determine the paths for successful operation. It is often assumed that the subsystems that make up a system are independent and can be described in discrete terms as either failed or operational.[1] Probability theory is applied to each subsystem, until the whole system is analyzed. Failure information is then introduced to estimate the system's reliability.

8.11.1   Series Reliability

If subsystems in a system are connected in a manner (not necessarily physically) such that system failure would occur as a result of the failure of any of the subsystem, then those subsystems are considered to be connected in series. The success of the system can thus be no better than that given by the lowest probability of successful operation of any subsystem within the series system.

A system of n subsystems is assumed to be in series if failure of the system occurs when any subsystem fails. The reliability of the series system, R_{ss}, is therefore the product of the reliabilities of the individual subsystems.

$$R_{ss} = \prod_{i=1}^{n} R_i \tag{8.2}$$

where R_i is the reliability of ith subsystem. This is another way to state that system failure is a function of the weakest link in the system (i.e., $R_{ss} \leq \min\{R_i\}$).

8.11.2   Parallel Reliability

Redundancy implies that if one subsystem fails, there is a parallel or redundant subsystem that will continue to do the work. That is, redundancy suggests that there exist backup paths for system success. Reliability for such a system is defined as the probability that at least one subsystem is operable between the input and output, although some systems may require that more that one path is operable.

For a basic parallel system in which all subsystems are assumed to be active, n paths for success are possible. For the system to operate successfully, at least one subsystem must operate without failure for the duration of the mission. Conversely, such a system

---

[1]It must be assumed that the reliability of a system is solely a function of the reliability of the subsystems that make up the system. However, reliability-limiting items are often a result of other factors, such as misapplication of a component, inadequate timing analysis, lack of transient control, improper interconnections between systems and subsystems, and poor handling.

fails only when all the subsystems fail. This gives rise to the mathematical expression for the system unreliability Q_{ps} in terms of the subsystem unreliabilities Q_i.

$$Q_{ps} = \prod_{i=1}^{n} Q_i = \prod_{i=1}^{n} (1 - R_i) \qquad (8.3)$$

The system reliability can thus be expressed as

$$R_{ps} = 1 - Q_{ps} = 1 - \prod_{i=1}^{n} (1 - R_i) \qquad (8.4)$$

Parallel systems will always generate better reliability than series systems because parallel systems provide more than one subsystem to do the job, and if one fails there is another to take its place. In fact, the reliability of an active parallel system will always be greater than that of the subsystem with the highest reliability.

The reliability R_s of a system in which at least r out of n possible subsystems must operate if the system is to operate is given by

$$R_s = \sum_{i=1}^{n} \binom{n}{i} R^i (1 - R)^{n-i} \qquad (8.5)$$

where R is the subsystem reliability and is assumed equal for all subsystems, and

$$\binom{n}{i} = \frac{n!}{i!(n-i)!} \qquad (8.6)$$

Examples of r out of n redundancy occur in computer memory subsystems in which a certain minimum number of bytes are necessary to support an outcome, and in systems such as a space shuttle, where a minimum number of similar computer outcomes are required for a safe flight.

8.11.3 Mixed Series-parallel Reliability

For a series-parallel system, such as that shown in Figure 8.2 in block diagram form, the reliability for each parallel unit i is

$$R_i = 1 - \prod_{j=1}^{n} Q_{ij} \qquad (8.7)$$

where n is the number of parallel systems j in the unit. For m subsystem in series, the system reliability R_T is thus

$$R_T = \prod_{i=1}^{m} R_i = \prod_{i=1}^{m} \left(1 - \prod_{j=1}^{n} Q_{ij} \right) \qquad (8.8)$$

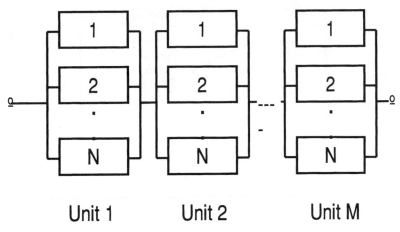

Unit 1   Unit 2   Unit M

Figure 8.2   Series-parallel system.

If all the systems are identical such that $Q_{ij} = Q$, then

$$RT = (1 - Q^n)^m \qquad (8.9)$$

For the case of a parallel-series configuration (diagramed in Figure 8.3), the reliability of the series systems for each path i is calculated first

$$R_i = \prod_{j=1}^{n} R_{ij} \qquad (8.10)$$

where n is the total number of systems j in path i. Then taking the reliability for each of the m paths in parallel gives a total system reliability R_T of

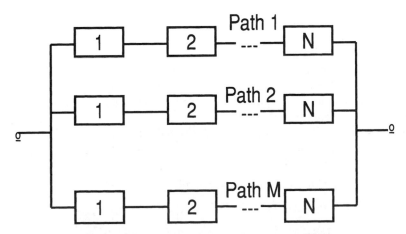

Figure 8.3   General parallel-series configuration.

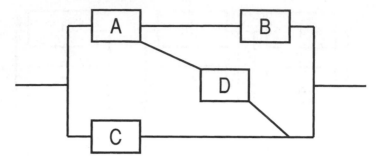

Figure 8.4 Complex system.

$$R_T = 1 - \prod_{i=1}^{m} (1 - R_i) = 1 - \prod_{i=1}^{m} \left(1 - \prod_{j=1}^{n} R_{ij} \right) \tag{8.11}$$

If all paths are identical with identical systems such as $R_{ij} = R$, then

$$R_T = 1 - (1 - R^n)^m \tag{8.12}$$

Another type of parallel redundant system is the voting redundant system. A voting redundant system consists of three or more systems in active redundancy. A comparator samples the outputs of the systems and switches off systems that do not agree with the majority. The system can operate successfully if at least two (i.e., a majority) of the systems are operating properly. The reliability of the three-part voting redundant system is calculated as

$$R_{sy} = R_A R_B + R_A R_C + R_B R_C - 2 R_A R_B R_C \tag{8.13}$$

8.11.4 Complex System Analysis

System architectures that cannot be resolved into combinations of parallel and series configurations are called *complex systems*. An example of a complex system is given in Figure 8.4. Although there are many approaches to solving complex configurations, there is no generally efficient method. One method of determining the reliability of a complex system is by enumeration, whereby all possible configurations are examined to see whether the system can operate in those configurations. Table 8.1 lists all possible states for Figure 8.4 along with those that are operable.

Using the data from Table 8.1 and letting $a = p(A)$, $1 - a = p(\bar{A})$, $b = p(B)$, and so on, the system reliability, R, is

$$
\begin{aligned}
R = {} & abcd + abc(1 - d) + ab(1 - c)d + ab(1 - c)(1 - d) \\
& + a(1 - b)cd + a(1 - b)c(1 - d) + a(1 - b)(1 - c)d \\
& + (1 - a)bcd + (1 - a)(1 - b)cd + (1 - a)(1 - b)c(1 - d) \\
& + (1 - a)bc(1 - d) = c + ab + ad - abc - acd - abd + abcd
\end{aligned}
\tag{8.14}
$$

TABLE 8.1 Complex System Analysis Chart

Operating States		Nonoperating States
$A\,B\,C\,D$	$\bar{A}\,B\,C\,D$	$A\,\bar{B}\,\bar{C}\,\bar{D}$
$A\,B\,C\,\bar{D}$	$\bar{A}\,B\,C\,\bar{D}$	$\bar{A}\,B\,\bar{C}\,D$
$A\,B\,\bar{C}\,D$	$\bar{A}\,\bar{B}\,C\,D$	$\bar{A}\,B\,\bar{C}\,D$
$A\,B\,\bar{C}\,\bar{D}$	$\bar{A}\,\bar{B}\,C\,\bar{D}$	$\bar{A}\,\bar{B}\,\bar{C}\,D$
$A\,\bar{B}\,C\,D$		$\bar{A}\,\bar{B}\,\bar{C}\,\bar{D}$
$A\,\bar{B}\,C\,\bar{D}$		
$A\,\bar{B}\,\bar{C}\,D$		

In general, calculation of system reliability for even a moderately large complex system is extremely time-consuming. As a result, a number of researchers have proposed techniques to determine the bounds on reliability using minimal paths and minimal-cut methods. Basically, a minimal path is the minimal set of operating systems that ensures system operation and a minimal cut is a minimal set of failed systems that ensures system failure.

8.11.5 Reliability of Standby Systems

A standby system consists of parallel redundancies, except that the standby subsytems wait idle until required to function. A monitor detects when the active system has failed and singles the switch to activate or connect a standby subsystem. A switch controls which one of the redundant subsystems is active.

To incorporate time dependencies in the model, we present the case of two subsytems in a standby redundancy and neglect the switch and monitor. The probability that the primary subsystem fails during time t_1 is

$$P(E_1) = f_1(t_1)\,dt_1 \tag{8.15}$$

The probability that the secondary subsystem fails after time t_1 is

$$P(E_2) = f_2(t - t_1)\,dt \tag{8.16}$$

The system failure in time dt is

$$f_s(t)dt = \int_{t=0}^{t} f_1(t_1)\,dt_1 f_2(t - t_1)\,dt \tag{8.17}$$

The system reliability is thus

$$R_s = \int_{t}^{\infty} f_s(\tau) = \int_{t}^{\infty} \int_{0}^{t} f_1(t_1)f_2(\tau - t_1)\,dt_1\,d\tau \tag{8.18}$$

8.11.6 Redundancy Tradeoffs and Optimization

Although redundancy methods can be effective in increasing overall system reliability, implementing redundancy involves designing for alternative paths to connect the redundant subsystems, while meeting space, weight, and any other requirements. Thus, it is necessary to trade off the "cost" of the improvement in reliability with the "cost" of the constraints. Tillman et al. (1977) present a review of solution procedures to maximize system reliability subject to resource constraints, such as cost, weight, or power.

Redundancy may be justified when the increased cost due to additional systems and connections is compensated for by increased reliability. Redundancy may not be justified when one cannot satisfactorily answer questions such as how reliable the new interconnections are, whether there is any interaction among the redundant subsystems, whether if one subsystem fails the other(s) can carry the load properly, how the systems will be switched in a standby system, and whether spurious results can occur. Achieving the desired reliability while minimizing the forementioned factors is of paramount importance.

8.12 FAULT TREES

Fault-tree analysis consists of determining the probability of occurrence of a system event, given the probability of occurrence of the subsystem events (the usual assumption is that the subsystem events are statistically independent). Since the fault tree represents failure criteria (as contrasted with the reliability block diagram, which represents success criteria using path-sets), fault trees are analyzed by generating the cut-sets of the system. Thus, the path-sets establish the success criteria for a system, while the cut-sets establish the failure criteria. A *minimal cut-set* contains the minimum number of elements whose failure ensures system failure.

A *fault-tree* model is a logical representation of the failure criteria of a system. The system event of a fault tree generally represents the failure of the system being analyzed, and is broken down into its constituent causes. These contributing causes are connected by logical gates, including, for example, the *AND*, *OR* and *m-out-of-n* gates (An *m-out-of-n* gate is true when *m* of the *n* inputs have occurred.) The system event *(system failure)* is caused by A_1 and A_2 failing or by B_1 and B_2 failing. The subsystem events of a fault tree (i.e., A_1 and B_1) usually represent subsystem failures.

8.13 FAILURE MODES AND EFFECTS ANALYSIS

A FMEA is an evaluation process for analyzing and assessing the potential failures in a system. The objective is to determine the effect of failures on system operation, identify the failures critical to operational success and personnel safety, and assess each potential failure according to the effects on other portions of the system. In general, these objectives are accomplished by itemizing and evaluating system composition and functions.

FMEAs were first utilized in the 1960s and were used in the Apollo space program to evaluate the reliability and criticality of the complex systems required for spaceflight. Today, FMEAs are used early in the design and testing phases to avoid costly modifications by ferreting out latent design and operational deficiencies to ensure a high level of achieved reliability before the initiation of full-scale production. FMEA is also a system of the logistic support analysis used to help determine the maintenance tasks required for each failure, the repair systems material list, and the personnel requirements for support of a failure.

There are two approaches to FMEA: functional and hardware. The functional approach is normally used when the system definition has been identified, making a system level-down analysis the more practical approach. However, the function-oriented approach is not as detailed as the hardware approach, and as a result certain failure modes can be overlooked.

The functional approach to FMEA begins with the initial system indenture level and proceeds downward through the lower indenture levels. The indenture levels are determined from the system information flow diagrams. Flow diagrams are employed as a mechanism for portraying system design requirements, illustrating series-parallel relationships, the hierarchy of system functions, and functional and hardware interfaces. The flow diagrams are designated as top level, first level, second level, and so on. The top level shows gross operational and mission requirements, while the lower level diagrams represent progressive expansion of the individual functions of the preceding level. Normally this documentation is prepared down to the level necessary to establish the hardware, software, facilities, personnel, and data requirements of the system. This system identification methodology permits a natural extension to the hardware approach.

The hardware approach to FMEA is practical only when the RMA engineering team has access to schematics, drawings, and other engineering and design data normally available once the system has matured beyond the functional design stage. This approach is initiated in a part level-up fashion. For example, in electronic systems the lowest level may be associated with discrete electronic components making up printed wiring-board assemblies (PWAs). PWAs can be line replaceable units (LRUs) or subassemblies of a chassis that has been designated an LRU in place of the PWAs. A chassis-level LRU usually represents a subsystem.

Components can also refer to building blocks for mechanical systems. Consider, for example, a ground-based radar system. Here the antenna is a subassembly, but it is highly mechanical in nature. The antenna is built up from waveguides, gimbals, servomotors, gear assemblies, and the like. These would all represent subassemblies and finally components of the antenna.

The level of detail refers to the system hardware level at which the failures are postulated. In relation to the functional approach, the failures are associated with the input and output requirements of the particular system under evaluation. The hardware approach level of detail includes failures that could occur at the individual component level. During the conceptual stage the level of detail will be driven by the design status maturity. The benefit of a lower-level-of-detail FMEA performed early in the design development cycle is the uncovering of design mistakes.

To be effective, the engineering team needs to determine the failure modes and effects accurately. A poor understanding of the system design and operation can lead to incorrect failure-mode classifications, resulting in inaccurate FMEAs. The general guidelines that should be followed in performing a FMEA are:

- Select the approach for the analysis, either functional or hardware.
- Define the system utilizing block diagrams and indenture-level numbering.
- Define the system or mission requirements.
- Define the failure modes and the failure mechanisms, indicating any environmental stress factors if known. Failure modes include premature operation, failure to operate at a prescribed time, failure to cease operation at a prescribed time, and failure during operation.

• Generate required corrective action recommendations for the design or manufacturing process and evaluate new procedures and their effectiveness, including the recommendations for system reliability improvements.

8.14 DESIGN FOR MAINTAINABILITY

Maintainability refers to the ability to keep an existing level of system performance through the use of preventive or corrective repairs. Due to the expense of labor, design for maintainability is generally based on system cost vs. repair cost basis. As a result, certain systems may be disposable while others are repairable. Thus, design for maintainability is generally reserved for high-cost modular systems.

The purpose of design for maintainability is to facilitate easy maintenance and repair. The ease of maintenance depends on the architecture of the system (see Section 8.4). Maintainability activities are generally included with field support of the equipment, including overall maintenance concepts, spare parts provisioning, and in many cases safety analysis.

Design features that will expedite maintenance will enhance maintainability. Thus, in design for maintainability, the system should be fashioned so that the primary maintenance consists of subsystem replacement. Design for maintainability also dictates that these replaceable subsytems be easily accessible and removable. For example, a unit designed for socket insertion and manual latch-lock retention is preferable to one requiring the use of screws and fasteners or the reflowing and reformation of solder joints because it reduces the design complexity, reduces the service time requirements, reduces the service interval sophistication, reduces the service technician training requirements, and ultimately reduces the costs transferred to the customer.

The design team needs to consider how the equipment actually will be repaired in the field, perhaps under the pressures of inexperienced maintenance crews, or by field service personnel with limited knowledge about specific system failure modes. The design team must always keep in mind that equipment may be used and repaired by people who have other things in mind than "babying" the equipment. The design team must also realize the difference between what people will probably do and what the design team thinks they ought to do.

The design team, in acknowledging that the system must be maintained and can fail, should provide means for ease of maintenance, ease of removal of a failed unit, troubleshooting, access to the failed unit, and repair or replacement. Such means, in addition to improving maintainability, will also improve reliability through the averting of subsequent failures due to human errors during maintenance. Many design details, often overlooked, are important to the maintenance technician. A list of "Don'ts" in designing for maintainability is presented below (Pecht, 1995).

1. Don't place systems or maintenance structures (e.g., oil filter) where they cannot be removed without removal of the whole unit from its case or without first removing other systems.
2. Don't put an adjustment out of arm's reach.
3. Don't neglect to provide enough room for the technician to get a gloved hand into the unit to make an adjustment.

4. Don't put an adjusting system in such a place that the technician has difficulty in locating it.

5. Don't screw subassemblies together in such a way that the maintenance technician cannot tell which screw holds what.

6. Don't use chassis and cover plates that fall when the last screw is removed.

7. Don't make sockets and connectors for modules of the same configuration so that the wrong unit can be installed.

8. Don't provide access doors with numerous small screws or attachments.

9. Don't use air filters that must be replaced frequently; don't place air filters so that it is necessary to shut down power and disassemble the equipment to reach them.

10. Don't omit the guide from a screwdriver adjustment, so that while the technician is adjusting and watching a meter, the screwdriver slips out of the slot.

11. Don't design frequent failure or highly adverse impact failure items in least accessible locations.

12. Don't unnecessarily subject nonwear components to failure stress caused by maintenance actions.

Design mistakes are often the bane of the maintenance technician. The design team is not always made aware of these mistakes. In the following list are some design errors that should be avoided (Pecht et al., 1994).

1. Design for minimum maintenance skills; some technicians are neither well-trained nor well-motivated.

2. Design for minimum tools; special tools and laboratory test equipment may not always be available.

3. Design for minimum adjusting, adjustment for shift, drift, and degradation should not be necessary in most cases.

4. Use standard interchangeable systems wherever possible; special systems create problems.

5. Group subsystems so they can be easily located and identified.

6. Provide for visual inspection.

7. Provide troubleshooting techniques including panel lights, telltale indicators, built-in test panels, and the like. Use overload indicators, alarms, and lighted fuses to assist in locating a failure.

8. Provide test points. Use plain marking and adequate spacing and accessibility.

9. Label units; labels on top of components systems and structures (or from the direction of normal human access) should agree with instruction manuals to aid in locating suspected problems.

10. Use plug-in modules when possible; ease in replacement avoids errors and mutilation of harnesses.

11. Orient plug-ins all in the same direction to avoid the need to look at each key position.

12. Provide handles or handholds on heavy components for easy handling.

13. Use BAD-GOOD meters, red-lined meters, or tolerance bands. Where possible, avoid using meters that must be read and evaluated from a document or table.

14. Design for safety; use interlocks, safety covers, and guarded switches, and do not permit exposed high voltages.

15. Minimize the number of removable bolts or screws in access panels or doors.

8.15 DATA COLLECTION, CLASSIFICATION, AND REPORTING

RMA data collection, classification, and reporting must be complete and consistent with the development of the system. RMA data come from manufacturing, assembly, yields, tests, and field characterizations. The RMA team must be aware of the source of data, the specific failure modes uncovered in failure analysis, the manufacturing and handling conditions, the application of the system in the field, and the applied stress history. Test data often neglect the kinds of stresses that actually arise in the field while field data collection can be inadequate and inaccurate in the sense that a failure is often associated with a removal rather than with a well-identified failure mechanism (Pecht and Ramappan, 1992). In many cases, field removals are retested as operational (called RTOK for retest OK) and the true cause of failure is never determined. In other cases, the cause of failure is not reported or is inaccurately reported, especially if the failure was caused by an incorrect installation, a mishandling, or an incorrect use (Pecht and Ramappan, 1992).

Failure diagnosis and corrective actions must be part of a continuous system improvement program. When the only goal is to meet warranty requirements, there is seldom any interest in further diagnosis and corrective action after meeting goals. This can provide the basis for a hindrance to continued improvements in reliability.

Failures are typically grouped into one of three classifications, depending on when the failure occurs during a system's operating life. The classifications are infant mortality, useful life, and wearout failure.

Infant mortality failures occur early in a system's operating life. The definition is that the rate of failures decreases with age. The causes of early failures include:

- Poor workmanship
- Poor manufacture, assembly, storage, and transportation
- Poor quality control (flawed materials, contamination)
- Insufficient burn-in debugging
- Improper start-up

The associated rate of failure during the "useful" life of the system is assumed to be constant. The causes of failures include:

- Inadequate derating, higher than expected random loads, and/or lower than expected strength characteristics
- Acts of humans, such as abuse or misapplication
- Acts of God, such as lightening or flood

Wear-out failures occur late in the operating life of a system. The definition is that the rate of failure increase with age. The causes of wear-out failures include:

- Aging
- Wear
- Creep
- Fatigue
- Poor service, maintenance, or repair

8.16 WARRANTIES AND LIFE-CYCLE COSTS

The expectations of RMA often affect the warranty terms. In some cases, suppliers may only be required to meet contractual goals without incentive for, or interest in, continued improvement. That is, the concept of "attainable maximum" often provides an easily achieved cap on expectations.

There are many other warranty arrangements that are often intended to encourage suppliers to treat system RMA seriously. For example, the desired reliability goals bear economic considerations that affect life-cycle costs. Those costs are usually included in the fundamental economic analysis to determine economic feasibility of the total program and, in some cases, can be an important item in the total cost of ownership.

Variables that affect life-cycle cost include maintenance, support, personnel, spares, and a myriad of communications factors. Cost studies can be quite complex. For example, when RMA are major elements, as is the case with aviation equipment, initial dollar cost can be less significant than factors such as weight, volume, and power consumption, but must be defensible in terms of long-term system value.

8.17 OPERATIONAL READINESS AND AVAILABILITY

The capability of a system to perform its intended function when called upon is its operational readiness or its operational availability. The difference between readiness and availability is that the latter includes only operation and down times, while the former also includes free and storage times when the system is not needed. Operational readiness and availability emphasize the "when called upon" aspect of the task or mission. The emphasis is on a probability at a point in time rather than over an interval, as is the case with the mission success rate (the percentage of successfully completed missions). This interval of time can be extremely long, as in the case of a satellite on a long-term mission to another planet; the satellite may be operationally available at launch time, but that does not ensure that it will operate successfully for the duration of its mission. For systems that are continually used and are providing useful output, availability is often estimated by calculating the fraction of total "need time" that the system is operational or capable of providing useful output.

Operational availability addresses the system's readiness to perform its intended function at a particular point in time. The difference between a system's being up or down, however, is often a function of the customer's definition of failure, which depends on the use of the system. Thus, if the performance related to a critical attribute is not satisfactory, the customer may consider the system to be "down," and readiness or availability from that point until the need ends, or until the deficiency is corrected, is zero.

For example, if a radar has a specified range of 50 miles, is the radar considered down if it is effective only to 45 miles? If the 50-mile range is the absolute minimum needed to avoid midair collisions, the aircraft on which the radar is installed would be considered unflyable, and the radar would be considered unavailable for the mission. If the 50-mile range is a goal value and 20 miles is the absolute minimum, a 45-mile range might be acceptable. An availability calculation could be based on a definition that includes as uptime all periods for which the range is at least 20 miles.

In summary, reliability translates into a demand for support resources, while maintainability translates into the range of resources (i.e., people, space) and the time required to support the system operation. The interaction of reliability and maintainability results in the need for support assets to maintain a level of operational readiness or availability over the time desired by the system user.

REFERENCES

Coudert, O., and Madre, J. C. (1995). Fault tree analysis: 10^{20} prime implicants and beyond. *Proc. the Reliab. Maintainab. Symp.*

Dugan, G. B., Bavuso, S., and Boyd, M. (1992). Dynamic fault tree models for fault tolerant computer systems. *IEEE Trans. Reliab.*

Kelkar, N., Dasgupta, A., Pecht, M. G., Knowles, I., and Hawley, M. (1996) "Smart" electronic systems for condition-based health management. In *Condition Monitoring and Diagnostic Engineering Management, (COMADEM)*, pp. 591–601, London: Academic Press.

Lass, R. (1975). Optimization of system reliability by a new nonlinear integer programming procedure. *IEEE Trans. Reliab.* **R-24**(1), 14–16.

Martz, H.F., and Waller, R. A. (1982). *Bayesian Reliability Analysis*. New York: Wiley.

Pecht, M. G. (1995). *Product Reliability, Maintainability and Supportability Handbook*. Boca Raton, FL: CRC Press.

Pecht, M., and Dasgupta, A. (1995). Physics-of-failure: An approach to reliable product development. *J Inst. Environ. Sci.,* September/October, pp. 30–34.

Pecht, M. G., and Ramappan, V. (1992). Are components still the major problem: A review of electronic system and device field failure returns. *IEEE Trans. Components, Hybrids, Manuf Techno.* **15**(6), 1160–1164.

Pecht, M. G., Dasgupta, A., Evans, J., and Evans, J. (1994). *Quality Conformance and Qualification of Microelectronic Packages and Interconnects*. New York: Wiley.

Smith, C. O. (1976). *Introduction to Reliability in Design*. New York: McGraw-Hill.

Tillman, F. A., Hwang, C. L., and Kou, W. (1977) Optimization techniques for system reliability with redundancy—a review. *IEEE Trans. Reliab.* **R-26,** 148–155.

9 Concurrent Engineering

ANDREW KUSIAK and NICK LARSON

9.1. INTRODUCTION

The process of bringing new products to market may vary dramatically across cultures, industries, and firms; however, the following core phases are essential: marketing, design, manufacturing, and sales. Each of these phases can be decomposed into lower-level activities that also remain fairly common. For example, the design process typically includes recognition of needs, development of requirements, conceptual design, embodiment design, and detailed design. At lower levels of detail, firms begin to adopt distinct methods for executing and managing the function of design. However, the process of designing different products within the firm usually remains the same.

Trends in business and society have contributed to the increasing complexity of the design process. The customer continues to demand variety, quality, and service in an economical product. In an attempt to address the needs of the customer, competition has become fierce in all sectors of the economy. As a result, companies have tightened operations to become more efficient. Popular terms such as concurrent engineering and business process reengineering have loosely labeled very stern efforts to improve the product development process. The organizational structure of firms has flattened out by eliminating the functions of middle-level management. The capability to perform such restructuring is a result of the information revolution. Computer hardware and software have enabled rapid completion of complex technical tasks, but perhaps the most dramatic illustration of the new reliance on technology is e-mail. Many traditional physical boundaries have been eliminated by mechanisms that provide instant transfer of information between the desktops of engineers, managers, sales personnel, and all others involved in the process. Therefore, the concept of concurrent engineering (see Section 9.2), which requires the participation of many functional disciplines, has become a reality.

Before providing a complete definition of concurrent engineering, it is helpful to highlight some of the drawbacks to the traditional "throw it over the wall" design process. In this approach, the design is passed from the design department to the manufacturing department, where manufacturing experts determine how to make the product. The design process is managed internally and there is little or no communication with experts in other functions, such as manufacturing, assembly, quality control, finance, marketing, and customer service. This lack of communication often results in redesign, which increases product cost and delays the release of the product to market. Furthermore, design seldom has access to the data needed to evaluate manufacturability, assemblability, serviceability, and quality. As a result, design engineers make uninformed decisions during the first iteration of the process, once again leading to redesign.

Handbook of Systems Engineering and Management, Edited by A. P. Sage and W. B. Rouse
ISBN 0471-15405-9 ©1999 John Wiley and Sons, Inc.

This chapter presents a systems engineering perspective of concurrent engineering that focuses on integrating people, processes, problem-solving mechanisms, and information. Several of the most common problems facing systems engineers are explored. These problems are illustrated with examples and when additional information is needed on specific techniques, the appropriate references are provided. Section 9.2 takes a closer look at concurrent engineering and the product life cycle. For more information on engineering design methodologies and management of product design, see Kannapan and Marshek (1992) and Mallick and Kouvelis (1992), respectively.

9.2. CONCURRENT ENGINEERING AND THE PRODUCT LIFE CYCLE

Concurrent engineering (CE) can actually be defined in two ways: (1) the practice of considering the entire functionality of the product, as well as its assembly and manufacture, in an integrated design process; or (2) the practice of considering the entire product life cycle, from design to disposal, in an integrated design process. The first definition was the original vision for concurrent engineering initiatives that focused on complex products. Thus, concurrent engineering was considered a response to the problem of product designs being incompatible with existing manufacturing and assembly processes. As the concept of concurrent engineering evolved, technological advances, societal concerns, and the increasing demand for smaller, simpler products broadened the scope of concurrent engineering to include the entire product life cycle. For a complete description of concurrent engineering and its history see, for example, Carter and Baker (1992).

The foundation for concurrent engineering is in the design for manufacture and design for assembly concepts (Boothroyd and Dewhurst, 1988). Design for manufacture and assembly (DFMA) has been widely implemented as both a philosophy of design and design support software. DFMA methods establish guidelines for manufacturing and assembling components and store the information in vast data repositories that can be queried by engineers during design. With the advances in computer technology over the last decade, powerful DFMA software systems have become readily available.

The ability to clearly link customer requirements to the design of the product and obtain the highest level of quality are also essential in concurrent engineering. Quality function deployment (QFD), also known as the "House of Quality" (Hauser and Clausing, 1988), has become popular for relating customer requirements to specific components of the design. To improve quality in concurrent design, Taguchi's concept of robust design is used to establish acceptable tolerances that achieve both part function and realistic manufacturability (Byrne and Taguchi, 1986). DFMA, QFD, and robust design are examples of specific methods. However, the need to integrate experts from all functions in the product development process remains at the heart of concurrent engineering. This demands a cultural change that is often pursued very differently in different firms.

Concurrent engineering initiatives have produced impressive results in such U.S. companies as Hewlett-Packard, Raytheon, John Deere & Co., AT&T, and Boeing (Rosenblatt and Watson, 1991). The U.S. government has also encouraged the practice of concurrent engineering by supporting the DARPA Initiative in Concurrent Engineering (DICE) in the U.S. military, industry, and academic research (Reddy et al., 1991). A recent survey reported that various-sized companies have realized dramatic benefits from concurrent engineering in a broad range of industries (see Table 9.1).

The latest trend in concurrent engineering is to broaden the scope to include the entire product life cycle (Alting, 1993), rather than simply manufacturing, assembly, and quality.

TABLE 9.1   Benefits of Concurrent Engineering

Performance Measure	Benefit
Development time	30–50% less
Engineering changes	60–95% less
Scrap and rework	75% reduction
Defects	30–85% fewer
Time to market	20–90% less
Field failure rate	60% less
Service life	100% increase
Overall quality	100–600% higher
White-collar productivity	20–110% higher
Return on assets	20–120% higher

Source: Lawson and Karandikar, 1994.

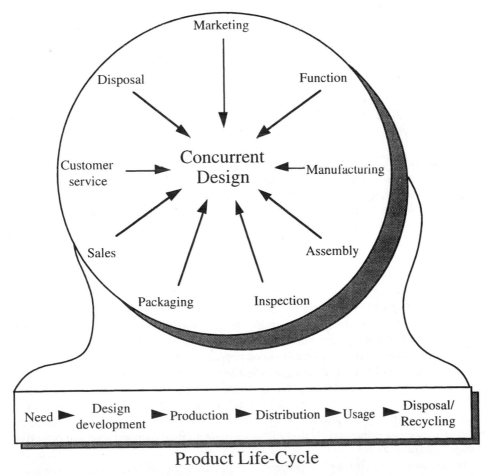

Figure 9.1   The concurrent engineering environment.

Initiatives such as design for serviceability and design for the environment bring focus to issues in previously overlooked phases of the product life cycle. In keeping with the traditional terminology, design for X (DFX) has become a popular term, where X can now represent any number of "ilities" associated with the product life cycle (i.e., manufacturability, quality, serviceability, disposability). Figure 9.1 illustrates the resulting structure of the concurrent engineering environment. With the increasing number of perspectives involved with design and the tremendous amount of information available, the need for systems engineers in concurrent engineering has never been greater.

9.3 BUILDING A CONCURRENT ENGINEERING ENVIRONMENT: A SYSTEMS ENGINEERING PERSPECTIVE

Concurrent engineering is a much more complex approach to product development than traditional methods of design. Therefore, installing a concurrent engineering environment requires the expertise of systems engineers. This section presents a systems engineering perspective of concurrent engineering that emphasizes four components: (1) people, (2) processes, (3) problem-solving mechanisms, and (4) information. The objective of the systems engineer is to determine an appropriate strategy for integrating these components of product development to achieve a concurrent engineering environment.

9.3.1 Integrating People

Integrating the people that perform the vast array of functions in the product development process is at the heart of concurrent engineering. This section focuses on two specific concurrent engineering tasks that require the expertise of systems engineers: building concurrent engineering teams and negotiation in engineering design.

9.3.1.1 Building Concurrent Engineering Teams. The concept of multifunctional teams is one of the key aspects in concurrent engineering. Specialists from various disciplines, such as design, manufacturing, quality control, and marketing, work in a group rather than individually in order to design the product. The team approach has been used, for example, by the Ford Motor Company in the development of the Ford Taurus (Belson, 1994).

The membership of a concurrent engineering team depends on the type of product to be developed, customer requirements, engineering and product characteristics, and so on. Numerous teams are often organized in a hierarchy, where each team is concerned with design of a subsystem of the product. Askin and Sodhi (1994) presented an approach to organizing teams in concurrent engineering. They developed five different criteria for team formation and discussed team training, leadership, and computer support issues. A labor assignment heuristic was developed for team formation. Zakarian and Kusiak (1995) presented a conceptual framework for prioritizing team members based on customer requirements and product characteristics. This approach combines design and decision analysis tools, such as quality function deployment (QFD) and the analytic hierarchy process (AHP), with mathematical programming to allocate personnel to concurrent engineering teams. This section elaborates on this approach to illustrate the importance and objectives of team selection.

The use of QFD facilitates the process of concurrent engineering and promotes teamwork toward the common goal of assuring customer satisfaction (Smith, 1991). QFD was first introduced in 1972 by Akao at Mitsubishi Industries Kobe Shipyard and was then popularized as the House of Quality in the United States (Hauser and Clausing, 1988). In the team selection process, the House of Quality is used to collect and represent the data for the multifunctional team selection model. Figure 9.2 illustrates the planning matrices used in the house of quality. The customer's requirements are propagated through the matrices to eventually relate the components of the product to the potential team members who are responsible for producing those components (Fig. 9.2c).

After the components—team members matrix is constructed, the AHP methodology (Saaty, 1981) may be deployed for prioritizing team members with respect to each component of the product. The AHP is a multicriteria decision making method that uses a hierarchy to represent a decision problem. At the top of the hierarchy, the goal (objective) upon which the best decision should be made is placed. The next level of the hierarchy contains attributes or criteria that contribute to the quality of the decisions. Each attribute can be decomposed into more detailed attributes. The lowest level of the hierarchy contains decision alternatives.

After the hierarchical network is constructed, one can determine the priorities of the elements at each level of the decision hierarchy and synthesize these measures to determine the priorities of decision alternatives. Pairwise comparisons are made to first determine the relative importance of each criterion with respect to the overall goal, and then to determine the relative importance of each alternative with respect to each criterion. The pairwise comparisons are made using the nine-point scale developed by Saaty (1986). Table 9.2 summarizes Saaty's relative-importance scale modified for the concurrent engineering team selection problem. Typically, the decision maker has to evaluate the upper triangular part of the comparison matrix, while reciprocals are placed in the lower triangular part. Therefore, the diagonal elements of the matrix are always one.

To select a concurrent engineering team, one must elicit all possible characteristics or attributes desired, and then develop a method for assessing their importance to the overall goal. AHP is used as a framework to develop the measure of priority for components of the product, as well as the measure of priority of the team members with respect to each distinct component. The priority measures can then be used in a mathematical model to allocate personnel to concurrent engineering teams.

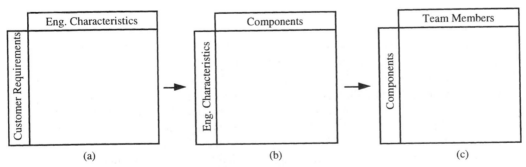

Figure 9.2   The basic House of Quality matrices: (*a*) customer requirements–engineering characteristics planing matrix; (*b*) engineering characteristics–components of the product planning matrix; and (*c*) components of the product–team members planing matrix.

TABLE 9.2 Scale of Relative Importance

Scale Value	Definition	Explanation
1	Equal importance	Two members contribute equally to the goal
3	Moderate importance one over another	Experience and judgment slightly favor one team member over another
5	Essential or strong importance	Strongly favor one team member over another
7	Demonstrated importance	A team member is strongly favored and its dominance is demonstrated in practice
9	Extreme importance	The evidence of favorite one team member over another is of the highest possible order of affirmation
2, 4, 6, 8	Intermediate values between two adjacent judgments	When compromise is needed
	Reciprocal	For inverse comparison

Source: Zakarian and Kusiak, 1995.

Example 9.1 *Automobile Design.* Consider an automotive company that intends to select a concurrent engineering team to develop a car that satisfies the customer requirements. Therefore, a pairwise comparison of the importance of possible team members with respect to the overall goal (development of a car) is required. Before applying the AHP, QFD planing matrices are constructed. First, the customer requirements are determined and related to the engineering characteristics (see Fig. 9.2*a*). Next, the desired engineering characteristics are related to the components (subsystems) of the car (see Fig. 9.2*b*). Finally, the subsystems of the car–potential team members planing matrix is constructed (see Fig. 9.3).

The matrix in Figure 9.3 is used to organize the factors considered in the decision-making process into a hierarchical structure. The highest level in the structure is the goal (i.e., build a car). The factors contributing to the goal (i.e., subsystems of the car) are placed on the next level. The decision alternatives (i.e., team members) are placed at the lowest level of the hierarchy. The hierarchical structure is shown in Figure 9.4. (Note that the figure has been abbreviated for simplicity. In practice, every element on a given level of the hierarchy is connected to every element on the next lowest level.) In order to select the best team for each particular subsystem of the car, the team members included in the hierarchy have to be prioritized.

Table 9.2 is then used to make pairwise comparison matrices for the goal with respect to the subsystems, and for the subsystems with respect to the team members. The weight vectors for these matrices are then calculated and aggregated to determine the priority measures for each team member with respect to each subsystem and normalized subject to the overall goal (see Table 9.3). Higher values in the table denote higher priority for team selection For details on performing the calculations, see Zakarian and Kusiak (1995).

The multifunctional teams selection problem is formulated as an integer programming problem (Zakarian and Kusiak, 1995). The model is based on the product com-

Team Members \ Subsystems	Power brakes	Transmission	Engine	Design	Lighting system
Mechanical engineer (ME)					
Manufacturing engineer (MF)					
Design engineer (DE)					
Quality engineer (QE)					
Finance expert (FE)					
Electrical engineer (EE)					
Reliability engineer (RE)					

Figure 9.3   Team members–components (subsystems) of the car planning matrix.

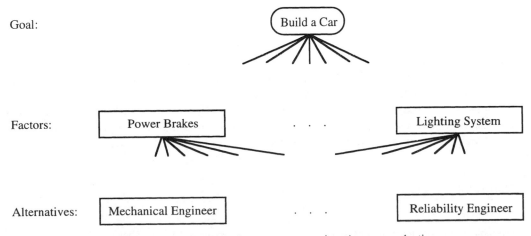

Figure 9.4   Hierarchy for concurrent engineering team selection.

TABLE 9.3   Normalized Importance Measure of Each Team Member with Respect to Each Subsystem

Subsystem	ME	MF	DE	QE	FE	EE	RE
Power brakes	0.0413	0.0245	0.0158	0.0120	0.0030	0.0185	0.0300
Transmission	0.0206	0.0081	0.0135	0.0090	0.0020	0.0199	0.0243
Engine	0.0804	0.0514	0.0574	0.0260	0.0066	0.0254	0.0574
Design	0.0168	0.0727	0.1302	0.0466	0.0668	0.0277	0.0584
Lighting system	0.0024	0.0014	0.0011	0.0043	0.0060	0.0138	0.0086

ponent–team-member-type priority incidence matrix (i.e., Table 9.3). Each row of the matrix corresponds to a distinct component of the product and each column denotes a type of a team member. Each entry in the incidence matrix indicates the priority weight of a team member with respect to a product component. The mathematical programming model is used to group potential team members into concurrent engineering teams. For example, suppose that the available number of engineers are as defined in Table 9.4. Solving the mathematical model defines the teams in Table 9.5.

9.3.1.2   Negotiation in Engineering Design.

The objective of concurrent engineering is to simultaneously consider all life-cycle perspectives, including functionality, manufacturability, reliability, and serviceability. During the design process, different perspectives impose certain relationships, or constraints, among features of the design. A constraint can be expressed in terms of equations, tables, charts, curves, or qualitative relationships. In constraint-based design, decision making involves retraction, creation, and/or refinement of constraints. The design is completed when a set of constraints becomes stable and all of the constraints are satisfied.

Different perspectives rely on varying design models that often lead to conflicts during the design process. A design conflict occurs when a constraint is violated in one perspective as the result of a decision made in another perspective. To maintain the consistency of design solutions and produce an acceptable design, conflicts must be resolved. Since negotiation is costly and difficult, it is important to have a methodology that reduces the negotiation time and improves the quality of design decisions.

Many methods exist that use negotiation models for conflict detection and constraint-based resolution (Mayer and Lu, 1988; Premkumar and Kramer, 1989; Lander et al., 1989; Klein, 1991; Werkman, 1991). Klein (1991) proposed a knowledge-based framework that represented conflict resolution expertise explicitly. Bowen and Bahler (1991) developed a constraint-based approach for conflict detection among designers in different perspectives. Kannapan and Marshek (1993) applied an axiomatic approach based on game theory to determine a compromise solution. Utility theory was used to rate the satisfaction level of the design perspectives. Sycara and Lewis (1991) used multiattribute utility theory to model the preference structure of different perspectives and applied case-based reasoning

TABLE 9.4   Available Engineers for Team Selection

Team Member Type	ME	MF	DE	QE	FE	EE	RE
Available number of team members	4	3	5	3	3	5	4

TABLE 9.5   Concurrent Engineering Teams Defined by the Model

Subsystem	Concurrent Engineering Team
Power brakes	ME, MF, DE, EE, RE
Transmission	ME, DE, QE, EE, RE
Engine	ME, MF, DE, QE, FE, EE, RE
Design	ME, MF, DE, QE, FE, EE, RE
Lighting system	DE, FE, EE

to generate design alternatives as conflicts occur. Several domain-independent negotiation issues, such as constraint relaxation and constraint removal were, introduced.

To illustrate the concept of negotiation in concurrent design, this section focuses on a method introduced by Kusiak and Wang (1995). A qualitative constraint network is used to characterize the qualitative and quantitative dependencies among design constraints for each perspective. When conflicts occur, qualitative reasoning (deKleer and Brown, 1984; Kuipers, 1986) is used to derive the dependencies among design goals, decisions, and conflict variables. This provides the necessary information to generate strategies for negotiation among different perspectives.

As stated earlier, each design perspective has its own set of features that can be represented with a set of constraints and variables. Features are the attributes that characterize a design from a particular perspective. Consequently, each perspective may have its own set of constraints that interact with each other, as illustrated in Figure 9.5. The variables used by each perspective are classified into three types:

1. *Decision Variable.* Each design perspective is allowed to make independent decisions on the variable. The subjective knowledge, experience, and desires of the designer are used to imprecisely determine the decision variable.
2. *Intermediate Variable.* The variable is determined by processing decision variables.
3. *Performance Variable.* The variable is dependent on the decision variables and is used to measure the performance of a system.

The interacting constraints (concerning shared design features) describe the relationships among different perspectives. A conflict occurs when an interacting constraint is violated. The negotiation process involves resolving conflicts through three main activities:

1. Generation of a proposal (initial value for each variable in conflict) by a design coordinator.
2. Evaluation of each proposal by all relevant perspectives.
3. Modification of a solution generated by each perspective.

When a conflict occurs, a design coordinator (see Fig. 9.5) suggests an initial proposal based on the partial solutions presented by different perspectives. Each perspective evaluates the proposal based on its expertise and goal and determines if it is acceptable. If the proposal is not acceptable, the perspectives involved in the conflict propose modified solutions. This process iterates until an agreement is reached.

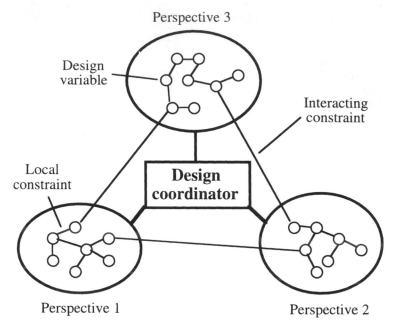

Figure 9.5 The sets of constraints owned by different perspectives.

The power of the method proposed by Kusiak and Wang (1995), as well as similar methods, is in its ability to capture the negotiation problem in a dependency network and perform qualitative reasoning on the effects of changes in design variables. The following example highlights the activities performed in the negotiation process, which is summarized later in Figure 9.10.

Example 9.2 *Valve Design.* The negotiation problem is illustrated with the example of a poppet relief valve. Figure 9.6 shows the schematic of a poppet relief valve that

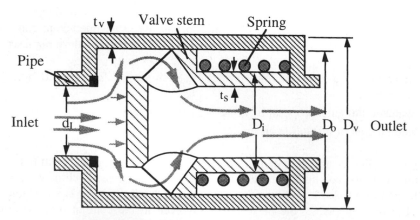

Figure 9.6 Schematic of a poppet relief valve.

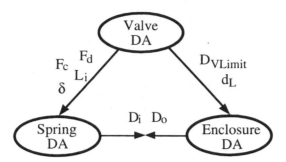

Figure 9.7 Collaboration among valve DP, spring DP, and enclosure DP.

includes a poppet enclosure, poppet valve stem, and helical compression spring enclosed in a pipe (Kannapan and Marshek, 1993). The variables and constraints used are defined in (Lyons, 1982).

This example considers the valve design problem as a constraint-based design problem. Assume that three design perspectives (DPs) are cooperating to design the valve (see Fig. 9.7). The valve DP is responsible for determining the configuration of the valve based on the design requirements. After completing the configuration design, the valve DP passes some parameters to the spring DP and the enclosure DP. The spring DP determines the cracking pressure, the distance the poppet will move, and the stability of the seal in the closed position. The size and the thickness of stem and enclosure are determined by the enclosure DP. Note that the decisions of valve DP dominate the entire design process, because the valve DP makes decisions prior to the other two DPs. All perspectives are assumed to be benevolent, that is, they assist each other in reaching a solution that is acceptable to all perspectives. The design variables for each perspective are listed in Table 9.6.

In Figure 9.8, the sets of constraints used by the spring DP and enclosure DP are represented with the qualitative constraint network. By applying qualitative reasoning, the entire network can be reduced to a simpler network shown in Figure 9.9. For details on constructing and reducing the network, see Kusiak and Wang (1995). The square boxes represent decision variables, while octagons represent performance variables. The qualitative and quantitative dependencies among decision variables, performance variables, and interacting variables are listed in Table 9.7. Once again, for details on calculating the qualitative dependency (δ_{ab}) and quantitative dependency (ψ_{ab}), see Kusiak and Wang (1995). As an example, if the value of variable d increases by 1

TABLE 9.6 Design Variables for Each Perspective

	Valve DP	Enclosure DP	Spring DP
Decision variables	d_L, C_v, C_f, K, L_i	D_V, S_p, A_1, A_2, t_s	D, d, G, r_{s1}, r_{s2}
Intermediate variables	$d_o, F_c, d_{eo}, \delta, F_d$	t_v, D_i, D_o	$K, C_s, K_s, S_s, N, L_s,$ $L_f, D_o, D_i, C_{11}, C_{12}$
Performance variables	$L_i, D_{v\text{Limit}}$	V_V	V_S

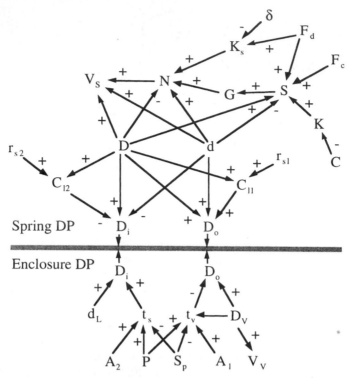

Figure 9.8 Qualitative constraint network for the spring DP and enclosure DP.

percent, then variable D_o increases by 0.0435 percent, and variable D_i decreases by 0.0588 percent.

The set of qualitative strategies is generated by reasoning with the qualitative constraint network. For example, when variable D_o is in conflict, the spring DP has to decrease the value of variable D_o to reach the proposal. Figure 9.9 shows that to decrease the value of variable D_o, one can decrease the values of variables D, d, or r_{s1}. The selection of a strategy usually depends on its impact on the performance variables. If the action to decrease variable D is selected, then the performance variable V_S increases. In contrast to variable D_o, decreasing variable d decreases the performance variable V_S. The variable r_{s1} is independent from V_S. Thus the spring DP prefers to change variable d or r_{s1} rather than D_o.

Once a qualitative strategy is selected, the corresponding quantitative analysis is performed by a perspective. The quantitative analysis is based on the quantitative dependency shown in Table 9.7. If the current strategy cannot make the proposal acceptable, then select another strategy and repeat the process. If no strategy is feasible, then one may either modify the current solution or maintain the original solution. A new solution is obtained by setting a conflict variable at an extreme value or restructuring the problem. Figure 9.10 illustrates the negotiation process.

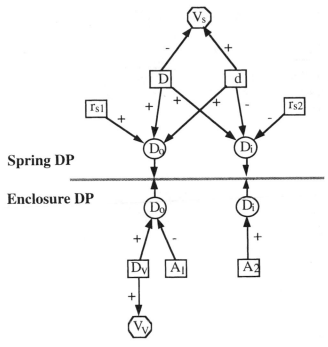

Figure 9.9   The reduced qualitative constraint network for the spring and enclosure DPs.

9.3.2   Integrating Processes

Developing a concurrent engineering environment does not necessarily require the scrapping of existing design processes. Rather, systems engineers can typically reengineer and supplement existing design processes to shorten design leadtime and improve product quality. This section addresses the issues of process modeling and design process reengineering. Formal methods of model analysis are discussed and illustrated with industrial examples.

TABLE 9.7   Qualitative and Quantitative Dependencies in the Network in Figure 9.9

Variable a to Variable b ($a \rightarrow b$)	Qualitative Dependency (δ_{ab})	Quantitative Dependency (ψ_{ab})	Variable a to Variable b ($a \rightarrow b$)	Qualitative Dependency (δ_{ab})	Quantitative Dependency (ψ_{ab})
$D \rightarrow D_o$	+	0.9558	$A_1 \rightarrow D_o$	−	0.0102
$D \rightarrow D_i$	+	1.0588	$A_1 \rightarrow D_i$	+	0.0113
$d \rightarrow D_o$	+	0.0435	$D_v \rightarrow D_o$	+	1.0102
$d \rightarrow D_i$	−	0.0588	$d \rightarrow V_S$	+	0.0674
$r_{s1} \rightarrow D_o$	+	0.0862	$D \rightarrow V_S$	−	1.4869
$r_{s2} \rightarrow D_i$	−	0.1176	$D_v \rightarrow V_V$	+	0.0210

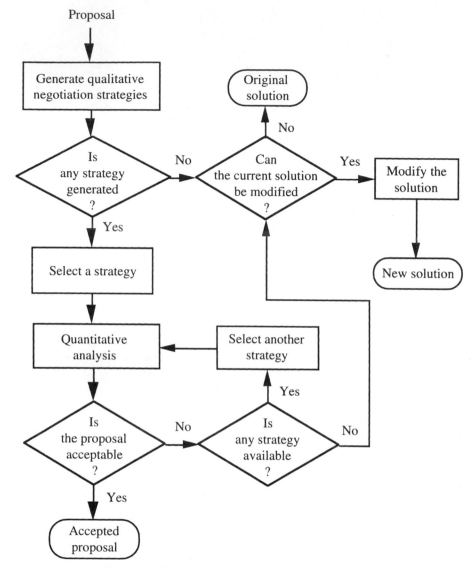

Figure 9.10 The procedure for evaluation of a proposal.

9.3.2.1 Process Modeling.   A model represents a system or an object and is typically constructed for the purpose of analysis. Models are needed to describe existing systems, as well as evaluate the feasibility and anticipated performance of proposed systems. Although models must capture enough detail to facilitate reliable experimentation, the purpose of modeling must not be violated by including unnecessary information that results in an investment exceeding that required to build and/or experiment on the actual system. The motivation and potential drawbacks of modeling efforts vary considerably between applications and methods. To develop a concurrent engineering environment, it is essential to model the processes involved not only in design, but in the entire product life cycle.

There are several factors that motivate model development in concurrent engineering. Developing an infrastructure to support concurrent engineering requires a substantial capital investment. Therefore, a thorough understanding of design and manufacturing functions, data, resources, and the organizational structure is essential. A model of the system can provide this understanding without disturbing the actual environment. Furthermore, the model can be used to analyze the system's ability to respond to anticipated market changes. This enables rapid and accurate reconfiguration when new products are demanded and new technology becomes available. Ultimately, an executable version of the model can simulate and even control the product development process.

Many effective methodologies exist for modeling design and manufacturing processes (see Section 9.4.1). Development of the IDEF (ICAM Definition) methods began with the Air Force program for integrated computer-aided manufacturing (ICAM). Through this work, the need for a family of mutually supportive methods for enterprise integration was realized and development was continued as part of the Air Force Information Integration for Concurrent Engineering (IICE) program. "The IICE program was chartered with developing the theoretical foundations, methods, and tools to successfully implement and evolve towards an information-integrated enterprise" (Mayer et al., 1992a). As the IDEF methods have become widely used as part of concurrent engineering, total quality management (TQM), and business process reengineering (BPR) initiatives, the IDEF acronym has come to stand for a family of integration definition methods.

The IDEF methods (see Chapter 25) provide several architectural representations that can effectively model the engineering design process. IDEF0 was developed for modeling a wide variety of systems that use hardware, software, and people to perform activities (U.S. Air Force, 1981). An IDEF0 model consists of three components, diagrams, text, and a glossary, all cross-referenced to each other. The box and arrow diagrams are the major components of the model. In a diagram, a box represents a function and an arrow represents an interface. A box is assigned an active verb phrase to represent the function. An interface may be an input, an output, a control, or a mechanism, and is assigned a descriptive noun phrase. Inputs (I) enter the box from the left, are transformed by the function, and exit the box to the right as an output (O). A control (C) enters the top of the box and influences or determines the function performed. A mechanism (M) is a tool or resource that performs the function. The interfaces are generally referred to as the ICOMs (see Fig. 9.11).

IDEF0 provides a structured representation of the functions, information, and objects that are interrelated in a manufacturing system. IDEF3 was created specifically to model the sequence of activities performed in a manufacturing system. An IDEF3 model enables an expert to communicate the process flow of a system through defining a sequence of activities and the relationships between those activities. There are two basic components of the IDEF3 process description language, the process flow description and the object state transition network description. The two components are cross-referenced to build IDEF3 diagrams (Mayer et al., 1992a).

The IDEF3 process flow description is made up of units of behavior (UOBs), links, and junction boxes. A UOB represents a function or activity occurring in the process. Relationships between UOBs are modeled with three types of links: precedence links, relational links, and object flow links. Precedence links express simple temporal precedence between UOBs. Relational links highlight the existence of a relationship between two or more UOBs, however, no temporal constraint is implied. Object flow links provide a mechanism for capturing object related constraints between UOBs and carry the same temporal se-

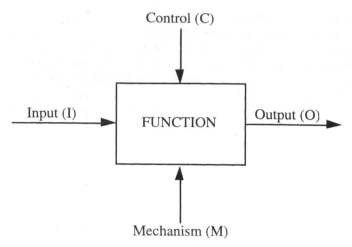

Figure 9.11   IDEF0 function box and interface arrows.

mantics as a precedence link. The logic of branching within a process is modeled using junctions. Several classifications are used to define junction boxes. Junctions are classified according to logical semantics as *and* (&), *or* (O), and *exclusive or* (X). Multiple process paths are classified as *fan-in* or *fan-out* corresponding to converging and diverging paths, respectively. The relative timing of process paths that converge or diverge at a junction are classified as *synchronous* or *asynchronous*. An example of an IDEF3 process flow diagram is shown in Figure 9.12. For more information on the IDEF methodologies, see Chapter 25.

9.3.2.2   Process Reengineering.   To develop a concurrent engineering environment, it is not enough to simply model the existing design process. It is essential to perform formal model analysis to identify the underlying structure of the process. Through model analysis, the systems engineer identifies the existing cycles, resource requirements, information

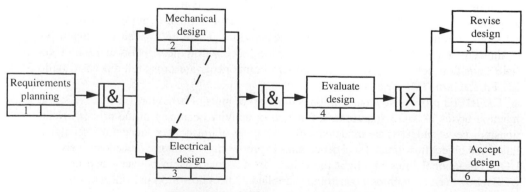

Figure 9.12   IDEF3 process flow diagram

flow, and value-added activities in the process. The concurrent engineering environment is developed by restructuring the design process to shorten lead times, ensure producablity, and improve quality. In this section we discuss formal methods for model analysis based on the triangularization and decomposition algorithms. For more information on reengineering, see Chapter 23.

Kusiak et al. (1994) presented the triangularization algorithm as a formal method for reengineering design activities. The algorithm identifies groups of activities and arranges them concurrently. An activity–activity incidence matrix and the corresponding directed graph is used to represent the relationships between activities in a process. A nonempty element in an incidence matrix represents a relationship between the corresponding activities. The objective of the algorithm is to order rows and columns of the incidence matrix into a lower triangular form with the minimum number of nonempty elements in the upper triangular matrix. A nonempty element in the upper triangular matrix verifies the existence of a cycle in the process.

The algorithm is very effective at identifying the structure of IDEF0 and IDEF3 models. The IDEF model may be represented as a process graph and corresponding activity–activity incidence matrix. The triangularized activity–activity incidence matrix identifies cycles, concurrent activities, and a dependency path.

Although the algorithm seems best suited for analysis of IDEF3 process flow diagrams, it may be applied to IDEF0 models as well. The information contained in the ICOM structure of an IDEF0 model must be preserved during analysis of the model. Therefore, identifying the structure of the model with the triangularization algorithm requires consideration of three types of constraints: (1) output–input constraints, (2) output–control constraints, and (3) output–mechanism constraints. Three subproblems can be solved, one for each type of constraint. Each subproblem will reveal the structure of the system from a different perspective: the product flow perspective (output–input constraints); the information flow perspective (output–control constraints), and the resource perspective (output–mechanism constraints). The following example of IDEF0 model analysis using the triangularization algorithm illustrates process reengineering to achieve concurrency.

Example 9.3. *Reengineering the Conceptual Design Phase.* Conceptual design occurs at the front end of the design process as functional requirements are transformed into a physical description of the product. To reengineer the conceptual design phase, it is necessary to consider the tasks (activities) in the process, as well as the interfaces (inputs, outputs, etc.) between tasks. The process graph shown in Figure 9.13 is based on an IDEF0 model of the conceptual design process at an industrial company. The model has been simplified to include 16 activities on a single level of abstraction. Inputs, outputs, controls, and mechanisms are shown on the process graph using the modeling conventions of IDEF0 (i.e., controls enter the top of the activity). The model includes 20 mechanisms (numbered 1, . . . , 20 on the process graph) that are required to perform the tasks in the conceptual design phase.

The conceptual design process is decomposed by considering three perspectives: (1) the product flow perspective, (2) the information flow perspective, and (3) the resource perspective. Output–input constraints in the IDEF0 model define the product flow perspective, and output–control constraints define the information flow perspective. In this process, there are no output–mechanism constraints. In other words, all of the mechanisms used come from external sources, rather than other activities in the process. This is common in most practical applications, where typical mechanisms include en-

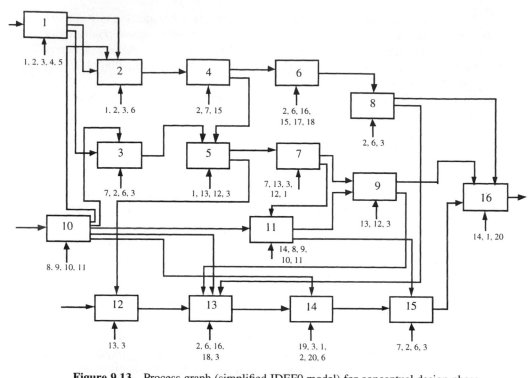

Figure 9.13 Process graph (simplified IDEF0 model) for conceptual design phase.

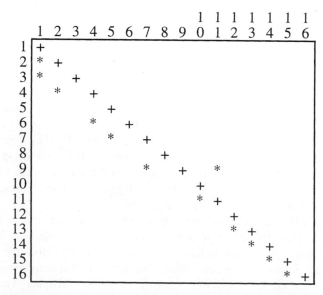

Figure 9.14 Activity–activity incidence matrix for product flow perspective prior to triangularization.

gineers, computers, and design software. Therefore, the resource perspective is not considered in this example. For each perspective, an extended triangularization algorithm (Kusiak et al., 1994) is used to identify cycles and activities that can be performed concurrently.

Figure 9.14 shows the activity–activity incidence matrix for the product flow perspective prior to applying the triangularization algorithm. The activity–activity incidence matrix following triangularization and the corresponding process graph for the product flow perspective are shown in Figures 9.15 and Figure 9.16, respectively.

The same procedure is used to analyze information flow in the conceptual design process. Figure 9.17 shows the activity–activity incidence matrix for the information flow perspective following triangularization Figure 9.18 shows the corresponding process graph for the information flow perspective. The resulting process graphs clearly identify the underlying structure of the process. The systems engineer can now reengineer the process by identifying non-value-added activities, cycles, and information-intensive processes that prevent concurrency.

Although more complex, the information flow perspective exhibits a greater degree of concurrency, that is, the process reduces to fewer concurrent levels. One reason for this might be redundancy (or duplication) of controls in the process. This is not to be confused with data redundancy, which is often viewed negatively. A redundant control, such as a procedure or statement of work, is fixed (i.e., not altered during the process) and simultaneously available to multiple activities in the process. Therefore, all of the activities that require the control can proceed concurrently. Frequently used controls should be fixed as early in the process as possible to maximize concurrency of activities in the information flow perspective. Information-intensive processes are characterized by long sequences of activities that cannot be broken down into fewer concurrent levels

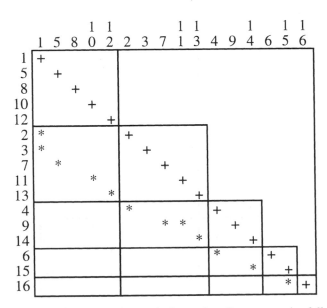

Figure 9.15 Activity–activity incidence matrix for product flow perspective following triangularization.

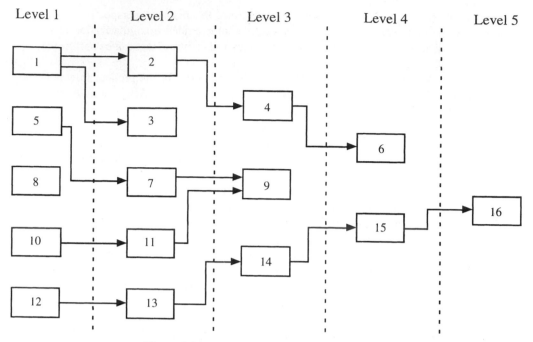

Figure 9.16 Product flow perspective process graph.

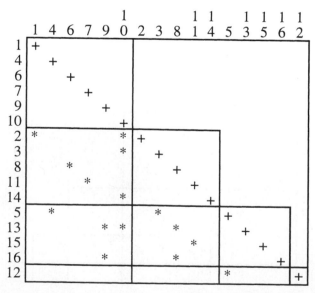

Figure 9.17 Activity–activity incidence matrix for information flow perspective following triangularization.

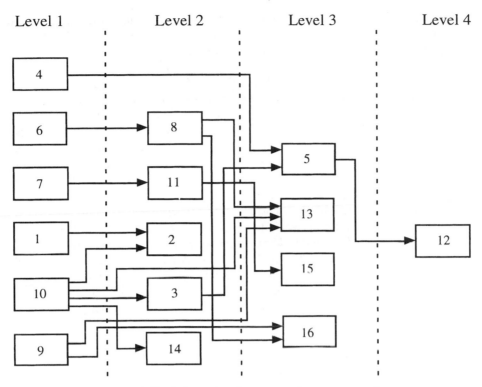

Figure 9.18 Information flow perspective process graph.

in the information flow perspective. This does not appear to be a problem in the existing conceptual design process. Since the process does decompose into many concurrent activities in each level, however, the systems engineer should verify that each activity has the appropriate access and modification rights to the controls. This will help assure that any changes in controls are visible to all appropriate activities.

The product flow perspective decomposes to five concurrent levels, with three completely separable groups of activities: group G1 (activities 1, 2, 3, 4, and 6), group G2 (activities 5, 7, 9, 10, and 11), and group G3 (activities 12, 13, 14, 15, and 16). Activity 8 does not affect product flow, and there are no cycles. Recall that the product flow perspective is defined by the output–input constraints in the model and that, by definition, an input to an activity undergoes a state change. The product is ultimately realized by changes in the state of the design, hence, the definition of the product flow perspective. When an activity does not affect product flow, the value added to the product by that activity comes into question. In other words, if an activity does not participate in product flow, it does not change the state of the design. Thus, the activity can usually be combined with another activity to simplify the process. This is the case with activity 8, which may be combined with activity 6. It is important to note that modeling and domain experts may overlook inputs when defining the model, or may consider something a control when it is actually an input (i.e., undergoes a state change). Even if the model is not completely accurate, however, it can still clearly identify non-value-added or, at the very least, low-value-added activities. The debate about whether something is an input (which undergoes a state change) or a control (which does not change state) is, in itself, an important step in the reengineering process.

The decomposition algorithm (Kusiak and Wang, 1993b) is based on the premise that in order to significantly reduce cycle time, the process has to decompose. Once again, the process is represented with a process graph and corresponding task-parameter incidence matrix. A nonempty element in the incidence matrix identifies the relationship between a task and a parameter. The algorithm decomposes the task-parameter incidence matrix into mutually separable submatrices (groups of tasks and groups of parameters) with the minimum number of overlapping parameters. Grouping identifies tasks that may be performed concurrently and critical parameters in the process.

The algorithm is effective in decomposing IDEF0 models based on the ICOM structure of the model. Once again, it is necessary to represent all activities and ICOMs on a single level in a process graph of the model. The corresponding activity–ICOM incidence matrix may then be constructed. Each ICOM in the model should correspond to a column in the activity–ICOM incidence matrix. A nonempty element in a row of the matrix identifies a relationship between the activity and an ICOM. The decomposition algorithm identifies groups of activities and ICOMs, as well as critical ICOMs that link groups.

Unlike the triangularization algorithm, the decomposition algorithm considers specific ICOMs (in IDEF0 models) and links (in IDEF3 models). Critical relationships between groups of activities are extracted from the complex structure of the model. In Figures 9.15 and 9.17, the nonempty elements of the matrix are forced below the diagonal, illustrating the objective of the triangularization algorithm. In contrast, the decomposition algorithm forces nonempty elements into the diagonal (see Fig. 9.21). Groups of activities are formed in submatrices along the diagonal, and overlapping ICOMs are columns on the right of the matrix.

Example 9.4. *Decomposing the Design Process.* Figure 9.19 is a process graph of an IDEF0 model constructed for an electronics manufacturer. For simplification, the model has been reduced to 12 activities with only input and output ICOMs. Figure 9.20 is the corresponding activity–ICOM incidence matrix. The decomposition algorithm (Kusiak and Wang, 1993b) may be used to analyze the impact of ICOMs on the structure of the model.

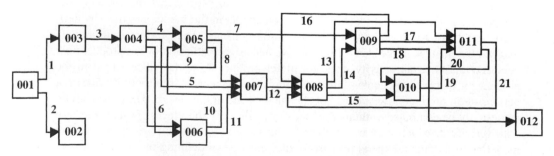

Figure 9.19 Process graph for ICOM-based analysis.

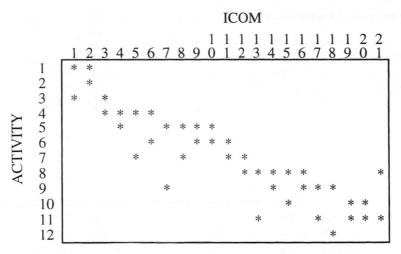

Figure 9.20 Activity–ICOM incidence matrix for process graph.

The decomposed activity–ICOM incidence matrix is shown in Figure 9.21. The model decomposes into two groups of activities with two overlapping ICOMs. Group 1 (activities 1, 3, 2, 4, 5, 7, and 6) and group 2 (activities 8, 11, 9, 10, and 12) are joined by ICOMs 7 and 12. The decomposition algorithm identifies overlapping activities that are critical to process completion. If an overlapping ICOM is an input, the analysis should focus on the quality and/or reliability of the ICOM. If the ICOM is a control or mechanism, alternatives to the current information or resources should be considered. These approaches relax constraints that are imposed by overlapping ICOMs, allowing the groups of activities to be performed concurrently.

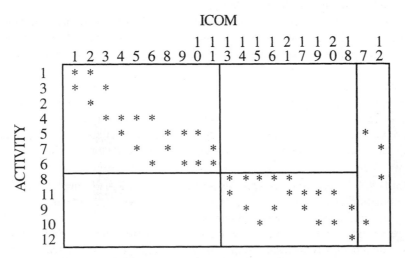

Figure 9.21 Decomposed activity–ICOM incidence matrix.

9.3.3 Integrating Problem-solving Mechanisms

Many approaches to solving specific design problems have been proposed, including strategies ranging from classic problem-solving methodologies to artificial intelligence. Morphological analysis can be used to decompose complex engineering design problems into a manageable number of subproblems that can be solved independently (Pahl and Beitz, 1984). Chandrasekaran et al. (1993) discussed the use of functional representation to capture information in the design rationale (i.e., the problem- solving process) by focusing on the functionality of the artifact. Brown and Chandrasekaran (1986) developed a design language (DSPL) for modeling the hierarchical structure of problem solving in general design.

Specification of requirements is performed early in the conceptual design stage to define the functionality of the product. Kusiak and Szczerbicki (1992) proposed a method of decomposing requirements into various levels of abstraction. Upon decomposing product requirements, design parameters and constraints are identified. Several authors, including Steward (1981, 1993), Rinderle (1986), and Kusiak and Wang (1993a,b) have used network models to represent design parameters and constraints as nodes and arcs, respectively. The incidence matrix corresponding to the network is reordered to decompose the design problem into a set of subproblems. Once a design problem is decomposed, mathematical methods (i.e., nonlinear programming) become more efficient for design optimization. Many authors, including Sobieszczanski-Sobieski (1982), Azarm and Li (1989), and Rogers and Bloebaum (1994) have applied decomposition-based approaches to solve design optimization problems.

Several specific problem-solving methodologies are discussed in this section. Examples are provided to illustrate requirements decomposition and constraint–parameter decomposition. The foundation for a growing body of literature in the area of decomposition-based design optimization is also presented.

9.3.3.1 Requirements Decomposition. Design *requirements* provide an abstraction of the design task from the most general demand (the overall requirement) to more specific demands (subrequirements). The number of levels in the decomposition of requirements depends on the complexity of the design task. The capability to satisfy a given requirement is provided by a design *function*. Each requirement may be satisfied by more than one function, and each function may correspond to more than one requirement. A requirement–function incidence matrix can be used to represent the interaction between functions and requirements (Kusiak and Szczerbicki, 1992). The design engineer can identify sets of functions that satisfy a given requirement by constructing a logical tree to decompose an overall requirement and all higher-level requirements into subrequirements. Each node in the tree represents a requirement, and leaf nodes contain functions that satisfy the corresponding requirement. Example 9.5 illustrates requirements decomposition.

Example 9.5. *Decomposition of a Design Requirement R1.* Figure 9.22 illustrates the decomposition of an overall requirement $R1$ into subrequirements and functions that satisfy the corresponding subrequirements. An arc between nodes of the tree in Figure 9.22 represents a conjunction (AND relationship), while a node without an arc represents a disjunction (OR relationship). In the example, the overall requirement $R1$ is satisfied by each of the following four sets of subrequirements: $\{R11, R8\}$, $\{R12, R8\}$, $\{R10, R5, R3\}$, and $\{R9, R5, R3\}$. Each of the sets may lead to a different design. When

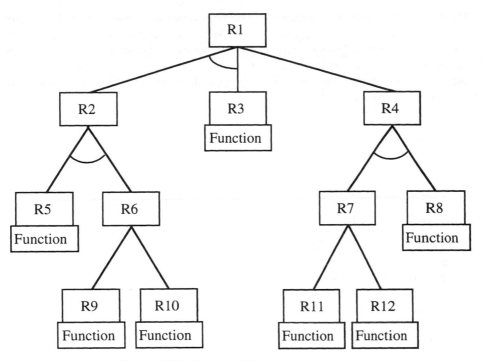

Figure 9.22 Decomposition of requirement $R1$.

a subrequirement cannot be further decomposed, a function is selected or specified to satisfy the requirement. In Figure 9.22, there are three different levels of complexity at which functions have been specified. The first level contains requirements $R2$, $R3$, and $R4$; the second level contains requirements $R5$, $R6$, $R7$, and $R8$; the third level contains requirements $R9$, $R10$, $R11$, and $R12$. Kusiak and Szczerbicki (1992) provide examples of decomposition in requirement space, as well as functional space.

9.3.3.2 Constraint–Parameter Decomposition.

Parameters describe the features of a product, and *constraints* define the range of values that are assigned to parameters. A parameter may be assigned a numerical value, a binary value, or a string of characters (i.e., color = red, material type = aluminum). Constraints are defined by product requirements. In constraint–parameter decomposition, the structure of the problem is represented by an incidence matrix. The problem is decomposed by transforming the matrix so that all nonempty elements are grouped into the diagonal. It is desirable that the diagonal blocks of nonempty elements would be mutually separable. A block of nonempty elements identifies constraints that must be evaluated simultaneously.

An advantage to this method of decomposing constrained design problems is the ability to handle qualitative, as well as quantitative constraints. Many "real-world" design problems contain nonnumerical constraints (i.e., color, material type). The method identifies *relationships* between parameters and constraints. However, since values are not assigned to parameters, the design engineer selects the procedure that is most appropriate for solving the subproblem. Example 9.6 illustrates constraint–parameter decomposition.

Example 9.6. *Constraint–Parameter Decomposition for Design of a Ball Bearing.* The parameters in the design of a ball bearing are listed in Table 9.8 (Hamrock and Anderson, 1985). The geometric constraints for the design of the component follow the table. The design problem is to assign values to the parameters without violating the constraints. Problem decomposition is applied to identify critical parameters and groups of constraints.

Constraint $C1$: $d_e = \dfrac{1}{2}(d_o + d_i)$

Constraint $C2$: $P_d = d_o - d_i - 2d$

Constraint $C3$: $f = r/d$

Constraint $C4$: $B = f_o + f_i - 1$

Constraint $C5$: $D = Bd$

Constraint $C6$: $\beta_f = \arccos\left\{\left[r_o + r_i - \dfrac{1}{2}(d_o - d_i)\right]\middle/(r_o + r_i - d)\right\}$

Constraint $C7$: $P_e = 2D \sin \beta_f$

Constraint $C8$: $s = r(1 - \cos \theta)$

Constraint $C9$: $\dfrac{1}{R} = \dfrac{1}{R_x} + \dfrac{1}{R_y}$

Constraint $C10$: $\Gamma = R\left(\dfrac{1}{R_x} - \dfrac{1}{R_y}\right)$

Constraint $C11$: $R_x = d(d_e - d \cos \beta)/2d_e$

Constraint $C12$: $R_y = f_i d/(2f_i - 1)$

Figure 9.23 shows the constraint–parameter incidence matrix for the ball bearing design problem prior to decomposition. The decomposed constraint–parameter inci-

TABLE 9.8   Parameters in the Design of a Ball Bearing

Parameter	Description	Parameter	Description
d_e	Pitch diameter	D	Race curvature distance
d_o	Outer-race diameter	r_o	Outer-race curvature
d_i	Inner-race diameter	r_i	Inner-race curvature
P_d	Diametral clearance	P_e	Free endplay
d	Rolling-element diameter	s	Shoulder height
f	Race conformity ratio	R	Curvature sum
r	Race curvature radius	R_x	x-Direction effective rad.
B	Total conformity	R_y	y-Direction effective rad.
f_o	Outer-race conformity	f_i	Inner-race conformity
Γ	Curvature difference	β	Contact angle
β_f	Free contact angle	θ	Shoulder-height angle

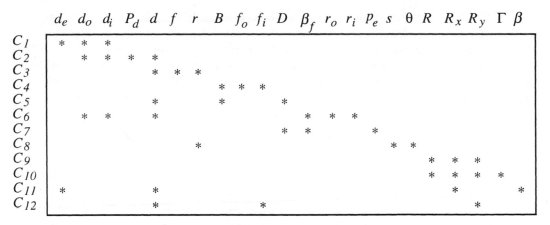

Figure 9.23   Constraint–parameter incidence matrix for ball bearing design.

dence matrix is shown in Figure 9.24; see Kusiak and Wang (1993b) for a constraint–parameter decomposition algorithm. Four critical parameters are identified: d, β_f, R_x, and B. Once values are determined (i.e., assumed) for the critical parameters, the problem decomposes into four subproblems, corresponding to the constraint–parameter blocks along the diagonal of the matrix in Figure 9.24.

9.3.3.3   Decomposition–based Design Optimization.

If a complex design problem is decomposed into subproblems, mathematical programming techniques become much more efficient for design optimization. Furthermore, subproblems can be treated independently with methods that exploit the structure of the problem (i.e., integer programming). Whereas a large scale design problem may be difficult, if not impossible, to solve, subproblems (sometimes called modules or modular subproblems) may be solved and integrated to provide an overall solution.

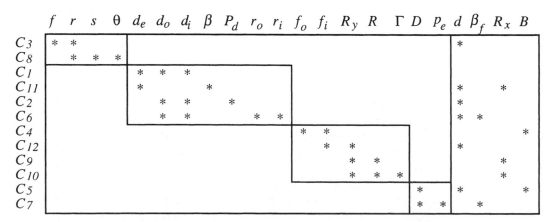

Figure 9.24   Decomposed constraint–parameter incidence matrix.

The literature in decomposition-based design optimization is extensive. Kirsch (1981) identified two types of decomposition-based methods for solving design optimization problems: (1) the model coordination method, and (2) the goal coordination method. Sobieszczanski-Sobieski (1982, 1989) used decomposition-based methods to formulate multilevel linear and nonlinear design optimization problems. Johnson and Benson (1984) developed a two-stage decomposition strategy for design optimization. The strategy assumes the design problem can be decomposed into independent subproblems. The concept of monotonicity analysis (Papalambros and Wilde, 1988) was used to formulate two-level (Azarm and Li, 1988) and three-level (Azarm and Li, 1989) design optimization problems. Steward's design structure matrix (1993) was used by Rogers and Padula (1989) to develop a knowledge-based system for multilevel decomposition of complex design problems. The system was recently enhanced to aid in tracking the flow of information during the solution procedure and performing sensitivity analysis (Rogers and Bloebaum, 1994). Wagner and Papalambros (1993a,b) suggested "decomposition analysis" as a method of identifying linking (or coordinating) variables and local variables that define the master problem and subproblems.

9.3.4 Integrating Information

Perhaps the most complex and dynamic component of a concurrent engineering environment is its vast information network. Very challenging problems face systems engineers attempting to integrate information from numerous databases that are continually accessed by various design, scheduling, and control systems. This section provides a brief overview of two key topics that impact information integration in concurrent engineering: database management systems, and information and data modeling.

9.3.4.1 Database Management Systems. The volume of engineering data can be overwhelming to systems engineers attempting to implement concurrent engineering. Although the amount of data generated during product development is staggering, however, it is the very existence of these data that makes concurrent engineering a reality. Unfortunately, the data are often generated by different computer programs that may even be operating on different platforms, and use different file formats to store data independently. Thus, engineering design environments cannot regularly benefit from central databases similar to those used in financial and other commercial applications. Individual users may develop file-naming strategies to manage data locally and may use their system's password structure to restrict access, however, this solution is inefficient even in small organizations. In large organizations it is virtually impossible to manage file-based data efficiently, but many existing engineering design systems remain file-based islands of automation.

The database management system (DBMS) takes a top-down, global approach to data and is oriented toward data rather than toward individual programs (Stark, 1992). DBMS software manages the database by providing multiple users with efficient, secure, and convenient access to data. Components of the DBMS include the data model, the data dictionary, the data definition language, the data manipulation language, data management functions, a query language, and query programs (i.e., search engines). Rather than storing data in individual files, the user stores data in a global database that is queried when data are needed. Depending on the user's access privileges, the data can then be modified and returned to the database. The changes are then immediately available to all other users. The use of a central data repository also reduces data redundancy.

Although a DBMS is much more efficient than a file-based system, there are still several disadvantages to using a DBMS in engineering environments. The most common disadvantage is the inability of a DBMS to handle complex graphic data, which is typical of engineering design. A DBMS is also structured to handle brief transactions that require only a few seconds to a few minutes. In the engineering environment, a transaction might last several hours, days, or even weeks (i.e., designing a part). This results in version-control complications that are not always handled effectively by a DBMS. In commercial databases, the programs that use the data are often simple and output data directly to the database. However, engineering design programs, such as computer-aided design (CAD) systems, usually output data to a separate file. Finally, engineering data are typically dynamic and linked by complex, or even unclear, relationships.

Since conventional DBMS tools are usually not effective in engineering environments, engineering-oriented DBMS solutions are becoming available. "Metadata" is a concept that seems to be a compromise between strictly file-based systems and the central data repository inherent to the DBMS. In this approach, the data are left in the files in which they were created, and a limited amount of information about each file is stored in the DBMS. In essence, this approach relies on storing "data about data" (Stark, 1992). Information about the file can include file location, file name, program creating, creator, owner, status, creation date, release date, release procedure, reviewer, releaser, part name, data type, data structure, data format, rights, version number, project, references, log information, and so on.

In building a concurrent engineering environment, one role of systems engineers in the information integration task is to determine the metadata to be included in the DBMS. This requires systems engineers to be familiar with all aspects of the product development process, including information systems, engineering design software, and engineering personnel. To achieve and maintain the necessary understanding of the system, information and data modeling techniques, as discussed in the following section, are essential.

9.3.4.2 *Information and Data Modeling.*

To achieve concurrent engineering, it is necessary to store a large volume of data for various applications, such as manufacturing, sales, and marketing. Furthermore, this data must be accessible to engineers in design, as well as manufacturing. Companies realize that data are a resource, and must be managed in a flexible manner. Data models semantically model the relationships between various pieces of data while maintaining this flexibility. There are four common types of data models: (1) hierarchical data models, (2) network data models, (3) relational data models, and (4) object-oriented data models. Although object-oriented models have gained recent popularity, this section illustrates the concepts of information and data modeling with a relational data modeling methodology.

IDEF1x (Mayer, 1990) is based on the entity-relationship approach to semantic data modeling developed by Chen (1976) and the relational database theories of Codd (1970). The basic constructs of an IDEF1x model include entities, attributes, and relationships.

An entity is a class of real or abstract things that share common attributes or characteristics. An individual member of this class is called an *entity instance*. If the entity class is "People," an entity instance might be the person John Doe. An entity instance can belong to more than one entity class. For example, if another class is "Managers," and John Doe is a manager, then this instance would be a member of the manager class as well.

In data modeling methodologies such as IDEF1x, an entity is represented by a box. Each entity is assigned a unique name and number. The name is typically a noun with adjectives or prepositions that describe what the entity represents, and the number is a positive integer. A formal definition of the entity is maintained in the model glossary.

A connection relation is an association between one entity and another. Each instance of the first entity (the parent entity) is related to zero, one, or more than one instance of the second entity (the *child entity*). Each instance of the child entity is associated with exactly one instance of the parent entity. In other words, a child entity can only exist if there is a corresponding parent entity. The connection relationship can be further defined by stating the cardinality of the relationship, that is, the number of child entities for each parent entity.

A specific connection relationship is shown as a line from the parent entity to a child entity, with a dot on the child end of the line. The default cardinality for such a relationship is zero, one, or more. A relationship is given a name, usually a verb phrase, consisting of a verb with possible adverbs and prepositions. The names of relationships between entities must be unique for each entity pair. The relationship names, however, may not necessarily be unique within the model. The relationship is always expressed in the parent to child direction, so that a sentence may be formed from the parent entity name, the relationship name, relationship cardinality, and the child entity name. For example, the phrase "Managers manage one project" could be derived from a relationship where the parent entity is Managers, the relationship name is Manage, the relationship is one-to-one, and the child entity is Projects. The relationship must hold from the opposite direction as well, that is, "Projects are managed by exactly one manager." Figure 9.25 illustrates entities and relationships.

An *attribute* is a characteristic of an entity. For the Employee entity, one attribute might be "Employee number." An attribute instance is a set of specific characteristic values. An attribute value is the contents of the attribute instance. In this example, an attribute instance might have "11234" as the "Employee number" value. An entity must have an attribute or group of attributes whose values uniquely identify each instance of the entity. This group of attributes is known as the *primary key* of the entity. If the Employee entity has attributes "Employee number," "Employee name," and "Employee salary," the "Employee number" attribute is the primary key, since it would be possible to have more than one employee with the same name or the same salary.

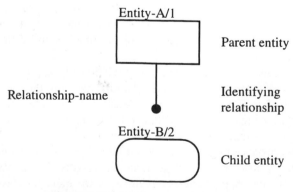

Figure 9.25 Entity relationship syntax.

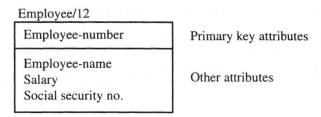

Figure 9.26   Attribute and primary key syntax.

In an IDEF1x model, each attribute is used by only one entity (the single-owner rule) and each instance of the entity must have a value for each attribute of the entity, that is, the attribute must be applicable to every instance associated with the entity (the no-null rule). No entity instance can have more than one value for any attribute of the entity (the no-repeat rule). Attributes are identified with unique, singular names. They are shown by placing each attribute name on one line inside the entity box. Primary key attributes are placed at the top of the list, separated from the other attributes by a line across the entity box (see Fig. 9.26).

IDEF1x provides a rich syntax that extends far beyond what is briefly introduced in this section. In practice, systems engineers use the syntax provided by a specific data modeling methodology to derive a complete description of the data elements in the system. Object-oriented methods are similar in that a rich syntax is provided for data modeling; however, the features of the object-oriented paradigm provide the foundation. The object, which includes both data structure and behavior, is the foundation of the object-oriented paradigm. *Objects* are defined as members of a class. Instance variables are used to define the state of an object. Objects communicate with one another through messages. Other features of the object-oriented paradigm include encapsulation, generalization, inheritance, and polymorphism. An example of an object-oriented data modeling methodology is IDEF4 (Mayer et al., 1992b).

To facilitate information integration in concurrent engineering, systems engineers must select and utilize a suitable data modeling methodology. Furthermore, as with process modeling, data modeling must include the perspectives of all relevant domain experts. The result will be valid data models that can be used to identify metadata for engineering DBMS tools, reengineer information flow, and control data access and modification.

9.4   MANAGING A CONCURRENT ENGINEERING ENVIRONMENT: TOOLS AND TECHNIQUES

Many different tools and techniques have been developed for managing concurrent engineering initiatives in specific domains and/or firms. However, nearly all of these methods exhibit the features of a structured process modeling language and techniques for evaluating the performance of a concurrent engineering environment. In this section we outline some useful enterprise modeling methodologies that have been used to study concurrent engineering systems. In addition, a product life-cycle-oriented risk assessment strategy is presented for managing concurrent engineering projects.

9.4.1 Methodologies for Modeling Concurrent Engineering Processes

In addition to being an essential tool for building concurrent engineering environments, process models are also useful for managing and monitoring concurrent engineering projects. Many methodologies exist for modeling enterprises. Although the methods vary in scope, appearance, and theoretical foundation, they all provide insight to the problem of concurrent engineering process model development. This section briefly describes several of the existing methodologies.

9.4.1.1 CIM-OSA. Computer Integrated Manufacturing–Open Systems Architecture (CIM-OSA) is the subject of on-going development by the ESPRIT Consortium AMICE. The methodology facilitates total enterprise modeling through a model construction process that includes enterprise requirements definitions, enterprise design specifications, and an enterprise implementation description. Four enterprise views, or perspectives, are considered: function, information, resource, and organization. Within each view, generic building blocks describe the functions, information, and resources in the system. Relations between building blocks define the total enterprise (Beekman, 1989).

CIM-OSA recognizes the functional, information, resource, and organizational perspectives that must be considered explicitly in modeling concurrent engineering environments. Furthermore, abstraction concepts such as encapsulation, classification, and inheritance are supported.

9.4.1.2 EXPRESS. In 1980 ISO TC184/SC4/WG5 initiated work on EXPRESS and Version 1.0 was approved in 1991. The PDES consortium uses EXPRESS systematically in its work on STEP. A graphical representation of the language is available as EXPRESS-G. The methodology provides a syntax for defining classes of entities (which may be information, resources, material, products, etc.) that support abstraction. Dynamic behavior, however, cannot be modeled. Although not as explicit as CIM-OSA, EXPRESS can capture the functional, information, resource, and organizational perspectives of concurrent engineering environments (European Committee for Standardization, 1994).

9.4.1.3 GRAI Method. The GRAI method was developed by the GRAI Laboratory in the early 1980s and has been used internationally in manufacturing applications (Doumeingts et al., 1987). The GRAI method is built around a conceptual reference model that is based on the theories of complex system, hierarchical system, organization system, and discrete activities theory. The manufacturing system is structured in three subsystems: a physical system, an information system, and a decision system. The GRAI formalism focuses on the decision subsystem and relies on other methods, such as IDEF0 and entity relationship attribute (ERA), to model the physical and information systems. The GRAI formalism is supported by two graphical representations: the GRAI grid, and the GRAI net.

Although the GRAI method explicitly focuses on decomposition of the organizational perspective, it does not, in itself, cover the functional, information, and resource perspectives. Through decomposition, the method supports encapsulation. However, classification and inheritance are not supported.

9.4.1.4 IDEF Methods. As discussed in Section 9.3.2.1, development of the IDEF methods began with the Air Force program for ICAM and continued as part of the Air Force

IICE program. Although IDEF0 and IDEF3 are the most popular, additional IDEF methods have been, and continue to be developed.

The IDEF1x model is used to semantically model the relationships between various pieces of data. The basic constructs of an IDEF1x model include entities, attributes, and relationships (Mayer, 1990). The IDEF4 methodology provides syntax and semantics for capturing the thought process that is required to develop modular, maintainable, and reusable applications programmed in object-oriented languages such as C++, Object Pascal, Common Lisp Object System (CLOS), and Smalltalk (Mayer et al., 1992b). The dynamic behavior of a system can be captured using IDEF2. Methods for modeling domain ontologies (IDEF5) and defining the motives that drive the decision-making process (IDEF6) are in development. As a whole, the IDEF family of methods facilitates model construction for a variety of architectural perspectives.

9.4.1.5 IEM.
Integrated Enterprise Modeling (IEM) is a public domain methodology developed by IPK Berlin (European Committee for Standardization, 1994). Unlike the previous methods (with the exception of IDEF4), IEM is designed around the object-oriented paradigm. Objects in manufacturing and design systems are described by data and functions that change the objects. Objects are distinguished by purpose into three classes: products, orders, and resources. All real objects in the manufacturing environment belong to one of the three IEM classes and subclasses. A generic activity model is defined for operating on objects.

The object-oriented paradigm allows for the simultaneous modeling of the functional and information perspectives through a single construct class. Although not explicitly considered in IEM, the organizational perspective can be added and integrated using the class concept. This methodology illustrates the robust, generic modeling capabilities provided by the object-oriented paradigm that are useful for modeling the concurrent engineering environment.

9.4.1.6 PSL/PSA.
Problem Statement Language/Problem Statement Analyzer (PSL/PSA) was commercially developed by META Systems (European Committee for Standardization, 1994). The PSL component is a language that can be used to describe information systems in terms of objects, properties, and relationships. As with IDEF1x, PSL/PSA is based on the concepts of relational database theory. Formal and graphical representations are provided, and reports can be generated from the commercially available software. The relational approach to modeling provided by PSL/PSA is less desirable in comparison to object-oriented methodologies that are available.

9.4.1.7 SSADM.
Structured Systems Analysis and Design Method (SSADM) is the standard method of systems analysis and design used by the UK government. SSADM was developed by the Central Computer and Telecommunications Agency in the early 1980s. The method focuses on analysis of business requirements, design, and specification of application databases and software. A project is broken down into modules that contain activities that must be completed to deliver the product. Each step has a list of tasks, inputs, and outputs. SSADM has modules for feasibility study, requirements analysis, requirements specification, logical system specification, and physical design (Ashworth, 1988).

The clear focus of SSADM is on the information perspective. Although many information modeling techniques are incorporated (i.e., data flow modeling, entity event modeling, and relational data analysis), the method does not employ the object-oriented paradigm.

9.4.1.8 OOMIS. The Object-Oriented Modeling Methodology for Manufacturing Information Systems (OOMIS) consists of two phases, an analysis phase and a design phase (Kim et al., 1993). The first task of the analysis phase is to decompose the manufacturing functions into component functions using an approach similar to IDEF0. After constructing a functional model, function tables, data tables, and operation tables are generated. In the design phase, the object-oriented paradigm is used to translate the function tables, data tables, and operation tables into an integrated information model. Classes consisting of an identifier, attributes, and methods are defined for components of the manufacturing system. Two specific class types are used, function class and entity class. Semantic design is facilitated by relationship diagrams.

OOMIS displays many features that are desirable for modeling the concurrent engineering environment. Unlike other methods that treat the functional and information perspectives independently (i.e., IDEF), the object-oriented paradigm is employed to form an integrated model. In fact, the information perspective is derived directly from the functional model.

9.4.2 Risk Assessment in Concurrent Engineering

This section presents a strategy for risk assessment in concurrent engineering environments. The proposed strategy is based on the premise that a holistic model of the product life cycle can be used to evaluate the design of any product in the domain of the firm. Therefore, the product can be defined in the context of the activities that must be performed to result in a successful product, rather than traditional methods of modeling based on the design artifact. To capture the necessary level of detail in a model of the product life cycle, a comprehensive modeling methodology must be employed. The IDEF family of methods has been adopted for this purpose.

Once the model has been developed, it can be used repeatedly to evaluate the design of different products. Customer requirements provide an initial summary of the activities that must be performed; however, the entire design scenario (and ensuing path through the product life cycle) may seldom, if ever, be realized. Therefore, the project management challenge is that of determining the remaining activities in a project plan that result in a successful design. The determination of a final design scenario is based on a variety of concurrent engineering risk factors that impact the remainder of the product life cycle. As a result, the overall risk, considering the perspectives of many different functional areas, can be mitigated.

Risk assessment is a process that attempts to answer three questions: (1) What can go wrong? (2) What is the likelihood that it will go wrong? (3) What are the consequences? (See Chapter 3.) Based on these questions, Equation (9.1) is a quantitative definition of risk, where $\hat{R}$ is risk, $\tilde{S}$ is a scenario of events that leads to a problem, $\tilde{P}$ is the likelihood of the scenario, and $\tilde{C}$ is the consequence of the scenario:

$$\hat{R} = \{\tilde{S}, \tilde{P}, \tilde{C}\} \tag{9.1}$$

A scenario captures all of the activities in the product life cycle, including everything from design to disposal. The probability of the scenario is determined by the likelihood that the activities in the scenario do not accomplish the objectives of the product. Adverse consequences for a scenario, or product/process design (e.g., violated due dates, low potential recyclability), can result from activities throughout the product life cycle. Risk $\tilde{R}$ is calculated by Equation (9.2), where $\tilde{P}_k$ is the probability of scenario k, and $\tilde{C}_k$ is the consequence of scenario k:

$$\tilde{R} = \sum_{k=1}^{K} (\tilde{P}_k \times \tilde{C}_k) \tag{9.2}$$

The life-cycle-oriented risk assessment procedure is summarized below. In sections 9.4.2.1–9.4.2.4 we discuss each step in greater detail.

Risk Assessment Procedure

1. Identify scenarios by determining all possible path-sets in the IDEF3 model of the product life cycle.
2. Estimate the probability of each scenario using a formal weighting scheme.
3. Determine the consequences of each scenario based on the relevant performance measures and domain experts' perceptions of risk.
4. Calculate risk using Equation (9.2).

9.4.2.1 Identifying Product Life-cycle Scenarios. The concept of path-sets in IDEF3 models is used to identify scenarios. A path-set $p_k(k = 1, \ldots, K)$ in an IDEF3 model is a set of all UOBs that define a path from source to sink in the model. The path-set is determined by junctions in the model and only one path set is taken for each execution of the model. Thus, a path-set identifies a scenario in the product life cycle. Due to the semantics of IDEF3 junction boxes, path-sets are difficult to identify by inspection, even in relatively simple models. The IDEF3 path-set algorithm (Larson and Kusiak, 1996) uses an activity–activity precedence matrix to represent the model and determine all possible path sets.

> **Example 9.7.** *Identifying Product Life-cycle Scenarios.* This example is based on a portion of the product life cycle corresponding to the manufacturing process at an electronics manufacturer. Figure 9.27 shows a simplified version of the IDEF3 model for the process with 27 UOBs and various junction boxes. The IDEF3 path-set algorithm is applied to obtain the path sets listed in Table 9.9. As shown in this example, for even small models, it is not feasible to determine all possible path-sets by inspection.
>
> This example illustrates the large number of scenarios defined for even one phase of the product life cycle. Thus, it is obvious that as the design, manufacturing, sales, service, and disposal functions are aggregated, the model becomes extremely complex and many possible product life-cycle scenarios must be evaluated (one corresponding to each path-set). As customer requirements become known for new products, perhaps through the use of QFD (Hauser and Clausing, 1988), infeasible scenarios can be eliminated from the product life cycle.

9.4.2.2 Weighting and Aggregating Scenario Probabilities. Recall that the probability of a scenario is the likelihood that the UOBs in the path-set are not successfully completed.

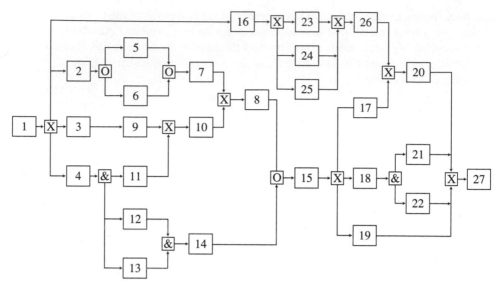

Figure 9.27 IDEF3 model with 27 UOBs (adapted from the manufacturing process at an industrial corporation).

The first step in calculating scenario probability is to determine the risk factors that will be considered and weight them accordingly. This information can be obtained by asking domain experts the following questions: (1) What key characteristics of products and processes are important to complete a successful design? (2) How can these characteristics be measured, monitored, and evaluated? The key characteristics provide valuable infor-

TABLE 9.9 Path-Sets for the IDEF3 Model in Figure 9.27

Path-Set	UOBs in Path-Set
1	1, 2, 5, 7, 8, 15, 19, 27
2	1, 3, 9, 10, 8, 15, 19, 27
3	1, 4, 11, 12, 13, 10, 8, 14, 15, 19, 27
4	1, 16, 23, 26, 20, 27
5	1, 2, 6, 7, 8, 15, 19, 27
6	1, 2, 5, 6, 7, 8, 15, 19, 27
7	1, 2, 5, 7, 8, 15, 17, 20, 27
8	1, 3, 9, 10, 8, 15, 17, 20, 27
9	1, 4, 11, 12, 13, 10, 8, 14, 15, 17, 20, 27
10	1, 2, 6, 7, 8, 15, 17, 20, 27
11	1, 2, 5, 6, 7, 8, 15, 17, 20, 27
12	1, 2, 5, 7, 8, 15, 18, 21, 22, 27
13	1, 3, 9, 10, 8, 15, 18, 21, 22, 27
14	1, 4, 11, 12, 13, 10, 8, 14, 15, 18, 21, 22, 27
15	1, 2, 6, 7, 8, 15, 18, 21, 22, 27
16	1, 2, 5, 6, 7, 8, 15, 18, 21, 22, 27
17	1, 16, 24, 26, 20, 27
18	1, 16, 25, 26, 20, 27

mation about risk factors early in the design process. For example, on-time delivery to the customer may be cited as a key characteristic. Therefore, schedule risk may be identified as a risk factor. A possible measure of this characteristic may be the percent on-time delivery for similar products. After collecting this information, a hierarchy similar to that shown in Figure 9.28 can be constructed to apply weights to the relevant risk factors. A method such as the AHP can be used for weighting and aggregating risk factors.

The AHP is a method for multicriteria decision making proposed by Saaty (1981). It is based on the premise that if a complex problem is decomposed into a hierarchy, pairwise comparisons can be made between the attributes of the problem to determine the best alternative. The goal (objective) is placed at the top of the hierarchy. The next level of the hierarchy contains attributes, or criteria, which contribute to the quality of the decisions. Each attribute can be decomposed into more detailed attributes, with the lowest level of the hierarchy containing the decision alternatives. After the hierarchical network is constructed, priorities of the elements are determined at each level of the decision hierarchy, and the elementary priorities are synthesized to rank the decision alternatives.

The hierarchy can focus the analysis on specific risk factors that are relevant to the overall goal of achieving a successful design. Domain experts can then estimate the likelihood that adverse effects will result from a risk factor. For example, failure rates of machines can be used to estimate the probability that a production schedule will not be met. In practice, many factors can contribute to the probability of adverse effects (i.e., the probability a scenario results in a flawed design). Ongoing research is exploring additional methods, such as fuzzy logic, for obtaining accurate measures for scenario probability (Larson and Kusiak, 1996).

9.4.2.3   *Identifying and Evaluating Consequences of Scenarios.*   The consequences of a scenario are the adverse effects realized by failing to successfully complete the UOBs in the corresponding path-set. Table 9.10 lists some common consequences of various

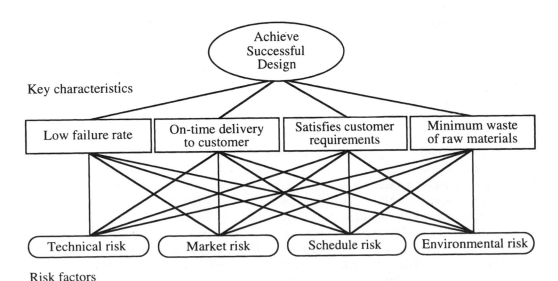

Figure 9.28   Hierarchy for calculating risk factor weights.

TABLE 9.10   Consequences of Concurrent Engineering Risk Factors

Risk Factor	Consequences	Measures of Consequence
Requirements risk	Loss of customer base	Number of customer complaints
	Due date violation	Days past deadline
Technical risk	Poor quality	Number of rejects
		Rework cost
Schedule risk	Additional resource requirement	Days past deadline
	Due date violation	Personnel cost
		Overhead cost
Cost risk	Higher product cost	Sale price of product
		Loss of market share
Network risk	Due date violation	Capital cost
	Information loss	Days past deadline
Redesign risk	Additional design iterations	Personnel cost
	Due date violation	Overhead cost
	Additional resource requirement	Days past deadline
Resource risk	Due date violation	Capital cost
	Additional resource requirement	Personnel cost
		Overhead cost
		Days past deadline
Environmental risk	Pollution	Cleanup expenses
	Negative public perception	Product disposal costs

concurrent engineering risk factors. Obviously, each of these consequences results in higher product development cost. An aggregate measure of this cost is typically domain dependent and reflects management objectives and customer concerns.

Once again, the relationships between a risk factor, consequences, and alternative design scenarios can be captured in a hierarchy, as shown in Figure 9.29. The overall goal is to mitigate risks. The consequences on the second level can be evaluated to determine those that are most important to achieving that goal. The importance of each scenario in contributing to the consequences is evaluated at the lowest level. The scenarios at the lowest level correspond to product life-cycle scenarios, such as those listed in Table 9.9.

The weights obtained at the second level in Figure 9.29 can be used to derive a cost function for a given risk factor. Formal procedures for determining consequences must address the issue of building valid hierarchies, including protocols for conducting group meetings (Saaty, 1981; Larson and Kusiak, 1996).

9.4.2.4   *Calculating Risk.* Desirable product life-cycle scenarios can be identified using the AHP and hierarchies such as the one shown in Figure 9.29. However, it is also informative to calculate a value for process risk and compare it to other scenarios. Recall that if the scenario probability can be estimated and a cost function is derived for the consequences of a scenario, process risk is calculated using Equation (9.2).

All activities in the life-cycle-oriented risk assessment procedure are performed by concurrent engineering teams (see Section 9.3.1.1). This allows a broad range of functional perspectives to contribute to the risk assessment process. In fact, any successful life-cycle risk assessment program must be constructed around the concurrent engineering design

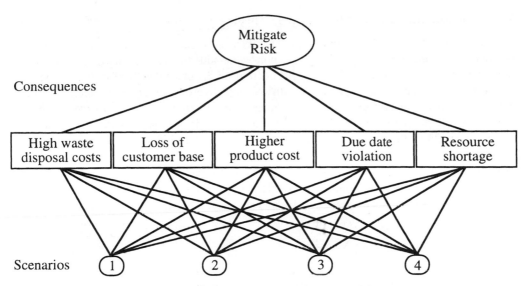

Figure 9.29   Hierarchy for mitigating risks.

team concept. Furthermore, risk assessment must serve as a key component to managing the concurrent engineering environment.

9.5 IMPLEMENTATION

Sections 9.3 and 9.4 discuss specific issues in building and managing concurrent engineering environments, respectively. Implementing concurrent engineering is obviously an enormous task that varies considerably between firms. The general steps discussed in this section, however, are consistently visible in successful concurrent engineering initiatives.

9.5.1   Develop an "As-Is" Model of the Existing Design Process

Developing a concurrent engineering environment requires significant reengineering of existing design processes. To facilitate the reengineering task, it is first necessary to develop an "as-is" model of the product development process. Section 9.4.1 introduced several of the most popular enterprise modeling methodologies. Although the techniques discussed in this chapter employ the IDEF modeling methodologies, all of the methods share some common features that are useful for modeling the product development process. It is essential to capture several perspectives of the process, including the functional perspective, the information (or data) perspective, and the temporal perspective. Capturing these perspectives requires systems engineers to construct IDEF0, IDEF1x (or IDEF4), and IDEF3 models, respectively. All domain experts must participate in the model construction process, either through conducting team meetings or interviews. The resulting models must then be analyzed using observational, as well as formal analysis techniques.

9.5.2 Develop the "To-Be" Model of the Concurrent Engineering Process

Developing the "to-be" model can also be described as the reengineering phase. The "as-is" models are analyzed to identify cycles, non-value-added activities, shared resources, redundant information, and alternative logic. Section 9.3.2.2 presented two formal methods for performing process model analysis with the objective of reengineering design processes. Systems engineers can also utilize rules of observational analysis to reengineer process flow logic (Kusiak et al., 1994). Several techniques include shortening the duration of activities, eliminating redundant activities, partitioning activities, combining serial activities, and eliminating cycles. Although useful methods exist, few, if any, software tools provide functionality beyond model construction. Intelligent systems that automatically derive the "to-be" model from an "as-is" model are needed to assist systems engineers in the reengineering task.

9.5.3 Identify Performance Measures

The motivation for moving to concurrent engineering is the desire to dramatically improve the product development process. In order to evaluate the success of concurrent engineering initiatives, it is necessary to establish a meaningful set of performance measures that match the objectives of the firm. In some industries, such as competitive, high-tech electronics, it might be more desirable to reduce time-to-market. However, in some consumer-based industries, it might be more desirable to minimize total product cost. Depending on the firm's objectives, the key performance measures may vary. As a good starting point, refer to Table 9.1.

9.5.4 Monitor the Concurrent Engineering System

Once the firm is operating a concurrent engineering environment and the key performance measures have been established, it is essential to continually monitor the system. Continuous process improvement should remain a focus, and in environments as dynamic as concurrent product development, there will be many opportunities. Once again, enterprise modeling will establish a medium for system monitoring. Furthermore, management techniques that emphasize key performance measures, such as the life-cycle-oriented risk assessment procedure presented in Section 9.4.2, must be implemented.

9.6 CONCLUSION

Many companies are making the move to concurrent engineering in an attempt to realize the benefits of shorter lead times, higher quality, and lower cost. To be successful, systems engineers must be deeply involved with efforts to build and manage concurrent engineering environments. There are four core tasks that require the expertise of systems engineers: (1) integrating people, (2) integrating processes, (3) integrating problem- solving mechanisms, and (4) integrating information. This chapter presented several challenging problems that face systems engineers and illustrated these problems with specific examples. Some specific methods were cited to provide a foundation for attacking very general problems, such as team selection, negotiation, process modeling, process reengineering, problem decomposition, and information management.

Managing a concurrent engineering environment requires achieving and maintaining an understanding of the entire enterprise. A variety of enterprise modeling methodologies were briefly described and the common, essential features were noted. A life-cycle-oriented risk assessment strategy was introduced for managing concurrent design. Finally, key steps in the implementation process were discussed. The need for systems engineering professionals in concurrent engineering initiatives is fairly obvious. However, the role of systems engineers is not always as clear. This chapter provides a reference to key problems that should be addressed by systems engineers seeking to build and manage concurrent engineering environments. Chapter 23 provides additional discussions concerning reengineering and the closely related subject of Integrated Product and Process Development (IPPD).

ACKNOWLEDGMENT

This research has been partially supported by contracts from the U.S. Army Tank Automotive Command (Grant No. DAAE07-93-C-R080), the National Science Foundation (Grant No. DDM-9215259), and Rockwell International Corporation. The authors would also like to acknowledge Armen Zakarian, Juite Wang, and Terry Letsche of the University of Iowa for their contributions.

REFERENCES

Alting, L. (1993). Life-cycle design of products: A new opportunity for manufacturing enterprises. In *Concurrent Engineering: Automation, Tools, and Techniques* (A. Kusiak, ed.), pp. 1–17. New York: Wiley.

Ashworth, C. M. (1988). Structured systems analysis and design method. *Inf. Software Technol.*, **30**, 153–163.

Askin, R. G., and Sodhi, M. (1994). Organization of teams in concurrent engineering. In *Handbook of Design, Manufacturing, and Automation* (R. D. Dorf and A. Kusiak, eds.), 85–105. New York: Wiley.

Azarm, S., and Li, W.-C. (1988). A two-level decomposition method for design optimization. *Eng. Optim.* **13**, 211–224.

Azarm, S., and Li, W.-C. (1989). Multilevel design optimization using global monotonicity analysis. *J. Mech., Transm., Autom. Des.* **111**, 259–263.

Beekman, D. (1989). CIMOSA: Computer integrated manufacturing—open system architecture. *Int. J. Comput.-Integr. Manuf.* **2**(2), 94–105.

Belson, D. (1994). Concurrent engineering. In *Handbook of Design, Manufacturing, and Automation* (R. D. Dorf and A. Kusiak, eds.), 25–33. New York: Wiley.

Boothroyd, G., and Dewhurst, P. (1988). Product design for manufacture and assembly. *Manuf. Eng.*, April, pp. 42–46.

Bowen, J., and Bahler, D. (1991). Supporting cooperation between multiple perspectives in a constraint-based approach to concurrent engineering. *J. Des. Manuf.* **1**, 89-105.

Brown, D. C., and Chandrasekaran, B. (1986). Knowledge and control for a mechanical design expert system. *Computer* **19**, 92–100.

Byrne, D., and Taguchi, S. (1986). The Taguchi approach to parameter design. *ASQC Qual. Congr. Trans.*, Anaheim, CA.

Carter, D. E., and Baker, B. S. (1992). *Concurrent Engineering: The Product Development Environment for the 1990s.* Reading, MA: Addison-Wesley.

Chandrasekaran, B., Goel, A. K., and Iwasaki, Y. (1993). Functional representation as design rationale. *Computer* **26,** 28–37.

Chen, P. P. S. (1976). The entity-relationship model: Toward a unified view of data. *ACM Trans. Database Sys.* **1**(1), 9–36.

Codd, E. F. (1970). A relational model for large shared data banks. *Commun. ACM* **13**(6), 377–387.

deKleer, J., and Brown, J. S. (1984). A qualitative physics based on confluences. *Artif. Intell.* **24**(1–3), 7–84.

Doumeingts, G., Vallespir, B., Darricar, D., and Roboam, M. (1987). Design methodology for advanced manufacturing systems. *Comput. Indu.* **9**(4), 271–296.

European Committee for Standardization (1994). An evaluation of CIM modeling constructs: Evaluation report of constructs for views according to ENV 40 003. *Comput. Ind.* **24**(2–3), 159–236.

Hamrock, B. J. and Anderson, W. J. (1985). Rolling-element bearings. In *Mechanical Design and Systems Handbook,* (H. A. Rothbart, ed.), 2nd ed., pp. 29.29–29.24. New York: McGraw-Hill.

Hauser, J. R., and Clausing, D. (1988). The house of quality. *Harv. Bus. Rev.,* May–June, pp. 63–73.

Johnson, R. C., and Benson, R. C. (1984). A basic two-stage decomposition strategy for design optimization. *J. Mech. Des.* **106,** 380–386.

Kannapan, S. M., and Marshek, K. M. (1992). Engineering design methodologies: A new perspective. In *Intelligent Design and Manufacturing* (A. Kusiak, ed.), pp. 3–38. New York: Wiley.

Kannapan, S. M., and Marshek, K. M. (1993). An approach to parametric machine design and negotiation in concurrent engineering. In *Concurrent Engineering: Automation, Tools, and Techniques* (A. Kusiak, ed.), pp. 509–533. New York: Wiley.

Kim, C., Kim, K., and Choi, I. (1993). An object-oriented information modeling methodology for manufacturing information systems. *Comput. Ind. Eng.* **24**(3), 337–353.

Kirsch, U. (1981). *Optimal Structural Design.* New York: McGraw-Hill.

Klein, M. (1991). Supporting conflict resolution in cooperative design systems. *IEEE Trans. Syst. Man, Cybernet.* **21**(6), 1379–1390.

Kuipers, B. (1986). Qualitative simulation. *Artifi. Intell.* **29**(3), 289–338.

Kusiak, A., and Szczerbicki, E. (1992). A formal approach to specifications in conceptual design. *J. Mech. Des.* **114,** 659–666.

Kusiak, A., and Wang, J. (1993a). Qualitative analysis of the design process. *Proc. Winter Annu. Meet. ASME,* New Orleans, LA, pp. 21–32.

Kusiak, A., and Wang, J. (1993b). Decomposition in concurrent design. In *Concurrent Engineering: Automation, Tools, and Techniques,* (A. Kusiak, ed.), 481–508. New York: Wiley.

Kusiak, A., and Wang, J. (1995). Dependency analysis in constraint negotiation. *IEEE Trans. Syst., Man, Cybernet.* **25**(9), 1301–1313.

Kusiak, A., Larson, T. N., and Wang, J. (1994). Reengineering of design and manufacturing processes. *Comput. Ind. Eng.* **26**(3), 521–536.

Larson, T. N., and Kusiak, A. (1995). *A Hybrid System for Risk Assessment in Product Development,* Working Paper, Iowa City: University of Iowa, Department of Industrial Engineering.

Larson, T. N., and Kusiak, A. (1996). Managing design processes: A risk assessment approach. *IEEE Trans. Sys., Man, Cybernet.* **26**(11), 749–759.

Lawson, M., and Karandikar, H. M. (1994). A survey of concurrent engineering. *Concurr. Eng.: Res. Appl.* **2,** 1–6.

Lyons, J. L. (1982). *Lyons' Valve Designer's Handbook,* New York: Van Nostrand-Reinhold.

Mallik, D. N., and Kouvelis, P. (1992). Management of product design: A strategic approach. In *Intelligent Design and Manufacturing* (A. Kusiak, ed.), pp. 157-177. New York: Wiley.

Mayer, A. K., and Lu, S. C.-Y. (1988). An AI-based approach for the integration of multiple sources of knowledge to aid engineering design. *J. Mech., Transmi., Autom. Des.* **110**(3), 316–323.

Mayer, R. J. (ed.) (1990). *IDEF1x Data Modeling: A Reconstruction of the Original Air Force Report.* College Station, TX: Knowledge Based Systems.

Mayer, R. J., Cullinane, T. P., deWitte, P. S., Knappenberger, W. B., Perakath, B., and Wells, M. S. (1992a). *Information Integration for Concurrent Engineering (IICE) IDEF3 Process Description Capture Method Report* AL-TR-1992-0057. Wright-Patterson AFB, OH: Armstrong Laboratory.

Mayer, R. J., Keen, A., and Wells, M. S. (1992b). *Information Integration for Concurrent Engineering (IICE) IDEF4 Object-Oriented Design Method Report* AL-TR-1992-0056. Wright-Patterson AFB, OH: Armstrong Laboratory.

Pahl, G., and Beitz, W. (1984). *Engineering Design.* London: The Design Council, Springer-Verlag.

Papalambros, P., and Wilde, D. J., (1988). *Principles of Optimal Design: Modeling and Computation.* Cambridge, UK: Cambridge University Press.

Premkumar, P., and Kramer, S. (1989). A generalized expert system shell for implementing mechanical design applications: Review, introduction, and fundamental concepts. *J. Mech. Des.* **111**(3), 616–625.

Reddy, R., Wood, R. T., and Cleetus, K. J. (1991). The DARPA initiative: Encouraging new industrial practices. *IEEE Spectrum,* July, pp. 26–30.

Rinderle, J. R. (1986). Implications of function-form-fabrication relations on design decomposition strategies. *Proc. ASME Conf. Comput. Eng.,* Chicago, *1986,* pp. 193–198.

Rogers, J. L., and Bloebaum, C. L. (1994). *Ordering Design Tasks Based on Coupling Strengths.* NASA Tech. Memo. 109137. Hampton, VA: Langley Research Center.

Rogers, J. L., and Padula, S. L. (1989). *An Intelligent Advisor for the Design Manager,* NASA Tech. Memo. 101558. Hampton, VA: Langley Research Center.

Rosenblatt, A., and Watson, G. F. (eds.) (1991). Special report: Concurrent engineering. *IEEE Spectrum,* July, pp. 22–37.

Saaty, T. L. (1981). *The Analytical Hierarchy Process.* New York: McGraw-Hill.

Saaty, T. L. (1986). Axiomatic foundation of the analytical hierarchy process. *Manage. Sci.* **32**(7), 841–855.

Smith, L. R. (1991). QFD and its application in concurrent engineering. *Des. Prod. Int. Conf.,* Honolulu, Hawaii, pp. 369–372.

Sobieszczanski-Sobieski, J. (1982). *A Linear Decomposition Method for Large Optimization Problems: Blueprint for Development.* NASA Tech. Memo. 83248. Hampton, VA: Langley Research Center.

Sobieszczanski-Sobieski, J. (1989). *Multidisciplinary Optimization for Engineering Systems: Achievements and Potential.* NASA Tech. Memo. 101566. Hampton, VA: Langley Research Center.

Stark, J. (1992). *Engineering Information Management Systems: Beyond CAD/CAM to Concurrent Engineering Support.* New York: Van Nostrand-Reinhold.

Steward, D. V. (1981). *Systems Analysis and Management: Structure, Strategy and Design.* New York: Petrocelli Books.

Steward, D. V. (1993). *NSF Report: Using the Design Structure Method.* Sacramento: California State University, Computer Science Department.

Sycara, K., and Lewis, C. M. (1991). Modeling group decision making and negotiation in concurrent product design. *Int. J. Sys. Autom.: Res. Appl.,* **1**(3), 217–238.

U.S. Air Force. (1981). *Integrated Computer Aided Manufacturing (ICAM) Architecture. Part II. Vol. IV. Functional Modeling Manual (IDEF0)* (AFWAL-tr-81-4023). Wright-Patterson AFB, OH: Air Force Materials Laboratory.

Wagner, T. C., and Papalambros, P. Y. (1993a). A general framework for decomposition analysis in optimal design. *Proc. ASME Adv. Des. Autom. Conf.,* Vol. 65-2, pp. 315–325.

Wagner, T. C., and Papalambros, P. Y. (1993b). Implementation of decomposition analysis in optimal design. *Proc. ASME Adv. Des. Autom. Conf.,* Vol. 65-2, pp. 327–335.

Werkman, K. J. (1991). Negotiation as an aid in concurrent engineering. In *Issues in Design/Manufacture Integration* (A. Sharon et al., eds.), pp. 23–30. Atlanta, GA: ASME.

Zakarian, A., and Kusiak, A. (1995). *Selection of Teams in Concurrent Engineering,* Working Paper, Iowa City: University of Iowa, Department of Industrial Engineering.

10 The Logical Framework in a Project Management System

JOSEPH D. BEN-DAK

10.1 INTRODUCTION

The *logical framework* is a conceptual tool, or systematic process, that helps translate common sense into a logical connection. It is a process that helps formalize problems in order to make it possible for decision makers to see what needs to be changed. It clarifies situations. It converts and/or corrects an intuition into a rational logical decision. And, it helps both to seize the dynamics involved in any form of organization and transform these dynamics into an orderly, logical sequence of steps. In other words, the logical framework is a management tool intended to provide greater insight into organizational thinking, the allocation of responsibilities, and unambiguous communication.

Rooted in principles similar to those of contract law, the logical framework consists of agreements that bind different parties on the basis of the achievement of mutual expectations. More precisely, the source of inspiration in both the logical framework and contract law is the necessity to ensure that the agreement is enforced at any time. This task requires a precise assessment of the extent to which the commitments or expectations have been fulfilled. This is the concept of performance measurement, which is one of the components of evaluation.

The logical framework has also been inspired by the concept of *management by objectives,* which takes objectives and goals as a starting point, integrates them into a hierarchy, and helps define a hierarchy of responsibilities. The logical framework provides an understanding of how things function and prepares the ground for explaining why things do not function as they are expected to. This is the concept of *explanation,* which is the other component of evaluation.

As a result, the logical framework is a foundation to allow defining evaluation essentially as two activities:

- Comparing expectations and observations.
- Explaining the differences between expectations and observations if expectations have not been fulfilled.

Because the logical framework focuses on evaluation feedback, it can almost be envisioned as a *dashboard,* that is, as a control panel that informs the project stakeholders of the state of the project, how well the project is performing given the actual resources, and what should be done in order to improve the project performance within the limited amount

Handbook of Systems Engineering and Management, Edited by A. P. Sage and W. B. Rouse
ISBN 0471-15405-9 ©1999 John Wiley and Sons, Inc.

of resources. What is most useful is that this evaluation is applied and considered at a time (or during a period) when it is still possible to effect project outcomes. This is the principle of *formative evaluation*.

The logical framework (or LogFrame) can be applied to any organizational structure. It can be effectively applied to projects in technology transfer, military procurement, and business training. Thus far, it has been most prominently applied in projects for developing nations and has been put to use in various projects at the U.S. Agency for International Development (US-AID) and the United Nations Development Programme (UNDP). In this context, the LogFrame provides a foundation for formative evaluation, illustrating the continuous effort of the international organizations to prove the value and effectiveness of their intervention in order to both eliminate poverty and promote self-sustaining development. The examples presented in this chapter are taken from projects planned and executed from 1993 to 1996 by a group in the UNDP, the Global Technology Group, devoted to private-sector projects in less developed countries in the areas of science and technology.

In the sections that follow, the LogFrame is successively presented as four tools: (1) to break down a project concept; (2) to measure performance; (3) to evaluate projects; and (4) to design projects. At the end of these discussions, a comparative assessment of the tool is presented. Chapter 29 presents a number of other tools and approaches for project management.

10.2 THE LOGICAL FRAMEWORK: A TOOL FOR BREAKING DOWN THE PROJECT CONCEPT

The goal of this section is to show to what extent the LogFrame helps establish a diagnosis and provides a clear picture of the project status. The LogFrame breaks down the project concept, which helps decision makers convert the project from what they see into a logical structure.

10.2.1 Definition of the LogFrame Concepts

The LogFrame is represented as a 4×4 matrix (Table 10.1). The matrix shows that a project is composed of goals, purposes, outputs, and inputs. This breakdown of the project concept is also done in terms of narrative summary, objectively verifiable indicators (OVIs), means of verification (MOV), and assumptions.

TABLE 10.1

	Narrative Summary	Objectively Verifiable Indicators	Means of Verification	Assumptions
Goals				
Purposes				
Outputs				
Inputs				

It is important to note at this stage that a program differs from a project in the sense that one program includes several projects. A program corresponds to an entity pursuing an objective and is composed of projects. Each of these projects is itself an organized entity pursuing an intermediate objective, which in turn contributes to achieving the overall objective of the program. This description of programs and projects explains why it is possible to conceive of projects and programs as organizations.

A *Goal* is a term characterizing an objective that goes beyond the project level. It is created at the program level. Goal denotes a desired result to which an entire program of development may be directed. It provides the reason for dealing with the problem that the project is intended to solve. For example, "Reach self-sufficiency" could be a goal for a project that is dealing with the development of agrobusiness in a country X. Indeed, the desired result, which is to attain self-sufficiency, is not restricted to the agrobusiness domain encompassed by the project. Goals are more often than not stated in general terms and refer to long term notions that are included in a large scope.

A *purpose* corresponds to what is hoped to be achieved by undertaking the project. The purpose of a project represents the solution to a specific development problem and may be derived by inverting the statement of the problem into a statement of the appropriate solution. For instance, a purpose could be stated in the following form "Increase the competitiveness of the food-processing industry of the country" for the project dealing with the development of agrobusiness in a country X previously referred to. Here, the desired result is restricted to the domain encompassed by the project, which is agrobusiness. Purposes are usually exemplified by the significant change in people or organizations thought to be required in order to provide important social or economic benefits for the target beneficiaries. They refer to midterm or short-term notions that are included in a more limited scope.

An *Output* is the result of a process. It represents what has been produced or created. For the same agrobusiness project, for instance, an output could be stated in the following form "Training of farmers to new technologies used in the field of agrobusiness.". Here, the desired result is a punctual action that concerns a particular segment of the project, which is the technologies used in agrobusiness. Outputs refer to punctual notions that are included in a very specific scope.

An *Input* is a resource that is to be used and transformed. An input in the same project could be "The farmers who serve as a task force to develop the agrobusiness in the country." Inputs generally refer to countable or measurable concepts in a 0-dimension scope.

Goals, purposes, outputs, and inputs will be henceforth referred to as the project levels. They can be distinguished with regards to two dimensions: geographical scope (that is to say, the domain encompassed by the goal, purpose, output, or input), and time scope. This is summarized in Table 10.2.

The basic logic connecting the goals, purposes, outputs, and inputs is shown in the following example from an agriculture project (Fig. 10.1). The goal and purpose answer the question, Why? while the purpose, outputs, and inputs are combined to answer, How? The purpose is the pivotal point that connects the Why? and the How? of the project.

A *narrative summary* is the brief description, or statement of the idea, that is to be categorized either as a goal, purpose, output, or input. The following is an example of narrative from a tourism project:

TABLE 10.2

Project Steps	Geographical Scope	Time Scope
Goal	Broad	Long or midterm
Purpose	Limited	Mid- or short term
Output	More limited	Short term or punctual
Input	Most limited	Punctual

Supergoal: Health and Education of Population Increased

Goal: Government Revenue Increased

Purpose: Active Tourist Industry Established

Outputs: 1. Hotels Completed
2. Airport Upgraded
3. Publicity Campaign Completed
4. Personnel Trained

Inputs: 1.1 Select Sites   2.1 Determine Performance
1.2 Design Hotels   Requirements
1.3 Construct Hotels   2.2 Solicit Bids
1.4 Purchase and   2.3 Select Contractor
        Install Furniture   2.4 Monitor Construction

An *objectively verifiable indicator* (OVI) is a specific statement on either the goal, purpose, output, or input. It aims at characterizing the project level. It measures the characteristics of the project level with adequate precision. An OVI is a measure designed to verify an accomplishment of a project level, and provides an objective basis for monitoring and evaluation. Many OVIs can help define one level of the project. For example, for the goal of the agrobusiness project, an OVI can be "the trade balance of the country X." Indeed, this concept helps visualize whether or not country X has reached self-sufficiency because it offers one alternative:

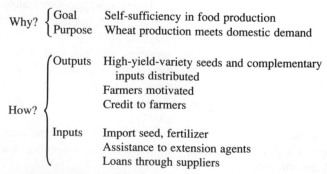

Figure 10.1 Narrative summary of a basic logic example in agriculture.

- If the trade balance of country X < 0, then country X is still dependent on other countries.
- If the trade balance of country X > 0, then country X has reached self-sufficiency.

For the purpose of the same project, an OVI could be the "Relative Market Share of the agrobusiness sector of country X in the world." For the output of the same project, an OVI could be a "quality index for the training of the farmers," while for the input of the same project, an OVI could be the "number of farmers involved in the project." OVIs are generally stated in terms related to quantity, quality, and time.

A *means of verification* (MOV) is the source of data needed to verify the status of the project-level indicator. For the OVI just mentioned, the trade balance of country X, an MOV could be any published statistics, such as reports issued by International Organizations (IMF, UN, and others). An MOV also pinpoints the choice of the source of the data used. There must be a consistency among the data sources.

An *assumption* is a supposition that something is true; a factor or statement that is taken for granted. The LogFrame defines assumptions as expectations about external factors that influence the success or failure of a project. A condition of no civil war in country X is an example of an assumption.

All of the preceding discussion is summarized in Table 10.3.

10.2.2   How These Concepts Interact

After having laid down the basics of the LogFrame, it is necessary to understand how the concepts are related to each other, and to determine the nature of their interactions. These interactions are of two sorts: quantitative—the elements are characterized by a singular to plural relationship; and qualitative—the elements are ordered in a hierarchy, which displays both a vertical and a horizontal logic.

10.2.2.1   The Singular-to-Plural Relationships. First, it is noticeable that the relationships between the LogFrame concepts are many to one, one to many, one to one, or many to many. Indeed, one program has many projects and one project must have one goal and one purpose but may have many outputs and inputs. Each of the goals, purposes, outputs, and inputs may have many OVIs, many MOVs, and many assumptions. This is summarized in Figure 10.2, where the cardinalities

$(1, n)$ means that the relationship is from 1 to many

$(n, 1)$ means that the relationship is from many to 1

$(1, 1)$ means that the relationship is from 1 to 1

10.2.2.2   The Hierarchy Between the Logical Framework Concepts

The Vertical Logic.   In addition to this singular-to-plural relationship, there is a hierarchy, or an order between these concepts. The order is conditioned by the means—ends or cause—consequence relationship. It relies on a set of linked hypotheses, that is to say an "if-then"

TABLE 10.3

	Narrative Summary	Objectively Verifiable Indicators	Means of Verification	Assumptions
Goals	The higher-order objective to which the project contributes	Measures to verify accomplishment of goal	Sources of data needed to verify status of goal-level indicators	Important external factors necessary for sustaining objectives in the long run
	Example: Reach self-sufficiency	Example: Trade balance of country X	Example: World Bank Report	Example: No civil war in country X
Purposes	The effect or impact of the project	Measures to verify accomplishment of purpose	Sources of data needed to verify status of purpose-level indicators	Important external factors needed in order to attain the goal
	Example: Increase competitiveness of agrobusiness products of country X	Example: World relative market share of agrobusiness sector of country X	Example: Market survey published by Lehman Brothers	Example: No significant change in the market outlook
Outputs	The deliverables of the project	Measures to verify accomplishment of outputs	Sources of data needed to verify status of output-level indicators	Important external factors needed in order to attain the purpose
	Example: Training of the farmers to new technologies	Example: Quality index of the training of the farmers	Example: Reports made by the consultants	Example: No culture barriers
Inputs	The resources that are to be used and transformed	Inventory of the used and transformed resources	Sources of data needed to verify status of input-level indicators	Important external factors needed in order to produce outputs
	Example: Farmers involved	Example: Number of farmers	Example: Report made by the consultants	Example: Farmers are in good shape

relationship. It is considered hypothetical because in many cases the linkage between the "if" statement and the "then" statement is uncertain. Implementation of the project is in fact a way of testing the hypotheses.

From this perspective, a goal can be defined as the higher-order objective for which the project purpose is one logical precondition. A purpose can be defined as what we hope

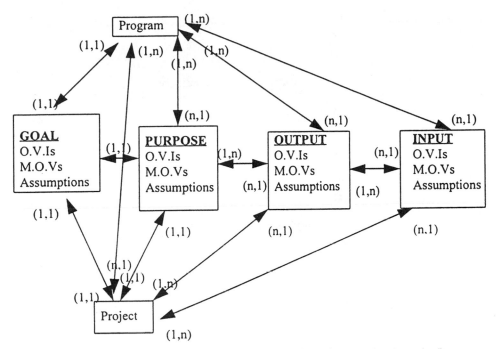

Figure 10.2   The four levels in project conceptualization when translated to a logframe.

to achieve *in order to* accomplish the goal. A downstream definition of purpose is the result aspired to *if* the required outputs are produced. A downstream definition of an output could be the specifically intended result that is expected from good management of the inputs.

The following logic can be derived:

- *If* a goal is desirable, *then* the project purpose will be necessary to achieve the goal.
- *If* a project purpose will assist goal achievement, *then* outputs will be necessary to achieve the purpose.
- *If* the outputs will enable purpose achievement, *then* inputs will be necessary to produce outputs.

In the frame of a means-to-ends thinking process, the end-of-project and end-of-program status should be defined. The *end-of-program status* is the condition or situation that will exist when the program achieves its goal. Similarly, the *end-of-project status* is the condition or situation that will exist when the project achieves its purpose; the means for achieving the goal is the purpose; the means for achieving the purpose is the outputs; and the means for producing the outputs is the inputs.

The project levels can be ordered vertically, as shown in Figure 10.3.

It is important to note, however, that in most cases one project level is a necessary, but not a sufficient condition to reach the upper level. This is where the concept of assumptions intervenes. Indeed, an assumption is an event or action that must take place, or a condition that must exist for a project to succeed. An assumption is an exogenous factor that belongs to the external environment of the project. It is an element over which the project team

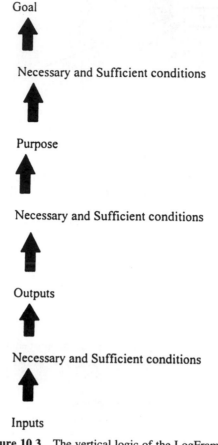

Figure 10.3 The vertical logic of the LogFrame.

has usually little or no control. Therefore, assumptions associated with project levels constitute sufficient and necessary conditions to reach the upper project level.

Therefore, we have the following relationship:

Project level + assumptions
= Necessary and sufficient conditions to reach the upper project level

At this point, it is possible to have a complete view of the vertical logic of the LogFrame. It can be illustrated as shown in Figure 10.4.

The Horizontal Logic. In addition to the vertical logic, the LogFrame contains a horizontal logic, which links program and project levels to OVIs, MOVs, and assumptions. Here, the

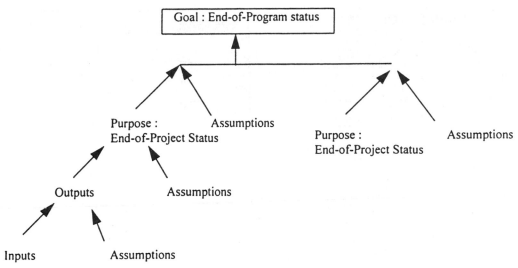

Figure 10.4   LogFrame vertical logic when influencing assumptions.

connection cannot be formalized in if–then terms, because it is of a different sort. Three types of relationships must be distinguished: the program–project-level OVI relationship; the OVI–MOV relationship; and the program–project-level assumption relationship.

The program–project-level OVI relationship can be compared in grammatical terms to that of an adjective[1] to a noun, that of an adverb[2] to an adjective, or that of an adverb to a verb. Indeed, as it has been previously mentioned in the definition section, an OVI aims at characterizing the project level. It emphasizes a selected part of the project or program level, focusing on a specific part that is considered to be the important part of the program or project level. Like all indicators, OVIs help us visualize the project level in terms of conditions or situations for goals and purposes, activities for outputs, and resources for inputs. OVIs represent the conceptual stage of the visualization process of the project or program level. For example, the trade balance of country X gives an idea of the extent to which country X is self-sufficient. OVIs provide a benchmark and a ground for comparison.

The OVI–MOV relationship represents the practical stage of the visualization process. MOVs contain information that either serves as a source of inspiration for the generation of OVIs, or provides a concrete expression of the OVI. For instance, the World Bank Report contains data that reveal whether the trade balance of country X is positive or negative. There also needs to be consistency between MOVs and OVIs in the sense that for a particular OVI there may not exist an appropriate MOV. In such cases, an independent study must be undertaken.

The program–project-level assumption relationship can be formulated in terms of inclusion. Assumptions are external factors that, when mixed with project levels, constitute necessary and sufficient conditions for the project's success. Assumptions belong to the

---

[1]Adjective: "Part of speech used to qualify, define, or limit a substantive [noun]," *New Webster Dictionary and Thesaurus of the English Language,* Lexicon Publications, Lexington, Mass. 1993.
[2]Adverb: "A part of speech which modifies or limits a verb, an adjective, or another adverb," *New Webster Dictionary and Thesaurus of the English Language,* Lexicon Publications, Lexington, Mass. 1993.

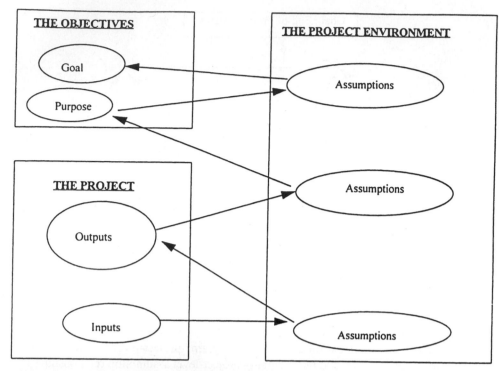

Figure 10.5   Different influencing entities involved in the program–project-level assumption relationship

environment of the project. The different entities involved can be illustrated as shown in Figure 10.5.

Definitions of the LogFrame concepts and of all the interactions between these concepts make it possible to build a logic, that is, to draw an organizational chart (Fig. 10.6). This chart illustrates how the LogFrame provides an in-depth analysis of the elements involved in organizational thinking.

10.3   THE LOGFRAME: A TOOL FOR MEASURING PERFORMANCE

It is important to note that the LogFrame contains certain limits. Indeed, this methodology does not assure that the project is optimal, that is to say, that the project directly addresses the most critical constraint of achievement, and is the most effective means for overcoming that critical constraint unless the planners and/or evaluators choose to explore alternative approaches. This is where the concept of performance measurement is of great importance. Performance measurement is the LogFrame pushed one step further, from observation to logical connection and from logical connection to optimization.

10.3.1   Comparing Expectations and Observations

The basic step for performance measurement is to compare planned achievements to actual achievements and see to what extent they match. This comparison must be done at each

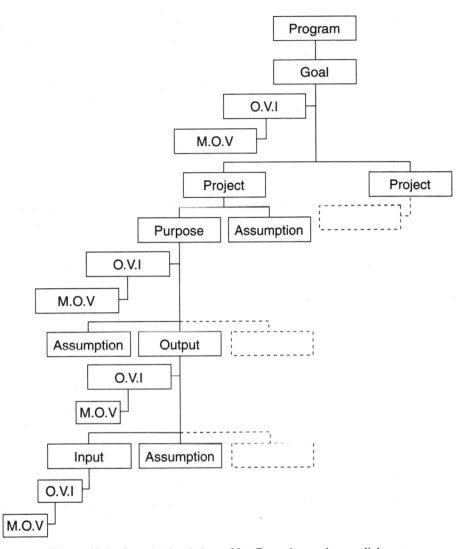

Figure 10.6   Organizational chart of LogFrame interactions → linkages.

project level. Here, OVIs are used to determine the accurate observed outcomes and to decide whether or not they agree with the expected outcomes. The OVIs basically work as standards for evaluating the success of the program and its projects. If the expected results of the projects were not achieved even though the projects were carried out according to the guidelines of the project documents, an analysis can be made to identify the possible reasons for this, and suggestions of how the project could have been formulated differently can be made. Therefore, the LogFrame can be transformed into the matrix shown in Table 10.4.

The LogFrame has been used at the UNDP to evaluate several individual programs and projects. For example, in early 1996 the report "Mid-Term Evaluation," which was an extensive evaluation of the effectiveness of projects conducted by the Global Technology Group/UNDP, was issued. Preparers of the evaluation used the LogFrame to look at the

TABLE 10.4

Project Level	OVIs	Expectations	Observations	Explanations for Differences Between Expectations and Observations
Goal				
Purpose				
Output				
Input				

Global Technology Group as a whole and to examine nine individual projects as a representative sampling of the group's work. One of the projects evaluated concerned the African Mathematical Union Women's Conference. The matrix was applied to this project as is shown in the following section.

10.3.1.1 Project Strategy and Brief Description. This project focused on advancing the careers of women scientists and mathematicians living and working in Africa. It was intended to create an awareness of their contributions and potential contributions, not only to the fields of advanced S&T and mathematics, but most importantly, to furthering sustainable human development in various African countries through applications of advanced mathematical modeling to public policy formulation, and to the measuring and monitoring policy outcomes.

The project selected six highly talented women scientists and financed their research, and in 1994 paid their travel expenses to the annual African Mathematical Union Conference held in Ibadan, Nigeria, to present their research. The conference was expected to give them greater exposure and an opportunity to tap into the network of African intelligentsia committed to solving developmental problems through their expertise in mathematical modeling and technology applications.

Follow-up on the impact of this project showed that two of the women sponsored used the opportunity to develop their careers as well as their interest in sustainable human development (SHD) activities. Another participant was promoted to the Head to the Faculty of Mathematics at the Chellah University in Rabat, Morocco, and a fourth developed a business in mapping and commercializing herbal medicines in Cameroon for the export market (see Table 10.5).

In addition to the African Mathematical Union Women's Conference project, the LogFrame was used to evaluate eight other projects in the Mid-Term Evaluation report. The variety of projects evaluated included small- and middle-size enterprise diamond mining in Southern Africa; national science and technology policy in Mongolia; long-distance and regional science education in Brazil; and military plant conversion in China. The diversity of subjects in this list clearly shows that the LogFrame can be a versatile tool applicable to a variety of project types. In all of these evaluations, the LogFrame was used effectively to analyze the goals, purposes, outputs, and inputs of the projects, as well as to highlight and explain cases where the project outcomes did not meet expectations.

10.3.2 Relating Achievements to the Amount of Resources Used

Comparing expectations and observations provides a good understanding of how well the project went. However, it does not tell whether the way the project was designed or implemented was optimal. This is why it is helpful to establish a relationship between the amount of resources used or the way the activities have been conducted, and the results obtained. In other terms, the idea is to determine whether the same result could have been achieved with fewer resources or whether the result could have been better with the same amount of resources.

Two main types of methodologies can be depicted: one focuses on the relationship between inputs management and the results obtained at the purpose level for a project/goal level for a program; and the other emphasizes the way the activities were conducted at the output level and its impact on the results.

10.3.2.1 Relating Inputs to Results. The most common method is the cost/benefit ratio, which compares the benefits generated by the project with the explicit and implicit costs generated by the project. By implicit costs is meant costs that are created by a situation in which an element is missing. For instance, the absence of a salesperson in the salesforce enables a company to save $50,000 a year on salaries expenditures, but if it is estimated that it cuts the turnover by $100,000, then it becomes an implicit cost of $50,000. Implicit costs more often than not correspond to the loss of a potential gain. The estimation of all these implicit costs is a difficult part of the cost/benefit analysis.

The estimation of benefits is also far from being self-evident for projects in which the benefits are qualitatively stated. Indeed, for a project dealing with the creation of a park in an urban area, the benefits for the community and the return on investment for the company are difficult to quantify. Benefits can be stated in terms of social impact, the creation of jobs, for instance, and the translations of this benefit into financial terms is a difficult task. This translation is nevertheless necessary in order to compare costs, which are expressed in financial terms, with benefits.

On the whole, the main difficulty in establishing a relationship between the management of inputs and the achievement of the project purpose is the creation of a linked hypothesis from the input level to the purpose level. Sometimes it is impossible to formalize the optimization of the use of the resources in cost/benefit terms. Indeed, in the case of a $50,000 UNDP project dealing with the development of the stone industry in Lesotho, South Africa, it is difficult to establish a logical link between the $50,000 invested in the project in 1996 and the industry reaching the growth stage in the life cycle in 1999. In this case, the government's incentives and investments to make this happen may have mounted up to $10 million from 1996 to 1999. Therefore, it is difficult to assess the contribution of the initial UNDP investment to the project. Also, from a UNDP point of view, it is difficult to determine the benefits obtained, the nature of the return on investment (e.g., justifying allocation of more funds for the next fiscal year). In development studies it appears more often than not, however, that if an early input redefines the organizational nature of relationships of partners-to-be or defines the very existence of entrepreneurial actors, a context for formulating a "contribution linkage" exits.

On the whole, the cost/benefit analysis seems appropriate when the benefits provided by the project and the resources used can easily be stated in the same terms or units. This can be demonstrated by the purchase of a share on the stock market, where the benefit is

TABLE 10.5

Project Level	OVIs	Expectations	Observations	Explanations for Differences Between Expectations and Observations
Goal		Advancement of women's careers in S&T Increasing awareness of the interconnectedness of SHD to even the most advanced S&T research Encouraging applications of advanced sciences, including mathematics in economic and social modeling, especially to measurement of SHD activities and outputs	Provided the opportunity for several skilled and capable female mathematicians to gain recognition in African science and development forums Research for and after the conference became the basis for identifying and developing unique processes/products	All expected goals achieved to the maximum possible despite resource constraints and the brevity of project lifecycle. Given the degree of success from the project implementation, follow-up on activities linking advanced science to SHD should be pursued
Purpose	OVI-1: Research applications to SHD developed OVI-2: Training local scientists	To empower a group of extremely promising, talented women scientists and mathematicians to develop their research, and through its applications, find solutions for problems of development in their countries Commitment to develop high-level capacity building and human resources in the mathematical sciences, necessary for locally driven development in Africa	Conference gave the sponsored participants the opportunity to develop strategies for applying their previously purely scientific interests in broader areas of economic development UNDP's interest in the AMU conference was a serious attempt at integrating S&T with field-level activity	All expected purposes were achieved

Output	OVI-3: Women scientists sponsored OVI-4: International Experts OVI-5: Quality Index for Conference	Sponsor the participation of 6 African female mathematicians to prepare and present their research and ideas at the Annual African Mathematical Union Conference, Ibadan, Nigeria, January 17–21, 1994 Sponsor an international expert to prepare and present an overall strategy to utilize African expertise in mathematical applications to development of new technologies Partial sponsorship of the Proceedings of the AMU conference in order to direct the focus of the conference to solving problems of SHD in Africa through S&T applications	Six mathematicians given tickets and DSA to attend the conference to present research and ideas One international expert sponsored to give presentation on "Applied Mathematics and Local Manufacturing." AMU subcontracted for $5,500 to cover expenditure on sponsoring special lectures on SHD and S&T applications in Africa. *Rating on Quality Index for Conference = 4* Follow-up efforts included contacting all 6 sponsorees, administering survey questionnaire on the impact of the conference on their productivity and research applications	All expected outputs realized
Input	OVI-6: Budget OVI-7: Number of Missions OVI-8: Quality Index of Women Participants	UNDP Finances: $40,000 UNDP Mission to Ibadan Selection of a delegation of 6 female mathematicians from Morocco, Cameroon, Tanzania, Ivory Coast, Senegal, and Rwanda Recruitment of international expert with considerable experience in applied mathematics and developing countries	UNDP Actual Inputs: $40,000 One UNDP mission to discuss SHD at the conference Participants were all Ph.D.'s holding important university positions in the mathematics, engineering, and science faculties in their countries. They all had extensive publication and teaching records. *Rating on Quality Index = 4* International expert was world renowned automated manufacturing technology expert and the founder of Ennex Technologies	All expected inputs realized

measured in terms of monetary units and so is the cost. There needs to be a consistency in the units used to determine the resources/costs and the purpose/benefits.

10.3.2.2 *Relating Outputs to Results.*

The linkage of outputs to results is between the output and the purpose levels of the project. It provides an interesting basis for formative evaluation in cases where the cost/benefit analysis proves to be insufficient because the linkages between the input and the purpose levels cannot be made.

The idea is to focus on minimizing human error. Human error can be represented by people who were not working while being kept waiting in the field for instruction. The time loss can be converted in monetary units. Another type of error can be having to redo certain tasks because the job was not done properly. The time loss can again be converted into monetary units. The evaluation of human error can help improve the project or other projects of the same type by not repeating the same mistakes. This is the essence of *summative* evaluation, which is concerned with "grading" projects that have reached completion and drawing lessons from those projects in order to apply them elsewhere.

On the whole, relating outputs to results relies on analyzing the errors made at either the planning or implementation stage of the project, converting these errors into time loss, and translating this time loss into monetary units. At this point, it is possible to determine the extent to which the project could have been carried out in a more efficient way.

10.4 THE LOGFRAME: A TOOL FOR EVALUATING PROJECTS

As has been previously mentioned, the other component of evaluation consists of providing explanations for the differences between the planned and the actual achievements in order to provide decision makers with recommendations. Because explanations may consist of formulating other hypotheses, it is necessary to ensure the exactitude of those "linked hypotheses" defined by the LogFrame. Indeed, the set of linked hypotheses is what defines the hierarchy, and therefore the coherence and structure of the organization. At this point, formulation of hypotheses that have not been verified during the implementation of the project may prove to be misleading for the entire organization. Therefore, evaluation involves testing hypotheses and formulating recommendations.

10.4.1 Formulating Relationships Between Expectations About Project Impacts and Field Examinations of Actual Impacts

The formulation of hypotheses stems from the explanations for the differences between expected happenings and observed happenings. Those hypotheses need further investigation in order to be verified in the field. An example of testing hypotheses may be the following. Three options are considered:

A. Fourteen officers graduated and no books were provided.
B. Fourteen officers graduated and books were provided to only four people.
C. Six out of 14 officers graduated and books were provided to seven people (including the six).

TABLE 10.6

Level	Expected	Observed	Difference Statement	Statement of Hypothesis to Explain Difference	Verification of Hypothesis in the Field	Findings	Conclusion
Goal							
Happening							
Assumption							
Purpose							
Happening							
Assumption							
Output							
Happening							
Assumption							
Input							
Happening							
Assumption							

Here, option A shows that there is no causal linkage between books and the graduation of the 14 officers. Therefore, the conclusion may be that there exist other elements, such as practical training, that help the officers prepare for their graduation. Option B shows that the impact of the books on the education of the 14 officers is very limited and that other elements intervene in the preparation of those 14 officers. Option C clearly shows that officers cannot graduate without the books.

The validation of hypotheses leads to further explanations, which lead to findings. The same procedure should to be done for the project assumptions. Once the findings concerning assumptions are formalized, they can be compared to the findings of the project happenings. Confrontation of both findings (assumptions and facts) then leads to conclusions, which lead to recommendations.

The whole process of converting hypotheses to recommendations can be seen in the overall view of the LogFrame shown in Table 10.6. The LogFrame can be applied to the case of the 14 officers who were expected to graduate at some point in the development of a project. For the purpose of our example, it will be assumed that option C occurred, and that graduation of the 14 officers should be accomplished at the output level of a project. The matrix would be filled as shown in Table 10.7. At this point, the formulated recommendation is that the persons in charge of training the 14 officers provide those 14 officers with 14 books and additional study material. This should be done at all project levels (goal, purpose, output, and input).

10.4.2   The Need to Develop Objectively Verifiable Indicators

All the project examples in this chapter thus far have provided objectives that were easily measurable, and therefore the accomplishment of project levels could be verified and measured. Also, the objectively verifiable indicators could easily be formalized in quantitative terms ("14 officers graduated"). This is not always the case. There is no denying

TABLE 10.7

Level	Expected	Observed	Difference Statement	Statement of Hypothesis to Explain Difference	Verification of Hypothesis in the Field	Findings	Conclusion
Goal							
Purpose							
Output Happening	14 officers graduated	6 officers graduated	8 officers did not graduate	Lack of books 1 book used by 2 persons and no other material used	7 officers used one book each and the other officers used nothing 1 book used by 1 person and no other material used	Out of the 7 officers equipped, only 6 graduate The only material used per officer is 1 book	The book that is used by only 1 officer at a time is a necessary but insufficient material for the officers to graduate
Assumption	7 books available	7 books available	OK				
Input							

that performance measurement assumes an accurate definition of OVIs. This subsection encompasses the various types of OVIs in order to provide a methodology on which to build them.

The project objectives can be quantitative, a binary event, or qualitative. Each of these objectives has different natures of measurement, that is to say, different sorts of OVIs.

A *quantitative* objective represents a quantifiably verifiable end or result. Quantitative objectives can be measured in either deterministic or probablistic terms. A deterministic measurement of an objective is where the realization of the objective is unequivocally determined from numerical data. A probabilistic measurement of an objective occurs when the attainment of the objective cannot be determined with certainty. Examples of OVIs can be "to build 150 low-income houses" for a deterministic measurement and "to reduce crime rate by 50 percent" for a probabilistic measurement.

A *binary-event* objective either clearly occurs or does not occur. Binary-event objectives are measured in logical terms. A logical measurement determines whether the objective has occurred or not occurred. An example of an OVI for an axiological measurement can be "to find the lost ship." Since the OVI is to help visualize the advancement of the project/program, it is necessary to convert the statement "to find the lost ship" into terms that help illustrate whether or not this objective has been met. The most common way of doing it is to assign a value of 1 if the event did occur and a value of 0 if the event did not occur. Therefore, the possible values of the OVI (the units) are defined (1 or 0).

A *qualitative* objective is judged subjectively to determine whether or not the objective has been accomplished. Qualitative objectives are measured in axiological terms. An axiological measurement involves value judgments, where the data necessary to determine accomplishment of an objective are gathered via subjective methods such as interviews or surveys. An example can be "to build a beautiful library." Here, this statement as it exists cannot be termed an OVI, because the concept of "beautiful" is subjective while an OVI is objective. Therefore, the term "beautiful" has to be translated first into objective terms and then into a measurement that helps illustrate whether or not the objective has been met. The way of doing this is to break down the idea of "beautiful" into a set of criteria, which can be stated in binary-event terms. Then, each of these criteria can be measured in logical terms. Afterwards, a scale can be determined by a maximal value equal to the number of criteria and a minimal value of 0, meaning that none of the criteria have been met. Therefore, OVIs for qualitative objectives can be represented by composite indexes.

To show this, we can define three criteria that the library stakeholders agree on for a "beautiful library." These can be that the shelves have to be made of mahogany, the seats have to be in leather, and the color of the tables has to be black. The composite index would be as follows:

Are the shelves made of mahogany?   If yes, then $OVI_1 = 1$;   if not, then $OVI_1 = 0$
Are the seats in leather?   If yes, then $OVI_2 = 1$;   if not, then $OVI_2 = 0$
Is the color of the tables black?   If yes, then $OVI_3 = 1$;   if not, then $OVI_3 = 0$

The final index is the composite index $OVI = \Sigma_k OVI_k$.

In this case, the maximal value of the OVI is 3 and the minimal value is 0. A value of 2 would mean that two of the three criteria have been met. However, it must be determined if the two criteria that have been met are very important or meaningless in order to draw a general conclusion on whether the library is beautiful. In other terms, if the most important criterion of beauty determined by the stakeholders is that the tables have to be black and this criterion has not been respected, then it cannot be considered that the library is beautiful even though the value of the OVI is 2 on a scale of 3. Therefore, the criteria must be weighted and then the composite index be calculated as follows:

$$OVI = (\Sigma_k((OVI_k) * W_k)$$

where W_k is the weight of the OVI_k.

For example, the concept of "beautiful" could be broken down as follows: shelves made of mahogany (20 percent), seats in leather (30 percent), black tables (50 percent). The maximal value of the OVI is now 1 and the minimal value is now 0. Then if only the criterion of black tables has not been met, the value of the OVI is 0.5 out of 1. It can then be interpreted by the stakeholders whether a grade of 0.5 out of 1 is satisfactory, sufficient, or not satisfactory.

On the whole, it appears that the creation of OVIs is problematic for qualitative objectives. The difficulty lies mainly in the conversion of an indicator from a subjective concept of, for example, beauty into numerical data. Therefore, it seems that the common difficulty of qualitative objectivity is the actual creation of numerical data. Indeed, numerical data can be broken down into units, expressed in units, and integrated into a scale or a continuum. A scale is necessary for OVI building in the sense that it provides a measurement that includes a maximal and minimal value for the OVI. A scale or a stated continuum helps to illustrate where the concept, which is to be measured and translated into an OVI, stands as compared to the best possible value. Hence, OVIs should be expressed in quantifiable terms, in order to be integrated into a scale. They should focus on the characteristic of the concept that is to be measured.

It is nonetheless important to note that breaking down qualitative concepts into criteria poses a practical problem of how to collect the information and how to objectively determine the criteria that are involved and their relative importance according to the stakeholders, who do not necessarily share the same viewpoint. Perhaps the simplest way to handle such situations is to use the services of a rating or judging panel, including equal numbers of a priori proponents and antagonists of the project purpose. This may be a means of translating subjective concepts to near-objective OVIs, at least in terms of ordinal categorization. Needless to say, however, this is only practically the simplest way to proceed.

10.5 THE LOGFRAME: A TOOL FOR DESIGNING PROJECTS

The LogFrame can also be utilized as a tool for designing projects. Indeed, evaluation is the foundation for feedback, learning, and strategy formulation. As is stated in "UNDP Evaluation Findings in 1994" (Capeling-Alajika et al., 1995), "evaluation in UNDP has generally tended to serve the functions of impact and budget review; monitoring and assessing project performance; determining whether to continue or terminate the project; or identifying short-term corrective measures to meet objectives." This corroborates the fact that the LogFrame, while breaking down the organization of a project management system into a logical connection and thereby providing a local and almost objective basis for decision making, helps determine the steps that need to be achieved and the resources that need to be allocated to carry out a project.

The purpose of the next section is to introduce the issues of formulating objectives and understanding the project environment, both of which are necessary for project design. The notion of project and programs as components of an overall effort are discussed at the end.

10.5.1   Formulating Objectives

The issue of formulating objectives is critical insofar as it determines the vision of what is to be achieved. In order to facilitate the implementation, and particularly the evaluation of the project, performance measures should be addressed at the design stage as well as in the implementation stage of the project. There is no denying that goals and purposes stem from the identification of problems. The formulation of goals and purposes is the translation of problems into a need, which is itself at first general and theoretical, and then expressed in concrete terms. Therefore, we address matters of identification of (new) problems and the determination of objectives.

10.5.1.1   How to Identify the (New) Problems to be Addressed. A problem analysis starts from an analysis of the existing situation. A problem corresponds to a proposition in which the logic has yet to be constructed. Problems derive from a missing element in the logical connection. Therefore, a problem can be seen as an existing negative state.

Once the existing negative state is made clear, it is necessary to identify the major problem conditions in the existing situation or environment. Eventually, it is necessary to determine the causes and the effects of the problem at hand. The causes and effects of a problem can be visualized in a *problem tree.*

The problem tree is composed of five steps that can be easily figured out by intuition:

1. Identification of major problems existing within the stated problem situation.
2. Statement of tentative core problem.
3. Identification of other problems that cause core problem.
4. Identification of the effects caused by the core problem.
5. Construction of a diagram showing the cause and effect relationships in the form of a problem tree.

One example of an application is the observation of frequent bus accidents in a particular area. This existing negative state can be selected as the core problem.

The causes of this problem can be the following: bad condition of buses, drivers not being careful enough, and bad road conditions. The bad condition of buses can be explained by the aging of the vehicles and/or by the poor or nonexistent maintenance system. The fact that the drivers are not careful enough can itself be explained by their excessive accumulation of working hours.

The effects of the core problem can be the following: a large number of passengers and/or drivers hurt or killed and schedules not being respected. Another consequence of the fact that the schedules are not respected is that people who take the bus are late. The major consequence of this is that people will no longer trust the bus company. If nothing is done to solve this problem, the company will eventually lose its customers and close down. The problem tree will then take the form shown in Figure 10.7.

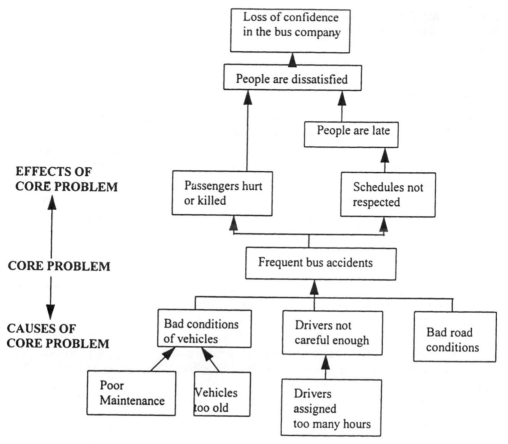

Figure 10.7 Problem tree for the case of frequent bus accidents.

10.5.1.2 How to Define Objectives. At this point, it is possible to convert the problem into a need, formulate this need as a goal or purpose according to the dimension of the problem, and determine whether it should lead to a program or a project. Objectives describe future situations that will be achieved by solving problems. Since problems are caused by problems and result in problems, objectives stem from objectives and help identify potential objectives from the project or the organization. It is important to note that the objectives defined often serve as a basis for developing OVIs. Therefore, reflections on objectives help clarify the way to generate OVIs.

The example of the frequent number of bus accidents can be transformed into a set of objectives. Indeed, the core problem can be stated as the following objective: reduction of the frequency of bus accidents. Causes can also be turned into the following objectives: vehicles well maintained and renewed every three years, drivers working fewer hours, road conditions improved. Consequences can be transformed into objectives as well: fewer passengers hurt or killed, schedules are respected, people are satisfied, and eventually confidence in the bus company regained.

Therefore, problems can easily be translated into objectives. Nevertheless, it is necessary to prioritize these various objectives. This is all the more necessary as some actions

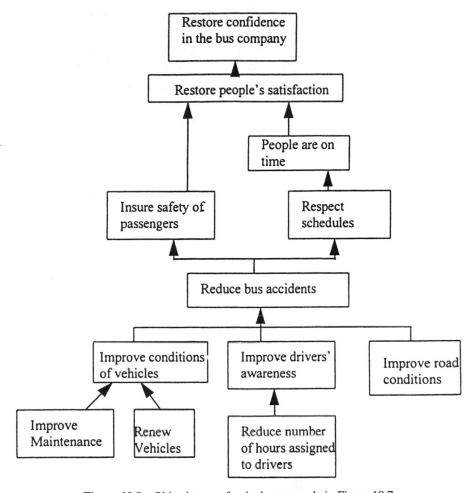

Figure 10.8   Objective tree for the bus example in Figure 10.7.

undertaken in order to achieve the objectives cannot be performed at the same time. A hierarchy of objectives helps organize the set of actions in order to meet the objectives. Using the LogFrame lexicon, it can be said that ordering the goals and the purposes help define the outputs and select the inputs accordingly.

The methodology used to prioritize the objectives is the *objective tree,* which is similar in many respects to the problem tree. An objective tree indicates how the attainment of subobjectives contributes to the accomplishment of higher-level objectives. It also reveals interconnections between the various objectives. To a certain extent, it can be regarded as the graphical representation of the LogFrame.

The objective tree can be applied to the same bus company (see Fig. 10.8). The 4 × 4 LogFrame matrix seen in Table 10.8 can be derived from this objective tree. The main limitation of the objective tree is that there is no single objective tree valid for a particular project or program because of the linked hypotheses. Therefore, like the LogFrame, an

TABLE 10.8

	Narrative Summary	Objectively Verifiable Indicators	Means of Verification	Assumptions
Goal	Example: Restore confidence in the bus company	Example: (1) Increase in number of customers by at least 20% (2) Ratio of customers loyalty to the bus company >75%	Example: Surveys conducted by a marketing firm	Example: No significant increase in competition intensity
Purpose	Example: Ensure safety of passengers	Example: Number of bus accidents = 0	Example: Annual report of the bus company	Example: Improvement of road conditions
Outputs	Example: (1) Additional training of bus drivers (2) Better maintenance of vehicles (3) Reduction of number of hours assigned to drivers	Example: (1) Quality index of the training of the bus drivers (2) Renew vehicles every three years (3) Reduce number of hours assigned to bus drivers by 15%	Example: (1) Report made by the Human Resources Department (2) Report made by the Stock Unit (3) Report made by the Human Resources Department	Example: (1) Bus drivers are willing to participate and change their habits (2) The company will have the sufficient resources to do so (3) Competition will make it possible for the company to reduce the number of working hours
Inputs	Example: (1) Recruit additional bus drivers (2) Invest in training and recruitment and vehicles maintenance	Example: (1) 15 additional bus drivers recruited (2) Investment of $2 million	Example: (1) Report made by the Human Resources Department (2) Report made by the Finance Department	Example: New bus drivers will integrate quickly into the company

objective tree does not necessarily represent the optimal way of organizing the objectives in order to build a strategy.

10.5.2 Assumptions About Project Impacts: Sizing the Project Environment

The analysis of a project management system assumes an accurate definition of the project system and the project environment system. In order to define these systems, it is necessary to determine the factors that affect the project concept: internal factors, which will help define the project system, and external factors, which will help define the project environment system. Internal factors can also be regarded as factors that can be influenced by the behavior of the project team, whereas external factors are influenced by other factors. Identification of the factors that affect the project may provide a picture of the project in terms of strengths, weaknesses, opportunities, and threats.

10.5.2.1 Internal Factors. The diagram of the project and the project environment shows that the internal factors mainly concern the input and the output project level.

At the Input Level. As regards the input level, various types of resources may be characterized in either financial, natural, human, or infrastructural terms. The appropriate choice of resources to conduct the project represents the key success factor at the input level.

In financial terms, the variety of resources used is very wide, ranging from loans to clearing, to borrowing from stock markets or governments, and to borrowing in soft or hard currency. The financial tool choice will condition the number and the nature of the creditors. For instance, for a French company starting to build a highway in Africa, the financing of initial expenditures by the government of the beneficiary country will restrict the degree of freedom of the French company in the way it wants to conduct the construction. On the other hand, if the French company pumps money out of its own reserves, it will be granted a much larger degree of freedom. Therefore, the choice of the financial resources affects the degree of freedom in conducting the project. Financial resources may be chosen according to their cost, liquidity (the ability to convert assets into cash), and time.

In terms of human resources, the choice of the team is conditioned by the complementarity and competency of the people constituting the task force. Human resources mostly include skills, which can be separated into technical skills required for engineers, managers, and technicians, and general skills required for teamwork and administrative disciplines. The compatibility of personalities involved is also a factor that needs to be taken into account. Human resources are chosen according to cost, skills, and compatibility with the team.

As regards natural resources, they must combine with human resources to represent the basics for undertaking projects in developing countries. As was mentioned in the United Nations Development Programme Mid-Term Evaluation Report, it is necessary to

first pay attention to indigenous capacity, especially that which has been dormant, fledgling or inconspicuous; then transfer technology and know-how to maximize its productivity. The latter is achieved by helping people organize, manage, and market their knowledge and by mapping new environmental and technical possibilities to induce competition and sustainability.

Therefore, natural and human resources are envisioned as a starting point of any development strategy. The UNDP Global Technology Group defined seven priority areas according to the aforementioned principle, among which is the "Development of Unique Products." "Unique Products" are defined as

> those that are available (or can be manufactured, grown, or cultivated) in abundance in a country/regional geographic boundary, that can be produced with relatively simple or easily acquired extent technologies involving local men and women, and those that have a distinctive market advantage in local and/or North countries.

The terms "easily acquired" and "distinctive advantage" are not simple, static concepts and must be nurtured strategically in order to make a unique product have a true survival value. This priority area includes the essential mapping of natural resources as a major type of project. It represents a significant percentage of the projects supervised or initiated by the Global Technology Group. This shows the growing importance of natural resources in a development strategy. Natural resources are chosen with regards to cost, physical properties, and environmental sustainability.

As far as infrastructure or equipment is concerned, they will condition productivity. For instance, the purchase of high-performance computers for a World Wide Web site in the frame of an educational program enables information to be carried faster and costs to be shared, costs that would otherwise have been incurred by each group purchasing the data individually. For a virtual school, it means that books made available on the internet by one school do not have to be purchased by other schools who can have access to the internet. Equipment is chosen in terms of cost, productivity, and reliability. Time can also be regarded as a resource. It is a unit in itself and units chosen depend on the underlying costs. All this is summarized in Table 10.9.

At the Output Level. Here, the factors cannot be categorized because the types of outputs produced and the types of processes are infinite. Minimization of human error is where the project team places its greatest effort. This is usually accomplished through the minimization of human intervention in conducting the project. The cost of automation is

TABLE 10.9

Type of Resource	Minimal Parameters that Condition the Choice of Resource
Financial	Cost
	Liquidity
	Time
Human	Cost
	Skills
	Compatibility with the team
Equipment	Cost
	Productivity
	Reliability
Natural	Cost
	Physical properties
	Environmental sustainability
Time	Cost

TABLE 10.10

Type of Output Produced	Parameters on which the Team Focuses
Any	Minimization

compared with decreases in the probability of human error, as well as its compatibility with the values of the organization, or the project team. This is summarized in Table 10.10. On the whole, though, internal factors that affect the project help provide a full picture of the project in terms of strengths and weaknesses.

10.5.2.2 External Factors. External factors or assumptions that affect the project are infinite. It seems critical for any assessment of *actual project impact* to determine what is expected to happen externally to project effort, so as to know why project results are achieved. Project planners and evaluators alike must be able to relate effort to results and not assign the success of internal efforts to cases where external factors affected the results more significantly. It seems possible to divide the project environment, which includes all the assumptions, into seven domains:

- Economic
- Political
- Socioeconomic
- Sociocultural
- Institutional or legal
- Environmental
- Technological

The economic domain encompasses issues related to economic matters, that is to say, matters concerned with profit making. It may be illustrated by the state of the coffee market for a project dealing with the development of new food-processing technologies in the field of coffee. Indeed, if the coffee market is bullish, then there is hope for making a profit by investing in it.

The political domain covers issues dealing with policy making and diplomatic affairs. This can be illustrated by the relationship between Taiwan and China, which affects a project dealing with the military industry conversion in China. Here, the political tensions between the two countries is not likely to ease the process of military conversion on the mainland.

The socioeconomic domain includes issues that are of economic importance but focus on the social impact. It can be illustrated by the unemployment rate for any type of project. Indeed, conducting a project in a country where the unemployment rate is high will make it easy for the project team to recruit local people because they are available.

The sociocultural domain includes issues dealing with social phenomena, with a focus on the cultural impact. This can be illustrated by the power of a country's unions in a project dealing with the development of small and medium enterprises. If the unions wield

TABLE 10.11

Type of Factor that Affects the Project	The Picture of the Project it Provides
Internal	Strengths/Weaknesses
External/Assumptions	Opportunities/Threats

great power, then the project team will have a lower degree of freedom, as compared to a situation where the unions have less power.

The institutional or legal context includes issues dealing with regulations that are of either legal or institutional origin. This can be illustrated by the constitution of Japan, which does not allow the formation of an army. This fact would compromise a project dealing with the introduction of Japan into the UN Security Council in exchange for a higher financial contribution, given that all the members of the Security Council must be ready to provide military assistance.

The environmental context deals with environmental issues. An example of an environmental external factor can be a project to install a soda plant in a particular area and the growing concern over reducing phosphate in the air once the plant becomes operational.

On the whole, external factors that affect the project help provide a full picture of the project in terms of opportunities and threats. This is summarized in Table 10.11.

10.5.3 Project and Program: Efforts as Components of a Whole

The standards of living in some countries and segments of populations in many other countries have improved in the developing world. Regarding education, primary and secondary enrollment has significantly increased. The access to health and sanitation services has been facilitated and consequently mortality is decreasing. Average per capita income has also increased.

It is nevertheless important to note that these results are not equally shared among all developing countries. Poverty continues in a significant part of the global population, illiteracy still concerns 35 percent of the global adult population, 30 percent of children do not complete primary school, and the lack of basic services affects 40 percent of developing countries. In most of sub-Saharan Africa, per capita income has decreased to the level of the 1970s.

At a time when donor countries experience budget deficits, the "transfer of resources" becomes a critical issue. Indeed, their donations are restricted insofar as they have to face an increasing domestic demand for economic support. Consequently, donor countries are concerned about how efficiently development organizations use their donations. At this point, development organizations face a major challenge in proving that their assistance to support and carry out projects and programs in developing countries is "an effective instrument to eliminate poverty and promote self-sustaining development."

It appears that the overall effort made by development organizations is to diminish the gap between developing and developed countries, as well as among developing countries themselves. In "Planning Projects that Breed Equality" (Ben-Dak, 1975) there is the problem of identifying the "most basic values that can be considered globally common." In other words, the project must not impress cultural values on other populations.

This is why the LogFrame is so useful for a development organization such as the United Nations Development Programme. Indeed, it helps determine and categorize the objectives and outputs while trying not to model values that are not shared by the culture of the recipient country. Examples for value orientation (goals), objectives, outputs, and inputs can be categorized as follows:

10.5.3.1   Goals. Some examples of goals are listed below:

1. Poverty elimination
2. Employment, sustainable livelihood
3. Advancement of women and other disadvantaged groups
4. Environmental protection and regeneration

10.5.3.2   Purposes. The following paragraphs discuss some examples of purposes.

1. *Development of Unique Products.* The Global Technology Group of the UNDP defines unique products as "those that are (or can become) available in abundance in a country/regional geographic boundary, that can be produced with simple extent technologies involving local men and women, and those that have a market base in local and/or North countries." A typical example of a project geared to developing unique products is the mapping of the animal and plant resources of the Gobi Desert (project MON/95/309).

2. *Development of Business Catalysis.* The Global Technology Group defines the catalytic process as "experimentation with the numerous modes and strategies of *mixed* local and foreign decision groups and increasingly with joint ownership that are fair, participatory, and based on realistic learning curves in local and foreign industrial plants." A typical example of a project geared to developing business catalysis is the identification of 100 innovative businesses in Latin America to further their export promotion activities (project RLA/94/323).

3. *Development of Environmentally Sound Technologies.* The Global Technology Group focuses its effort on ensuring the sustainable use of land and sea-based plant and animal resources. A typical example of a project geared to developing environmentally sound technologies (EST) is the establishment of a database of locally available and applied ESTs to engage in South–South transfer for "greening" enterprises in Zimbabwe (project ZIM/93/340).

4. *Introduction and Effective Enforcement of Intellectual Property Rights.* This system is not only significant for the people of the South in that it is *their* way of making knowledge count, but it is also significant in contributing to the preservation of agricultural and pharmaceutical biodiversity, new industrial bases, and human survival in the fuller sense of both good health and productivity. A typical example of an intellectual property rights (IPR) project is the development of creative interfacing between the Western IPR and the Southern Cooperative systems of innovation in Mongolia (project MON/93/010).

5. *Mobilization of Diverse Groups.* The Global Technology Group concentrates on gathering men and women from all regions and various ethnic groups of a country in a manner that makes certain desired (collectively or otherwise) goods or ends possible. The Group has also consistently advocated cooperation with nontraditional development partners, such as the military, in national and regional environmental care. An example of a project dealing with the mobilization of diverse groups is the military industry conversion in China (project CPR/94/319).

All of the preceding purposes are suitable for attaining any of the goals, and there is no particular goal for one purpose. Indeed, the development of unique products is a purpose that can be integrated into a program to achieve any of the just-mentioned goals. In the frame of the program, however, this purpose will serve only one of the four goals.

10.5.3.3 Outputs. Many of the UNDP outputs can be categorized as follows:

1. Feasibility reports
2. Training workshops
3. Limited sponsorship demonstration programs
4. Conferences
5. Expert advice
6. Round tables involving government, financial institutions, and the private sector (local and international)
7. Technology transfer actions, in-licensing, and licensing out

10.5.3.4 Inputs. The UNDP inputs can be categorized as follows:

1. Financial resources: UNDP financial inputs
2. Human resources: Consultants, task forces, UNDP personnel
3. Equipment: Computers mostly
4. Time

10.6 FIRST-ORDER CONCLUSIONS: THE LOGFRAME, A FORMATIVE EVALUATION OF THE CONCEPT OF EVALUATION

At this point, the LogFrame can be envisioned as a conceptual tool that provides rational thinking on the concept of evaluation in order to improve the feedback system within an organization (i.e., the project or program). This is why the LogFrame can be looked upon as a formative evaluation of evaluation.

Evaluation has been defined as the combination of a comparison of actual performance to project plans and a test of causal linkages. Evaluation also focuses on redesigning, replanning, and reimplementing. It aims at improving the management of an organization in terms of communication, clarification of objectives, candid reassessment of plans and alternatives, and introduction of change in the organization.

Evaluation must be performed when decisions are to be made, preferably on a periodical basis. It is also needed in cases where there is a transfer of experience from one project to another. It must be envisioned on a service–provider–client basis, that is, the evaluation must start with the stakeholders and reflect their views, while trying to remain objective.

Regarding the evaluation team, the problem of insiders and outsiders was raised earlier in this chapter. This issue is critical for ensuring that the evaluation is done on an objective basis and at a reasonable cost. *In-house evaluators* have a greater knowledge of the environment of the organization and of the specific operations of that organization. They can provide immediate and direct feedback. Their cost is lower and their coverage is broader. Their main drawback is that they have a preconception of what the problems of their organization are. They also have an interest insofar as they are stakeholders. *Outside experts* are, on the other hand, disinterested experts whose technical knowledge is broader since they are not limited to that of the organization they are evaluating, and they have a broader view in terms of problem formulation and identification. Therefore, a joint effort of insiders and outsiders may be best, depending, of course, on the evaluation needs. Evaluators are to provide evidence to decision makers, involve all the stakeholders, remain objective and help others remain so, and impartially compare present projects to possible alternatives.

As regards the investment in evaluation, the basic condition of performing an evaluation is that the benefit brought by this evaluation must exceed its cost. This benefit must also exceed that of an alternative investment. A method for determining the cost of an evaluation can be the expected increased value of the project enabled by the conduction of the evaluation. For example, if a project worth $5 million has a 50 percent probability of success, and it has been found that conducting an evaluation would increase its success potential to 60 percent, then the cost of the evaluation should not exceed $5 million $\times$ 10 percent, or $0.5 million.

At this point, an evaluation process can be envisioned. It can be depicted as a systematic series of actions directed toward a specific end and a managed sequence of activities. These activities can be organized as follows.

The design clarification is to both clarify the project's or organization's intent and to reexamine the need to evaluate at a particular time. Once clarified, the evaluation process goes through the planning phase, the conduction or implementation phase, and the utilization of the evaluation results.

The planning phase involves specifying the required data. Prerequisites for the identification of those data are: understanding the target users of the evaluation, knowledge of the available information sources, and insightful prediction of the findings. The evaluation process may lead to five types of decisions: modification of the organizational or project design (or not), expansion of the scope of the project or organization, termination of the project or organization, replication of the project or organization elsewhere; replanning of the program or larger system to which the project/organization contributes. The types of data required may vary from data on changes in the project/organization environment to data on unplanned results or possible alternative explanations for actual results.

Conducting the evaluation process includes data collection and transforming data into organized information. The issue of data collection poses problems of determining the size of the sample from which the data will be collected, the way the data will be collected (interviews, questionnaires, direct observation), identification of the data collector, and the timing of data collection. The underlying issue that has been previously raised is the need for objective data collection. The test of causal linkages will constitute the transformation of data into information. Eventually, the evaluation must lead to findings, which must be utilized.

The utilization phase consists of transforming findings, which are facts, into conclusions, which are interpretations of those facts. This is the extrapolation component of the utilization phase. Then, those conclusions must lead to concrete recommendations, that is to say, the definition of alternative courses of action. The evaluation review, which corresponds to the evaluation of the evaluation, consists of collaborative discussions of evaluation findings, issues, alternatives, and confirming no change in action decisions. It ends the evaluation process.

The end product of the evaluation process can be either a plan, a report, or an input for the design of other projects.

10.7 SECOND-ORDER CONCLUSIONS: THE LOGFRAME IN THE CONTEXT OF DEVELOPMENT STUDIES AND BEYOND

10.7.1 The Reality of Nonapplied LogFrame

If first-order consequences of the LogFrame (i.e., the *intended* consequences of this tool) are to capture conceptualization that helps the statement of logical connections between levels of effort, then there are numerous unintended consequences when the tool is applied correctly.

Experience with the LogFrame for the past 20 years has suggested that the tool has not had a major impact in the development field, or at the more specific project level, simply because it is not applied correctly or not applied at all. Correct application in this case refers directly to the original thinking at US-AID and Practical Concepts Incorporated (see also Maclean, 1988). Given that systems thinking encompasses a large and fairly amorphous body of methods, tools, and principles "all oriented to looking at the interrelatedness of forces and seeing them as part of a common process" (Senge et al., 1994, p. 83), how can a practical technique for process mapping evolve? The challenge was, and still is, not necessarily to go toward the complexity of comprehensive system dynamics *á la* Jay Forrester and tools captured in other chapters of this handbook, but to *create a development worker's perspective of a useful, sufficiently simple, clear set of directives*. Clearly, in development thinking one must capture the essentials of the internal relatedness of project efforts and relate the internal set of expectations to the minimal "real world" constraints. The original LogFrame thinking was to capture this imperative (Ben-Dak, 1975; Practical Concepts, 1971, Ben-Dak, 1979). This way of thinking was the key reason for US-AID to adopt the tool (AID Bureau, 1983; Adoum and Ingle, 1991; Solem, 1987) and the motivation of many commercial vendors, such as Team Technologies of Virginia, to adapt the tool for their purpose.

A closer look at evaluations and related project designs suggests that even when the comparisons between expected-level performance and actual achievements could have benefitted from LogFrame application in an early stage, or from postproject reanalysis (i.e., rewrite from records the "project LogFramed history"), much insight was lost because the tool was not applied (Rivera, 1990; Bartone, 1990; Kulaba, 1990; Khandker, 1989). These projects involving UNDP and the World Bank work on private- and urban-sector projects perhaps duplicate other evaluation realities of less aware areas of project conceptualization (aware in the sense of being cognizant of LogFrame advantages). Major international assistance projects that could have benefitted from applying the LogFrame in project design and evaluation are efforts concerning the feeding and greening of the world, in particular international agricultural research for development (for example, Hazell and Ramasamy, 1991; Hoggart, 1992; National Planning Association, 1987; National Research Council, 1993; Persley, 1990; Ryan, 1987; Vosti et al., 1991). Rarely do decisions to cut or increase financial allocation for agricultural development rely on *disciplined comparisons* of the sort insisted on by the LogFrame directives (Tribe, 1994). Topping this list of unfortunate evaluations efforts are the pilot studies of capacity building done for UNDP by the Harvard Institute for International Development. These studies of Bolivia, Central

African Republic, Morocco, Sri Lanka, and Ghana (Project INT/92/676) produce a great deal of insight regarding completed and expected financial reforms, environmental and public-sector changes and reformulations, and a wealth of case studies, but there is no benchmark to begin a within-country comparison and a dashboard as explained in Section 10.1.

UNDP's and US-AID's use of the LogFrame has been mainly to apply the concepts of level breakdown and objectively verifiable indicators. In a variety of project manuals and instructions for project documents there has not been real insistence on working with all cells, even though US-AID has been a bit more explicit in recommending the play according to the original script. In addition, evaluations have very rarely used the full impact analysis of relating a 4×4 project design to field observation/hypothesis testing.

Thus, it appears that the LogFrame has not been applied to a large body of development work when it would have been appropriate. In fact, not withstanding cursory, anectodal references, no systematic general evaluations were attempted when insufficient bodies of evidence were accumulated in both UNDP and US-AID. Because the cost of not applying the LogFrame has been known from the beginning, especially in terms of second-order consequences, it is extremely important to review its particular new costs since the ideology of sustainable development has become fashionable in current thinking about development.

As demonstrated in this chapter, sustainable development, has come to mainstream UN work; for example, the Commission on Sustainable Development was created to move with this old-new thinking so that new realities can be effected in the less developed world or anywhere on this planet (Brown et al., 1995a,b; Callenbach, 1993; Daly and Cobb, 1994). The conceptual work demonstrated in this chapter belongs perhaps to a specific effort that can be labeled "sustainable livelihood" projects.

10.7.2 Sustainable Livelihood as a Guiding Concept

Use of the concept of sustainable livelihoods has evolved in response to the challenges of development in today's world and to the need for new approaches and concepts that are integrative, can accommodate diverse constraints, and do not attempt to be so all-encompassing that they are too vague and ineffective to lead to action. Such action must still flow from institutional arrangements or rearrangements, policy making, resource mobilization, and initiatives by society's major actors, such as governments (at all levels), nongovernment organizations, business, and civil society leaders/groups. The flow must, however, be within an interactive and dynamic framework that has as its goal the promotion of sustainable livelihoods for identifiable groups of poor people. Systematic logic is clearly the necessary direction that relevant project design and evaluation must take (see also Sideman and Ben-Dak, 1997).

More specifically, sustainable livelihood seeks to offer an alternative approach to the essential failure of development strategies over the last four decades to stem the tide of growing poverty. It builds on the lessons learned in "delivering development," and the challenges presented by today's world in the changing science and technology, nature of work, employment, jobs, education and training modes, multisources of incomes, the desire for sustainable well-being, and livelihood security. The concept of job security is gradually yielding to livelihood security. The sustainable livelihood concept allows the integration of social, gender, and environmental equity issues with the foregoing "employment" issues, in the search for ways to help poor people make a better living in a sustainable manner (Gore, 1992).

Within the UN intergovernmental system, the concept was highlighted at both Rio and post Rio Copenhagen. Agenda 21 used the term "livelihoods" in the context of combatting poverty. Not only did Agenda 21 objectives include providing "all persons urgently with the opportunities to earn a sustainable livelihood," but "alternative livelihoods" were singled out as a priority for ecologically fragile areas. Livelihoods of disadvantaged groups and resource-poor populations in both urban and rural areas were also targeted for improvement (Purcell and Morrison, 1987). The environmental and natural resource aspects of sustainable livelihoods have thus been established as critical to the concept.

In the current literature on development, sustainable livelihoods are viewed in a variety of ways that can be summarized in two broad dimensions. The sustainability dimension implies both (1) time/durability and (2) continued environmental conservation, while the livelihoods dimension implies a variety of means for making a living (gainful occupational activities) that are at least adequate for satisfying basic needs, and secure against shocks/ stresses.

The goals of full-employment/livelihoods policy approaches are to move toward more sustainable methods of production and consumption, underpinning modern science technology and economies, as well as to reduce vulnerability and insecurity in the occupational affiliations of those living in poverty, and especially of female-headed households. How to make these approaches operational in ways that are most useful to UNDP regional and country programs, or US-AID, or any other development and international assistance programs, is the major purpose of the LogFrame conceptualization presented in this chapter.

UNDP's Executive Board has shown vision, boldness, and leadership in confronting the failures of past development strategies, including the lack of systematic feedback, and in coming to grips with the challenges of current priorities. The agency's mission, or "supergoal," is sustainable human development and the eradication of poverty. Being aware that such a mission statement has historical and political roots while promising practical fruits, the next step is to identify the stems and leaves, that is, the operational steps that lead to effective programming and action toward achieving the goal. We know from the analysis developed in this chapter that program goals depend on achieving project purpose, outputs, and inputs. We know that the emphasis on measuring success is critical. It is also important to accept that goal achievement by UNDP alone is not a plausible objective. The operational objective must be progress toward the goal as a measure by the agency's contribution to that effort. Measurement and monitoring of progress toward the goal thus become the critical management tasks involving a system examination of other agencies as well. For this purpose, assessment procedures for sustainable livelihood outcomes of UNDP programs, as well as indicators for measuring progress toward sustainable livelihood, depend on the LogFrame, or similar systems that will be or can be developed.

To move toward operational steps, planners should accept poor people (however defined, typified, or located) as their primary constituents. Project stakeholders' work begins with a focus not on poverty but on poor people (in households and communities) and their livelihood activities (their stores, assets, entitlements, and access to and control over resources required for their livelihood). Such activities and access are considered in relation to both their interaction with other (social, cultural, political, economic relationships) and with their natural and human-made environment (environmental sustainability considerations). Agency interventions should include either directly or indirectly all those critical areas that will synergistically contribute to enhancing existing or potential opportunities for helping the poor achieve sustainable livelihoods. Most poignantly, perhaps, is the observation that beyond greening and traditional acting modes available to UNDP and similar organizations, private-sector small, middle-size, and large corporations of both local and international origin will have to play a major role (see Hart, 1997). This has been the Global Technology Group's main conclusion from its LogFrame work from 1993 to 1996.

Given this perspective, one can appreciate the degrees of fit between the level of sophistication of the LogFrame tool and its prescribed utility by comparing its conceptual deliverables in practice (when it is applied correctly) to other instruments of similar promise but of less capability.

One such promising concept system has been identified by the environmental management area as part of the new callings for sustainable environment. To address this calling there was a need for strategic vision and planning that can be translated into effective operational implementation. This multifaceted area has introduced the idea that one of its main orientations is the integration of neutralized wastes as a resource-deployment strategy (Ben-Dak, 1995, 1996). Such a strategy deals with such questions as: How wide is the net of issues to be included? What enters the realm of acceptable and needed impacts? How should uncertainty be handled? What is the most appropriate policy and regulatory framework? What is the most appropriate value-adding business? While these questions may yet merge to common management principles (following the example of ISO 9000), they do not provide an answer to most business or capability evaluations *in the developing world*. Many of the ideas of waste-management systems, including EMS's, and EIA, touch areas external to the direct impact of a project in contexts such as legislation, social, political, and ideological constraints (Gilroy, 1992; Godet, 1994; IUCN/UNEP/WWF, 1991; Janssen, 1992; Lave and Giuenspecht, 1991; White and Becker, 1992; Winn and Roome, 1993) with essentially two ad hoc observations, as follows.

1. Conceptually, the field is still evolving and has not yet produced a single agreed-upon frame of reference that allows uniformity in, say, having two groups working on a formative evaluation of the same project. The fact that external factors figure so prominently that they differ across the cultural, language, and development continuum, makes the wish list of these technology/resource management systems much more obvious than the directives to project stakeholders. This is true, to varying degrees, for both developed and less developed country contexts.

2. The cumulative, environmental-management, empirical evidence suggests that the level of analysis of the LogFrame—*the manageable interests in a given project*—is close to optimal. The moment project or impact conceptualizers go beyond their stakeholders' "internalized" concepts of the whys and hows of a given project, they enter areas of possible and growing disagreements and information imbalances. Clearly, in the long run and for true optimal project handling one must deal effectively with disagreement and information about the external environment of the project simply because it may be the final judge and context for project success. Currently, *participation* and *involvement* are key factors in development studies. Because enough experience has been documented that the LogFrame captures group and immediate participant conceptualization of project terms in less developed countries handling external factors qua project constraints, the LogFrame approach emerges as both sufficiently simple and complex for the purposes at hand.

10.8 CONCLUSIONS

It is precisely in this context that some of UNDP's work has already attempted to tease out the relatedness of in-project levels and out-of-project realities so that lessons can be learned. The Global Technology Group has used the *LogFrame for this very reason*. The approach taken was, essentially guided by a determination to completely "fill" and carefully execute 16-fold cells that must be properly considered and reconsidered for improving project design. It then, and only then, follows that formative and summative evaluations can be properly executed, and that project planners and evaluators can agree on findings about project performance, then conclusions, and then recommendations. There will be more speculation and less disciplined insight as one moves to the creative interpretation of findings.

The fairly sketchy and telegraphic-simplistic nature of some of the examples elaborated earlier can still prove to the discerning reader that there is reason and logic in following up on all 16 cells for basic, required detail. One can do with less for some purposes, for example working with narrative summary alone or eliminating altogether the external world of impacts (i.e., the assumptions), but there will be a corresponding reduction in policy and learning value in the execution and evaluation.

The recommendation here is that for both practicality and inclusion of key data the LogFrame is a very appropriate system for policy studies in development. The unintended consequence of not applying it at all, or applying only part of it, is perhaps to miss what common sense can do for sustainable livelihood or project success in many fields of development or for that matter, *any* project management that has a future.

APPENDIX 10A

Table 10A.1 is a checklist for ensuring that a new LogFrame being developed is based on the correct conceptual thinking. Each of these items, one through sixteen is to be examined first by the individual structuring the evaluation, then discussed by the group or team for the project. The final column represents the combined opinion of the individual and the group, which is the final verdict on whether or not the devised LogFrame conforms to the correct conceptual thinking.

GLOSSARY OF TERMS WITH SPECIAL MEANING IN THE LOGICAL FRAMEWORK

Assumption   A supposition that something is true; a factor or statement that is taken for granted. **LogFrame** definition: Expectations about factors that influence the success or failure of a project, but over which the **project manager** has no control.

Achievement Reporting   **Reporting** from one level of **management** to the next higher level on the accomplishment (in terms of quality, quantity, and timeliness) of previously determined performance targets.

Congruence Diagrams   A graphic tool used in evaluating a project. It is useful for drawing conclusions and making recommendations about a project. The congruence diagram

TABLE 10A.1   Checklist for Ensuring Project LogFrame Based on Correct Conceptual Thinking

Characteristics of "Good" LogFrame	Personal	Group	Final
1. The project has a single purpose.			
2. The purpose is not a restatement of outputs.			
3. The purpose is above the manageable interest of the project manager.			
4. The purpose is clearly stated.			
5. All outputs listed are necessary for achieving the purpose.			
6. Outputs are clearly stated.			
7. Outputs are stated as results.			
8. Input statements identify activities as well as resources.			
9. The goal is clearly stated.			
10. The if–then relationship between purpose and goal is the next logical step in the project hierarchy—two or more linkages have not been omitted.			
11. Assumptions stated at the input level do not include project preconditions.			
12. The outputs, together with the output-level assumptions, create the set of necessary and sufficient conditions to achieve purpose.			
13. The purpose-level assumptions provide a functional statement of the critical conditions (including other project purpose conditions) that are required to achieve the goal.			
14. The input-to-output linkage is plausible.			
15. The output-to-purpose linkage is plausible.			
16. The vertical logic of the project is sound; all three linkages (I-O-P-G), if reviewed simultaneously, are plausible.			

compares project results in two ways at the same time: (1) obtained or actual results against planned; (2) among the four levels of the **logical framework.**

Baseline Data   Data characterizing the state of affairs at the time a project is started, hence providing a "baseline" against which to assess the nature and extent of change effected by the project. Baseline data should be captured or reconstructed for all OVIs set forth on the **logical framework** before the project is started.

Development Hypothesis   A project or program hypothesis where the expected result is development effect. If outputs, the **purpose** is called the project hypothesis. The hypothesis that purpose will lead to goal is called the program hypothesis. We state these relationships as hypotheses because we are not certain of the causal relationships between the "if" statement and the "then" statement. Implementing the project is in fact a way of testing the hypotheses.

End-of-Project Status (EOPS)   The set of objectively verifiable indicators that will signal the successful achievement of the project **purpose.**

Exception Reporting   Reporting to the next higher level of **management** that a planned event did not occur or is in danger of not occurring on time and/or in the same manner as was planned (e.g., with respect to the quantity and quality of performance).

Evaluation   An orderly examination of progress at each level of objectives (GPOI). Examines validity of hypotheses, challenges relevance of objectives, assesses **project design,** and results in redesign and replanning actions. Evaluation is oriented more to assessing output-to-purpose and purpose-to-goal linkages. This contrasts with **monitoring,** which is oriented more to the input-to-output linkage.

Manageable Interest   Defines the responsibility of the **project manager.** Project managers commit to delivering outputs if the requested resources are put at their disposal. It is within their manageable interest to reallocate or otherwise modify inputs and do whatever else is necessary to produce outputs aimed at achieving an agreed-upon **purpose.**

Management   Functionally defined, management has five operational criteria: (1) organizational activity; (2) objectives; (3) relationships among resources; (4) working through others; (5) decisions (from Cleland and King, 1983). "Management is working with and through individuals and groups to accomplish organizational goals" (from Hersey and Blanchard, 1977, 1996).

Matrix for the Logical Framework (LogFrame)   A 4×4 matrix that displays the interrelationships of the design and evaluation components of a development project. The matrix is displayed on a worksheet divided into four rows (for goal, purpose, outputs, and inputs) and four columns (for narrative, objectively verifiable indicators, means of verification, and important assumptions).

Means of Verification   The actual type, source, and means of obtaining data that will be used to verify an indicator (e.g., change in birth rate as evidenced from records of the Ministry of Health).

Monitoring   The **management** function of following the progress and overseeing the operations of a project. Monitoring is oriented more to the input-to-output linkage of the **logical framework** in contrast to **evaluation,** which is oriented more to the output-to-purpose and purpose-to-goal linkages. Monitoring is concerned with work activities and the procurement and use of resources.

Network   A graphic representation of the logical sequence of activities and events required to reach a specified objective.

Project Design   A summary of what the project is expected to achieve (**purpose**) and how it will be achieved with the inputs and time available. The key elements of project design can be summarized in the **logical framework** format.

Project Manager   The individual who is personally accountable for the success of the project. More specifically, the individual who is charged with producing those agreed-upon outputs within the specified time and cost constraints.

Purpose   What is hoped to be achieved by undertaking the project. The result aspired to if the required outputs are produced. Usually the significant change in people or organizations thought to be required to effect important social or economic benefits for the target population.

Reporting   Providing the necessary information to appropriate people for timely decision making regarding the successful implementation of the project. Includes both formal and informal communications; for example, a formal (fixed-format) report may be the stimulus for personal discussions.

Scientific Method   "A method of research in which a problem is identified, relevant data collected, a hypothesis formulated, and the hypothesis empirically tested" (from the *Random House College Dictionary of the English Language,* 1975).

System   Interrelated activities and events organized to perform a specific function(s), such as, produce certain outputs(s). A system may comprise any number of elements, but the interrelatedness of those elements is that required to perform the system functions(s) or to achieve its output(s). No system exists without connections to other systems.

Tree Analysis   A group of graphic tools used in problem diagnosis and objective setting for an organization. This group includes problem trees, objective trees, decision trees, and alternative trees. Implicit within tree analysis is the concept of cause-and-effect relationships.

Vertical Logic   Represent a prediction that *if* the expected results at one level of the **LogFrame** hierarchy are achieved, and *if* the assumptions at the level are valid, *then* the expected results at the next higher level will be possible (and will be achieved with the proper effort at the next higher level).

REFERENCES

Adoum, C. M., and Ingle, M. D. (1991). *Logical Framework and Benefit Sustainability.* Washington, DC: US-AID.

A.I.D. Bureau for Program and Policy Coordination. (1983). *Logical Framework: Modifications Based on Experience.* Washington, DC: US-AID.

Bartone, C. R. (1990). *Sustainable Responses to Growing Urban Environmental Crises.* Washington, DC: World Bank.

Ben-Dak, J. D. (1975). Planning projects that breed equality. *Manage. Constraints Less Dev. Countries, ICUS Conf. Proc.*

Ben-Dak, J. D. (1995). *Integrating Neutralized Wastes as a Resource Deployment Strategy (INWARDS).* UNDP: New York.

Ben-Dak, J. D. (1996). Managing and INWARDS strategy. *World Recycl. Congr.,* Berlin.

Ben-Dak, J. D., Posner, L., and Rosenberg, L. (1979). *Logical Framework: A Manager's Guide to a Scientific Approach to Design and Evaluation.* Washington, DC: Practical Concepts.

Brooks, B. (1980). *Logical Framework.* Project Plan. Manage. Ser., Module 2. Kingston, Jamaica: Project Analysis and Monitoring Co.

Brown, A. L. (1977). *Program Design Guidelines Using a Logical Framework—Goal Hierarchy Combination.* Washington, DC: American Technical Assistance Corp.

Brown, L., and Renner, M. (1995a). *State of the World.* Washington, DC: Norton.

Brown, L. et al. (1995b). *Vital Signs.* Washington, DC: Norton.

Callenbach, E. et al. (1993). *EcoManagement.* San Francisco: Berrett-Koehler Publishers.

Capeling-Alajika, S., Lopez, C., Benbouali A., and Jamal, Z. (1995). *UNDP Evaluation Findings in 1994,* OESP Ser. Lessons Learned. New York: UNDP. Office of Evaluation and Strategic Planning.

Cleland, D. I., and King, W. R. (1983). *Management: A Systems Approach,* p. 7. New York: McGraw-Hill.

Daly, H., and Cobb, J. (1994). *For the Common Good.* Boston: Beacon.

Global Technology Group, Bureau for Policy and Programme Support. (1996). *Mid-Term Evaluation, Special Programme Resources B.5 Technology Transfer and Adaptation.* New York: UNDP.

Godet, M. (1994) *From Anticipation to Action: A Handbook of Strategic Prospective.* Tokyo: Unipub.

Gore, A. (1992). *Earth in the Balance: Ecology and the Human Spirit.* New York: Houghton Mifflin.

Hart, S. L. (1997). Beyond greening: Strategies for a sustainable world. *Harv. Bus. Rev.,* January–February, pp. 66–76.

Harvard Institute for International Development. (1994). *Pilot Studies of Capacity Building: Morocco, Central African Republic, Bolivia, Ghana, Sri Lanka,* Prepared for UNDP Proj. INT/92/676. Boston: HIID.

Hazell, P. B. R., and Ramasamy, C. (1991). *The Green Revolution Reconsidered.* Baltimore, MD: Johns Hopkins University Press.

Hersey, D. D., and Blanchard, K. H. (1996). *Management of Organizational Behaviour,* p. 3. Englewood Cliffs, NJ: Prentice-Hall.

Hoggart, K. (ed.). (1992). *Agricultural Change, Environment and Economy.* London: Mansell Publishing.

Janssen, R. (1992). *Multiobjective Decision Support for Environmental Management.* Dordrecht, The Netherlands: Kluwer Academic Press.

Khandker, R. H. (1989). *UNDP Activities in the Realm of Private Sector Development During 1987–1988.* New York: UNDP.

Kulaba, S. M. (1990). *Thematic Evaluation of UNDP Urban Projects in Tanzania.* New York: UNDP.

Lave, L., and Gruenspecht, H. (1991). Increasing the efficiency and effectiveness of environmental decisions: Benefit-cost analysis and effluent fees. *J. Air Waste Manage. Assoc.* **45**(5), 680–693.

McLean, D. (1988). *Logical Framework in Research Planning and Evaluation.* Working Paper, No. 12. The Hague: International Service for Agricultural Research.

Mellor, J. W. (1966). *The Economics of Agricultural Development.* Ithaca, NY: Cornell University Press.

National Planning Association. (1987). *US Agriculture and Third World Economic Development: Critical Dependency.* Washington, DC: National Planning Association.

National Research Council. (1993). *Sustainable Agriculture and the Environment in the Humid Tropics.* Washington, DC: National Academy Press.

Persley, G. J. (1990). *Beyond Mendel's Garden: Biotechnology in the Service of World Agriculture.* Wallingford, UK: CAB International.

Practical Concepts. (1971). *Logical Framework.* Washington, DC: Practical Concepts.

Purcell, R. B., and Morrison, E. (1987). *US Agriculture and Third World Development: The Critical Linkage.* Boulder, CO: Lynne Reinner Publishers.

Rivera, M. C. (1990). *Ex-post Evaluation of Cooperation Projects for Integral Development of Popular Settlements in Colombia.* New York: UNDP.

Ryan, J. G. (ed.). (1987). *Building on Success: Agricultural Research, Technology and Policy for Development.* Canberra: ACIAR.

Senge, P. M., Kleiner, A., Roberts, C., Ross, R. B., and Smith, G. J.. (1994). *The Fifth Discipline Fieldbook: Strategies and Tools for Building a Learning Organization.* New York: Currency/Doubleday.

Solem, R. R. (1987). *Logical Framework Approach to Project Design, Review and Evaluation in*

A.I.D.: Genesis, Impact, Problems, and Opportunities, AID Working Paper, No. 99. Washington, DC: AID Center for Development Information and Evaluation.

Sideman, S., and Ben-Dak, J. D., (1997). Assessing medical technology in less-developed countries, Int. J. Tech. Assess. Health Care **13**(3), 463–472.

Tribe, D. (1994). *Feeding and Greening the World.* Wallingford, UK: CAB International and the Crawford Fund for International Agricultural Research.

Vosti, S. A., Reardon, T., and von Urff, W. (1991). *Agricultural Sustainability, Growth and Poverty Alleviation: Issues and Policies.* Feldafing: Deutsche Stiftung fur Internationale Entwicklung.

White, A. L., and Becker, M. (1992). Total cost assessment: Catalyzing corporate self-interest in pollution prevention. *New Solutions* **2**(3), 34–40.

Winn, S. F., and Roome, N. J. (1993). R&D management eesponses to the environment: Current theory and implications for practice and research. *R&D Manage.* **23**(2), 147–160.

11 Standards in Systems Engineering

STEPHEN C. LOWELL

11.1 INTRODUCTION

In this chapter we present an overview of the role of standards, and discuss key standards that can be used to develop and assess the major elements of the technical development process that must be integrated as part of the systems engineering effort. In addition, we give information on tailoring of requirements, proper application of standards, and how to locate and obtain standards.

11.2 DEFINITION

While there are many different definitions for specifications and standards, in the context of systems engineering, specifications and standards are technical documents used to define the requirements for the development, manufacture, and integration of products, processes, and services. *Specifications* describe the technical requirements for products or services, and the verification procedures to determine that the requirements have been met. Specifications are sometimes referred to as product standards. *Standards* are process-oriented technical documents. They are often referenced in specifications to establish or limit engineering processes and practices in the preparation, manufacture, and testing of materials, designs, and products. Standards establish uniform methods for testing procedures; define the engineering characteristics of parts and materials; provide for physical and functional interchangeability and compatibility of items; and establish quality, reliability, maintainability, and safety requirements.

11.3 HISTORICAL HIGHLIGHTS OF STANDARDS IN THE UNITED STATES

Specifications and standards have existed for thousands of years. The Bible has the specifications for Noah's ark, and the building code for Solomon's temple in Jerusalem. The Chinese used standards for the manufacture of pottery as early as 1600 B.C.. The precision with which the ancient Egyptians built the great pyramids would not have been possible without standards.

In the United States, the Founding Fathers placed such importance on standards that the Constitution charged the Congress with establishing standard weights and measures to provide an equitable basis for commerce between the states and for the collection of tariffs

Handbook of Systems Engineering and Management, Edited by A. P. Sage and W. B. Rouse
ISBN 0471-15405-9 ©1999 John Wiley and Sons, Inc.

at custom houses. Despite the commercial need for standards and the Constitutional mandate, political disagreements kept the Congress from acting quickly. The official standard measures were not finally adopted until 1856.

From the outset of the Industrial Revolution in the 19th century, specifications and standards began to take on new importance and different roles. The mass production of the industrial era was made possible largely because of the use of standard, interchangeable parts. In the early 1800s, Eli Whitney developed what he called the "uniformity system" in order to fill a government contract to manufacture 10,000 muskets. Up until this time, weapons manufacture relied more on the individual talents of highly skilled gunsmiths than mass production. The newly formed United States, however, had very few trained gunsmiths. Faced with a shortage of skilled labor, Whitney was able to fill this huge order only by using special machines to produce standard parts, which could then be assembled by lesser-skilled workers. Perhaps most importantly, the standards that Whitney developed to mass produce these muskets were subsequently used at other factories to make guns for the government, which marked the beginning of national standardization in modern American industry.

It is sometimes said that standards are born from disasters. The Industrial Revolution certainly contributed to many safety-related disasters, which eventually led to a demand for safety standards in the United States. Throughout the 19th and early 20th centuries, boiler explosions were a national nightmare. From 1870 to 1910, there were at least 10,000 recorded boiler explosions in the United States. In 1894, the simultaneous explosion of 27 boilers at the Henry Clay mine in Shamokin, Pennsylvania, leveled the town and killed thousands of people. With explosions totaling 1,400 a year by 1910, the American Society of Mechanical Engineers reacted by developing the first comprehensive boiler and pressure vessel code. Quickly adopted by states and cities, it virtually eliminated boiler explosions. In fact, between 1974 and 1984, there was not a single boiler explosion in the United States.

As science and technology advanced, the need for new and different types of standards became apparent. In 1901, the National Bureau of Standards was created to help identify these standards on a national level. One of the first efforts of the new bureau was in the area of fire safety. The great Baltimore fire of 1904 burned down a 70-block area in the business district. Scores of fire companies within a 100-mile radius of Baltimore responded. The devastation of the fire would have been significantly less, but unfortunately, the firefighters found that their hose coupling could not attach to the hydrants nor to other hoses. There were no standard couplings. A study by the National Bureau of Standards found that there were at least 600 different sizes and varieties of hose couplings being used in the United States. A year after the fire, the first national standards for fire hose couplings were adopted.

Wars have also highlighted the need for standards and fostered their rapid development. World War I was the first truly industrial war where the new technologies of war required mass production on an unprecedented level. When the United States entered the war in 1917, there was a critical need to standardize production and products to maximize industrial output. At the request of the federal government, professional societies of engineers and the American Society for Testing and Materials organized the American Engineering Standards Committee. This committee did not develop standards, but coordinated the efforts of those groups that did. The committee worked so well that after the war, it continued to function as a peacetime, civilian organization called the American Standards

Association, which eventually became the American National Standards Institute that exists today.

The 1920s are considered the "Golden Era of Standards" in the United States. The rapid expansion of American industry during and after the war emphasized the need for standards. A study initiated by Herbert Hoover in 1920, when he was the president of the Federated American Engineering Societies, estimated that American industry could reduce its production and distribution costs by 49 percent through improved standardization. New or more widely available consumer products also contributed to the demand for more standards. Consumer demand for radios, phonographs, and other electronic products necessitated the creation of new standards. In 1924, the Electronic Industries Association was formed to meet this demand for electronic standards. On the national scene, some 200 organizations became involved in developing standards.

The "Golden Era of Standards," however, came to an abrupt end in 1929 with the Great Depression. Companies did not have the funds to support standards development, and with the closing of many factories, there was not the same need to have standards to support mass production. With the start of World War II, there was a resurgence of interest in industry and government developing standards.

Today, environmental concerns and consumer protection provide a basis for much of the new standards development. But it is the globalization of the marketplace that is the primary driver behind standards development. Organizations that participate in the global market are emphasizing international standards instead of national ones for the following reasons:

- *Trade.* Expansion of free-market economies have allowed for more diverse sources for products and services. International standards serve as the language of trade to provide a common frame of reference to compare the products or services from one country to the next. As companies become increasingly multinational, standards provide a common set of requirements for manufacturers to acquire parts and services from a variety of suppliers from around the world.
- *Global Communications Systems.* Having interoperable computers, telephones, fax machines, and other communication systems is essential for doing business in the global market. International standards have played a key role in establishing an open systems framework for communications interface, while fostering innovation, productivity, and reduced costs.
- *Access to Developing Nations.* Developing nations rely increasingly on international standards. For an organization to have competitive access to these markets, they must be aware of and comply with international standards.

11.4   REASONS FOR USING SPECIFICATIONS AND STANDARDS

Specifications and standards are used by both industry and government throughout the world to conserve money, manpower, time, production facilities, and natural resources. Because specifications and standards usually represent known solutions to technical problems, and are the result of much experience, trial, and study, they benefit both the customer and the supplier. Some of the benefits include the following:

1. *Safety.* Safety considerations may be the most important reason to use standards. Failure of a system may result in death, injury, and property damage. Standards contribute to safety management by promoting the use of proven products, materials, processes, and practices. In general, safety standards do not dictate a specific design solution, but provide fundamental design data to the systems engineer to avoid "building in" catastrophic failure.

2. *Save Time and Money.* Some ways standards save time and money follow:
   a. Standards save the design engineer from having to reinvent the wheel each time for each part and process. The design engineer is able to benefit from the experience of other experts who developed the standard. The time not spent redesigning proven standards parts is a major cost avoidance in the development of any system.
   b. Standards identify parts with known quality and reliability. This shortens the time for testing and debugging the final product and reduces the cost associated with scrap and rework.
   c. Standards lower unit costs and shorten delivery times due to mass production. Custom-designed items cost considerably more to manufacture than do standard materials, parts, and components. It also takes much longer to have nonstandard items built and delivered. Because standard items usually have several manufacturers or distributors, costs are also less because of competitive bidding.
   d. Standards reduce the duplication and proliferation of parts, which cuts inventory control costs by reducing storage space, handling, incoming inspection of parts, and the recordkeeping associated with tracking parts and accounting.
   e. Standards reduce training time and costs. The greater the variety of parts, the more time is needed to train design engineers about the different characteristics of the parts; production staff about the installation of the different parts in the assembly line; and support personnel about the maintenance and repair of different parts.

3. *Quality and Reliability.* Standards enable quality management to be based on accepted, explicit requirements and the application of proven products, processes, and practices. The risk associated with designing new, untried parts is reduced significantly.

4. *Avoid Production Delays.* Standard parts and materials are more readily available from multiple sources at competitive prices.

5. *Improved Communications Between Buyer and Seller.* Standards help reduce misunderstandings between buyers and sellers, which reduces changes to the order, setup time, production lead time, contract disputes, and litigation.

6. *Market Acceptance and Customer Confidence.* Standards provide a measure for customers to buy a product with reasonable confidence that it will perform as advertised, meet an assured level of quality, or even exceed the standard level of performance. In the global marketplace, customers are demanding products that conform to certain standards or have standard interfaces. In many instances, local or national regulations or codes require that a product conform to certain standards before it can be sold.

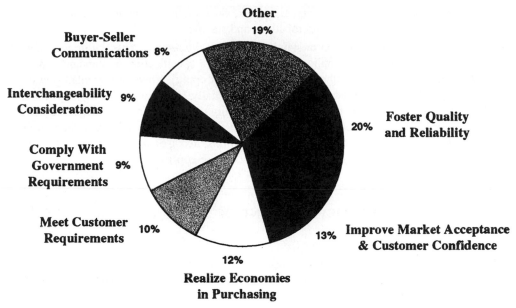

Figure 11.1   The reasons companies use standards shown as percentages.

In a 1987 industry survey of companies, ranging in type from building hardware, aerospace, petroleum refiners, manufacturers of mechanical equipment, to public service, 177 respondents gave the reasons in Figure 11.1 for why they use standards.[1]

11.5   PROPER APPLICATION OF SPECIFICATIONS AND STANDARDS

The primary problem associated with specifications and standards in the systems engineering process is misuse. Specifications and standards are tools that can provide solutions to technical problems. Unfortunately, sometimes they are used as substitutes for good engineering analysis, tradeoffs, and risk management. Here are some of the common pitfalls to avoid:

- Locking yourself into a set of detailed specifications and standards too early in the design and development process. Normally, customers will state their needs in terms of performance or functional requirements. Determining the exact specifications and standards used to establish product and process requirements should evolve through the requirements analysis and design process. Selecting specifications and standards prematurely can stifle innovation and limit the benefits of tradeoffs, risk management, and other alternatives typically considered under systems engineering analysis.
- Overspecifying. There is tendency to cite more specifications and standards than are needed in order to meet a requirement or to reduce risk. Such a practice can increase

[1]Source: October 1990 Press Release Package from the Standards Engineering Society for National Standards Week.

cost and time, while perhaps increasing risk, since the more important requirements can become lost in a sea of specifications and standards.

• Citing an entire specification or standard when all the requirements within the referenced document are not applicable. Often times, a specification or standard can be tailored to focus on only those requirements that apply. When only a small portion of a specification or standard is needed, it is usually better to extract the applicable sentences or paragraphs rather than cite an entire document unnecessarily. If the requirements are too extensive to extract, limit the applicability of a specification or standard by citing only those paragraphs, sections, tasks, methods, and so forth, that do apply or deleting those requirements that do not apply.

11.6 SELECTION AND DEVELOPMENT OF SPECIFICATIONS AND STANDARDS

The selection and use of specifications and standards is an iterative process. At the outset of system design, there should be relatively few fixed specifications and standards, except

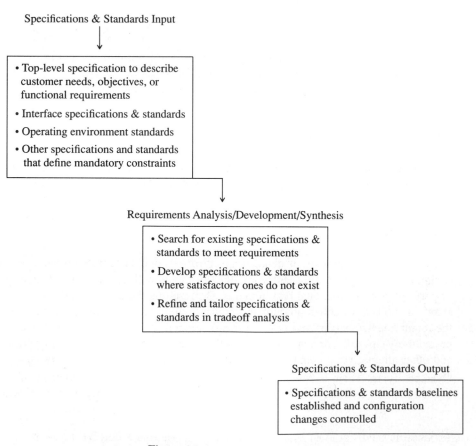

Figure 11.2 The iterative process.

TABLE 11.1   Several Standards Important to Systems Engineering

Technical Discipline	Standard Number and Title	Description of Content
Systems Engineering	EIA Standard IS-632, "Systems Engineering"[a]	This standard describes the systems engineering process, and provides guidance on how to perform, manage, and evaluate systems engineering efforts. It includes information on the systems engineering management plan, technical reviews, and measuring system and cost-effectiveness.
Configuration Management	EIA Standard IS-649, "National Consensus Standard for Configuration Management"	This standard describes basic configuration management principles and the best practices to implement them. It provides guidance on the establishment, documentation, and maintenance of a product's functional, performance, and physical attributes, and how to manage changes to those attributes.
Quality	ISO 9000, "Quality Management and Quality Assurance Standards—Guidelines for Selection and Use"[b] ISO 9001, "Quality Systems—Model for Quality Assurance in Design, Development, Production, Installation, and Servicing" ISO 9002, "Quality Systems—Model for Quality Assurance in Production and Installation" ISO 9003, "Quality Systems—Model for Quality Assurance in Final Inspection and Test" ISO 9004, "Quality Management and Quality System Elements—Guidelines"	ISO 9000 is often referred to as a single standard, but it is actually a series of different standards. It is an international standard, which has broad use throughout the world. While the different ISO 9000 series standards describe distinct quality models of varying stringency for use in different applications, they do share common elements for ensuring valid measurements; test equipment calibration; use of appropriate statistical techniques; training; product identification and traceability; maintenance of an adequate recordkeeping system; an adequate inspection process, including how to deal with nonconforming items; and an adequate system for product handling, storage, and delivery.

TABLE 11.1 **Several Standards Important to Systems Engineering** *(Continued)*

Technical Discipline	Standard Number and Title	Description of Content
Quality	ASQC 9000, "Quality Management and Quality Assurance Standards—Guidelines for Selection and Use"[c] ASQC 9001, "Quality Systems—Model for Quality Assurance in Design, Development, Production, Installation, and Servicing" ASQC 9002, "Quality Systems—Model for Quality Assurance in Production and Installation" ASQC 9003, "Quality Systems—Model for Quality Assurance in Final Inspection and Test" ASQC 9004, "Quality Management and Quality System Elements—Guidelines"	The ASQC 9000 series of quality standards is the U.S. national standard equivalent for the ISO 9000 standards. It is identical to the ISO 9000 series, except for some spelling differences.
	MIL-Q-9858, "Quality Program Requirements"[d]	This military specification describes the criteria for the manufacturer to establish and document procedures to control quality in the areas of design, development, fabrication, processing, assembly, testing, maintenance, packaging, shipping, storage, and installation. This military specification was recently canceled, but continues to be widely used. ISO 9000 is often used as a replacement for this document.
	SAE AS7106, "National Aerospace and Defense Contractors Accreditation Program Quality Program Requirements"[e]	This Society of Automotive Engineers standard was produced for the aerospace industry in an effort to harmonize differences between a variety of quality standards, including the ISO 9000 series and MIL-Q-9858. While the intent of the standard is to establish the minimum quality program requirements necessary for certification under the National Aerospace and Defense Contractors Accreditation Program, the quality program criteria could be applied in nonaerospace efforts as well.

TABLE 11.1 *(Continued)*

Technical Discipline	Standard Number and Title	Description of Content
Reliability	Military Standard 785, "Reliability Program for Systems and Equipment Development and Production"	This military standard provides general requirements and specific tasks for reliability programs during the development, production, and deployment of systems and equipment. While this is a military standard, it is also used in commercial applications. The Society of Automotive Engineers and the Institute of Electrical and Electronic Engineers both have efforts underway to develop the first U.S. national reliability standard.
Maintainability	Military Standard 470, "Maintainability Program for Systems and Equipment" Military Standard 471, "Maintainability Verification, Demonstration, and Evaluation"	This military standard provides task descriptions to establish maintainability programs. This military standard is a companion document to Military Standard 470; it provides the procedures and tests methods to verify and evaluate maintainability.
Supportability	Military Handbook 502 "Acquisition Logistics"	This military handbook identifies management objectives, tasks, and events associated with the logistics support development and implementation.
Human Factors	Military Standard 1472, "Human Engineering Design Criteria"	This military standard establishes general human engineering design criteria, principles, and practices to improve the effectiveness, simplicity, reliability, and safety of system operation, training, and maintenance.
Producibility	Military Handbook 727, "Design Guidance for Producibility"	This military handbook provides information to help the engineer detect, and then reduce or eliminate, design features that would make production difficult to achieve.

TABLE 11.1   Several Standards Important to Systems Engineering *(Continued)*

Technical Discipline	Standard Number and Title	Description of Content
Software Development	ISO/IEC 12207, "Information Technology Software Life Cycle Process"	This international standard establishes uniform requirements for software development activities, including requirements definition, documentation, product evaluation, quality assurance, configuration management, testing, and software reuse.
	EIA/IEEE J-STD-016, "Standard for Information Technology Software Life Cycle Processes Software Development Acquirer-Supplier Agreement"[f]	This standard is based on the major technical content of the ISO/IEC 12207.
Parts Management	AIAA R-100, "Recommended Practice for Parts Management"[g]	This standard establishes the criteria for a parts management program. While it primarily reflects the practices of the aerospace industry, the principles can be applied in any segment of industry. This standard integrates ten key elements into the parts management approach: part obsolescence management; supplier management; supplier assessments; cost; technology insertion; information exchange; process control; oversight; concurrent engineering; and training.
System Safety	Military Standard 882, "System Safety Program Requirements"	This standard provides uniform requirements for developing and implementing a system safety program to identify the hazards of a system and to impose design requirements and management controls to prevent mishaps.
Work Breakdown Structure	Military Handbook 881, "Work Breakdown Structures"	This military handbook provides criteria for the preparation and use of work breakdown structures. While the specific criteria relates to defense items, the general principles can be applied to commercial items as well.

TABLE 11.1 *(Continued)*

Technical Discipline	Standard Number and Title	Description of Content
Environmental Management	NAS 411, "Hazardous Materials Management Program"[h]	This standard establishes the requirements for a hazardous materials management program to ensure that appropriate consideration is given in the system design process to eliminating, reducing or minimizing hazardous materials, and controlling hazardous materials for the protection of human health and the environment.
	ISO 14001, "Environmental Management System"	A family of environmental management standards is being developed by the International Standards Organization under the 14000 number series. ISO 14001 is the core document in the 14000 series. ISO 14001 defines terms and processes for the management of environmental requirements; provides guidance on creating organizational environmental plans; gives direction for establishing roles and responsibilities for implementing plans; and establishes criteria for monitoring and measuring operations and activities, and providing management reviews.

[a] Copies of Electronic Industries Association (EIA) standards can be purchased from Global Engineering Documents, 15 Inverness Way East, Englewood, CO 80112-5704 or call 1-800-854-7179.

[b] Copies of ISO standards can be purchased from the American National Standards Institute (ANSI), 11 West 42nd Street, New York, NY 10036, or send a sales fax to 212-302-1286.

[c] Copies of ASQC standards can be purchased from the American Society for Quality Control (ASQC), P.O. Box 3005, 611 E. Wisconsin Avenue, Milwaukee, WI 53201-4606.

[d] Copies of military specifications and standards can be purchased from the Department of Defense Single Stock Point, Standardization Document Order Desk, Building 4D, 700 Robbins Avenue, Philadelphia, PA 19111-5904.

[e] Copies of SAE standards can be purchased from the Society of Automotive Engineers (SAE), 400 Commonwealth Drive, Warendale, PA 15096-0001.

[f] Copies of this joint EIA/IEEE standard can be purchased either from Global Engineering Documents, 15 Inverness Way East, Englewood, CO 80112 or the Institute of Electrical and Electronic Engineers (IEEE), 445 Hoes Lane, P.O. Box 1331, Piscataway, NJ 08855-1331.

[g] Copies of this AIAA standard can be purchased from the American Institute of Aeronautics and Astronautics, 1801 Alexander Bell Drive, Reston, VA 22091.

[h] Copies of National Aerospace Standards (NAS) can be purchased from Global Engineering Documents, 15 Inverness Way East, Englewood, CO 80112 or the Aerospace Industries Association, 1250 Eye Street, NW, Washington, DC 20005.

those that define the customer's requirements in terms of performance, function, need, or objective; any mandatory physical or operational interfaces; the environment in which the system must operate; and any other constraints that must be considered during system design. During the development process, it should be determined if specifications and standards already exist that would define the requirements (see Sections 11.7 and 11.8 for guidance). If satisfactory specifications and standards do not already exist, they will have to be developed. The important thing to keep in mind when selecting or developing the needed specifications and standards is that these specific technical requirements must be balanced against cost, schedule, risk, performance, and the overall customer need. At the end of the process, there should be a set of defined, tailored specifications and standards that establish the product baseline whose configuration must be managed. Figure 11.2 is a graphic representation of the iterative process described above.

11.7 USEFUL STANDARDS IN THE SYSTEMS ENGINEERING PROCESS

The standards given in Table 11.1 provide guidance on the various processes that are an integral part of systems engineering. These standards represent those that are most commonly used by industry and government today; however, systems engineering is a robust discipline and new standards continue to be developed. These are general standards that can apply to any or most systems, but there are also numerous standards not cited here that only apply to specific product lines or processes. See Section 11.8 for a discussion of how to locate those documents.

11.8 LOCATING AND OBTAINING SPECIFICATIONS AND STANDARDS

There are hundreds of thousands of international, foreign, U.S. national, and U.S. government specifications and standards from which the systems engineer can select when designing a system, and thousands more documents are created every year. The key is knowing where to search for these standards, how to limit the search, and how to obtain copies of documents. Every standards developing organization has an index of their documents, and many have their indexes available on the Internet. However, rather than search through hundreds of different indexes, the following sources provide reasonably complete indexes of large segments of specifications and standards, and many can provide copies as well.

Organization	Information Provided
American National Standards Institute (ANSI) 11 West 42nd Street New York, NY 10036 Telephone: (212) 642-4900 Web Site: http://www.ansi.org	The American National Standards Institute (ANSI) can provide indexes and copies of ANSI-approved and draft U.S. industry standards, and international standards. ANSI has an on-line Standards Information Database (SID), which provides a complete list of approved American National Standards originating from over 250 standards organizations, a list of proposed American National Standards under development, and a list of contacts within the United States Accredited Standards Developer and U.S. Technical Advisory Group. This on-line Standards Information Database can be accessed through the ANSI Web site. This database only provides lists of documents. Copies must be obtained from the issuing standards organization.

Global Engineering Documents
15 Inverness Way East
Englewood, CO 80112-5704
Telephone: (800) 854-7179

Provides search services and copies of U.S. industry standards, international and foreign standards, and federal and military specifications and standards.

Information Handling Services
P.O. Box 1154
Inverness Way East
Englewood, CO 80150
Telephone: (800) 241-7824

Affiliated with Global Engineering Documents, Information Handling Services provides subscription services to indexes and copies of U.S. industry standards, international and foreign standards, and federal and military specifications and standards on CD-ROM.

ILI Infodisk, Inc.
14-25 Plaza Road
Fair Lawn, NJ 07410
Telephone: (201) 703-8418

Offers index services with information on over 235,000 standards from all over the world. Is primarily an index resource, but can also help customers obtain copies.

Custom Standards Services, Inc.
Web Site: http://www.cssinfo.com

Provides on-line search and document ordering capability for U.S. industry standards, international and foreign standards, and federal and military specifications and standards. Offers a number of customized services, including a free notification service to alert customers when a document they have ordered in the past is updated.

Document Center
1504 Industrial Way, Unit 9
Belmont, CA 94002
Telephone: (415) 591-7600

Provides search services and copies of U.S. industry standards, international and foreign standards, and federal and military specifications and standards.

Department of Defense Single
 Stock Point
Standardization Document
 Order Desk
700 Robbins Avenue
Building 4D
Philadelphia, PA 19111-5094
Telephone:
 (215) 697-2179 or 2667
Web Site:
 http://www.dodssp.daps.mil

Provides the DoD Index of Specifications and Standards, which lists nearly 50,000 government specifications and standards, and industry standards adopted for defense use. Copies of the federal and military specifications and standards are available. The web site provide document ordering information and subscription services.

National Standards System
 Network (NSSN)
Web Site: http://www.nssn.org

The National Standards System Network (NSSN) provides on-line search capability for more than 250,000 private sector and government specifications and standards. The ultimate goal of the NSSN is to make these documents available over the Internet.

International Organization for
 Standardization (ISO)
Web Site: http://www.iso.ch

ISO is the leading international standards organization in the world. It is a federation of national standards bodies from about 100 countries. ANSI is the U.S. member to ISO. The ISO Web site provides an on-line index and ordering capability for all ISO standards.

11.9 OTHER USEFUL WEB SITES FOR STANDARDS

Aerospace Industries Association (AIA)—http://www.aia-aerospace.org

American Society of Mechanical Engineers (ASME)—http://www.asme.org

American Society for Testing and Materials (ASTM)—http://www.astm.org

Electronic Industries Association (EIA)—http://www.eia.org

Institute of Electrical and Electronic Engineers (IEEE)—http://www.ieee.org

National Institute of Standards and Technology (NIST)—http://www.nist.gov

Society of Automotive Engineers (SAE)—http://www.sae.org

12 System Architectures

ALEXANDER H. LEVIS

12.1 INTRODUCTION

Systems architecting has been defined as the process of creating complex, unprecedented systems (Rechtin, 1991; Rechtin and Maier, 1996). This description fits well many of the systems that are being created or planned today, whether in industry, government, or academia. The requirements of the marketplace are ill-defined; the rapidly evolving technology has been making possible the offering of new services at a global level. At the same time, there is increasing uncertainty about the way in which they will be used, what components will be incorporated, and the interconnections that will be made. Generating a system architecture as part of the systems engineering process can be seen as a deliberate approach to cope with the uncertainty that characterizes these complex, unprecedented systems.

The word architecture derives from the Greek word architecton, which means master mason or master builder. The architect, now as then, is a member of the team that is responsible for designing and building a system; then the systems were edifices, now they are computer-based and software intensive. Indeed, the system architect's contribution comes in the very early stages of the systems engineering process, at the time when the operational concept is defined and the basic structure of the system is conceptualized. Consequently, the design of a system's architecture is a top-down process, going from the abstract and general to the concrete and specific. Furthermore, it is an iterative process. The process of developing an architecture in response to requirements (that are ill-structured because of uncertainty) forces their re-examination. Ambiguities are identified and resolved and, when inconsistencies are discovered, the requirements themselves are reformulated.

One thinks of system architectures when the system in question consists of many diverse components. A system architect, while knowledgeable about the individual components, needs to have a good understanding of the interrelationships among the components. While there are many tools and techniques to aid the architect, and there is a well-defined architecture development process, architecting requires creativity and vision because of the unprecedented nature of the systems being developed and the ill-defined requirements. For detailed discussions on the need for systems architecting, see Rechtin (1991, 1992, 1994), Chorafas (1989), Sage (1992, 1995), and Sage and Lynch (1998).

Many of the methodologies for systems engineering have been designed to support a traditional system development model. A set of requirements is defined; several options are considered; and through a process of elimination a final design emerges that is well-defined. A schematic representation of this model is shown in Figure 12.1. This approach, based on structured analysis and design, has served well the needs of systems engineers

Handbook of Systems Engineering and Management, Edited by A. P. Sage and W. B. Rouse
ISBN 0471-15405-9 ©1999 John Wiley and Sons, Inc.

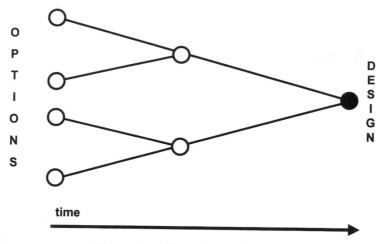

Figure 12.1 The traditional approach.

and has produced many of the complex systems in use today. It is effective when the requirements are well-defined and remain essentially constant during the system development period. However, when the time between requirements definition and operational implementation of the system is long—of the order of years—two things happen: the requirements change because of changing needs and new technology offers different alternatives and opportunities. This well-focused approach cannot handle change well; its strength lies in its efficiency in designing a system that meets a set of fixed requirements.

More recently, a new approach, with roots in software systems engineering is emerging to deal better with uncertainty in requirements and in technology, especially for systems with a long development time and expected long life cycle. This approach is shown schematically in Figure 12.2 and it encompasses such notions as evolutionary acquisition, or

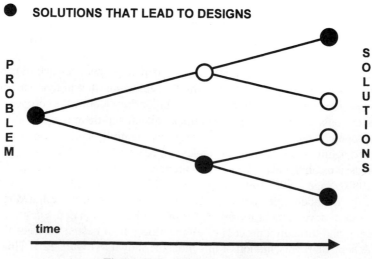

Figure 12.2 The new paradigm.

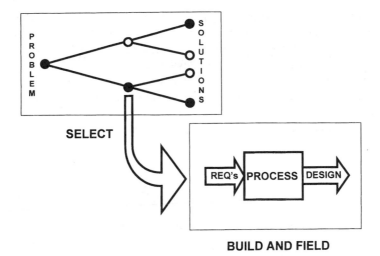

Figure 12.3   The assumed approach.

build-a-little, test-a-little. The problem is formulated in general terms and the requirements are more abstract, and therefore subject to interpretation. Alternative solutions to the problem are explored and pursued further as new technology options become available. Intermediate designs are saved: some are implemented in prototype versions but are not operationally implemented, while others represent transitional designs. The advantage of the

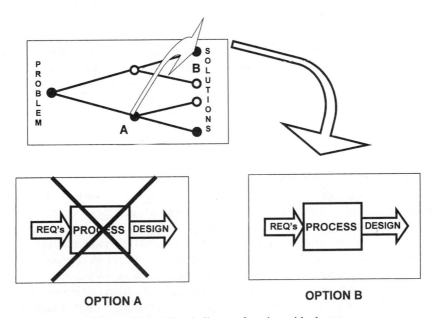

Figure 12.4   The challenge of coping with change.

object-oriented approach is that it allows flexibility in the design as it evolves over time; the disadvantage is that the number of options may increase substantially, requiring some early elimination of technology alternatives in the absence of reliable information.

The basic premise of this paradigm is illustrated in Figure 12.3. At any time in the development process, when there is need to build a system, the available solution that best meets the current requirements is selected and implemented using any systems engineering approach. If however, the implementation time is long (long enough to require a major change in the solution, either because requirements have changed or a new technology has become available and is cost-effective), then the situation shown in Figure 12.4 prevails, with the unfortunate consequence that very little, if any, from the work done on Option A can be used for Option B. The need to migrate gracefully, quickly, and at a low cost from one solution to another, especially in information systems, is the major driving force for system architectures.

There are several kinds of architectures that have existed in a variety of forms for a long time: technical, functional, and physical architectures. More recently, the Department of Defense (DoD) has defined three constructs: technical, operational, and systems architectures.

12.2 DEFINITION OF ARCHITECTURES

In defining an architecture, especially of an information system, the following items need to be described. First, there are processes that need to take place in order that the system accomplish its intended functions; the individual processes transform either data or materials that "flow" between them. These processes or activities or operations follow some rules that establish the conditions under which they occur; furthermore, they occur in some order that need not be deterministic and depends on the initial conditions. It is also necessary to describe the components that will implement the design: the hardware, software, personnel, and facilities that will be the system.

This fundamental notion leads to the definition of two architectural constructs: the functional architecture and the physical architecture. A *functional architecture* is a set of activities or functions, arranged in a specified partial order that, when activated, achieves a set of requirements. Similarly, a *physical architecture* is a representation of the physical resources, expressed as nodes, that constitute the system and their connectivity, expressed in the form of links. Both definitions should be interpreted broadly to cover a wide range of applications; furthermore, each may require multiple representations or views to describe all aspects.

Before even attempting to develop these representations, the operational concept must be defined. This is the first step in the architecture development process. An *operational concept* is a concise statement that describes how the goal will be met. There are no analytical procedures for deriving an operational concept for complex, unprecedented systems. On the contrary, given a set of goals, experience, and expertise, humans invent operational concepts. It has often been stated (Rechtin, 1991) that the development of an architecture is both an art and a science. The conceptualization of an operational concept falls clearly on the art side. A good operational concept is based on a simple idea of how the overriding goal is to be met. For example, "centralized decision making and distributed execution" represents a very abstract operational concept that lends itself to many possible implementations, while an operational concept such as the "client-server" is much more limiting. As the architecture development process unfolds, it becomes necessary to elaborate on the operational concept and make it more specific. The clear definition and understanding of the operational concept is central to the development of compatible, functional, and physical architectures.

Analogous to the close relationship between the operational concept and the functional architecture (to the extent that often a graphical description of the operational concept is improperly presented as the functional architecture) is the relationship between the technical architecture and the physical one. A *technical architecture* is a minimal set of rules governing the arrangement, interaction, and interdependence of the parts or elements whose purpose is to ensure that a conformant system satisfies a specified set of requirements. It provides the framework upon which engineering specifications can be derived, guiding the implementation of the system. It often has been compared to the building code that provides guidance for new buildings to be able to connect to the existing infrastructure by characterizing the attributes of that infrastructure.

All these representations, even when they describe dynamic behavior of the architecture, are static representations; they consist of diagrams. In order to analyze the behavior of the architecture and evaluate the performance characteristics, an executable model is needed. After all, the systems that are to be designed are dynamic systems. An executable model is a dynamic model; it can be used to analyze the properties of the architecture and it can also be used to carry out simulations. But it also serves in a more subtle, but very important role. It becomes the litmus test by which one can determine whether the description of the system architecture, as given by a set of static representations or models, is complete. Indeed, the methodologies, whether structured analysis based or object-oriented based, become rigorous when an executable model is derived and the condition is imposed that all information contained in that model must be traced back to one or more of the static

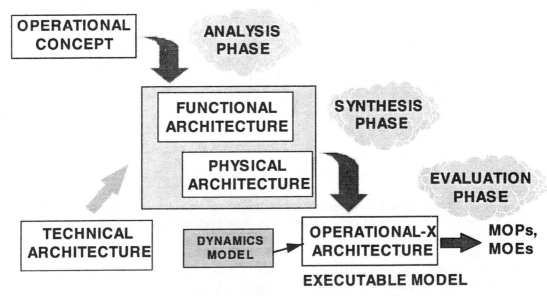

Figure 12.5 The three-phase process of architecture development.

diagrams. This dynamic model of the architecture is called the *operational-X architecture*, where the X stands for the executable property.

The architecture development process can be characterized as consisting of three phases: the *analysis* phase, in which the static representations of the functional and physical architectures are obtained using the operational concept to drive the process and the technical architecture to guide it; the *synthesis* phase, in which these static constructs are used, together with descriptions of the dynamic behavior of the architecture (often referred to as the *dynamics* model), to obtain the executable operational-X architecture; and the *evaluation* phase, in which measures of performance (MOPs) and measures of effectiveness (MOEs) are obtained. This three-phase process is shown schematically in Figure 12.5.

In the next two sections, the two paradigms for architecture development are described. First, the structured analysis approach and a set of tools for implementing it are given; second, one of the several object-oriented approaches is described. It will be observed that, even though the two approaches are conceptually different, many of the same tools can be used to construct the various representations that support each approach. Furthermore, once the executable model is obtained, whether through structured analysis or object-oriented constructs, the evaluation phase is the same.

12.3 STRUCTURED ANALYSIS APPROACH

The structured analysis approach has its roots in the structured analysis and design technique (SADT) that originated in the 1950s (Marca and McGowan, 1987) and encompasses structured design (Yourdon and Constantine, 1975), structured development (Ward and Mellor, 1986), the structured analysis approach of DeMarco (1979), structured systems analysis (Gane and Sarson, 1978), and the many variants that have appeared since then, often embodied in software packages for computer-aided requirements generation, and analysis. This approach can be characterized as a process-oriented one (Solvberg and Kung, 1993), in that it considers as the starting point the functions or activities that the system must perform. A second characterizing feature is the use of functional decompositions and the resulting hierarchically structured diagrams. However, to obtain the complete specification of the architecture, as reflected in the executable model, in addition to the process or activity model, a data model, a rule model, and a dynamics model are required. Each one of these models contains interrelated aspects of the architecture description. For example, in the case of an information system, the activities or processes receive data as input, transform it, and produce data as output. The associated data model describes the relationships between these same data elements. The activities take place when some conditions are satisfied. These conditions are expressed as rules associated with the activities. But for the rules to be evaluated, they require data that must be available at that particular activity with which the rule is associated; the output of the rule also consists of data that control the execution of the process. Furthermore, given that the architecture is for a dynamic system, the states of the system need to be defined and the transitions between states identified to describe the dynamic behavior. State transition diagrams are but one way of representing this information. Underlying these four models is a data dictionary or, more properly, a system dictionary, in which all data elements, activities, and flows are defined. The construct that emerges from this description is that a set of interrelated views, or models, are needed to describe an architecture using the structured analysis approach.

In an ideal world, a tool would exist that would support all these models of an architecture and generate a consistent data dictionary. While the four types of models exist in many forms and software tools for their generation are available, they have been developed independently from a different starting point: it is possible to approach the problem by starting with a data model (data-oriented approach) or with a rule model (rule-oriented approach). At this time, the architect must use a suite of tools and, cognizant of the interrelationships among the four models and the features of the tools chosen to depict them, work across models to make the various views consistent and coherent. And the architect must obtain a single, integrated system dictionary from the individual dictionaries generated by the various tools.

The activity model, the data model, the rule model, and the supporting system dictionary, taken together, constitute the functional architecture of the system. The term functional architecture has been used to describe a range of representations, from a simple activity model to the set of models defined here. What a functional architecture does not contain is the specification of the physical resources that will be used to implement the functions or the structure of the human organization that is supported by the information system. These descriptions are contained in the physical architecture. The structure of the functional architecture is shown in Figure 12.6.

12.3.1 Functional Decomposition and Activity Model

In a process-oriented approach such as structured analysis, the architecture development process can start with a very abstract operational concept. As the analysis evolves, the operational concept becomes more specific. The operational concept is often described pictorially with an associated narrative that explains how the operation is to take place. Cartoons and clip art are often used in the graphical representation. This depiction is often inappropriately referred to as an architecture. It is not. It is equivalent to a sketch that an architect may make of a house, how it sits on the land, and where the main functional areas are. It is not the model of the house itself or the schematics that would allow someone to build this house.

Given an operational concept at some level of abstraction, the first step in the development of the functional architecture is the *functional decomposition*. Starting with a verb or verb phrase that articulates the function of the system, a first-level decomposition is done, separating into functions that *are part of* the top-level function. These first-level

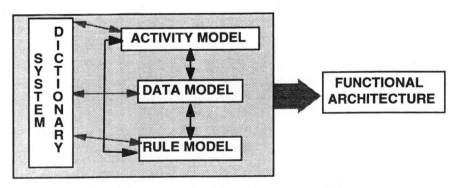

Figure 12.6 Components of the functional architecture.

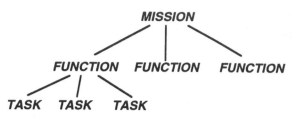

Figure 12.7 A functional decomposition.

functions are mutually exclusive and could be totally exhaustive. Each of these functions can be decomposed further into level-two functions that are parts of it, and so forth. This decomposition can be shown in outline form or graphically as a tree structure. In various approaches, specific names are given to the decomposition levels: for example "mission—function—task" is one set of labels. A schematic of this tree structure is shown in Figure 12.7. The decomposition is carried out to as many levels as is necessary, always guided by the operational concept. However, keeping the levels as few as possible is recommended for two reasons: each additional level increases substantially the complexity of the problem (and may not be supportable by the other parts of the architecture process), and because a multilevel decomposition may introduce implicitly a physical architecture—the way the functions are partitioned may prematurely specify implementation solutions. One useful rule is to decompose until each function can be assigned to a single physical resource. To achieve that, it is implied that the physical architecture is available. This is usually not the case; the two should be developed in parallel with much interaction between them. Indeed, the functional decomposition is an iterative process and should be done with care, because it is difficult to go back to the higher levels and make changes to them—the lower levels and everything related to them will have to be reexamined.

There are two primary methods in wide use for representing an activity model. The first, IDEF0, has systems engineering roots, while the second, data flow diagrams, has its roots in software systems engineering. Recently, the National Institute of Standards and Technology (NIST) published Draft Federal Information Processing Standard #183 for IDEF0. There are various variants and extensions of the data flow diagramming approach. Each of the approaches has advantages and disadvantages; choosing one of them depends on the features of the problem to be addressed. For the history of IDEF0, see Marca and McGowan (1987), and for data flow diagrams, see Yourdon (1989) or Rumbaugh et al. (1991). IDEF0 is described briefly in this section; the data flow diagrams approach is described in Section 12.7.2.

IDEF0 is a subset of the SADT. It is a modeling language for developing structured graphical representations of the activities or functions of a system that has been designed to describe and aid in understanding complex systems. It is a static representation, designed to address a specific question from a single point of view, and it has two graphical elements: a box, which represents an activity, and a directed arc, which represents the conveyance of data or objects related to the activity. A distinguishing characteristic of IDEF0 is that the sides of the activity box have a standard meaning, as shown in Figure 12.8. Arcs entering the left side of the activity box are inputs, those entering the top side are controls, and those entering the bottom side are mechanisms or resources used to perform the activity. Arcs leaving the right side are outputs—the data or objects generated by the

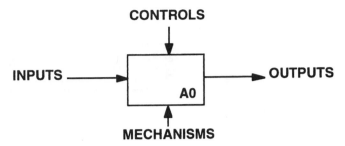

Figure 12.8 Box and arrow semantics in IDEF0.

activity. When IDEF0 is used to represent the process model in a functional architecture, mechanisms are not needed; they are part of the physical architecture.

Verbs or verb phrases are inscribed in the activity boxes to define the function represented. Similarly, arc inscriptions are used to identify the data or objects represented by the arcs. There are detailed rules for handling the branching and the joining of the arcs.

A key feature of IDEF0 is that it supports hierarchical decomposition. At the highest level, the A-0 level, there is a single activity that contains the root verb of the functional decomposition. This is called the context diagram, and also includes a statement of the purpose of the model and the point of view taken. The next level down, the A0 level, contains the first-level decomposition of the system function and the interrelationships between these functions. It is a single page. Each one of the activity boxes on the A0 page can be further decomposed into the A1, A2, A3, . . . page, respectively. A typical IDEF0 diagram of the first two levels—A-0 and A0—is shown in Figure 12.9. There are two inputs, one control, and three outputs.

Associated with IDEF0 is a data dictionary that includes the definitions and descriptions of the activities, listing and description of the inputs, controls, and outputs, and, if entered, a set of activation rules of the form "preconditions → postconditions." These are the rules that indicate the conditions under which the associated function can be carried out.

12.3.2 Data Model

The arcs in the activity model represent data or objects. The purpose of a data model is to analyze the data structures and their relationships independently of the processing that takes place, already depicted in the activity model. There are two main approaches with associated tools for data modeling: IDEF1x (IDEF1 extended) and entity-relationship (E-R) diagrams. Both approaches are used widely. The NIST has published Draft Federal Information Processing Standard #184 in which IDEF1x is specified. There are many books that describe E-R diagrams: Sanders (1995), Yourdon (1989), McLeod (1994).

IDEF1x is a modeling language for representing the structure and semantics of the information in a system. The elements of IDEF1x are the entities, their relationships or associations, and their attributes or keys. An IDEF1x model comprises one or more views, definitions of the entities, and the domains of the attributes used in the views.

An entity is the representation of a set of real or abstract objects that share the same characteristics and can participate in the same relationships. An individual member of the set is called an *entity instance*. An entity is depicted by a box; it has a unique name and

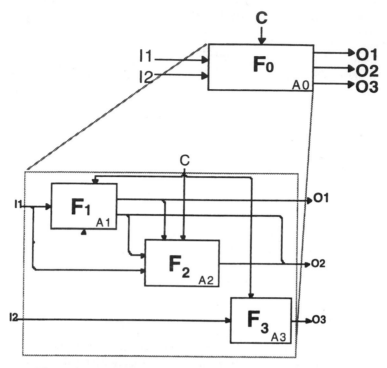

Figure 12.9 The first two levels of an IDEF0 activity diagram.

a unique identifier. If an instance of an entity is identifiable with reference to its relationship to other entities, it is said to be *identifier dependent*. A slightly different form of the box is used to distinguish identifier independent and dependent entities. The box depicting the entity instance is divided into two parts, the top part containing the primary key attributes; the lower one the nonprimary key attributes. Every attribute must have a name (expressed as a noun or noun phrase) that is unique among all attributes across the entities in the model. The attributes take values from their specified domains. This formalism is shown in Figure 12.10.

Relationships between entities are shown in the form of lines that connect entities; a verb or verb phrase is placed beside the relationship line. The connection relationship is directed—it establishes a parent–child association—and has cardinality. Special symbols are used at the ends of the lines to indicate the cardinality. The relationships can be classified into types such as identifying or nonidentifying, specific and nonspecific, and categorization relationships. The latter, for example, is a generalization/specialization relationship in which an attribute of the generic entity is used as the discriminator for the categories. A simple example is shown in Figure 12.11.

12.3.3 Rule Model

In a rule-oriented model, knowledge about the behavior of the architecture is represented by a set of assertions that describe what is to be done when a set of conditions evaluates

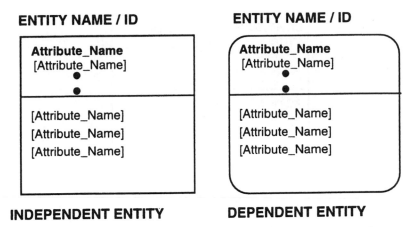

Figure 12.10 Independent and dependent entities with key and nonkey attributes.

as true. These assertions, or rules, apply to specific functions defined in the activity model and are formulated as relationships among data elements. There are several specification methods that are used depending on the application. They include decision trees, decision tables, structured English, and mathematical logic. Each one has advantages and disadvantages; the choice often depends on the way that knowledge about rules has been elicited and on the complexity of the rules themselves.

A decision tree is most appropriate when each rule has as a consequent a single action, the execution of an activity. A decision tree has a single root that represents the first

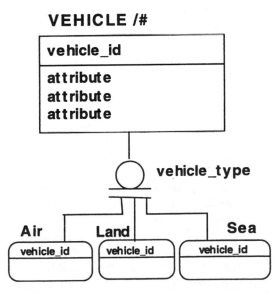

Figure 12.11 Example of generic and category entity specification using a discriminator.

CONDITION STUB	**CONDITION MATRIX**
ACTION STUB	**ACTION MATRIX**

Figure 12.12   The four parts of a decision table.

decision. Subsequent decisions are depicted as the branches and leaves of the tree. Each node of the tree represents a decision, while the leaves indicate the resulting actions. A decision table is more useful when a set of conditions that evaluates as true results in multiple actions. The table is partitioned in four sections, as shown in Figure 12.12. The Condition stub contains the list of the decision variables, while the Action stub contains the list of actions. The Condition matrix consists of columns, with each column depicting a combination of conditions. The columns of the Action matrix show which actions are to take place when the conditions in the corresponding Condition column evaluate as true.

Both decision trees and decision tables show only selection or decision constructs; they do not show sequencing or repetition and iteration. Structured English shows all three control structures. In structured English, the rules are expressed using nested patterns of the form "if-then-else-so." The action that results when the conditions evaluate as true is expressed in the form of a command: then do this.

Of course, mathematical logic can be used (whether symbolic logic or predicate logic) to represent the set of rules. This is a very general representation that allows for the modeling of very complex rules.

12.3.4   Dynamics Model

The fourth type of model that is needed is one that characterizes the dynamic behavior of the architecture. This is not an executable model, but one that shows the transition of the system state as a result of events that take place. The state of a system can be defined as all the information that is needed at some time t_o so that knowledge of the system and its inputs from that time on determines the outputs. The state space is the set of all possible values that the state can take.

There is a wide variety of tools for depicting the dynamics, with some tools being more formal than others: state transition diagrams, state charts, event traces, key threads, among others. Each one serves a particular purpose and has unique advantages.

A *state transition diagram* is a representation of a sequence of transitions from one state to another—as a result of the occurrence of a set of events—when starting from a particular initial state or condition. The states are represented by nodes (e.g., a box), while the transitions are shown as directed arcs. The event that causes the transition is shown as

Figure 12.13 A transition from state 1 to state 2.

an arc annotation, while the name of the state is inscribed in the node symbol. If an action is associated with the change of state, then this is shown on the connecting arc, next to the event. An example is shown in Figure 12.13.

Note that the dynamics model is not a dynamical model; it is not itself executable. It characterizes in a static manner aspects of the dynamic behavior of the model. Furthermore, since a state transition diagram represents the transitions from an initial condition and a sequence of events, it follows that many such diagrams are needed to characterize a system's behavior.

12.3.5 System Dictionary and Concordance of Models

Underlying all these four models is the system dictionary. Since the individual models contain overlapping information, it becomes necessary to integrate the dictionaries developed for each one of them. Such a dictionary must contain descriptions of all the functions or activities including what inputs they require and what outputs they produce. These functions appear in the activity model (IDEF0), the rule model (as actions), and the state transition diagrams. The rules, in turn, are associate with activities; they specify the conditions that must hold for the activity to take place. For the conditions to be evaluated, the corresponding data must be available at the specific activity—there must be an input or control in the IDEF0 diagram that makes these data available to the corresponding activity. Of course, the system dictionary contains definitions of all the data elements as well as the data flows that appear in the activity model.

The process of developing a consistent and comprehensive dictionary provides the best opportunity for ensuring concordance among the four models. Since each model has different roots and was developed to serve a different purpose, together they do not constitute a well-integrated set. Rather, they can be seen as a suite of tools that collectively contain sufficient information to specify the architecture. The interrelationships among models are complex. For example, rules should be associated with the functions at the leaves of the functional decomposition tree. This implies that, if changes are made in the IDEF0 diagram, then the rule model should be examined to determine whether rules should be reallocated and whether they need to be restructured to reflect the availability of data in the revised activity model. A further implication is that the four models cannot be developed in sequence. Rather, the development of all four should be planned at the beginning with ample opportunity provided for iteration, because if changes are made in one, they need to be reflected in the other models.

Once concordance of these models has been achieved, it is possible to construct an executable model. Since the physical architecture has not been constructed yet, the executable model can only be used to address logical and behavioral issues, but not performance issues.

12.4 THE EXECUTABLE MODEL

Information systems are dynamic in nature. Events occur that trigger the execution of functions, and many functions can be executed concurrently. Therefore, an accurate representation of an information architecture should be executable. There exist some graphical modeling approaches that allow a dynamic representation of a system. Behavior diagrams and colored petri nets are examples of such approaches. They can be used directly to model a discrete event dynamical system representation of an information architecture. The problem, however, is that they are much more complex than the four models already described and cannot be easily understood by the user. The solution, therefore, is to derive from the static representations a dynamic representation of the system. A methodology has been developed that allows the development of a colored petri net model of an architecture that can be traced back to the four models.

Colored petri nets (Jensen, 1992) are a generalization of ordinary petri nets. The latter are bipartite directed graphs that are executable. In ordinary Petri Nets, two types of nodes are defined: places and transitions. The arcs that join two nodes are directed; furthermore, arcs can connect only nodes of different types. Directed arcs connecting places to transitions establish the inputs to that transition, while arcs connecting transitions to places establish the outputs. The arcs can have inscriptions that define the degree of multiplicity of that arc. An illustrative petri net is shown in Figure 12.14.

In order for the petri net to execute and be a dynamical model, another element needs to be introduced. This is the token. A *token* is an indistinguishable marker that resides in places. The distribution of tokens in the places of a petri net is called a *marking* and defines the state of the system or net. Markings enable transitions that can then fire. The execution rule is as follows: A transition is enabled if every one of its input places has at least as many tokens as the multiplicity inscribed on the arc connecting the place to the transition. An enabled transition can fire. When it fires, the tokens used to satisfy the enablement condition are removed from the net; new indistinguishable tokens are generated in the output places of the transition. The number of tokens generated depends on the multiplicity of the outgoing arcs. Figure 12.15 shows an initial marking for the net of Figure 12.14 that enables the transition, and the results of the transition firing.

In colored petri nets, the tokens are distinguishable; they are characterized by their color. An attribute vector is associated with the token. The assignment of values to the attributes from their respective domains specifies the color of the token. Color sets are associated with places; they specify which token can reside in that place. Instead of a

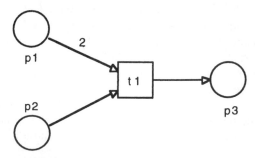

Figure 12.14 An ordinary petri net.

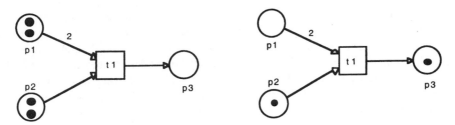

Figure 12.15 Enablement and firing of transition.

simple multiplicity number inscription that was allowed on the arcs, complex enablement conditions can now be specified. Each input arc inscription specifies the number and type of tokens that need to be in place for the transition to be enabled. The output arc inscriptions indicate what tokens will be generated in an output place when the transitions fires. Furthermore, guard functions associated with transitions are allowed. These guard functions specify additional conditions that must be satisfied, that is, in addition to those inscribed on the arcs, for a transition to be enabled. Code segments can be associated with transitions. These code segments can represent the function modeled by the transition and complement the output arc inscriptions.

Each colored petri net model has associated with it a global declaration node that contains the definitions of all color sets and their associated domains and the definition of variables. It becomes apparent then that much of the data in the data dictionary appear in the global declaration node of the colored petri net model.

The use of colored petri nets to develop an executable model from the structured analysis models can be described as follows. One starts with the activity model. Each IDEF0 activity is converted into a transition; each IDEF0 arrow connecting two boxes is replaced by a set arc-place-arc (Fig. 12.16), and the label of the IDEF0 arc becomes the color set associated with the place. All these derived names of color sets are gathered in the global declaration node. From this point on, a substantial modeling effort is required to make the colored petri net model a dynamic representation of the system. The information contained in the data model is used to specify the color sets and their respective domains, while the rules in the rule model result in arc inscriptions, guard functions, and code segments.

The process of deriving the executable model invariably leads to some revision of the static models. It is most important that discipline be exercised so that any change introduced at the executable model level is reflected back in the static models. Only this way can a documented and easily reviewed representation of the architecture be maintained.

The executable model becomes the integrator of all the information; its ability to execute tests some of the logic of the model. Given the colored petri net model, a number of

Figure 12.16 From IDEF0 to a colored petri net representation.

analytical tools from petri net theory can be used to evaluate the structure of the model, for example, determine the presence of deadlocks, or obtain its occurrence graph. The occurrence graph represents a generalization of the state transition diagram model. By obtaining the occurrence graph of the petri net model, which depicts the sequence of states that can be reached from an initial marking (state) with feasible firing sequences, one has obtained a representation of a set of state transition diagrams. This can be thought of as a first step in the validation of the model at the behavioral level. Of course, the model can be executed to check its logical consistency, that is, to check whether the functions are executed in the appropriate sequence and that the data needed by each function are appropriately provided. Performance measures cannot be obtained until the physical architecture is introduced; this architecture provides the information needed to compute performance measures.

Since colored petri nets with their dense annotation are not very easily understood by the information system users, all the information gathered in the design and exploitation of the executable model needs to be brought back into the static models. This annotated and validated representation now constitutes a sound basis for system development.

12.5 PHYSICAL ARCHITECTURE

To complete the analysis phase of the procedure (Fig. 12.17), the physical architecture needs to be developed. There is no standardized way to represent the physical systems, existing ones as well as planned ones, that will be used to implement the architecture. They range from wiring diagrams of systems to block diagram representations to node models to organization charts. While there is not much difficulty in describing precisely physical subsystems using the terminology and notation of the particular domain (communication systems, computers, displays, databases), a problem arises on how to depict the human organization that is an integral part of the information system. The humans in the organization cannot be thought of simply as users; they are active participants in the workings of the information system and their organizational structure that includes task allocations, authority, responsibility, reporting requirements, and so on, must be taken into account and be a part of the physical model description. This is an issue of current research,

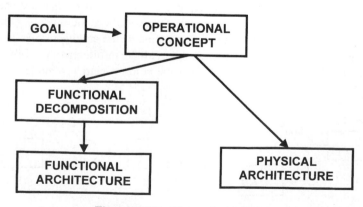

Figure 12.17   The analysis phase.

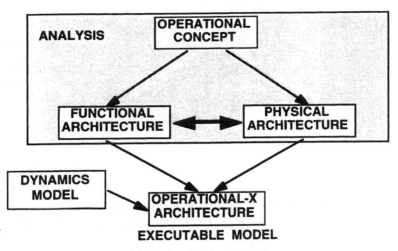

Figure 12.18 The synthesis phase.

since traditional organizational models do not address explicitly the need to include the human organization as part of the physical system description.

Once the physical architecture is available, then the operational-X architecture shown in Figure 12.5 can be obtained. The process is described in Figure 12.18; it is the synthesis phase. The required interrelationship between the functional and the physical architectures is shown by the boldface two-way arrow. It is critical that the granularity of the two architectures be comparable and that partitions have taken place in the hierarchical decompositions in a manner that allows functions or activities to be assigned unambiguously to resources, and vice versa. Once the parameter values and properties of the physical systems have become part of the database of the executable model, performance evaluation can take place.

12.6 PERFORMANCE EVALUATION

A detailed view of all the models and their interrelationships is shown in Figure 12.19. MOPs are obtained either analytically or by executing the model in simulation mode. For example, if deterministic or stochastic time delays are associated with the various activities, it is possible to compute the overall delay or to obtain it through simulation. Depending on the questions to be answered, realistic scenarios of inputs that are consistent with the operational concept need to be defined. This phase allows for functional and performance requirements to be validated, if the results obtained from the simulations show that the measures of performance are within the required range. If not, the systems may need to be modified to address the issues that account for the encountered problems.

This is actually an iterative process, as shown in Figure 12.20. The executable model can be used both at the logical and behavioral level, as well as the performance level. The latter requires the inclusion of the physical architecture. In one consistent architectural framework supported by a set of models, both requirements analysis, design, and evalu-

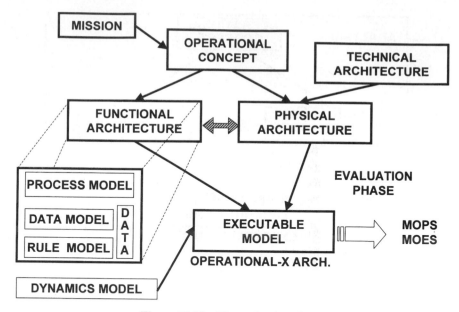

Figure 12.19   The evaluation phase.

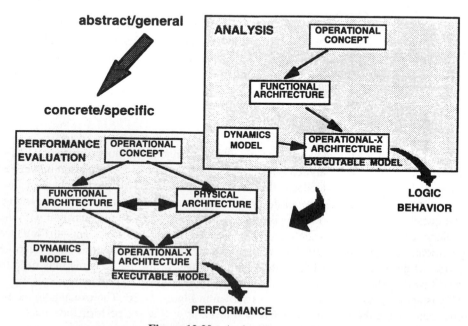

Figure 12.20   An iterative process.

ation can be performed. Furthermore, the process provides a documented set of models that collectively contain all the necessary information.

However, this set of tools does not provide for a flexible approach if major changes are expected during the development and implementation process. An alternative approach, which uses many of the same tools, has begun to be used in an exploratory manner.

12.7   OBJECT-ORIENTED APPROACH

The approach that allows for the graceful migration from one option to another, in a rapid and low-cost manner (see Fig. 12.4), is to look for the things that do not change across the options, or that change in well-understood ways. First, an information architecture can be seen as a repository of information about a system, because it contains a data dictionary, depicts data and control flows, and describes system behavior. An information architecture (IA) consists of components (computer programs, data ensembles, hardware elements, human elements), connections (interrelationships between pairs of components), and constraints (governing principles and shared assumptions across the architecture) (Druffel et al., 1994). Furthermore, an IA can be shown at different levels of abstraction, which is not the same as different levels of detail. This is illustrated in Figure 12.21, where the same number of objects connected in the same manner is used in all three representations, but the level of abstraction of each is clearly different.

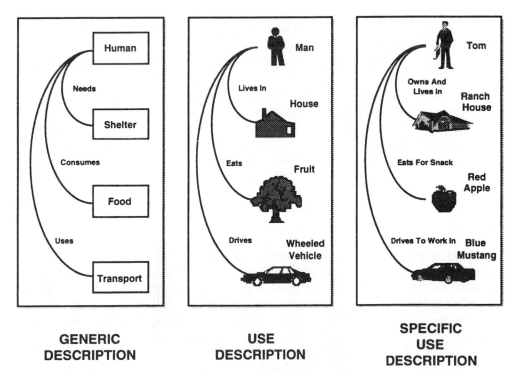

Figure 12.21   Different levels of abstraction.

The most abstract level of the information architecture is the Domain IA. A domain defines what information is to be used, by what applications, and by which users. It establishes the universe within which a certain level of interoperability is desired. Typical domains can be discrete—part flexible manufacturing, payroll systems, health services delivery systems, personnel systems, financial systems, defense systems. The domain IA does not include an operational concept, that is, a description of how the tasks are to be performed. The inclusion of an operational concept reduces the level of abstraction and leads to the concept of operations (or ConOps) IA. For example, while different banks may have the same domain IA, they may have different ConOps IAs to reflect differences in corporate culture and operational philosophies.

When the generic entities or objects in a ConOps IA are associated with system components—generic ones—then an instantiated IA is obtained. It should be clear that there can be many instantiated IAs for the same ConOps IA, while there will be only a few ConOps IAs for the same domain IA. Thus, a particular solution is a design based on an instantiated IA. When there is need to move to another design, much is preserved, because all instantiated IAs share the same domain IA and often the same concept of operations, though different resources. This leads to reuse of software components, and places emphasis on system integration rather than on doing one-of-a-kind designs.

The system engineering paradigm that is well suited for such an approach, that is, for the use of information architectures in systems design, is object-oriented analysis and design. The basic elements of the approach, from the point of view of systems engineering, are described in this section.

There is extensive literature in object-oriented programming; more recently, the literature has begun to address problems in object-oriented analysis and design, as applied to software systems engineering (Booch, 1994; Rumbaugh et al., 1991; Berard, 1993). Sage (1993) reviewed the basic concepts of object-oriented design and formulated a methodology for the design of decision support systems. There are several major ongoing efforts by industry to extend the software systems engineering constructs to a full-scale systems engineering approach, particularly of software-intensive systems, but they have not been documented yet in the open literature.

The fundamental notion in object-oriented design is that of an object: an *object* is an abstraction of the real world that captures a set of variables that correspond to actual real-world behavior (Sage, 1993). The boundary of the object is clearly defined. This boundary separates or hides the inner workings of the object from other objects. Interactions between objects occur only at the boundary through the clearly stated relationships with the other objects. The selection of objects is domain-specific. In information system design, candidates for objects are decision makers or users, communication systems, databases, workstations, and the like.

A class is a template, description, or pattern for a group of very similar objects, namely, objects that have similar attributes, common behavior, common semantics, and common relationship to other objects. In that sense, an object is an instance of a class. For example, "air traffic controller" (ATC) is an object class; the specific individual that controls air traffic during a particular shift at an ATC center is an object, that is, an instantiation of the abstraction "air traffic controller." The concept of object class is particularly relevant in the design of information systems, where it is possible to have hardware, software, or humans perform some tasks. At the higher levels of abstraction, it is not necessary to specify whether some tasks will be performed by humans or by software running on a particular platform.

Encapsulation is the process of separating the external aspects of an object from the internal ones. In engineering terms, this is defining the boundary and the interactions that cross the boundary—the old black-box paradigm. This is a very natural concept in information system design; it allows the separation of the internal processes from the interactions with other objects, either directly or through communication systems. One can change the algorithm for resource allocation, but require the same inputs and produce the same outputs.

Modularity is another key concept in object-oriented design that has a direct, intuitive meaning. *Modularity,* according to Booch (1994), is the property of a system that has been decomposed into a set of cohesive and loosely coupled modules. Consider, for example, the corporate staff, the line organization, and the marketing organization of a company. Each module consists of objects and their interactions, the assumption here being that the objects within a module have a higher level of interaction than there is across the modules.

It is fairly clear that encapsulation, modularity, and object class are closely related concepts. Encapsulation is the process by which the boundary characterizing an object class is defined, and the nature of the object classes leads to the identification of modules.

The concept of *hierarchy* has its roots in organization theory. In the context of object-oriented design, hierarchy refers to the ranking or ordering of abstractions, with the more general one at the top and the more specific one at the bottom. An ordering is induced by a relation, and the ordering can be strict or partial. In the object-oriented paradigm, two types of relations are recognized: aggregation and inheritance.

Aggregation refers to the ability to create objects composed of other objects with each a *part* of the aggregate object. This concept is well known and understood in organization design and has been exploited as the means to design large organizations. The "line" organizations of the military are constructed in this way: the individual infantry man, a squad, a platoon, a company, a battalion, a regiment, a division, and so forth. This is a useful construct, even though it tends to oversimplify the organizational structure. As one moves up the hierarchy, the organizational units need to add components or sections that provide services across the organization (staff functions). The concept of aggregation provides the means of incorporating functional decompositions from structured analysis in the object-oriented approach.

The second concept, *inheritance,* has some interesting implications for information system design. Inheritance is the means by which an object acquires characteristics (attributes, behaviors, semantics, and relationships) from one or more other objects (Berard, 1993). In single inheritance, an object inherits characteristics from only one other object; in multiple inheritance, it inherits from more than one object. Inheritance is a way of representing generalization and specialization. The "navigator" in an air crew inherits all the attributes of the "air crew member" object class, but has additional attributes that specialize the object class. The "pilot" and the "copilot" are different *siblings* of the air crew object class. The inheritance concept has not been explored fully in system design where specialization has been thought of more in terms of functional decompositions that are partitions of a function.

There are additional concepts in object-oriented design, such as polymorphism, persistence, reuse, message passing, and dynamic binding. These concepts refer mostly to the software implementation aspects, and have not been explored fully yet for the development of systems architectures.

The object modeling technique (OMT) of Rumbaugh et al. (1991) requires three views of the system: the object view, the functional view, and the dynamic view (Fig. 12.22).

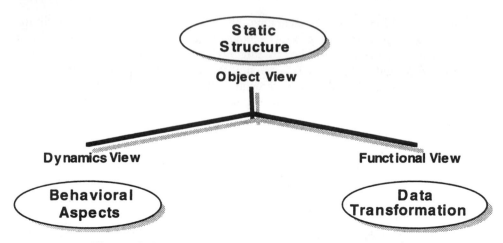

Figure 12.22   The three views in the object modeling technique.

The object view is represented by the object model that describes the structure of the system. This view is a static description of the objects; it shows the various object classes and their hierarchical relationships. The functional view is represented in terms of data flow diagrams that depict the dependencies between input data and computed values in the system. The dynamic view is represented in terms of state diagrams, and shows the sequence of events that occur. All three views, which are interrelated, are needed to describe the system. As mentioned in Section 12.1, this representation, while adequate for object-oriented software system design, is not sufficient for systems engineering. As in the structured analysis approach, an executable model is needed to bring them all together and to provide a means for performance evaluation. This extended construct is shown in Figure 12.23. Note the feedback loop and the need to compare predicted system behavior with the requirements. This is a two-way process that may lead to some changes in requirements by the user.

12.7.1   The Object View

The *object view* presents the static structure of the object classes and their relationships. The object view is a diagram that is similar to the data model described in Section 12.3.2, only here there are object classes in place of the data entities. An object class is depicted by a box divided into three parts. As shown in Figure 12.24, the top part contains the name of the class. The second part contains the attributes, such as the data values held by all the objects in the class. The third part contains the class operations, which are the functions or transformations of the class that can be applied to the class or by it (Rumbaugh et al., 1991).

The lines connecting the object classes represent relationships between classes. These relationships have cardinality (one-to-one, one-to-many, etc.) as already described for the data model. The aggregation relationship shows how one class is composed of other classes; the other classes are *part* of the superclass. For example, the computer keyboard is part of a workstation. The generalization relationship shows how subclasses inherit the properties of the superclass; for example, a workstation is a subclass of the class computer

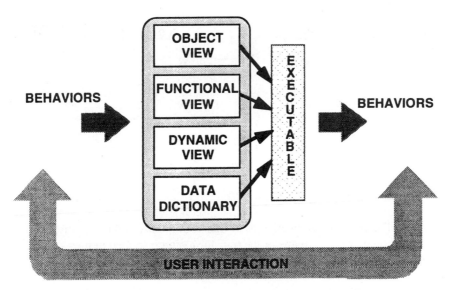

Figure 12.23   Architecture development using the object-oriented approach.

and, clearly, it has the general attributes of a computer and its functionality. It also has attributes and functionality specific to a workstation. The third type of relationship is called an *association*, and it shows how one class accesses the attributes or invokes the operations of another.

The definition of the classes for a particular domain should remain invariant over time, and the relationships at the domain level will also not change. The object view is the basic representation of the Domain IA. However, it is not a complete one.

12.7.2   The Functional View

The *functional view* consists of a set of data flow diagrams that are analogous to the activity models in structured analysis. A data flow diagram, as used in the object modeling technique, depicts the functional relationships of the values computed by the system; it specifies the meaning of the operations defined in the object model without indicating where these operations reside or how they are implemented. A typical data flow diagram is shown in

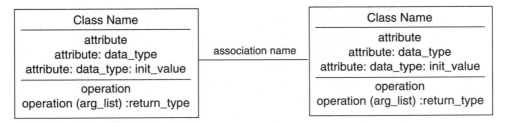

Figure 12.24   An object view.

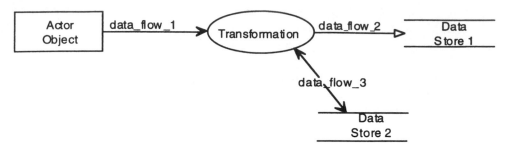

Figure 12.25   A functional view: data flow diagram.

Figure 12.25. The functions or operations or transformations, as they are often called, are represented by ovals with the name of the transformation inscribed in them, preferably as a verb phrase. The directed arcs connecting transformations represent data flows, while the arc inscriptions define what flows between the transformations. Flows can converge (join) and diverge (branch). A unique feature of data flow diagrams is the inclusion of data stores that represent data at rest — a database or a buffer. Stores are connected by data flows to transformations, with an arc from a store to a transformation denoting that the data in the store are accessible to the transformation, while an arc from a transformation to a store indicates an operation (write, update, delete) on the data contained by the store. Entities that are external to the system, but with which the system interacts, are called *terminators* or *actors*. The arcs connecting the actors to the transformations in the data flow diagram represent the interfaces of the system with the external world. Clearly, data flow diagrams can be decomposed hierarchically, in the same manner that the IDEF0 diagram was multileveled.

While data flow diagrams have many strengths, such as simplicity of representation, ease of use, hierarchical decomposition, the use of stores and actors, they also have weaknesses. The most important of these latter is the inability to show the flow of control. For this reason, enhancements exist that include the flow of control, but at the cost of reducing the clarity and simplicity of the approach.

There are several notations used to describe data flow diagrams; the one used in Figure 12.25 is the DeMarco notation (De Marco, 1979). An alternative one is the Gane and Sarson (1978) notation. Commercial packages that support the data flow diagram models offer a choice of notation through the use of alternate palettes.

12.7.3   The Dynamics View

The dynamics view in OMT is similar to the one in structured analysis — state transition diagrams are used to show how events change the state of the system (see Section 12.3.4). In addition to showing the events and the actions on the arcs, however, conditions are also shown in brackets. These conditions are guards on the transitions in that they specify what condition must evaluate as true in order for the transition to take place. Thus, the rules that govern the operations of the system are not shown as an independent model, but are integrated in the dynamics model, as shown in Figure 12.26.

A final construct that describes the "trajectories" of the system using events and objects is the *event trace* (Fig. 12.27). In this diagram, each object in the object view is depicted as a vertical line and each event as a directed line from one object to another. The se-

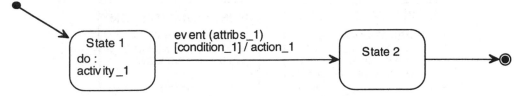

Figure 12.26   A state transition diagram with rules.

quencing of the events is depicted from top to bottom, with the initiating event as the topmost one. The event traces characterize behaviors of the architecture: if given, they provide behavioral requirements; if obtained from the executable model, they indicate behavior.

12.7.4   The Executable Model

The three views, when enhanced by the rule model embedded in the state transition diagrams, provide sufficient information for the generation of an executable model. Colored petri nets can be used to implement the executable model, although the procedure this time is not based on the functional model. Instead, the object classes are represented by subnets that are contained in "pages" of the hierarchical colored petri net software implementation. These pages have port nodes for connecting the object classes with other classes. Data read through those ports can instantiate a particular class to a specific object. The operations of the pages/object subnets are activated in accordance with the rules, and again the marking of the net denotes the state of the system.

Once the colored petri net is obtained, the evaluation phase is identical to that of structured analysis. The same analytical tools (invariants, deadlocks, occurrence graphs) and the same simulations can be run to assess the performance of the architecture.

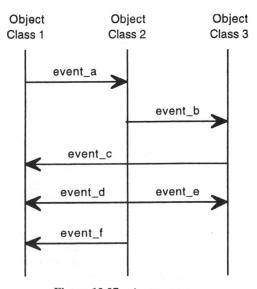

Figure 12.27   An event trace.

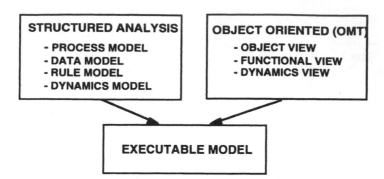

Figure 12.28   Both approaches lead to an executable model.

12.8   CONCLUSION

The problem of developing system architectures, particularly for information systems, has been discussed. Two main approaches, structured analysis, with roots in systems engineering, and the object-oriented one, with roots in software system engineering, have been described. Both of them are shown to lead to an executable model, if a coherent set of models or views is used. This concept is shown in Figure 12.28. Note that the executable model, whether obtained from the structured analysis approach or the object-oriented one, should exhibit the same behavior and lead to the same performance measures. This does not imply that the structure of the colored petri net will be the same. Indeed, the net obtained from structured analysis has a strong structural resemblance to the IDEF0 (functional) diagram, while the one obtained from the object-oriented approach has a structure similar to the object view. The difference in the structure of the two models is the basis for the observations that the two approaches are significantly different in effectiveness, depending on the nature of the problem being addressed. When the requirements are well-defined and stable, the structured analysis approach is direct and efficient. The object-oriented approach requires that a library of object classes be defined and implemented prior to the actual design. If this is a new domain, there may not be earlier libraries populated with suitable object classes. Instead, these libraries may have to be defined. Of course, as time passes, more and more object class implementations will become available, but at this time, the start-up cost is not insignificant. On the other hand, if the requirements are expected to change and new technology insertions are anticipated, it may be more effective to create the class library in order to give the systems engineering team the requisite flexibility in modifying the system architecture.

ACKNOWLEDGMENT

The work was funded in part by the Office of Naval Research under Grant No. N00014-93-1-1902, and by the U.S. Air Force Office of Scientific Research under Grant No. F49620-95-0134. The contributions of Didier M. Perdu and Robert F. Phelps in the development and testing of the concepts related to the executable models are gratefully appreciated.

REFERENCES

Berard, E. V. (1993). *Essays on Object-Oriented Software Engineering.* Englewood Cliffs, NJ: Prentice-Hall.

Booch, G. (1994). *Object-Oriented Analysis and Design.* Redwood City, CA: Benjamin/Cummings.

Chorafas, D. N. (1989). *Systems Architecture and Systems Design.* New York: McGraw-Hill.

DeMarco, T. (1979). *Structured Analysis and Systems Specification.* Englewood Cliffs, NJ: Prentice-Hall.

Druffel, L., Loy, N. E., Rosemberg, R. A., Saunders, T. F., Sylvester, R., and Volz, D. (1994). *Information Architectures that Enhance Operational Capability in Peacetime and Wartime* (Rep. SAB 94-002). Washington, DC: U.S. Air Force.

Gane, C., and Sarson, T. (1978). *Structured Systems Analysis: Tools and Techniques.* Englewood Cliffs, NJ: Prentice-Hall.

Jensen, K. (1992). *Coloured Petri Nets.* Berlin: Springer-Verlag.

Marca, D. A., and McGowan, C. L. (1987). *Structured Analysis and Design Technique.* New York: McGraw-Hill.

McLeod, R., Jr. (1994). *Systems Analysis and Design.* Fort Worth, TX: Dryden.

Rechtin, E. (1991). *Systems Architecting, Creating and Building Complex Systems.* Englewood Cliffs, NJ: Prentice-Hall.

Rechtin, E. (1992). The art of systems architecting. *IEEE Spectrum,* **42**(10), 66–69.

Rechtin, E. (1994). Foundations of systems architecting. *J. NCOSE* **1**(1), 35–42.

Rechtin, E., and Maier, M. (1997). *The Art of Systems Architecting.* Boca Raton, FL: CRC Press.

Rumbaugh, J., Blaha, M., Premerlani, W., Eddy, F., and Lorensen, W. (1991). *Object-Oriented Modeling and Design.* Englewood Cliffs, NJ: Prentice-Hall.

Sage, A. P. (1992). *Systems Engineering.* New York: Wiley.

Sage, A. P. (1993). Object oriented methodologies in decision and information technologies. *Infor. Dec. Techn.* **19**(1), 31–54.

Sage, A. P. (1995). *Systems Management for Information Technology and Software Engineering.* New York: Wiley.

Sage, A. P., and Lynch, C. L. (1998). Systems Integration and Architecting: An Overview of Principles, Practices, and Perspectives, *Systems Engineering* **1**(3), 176–227.

Sanders, G. L. (1995). *Data Modeling.* Danvers, MA: Boyd & Fraser.

Solvberg, A., and Kung, D. C. (1993). *Information Systems Engineering.* Berlin: Springer-Verlag.

Ward, P., and Mellor, S. (1986). *Structured Development of Real-Time Systems.* New York: Yourdon Press.

Yourdon, E. (1989). *Modern Structured Analysis.* Englewood Cliffs, NJ: Yourdon Press.

Yourdon, E., and Constantine, L. (1975). *Structured Design.* New York: Yourdon Press.

13 Systems Design

K. PRESTON WHITE, JR.

13.1 INTRODUCTION: WHAT IS SYSTEMS DESIGN?

Design is the essence of engineering. Broadly defined, *design* is the creative process by which our understanding of logic and science is joined with our understanding of human needs and wants to conceive and refine artifacts that serve specific human purposes. Such a sweeping definition clearly might serve as well to define the whole of engineering as a human endeavor. As a profession, therefore, engineering is concerned primarily with design—the design of processes, structures, machines, circuits, and software—and with the purposeful combinations of these elements to result in what we call systems.

13.2 STEPS IN THE DESIGN PROCESS

The result of the engineering design process typically is a set of drawings [now most typically in a digital, computer-aided design (CAD) format] that detail the structure of the system—including the size, shape, materials, and quantities of components and the inter-relationships among these design elements—together with the calculations, analyses, surveys, and reports required to support the selection of a particular design and to enable its eventual fabrication. To achieve this result means bridging the gap between the realm of human needs (tangible and intangible) and the realm of concrete expression. This effort is generally understood to entail the specification of measurable goals, objectives, and constraints for the design; the conceptualization and parameterization of alternative candidate designs that meet or surpass specifications; the analysis and ranking of design alternatives; and, finally, the selection, implementation, and testing of the most preferred alternative (Reilly, 1993; Sage, 1977).

There is no universal theory describing exactly how all of these activities do or should come together in the design process. Indeed, since design is practiced both by individuals and by teams, and in all branches of engineering and in other disciplines, as well, it may be that there is no single realization of the design process that is just the right style for all. However, because the translation from human needs to final design invariably involves the experience, intuition, skill, and creativity of the designer or design team, and because there is no unique solution to a given design problem, design is universally understood to be a *creative, iterative, decision-making process.* There is enough commonalty among the different views of design to propose the general organization of the process depicted in Figure 13.1 and outlined below.

Handbook of Systems Engineering and Management, Edited by A. P. Sage and W. B. Rouse
ISBN 0471-15405-9   ©1999 John Wiley and Sons, Inc.

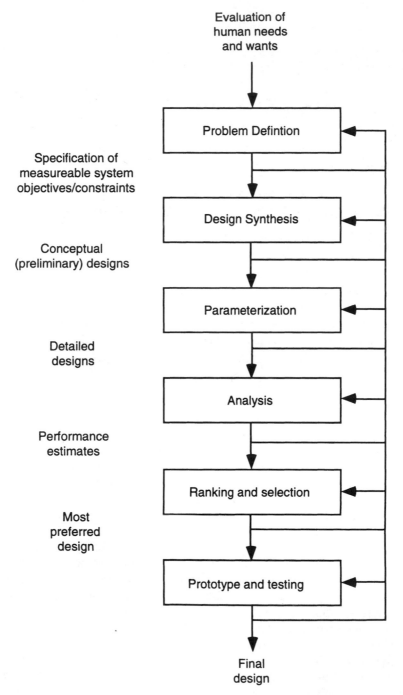

Figure 13.1 Design as an iterative process.

13.2.1   Problem Definition

The design process begins with the identification of a human need and the reduction of this need to a precise set of specifications for the system to be built. The definition of *system specifications* for a given design problem usually requires considerable interaction with the client or sponsor in order to ensure that the design problem is well understood, that goals and objectives of the design are explicit, and that the success of alternate designs can be measured quantitatively. Defining the problem is as much a part of design as is defining the artifact itself. The most elegant and efficient solution to the wrong problem is worse than useless.

A principal contribution of the systems method to the design process is the attention and emphasis afforded to proper problem definition. Gibson (1991) observes that inter-action with the client (sponsor or customer) is critical to appreciate the environment in which the design problem arises and in which the design solution will operate and be judged. It is almost always the case that the needs being addressed are poorly specified at the outset.

> It may seem unlikely to the novice systems analyst that a client would not understand his own problem, but this is often the case. The fact that the client cannot define his own problem is particularly exasperating to a theorist who has been taught that one cannot handle a problem until it is completely defined. In practice the converse is true. The systems analyst must expect to engage in a dialogue with his client to arrive at a suitable statement of a large-scale system problem.

Moran and Carroll (1996) reiterate this observation specifically in the context of design:

> needs are vaguely stated, usually tacit, often latent, and sometimes wrong (people aren't sure want they want). Requirements are wishfully overstated and it is in the process of design that these are cut back to what is realistic.

At the same time, the designer must guard against hidden agenda and premature closure on preconceived options, through the imposition of unnecessary constraints on the system. Gibson (1991) is succinct: "The client always lies." Hazelrigg (1996) argues that we want

> to instill in the design engineer a sense of freedom, not one of constraint. We want the engineer to think of the design process not as a process of seeking to satisfy a set of [functional] constraints at minimum cost but rather as a process of freely taking advantage of an opportunity through the ability to manipulate nature.

Gibson (1991) recommends seven steps for the systematic development of goals. These are:

1. *Generalize* the question in order to provide a correct problem statement placing the problem in its proper context.
2. Develop a *descriptive scenario,* honestly assessing where you are now.
3. Develop a *normative scenario* describing where you want to be at some time in the future.
4. Develop the *axiological component,* setting out the values of the sponsor.

5. Prepare an *objectives tree* that contains the most general goals at the top, and successively more specific and measurable objectives in the branches below.
6. *Validate* the goals and objectives developed in the preceding steps.
7. *Iterate* through several passes to refine and perfect the objectives tree.

In the nonlinear model of Figure 13.2, Hazelrigg (1996) underscores the relationship between systems engineering and strategic planning, emphasizing the role of Gibson's normative and descriptive scenarios and plan for action in systems design:

> In systems engineering the extant technology is where we are now. The system objectives are our statement of where we want to be at some time in the future. And the systems engineering process is how we get from here to there.

Figure 13.3 provides an example of an objectives tree that was used in the redesign of a congested traffic intersection. The top-level goal captures the context and underlying purpose of the design problem, generally, which is to enhance the quality of life for people living in the region served by the intersection. While the quality of life certainly has many components, the second-level objective suggests that one contributing factor is mobility, which can be improved by reducing congestion along the principal traffic corridor within the region. A third-level objective indicates the belief that congestion along the corridor can be reduced by reducing congestion at the particular intersection under study. At the bottom of the tree are specific design objectives, the satisfaction of which can be quantified and ultimately measured.

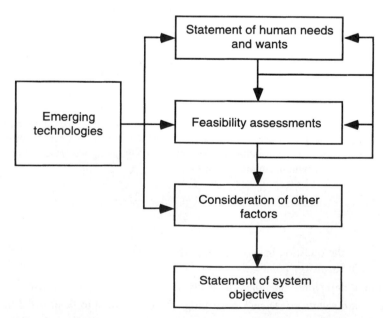

Figure 13.2   A nonlinear model for determining system objectives (after Hazelrigg, 1996).

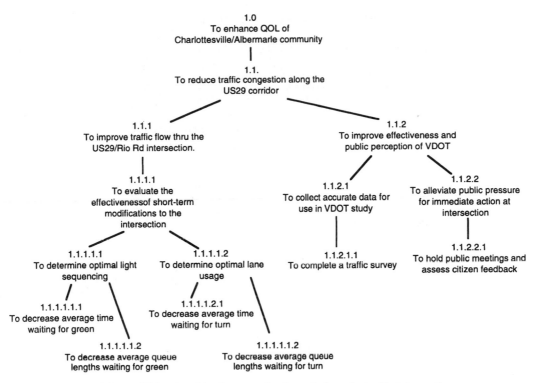

Figure 13.3   An objectives tree for the redesign of a traffic intersection.

13.2.2   Design Synthesis

The next task in the design process is to generate alternative designs, or *design options,* that might reasonably satisfy system specifications. Because these specifications can be fulfilled in a great many ways, it is characteristic that there is no unique solution to a design problem. There are usually a host of adequate solutions, some of which can be identified as better than others. Moreover, because the objectives of design are many (and conflicting and noncommensurate), some solutions may be preferred with respect to certain objectives, while other solutions are preferable with respect to different objectives. Various concepts for the system must be studied.

Systems engineers typically think of alternative designs as points in an *options* or *design space.* The options space is the set of possible system configurations and decisions, and represents the full range of choices available to the designer. For complex problems, defining (and paring down) the options space can be a very difficult task. This task can be eased by decomposing it into two component subtasks—*design synthesis* and *parameterization*—which can be addressed sequentially, but iteratively.

Hazelrigg (1996) defines design synthesis as the task of eliciting feasible system configurations, including both physical and nonphysical elements of a system.

TABLE 13.1   How to Develop Design Options

Modern Tools for Creativity

Phase	Tools
Unstructured search (out-scoping)	Brainstorming
	Brainwriting
	Dynamic confrontation
Examine various combinations of elements	Morphological box
	Options field
Assemble elements into complete candidate solutions	Options profile
	Computer simulation
	Delphi

After Gibson, 1991.

In design synthesis, we are not concerned with dimensions of the physical elements of the system, only with the configuration of the parts that comprise a system, how these are arranged, and how these are used, but not with their individual specifications.

Arora (1989) concurs in his earlier statement:

Since design concepts must be studied at a high level in a relatively short time, highly idealized models are used. Various subsystems are identified and their preliminary designs estimated. Decisions made at this stage generally effect the final appearance, performance, and cost of the system. At the end of the preliminary design phase, a few promising concepts needing further analysis are identified."

Eliciting feasible system configurations clearly demands *creativity* on the part of the designers. For this reason, theories of human creativity and tools to stimulate creativity are of interest to systems designers. Evans (1991) defines creativity as "the ability to discover new relationships, to look at subjects from new perspectives, and to form new combinations from two or more concepts already in the mind." Following the work of psychologist J. P. Guilford, Evans argues that creativity is the product of *divergent thinking,* which is nothing more than the ability to seek and discover many alternatives, as opposed to *convergent thinking,* which follows a determined path to a unique answer. In addition to knowledge and imagination, Evans lists the following characteristics of the individual that support creative behavior and offers insights on how to cultivate these: awareness and sensitivity to problems, memory, fluency, flexibility, originality, self-discipline and persistence, adaptability, intellectual playfulness, humor, nonconformity, tolerance for ambiguity, self-confidence, skepticism, and intelligence.

Like Evans, Gibson (1991) reviews classic theories of creativity. He then presents three modes of thought that are involved in the modern approach to developing design alternatives, as well as a selection of tools appropriate to each mode. These are summarized in Table 13.1. In the *unstructured search mode,* the designers range freely in an exercise intended to augment more conventional ideas by bringing into play new and unconventional concepts. Tools appropriate to this first mode of thought include brainstorming, brainwriting, and so-called dynamic confrontation. Each of these group-stimulation tools provides a structured approach that makes effective use of the interplay of ideas among the design team.

In the *examination of various combinations* and subsystem elements, the designers lay out options for various components of a design and seek to examine both conventional and unconventional combinations of these options. Tools appropriate to this second mode of thought include the morphological box and its modification into the so-called options field approach. In Gibson's final mode of thought, the designers want to assemble the elements into complete candidate solutions. Tools appropriate to this second mode of thought include dynamic interactive computer simulation and so-called options profiles. Gibson emphasizes the need for out-scoping in design synthesis (Guilford's divergent thinking) and cautions against skipping the first two modes of creative thought and jumping directly to the third:

> we find that creating "complete solutions" initially often results in premature in-scoping [Guilford's convergent thinking]. "Complete" solutions offered at this point tend to be old and tired ideas. They seldom, if ever, are imaginative or take full advantage of the situation.

Brainstorming is a group process designed to stimulate the discovery of new solutions to problems. In essence, it is a structured "bull session" under the supervision of a trained facilitator in which participants throw out ideas and build on and add to the ideas of one another. Under the right set of conditions, brainstorming has proven successful in applications spanning the past three decades. The rules are few:

- A trained leader, called a "facilitator," is absolutely necessary
- Sessions are kept informal, with no bosses; use first names
- Sessions are held away from the usual place of work; dress is casual
- Interruptions are not allowed
- Ideas are not criticized, categorized, or organized during sessions
- Building on "main ideas" with "helpers" is encouraged
- Everybody talks as they please, but positively
- A break is taken after an hour or so, when idea creation slows
- After one or two sessions, move on to an evaluation and categorization mode

Gibson also cautions about the known drawbacks of the technique:

- Brainstorming can be emotional and stressful and the possibilities for interpersonal conflict are real
- Individuals uncomfortable with verbal free-for-all can resent being drawn into the process
- Slow-talking individuals risk being shut out by more vocal individuals
- More vocal individuals may enjoy the process, but are difficult to cut off without dampening the process
- Only one person can talk at a time and other participants must idle while waiting their turn

Given these drawbacks, *brainwriting* is an attractive alternative that reduces the potential for domination by vocal or empowered participants and minimizes the likelihood of interpersonal conflicts. Instead of loud talking and laughter, a brainwriting session is conducted in silence. Five or six participants sit around a table. In the center of the table are sheets of paper, each headed by a so-called trigger question. Each individual takes one of the sheets, writes a sentence that responds to the corresponding trigger question, returns the sheet to the center, and takes another sheet. After reading the trigger question and prior comments by other participants, the individual adds his own positive comment. The process continues until useful comments are exhausted, then the ideas are edited, classified, and compiled into a narrative by the group.

Dynamic confrontation is an adversarial group process. In stark contrast to brainstorming and brainwriting, in which no criticism is allowed, the essence of dynamic confrontation is to criticize every idea. A presentation is made and then each main claim or assumption presented is deliberately and intensely challenged. The idea is to test conventional wisdom and require each participant to think through every claim.

Zwicky and Wilson (1967) proposed a method for examining unusual and original combinations of elements that he calls the *morphological box*. The underlying idea is the basis for the *options field/options profile* approach propounded by Warfield (1976). The designer or design team forms a table or box in which each column defines the functional classes that make up the basic subsystems of the design. Under each functional category, an exhaustive list of technological alternatives for achieving the corresponding function is developed. Finally, conceptual designs are enumerated by picking one option from each column, until all combinations are considered. It is understood that most combinations will be impractical, but the idea is to stimulate original concepts that might otherwise be overlooked. Table 13.2 illustrates the application of a morphological box to the design of a passenger vehicle.

13.2.3 Parameterization and Analysis

The result of design synthesis is the selection of a promising set of *preliminary* or *conceptual designs* for the system. These design concepts need to be converted into *detailed designs* for all components and subsystems. This requires the identification of system parameters and selection of parameter values. Hazelrigg (1996) defines this as the task of design *parameterization:*

> Parameterization consists of identifying the numerical quantities that specify each element in the system and their permissible ranges. A system specification consists of a set of values for the system parameters together with the system configuration. Alternatively, we could view this as the specification of a set of parts and the instructions to assemble and put these to use.

Parameterization of a conceptual design and analysis of the performance of the resulting detailed design typically are tightly coupled in an iterative loop. The purpose of such iteration is, first, to find a detailed design that meets or exceeds all design specifications (if such a design exists) and, subsequently, to find alternate parameterizations that improve on this performance. For each design concept, an initial parameterization is proposed based on experience and heuristics. Using the appropriate analysis methods, the performance of the corresponding detailed design is estimated with respect to the design specifications. Based on this performance estimate, a new parameterization is determined. The iterative process of parameterization and analysis of performance continues until the designer is satisfied that the parameter space has been explored sufficiently.

TABLE 13.2   Morphological Box Showing Possible Options Under Three Functional Categories for the Conceptual Design of a Passenger Vehicle

Suspension	Propulsion	Guideway
Air cushion	Jet engine	All terrain
Magnetic field	Traction	Steel rail
Steel wheels	Paddle wheel	Wire
Rubber tires	Sail	Side bumpers
Legs	Water jet	Road
Flotation	Muscle	Air
Hot-air balloon	Linear induction	Waterway
Crawler treads	Magnetic field	Tube
Skids/skis	Propeller	Laser beam
Antigravity	Air jet	GPS
Overhead cable	Water jet	Other
Overhead rotator	Suction	
Wings	Other	
Other		

Arora (1989) contrasts the conventional process of parameterization with optimal design, as illustrated in Figure 13.4:

> The design parameters for subsystems must be identified. The parameters must be such that once their numerical values are specified, the subsystem can be fabricated. The design parameters must also satisfy technological and system performance requirements. Various subsystems must be designed to maximize system worth, or minimize a measure of cost. Systematic optimization methods can aid the designer in accelerating the detailed design process. At the end of the process, a description of the system is available in the form of reports or drawings.

13.2.4   Ranking and Selection

The ranking of design alternatives and the ultimate selection of the most preferred design involves the selection of the best parameterization of the best conceptual design. While the best parameterization of a given design concept can sometimes be determined by the solution of a deterministic optimization problem, as suggested by Arora (1989), in general this step must be caste as a multiple-objective decision problem under uncertainty (Sage, 1977, 1992; Changkong and Haimes, 1983; Goicoechea et al., 1982; Hazelrigg, 1996; White, 1990).

13.2.5   Prototype and Testing

The ultimate step in the design process involves the fabrication and testing of a prototype or system. For mass-produced items, the opportunity exists to revise specifications and design details to improve performance, based on what is learned by actually making and using the artifact, either in a test setting, or in the field, or both. This may result in the fabrication of several generations of prototypes and actual products, each intended to be an improvement over the original design.

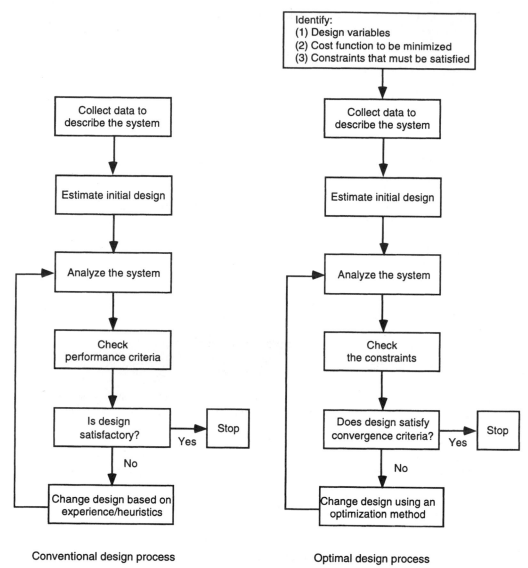

Conventional design process   Optimal design process

Figure 13.4 Conventional parameterization versus optimal design. (After Arora, 1989.)

For large and costly or one-of-a-kind systems, it may not be possible or feasible to construct a prototype. Nevertheless, it is not unusual that specifications and design elements are revised during fabrication and operational tests. Such revisions may be made to improve producibility, operability, or functionality; to reduce unenvisioned production, operating, or maintenance costs; or to remedy unwanted legal, societal, or environmental impacts.

Of course, design changes made at this last step typically are far more expensive and time-consuming than changes made earlier in the design cycle. These may imply the need and associated cost of scrapping units or production components and costly retooling of the production process. Significant changes can further imply schedule delays and increased time to market. Moreover, costly changes can lead to intraorganization stress and conflict between design and manufacturing functions and individuals. If the design process has been successful, major problems will have been anticipated during analysis and the number of postdesign changes will be few and minor. The introduction of CAD tools and computer-based simulation has contributed greatly to the ease of discovering functional design problems "on the drawing board," while design for manufacture (DFM) and concurrent engineering concepts have helped to anticipate the number of conceptual, production-related, and postproduction design flaws.

13.3   DESIGN TOOLS

13.3.1   CAD, CAE, CAM, and CIM

Computer-aided design (CAD) refers to the use of modern computing hardware and software in converting the initial idea for a product into a detailed engineering design. The evolution of a design typically involves the creation of geometric models of the product, which can be manipulated, analyzed, and refined. In CAD, computer graphics replace the traditional sketches and engineering drawings used to visualize products and communicate design information.

Engineers also use specialized computer-based analysis programs to estimate the performance and cost of design prototypes and to calculate optimal values for design parameters. When combined with CAD, these automated analysis and optimization capabilities are called computer-aided engineering (CAE).

Computer-aided manufacturing (CAM) refers to the use of computers in converting engineering designs into finished products. Production requires the creation of process plans and production schedules that tell how the product will be made, what resources will be required to make it, and when and where these resources will be deployed. Production also requires control and coordination of the physical processes, equipment, materials, and labor needed to implement the plan and schedule. In CAM, computers assist managers, manufacturing engineers, and production workers by automating many of these production tasks. Computers help to develop process plans, order and track materials, and monitor production schedules. Computers also help to control the operation of machines, industrial robots, test equipment, and the systems that move and store materials in the factory.

The deployment of CAD/CAM in industry is helping to shape new ideas about how better to organize and manage production. When CAD/CAM technology is introduced into traditional, functionally organized firms—with separately managed units for marketing, design, engineering, and production—the result has been called "islands of automation." But CAD, CAE, and CAM work best together. This has made clear the need to break down many of the traditional barriers between different functional units and manufacturing shops. The goal of computer-integrated manufacturing (CIM) is a common database, created and maintained on a factorywide computer network, that will be used for design, engineering analysis and optimization, process planning and production scheduling, part and robot programming, material handling, inventory control, maintenance, and marketing. While there are still many technical and managerial obstacles to achieving this vision, CIM appears to be the future of CAD/CAM.

13.3.2 CAD Workstations

A typical CAD workstation consists of a graphics display terminal and various input devices, such as a keyboard, mouse, light pen, and digitizing tablet. One or more workstations are linked to a computer, which runs the CAD software. Also linked to the computer are peripheral devices, such as printers, plotters, modems, and mass storage devices. The CAD system allows the designer to create a model of the product in computer memory, display different views of the model on the terminal, modify the model as desired, save different models in a database for later recall, and make hard copies of design notes and engineering drawings as needed.

The type of computer model used depends on the product being designed. In the design of printed wiring assemblies, for example, two-dimensional models are used to lay out electronic components and interconnections on the face of different layers of the circuit board and to test the results against design rules and performance specifications. In mechanical and civil engineering design, three-dimensional geometric models are used to represent discrete parts, mechanical assemblies, and engineering structures.

Several types of geometric models are possible, including wire-frame (line) drawings, surface representations (planes, patches, and sculpted contours), and solids models. Solids modeling is the basis for most current CAD systems, because the information in the model database represents a distinct solid object, completely and unambiguously. Complicated solid objects are built-up by adding and subtracting simpler geometric shapes—cubes, slabs, cylinders, and cones—called *primitives.*

Many CAD systems currently are used as sophisticated electronic drafting devices. This has many time- and labor-saving advantages. While a traditional drafter has to make several views of a single object to capture its geometry, the electronic drafter creates a single solids model from which any particular view can be derived automatically. As the design evolves, each change does not require reentering an entirely new model. In the future, as CAD, CAE, and CAM are integrated into CIM, the CAD model will become part of a common database for the entire design and manufacturing product cycle.

13.3.3 Systems Engineering Design Tools

The characteristic activities and tools of the systems engineer can be interpreted in terms of how these activities support the central mission of systems design. Indeed, most of the systems engineering tools, techniques, and methods described elsewhere in this handbook are important in systems design. For this reason, the present chapter focuses on systems engineering as seen from the perspective of the designer and leaves further discussion of specific design tools and techniques to the state-of-the-art expositions in the surrounding chapters.

For example, requirements management (see Chapter 4), issue formulation (see Chapter 24), and functional analysis (see Chapter 25) exist to probe and clarify human needs and wants in defined problem environments and to refine vaguely expressed ambitions and goals into explicit and coherent design requirements and specifications. Analysis and modeling methods (see Chapter 26) seek to predict the performance and cost of alternative designs under specified operating and market conditions; optimization techniques (see Chapter 27) are used to determine the best parametric specifications for alternative design concepts; and benchmarking (see Chapters 10 and 15) and standards (see Chapters 11 and

15) provide a frame of reference for the evaluation (see Chapter 22) of designs and design prototypes. Design is informed by the engineer's understanding of risks (see Chapter 3), requirements (see Chapter 4), costs (see Chapter 5), quality (see Chapter 7), operations (see Chapter 8), and human factors (see Chapters 19 and 21).

Blanchard (1986) notes that "Effective system design is basically realized though the system engineering process" (see Chapter 1), and Hazelrigg (1996) defines systems engineering as *the treatment of engineering design as a decision-making process* (see Chapter 28). Meredith et al. (1985) and Chadwick (1971) view design and planning (see Chapter 29) as an integrated process in engineering systems. Reengineering (see Chapter 23) concerns the design of processes and integrated product development, and concurrent engineering (see Chapter 9) has its roots in design for manufacture (DFM).

13.4   A BRIEF HISTORY OF RECENT DESIGN THEORY

Researchers from many backgrounds and traditions have attempted to elicit and capture the fundamental nature of design. In spite of its central role in engineering (and other disciplines as well), however, there is no unifying or universally accepted theory of design. Moran and Carrol (1996) briefly review some of the major influences on current design research.

The early work of Christopher Alexander (1964) characterized design as a decomposition and resynthesis problem—the system engineer's $\Pi-\Sigma$ process later described by Warfield (1976). Alexander suggests that the ultimate object of design is "form" and that every design problem is an effort to achieve fitness between the form in question, on the one hand, and its context, on the other. "The form is the solution to the problem, the context defines the problem." Context is often obscure, however, so that the designer can not give a fully coherent criterion for the fit that is to be achieved. Context must therefore be defined in appropriate ways to establish the kind of fit required. While Alexander et al. (1977) later rejected this mathematical line of thinking in favor of a notion of interlocking patterns, Alexander's original perspective fueled considerable related research on design methodology in the 1960s.

Noble laureate Herbert Simon (1969) proposed a view of the design process as involving search in a design space. A design alternative is satisfactory if there exists a set of criteria that describes minimally acceptable alternatives and if that alternative meets or exceeds all criteria. Because of the limits of human understanding and cognitive processing power, however, design criteria are not absolute, and expectations can rise or fall as the search for alternatives proceeds and more is learned about the design space. Simon's account of the design process continues to hold sway, particularly in the cognitive-science and AI communities.

Rittel and Weber (1983) observe that design is a process of negotiation and deliberation, dealing inherently in the resolution of uncertainty and conflict, while Schon (1983) sees design from a social and ethnographic perspective, emphasizing the reflective nature of design. Based on these earlier ideas, Moran and Carrol (1996) observed that

> like other professionals, designers are well aware that they are coping with elusive problems, searching, analyzing, synthesizing, negotiating, deliberating, and reflecting. Indeed, to be effective designers they must engage in and manage all of these modes of activities.

Moran and Carrol, among others, argue the necessity of a *design rationale*. This notion goes beyond merely the accurate representation of concrete artifacts, or even the written documentation of specifications for that artifact, and articulates the reasons and the reasoning process behind a given design. This legacy "makes the decisions more understandable to designers at a later time and exposes them to reflection and reconsideration." Recently, design rational has become an organizing principle in engineering design and related design disciplines. It appears to be of special importance in software design, where the structure, function, and intent of large codes is difficult to assess independently by subsequent generations of programmers charged with maintaining and enhancing applications.

Approaching design from the manufacturing side, Suh and his associates at MIT (Suh et al., 1978) hypothesized an axiomatic approach to design in which a very few global principles, or axioms, guide the entire design process. These axioms cannot be proven, but in the absence of counterexamples are accepted as general truths, applicable to the full range of design and production decisions. Most have been learned empirically and are deeply rooted in the long history of design and manufacturing. Understanding these axioms (if only implicitly) and applying them correctly in a specific context is the hallmark of an experienced systems designer.

Although several axioms were originally proposed, Yasuhara and Suh (1980) reduce these to two:

Axiom 1: In good design the independence of functional requirements is maintained.
Axiom 2: Among the designs that satisfy Axiom 1, the best design is the one that has the minimum information content.

These axioms convey the basic idea that specification of fewer functional requirements than necessary to achieve design objectives leads to unacceptable solutions, while the specification of more functional requirements than necessary leads to overdesign and attendant costs. Typical of the corollaries derived form these axioms (Yasuhara and Suh, 1980) are:

1. Decouple or separate parts or aspects of a solution if functional requirements are coupled or become coupled in the design of products and processes.
2. Integrate functional requirements into a single physical part or solution if they can be independently satisfied in the proposed solution.
3. Minimize the number of functional requirements and constraints.
4. Use standardized or interchangeable parts whenever possible.
5. Make use of symmetry to reduce the information content.
6. Conserve materials and energy.
7. A part should be a continuum if energy conduction is important.

Stoll (1990) contends that these axioms define a two-step process for design. First, the functional requirements and constraints of a product, device, or system should be satisfied independently by some aspect, feature, or component within the design. Second, the axioms should be applied individually to each decision as the design process progresses.

13.5 DESIGN AND CONCURRENT ENGINEERING

DFM applies the systems method to the manufacturing enterprise by simultaneously considering all design goals and constraints for the products and systems that will be produced. In many industries, DFM has become a central concept for survival in increasingly competitive global markets. One of the important lessons of DFM is that consideration of manufacturing issues early in the design phase can reap substantial benefits. These include savings in setup and production costs, reduction of lead times required to bring a new product to market, reduction of parts inventories and associated overhead, and improvements in overall product quality and reliability.

There is now a large body of literature on DFM and the subject continues to be an active area of research. While there are some general principles that can be distilled from case studies, application of DFM to a specific product appears to be as much art as science. Successful applications clearly demand a great deal of domain-specific expertise.

13.5.1 DFM Philosophy

DFM was born from the recognition that the delivery of a product to market typically requires complex interactions among a number of distinct activities, including

- Market analysis and product selection
- Product and component design
- Material selection
- Component and material purchasing and inventory
- Fabrication technology and tool selection
- Material handling and process control
- Assembly
- Test and rework
- Marketing, sales, and distribution

These activities, individually and in combination, ultimately determine the quality and cost of a product, as well as the time-to-market of new or improved products.

In its broadest sense, DFM seeks to understand the interaction of all of these activities and to leverage this understanding in order to optimize the product development/production/delivery process as a whole. In this sense, DFM is synonymous with so-called *concurrent* or *simultaneous engineering* (Salomone, 1995), or *integrated product development*. More narrowly, DFM focuses on the specific subset of interactions between the product design process and each of the various manufacturing subsystems, in order to develop products that are easier, faster, and less expensive to make, service, and retire, while maintaining required standards of quality and functionality.

As a production philosophy, DFM would seem to rank with Truth, Beauty, Liberty, and Justice as an intuitively correct and intrinsically desirable norm. Indeed, in craft production DFM is inherent in the practice of skilled artisans, where the product designer and fabricator are intimate and most often the same individual. Industrialization was attended by increasing specialization of function, however, which permitted the mass production of sophisticated products using machines and abundant unskilled or semiskilled labor. With functional specialization, design and manufacture gradually came to be viewed and practiced as distinct, sequential activities. Product designers were increasingly distanced from manufacturing issues and details. Only in the wake of rising global competition in international markets for manufactured goods that began in the 1970s did the reintegration of design and manufacturing functions crystallize as a potentially advantageous strategy for developing products.

One of the primary motivations behind DFM is the empirical observation that, for many different kinds of products, as much as 70 percent of the ultimate manufacturing cost is established during the design phase (Corbett et al., 1991; Boothroyd et al., 1994). By designing a product in such a way as to minimize the effort required for its manufacture, very often it is possible to reduce costs while improving the quality of the product, without loss in functionality or performance. In contrast, reducing labor costs or improving productivity and efficiency in the manufacturing processes per se has relatively little impact on the overall cost, if manufacturing issues were not properly taken into account early in the design phase.

Historically, product design considers three major factors: product function, product life, and component cost (Corbett et al., 1991). Manufacturability issues typically arise in so far as these issues contribute to component cost. Typical approaches to controlling or reducing component cost are value engineering and producibility engineering.

The basic goal of value engineering is maximum performance per unit cost. This goal is pursued, first, by altering the detailed design in ways that reduce cost while maintaining performance and, second, by identifying and modifying manufacturing methods that realize additional savings. Stoll (1990) identifies two kinds of value that are associated with a product: use or functional value, and esteem or prestige value. To achieve maximum use value is to provide the greatest ratio of utility and reliability value per unit cost, whereas to achieve maximum esteem value is to provide the greatest ratio of appearance, attractiveness, and feature value per unit cost.

As early as the 1960s, several companies developed producibility guidelines for use by designers. General Electric, for example, complied manufacturing data into one large reference volume, the *Manufacturing Producibility Handbook* (1960). The goal of producibility engineering is to ensure that parts can be manufactured, assembled, and tested using current or readily available techniques and processes, while meeting performance requirements. This usually translates into protecting manufacturing from unreasonable demands resulting from design engineering's desire to optimize performance. Little attention, however, is given directly to the concurrent design of products and the processes used to make these products.

Value engineering and producibility engineering are artifacts of a serial view of the

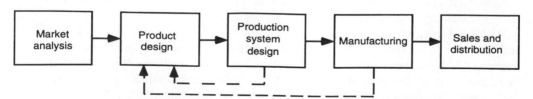

Figure 13.5 Classic (linear) model of product development. (Stoll, 1986.)

design/manufacturing product cycle illustrated in Figure 13.5. Issues of product function and life are addressed by the design section and result in detailed design concepts. Issues of manufacture are addressed by the manufacturing section and usually come into play *after* the major product concept decisions have been made. Manufacturability is thus an ancillary concern. The unidirectional flow of decision-making authority embodied in this serial model has led to the now standard metaphor in which designs are thrown "over the wall" to be implemented by manufacturing with little or no feedback regarding manufacturing difficulties.

This serial view belies the interactions among design and manufacturing activities illustrated in Figure 13.6, and contrasts markedly with the modern, integrated DFM philosophy illustrated in Figure 13.7. This integrated model recognizes that consideration of manufacturing issues early in the design phase, before detailed design decisions are made, can reap substantial benefits. DFM recognizes that, while valid and worthwhile, value engineering and producibility engineering too often seek to mitigate problems that could have been avoided altogether if manufacturing considerations had been taken into account early in the design phase. In Peter Drucker's words, "trying to do better that which should not have been done at all."

The integrated view of design and manufacturing issues embodied in DFM has been inspired at least in part by industry's attempts to achieve short-term gains by implementing robotics, flexible manufacturing, and CIM. These efforts led to the recognition that the interface between design and manufacturing was much more complex and pervasive in its

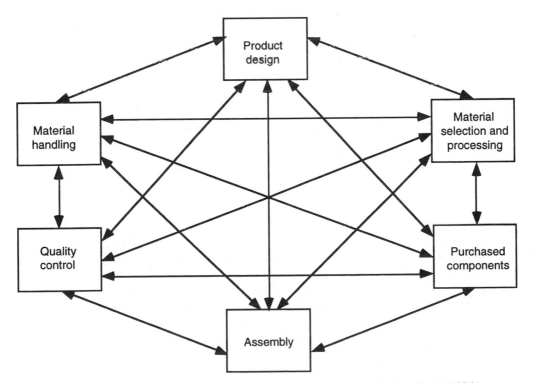

Figure 13.6   Interactions among design and manufacturing activities. (Stoll, 1986.)

effects than previously understood. In order to realize the benefits promised by advanced manufacturing techniques, it has become apparent that product design and process design have to be conducted in parallel rather than in an essentially serial manner.

Indeed, the desired integration extends *beyond* the design and manufacturing sections to concurrent or simultaneous engineering, as illustrated in Figure 13.7. The basic paradigm for product design must change to parallel, coupled activities between market analysis, product design, production system design, manufacturing, and sales and distribution. Clearly, this concurrent approach is much more complex and sophisticated than the serial model.

Salomone (1995) discusses key organizational issues in the implementation of concurrent engineering:

- *Collaboration Among Team Members.* Collaboration—literally, "working together"—is usually expressed in terms of teams and teamwork. Cross-functional product-development teams, including members representing marketing, design, process design, manufacturing, and sales, are often considered to be the central enabler in implementing DFM. It is well known, however, that cross-functional industry teams can operate without deliberate collaboration between team members and can lead to disastrous product failures. This occurs when a team is simply a collection of individuals with a common cause, but in which individual success is dictated by the agenda and criteria established by the individual team members' differing functions. "Since new product introductions often negatively affect traditional business measurements aimed at manufacturing efficiency or controlling marketing expense . . . trade-offs that are good for product development but negatively affect immediate business measurements have the potential of producing disastrous results for the . . . team member's career." This and other problems inherent in so-called

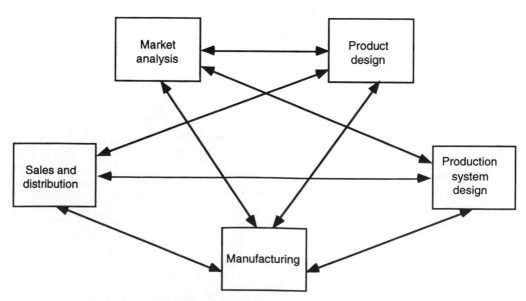

Figure 13.7 Integrated (concurrent engineering) model of product development. (Stoll, 1986.)

"matrix organization" are discussed at length in the literature on project management (Kerzner, 1995; Meredith and Mantel, 1989; Obradovitch and Stephanou, 1990). For cross-functional teams to succeed, new metrics need to be established that recognize and reward teams and individual team members in order to promote successful collaboration.

- *Virtual Teams.* Effective collaboration should occur beyond the team, including the full function of organizations within the firm, as well as suppliers, customers, consultants, resellers, distributors, and in some cases collaboration with other companies on the development of a single product or technology. This concept has come to be called the "supply chain." A recent special issue of the *Texas Instruments Technical Journal* (1996) reports such a successful collaboration between TI and Sun Microsystems, Inc., in the rapid manufacturing ramp of new technologies for the Ultra-SPARC-I family of computer workstations. As in building product design teams within the organization, individuals and their willingness to collaborate outside the company are a major factor in determining success. The organizational infrastructure that drives behavioral changes of individual toward collaborative activities—reward systems, goal setting, individual recognition, and functional measurements—again is key.

- *Implementation of Information Technology.* "Information technology has become the foundation enabler that allows concurrent engineering to happen." This extends beyond the use of shared CAD, CAM, and CAE systems, which provide automated design, drafting, and analysis functions, to include:

  Simulation technologies that allow design and manufacturing problems to be predicted and addressed early in the design cycle, when corrections are simple and easy to make;

  Rapid modeling and prototyping, which allows early testing and review of products using physical models constructed directly from CAD drawings and databases;

  Data libraries, which allow complex accumulations of detailed design data form one product to the next, leveraging investments across products and product generations;

  Communication technologies that allow functions to share information on a regular basis, review data, run independent analyses, interpret results, and feedback findings and concerns across functions;

  Network technologies linking external collaborators, which allow suppliers to provide rapid feedback on manufacturability and lower-cost alternatives up and down the supply chain;

  Network technologies that also link customers, allowing direct market analysis, collaboration and feedback on product designs, and improved customer relationships.

- *Establishment of Formal Concurrent Processes.* "The fundamental thought process of concurrent engineering is a process of convergence, where large scale collaborative thinking leads to innovative ideas, market understanding, and product and process knowledge [ultimately resulting in] an engineered product and processes for manufacturing and marketing [that product]." This convergence of ideas is illustrated in Figure 13.8. During the process of product development, many inputs are given, many sources of knowledge and information are available, and many tools and methodol-

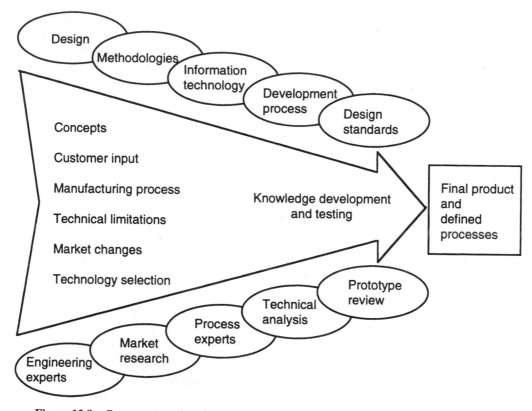

Figure 13.8 Concurrent engineering as the convergence of ideas from many sources over time on a final product and defined processes for that product. (After Salomone, 1995.)

ogies are provided to the product development team. In order to use these inputs constructively, the team expands and refines its knowledge about the product and the component processes, constantly changing and improving the design, while converging on a final design and well-defined processes for engineering, marketing, manufacturing, supply, and distribution.

Ettlie and Reifus (cited by Stoll, 1986) report that the following innovations characterize the organizational shift to DFM in many industries:

- Design-manufacturing teams
- Shared CAD systems for design and tooling
- Common reporting positions for computerization
- Philosophical shift to DFM
- Development and promotion of the engineering generalist or systems engineer

These innovations largely echo key organizational issues in the implementation of concurrent engineering discussed by Salomone.

DFM principles and rules enable product designers to consider manufacturing issues early in the design process. Following Stoll (1986), use "of this human-oriented, largely heuristic body of knowledge helps narrow the range of possibilities so the mass of detail which must be considered is within the capacity of the designer and planner." While much of this general knowledge is not new, and while "the principles themselves are largely invariant and universal, their organization and statement in the context of a DFM philosophy of global manufacturing system optimization give them new value and usefulness." In addition to the codification and dissemination of DFM principles, specific DFM rules governing detailed design are being incorporated in many CAD systems.

Given their common origins in the observations of experienced designers, the general principles for DFM most often cited are tellingly similar to the design axioms and corollaries outlined in the preceding section. These include (Boothroyd et al., 1994; Chow, 1978; Corbett et al., 1991; Stoll, 1986):

- *Minimize Total Number of Parts.* A part, component, or subsystem eliminated from a system costs nothing to design, order, buy, fabricate, transport, inventory, handle, orient, assemble, clean, inspect, rework, service, deliver, or retire. It is always in stock, weighs nothing, is highly producible, and never malfunctions, fails, or requires adjustment in operation.
- *Develop a Modular Design.* A module is a self-contained component with a standard interface to other components within a system. Modular design simplifies final assembly because there are fewer modules than subcomponents and because standard interfaces typically are designed for ease of fit. Each module can be tested prior to assembly and, in the field, repairs can be made by replacing defective modules. Custom systems can be realized by different combinations of standard components; existing systems can be upgraded with improved modules; and new systems can be realized by new combinations of existing and improved modules.
- *Use Standard Components.* Stock items are made in volume by a supplier and typically are less expensive, more reliable, and more easily obtained and replaced than custom-made components.
- *Design Parts to be Multifunctional.* Part counts typically can be reduced by designing components that serve more than one physical (structural, electrical, thermal, etc.) requirement. Manufacturing operations can be reduced or simplified by designing components that serve production as well as physical functions (self-alignment, built-in test, etc.).
- *Design Parts for Multiuse.* Group technology sorts parts, first, into groups that are unique to a particular system and groups that are used in many systems, and, second, into families of similar parts. Multiuse parts can then be designed by standardizing similar parts, minimizing the number of families, the number of variations within families, and the number of design features within variants. Manufacturing processes then can be developed for entire families, with minor process variants corresponding to design variants. Existing systems can be redesigned to make use of standardized parts where feasible and new systems designed using standard parts almost exclusively.
- *Design Parts for Ease of Fabrication.* Higher component costs in many instances can be more than compensated by lower overall production costs. Designs should use materials and primary production processes that minimize or eliminate material waste

(e.g., near net shape) and the need for secondary processing (e.g., surface treating).

- *Avoid Separate Fasteners.* Part counts and assembly times can be reduced dramatically by building fastening functions into the components to be fastened, using tabs, snap fittings, or unitized components and housings. Where separate fasteners are required, these should be standardized and self-positioning/self-guiding wherever possible.

- *Minimize Assembly Directions.* Ideally all systems should resemble a sandwich, with components added from the top and positively located. This reduces the number of directions and the time required for assembly moves, assembly stations, transfers, fixtures, and setups, leading to improved quality and reduced costs.

- *Maximize Compliance.* Small variations in stock parts and in fabrication and assembly processes can accumulate during assembly, resulting in misalignment, outright production failures, or system malfunctions. Compliance can be achieved by using highly accurate (consistent) parts, "worn-in" production equipment, tactile sensing, and vision systems, as well such built-in compliance features as accurate registration, positive location, generous clearances, and exaggerated guiding features (tapers, chamfers, and radii). External effects providing compliance include gravity and vibration feeds.

- *Minimize Handling.* Orienting components for fabrication or assembly can be expensive, risks damaging the component, and does not add value to the product or system. Parts therefore should be designed to make positioning easy both to achieve and maintain. The use of symmetry can obviate the need for orientation, whereas exaggerated asymmetry aids orientation when polarity is important. Parts should be designed to avoid tangling or nesting in storage, and part strips, tube feeders, and magazines should be used to preserve orientation. Rigid components are easier to handle (especially robotically), which suggests design rules such as the use of connectors rather than wire leads and circuit boards instead of wires and cables. Proper component and system design can also eliminate or simplify flows of material, people, and information within the plant. Standardizing the units of material flow, for example, using kits, standard outer-dimension packaging, or standard-dimension panels of circuit boards, is a proven means of simplifying material flows by easing handling.

13.5.2 DFM Methodology

DFM includes both design for fabrication (DFF), which is concerned with making individual parts of the product, and design for assembly (DFA), which is concerned with how and why the parts go together. According to the Stoll (1986), DFF has always been part of competent product design. DFA has received a great deal of comparatively recent emphasis because, for many products, assembly is a major part of the overall cost. More importantly, however, taking ease of assembly into account as an early part of product design usually results in substantially reduced part counts, which translates into savings in overhead costs, assembly costs, and net component manufacturing costs.

The *quantitative* evaluation method of Boothroyd and Dewhurst, "Design for Manufacture and Assembly" (DFMA, a registered trademark of Boothroyd Dewhurst Inc.), is based on this idea (Boothroyd et al., 1994). This method is illustrated schematically in Figure 13.9. The first step in DFMA is to design for assembly, that is, to derive a design

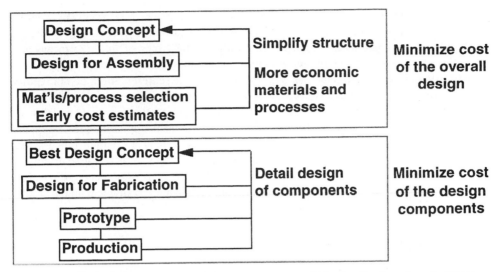

Figure 13.9   The DFMA process. (White and Trybula, 1996, after Boothroyd et al., 1994.)

that minimizes assembly costs and times. Boothroyd notes that such designs usually have substantially fewer parts and use manufacturing techniques that produce near-net parts, which need less machining and finish work. This often translates into considerable savings downstream. The second step is to refine the design of the parts identified in the first step, so as to minimize the effort required to fabricate these. This is what Boothroyd usually means by design for manufacture.

The key to the Boothroyd Dewhurst method (and to DFM in general) is to have suitably accurate cost models available for each iteration of design. At the first stage, these models need only be approximate. Since the purpose of initial cost estimates is to compare different design concepts, materials, and processes, the designers should be looking for relatively large differences in cost between competing design concepts. Boothroyd et al. (1994) have assembled approximate cost data for an impressively large number of manufacturing processes and materials. These include

- Mechanical assembly (manual and robotic)
- Electronics board assembly (manual and robotic)
- Wire harness assembly
- Machining
- Injection molding
- Sheet metalworking
- Die casting
- Powder metal processing

The ability to couple these data with a consistent, straightforward procedure for deriving cost estimates is the primary value in their proprietary method and supporting software.

Coupling of DFM methodology with CAD or CIM software is a promising area of ongoing research, as is the possibility of adapting AI tools. Boothroyd et al. (1994) discuss CAD and CIM at some length, focusing primarily on the issues of linking process and material cost information with the representation of a part's geometry. In particular, some geometry modeling methods, such as wire frame or surface representation, do not readily admit computation of properties such as volume or center of mass. At the same time, producing a part design on many CAD systems can be too time-consuming to make these attractive for preliminary design activities. With respect to artificial intelligence, while the conceptual design of a new or improved product is essentially a creative process that does not lend itself to application of AI techniques per se, application of DFM principles and techniques to a rough design can be greatly *facilitated* using expert systems and automatically generated cost estimates, derived from information available with CAD software.

13.5.3 Design for X

Product design can be viewed as an optimization problem, with multiple design objectives and constraints. These objectives very often conflict (improvement with respect to one objective is not possible without deterioration with respect to another objective), and in many cases are noncommensurate (performance cannot be measured in the same units). In traditional practice, the designer seeks to optimize the functional performance of a product within constraints imposed by cost and manufacturing technologies. These constraints typically and incorrectly are applied after the fact, to the detailed design. Narrowly interpreted, DFM can be viewed as a reformation of the design problem, where the designer seeks to minimize manufacturing costs, subject to constraints imposed by performance requirements, as well as by the available materials, purchased parts and components, and manufacturing technologies.

Viewed in this way, it should be clear that the objectives of product design should not be limited just to functional and manufacturing issues. *All* phases of the product life cycle—from product concept to retirement—should be considered in the early design phase and corresponding methodologies developed to facilitate this end. This more inclusive interpretation of DFM is sometimes called Design for X (DFX), where X can be taken from any of a long list of relevant design objectives, with all other objectives (perhaps) treated as constraints.

Recent work includes component (if still isolated) aspects of a total DFX vision:

- Design for *disassembly* (product disposability) (Beasley and Martin, 1993; Jovane et al., 1993).
- Design for *green,* or design for *environmentally conscious manufacture* (Ishii et al., 1994; Navinchandra, 1991; Girard and Boothroyd, 1995; Zust and Wagner, 1992).
- Design for *maintainability* and *service evaluation* (Moss, 1985; Abbatiello and Dewhurst, 1994, 1995a,b)
- Design for *inspection* (Feenstra and Knight, 1993).
- Design for *reliability* (Priest, 1988).
- Design for *quality* (Crow, 1983; Ehresman, 1988; Hubka and Edor, 1988; Hubka, 1989; Phadke, 1989; Taguchi, 1993; Unal and Lance, 1992; Porter and Knight, 1994).
- *Life-cycle* design (Fabrycky, 1987; Hedge, 1994; Hazelrigg, 1996).

Specializations can also be developed, such as design for electronics assembly (Stanziano and Knight, 1995), design for photonic systems (Leclerc and Subbarayan, 1996, 2nd design for semiconductor manufacturing (White et al., 1997).

Some of these design objectives are compatible and some are not. For example, design for disassembly is clearly compatible with design for green and design for maintainability, since recycling and maintenance typically involve disassembly, albeit to differing degrees. Similarly, design for disassembly does not necessarily present constraints on design manufacturability analysis (Girard and Boothroyd, 1995). On the other hand, designing for ease of service can adversely impact ease of assembly, since assemblies that are easy to put together are not always easy to take apart. In many such cases, different aspects of the product's life cycle cannot be optimized simultaneously and the designers must make tradeoffs.

REFERENCES

Abbatiello, N., and Dewhurst, P. (1994). *Design for Service Case Study: Hewlett Packard Deskjet Printer: Phase I*. Kingston: University of Rhode Island, Department of Industrial and Manufacturing Engineering (unpublished manuscript).

Abbatiello, N., and Dewhurst, P. (1995a). *A Strategy for Design for Service Evaluation*. Kingston: University of Rhode Island, Department of Industrial and Manufacturing Engineering (unpublished manuscript).

Abbatiello, N., and Dewhurst, P. (1995b). *Extensions of the DFS Methodology*. Kingston: University of Rhode Island, Department of Industrial and Manufacturing Engineering (unpublished manuscript).

Alexander, C. A. (1964). *Notes on the Synthesis of Form*. Cambridge, MA: Harvard University Press.

Alexander, C. A., Ishikawa, S., Silverstein, M., Jacobson, M., Fiksdahl-King, I., and Angel, S. (1977). *A Pattern Language: Towns, Building, Construction*. Cambridge, MA: Harvard University Press.

Arora, J. S. (1989). *Introduction to Optimum Design*. New York: McGraw-Hill.

Beasley, D., and Martin, R. R. (1993). Disassembly sequences for an object built from unit cubes. *Comput.-Aided Des.* **25,** 751–761.

Blanchard, B. S. (1986). *Logistics Engineering and Management*. Englewood Cliffs, NJ: Prentice-Hall.

Boothroyd, G., Dewhurst, P., and Knight, W. (1994). *Product Design for Manufacture and Assembly*. New York: Dekker.

Chadwick, G. (1971). *A Systems View of Planning*. New York: Pergamon.

Changkong, V., and Haimes, Y. Y. (1983). *Multiobjective Decision Making: Theory and Methodology*. New York: Elsevier.

Chow, W. W. C. (1978). *Cost Reduction in Product Design*. New York: Van Nostrand-Reinhold.

Corbett, J., Dooner, M., Melela, J., and Pym, C. (1991). *Design for Manufacture: Strategies, Principles, and Techniques*. Reading, MA: Addison-Wesley.

Crow, K. A. (1983). Concurrent engineering. In *Tool and Manufacturing Engineers Handbook: Design for Manufacturability* (R. Bakerjian, ed.), Vol. 6. Dearborn, MI: SME Press.

Ehresman, T. (1988). Using quality tools in DFM. In *Tool and Manufacturing Engineers Handbook: Design for Manufacturability* (R. Bakerjian, ed.), Vol. 5. Dearborn, MI: SME Press.

Evans, J. R. (1991). *Creative Thinking in the Decision and Management Sciences*. Cincinnati, OH: South-Western Publishing.

Fabrycky, W. J. (1987). Designing for the life-cycle. *Mech. Eng.* **109,** 72–75.

Feenstra, T. J., and Knight, W. A. (1993). *Design for Inspection.* Kingston: University of Rhode Island, Department of Industrial and Manufacturing Engineering (unpublished manuscript).

General Electric. (1960). *Manufacturing Producibility Handbook (MPH).* Schenectady, NY: GE Manufacturing Services.

Gibson, J. E. (1991). *How to Do Systems Analysis.* Charlottesville: University of Virginia.

Girard, A., and Boothroyd, G. (1995). *Design for Disassembly.* Kingston: University of Rhode Island, Department of Industrial and Manufacturing Engineering (unpublished manuscript).

Goicoechea, A., Hansen, D. R., and Duckstein, L. (1982). *Multiple Objective Decision Analysis with Engineering and Business Applications.* New York: Wiley.

Hazelrigg, G. A. (1996). *Systems Engineering: An Approach to Information-Based Design.* Upper Saddle River, NJ: Prentice Hall.

Hegde, G. G. (1994). Life cycle cost: A model and applications. *IIE Trans.* **26,** 56–62.

Hubka, V. (1989). Design for quality. *Proc. Int. Conf. Eng. De.,* Harrogate, UK, pp. 1321–1333.

Hubka, V., and Eder, W. E. (1988). *Theory of Technical Systems: A Total Concept Theory for Engineering Design.* New York: Springer-Verlag.

Ishii, K., Eubanks, C. F., and Marco, P. D. (1994). Design for product retirement and material life-cycle. *Mater. and Des.* **15,** 225–233.

Jovane, F., Alting, L., Armillotta, A., Eversheim, F. Feldman, K. Seliger, G., and Roth, N. (1993). A key issue in product life cycle: Disassembly. *Ann. CIRP* **42**(1), 1–13.

Kerzner, H. (1995). *Project Management: A Systems Approach to Planning, Scheduling, and Controlling,* 5th ed. New York: Van Nostrand-Reinhold.

Leclerc, S., and Subbarayan, G. (1996). A design for assembly evaluation methodology for photonic systems. *IEEE Trans. Components, Packag., and Manuf. Technol. Part C: Manuf* **19**(3), 189–200.

Meredith, D. D., Wong, K. W., Woodhead, R. W., and Wortman, R. H. (1985). *Design and Planning of Engineering Systems.* Englewood Cliffs, NJ: Prentice-Hall.

Meredith, J. R., and Mantel, S. J., Jr. (1989). *Project Management: A Managerial Approach,* 2nd ed. New York: Wiley.

Moran, T. P., and Carroll, J. M. (eds.) (1996). *Design Rationale: Concepts, Techniques, and Use.* Mahwah, NJ: Erlbaum.

Moss, M. A. (1985). *Design for Minimal Maintenance Expense.* New York: Dekker.

Navinchandra, D. (1991). Design for environmentalability. *Des. Theory Methodol.* **31,** 119–124.

Obradovitch, M. M., and Stephanou, S. E. (1990). *Project Management: Risks and Productivity.* Bend, OR: Daniel Spencer.

Phadke, S. M. (1989). *Quality Engineering Using Robust Design.* Englewood Cliffs, NJ: Prentice-Hall.

Porter, C. A., and Knight, W. A. (1994). *Design for Quality.* Kingston: University of Rhode Island, Department of Industrial and Manufacturing Engineering (unpublished manuscript).

Priest, J. W. (1988). *Engineering Design for Producibility and Reliability.* New York: Dekker.

Reilly, N. B. (1993). *Successful Systems Engineering for Engineers and Managers.* New York: Van Nostrand-Reinhold.

Rittel, H., and Weber, M. (1983). Dilemmas in a general theory of planning. *Policy Sci.* **4,** 155–169.

Rychener, M. D. (ed.), (1988). *Expert Systems for Engineering Design.* San Diego, CA: Academic Press.

Sage, A. P. (1977). *Methodology for Large Scale Systems.* New York: McGraw-Hill.

Sage, A. P. (1992). *Systems Engineering,* New York: Wiley.

Salomone, T. A. (1995). *What Every Engineer Should Know About Concurrent Engineering.* New York: Dekker.

Schon, D. A. (1983). *The Reflective Practitioner: How Professionals Think in Action.* New York: Basic Books.

Simon, H. A. (1969). *The Sciences of the Artificial.* Cambridge, MA: MIT Press.

Stanziano, M., and Knight, W. A. (1995). *Design for Electronics Assembly.* Kingston: University of Rhode Island, Department of Industrial and Manufacturing Engineering (unpublished manuscript).

Stoll, H. W. (1986). Design for manufacture: An overview. *Appl. Mech. Rev.* **39,** 1356–1364.

Stoll, H. W. (1988). Design for manufacturing. *Manuf. Eng.* **100,** 67–73.

Stoll, H. W. (1990). Design for manufacturing. In *Simultaneous Engineering* (C. W. Allen, ed.). Dearborn, MI: SME Press.

Suh, N. P., Bell, A. C., and Gossard, D. C. (1978). On an axiomatic approach to manufacturing and manufacturing systems. *ASME J. Eng. Ind.,* **100**(2), 127–130.

Taguchi, G. (1993). *Taguchi on Robust Technology Development: Bringing Quality Engineering Upstream.* New York: ASME Press.

Texas Instruments. (1996). *Texas Instruments Technical Journal,* Spec. Issue.

Unal, R., and Lance, B. (1992). Engineering design for quality using the Taguchi approach. *Eng. Manage. J.* **4,** 37–47.

Walton, J. (1991). *Engineering Design: From Art to Practice.* St. Paul, MN: West Publ. Co.

Warfield, J. N. (1976). Societal Systems: Planning, Policy, and Complexity. New York: Wiley.

White, D. J. (1990). A bibliography of applications of mathematical programming multiple-objective methods. *J. Oper. Res. Soc.* **41,** 669–691.

White, K. P., Jr. and Trybula, W. (1996). DFM for the next generation. *Proc. 17th Int. Electron. Manuf. Technol. Symp.,* Austin, TX.

White, K. P., Jr., Trybuk, W., and Athay, R. (1997). Design for semiconductor manufacturing: A perspective. *IEEE Trans. Components, Packag., 2nd Manuf. Technol. Part C: Manuf.* **20**(1), 58–72.

Yasuhara, M., and Suh, N. P. (1980). A quantitative analysis of design based on the axiomatic approach. *Comput. Appl. Manuf. Syst.,* pp. 1–20.

Zust, R., and Wagner, R. (1992). Approach to the identification and quantification of environmental effects during product life. *Ann. of CIRP* **41,** 473–476.

Zwicky, F., and Wilson, A. G. (1967). *New Methods of Thought and Procedure.* Berlin: Springer.

14 Systems Integration

JAMES D. PALMER

14.1 INTRODUCTION

Systems integration (SI) is essential to the development of large, complex engineered systems. It combines the practice and application of systems engineering and systems management, involving the organization, management, and technical skills necessary for success in systems integration programs, to large complex engineered systems. These concepts are utilized to address latent opportunities and problems associated with melding existing systems and new technologies to form more capable systems that are intended to take on additional tasks, exhibit improved performance, and/or enhance existing systems.

Typically, SI requires the coordination of preexisting and coexisting system components with newly developed ones. The outcomes may result in the significant growth of these systems or the merging of two or more systems or the combining of existing system fragments with commercial off-the-shelf components or various combinations of these. At a tactical level, SI is involved with ensuring that specific hardware/software components fit together smoothly in a stated configuration. Indeed, at this level SI is often referred to as *configuration management*. But at a broader, more strategic level, SI is concerned with interpreting the overall performance needs of a sponsor into technical performance specifications and ensuring that these system requirements are met.

14.1.1 Objectives for a Systems Integration Methodology

The objectives for a systems integration engineering methodology can be stated as follows:

1. To provide a suitable methodology that encompasses the entire integration program, from requirements through design, to construction, to test, and finally to deployment and maintenance.
2. To support problem understanding and communication between all parties at all stages of development.
3. To enable capture of design and implementation needs early, especially interface and interactive needs associated with bringing together new and existing equipment and software.
4. To support both a top-down and a bottom-up design philosophy.
5. To support full compliance with audit trail needs, system-level quality assurance, and risk assessment and evaluation.
6. To support definition and documentation of all aspects of the program.

Handbook of Systems Engineering and Management, Edited by A. P. Sage and W. B. Rouse
ISBN 0471-15405-9   ©1999 John Wiley and Sons, Inc.

7. To provide a framework for appropriate systems management application to all aspects of the program.

14.1.2 Systems Integration Personnel Needs and Technical Areas

Integration activities may be viewed from the perspective of the contracting corporation or agency (client), or the organization engaged to carry out the SI contract. The client must define the SI requirements, specifications, constraints, and variables in a manner so as to provide the means for a SI organization to deliver the necessary systems and services required to fulfill the client mission under the specific contract. The organization engaged to carry-on the SI program must be able to understand the requirements documents, comprehend the issues associated with bringing together existing hardware and software with new hardware and software, and do so in an effective and efficient way so as to deliver the integrated product on-time and within budget. SI programs may be hardware only (unlikely) or software only (likely) as well as the combinations of these (most likely).

Systems engineering and systems management combine hardware, software, facilities, personnel, procedures, and training to achieve specific goals and objectives. SI requires personnel who possess sound technical and management skills that combine to provide the ability to integrate technology and operations with technical and managerial direction. Systems engineering addresses tasks that are essential for the design, development, and operation of large complex engineered systems. Systems management covers the tasks and related activities associated with planning, control, and operations to achieve the goals and objectives of a specific systems integration program. For a detailed exposition of systems engineering principles and systems management processes, please refer to Chapters 1 and 2.

SI programs run the gamut from procurement of major weapons systems, to business management systems, to life-critical support systems. Typical SI applications are for large-scale programs that include hardware, software, facilities, personnel, procedures, and training, as these relate to the integration of existing and new capabilities to achieve specific goals and objectives. SI programs have come to include the procurement of major ADP systems such as those envisioned under the Department of Defense Computer-Aided Acquisition and Logistic Support (CALS)[1] program, MIS programs for the Environmental Monitoring and Assessment Program (EMAP),[2] and the Defense Information Systems Agency (DISA)[3] business process reengineering programs (see Chapter 23 for details on reengineering).

14.1.3 Definition of Systems Integration

SI is a logical, objective procedure for applying new and/or expanded performance requirements in an efficient, timely manner to the design, procurement, installation, and operation of an operational configuration consisting of distinct modules (or subsystems), each of which may embody inherent constraints or limitations.

---

[1] CALS is a program of the United States Department of Defense.
[2] EMAP is a program of the United States Environmental Protection Agency.
[3] DISA is an agency of the United States Department of Defense.

This definition of SI contains a number of key terms that require further explanation within the context of SI:

Logical, Objective Procedure. The SI process is clear to external observers and all steps have a built-in audit trail.

Efficient and Timely. The SI process will not be unduly burdened with delays and bureaucratic procedures that increase cost to the client and delay deployment of the system.

Design, Procurement, Installation, and Operation. The SI process will be employed throughout the entire process. Life-cycle costing will be considered together with retrofits, extension of system capability, and the like.

Distinct Modules with Inherent Limits or Constraints. The concept of distinct modules with inherent limits or constraints is central to the concept of SI. SI is necessary when the configuration to be deployed includes devices with intimate connections to other devices previously deployed or to be deployed under a later procurement, particularly if these devices were designed and constructed *de novo* by subcontractors with only partial design responsibility for the overall system.

14.1.4 Role of Systems Integration in Large, Complex Engineered Systems

The rationale behind the use of SI in the development and procurement of large complex engineered systems is to provide an organized, sensible, accountable, and workable approach to otherwise seemingly incomprehensible programs. Some of the attributes emphasized through the use of SI for large complex engineered systems include:

- Development and utilization of a strategic plan for management and technical aspects of the program;
- Establishment of a complete audit trail;
- Assistance in meeting initially unrecognized needs (including changes in system requirements);
- Avoidance of under- and overprocurement;
- Development and utilization of risk management plans;
- Management of subcontractors to the same specifications as employed on the prime contract; and
- Provisions for future modification and expansion.

Thus, application of SI provides an organized approach to carry out the design, management, and deployment of large, complex engineered systems.

In this chapter we define SI, examine the role of SI in large complex engineered systems, develop an SI life cycle and indicate the tasks that must be completed, look at personnel requirements, explore an SI strategy for success, investigate some of the application methodologies for SI, review the relevant management characteristics, see how disparate subsystems are brought together and tested, see to delivery, and examine risk management strategies. We use actual examples of SI projects throughout.

14.2 SYSTEMS INTEGRATION IN LARGE, COMPLEX ENGINEERED SYSTEMS AND A SYSTEMS INTEGRATION LIFE CYCLE

The methodologies of systems engineering and systems management are incorporated to form a formal approach to carry out SI programs, including the objectives for SI listed earlier. This organized approach supports and sustains the use of an SI life cycle to provide for the smooth rational development of an SI program. A typical SI life cycle and associated activities are addressed in the following section.

14.2.1 Typical Systems Integration Life-cycle Phases

An organized approach to SI is to view the program from the perspective of a systems development life cycle. This systems life cycle generally comprises a number of phases, usually seven (the minimum number is three and the maximum number unspecified) that range from the identification of requirements and the statement of specifications, to operational deployment of the system. Minimally, the system life-cycle phases are requirements definition, design and development, and operations and maintenance.[4] The seven-phase life cycle that is most commonly used in SI programs is as follows:

- Requirements definition and specification
- Feasibility analysis
- System architecture development
- Management plan: program and project plan
- Systems design: logical and physical design
- Implementation: design implementation, system tests, and operational deployment
- Evaluation: system review and plan for replacement/retirement

The SI life cycle is both interactive and iterative; however, for simplicity of presentation, it is depicted as a waterfall model, as shown in Figure 14.1.

14.2.2 Primary Activities Conducted for the Life-cycle Phases

Some of the primary activities that are conducted during each of the typical life-cycle phases are noted below.

Requirements Definition and Specifications

- Definition of requirements by use
- Review of requirements for ambiguity, conflict, and other issues
- Development of systems specifications

The goal for requirements definition and specification is to completely define and correctly interpret the client's real needs. It sometimes surprises the systems integrator when the

---

[4]While the latter two functions may be divided into separate parts, these are generally lumped together for purposes of establishing the minimal set of system development activities.

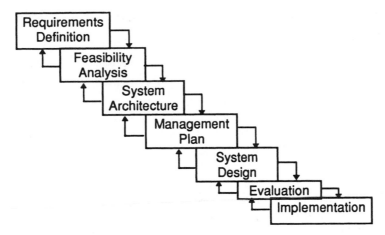

Figure 14.1 Systems integration life cycle.

client appears not to understand completely the system requirements being proposed, however, this should come as no shock. The client is undoubtedly an expert in the area in which the system is to be constructed (or certainly should be) and there is no reason to expect that an expert in the particular technical area should also be an expert in system component design and integration, especially when the need is for new functionality to be incorporated in an existing system, or when several systems are to be coupled together with added functionality. As an analogy, consider an individual who wishes to build a new house but is not trained in architecture. The person may well know what functions are most desired, but this does not an architect make. Many costly mistakes have been made by overlooking this apparently simple point.

Feasibility analysis

- Determine the likelihood of successful system development and deployment
- Examine new technologies
- Assess risk and develop risk strategies

It is necessary to establish complete and consistent system performance criteria to determine feasibility as well as assess the risk potential. The criteria must be developed on a *functional basis* and not by reference to existing commercial configurations. This may seem unworldly and unrealistic. Why should not the system integrator examine the configurations available in the marketplace and select the one that seems best to fit the client's needs? This is certainly the approach one may instinctively think of using. But there are many arguments against this untutored approach. Chief among them is that in procurement of large complex engineered systems, there is the almost certain risk of not taking advantage of technology growth and change, and the lack of standard interfaces across commercial systems. It will be impossible to develop an audit trail that leads from the functional performance requirements of the client to the final procurement choice, if in fact the choice was not functionally based, but rather was based on a survey of available configurations and a subjective decision made by a procurement agent. Assessment of risk potential based on performance and technical characteristics constitutes one of the major activities during feasibility analysis.

System Architecture Development

- Describe functional system architecture
- Specify required technical capabilities

Just as in the instance with feasibility analysis concerning performance, it is necessary to specify the system architecture and to establish the required technical capabilities for the system to be deployed. This is an especially critical activity for SI in that the architecture selected will most certainly determine the ease or difficulty with which the ensuing system may be put together and made operational. Good systems architecture can also provide significant cost savings, especially during maintenance activities, as well as greatly increased system flexibility. We must be certain that we put architectural requirements in system specifications, emphasize the early satisfaction of these specifications, give incentives to use proven architectural concepts, control architectural configuration over the life cycle of the system, and accommodate subsequent systems integration (Sage and Lynch, 1998).

Management Plan: Program and Project Plan

- Identify technical architecture alternatives
- Specify required configuration categories
- Prepare program and project plans (e.g., work breakdown structure)
- Prepare subcontractor management plan
- Prepare risk management plan

The development and utilization of a management plan is the key to success for the SI program. Architectural and configuration alternatives must be examined and explicit program and project plans developed. One of the most important activities for the success of development of large, complex engineered systems is the identification and assessment of risk factors and the development of a successful risk management strategy. This development includes assessment of technical and nonfunctional risk factors and the development of a risk management plan.

Systems Design: Logical and Physical Design

- Design approaches (e.g., top-down, bottom-up, etc.)
- Use of CASE tools and other automated aids

A major activity is to define and explore all options that could meet the performance criteria established earlier. There are two completely different approaches to this step depending on whether the design will lead to in-house development or external procurement or combinations thereof. Beyond this, the procurement can be through a free and open bidding process or by a directed procurement, depending on the nature of the contract.

Directed Procurement. In this approach it is the duty of the system integrator to survey all available combinations and configurations, probably by use of an options field and options profile, and establish specific functional parameters.

Free and Open Bidding Process. Most major government procurements fall within this category. Here the system integrator will act to ensure that the performance requirements established earlier are widely circulated to all qualified bidders.

For any of the several alternatives, it is desirable to rate and weight the available options, prior to making specific selections. It is the responsibility of the system integrator to apply the weightings to the separate performance characteristics of the most likely design configurations or the qualified responses to a request for proposal (RFP). These weights should have been established by the systems integrator in consultation with the client, prior to design or solicitation of the RFP. It must be possible to demonstrate that the weights recommended by the systems integrator and approved by the client are consistent with the importance of the functional requirements of the system. This is another way in which the audit trail that leads back to the functional performance requirements is maintained and utilized.

Implementation: Design Implementation, System Tests, and Operational Deployment

- Identify technical configuration
- Specify required configuration component items
- Procurement from subcontractors
- Perform system tests
- System deployment

Select and deploy the preferred option(s). Included in this step are such widely diverse matters as ensuring that nonoperating questions such as operator training, systems shakedown, and client acceptance bench marking are addressed, and in addition, ensuring lifetime system maintainability.

Evaluation: System Review and Plans for Replacement/retirement

- Review and evaluate system functioning
- Obtain, install, test, and accept modified components
- Maintain, modify, augment, and enhance systems
- Plan for system retirement/replacement.

Conduct various system performance tests and evaluate the outcomes in light of the client requirements. Once the system has been deployed and is operational, it is necessary to conduct total SI tests and evaluate the outcomes to ascertain that the functional and performance requirements have been met.

The phases in the system life cycle are interactive and iterative throughout the program development activity. Depending on the state of development and understanding of a specific project, it is possible to initiate the life cycle at any one of the specific phases. However, it is recommended that the life cycle be used from the beginning of the program and followed through until completion. Change management (and for large, complex en-

TABLE 14.1   Total Life-cycle Cost of Phases of Digital Computer Software Production[a]

Life-cycle Phase	Share of Life-cycle Cost	
Establish scope of project	3%	⎫
Requirements specification	9%	⎬ 20%
System design	8%	⎭
Coding	14%	⎫
Integration testing	15%	⎬ 30%
Acceptance Testing	5%	⎭
Operations, maintenance, and version upgrades	50%	50%

[a] Note that about half of the total system cost occurs *after* the system has been accepted by the client.
Source: MITRE Corporation

gineered systems change is inevitable) will involve the utilization of this same life cycle; thus, it is prudent to have such a system in place from the onset of the program, as changes are a fact of life when dealing with large, complex engineered systems.

14.2.3   Effort Expended as a Function of Life-Cycle Phase

While the actual records of the amount of time expended in each of the several phases of the SI life cycle is most difficult to come by—most corporations either do not maintain such records or do not make them public—it is helpful to examine data that are available. One such study of the SI life cycle from the perspective of the percentage of time estimated for each of the activities has been conducted by MITRE, using a variation of the SI life cycle, just presented. The particular SI program development efforts were for information systems, and the percentage of time generally allocated to each of the activities is noted in Table 14.1.

Table 14.1 shows the results for the purchase and operation of digital computer software. We notice several things. First, that establishment of the goals or scope of the project and of functional specifications together account for only about 10 percent of the total project cost. Notice further that about half of the total cost of the system is incurred *after* the system has been tested, accepted, and put into operation. If this portion of the life cycle is ignored in the specification process, serious inefficiencies can result.

Thus we see that the life cycle for SI is quite similar to a typical systems engineering life cycle, which is to be expected, as much of the technical aspect of SI is drawn from systems engineering. The differences are significant in that SI places much more emphasis on systems architecture, systems management, and risk management than does the typical systems engineering project.

14.3   SYSTEMS INTEGRATION MANAGEMENT AND TECHNICAL SKILLS AND TRAINING REQUIREMENTS

SI has come to impact all markets in virtually every business sector in the world: ADP systems, life-support systems, defense systems, information management systems, and so on. Technological advances in hardware and software virtually assure that every entity will find it necessary to migrate to new technologies, not once, but two or three times in the course of the next decade simply to maintain a competitive position in their own sector. This means that the entire technology industry infrastructure will be remodeled in the next 10 years as new technologies, advanced processes, and new approaches are introduced. Traditional SI programs of the 1980s have given way to microsystems integration of the early 1990s, which will give way to cooperative processing by the end of the decade. Technological changes are occurring on a grand scale with ever shrinking times to bring new technologies to bear on the market. A comparison of the activities of a system integrator in contemporary applications with those of the future is depicted in Table 14.2.

14.3.1 Functional Activities of Systems Integrators

Systems integration, as a professional practice, requires the practitioner to merge information from several academic disciplines, as SI involves cross-disciplinary activities. Some of the functional activities that are done by SI professionals involve the following:

- Conduct general studies of needs to realize improved system performance.
- Develop detailed specifications and designs.
- Conduct risk studies and implement risk minimization strategies.
- Perform system analysis and design.
- Develop hardware and software design.
- Employ project planning and control.
- Perform business management and accounting.
- Develop and nurture relationships with customers and subcontractors.
- Develop hardware design and specification.
- Carry out configuration management.
- Accomplish testing.
- Implement technology based solutions to business needs.
- Train users of new systems.

14.3.2 Systems Integrator Knowledge Requirements

To accomplish these duties and the many other responsibilities of the systems integrator, many different kinds of knowledge are required. Some of the more important knowledge areas that are essential for professionals working in SI are shown in Table 14.3.

TABLE 14.2 Comparison of Systems Integration Activities

Current SI Activities	Future SI Activities
Batch	Client server (distributed Servers)
On-line	Human metaphor
Homogeneous	Heterogeneous
Long life cycles	Short lead times
Instructor-led training	User controlled training
Paper-based self-studies	Computer-based point of need
	Integrated performance support

TABLE 14.3 Knowledge Areas Needed for SI Professionals Working in Systems Integration Programs

Engineering Skills	Management Skills
Computers	Leadership
Networks	Conceptual design
Software	Information engineering
Industry and application familiarity	Staff supervision
Communications—in all forms	Planning and tracking
General business concepts	Technologies
Strong learning skills	Industry and applications
Risk analysis	Risk management plan
	Accounting and finance
	Contract administration

In the present time-frame, 1990 to 2000, we find that businesses are in a basic defensive posture, as competition has continued to shrink margins on manufactured products (cf. Apple Computer, IBM, AT&T, Kodak, and other companies' reasons for downsizing). At the same time, margins on services have not been impacted in the same way. Competitiveness in service industries is greater than ever; however, profit margins have been only slightly impacted. Several trends are apparent for companies that are dependent on technology and information processing either as a service or as a product. These trends indicate that these areas are undergoing a significant downsizing, with reductions of up to two-thirds in many instances. This leads to significant reengineering of existing systems and the eventual replacement of systems that are significantly out-of-date and beyond recovery. These trends have implications beyond user industries and impact vendors and the entire technology infrastructure. This infrastructure will require remodeling to accommodate new technologies and the subsequent ways in which industries view them. Another significant trend is the globalization of the technology service industry. Global competition is the rule rather than the exception. SI and network integration have a worldwide impact and are of concern for all markets. The implications of these trends on the knowledge skills of SI professionals are great. It will no longer be feasible for the systems integrator to "bootstrap" personnel, as formal training will be necessary to remain competitive. The range of the knowledge areas continues to increase, thus leading the way to requiring teams of SI personnel to do these large, complex engineered systems jobs, rather than have one or two "experts" available for consultation and the remainder of the work done by engineers and managers not trained in SI.

14.3.3 Systems Integration Strategies for Success

For SI management, some of the strategies that will be necessary to compete in the SI world of the future include:

- Strict control of overhead
- Increased knowledge of client businesses
- Proactive management to control costs and increase market share

Overhead Must be Controlled and Reduced by a Significant Factor. Up to a one-half reduction is required in many sectors, to enable SI corporations to meet competitive challenges from worldwide competition.

Selective Targeting of Opportunities and of Systems Will Become Necessary. As companies are increasingly obligated to have intimate knowledge of their clients' businesses and needs for technology and information processing, it is essential that they know and understand the client domain and markets.

Proactive Interactions with Clients Are the Required Norm, Rather than Reactive Postures to Market Systems Integration Programs. It is no longer acceptable to be known as the "best" in your field and wait for business opportunities to come to your door. In most situations, this will require retraining of personnel to acquire new skills in SI activities of all kinds.

The question for all systems integrators, buyers or sellers, is how clients are looking at SI solutions to make decisions as to what to do, when to do it, and who will perform the necessary activities. Every technology program must justify its need based on meeting specific user business objectives and knowing the connection between investment and business objectives. A strategic evaluation of the investment must show the potential benefits for making the commitment. Risk must be examined and evaluated and an independent assessment must be made. Finally, the benefits to the business before, during, and after program implementation must be made. Ideally, the out-of-pocket costs will be less than the accrued benefits of the SI program.

Competitive pressures, the reduction in the number of large-scale programs, and increasing interest in projects of this type by firms redirecting their efforts from traditional work requires that we examine the effectiveness of our processes and procedures for carrying out SI programs. This includes activities such as examination of the ways in which we develop systems and ways in which we state improvement goals and measure progress toward these goals. TQM is covered in much detail in Chapter 7.

14.4   SYSTEMS INTEGRATION STRATEGY FOR SUCCESS

To be successful in an SI effort it is necessary that the client and the corporation have a well-articulated strategy for development and implementation of the SI program. This strategy must be business-oriented, cost-effective, and provide the best return to the client for the investment made. It is highly desirable to have a strategic plan in place prior to the development or acceptance of an SI program. From the onset it is necessary to know how to proceed with SI development beginning with requirements specifications for the client and with the proposal response on the part of the corporation. In general, the position of the client may be reflected as the mirror image of the corporate position in that the client must be prepared to administer and monitor the contract once it has been let, to manage and execute the effort. The client and the corporation must have a strategy that is as well developed and managed as the corporate program for success in the SI program.

The initial need, on behalf of the client, is for the preparation of proper specifications that accurately reflect the needs of the user as known at the time of concept formation. This is followed by the preparation of an RFP based on these requirements. Upon receipt of the RFP, the corporation interprets the requirements and prepares a proposal response to the user specifications as provided by the client. When responses to the RFP are received, the client must have proper evaluative criteria and processes by which to assess the validity of proposals to be able to make an award. Both client and contractor must be skilled in SI for an effective program to be completed.

The basic activities that are undertaken for implementation of SI are issue identification, issue formulation, and issue resolution. For an in-depth treatment of these topics, please refer to Chapter 24. These three activities are to be completed for each of the phases of the life cycle. Since requirements assessment is the initial phase, we examine the application of this process to this phase.

- *Issue Identification.* The initial task of requirements review has as its purpose a deep analysis of all of the requirements to determine if there are latent problems within any of the individual requirements, with combinations of requirements, and finally the entire set of requirements. These problems are identified, cataloged, documented, and archived in a database that is maintained intact as version 0 for this SI project. This issue identification process is the first essential step in establishing and embedding an audit trail for our SI project.

- *Issue Formulation.* The next step is issue formulation. Having identified the problems with the requirements as received from the corporation and classifying these as issues, we begin the process of issue formulation. Issue formulation consists of examining the issue in detail and coming to an understanding of the rationale behind the issue and formulating this understanding into a series of pertinent questions, determining potential impact on the SI project, and establishing the level of risk related to problem resolution and also in the event we are not able to resolve the issue. A record of the issue formulation is maintained in our same database as for issue identification, and these two aspects are explicitly coupled one to the other for the life of the project.

- *Issue Resolution.* The third step is to move to issue resolution. Issue resolution means that we must examine each of the issues that we have identified and formulated and seek a means to resolve each of these. Issue resolution has implicit risk factors tied to each and every possible path that is open for solution. Where there are multiple paths to effect a solution or where there is apparently no solution available at the time, an options field matrix is developed that becomes part and parcel of the audit trail. This matrix provides us with a quick review of the various alternatives, gives us a sense of the degree of risk involved, and establishes a sense of the probability that the solution path will lead to an approved resolution of the issue by the client. Issues are then resolved.

What is required is a means of achieving these basic activities during the SI program. The embedding of these activities is essential to the successful strategy for SI implementation.

14.4.1 Rules for Success

A successful SI strategy may be developed by following the 12 "rules for success" given below:

1. *Have a Commitment by Client and SI Organization.* A commitment by the client and the SI organization to the SI project and the identification of "champions" in both groups to administer the project is needed.
2. *Understand the Objectives of the Program and How to Measure Progress Toward these Objectives.* The capability to assess progress and determine and assign appropriate metrics to ascertain SI project status is needed.
3. *Have an Overall Strategic Design Plan.* An overall technical strategy for design development and implementation is needed.
4. *Have an Overall Management Strategy.* An overall **management** strategy for SI implementation is needed.
5. *Have a Quality Assurance Strategy in Place.* A methodology that provides quality assurance and quality control to all aspects of the program is needed.
6. *Provide for Personnel Preparation.* Adequate preparation and training of personnel responsible for all aspects of the SI project is needed.
7. *Have In-place Cost Management and Control Procedures.* A well-defined sense of cost awareness and control structures that have been tested and are operational is needed.
8. *Have a Plan to Manage Subcontractors.* A plan to manage subcontractors and provide interfaces to other major players in the project is needed.
9. *Have a Plan for Handling Exigencies.* A plan that determines the way to give consideration as to how to handle constraints and variables related to requirements that interface with existing systems, and conduct impact assessments of costs related to these requirements is needed.
10. *Have a Risk Assessment Plan.* A plan is needed that enables identification and management of risk from the beginning of the project that continues through to delivery.
11. *Have an Action Plan for Development of an Audit Trail.* An audit trail is essential to the success of an SI program in that it provides the necessary linkages across all aspects of the program from specifications to final test and implementation.
12. *Have a Performance Evaluation Plan.* The final success of any SI program is the way the installed system performs. A performance evaluation plan is necessary from the receipt of the initial requirements to assure that the architecture, design, and implementation meet client needs.

These 12 rules for success represent a commonsense approach to implementation of an SI program. The prescription is relatively easy to give, but difficult to follow. Shortcuts appear easy to take. However, if we do not address each of these areas, we increase the vulnerability of the program to a variety of difficulties ranging from not meeting performance requirements, to increased risk exposure, to simply not doing the job properly. Corporate strategy must be sensitive also to the needs of the client and to the activities of subcontractors.

14.4.2   Strategy Implementation

Both a top-down and a bottom-up approach will generally be needed and used for SI projects. The top-down approach will be primarily concerned with long-term issues that relate to the structure and architecture of the overall system. The bottom-up approach will be concerned with detailed designs and with parallel efforts to make existing systems more efficient and effective such that they can be potentially incorporated into an overall integration concept for the newly implemented system. The overall integration design must be such as to take into consideration various existing hardware and software that are not subject to change (system-level constraints), and potentially other initiatives that modify the overall integration concept (system-level constraints and variables). This concept must be consistent with both existing and evolving systems.

One of the central concerns of a strategic plan for the SI effort is the system-level architecture for the overall concept. While this aspect is provided at a general level by the client, it must be subjected to careful and consistent development by the corporation. Conceptually, this might appear as in Figure 14.2. It should be explicitly noted that the operational system architecture will, more often than not, be distributed spatially and temporally; and what is depicted in Figure 14.2 is a simplistic representation of system-level architecture.

To be successful in the understanding and development of technical approaches to the resolution of SI issues, the team assigned to system implementation and integration should have a keen understanding of principles for the design of integrated systems that are often temporally and spatially distributed as well as multipurpose. They also need to have systems management capability to assist in the technical direction of the work of others, especially subcontractors, should this be required. Another of the primary systems engineering and technical direction activities of the SI team involves architectural design of the overall system and integration of new requirements into the overall system. The central activities in SI project implementation are shown in Figure 14.3. Of particular importance here are the many feedback loops that enable learning through experience, and the resulting improvement in the overall methodology for design implementation and evaluation as the experiential knowledge base grows.

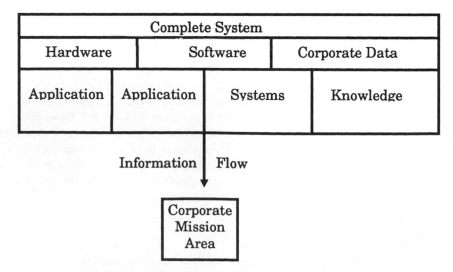

Figure 14.2   A conceptual perspective of the system architect.

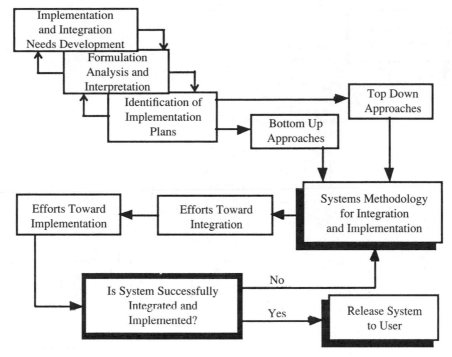

Figure 14.3 Typical effort flow for SI implementation.

14.4.3 Steps to Follow for Implementation

The technical direction of various projects and subcontractors involved in the program or other corporations with whom interfaces are needed requires the necessary interactions to assure successful integration of new capabilities into existing systems. The efforts associated with these activities are shown in Figure 14.4.

The development of an appropriate strategy by the client and the corporation carries with it benefits that accrue to each, plus it can be used to conduct useful benefit/cost assessments related to progress on the SI project. Please refer to Chapter 6 for a discussion of cost management.

There are a number of technology and management issues involved in systems integration. Technical issues generally involve the conduct of impact assessments and the preparation of systems engineering reports on architectural changes and associated integration concerns, by the SI team. The SI team should also be capable of providing systems management support relative to technical and scheduling decisions. These will generally involve life-cycle costing studies of possible acquisition strategies and configuration management studies as they specifically relate to implementation needs. From this information, early benefit/cost studies of appropriate strategies should be conducted.

14.4.4 Implementation and Integration Activities

The best strategy is to have an action plan in place prior to a positive "go ahead" given by the client. The SI organization will generally be tasked to prepare design specifications

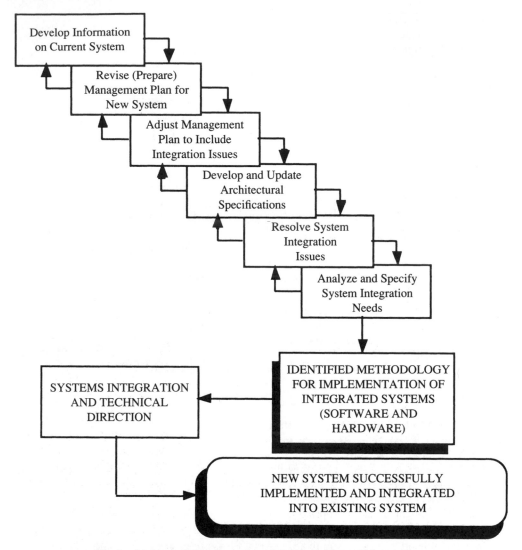

Figure 14.4 Activities in systems implementation and integration.

and to make a recommendation concerning the methodology based upon the client-developed requirements that will be used to accomplish implementation. This will include operational test and evaluation according to client-prepared and -approved operational test and evaluation plans. Figure 14.5 illustrates these implementation and integration activities from the perspectives of technology and management.

Corporate strategic plan approaches should include situation assessment that results in definition of problem areas in terms of the needs, constraints, and variables specifically related to system implementation. With respect to implementation, there will generally exist a baseline configuration that provides both a reference point and a set of constraints for the existing system that is to be modified through augmentation and installation of the

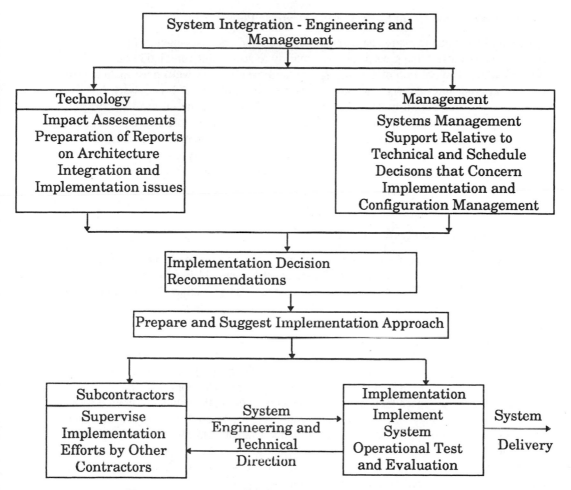

Figure 14.5 Technological and management issues in implementation.

new system. Thus, another necessary first task must be to identify potential alternative augmentations, additions, and modifications of the existing system.

This leads to the next aspect of our overall strategic plan for SI: impact assessment related to constraints from existing systems. The analysis and impact assessment activities are those in which the impacts of identified alternatives on the performance of the resulting operational implementation will be determined. This step should allow for some refinement of parameters within the assumed architectural structures for the alternative implementations, such that each of the alternatives is optimized for best performance. These alternative systems and processes should be capable of being interfaced and interoperable with the appropriate existing systems, integrated with other systems, and compatible with still other systems. There is a major need for an evaluation and review methodology to verify and validate the software, the hardware, the human interfaces, and the trustworthiness of the resulting system.

14.4.5 Risk Management as Part of the Strategic Plan

The overall SI strategic plan must include provisions for risk assessment. The greater the risk for the SI project, the greater the vulnerability to successful completion on time and within budget. The position of the corporation (and the client, as well) must include processes and procedures that examine and assess the impact of risk. It is necessary that risk assessment potential from major subcontracts be built in from the onset of the project. Otherwise, we are left to the business of retroactively attempting to create an audit trail, an awkward position. Thus, a major part of our overall SI position must be concerned with risk impact assessment and the ability to complete the necessary program on time and within budget. Risk management is such a major activity that it is covered more completely in Section 14.9.

14.4.6 A Systems Integration Example

Let us examine a specific instance to illustrate the importance of an appropriate client and SI organization position with regard to SI. Particularly, we will look at the case of a SI problem in which the client failed to provide the necessary guidelines for the SI program and the SI organization did not have a strategic plan for dealing with SI programs. The client developed a "felt" need to replace an existing computer information system that was critical to the day-to-day operations of the client. The client provided a basic description of the desired system improvements and placed the contract with the SI organization and proceeded to distance itself from any interaction with the development process. As the SI organization had no strategic plan formulated to guide the development of SI programs, they accepted the proposed "felt" need requirements as being the actual considered needs of the client. Further, the client mode of operation required no interaction as a condition of the contract. Here we see all of the ingredients of a formula for failure: unstated requirements on the part of the client, no follow-up process on the part of the client, and no SI strategic planning on the part of the SI organization. As the project developed, many other indicators were present to presage failure.

Following a difficult development program that necessitated the resolution of many knotty problems, such as conflict across requirements, redundancy of requirements, requirements ambiguity, and requirements imprecision; the SI product was completed and delivered. Typical of the requirements that were included was that the system interfaces be "user-friendly." Since there was little contact between client and SI organization, the systems integrators took it upon themselves to define what the client intended for the system interfaces to be user-friendly. In this instance they provided for the use of pull-down menus with help screen availability from the pull-down menu. As might be expected, the client had no capability for inclusion of a mouse on the system and was totally put off by this solution. This was only one small example of the kinds of dilemma found by both client and SI organization. Other major problems related to conflict between requirements such as the client requiring high maintainability while also giving equal weight to the necessity of operation on multiple hardware sets. Imprecision was included by the use of terms such as "high" for maintainability and "ease" of use. These, and terms of a similar nature, would have been routinely resolved with a well-defined client and corporate strategic plan with regard to SI programs.

The result of the total effort was a program that was not wanted by the client and a totally frustrated SI organization that was told not to bother to bid on other client programs. The bottom line for the participants in this example were that with neither the client nor the SI organization having an SI strategic plan, both were unhappy with the outcome. Neither the client nor the SI organization had considered SI as being other than a common ordinary project, with the resulting consequence of a program that did not meet the need of the client and a frustrated SI organization that would never again be permitted to bid on work from this client.

We all are aware of the successes, as well as the horror stories, that accompany the folk lore surrounding SI programs. If we know of a success and the SI organization did not have a good strategic plan in place to carry out the SI program, we either know a situation where the instincts of the principals were excellent, or we have a case of blind dumb luck. If we know of horror stories, the likelihood of the SI program being conducted in accordance with a well-defined strategic plan that was followed is minuscule. There is no sense in leaving the process to chance. It is far better to make the initial investment in the development of a strategic plan than to expose the program to the risks involved and increase the possibility of failure.

14.5   THE AUDIT TRAIL

Implementation of an audit trail and following internal rules, processes, and procedures are essential for good design decisions, actions related to procurements, and definition of constraints. Embedding an audit trail as part of normal procedures will enable realization of the potential benefits that accrue as a result of incorporating this process. What follows is a step-by-step process for development of an audit trail that will enable assurance that it has been properly done.

The first step in an ideal process for embedding an audit trail into an SI project would be to define and develop a strategic plan for implementing an SI business. Since the ideal seldom takes place, it is necessary to describe how to prepare and utilize an audit trail in light of real-world working conditions. We show how to embed the audit trail within the existing conditions that typically prevail at either the client or SI organization.

14.5.1   Audit Trail Process

Establishing an audit trail and embedding this information in the SI process begins immediately upon receipt of the SI contract. Several tasks must be accomplished at the onset of the SI program to ensure that an audit trail is properly developed and embedded in our SI program plan. These essential tasks are to:

1. Review the requirements in detail that we have accepted from the client.
2. Track program progress.
3. Resolve problems that occur.
4. Assign metrics to determine when these problems have been satisfactorily resolved.
5. Establish requirements for subcontractors to procure, deliver, and install the items called for in the subcontract.
6. Document all of this in a database.

The process of embedding the audit trail begins with the three-step process presented earlier that is conducted for each of the activities listed in the SI life cycle:

1. Issue identification
2. Issue formulation
3. Issue resolution

These three steps form the activities in which we must engage to establish and maintain an audit trail of significant events, issues, and decision paths. This sequence of three steps is then repeated for each of the phases of the SI project life cycle with appropriate entries in the audit database. For example, if the next phase in the SI life cycle is design of the intended system solution, we again establish the set of issues that confront us, formulate these issues, and propose an options field matrix for alternative paths to successful issue resolution. A unique database entry is made for each of the issues as it is generated, formulated, and finally disposed of by resolution or by the issue remaining an unsolved problem. Metrics generated for each of the issues are identified so that we can be certain that we know when we have reached a successful resolution and how we know that we have succeeded.

In a recent study performed on requirements generated by several government agencies, it was found that persons responsible for the development of system-level requirements frequently utilize words that are imprecise and ambiguous, provide requirements that are incomplete, and generate conflicting requirements.[5] Problems such as these may be found and resolved early on if we have provided for the implementation of a process that includes issue identification, formulation, and resolution. There are many examples of successful application of this approach to resolve fundamental difficulties with system-level requirements, just as there are many examples of failed SI programs that did not pay adequate attention to resolution of issues until it was too late. The best approach is to examine each of the requirements to identify issues, follow this with issue formulation, review resolution alternatives, and work with the client for final satisfactory resolution of latent issues. If the requirements are correct, the likelihood that the delivered system will be adequate rises by orders of magnitude. This should be continued throughout the development of the program.

14.5.2 Documentation and Indexing

The next audit trail activities are to document significant events that take place during the development process, specifications that we have developed for procurements, and special problems that have arisen due to client constraints that appear to impact the procurement processes. For example, if the activity deals with procurement of software and hardware from subcontractors, we establish the potential issues that might confront the subcontractors, formulate our understanding of them, and provide an options matrix for solution fields for our use in evaluating the various alternative solutions proposed by subcontractors. In this way we have built-in measures as to how to evaluate subcontractor submissions on predefined parameters that are other than price and performance. As before, each of the

[5]This work was done by the Center for Software Systems Engineering of George Mason University.

entries into our audit trail database carries a unique identifier that is retained by this item through the life of the SI project. This procedure provides us with total traceability for each of the items through the entire life of the program.

In this way we are able to index and sort the items in the audit trail database for reference, support of our actions, and in defense of options from which we made choices relative to solution paths. This database may also be exercised to assist in the determination of the degree of risk and the probability of success for any of the alternative paths that we have selected for our approach. This is accomplished by examination of the alternatives and by placing a probability of success on each for the options. Thus, it is a valuable tool for audit trail purposes, and it becomes an invaluable tool for management functions relative to understanding why and how we have arrived at various conclusions relative to alternative paths in design, procurement, and issue resolution.

14.5.3 Steps to Take to Embed an Audit Trail

The way to embed an audit trail in an SI program, either by the client or the SI organization, is to

1. Document system-level requirements into a database;
2. Establish whether or not issues are present and, if so, resolve them;
3. Assign appropriate validation and verification metrics;
4. Establish guidelines for subcontractors;
5. Provide solid unambiguous procurement instruments; and
6. Track these activities from specifications through installation and operation.

Effective traceability of all SI activities requires that we know with great confidence and accuracy the actions and consequences of such actions that we have taken during the course of conducting the SI project. This means that we must have an audit trail to document and describe the fundamental soundness of our activities. The audit trail is perhaps the most critical activity in which we engage in developing our process control document. The audit trail is no less than the entries in our SI "bookkeeping general ledger" that enable us to declare what, how, when, and why we have performed a specific act. Without this we are always in the difficult circumstance of trying to recreate what has been done to whom, when, for what reason, how we did it, and why.

This information should prove invaluable to proactive clients and SI organizations interested in much better performance the very next time that we are called upon to do an SI (or other related job) project. See Chapter 2 for a discussion of proactive management.

14.6 QUALITY ASSURANCE IN SYSTEMS INTEGRATION

To be able to indicate how to perform and document the quality assurance process, we first must be able to define quality attributes for the system, assess the affect of quality on system performance, and finally determine the metrics that are necessary for evaluation of quality. Quality is a very subjective term, a multiattributed one as well, and one that means different things to different observers. For example, the client views quality assurance from a very different perspective from that of the SI organization for the very simple reason that the client seeks the very highest indicators for performance, while the SI organization may well be satisfied by the system being adequate.

14.6.1   Definitions of Quality Assurance

There are many definitions of quality and quality assurance, and we will examine only two of the more popular ones that are in current use. For example, DoD Standard 2168 (U.S. Department of Defense, 1987) provides a very simple definition of software quality:

> Software quality is the degree to which the attributes of the software enable it to perform its specified end-item use.

A companion definition for software quality assurance plans has been developed by the IEEE; it is used in their Standard for Software Quality Assurance Plans (1986):

> Quality Assurance is a planned and systematic pattern of all actions necessary to provide adequate confidence that the item or product conforms to established technical requirements.

For the most part, these are very reasonable definitions. Each contains the notion of a metric to indicate the degree of quality or degree of conformance to the requirements of the user or client. Neither is as specific as might be desired relative to the need to measure the quality of performance and processes for the SI project. As systems engineers on the SI project, we must be concerned with an engineering design interpretation of quality assurance.

For example, we need appropriate metrics or indicators of systems quality to be able to

- Obtain an early-warning indicator of potential difficulties.
- Make appropriate design changes early in the SI development life cycle.
- Monitor the progress of the risk management plan.

Although not explicitly mentioned, it is our belief that the notions of conformity with standards of good practice, legality, and so forth, are intrinsic to these definitions. Also inherent in these definitions of quality assurance is the notion of *testing* as the primary tool of systems quality assurance.

For example in the development of software, this includes, as has been noted by Beizer (1982), the notion of *software design for testability*. Similar statements have been made for hardware and for the entire system integration activity. To do this requires appropriate metrics. It is not good enough to identify attributes of subsystems or hardware or software functionality; we must have attribute measures or metrics of total functionality as well. While an attribute need not necessarily be quantifiable, an attribute measure should be. Thus the term attribute measure is synonymous with metric.

Quality assurance indicators should lead to the detection of errors, if any, and the diagnosis of these, such as identifying a means of eliminating the problem. Correction of specification errors is, however, an activity that should be performed by a very different group of people than those testing for quality assurance.

Just as is the case for embedding the audit trail, a quality assurance plan is a critical pacing element for performance and documentation of the quality assurance process. This leads to the notion of a three-stage process in quality assurance, and subsequent maintenance efforts. Quality assurance will be primarily concerned with *detection* of the existence of faults, *diagnosis* of the type and location of the fault, and *correction* of the fault.

14.6.2 Quality Assurance and Testing

Quality assurance and associated testing can be conducted from either a structural, a functional, or a purposeful perspective. From a structural perspective, the system would be tested in terms of microlevel details that involve attributes such as hardware processor input–output (I/O) lines, throughput capability, cycle time, programming language, control, and coding particulars. Other quality assurance indicators would include system reliability, availability, and maintainability as well as factors such as interoperability or usability. From a functional perspective, quality assurance and testing involves treating the system as a black box and determining whether the performance conforms to the technical requirements specifications. From a purposeful perspective, the system must be tested to determine whether it does what the client really wishes it to do. This is generally known as validation testing.

These perspectives are surely not mutually exclusive, and each needs to be employed in a typical quality assurance effort. There are problems in implementing each of these. Complete functional testing will often, perhaps almost always, be impossible in practice because this would require subjecting the system to all possible inputs and verifying that the appropriate output is obtained from each of them. There will not generally be sufficient time to allow this, and there will often be other difficulties also. One of them is that we will generally not have sufficient ability and experiential familiarity with a particular application to identify all possible inputs.

Notions of quality indicators are quite closely related to notions of process management and therefore require the development of sets of plans to be executed to ascertain system quality. There are several dimensions on which to develop a taxonomy of system quality indicators. We have already mentioned the notions of structural, functional, and purposeful assurance. External vs. internal quality indicators is another related dimension. The two are not independent concepts, as the external quality of a system is generally quite dependent upon internal quality as well as knowledge of the purpose to which the system will be put.

Test plans are required if we are to have any hope of measuring many of the important dimensions noted. Two of these are consistency and completeness. These measures are certainly internal quality indicators, and they influence functionality, which is an external indicator. Even these internal notions are subtle in that it is not at all always clear what many of them, such as completeness, mean. When we consider the great variety of ways in which systems can be incomplete, then we fully realize that the development of a *testable metric* for completeness will generally not be an easy task. What we do infer is that, for a complex system, it does not appear possible to separate internal from external factors at all. The external factors are difficult to establish since they involve client or user needs, and the environment in which the system is to function. Once the external factors have been determined, these must somehow be translated into internal factors which, in turn, influence the external factors.

14.6.3 The Role of Software Processes in Quality Assurance for Systems Integration

Generally, the interest is going to be initially concerned only with the *quality* of the system delivered in terms of usability of the product for an assumed set of purposes, which are often difficult to specify in advance. But, the quality of the end product is a direct function of the quality of the process that produced it. Given that the fundamental quality of most of the hardware components is quite high, it will often be the case that the quality of an overall system is much more dependent upon the quality of the software product than anything else. Since software development is primarily an information technology development, and since this is an intellectual activity with no real physical or material component, we see that *software processes may be the most important software product from design or management perspectives*. Thus, the plan must be tailored to address the outcomes of the software development processes, and this is dependent upon the SI corporation processes that are in place for this activity.

We also note that the interests change over time and this adds yet another dimension to our determination of performance and documentation of quality assurance processes. As time progresses, the interest in quality assurance may expand to include not only software product operation, but also software revision to meet evolving needs, and the transition of the software product to meet new needs in an efficient and effective way.

14.6.4 Infusion of Quality in Systems Integration

When we put all of these dimensions together and include the introduction of time as an important dimension, it leads to a hierarchically structured notion of quality assurance attributes. Useful rudimentary efforts toward quality infusion into systems development efforts provides us with a plan that involves three steps:

1. Identifying quality attributes important for a specific situation.
2. Determining importance weights for these attributes.
3. Defining and instrumenting operational methods of determining the attribute scores for specific development approaches.

Quality assurance involves those systems management processes, systems design methodologies, and development techniques and tools that act to ensure that the resulting product meets or exceeds a set of multiattributed standards of excellence. Since we have determined that our major problems stem from software development, we will examine software quality assurance approaches for performance and documentation of our quality assurance process.

14.6.5 The Role of Verification and Validation

Software quality assurance activities are generally related to those software *verification* and *validation* activities that are conducted throughout all stages of the software development life cycle. Software *verification* is the activity of comparing the software product produced at the output of each phase of the life cycle with the product produced at the output of the preceding phase. It is this latter output that serves both as the input to the next phase and as a specification for it. Software *validation* compares the output product at each stage of the software life cycle, occasionally only the final product phase, to the initial system requirements. Often these activities are performed by people outside of the software development organization, and the prefix *independent* is sometimes used in these cases.

Thus, verification seeks to determine whether the software product is being built correctly, while validation seeks to determine whether the right product has been produced from an assumed set of correct specifications. So, verification and validation are quality control techniques and a part of quality assurance. An external assessment of the developed and implemented software would generally need to be made to ensure that concerns of software product revision and software product transition are addressed, as well as that of software operational functionality.

14.6.6 A Process for Fusion of Quality Assurance in Systems Integration

Over the course of the development life cycle, SI programs typically undergo the steps listed below:

- Identification of need and specification of requirements
- Initial design and development
- Controlled introduction to a client
- Release to client
- Modification to meet evolving needs
- Transition to a new environment

To assure integration of the quality assurance (QA) activities in the SI program, the following steps should be taken.

1. Organize a QA team as a separate entity, reporting to the top management.
2. Establish a QA process that is to be utilized for all technical activities.
3. Validate the requirements and specifications with the client.
4. Monitor the progress of all aspects of the program.
5. Conduct the necessary tests to assure that QA is being met.
6. Provide a test regimen for performance evaluation and final delivery.

Attributes, and attribute measures or metrics, are needed that serve project management needs during initial design through controlled introduction of the system to the client. There are also needs at the corporate level of management to ensure that the product is ready to be released and transitioned to the new environment.

Upper-level management will often need quality indicators to be able to quantify the product development quality, and to compare different and potentially competing development options on the basis of cost and effectiveness. Metrics allow management to select appropriately from among the potentially competing approaches. The use of appropriate metrics allows for and supports the goals of better responsiveness to client needs.

This leaves us with the final task of documentation. Documentation should be entered and maintained in a quality assurance database that is constructed just for this purpose. This area is sufficiently important to provide for this extra measure of effort. This information could be developed and maintained together with the development process and recorded in the audit trail database. However, since these factors taken together often determine the success or failure of the SI project in meeting client requirements, it is often worth the extra effort also to maintain this database as a separate entity.

14.7 SUBCONTRACTOR MANAGEMENT FOR SYSTEMS INTEGRATION

One of the major problems that confronts SI organizations is how to deal with subcontractors to assure that these will perform the right job in a correct and proper way. The best way to assure this is to have a set of processes and procedures that relate directly to the activities of subcontractors. These processes and procedures derive directly from the strategic plan that serves as a guide on how to conduct business. We should expect and demand no less from our subcontractors. We have acted with due diligence to work with the client to understand the needs and requirements of the SI project, the constraints and variables that are established, and the limits of business conduct that are acceptable for the particular job under contract. We should impose nothing less than these requirements on all those who work with us. That is, we pass through all appropriate requirements that have been assigned us in performing on the SI contract.

14.7.1 Subcontract Management

So what does all of this mean to the office responsible for issuing contracts and subcontracts that enable us to meet the requirements of the SI project? We must line out careful unambiguous instructions as part of the subcontracts issued to subcontractors with metrics spelled out by which we are able to gauge both technical and on-time performance.

We incorporate this into the terms and conditions of the subcontracts. We must be careful to avoid the pass-through of undocumented risk factors, as we lose control of these once they are in the hands of others. Pass-through of known risk elements is natural, and a review process must be in place such that we are able to keep track of the progress in the resolution of risk items.

We have discussed how to implement and maintain an audit trail throughout the SI process and how to perform and document the quality assurance process. Each of these activities bears special significance on how we implement the SI approach with subcontractors that we engage for either assistance with the project or for procurement of hardware and software. Just as the client provides the corporation with a set of requirements that it believes to be representative of the actual needs of the user, the corporation must prepare a detailed set of valid requirements for subcontractors. Absence of a strategic plan on the part of a subcontractor should result in imposition of the SI organization strategic plan, especially those parts that relate to audit trail maintenance; risk identification, formulation, and resolution; and such management processes and procedures as we feel are essential for satisfactory performance on the contract or subcontract. We should in no way feel that we are imposing on subcontractors in assigning these responsibilities. It is in this way that we manage and reduce the potential for risk and increase the potential for success.

14.7.2 How to Deal with Subcontractors

The systems integrator has the responsibility of helping the subcontractor take a functional point of view of the organization and of all procurement efforts. It is the responsibility of the systems integrator to aid the subcontractor in erecting a parallel technical auditing process. In addition to interaction matters already discussed there are several other points to be made. These include the following.

1. *No Favored Treatment for Specific Vendors.* It is only human perhaps for clients and system integrators to have favorite vendors. These are companies or individuals within certain companies that have provided excellent service in the past. Perhaps a previous acquisition more than met specifications or was unusually trouble-free. Sometimes a particular marketing organization has gone the extra mile to be of assistance in an emergency. It is only natural under such circumstances that this favorable past impression might bias the client or systems integrator. Indeed, the concept of the favored client is a common one in the private sector. But this attitude is illegal and improper in government procurements. We want to emphasize here that we are *not* talking about collusion or conspiracy to defraud the government. It is entirely possible that on occasion that biased behavior could *benefit* the government. That is of no matter. It is illegal and not to be condoned.

2. *Timely, Accurate Client Reports.* Technical personnel, engineers, computer scientists, and the like, tend not to support active, timely reporting on progress to clients. They follow the mushroom growers approach to client interactions — "keep 'em in the dark and cover 'em with manure." That approach may work when things are moving well, but it runs the risk of forfeiting client confidence in troubled times. It seems better to report progress accurately and in a timely fashion, so that if slippages occur they are minor when first mentioned. Naturally the systems integrator should make every effort to stay on schedule, and if the schedule slips or a problem surfaces, the systems integrator should present the recommended solution at the same time the problem is first mentioned.

3. *Prudential Judgment.* Suppose the systems integrator has reason to believe that the client is unable or unwilling to handle setbacks in an objective manner. The parable of the king who "killed messengers who brought him bad news" would not remain current in our folklore if it did not have a basis in reality. Thus reports of delays and difficulties should be brought to the attention of top management rather than directly to the client. This is the sort of prudential judgment call that should be handled by the top management within your organization rather than someone at the operating level. It is suggested that the matter be brought to the attention of top management within the organization as soon as possible and in a calm, factual manner.

Management of subcontractors is of special importance for systems integration involving large complex engineered systems. It is highly likely[6] that multiple subcontractors will be employed by the prime contractor. Prudent management of these subcontracts is critical to the success of the SI program.

14.8   SUBSYSTEM INTEGRATION AND DELIVERY

There are a number of key activities that must be completed by the systems integrator to assure integration of the products provided by the subcontractors prior to test and delivery

---

[6]In the case of most procurements for such systems, it is mandated by the client that there be several subcontractors.

of the final configuration. Some of the more important activities that must be accomplished include:

1. Organization of overall team support for the subsystem integration and test activity, including personnel from various subcontractors.
2. Validate incremental deliveries as these are made by a subcontractor.
3. Preparation of the various subsystems for test and evaluation prior to integration to assure performance meets the stated specifications.
4. Integration of HW/SW subsystems from subcontractors with systems developed by the corporation and legacy systems of the client.
5. Monitor test activity and assure that all tests conform to the system testing regimens agreed to by the client.
6. Provide for both Alpha and Beta site tests.
7. Conduct necessary posttest activities to review outcomes with all concerned parties.
8. Conduct formal reviews and review documentation.
9. Provide for failure recovery and error correction in the event subcontractors are unable to meet design specifications.

The corporation must be able to demonstrate that it has gone about its business in a legal, objective, unbiased fashion. In large procurements it is often the case that outside contracts will be let for validation and verification and to develop and administer an audit trail relative to the prime contractor. The necessity for an external enterprise to create and follow a technical audit trail arises not so much from the need to respond to potential procurement difficulties as it does from a need to be able to demonstrate that an objective and unbiased procurement process was utilized. The systems integration acquisition strategy is shown in Table 14.4.

This Validation Test Document will contain a conceptual discussion of items such as the following.

- Traceability
- Potential conflicts and resolution procedures
- Risk analysis and management
- Consistency of requirements
- Potential ambiguities in evaluation procedures
- Testability

We discuss these items next.

1. *Traceability.* The fundamental requirement for the auditing component is traceability. This is a classic requirement in all of engineering and in scientific efforts. All work must be written up on a regular basis in a laboratory notebook, dated, signed, and witnessed. All engineering drawings must be inspected, dated, witnessed, and signed. In engineering construction only registered professional engineers inspect and approve drawings. This seems to be a reasonable precaution when lives may be at stake when using the finished product. While the traceability and validation aspect of computer software is not

TABLE 14.4   The Generic Technical Acquisition Strategy for a Systems Integration Viewpoint

Systems Integration Acquisition Strategy	
Specification Component	Auditing Component
I *Functional Architecture Concept* Establish the general technical capabilities (i.e., the client functional needs)	*Validation Test Document* *Conceptual Discussion of:* 1. Traceability 2. Conflict resolution 3. Risk analysis and management 4. Consistency 5. Ambiguity evaluation 6. Testability 7. Constraints 8. Feasibility
II *Technical Architecture Plan* Define the configuration categories	*Validation Test and Audit Plan* For each configuration category, name and describe the relevant characteristics that delimit the requirement
III *Technical Component Specifications* Define and select the configuration components	*Validation Test and Audit Implementation* For each configuration component, set down explicit functional and quantitative tests
IV *Contract(s)*	*Establish the operational requirements for validation and audit*

as formal and rigid as in conventional engineering, the trend is undoubtedly in that direction.

2. *Potential Conflicts and Resolution Procedures.* At the Validation Test Document level we do not identify specific technical conflicts and their solutions. At this highest level we expect only to see outlined the recommended procedure for resolving technical conflicts. This procedure should be formal, with a special form to be filled out if the conflict is not resolved at the first level discovered.

Informal resolution of potential conflicts is the purpose of frequent peer reviews of the system while it is under construction. Yourdon (1988) recommends this in his *data flow method* of design. But the idea of frequent peer reviews is a general tool and should be adopted in some form of team design or analysis. Peer review meetings should probably occur at least weekly, with any conflicts not resolved at that time being written up and forwarded to the first level of management. This should not be viewed as an additional burdensome administrative load; rather it is simply what a group leader would do automatically in a management-by-exception environment.

3. *Risk Analysis and Management.* Risk analysis and management is covered in Section 14.9.

4. *Consistency of Requirements. Consistency of requirements* would seem to be essentially similar to the previous issue of *conflict resolution procedures* and it may be taken

as so if convenient. We separate the two simply to indicate that consistency of requirements can be checked at the general level, whereas conflicts sometimes occur in an unfortunate application of requirements that are not of necessity inconsistent in themselves.

(a) *Potential ambiguities in evaluation procedures.* In effect a *conflict* is an error of commission, while an *ambiguity* is an error of omission. It is almost impossible to write a set of specifications for a complex system that is totally without conflict and ambiguity. Be that as it may, it is the job of system integrators to produce a set of specs that reduce ambiguity to a minimum, while at the same time remaining within the bounds of reasonableness as far as complexity goes.

(b) *Testability.* Testability is an absolutely necessary component. If the specification is not testable, it does not have scientific reality. Thus it is the job of the installation team or the validation component of the SI effort to require a feasible test scheme for each proposed specification. Some specs can be validated or tested by simple observation. One can count the entry ports or disk drives or what have you. But other specs are intrinsically impossible to complete until after final installation and break-in of the system.

The second level of the audit component is the Validation and Audit Plan. At this level the generic Validation Test Document produced in the first phase is refined and sharpened. For each configuration category, name and describe the relevant characteristics that delimit the requirement.

Then in the third audit component, Validation and Test Audit Implemenation, for each configuration component set down explicit functional and quantitative tests.

At the fourth and final audit level, within the contract RFP, establish the Operational Requirements for Validation and Audit.

(c) *Audit Reports and Sign-off.* In a previous section we set out the general auditing procedure to be followed, but in this area we wish to be more specific. The procedures just discussed establish the requirements for a complete audit trail, but only if the requirements are actually followed. Often in practice, however, reality is far from the theoretical ideal. For example, PERT and CPM charts are merely useless impedimenta if not maintained on a timely basis. We also know that documentation sometimes lags production by several cycles. Similarly, audit reports and sign-off will not be kept up to date and functional unless management insists. This is especially so in dealing with subcontractors and one can see why this is so.

A subcontractor is paid to produce one or more deliverables. Paper records of any kind seem to some subcontractors to be a nonfunctional and unnecessary expense. It may be difficult for a minor subcontractor to comprehend the overall SI strategy; besides, why worry about the big picture as such a minor player. Probably the simplest answer to this issue is implicit in the first sentence of this paragraph. Simply make the audit reports and sign-off part of the deliverable.

Beyond making audit records and other documentation part of the deliverables, however, we need to do more. We need to enlist the voluntary and willing cooperation of subcontractors, even the most anti-systems-oriented ones. We can do this by better communications. More complete and timely access by subcontractors to the current state of the overall contract should make them more aware of the importance of their cooperation.

14.9 RISK MANAGEMENT

In general, risk to a SI program originates from functional and nonfunctional requirements. Functional sources are usually determined by technical aspects of the project, such as architecture, design, performance, and test. Nonfunctional sources include parameters such as management, political climate, team capability, and cost. The assessment of risk is typically obtained by taking the product of two parameters: (1) the probability of occurrence; and (2) the severity of impact. Two of the four possible high–low value combinations of these two parameters are then treated as high risk: (1) high impact with high probability, and (2) high impact, even with low probability. Since both probability of occurrence and severity of impact are subjective evaluations in most systems integration activities, formal methods of risk analysis, for which significant amounts of data are available, are generally not applicable. In practice, what is substituted are decision trees, lists of items at risk, or no approach at all. However, many of the techniques that are developed in Chapter 3 are applicable to SI programs. The degree to which these may be utilized depends on the nature of the particular form of risk and the amount of information that is available at the time of assessment. The reader is referred to Chapter 3 for various methods of risk evaluation and management. In this section, we concentrate on risk and SI programs and ways with which such risk has been handled.

14.9.1 Approaches to Risk Management for Systems Integration

Several approaches have been utilized to address the detection and identification of program elements at risk. These approaches are all of a qualitative nature. For example, program elements at risk are identified through administration of questionnaires to program personnel or by personal experience of program personnel comparing another program to the current one or by utilization of a list of common elements that nearly always lead to risk as determined through a catalog of past risk experiences.

The use of the questionnaire is the most commonly used "formal" approach to the detection and identification of program elements at risk. A typical procedure would be similar to that developed by Boehm (1991). Program elements thought to be at risk are identified either by individuals who compile and rate such elements or through the administration of a questionnaire or both. Functional and nonfunctional sources are considered by the SI team for potential risk. Following this review, the top ten risk elements are placed on a list and a risk management team develops a risk management plan. Once the program elements at risk are determined and identified and the risk management plan implemented, risk amelioration commences. This is an iterative process: as additional program elements at risk are discovered, they are added to the list, the risk management plan is modified, risk amelioration commences, and risk discovery is continued. As the program elements that were determined to be at risk are resolved, these elements are removed from the list and other program elements added. This process continues until the program managers are satisfied that the degree of risk to the success of the program has become manageable or all risk issues that could be discovered through this process have been discovered and resolved.

While this approach, and others that use questionnaires to elicit program elements at risk, are capable of detection and identification of program elements determined to be at risk, it is often the case that program elements at risk will be missed and that some program elements determined to be at risk will be incorrectly identified. Thus, there is no assurance that all program risks have been detected and identified or that major high-impact risk elements have been identified. This approach also ignores the potential major impact of incorrect or improper resolution of risk on downstream development procedures. This approach is exclusively dependent on the knowledge of individuals who conduct the surveys of potential risk situations or on the knowledge of other persons who have a "feel" for the risk involved in the situation. The same assertion can be made for processes that employ questionnaires; that is, the process is no better than the individuals involved in the process. There also is no way to determine the ripple affect of high-risk program elements on other program elements with which they may interact nor are they repeatable.

These approaches offer a marked contrast to program risk detection and identification approaches such as those used in manufacturing or in hardware development. Here statistical analysis, sample data techniques, and random sampling techniques provide quantitative measures to the detection and identification of program elements at risk as well as measures of the potential ripple affect and impact on interacting components.

14.9.2 Components of a Risk Management Plan for Systems Integration

The potential sources of risk to an SI program are many and varied and run the gamut from deficiencies in requirements to failure to provide adequate controls on subcontractors. The approach that provides the most opportunity to ferret out risk and resolve these problems is one that addresses the topic at the beginning of the program. Thus, a major part of the strategic plan is the development of the means to address risk. The major components of the risk management plan are:

1. Identification of "at-risk" system components
2. Risk analysis
3. Risk avoidance measures
4. Risk management;
5. Internal processes and procedures to handle risk

The need for risk management techniques over the entire SI life cycle is illustrated by the conceptual diagram below in Figure 14.6.

For each of the activities, components "at risk" are identified, the risk aspects analyzed, the steps to avoid the risk and the ensuing consequences are taken, management of the risk is initiated, and internal processes and procedures are developed to address components at risk. In addition, the risk detection and identification plan is modified to incorporate similar occurrences of such risk, if these are not already included in the plan. The risk management plan, as part of the overall strategic plan for the SI program, begins with an analysis of the requirements at the onset of the program to ascertain if there are requirements statements that could jeopardize successful completion of the program. The risk management plan continues with risk assessment for each of the phases of the SI life cycle, as depicted earlier in Figure 14.1.

One of the most vexing problems in risk management is the early identification of potential causes of risk. This is especially true in the development of large complex life-support systems and for large systems integration programs that are heavily dependent on the integration of legacy systems and newly developed requirements. What has made this problem particularly difficult has been the necessity of using qualitative processes to at-

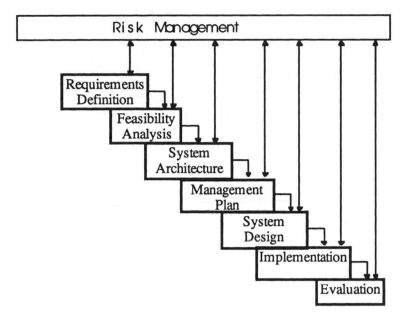

Figure 14.6 Risk management applied to the SI life cycle.

tempt to identify risk areas and risk situations. Risk detection and identification should commence with the issuance of requirements and the development of specifications. It is often the case that the risk assessment process is delayed until development of system designs or even until procurements of major subsystems. This is fundamentally an untenable situation, since by this point in a program, investments of resources and personnel have been made, designs have been developed, and it is much too late to achieve an economical and efficient recovery without significant rework. The impact and ripple affect due to program elements at risk becomes known only after the discovery of the nature and character of risk, thus jeopardizing the entire development program.

Consider the instance of system and hardware and software requirements that may be at risk. If these requirements are found to be ambiguous, in conflict, incomplete, or changing too much (requirements volatility), they may be considered to be a cause of risk to successful completion of the program. Any one of these sources may, in and of itself, be sufficient to jeopardize the entire program if not resolved. Thus, it is necessary to examine each of the risk categories in some detail to determine the level of risk involved. Following discovery of any statements deemed to be at risk, these statements would be analyzed, in consort with related statements, to determine the degree of threat to the program. Once this is complete, a risk avoidance activity is initiated and those statements that are considered to be a threat are modified to assure risk reduction to an acceptable level. These statements are then reviewed and the appropriate actions taken to either eliminate or reduce the risk potential. Finally, the types of errors detected are reviewed and if necessary, the risk plan is modified so as to assure detection of similar risk elements in other program activities. Prior to proceeding to the next phase, the modified statements are reviewed in the context of all other requirements (ripple effect) to determine if additional errors have been introduced as a consequence of the changes (which is very often the case). Thus, it is also necessary to provide a methodology to determine traceability between requirements and risk topics.

14.9.3 Traceability as Part of Risk Management

That traceability is clearly an essential capacity in SI to observe the relative interrelationships becomes quite apparent. For example, if a contention for resources was the detected conflict, a solution that eliminates access to the resource for one of the activities is no solution at all if there exists a need for both activities to access the resource. What is utilized to address this is a methodology to assess requirements in the context of *clusters* of similar statements for risk detection and traceability to related statements, which provides a trace within clusters.[7] This enables utilization of metrics in terms of a variety of attributes as ratios within clusters, across clusters, and with the set of requirements as a whole. Typical of the specific metrics are those that include the ratio of the number of statements in conflict to the entire number in the cluster, the number of ambiguous statements to the number of statements in the cluster, the number of statements undergoing changes compared to the number of total statements in the cluster, and the degree of cohesion within a cluster, and other ratios. These ratios are subsequently used to indicate the level of potential problems within and across clusters and relative to the entire set of requirements, as well. These clusters typically contain large groups of requirements statements for which intercluster coupling is typically low, but the cohesion within a given cluster is generally high.

For example, Client A had issued a set of requirements statements for bid purposes without first establishing whether or not there were any potential problems. Upon receipt of a large number of requests for clarification of the ambiguities, conflicts, and incomplete statements, the client determined that it should withdraw the bid documents and perform an analysis of the requirements for potential problems. The client had determined that the risk involved with continuing the RFP process was simply too great and the entire effort was in jeopardy even before work commenced by a corporation. Client A subsequently conducted a study of the requirements statements using the cluster-by-similarity approach and metrics noted previously and determined that of the seven hundred (700) individual requirements statements, some one hundred twenty-five (125) had problems of one kind or another. Using the metrics noted earlier, Client A calculated the ratios of errors to the set as a whole, determined the degree of severity of the problem, sampled individual statements, and decided to assign "RED Teams" to ferret out and correct the statements with errors. Following this work, the bid package was modified and the RFPs once again were sent out for response, this time resulting in several acceptable bids.

Clearly, risk management is a critical activity for SI programs. The very size and nature of the activities associated with SI lead to error-prone situations. Thus, a risk management plan is an essential part of the overall SI strategic plan.

REFERENCES

Air Force Systems Command. (1989). *Software Risk Abatement,* Pamphlet 800-45. Washington, DC: Andrews Air Force Base.

[7]Clusters are the groupings of requirements based on stated similarity criteria determined by the user.

Arthur, L. J. (1988). *Software Evolution: The Software Maintenance Challenge*. New York: Wiley.

Arthur, L. J. (1993). *Improving Software Quality: An Insider's View of TQM*. New York: Wiley.

Beam, W. R. (1990). *Systems Engineering: Architecture and Design*. New York: McGraw-Hill.

Beizer, B. (1982). *Software Testing Techniques*. New York: Van Nostrand-Reinhold.

Billington, R., and Allan, R. (1983). *Reliability Evaluation of Engineering Systems*. New York: Plenum.

Blanchard, B. S. (1991). *Systems Engineering Management*. New York: Wiley.

Boar, B. H. (1993). *The Art of Strategic Planning for Information Technology*. New York: Wiley.

Boehm, B. W. (1981). *Software Engineering Economics*. Englewood Cliffs, NJ: Prentice-Hall.

Boehm, B. W. (1989). *Software Risk Management: A Tutorial*. Los Alamitos, CA: IEEE Computer Society Press.

Boehm, B. W. (1991). *Software Risk Management: Principles and Practices*. New York: IEEE Software.

Charatte, R. N. (1989). *Software Engineering Risk Analysis and Management*. New York: McGraw-Hill.

Chittester, C., and Haimes, Y. Y. (1994). Assessment and management of software technical risk. *IEEE Trans. Syst., Man, Cybernet.* **24**(2), 187–202.

Conte, S. D., Dunsmore, H. E., and Shen, V. Y. (1986). *Software Engineering Metrics and Models*. Menlo Park, CA: Benjamin/Cummings.

Defense Systems Management College. (1989). *Risk Management: Concepts and Guidance*. Ft. Belvoir, VA: Def. Syst. Manage. Coll.

Fairley, R. (1994). Risk management for software project. *IEEE Software* **11**(3), 57–67.

IEEE Standard 983-1986. (1986). *IEEE Guide for Software Quality Assurance Planning*. New York: IEEE.

Jenkins, A. M., Naumann, J. D., and Wetherbee, J. C. (1984). Empirical investigation of systems development practices and results. *Info. Manage.* **7**, 73–82.

Kitchenham, B., Pfleeger, S. L., and Fenton, N. (1995). Towards a framework for software measurement validation. *IEEE Trans. Software Eng.* **21**(12), 929–944.

Meyer, B. (1988). *Object-Oriented Software Construction*. Englewood Cliffs, NJ: Prentice-Hall.

Nakajo, T., and Kume, H. (1991). A case history analysis of software cause-effect relationships. *IEEE Trans. Software Eng.* **17**(8), 830–837.

Palmer, J. D., and Myers, M. (1988). Knowledge-based systems application to reduce risk in software requirements. *Proc. Uncertainty Intellig. Syst., 2nd Int. Conf. Inf. Process. Manage. Uncertainty in Knowl. Based Syst.,* Urbino, Italy, *1988,* pp. 351–358.

Phan, D., Vogel, D., and Nunamaker, J. (1988). The search for perfect project management. *Computerworld,* September 26, pp. 95–100.

Rezpka, W. E. (1988). *A Requirements Engineering Testbed: Concept, Status and First Results*. Griffis Air Force Base, NY: Rome Air Development Center.

Rowe, W. D. (1988). *An Anatomy of Risk*. Malabar, FL: Krieger Publ. Co.

Sage, A. P. (1990). *Concise Encyclopedia of Information Processing in Systems and Organizations*. pp. 466–472. Oxford: Pergamon.

Sage, A. P. (1995). *Systems Management for Information Technology and Software Engineering*. New York: Wiley.

Sage, A. P., and Palmer, J. D. (1990). *Software Systems Engineering*. New York: Wiley.

Sage, A. P., and Lynch, C. L. (1998). Systems integration and architecting: an overview of principles, practices, and perspectives. *Systems Engineering,* **1**(3), 176–227.

Thayer, R. H. (ed.). (1987). *Software Engineering Project Management.* Silver Spring, MD: IEEE Computer Society Press.

Thayer, R. H., and Dorfman, M. (eds.). (1990). *System and Software Requirements Engineering.* Los Alamitos, CA: IEEE Computer Society Press.

U.S. Department of Defense. (1987). *Defense Systems Software Quality Program (DoD STD 2177A).* Washington, DC: USDoD.

U.S. Department of Defense. (1987). *Defense System Software Development.* Washington, DC: USDoD.

Weintraub, R. M., and Burgess, J. (1994). Drastic revisions are likely for air traffic control contract. *Washington Post,* April 11, p. A6.

Yau, S. S., Collofello, J. S., and MacGregor, T. (1978). Ripple effect analysis of software maintenance. *COMPSAC Proc.*

Yourdon, E. N. (1988). *Managing the System Life Cycle: A Software Development Methodology Overview,* 2nd ed. New York: Yourdon Press.

Zimmer, J. A. (1989). Altering with high changes. *J. Software Maint.* **1**(1), 36–42.

15 Systematic Measurements

ANDREW P. SAGE and ANNE J. JENSEN

15.1 INTRODUCTION

One major need in all of systems engineering and systems management, at all levels of the organization, is to obtain the information and the knowledge that is necessary to organize and direct individual programs associated with the production of products and services. These products and services result from the use of systems engineering process life cycles that are associated with research, development, test, and evaluation (RDT&E); systems acquisition, production, procurement, or manufacturing; and systems planning and marketing. This information can only be obtained through an appropriate program of systematic measurements and the development of appropriate models for use in processing this information. We discuss frameworks for measurements at the level of product, process, and systems management of process and product. Associated with each of these levels are a variety of methods of systems analysis and operations research, and metrics. There is not an abrupt transition from activity at one level to activity at another, nor should there be. Measurements are needed at each of these levels. In this chapter, we discuss approaches and frameworks for measurements and models that support organizational success through effective use of innovations brought about in systems engineering and management and the related subjects of information technology and knowledge management.

Success in implementation of systems engineering for organizational support, or systems management efforts generally, is critically dependent upon the availability of appropriate measurements. *Management and measurements are irretrievably interconnected* since:

1. Organizational success is dependent upon management quality;
2. Management quality depends upon decision quality and organizational understanding;
3. Decision quality and organizational understanding depend upon information and knowledge quality and appropriateness;
4. Information quality depends upon measurement quality and appropriateness; and
5. Knowledge quality depends upon information quality and appropriate experiential familiarity with the situation and task at hand to enable formation of an appropriate context for the information that is obtained.

If we believe each of these five assertions, then it is hard to escape the conclusion that appropriate measurements are very important.

This chapter, derived in part from the discussions in Sage (1995b), provides a number

Handbook of Systems Engineering and Management, Edited by A. P. Sage and W. B. Rouse
ISBN 0471-15405-9 ©1999 John Wiley and Sons, Inc.

of foundations for instituting a program of systematic measurement that is supportive of enhanced organizational success. This success will occur because of

- The improved quality of the delivered systems and products.
- The improved quality of the processes used to develop these systems and products.
- The improved predictability of product and system performance.
- The improved schedule and cost accuracy and predictability.
- The much enhanced customer satisfaction that is thereby made possible.

While systematic measurements and associated metrics have a great many uses, there are abuses as well. There is a danger of allowing the medium of measurements to become the message itself. When the measurements or metrics, or objectives measures, become sublimated for purposeful objectives, all sort of horrors can result. When this happens, for example, a quality software product is deemed to be one that scores high on verification and validation tests. Verification is an interactive approach designed to ensure that the software product as it evolves through various phases of the production life cycle is as faithful as possible to the specifications for that phase as determined by the planning for that phase that resulted from the previous phase. Validation is the activity of comparing the output of each phase of the life cycle, sometimes only the result of the final development phase, with the initial system specifications. Unless the technical system specifications are a faithful translation of user requirements, however, there is no assurance that the system produced will satisfy user requirements. Focus on the metrics as the objective, rather than on the objective of fulfilling user needs, results in a sublimated objective that is not necessarily associated with production of a software product that satisfies the customer, but one that will score highly on the inspection-oriented test. Many illustrations of sublimation of objectives by objectives measures could be provided.

Many have addressed the need for systematic measurements in organizational settings. Harrington (1991), for example, indicates well the case for measurement at the level of systems management when he states that "to measure is to understand, to understand is to gain knowledge, to have knowledge is to have power." A six-phase process called an *opportunity cycle* is suggested to allow for corrective action through measurement. This involves assessment and detection of difficulties, analysis and diagnosis of the causes of the difficulty and potential corrective actions, implementation of a chosen corrective action, action to prevent recurrence, and measurement throughout. These phases are not atypical of the general life-cycle phases associated with systems engineering and management efforts described throughout this handbook, especially in Chapter 1 and in Sage (1992a,b, 1995a,b). In this chapter, we first describe some needs for, and the associated purposes and objectives for, systematic measurements that support the use of information technology innovations in enhancing organizational success. We discuss the central role that metrics play in organizational performance assessment. These measurements are necessarily associated with models, and some of our efforts relate to models for effective measurements.

15.2 ORGANIZATIONAL NEEDS FOR SYSTEMATIC MEASUREMENT

Managers have long recognized the need to measure the performance of their organizations. Often, there is a major concern relative to what to measure and for what purpose.

More recently, in light of increased competition and the global economy, managers have also recognized the need to measure relevant performance facets of not only their own organizations but also those of their suppliers, customers, competitors, and the business environment. The same questions relative to measurement purpose arise here. Much organizational research addresses these needs (Christopher and Thor, 1993; Drucker, 1954, 1966; Kaplan and Norton, 1993, 1996a, b; Peters and Waterman, 1982; Sink, 1985; Sink and Tuttle, 1989; Struebing, 1996). The needs appear simple. To address competitive concerns, many prescriptions suggest that we should gather needed performance information, and use that information to improve performance through such programs as total quality management (TQM). These programs are intended, in part, to transform organizations into entities that continually, systematically, and proactively use measurement to assess their performance, identify areas for improvement, and evaluate the results of improvement interventions.

Development of an organizational performance measurement and reporting system that provides appropriate information requires thorough analysis of the organization, its performance goals and objectives, and its measurement needs and capabilities. Practitioners must identify performance improvement goals, determine how the performance of each organizational component should be measured, and decide how the performance information should be managed, what should be reported, how it should be reported, and to whom it should be reported. They must then decide what measurement and reporting capabilities can be implemented in a feasible manner within existing organizational constraints, and which provide the information necessary to support continual organizational performance improvement. This process can be unwieldy due to the volume of information that must be collected and analyzed, and can be further complicated by continual changes in the internal environment of the organization and the external environment in which it operates. Information is not necessarily the equivalent of knowledge, as is stressed in Chapter 30, and it is important that information accessed be that most relevant to the production of useful and applicable knowledge.

Many organizational specialists point to the need for collecting, reporting, and utilizing organizational performance information; yet little guidance has been found to assist managers in determining what information is needed, for what purpose, and how it should be obtained and used to support knowledge management. Subsystem-level measurement and improvement approaches are useful, but may result in subsystem-level optimization at the expense of the larger system. Organization-wide performance improvement approaches, awards, and certifications such as TQM, the Deming Award, the Baldrige Award, ISO 9000, and the Software Engineering Institute's Capability Maturity Model are potentially also very useful. These are primarily prescriptive frameworks, however, and do not specifically provide guidance for the tailoring and inevitable tradeoffs that must be performed. Few organizations have unlimited budgets with which to plan, implement, and manage the ultimate organizational performance measurement and reporting system. Instead, appropriate solutions—ideally, optimum solutions—must be developed that leverage existing resources and that allow for tradeoffs between competing organizational performance measurement needs.

Great opportunities for enhancement of organizational performance measurement are provided by recent advances in information technology that enable effective management of the environment for knowledge production and integration, or knowledge management. Many organizations now have extensive information technology infrastructures that include client–server systems, integrated software systems, databases, and communications capabilities. The databases in these infrastructures often contain vast amounts of accurate and readily available operational, financial, and personnel information that can often be easily and inexpensively leveraged for organizational performance measurement purposes. The communications infrastructure can also often be leveraged for the collection of additional performance data and the ultimate distribution of performance measurements. To leverage these capabilities and facilitate the metrics identification and analysis process, a methodology is needed that incorporates information technology and knowledge management considerations.

15.3 MEASUREMENT NEEDS

It is appropriate to ask several questions concerning a proposed set of metrics. The answers to these questions should enable us to determine if the proposed set of metrics is efficient and effective for the intended purpose. We generally seek to determine whether identification of near-optimal organizational performance metrics sets is enhanced by the development of an information technology and knowledge-management-oriented methodology that incorporates analysis of information technology factors and that facilitates the analysis of metrics needs vs. information considerations. Two prototypical question and an associated hypothesis for each are as follows.

Question 1: Will the use of a methodology that emphasizes information and knowledge oriented metrics considerations and use of the needed and existing information technology resources result in the selection of near-optimal metrics sets?

Hypothesis 1. Use of a methodology that requires users to analyze information-oriented considerations and existing information technology resources will result in the selection of metrics sets that are near optimal (complete, balanced, implementable, and cost-effective), and that will result in measurements that are accurate and reliable.

Question 2: Will the use of a methodology that incorporates information technology and knowledge infrastructures in the selection of metrics result in recommendations for replacing the computerization of steep hierarchies with more efficient, effective, and flatter networked organizations?

Hypothesis 2. The systematic measurement methodology used requires users to identify the existing information technology and knowledge management infrastructures, and to specify how data for each metric will be collected. This will result in increased recognition of the need to move to a networked organization where information and knowledge is readily available to those who need it.

These questions and the associated hypotheses need to be specifically tailored to the purposes to be achieved by the systematic measurements and, generally, a number of other relevant questions and hypotheses will be needed as well. Hypothesis testing is one of the activities necessary for answering the questions associated with evaluation of the effectiveness and efficiency of a proposed set of systematic measurements. Definition, development, and deployment of an information technology and knowledge-management-oriented methodology for identifying, analyzing, and suggesting improvements in appropriate and useful performance metrics is a need that, when fulfilled, greatly enhance the value of systems engineering and management efforts. Such a methodology will include generic metrics categories, lists of metrics, and such measurement system considerations as the means by which performance data are collected, verified and validated, and converted into information and knowledge. It will also include information management considerations such as the physical and/or electronic means used to archive and retrieve information, and information distribution and presentation considerations such as the physical, verbal, electronic, or other means by which performance information is disseminated. It will also include the knowledge management considerations discussed in Chapter 30.

To manage the many factors involved in identifying an appropriate and useful, ideally optimum, organizational performance measurement system, and to accommodate the continual changes in many of these factors; a methodology is needed that portrays organizational performance measurement in a measurement process maturity context. As an organization gains performance measurement experience and develops insights into its performance drivers, its measurement processes will mature. The purpose of its management efforts, the performance facets it measures, and how it measures performance, will all change over time, and the processes by which it evaluates and refines these factors will mature. Over time, measurement efforts should evolve from reactive to interactive to proactive, which should result in increased effectiveness and possibly decreased costs. They should also evolve from the more obvious measures associated with the organization's products, processes, and finances to more sophisticated measures that address the key performance indicators that directly impact long-term organizational success. Incorporation of such a process maturity context in the methodology in the form of a *systematic measurement advisor* (SMA) will assist users in

- Addressing requirements resulting from formal measurement process documentation.
- Periodic review of long-term organizational goals and objectives.
- Continual review and revision of the current optimal metrics set.
- Management of the evolution of the entire organizational performance measurement process.

Emphasis will be placed on identifying opportunities to leverage existing information technologies such as networks and corporate databases, and such newer information technologies as executive information systems and client–server systems. There is a substantial amount of information that must often be collected as data, and processed as information and knowledge, in order to measure broad-scope organizational performance. Leveraging information technology and knowledge management capabilities for performance measurement efforts, and coping with these organizational realities is a critical requirement for the success of large-scale, organization-wide measurement systems.

The contemporary organizational performance improvement literature indicates that only systematic, well-planned, well-implemented, and well-managed organizational performance measurement and improvement efforts will succeed. Organizations that implement TQM, reengineering, or similar approaches only as "quick fixes" usually fail to improve their performance. Much more than a quick fix is needed. The literature also indicates that economic globalization, geographic dispersion, the decreasing duration of business cycle times, reductions in personnel and other resources, and the proliferation of information technology necessitate modifications to the more traditional organizational performance measurement and reporting approaches in use today (Davenport, 1997). It appears that much effort should be placed on systematically identifying those metrics that

- Are strongly associated with the organization's mission and goals.
- Can support decisions that result in the greatest improvements at the lowest cost.
- Make optimal use of performance data already available within, and outside of, the organization.
- Reuse existing information technology infrastructures where feasible.
- Strongly associate information needs with knowledge management objectives.
- Help organizations be responsive and innovative in light of the many rapid changes in the business environment.

There are four needs to be addressed in defining, developing, and deploying an information technology and knowledge-management-oriented methodology for the analysis and selection of organizational performance metrics.

1. A measurement-process maturity-oriented life cycle should facilitate the evolution of the metrics set and ensure that it is focused on the organization's goals and objectives, and critical core capabilities.
2. An organizational performance metrics taxonomy should facilitate the process of metrics definition, development, and deployment.
3. A metrics-set refinement process should facilitate the identification and acquisition of a complete and well-balanced metrics set that represents an effective and efficient transition from the existing set of metrics.
4. An automated tool may be developed to facilitate use of this process and associated algorithms.

The first need is for a life cycle for metrics-set definition, refinement, and implementation that supports an evolutionary, goal-oriented approach to the development and continual refinement of metrics and measurement systems. A key critical success factor for large-scale organizational performance measurement systems is that the measurements be capable of being continually refined in an adaptive manner as organizational characteristics change. A structured life cycle for metrics-set refinement would facilitate the metrics identification and acquisition processes and support the evolution of the documentation describing them. The second need is for taxonomies of organizational performance metrics. Many organizations use the measurement requirements from awards or certifications such as ISO 9000 and 14000, the Deming Prize, the Baldrige Award, and the Capability Maturity Model. These metrics and the associated measurement requirements provide good guidance and encourage measurement of a broad array of performance facets in a systematic manner. However, organizations must appropriately tailor these generic measures for their unique measurement needs and must focus attention on a limited set of metrics that can provide the most beneficial and cost-effective information. Generic taxonomies of metrics should facilitate the metrics identification process. The third need is for algorithms that can be used to assess the degree of optimality and acquirability of the metrics set as it is evolved. Here optimality is defined as the completeness and appropriate balance of many metric characteristics including organizational level, purpose, diversity, cost-effectiveness, reliability, and ease of implementation. Acquirability is defined as the ability to gain the required data, processing and reporting capabilities required to implement the metrics set. These algorithms should emphasize parameters related to desirable measurement and reporting system characteristics such as the reuse of existing information technology (IT) repositories and capabilities. The fourth need is for an automated tool, such as the SMA, as described earlier in a conceptual manner, to facilitate use of the methodology and management of measurement system evolution.

There are a number of needs for, and approaches to, systematic measurements. An individual or an organization can approach systematic measurements from at least four perspectives:

1. *Inactive.* Inactive denotes an organization that does not use metrics, or that does not measure at all except perhaps in an intuitive and qualitative manner.
2. *Reactive.* Reactive denotes an organization that will perform an outcomes assessment, and after it has detected a problem, or failure, will diagnose the cause of the problem and, will often get rid of the symptoms that result from the problem.
3. *Interactive.* Inferative denotes an organization that will measure an evolving produce as it moves through various phases of the life-cycle process in order to detect problems as soon as they occur, diagnose their causes, and correct the difficulty through recycling, feedback, and retrofit to and through that portion of the life-cycle process in which the problem occurred.
4. *Proactive.* Proactive measurements are those that are designed to predict the potential for errors and synthesis of an appropriate life-cycle process that is sufficiently mature such that the potential for errors is minimized.

All of these perspectives on measurement purpose are needed. All but the first are appropriate, and there are perhaps very rare occasions when one might even justify an inactive approach that involves no measurements at all. Such might be the case, for example, in an organizational environment dominated by opportunistic political considerations (Mintzberg, 1989; Sage, 1992a).

Inactive and reactive measurements are associated with organizations that have a low level of process maturity (Paulk et al., 1995). As one moves to higher and higher levels of process maturity, the lower level forms of measurement purpose become less and less used. In part, this is so because a high level of process maturity results in such appropriate metrics for systems management that final product errors, which can be detected through a reactive measurement approach, tend to occur very infrequently. While reactive measurement approaches are also used at the higher levels of process maturity, they are not at all the dominant focus of measurement. In a very highly mature organization, they might be only needed on the rarest of occasions when occasional problems manifest themselves and corrective efforts are needed.

The metrics needed in any given situation, as well as the needed methods for evaluation and assessment may correspond to any of three levels for systems engineering—product, process, and systems management—as suggested in Figure 15.1. These metrics may be very

- *Product oriented,* in terms of inspections and quality control of the product itself.
- *Process oriented,* in terms of life-cycle evolution and configuration management of

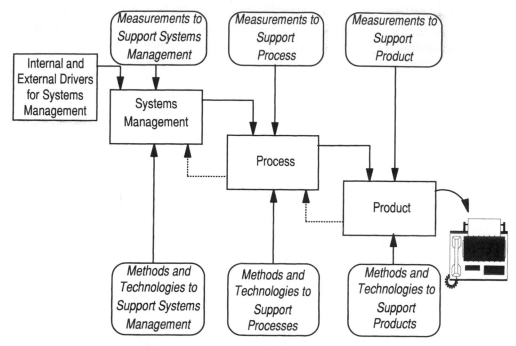

Figure 15.1 A three-level structure of systems engineering.

the product, such as to lead to operational-level task quality assurance for the process and the evolving product.

• *Systems management oriented,* in terms of strategic-level quality assurance and technical direction of the process.

It should be explicitly noted here that the use of the word "product' is intended to imply both "hard physical product" as well as "soft intangible and possible intellectual product or service." Thus the term product is intended to infer both conventional products and services. Note also that the metrics at these three levels of product, process, and systems management, are generally also:

• *Reactive*—product-oriented metrics that catch such defects as bugs in the programs code, and dysfunctional microelectronics chips.

• *Interactive*—process-oriented metrics that attempt to assure real-time control of quality through operational-level assurance at the process level, such as, for example, through verification at each life-cycle phase of the system development process.

• *Proactive*—in order to yield prospective and predictive control of quality such as to result in the engineering of a process that can be expected to result in delivery of a trustworthy and high-quality operational product with high effectiveness, high product differentiation, and minimal development cost.

It is quite possible and desirable to comment briefly on connections between this three-level view and the notions of TQM, such are as discussed in Chapter 7 of this handbook.

This is important, as system or product quality is a major concern today. Most of the approaches advocated in the past have been at the levels of operational and task management for product improvement. While necessary, these approaches will often not be sufficient. Also required will be efforts at the level of strategic systems management. This will generally result in a set of strategic quality assurance and management plans and these, in turn, will enhance process quality at the level of systems management, or management control (Anthony et al., 1992). The term *total quality management* is generally understood to refer to the continuous improvement of processes and products through use of objectives and associated measures with which to determine objective satisfaction, an integrated systems management approach to a life-cycle process of product development through process-oriented organizational teamwork and associated process-related methods to achieve customer satisfaction (Sage, 1992a, b, 1995b).

Virtually all TQM efforts, which are discussed in Chapter 7, suggest several major interrelated efforts that make up a systems approach to strategic quality enhancement. We might summarize them here as follows:

1. Know and understand customers, consumers, and clients and their motivations and needs.
2. Know and understand the internal and external environment of the organization.
3. Concentrate on continuously improving systems products and processes.
4. Identify appropriate process and product metrics for quality assurance measurements.
5. Do the right things right in the sense of focusing on doing the important things that help achieve meaningful organizational objectives.
6. Then and only then, after these have been identified and the issues formulated, do these "smart" things efficiently and effectively.
7. Exercise administration, management, leadership, and governance in a professional systems management setting that emphasizes equity and explicability, and which is attuned to the cultural realities extant.
8. Develop and use appropriate proactive systems management and measurement approaches, such as to empower people for maximum organizational achievement.

We will find a great deal of use for systematic measurement in virtually all systems engineering and management efforts.

One essential feature of TQM is the notion of the *totality* of the approach. An additional and essential feature is that it is fundamentally related to process improvement, and though this to product and service improvement. It involves efforts at the level of strategic planning, management control, and operational and task control. It involves approaches to process and product, and major focus on problems and issues addressed by the product. It is important to note that measurements, or metrics, are an essential feature of TQM.

This advocacy of measurement might and should be contrasted with the often stated view of Deming (1986) that management by current statistics and associated measurements is devastating to the goals of TQM. He strengthens this further with the admonition to eliminate numerical production quotas for the workforce and numerical financial goals for management. This seeming contradiction is not one at all, we believe. Deming is, we believe, strongly cautioning against reactive- and inspection-oriented metrics. Instead, proactive metrics at the level of systems management are much to be encouraged. These proactive metrics plan for quality and develop appropriate metrics for this, as contrasted with the reactive inspections that ultimately serve to eliminate the unfit. Further problems with the reactive measurement approaches, especially when used exclusively, is that they impose barriers that ultimately rob people of pride of work quality. Further, they encourage the sublimation of objectives to objectives measures, and they encourage us to seek approaches that work only in the sense of maximizing objective measurement attainment, rather than the desired maximization of attainment of objectives.

Thus, we get to the nexus of many difficulties relative to innovation and productivity in general, and innovation and productivity relative to systems engineering products, processes, and management in particular. These processes are strongly related to group and organizational issues, and they primarily concern very important and often neglected issues that relate to organizational culture, and leadership. They relate to such important behavioral issues as how leaders create, imbed, and transmit organizational culture, and thereby bring about process-related changes that lead to development of an intelligent, knowledge-based enterprise that will enable successful coping with the challenge of organizational change for productivity enhancement. This relates strongly to organizational architectures that result from reengineering the organization. The resulting corporate culture and performance enables the reengineering of work through information technology to enable process innovation and profiting from innovation. Thus, we should necessarily be very concerned with this subject of organizational culture and leadership for process-related improvements through systems management. This subject also relates well to managing complexity in high-technology organizations through the strategic use of systems-engineering- and systems-management-based approaches. The enabler for much, if not all, of this is the development of systemic measurement approaches to enhance organizational development.

One of the difficulties, we believe, with much past and present measurement practice is that there is a separate measurement department that is in itself a closed system. In effect, these people become measurement czars. Measurements, in such an environment, are often used for punitive purposes rather than to encourage teamwork and cooperation. In addition, there is often a paucity of models supporting even the data that are obtained. So, there is no meaningful way to provide the model-based management (Blanning, 1992) of data necessary to convert it into meaningful, or actionable, management information in the form of organizational knowledge. This suggests five realities:

1. Context-free data collection generally serves no purpose in that much of the data collected will be useless, and needed data will often not be obtained.
2. This leads to great disrespect for the data collectors and their wasteful, context-free, efforts.
3. Model-directed data collection is a real need, and model-based management of this data collection is needed as well.
4. There is a major need for effective systems management of the organization to obtain relevant, useful, and economic collection of measurement data such that these data can become similarly useful and appropriate information and knowledge.
5. There also needs to be appropriate dialog generation and management such that

management has available to it the sort of presentations that are supportive of organizational knowledge that will lead to high-quality decisions.

This suggests ultimate development of something like a decision support system to aid management in the best use of systematic measurements in the form of an SMA, and this is what was described conceptually earlier in this section.

There are a number of purposes for which measurements are critical:

- Enhancing our understanding of the functional performance of the organization;
- Identifying external threats and opportunities;
- Identifying internal strengths and weaknesses of the organization;
- Identifying the need for changes at the levels of systems management, processes, or product;
- Improved quality of delivered systems and products;
- Improved quality of the processes used to develop these systems and products;
- Improved predictability of product and system performance;
- Improved schedule and cost accuracy and predictability; and
- Maximized customer satisfaction

There is a major need for models and model management of data to provide the actionable information for management decision and control, priority setting, and organizational steerage.

There are a number of objectives for systematic measurement. A rather incomplete set of these include:

1. To Identify potentially successful products, systems, and services, and the associated needs for RDT&E and acquisition.
2. To translate these needs into requirements for RDT&E.
3. To translate these needs into requirements for systems acquisition and production.
4. To transfer or transition successful emerging technologies from the technology base to the systems acquisition base.
5. To establish a risk management program for strategic planning and systems management, as well as for process and product development.
6. To determine the extent of customer and stakeholder satisfaction with these products.
7. To verify that the products of systems acquisition meet the requirements set forth in systems planning and marketing.

The following two precepts support quality and quality enhancements through systematic measurements:

1. Quality should never be "inspected in" at the end through reactive-based testing only, except as a last resort when all else has failed, as this is a very expensive and ultimately debilitating approach.

2. Quality is built in from the start through a program of proactive and interactive systematic measurements.

On the basis of these objectives and a preference for a balanced measurement approach that utilizes appropriate proactive, interactive, and reactive measurements, we know *why* we should measure. We are also in a position to discuss *where, when,* and *what* should be measured and *who* should accomplish the measurements.

We can take ordinal or qualitative measurements, such as observing that a software product functions with no errors, or we can make cardinal or quantitative measurements, such as measuring the average number of errors per thousand lines of code. Generally, and for most uses, cardinal measurements are preferred. We can measure efficiency, effectiveness, quality, and risk-related facets of efforts at the level of systems management, process, or product.

We measure because we wish to make continuous improvements. We measure all along the various organizational life cycles. We measure as soon as it becomes meaningfully possible to do so. We measure all aspects of systems-management-, process-, and product-related activities. Systematic measurements are best made by the individuals who have major reasons to know the results of measurements and who can utilize them to enable continual improvements.

Systematic measurements are not cost free. Thus, there is a need to select a systematic measurement program that is effective and efficient, and explicable as well as one that yields results that are viewed as equitable by those whose products, processes, and systems management efforts are subject to measurement. These may be measured through reactive, interactive, and proactive approaches. In addition, we may have metrics that are associated with structure, function, and purpose. So, we can imagine a three-dimensional representation for metrics, such as shown in Figure 15.2. This should be viewed as a conceptual

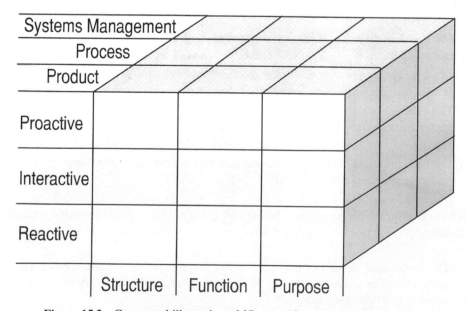

Figure 15.2 Conceptual illustration of 27 types of systematic measurements.

picture only in that the boundaries between the 27 different measurement types shown in the figure are not at all sharp, as they appear to be in this figure.

We believe that there is a natural flowthrough from improvements at the level of systems management to improvements at the levels of processes. There is also a natural flowthrough from improvement efforts at the level of process to improvements at the level of product. There are more major advantages to improvements at the level of product that result from generally proactive improvements at the level of systems management than there are from generally reactive efforts that are implemented direct at the product level. As represented in the illustration, enhancements to cost-effectiveness result from improvements at the level at which the metrics improvements are implemented, and at lower levels.

There are a large number of competitive advantages that an organization can possess and a large number of associated competitive strategies that it can pursue. Metrics need to be associated with relevant facets of both competitive structure and with competitive strategy. It is the blend of competitive advantage with the chosen competitive scope that leads to a competitive strategy. Fundamentally, an organization can possess competitive advantage (Porter, 1985) because of being a *low-cost* or a *high-product differentiation* producer. It may adopt a competitive scope, through market strategy, based on either a *broad focus* or a narrow, or *niche focus*. In most situations, an organization should select a single competitive advantage and a single competitive scope. If the organization is a low-cost producer, it will be especially concerned with maximum efficiency, or minimum production effort and scheduled production time. On the other hand, if it gains competitive advantage through high product differentiation, it will be concerned with maximum product effectiveness and quality. If an organization's competitive scope is in a broadly focused market arena, it will be very concerned with minimum product defects, since the large production volume would make it extraordinarily costly to accomplish broad scale repairs. If the organization is concerned with a narrow-focused market, it should be very concerned with maximum customer satisfaction in the specialized niche market area in which the product competes.

15.4   ORGANIZATIONAL MEASUREMENTS

Obviously, a mature organization is concerned with each of the approaches to success just delineated. Yet, it should be anticipated that one of the first two competitive advantage approaches, and one of the latter two competitive scope approaches will represent the competitive strategy adopted by a given organization. Even within this set of four possible strategy pairs, one of the approaches within each pair may dominate the other for any given organization. Thus we suggest that a given organization can focus on one of these product-quality-related measurement efforts as reflective and supportive of its prevalent approach to competitive strategy. This does not at all suggest that it neglects the other approaches, but that the prevalent focus is upon a single one of these four approaches. Nor does this suggest that it neglects attention to leadership, organizational processes, fast-paced innovation, or the development of a high-quality workforce.

These are not dissimilar from the observations of Grady (1992) in an excellent work concerned with software metrics at the levels of management and processes, and an earlier work concerned primarily with process and product metrics (Grady and Caswell, 1987). In the former work, Grady identifies three competitive strategies: maximize customer satisfaction, minimize engineering effort and schedule, and minimize defects. These strategies are associated with user satisfaction, productivity in terms of labor efficiency, and quality in terms of defect minimization. We can also identify cost-effectiveness as an additional competitive strategy. This can be viewed as a composite and blend of the other three strategies. Neither of the three, or four, strategies are independent, and there are dangers in attempting to focus on one to the exclusion of the other.

The seven major characteristics of each of these strategies are identified by Grady as:

1. The major organizational factors driving each strategy.
2. The time in the overall product life cycle when the strategy is most effective.
3. The essential characteristics and features of the strategy.
4. The most visible and useful metrics.
5. The organizational entity most likely to be responsible for the strategy.
6. The organizational entity most likely to be in contact with the customer.
7. The potential risks if the strategic focus is too restricted to only the selected strategy.

Associated with these strategies there should also be an identification of complementary but secondary developments of strategies to ensure that the risks of having too restrictive a metrics strategy do not materialize.

In this work, a goal–question–metric (GQM) paradigm, initially due to Basili and Weiss (1984), and which has been applied to develop software maintenance metrics (Rombach and Ulery, 1989), was suggested and used to develop metrics for the specific single competitive strategy that has been adopted. It was also suggested that metrics from the ancillary competitive strategies be incorporated to avoid the risks associated with too narrow a focus. The GQM exemplar indicates who needs to know what and when, and why this knowledge is needed. The three principal steps involved in the GQM approach are suggested in Figure 15.3. The goals, questions, and metrics considered may be associated with the levels of systems management, process, or product. Three prototypical goal statements for each of these three levels might be to stay within budget; to maximize

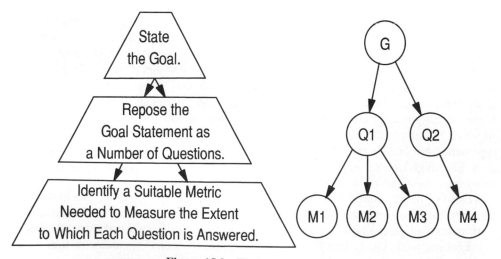

Figure 15.3 The GQM approach.

requirements stability and minimize requirements creep; and to meet product performance objectives. While it might seem, and correctly so, that these three goals are primarily associated with systems management, process, and product, we can attempt metrics at any level to improve upon goal satisfaction. Thus, we should determine the highest level at which we are concerned with improvements—management, process, or product. We then need to identify and to implement metrics that determine goal satisfaction at that level and at lower levels. The GQM approach would be implemented by associating these goals with an appropriate number of questions. Each of these questions would lead to a number of further questions. These questions can be posed at the level at which each goal statement is made, and at lower levels. Finally, each of the identified questions is associated with a number of metrics, or goal measures, such that it then becomes possible to measure the extent to which the goal statement is achieved.

One of the greatest challenges to developing a systemic approach for organizational performance measurement is identification of the specific purpose of measurement and, from this, the data that should be collected to fulfill this purpose. The historical evolution of organizational performance measurement systems provides some insights into how an appropriate set of metrics can be identified and acquired.

Organizational performance measurement has evolved dramatically during the 20th century. The types of metrics utilized, their complexity, and such considerations as time-liness and security have increased considerably. This information demonstrates how rapidly the amount and different types of information required by managers has increased. It also demonstrates that the major performance improvement standards require diverse sets of well-documented performance information and transition of this information into knowledge that serves performance improvement.

The types of metrics needed to assess organizational performance have changed dramatically during this century. Fredcrick Taylor's time studies and the Gilbreth's time-and-motion studies at the turn of the century are believed to be the first scientific organizational performance improvement efforts (Emerson and Naehring, 1988). These control-oriented efforts focused almost exclusively on the human as a machine, and used information to identify the most efficient movements each worker should perform. At that time, quantitative data that existed internal to the organization was used to calculate such efficiency measures as units assembled per hour. Subsequent organizational performance improvement advancements, such as Shewart's strategies for identifying and minimizing product variation, occurred in the 1920s. These strategies are based on statistical quality-control (SQC) methods that require:

- Collection of internal quantitative data for identifying process variation.
- Collection of external (from outside organizations such as suppliers) quantitative data for identifying variations in raw materials or equipment used in the internal processes.

Statistical limits are then calculated. When excessive variation occurs in the process and the limits are exceeded, efforts are initiated to identify and resolve the source(s) of the variation.

In the 1950s, performance improvement efforts began to include qualitative management issues such as worker motivation. This change is at least partially attributable to the growing size and complexity of organizations and recognition of the applicability of Maslow's Hierarchy of Human Needs (Barnard, 1938), which were discussed in the Introduction to this handbook, and the results of the Hawthorne studies and other psychological models. Performance improvement strategies emerging during this period heavily emphasized the insights and improvements made possible by including employee quality-of-work-life measures, as well as employee ideas and suggestions, in performance improvement efforts. Representative techniques include quality circles, process action teams, and brainstorming. These techniques emphasize the use of qualitative information in conjunction with quantitative internal and external information. During that period W. Edwards Deming developed his philosophy of total quality management, which mandates that every employee continually look for ways to improve performance (Deming, 1986).

The current business environment has resulted in a dramatic evolution in organizational performance objectives and associated measurement. This has involved not only TQM-related efforts, but also efforts that involve reengineering (see Chapter 23) and learning and knowledge management in organizations (see Chapter 30). One of the most fundamental advances is the recognition that performance improvement must be a highly structured, systematic effort that actively involves the entire organization. Practitioners also recognize that every organization must gather not only its own internal quantitative and qualitative performance information, but also external quantitative and qualitative performance information about its customers, suppliers, industry, and competitors. In addition, every organization must be able to use that information to assess its own performance, benchmark against other organizations, identify areas for improvement, formulate integrated improvement plans, and assess improvement progress. To do this effectively requires that information be capable of being interpreted as knowledge and, as a result, knowledge management is a major issue for the latter part of the current decade and (at least) the early part of the 21st century. Information technology and knowledge management approaches have tremendous potential for facilitating these efforts, as discussed in Chapter 30. This brief summary of information utilization for performance improvement efforts during this century illustrates:

- How the breadth and depth of information required for management of organizational performance has increased,
- Why the focus has shifted to the utilization of information as knowledge, and
- The resulting needs for effective knowledge management.

15.5 METRICS FROM WIDELY ACCEPTED STANDARDS, AWARDS, AND GOVERNMENT REQUIREMENTS

Many organizations design their performance measurement systems based on the organizational performance criteria that have been specified in national and international organizational performance certifications and awards (Christopher and Thor, 1993; Sage, 1992a, b; 1995b; Zairi, 1994). These criteria often reflect current knowledge regarding the facets of organizational behavior that are most important, and what approaches to improving these facets are most desirable from the perspective of the standards and awards.

There has been some debate over the benefits organizations reap while trying to win organizational performance improvement awards (Sashkin and Kiser, 1993). Some experts such as Crosby (1979) believe that companies that strive to win such awards are totally misguided. Some of the more visible performance improvement failures would seem to support this belief. In 1990, when Florida Power & Light (FP&L) became the first company outside of Japan to win the Deming Prize, it was used as an example of what can be accomplished with structured organizational performance measurement and improvement. During the years they spent trying to win the award, however, too much emphasis was placed on trying to win the award and not enough was placed on implementing sustainable performance improvements. By 1993, FP&L was in financial trouble and had sharply declining customer satisfaction ratings (Main, 1991). The Wallace company, a 1990 winner of the Baldrige Award, had to file for bankruptcy protection shortly after receiving its award. Wallace's chief executive officer reported that their emphasis on winning and then marketing their award contributed to their financial problems (Sashkin and Kiser, 1993). Senior managers also cited Wallace's overexuberance in fulfilling the Baldrige Award requirement that they share their knowledge with others, and the Texas oil-field economic problems of the early 1980s, as contributors (Clark, 1993). Many other experts believe that the rationale behind the national and international organizational performance certifications and awards is sound. They cite shortsighted emphasis on winning the awards, rather than improving performance, as a primary cause of many award-winning companies' failures (Mahoney and Thor, 1994; Sink, 1985). Organizations that implement well-planned performance measurement and improvement programs and manage them appropriately, regardless of whether or not they might lead to an award, often report significant and sustained performance improvements (Drucker, 1954; Kaplan and Norton, 1996a, b; Peters, 1987; Sink and Tuttle, 1989). The chief executive of Globe Metallurgical, Inc., the first small company to win a Baldrige Award, estimates that the company's quality efforts have netted a 40-to-1 return on investment (Clark, 1993).

To help increase the success of organizations trying to base their performance measurement and improvement programs on widely used awards or certifications, it is important (Sage, 1995) to tailor the evaluation criteria. This is so since it is possible for organizations to focus their improvement efforts in such a way that they receive outstanding evaluations, but do not actually improve their overall performance. To address this problem, it is suggested that organizations must identify their unique goals and objectives, and then tailor performance criteria to support them. While an organization's chosen set of performance criteria may be based on an award or certification, it must be of "a sufficiently broad scope and scale" and be tailored to such a degree that "it becomes increasingly difficult for a high score on an evaluation or audit to be non-indicative of high performance." This suggests that we must be very careful to not sublimate objectives to objectives measures.

Four widely implemented performance certification and improvement awards are discussed here, in terms of suggested metrics: the Deming Prize, the Baldrige Award, ISO 9000, and ISO 14000. All four of these awards establish organizational performance requirements that either specifically include, or imply, metrics requirements. The Deming Prize and the Baldrige Award are performance improvement awards; ISO 9000 is a quality performance certification standard; and ISO 14000 is a new environmental standard. The following subsections contain reviews of these four efforts and their metrics requirements. We also discuss some recent U. S. government requirements that call for metrics.

15.5.1 The Deming Prize

The Union of Japanese Scientists and Engineers initiated the Deming Prize in 1951 to honor W. Edwards Deming's contributions to their country (Walton, 1986). Its purpose is to honor "those companies recognized as having applied CWQC (Company-Wide Quality Control) based on statistical quality control," as well as individuals who have contributed to the advancement of statistical theory. While the award's primary emphasis is on statistical process control, awards are made on the basis of a number of criteria that fall within the ten categories, and the associated criteria, suggested in Table 15.1 (Mahoney and Thor, 1994).

As indicated in Table 15.1, the Deming Prize focuses primarily on the traditional quality program elements of documentation, records, control procedures, policies, and traceability; transformation and added value (production, service process activities); and process quality control, standards, quality results, benchmarking, and auditing. Deming Prize examiners can, and sometimes do, include the following topics in their evaluations: after-sale service; complaint handling; cost controls; customer opinion utilization; delivery performance; education and training; equipment maintenance; instrumentation and inspection; inventories; manufacturing processes; personnel and labor relations; product development and design; profits; quality assurance coordination; relationships with associates, subcontractors, suppliers, and customer companies; research; and safety (Mahoney and Thor, 1994). When company evaluations are performed, about 60 percent of the award criteria weight is placed on the organization's quality control processes, while the remaining 40 percent is placed on the results obtained as a result of such processes. Mahoney and Thor (1994) also note that the Deming Prize criteria do not include metrics for purchasing–procurement proficiency, contracting methods, supplier performance; handling, labeling, storage, and safety; packaging, handling, inventory procedures; marketing, distribution, delivery, installation; or customer service, customer satisfaction, guarantees, and warranties. However, many of these metrics categories are included in the Baldrige Award and ISO 9000.

Zairi's (1994) review of the metrics actually used by Deming Prize winners indicated metrics that relate to functional performance, presented in Table 15.2, and metrics that relate to management, presented in Table 15.3. These lists indicate that a wide variety of quality control metrics, and other metrics, are used to satisfy the Deming Prize criteria.

15.5.2 The Malcolm Baldrige National Quality Award

The United States' Malcolm Baldrige National Quality Award (MBNQA) was established by act of Congress through the signing of Public Law 100-107, the Malcolm Baldrige National Quality Improvement Act of 1987 and is named after former Secretary of Commerce Malcolm Baldrige, who, before his untimely death, lobbied extensively for the creation of a national quality award as a means of increasing U.S. competitiveness. The award is managed by the National Institute of Standards and Technology (NIST) and is intended "to promote awareness of quality excellence, to recognize quality achievements of U.S. companies, and publicize successful quality strategies" (NIST, 1995). Up to two awards are made annually in each of three categories: manufacturing companies, service companies, and small businesses. The MBNQA's criteria and weightings are revised annually as it evolves "toward comprehensive coverage of performance, addressing the needs and expectations of all stakeholders—customers, employees, stockholders, suppliers, and the public" (NIST 1995). These criteria and weightings have been criticized both for being too focused on processes, and for being too focused on results, and also as being more appropriate for organizations at lower levels of quality process maturity. Due to increased emphasis on the financial viability of award winners, and some sad experiences with failure of previous award recipients, the weighting has been changed to place greater emphasis

TABLE 15.1 Deming Prize Criteria

Category	Criteria
Analysis	Assertiveness of improvement suggestions Linkage with proper technology Propriety of analytical research Quality analysis, process analysis Selection of key problems and themes Utilization of analytical results Utilization of statistical methods
Collection, dissemination, and use of information on quality	Collection of external information Data processing, statistical analysis of information, and utilization of results Speed of information transmission (use of computers) Transmission of information between divisions
Control	Actual conditions of control activities Contribution to performance of quality control circle activities Control items and control points State of matters under control Systems for the control of quality and such related matters as cost and quantity Utilization of statistical control methods as control charts, other statistical concepts
Education and dissemination	Education of related companies Education programs and results Grasp of the effectiveness of quality control Quality control circle activities Quality and control consciousness and degree of understanding of quality control System for suggesting improvements Teaching of statistical concepts and methods and the extent of their dissemination
Effects (results)	Intangible results Measurement of results Measures for overcoming defects Substantive results in such matters as quality, services, delivery, time, cost, profits, safety, and environment
Future plans	Grasp of the present state of affairs and concreteness of the plan Linkage with long-term plans Measures for overcoming defects Plans for further advances
Organization and its management	Appropriateness of delegations of authority Committees and their activities Explicitness of the scopes of authority and responsibility Interdivisional cooperation Quality control diagnosis Utilization of quality control circle activities Utilization of staff

Table continues on following page

TABLE 15.1   Deming Prize Criteria *(Continued)*

Category	Criteria
Policies	Justifiability and consistency of policies
	Method of establishing policies
	Policies pursued for management, quality, and quality control
	Relationship between policies and long- and short-term planning
	Review of policies and the results achieved
	Transmission and diffusion of policies
	Utilization of statistical methods
Quality assurance	Actual state of quality assurance
	Equipment maintenance and control of subcontracting, purchasing, and services
	Evaluation and audit of quality
	Instrumentation, gauging, testing, and inspecting
	Procedures for the development of new products and services
	Process capability
	Process design, process analysis, and process control and improvement
	Quality assurance systems and its audit
	Safety and immunity from product liability
	Utilization of statistical methods
Standardization	Accumulation of technology
	Content of the standards
	Method of establishing, revising, and abolishing standards
	Outcome of the establishment, revision, or abolition of standards
	Systematization of standards
	Utilization of standards
	Utilization of statistical methods

on results. Of course, results is the response variable for such drivers as process. The 1998 MBNQA criteria and weightings are presented in Table 15.4.

These attribute weights can be used in a standard multiple-attribute utility assessment, such as discussed in Chapter 28. These Baldrige award criteria have sometimes been criticized for some of the chosen weights, in particular for the low weight assigned to the Information and Analysis attribute. A previous year's Baldrige award criteria descriptions have referred to Information and Analysis as "the focal point within the criteria for all key information to drive improvement of quality and operational performance. In simplest terms, it is the brain center for the alignment of an organization's information system with its strategic directions." In the 1996 criteria, however, Information and Analysis received only 75 of the 1,000 points, reduced from 80 in the previous year. Clearly internal and external information gathering and analysis, automated information management systems, and information utilization are key drivers that impact the success each organization experiences in the other six evaluation categories. This suggests that the Baldrige Award attributes are not mutually exclusive. Nor are they collectively exhaustive. In the 1998 version of the Baldrige Award criteria, Information and Analysis is appropriately depicted as part of the framework that integrates the other criteria activities. However, the weighting

TABLE 15.2 Functional Metrics for the Deming Prize

Indicator	Subindicator	Metrics
Cost	Cost reduction	Cost reduction
		Customer cost per defect amount
		Failure rate
		New product mass production start-up cost
		Target cost attainment condition
	Rationalization	Delivery price reduction
		Inventory turnover
		Rationalization amount
Departmental activities	Marketing capability	Sales target attainment rate
		Scheduled contract securing rate
Departmental activities	Product development capability	Enrichment of model assortment
		Improvement in new product mass-production start-up
		New product sales (amount, rate)
		Number of design alterations
		Number of patent applications
		Period for new product development
Human resources aspect	Manpower resources aspect	Attendance rate
		Industrial accident occurrence rate
		Number of complete quality control circle themes
		Number of improvement proposals
		Number of newly acquired qualifications
		Number of training course participants
Production/ delivery time	Delivery time	Delivery time
	Inventory	Inventory reduction
		Inventory turnover
		Number of days held
Production and delivery time	Production quality	Production quantity
Quality	Finished goods inspection	Delivery inspection rejection rate
		Quality assurance department inspection rejection rate
	Manufacturing failure	Failure cost
		Process material percent defective per unit
		Process percent defective
		Yield
	Marketability	Market share
	User advantages	Changes in quality problems
		Comparison with international standards
		Extension of warranty period
		Market quality evaluation data
	User disadvantages	Claims (amount, rate, number of cases)
		Compensation work cost
		Customer acceptance inspection rejection rate
		Customer production line claim rate
		Market claims (amount rate, number of cases)

TABLE 15.3   Management Metrics for Deming Prize

Earning rate	Break-even point ratio
	Net profit
	Ordinary profit (amount, rate)
	Per capita ordinary profit
	Profit
	Profit after turnover tax
	Profit before tax (amount, rate)
	Profit ratio of total liabilities and net worth
Growth rate	Exports
	Growth rate of sales
	Sales
Indicators	Management metrics
Productivity	Break-even point operating rate
	Per capita sales
	Per capita value-added productivity
	Value-added productivity
Safety	Net worth rate
	Debt ratio
	Ratio of financial expense to sales

for this category has only been retained at the value of 80 points, to which it was restored from 75 in 1997.

15.5.3   ISO 9000

ISO 9000 is a series of quality management standards developed by the International Standards Organization in Geneva, Switzerland, in 1987. It has been incorporated in many European Community (EC) product safety and liability laws, which make it a virtual necessity for companies doing business in EC countries. The purpose of ISO 9000 is to provide third-party assurance of an organization's ability to meet specifications and perform according to standards. It is also used to certify that organizations have consistent, well-documented processes, and mechanisms that ensure compliance with these processes in order to ensure their ability to meet contractual specifications. ISO 9000 is also intended to be industry independent and require interpretation for specific industries.

The purpose of each of the ISO 9000 series standards is suggested in Figure 15.4, and is as follows:

1. ISO 9004 is really more of a document of guidelines than of standards. The term "should" is used, rather than the "shall" that is used in the other ISO 9000 series standards. It relates specifically to system elements rather than life-cycle process elements. There will be a migration of some of the guidelines in ISO 9004 into one of the ISO standards. For example, a quality plan is not specified as part of the specifications, but is included as part of the ISO 9004 guidelines. In all likelihood, this will become a part of the standards in future versions.

TABLE 15.4 The 1996 Baldrige Award Criteria Weightings

Evaluation Category	Weight
1 Leadership	**110**
1.1 Leadership System	80
1.2 Company Responsibility and Citizenship	30
2 Strategic Planning	**80**
2.1 Strategy Development Process	40
2.2 Company Strategy	40
3 Customer and Market Focus	**80**
3.1 Customer and Market Knowledge	40
3.2 Customer Satisfaction and Relationship Enhancement	40
4 Information and Analysis	**80**
4.1 Selection and Use of Information and Data	25
4.2 Selection and Use of Comparative Information and Data	15
4.3 Analysis and Review of Company Performance	40
5 Human Resource Focus	**100**
5.1 Work Systems	40
5.2 Employee Education, Training, and Development	30
5.3 Employee Wellbeing and Satisfaction	30
6 Process Management	**100**
6.1 Management of Product and Service Processes	60
6.2 Management of Support Processes	20
6.3 Management of Supplier and Partnering Processes	20
7 Business Results	**450**
7.1 Customer Satisfaction Results	125
7.2 Financial and Market Results	125
7.3 Human Resource Results	50
7.4 Supplier and Partner Results	25
7.5 Company Specific Results	125
Total Points	**1000**

2. ISO 9003 represents the standards for final inspection and test. These standards only might be sufficient for an organization whose effort involved not RDT&E or production, but only the supply of already available products. The vendor of software products represents an example of a commodity supplier that might need to become familiar with only this standard.

3. ISO 9002 deals explicitly with production and installation of products and systems. The focus here is on system-acquisition-type efforts for products that are not new and that do not involve RDT&E.

4. ISO 9001 is for use by organizations that need to be engaged in RDT&E, new-product planning and marketing, and any other efforts needed to a new and emerging technology into actual functional use in a quality-conscious manner. While it encompasses system definition through deployment (and disposal), it does not appear to encompass all of the definitional activities needed in the first generic phase of systems engineering life cycles to identify user requirements and translate these over to technological specifications.

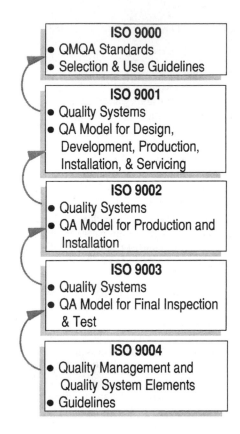

Figure 15.4 ISO 9000 series hierarchy.

5. ISO 9000 is an overview document that provides suggestions and guidelines for implementing the more detailed standards: ISO 9001, ISO 9002, and ISO 9003. A principal focus is upon documentation standards.

Many of the major industrialized nations have translated and published, or republished, ISO standards in terminology and language more appropriate for internal use. These include a European version, a United Kingdom version, a Unites States version, a Japanese version, and a version for The Netherlands. In the United States, the ISO Standards 9000,1,2,3,4 have been published by ANSI and ASQC efforts and bear the numbers Q91, Q92, Q93, and Q94.

There are 12 quality system requirements in ISO 9003, 18 in ISO 9002, and 20 in ISO 9001. Figure 15.5 illustrates these requirements and their hierarchical relationships within the various standards. Also shown in this figure is the section number in which discussion of the standards and efforts at their compliance is to appear in a Quality Manual (QM) that is prepared by individual organizations. For organizations not seeking approval at the highest level, ISO 9001, certain sections of the quality manual may be omitted. In the most general case, where ISO 9001 is used, the 20 sections of the ISO 9001 standard also define the 20 major requirements for a quality management system (QMS). Each of these is associated with appropriate metrics. They are as follows:

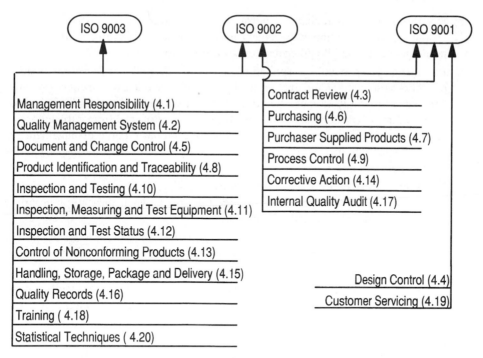

Figure 15.5 Illustration of the criteria making up ISO 9001, 9002, and 9003.

1. *Management responsibility* is a need. The organization is required to define and document management policy, objectives, and processes that indicate a commitment to quality. It must ensure that these policies are understood and implemented at all organizational levels, from the top down, and especially by those who verify and maintain quality assurance practices.

2. A *documented quality management system* must be established. This system must address all quality facets and must document these in the form of a QM that identifies the QMS in force in the organization. Ideally, an appropriate body recognized by the ISO accredits the QMS.

3. A *contract review* procedure must be established to ensure that all contracts begin with a mutually agreed upon set of customer requirements for a system that will satisfy the system user. The procedure must also ensure that the system developer, or contractor, is capable of delivery of an acceptable product to the end user.

4. *Design control* must be established in order to ensure that the system developer utilizes the procedures identified in the contract review in order to verify and control the design quality of the system to be produced, so as to ensure that it meets stated system requirements. These quality control procedures should cover the requirements that are input to the design process; design outputs, to ensure customer satisfaction with the intended final product; design verification in terms of reviews that ensure compliance with customer requirements, and government and industry standards; and design changes and modifications to accommodate potentially changed customer needs and requirements.

5. *Document control,* and change control, procedures must be established in order to ensure that configuration management information is made available to all appropriate people and at all appropriate organizational locations.

6. *Purchasing* plans must be addressed whenever it is desired to incorporate products of others as subunits in the final product. This must include assessment of potential subcontractor ability to meet quality and other performance standards. This must include assessment approaches to ensure that items procured from subcontractors satisfy stated requirements. It must also include proper records management for subcontractor-supplied items. This is especially important when third-party items are incorporated into a product.

7. Procedures must be established to specify, identify, verify, store, and maintain third-party products, or *purchaser-supplied products,* when they are appropriate for inclusion in an organization's final product.

8. *Product identification and tractability* must be addressed in order to ensure proper configuration management of an evolving system throughout the system acquisition life cycle.

9. *Process control* procedures must be established in order to ensure a well-planned and executed product line. This necessarily includes monitoring in terms of interactive systematic measurements throughout the life cycle.

10. *Inspection and testing* is required throughout the life cycle, from initially received materials and products, to in-production efforts, to efforts on the final product at the time of deployment. This includes appropriate recordkeeping.

11. *Inspection, measuring, and test equipment* must also be properly controlled in terms of appropriateness to the tasks for which they are used, and in terms of quality as well.

12. The *inspection and test status* of the evolving product or system must be identified throughout the life cycle. For example, design specifications should indicate whether they have been subject to final review and approval, or whether they are initial-draft specifications for the design. The appropriate release authority should identify conforming products that have been approved and released for the next phase of effort.

13. *Control of nonconforming products* must be established to ensure that products that do not meet standards are identified such that they cannot be inadvertently used. An appropriate detection, diagnosis, and correction procedure is needed such that nonconforming products may be scrapped, reworked into conforming products, returned to their supplier, or otherwise disposed of appropriately.

14. *Corrective action* procedures must be established. It must be possible to do this at the reactive level of the product, and at the interactive level of the process. In other words, faults must be removed from products, or they must be otherwise disposed of, and the process must be examined and potentially modified in order to prevent recurrence of the nonconformity. This may well require a considerable amount of formulation, analysis, and interpretation of the process itself in order to prioritize remediation efforts in order to make the most cost-effective improvements in the process.

15. *Handling, storage, package, and delivery* procedures must be established such that there is an audit trail of the evolving product from its initial conception through deployment.

16. *Quality records* must be prepared, kept, and maintained. These must be such as to enable demonstration that the requisite quality has been achieved and, also, that the QMS and associated processes are appropriate and that the QMS is used in practice.

17. *Internal quality audits* must be established, including independent-party evaluations of products, processes, and the QMS. The findings of these audits must be reported to organizational management for corrective actions.

18. *Training,* including the identification of training needs, must be such as to enable people to perform their work with the highest standards of quality.

19. *Servicing of customers,* both internal and external, must be addressed. This must include identification of customer-service responsibilities and record-keeping.

20. *Statistical techniques,* where appropriate, must be established in order to verify the appropriateness and reliability of products, processes, and the QMS.

It is very clear that the ISO 9001 standards, or the ISO 9002 or ISO 9003 subsets thereof, are exhaustive. Following them will require much effort, but the effort should result in considerable productivity and quality dividends for the organization and its members.

ISO 9004 represents a set of guidelines that promulgate through all of the other (ISO 9003, 9002, 9001, and 9000) documents. First issued in 1987, the ISO 9000 series standards are subject to continued revision and refinement, by ISO Technical Committee 176, for which the ASQC is the U.S. national representative. The standards are reviewed and revised every five years, and the latest revision was in 1997. An ultimate goal for the ISO 9000 series is stated in Vision 2000 (Marquadt, et al., 1991):

- *Universal acceptance,* such that the standards will continue to meet wise-scope customer needs in a useful and easy-to-use manner.
- *Current compatibility,* such that all future documentation is consistent with and supportive of present documentation.
- *Forward compatibility,* such that the necessary evolutionary revisions will not constrain efforts through the imposition of inappropriate requirements.
- *Forward flexibility,* such that the revisions are broadly applicable and continually supportive of quality progress.

The ISO standards are formally voluntary and compliance with them is formally optional. But, development may be such that they, in effect, become mandatory options for one who wishes to be competitive in an increasingly competitive world.

The significant level of interest in the ISO 9000 series standards is due primarily to marketplace considerations and, to a somewhat lesser extent, by government regulations. European purchasers often desire ISO registration of suppliers. Product safety and liability laws, especially in Europe, often place a burden upon producers to furnish a documented set of quality assurance procedures. An ISO 9000 registered supplier will be better able to demonstrate that their products have no defects. We examine issues of registration and audits in our next subsection.

There has also been a considerable move, primarily by government agencies, in the United States to utilize ISO 9000 standards. This includes efforts by the Food and Drug Administration (FDA), the Federal Aviation Administration (FAA), and the National Aeronautics Administration (NASA). The Department of Defense plans to replace several MIL quality system standards by ISO 9000, and contemplated changes in the Defense Federal Acquisition Regulations would allow use of ISO 9000 by contractors.

It is, we feel, important to issue a cautionary note here. Most of the ISO 9000 standards can be interpreted as objectives measures, and not as objectives. If this is done, then it is entirely possible that a given organization may configure itself in such a way that it can achieve a "high score" on a formal evaluation of the QMS. This does not immediately guarantee that the ultimate product produced will be of high quality. This potential problem, of sublimating objectives with objectives measures, is ubiquitous across many areas of effort and not uniquely related to quality management. Of course, a goal should be to establish a sufficiently broad scope and scale for objectives measures, or attributes, such that it becomes increasingly difficult for a high score on an evaluation or audit, to not be truly indicative of high performance in achieving objectives.

ISO 9004 is the standard most applicable to designing and managing performance improvement efforts, and is therefore very applicable to this systematic measurements effort. Summaries of the ISO 9004 guidelines, which should be used in addition to the requirements specified in the selected ISO 9001, 9002, or 9003 standard, are described in Table 15.5, which contains management and leadership issues, and Table 15.6, which contains human resources development and management issues.

Missing from the ISO 9000 certification criteria are explicit organizational management and customer satisfaction criteria. Therefore, organizations cannot claim to be world-class simply on the basis of their ISO 9000 certification, as ISO 9000 only certifies that an organization can comply with the requirements specified in a negotiated contract (Mahoney and Thor, 1994). It appears that the ISO 9000 criteria only address 40 percent of the Baldrige Award criteria, and some organizations have encouraged ISO to incorporate benchmarking, customer relations, and cost-of-quality and poor-quality criteria in future guidelines.

15.5.4 ISO 14000

ISO 14000 (Puri, 1996) defines a voluntary environmental management system and is intended to be a significant international initiatives for sustainable development. Like other standards, it is intended to be used in conjunction with appropriate objectives and with management commitment to their attainment in order to improve organizational performance and to support sustainable development (International Institute for Sustainable Development, 1996). The standards are intended to provide an objective basis for verification of an organization's claims about its environmental performance. Consumers, governments, and organizations are now seeking ways to reduce their environmental impact and increase long-term sustainability by improving performance all along the value chain (Desimone and Popoff, 1997; National Research Council, 1996). This results in organizations adopting performance improvement criteria, in the sense of becoming more efficient, maintaining trustworthiness, and earning a profit. At the same time, they reduce the risk of costly fines and environmental restoration mandates, and promote their reputation as a good environmental neighbor and custodian of our natural resources.

The ISO 14000 voluntary standards do not specify specific environmental performance objectives. These are to be prescribed by the organization as a part of its strategic planning, and action efforts and should take into consideration its impact on the environment and its relationships with stakeholders. Performing according to the management-system-based

TABLE 15.5   ISO 9004 Management and Leadership Issues

Categories	Management and Leadership Issues
General	Highest management levels have responsibility for, and must be committed to, a basic quality management system. Management must appropriately resource the system.
Quality Policy	Management should announce its corporate quality management policy, which should be consistent with other policies and company values. Management should take all appropriate steps to ensure that its quality management approach is communicated, implemented, and sustained over time.
Specific Quality Objectives	Management should define objectives for the key elements of quality (e.g., performance, reliability, consistency, value added) in the quality policy. Objectives should be consistent with other corporate objectives and the quality policy. Quality costs should always be an important consideration; the objective is to minimize quality losses at an appropriate quality cost level.
Quality System	The elements are organizational structure, roles and responsibilities, procedures, and processes and feedback requirements for implementing quality management. The quality management system is the means by which stated management and quality policies and objectives are to be accomplished. The quality system should be tailored to the particular type of business and the business situation. The quality management system, to be in tune with today's approach to quality, should function so as to provide confidence that it is easily understood and effective; internal and external customer expectations are satisfied; and prevention is emphasized rather than only fault detection. Responsibilities (general, specific), accountabilities, and reporting and communication channels should be explicitly defined. Delegation (responsibility, authority) should be clearly established and sufficient to attain assigned objectives with desired efficiency. Interface control and coordination measures between different activities should be defined; broadly accepted protocols should be in place for problem resolution and resource distribution disputes. Emphasis should be placed on identification of actual or potential problems and the initiation of effective "fixes" or preventive measures. Provision must be made for sufficient and appropriate resources (e.g., human resources personnel having specialized skills and experiences; training; inspection, test, and examination equipment; computer hardware and software).

approach of ISO 9000, ISO 14000 is intended to assist organizations in focusing attention on environmental issues. The ISO 14000 is intended to provide customer assurance that the environmental performance claims of an organization are factual. ISO 14000 is intended to help integrate environmental management systems of organizations throughout the world. To date however, ISO 14000 has not been adopted or fully accepted by all nations or by all organizations throughout the world, in part due to skepticism that has

TABLE 15.6   ISO 9004 Human Resources Development and Management Issues

Category	Issues
Training	In general, it is essential to identify quality training needs and provide mechanisms to deliver at all levels and to newly hired and transferred personnel. Executive and management personnel must: • Understand the quality management system. • Have tools and techniques to operate the quality system. • Accept and use evaluation criteria for system effectiveness. Professional and technical personnel must: • Have broad-gauge training to enhance management system contribution. • Be trained in statistical techniques at appropriate levels of complexity. Production supervisors and workers must have: • Methods and skills to perform tasks (e.g., operation of instruments, tools, and machinery; reading and understanding documentation; relationship of work to quality objectives; safe operations). • Training in basic statistical technical techniques, as needed. • Operator certification, as appropriate.
Performance Management	Orientation and awareness; create understanding of tasks expected to be performed and their relationship to overall quality activities; establish the necessity for proper job performance at all levels; clarify how poor job performance affects other employees, customer satisfaction, operating costs, and the organization's profitability. Certification. Establish a competency-based system tied to recognition and rewards. Application. Efforts should cover all who can impact quality, both by assignment and by level. Feedback. Everyone involved must know how things are going, individually and organizationally.
Quality Measurement	Publicize definitive measures of quality achievement by individuals and teams to validate the system and encourage continuing creation of appropriate quality. Recognize performance when satisfactory quality levels are attained.

ensued, perhaps because the standard only addresses environmental issues and not others of sustainable development in a more generic sense. Yet acceptance is needed in order for in order for ISO 14000 to work well. Sustainable development is based both on human and social development, as well as natural resource preservation and environmental conservation. This may require a refocusing of ISO 14000 standards to deal with other than only environmental issues. Perhaps, this need will result in an aggregation of the series 9000 and series 14000 standards.

Historically, the ISO 14000 series of standards resulted primarily from the Uruguay round of the GATT negotiations and the UN Rio Summit on the Environment held in 1992. The GATT effort focused on the need to reduce nontariff barriers to trade. The Rio Summit resulted in a commitment to environmental protection and restoration throughout the world. Following acceptance of ISO 9000, the, ISO assessed the need for international environmental management standards through the 1991 creation of the Strategic Advisory

Group on the Environment (SAGE) in 1991. This group considered whether standards could serve

- To advance a shared approach to environmental management comparable to quality management and ISO 9000.
- To enhance the ability of organizations to measure and improve environmental performance.
- To facilitate trade and remove trade barriers.

The result of this effort was the series of ISO14000 standards. They are intended to embrace:

- Environmental aspects in product standards
- Environmental auditing
- Environmental labeling
- Environmental management systems
- Environmental performance evaluation
- Life-cycle assessment

The Environmental Management System (EMS) is the central part of the ISO 14000 standard. An EMS is a continual cycle of planning, implementing, reviewing, and improving actions that an organization implements in order to meet its environmental responsibility. Figure 15.6 illustrates these steps. All responsible organizations have operational management systems. These include financial systems, which define how decisions on budgeting, cash management, and accounting are made. Organizations also have human resource departments, sales and marketing departments, and manufacturing departments. Each of these departments is generally associated with a management system. ISO 14000 is being developed to satisfy the need for an environmental management system. The ISO 14000 standards, which are more realistically thought of as guidelines, define a core environmental management system and the measurement procedures necessary for its implementation. The management system also defines three sets of tools that are important in implementing an EMS: life-cycle assessment, environmental performance evaluation, and environmental labeling.

The ISO 14000 is intended to serve two major needs of an organization. There is an internal need for a standard that will help the organization address legal, commercial, and other challenges related to the environment. There is a need to assure those outside of the organization that it is meeting its stated environmental objectives. The internal benefits to the organization of ISO 14000 compliance include reduced incidents and liability through use of systematic measurements to establish verifiable performance. Such a systematic approach to managing environmental issues will also result in the identification of opportunities to conserve material and energy inputs, reduce wastes, and improve process efficiency. This is the basis for industrial ecology as an important set of systems engineering efforts (Graedel and Allenby, 1995; Allenby, 1999) that will lead to improved environmental performance and to improved product performance and profit. There are external benefits in terms of third-party assurance and verification that the products and services of an organization meet prescribed standards. This eliminates the need of organizations to

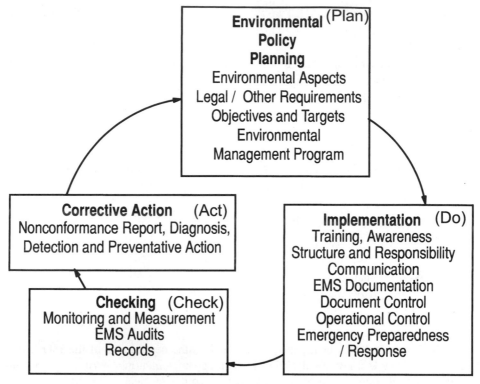

Figure 15.6 The plan–do–check–act life cycle for EMS establishment.

individually inspect the products and services of each supplier through this assurance. Thus, compliance with ISO 14000 may become a required prerequisite to operation of an organization.

Perhaps the best way to describe what the ISO 14000 standards cover is to identify the eight major component of these standards.

1. *Environmental Aspects of Product Standards.* This standard is intended to raise awareness that product design provisions can affect the environment in both negative and positive ways, and recommends the use of life-cycle thinking and recognized systems engineering methodologies in developing product standards that incorporate environmental aspects. This standard incorporates "design for the environment" or "green design" methodologies (Fiskel, 1996).

2. *Environmental Auditing Requirements.* The environmental auditing standards suggest requirements for environmental auditing, guidelines for auditing the EMS, and qualification criteria for auditors.

3. *Environmental Labeling Requirements.* The environmental labeling standard provides requirements for three types of labels: a "seal of approval" for products and services that meet specified requirements; a "single-claim labels" in terms of measurable criteria for such things as recycled content and energy efficiency; and an "environmental report card" that uses a life-cycle approach that provides for com-

parison of environmental effects of the manufacturing and use of products and services.

4. *Environmental Management System Certification Requirements.* The EMS certification standard contains the fundamental requirements for developing and implementing an EMS; it also can be certified by an external third party in the same way as for the ISO 9000 series standards. This standard is intended to integrate the environmental management system with existing and evolving organizational management policies and operations.

5. *Environmental Management System Guidance Document.* The EMS guidance document is intended to provide assistance to organizations initiating, implementing, or improving an EMS. It outlines the elements of an EMS and provides practical advice on implementing or enhancing a system. The EMS objectives include identification of applicable regulatory requirements, commitment to continual improvement, and regular periodic evaluation of environmental performance.

6. *Environmental Performance Evaluation Requirements.* Environmental performance evaluation (EPE) is a process for measuring, analyzing, assessing, and describing an organization's environmental performance according to agreed-upon objectives. This is a measurement-intensive effort for gathering, sorting, and aggregating data in order to provide information and knowledge concerning how well targets and performance objectives were satisfied. EPE is an EMS tool that provides for measuring environmental impacts for purposes of control.

7. *Life-Cycle Assessment Requirements.* Life-cycle assessment (LCA) (Graedel, 1998) is a tool for evaluating environmental attributes associated with a product, service, or process.

8. *Terms and Definitions.* This standard attempts to provide a clear and unambiguous definition of terms used throughout the ISO 14000 standard. This is important since many terms can be easily misunderstood and can vary significantly in meaning when translated into different languages.

The ISO 14000 international standard is intended to be applicable to any organization wishing to

- Define, develop, and deploy an environmental management system.
- Assure concordance with organizationally established environmental policies and strategies.
- Demonstrate this conformity to others.
- Obtain certification or registration of an environmental management system by an external organization.
- Self-determine and self-proclaim conformance with ISO 14000.

At this writing (1998), ISO 14000 and the EMS are a working draft. The standard has been written to apply to all types and all sizes of organizations. It is designed to be responsive to diverse geographical, cultural, and social conditions. The standard speaks to commitment to continual improvement and compliance with applicable legislation; however, it does not establish absolute requirements. Thus, two organizations engaging in similar activities, but having different environmental performance and requirements, may each comply with the standard. The standard applies to plant operations and processes. There is a current question as to whether it applies to products as well. Some feel that products should not be included because of possible creation of a nontariff trade barrier. Many feel that products can be included in the standard if this inclusion issue is approached in a flexible manner. Figure 15.6 presents a plan–do–check–act phased life cycle for establishment of an EMS for ISO 14000.

15.5.5 Government Performance and Results Act of 1993

The Government Performance and Results Act of 1993 (GPRA) was passed by Congress "to provide for the establishment, testing, and evaluation of strategic planning and performance measurement in the federal government." Concerned with rising costs and constituents' complaints about federal organizations' efficiency and effectiveness, and encouraged by General Accounting Office (GAO) findings and private-sector successes, Congress passed the GPRA in order to force federal organizations to set clear goals, define and implement performance metrics, and regularly report on their performance. Congress plans to use these goals, metrics, and performance reports in conjunction with policy- and decision-making activities to ensure that funding is focused on well-managed, efficient, and effective programs that are making progress toward their strategic goals.

The GPRA required that all federal organizations submit their first five-year strategic plan by September 30, 1997, and updated plans every three years thereafter. By March 31, 2000, organizations must begin submitting annual performance reports based on their strategic plans. To help clarify and standardize what is required, the GPRA provides descriptions of the processes and results that are required to formulate the reports.

1. Establish performance goals to define the level of performance to be achieved by a program activity.
2. Express these goals in an objective, quantifiable, and measurable form, unless they are specifically authorized to be in an alternative form.
3. Briefly describe the operational processes, skills and technology, and the human, capital, information, or other resources required to meet performance goals.
4. Establish performance indicators to be used in measuring or assessing the relevant outputs, service levels, and outcomes of each program activity.
5. Provide a basis for comparing actual program results with the established performance goals.
6. Describe the means to be used to verify and validate measured values.

In addition, except for the first plans submitted in 1997, all plans must:

- Review the success of achieving the performance goals of the fiscal year.
- Evaluate the performance plan for the current fiscal year relative to the performance achieved.
- Explain and describe, where a performance goals has not been met, why the goal was not met, those plans and schedules for achieving the established performance goal, and if the performance goal is impractical or infeasible, why that is the case and what action is recommend.

• Include the summary findings of those program evaluations completed during the fiscal year covered by the report.

Descriptions of the measurement and reporting process terminology to be used are also provided in the GPRA. These descriptions are consistent with most standard performance measurement and improvement terminologies. However, there is considerable emphasis placed on the difference between what GPRA called "output measures" and "outcome measures." Output measures are intended to measure the direct impact of a program or activity, as well as the impact that program has on overall performance. For example, for an immunization program, an output measure might be the percentage of people immunized, while an outcome measure might be the percent decrease in the number of people who contract the disease. This distinction highlights the need to ensure that lower-level improvement activities positively impact higher-level performance improvements, and ensure that undesired sublimation of the two measures does not occur.

In summary, the GPRA provides thorough requirements for linking actionable objectives in the one-year plans to long-term strategic objectives, and provides some definitions to standardize performance assessment and reporting. Like other frameworks, however, it does not provide guidance on metrics prioritization and selection, metrics formulation, data collection, information validation and consistency, or many of the other issues associated with implementing and managing an organizational performance management system. One of many discussions of implementation of the GPRA may be found in the GAO report by Stevens and Rezendes (1996).

15.6 SELECTED MEASUREMENT APPROACHES

In this section, we provide an overview of several organizational performance approaches that have been suggested in recent times. Each of these approaches provides valuable guidance concerning selection of appropriate metrics.

15.6.1 Sink and Tuttle's Organizational Performance Measurement Approach

One subject missing in many discussions of such subjects as quality management is how performance measurement should actually be planned, analyzed, and performed. Identifying categories of performance metrics, and determining how performance data will be collected, converted into information, and then measured, requires analysis of many factors. Sink and Tuttle (1989) have endeavored to provide a systematic approach to the selection, formulation, implementation, and execution of organizational performance improvement metrics. They state that the most important reason to measure is to assess performance improvement progress, and note that there are three counterproductive measurement traps into which managers often fall instead. These traps are:

1. Measuring A while hoping for B. We measure the easy things, the most pressing things, the wrong things; we hope for quality while measuring and controlling only production schedules.
2. Measuring to control in such a way as to make improvement more difficult. In so

doing, one focuses on control of excess, creating a compliance mentality rather than an improvement orientation.

3. Measuring to find those who have performed poorly in order to punish them while ignoring the good performers.

Sink and Tuttle discuss performance measurement in the context of an eight-step performance improvement planning process. Our representation of this is shown in Figure 15.7. As depicted in this figure, it is imperative that performance measurement and feedback be incorporated in order to achieve two objectives:

- To continually assess and evaluate operational performance improvement interventions.
- To obtain periodic reassessment and refinement of the integrated, organization-wide performance improvement effort.

Sink and Tuttle use an input–output analysis diagram to model the organization and its suppliers, inputs, outputs, and customers. To help identify the relationships between management, performance improvement interventions, and performance measurement that must also be considered when developing a measurement system, Sink and Tuttle incorporated Kurstedt's Management Systems Model (1985), represented in Figure 15.8, into

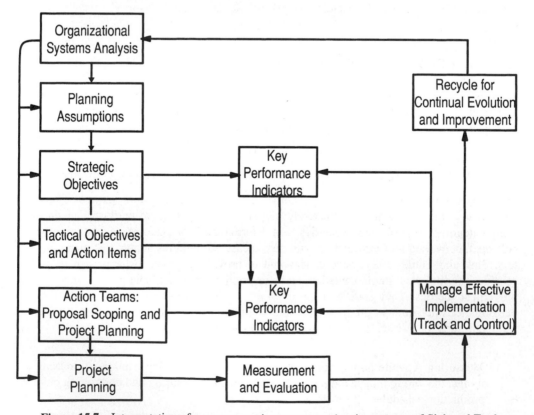

Figure 15.7 Interpretation of measurement improvement planning process of Sink and Tuttle.

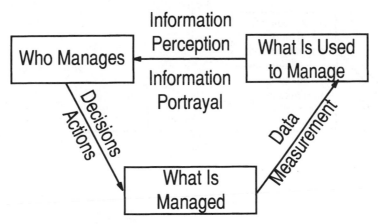

Figure 15.8 Interpretation of the management system model of Kurstedt.

their approach and obtain the representation shown in Figure 15.9. This model shows that management is a cyclic process that relies heavily on assessment of organizational performance information and, in turn, makes decisions that directly impact the performance of the organization. Every step in this process must be performed appropriately or organizational performance improvement will generally not be obtained. To obtain a comprehensive, standardized approach to metrics identification, Sink (1985) defined seven operational organizational performance metrics categories, and Sink and Tuttle (1989) elaborated upon and enhanced these categories to develop metrics for effectiveness, efficiency, productivity, quality of life, innovation, profitability and budgetability, and quality. The following operational definitions are associated with these categories.

- *Effectiveness* or accomplishment of the right things. This refers to actual accomplishment of the "right" things, on time, and within the specified quality requirements specified. Example metrics include percentage of sales quota met by salesperson; customer turnover per district and product; slippage of schedules; actual and potential market share, reputation, and trustworthiness as perceived by customers; and percentage of product shipped on time.
- *Efficiency* relates to the resources expected or predicted or forecasted or estimated to be consumed divided by the resources actually consumed. Example metrics include maximum, minimum, or low-limit target levels; percentage of shortages of scheduled production material; change in the average total requisition handling cost; inventory reports of items below minimum point; rates of spoilage or waste; average lead time; machine utilization ratio; and space utilization.
- *Productivity* is the ratio of output over input. It is the relationship between what comes out of an organizational system divided by what goes into that organizational system. Example metrics include sales per employee; production rate per employee; and production lead time from raw material to finished product.
- *Quality of work life* represents how organizational employees feel about various aspects of work life. It represents the affective or emotional response or reaction of the people in the organizational system to any number of factors, such as pay, working

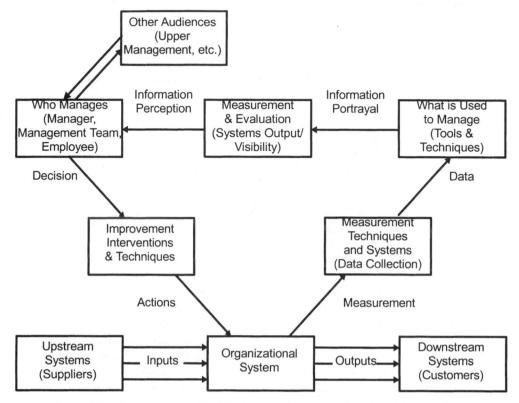

Figure 15.9 Interpretation of modified management system model of Sink and Tuttle.

conditions, culture, leadership, coworker relations, feedback, autonomy, skill variety, task identity, task significance, the boss, amount of involvement in planning, problem solving, and decision making. Example metrics include employee absenteeism and turnover rates; number of employee grievances; number of accidents; and employee hours worked exceeding target levels.

- *Innovation* represents the creative process of changing. It can refer to what we are doing and how we are doing things. It is concerned with structure, technology, products, services, methods, procedures, and policies that successfully respond to internal strengths and weaknesses, and external opportunities and threats. Example metrics include number of new production methods adopted; time and cost savings by employing new methods or technologies; and number of patents received per time period.

- *Profitability/budgetability*. Profitability relates to profit centers and involves measures that relate revenues to costs. Budgetability relates to cost centers and involves measures of the relationship between budgets and agreed-upon goals, deliverables, and timeliness with the actually obtained values. Example metrics include actual product sales as opposed to budgeted sales; budget variances exceeding target levels; profits as percent of capital employed or return on investment (ROI); profits as percent of sales; profits per employee; percentage of increase in dividends; and debt ratios to total assets.

• *Quality* is defined relative to five factors: upstream systems of vendors and suppliers, inputs from these, transformation processes that result in products or services, outputs in terms of the product or service produced, and downstream systems in terms of customer responses.

Each of these seven categories is an important source of measurement information concerning organizational productivity measurement. These need to be associated with measurements of the 27 generic types illustrated in Figure 15.2. Of particular concern in use of this approach is that we represent learning and knowledge management efforts as they apply to the three integrated levels of product, process, and systems management.

15.6.2   Kaplan and Norton's Balanced Scorecard Approach

Kaplan and Norton (1996a, b) developed what they have denoted as a "balanced scorecard" approach to metrics-set identification. This was accomplished in order to recognize and make explicit the need to complement traditional financial metrics with metrics focusing on customer satisfaction, internal business processes, and learning, innovation, improvement, and growth. Four "scorecards" are developed, one for each of these, as illustrated in Figure 15.10. In developing this approach, emphasis is placed on using a top-down approach to identify a number of most appropriate metrics on which the organization should focus. These metrics are conveyed in a graphical framework that shows the metrics

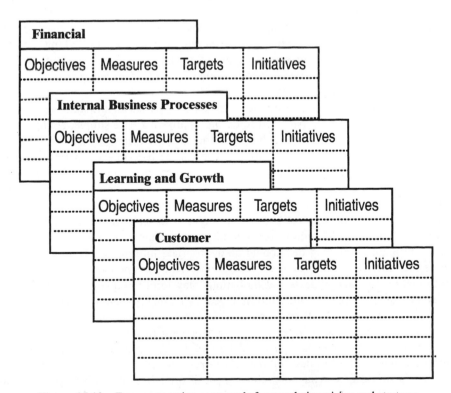

Figure 15.10   Four perspective scorecards for translating vision and strategy.

in each group, or scorecard, and facilitates the identification of linkages and complimentary interests of the four metrics sets.

Four phases of effort are involved in managing a balanced scorecard, and they follow the plan−do−check−act sequence in an iterative manner.

- Business planning.
- Feedback and learning.
- Clarifying and translating the vision.
- Communicating and linking.

These four life-cycle ingredients become the long-term drivers of success across each of the four major processes of the organization: financial processes; internal business processes; learning and growth processes; and customer interaction processes. In business planning, operational and financial plans are integrated by giving those metrics included in the balanced scorecard the highest priorities and allocating time and resources to them accordingly. This enables the setting of targets, aligning strategic initiatives, resource allocations, and the establishment of milestones. In the feedback and learning phase, progress achieved in obtaining scorecard objectives is reviewed, new relationships and tradeoffs are identified, and the scorecard is modified as required. This enables articulation of the shared vision, supplying needed strategic feedback and facilitating strategy analysis and learning. In translating the vision, actionable objectives and measures are developed that are in alignment with the organization's vision, mission, and strategy. Targets and initiatives are also developed. In the communication and linking phase, management strategy is communicated throughout the organization via the concise, easily interpreted, graphic scorecard that enables communicating and educating, goal setting, and linking organizational rewards to performance measures. Figure 15.11 illustrates four scorecards and their use in a plan−do−check−act type of iterative life cycle to obtain measurements and associated organizational improvements.

When compared to other organizational performance metrics identification approaches, the primary advantages of the balanced scorecard approach are its simple, readily understandable four-part structure; its graphical format; and the clear way that this leads to identification of the metrics that organizations should include in their performance improvement efforts. The approach's potential drawbacks are that it provides little guidance on the prioritization and selection, formulation, and implementation of specific metrics. Prioritization and selection may be left to senior managers and facilitators who may not always have the operational-level knowledge required to make those decisions. Implementation is also left to the senior executive team. This team should agree on an implementation program, including developing an appropriate information system to support the scorecard. Incomplete knowledge of operations, data validity and consistency issues, and information technology costs and limitations may limit the quality of the decisions made by a nonrepresentative team. Thus, it appears vital that the senior executive team be experientially familiar with all environmental issues surrounding the organization such that they are fully capable of interpreting the scorecard information in an appropriate context such that it becomes knowledge. As an alternative, people from throughout the organization may be used to develop the scorecards, such that "everyone in the organization has a scorecard," and support achievement of organizational objectives. Thus, ap-

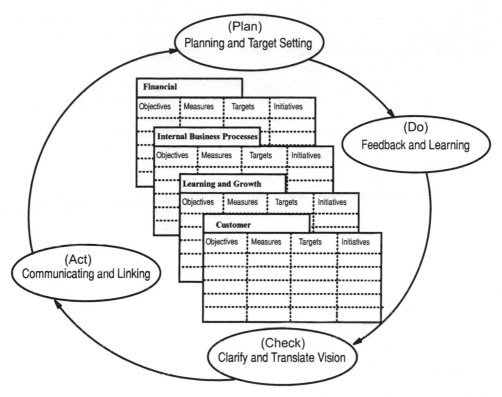

Figure 15.11 The four processes for managing strategy in the balanced scorecard of Kaplin and Norton.

propriate use of a scorecard could well represent a critical capability evaluated at review time.

When the feedback and learning scorecard is augmented to include acquisition of intellectual capital and associated knowledge management, use of this scorecard should be particularly appropriate. In a similar way the customer scorecard might be augmented to represent product attributes as well, or an additional scorecard developed for this purpose. Some of the scorecards could become quite sizable, especially the one for the organizational or internal business processes, as these include RDT&E, acquisition, and planning and marketing.

Kaplin and Norton (1996a,b) provide a wealth of information concerning use of this approach to strategy and measurement. They are particularly concerned with three principles to enable an organization to develop a balanced scorecard that is tightly linked to organizational strategy.

1. Cause and effect relationships are very important, as a strategy is little more than a set of approaches based on hypotheses associated with assumed, and ideally demonstrated, cause-and-effect relationships. The balanced scorecards should both identify and make explicit the cause-and-effect relationships between objectives mea-

sures and the performance drivers associated with the correspondence objectives. Taken together, these objectives measures should represent a chain of the cause-and-effect relationships that will communicate the organization's strategy, and its meaning, throughout the organization.

2. Each balanced score card will contain generic objectives measures that include such lag indicators as customer satisfaction and retention, and profitability. The performance drivers are the lead indicators that represent critical core capabilities for each organization, such as intellectual capital. A good balanced scorecard should have a mix of these lead and lag indicators, or a mix of objectives measures and performance drivers. Objectives measures alone do not generally indicate how objectives are to be achieved. Performance drivers indicate potential in an organization, but not necessarily whether these have been translated into results. An appropriate balanced scorecard is a measurement system that has an appropriate mix of objectives measures and performance drivers that have been customized to the specific objectives and strategies of the organization.

3. A balanced scorecard should link all strategic efforts and operational improvements to economic returns, or whatever performance measure represents the bottom line for a given organization. To not do this is to run the risk that the improvement programs themselves, and perhaps even achievements of the objectives measures associated with them, are sensed as the major objectives.

The major purpose in making scorecards is to develop a set of strategic measures that are critical for successful performance and then a set of diagnostic measures that support these strategic measures.

15.6.3 Thor's Family of Measures Approach

Thor's approach (1994) to the development of organizational performance metrics sets focuses on the identification of five categories of metrics:

- Profitability
- Productivity
- External quality (customer focused)
- Internal quality (waste reduction)
- Such other metrics as innovation, safety, and organizational culture

Thor points out that a family of metrics is needed at every step in the process, not just at the strategic level, with some metrics being used only at the lower levels and some metrics being rolled up into organization-wide metrics.

The Family of Measures approach is based on the use of a brainstorming technique with a representative group of organizational personnel to identify, analyze, and ultimately select metrics. Once a family of measures, or metrics set, has been identified, it should be evaluated using whatever appropriate data are available. After the measurement system has been created, a formal or informal break-in period should be established for the purpose of identifying problems with communication of the measures or interpretation of their quantifications or instantiations. The break-in period is essential because of the problems that can result if managers do not receive current and complete measurements or if they inaccurately interpret the performance aspects these measurements are designed to reflect.

Thor identifies seven characteristics that must exist if a family of measures is to be capable of being used successfully:

1. The family of measures must be linked to the appropriate level of the strategic plan and expressed in that level's language.
2. The family of measures must be well communicated throughout the organization.
3. The family of measures must be made up of enough members to ensure completeness, but not so many that the organization loses its focus.
4. The family of measures must be technically sound.
5. The family of measures must be reviewed as often as appropriate.
6. The family of measures must provide information on both level and trend.
7. The family of measures must be consistent with rewards, recognition, and management style.

He also suggests that the following questions must be answered with regard to the effectiveness and thoroughness of the family of measures.

1. Are we leaving out something significant?
2. Are the measures inconsistent with individual motivations?
3. Are the proper authority, tools, and training available to ensure accurate measurement?
4. Can the measures be updated as needed?
5. Are the measures consistent with group rewards and recognition?
6. Do the measures foster good customer and supplier relations?

The Family of Measures approach provides a useful framework (categories) for metrics and associated guidelines for their selection. It differs from some other approaches in that it requires the involvement of personnel from every part of the organization, and specifies the development of a family of metrics for every organizational unit and level.

15.7 SYSTEMATIC MEASUREMENTS OF CUSTOMER SATISFACTION

Grady (1992) suggests that the major organizational purpose associated with customer satisfaction strategies is that of capturing market share through understanding of customer needs and beliefs and responding to these in an effective manner. While this is an effective objective at any time, it is most effective when initially entering a market. It is often very difficult to establish when the initial customer satisfaction level is low. Thus, the most effective implementation of a customer satisfaction strategy begins at the onset of product introduction. Customer communication and quick responses to concerns and inquiries are characteristic features of this strategic approach. As a consequence of this, surveys and interview data are important metrics, as are metrics associated with product performance and defects.

These measurement strategies may be pursued at the level of the product only, and most of the direct metrics associated with customer satisfaction are at this level. Many have developed attribute trees for product quality that are based upon customer satisfaction notions. Figure 15.12 presents an attribute tree used in Sage and Palmer (1990) for product quality evaluation and that is based on previous efforts (Boehm et al., 1978) that also deal with product quality and related customer satisfaction issues.

Grady (1992) also suggests using a set of product attributes based on functionality, usability, reliability, performance, and supportability (FURPS) (also see Grady and Caswell, 1987). This approach, denoted by the Hewlett-Packard Corporation as the FURPS+ model, where + denotes any of several extensions that have been considered, is given in Figure 15.13 in terms of the three generic phases of the system acquisition life cycle. This figure suggests that a number of objectives and objectives measures be identified for each FURPS attribute. The GQM methodology is a potential way of accomplishing this.

Figure 15.14 represents these same FURPS attributes and associated evaluation of them by the customer. Each of the lowest-level attributes is associated with an importance weight, as in standard multiple-attribute utility-assessment-based approaches, such as discussed in Chapter 28. Rather than the generic FURPS attributes shown in the illustration, those importance weights associated with the specific FURPS attributes for the product in question should be included here. The customer evaluation of the current product and the new product being fielded may be scored and an overall evaluation of the performance of the current system and the new system obtained by multiplying the importance weights by the corresponding system or product scores and summing these. There are a number

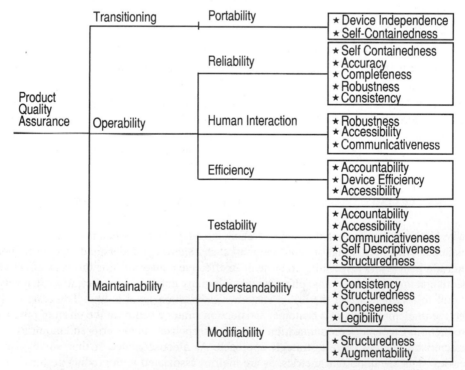

Figure 15.12 A possible attribute tree for software product quality assurance.

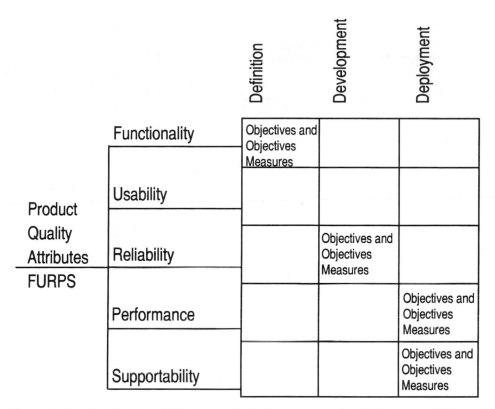

Figure 15.13 Identification of objectives and objectives measures for FURPS+ at each life-cycle phase.

of normalization and anchoring issues involved (Sage, 1992a), but the procedure is generally a very straightforward one, and many have applied it with good results in a variety of application settings.

Grady suggests use of a quality function deployment-based approach to enable tracking of the various objectives, attributes, and metrics associated with determination of customer satisfaction. A quality function deployment matrix represents an actual or conceptual collection of interaction matrices that provides the means for transitional and functional planning and communication across groups. Basically, it is an approach suggested in the total quality management literature that attempts to encourage an early identification of potential difficulties at an early stage in the life cycle of a system. It encourages those responsible for fielding a large system to focus on customer requirements and to develop a customer orientation and customer-motivated attitude toward everything. Thus, the traditional focus on satisfying technical system specifications is sublimated to, but not replaced by, the notion of total satisfaction of customer requirements.

Quality function deployment (QFD) has been suggested as one successful approach to use in conjunction with implementation of TQM. A QFD interaction matrix is one approach for representation and communication of QFD results. The purpose of QFD is to promote integration of organizational functions to facilitate responsiveness to customer

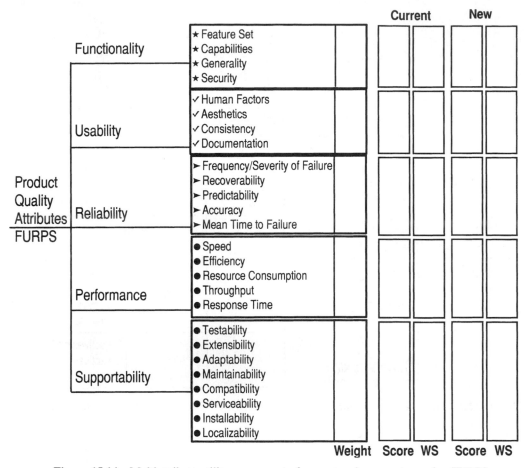

Figure 15.14 Multiattribute utility assessment of current and new system using FURPS.

requirements. As described by Clausing (1994), QFD comprises structured relationships and multifunctional teams. The members of the multifunctional teams attempt to ensure that

- All information regarding customer requirements and how best to satisfy these requirements is identified and used.
- There exists a common understanding of decisions.
- There is a consensual commitment to carry out all decisions.

These are, of course, especially important for an organization that attempts customer satisfaction as the primary component of its competitive strategy.

Two applications to QFD are often suggested. One is a "house of quality" effort matrix, and the second is a "policy assessment" matrix. Generally, the associated interaction matrices might appear as illustrated conceptually in Figures 15.15, 15.16, and 15.17. Figure 15.15 illustrates some generic interactions among potential elements in the matrices; Figure

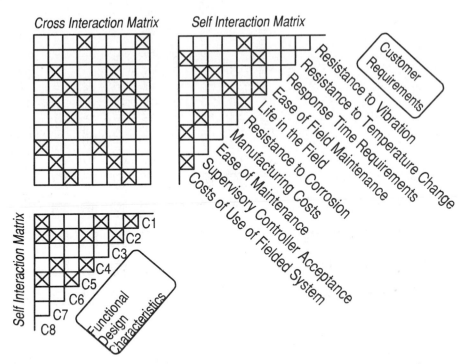

Figure 15.15   Typical self- and cross-interaction matrices for quality function deployment.

15.16 illustrates a more complete set of QFD matrices, and Figure 15.17 presents an illustrative picture of these, as a house of quality and house of policy, that follows an illustrative house-of-quality depiction of these interactions. In this illustration, the system production output, which represents the product delivered to the customer, would be subject to the FURPS measurement represented in Figure 15.14 or through use of the attributes described in Figure 15.12.

The generic efforts involved in establishing the QFD deployment matrices would be as follows.

1. Identify customer requirements and needs. Establish these needs in the form of weighted requirements expressing their importance.
2. Determine the systems engineering architectural and functional design characteristics that correspond to these requirements.
3. Determine the manner and extent of influence among customer requirements and functional design characteristics and how potential changes in one functional design specification will affect other functional design specifications.

As a related and more strategic matter, it would be necessary to determine the extent to which the specific systems acquisition effort, or RDT&E, or marketing effort, under consideration is supportive of the long-term objectives of the organization, and the competitive position and advantage to be obtain through undertaking the development. We see that this QFD approach has uses in strategic management to see if the management controls

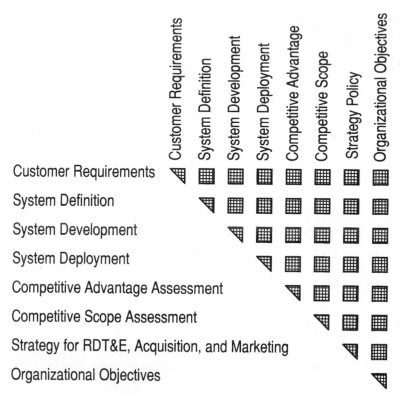

Figure 15.16 Interaction matrices for organizational policy assessment and quality function deployment.

or systems management effort associated with the development of the system under consideration is supportive of the overall development strategies of the organization.

Hypothetically, at least, the QFD matrices provide a link between the semantic prose that often represents customer requirements, and the functional requirements that would be associated with technological system specifications. In a similar manner, an interaction matrix can be used to portray the transitioning from functional requirements to detailed design requirements. Thus, they provide one mechanism for transitioning between these two phases of the systems life cycle. One goal in this would be to maintain independence of the various functional requirements, to the extent possible. A measure of this independence can be obtained from the density and location of the interactions among the functional requirements.

Peters (1987) has identified customer responsiveness prescriptions as a major driver for future organizational success and as one of 5 major prescriptions for excellence shown in Figure 15.18. He associates ten important activities with attaining customer responsiveness, and these are represented in Figure 15.19. Each of these ten activities can easily be written as goals or objectives and the GQM methodology used to obtain relevant metrics with which to measure customer responsiveness satisfaction.

Peters takes special care in identifying measures for customer satisfaction. He identifies

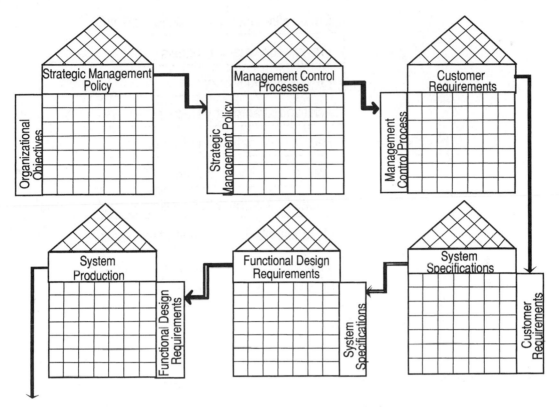

Figure 15.17   Interpretation of quality function deployment concepts.

Prescriptions for Excellence	Customer Responsiveness
	Fast Paced Innovation
	Flexibility Through Empowered People
	Leadership
	Organizational Processes

Figure 15.18   Peters' five first-level prescriptions for excellence.

	Create New Specialized Niche Markets
	Provide Top Quality Products
	Provide Superior Service
	Achieve Extraordinary Responsiveness
Customer Responsiveness	Become International and Global
	Create Uniqueness of Product /Service
	Create Obsession with Listening
	Make Manufacturing a Marketing Weapon
	Make Heroes of Sales and Service Forces
	Pursue Fast Paced Innovation

Figure 15.19   Peters' ten prescriptions for customer responsiveness.

ten major considerations that are suggested as major facets of measurement of customer satisfaction.

1. The frequency of measurements must be appropriate and should invariably include informal monthly surveys, and more informal semiannual and annual surveys.

2. The survey format should include both informal customer focus groups and assessments made by the organization itself, as well as third-party or external surveys that are conducted by an independent organization.

3. The content of the surveys should include and call for important quantifiable and nonquantifiable responses. It should recognize that no single survey instrument will be best or sufficient by itself.

4. The design of the surveys should include both systematic approaches and less formal approaches and should emphasize both realities and perceptions.

5. Surveys should involve all stakeholders with the customer responsiveness issues being measured.

6. Surveys should measure the satisfaction of all stakeholders.

7. Combinations of measures should be employed to provide an index of performance and responsiveness.

8. The results of these systematic measurements should be used to affect compensation and other rewards.

9. The obtained customer satisfaction measures should be posted throughout the organization.

10. All job descriptions should include a description of connectivity to the customer, and performance evaluations should include assessments of customer orientation.

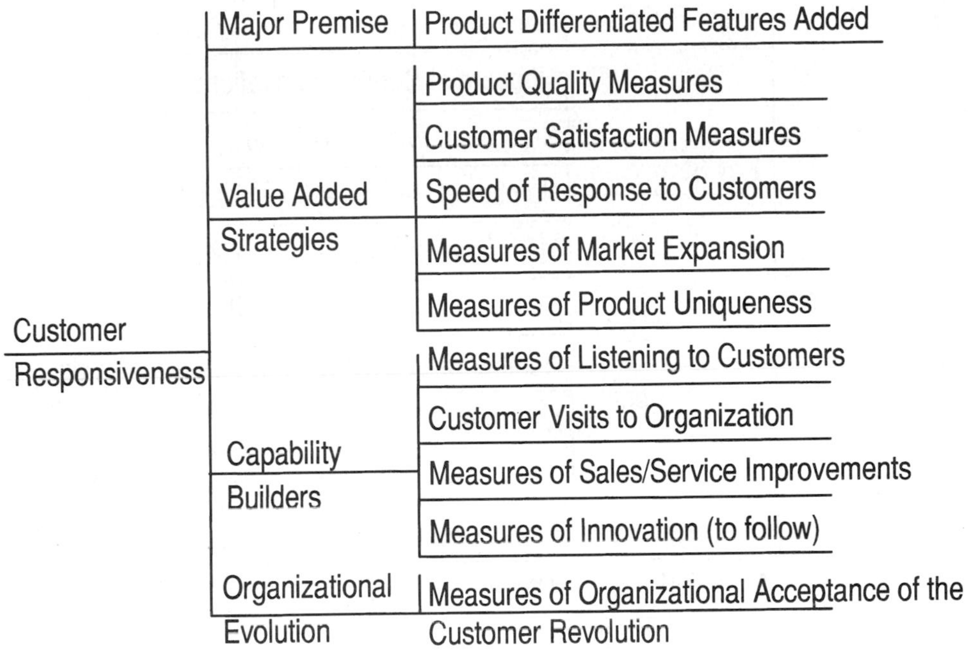

Major Premise	Product Differentiated Features Added

Customer Responsiveness	Value Added Strategies	Product Quality Measures
		Customer Satisfaction Measures
		Speed of Response to Customers
		Measures of Market Expansion
		Measures of Product Uniqueness
		Measures of Listening to Customers
	Capability Builders	Customer Visits to Organization
		Measures of Sales/Service Improvements
		Measures of Innovation (to follow)
	Organizational Evolution	Measures of Organizational Acceptance of the Customer Revolution

Figure 15.20 Peters' eleven measures of customer responsiveness.

Peters is especially cogent in pointing out the need for subjective and objective, and systematic and informal, approaches to measurement.

Many, if not most, of the suggested metrics for customer satisfaction measures are at the level of product. Product quality and appeal to customers is, of course, very important. But attention to this *only* at the level of product is likely to yield disappointing results. The product line, or process, that results in the product represents a major opportunity for improvement in productivity, as do efforts at systems management.

In his management handbook, Peters (1987) identified first level prescriptions for excellence, as shown in Figure 15.18. Measurements are also discussed. Figure 15.20 illustrates suggested measurements for the customer responsiveness prescriptions for excellence, and Figure 15.21 illustrates measures for organizational innovation. Many attributes and measurements suggested here need to be used in determining the overall worth of proposed alternatives for organizational advancement. Cost–benefit and cost-effectiveness analysis have been traditional measures used to determine this worth and are suggested for this purpose. This form of analysis is supportive of organizational effectiveness and requires systematic measurements for success. Cost-effectiveness analysis and related approaches for economic systems analysis are discussed in Chapter 7 and Sage (1983).

15.8 SYSTEMATIC MEASUREMENTS OF EFFORT AND SCHEDULE

At first glance, it might appear that efforts to minimize the effort and time required to produce a product or service are exclusively focused on the desire to become a low-cost producer of a potentially mediocre product. While this may well be the case, it is not at

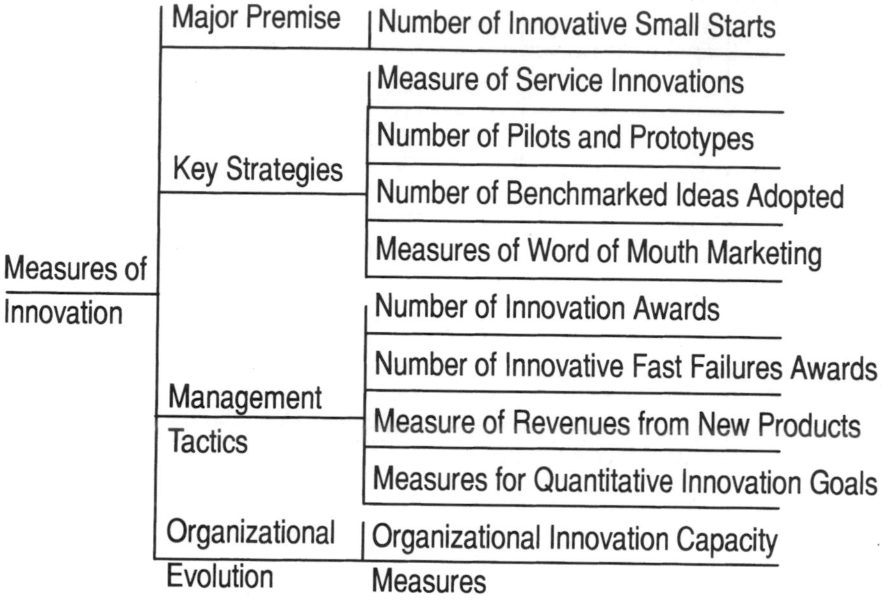

Major Premise	Number of Innovative Small Starts

Measures of Innovation

Key Strategies
- Measure of Service Innovations
- Number of Pilots and Prototypes
- Number of Benchmarked Ideas Adopted
- Measures of Word of Mouth Marketing
- Number of Innovation Awards

Management Tactics
- Number of Innovative Fast Failures Awards
- Measure of Revenues from New Products
- Measures for Quantitative Innovation Goals

Organizational Evolution
- Organizational Innovation Capacity Measures

Figure 15.21   Measures for organizational innovation.

all necessary that this be the focus of efforts in this direction. An organization desirous of high product differentiation, and perhaps such other features as rapid production cycle time, is necessarily concerned with knowing the cost and time required to produce a product or service.

Concerns with productivity at the level of processes naturally turn to the desire to maximize the benefit–cost ratio for a product. This allows an organization to cope with competitive pressures through the development of new and improved products, with simultaneous control of costs and schedule needed to produce a superior product in a minimal amount of time that can be marketed at a lower price than might otherwise be the case. An exclusive focus on cost and schedule reduction can lead to poor results if other efforts are not also included. These include attention to product and process quality and defects, and attention to these effectiveness issues at the level of systems management as well. Thus, when one wishes to minimize effort, or cost, and schedule, it is appropriate to consider metrics-based approaches. These process-related approaches are closely related to approaches to maximize quality and minimize defects. So, we will find a great deal of related commentary on metrics in other chapters in this handbook. Our primary purpose here is to provide a measurement framework and perspective for these related discussions.

15.9   SYSTEMATIC MEASUREMENTS OF DEFECTS

Clearly, neither the customer nor the organization responsible for a product wishes to have defective products. Some organizations will place defect minimization as their most important objective. Product defects may be omnipresent. Defects may be found in a product prior to its deployment and corrected at the development phase of the life cycle. Defects may not be found until just after a product or system is initially delivered to the customer. Defects may remain long after a product has been released to a customer. Defects may fall into any of a large number of categories. Systematic measurements are needed to determine defects at each of these times in the life cycle, and for a number of defect types. Defect removal involves detection, diagnosis, and corrective action. Generally, the measurements associated with defect removal are reactive, and possibly interactive in nature. Defect prevention is also very important, in many ways more important than defect removal. Obviously, the defects that do occur should be detected, diagnosed, and corrected. Defect correction is very much a function of operational-level quality control and quality assurance. There are a number of approaches for test and inspection, verification and validation, and the many other approaches that may be taken to defect prevention and defect correction. The discussion of systems engineering life cycles in Chapter 1 included some of the configuration management issues associated with defect correction, as configuration management, discussed in Chapter 5, is very necessary as an interactive approach to defect prevention and removal. Important works that discuss a number of important approaches in this area, especially as applied to software, include Conte et al. (1986), Moller and Paulish (1993), Card (1990), and Fenton (1991). Recent reprint books include Wheeler et al. (1996) and Baumert et al. (1994). A relatively extensive discussion of systems management use of defect metrics, and the relationship between this approach as an organizational driver and other approaches is provided by Grady (1992).

15.10 METRICS PROCESS MATURITY

Development of an optimal set of metrics for an organization requires a well-structured, well-documented, evolutionary approach. Many researchers have noted that metrics sets must evolve as the organization and its environment evolve; others have noted that metrics sets will improve as the staff involved gain measurement experience. The Capability Maturity Model (CMM), which was initially designed to assess software development organization's process maturity, provides insights into how process maturity might be used to guide the evolution of an organization's performance measurement activities and metrics set.

The CMM was developed by the Software Engineering Institute (SEI) at Carnegie Mellon with DoD funding (Humphey, 1995; Paulk et al., 1995). The initial objective of CMM development was to promote software engineering technology transfer within the defense industry. As knowledge of the Institute's work spread, along with recognition of the tremendous risks and cost overruns afflicting large software development efforts, non-DoD industries also became interested in the SEI's work. In 1992 the SEI invited non-DoD organizations to participate in its activities. The result has been a surge of interest, particularly in software and systems engineering process standardization and improvement, and in the CMM that was developed as a framework for assessing process maturity.

The CMM comprises five "maturity" levels. Organizations are expected to identify their current maturity level, and then use the CMM to help them identify and implement the actions needed to advance to the next level. A summary of the characteristics of the five maturity levels follows.

1. *Level 1: Initial.* The process is not under statistical measurement control at even the operational level. No systematic process improvement is possible. The process is

undisciplined and ad hoc. Success is very much a function of individual efforts, and those who achieve success under these circumstances are organizational heroes. When the hero leaves, the organization suffers greatly. Product quality and trustworthiness are unpredictable. There is little or no effort at risk management. Risk management is, at best, reactive in nature.

2. *Level 2: Repeatable.* A measure of thorough operational-level product control is achieved through metrics associated with cost, schedules, and product configuration changes. Thus, basic program management processes are established. Earlier successes may be repeated for very similar applications. The beginnings of an interactive approach to risk management have been established.

3. *Level 3: Defined.* The process has been adequately understood and specified such that operational quality control is able to yield products of specified trustworthiness with predictable costs and performance schedules. The organization has a set of standardized, consistent, and repeatable processes. Risk management is interactive. These are well integrated across the organization.

4. *Level 4: Managed.* Comprehensive process-related measurements and significant improvements in product quality are possible through the understanding and control thereby made possible. Interactive risk management processes are well in place, and some proactive approaches are in evidence.

5. *Level 5: Optimized.* The zenith of maturity. The organization is then able to make continuous improvements to products and services through continuous process improvements. Innovative ideas and leadership abound. There is much double-loop learning, and this further supports this highest level of process maturity. Risk management is highly proactive, and there are interactive and reactive controls and measurements.

Increasing levels of process maturity are characterized by improvements in systematic performance measurement. As an organization matures in its performance measurement endeavors, the cost of measurement should fluctuate and eventually decrease, while performance measurement effectiveness should increase significantly.

There are a number of related process maturity models. The SEI and others are developing a systems engineering capability maturity model (SE-CMM) (Kuhn et al., 1996). There are five maturity levels.

1. *Level 1: Performed Informally.* The base practices of the process area are generally performed. However, the performance of these base practices are not, in general, rigorously planned and tracked. Thus, performance depends on individual knowledge and effort, and may vary widely across individuals. Individuals within the organization recognize that an action should be performed, and there is general agreement that this action is performed as and when required.

2. *Level 2: Planned and Tracked.* The base practices of the process area are planned and tracked, as well as performed. Performance is according to specified procedures and is verified. The various work products conform to specified standards and requirements. The organization uses measurement to track process area performance. This enables the organization to manage its activities as based on actual performance. The major distinction from the Performed Informally level is that the performance of the process is planned and tracked.

3. *Level 3: Well Defined.* The base practices are performed according to a well-defined

process using approved, tailored versions of standard, documented processes. Measurements are a very important aid to effort at this level. The primary distinction from the planned and tracked level is that the process is planned and managed using an organization-wide standard process.

4. *Level 4: Quantitatively Controlled.* At this level, detailed measures of performance are collected and analyzed, and this leads to a quantitative understanding of process capability and improved abilities at performance prediction. Performance is now objectively managed, and the quality of work products is quantitatively known. The primary distinction from the Well Defined level is that the defined process is quantitatively understood and controlled.

5. *Level 5: Continuously Improving.* The organization now is able to establish quantitative performance goals, or targets, for process effectiveness and efficiency. These are based upon organizational objectives. The organization continuously improves its processes by using quantitative data and information obtained from performing the defined processes and from the use of innovative ideas and innovative technologies. The primary distinction from the quantitatively controlled level is that the defined process and the standard process undergo continuous refinement and improvement. These are based on quantitative understanding of change impacts on organizational processes.

The SE-CMM model is composed of two major parts. The capability-level part comprises the five maturity levels, as just described. There are also three major process area categories: engineering, project, and organization. There are a total of 18 process areas assigned to the three major process areas as now described.

1. Engineering Process Areas
   - (a) Analyze candidate solutions
   - (b) Derive and allocate requirements
   - (c) Evolve system architecture
   - (d) Integrate disciplines
   - (e) Integrate system
   - (f) Understand customer needs and expectations
   - (g) Verify and validate system
   - (h) Evolve system architecture
2. Project Process Areas
   - (i) Ensure quality
   - (j) Manage configurations
   - (k) Manage risk
   - (l) Monitor and control technical effort
   - (m) Plan technical effort
3. Organizational Process Areas
   - (n) Define organizational systems engineering process
   - (o) Improve organizational engineering processes
   - (p) Manage product line evolution
   - (q) Manage systems engineering support environment

(r) Provide ongoing knowledge and skills

(s) Coordinate with external suppliers

Measurements are needed to be associated with each of these. In evaluating systems engineering organizations according to the SE-CMM, an organization is evaluated on each of the 18 process area categories and a maturity level is established for each. This profile is presented and, if desired, an aggregate score can be obtained. Thus, this is a much more detailed model than the software CMM.

In a related effort at describing process modeling efforts, the plan–do–check–act generic life cycle is used to obtain a model for process engineering in the form of initiating, diagnosing, establishing, acting, and leveraging (IDEAL) (McFeeley, 1996). Figure 15.22 represents the life-cycle flow of the various process phases in the IDEAL model. There are a number of related capability maturity models, many of which are discussed in Sage and Lynch (1998).

Venkatraman (1994) has developed a five-level model of how organizational transformation can be enabled and inhibited by information technology. Our interpretation of these levels is as follows.

1. *Level 1: Product Improvement.* Redeploy improved systems and products with minimal changes in organizational structure, functions, purpose, or processes. Leverage IT for redesign of products and systems to increase organizational functionality.

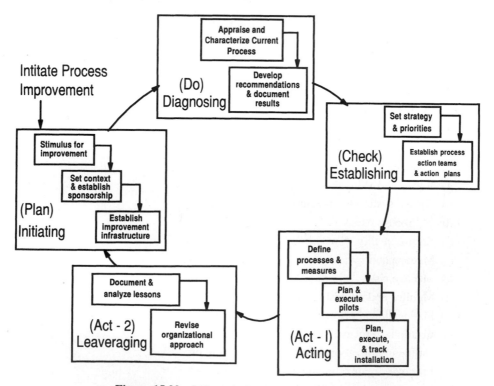

Figure 15.22 Life cycle for the SEI IDEAL process.

Lack of integratibility and duplication of functions may significantly limit ultimate performance benefits. There will generally be very little internal resistance to change.

2. *Level 2: Product Integration.* Systematically attempt to leverage IT for improvement of total organizational capability. Place more focus on technological connectivity and informational interdependence of organizational units than on interdependence of organizational functions. Require system improvement and integration of new system with legacy or heritage systems. Focus generally on improvement of present functionality, as contrasted with future organizational performance needs.

3. *Level 3: Process Reengineering.* Systematically attempt to redesign new innovative processes for the production of more cost-effective, higher quality, and customer responsive products. This may be implemented in response to competitor challenges and/or to better satisfy customers. This may be restricted to a single process, implemented in response to operational crises, and without potential need for organizational culture and leadership change if implemented for product improvement reasons only. This may be responsive to planned organizational change issues.

4. *Level 4: Organizational Networked Reengineering.* Redeploy improved IT systems for enhanced management effectiveness and operational efficiency. Redeploy improved IT systems for enhanced organizational learning and knowledge leveraging across the organization, effectiveness in the marketplace, and in coping with competitors. This involves enhanced strategic scoping decisions and results. It involves enhanced capabilities for cross-functional work teams.

5. *Level 5: Organizational Scope Reengineering.* Systematically attempt to strategically position the core competencies of the organization on critical performance measures. Make a fundamental effort to challenge the present structure, function, and purpose of the organization. Focus major effort at top-down redesign of efforts at this and lower organizational levels for enhanced effectiveness and efficiency.

These levels do not specifically indicate maturity of approach; rather they indicate levels of organizational effort at improvement. The first two relate to product, the third to process, and the last two to change at organizational levels of systems management. Clearly, a maturity score or level can be established with respect to the capabilities of an organization at each of these improvement levels. Efforts such as this can always be extended. For example, process integration and organizational network and scope integration can be added to this model to augment levels 3, 4, and 5. This would emphasize systems integration, as discussed in Chapter 14 and Sage and Lynch (1998), as one of the major concerns in systems engineering and management. Each of the resulting levels could then be associated with a capability maturity achievement level for a given organization.

In 1996, the American Productivity & Quality Center's Corporate Performance Measures System Consortium performed a benchmarking study of 31 best-practices organizations. One area of investigation was the measures these mature, best-practices organizations use to assess and improve their performance. While the detailed results of this study are not yet available to the general public, high-level results have been reported. According to the reports, the best-practice organizations trade off between using financial measures and using quality and productivity measures. Product manufacturing organizations tend to use more financial measures than support service organizations. Support service organizations use more quality measures, and are generally moving toward performance measurement systems that contain few financial measures. The six measures that are most important to the chief executive officer of these organizations have been identified as follows (Struebing, 1996):

- 36 percent of the top measures reported were financial measures
- 13 percent were quality measures
- 12 percent were customer measures
- 11 percent were productivity measures
- 12 percent were workforce measures
- 9 percent were market measures
- 7 percent were other measures

Approaches to the transfer of internal knowledge and best practices are discussed in O'Dell and Grayson (1998).

These latter two discussions lead us to the subject of use of information technology for organizational performance measurement.

15.11 INFORMATION TECHNOLOGY AND ORGANIZATIONAL PERFORMANCE MEASUREMENT

Information technology has evolved considerably over the past two decades. Database management systems (DBMS) were the first systems that supported organizational performance measurement by facilitating the storage and retrieval of performance data. Soon after, management information systems (MIS) were developed to provide managers with the information they needed, along with querying and reporting capabilities. Decision support systems (DSS) followed, which enhanced existing capabilities by providing managers with analytic models and other tools to help them analyze performance data. Most recently, executive information systems (EIS) were developed, they add even more manager-friendly capabilities, such as diverse information sources, advanced data querying, and graphical information display capabilities. Chapter 30 provides a discussion of this evolution and the development of intellectual and knowledge management efforts as a natural outgrowth of the information technology revolution.

The early work of Mintzberg (1975) on MIS provides insights into information needs and usage considerations from the manager's perspective. He makes the following observations, which are intended to be used by MIS developers to reduce the impediments that can reduce a manager's ability to use MISs.

1. Managers need broad-based formal information systems, in large part independent of the computer.
2. In an ideal MIS, the rate of information bombarding the manager would be carefully controlled.
3. Concentration on intelligent filtering of information is a key responsibility of the MIS.
4. Careful determination of channels is necessary in MIS design.

5. The formal information system should encourage the use of alternative and in-depth sources of information.

6. Stored information must be conveniently available to the manager.

7. The information specialist must be sensitive to the manager's personal and organizational needs.

8. The MIS should be designed to minimize some of its disruptive behavioral effects (politics).

Several characteristics distinguish EISs, designed specifically for managers, from other computer-based information systems (CBISs) such as transaction-based systems. First and foremost, EISs collect and utilize information from more than one source. It is not uncommon for an EIS to include data that have been gathered from an organization's business processes and operations, personnel, accounting, finance, and marketing CBISs. Also, data may come from such external commercial sources as stock market data services, industry statistical and analytical services, and news. EISs are capable of receiving data from diverse platforms (e.g., different types of computers, databases, file formats, and communication protocols), that are updated on different calendar cycles, that contain different types of data (e.g., hard, soft, graphical), and that may have different data element definitions services (Belcher and Watson, 1993; Watson et al., 1992). Finally, EISs provide the capabilities required to formulate, analyze, and interpret the diverse subject and format information received and to define, develop, and deploy the metrics needed by senior managers, and to store the resulting information for historical-trend analysis purposes.

While information systems have been used for several decades to automatically collect data about manufacturing processes, material movement, and inventories, they are just beginning to be used to collect and disseminate the non-process-oriented information required to support organizational decision making. In the early 1990s, a number of researchers began specifically identifying ways to enhance organizational performance measurement and improvement efforts through information technology (Boynton et al., 1993; Henderson and Venkatraman, 1993; Ives et al., 1993; Luftman et al., 1993). The impetus for this research was increasing information technology capabilities, as well as increasing awareness of the need to continually and systematically improve organizational performance and make performance information available to decision makers.

Recent advances in information technology provide much more efficient means for collecting, processing, storing, and communicating organizational performance information than the early information systems approach, which is based on database management. Information gateways form a general-purpose data bridge between various workstation networks and allows for information delivery and user interfaces to be constructed at workstations and at mainframe-based legacy systems that may be home to the central data resources of the enterprise (Volonino et al., 1995). Hardware and software standards support such integration. Distributed databases such as Oracle, Informix, and Sybase can be used with information gateways, as well as telecommunications capabilities and network technologies, to create comprehensive, accurate, real-time views of data throughout even geographically dispersed organizations. High-performance processors such as 32-bit multitasking operating systems can be used to rapidly and seamlessly integrate data collection, processing, communications, storage, and information retrieval and presentation. Rapid application development (RAD) and other development tools and methodologies, such as object-oriented programming (OOP), facilitate the development and efficient modification of large-scale systems. High-performance, cost-effective workstations, coupled with managers' increasing computer skills, enable complex but user-friendly applications to be developed and implemented. Finally, the Internet and World Wide Web further increase the number of available cost-effective information collection and sharing capabilities.

For years, even if extensive performance measurement data was collected, it was often not analyzed or used in decision making (Goodman et al., 1996). As a result of recent advances in information technology and its utilization, much computerized data, regardless of its location, can now be collected, processed, and presented to management in a timely manner as information in order that it can be used as knowledge in support of day-to-day decision making. With the enhanced data-processing, storage, data-analysis, and drill-down capabilities that allow high-level aggregates to be broken down into lower-level inputs upon demand, as well as the increasing computer competence of senior managers, the time and number of people required to massage performance data before it can be delivered to decision makers in the form of useful information can be reduced. Information security is still a concern, however, especially given the competitive value of business information, the immature current state of computer data security, and the mere seconds required to send confidential business information to virtually anyone and everyone on the Internet (Miller, 1996).

Volonino et al., (1995), as an example of the new organizational performance measurement and reporting capabilities made possible by information technology advances, describe a newly possible computer application developed by a Fortune 500 packaged-goods company. This application was represented to allow marketing executives to track all product orders and shipments worldwide by market and geographic region, and to obtain information about this in real, or near-real time. This enabled users to make more informed and timely decisions that relate to production, promotion, pricing, and advertising in response to changing conditions in the marketplace. It facilitated analysis of the data and presentation of the results of this analysis and interpretation far more rapidly than was previously possible. Many other such examples could be cited.

There are also public-sector illustrations of this use of information technology as a source of metrics. Even the federal government, which, like many large bureaucracies, is often criticized for lagging in the application of new capabilities, has recognized the importance of using information technology to enhance organizational performance. A report titled "Creating a Government That Works Better & Costs Less: Reengineering Through Information Technology" (Gore, 1993), stresses the need to leverage information technology to enhance the service provided to U.S. citizens by federal agencies. In this report, information technology is presented as the essential infrastructure for government of the 21st Century and as a modernized "electronic government" to give citizens broader, more timely access to information and services through efficient, customer-responsive processes. Eight initiatives for achieving that electronic government are presented below.

1. Integrated electronic benefit transfer
2. Integrated electronic access to government information and services
3. National law enforcement/public safety network
4. Intergovernmental tax filing, reporting, and payments processing
5. International trade data system
6. National environmental data index

7. Government-wide electronic mail
8. Establish support mechanisms for electronic government (training)

Using information technology as the means by which these services are provided will also service performance information to be collected and used to further improve service.

Using information technology to enhance organizational performance measurement and improvement is also addressed in the U.S. General Accounting Office (USGAO), (1994) report titled *Executive Guide: Improving Mission Performance Through Strategic Information Management and Technology*. In the preface to this report, it is noted that government agencies can use information technology to enhance the quality and accessibility of important information and knowledge for federal managers. The eleven best "practices" gleaned from leading organizations and recommended in this report are presented below. These practices are categorized into "three key functions critical to building a modern information management infrastructure: (1) *deciding* to work differently, (2) *directing* resources toward high-value uses, and (3) *supporting* improvement with the right skills, roles, and responsibilities." These functions and the eleven associated practices are based on lessons learned from leading organizations and are intended to both implement revamped services and to provide the performance information necessary to continuously monitor and improve performance.

1. Decide to Change
   (a) Recognize and communicate the urgency to change information management practices
   (b) Get line management involved and create ownership
   (c) Take action and maintain momentum
2. Direct Change
   (d) Anchor strategic planning in customer needs and mission goals
   (e) Measure the performance of key mission delivery processes
   (f) Focus on process improvement in the context of an architecture
   (g) Manage information systems projects as investments
   (h) Integrate the planning, budgeting, and evaluation processes
3. Support Change
   (i) Establish customer/supplier relationships between line and information management professionals
   (j) Position a Chief Information Officer as a senior management partner
   (k) Upgrade skills and knowledge of line and information management professionals

As a result of efforts such as these there have been major incentives to strategically align information technology capabilities and business strategy in support of better systematic measurements, and quality information and knowledge for improved decision making. The business community is making major strides in the use of information technology as a means of enhancing business performance. Numerous articles have been published in which researchers identify the need to align organizational information technology capabilities with business needs (Dertouzos, 1997). For example, works address the need to align and integrate business strategy, information technology strategy, information technology infrastructure, and organizational infrastructure from four perspectives: competitive potential; technology potential; service level; and strategy execution (Luftman, et al., 1993; Henderson and Venkatraman, 1993). Clark (1993) addresses the need for the information chain to support and integrate with the value chain. He notes that a key benefit of value-chain logic is the clarification of cause-and-effect relationships that relate the internal operations and activities in an organization as drivers of outcome events that are externally visible to the organization's stakeholders. The hypothesis here is that information technology must be aligned to an organization's value chain to support the association of specific plans and results and to provide cause-and-effect traceability. Boynton et al., (1993) examine information technology and its role in four combinations of product change and process change:

- *Mass production,* which leads to automation of manual processes to achieve cost-justified efficiency enhancements.
- *Mass customization,* which leads to integration of constantly changing network information processing and communication requirements, interoperability, data communication, and coprocessing critical to network efficiency.
- *Invention,* which leads to development and distribution of customized systems.
- *Continuous improvement,* which leads to the design of cross-functional information and communication systems that support organizational transformations.

In addition to determining what metric set formulations are required and what information technology capabilities might be leveraged to provide them, organizations must also be able to develop integrated organizational performance measurement and reporting *systems* that provide:

1. The ability to formulate the data and metrics into meaningful measures.
2. Timely access to relevant, accurate, current, complete measures of performance.
3. The capability to track these measures and identify historical-performance trends.
4. The capability to translate the resulting information into intellectual capital for performance improvement.

The ability of an organization to develop such a system cost-effectively, particularly if it is to be partially or completely computerized, should be a consideration in the selection of specific metrics to be included in the system.

One of the most promising information technologies for supporting organizational performance measurement activities is the EIS. EIS technology was introduced in the late 1970s (Watson et al., 1992) as a means of providing senior managers with the information they need to make decisions. Typical EISs include information about an organization's performance, the business environment in which the organization competes, its suppliers, and its competitors. EIS technology facilitates the delivery of this information via easy-to-use graphical user interfaces, intuitive information presentation approaches, exception reporting, and drill-down information analysis capabilities. Approximately one dozen established, commercial EIS tools exist, such as Comshare's Commander and Pilot's Executive Software, although many EISs are custom-built in-house. One purpose of an EIS is to support the intellectual capital of an organization, although much more than just an

EIS is needed to provide this. However, problems have plagued early EISs and failures were not uncommon (Rainer and Watson, 1995). In a survey of 50 large ($6.6 billion average total assets) companies with EISs, 21 of the companies reported that their EISs had failed at some time (Glover et al., 1992). As part of that study, these researchers identified the characteristics that caused the EIS failures. These failure characteristics and the number and percentage of the 21 companies that experienced them were as follows:

- Inadequate technology, 12, 57 percent
- Lack of sponsorship, 12, 57 percent
- EIS too complicated, 9, 43 percent
- EIS perceived as unimportant, 9, 43 percent
- EIS failed to meet objectives, 7, 33 percent
- Insufficient IS resources, 7, 33 percent
- Management not committed, 6, 29 percent
- Attempt to cost-justify EIS, 5, 24 percent
- Corporate culture not ready for EIS, 5, 24 percent
- Unknown objectives for EIS, 4, 19 percent
- Executives lost interest in EIS, 4, 19 percent
- Inability to define information requirements, 4, 19 percent
- Data integrity in doubt, 3, 14 percent
- Organizational resistance, 3, 14 percent
- Insufficient depth of information, 3, 14 percent
- EIS not linked to critical success factors, 3, 14 percent
- Sponsor turnover, 2, 10 percent
- Too much time required to develop applications, 1, 5 percent
- Information requirements were too detailed, 1, 5 percent
- Vendor support for EIS was discontinued, 1, 5 percent

Of the 20 EIS failure characteristics just described, 8 of the 20 characteristics (EIS perceived as unimportant, EIS failed to meet objectives, Unknown objectives, Inability to define information requirements, Data integrity in doubt, Insufficient depth of information, EIS not linked to critical success factors, and Information requirements too detailed) relate to the perception of, or value of, the information contained in the EISs. As stated previously, the primary goal of EISs is to provide executives with the information they need to make better decisions. Ideally, therefore, the management activities supported by an EIS should have resulted from better decisions enhanced and enabled through use of the EIS, and which have significant beneficial impacts on the organization's performance, and the result of which will be measurable via the organization's performance metrics. If the objectives for which the objectives measure information is needed are not clearly defined, appropriate information in the form of objectives measures cannot be identified for inclusion in the EIS. Even when appropriate information is selected, failure can result if the objective measure information in the EIS is (or is believed to be) inaccurate, out-of-date, incomplete, at too high a level, or if there is simply too much of it (prioritization, filtering, and/or higher-level aggregation are needed). Another information-related reason EISs fail is cost justification. In that study, five of the participating companies scrapped their EIS efforts after traditional cost–benefit analyses were performed. Some of these failures were attributed to EISs providing existing information that was already available. The fact that the information was available faster and in easier to interpret formats did not justify use of the EIS. Many other studies have identified similar difficulties. Even when an EIS does provide highly beneficial, nontraditional information to senior executives, its benefits are still extremely difficult to quantify (Belcher and Watson, 1993). Often the most valuable information is the most difficult and costly to collect, process, and protect from unwanted distribution. If efforts are not made to distinguish between the essential, expensive information and the less valuable, less expensive information, the EIS may fall victim to cost-cutting, despite its provision of useful information. For these reasons, information value should be considered, as well as logistical costs (data collection, processing, security, dissemination), during selection of information content for an organizational performance measurement and reporting system such as an EIS.

In another study performed to identify the factors leading to successful EIS development and ongoing operations, Rainer and Watson (1995) interviewed 48 EIS developers or operators from seven organizations. They asked the developers/operators to identify the most important factors in successful EIS development, as well as the most important factors in EIS on-going operations. In regard to the success of EIS development efforts, the interviewees identified the factors they considered important. Two of the four most often identified factors were "define information requirements" and "manage data." Rainer and Watson quote one executive interviewee as saying, "'This process [determining information requirements] is the most important part of the development process . . . it is, in essence, a complete corporate planning meeting.'" Another person indicated that "'The quickest way to get an executive to realize the value of an EIS is to make sure the system directly addresses a business problem he or she has. If the EIS does not relate to a specific business problem, it is much more difficult to get their information requirements. What we did not want to have was a solution in search of a problem.'" Other consultants who participated in the survey noted, "'Data comes from many sources into our EIS, both internal and external. The data must have consistent formats, but, more importantly, they must be accessible to us. We had to work very hard to get the data we needed to support our executive's information requirements'" and "'Our EIS stayed in the evolutionary stage for several months while the executives settled on their information requirements. Our EIS team experienced a lot of frustration during this process, but they stayed close to the executives and made our EIS a success.'"

Some other information-related factors identified as important to on-going EIS operations were:

- Timely information
- Accurate information
- Relevant information
- Adapting to changing information requirements
- Convenient information
- Standard (information) definitions in the enterprise
- Access external data
- Drill down (into data)
- Access soft, human data

- Access internal data
- Concise information
- New, unique information
- Comprehensive information
- Multiple methods to find information
- Integrate all types of data

One finding that apparently surprised Rainer and Watson in their survey of EIS success factors was that "higher quality decisions" was rarely cited as a success factor. They found that many managers felt that "'My EIS doesn't give me anything more than I was getting already. I'm just getting it faster with less effort. I think my EIS is increasing my efficiency, but I still do my job the same way as before I got the computer.'" One of their major conclusions—that executives must be very active in the development of an EIS if the resulting tool is to prove beneficial—is very important for those who engineer these systems. According to these authors, executives must "take an active role in EIS development, an area in which they are, for the most part, unfamiliar," and they "must champion the system and devote time to clarify their information requirements." Also important is that the things executives want from their EISs include "high quality information that they can obtain more quickly and with less effort than they could prior to having their EISs, thus improving their efficiency"; access to "information quickly and easily, requiring EISs to be easy to use and require little training"; "status access and exception reporting functions to enable them to assimilate information efficiently and effectively"; and improved communications. Of course, this study may reflect an inadvertent bias on the part of managers, who might feel that acknowledging the effectiveness of their EIS might also be an acknowledgement of their prior lack of management success. While this possibility cannot be discounted, it does not appear a very likely possibility. One important conclusion to this, and other studies of this issue, is that a multidisciplinary team is required continuously throughout the definition, development, and deployment of an EISs if it is to support executive functions, including the provision of relevant data and information that supports knowledge and associated increases in productivity.

How information content requirements are identified is crucial to an EIS's usefulness. To achieve its potential and increase its value to senior executives, an EIS must provide not only traditionally reported information, but also nontraditional information. Two problems often associated with identifying what information to include in an EIS are:

1. How to get senior executives to specify what they need.
2. How to keep up with senior executives' changing information needs (Watson and Frolick, 1993).

As Drucker (1995) points out, "Few executives yet know how to ask: What information do I need to do my job? When do I need it? And from whom should I be getting it?" He also emphasizes the great need to measure and not just to count. Mintzberg (1975) cautions that even when the essential information is identified and provided, managers may still be skeptical of its timeliness and reliability, and may "turn to ad hoc, informal information systems that they design and prove for themselves."

To facilitate identification of the information an EIS should contain, numerous approaches have been developed. In traditional software development efforts, a linear waterfall or system development life-cycle methodology is used in which the information requirements definition is a formal process involving observation of users' behaviors and analysis of their information needs that is completed before coding begins. As discussed in Chapter 1, however, many iterative approaches are now being used, including prototyping; spiral, operational-transformation; and reusability models. For an EIS information requirements definition, iterative approaches such as prototyping are quite prevalent, as they enable senior executives who may have trouble initially defining their information needs to refine their information requirements multiple times. It is important to note that multiple perspectives beyond information technology approaches must be adopted when designing an EIS, including economic, social, organizational, cultural, political, and relevant, security perspectives. Components of these approaches and perspectives need to be integrated with organizational performance measurement concepts to form an appropriate basis for a program of systematic measurements.

15.12 SUMMARY

In this chapter, we have described a number of perspectives from which we might view systematic measurements. In particular, we discussed systematic measurements from the vantage point of an organization that wished to obtain competitive advantage through customer satisfaction; effort and schedule; defect minimization; or systems management, especially for product differentiation. Clearly, a combination of these perspectives is possible, and desirable as well. The first and third approach is mostly concerned with product. The second approach is mostly concerned with process. The last perspective is based on systems management use of metrics. All of these are related and, generally, all perspectives should be taken.

By including elements such as these, we could then discuss how a systematic measurement approach would differ across the various maturity levels. For example, we might have the following sort of descriptors across a problem-handling strategy:

1. *Level 1.* Problems are fought over when they are first diagnosed, and usually with an improper and inadequate formulation of the problem. There is no measurement of anything. There is much screaming, accusations, and a confrontational atmosphere.
2. *Level 2.* Crisis management teams are set up to attack major problems as they occur through the development of short-range solutions. The reactive inspections of products creates the need for hypervigilance teams that create intraorganizational conflict. Long-term solutions are not sought and there is little organizational learning. This represents the beginning of measurement awareness, at the level of reactive measurements.
3. *Level 3.* Process measurements now occur, and there is a focus on measurement throughout the life cycle. Problems are diagnosed early through the use of interactive measurements at the level of process. Problems are resolved in an orderly manner, generally due to internal verification and validation of efforts throughout the life cycle. Learning is slow, however.
4. *Level 4.* Problems are identified at the planning stage, and a high-quality process is implemented that will result in doing things right the first time. This represents a high degree of interactivity and the beginnings of proactivity.

5. *Level 5*. Organizational learning has become very mature through experiential familiarity with proactive problem handling at Level 4 and, as a result, problems are prevented from occurring through use of proactive and interactive measurements. Reactive measurements are used, but there is less reliance on them at this highest measurement maturity level.

It would be highly desirable to characterize an organization's measurement maturity according to a scale, such as provided by this description. Figure 15.23 represents a hypothetical mix of measurement costs and the resulting cost and effectiveness that might be associated with systematic measurements at each of these five levels. Thus, we have suggested a measurement maturity model in our efforts here.

It is very important not to be deluded by poor measurements, or poor strategies for the definition, development, and deployment of metrics. In an insightful paper, Schiemann and Lingle (1997) suggest what they consider to be the seven greatest myths of measurement. These provide a fitting conclusion for this chapter. Our interpretation of these seven myths is as follows:

1. If we measure all the hard quantitative results, the soft behavioral stuff will naturally follow as a consequence.
2. Measurement is for bean counters, as the resulting data cannot be translated into bottom-line results that affect markets, people, operations, innovation, and the organizational environment.
3. Measurement is too hindsight oriented to be used for foresight-oriented efforts.

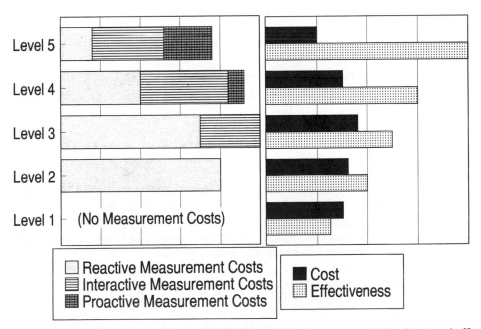

Figure 15.23 Relative measurement costs and the resulting system organizational cost and effectiveness at each of five measurement maturity levels.

4. The measurement may become the message and may impose attitudes rather than yielding information that aids understanding, knowledge acquisition, and problem solving.

5. Measurement will necessarily stifle creativity.

6. Measurement necessarily thwarts productive human activities and is unfriendly to humans.

7. The more we measure, the better we will be able to engineer quality systems and manage.

These authors also suggest four positive actions that may help avoid the pitfalls of inappropriate measurements and measurement processes:

1. Become an advocate of appropriate measurements.
2. Use measurements to promote understanding and not to blame people.
3. Set strategic priorities relative to the appropriate and wise use of measurements.
4. Share measures, and the foundations that lead to selection of metrics, with key stakeholders.

We believe that the approaches described here support attainment of these objectives.

REFERENCES

Allenby, B. R., and Richards, D. J. (eds.). (1994). *The Greening of Industrial Ecosystems.* Washington, DC: National Academy Press.

Allenby, B. R. (1999) *Industrial Ecology: Policy framework and Implementation,* Upper Saddle River, NJ: Prentice Hall.

Anthony, R. N., Dearden, N. J., and Govindarajan, V. (1992). *Management Control Systems.* Homewood, IL: R. D. Irwin.

Barnard, C. I. (1938). *The Functions of the Executive.* Cambridge, MA: Harvard University Press.

Basili, V., and Weiss, D. M. (1984) A methodology for collecting valid software engineering data. *IEEE Trans. Software Eng.* **10**(6), 728–738.

Baumert, J. H., Fendrich, J. W., and Tripp, L. L. (1994). *Achieving Quality Software through Standards, Metrics, and Process Improvements.* Los Altos, CA: IEEE Computer Society Press.

Belcher, L. W., and Watson, H. J. (1993). Assessing the value of Conoco's EIS. *MIS Q.* **17**(3), 239–253.

Blanning, R. W. (1992). Model management systems. (E. A. Srohr, and B. R. Konsynski, eds.). *Information Systems and Decision Processes,* Los Alamitos, CA: IEEE Computer Society Press pp. 36–48.

Boynton, A. C., Victor, B., and Pine, B. J., II. (1993). New competitive strategies: Challenges to organizations and information technology. *IBM Syst. J.* **32**(1), 40–63.

Christopher, W. F., and Thor, C. G. (eds.). (1993). *Handbook for Productivity Measurement and Improvement.* Portland, OR: Productivity Press.

Clark, M. (1993). Creating customer value: Information-chain-based management. *Inf. Strategy: Exec. J.,* Fall, pp. 13–18.

Clausing, D. (1994). *Total Quality Development: A Step by Step Guide to World-Class Concurrent Engineering.* New York: ASME Press.

Crosby, P. B. (1979). *Quality is Free.* New York: American Library.

Davenport, T. H. (1997). *Information Ecology.* Oxford: Oxford University Press.

Deming, W. E. (1986). *Out of the Crisis.* Cambridge, MA: MIT Press.

Dertouzos, M. (1997). *What Will Be: How the New World of Information Will Change Our Lives.* New York: HarperCollins.

Desimone, L. D., and Popoff, F. (1997). *Eco-Efficiency: The Business Link to Sustainable Development.* Cambridge MA: MIT Press.

Drucker, P. F. (1954). *The Practice of Management.* New York: HarperCollins.

Drucker, P. F. (1966). *The Effective Executive.* New York: Harper & Row.

Drucker, P. F. (1995). *Managing in a Time of Great Change.* New York: Dutton.

Emerson, H. P., and Naehring, D. C. E. (1988). *Origins of Industrial Engineering.* Atlanta, GA: Industrial Engineering and Management Press.

Ewusi-Mensah, K. K. (1989). Evaluating information systems programs: A perspective on cost-benefit analysis. *Inf. Syst.* **14**(3), 205–217.

Fiskel, J. (ed.). (1996). *Design for Environment: Creating Eco-Efficient Products and Processes.* New York: McGraw Hill.

Glover, H., Watson, H., and Rainer R., Jr. (1992). 20 ways to waste an EIS investment. *Inf. Strategy: Exec. J.,* Winter, pp. 11–17.

Goodman, J., DePalma, D., and Broetzmann, S. (1996). Maximizing the value of customer feedback. *Qual. Prog.,* December, pp. 35–39.

Gore, A. (1993), *Creating a government that works better and costs less: Reengineering through information technology.* Accompanying Report of the *Nat. Perform. Rev.,* U.S. Government Printing Office, September.

Grady, R. B. (1992). *Practical Software Metrics for Program Management and Process Improvement.* Englewood Cliffs, NJ: Prentice-Hall.

Grady, R. B., and Caswell, D. L. (1987). *Software Metrics: Establishing a Company-Wide Program.* Englewood Cliffs, NJ: Prentice-Hall.

Graedel, T. E., and Allenby, B. R. (1995). *Industrial Ecology.* Upper Saddle River, NJ: Prentice Hall.

Graedel, T. E. (1998), *Streamlined Life Cycle Assessment.* Upper Saddle River, NJ: Prentice Hall.

Harrington, H. J. (1991). *Business Process Improvement: The Breakthrough Strategy for Total Quality, Productivity, and Competitiveness.* (1993). New York: McGraw-Hill.

Henderson, J. C., and Venkatraman, N. (1993). Strategic alignment: Leveraging information technology for transforming organizations. *IBM Syst. J.* **32**(1), 4–15.

Humphrey, W. S. (1995). *A Discipline for Software Engineering.* Reading, MA: Addison-Wesley.

International Institute for Sustainable Development. (1996). *Global Green Standards: ISO 14000 and Sustainable Development.* Winnipeg, Manitoba, Canada: IISD.

Ives, B., Jarvenpaa, S. L., and Mason, R. O. (1993). Global business drivers: Aligning information technology to global business strategy. *IBM Syst. J.,* **32**(1), 143–160.

Jensen, A. J., and Sage, A. P. (1995). The role of information in organizational performance improvement strategies. *Inf. Syst. Eng.* **1**(3/4), 193–206.

Kaplan, R. S., and Norton, D. P. (1993). Putting the balanced scorecard to work. *Harv. Bus. Rev.,* September-October, pp. 134–142.

Kaplan, R. S., and Norton, D. P. (1996a). Using the balanced scorecard as a strategic management system. *Harv. Bus. Rev.,* January-February, pp. 75–85.

Kaplan, R. S., and Norton, D. P. (1996b). *The Balanced Scorecard.* Boston: Harvard Business School Press.

King, J. L., and Schrems, E. L. (1978). Cost-benefit analysis in information systems development and operation. *Comput. Surv.* **10**(1), 20–34.

Kuhn, D. A., et al. (1978). *A Description of the Systems Engineering Capability Maturity Model Appraisal Method Version 1.1. Carnegie-Mellon University, Pittsburgh, PA: Software Engineering Institute. (CMU/SEI-96-HB-004).*

Kurstedt, H. A. (1985). *The Management System Model Helps Your Tools Work for You,* Working Draft. Virginia Tech. Management Systems Laboratory. Blacksburg: Virginia Tech.

Luftman, J. N., Lewis, P. R., and Oldach, S. H. (1993). Transforming the enterprise: The alignment of business and information technology strategies. *IBM Syst. J.* **32**(1), 198–221.

Mahoney, F. X., and Thor, C. G. (1994). *The TQM Trilogy.* New York: American Management Association.

Main, J. (1991). Is the Baldrige overblown? *Fortune,* July 1, pp. 62–65.

Makridakis, S. G. (1990). *Forecasting, Planning, and Strategy for the 21st Century,* New York: Free Press.

Marquadt, D., Chove, J., Jensen, K., Pyle, J., and Strahle, D. (1991). Vision 2000: The strategy for ISO series standards in the 1990s. *Qual. Prog.* **10**(5), 25–31.

McFeeley, B. (1996). *IDEAL: A Users Guide for Software Process Improvement* (Handbook CMU/SEI-96-HB-001). Pittsburgh, PA: Carnegie-Mellon University, Software Engineering Institute.

Merkhofer, M. W. (1987). *Decision Science and Social Risk Management.* Dordrecht, the Netherlands: Reidel Publ.

Miller, J. P. (1996). Information science and competitive intelligence: Possible collaborators? *Bull. Am. Soc. Inf. Sci.* **23**(1), 11–13.

Mintzberg, H. (1975). *Impediments to the Use of Management Information.* New York: National Association of Accountants.

Mintzberg, H. (1989). *Mintzberg on Management: Inside our Strange World of Organizations.* New York: Free Press.

Mishan, E. J. (1976). *Cost-Benefit Analysis.* New York: Praeger.

Moller, K. H. and Paulish, D. J. (1993). *Software Metrics: A Practitioners Guide to Improved Product Development,* New York: IEEE Press.

National Institute of Standards and Technology (NIST), Technology Administration. (1995). *Malcolm Baldrige National Quality Award 1996 Fact Sheet.* Washington, DC: U.S. Department of Commerce.

National Research Council. (1996). *Linking Science and Technology to Society's Goals.* Washington, DC: National Academy Press.

O'Dell, C. and Grayson, Jr., C. J. (1998). *If Only We Knew What We Know: The Transfer of Internal Knowledge and Best Practice.* New York: Free Press.

Paulk, M. C., Weber, C. V., Curtis, B., and Chrissis, M. B. (eds.). (1995). *The Capability Maturity Model: Guidelines for Improving the Software Process.* Reading, MA: Addison Wesley.

Peters, T. J. (1987). *Thriving on Chaos: Handbook for a Management Revolution.* New York: Knopf.

Peters, T. J., and Waterman, R. H. (1982). *In Search of Excellence.* New York: Warner Books.

Porter, M. E. (1985). *Competitive Advantage: Creating and Sustaining Superior Performance.* New York: Free Press.

Puri, S. C. (1996). *Stepping Up to ISO 14000.* Portland, OR: Productivity Press.

Quinee, J. B. (1961). Long range planning of industrial research. *Harv. Bus. Rev.* **38**(4), 88–102.

Rainer, R., Jr., and Watson, H. (1995). What does it take for successful executive information systems. *Decis. Support Syst.* **14**(2), 147–156.

Rombach, H. D., and Ulery, B. (1989). Improving software maintenance through measurement. *Proc. IEEE* **77**(4), 581–595.

Sage, A. P. (1983). *Economic Systems Analysis: Microeconomics for Systems Engineering, Engineering Management, and Project Selection,* New York: Elsevier North-Holland.

Sage, A. P. (1992a). *Systems Engineering.* New York: Wiley.

Sage, A. P. (1992b). Systems engineering and information technology: Catalysts for total quality in industry and education. *IEEE Transa. Syst., Man, Cyberne.* **22**(5), 833–864.

Sage, A. P. (1995a). Systems engineering and systems management for reengineering. *J. Sys. Software* **30**, 3–25.

Sage, A. P. (1995b). *Systems Management for Information Technology and Software Engineering.* New York: Wiley.

Sage, A. P., and Lynch, C. L. (1998). Systems integration and architecting: an overview of principles, practices, and perspectives. *Syst. Eng.* **1**(3), 176–227.

Sashkin, M., and Kiser, K. J. (1993). *Putting Total Quality Management to Work.* San Francisco: Berrett-Koehler.

Sassone, P. G., and Schaffer, W. A. (1978). *Cost-Benefit Analysis: A Handbook.* New York: Academic Press.

Schiemann, W. A., and Lingle, J. H. (1997). Seven greatest myths of measurement. *Manage. Rev.* May 1997. Reprinted in *IEEE Engineering Management Review* **26**(1), 1988, 114–116.

Sink, D. S. (1985). *Productivity Management: Planning, Measurement and Evaluation, Control and Improvement.* New York: Wiley.

Sink, D. S. (1991). The role of measurement in achieving world class quality and productivity management. *Ind. Eng,* **23,** 23–28.

Sink, D. S., and Tuttle, T. C. (1989). *Planning and Measurement in Your Organization of the Future.* Norcross, GA: Industrial Engineering and Management Press.

Stevens, L. N., and Rezendes, V. S. (1996). Managing for results: Key Steps and challenges in implementing GPRA in science agencies, GAO Report T-GGD/RCED-96-214, July 10, 1996, 16 pp.

Struebing, L. (1996). Measuring for excellence. *Qual. Prog.* **29**(12), 25–28.

Sugden, R., and Willliams, A. (1978). *The Principles of Practical Cost-Benefit Analysis.* Oxford: Oxford University Press.

Thor, C. G. (1994). *The Measures of Success.* New York: Wiley.

United Nations Development Program. (1996). *Human Development Report 1996.*Cary, NC: Oxford University Press.

U.S. Congress. (1993). *Government Performance and Results Act of 1993,* Public Law 103-62, 103d Congress, 107 Stat. 285. Washington, DC: U.S. Congress.

U.S. General Accounting Office (USGAO). *Executive Guide: Improving Mission Performance Through Strategic Information Management and Technology.* (GAO/AMID-94-115). Washington, DC: USGAO, Comptroller General of the United States.

Venkatraman, N. (1994). IT-enabled business transformation: From automation to business scope redefinition. *Sloan Manage. Rev.* **35**(2), 73–88.

Volonino, L., Watson, H., and Robinson, S. (1995). Using EIS to respond to dynamic business conditions. *Decis. Support Syst.* **14**(2), 105–116.

Walton, M. (1986). *The Deming Management Method.* New York: Perigee.

Watson, H. (1992). How to fit an EIS into a competitive Context. *Inf. Strategy: Exec. J.,* Winter, pp. 5–10.

Watson, H., Rainer, R. K, and Houdeshel, G. (1992). *Executive Information Systems.* New York: Wiley.

Watson, H. J., and Frolick, M. N. (1993). Determining information requirements for an EIS. *MIS Q.* **17**(3), 255–269.

Zairi, M. (1994). *Measuring Performance for Business Results.* London: Chapman & Hall.

16 Human Supervisory Control

THOMAS B. SHERIDAN

16.1 INTRODUCTION

16.1.1 Definition of Supervisory Control

Computers and human users of computers are coming to interact in a broad range of tasks or applications in what has come be called *human supervisory control*. The term supervisory control can apply to any task in which the computer receives information about the ongoing state of a physical process (e.g., vehicle, robot hand, chemical plant) and, based upon such sensed information as well as information programmed into it by a human supervisor, directs actuators that act on that process. The human supervisor interacts with the computer to acquire information, make decisions, and give instructions. The computer is thus a mediator—communicating in one direction to the supervisor and in the other direction to the physical process.

Supervisory control is analogous to supervision of subordinate staff in an organization of people. The supervisor gives human subordinates general instructions that they in turn may translate into action. The supervisor of a computer-controlled system does the same.

Other popular terms such as *human-centered automation* imply supervisory control. Defined strictly, supervisory control means that one or more humans set initial conditions for, intermittently adjust, and receive information from, a computer that itself closes a control loop through a well-defined controlled process by means of artificial sensors and effectors.

Defined somewhat more loosely, supervisory control means a computer transforms human operator commands to generate detailed control actions, or makes significant transformations of measured data to produce integrated summary displays. In the looser definition the computer need not be able to exert closed-loop control on the process by itself, whereas in the stricter definition it must. The two cases may appear similar to the human supervisor, since the computer mediates both human outputs and inputs, and the supervisor is thus removed from detailed events at the low level.

A supervisory control system is represented in Figure 16.1. The human operator issues commands **c** to a *human-interactive* computer capable of understanding high-level language and providing integrated summary displays of process state information **y** back to the operator. This computer, typically located in a control room or cockpit or office near to the supervisor, in turn communicates with at least one, probably many (only three are shown), *task-interactive* computers, located with the processes they are controlling. The task-interactive computers thus receive subgoal and conditional branching information from the human-interactive computer. Using such information as reference inputs, the task-

Handbook of Systems Engineering and Management, Edited by A. P. Sage and W. B. Rouse
ISBN 0471 15405-9 ©1999 John Wiley and Sons, Inc.

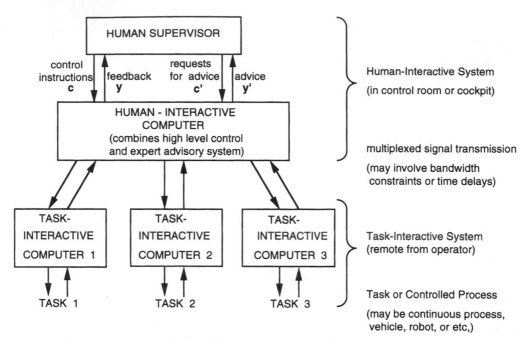

Figure 16.1 Supervisory control system.

interactive computers serve to close low-level control loops between artificial sensors and mechanical actuators, that is, they accomplish the low-level automatic control.

The task at the lower level usually is at some physical distance from the human supervisor and the human-friendly display-control computer. Therefore the communication channels between computers may be constrained by multiplexing, time delay, or limited bandwidth. The task-interactive computer sends control signals to and receives feedback signals from the controlled process.

In Figure 16.1 the supervisory command and feedback channels for process state information are shown to pass to and from the left side of the human-interactive computer. On the right side we represent the decision-aid functions, with displayed output y' being advice relevant to operator requests for auxiliary information c'. Earlier the latter interaction was not recognized explicitly as part of a supervisory control system. New developments in computer-based "expert systems" and other decision aids for planning, editing, monitoring, and failure detection have changed that. There is a useful analogy to the nervous system of higher animals, wherein commands are sent from the brain to local ganglia, and peripheral motor-control loops are then closed locally through receptors in the muscles, tendons, or skin. The brain, presumably, does higher-level planning based on its own stored data and "mental models," an internalized expert system available to provide advice and permit trial responses before commitment to actual response.

16.1.2 A Brief History of the Supervisory Control Concept

There is nothing new about supervisory control. Early forms of "automation" required humans to program or set in initial conditions for "automatic" machine tools, elevators, washing machines, and other such devices. Theorizing about supervisory control began as it became evident that direct manual control by the human operator was being replaced by the computer in aircraft. Yet these systems still required the human operator in a new role of monitor and goal-constraint setter (Sheridan, 1960). An added incentive was the U. S. space program, which posed the problem of how a human operator on earth could control a manipulator arm or vehicle on the moon through a 3-second communication round-trip time delay. The only solution that avoided instability was to make the operator a supervisory controller, communicating intermittently with a computer on the moon, which in turn closed the control loop there (Ferrell and Sheridan, 1967).

The rapid development of microcomputers was clearly forcing a transition from manual control to supervisory control in a variety of industrial and military applications. The design of display and control interfaces was seen to require a number of new developments (Sheridan and Johannsen, 1976; Sheridan, 1984). The National Research Council (1983, 1984) recognized supervisory control as one of the most important areas in which new human factors research was needed. Supervisory control has been reviewed previously by Moray (1986) and Sheridan (1987).

16.1.3 Current Salience of Supervisory Control

Automation and *smart machines* (those that include some form of computation responsive to their external environments) are everywhere, not only in the workplace but also in public transportation, schools, hospitals, government and military operations, and the home. However, seldom is the automation "stand alone" for any period of time; there almost always are human operators or users interacting with it in one mode or another. This mode (or these modes) almost always take the form of what is here described as supervisory control. Thus we see supervisory control coming to be the dominant mode of human interaction with automation or smart machines. Applications include:

- *Transportation,* including aircraft piloting and air-traffic control, rail system driving and dispatching, automobile driving and traffic control, loading and unloading of trucks, ship control and loading and unloading of cargo containers, control of undersea vehicles, control of manned and unmanned space craft.

- *Robotics* as applied to welding, cutting and other materials processing, painting, materials handling, machine assembly, computer-chip insertion, labeling of packages, product inspection, machine inspection and repair, building and highway construction, building cleaning and maintenance, mail delivery, warehousing, trash pickup, farming, mining, ordnance disposal, police sentry, and a host of military applications.

- *Process control,* including oil and gas extraction, pipeline control, petroleum refining, manufacture of paper, paint, pharmaceuticals, and chemicals generally.

- *Health care,* including telesurgery and semiautomated machining of bones for fitting prostheses, anesthesia dispensing and monitoring in operating rooms, patient monitoring in intensive care units and nursing homes and private homes, patient record handling, hospital laundries.

- *Other applications,* including automatic teller machines, supermarket and retail checkout, various home appliances, health monitoring and security devices carried on peoples' bodies.

• *Or just about any other aspect of life* where technology is helping people to acquire needed information, make decisions, and execute control over variables in their environments.

16.2 TASK ANALYSIS AND FUNCTION ALLOCATION

Task analysis is commonly regarded to be the starting point for design or improvement of any human–machine system, where *task* refers to *result* to be achieved, as defined by given criteria of completion or satisfaction. *Function allocation* is the allocation of functions between human(s) and machine(s), where *function* refers to the *means,* or the *operations*, by which a task is done. Task analysis is an act of analysis of the given specifications and criteria. Function allocation is an act of synthesis performed by the designer. These components of design are particularly important for supervisory control systems. It therefore is important to consider task analysis and function allocation at the outset of this chapter.

Task analysis and function allocation are highly interrelated. It is sometimes asserted that task analysis is the "what" and function allocation is the "how." This distinction is helpful only to a point, because what and how are not easily separable. It can be said that the main difference between task analysis and function allocation is that task analysis is specification and clarification of the *constraints that are given initially*, while function allocation is the specification of additional constraints that *determine a demonstrably best (or acceptable) means to achieve solution or design to the problem (task) in terms of available resources* (the principal categories being human and computer).

16.2.1 Task Analysis

Task analysis requires the breakdown of a given overall task into elements, and the specification of how these elements relate to one another in space and time. By task is meant the achievement of a specified result, the execution of a given set of actions, the production a given thing, the operation of a given system, or diagnosis and solution of a given problem. Sometimes *mission* is used to connote the full task or end state to be achieved, where task or subtask are used to refer to some component. Terminology for such breakdown of tasks is of no particular concern here except to note that some hierarchical breakdown is usually helpful or even necessary.

In defining task analysis, the term *given* was used. What is given are the task constraints, and task analysis is really a matter of articulating these constraints and making their implications visible. The task analyst's first duty is to bound the system and to list the salient independent variables (inputs) that must be considered when doing the task, and the dependent variables (outputs), the measures of which constitute task performance. This also includes specification, in as objective a form as possible, of system objectives (or overall mission).

The constraints are properties of both the independent and dependent variables. Specifying constraints on the independent variables is a matter of observing and estimating the properties of the system inputs in actual context. Specifying system objectives is particularly difficult because what seems intuitively important to the analyst may not be important to the operators, users, or other humans affected by the system. (Further discussion of utility theory and assessment is found elsewhere in this volume.) In contrast, specifying constraints on the dependent variables is a matter of deciding what limits there must be on the system outputs because of laws of nature, resource limitations, or other design constraints or tradeoffs that limit how well one can satisfy system objectives for expected inputs.

Constraint properties can take several forms: (1) they can be fixed numbers, which might, for example, set the mean values or ranges of the variables; (2) they can be functional relations between two or more variables, for example, laws of nature such as those of Newton (force = mass times acceleration) or Ohm (voltage = current times resistance), or economic statements of the dollar or time cost of certain objects or events; or (3) they can be objective functions of two or more variables that define the relative worth (goodness, utility) of particular states (combinations of variables).

Task analysis for human–machine systems means analysis into elements of seeking information, sensing various displays, perceiving patterns of information, abstracting, re-membering, and recalling parts of that information, making decisions, and taking control actions. These are all steps that can be done by either human or machine. It is important at the task analysis stage to stay relatively abstract and refrain from implying the detailed behavioral *hows* of either human or machine until the constraints of *what* is to be done are clearly specified.

While there are many techniques now used for task analysis, it is still very much an art. Task analyses end up being verbal statements of missions with qualifications, mathematical equations, block diagrams defining elements with arrows indicating what influences what, lists of variables with ranges and/or statistical properties, timelines, flowcharts, and so forth.

When performing a task analysis it is all too easy to assume a given specification to be more constrained than necessary. For example, let us assume one seeks to improve a particular system that currently consists of various displays and controls. The redesign can include some machine sensing and some automatic control, or at least a redesign of the displays and controls. There is a tendency to fall into the trap of analyzing the task as a series of steps for a human to look at particular existing displays and operate particular existing controls, the very displays and controls that constitute the present interface one seeks to improve upon or automate. The correct way to analyze the task is, at each step, to specify the information required, the decisions to be made, the control actions to be taken (at the level of the controlled process), and the criteria of satisfactory completion of each step, independent of the human or machine means to achieve those steps or particular displays or controls that might be used (Figure 16.2).

It is not easy to perform a task analysis without starting into the function allocation. But the effort must be made to determine what is really a given and what may be left as degrees of freedom for synthesis (function allocation). Sometimes what appears to be given when the designer first faces the problem or system is not necessarily given.

16.2.2   Function Allocation

Function allocation requires the designer to determine (i.e., to allocate) the various functions (means, techniques, roles, behavior) of the available resources (agents, instruments, tools, be they human or machine) of the required tasks and subtasks. Conversely, one can think of it as allocating the required tasks to the functions and their associated resources.

Task Step	Operator or Machine Identification	Information Required	Decision(s) to be Made	Control Action Required	Criterion of Satisfactory Completion

Figure 16.2 Task analysis column headings.

Issues of task analysis and function allocation to human vs. machine have continued to be a challenge for many years. As of now, however, while there are many accepted techniques for doing task analysis, there is no commonly accepted means of performing function allocation (and in particular to *optimize* this allocation, a bit of nonsense that one sometimes sees in print). The reasons are several: (1) while tasks/functions (the whats and the hows) may indeed be broken into elements, those elements are seldom independent of one another, and the task components may interact in different ways, depending upon the functions/resources chosen for doing them; (2) there is an infinity of ways the human and computer can interact, resulting in an infinite spectrum of allocation possibilities from which to choose; and (3) criteria for judging the suitability of various human–machine mixes are usually difficult to quantify and often implicit. Though the chapter suggests some things to consider when allocating tasks to various functions of people and machines (especially machines with "intelligent" capabilities), no attempt is made to offer a general procedure for synthesizing the allocation.

Fitts (1951) proposed a list of what "men are better at" and what "machines are better at" (see below). This is sometimes called the Fitts MABA-MABA List, or simply the Fitts List. It is often considered to be the first well-known basis for function allocation.

Men Are Better At

- Detecting small amounts of visual, auditory, or chemical energy
- Perceiving patterns of light or sound
- Improvising and using flexible procedures
- Storing information for long periods of time, and recalling appropriate parts
- Reasoning inductively
- Exercising judgment

Machines Are Better At

- Responsing quickly to control signals
- Applying great force smoothly and precisely
- Storing information briefly, erasing it completely
- Reasoning deductively

Jordan (1963) is concerned that the Fitts List is used by people who assume that the goal is to *compare* men and machines, then decide which is best for each function and for each task element. He quotes Craik (1947a,b), who had earlier pointed out that to the extent that man is understood as a machine, we know how to replace man with a machine. Early studies by Birmingham and Taylor (1954) revealed that in simple manual control loops performance can be improved by quickening, wherein visual feedback signals are biased by derivatives of those signals, thereby adding artificial anticipation (what the control engineer would call proportional-plus-derivative control) and saving the human operator the trouble of performing this computation cognitively. Birmingham and Taylor concluded that "man is best when doing least" in this case. Jordan suggests that this is a perfect example of Craik's tenet. He also quotes Einstein and Infeld (1942), who discuss the development and then the demise of the concept of *ether* in physics, and how when empirical facts do not agree with accepted concepts it is time to throw out the concepts (but retain the empirical facts). Jordan's point is that we should throw out the idea of comparing man and machine but keep the facts about what people do best and what machines do best. Jordan's idea of what we should espouse, and the main point of retaining the Fitts List, is that people and machines are *complementary*.

A more straightforward procedure for doing funtion allocation was suggested by Meister (1971): (1) write down all the salient mixes of allocation; (2) write down all the applicable criteria. Following this, one should rank order all combinations of allocation mix and criteria, thus determining a rank-order score. Alternatively one could weight the relative importance of each criterion, rate each mix by each criterion, multiply by the weight, and add up the scores for each allocation mix. However, there are difficulties with any such direct method, such as hidden assumptions, unanticipated criteria situations, nonindependence of tasks, nonindependence of criteria, nonlinearities that invalidate simple multiplication of weight by rating and addition of products, and most of all the fact that a very large number of possible interactions between human and computer compete for consideration, not simply "human vs. computer."

Price (1985) asserts that in order to make use of the Fitts MABA-MABA list, one needs data which are context dependent, but these data are mostly not available. Acquisition of these data is exacerbated by the fact that the machine is not static, that the capabilities of machines to perform "intelligent" acts such as automation and decision-support are ever improving. But, claims Price, automation can "starve cognition" if the human is not kept in sufficient communication with what the automation is doing or intending. He seems to agree with Jordan when he points out that human performance and machine performance are not a zero-sum game, implying that the combination can be much better than either by itself. Kantowitz and Sorkin (1987) and Price (1990) provide recent reviews of the literature in function allocation.

The public (and unfortunately too many political and industrial decision makers) have been slow to realize that function allocation does not usually mean allocation of a whole task to either human or machine, exclusive of the other. For example, in the space program, it has been common for the layperson to consider that a task must be done by either an astronaut or a "robot," that if a spacecraft is manned, then astronauts must do almost everything, and that if a spacecraft is unmanned, every task must be automated. In fact, on manned spacecraft many functions are automatic, and on unmanned spacecraft many functions are performed by human remote control from the ground. In this way alternative functions can mediate between the task and the resource (human or machine).

The Fitts generalizations remain a useful starting point for allocation of functions between human and computer. No other allocation model has replaced it in terms of simplicity and understandability. It is, however, a qualitative statement, subject to interpretation. As computers become "smarter," they creep closer to bettering some human properties. We can say for sure from lots of evidence that human memory prefers large data chunks with interconnected or associated elements and relatively complete pictures and patterns, and tends to be less good at details. These empirical facts strongly suggest that insofar as it is feasible and practical, the human should be left to deal with the "big picture" while the computer copes with the details.

It should also be said that human and computer can either function simultaneously (called *sharing* the task), or they can hand control back and forth (called *trading*). Both trading and sharing may occur as parts of the same overall task.

We now turn to the five supervisory phases of the human in supervisory control, discussing each phase in order, in terms of what the designer needs to consider for task analysis and function allocation.

16.3   THE PHASES OF SUPERVISORY CONTROL

There are five phases of any supervisory control effort, which occur necessarily in order, but with frequent looping back: this structure of supervisory phases is diagrammed in Figure 16.3. The supervisor must attend to these phases in performing the assigned task. In some control tasks one or another phase may be implicit (e.g., there is no time explicitly set aside for planning or learning, or the programmimg is a matter of activating one or another stored programs). The time duration of remaining in one or another phase before sequencing to the next phase or looping back may vary considerably. In order to perform the required task analysis and make the proper allocation of functions, the system designer must consider these phases, explicit or implicit, and understand what the supervisor must think and do on what time scales.

The first phase is for the supervisor to *plan* what needs to be done over some period of time, before any automation is turned on. The second phase is to *program (command)* the computer with what it needs to know to perform its assigned function for that time period. The third phase is to turn on the automation and *monitor* the automatic action, detecting any failures. The fourth phase for the supervisor is to *intervene* into the automatic action as necessary, deciding on any necessary changes manually (or cycling back to the instruction phase and making program modifications) before automatic control is resumed. This is the inner feedback loop in Figure 16.3. The fifth phase is to evaluate performance and *learn* from experience, and cycle back to the planning phase, the outer feedback loop.

Associated with each phase, there may be one or more ways the computer can serve some advisory function. Computer advice-giving has been particularly prevalent in planning, stemming from an appreciation of simulation and management decision aids; it is often inherent in programming as editing prompts built into the programming language; it is inherent in alarming and failure diagnosis during monitoring; it is increasingly emerging for the intervention phase, and except for simple data logging, it has been conspicuously absent for learning.

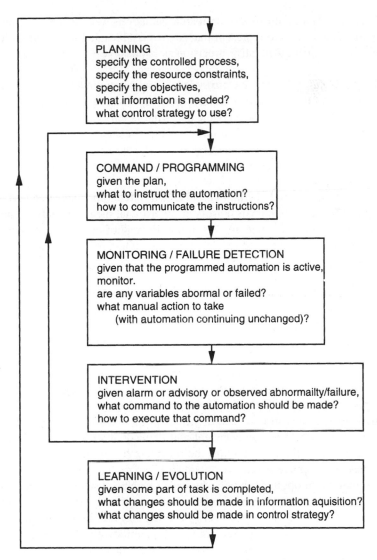

Figure 16.3 The five sequential phases of supervisory control.

16.3.1 Planning

16.3.1.1 Understanding the Given Controlled Process. The first order of planning business for the supervisor to know about is to know what physical thing (or mix of people and machines) is to be controlled, that is, the *controlled process*. What are the independent variables (inputs) that cannot be affected but must be expected? What are the independent variables that can be manipulated manually by the operator(s) or through mechanical or electronic actuators? And what are the dependent or output variables of interest?

The designer, when thinking through the supervisor's role, must limit the variables: physical space, items of hardware of software, affected parties, expected training of operators, degree of abnormality of the contextual situations, and so on, that one can be concerned with. The ideal is to represent that set of variables and their known relation to one another with as much quantitative precision as is possible.

16.3.1.2 Understanding Resource Constraints.

Control requires energy in some appropriate form, time, and money. It also requires personnel with qualifications and training. The supervisor must know what is available and what is reasonable.

System operation must obey the laws of physics. The supervisor must understand, at least at an intuitive level, what the salient laws are. The system must abide by reasonable economic constraints. And any humans must conform to what is known of ergonomic and cognitive limitations.

In order for control to occur, certain variables will have to be measured and presented to the supervisor, whether by human sensors or by artificial means. It must be understood what the required magnitude range and resolution in time of each variable are, and how well the salient variables (range and resolution in magnitude, resolution in time, degree of expectation and predictability) can be measured.

16.3.1.3 Understanding Control Objectives.

The human–machine control systems cannot be all things to all people and/or perform perfectly at zero cost, so the goal and the "how well" must always be some type of tradeoff between goodness of performance and cost in resources of various kinds. The system designer would like to have specified the goal(s) or objectives of the system in a quantitative way, stated as relative goodness of performance (a scalar) as a function of system inputs and outputs and resources used. This is called an *objective function* or *utility function* (Keeney and Raiffa, 1976; see other chapters in this volume regarding *utility* in decision making).

In the engineer's ideal world one could write explicit equations for (a) controlled process (i.e., the relations between independent and dependent variables); (b) the physical, economic, and human constraints; and (c) the objective function. Then, by simultaneous solution of these equations to maximize "goodness," an *optimal* control strategy is determined. In the real world optimization of a human–machine system is not possible.

Planning is a quintessential human function, so it is normally left mostly to the human supervisor (and the designer before the fact). Planning includes integrating the three steps just cited, even though the elements are not stated quantitatively. Indeed, if the elements were available quantitatively, the human could not do much with them. But human operators, especially those with experience, have a remarkable capacity for putting the three elements together to perform what is called "satisficing," namely coming up with plans, designs, or control strategies that are not globally optimal, but that are acceptable, or satisfactory (March and Simon, 1958; Charny and Sheridan, 1986).

16.3.1.4 Information Acquisition and Integration for Planning of Control.

People need not do all the planning by themselves. These days computers can store data based on past experience, can *fuse* data from current measurements, can apply a large number of crisp or fuzzy rules from a preprogrammed rule base, can run "what would happen if—" simulations, and can display the results to the human in whatever format is desired, including fancy computer graphics and virtual reality. Call this *automation of information acquisition and integration.*

These new aids for acquiring and integrating information and planning pose many new problems of "how much." Following is a list of ten levels of automation of information acquisition and integration. For most supervisory tasks both the supervisor (and the designer) would prefer some but by no means the maximum "help" from the computer.

1. The computer offers no assistance: the human must get all information.
2. The computer suggests many sources of information, or
3. Narrows the sources to a few, or
4. Guides the human to particular information, and
5. Responds to questions posed in restricted syntax, or
6. Responds to questions posed without restricted syntax, and
7. Integrates the information into a coherent presentation, or
8. Integrates the information into a coherent presentation, with an indication of confidence about each aspect, and
9. Passes it to the human or automation for action, but allows for other considerations.
10. The computer collects information as it sees fit, packages it, and presents it to human or automation for action with no opportunity to consider alternatives.

16.3.1.5   Decision and Control.   The final step of the supervisor in phase 1 is to make a decision and plan for control action. The final step for the designer is the corresponding allocation of human and/or computer for these functions. The following list shows a ten-level scale, similar to that of the preceding list, of automation for decision and control action. While there are some differences between the information acquisition scale and the control action scale, the idea is the same.

1. The computer offers no assistance: the human must make all decisions and actions
2. The computer offers a complete set of decision/action alternatives, or
3. Narrows the selection down to a few, or
4. Suggests one alternative, and
5. Executes that suggestion if the human approves, or
6. Allows the human a restricted time to veto before automatic execution, or
7. Executes automatically, then necessarily informs the human, and
8. Informs the human only if asked, or
9. Informs the human only if it, the computer, decides to.
10. The computer decides everything and acts autonomously, ignoring the human.

16.3.2   Command Programming

Given the plan, the supervisor must now decide (1) what to instruct the computer, and (2) how to communicate those instructions. These are two separate tasks, neither is trivial, and both are common sources of error. The first requires translation of the plan into functional steps, and the second requires the translation of the functional step into some formal command language that the computer can understand.

16.3.2.1   Forms of Command Language.   Command languages can take many forms: dedicated analog controls like joysticks and knobs, dedicated symbolic controls like discrete push buttons, concatenations of keypresses of alphanumeric characters to form computer code, interaction with computer-graphic touch screens, speech in limited vocabulary that is recognized by the computer.

Experimental studies by Brooks (1979) and Yoerger (1982) clearly showed that teaching a robot by giving symbolic commands relative to environmental objects is better than demonstrating desired movements in analogic fashion.

Mostly supervisor command or programming is proceduralized and intentionally not flexible (for fear that flexibility will lead to more errors), though there is controversy about how rigid or flexible command languages should be. As with desktop computer applications, it may well be desirable to have menu-driven commands available as well as shortcut key commands for those who are more experienced or happen to remember the latter.

Different procedures apply to different phases of the task and different contingencies, and the supervisor must either have these memorized or have ready access to information if a reminder of the procedures is needed. One new development is the *smart check list,* in which the computer keeps track of whether the operator performed the required step and if not displays an alarm or alert.

A serious problem occurs when the computer has inadequate capability to decide if it is getting appropriate commands for the current situation and to give feedback to the supervisor of what it understands it has been commanded to do. In normal discourse between two humans there are subtle cues by which the speaker gets to know if the listener has understood correctly, and/or the listener can ask for clarification. One would like to enhance supervisory command language interfaces with such a capability, but there remains a long way to go.

16.3.2.2   Operator Actions that Do Not Affect the Automation. Within command programming we mean to include any small changes that modify the way the information acquisition occurs (reprogramming the information agents) or the control action occurs. If, however, there are manual actions that operate directly on the controlled process and that do not affect the automatic control logic, we do not include them here. Examples of the latter would be direct manual manipulation of objects that are otherwise manipulated by robots, direct joystick, or steering wheel control of the aircraft or automobile rather than command through the autopilot.

16.3.3   Monitoring and Failure Detection

At this phase, with the automation turned on, the human becomes a monitor. Monitoring means continuously (or continually) observing many different variables, trying to be aware of the general "situation," and being on the lookout for abnormalities or failures. Direct actions upon the controlled process that do not amount to parameter or program changes to the automation itself (the task-interactive computer) are considered part of the monitoring function.

16.3.3.1   Attention Allocation in Monitoring. By what strategies should the human supervisor share attention among many variables, assuming it is possible to attend to but one at a time? Operations can be performed in parallel by essentially different computers, and there is plenty of time for a single computer to time-share. But the same cannot be said about the human operator. The human, relatively, is very slow, and cannot shift attention rapidly from one function to another.

It might be expected that the supervisory operator would rotate attention in a regular round-robin pattern among the functions. But that would not be efficient if some variables changed very rapidly and others changed very slowly. Proper allocation of attention obviously depends upon what is most urgent in terms of importance as well as window of time to take action and required human and physical resources (Wickens, 1992). One can model the problem as represented in Figure 16.4, where blocks represent different tasks that occur at unpredictable times and positions on the screen and move at uniform speed toward a *deadline* at the right. The width of each block represents effort or time required of that task, block height represents relative reward, and distance to the deadline represents time available. The problem is how to allocate attention among the blocks (sensory or motor tasks).

Tulga and Sheridan (1980) ran experiments using such a display and compared subjects' decision behavior to what was optimal (an algorithm that itself took considerable effort for the authors to figure out in this case). Results indicated that subjects' choices were not far from optimal when they had plenty of time to plan ahead. As the task demands became heavier and the *mental workload* grew, there came a point where the strategy switched from "plan ahead" to "put out bonfires," that is, do whatever needs to be done that there is still time to do. Interestingly, the subjects found that as pace and loading increased, and "planning ahead" finally had to be abandoned in favor of "do what has an immediate deadline," subjective mental workload actually *decreased*!

Some kinds of attention demands are simply to observe, and insofar as the resource constraint is the effort of switching attention (including focusing and reading) from one relatively simple display to another, it may be assumed that observation time is constant

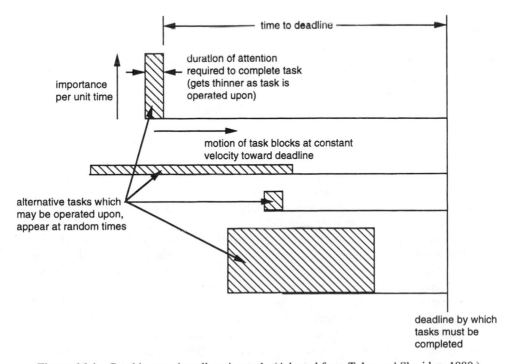

Figure 16.4   Graphic attention allocation task. (Adapted from Tulga and Sheridan, 1980.)

across displays. In this special case the strategy for attending becomes one of relative sampling frequency, that is, checking each of many displays often enough to ensure that no new demand is missed and/or sampling a continuously changing signal often enough to be able to reproduce the highest frequency variations present (the bandwidth). This latter is called the Nyquist frequency and is precisely twice the highest frequency component in the signal. Senders et al. (1964) showed that experienced visual observers do tend to fixate on each of the multiple displays (of varying frequency) in direct proportion to each maximum signal frequency. However, subjects tend to oversample the lowest frequency signals and undersample the highest frequency signals—the same kind of "hedging toward the mean" that is found in other forms of operator decision behavior.

At the instant ($t = 0$) an operator observes a display of input variable x, that is, x_0, the information may be regarded as perfect (see horizontal spike representing uncertainty as a density function in Figure 16.5). As time passes after one observation and prior to a new one, the information becomes stale, eventually converging to some a priori statistical expectation $p(x)$. This is represented by a series of three spreading uncertainty functions in Figure 16.5. Sheridan (1970) showed that one can optimize sampling rate T and control action u, given the cost of sampling C, $p(x)$, some model of how rapidly after a sample the knowledge of the input converges to $p(x)$, and the gain $V(x, u)$ associated with each combination of input x and action u. Figure 16.5b shows the declining expected value of greatest possible expected gain for control action based on inference from the sample x_0 (upper light line, see reference for derivation) and the average cost of sampling (lower light line); the difference is the net expected gain (heavy line), which has a maximum at the optimal T. This is equivalent to finding the best tradeoff between (1) sampling too often, to the point of diminishing returns (nothing much has changed in the situation), while incurring some cost per sample, and (2) sampling too seldom, having only a rough idea of what the input x is doing, and therefore not selecting a control action u at each instant that gives a very good return. The experienced human supervisor mostly allocates attention to various monitoring needs on an intuitive basis. If suitable value functions (V) were available, the monitoring strategy might be improved, but in practice making $V(x,u)$ explicit is a problem.

16.3.3.2 Situation Awareness and Mental Workload.

Situation awareness is usually defined in terms of missed signals or questions posed to an experimental subject when a simulator exercise is suddenly stopped (Endsley, 1996). Mental workload (Moray, 1979; Wierwille et al., 1985; Sarter and Woods, 1991) is usually measured by a subjective scale, since physiological indices show great inter- and intrasubject variability, and in real tasks operators are likely to refuse experimenter requests to perform *secondary tasks* (where decreased performance on the secondary task supposedly measures increased fraction of attention demanded by the primary task).

In the past operators have had to assess the current situation by keeping track of myriad individual displays, remembering the past history of these signals, and mentally extrapolating into the future to judge whether the trend is bad. Computers allow the integration of the formerly separate and discrete signals into interrelated graphical "pictures" (e.g., in aircraft the combination display for pitch–roll–bank–attitude, ground and absolute altitude, altitude rate, with flight director "bugs" appended to these) and the horizontal situation (map) display showing heading, waypoints, weather, traffic information, etc.). Increasingly displays are combining history (trends) and predictions with present status, so

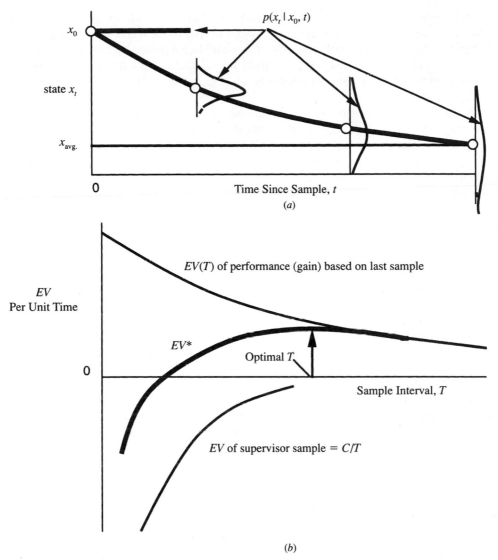

Figure 16.5 Optimization of sampling interval T. (a) Convergence of expected input x to prior expectation following sample x_0. (b) Decline of expected value of gain after sample at x_0 (upper light curve), average cost of sampling (lower light curve), and net expected gain (heavy curve).

as to give overviews with respect to time as well as overviews with respect to interrelated variables.

*16.3.3.3 **Abnormality Detection.*** Much effort has been given to the problem of how best to detect failures and get the operator's attention (Curry and Gai, 1976). Auditory alarms, either speech or tonal signals, are generally regarded as best, and visual indications are good for providing the details. Not more than seven different auditory alarms should be used to ensure easy recognizability. Visual lights or lists or messages should be arranged hierarchically and with as little redundancy as possible, so if the key information is provided at a high level (e.g., 24 V power supply failed), there is usually no need to present lower-level information that is an obvious result (e.g., failure of individual devices known to depend on 24 V).

In complex control rooms there is sometimes a special computer system devoted to alarming and diagnosing. An example is the safety parameter display system in nuclear plant control rooms. A common problem with such systems is what is called the *keyhole problem,* namely that the system becomes disassociated from the actual equipment and its actual location (the operator sees the world through the limiting "keyhole" of the one computer).

In any automatic alarming system or human alarming strategy there is the problem of trading off between responding to false alarms and associated costs (probability times cost) and ignored true alarms and that cost (probability times cost). Correct action for alarms and correct action for no alarm are sometimes assumed to incur zero cost. Because false alarms may be more frequent than true alarms, there is a danger of ignoring alarms, since the cost of an ignored true alarm is far greater than that of responding to a false alarm.

Alarm response is very situation dependent, especially with regard to whether the computer or human should be given the authority to respond to the alarm. For example, Inagaki (1997) analyzes the situation of the aborted takeoff of an aircraft, and the dependence of the ideal allocation on when the alarm occurs relative to rotation speed, and the relative trust of the pilot in the alarm.

There is also the chance that the very mechanism (circuitry) that activates the alarm will be disabled by the events that cause the alarm (e.g., fire, explosion, etc.). As pointed out by Inagaki and Ikebe (1988), there is a need for positive knowledge that the alarm system is working, and if it goes out for any reason, then the alarm itself should signal the operator in some way.

16.3.4 Intervention

Intervention occurs when the human supervisor makes any parameter or program change to the automation (task-interactive computer). This occurs when the automation completes a planned operation and it is time to invoke or program the next phase of operation. It also occurs when an abnormality or failure occurs and either the task-interactive computer must be quickly reprogrammed or the supervisor bypasses the automation and manually takes over control of the process.

16.3.4.1 When to Intervene? This is the phase of supervisory control where the human operator is deemed most essential, since by definition if the automation fails, it must be revised or replaced. The decision as to whether and when to intervene is a critical one. On the one hand the supervisor, upon getting an alarm signal or diagnosing a failure from nonalarmed displays, can react immediately; on the other hand, it may be wise to collect more observations, with hands in pockets, so to speak. There is an obvious tradeoff here: the more data one gets, the more sure one is of what action to take. But the longer one waits, the worse the potential consequences. A rational decision depends on the costs in each case, both functions of time. How long the aircraft pilot should wait after hearing the ground proximity warning, "whoop, whoop, pull up, pull up" is likely to be very different from how long the automatic machine tool operator should wait after an indication of tool wear or the process plant operator after an indication of low feedwater.

16.3.4.2 Do Operators Under Stress Really Make Decisions?
There is a debate about whether human operators under stress make considered rational decisions, just react as they have been conditioned from past experience, or something in between. We are quite ignorant here. The stress and confusion is compounded when the alarm or other indications of failure are believed to be unreliable, but the consequences of not responding to real failure in time are great.

16.3.4.3 Computer Aids to Intervention.
As mentioned earlier, an inherent part of modern supervisory control systems are computer-based decision aids (part of the human-interactive computer) that not only indicate failures but also perform diagnoses as to the cause and recommend remedial actions for the human to take. If such systems were reliable, of course, one could argue that the recommended action should be taken automatically, with the human acting only as a supervisory monitor (in this case with both automation and supervision escalating to a higher level). If such advisory systems are somewhat less reliable, one must argue that the human should consider the advice and weigh it together with that person's own independent judgment.

But then the question arises as to whether the human will put too much faith in the advisor and neglect making independent observations and judgments, or will have too little faith, stubbornly believing in personal judgment when in fact the computer is both quicker and more often correct. Pritchett (1997) found that experienced pilots in a collision-avoidance simulation had a tendency not to follow a new system's recommended avoidance maneuvers, probably in part due to disbelief that a real emergency existed, in part due to lack of understanding of the system's basis for its recommendations, and in part due to habituated responses that were inappropriate in this case. The correct level of trust can be approached by designing the advisor to give not only recommendations but also reasons for same. This enables the human to better calibrate the advisory system.

When there is an advisory system the tradeoff between collecting more observations and responding more quickly becomes more complex. It becomes a matter of executing the advisor's recommendation, deciding on and executing another action, or taking more observations before any response. Kim (1997) showed that if the costs and probabilities are known in each case one can derive the optimal decision. System reliability can often be estimated from available data; the costs of consequences are the more difficult factors to get.

Computer-based aids to intervention decisions by the operator are rapidly evolving, but we continue to lack sufficient understanding of human decision making under stress, trust, situation awareness, and mental workload. This is further exacerbated by the fact that in many of the most critical cases there are multiple humans and multiple independent computers interacting within the same decision process: for example, in aircraft piloting and air-traffic control, in automobile traffic, in nuclear power plants, and in surgical operating rooms.

16.3.5 Learning and Evolution

The final phase of supervisory control is that of learning, by the human(s) and the computer(s), and the consequent evolution of the parameters of the automation (both information acquiring and control executing).

In order for learning to take place, there must be some measures of performance that are evident (to the human or the computer, whichever is doing the learning) that can be associated with actions taken under various inputs or circumstances or system states. This is the $V(x,u)$ function referred to earlier in discussing attention allocation, but here it is meant at a higher level of aggregation, where x is a task context, or set of inputs, and u is a program or parameter set that determines actions.

Learning also requires that the human supervisor be trained to attend to the performance measures and to make associations with programs (plans, strategies) so as to be able to decide what worked better and what worse. It may demand that the supervisor be a bit experimental, and not just follow procedures like a mechanized zombie, even if it means making a few small but tolerable errors.

Obviously there is a big role for the computer here, one of measuring performance, making the correlations with programs and parameters set in by the human, as well as the exogenous variables, and plotting trends, and making recommendations to the supervisor for future plans.

16.4 EXAMPLES OF SUPERVISORY CONTROL APPLICATIONS AND PROBLEMS

The concept of supervisory control has found most currency in control of aircraft, ships, automobiles, and other vehicles; control of telerobots for earth, undersea, and space applications; control of nuclear, chemical, and other process control systems; and control tasks in hospitals and health-care systems. Computer word processing and "number crunching" are not normally considered supervisory control, since the physical process beyond the computer itself is relatively trivial. There is little doubt that the paradigm could be extended to just about all computer-interactive tasks.

16.4.1 Aviation

New technologies for commercial aviation (Wiener, 1988; Billings, 1991; Sarter and Woods, 1994) include the Traffic Alert and Collision Avoidance System (TCAS), the Global Positioning System (GPS) to take fixes of latitude and longitude from orbiting satellites, and various aids to detect windshear (in which neighboring air masses have significantly different velocities). The "glass cockpit" came in a decade ago with Boeing's 757 and 767, in which information, which before that was presented on separate displays, is now integrated by computers and displayed on several CRTs. Autopilots have been provided with multiple control modes, for example, for going to and holding a new altitude, flying to a set of latitude–longitude coordinates, or making an automatic landing when the airport has the supporting equipment. In the recent Airbus aircraft the primary flight mode is fly-by-wire through small sidesticks, in dramatic contrast to the old control yokes. In the cockpit computer-based expert systems give the pilot advice on engine conditions, how to save fuel, and other topics. Performance management systems are now available to optimize fuel and time.

The Flight Management System (FMS) is a recent addition to the flight deck. The FMS is the aircraft embodiment of the human-interactive computer. It permits pilot selection among many levels of automation, provides the pilot advice on navigation and other subjects, and detects and diagnoses abnormalities. The typical FMS control and display unit (CDU) has a CRT display and both generic and dedicated keysets. More than 1,000 modules provide maps for terrain and navigational aids, and procedures. A closely associated engine indicating and crew alerting system (EICAS) presents synoptic diagrams of various electrical and hydraulic subsystems. A proposed electronic map that would show flight plan route, weather, and other navigational aids is illustrated in Figure 16.6. When the pilot enters a certain flight plan, the FMS can automatically visualize the trajectory and call attention to any way points that appear to be erroneous on the basis of a set of reasonable assumptions. (This might have prevented the programmed trajectory that allegedly took the commercial Korean jetliner, KAL 007, into airspace of the former Soviet Union.)

The FMS extends a trend in aircraft control, which began with purely manual control

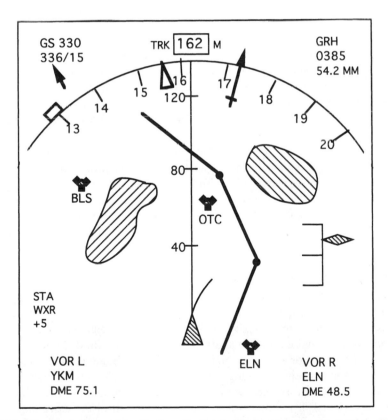

Figure 16.6 Proposed horizontal situation display for flight management system. (Courtesy of Billings, 1991.)

through a joystick (and later yoke) to control ailerons and elevators and pedals to control the rudder (and brakes). That control became power-aided (like power steering in a car, and later still became fly-by-wire (with no mechanical connection to transmit forces and displacements). The purely manual control began to be replaced, first by simple autopilots to hold an altitude and/or heading, later by autopilot modes to achieve an altitude or heading in a transient maneuver, and still later to automatically takeoff or land the aircraft. At first there were essentially no buttons to push, but gradually piloting has become to a far greater extent a matter of button-pushing. This trend started as a matter of pushing buttons on the panel that are uniquely dedicated to specific functions, and many of those are still there, collected into the mode control panel (MCP). However, the pilot has increasingly come to use the general alphanumeric keyset of the FMS CDU. Typically one CDU is at the most convenient location for the captain's right hand, the other is similarly located for the first officer's left hand, in such a way that either can back up the other.

There has been a move not only away from sticks and control yokes to button-pushing, but also a move away from direct body image control (which the psychologist might call "psychomotor skill") to supervisory control (which the psychologist might call "cognitive skill"). It is the FMS that more than any other technology forces this move, since the FMS forces the language of communication (and probably more critical, the language of the pilot's thinking) to include the alphanumeric streams (words, mnemonic codes). The pilot's mental models still include the physical space of the aircraft, and the lift, drag, thrust, and wind-vector components of force that force the aircraft in its Newtonian trajectory, but must now also include some of the "if– then–" logic in which computer technicians think.

Aircraft control is determined not only by the pilot and the FMS, but is also by air-traffic control (ATC). The same kinds of technology changes that have been taking place in the aircraft itself have been happening in ATC. Traditionally ATC is accomplished through a network of stations around the world that are connected to the aircraft by two-way radio and "see" the aircraft by means of radar. In most countries not only commercial carriers but also general aviation (GA) aircraft are required to carry transponders that identify them to ATC with a simple code, and if they plan to fly into the airspace of major airports, these transponders must have the capability of transmitting aircraft altitude. This provides an identification tag next to the blip seen on the ATC operator's radar display.

Airports with control towers (the smaller GA airports are an exception) typically have one ATC operator for takeoffs and landings and one for ground (taxiway) operations, both of whom operate from visual observation. Somewhat larger airports have an additional operator for approach and departure operations who works primarily from a radar screen. The most critical operations are usually found in the terminal area radar control rooms (TRACONS) associated with major airports. These are the ATC operators who must ensure that aircraft arriving from various directions are ordered with proper spacing into the landing pattern and out of the way of other aircraft that are leaving that airspace or that are just passing through. A final category of ATC operators is found in so-called "enroute" facilities (one for each major sector in a large country like the United States) that monitor commercial traffic on major airways ("highways in the sky") laid out between radio navigation beacons. Some recent technology permits every aircraft in the United States to be displayed on a single large screen. During major weather problems when, for example, airports must be closed, it is important to monitor and control the cascading effects of backups and delays from one sector to another.

A major innovation in ATC is "datalink," two-way communication with aircraft by digitally encoded messages. Datalink allows much more data communication with the ground than was possible through traditional voice channels. This enables new capabilities, but also poses new problems. Such capabilities include sending a variety of information about aircraft conditions to the ground for recording and analysis in real time, but there are commercial rights to privacy that may inhibit that possibility. Such capabilities also include making available to the pilot any or all of the radar information about other aircraft in that pilot's area, which might seem to be only an advantage. But the problem is that if the pilot sees all of what the ATC operator sees, and given that the pilot is ultimately in charge of the aircraft, ATC thereby loses control.

Perhaps even more serious is the question of how communication between ATC and the pilot is accomplished. With datalink, ATC instructions, landing clearances, and so forth, after being sent to the aircraft digitally can be put on visual displays. So too can pilot queries and responses, thereby (perhaps) obviating the need for voice communication. But pilots like the voice communications and have confidence in it (it is immediate, flexible, can be as informal as necessary, one gets immediate confirmation that the other party is paying attention, and by listening to the "party line" one can get a sense of what is going on in the surrounding air-traffic situation). Communicating by datalink may be more reliable in a narrow technical sense (voice messages get cut off and essential information can be lost, or require repeating, etc.). But datalink communication could turn out to be less reliable in a broader sense. What about having both? That may increase workload. So the form in which datalink will be implemented remains under active discussion.

A major ATC policy change now being seriously considered in the United States is "free flight," under which aircraft would be free to fly enroute by direct great circle routes and deviate from those as they wished, ether vertically or horizontally, depending on winds aloft, weather conditions, and other air traffic, all of which they now can observe for themselves. (They still would have to abide by ATC instructions as they converged on their destination airport.) Current efforts in the author's laboratory at MIT are focused on evaluating pilot decision aids for vertical free flight, and suggest that significant fuel savings are possible in this manner (Patrick, 1996).

Unmanned aerospace vehicles (UAVs) are now being developed by the Air Force. These are aerodynamic telerobots (see discussion on telerobots below) that fly autonomously like cruise missiles for periods of time, but require supervisory monitoring, corrections, and course updates intermittently. Initial development of UAVs neglected the human factors of supervisory control, which resulted in a major crash, so there is renewed interest on human supervisory control.

16.4.2 Rail Transport Systems

Although aviation is surely the most "high-tech" mode of transportation (and also the newest, the one with tightest selection and training requirements for human operators, and the most regulated in other respects), it is interesting to consider how similar changes are occurring in other transport modes. Consider railroad trains, the oldest of the three modes. Although slow to follow, railway systems are copying much of aircraft technology, such as the use of GPS for determining latitude and longitude, use of new sensing technology to measure critical variables (in this case, for example, the temperature of wheel bearings). Locomotives are even using "fly-by-wire" controls, which in the case of trains combine the brake and throttle adjustments into a single control, since both should not be operated at the same time. (Existing systems have separate controls for brake and throttle, evolved from air pipe valves and locomotive steam valves, which the driver operated directly.)

The primary task of a train driver is to control speed, and as part of this to anticipate the uphills and downhills, which tend to decelerate or accelerate the train by gravity, to know the fixed speed limits at different locations along the track where there are curves or switches or grade crossings or heavily populated areas, and to stay current with information about temporary speed limits due to weather or problems with the track or roadbed, or because maintenance crews are currently working at certain locations. The driver must know the train momentum and friction effects of wind and track as a function of speed for different consists (number of cars and type of load), and know the limits on train thrusting and braking (both normal and emergency). The driver is responsible for reading and comprehending the wayside signals at entry to each block or section of track (which becomes more and more difficult as speeds increase), communicate with the central dispatcher as necessary, know or be able to access the scheduled time of arrival at each station, and brake the train to a stop at precisely the correct point at each station. The driver must also monitor brake pressure and other variables of train and track condition, change the pantograph connection to the overhead electrical power cable from DC to AC and reverse, or operate Diesel engines, as required. The driver must integrate all of this so as to maintain safety, arrive at the next station on time, and minimize use of fossil fuel or electrical energy.

In some sense the train diver's job looks easier than that of the airplane pilot, since one might think of a train as a one-dimensional airplane. The special problem of the train is that it has very large momentum (it takes up to 3 km to stop a modern high-speed train traveling at 300 km/h, even under emergency braking). Yet, if an obstacle lies ahead—a truck stalled on a grade crossing, a bridge or section of track that is not in good condition, or a large rock or rock slide fallen from a mountain—the train does not have the lateral maneuvering room that the aircraft does; it must stop. Some trains are equipped with automatic speed controllers that can be set much the same as "cruise control" systems in automobiles (a primitive form of supervisory control). But, much like the automobile's cruise control, this does not take care of necessary braking. Some newer trains do have so-called "automatic protection systems" to signal the driver to put on the brakes if a train is detected immediately ahead (trains have been easier to detect than other obstacles by means of electromagnetic sensors). But the existing infrastructure of slow and stop signals are on fixed posts along the side of the trackbed, and not only can these not be seen from a distance, they are very difficult to read as the trains reach higher and higher speeds.

Flexible, computer-driven displays for monitoring, which are located on the operator's console (the "glass cockpit"), are gradually finding their way into railway systems. What might such displays be used for, and how might they assist the driver in speed control? One obvious use is for displaying in the cab the same information that now must be gleaned from the outside signals, providing this in an easy-to-read manner right in front of the driver. This, of course, depends on digital communication between the wayside and the locomotive cab, which means an expensive overhaul of the present signaling system.

An advanced version of such an in-cab display has recently been proposed by our group (Askey, 1995). Its purpose is to help the driver anticipate by (1) previewing speed constraints, (2) predicting the effects of alternative throttle and brake settings that might be set momentarily, and (3) providing an optimal throttle setting. Currently, even though speed signal information from the wayside is being transferred into the cab itself, this does not help if the operator who cannot see trouble far enough ahead. The driver certainly cannot see through the windscreen ahead for more than 1 km, and not even that at night (headlights on trains are more of a signal to outsiders than to provide illumination for the driver). Experienced drivers use a mental model of speed constraints set by curves in the track, grade crossings, or population-densities that are fixed and can be learned, but as noted earlier, because of track maintenance, rock slides, snow, and so on, there may be other speed constraints that are not so easily anticipated. For these reasons our proposed display previews the track for several kilometers, showing curves, speed limits, and other features, both fixed and changing (Figure 16.7).

It also shows prediction curves of the speed, determined from a computer-based dynamic model. These curves show how the speed will change as a function of track distance ahead if the current throttle setting is maintained. There are also predicted speed curves for maximum service braking and maximum emergency braking (a different braking system). Finally, there is a continuous indication of throttle settings that will get the train to the next station on schedule (assuming the train is at or near schedule currently and the winds are known), meeting the known speed limits, and under these constraints minimizing fuel. This indication can be updated iteratively by a dynamic programming algorithm. The latter display is akin to the "flight director" in an aircraft. This system was tested in a dynamic human-in-the-loop train simulator with a number of trained driver subjects and shown to improve performance significantly over driving with the conventional displays.

Rail systems have their equivalent of air-traffic control in the dispatch or traffic control centers usually located in major rail junctions. Here a crew of several people set schedules, set switches, communicate with trains by telephone as necessary, and monitor rail traffic for large areas. Up to now their principal tools have been the telephone, large sheets of paper with schedule tables and/or plots of location vs. time for every train for the main

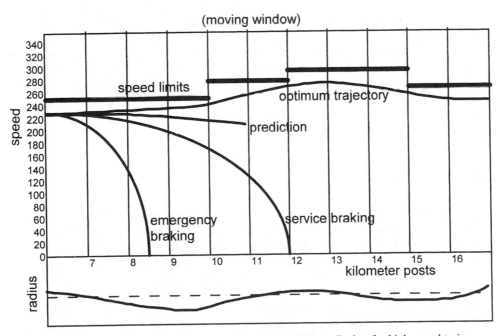

Figure 16.7   Combined preview, prediction, and advisory display for high-speed train.

line sections of track, and wall-mounted diagrams or maps showing (usually with a small light bulb for each block of track) which blocks are occupied by trains. They have few if any computer-based tools to predict the cascading effects of late trains or accidents.

16.4.3 Highway Vehicles

Highway vehicles, including trucks, busses, taxicabs, special vehicles, such as fire-trucks, police cars and ambulances, as well as private automobiles have also followed the trend toward use of new sensor, communication, computer, and supervisory control technology. The European Community countries were probably the first to organize a major development project. One called Prometheus began roughly a decade ago and has recently been completed. Its aims were to introduce new computer-based navigation and control systems; to look ahead to common signing, speed control, and other infrastructure developments for European countries; and to enhance the position of European automotive manufacturers in world markets. A second program, called Drive, with similar aims but a different organizational structure, is still ongoing. In Japan similar national projects instituted cooperation between vehicle manufacturers, suppliers, and government agencies. In about 1990 the U.S. Department of Transportation (USDoT), in concert with state agencies, vehicle manufacturers and suppliers, insurance companies, and universities, founded a unique advisory and educational organization initially called Intelligent Vehicle-Highway Society of America. Its name was later changed to Intelligent Transportation Society of America (ITS) after metropolitan rail interests complained of being left out, but it is still largely highway vehicle oriented. From the outset human factors considerations were seen as important to the success of ITS.

A first major demonstration project of passenger vehicle navigation was undertaken in the United States by General Motors, Avis, the American Automobile Association, the State of Florida, and the City of Orlando. One hundred rentable Oldsmobiles were outfitted with GPS transducers, simple inertial transducers, 386PCs, CD ROMs to hold street map information for Orlando, speech-generation devices, and CRT graphical display units. The Rental agency customers could indicate their destinations to the computer, and the system would then guide them there by a combination of voice and graphical displays. This included, for example, advice to move to one side of the road or another in preparation for a left or right turn at a forthcoming named intersection. (In some sense this is only the input side of supervisory control: the driver is aided in sensing and decision-making, but must still perform vehicle control by manual means.) The GPS was accurate enough in most cases to detect whether the driver stayed on the correct road, and if not would give instructions on how to get back on the correct course. A similar project by Ford and Motorola in the Chicago area is in progress at this writing.

Controversy over putting such navigation technology in passenger cars mostly centers on safety. Does use of such a system detract significantly from the driver's attention to the road, or does it result in greater confidence and greater safety? Can older drivers accommodate their visual focus between the road outside and the internal graphics or other displays sufficiently rapidly? Is this yet another addition to the driver's workload, which in the future may include not only cellular telephone but also fax machines and, e-mail. Are map displays too distracting (currently some states in the United States claim yes and legally prohibit such front-seat map displays, classifying them the same as television sets. If maps are used, should the map direction always correspond to the vehicle direction (the vehicle then becoming a fixed symbol), or should the map remain fixed with north (or south) up, one's own vehicle becoming a moving symbol? The answers to these questions may be settled to some extent by research, but the final arbiter will be the marketplace.

The major safety objective is to avoid collision. New technology provides a different approach to collision avoidance. Instead of accepting the "first crash" (of the vehicle with environmental obstacles) and focusing only on technological means to ameliorate the "second crash" (of the driver and passengers with the interior of the vehicle), there is now the opportunity for technology to prevent the first crash. In this regard perhaps the second major new development for "intelligent vehicles" is so-called "intelligent cruise control." Current cruise control systems do not know when the vehicle is about to collide with another. Intelligent cruise control systems make use of microwave and optical sensors to detect the presence of a vehicle in front, and the resulting signal can be used to warn the driver or to brake automatically. Interestingly current demonstration projects have mostly stayed away from actually applying the brakes, choosing instead to decelerate and downshift. The reason for this seemingly irrational choice is a concern that once claims are made that an automatic system will brake, the driver may become less inclined to apply the brakes personally, and the manufacturer will be wide open to tort litigation.

Antilock brake systems (ABSs) are known to be hard to use correctly by many or even most drivers. Instead of stepping hard on the brake pedal, drivers have a tendency to ease them on and pump the brakes, thus making them less effective than they were designed to be. Means to "program" the brakes for different road conditions, different drivers, and the like, are now being evaluated using a driving simulator in the writer's laboratory.

Currently the most ambitious U.S. ITS project is the "automated highway system" (AHS). This project seeks to demonstrate how vehicles can be driven into special lanes and turned over to computer control. A variety of automatic steering and longitudinal control techniques have been demonstrated, including platooning of vehicles at high speeds with but one meter headway, controlled electronically. Allegedly, according to the University of California (Berkeley) PATH Program researchers (Hedrick, personal communication), it can be shown that it is actually safer to have high-speed vehicles in close proximity to one another than spaced at longer distances, since in this way large velocity differences can never build up and hence large-impact collisions cannot occur.

The "intelligent highway system" also has its counterpart of air-traffic control, called an "advanced traffic management system." This is not a unique concept, for in fact many new highway and tunnel projects have built accompanying traffic management centers. The newest such system is associated with the largest current construction project in the United States, Boston's $8 billion Central Artery/Tunnel. In its control room several operators will monitor traffic by means of roughly 400 video cameras, and a comparable number of magnetic and other sensors (of not only traffic but also heat, smoke, carbon monoxide, and toxic chemicals of various kinds. They will use a sophisticated computer system that infers where and how serious any abnormality is and make a recommendation for intervention action (e.g., send in a tow truck from a particular location, or certain fire fighting facilities, or police, etc.). They can change variable message signs, block traffic lanes, initiate visual and auditory alarms, or activate fire extinguishers.

Our laboratory is currently participating in the design of this "incident response system," examining in particular the intervention phase of supervisory control, which has received all too little research attention to date (or as much as teaching and monitoring). In the particular case of the Central Artery/Tunnel, the supervisory operator is expected to decide within seconds whether to accept the computer's advice (in which case the communication to the proper responding agencies is automatically commanded), or reject the advice and generate commands personally (in effect intervene in an otherwise automatic chain of events), or do nothing while making more observations. An operator decision model of this three-way time-dependent decision has been developed by Kim (1997), that specifies an optimal decision based on costs of dispatching and not dispatching emergency services contingent upon probabilities of true and false information and the time variations in the quality of information. Kim also experimented with human subjects to evaluate what trained operators actually do under different conditions.

16.4.4 Telerobotics and Discrete Product Manufacturing

A *telerobot* is a system, supervised by a human from a distance, that is capable of sensing, deciding, and acting in a variety of task circumstances based on a stored program. In other words it is an automatic system operating under supervisory control, where in this case the "tele" prefix denotes remote supervision. Usually one thinks of a remote manipulator for space, undersea, mining, toxic or other hazardous environments. So-called industrial robots in factories are not telerobots because they are generally specialized to specific tasks such as welding or paint spraying, operate for long periods without human intervention, and human interaction is immediate. In some sense airplanes and other systems we do not normally think of as robots are really telerobots in the general sense, since they embody artificial sensors, actuators, computer decision making, and somewhat remote supervisory control by a human.

Figure 16.8 shows the undersea telerobot *Jason,* big brother of *Jason Junior* (when the ship *Titanic* was first discovered several thousand feet under water *Jason Junior* was guided manually inside). *Jason* has several propellers for mobility in any direction, a manipulator arm, and a variety of video, sonar, temperature, and other sensors. It can be

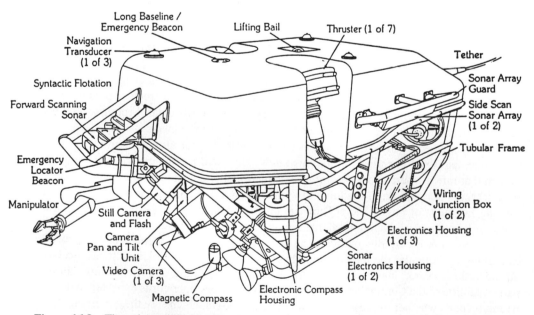

Figure 16.8 The submersible telerobot *Jason.* (Courtesy Woods Hole Oceanographic Institution.)

operated in a supervisory as well as a direct mode, using sonar triangulation to provide position feedback. See Sheridan (1992) for a more complete discussion of telerobot control.

A variety of control modes (Brooks, 1979; Yoerger, 1982) and graphical aids (Ellis, 1991) have been developed for the planning phase of supervisory control. Many of them construct in the computer a geometric model of the task and telerobot, allowing the supervisor to test out a set of commands before implementing them. Various graphical techniques help the user observe the robot arm/hand relative to various environmental objects from any arbitrary viewpoint, observe in the 2-dimensional display the orthogonal distances between objects and thereby avoid collisions (Das, 1989), and even get advice from the computer regarding the safest trajectory (Park,1991).

Command languages include both symbolic and analogic means of telling the telerobot where to go (e.g., some are equipped with wheels, tractor treads, or legs for mobility on land; some have propellers for mobility under water; some have thrusters for moving around in space) and what to do (e.g., pick up a designated object and place it in a new location). *Put that there* has become a common class of command, where the action command *put* is further modified by descriptors of how to grasp, how fast to move, how to make contact, how to assemble, and so forth; *that* designates the object to be handled; and *there* specifies the precise final configuration of object and environment.

The monitoring, intervention, and learning phases of telerobot supervision are not unlike those for other systems. The key problems have to do with providing good integrated displays that provide the supervisor with what is needed and when it is needed in synoptic or integrated form.

A special problem encountered in telerobotics (or any form of remote control) occurs when the communication channel contains a relatively long time delay, such as the 3-second round-trip delay that occurs when controlling a space telerobot from the earth. In this case any attempt to control with a closed-loop gain exceeding unity (normally imperative at frequencies where one wants faithful tracking and noise rejection) causes an instability. A human controller's natural tendency is to resort to a "move-and-wait" strategy, wherein the operator commands a short open-loop move of the remote telerobot arm, then waits for feedback, then makes another move, and so on. This procedure makes for extremely slow action, since the moves must be very short, especially as the telerobot hand approaches contact with an object. Ferrell (1965) first studied such human control experimentally, and constructed a model that predicted task completion time. This was precisely the problem that led the MIT Human-Machine Systems Lab to its efforts to develop supervisory control (Ferrell and Sheridan, 1967) as a viable alternative.

Industrial robots deserve at least some mention with regard to supervisory control, since they are very numerous in industry and are programmed and monitored by humans. However, the cycle of intervention is much less frequent than with the telerobots because the tasks they are called upon to perform are generally much more specialized and predictable.

16.4.5 Process Control

The term *process control* refers to control of any process that is continuous in time and space, where the product flows through the manufacturing or transportation process. These latter processes include electricity generation (nuclear, fossil, hydro, solar, etc.); electricity distribution; chemical manufacturing; petroleum cracking and processing; and fabrication of cloth, metal wire and strip, paper, rubber.

Process plants are increasingly run on supervisory control. In fact, they were among the first large-scale systems to be controlled this way, with large numbers of simple proportional-integral–derivative (PID) controllers to regulate flow, pressure, temperature, and so forth, scattered about the plant, and set points centrally adjusted by human operators who also monitor for many alarms, several thousand in some plants.

Beyond the simple PID controllers, computers have been used in process control mostly for recording data, performing off-line analyses on data, establishing trends and predictions, and generating displays. They have not been used much for control, in the sense of high level goal setting and programming, though that is changing in some plants where an entire batch run can be programmed for a series of operations as well as various abnormality contingencies.

With time and increase in the size of nuclear plants, the number of independent indicators and alarm lights in the control room grew into the thousands. But it became evident that after a serious malfunction too many displays changed state (in one large loss-of-coolant simulation the writer counted 500 displays that changed in the first minute, 800 more in the next minute). Operators commonly referred to the control board as a "Christmas tree." There was no way an operator could quickly determine what was going wrong. Experienced operators claimed that some obscure form of "pattern recognition" was the way they could gain insight. When the accident at Three Mile Island occurred, the Nuclear Regulatory Commission mandated a *safety parameter display system* (SPDS), a synoptic display of the health of the plant—a way to keep the operators informed regarding which major systems were functioning properly and which were not. However, one problem learned by the nuclear power community was that the SPDS, since it was forced on a community of operators who had little or nothing to do with its invention or use, at first was simply ignored by the operators. Technology that purports to be "helpful," especially in crisis, must be understood and accepted by the users!

16.4.6 Health Care

Supervisory control is coming to find application in all branches of health care. The hospital operating room and intensive care ward are coming to look more and more like control rooms, with many panels of displays and controls, all mediated by computers. For example, the anesthesiologist is really a process supervisory controller, where the process is the patient and the anesthesiologist plans a drug treatment, programs the machinery in the operating room to dispense the correct anesthetics in the correct amounts at the correct time, monitors the patient's vital signs, intervenes as necessary to change the treatment or directly handle the patient in other ways (or advise the surgeon to do the same), and learns from experience.

Miniaturization of sensors as well as radio communication devices has enabled a whole new prospect for health monitoring for patients with chronic problems. These advances enable automatic monitoring of patients in the hospital or chronic care facility, with alarms sounded in the nurse's station when variables exceed their set thresholds. There are exciting new prospects for instruments that are wearable by (even implantable within) patients living normally in their own homes, or working, shopping, and the like, that can signal the appropriate medical personnel by means of cellular telephone or pager communication infrastructure.

Programmed robots under supervisory control by the orthopedic surgeon are now being used to drill specially shaped holes in bones during operations to insert artificial joints.

The holes thus drilled can ensure much tighter and cleaner fits than previous manually drilled holes.

Minimally invasive surgery refers to operations performed on internal body structures through sealed puncture devices (trochars) by means of miniature fiber optics or video cameras *(endoscopes)* to view inside, and special tweezers, scissors, knives, staplers, and the like, that are handled from outside but inserted through the trochars to manipulate inside. While relatively few endoscopic operations are common thus far (such as *cholycystectomy* or gall bladder removal, or removal of torn cartilage from knee joints), there is much excitement about a broad range of new surgical procedures, and little doubt that supervisory control will soon be added to what is now essentially manual control. This is already happening in brain surgery, where computer-based images from prior magnetic resonance image (MRI) scans are superposed on the skull so that the surgeon knows where to open the skull and insert instruments.

Telemedicine refers to the delivery of health care over communication channels, and that can mean two-way video, audio, and data channels connecting the patient (and a nurse or paramedic accompanying the patient) in remote areas (or in space or on ships at sea) to the medical specialist at the big city hospital. Experiments with telemedicine have been promising, but the medical culture has been somewhat slow to adopt telemedicine to any great extent. The supervisory control here often means the more experienced specialist supervises the nurse/paramedic in doing examinations and performing therapeutic tasks (Ottensmeyer et al., 1996). The U.S. Army is even experimenting with telerobotic endoscopic surgery, thereby combining the two forms of remote supervision, to better treat soldiers on the battlefield and stop bleeding in time so they can be safely returned to field hospitals.

16.5 OVERVIEW CONSIDERATIONS OF SUPERVISORY CONTROL

16.5.1 What Is Special About Design of Human–Machine Systems

Design of human–machine systems is different from design of hardware or software in several ways:

1. Design engineers typically have some understanding of physics and can relate to mechanical and electrical forces and geometry. Seldom, however, do they have much understanding of psychology and physiology and empathy for a human user's perceptions. When involved with human–machine systems, they have to be taught that the object is not design of a thing, but design of a relationship between a human and a machine.

2. It is quite obvious, but important to emphasize here that the human is already "designed." Humans can be selected and trained, but otherwise their limits are to be reckoned with as they are.

3. The human brain does not work according to crisp rules. The rules are fuzzy, in the sense of fuzzy logic. Words and pictures are its normal medium of communication. Words and pictures have imprecise meanings and associations with many other words and pictures.

4. There are many possible modes and channels of communication between human and machine, and redundancy is typically a good thing. A single communication channel can be insufficiently reliable.

5. Humans are able to self-pace in performing tasks, speeding up on the easy parts, slowing down on the hard parts.

6. Humans are prone to making numerous errors, but are also able to discover their own errors and correct them in time and to a degree that they seldom have serious consequences. Humans must be free to make small errors, to experiment, and to adjust their own behavior in an effort to learn and improve. Humans should never be expected to perform tasks repetitively in the same way as before and with zero error.

7. Humans seem to exhibit creativity, free will, and indeterminism. To the degree that this is true, their behavior is unpredictable.

16.5.2   Who or What Is in Charge? The Difficult Problems of Authority

The problem of authority between human and computer is one of the most difficult in new transportation systems (Boehm-Davis et al., 1983). Popular mythology is that the human operator is (or should be) in charge at all times. In the past when a human turned control over to an automatic system, it was mostly with the expectation that circumstances were not critical, and that the person could do something else for a while (as in the case of setting one's alarm clock and going to sleep). Even when continuing to monitor the automation (as would occur in supervisory control) people are seldom inclined to "pull the plug" unless they receive signals clearly indicating that such action must be taken, and unless circumstances make it convenient for them to do so. Examples of some current questions being debated in the aviation field follow:

1. Should there be certain states, or a certain envelope of conditions, for which the automation will simply seize control from the operator? In the Airbus aircraft, for example, it is impossible to exceed critical boundaries of speed and attitude that will bring the aircraft into stall or other unsafe flight regimes. In the MD-11 the pilot can approach the boundaries of the safe flight envelope only by exerting much more than the normal force on the control stick.

2. Should the computer automatically force deviation from a programmed control strategy if critical unanticipated circumstances arise? The MD-11 will deviate from its programmed plan if it detects wind shear.

3. If the operator programs certain maneuvers ahead of time, should the aircraft automatically execute these at the designated time or location, or should the operator be called upon to provide further concurrence or approval? The A320 will not initiate a programmed descent unless it is reconfirmed by the pilot at the required time.

4. In the case of a subsystem abnormality, should the affected subsystem automatically be reconfigured, with after-the-fact display of what has failed and what has been done about it? Or should the automation wait to reconfigure until after the pilot has learned about the abnormality, perhaps been given some advice on the options, and had a chance to take initiative? The MD-11 goes a long way in automatic fault remediation.

5. One can further ask: if all the information required for full automation is available, why not automate fully? In the railway sector this question has been asked, and in some special cases the decision was to remove the driver entirely (e.g., trains on rails or rubber tired guided vehicles in some new airports, or trains on dedicated tracks such as those between Orly Airport and Paris). Many engineers assert that automatic control is essential for modern high-speed trains, and there is simply nothing to debate. However, history has shown with regard to automation that we are not always as smart as we may think. Though automation is now widely accepted in aviation by pilots, by airlines, and by regulators, there remain accidents that have been blamed on the automation itself. In the case of the Airbus there is now an active discussion about whether automation has gone too far. It is salient to note that Charles Stark Draper, the "father" of inertial guidance and the Apollo navigation system, which took the astronauts to the moon, was observed by the writer to proclaim at the outset of the Apollo Program that the astronauts were to be passive passengers and that all the essential control activities were to be performed by automation. It turned out that he was wrong. Many routine sensing, pattern-recognition, and control functions had to be performed by the astronauts, and in several instances they "saved" the mission.

Assuming sufficiently accurate models of vehicles and their interaction with their respective media, and sufficiently accurate state measurements, optimal automatic control is quite feasible. Alternatively, as noted earlier, an optimization calculation can be made continuously and used, not for automatic control, but to display to a human operator a best profile for human control action. Proponents of this view maintain that if the human kept precisely to such a profile, he could in any case do a better job at control than trying, in effect, to perform complex calculations in his head. In addition he would be there if some totally unpredictable events occurred, would allay public fear of riding in an unmanned vehicle, and would inhibit litigation on the basis that if an accident occurred both automation and human were present doing their best.

Automation is sure to improve and to relieve the human of more tasks. However, we must guard against unsubstantiated claims, as are now being made for how "intelligent automobiles" will increase safety and reduce congestion. Empirical systems demonstrations will speak louder than words.

16.5.3 Human Trust in Automation

Human operator trust in automation is coming to be a major topic of interest (Muir and Moray, 1987; Riley, 1994; Wickens, 1994). Parasuraman (1997) makes a compelling case about misuse, disuse, and abuse of automation, based on distrust and overtrust as well as workload and other factors. Engendering trust is often a desirable feature of a system, something the designer strives for. But not always. Below are listed seven aspects of trust (Sheridan, 1988), with both the positive and the negative aspects mentioned.

1. *Reliability* of the system, in the usual sense of repeated, consistent functioning. Probability of failure or mean-time-to-failure are standard measures of this. (But the technology may be reliably harmful, always performing as it was designed, but designed poorly in terms of human or other factors.)

2. *Robustness* of the system, meaning demonstrated or promised ability to perform under a variety of circumstances. Some measure of versatility can be generated. (But it may able to do a variety of things, some of which need not or should not be done.)

3. *Familiarity,* that is, the system employs procedures, terms, and cultural norms that are familiar, friendly, and natural to the trusting person. Training time would be a measure, or years of experience with a particular system by a particular operator. (But a person may be so familiar as to ignore certain pitfalls.)

4. *Understandability,* in the sense that the human supervisor or observer can form a mental model and predict future system behavior. Some kind of test score would be the appropriate measure here. (But ease of understanding parts of it may lead to overconfidence in one's understanding of other aspects.)

5. *Explication of intention,* meaning that the system explicitly displays or says that it will act in a particular way, as contrasted to its future actions having to be predicted from a model. The test here would be whether the system is able to communicate its intention and whether it carries out such a stated intention. (But knowing the computer's intention may also create a sense of overconfidence and a willingness to outwit the system and take inappropriate chances.)

6. *Usefulness,* or utility of the system to the trusting person in the formal theoretical sense. The measure here would have to be a subjective rating scale, or a Von Neumann utility measure (see Keeney and Raiffa). (It may surely be useful, but for unsafe purposes.)

7. *Dependence* of the trusting person on the system. For a measure one could either observe the operator's consistency of use, or use a subjective rating scale, or both. (Finally and obviously, overdependence may be fatal if the system fails.)

16.5.4   Social Implications of Supervisory Control

The relentless drive of new technology results in takeover of more and more functions from the human operator by the computer. Many decades ago the operator's muscles were relieved by mechanization, so that it was necessary only to provide guidance for the movements of aircraft, lathe, or whatever, and the machine followed without the operator having to exert much physical effort. Then, two decades ago, stored program automation allowed the operator to indicate movements only once, and the machine would dutifully repeat those movements forever, if necessary. Then the operator was bumped up to a higher level, so that the computer could be given goal statements, branching instructions, production rules, and "macro" information, and the computer would dutifully oblige. All these advances have elevated the stature of the human operator in "motor" activities.

This elevation has also been taking place in sensory and cognitive activities. Artificial sensors now feed data into computers, which process the data and give the operator graphic displays of summary assessments, formatted in whatever manner is requested. "Intelligent" sensory systems draw attention when "something new" has occurred, but otherwise may leave the operator alone. Newer computer systems build up "world models" and "situation assessments." "Expert systems" can now answer the operator's questions, much as does a human consultant, or give suggestions even if they are not requested.

How far will or should this "help" from the computer go? Are there points of diminishing returns? Already the aircraft pilots have an expression they use to refer to new computer cockpit automation and advisory aids: "killing us with kindness."

There is another problem of supervisory control, apart from the matter of trust per se. That is, even if the computer were quite competent and reliable, there may be other reasons for not having it assume so much subordinate responsibility. This problem may be summarized by the word "alienation." Elsewhere (Sheridan, 1980), seven factors contributing to alienation of people by computers are discussed in detail. They are as follows.

1. People worry that computers can do some tasks much better than they themselves can, such as those dealing with memory and calculation. Surely people should not try to compete in this arena.

2. Supervisory control tends to make people remote from the ultimate operations they are supposed to be overseeing—remote in space, desynchronized in time, and interacting with a computer instead of the end product or service itself.

3. People lose the perceptual-motor skills, which in many cases gave them their identity. They become "deskilled," and, if they were ever called upon to use their previous well-honed skills, they could not.

4. Increasingly, people who use computers in supervisory control or in other ways, whether intentionally or not, are denied access to the knowledge to understand what is going on inside the computer.

5. Partly as a result of the last factor, the computer becomes mysterious, and the untutored user comes to attribute to the computer more capability, wisdom, or blame than is appropriate.

6. Because computer-based systems are growing more complex, and people are being "elevated" to roles of supervising larger and larger aggregates of hardware and software, the stakes naturally become higher. Where a human error before might have gone unnoticed and been easily corrected, now such an error could precipitate a disaster.

7. The last factor in alienation is similar to the first, but all-encompassing, namely the fear that a "race" of machines is becoming more powerful than that of the human race.

These seven factors, and the fears they engender, whether justified or not, must be reckoned with. Computers must be made to be not only "human friendly" but also "human trustable." Users must become computer-literate at whatever level of sophistication they can deal with.

There is no question but that supervisory control puts more and more power into the hands of fewer people, and human engineering to make this kind of human–computer interaction safe and productive for all is a challenge.

16.5.5 Current Confusion About "Human-centeredness" of Automation

The topic of "human factors in automation" subsumes supervisory control. The classic Rasmussen (1986) treatise should be mentioned in this regard, as well as two meetings on the topic (Mouloua and Parasuraman, 1994; Parasuraman and Mouloua, 1996).

Currently it is fashionable to speak of *human-centered automation*. The difficulty is that this term connotes different meanings to different people. Below are listed ten alternative meanings, all surely somewhat idealistic, and for each a rather restrictive qualification is added, which to the writer seems necessary (Sheridan, 1996):

1. Allocate to the human the tasks best suited to the human; allocate to the automation the tasks best suited to it. (Automate only the tasks requiring intermediate instructional complexity, but not the one-of-a-kind tasks that are usually easier to do oneself rather than bother trying to teach the computer, and the very complex tasks no one has the slightest idea how to program.)

2. Keep the human operator in the decision and control loop. (The human can handle only control tasks of appropriate bandwidth, attentional demand, etc.)

3. Maintain the human operator as the final authority over the automation. (This is not always the safest way. It very much depends on task context.)

4. Make the human operator's job easier, more enjoyable, or more satisfying through friendly automation. (Operator ease and enjoyment do not necessarily correlate with operator responsibility and system performance.)

5. Empower the human operator to the greatest extent possible through automation. (Operator empowerment is not the same as system performance.)

6. Support trust by the human operator. (Too much trust is just as bad as not enough trust.)

7. Give the operator computer-based advice about everything that it should be necessary to know. (Too much advice will overwhelm.)

8. Engineer the automation to reduce human error and keep response variability to the minimum. (Acceptable levels of error and response variability enhance learning. Darwin taught us about requisite variety many years ago.)

9. Make the operator a supervisor of subordinate automatic control system(s). (Maybe not always. For some tasks, direct manual control may be best.)

10. Achieve the best combination of human and automatic control, where best is defined by explicit system objectives. (Do we not wish we always had explicit system objectives!)

16.5.6 Two Models, and Two Limits in Our Ability to Model Supervisory Control

Compared to models of discrete manual control there are relatively few models of supervisory control extant. Perhaps the best known is PROCRU (Baron et al., 1980), a modeling approach based on the earlier optimal control model of Kleinman et al. (1970), which adds to continuous manual control some assumptions of visual sampling/attention allocation in following a given procedure as to when to perform a series of discrete functions. PROCRU has been used successfully, for example, to discriminate high and normal workload aircraft landings. A recent manifestation (Levison and Baron, 1997) models automobile driving. This class of models assumes that the human operator is well trained and internalizes an accurate representation of the controlled process (which can be represented by linear differential equations), that a quadratic performance criterion obtains (a weighted sum of mean squared deviation from an ideal course and mean-squared control resources used), and that operator limitations can be represented by wide-band Gaussian "noise." Tasks that compete for attention are selected according to assumed relative importance, and only one task at a time can be attended to. The authors claim good results after adjusting model parameters to fit experimental path deviations.

Papenhuijzen (1997) presents models of the navigator (speed controller) and helmsman (rudder adjustor) that work together to control a large ocean vessel navigating a narrow channel. The helmsman model assumes a nonlinear internal model as well as prediction–decision element, while the navigator model incorporates a Kalman filter for state estimation, as well as a set of fuzzy rules for track planning. The model predicts actual ship maneuvering reasonably well.

The paucity of supervisory control models is not surprising. Our ability to model a less restricted class of supervisory control tasks is limited by two factors:

1. The first factor is the same as that which makes any cognition difficult to model. As compared to simpler manual control and other perceptual-motor skills, the essence of supervisory control is cognition, or mental activity, particularly including objectives and command/control decisions made at least partly by the free will of the supervisor. This includes problem solving, where there are multiple ways to solve the problem or complete the assigned task. The problem is that such mental events are *not directly observable;* they must be inferred. This is the basis for the ancient mind–body dilemma, as well as the basis for the rejection, by some behavioral scientists, of mental events as legitimate scientific concepts. Such "behaviorism," of course, is out of fashion now. Computer science motivated "cognitive science," and by indirect inference, made computer programs admissible representations of what people think. However, the limits on direct measurement of mental events do not seem to want to go away.

2. The second factor is more specialized to supervisory control. It is a result of wanting to combine mental models internal to the human operator with external computer decision aids. If the operator chooses to follow the aid's advice, what can be said about the exercise of the operator's own mental model? If the operator claims to "know" certain facts or have a "plan," which happens to correspond to the knowledge or plan communicated by the computer, can the normative decision aid be considered a convenient yardstick against which to measure inferred mental activity? Where there is free will and free interaction between the two it is difficult to determine what is mental and what is computer. Mental and computer models might more easily be treated as a combined entity that produces operator decisions.

We must continue efforts to model and predict human behavior in this increasingly prevalent form of control system in spite of these difficulties mentioned earlier. Obviously the supervisory control paradigm is but one of many perspectives that may be useful; others are described in this book. But useful for what? For predicting behavior of human or computer or both? Unfortunately there are few, if any, current models of supervisory control with much predictive value. Supervisory control models refined to that point exist now only for a very narrow range of tasks, such as well-defined aircraft piloting or industrial plant operations where given procedures are being followed and well-defined global criteria apply. In the present context the usefulness of the supervisory control perspective, it is argued, is for classifying human functions with respect to computer functions, and for gaining insight into ways in which computers can aid people in doing jobs.

16.6 CONCLUSIONS

Technical devices of various kinds, from vehicles to robots to factories are getting "smarter" through the embodiment of sensors and computers that enable automation. Supervisory control is coming to be the dominant mode by which humans interact with such devices. This chapter has defined supervisory control, made a case for the importance of task analysis and function allocation in designing such systems, described in detail the five sequential phases of supervisory control, presented a number of applications, and presented some generic caveats about supervisory control. The field is yet young and rapidly evolving, being driven hard by the changing technological capabilities. It is an understatement to assert that much research is needed.

REFERENCES

Askey, S. (1995). Design and evaluation of decision aids for control of high speed trains: Experiments and model. Ph.D. Thesis, MIT, Cambridge, MA.

Baron, S., Zacharias, G., Muralidharan, R., and Craft, I. (1980). PROCRU: A model for analyzing flight crew procedures in approach to landing. *Proc. 16th Annu. Conf. on Manual Control,* pp. 488–520. Cambridge, MA: MIT.

Billings, C. S. (1991). *Human-Centered Aircraft Automation: A Concept and Guidelines,* NASA Tech. Memo. 103885. Moffet Field, CA: NASA Ames Research Center.

Birmingham, H. P., and Taylor, F. V. (1954). A design philosophy for man-machine control systems. *Proc. IRE,* **42,** 1748–1758.

Boehm-Davis, D., Curry, R., Wiener, E., and Harrison, R. (1983). Human factors of flight deck automation: Report on a NASA-industry workshop. *Ergonomics* **26,** 953–961.

Brooks, T. L. (1979). Superman: A system for supervisory manipulation and the study of human-computer interactions. SM Thesis, MIT, Cambridge, MA.

Charny, L., and Sheridan, T. B. (1986). Satisficing decision-making in supervisory control. Unpublished paper, MIT, Man-Machine Systems Laboratory, Cambridge, MA.

Craik, K. J. W. (1947a). Theory of the human operator in control systems: 1. The operator as an engineering system. *Br. J. Psychology,* **38,** 56–61.

Craik, K. J. W. (1947b). Theory of the human operator in control systems: 2. Man as an element in a control system. *Br. J. Psychology,* **38,** 142–148.

Curry, R. E., and Gai, E. G. (1976). Detection of random process failures by human monitors. In *Monitoring Behavior and Supervisory Control,* (T. Sheridan, and G. Johannscn, eds.), pp. 205–220. New York: Wiley.

Das, H. (1989). Kinematic control and visual display of redundant teleoperators. Ph.D. Thesis, MIT, Cambridge, MA.

Einstein, A. and Infeld, L. (1942). *The Evolution of Physics.* New York: Simon & Schuster.

Ellis, S. R. (ed.)(1991). *Pictorial Communication in Virtual and Real Environments.* London: Taylor & Francis.

Endsley, M. R. (1996). Automation and situation awareness, In *Automation and Human Performance: Theory and Applications.* (R. Parasuaraman, and M. Mouloua, eds.). Hillsdale, NJ: Erlbaum, pp. 163–182.

Ferrell, W. R. (1965). Remote manipulation with transmission delay. *IEEE Trans. Hum. Factors Electron.* **HFE-6,** No.1.

Ferrell, W. R., and Sheridan, T. B. (1967). Supervisory control of remote manipulation. *IEEE Spectrum* **4**(10), 88.

Fitts, P. M. (1951). *Human Engineering for an Effective Air Navigation and Traffic Control System.* Columbus: Ohio State University Research Foundation.

Hedrick, K. (1994). Personal communication.

Inagaki, T. (1997). Situation-adaptive autonomy for safety of human-machine systems in critical phases. *Proc. 6th IFIP Int. Conf. Depend. Comput. Crit. Situations,* Garmisch-Partenkirchen, Germany, *1997.*

Inagaki, T., and Ikebe, Y. (1988). A mathematical analysis of human-machine interface configurations for a safety monitoring system. *IEEE Trans. Reliab.* **37**(1), 35–40.

Jordan, N. (1963). Allocation of functions between man and machines in automated systems. *J. Appl. Psychology,* **47,** 161–165.

Kantowitz, B., and Sorkin, R. (1987). Allocation of functions. In *Handbook of Human Factors.* (G. Salvendy, ed.), pp. 365–369. New York: Wiley.

Keeney, R. L., and Raiffa, H. (1976). *Decisions with Multiple Objectives: Preferences and Value Tradeoffs.* New York: Wiley.

Kim, S. (1997). Theory of human intervention and design of human-computer interfaces in supervisory control application to traffic incident management. Ph.D. Thesis, MIT, Cambridge, MA.

Kleinman, D. L., Baron, S., and Levison, W. H. (1970). An optimal control model of human response. Part I. *Automatica* **6**(3), 357–369

Levison, W., and Baron, S. (1997). In *Perspectives on the Human Controller,* (T. B. Sheridan, and A. Van Lunteren, eds.). Mahwah, NJ: Erlbaum, pp. 232–247.

March, J. G., and Simon, H. A. (1958). *Organizations.* New York: Wiley.

Meister, D. (1971). *Human Factors, Theory and Practice.* New York: Wiley.

Moray, N. (ed.), 1979. *Mental Workload: Its Theory and Measurement.* New York: Plenum.

Moray, N (1986). Monitoring behavior and supervisory control. In *Handbook of Perception and Human Performance.* (K. Boff, L. Kaufmann, and J. P. Thomas, eds.). New York: Wiley.

Mouloua, M., and Parasuraman, R., (eds.) (1994). *Human Performance in Automated Systems: Recent Research and Trends.* Hillsdale, NJ: Erlbaum.

Muir, B. M., and Moray, N. P. (1987). Operators' trust in relation to system faults, *Proc. IEEE Conf. Syst., Man Cyberne.,* Atlanta, GA, *1987.*

National Research Council (1983). *Research Needs for Human Factors.* Washington, DC: National Academy Press.

National Research Council. (1984). In *Research and Modeling of Supervisory Control Behavior.* (T. B. Sheridan, and R. Hennessy, eds). Washington, DC: National Academy Press.

Ottensmeyer, M. P., Thompson, J. M., and Sheridan, T. B. (1996). Telerobotic surgery: Experiments and demonstrations of telesurgeon/assistant cooperation under different time delays and tool assignments. *Proc. S.P.I.E. Conf.,* Boston, *1996.*

Papenhuijzen, B. (1997). Navigation simulation. In *Perspectives on the Human Controller*, (T. B. Sheridan, and A. Van Lunteren, eds.). Mahwah, NJ: Erlbaum.

Parasuraman, R. (1997). Humans and automation: Use, misuse, disuse, abuse. *Hum. Factors,* **39,** 230–253.

Parasuraman, R., and Mouloua, M., (eds.) (1996). *Automation and Human Performance: Theory and Applications.* Mahwah, NJ: Erlbaum.

Park, J. H. (1991). Supervisory control of robot manipulators for gross motions. Ph.D. Thesis, MIT, Cambridge, MA.

Patrick, N. J. M. (1996), *Decision-aiding and optimization for vertical navigation of long-haul aircraft.* Ph.D. Thesis, MIT, Cambridge, MA.

Price, H. E. (1985). The allocation of functions in systems. *Hum. Factors* **27**(1), 33–45.

Price, H. (1990). Conceptual system design and the human role. In (H. Booher, ed.), *Manprint.* New York: Van Nostrand.

Pritchett, A. (1997). Pilot non-conformance to alerting system commands during closely spaced parallel approaches. Sc.D. Thesis. MIT, Cambridge, MA.

Rasmussen, J. (1986). *Information Processing and Human-Machine Interaction.* New York: Elsevier/North-Holland.

Riley, V. (1994). A theory of operator reliance on automation. In *Human Performance in Automated Systems: Recent Research and Trends.* (M. Mouloua, and R. Parasuraman, eds.). Hillsdale, NJ: Erlbaum, pp. 8–14.

Sarter, N., and Woods, D. D. (1991). Situation awareness: A critical but ill defined phenomenon. *Int., J. Aviat. Psychol.,* **1,** 45–57.

Sarter, N., and Woods, D. D. (1994). Decomposing automation: Autonomy, authority, observability and perceived animacy. In M. Mouloua, and R. Parasuraman, eds.), *Human Performance in Automated Systems: Recent Research and Trends.* pp 22–27. Hillsdale, NJ: Erlbaum.

Senders, J. W., Elkind, J. E., Grignette, M. C., and Smallwood, R. P. (1964). *An Investigation of the Visual Sampling of Human Observers,* NASA Rep. CR-434. Cambridge, MA: Bolt, Beranek & Newman.

Sheridan, T. B. (1960). Human metacontrol. *Proc. Annu. Conf. Manual Control,* Wright Patterson, AFB, OH.

Sheridan, T. B. (1970). On how often the supervisor should sample. *IEEE Trans. Syst., Man Cybernet.* **SSC-6,** 140–145.

Sheridan, T. B. (1980). Computer control and human alienation. *Technol. Rev.* **83**(1), 60–75.

Sheridan, T. B. (1984). Supervisory control of remote manipulators, vehicles and dynamic processes. In *Advances in Man-Machine Systems Research,* (W. B. Rouse, ed), Vol. 1, pp. 49–137. New York: JAI Press.

Sheridan, T. B. (1987). Supervisory control. In *Handbook of Human Factors/Ergonomics.* (G. Salvendy, ed.). New York: Wiley, pp. 1243–1268.

Sheridan, T. B. (1988). Trustworthiness of command and control systems. *Proc. 2nd IFAC/IFIP/IFORS/IEA Symp. Man-Mach. Sys., Oulu Finland.*

Sheridan, T. B. (1992). *Telerobotics, Automation and Human Supervisory Control.* Cambridge, MA: MIT Press.

Sheridan, T. B. (1996). Human-centered automation: oxymoron or common sense? *Proc. IEEE Int. Conf. Syst., Man Cybernet.,* Vancouver, Canada.

Sheridan, T. B., and Johannsen, G. (eds.) (1976). *Monitoring Behavior and Supervisory Control:.* New York: Plenum.

Tulga, M. K. and Sheridan, T. B. (1980). Dynamic decisions and workload in multitask supervisory control. *IEEE* Trans. Syst., Man Cybernet. **SMC-10,** 217–232.

Wickens, C. D. (1992). *Engineering Psychology and Human Performance,* 2nd ed. New York: HarperCollins.

Wickens, C.D. (1994). Designing for situation and trust in automation. *Proc. IFAC Conf. Integr. Syst. Eng.* Baden-Baden, Germany.

Wiener, E. L. (1988). Cockpit automation. In *Human Factors in Aviation* (E. L. Wiener, and D. Nagel, eds.), pp. 433–461. San Antonio, TX: Academic Press.

Wierwille, W. W., Casali, J. G., Connor, S. A., and Rahimi, M. (1985). Evaluation of the sensitivity and intrusion of mental workload estimation techniques. In *Advances in Man–Machine Systems Research,* (W. B. Rouse, ed.), Vol. 2, pp. 51–127. New York: JAI Press.

Yoerger, D. R. (1982). Supervisory control of underwater telemanipulators: Design and experiment. Ph.D. Thesis, MIT, Cambridge, MA.

17 Designing for Cognitive Task Performance

JUDITH M. ORASANU and MICHAEL G. SHAFTO

17.1 OVERVIEW AND INTRODUCTION

This chapter is about cognitive engineering. It addresses the question of why designers should be concerned with the cognitive capabilities of potential users of their products and how designers might take these capabilities into account as part of the total system design process.

The history of technology development traces efforts to make human jobs easier and to enhance human potential by relieving people of tasks that are difficult, time-consuming, and subject to error. From the creation of the wheel to the development of modern "glass cockpit" aircraft, from the abacus to the workstation, the purpose of design has been to offload functions from the human to an enabling tool or machine. Sometimes this shift has been swift and smooth. More often it has taken generations and has created new problems even as it solved old ones.

Consider for example, the development of the automobile. The original Model T Ford represented a historical leap into low-cost mass-produced vehicles. Ease of use was not a requirement. It demanded that owners be their own mechanics, and to have a complete understanding of how the car worked in order to propel this noisy marvel forward:

> Its most remarkable quality was its rate of acceleration. In its palmy days the Model-T could take off faster than anything on the road. The reason was simple. To get under way, you simply hooked the third finger of the right hand around the lever on the steering column, pulled down hard, and shoved your left foot forcibly against the low-speed pedal. These were simple, positive motions; the car responded by lunging forward with a roar. After a few seconds of this turmoil, you took your toe off the pedal, eased up a mite on the throttle, and the car, possessed of only two forward speeds, catapulted directly into high with a series of ugly jerks and was off on its glorious errand. (White, 1936)

The advent of the manual transmission simplified the knowledge requirements, but the driver still needed to understand and to manage the clutch, the throttle, torque, and revolutions per minute (rpm). Automatic transmissions further reduced the knowledge needed to operate the vehicle, so that practically anyone could master it, even those with no interest in automotives.

The evolution of automobiles has included both a reduction in the knowledge and a simplification of the skills needed to operate the vehicle. In addition, information about the status of the car's components has been reduced to those most essential for a driver to

Handbook of Systems Engineering and Management, Edited by A. P. Sage and W. B. Rouse
ISBN 0471-15405-9 ©1999 John Wiley and Sons, Inc.

notice while operating the car—speed, fuel level, and a few caution lights that signal potential problems. Minimizing these functions means the driver can devote greater attention to higher-order strategic tasks, such as navigating to the destination and monitoring road and traffic conditions, while driving at ever-faster speeds on crowded roads.[1]

How did the designers of each new generation of cars decide what changes to make to enhance their latest model's appeal to the public (apart from addition of new fins and grillwork)? Designing the technology to simplify operation and reduce effort is the process known as *cognitive engineering* (Norman, 1986). Norman defines cognitive engineering as an attempt "to apply what is known from science to the design and construction of machines." This means that design must be grounded in principles of human information processing and the essence of interaction. Additionally, it must be rooted in deep knowledge of the domain and the specific tasks the technology is intended to support. In his discussion of human-centered automation principles, Billings (1997) emphasizes design of technologies that take advantage of the strengths and compensate for the weakness of both elements of a system (that is, the human user and the technology).

When we talk about systems here, we are referring to the human interacting with the machine, working together in a context to accomplish a shared task. One can argue that our opening example, the automobile, is itself a system composed of static and dynamic mechanical, chemical, electrical, and hydraulic subsystems. However, the meaning of the term *system* in this chapter includes not only the device, but also the people who develop and utilize it, the functions, intentions, and outcomes—that is, the entire context of the dynamic person–machine relationship in use.

Given the recent advances in computing power, design tools, computational algorithms, and heuristic processes, it is possible for design engineers to develop devices with enormous functionality. Is it not sufficient for them to design devices that are elegant, responsive to needs, and highly functional? Focusing on the technology rather than on the human user assumes that if a system is good enough, humans will appreciate its merits and work at learning to use it, despite some errors, frustration, or grumbling along the way. After all, machines are precise, logical, tireless, unemotional, and have unlimited memory capacity (Norman, 1988).

An alternate view, and one we adopt in this chapter, is that designers ignore the human element of the system at their peril. Technology-centered design can lead to at least three consequences. First, such systems may be impossible to use. Norman (1988) points to examples of elegant designs, such as glass doors on modern skyscrapers with no evident hinges or opening mechanisms. These may become glass barricades instead of entries because people cannot figure out how to use them. Many of us have rented cars, and despite great standardization of operating interfaces, have been frustrated in our efforts to remove the key from the ignition or to open the gas filler lid until discovering the carefully concealed release button.

Second, technology-centered rather than human-centered design can increase the effort or difficulty of operating a device. Simply adding functionality without concern for the user often results in user confusion. Ask anyone who has tried to program a VCR using a multifunction remote. Interacting with computers in the early days was an activity limited to a very small elite who had mastered crude and cumbersome coding programs. The

---

[1] A by-product of this reduction in workload appears to be an increasing number of drivers who eat and drink, talk on their cellphones, read memos, put on makeup, shave, and log on their computers.

development of the graphical user interface (GUI) known as "windows" and controlled by a "mouse" represents a leap forward in the design of computers intended to be "user-friendly."

Finally, technology-centered design may actually increase the likelihood of errors and compromise the effectiveness and safety of the system rather than increase it. Wiener (1989) has pointed out that some components of automation in modern aircraft have reduced the demands on pilots during low workload periods, while increasing the demands during high-tempo critical phases of flight, such as takeoffs and landings. So-called "clumsy automation" has been implicated in a number of aircraft accidents.

In this chapter we describe the capabilities and limitations of the human cognitive system and their implications for system design. Several examples of incidents and accidents in engineered high-risk domains, such as aviation, medicine, space, and transportation, illustrate what can happen when the human element is insufficiently considered during the design process. In the Section 17.3 we describe three approaches to cognitive engineering, or human-centered design. The focus here is on information systems, given our concern with cognitive rather than physical requirements. The chapter ends with a discussion of why design has not always taken the user's cognitive requirements into account and the fallacious assumptions behind technology-centered reasoning.

17.2   COGNITIVE CONSTRAINTS ON SYSTEM DESIGN

First, we need to specify the nature of human cognition and the constraints it imposes on system design and function. Equally important are the unique strengths of the human cognitive apparatus and the advantages they confer on system functioning. By cognition we mean the combination of knowledge structures and the processes that operate on them in the course of achieving specific goals.[2] Cognition is what the mind does. Underlying all adult human activity is a vast storehouse of organized knowledge.[3] This knowledge supports so-called lower-level processes, such as attention and perception, that are involved in scanning the environment, recognizing patterns, and interpreting those patterns in terms of their significance for one's goals. Knowledge also is the foundation for the higher-order skills of planning, problem solving, and decision making, functions that are increasingly supported by technology.

17.2.1   Three Cognitive Limitations

Extensive research on human cognition has identified several bottlenecks or limitations in human capability. The first involves *attention*. Attention works like a flashlight. It can focus on elements of the environment that are of interest, but like the beam of a flashlight, its focus is limited to a relatively small area. Attention is essential for scanning the envi-

---

[2]A good source on human cognition is Anderson (1995); for a focus on human factors grounded in the human information processing system, see Wickens (1992).

[3]Our concern here is primarily with the mature adult who brings experience and knowledge to tasks supported by the system, not with developmental processes. We will address the issue of the target users of new systems in a later section. A major issue in design is whether the target is a broad audience of users who are relative novices or a much more narrowly defined group of individuals who already have knowledge and skills relative to the use of the system.

ronment to maintain situation awareness. People are poor at prolonged monitoring, however, and over time their vigilance decreases (Wickens, 1992). This can be a problem when conditions are changing rapidly and it is important to update one's situation model.

Considerable research has addressed the question of whether we can attend to more than one thing at a time (Broadbent, 1979). While the primary finding is that we cannot, the more accurate answer is that it depends. It depends primarily on the similarity between two tasks or types of cues. If the tasks draw on common underlying cognitive resources (like trying to read while watching TV or listening to a conversation amidst other conversations at a cocktail party), then performance on one of the tasks is likely to suffer (Wickens, 1992).

Thus, a system that places competing demands on one sensory channel is likely to result in poor performance because the operator is not able to allocate attention appropriately to all of the tasks. In the laboratory when a person is engaged in a task that requires focused attention, a secondary stimulus may not be detected. Many warnings on commercial aviation flight decks are now auditory, since the task of flying a plane places heavy demands on visual attention. Under stress, some alarms may not be heard or seen, a phenomenon known as *tunnel vision*. The goal is to design alarms to be highly salient so they are readily perceived. On the other hand, people can learn to manage their attention, deliberately focusing on one source of input rather than another, or scanning the environment for information that is particularly relevant to their task.

Another factor that determines how well we can perform two simultaneous tasks is the level of skill at each. After high levels of practice, some routine skills become "automatic" (Shiffrin and Schneider, 1977), which means the resource demands are reduced. We essentially perform that skill (e.g., driving a car) on "automatic pilot" and can focus our attention on concurrent tasks with little interference. While we are learning the skill, considerable attention is required. One only has to experience driving a car in Great Britain with the steering wheel on the "wrong" side and the lane directions reversed to realize the attention demands on a novice.

The second limitation or bottleneck involves *working memory*. Working memory is the mental component in which we actively process information, rehearsing, transforming, combining, or evaluating it in some way. Working memory has limited capacity. We can only deal with a few pieces of information at a time (try remembering a new phone number including the area code). If information is unfamiliar, we can usually handle around five elements at a time (Miller, 1956). Experts, on the other hand, can manage enormous amounts of information by exploiting their highly structured knowledge from long-term memory (Ericsson and Smith, 1991). Experts typically deal with larger "chunks" of information, thereby expanding their functional capacity (Chase and Simon, 1973).

Working memory limitations can interfere with carrying out procedures that involve steps that have few external cues connecting them. After completing one step, we may not remember what the next step is, we forget what we just did, or where we are in the sequence. This is especially true if there is a distraction. For example, while preparing for takeoff from Detroit, the crew of a DC9-82 was interrupted during its predeparture checklist by an air-traffic control (ATC) call. As a result, they neglected to extend the trailing edge flaps and leading edge slats, which are essential for lift on takeoff. The improper flap setting resulted in a crash just after liftoff [National Transportation and Safety Board (NTSB), 1988]. Strategies for workload management are critical to managing working memory limitations.

Working memory limitations can also affect the ease with which people can reason about and evaluate information. Research on decision making under time constraints has shown that people simplify the task, reducing the amount of information they consider before reaching a decision (relative to an "optimal" evaluation of options according to relevant criteria) (Payne et al., 1988). They tend to "satisfice," or choose the first satisfactory option, rather than the optimal one when resources are constrained (Simon, 1955). People sometimes work around their limitations by establishing contingency plans. That is, they evaluate options and conditionalize actions on various trigger events. By setting up contingencies during low workload periods, they reduce the demands on their limited capacities during anticipated high-stress periods, when they know they will not function as well. Tasks that place heavy demands on working memory are subject to error and shortcut strategies.

A third major limitation is *retrieval from long-term memory*. Even when we want to remember something, when we know what we are trying to remember, and are making an effort to think of it, we may not be able to recall it. Thus, tasks that depend on retrieval of knowledge for their performance are likely to suffer degraded performance.

People in high-consequence situations may be aware of this limitation and take steps to overcome it. Perhaps most familiar is the use of daily planners, in which we note meetings and tasks that must be done, thereby reducing the need to remember them. All we need to remember is to look in the book. Pilots who are transferring fuel from one tank to another have been observed to place the takeoff and landing speed book between the throttles as a reminder to close the fuel valve when the transfer is complete (Wiener, 1993). It is the "reminder" string we tie around our finger.

Difficulty retrieving information from long-term memory is more likely if the information to be retrieved is not well connected to other information in memory. Memory limitations reflect the type of memory structures being tapped by a particular task. Information is not simply stored as isolated bits of information, but is highly organized. The type of structure reflects the kind of knowledge. A distinction is often made between knowing "what" and knowing "how." Facts about the world, knowing what, are stored as declarative knowledge. This kind of knowledge is stored as propositions in semantic networks that reflect the relations among the propositions (Anderson, 1995). Discrete bits of declarative knowledge, "facts," are most susceptible to memory retrieval failure, especially in a domain about which we know little.

Knowledge of how to do things is called *procedural knowledge* and is stored as schemas, or patterns of states and actions. Tasks that involve high levels of skill depend on procedural knowledge. Knowledge about how things work is stored as mental models, which are complex structures involving objects, variable states, and variable operations. These are dynamic and allow us to "run" the system in our minds, to do "what if" reasoning by plugging in particular values. Mental models allow us to understand, explain, and predict the behavior of systems (Gentner and Stevens, 1983; Rouse and Morris, 1986).

However, mental models are not always accurate representations of a system. They may contain errors, gaps, or "bugs." Repetitive or regular errors have been attributed to buggy mental models (Van Lehn, 1990). Also, different operators of the same system may have slightly different mental models for the system, resulting in different operating patterns (Vicente, 1997).

17.2.2 Human Strengths

If machines can enhance, accelerate, and extend our physical and mental capacities, what do they need us for? Is it not sufficient to create the machine that does our bidding, and then get out of the way? Is not the robotic welding machine the final technological step that replaces the human welder? Actually, no. Human tasks change as technology develops. Industrial production lines may require far fewer manual laborers whose tasks are performed by robotic devices, but these in turn are controlled by computers which must be managed by people who make decisions, plan, change their minds with evolving conditions, evaluate, and direct. And current operators must understand the metallurgy of welding better than their manual predecessors. Note that even in the futuristic science fiction visions of *Star Trek* and its TV offspring, medical practice is still physician-intensive, albeit scientifically and technologically enhanced. In present-day medical labs, X-ray technology now shares the stage with PET, MRI, and CAT scans, but all their images are still subject to human interpretation and analysis.

In contrast to the reliability, accuracy, and persistent character of most machine systems, humans offer creativity, flexibility, resourcefulness, and adaptability. These are grounded in our highly structured long-term memory, which supports higher-order reasoning about a system—how to operate it, troubleshoot it, repair it, and optimize its performance. Cognitive science research has sought to understand the nature of the knowledge and skills required to perform complex tasks involving sophisticated systems. What has become clear is that the knowledge of domain experts consists of highly integrated, organized knowledge. Knowledge about a nuclear power plant, for example, does not consist of a list of discrete isolated facts, but rather an integrated model of the component parts, their functions, and how the elements operate together (Roth et al., 1994). Embedded in this model of the system are individual facts, but they are closely tied to other related facts to make up a network. As a result, thinking about one element is likely to retrieve automatically other information (Anderson, 1995). Mental models allow us to make predictions about what will happen at one site in the device when certain events happen at another point. They also support troubleshooting of device failures because of knowledge of the connectivity in the system elements (Kieras and Bovair, 1984).

Highly integrated knowledge structures permit knowledge-based reasoning. Humans are adaptive and flexible. They can deal with uncertainty and ambiguity, situations that typically stump intelligent but "brittle" machines. For example, modern "glass cockpit" aircraft include many "intelligent" elements in their guidance systems. One recent crash resulted because the autoflight system detected that an aircraft was two knots over the target landing speed and changed the mode to a go-around. The pilot was unable to counter the aircraft's "plan." Had the pilot been flying in a manual mode, he easily would have managed the landing with the two-knot variation in speed. As Wiener (1989) has noted, computers are "dumb and dutiful." They cannot take into account special circumstances as humans can. Humans are flexible enough to develop novel approaches to reach a goal in the face of impediments, or to change their goals as circumstances warrant. Computers plod along, despite changes that require adaptation. [Stix (1994) has noted that the ATC system works *because* of human operators who work around limitations in the 25-year-old system.]

Underlying knowledge allows humans to reason analogically, that is, to respond to a novel situation "as if" it were an old situation (Forbus et al., 1997). While artificial intelligence systems may require a perfect match between a new situation and patterns stored in memory in order to trigger "recognition," humans respond readily to similarity of patterns. In problem-solving and decision-making situations, experts typically recognize a situation as resembling a certain class of problems, and retrieve solutions that worked in the past. Klein's (1989, 1993) Recognition-Primed Decision Theory is based on pattern matching to memory for prior similar events. Recent efforts to distinguish experts from novices within a domain find that the primary difference is in the extent of organized knowledge that characterizes the experts. It is not that experts and novices think differently, in principle, or have different capacities or processes. Because of the differences in knowledge, experts appear to have better memory for domain information and to reason more powerfully.

One caution — while expert knowledge confers flexibility in a wide range of situations, it can also lead to error. Because of experience people may make assumptions that a new situation is like an old one, when in fact it is not. For example, when pilots observed engine oil quantity indicators decreasing during a flight, they thought the indicators were faulty because they had prior experience with that malfunction. In fact, the oil caps were missing, and oil was leaking during the flight (Billings, 1997).

Another human strength is our pattern-recognition capability. How our visual perception system operates is an object of marvel. How we recognize faces that change over time as a person ages, how we recognize objects that are partially obscured, how we project movement of objects, like a ball in the outfield and can often manage to intersect its path, how we recognize objects regardless of their orientation, size, or ambient lighting, are profound questions that have intrigued and motivated scientists for years. This is a set of skills that machines cannot yet duplicate.

Designers, however, can capitalize on these skills. Recent advances in aircraft guidance systems are building on these perceptual skills. New "highways in the sky" present pilots with graphic "tunnels" to direct them to the landing path (Grunwald and Merhav, 1996). The magenta line on the horizontal situation display shows the lateral flight path programmed into the flight management computer, along with the point at which the pilot is to begin to descend. Perceptually guided action is an approach to reduce the cognitive demands of tasks, relying instead on visual perception skills that are so efficient that they demand little effort (Vicente, 1995).

17.2.3 Cognitive Requirements of System Design

How can knowledge about human limitations and strengths serve to guide design of new devices and systems? Knowledge of human limitations can point out where trouble might be expected in operating a device. Knowledge of human strengths can be helpful in deciding what tasks to allocate to the human and to the machine components of a system.

Norman (1986) has developed a theory of action that can serve as a framework for examining the adequacy of a device from a human user's perspective. The theory describes the path from the user's goal to an action to achieve that goal, to evaluation of the outcome of the action with respect to the goal (see Figure 17.1).

17.2.3.1 Norman's Theory of Action. As shown in Figure 17.1, the user begins with a goal to be achieved. The first step involves formulating an intention to achieve the goal. However, simply having an intention is not a sufficient basis for action. A plan must be developed: What actions will actually lead to goal achievement under the current circumstances? Then the actions are taken. Actions are designed to operate on the world, changing something within the device or the situation. In order to know whether or not you have moved toward your goal, the effect on the system must be evaluated. One must perceive changes in the situation, interpret the changes in light of one's intentions, and evaluate the changed situation in terms of the original goal.

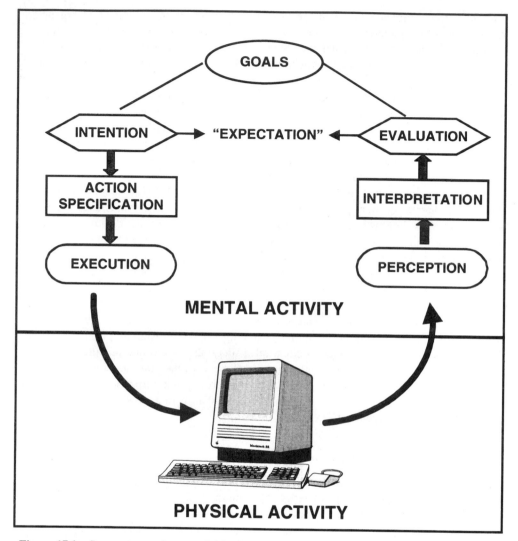

Figure 17.1 Seven stages of user activities involved in the performance of a task. (Reprinted with permission from D. A. Norman, "Cognitive Engineering." In *User Centered System Design,* D. A. Norman and S. W. Draper, Eds., Erlbaum, Hillsdale, NJ (1986).

This rough theory can aid a designer in answering the question: How do I know if I have designed an effective device? Good design should enable the user to bridge two gulfs—the gulf of execution and the gulf of evaluation (Norman, 1988). Bridging the gulf of execution depends on providing affordances in the system that support or invite the kinds of actions that lead to satisfaction of the user's goals. *Affordances* are "the perceived and actual properties of the thing, primarily those fundamental properties that determine just how the thing could possibly be used" (Norman, 1988, p. 8). Affordances provide strong cues to the operations of things. A good design should make obvious the conceptual model of the parts of the device and how they operate. Affordances support the user in going from intentions to actions, without putting an excessive burden on the user's cognitive system.

Bridging the gulf of evaluation depends on the degree to which a system makes the effects of operations visible to the user. Feedback is essential so that you can tell whether you have done the correct action or whether you have in fact achieved your goal. Without clearly visible feedback, users have no way of knowing whether their actions have "taken" and whether they were appropriate.

Consider the example of an infusion device designed to deliver measured doses of medicine to treat preterm labor in high-risk pregnant women (Obradovich and Woods, 1996). Since the device is designed for self-administration, women can remain at home rather than in a hospital and reduce the number of trips to a clinic. Users interact with the device through four multifunction buttons: select (SEL), activate (ACT), up arrow, and down arrow. To program the patient's dose profile requires the following sequence of steps: press the SEL button seven times, the ACT button twice, the SEL button once, and ACT button three times, and the SEL button once, and then the arrow key to set the dose level. A single screen displays a hierarchy of multiple modes: seven modes and under each are one to seven additional displays. Thus, one cannot display both the mode and variable level at the same time. The lack of clear feedback results in frequent confusions and errors. Improper programming of the device can result in under- or overadministration of the drug. Nurse practitioners have worked out special procedures to make sure that their patients administer the correct dosages, helping to overcome the gulf of evaluation built into the system.

17.2.3.2   *Four General Design Principles.*   Several principles of design emerge from Norman's analysis: build in appropriate affordances; use intuitive mappings; make important information visible; and rely on knowledge in the world rather than knowledge in the head.

1. *Affordances.*   Affordances signal to the user how a device might be used. Without these, considerable floundering and trial-and-error behavior may result, perhaps putting the system into hazardous states. An example from our laboratory illustrates their importance to bridging the gulf of execution. Recently, our research group tried to use a speakerphone for a conference call with colleagues across the country. The device consisted of three components connected by telephone wires: a standard Touch-Tone telephone, a flying-saucer-shaped device that appeared to be a speaker, and a rectangular box that appeared to be the control center. How to turn it on was not apparent. The telephone only had 12 standard buttons and no device for broadcasting. The flying saucer had three toggle switches each labeled "muting," but no on/off switch. The control box had a flat panel along one edge with five ovals printed on it. These were labeled Listen Only, Speak Only, Mute, Telephone, and Conference. No on/off switch.

Picking up the telephone gave us a dial tone, so we called our operator and asked her to place the conference call (two remote parties). All parties were on the line, but we still could not figure out how to broadcast. We pressed the oval labeled Telephone on the control panel of the rectangular box, but nothing happened. We tried pressing the Conference oval but nothing happened. In frustration we began pressing ovals in various combinations, until miraculously lights appeared on the flying saucer, a tiny light appeared in the oval labeled Telephone on the control panel, and voices filled the room. It appeared that it was necessary to hold down the Telephone oval for several seconds, which we did inadvertently while trying to combine oval presses. The affordances of this device were abysmal. No clues were provided to the actions required to make it work. However, the control box was very sleek, an elegant design, easy to keep clean with no protruding knobs.

2. *Mapping.*   A second factor in design is the *mapping* between intentions, the movement or operation of controls, and the results. A common example is the steering wheel of a car. When you want to turn right, you turn the wheel to the right. However, some mappings are not at all intuitive or are counterintuitive. For example, the throttle on a Cessna 152 aircraft is a knob that must be pulled all the way out to start the engine. Then, to increase engine speed one must push the knob in toward the control panel. To a novice pilot, this direction seems counterintuitive. Knobs with a variable range shaft seem like they should be pulled out more to increase the quantity. To remember this mapping, one can think of pressing down on the accelerator pedal on your car: more gas is pressing down; more throttle on the Cessna is pressing the knob in. Cultural norms clearly influence the adequacy of mappings.

Counterintuitive mappings can cause serious trouble. One of us recently rented a sailboat with an inboard motor. The transmission and throttle were combined, as in many boats. Unlike airplanes, boats can go forward and backward, so there is a transmission with three positions—neutral, forward, and reverse. Increasing power is accomplished by moving a lever further in the appropriate direction. However, the design of the controls made it difficult to remember which direction was correct. The transmission lever was mounted on the side of the steering column. To go forward one moved the lever down; to reverse one moved the lever up. The mapping of the controls to the desired direction was unintuitive and invited mistakes, especially in very close harbors that required precise maneuvering and quick changes of direction. Obviously, those controls were not designed by anyone who had ever operated a boat.

The preceding examples illustrate mapping of controls that involve physical movement. Often the controls are not physical, but electronic or digital, as in computers. Hutchins et al. (1986) have described "direct manipulation interfaces" for computers as one approach to providing devices with low cognitive costs. Direct manipulation interfaces revolutionized interaction with computers. Previous systems required meaningless sequences of commands that few could remember, such as IBM's Job Control Language. Now we have icons that are metaphors for actual desktops and control by mouse devices that map from intention to action in "natural" ways. Regardless of what the computer is doing behind the scenes, the relation between a user's actions and the results is direct and well mapped. When we move a mouse away from us, the cursor moves up on the screen; when we move the mouse closer to us, the cursor moves down. Natural mappings take advantage of physical analogs and cultural understandings to lead to immediate understanding of how to use a device. Three-year-olds quickly appreciate how to move a mouse to trace patterns on a computer screen.

3. *Visibility.*   A third element of Norman's design guidance is to make *visible* what is important for the user to know. Visibility pertains to advertising what the device can do and how (its affordances), feedback on what the user has done, and autonomous changes in the state of the system (those that do not result from a user's actions). As Norman put it, "The user needs help. Just the right things have to be visible: to indicate what parts

operate and how, to indicate how the user is to interact with the device" (1988, p. 8). Visibility can aid both execution and evaluation. On the execution side, in addition to telling what a device is for and how it should be operated, visibility can aid in remembering one's place in a lengthy sequence and remembering what to do next.

For diagnostic purposes, it is critical that a device articulate the sources and types of failures. For example, when *Apollo 13* experienced an explosion in the oxygen portion of its cryogenic system, it took 54 precious minutes for the fault to be identified (Murray and Cox, 1989). The pressure values that indicated the nature of the problem were embedded in 54 different digital data streams. To catch the problem immediately would have required an observer to notice a spike in one variable, followed by a precipitous drop. No graphical trend data were available to the technicians, which meant an exhaustive search through the possible problems.

4. *Rely on Knowledge in the World, Not Knowledge in the Head.*   Finally, in order to reduce the user's cognitive load (that is, not tax the limited capacities of attention, working memory, or long-term memory retrieval), Norman recommends relying on "knowledge in the world" rather than "knowledge in the head." Most often to perform a task we rely on a combination of knowledge in memory and supports from the environment. Reminding consists of a signal and the message: tying a string around your finger is a signal without a message; writing a date on your calendar is a message without a signal. Beeper alarms associated with calendars try to combine both. Checklists are good example of knowledge in the world—items and the proper order for doing them are listed on a piece of paper (or electronically). Paper flight strips enable air-traffic controllers to keep track of the aircraft currently under their control.

Relying on knowledge in the world to accomplish a task reduces cognitive demands, but it also may reduce what is in memory. Knowledge in the world reduces learning requirements. Getting information into memory is labor intensive, but once there, you can carry it with you. Knowledge in the world remains in its context; it is not transportable. Efficiency, flexibility, and generalizability may depend on knowledge in the head, but for users with little knowledge, put it in the world.

Hutchins (1995) has analyzed many aspects of flying airplanes, among them the task of remembering the appropriate speeds for the sequential stages during a descent and final approach. Certain critical actions can only be taken at or below specified speeds (e.g., extending flaps). While "speed cards" are available that illustrate takeoff and landing speeds at various aircraft weights, those are still numerical values that must be identified, processed, remembered, and retrieved when needed. Pilots need to coordinate their aircraft's location, altitude, and aircraft configuration to make sure they are taking appropriate actions at the correct points in the flight. By observing pilots in action, Hutchins has shown how they rely on knowledge in the world by using the technology available in the flight deck. Pilots typically use the "speed bugs"—arrow-shaped plastic markers that slide around the airspeed indicator dial—to mark target speeds for various flap settings (there are usually at least three speed bugs). By doing this, they do not need to remember their target speeds. Pilots rely instead on perceptual recognition rather than memory, thereby reducing their cognitive load during what is usually a busy phase of flight.

In the instance described by Hutchins it was the pilots who changed the nature of the task by establishing external supports for performance. Building in such supports during the design process is the focus of recent design efforts. Rasmussen (1997) captures this shift: "A visible trend is found from design focused on normative instruction and punishment, over focus on design to remove opportunities for human error, toward design aiming at control of behavior through selection of behavior-shaping system features, and making visible the boundaries of acceptable performance."

17.2.4 Designing to Prevent Error

At best a poor design results in inefficiency and frustration on the part of the user. At worst, it results in errors that may seriously interfere with safe operation in critical situations, like medicine, nuclear power, or aviation. Let us consider the types of errors people often make and how design may prevent or mitigate those errors.

17.2.4.1 Types of Error. There are at least two major types of errors: slips or lapses, and mistakes (Norman, 1981; Reason, 1990). *Slips or lapses* are failures to carry out an intention, but the goals or intentions are correct. In *mistakes,* the errors are actually in the goals or the intentions themselves. Slips are easy to detect providing there is feedback, because of obvious mismatches with intentions. Slips often occur while we are carrying out skilled actions because these require little attention; we are on automatic pilot. Hence, we are more subject to distractions. For example, we may find ourselves getting off the highway at our habitual exit after work when we had intended instead to go shopping; or dialed a phone number by picking up numbers off a pad sitting in front of us (not the intended phone number); or walked into a room and forgotten why we were there. People tend to monitor their actions much more closely when they are just learning a skill, and are less likely to make slips or lapses.

A particularly dangerous type of "slip" in the operation of automated systems is the "mode error." These errors occur when more than one action is associated with a particular display or control. Often high-tech devices have few controls (either to save "real estate" or for aesthetic reasons) and a single display window. These few controls serve multiple functions, usually by toggling between various modes. For example, clock radios, computer keyboards, telephones, and microwave ovens all have multiple modes. Mode errors occur when an action appropriate to one mode is entered when another mode is operative. These are especially insidious when there is no indication from the device of which mode is active. Mode errors appear to be quite common on modern flight decks and have been implicated in a number of aviation accidents.[4] One tragic case resulted when the flight crew entered 33 into a window to begin the descent of the aircraft. They thought they were entering a 3.3-degree descent angle, but because of the mode they were in, the aircraft's guidance system interpreted the entry as a 3,300 feet per minute rate of descent (Lenorovitz, 1992).

The other type of error, mistakes, results from choice of inappropriate goals or intentions. Their origin lies in the structure of knowledge in long-term memory or in the strategies or control processes used. People make poor decisions either because they misclassify or misdiagnose a situation, underestimate risk, fail to consider options, or to take relevant factors into account. When conditions are ambiguous or dynamically changing, mistakes are common. People may fail to update their assessment of the situation. Mistakes are much more difficult to detect, because goals or intentions are not always evident, and we often behave opportunistically, shifting our goals to match opportunities that present them-

---

[4]Sarter and Woods (1995) review mode-related errors in aviation and a series of experiments to assess their causes.

selves. More cognitive effort is required to prevent mistakes, or to correct them once recognized. Interpretations may seem reasonable at the time, but lead to poor outcomes because of inadequate assessment of the conditions or options. So-called metacognitive strategies may be needed to deal with these kinds of errors (Cohen and Freeman, 1997; Means et al., 1993).

17.2.4.2 Preventing Errors. What can be done at the system design level to prevent or mitigate errors? First, it is important for designers to understand the various types,of errors people are likely to make and what causes them (Lewis and Norman, 1986). The following steps can increase the error-tolerance of designs (Norman, 1988).

1. Make it easy to discover errors. The visibility principle applies here: feedback is essential so an error can be detected.
2. Make errors reversible. Slips or capture errors often are caught immediately after they occur. If reversible, they may do little harm. The UNDO command on my computer has saved me many times from inadvertent deletion of critical information.
3. Make it difficult to do what cannot be reversed. If you inadvertently hit the command to delete every file on your hard disk, a query will appear asking "Do you really want to do this?" Multiple steps are required to take this drastic step.
4. Use forcing functions, which prevent one step from being taken without an essential prior step. These can be of three types: *interlocks,* which force sequences; *lock-ins,* which prevent premature stopping of actions; and *lockouts,* which prevent one from getting into dangerous situations by mistake.

It is helpful to understand why errors occur in order to prevent them. For example, one type of error that has received considerable attention is the "post-completion error" (Byrne and Bovair, 1997). These are the kinds of errors where we leave our paper in the copying machine after making copies or our ATM card in the machine after getting cash. A cognitive analysis of the task showed that in these cases the forgotten step occurred after the main goal had been achieved (getting copies or cash). The solution is to put the usually forgotten step on the main goal path, so that it will not be forgotten. In fact, new ATM machines require that your card be removed before the money will be dispensed. So far no one has fixed the paper in the copier problem.

17.2.5 Issues in Applying Knowledge of Cognition to Design

While a few general principles have just been described, they are just that—general principles or design heuristics. What about designing a specific device to perform a specific task? How can a designer know the actual cognitive requirements that are relevant to a device? While this issue will be addressed in greater detail in Section 17.3, there are a few complexities that should be noted.

1. *Differences in User Skill Levels.* First, cognitive capabilities are not uniform and static. They depend on the experience and skill of operators. Who is the expected user of the new device? A large body of research shows enormous differences in the functional capacities between novices and experts in many fields (Ericsson and Smith, 1991, Chi et al., 1988). These differences are evident in their working-memory capacities (for content

VERY LOW ⟵ LEVEL OF AUTOMATION ⟶ VERY HIGH

VERY HIGH ⟵ LEVEL OF INVOLVMENT ⟶ VERY LOW

MANAGEMENT MODE	AUTOMATION FUNCTIONS	HUMAN FUNCTIONS
AUTONOMOUS OPERATION	Fully autonomous operation; Controller not usually informed. System may or may not be capable of being bypassed.	Controller has no active role in operation. Monitoring is limited to fault detection. Goals are self-defined; controller normally has no reason to intervene.
MANAGEMENT BY EXCEPTION	Essentially autonomous operation. Automatic decision selection. System informs controller and monitors responses.	Controller is informed of system intent; May intervene by reverting to lower level.
MANAGEMENT BY CONSENT	Decisions are made by automation. Controller must assent to decisions before implementation.	Controller must consent to decisions. Controller may select alternative decision options.
MANAGEMENT BY DELEGATION	Automation takes action only as directed by controller. Level of assistance is selectable.	Controller specifies strategy and may specify level of computer authority.
ASSISTED CONTROL	Control automation is not available. Processed radar imagery is available. Backup computer data is available.	Direct authority over all decisions; Voice control and coordination.
UNASSISTED CONTROL	Complete computer failure; No assistance is available.	Procedural control of all traffic; Unaided decision-making; Voice communications.

Figure 17.2 A continuum of system control and management for air traffic controllers. (Reprinted with permision from C. E. Billings, *Aviation Automation: The Search for a Human-Centered Approach (Human Factors in Transportation),* Erlbaum, Mahwah, NJ (1997).

within their domain of expertise), in the completeness of their mental models, and in their problem-solving strategies. These are all functions of knowledge in long-term memory. These knowledge differences are reflected in functional differences in performance, so a designer must have a particular type of user in mind.

2. *Function Allocation.* A second issue is at what level the device should support the human. How should tasks or functions be allocated to humans and machines? Billings (1997) presents a continuum of support (for an example, see Fig. 17.2) from devices as dumb tools through autonomous intelligent agents. The issue of function allocation depends on the nature of the problem and the generality of the solution. If the problem to be solved is very specific, and the tool is not to be used for a broad range of purposes, then it can be highly automated and very narrowly defined. Such devices may involve data-driven control. On the other hand, if it is to be a general-purpose tool to be used for a variety of tasks, then different principles apply. A personal computer falls into the latter category. Control in these cases is top-down; the user determines what the tool is to do at any time.

3. *Psychological vs. Engineering Specifications.* Third, it is difficult to go from general knowledge about cognitive processes, capabilities, and constraints to design because the two domains speak different languages. Different vocabulary and concepts characterize the two. Psychological language about cognitive capacities is not equivalent to the physical

language of designers. How to bridge this gulf? One approach, which has proven useful in well-defined situations, is to model the task. Card et al.'s (1983) goals, operations, methods, and selection (GOMS) technique was used by Gray et al. (1993) to evaluate the design of a new information system for a telephone company. GOMS is an approach to modeling the predicted sequences of steps, durations, and logical dependencies among them. Gray et al. applied this technique to the design of a new system for answering telephone information queries. Their analysis predicted that the new interface would in fact be less efficient than the old system, and even predicted where errors and delays would occur. Comparison of data from operators using the new interface with the model supported the prediction. One could use such an approach to evaluate candidate systems before they are fully designed and fielded, saving enormous costs.

4. *Organizational Factors.* A final issue is the impact of organizational context on the use of devices. No matter how well designed a device is, organizational pressures may work to undermine the overall efficiency of the human/machine system. Organizational norms, values, and goals influence performance of the task in many ways. Degani and Wiener (1994) have shown how organizational factors relate to policies and procedures on the flight deck, changing how pilots use automation in ways that differ from the assumptions and recommendations of the manufacturers and designers. Warnings or alerting devices that are built into the system are sometimes disabled because they interfere with practice. Such was the case in the crash of a DC-9-82 just after takeoff from Detroit. The flaps had not been configured for takeoff. A warning horn normally sounds if takeoff is attempted without proper flap configuration, but investigation of the accident revealed that the warning device had been disabled on this particular aircraft (NTSB, 1988). In a different context, Lee and Vicente (1998) report that management at a power plant instructed personnel to violate each of its Ten Commandments of Safety numerous times, thereby creating conflicts in operating principles.

Reason (1990) and others have analyzed the contribution of organizational factors in accidents and incidents. Faulty design per se may not be sufficient cause for an accident, but when certain organizational conditions exist, design failures become manifest. These "latent pathogens" are potential flaws that have their impact only when certain conditions are aligned (Reason, 1990). These flaws are easy to see in retrospect, but sometimes difficult to predict in advance.

Consider the case of the *Herald of Free Enterprise* (Sheen, 1987). This "roll-on/roll-off" passenger and freight ferry capsized only minutes after departing from Zeebrugge en route to Dover. The lower car deck flooded because the bow gates were still open when they departed. In addition, the ferry was riding low in the water because the ballast tanks had not been purged, due to schedule pressures. There was no alarm to indicate that the gates were still open when departing. We may ask why not. The company had decided against installing one because of cost, relying instead on human personnel to make sure the gate was closed. Unfortunately, personnel were in short supply on departure: one critical person was asleep, another was on a different deck. Redundancy in the system was inadequate. Schedule pressures to reach their destination on time contributed to not checking the gates prior to departure or to pumping the ballast tanks. This case illustrates the complexity of factors that contribute to accidents, a situation that is typical rather than unusual.

What is a designer to do? How is the designer to take into account expected users' cognitive requirements, the actual conditions of use, and organizational factors that influence the functional quality of the design? These issues are addressed in the following section of this chapter.

17.3 REDUCTION TO PRACTICE

17.3.1 Overview

Cognition, by definition, involves the access and utilization of knowledge. In the following discussion we will therefore limit our discussion to the design of information systems. The best design methods to support cognitive task performance are closely related to mainstream system and software engineering methods. Nevertheless, software engineering methods do not typically reflect cognitive factors.

Software design and engineering methods need to be extended to allow for characteristics of human cognition. In Section 17.2 we discussed cognitive invariants and bottlenecks that are well documented in the research literature. In addition to these facts about cognition, there are also emergent features that are dynamic, task-dependent, and are not as well understood as the invariant features.

We will discuss three questions about the relevance of cognitive research to design:

1. Is reliable information on cognition available?
2. If it is available, is it in a form that addresses design issues?
3. If cognitive design-support methods are available, is it cost-effective to apply them to real systems?

The first two of these questions are easily answered. We already have reviewed much of the research basis for cognitive design. In this section we mention some of the design-oriented sources that have been published during the past decade, and we describe cognitively oriented design-support methods that have links to software engineering methods. These techniques include the focused checklist, the cognitive walkthrough, human-system modeling, and iterative prototyping and operator testing. The third question, concerning cost-effectiveness, cannot be answered in a global sense.

The cost-effectiveness of different human factors methodologies depends upon the type of system, the costs of poor usability, the risks and costs of failure, and other factors. Sometimes, however, talk about "cost-effectiveness" can serve as an excuse for shortsighted thinking and bad design, rather than a careful consideration of the true life-cycle risks and costs of the to-be-designed system from the users' perspective.

17.3.2 Design Philosophy

We have limited our discussion to the design of information systems because these are the systems that highlight cognitive issues. Physical design issues concerned with anthropometry, ergonomics, and physical safety are important and are relatively well understood. The informational environment is less tangible, less stable, and harder to describe than the physical environment. It is often best described in discrete, not continuous, mathematics; this makes engineering analyses less tractable.

Software dynamics are thus harder to analyze than physical dynamics. Embedded systems, which were formerly associated with military or other exotic environments, are now the norm in everyday applications. Layered architectures and multilevel services define human-interactive systems ranging from the World Wide Web to the air-traffic management system. In such systems, dynamics become crucial. The relative timing and varying reliability of different levels create an unpredictable infrastructure for any particular target system. Subsystems, including the operator interface, must be designed to new standards of robustness.

Fortunately, the methods developed to support software engineering in these challenging environments can be adapted to the case of human-interactive systems. For example, iterative prototyping has gained acceptance as an alternative to top-down design (Curtis et al., 1988). It is widely acknowledged that top-down design forces important decisions to be made early, when ignorance is at its maximum. Iterative prototyping is a more sensible, learn-as-you-go strategy. A key element of iterative prototyping is the use of *threads* (Deutsch, 1996) or *scenarios* in the design process. (Because the term *threads* has several different meanings, we will use the term *scenarios* below.) Issues related to human factors and usability are best addressed in an iterative, scenario-based design process— the type of process that has proven effective for software engineering in general.

17.3.3 Cognitive Issues

Our claim that design for cognitive task performance is compatible with good software engineering practices does not imply that mainstream software engineering methods address cognitive factors. On the contrary, designers routinely ignore the operator's role and requirements. Our claim is that cognitive design entails no *fundamentally new* methods, but rather the extension and adaptation of mainstream software engineering methods to put the user back in the picture.

Cognitive task performance occurs in the context of physical and perceptual invariants. Invariant properties of human vision, audition, and motor behavior are well understood, as are aspects of fatigue, repetitive strain injury, eyestrain, and physical safety. In addition to physical and perceptual invariants, cognitive invariants are also well understood. As discussed in Section 17.2, human operators have known limitations of attention, vigilance, working-memory capacity, learning rate, and long-term memory reliability. These well-documented limitations impact higher-level cognitive processes such as decision making, problem solving, and communication.

Finally, there are emergent features of human–system interaction. These are not invariant, but depend upon features of tasks and goals. The environment of use, as well as levels of human expertise (designer, expert operator, novice operator, and trainee), can have significant impacts on cognitive performance. The easiest, fastest, and cheapest design approaches, such as generic checklists and context-free guidelines, fail to address these contextual issues. More robust design-support methods, such as modeling and experimental testing, are required.

17.3.4 Information Sources

Is reliable knowledge available about cognitive and other human factors? This question is easily answered. We simply cite some of the excellent sources of information that have been published since the 1980s. These are not research tomes, but are carefully organized and indexed to be relevant and useful to designers.

There are a number of textbooks that present factual information integrated with design principles, guidelines, and advice. Proctor and Van Zandt (1994) is an example, covering research on physical, perceptual, and cognitive factors, and relating this knowledge to design principles for workspace design, job analysis and design, communication within organizations, displays and controls, and human–computer interfaces.

In addition to textbooks, there are now extensive bibliographies and handbooks intended for use in the design process. The classic Smith and Mosier (1986) is 488 pages long and contains 944 proposed guidelines for designing the operator interface to computer-based information systems. It is organized into six functional areas: data entry, data display, sequence control, operator guidance, data transmission, and data protection. Some examples of guidelines from this source are:

> Require users to take explicit action to confirm doubtful and/or potentially destructive data change actions before they are accepted by the computer for execution.

> When display output is more than one page, annotate each page to indicate display continuation.

Boff et al. (1986) produced a comprehensive two-volume handbook on human perception and performance, consisting of 45 chapters and a 96-page index. There is a companion three-volume, 2,510-page *Engineering Data Compendium,* with a 134-page operator's guide and 58-page index (Boff and Lincoln, 1988). This work pulled together a large amount of design-relevant knowledge about human perception and cognition and organized it for use by designers and engineers.

NASA-STD-3000, the *Man-System Integration Standards for Space Flight* (National Aeronautics and Space Administration, 1989) is distributed through the Johnson Space Center and through CSERIAC (see below). It is roughly the same size as the Boff et al. *Engineering Data Compendium,* approximately 2,000 pages.

Helander et al. (1997) produced a 1,000-page handbook containing over 50 chapters on topics in human–computer interaction. Cardosi and Murphy (1995) contains comprehensive human factors guidelines just for the design and evaluation of ATC systems. Wagner et al. (1996) provides extensive human factors guidance for acquisition of systems by the FAA. It is specifically oriented to the contracting and acquisition process.

In addition to the thousands of pages of human factors design guidance just mentioned, there are also a number of on-line sources:

> HCIBIB, the *On-line HCI Bibliography* at http://www.hcibib.org/ is indexed by HCI keywords and by authors and is searchable.

> *CSERIAC* at http://www.dtic.dla.mil/iac/cseriac/cseriac.html, the Crew System Ergonomics Information Analysis Center, has links to current human factors information for designers, engineers, researchers, and human factors specialists. *CSERIAC* also offers a variety of products and services to government, industry, and academia to facilitate the use of human factors information in design.

> The *CLARE* Library at http://tommy.jsc.nasa.gov/~clare/index.html supports development of Control Center software for NASA's Johnson Space Center. *CLARE* makes applications inspectable so that designers can share ideas, techniques, operator-interface designs, and source code.

Given the depth and breadth of information in the sources just cited, it is no longer plausible to claim that we lack the knowledge base needed for design in support of cognitive task performance. Yet these sources may be too imposing, or they may not always be indexed in ways that a particular designer can easily use. In other words, a knowledge base may exist, but it may not be readily applicable to many design problems. Design-support methods are needed to bridge the gap between a static, disciplinary knowledge base and a dynamic, multidisciplinary design process. Human factors practitioners have heeded this criticism, and they have responded with methods like those described below.

17.3.5 Design-support Methods

This section addresses the second question: Can knowledge about cognition be applied during the design process?

The design-support methods discussed here range from low-cost to high-cost. All of these methods require significant human factors expertise on the design team. We believe that there are no valuable human factors design methods that can be implemented by nonprofessionals. Furthermore, the outcome of the design process, regardless of the methodology, will depend mainly on the technical and operational expertise of the human factors professionals on the design team. Again, the situation with respect to human factors is the same as with software engineering. No powerful software-engineering methodology could be successfully implemented by non-programmers. Analogous reasoning applies to human factors.

17.3.5.1 Focused Checklist. Remington and Johnston (1997) applied the knowledge just described, as well as their own knowledge of cognitive invariants, to a conflict-alert and resolution subsystem for advanced ATC. They produced a specialized checklist consisting of mixed physical, procedural, and design-process recommendations tailored to the target system. As a checklist, this method is not behavior-oriented. It summarizes certain perceptual and cognitive invariants as they apply to the target system and recommends analysis of dynamic contextual features by user testing.

The checklist is organized under five main topics, which correspond to five phases of operator activity in the task domain. A few excerpts from the checklist are given below for illustration. The full checklist contains about 100 items accompanied by a 50-page commentary discussing the reasoning behind and boundary conditions of each item.

Displaying Non-conflict Traffic Information
> *The representation of traffic information*
>> Redundant coding of information by text and spatial cues is used when both spatial and numeric judgments are required.
>
> *The use of color*
>> Colored text is distinguished from background by substantial differences in both hue and luminance; text is not separated from background by hue alone.
>
> *Highlighting to facilitate visual search*
>> Rapid blinking (>2 Hz) is used only to indicate the need for immediate controller action.

Detecting Conflicts
> *Division of responsibility*
>> The tasks assigned the [automated] conflict probe are a natural subset of the detection process and are recognized as such by controllers.

Controller acceptance

Controllers understand how the conflict probe *fails* and can predict situations that will be problematic for the conflict probe.

Displaying Non-urgent Conflict Information

Indicating the conflict pair

If more than one conflict is present, the most recent one is marked to distinguish it from already present conflicts.

Obtaining conflict information

Information needed by controllers to set action priorities, such as time-to-conflict, is easily obtainable.

Displaying Urgent Conflict Information (Alarms)

Summoning controller attention

An alarm signal is used only in cases requiring immediate attention by the controller; alarms should not be presented when the automated conflict probe is expected to solve the problem without controller intervention.

Alarm interference with conflict information processing

Controllers can cancel alarms by performing an action that indicates they are attending to the conflict (e.g. mouse click on conflict aircraft).

Identifying the conflict pair

Distinctive visual highlighting draws controller attention to the source of the imminent conflict that generated the alarm.

Resolving the Conflict

Division of responsibility

The role of the automation is conceptually clear to controllers and conforms to a natural division of labor.

Exploring resolutions

Controllers can examine the consequences of proposed resolutions by asking the automation to extrapolate trajectories, future states, and nearest points-of-approach to designated aircraft.

Performance and acceptance

The number of operating modes is small and the current mode prominently displayed.

17.3.5.2 Behavior-oriented Methods. The purpose of the focused checklist is to increase the utility of generic human factors design principles by selecting and tailoring them for a particular target system. The focused guidelines are expressed more in the language of the task domain (ATC, in this case) than in the language of perception and cognition. Nevertheless, the guidelines depend heavily upon perceptual and cognitive invariants. Behavior-oriented methods, as described below, are needed to examine the dynamics of human–machine interaction.

Deutsch (1996) discusses the concept of *scenarios* in software design. We believe that scenarios are fundamental to behavior-oriented design methods, and that behavior-oriented methods are necessary to address cognitive task performance issues. Scenario-based design forces the design team to confront the following three questions:

1. What can happen? In other words, in the intended operating environment, what events is the system supposed to detect and react to?

2. For each event that can happen, how should the system respond?

3. How can the system be designed so that it handles all the scenarios? That is, how can a single, integrated system be designed to cover all the relevant event sequences?

Finally, scenario-based design implies a test: Does the system handle the scenarios? By developing scenarios, the design team is simultaneously developing a test suite. Tests can be developed and applied iteratively in a cycle of incremental analysis, design, integration, and testing.

17.3.5.3 Cognitive Walkthrough. A *code inspection* or *code walkthrough* is sometimes used to detect and eliminate faults as early as possible in the software engineering process. The *cognitive walkthrough* (Lewis and Wharton, 1997; Polson et al., 1992; Wharton et al., 1994) is the human factors counterpart to the code walkthrough. It is focused on the human side of the human–computer system. Briefly, the elements of the cognitive walkthrough are the following:

1. The target system can be examined while it is still in the conceptual design phase; a working prototype is not required.
2. The analysts choose a particular scenario (event sequence) that the system is supposed to be able to handle.
3. The analysts determine how the human–computer system should respond to the scenario. There might be more than one correct response.
4. The analysts focus on each point in the scenario where the operators must decide on the correct action to take. The analysts ask whether the hypothetical operators would be able to make the correct decision and, if so, how; and if not, why not.

Like code-inspection methods, the cognitive walkthrough may be performed casually or rigorously. The cost and benefit of the method depend upon how carefully it is planned, executed, and documented.

Lewis and Wharton (1997) suggest that a cognitive walkthrough can catch about 40 percent of the design errors that would be caught by user testing. Cuomo and Bowen (1994) applied the cognitive walkthrough to a system with a graphical interface. They compared this method to two somewhat less complicated methods based on checklists and guidelines. "The walkthrough was best . . . and the guidelines worst at predicting problems that cause users noticeable difficulty (as observed during a usability study)."

It is difficult, however, to draw conclusions about the relative effectiveness of the cognitive walkthrough. The method and its competitors are not precisely defined. Also, the effectiveness of such design methods depend crucially on the breadth and depth of expertise of the analysts. It is more appropriate to note that the cognitive walkthrough is among the easiest behavior-oriented methods to apply, and that it can be applied early in the design process—exactly at the point where a design team should be developing a set of basic scenarios to support early prototyping. There is high payoff for design errors eliminated at this early stage. Later in the design process, when prototypes are more mature, user testing is recommended over the cognitive walkthrough (Lewis and Wharton, 1997, p. 730).

17.3.5.4 Modeling. The relation of modeling to scenarios is straightforward: an algebraic language can be used to describe the system and the scenarios. This approach facilitates communication among people with diverse backgrounds, such as pilots, engineers, programmers, and safety experts. It also facilitates integration of subsystems that are developed independently. Complementary logical and algebraic methods may be used — logic for global reasoning about system properties, and algebra for reasoning about dynamics.

Leveson (1995) has applied modeling to the analysis of safety. She notes that "most accidents are not the result of unknown scientific principles but rather of a failure to apply well-known, standard engineering practices." She shows how models can be used to assess both cost-effectiveness and safety. She has paid particular attention to the role of models in promoting clear communication among members of multidisciplinary design teams, and she has demonstrated how models help to solve real design problems in aviation (Leveson et al., 1997).

The AMODEUS (1998) Project is a research project of the ESPRIT program, funded by the Commission of the European Communities. The project supports interdisciplinary approaches to studying interactions between users and systems, with the goal of transferring advanced methods to operator-interface designers. AMODEUS is characterized by the development of cognitive models of users, formal representations of human and system behavior, detailed interactive-system architectures, and practical means of integrating operator and system representations.

Some software and human factors designers have advocated modeling to achieve simplicity in design (Griffeth and Lin, 1993; Segal, 1995; Zave, 1993). As Hoare (1980/1987, p. 157) puts it,

> Almost anything in software can be implemented, sold, and even used given enough determination. There is nothing a mere scientist can say that will stand against the flood of a hundred million dollars. But there is one quality that cannot be purchased in this way, and that is reliability. The price of reliability is the pursuit of the utmost simplicity. It is a price which the very rich find most hard to pay.

Actually, the goal is not simplicity, but rather a structured complexity. The point of using models in design is to make sure that systems can be understood in spite of their complexity. Hinchey and Jarvis (1995), and Mett et al. (1994) provide current summaries of the algebraic modeling approach advocated by Hoare, which is similar to that used by Leveson. The important characteristics of this style of modeling are the following:

1. *Consistent semantics,* so that any system describable by a model has a known behavior.
2. *Modularity,* so that different systems described by models can be combined, and the compound system will have known behavior predictable from its components.
3. *Factorability,* so that the qualitative behavior of a system can be described first, with continuous variables such as time and probability added later.
4. *Variable levels of detail,* so that some parts of a system can be modeled in great detail, but not all parts have to be modeled at the same level.
5. *Support for formal reasoning and for simulation,* so that the same model can be run as a simulation or analyzed by symbolic reasoning using an automated theorem-checker.

These characteristics are important for the use of modeling to support scenario-based design. Degani et al. (1998) show how algebraic modeling can be applied to a wide range of human-interactive systems, from simple household devices to a modern flight management system. Kitajima and Polson (1996) have applied model-based analysis to comprehension, action planning, and learnability. They have demonstrated that models can be used to clarify the background knowledge that must be assumed in the operator population, thus serving as a design evaluation technique.

Kirlik (1995) has discussed the kinds of extensions of current modeling frameworks that will be required to cover a broader range of cognitive factors. Although current models play a crucial role in design, modeling frameworks need to be improved to better support cognitive task analysis. We need better models of the interaction between perception and cognition, as well as models of how cognitive processes are modulated by fatigue, stress, and other factors. Better models are also required for learning and expert performance. At present, modeling must be considered complementary to the other design-support methods discussed here. User testing produces surprises for modelers as well as for designers.

17.3.5.5 Iterative Prototyping and User-testing. There is widespread agreement among human factors professionals that iterative prototyping and user testing are the best ways to design usable systems (Gould, 1988; Gould et al., 1987; Landauer, 1995; Lewis and Wharton, 1997; Small, 1994). When used in a scenario-based design process, user testing is much more effective than guidelines at highlighting the emergent dynamic features of real human–system interactions. It also exposes issues at levels of detail that could easily be missed by cognitive walkthroughs and modeling.

As documented by Gould et al. (1987), user testing can be used during the design process to examine design variants and to evaluate tradeoffs. (Do features really pay their way?) It is the only feasible method for estimating the true economic costs of poor usability (Landauer, 1995). To estimate users' error rates with a given system and a realistic set of tasks, the best approach is obviously to let real users try to do the task using a real or prototype system.

Documentation and procedures are notoriously weak spots in the design process. By conducting user tests at intermediate stages of development, procedures and documentation can be incrementally developed and tested.

For safety-critical systems, it is important to understand off-nominal and maintenance procedures, as well as nominal operations. Again, iterative prototyping and user testing are probably the most reliable ways of checking the adequacy of design and procedures. The thorough detection and correction of design problems also pays dividends by making training simpler and more cost-effective.

The most common reason given for *not* performing adequate user testing is that such testing is too time-consuming and is therefore not cost-effective. We are persuaded by the work of Gould and his colleagues that this excuse is without merit. If a reasonable analysis is performed, we are confident that iterative prototyping and user testing will be found to be cost-effective, especially if they are used in conjunction with modeling and quick-look techniques like the cognitive walkthrough. (See Landauer, 1995, for an extensive discussion of this topic.)

There are, however, some valid reasons for not placing too much confidence in user testing. In some cases, representative users may be unavailable. In other cases, due to the limited number of representative users, a small subset of the user population may become "professional test subjects." For example, certain air-traffic controllers or pilots may become honorary members of the design team for new air-traffic management or avionics systems. As such, they are no longer valuable as test subjects, although their expert opinion may still be useful. Test pilots and instructor pilots are notoriously unrepresentative of the pilot population.

Another caveat relating to experimental testing concerns safety-critical systems. It is well known in software engineering that no amount of experimental testing can prove that a system is fault-free. The same principle applies to user testing. Only a small fraction of the possible human–system–environment scenarios can be tested. Designers are well advised to use rigorous modeling in conjunction with user testing to minimize the chances of overlooking human–system failure modes. Although there is no such thing as a "provably fault-free" system, a reasonable investment in modeling and experimental testing can greatly improve the reliability and usability of information systems.

17.3.6 Cost-effectiveness

Cost-effectiveness can easily become a cover-term for shortsighted, narrow, and ultimately non-cost-effective thinking. Parnas (1997) takes programmers to task for not living up to the professional standards implied by the term *engineering*. Parnas writes, "Software is a major source of problems for those who own and use it. The problems are exactly those to be expected when products are built by people who are educated for other professions and believe that building things is not their 'real job'."

Humphrey (1989) describes how to manage large-scale software projects to a high standard of quality, but the processes and standards that he advocates have been met with industry opposition that is "loud and clear" (*Computer,* 1995). Birrell and Ould (1985), Deutsch (1996), and other sources document and recommend methods that have been shown to be cost-effective. Yet these design practices are not used because they are not *perceived* to be cost-effective.

Sims (1997) discusses the tradeoffs in software development projects. Rapid technology advances and intense market competition motivate continual product revisions. Technology change also drives changes in customer requirements. The popular media, trade journals, and the Internet focus intense customer scrutiny on each new product or new release. Media criticism may focus on limited or out-of-date functionality, as well as on poor usability. The implacable pressures of the marketplace drive vendors to release products that are buggy, poorly documented, and unusable owing to "reliance on aberrant development practices."

Thus, short-term thinking leads to bad engineering. Bad engineering increases life-cycle costs and risks (Landauer, 1995; Leveson, 1995). Landauer (1995) discusses the apparent costs of poorly designed systems where the costs are spread over years of use by perhaps millions of customers. Leveson (1995) discusses the complementary case where costs are concentrated in a few major disasters or a catastrophic failure. We suggest that these two kinds of failures (Landauer's and Leveson's) are two cells of a fourfold table crossing high and low probability of failure with high and low cost of failure. The case of high probability and high cost is likely to be detected even in a cursory design and test process. The case of low probability and low cost can be ignored.

Landauer's case, high probability and low cost of failure, distributes the large aggregate cost of bad design over a large number of low-cost failures. This is characteristic of software intended for home and office use, certain kinds of telecommunications systems, consumer electronics, and other mass-produced products. Leveson's case, low probability and high cost, concentrates the cost of bad design in a few accidents where there is significant loss of life, property loss, or health hazard. The combined work of Landauer and Leveson suggests that the true life-cycle risks and costs of bad design are rarely considered in shortsighted "cost-benefit analyses." In fact, the relevant data for a life-cycle cost-benefit study could not be obtained without extensive, realistic user testing.

17.3.7 Summary

There are strong connections between human factors engineering and software engineering. These connections have been recognized at least since the early 1980s.

In human factors, as in software engineering, the investment in design quality must be commensurate with the likely costs and risks of failures. A range of validated human factors design methods is now available. All require a high degree of human factors expertise on the design team, just as software engineering methods require computer science expertise. Taken together, these methods provide a cost-effective approach to human factors design for systems ranging from consumer electronics to ATC.

In human factors, as in software engineering, sound practices are available, but are often ignored in practice. The excuses for poor human factors engineering are the same as those for poor system engineering: short-sighted thinking leads to faulty cost-benefit assessments, mainly because life-cycle costs are not considered; rather, only the immediate costs of design, development, and testing are considered. Shortsighted thinking leads to bad engineering practices, which in turn lead to unnecessarily high life-cycle costs and risks.

The excuses for poor human factors engineering that may have been valid two or three decades ago—that human factors knowledge is lacking, irrelevant, too costly, or too time-consuming to apply—are out of date. The only valid reasons for ignoring human factors today are either that humans will not interact with the system, or that failure of the system will have no serious economic or safety impacts.

17.4 CONCLUSIONS

Stix (1994) describes a typical large-scale software engineering fiasco, the Advanced Automation System (AAS):

> In some respects the [AAS] undertaking seemed relatively straightforward. [...] Twelve years later much of the program has yet to be implemented. [...] The AAS project was structured like a typical defense contract. [...] Mesmerized by the flood of detail, both contractor and customer failed to pay attention to how a more automated system would change the way an individual controller performs. [...] What IBM devised, according to one controller, was a perfect electronic facsimile of the paper flight strip. The database contained the information that can be listed on a flight strip—about 140 separate units of information . . . enough to keep the controller tapping on a keyboard all day.

Based on his experience with the AAS project, Small (1994) offers the following list of recommendations to designers of human-interactive software systems:

More effective use of prototyping . . . [including appropriate involvement of human factors experts];

More appropriate use of controller [i.e., user, operator] teams, including more timely involvement, more appropriate roles, and more interaction with the developers;

Better integration of specialized human factors expertise throughout the development process, recognizing its limitations as well as its potential benefits;

Collection of objective data at appropriate times to validate the opinions of controller [user, operator] teams and the judgments of human factors experts;

Establishment of a mechanism for making the necessary trade-offs between an ideally human-oriented design and a design that is technically and financially feasible.

In the modern film classic *Monty Python and the Holy Grail*, the sorcerer Tim tries to warn King Arthur's band about the risks of confronting the deadly white rabbit. "It's a vicious killer!" Tim declares. Arthur and the knights are unimpressed by what they see as "a harmless bunny." Finally, indicating the remains of past victims, Tim shrieks, "Look at the bones!" But Arthur and his men can only learn from experience, and they lose several knights in a frontal assault on the rabbit before running away to a safe distance and lobbing the Holy Hand-Grenade of Antioch to subdue the beast.

Those of us, like Duane Small (cf. Curtis et al., 1988), who have encountered the killer rabbit of large-scale human-interactive systems, can do little more than point to the remnants of past failures and urge, "Look at the bones!"

ACKNOWLEDGMENTS

The authors express their appreciation to Christina VanAken for her assistance in preparation of this chapter. The opinions expressed here are the authors' and do not represent official opinion or policy of any government agency.

REFERENCES

AMODEUS. (1998). The AMODEUS Project home page: http://www.mrc-apu.cam.ac.uk/amodeus/

Anderson, J. R. (1995). *Cognitive Psychology and Its Implications*, 4th ed. New York: Freeman.

Billings, C. E. (1997). *Aviation Automation: The Search for a Human-Centered Approach (Human Factors in Transportation)*. Mahwah, NJ: Erlbaum.

Birrell, N. D., and Ould, M. A. (1985). *A Practical Handbook for Software Development*. Cambridge, UK: Cambridge University Press.

Boff, K. R., and Lincoln, J. E. (eds.). (1988). *Engineering Data Compendium: Human Perception and Performance*, Vols. 1–3. Wright-Patterson AFB, OH: Harry G. Armstrong Aerospace Medical Research Laboratory.

Boff, K. R., Kaufman, L., and Thomas, J. P. (eds.). (1986). *Human Perception and Performance*, Vols. 1 and 2. New York: Wiley.

Broadbent, D. E. (1979). Human performance and noise. In *Handbook of Noise Control* (C. M. Harris, ed.), pp. 17.1–17.20. New York: McGraw-Hill.

Byrne, M. D., and Bovair, S. (1997). A working memory model of a common procedural error. *Cogn. Sci.* **21**, 31–62.

Card, S., Moran, T., and Newell, A. (1983). *The Psychology of Human-Computer Interaction.* Hillsdale, NJ: Erlbaum.

Cardosi, K. M., and Murphy, E. D. (1995). *Human Factors in the Design and Evaluation of Air Traffic Control Systems.* Cambridge, MA: Volpe Center.

Chase, W., and Simon, H. (1973). Perception in chess. *Cogn. Psychol.* **4,** 55–81.

Chi, M. T. H., Glaser, R., and Farr, M. O. (1988). *The Nature of Expertise.* Hillsdale, NJ: Erlbaum.

Cohen, M. S., and Freeman, J. T. (1997). Improving critical thinking. In *Decision Making Under Stress: Emerging Themes and Applications.* (R. Flin, E. Salas, M. Strub, and L. Martin, eds.). Brookfield, VT: Ashgate Publ. Co.

Computer. (1995). The road to software maturity. *Computer* **28**(1).

Cuomo, D. L., and Bowen, C. D. (1994). Understanding usability issues addressed by three user-system interface evaluation techniques. *Interact. Comput.* **6**(1), 86–108.

Curtis, B., Krasner, H., and Iscoe, N. (1988). A field study of the software design process for large systems. *Commun. ACM* **31**(11), 1268–1287.

Degani, A. S., and Wiener, E. L. (1994). Philosophy, policies, procedures and practice: The four 'P's of flight-deck operations. In *Aviation Psychology in Practice* (N. Johnston, N. McDonald, and R. Fuller, eds.), pp. 44–67. Aldershot, Haunts: Avebury Technical/Ashgate Publishing Ltd.

Degani, A., Kirlik, A., and Shafto, M. (1999). Modes in human-machine systems: A review and classification. *Int. J. Avia. Psychol.*

Deutsch, M. (1996). *Systems Engineering of Software-Intensive Systems.* Berkeley: University of California, Berkeley Extension.

Ericsson, K. A., and Smith, J. (1991). Prospects and limits of the empirical study of expertise: An introduction. In *Toward a General Theory of Expertise: Prospects and Limits* (K. A. Ericsson and J. Smith, eds.), pp. 1–38. New York: Cambridge University Press.

Forbus, K. D., Gentner, D., Everett, J. O., and Wu, M. (1997). Towards a computational model of evaluating and using analogical inferences. In *Proceedings of the Nineteenth Annual Conference of the Cognitive Science Society* (M. G. Shafto and P. Langley, eds.), pp. 229–234. Mahwah, NJ: Erlbaum.

Gentner, D., and Stevens, A. L. (eds.). (1983). *Mental Models.* Hillsdale, NJ: Erlbaum.

Gould, J. D. (1988). How to design usable systems. In *Handbook of Human-Computer Interaction* (M. Helander, ed.), pp. 757–789. Amsterdam: Elsevier/North-Holland.

Gould, J. D., Boies, S. J., Levy, S., Richards, J. T., and Schoonard, J. (1987). The 1984 Olympic messaging system: A test of behavioral principles of system design. *Commun. ACM* **30**(9), 758–769.

Gray, W. D., John, B. E., and Atwood, M. E. (1993). Project Ernestine: Validating a GOMS analysis for predicting and explaining real-world task performance. *Hum.-Comput. Interact.* **8,** 237–309.

Griffeth, N. D., and Lin, Y.-J. (1993). Extending telecommunications systems: The feature-interaction problem. *Computer* **26**(8), 14–18.

Grunwald, A. J., and Merhav, S. J. (1996). Vehicular control by visual field cues: Analytical model and experimental verification. *IEEE Trans. Syst., Man, Cybernet.* **6,** 835–845.

Helander, M., Landauer, T. K., and Prabhu, P. (eds.). (1997). *Handbook of Human-Computer Interaction,* 2nd rev. ed. Amsterdam: Elsevier.

Hinchey, M. G., and Jarvis, S. A. (1995). *Concurrent Systems: Formal Development in CSP.* London: McGraw-Hill.

Hoare, C. A. R. (1980/1987). The emperor's old clothes. In *ACM Turing Award Lectures: The First Twenty Years, 1966–1985* (R. L. Ashenhurst, ed.). Reading, MA: Addision-Wesley (ACM Press Anthology Series), pp. 143–161.

Humphrey, W. S. (1989). *Managing the Software Process.* Reading, MA: Addison-Wesley.

Hutchins, E. (1995). How a cockpit remembers its speeds. *Cogn. Sci.* **19**(3), 265–288.

Hutchins, E. L.., Hollan, J. D., and Norman, D. A. (1986). Direct manipulation interfaces. In *User Centered System Design* (D. A. Norman and S. W. Draper, eds.), pp. 31–62. Hillsdale, NJ: Erlbaum.

Kieras, D. E., and Bovair, S. (1984). The role of a mental model in learning to operate a device. *Cogn. Sci.* **8**, 255–273.

Kirlik, A. (1995). Requirements for psychological models to support design: Toward ecological task analysis. In *Global Perspectives on the Ecology of Human-Machine Systems* (J. Flach, P. Hancock, J. Caird, and K. Vicente, eds.), pp. 1–13. Hillsdale, NJ: Erlbaum.

Kitajima, M., and Polson, P. G. (In press). A comprehension-based model of correct performance and errors in skilled, display-based human-computer interaction. *Intern. J. Hum. Comput. Stud.*

Klein, G. A. (1989). Recognition-primed decisions. In *Advances in Man-Machine Systems Research* (W. Rouse, ed.), Vol. 5., pp. 47–92. Greenwich, CT: JAI Press.

Klein, G. A. (1993). A recognition-primed decision (RPD) model of rapid decision making. In *Decision Making in Action: Models and Methods* (G. A. Klein, J. Orasanu, R. Calderwood, and C. E. Zsambok, eds.). Norwood, NJ: Ablex.

Landauer, T. K. (1995). *The Trouble with Computers: Usefulness, Usability, and Productivity.* Cambridge, MA: MIT Press (Bradford Books).

Lee, J. D., and Vicente, K. (1998). Safety concerns at Ontario Hydro: The need for safety management through incident analysis and safety assessment. *Proceedings of the Second Workshop on Human Error, Safety, and System Development* (Leveson, N., and Johnson, C., eds.), pp. 17–27. Seattle: University of Washington.

Lenorovitz, J. M. (1992). Confusion over flight mode may have role in A320 crash. *Avia. Week Space Technol.* **137**, 29–30.

Leveson, N. et al. (1997). *Safety Analysis of Air Traffic Control Upgrades.* Seattle: University of Washington.

Leveson, N. G. (1995). *Safeware: System Safety and Computers.* Reading, MA: Addison-Wesley.

Lewis, C., and Norman, D. A. (1986). Designing for error. In *User Centered System Design* (D. A. Norman and S. W. Draper, eds.), pp. 411–432. Hillsdale, NJ: Erlbaum.

Lewis, C., and Wharton, C. (1997). Cognitive walkthroughs. In *Handbook of Human-Computer Interaction* (M. Helander, T. K. Landauer, and P. Prabhu, eds.), 2nd rev. ed., pp. 717–732. Amsterdam: Elsevier.

Means, B., Salas, E., Crandall, B., and Jacobs, T. O. (1993). In *Decision Making in Action: Models and Methods* (G. A. Klein, J. Orasanu, R. Calderwood, and C. E. Zsambok, eds.), pp. 306–326. Norwood, NJ: Ablex.

Mett, P., Crowe, D., and Strain-Clark, P. (1994). *Specification and Design of Concurrent Systems.* London: McGraw-Hill.

Miller, G. A. (1956). The magical number seven plus or minus two: Some limits on our capacity to process information. *Psychol. Rev.* **63**, 81–97.

Murray, C., and Cox, C. B. (1989). *Apollo: The Race to the Moon.* New York: Simon & Schuster.

National Aeronautics and Space Administration. (1989). *NASA-STD-3000: Man-System Integration Standards.* Houston, TX: NASA-JSC, and Wright-Patterson AFB, OH: CSERIAC.

National Transportation and Safety Board. (NTSB). (1988). *Northwest Airlines, Inc., McDonnell Douglas DC-9-82, NC312RC, Detroit Metropolitan County Airport, Romulus, Michigan, August 16, 1987* (Aircraft Accident Report NTSB/AAR-8805). Washington, DC: NTSB.

Norman, D. A. (1981). Categorization of action slips. *Psychol. Rev.* **88**, 1–15.

Norman, D. A. (1986). Cognitive engineering. In *User Centered System Design* (D. A. Norman and S. W. Draper, eds.), pp. 87–124. Hillsdale, NJ: Erlbaum.

Norman, D. A. (1988). *The Design of Everyday Things*. New York: Doubleday.

Obradovich, J. H., and Woods, D. D. (1996). Users as designers: How people cope with poor HCI design in computer-based medical devices. *Hum. Factors* **38,** 574–592.

Parnas, D. L. (1997). Software engineering: An unconsummated marriage. *Commun. ACM* **40**(9), 128.

Payne, J. W. Bettman, J. R., and Johnson, E. J. (1988). Adaptive strategy selection in decision making. *J. Exp. Psychol.: Learn., Memory, Cogn.* **14**(3), 534–552.

Polson, P. G., Lewis, C., Rieman, J., and Wharton, C. (1992). Cognitive walkthroughs: A method for theory-based evaluation of operator interfaces. *Int. J. Man-Mach. Stud.* **36,** 741–773.

Proctor, R. W., and Van Zandt, T. (1994). *Human Factors in Simple and Complex Systems*. Boston, MA: Allyn & Bacon.

Rasmussen, J. (1997). Merging paradigms: Decision making, management, and cognitive control. In *Decision Making Under Stress: Emerging Themes and Applications*. (R. Flin, E. Salas, M. Strub, and L. Martin, eds.), pp. 67–81. Brookfield, VT: Ashgate Publ. Co.

Reason, J. (1990). *Human Error*. New York: Cambridge University Press.

Remington, R. W., and Johnston, J. C. (1997). *Human Factors Checklist for Air Traffic Control Automatic Conflict Detection Probes* (preliminary version). Mountain View, CA: NASA Ames Research Center. Available on-line at http://olias.arc.nasa.gov/hfas/indices/index.html

Roth, E. M., Mumaw, R. J., and Lewis, P. M. (1994). *An Empirical Investigation of Operator Performance in Cognitively Demanding Simulated Emergencies* (NUREG/CR-6208). Washington, DC: U.S. Nuclear Regulatory Commission.

Rouse, W. B., and Morris, N. M. (1986). On looking into the black box: Prospects and limits in the search for mental models. *Psychol. Bull.* **100,** 359–363.

Sarter, N. R., and Woods, D. D. (1995). How in the world did we get into that mode? Mode errors and awareness in supervisory control. *Hum. Factors* **37,** 5–19.

Segal, L. D. (1995). Designing team workstations: The choreography of teamwork. In *Local Applications of the Ecological Approach to Human-Machine Systems* (P. Hancock, J. Flach, J. Caird, and K. Vicente, eds.), Vol. 2. Hillsdale, NJ: Erlbaum.

Sheen, Mr. Justice. (1987). *MV Herald of Free Enterprise,* Report of Court No. 8074 Formal Investigation. London: Department of Transport.

Shiffrin, R. M., and Schneider, W. (1977). Controlled and automatic human information processing. II: Perceptual learning, automatic attending, and a general theory. *Psychol. Rev.* **84,** 127–190.

Simon, H. A. (1955). A behavioral model of rational choice. *Q. J. Econ.* **69,** 99–118.

Sims, D. (1997). Vendors struggle with costs, benefits of shrinking cycle times. *Computer* **30**(9), 12–14.

Small, D. W. (1994). *Lessons Learned: Human Factors in the AAS Procurement* (MP 94W0000088). McLean, VA: MITRE Corporation.

Smith, S. L., and Mosier, J. N. (1986). *Guidelines for Designing Operator Interface Software,* Tech. Rep. No. MTR-10090. McLean, VA: MITRE Corporation.

Stix, G. (1994). Aging airways. *Sci. Am.,* May, pp. 96–104.

Van Lehn, K. (1990). *Mind Bugs: The Origins of Procedural Misconceptions*. Cambridge, MA: MIT Press.

Vicente, K. J. (1995). A few implications of an ecological approach to human factors. In *Global Perspectives on the Ecology of Human-Machine Systems* (J. Flach, P. Hancock, J. Caird, and K. Vicente, eds.), pp. 54–67. Hillsdale, NJ: Erlbaum.

Vicente, K. J. (1997). Should an interface always match the operator's mental model? *CSERIAC Gateway* **7,** 1–4.

Wagner, D., Birt, J. A., Snyder, M. D., and Duncanson, J. P. (1996). *Human Factors Design Guide for Acquisition of Commercial Off-the-Shelf Subsystems, Non-Developmental Items, and Developmental Systems.* Washington, DC: Federal Aviation Administration, Office of the Chief Scientific and Technical Advisor for Human Factors.

Wharton, C., Rieman, J. R., Lewis, C., and Polson, P. (1994). The cognitive walkthrough method: A practitioner's guide. In *Usability Inspection Methods* (J. Nielsen and R. Mack, eds.), New York: Wiley.

White, L. S. (1936). Farewell my lady. *The New Yorker,* May.

Wickens, C. D. (1992). *Engineering Psychology and Human Performance,* 2nd ed. New York: HarperCollins.

Wiener, E. L. (1989). *Human Factors of Advanced Technology ("Glass Cockpit") Transport Aircraft,* NASA Contractor Rep. No. 177528. Moffett Field, CA: NASA Ames Research Center.

Wiener, E. L. (1993). *Intervention Strategies for the Management of Human Error,* NASA Contractor Rep. No. 4547. Moffett Field, CA: NASA Ames Research Center.

Zave, P. (1993). Feature interactions and formal specifications in telecommunications. *Computer* **26**(8), 20–30.

18 Organizational and Individual Decision Making

KATHLEEN M. CARLEY and DEAN M. BEHRENS

18.1 INTRODUCTION

Organizations do not make decisions, people do. This observation is a statement of both structural and operational fact: organizations (as physical realities, not accounting or legal entities) are made by, and are composed of, people. There may be transportation, transformation, technological computation, and communication infrastructures to support human decision makers. These infrastructures generally have differential impact on the individuals in question. These infrastructures affect what information people have access to, and so what decisions they make. Nevertheless, organizations are all created, supported, maintained, and operated by these individuals. Thus, the issue of socially constrained, but nevertheless indvidual, decision making, lies at the heart of research on organizational decision making.

That humans make up organizations is neither a questionable nor a key issue from an organizational decision-making perspective. What is important is whether any (or all) of individual behavior can affect the constructs theorized or measured at the organizational level. Some researchers have argued that human behavior is largely irrelevant. For example, Schelling (1978) presents the game of musical chairs as an example of a class of organizational behavior patterns that are realized in the aggregate independent of how the individuals who form the aggregate behave (within the rules of the game). No matter how the individuals play the game, one will always be left without a chair. Thus, if one has a theory of the game, the players are simply agents carrying out the rules and roles of the game, and general outcome can be predicted without models of the individual agents (again, as long as they play by the rules). Thus the form of the game itself makes the specific model of the individual agent irrelevant. Second, it has been argued that because of scale, the specific model of the individual agent is irrelevant. This would suggest that for markets, national economies, and social structures, it is important to measure the aggregate or collective behavior of players, but not the individual microprocesses underlying these individual behaviors. In this sense, individuals may be (succinctly) represented in an aggregate manner (e.g., a production function or a cost curve) that reflects their collective behaviors. Third, it has been argued that there are general principles of organizing that are true for any collection of entities and not peculiar to humans. Thus, these principles should be applicable to any collection of intelligent, adaptive agents, such as individuals, Webbots or robots, engaged in distributed and collaborative work.

Establishing a model of the individual agent requires making a series of simplifying assumptions. For Schelling (1978) these assumptions include that the game does not

Handbook of Systems Engineering and Management, Edited by A. P. Sage and W. B. Rouse
ISBN 0471-15405-9 ©1999 John Wiley and Sons, Inc.

change and that the agents follow the rules of the game. In neoclassical economics and political economy, on an aggregate level (for macroeconomics, the industry; for microeconomics, the firm), there are underlying assumptions of the participating agents' perfect knowledge and perfect choice. Making simplifying assumptions is an important step in science. For some types of questions these are acceptable simplifying assumptions. Nevertheless, it should be realized that these assumptions, as a representation of decision-making reality within organizations, are largely incorrect. Organizations, like games, are "artificial" in the sense that they are crafted by humans (Simon, 1981a). But, unlike many games, organizations are very volatile or fluid constructs. Within organizations it is the norm that the rules change, the players change, and the situations change (Cohen et al.,

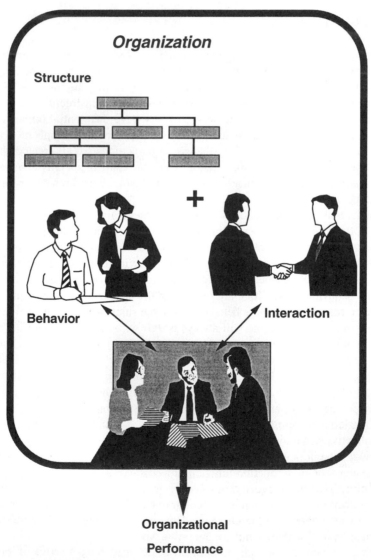

Figure 18.1 Factors afffecting organizational performance.

1972; March and Romelaer, 1976). This volatility is due in large part to the agents who compose them. Hence, within organizations, the form of the game depends on the agents and their personal history. From a managerial perspective, this strong interaction between cognition and task opens the avenue to strategies involving changing not just the task but the type of agents who engage in a task to achieve a particular level of performance.

Now consider the scale argument. The rational expectations model opposes much of what is known of human reasoning (Simon, 1979) and as a representation of decision making is largely incorrect (Simon, 1959, 1979). The principles of organizing argument is, from an organizational standpoint, the most intriguing. This argument cannot be easily wiped away by pointing to the interaction between agent cognition and task. Rather, the issue here forces the researcher to establish the general principles and then generate the conditions under which, and the ways in which, agent cognition matters.

For the most part, organizational theorists interested in individual and organizational decision making take this latter perspective and argue for the relevance of the agent model. In this case, organizational behavior is seen as an emergent property of the concurrent actions of the intelligent adaptive agents within the organization. This body of research has been informed by the work in distributed artificial intelligence, computational biology, information systems, social networks, behavioral decision theory, and human computer interaction and is influencing work in organizations, particularly that on organizational decision making and distributed work.

In summary, organizational performance is a function of both individual actions and the context in which individuals act (see Fig. 18.1). This context includes the web of affiliations in which the individual is embedded, the task being done, and the organization's structure and extant culture. Any model that does not include both individual cognition and context and the way in which these mutually coadapt will not be able to explain or predict behaviors associated with collaborative and distributed work.

18.2   THE INDIVIDUAL IN THE ORGANIZATION

Perhaps the individual who could best be described as the founder of the work on individual decision making within organizations would be Chester Barnard. In 1938, Barnard published the book, *The Functions of the Executive*. His analysis of individuals in organizations, particularly of individuals in cooperative systems, was the precursor for many future studies, as was the work by Roethlisberger and Dickson (1939). Barnard's work suggests that others' evaluations, particularly the manager's, directly affect concrete rewards such as pay. Feelings of fairness and equity in how one is treated in an organization stem from discrepancies between self and others' evaluations. Such discrepancies, therefore, should affect job satisfaction, organizational commitment, performance, and turnover. However, extensive studies of the relationships among job satisfaction, organizational commitment, individual and organizational performance, and personnel turnover have led to a set of complex and inconsistent results (Mowday et al., 1982; Mobley, 1982). Moving beyond this subarea, however, major advances in individual and organizational behavior have followed from alternative perspectives. Among these alternative perspectives are a predominantly psychologically and economically based behavioral perspective, an information-processing perspective, a cognitive perspective, and a structural or social network perspective.

18.2.1 The Individual as Behavioral Agent

Outside the field of organizations per se, there is an enormous body of research on individual decision making. Somh of this work lies in the field known as Behavioral Decision Theory (BDT). Depending on the perspective chosen by organizational researchers, BDT concepts have been applied to many levels: the individual in organizations, the individual in groups, or groups in organizations. Interestingly, not unlike the delineation between researchers of culture and those who research climate (Dennison, 1996), BDT seems to have two antecedent streams of research that can be grossly categorized as the psychological/descriptive approach and the economic/normative approach. Whereas both streams of research are considered predictive, the economic approach focused on the rational decision maker, and the approach that is somewhat more psychologically based attempts to describe and explain consistent deviations from rationality. It is this attempt by both psychologists and behavioral economists to explain fluctuations from rationality that can best be described as the field of Behavioral Decision Theory.

Although many individuals consider Bernoulli (1738) to be the forefather of modern day BDT, the major innovation to the concept of a rational decision process must be attributed to von Neumann and Morgenstern, with their publication of the book *Theory of Games and Economic Behavior* in 1947. This book, laid the framework for what was later to be referred to as Game Theory (see also Luce and Raiffa, 1957). Von Neumann and Morgenstern (1947) made explicit the assumptions and constraints that would provide for a rational (i.e., consistent and predictable) decision. This economic approach resulted in what is referred to as Expected Utility (EU) Theory. After von Neumann and Morgenstern (1947), researchers suggested variations on the strict interpretation of EU, still from the perspective of economics. Savage (1954) suggested that the actual process of decision making was modeled through a subjective expected utility. Moreover, researchers were (and still are) trying to develop methods to measure the difficult concept of utility (Marschak, 1950; Becker et al., 1964; Edwards, 1992). As Dawes (1988) wrote, "People, groups, organizations, and governments make choices. Sometimes the consequences of their decisions are desirable, sometimes not" (p. 2). Or, in a related vein, as others have argued, the choices made by individuals and groups are not rational (where rational is defined as making that decision predicted by EU theory). It was not until the 1970s and early 1980s that further major revisions to EU theory were published. Kahneman and Tversky (1979) broke ground with their Prospect Theory, which suggested that individuals have a different perception when considering losses vs. gains. Machina (1982) attempted to describe EU when one of the assumptions, called the independence axiom, is relaxed. Both Bell (1982) and Loomes and Sugden (1982) suggested that decisions were made on the basis of regret (i.e., what could have been) instead of the expected benefit (i.e., utility) of an outcome.

Essentially, this work has led to a wide range of findings concerning departures from rationality and biases common to social judgment processes (Ross et al., 1977; Kahneman et al., 1982). This research includes that on the framing effect (Tverksy and Kahneman, 1981), false consensus effect (Dawes and Mulford, 1996; Dawes, 1989, 1990; Orbell and Dawes, 1993), group think (Janis, 1982), and altruism (Orbell et al., 1988; Orbell and Dawes, 1993). The false consensus bias is premised on an individual's belief that everyone responds in the same manner as they do. In fact, we overestimate the degree to which our past behavior, as well as our expected behavior, is truly diagnostic of other individuals' future behavior. BDT and social psychology have examined this bias and have assessed that it is prevalent among individuals (Dawes and Mulford, 1996). Groupthink, on the other hand, is the tendency in groups for a convergence of ideas and a sanctioning of aberrant ideas to occur. Related to groupthink are the concepts of group polarization, and risky shifts (Pruitt, 1971a,b). However, this overdetermination of either the group's or an individual's future behavior is not seen when we examine how individuals compare themselves to others. In general, over 50 percent of the population, when asked to rate themselves on some mundane task, such as driving ability, see themselves as better than average. Of course, this is statistically impossible.

Biases also exist in the way individuals make judgements about individuals, future events, or causes. These biases are due in part to the personal characteristics of the individuals making the judgments (Fischhoff et al., 1981; MacCrimmon and Wehrung, 1986), as well as certain cognitive heuristics (i.e., mental short cuts or limitations) to which all of us are prone (Kahneman et al., 1982; Plous, 1993).

Kenneth MacCrimmon and Donald Wehrung (1986) provide a framework, as well as an assessment tool, which describe the risk propensity of a given individual. In addition, MacCrimmon and Wehrung describe the risk-taking behavior of 509 top-level executives and allow the readers to compare themselves with these managers. Borrowing from Fischoff et al. (1981), making choices under uncertainty predicates the prior judgment of (1) the uncertainty about the problem definition; (2) the difficulty in assessing the facts; (3) the difficulties in assessing the values; (4) the uncertainties about the human element; and (5) the difficulties in assessing the decision quality. It is this judgment process, affected by the risk propensity of various managers which MacCrimmon and Wehrung discuss. Of course, outside of the personality or characteristics of each manager (i.e., their risk propensities) there are also cognitive and perceptual biases that would need to be understood in order to understand theories of human action in organizations or in society. Amos Tversky and Daniel Kahneman (1974) discuss a number of the different types of biases inherent in the decision-making process that affect, if not all, the vast majority of us, at least unconsciously. Some of the heuristics that lead to biases discussed and elaborated in the book edited by Kahneman et al. (1982) are representativeness, availability, and adjustment and anchoring.

The representativeness heuristic suggests that individuals base judgements on similarity of characteristics and attributes. As Tversky and Kahneman (1974) suggest, people often make judgments based on "the degree to which A is representative of B, that is, by the degree to which A resembles B" (p. 1124). The representative heuristics can lead to the belief in "the law of small numbers," that is, that random samples of a population will resemble each other and the population more closely than statistical sampling theory would predict (Plous, 1993). Moreover, utilizing the representative heuristic can also result in people ignoring base-rate information (a base rate is the relative frequency an occurence is seen in the general population). The representative heuristic might be seen as a not-too-distant cousin to the availability heuristic.

The availability heuristic is the mental shortcut that allows individuals to "assess frequency of a class or the probability of an event by the ease with which instances or occurrences can be brought to mind" (Tversky and Kahneman, 1974, p. 1127). This heuristic does not necessarily result in a biased judgment. However, it can when the majority of available information is inaccurate because of recency or primacy effects. For example, the likelihood that your car is going to be stolen might very well be affected by the saliency of the information that your next door neighbors had their car broken into twice in the last two years. However, we rarely ask our other neighbours how often their cars have been broken into. Thus, the neighbor's information may become more salient only in our making the decision to purchase an antitheft device.

The heuristic of adjustment and anchoring causes extreme variations among judgments of individuals. This heuristic suggests that we take a piece of information, even a randomly chosen (i.e., noninformative) one, and then attempt to adjust our judgments around that piece of information. In other words, if I were to ask you the estimated income from a new sales project and told you that project alpha last year earned $40,000 your estimate for the expected income would be higher than if I told you that it only earned $4,000. Judgment makers tend to unconsciously anchor on a number and insufficiently adjust (either up or down) around that anchor. In fact, if I just had a wheel with dollar amounts ranging from $400 to $4,000,000 and spun a pointer so that it randomly landed on one value, your estimate would still be anchored and your judgment would be biased accordingly.

Thus, individual judgments of future events, outcomes, or processes are strongly affected by the information we perceive, we can remember, and the degree to which we are willing to expend energy on the judgment process. These judgment heuristics and the respective biases can be seen as limitations on the degree to which we can process information in a thorough and consistent manner (i.e., act rationally). They often lead to human and organizational error, a subject discussed in Chapters 17 and 19.

18.2.2 The Individual as Information Processor

The Carnegie School of Organizational Theory proposed an information-processing perspective in which individual and organizational decisions could be explained in terms of what information was available to whom, cognitive limits to information processing abilities, organizational (social and cultural) limits to access to information, the quality of the information, and so forth. Simon (1945), March and Simon (1958), and Cyert and March (1963) examined the decision-making components of organizational and firm action. Whether the decision was to restructure the organization or to outsource a given product, the firm was believed to follow a number of decision-making procedures prior to determining a solution. These procedures can be usefully represented using either formal logic or expert systems (Leblebici and Salancik, 1989; Salancik and Leblebici, 1988; Masuch and LaPotin, 1989). These procedures are both social and cognitive, are embedded in organizational routines and in individuals' mental models, and do not guarantee that the individual or the organization will locate the optimal solution for any particular task. Rather, individuals and organizations satisfice (Simon, 1959), that is, they make do with a decision that is satisfactory rather than one that is definitely optimal. Studies suggest that individuals in making decisions examine only a few alternatives, and even then do not consider all of the ramifications of those alternatives. As a result, decisions are more opportunistic than optimal.

This stream of research, which came to be known as part of the information-processing perspective, was later to include a rather well-known metaphor—the garbage can. Cohen et al. (1972) proposed a model of organizational choice that they entitled "A Garbage Can Model." Padgett (1980), Carley (1986a,b), and others went on to expand on this theory. According to this theory, organizational decision is a function of the flow of individuals, problems, and solutions. Individuals do not evaluate all possible solutions to a specific problem. Rather, in making a decision, individuals are prone to simply attach their favorite solution. Futher, whether or not a decision is actually made is a function of the effort that individuals expended on the problem and the number of individuals currently available to work on the problem. Researchers following Cohen et al. (1972) argued that the early model was insufficient to capture actual organizational behavior, as it ignored the role of organizational design and the limits on individual behavior dictated by organizational procedures such as those for data handling and personnel hiring. Recently, Carley and Prietula (1994b) demonstrated that to get interesting and detailed organizational predictions, one had to move beyond these models by incorporating a model of agents, organizational structure and situation, and task. In particular, task places an extremely strong constraint on individual and organizational behavior.

Information-processing theorists (March and Simon, 1958; Cyert and March, 1963; Galbraith, 1973, 1977) and social-information-processing theorists (Salancik and Pfeffer, 1978; Rice and Aydin, 1991) have argued that individual, and hence organizational, decisions depend on what information they have that in turn is constrained by the individual's position in the social structure. Structure influences individual decision making because it constrains access to information and because the decisions, attitudes, and actions of those to whom one is structurally connected have a strong influence on behavior. Further, the structure of the organization and the task limits access to information, determines the order of processing, and enables certain efficiencies. Moreover, the organizational structure can be viewed as a coordination scheme whose cost and performance depends on the network of connections and procedures within the organization (Malone, 1987; Krackhardt, 1994; Lin, 1994). Organizational slack as well as performance is thus a function of these information-processing constraints. This work is consistent with the arguments forwarded by, and is often carried out by, social network theorists.

18.2.3 Individuals as Intelligent Adaptive Agents

Organizations can be usefully characterized as complex systems composed of intelligent adaptive agents, each of which may act by following a set of relatively simple procedures or routines. However, if the agents coadapt, then the organization as a whole may exhibit complex patterns of behavior. In such systems, linear models cannot capture the complexities of behavior. Consequently, the level of prediction possible from the linear model is low.

Recently, computational organizational theorists and researchers in distributed artificial intelligence (DAI) have begun to study organizational adaptation, evolution, and learning using complex intelligent adaptive agent models. An intelligent adaptive agent is an agent (or set of agents) that makes decisions on the basis of information, but that information changes over time in response to the environment. Thus the agent (or set of agents) learns responses and may improve performance. An example of an intelligent adaptive agent would be an automated Web browser that searches for information on a particular topic, but it learns as it does so the preferences of the user for whom it is browsing. Models in this arena include those using simulated annealing, genetic programming, genetic algorithms, and neural networks. Some of these analyses focus on the evolution of industries and the sets of organizations within a market, rather than adaptation within a single organization (Axelrod, 1987; Axelrod and Dion, 1988; Crowston, 1994, 1998; Holland, 1975; Holland and Miller, 1991; Padget, 1998). Others explore issues of organizational performance and experiential learning (Carley, 1992; Lin and Carley, 1998; Verhagen and Masuch, 1994; Mihavics and Ouksel, 1996) or expectation-based learning (Carley and Svoboda, 1998). Another stream of research has occurred within DAI, one in which researchers have focused on the effect of coordination and communication among intelligent agents on performance (Durfee and Montgomery, 1991; Tambe et al., 1997).

These three streams of research collectively demonstrate the power of computational models and the intelligent adaptive agent approach for theorizing about organizational dynamics. These models employ the use of "artificial" agents, acting as humans. The agents in these complex intelligent adaptive multiagent models are nondeterministic and undergo a coevolutionary process. During their lifetimes, they may move through and interact with the environment, reproduce, consume resources, age, learn, and die. Although the agents are typically adaptive, they may vary in their intelligence and complexity. Using these models the researchers generate a series of predictions about the behavior of the system. Because the agents are artificial, the predictions may be equally applicable to organizations of humans and to organizations of "nonhumans" (Webbots, robots, etc.) Depenting on the assumptions built into the agent models, the results may be interpreted as predictions about organizing in general or about organizing in a particular context. Research is needed in this area to determine when artificial and human organizations are similar (Carley, 1996).

Most researchers in this area contend that organizational dynamics are due to, and may even emerge from, the adaptiveness of the agents within the organization. This process has been referred to by a variety of names, including colearning (Shoham and Tennenholtz, 1994), synchronization, and concurrent interaction (Carley, 1991b). For Carley (1991a) concurrent interaction and the coevolution of self and society is necessary for the emergence of social stability and consensus. For Shoham and Tennenholtz (1994) colearning is "a process in which several agents simultaneously try to adapt to one another's behavior so as to produce desirable global system properties." Collectively, the findings from these models indicate that emergent social phenomena (such as the emergence of hierarchy) and the evolutionary dynamics (patterns of change) depend on the rate at which the agents age, learn, and the constraints on their adaptation of interaction.

18.2.4 The Individuals' Mental Models

An alternative perspective on individual and organizational decision making has arisen out of the cognitive sciences. Here the focus is not on what decisions are made, or on rationality per se, but on how the individual and the team thinks about problems (Reger and Huff, 1993; Johnson-Laird, 1983; Klimoski and Mohammed, 1994; Eden et al., 1979; Carley, 1986c; Fauconnier, 1985; Weick and Roberts, 1993). As such, researchers draw from and make use of work on the coding and analysis of individual and team mental models. This research derives from work in cognitive psychology, philosophy, and artificial intelligence. Recent advances in textual analysis point to a future in which intelligent systems will exist for parsing and coding texts (Bechtel, 1998; Golumbic, 1990; Gazdar and Mellish, 1986; Winghart, 1988; Zock and Sabah, 1988). Since these techniques enable more automated coding of texts, they should make possible the analysis of larger quantities of texts, and thus make possible the empirical analysis of complex processes such as team mental model formation, negotiation, and the use of organizational rhetoric in establishing organizational effectiveness.

The expectation is that these systems will enable the researcher interested in individual and organizational decision making to move beyond content analysis to more relational modes of text analysis (Schank and Abelson, 1977; Sowa, 1984; Roberts, 1989, 1998) The expectastion is that by examining not just words, but the relations among concepts within the text, the researcher will be better able to analyze differences in meaning across individuals, groups, and organizations (Roberts, 1998; Carley and Palmquist, 1992; Carley, 1994; Carley and Kaufer, 1993; Kaufer and Carley, 1993). By focusing not just on words but on the relationships among the words present in the texts, researchers can examine texts for patterns of cognitive behavior, decision-making protocols, and can trace the logic of arguments. Texts as networks of concepts can then be analyzed using standard social network techniques (Carley, 1998a). Additionally, narratives or stories can be analyzed as event sequences (Heise, 1979, 1991; Ericsson and Simon, 1984), thus enabling organizational researchers to address the relationship between individual action and organizational behavior. An advantage of these techniques is that they allow the researcher to take rich verbal data and analyze them empirically. Empirical analysis makes it possible to statistically test hypotheses about the formation, maintenance, and change in team mental models over time and across teams.

Individuals' mental models can be characterized as the information known by the individual and the pattern of relationships among these pieces of information (Carley and Palmist, 1992). A mental model is not just all the information that in an individual's head. Rather, a mental model is a structure of information that gets created, and can be called up or used, in specific contexts. Individuals have many mental models about themselves, others, objects, the world, tasks, and so forth. These mental models include social, historical, cultural, environmental, personal, and task knowledge, and are specialized based on varying contexts and needs. From an organizational perspective it is useful to note that an individual's mental model includes the individual's perception of the sociocognitive structure—the sets of relations that an individual perceives as existing between other pairs of individuals (Krackhardt, 1987, 1990) and their understanding of others' knowledge and beliefs (Carley, 1986c). Problem solving involves searching through the set of salient mental models. As such, mental models influence not only what decisions individuals make, but their perceptions of others' decisions. According to this perspective, cognition mediates between structure and outcome, that is, it is the individual's perception of social structure (as encoded in the individual's mental model) that influences behavior, attitudes, evaluations, and decisions, and not the structure itself (Carley and Newell, 1994).

Individuals' mental models are thought to develop as they interact with other individuals. As a result, concurrent actions, interactions, and adaptations emerge more or less simultaneously at the individual, organizational, and social levels (Carley, 1990a, 1991a; Kaufer and Carley, 1993). In particular, individuals construct definitions of self that depend on their sociocultural–historical background and their interactions with others (Greenwald and Pratkanis, 1984; Higgins and Bargh, 1987; Markus, 1983). Individuals' mental models not only contain different information (as a result of their private history of interaction with others), but individuals may use the same information in different ways in making decisions. Attributions are a type of decision that has been widely studied from the mental mode perspective. Despite this research, how individuals make attributions about self and others remains unclear. For example, Heider suggested that a "person tends to attribute his own reactions to the object world, and those of another, when they differ from his own, to personal characteristics" (Heider, 1958, p. 157). This idea was extended by Jones and Nisbett (1972), who claimed that all attributions reflect the following bias: individuals tend to think that they themselves are responding to the situation or environmental demands, but they generally see others as behaving in particular ways because of their personality traits. In contrast, Bem (1965) argued that there is no difference in the factors individuals use to make attributions about their own and others' behavior. In their review of the literature Monson and Snyder (1977) concluded that there are systematic differences between attributions of self and others. However, the differences are not consistently in the direction predicted by either Heider or Jones and Nisbett. In fact, the differences appear to be a function of both cognitive and structural factors.

Often in individuals' mental models, people do not seem to discriminate between causality and correlation. People appear to construct correlations to confirm their prior expectations about what causes what (Chapman and Chapman, 1967, 1969). People appear to look for salient cues in suggesting causal links, rather than simply computing them from the statistical occurrences, as Heider had suggested. As such, individuals seek out obvious indicators of what they think should be causing some outcome and use such cues to make predictions about another's behavior or attitude. A possible explanation for this is that ambiguity in the organization may make evaluation of others difficult (Festinger, 1954). This forces individuals to use cues, as they are not privy to direct or statistical knowledge (Salancik and Pfeffer, 1978). Further, for some types of decisions, it may not be possible to determine the requisite information. For example, consider decisions that require evaluating an individual's contribution to a collaborative project. For this type of decision, it may not even be possible to evaluate the separate contribution of each organizational member (Larkey and Caulkins, 1991); thus, the decision maker must rely on organizational cues.

The relation of individual mental models to team mental models, and the value of team mental models to team and organizational performance are currently the subject of much debate. The construction of and change in team models is seen as integral to collaborative work. Differences in individual mental models are seen as potentially valuable in collaborative projects, as they enable the organization to learn from different individual's experiences (Knorr-Cetina, 1981; Latour and Woolgar, 1979). Team mental models are seen as critical for team learning and team performance, as they provide a shared understanding that enables action in the face of ambiguity and without making all information explicit (Hutchins, 1990, 1991a,b). Polanyi (1958, esp., pp. 216–219, 264–266) implicitly defined social knowledge and so team mental models as the articulation of "common experience." Thus, through articulation, a "tacit consensus" and understanding are developed. For Polanyi, social knowledge requires a transitivity of appraisal across a continuous network of individuals. What this means is that each piece of social knowledge in the team mental model is commonly, but not necessarily uniformly, shared. As such, the team mental model represents the tacit consensus to a set of information. It is not necessary for all members of a team to know that a piece of information is part of a team's mental model for it to be included. In contrast, Klimoski and Mohammed (1994, p. 422) suggest that a team mental model is an emergent group phenomenon. Since these team mental models facilitate group coordination and the allocation of resources "some level of awareness is necessary." In this case, there is a need for actual and not simply tacit agreement in order for a piece of information to be part of the team mental model. Among the issues being currently investigated in the area of distributed work is the extent to which individuals should share their mental models if they are to operate effectively as a team, and whether certain types of information more commonly appear in team mental models than others.

18.2.5 The Individual and the Social Network

From both an information-processing and a structural perspective has come a view that the organization, particularly its design or architecture, can be characterized as networks of people, tasks, and resources. In particular, attention has been paid to social network models of organizations and sets of organizations that are described in terms of the relationships or ties among individuals or organizations (for reviews, see Krackhardt and Brass, 1994). Researchers distinguish between the formal organizational structure (the organizational chart dictating who must report to whom) and the informal organizational structure (the emergent set of advisorial and friendship relations among the individuals in the organization). Social network models have successfully been used to examine issues such as organizational adaptation (Carley and Svoboda, 1998), power (Burt, 1976, 1992; Krackhardt, 1990), diffusion (Burt, 1973; Carley and Wendt, 1991; Carley, 1995a; Granovetter, 1973), changing jobs (Granovetter, 1974), structuration (DiMaggio, 1986), innovation (Burt, 1980), and turnover (Krackhardt, 1991; Krackhardt and Porter, 1985, 1986). These studies demonstrate that the structure of relations, both within and between organizations, in and of itself can affect individual and organizational behavior. Moreover, the informal structure often has as much or more influence on behavior than does the formal structure. This is probably particularly true for distributed teams. In teams and collaborative work groups, individuals are linked informally by many types of ties, including monetary, advisorial, and friendship (Boorman and White, 1976; White et al., 1976; Burt, 1976, 1977). The greater the overlap of different types of ties, the more effective the relationship and the more constraining on individual and group decision making.

Organizational learning is also intimately tied to the sharing or diffusion of information. As noted by Granovetter (1973, 1974), connections or ties among individuals determine what information is diffused and to whom. However, the strength of the ties among individuals may actually inhibit information diffusion. One reason for this is that in groups where the level of shared information is high, communication may tend to become ritualized and centered on repeating known information (Kaufer and Carley, 1993). In this case, the likelihood of new information diffusing along existing ties can actually decrease as individuals within the organization work together and become more similar in what they know. We can think of this as collaborative teams becoming stale over time as no new members are added. Both the level and pattern of ties among individuals in the group influences the speed with which information spreads and whom it spreads to (Becker, 1970; Burt, 1973, 1980; Coleman et al., 1966; Granovetter, 1973, 1974; Lin and Burt, 1975) and when it jumps organizational boundaries (Carley, 1990b). Individuals who are more tightly tied are less likely to be innovators (Burt, 1980), but may be more important in mobilizing others for change, which may be important for the development of coalitions such as unions or strikes (Krackhardt, 1990; Carley, 1990a). Further, Burt (1992) suggests that individuals can learn to control their corporate environment, their own career within the organization, and the organization's ability to respond to events by controlling the pattern of ties among individuals within the organization. Information technologies, however, may influence the pattern of these ties (Freeman, 1984) and their relative effectiveness for communicating different information (Carley and Wendt, 1991). Advances in the area of diffusion that are particularly relevant to organizations have been made by researchers using social network techniques. This work demonstrates that how integrated the individual is into the organization influences the likelihood that this person will diffuse new information and adopt innovations (Burt, 1973, 1980; Kaufer and Carley, 1993).

Finally, this research has demonstrated that there is no single adequate measure of structure (Krackhardt, 1994; Lin, 1994). This is true even if the focus is exclusively on the formal structure. The situation is compounded when one considers that the organization is really a composite of multiple structures such as the command structure, communication structure, task structure, and so forth. Thus, to understand the impact of structure on

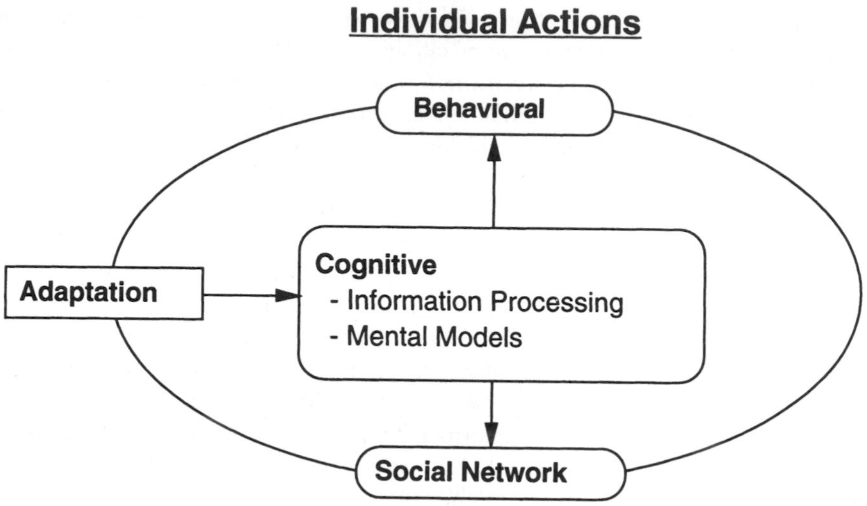

Individual Actions

Figure 18.2 Factors affecting individual actions.

collaborative work and distributed teams, it is necessary to simultaneously consider many measures.

18.2.6 The Situate Individual

Collectively these perspectives on the individual are leading to a broader understanding of the individual as a situated agent in the organization. Individual cognition mediates the individual's actions in, responses to, and interpretations of the context in which the individual is working (see Fig. 18.2). This cognition comprises both an architecture for processing information and a suite of mental models. The individual, however, is an intelligent adaptive agent. Through adaptation the individual changes the content, number, and type of mental models that are drawn on. Unlike some work on individual adaptation, however, within organizations the individuals' adaptation is constrained by their previous behaviors and their positions in the social network and organizational structure. Thus it is important to link a more macroperspective on the organization as a whole with the more microperspective on the individual.

18.3 THE ORGANIZATION

The vast majority of organizational decisions require attention from multiple individuals. That is, they are not the result of a single individual acting in isolation. Much of the research in organizational theory has been focused on examining how the organization's form, design, the task the individual is engaged in, or the environment in which it operates influences the decisions made by the organization. Decision makers may have access to different information, may have access to different technology for evaluating and gathering information, may have different criteria or goals for evaluating that information, may have different training or skills, and so forth. Thus, factors such as information flow, lack of resources, attention, timing, commitment, the degree to which consensus needs to be reached, and organizational design have as much influence on the organizational decision as the cognitive process by which individuals make decisions. Clearly both the social network and the information-processing approaches previously discussed point in this direction. Building on these traditions and other research on organizational design, researchers have begun to use computational models to address issues of organizational decision-making performance, learning, and adaptation.

18.3.1 Computational Organization Theory

Organizational decision-making theory has been strongly influenced by computational approaches (Ennals, 1991; Carley, 1995b). Early work was influenced by research in the areas of cybernetics and general systems (Ashby, 1956), system dynamics (Forester, 1961), economics and cognitive psychology (Cyert and March, 1963), information technology (Bonini, 1963), and social behavior and process (Dutton and Starbuck, 1971). Cyert and March's *A Behavioral Theory of the Firm* (1963) is a landmark text for organizational theorists interested in formal models. Cyert and March demonstrated the impact of bounded rationality on organizational decision making and the value of process models for understanding organizational decision making. With this work, a tradition began in which the organization is modeled as a collection of agents (who are at least boundedly rational), organizational behavior emerges from the concurrent interactions among these agents, and decisions are constrained by both agent capabilities and the social structure in which the agents are placed. In the past three decades there has been a tremendous growth in the use of mathematical and computational models for examining organizational decision making, particularly in complex or distributed settings. This area has come to be known as *computational organization theory*.

Computational organization theory focuses on understanding the general factors and nonlinear dynamics that affect individual and organizational behavior (Masuch and Warglien, 1992; Carley and Prietula, 1994a; Carley, 1995b) with special attention on decision making, learning, and adaptation. In these models, information, personnel, decision responsibility, tasks, resources, and opportunity are distributed geographically, temporally, or structurally within, and sometimes between, organizations. These models extend work in team theory by focusing on the nonlinear dynamics (Marschak, 1955; McGuire and Radner, 1986; Radner, 1993). Organizational decisions are seen to result from processes as diverse as problem resolution (rationally solving the problem), ignoring the problem, accident (as in a fortuitous result of solving a related problem), coordination of multiple decision-making units, and political negotiation among multiple decision makers. These models have been used to explore the way in which information technologies and tasks, individual, informational, cultural, environmental, demographic, and organizational characteristics impact the frequency, timeliness, accuracy, cost, complexity, effectiveness, and efficiency of organizational decisions, organizational learning, and organizational adaptation. Most of the current models come from either a neo-information-processing/social network perspective, or a DAI perspective (Bond and Gasser, 1988; Gasser and Huhns, 1989).

Models in this area range from simple intellective models of organizational decision-making behavior (Cohen et al., 1972; Carley, 1992) to detailed models of the decision processes and information flow that can emulate specific organizations (Levitt et al., 1994; Zweben and Fox, 1994) or follow specific management practices (Gasser and Majchrzak,

1992, 1994; Majchrzak and Gasser, 1991, 1992). These models vary in whether they characterize generic decision-making behavior (Cohen et al., 1972), make actual decisions in organizations (Zweben and Fox, 1994), make actual decisions given a stylized task (Durfee, 1988; Durfee and Montgomery, 1991; Carley and Prietula, 1994b; Lin and Carley, 1998; Carley and Lin, 1998), enable the researcher to examine the potential impact of general reengineering strategies (Gasser and Majchrzak, 1994; Carley and Svoboda, 1998), or enable the manager to examine the organizational implications of specific reengineering decisions (Levitt et al., 1994). These models typically characterize organizational decisions as the result of individual decisions, but they vary in the way in which they characterize the individual agent. Typical agent models include the agent as bundles of demographic and psychological parameters (Masuch and LaPotin, 1989), as simple information processors constrained by in–out boxes and message-passing rules (Levitt et al., 1994), or using some form of adaptive agent model (see following discussion). Further, most of these models characterize the organization as an aggregate of discrete and concurrently interacting complex adaptive agents (Prietula and Carley, 1994) or as a set of search procedures (Cohen, 1986) or productions (Fararo and Skvoretz, 1984).

Collectively, this work demonstrates that individual, task, environment, and design factors interact in complex and nonobvious ways to determine overall organizational performance. Task and environment are often the major determinants of organizational behavior. However, they interact with individual learning to the point that, depending on the complexity of the task and the quality of the feedback, the same people and the same organizations will in one circumstance overlearn, mislearn, and engage in otherwise maladaptive behavior and in another learn appropriate behavior. Moreover, as the level of detail with which task and organizational structure is modeled increases, the specificity and managerial value of the model's predictions increase. Finally, this work suggests that realistic organizational behavior, including errors, often emerge from processes of individual learning only when what the individual can learn is constrained by the organizational design, time, or the amount of information available.

18.3.2 Adaptive Organizations

Computational models of organizational decision making are particularly useful for examining issues of organizational learning and adaptation. Much of this work, particularly on the formal side, borrows from and is informed by work on adaptive architectures, more generally, and the work in computational biology and physics. Simon (1981a,b) has repeatedly argued that any physical symbol system has the necessary and sufficient means for intelligent action. The work by computational organizational theorists moves beyond this argument by arguing that a set of physical symbol systems that can communicate and interact with each other have the necessary and sufficient means for intelligent group means. Moreover, if the physical symbol systems can adapt in response to their own and other's actions, then the collection of such systems will exhibit emergent collective behavior.

While most of the organizational models share the perspective that organizational behavior emerges from the actions of intelligent adaptive agents, they differ in the way in which individual agents are characterized. A variety of agent models have been used, including traditional learning models (Carley, 1992; Lant and Mezias, 1990; Glance and Huberman, 1993), genetic algorithms (Holland, 1975, 1992; Holland and Miller, 1991;

Crowston, 1994), cognitive agent models like Soar[1] (Carley et al., 1992; Ye and Carley, 1995; Verhagen and Masuch, 1994; Carley and Prietula, 1994b), nodes in a neural network (Kontopoulos, 1993), and agents as strategic satisficers using simulated annealers (Carley and Svoboda, 1998). Regardless, individual learning is generally seen as one of the central keys to organizational learning (Lant, 1994), survival (Crowston, 1994, 1998), problem solving (Gasser and Toru, 1991), cultural transmission (Harrison and Carrol, 1991), emergent organizational behavior (Prietula and Carley, 1994), cooperation (Glance and Huberman, 1993, 1994; Macy, 1991a,b; Axelrod and Dion, 1988), and effective response to environmental uncertainty (Duncan, 1973). Most of the work in this area aggregate individual actions to generate organizational behavior. However, there is a smaller second tradition, a search procedure, in which the organization as a whole engages, and organizational behavior, and in particular learning, results (Lant and Mezias, 1992; Levinthal and March, 1981).

Organizational learning focuses on performance improvement and adaptation to the organization's external and internal environment. Thus, organizational researchers have found that modeling organizations as collections of intelligent adaptive agents acting more or less concurrently is critical for understanding issues of organizational learning and design. Currently, the four dominant methods used by organizational and social theorists to examine organizational adaptation are rule-based processors with detailed individualized models, neural networks (Rumelhart and McClelland, 1986; McClelland and Rumelhart, 1986; Wasserman, 1989, 1993), genetic algorithms and classifier systems (Holland, 1975, 1992; Holland et al., 1986), and simulated annealing (Kirkpatrick et al., 1983; Rutenbar, 1989).

The detailed rule-based models capture, using knowledge engineering and protocol analysis techniques, the detailed rules of behavior used by experts in performing some task. These rules or procedures are then placed in an artificial agent, who is given that and similar tasks to perform. In part, the goal here is emulation of an expert. Elofson and Konsynski (1993) apply AI and machine learning techniques to the analysis of organizational learning for the purpose of monitoring and analyzing decisions relative to organizational structure, and for monitoring organizational changes as part of the organizational learning and adaptation cycle. Their analysis demonstrates that increased flexibility is possible by knowledge caching, which provides a means of realizing an explicit organizational memory where information and processing capabilities arc distributed among the organizational members. Such distributed agents cannot act in a completely concurrent fashion, as one agent may not be able to begin a particular task until another agent has finished a different task. The key issue then is how to schedule and coordinate these intelligent agents. Coordination of these intelligent agents can be characterized as a search process through a hierarchical behavior space, in which case coordination emerges through a set of cultural- or task-based norms of behavior and response to other agents (Durfee, 1988; Durfee and Montgomery, 1991).

Neural networks are a computational analog of the biological nervous systems and represent the learning entity as a set of predefined nodes and relations in which the relations can change over time in response to inputs. Kontopoulos (1993) suggests that neural

---

[1]Soar can be characterized as a model of cognition in which all problem solving is search, and learning occurs through chunking.

networks are an appropriate metaphor for understanding social structure. In a neural network, information is stored in the relations between nodes that are typically arranged in sequential layers (often three layers) such that the relations are between nodes in contiguous layers but not within a layer. These systems learn slowly on the basis of feedback and tend to be good at classification tasks. For example, Carley (1991b, 1992) used a model, similar to a neural network, to examine how organizational structure constrains the ability of organizations to take advantage of the experiential lessons learned by the agents in the organization and demonstrated the resiliency of the hierarchical structure and not the team structure in the face of turnover. Carley demonstrated that when organizational learning was embedded in the relationships between agents and not just in the agents, the organization was more robust in the face of "crises" such as turnover and erroneous information.

Genetic algorithms are a computational analog of the evolutionary process. A genetic algorithm simulates evolution by allowing a population of entities to adapt over time through mutation and/or reproduction (crossover) in an environment in which only the most fit members of the population survive. These models require that the there is a fitness function against which each organization, or strategy, can be evaluated. The smoother the surface given the performance function, the more likely it is that this approach will locate the optimal solution. For example, Macy (1991a) utilizes evolutionary techniques to examine cooperation in social groups. One of the most promising uses of genetic algorithms is in the area of organizational evolution, in which the genetic algorithm is used to simulate the behavior of populations of organizations evolving their forms over time. Here, the concurrency across multiple organizations is key to determining the dynamics of organizational survival. Crowston (1994, 1998) has used this approach to examine Thompson's theory of organizational forms and the evolution of novel forms.

Simulated annealers are a computational analog of the process of metal or chemical annealing. Eccles and Crane (1988) suggest that annealing is an appropriate metaphor for organizational change. Simulated annealers search for the best solution by first proposing an alternative from a set of feasible and predefined options, seeing if this alternative's fit is better than the current system's, adopting the alternative if it is better, and otherwise adopting even the bad or risky move with some probability. The probability of accepting the bad move decreases over time as the temperature of the system cools. In organizational terms we might liken temperature to the organization's willingness to take risks. Like genetic algorithms a fitness function is needed in order to generate emergent behavior. Carley and Svoboda (1998) have used simulated annealing techniques to look at strategic change in organizations and suggest that such change may effect only a minimal change in performance over that made possible by simple individual learning.

The strategic management literature suggests that executives can and do actively restructure their organizations (Baird and Thomas, 1985; Miller et al., 1982; Staw, 1982; Staw and Ross, 1989). For these researchers, the outcome of the individual decision-making process is an organizational goal. The research on managerial decision making and its effects on structure and efficiency have been examined empirically by MacCrimmon and Wehrung (1986), as well as researched by March and Shapira (1987), and March (1981). Researchers using computational models are taking these empirical findings and using them as the basis for the computational models. In particular, when the organization is modeled as a simulated annealer, different strategies can be fruitfully modeled as the move set for changing states in the annealer.

Most of the computational work using adaptive agent techniques of neural networks and genetic algorithms have examined networks of individuals that are largely undifferentiated in terms of their structural position and their organizational roles, and are somewhat simple from a cognitive standpoint. Consequently, this work provides little insight into how to design, redesign, or reengineer organizations. An intriguing possibility is the combination of these models with models of organizational or social structure. Such combined models may provide insight into the relative impacts of, and interactions between, structural- and individual-based learning. For example, Collins (1992) demonstrates that spatial constraints on evolution can aid social learning. Early results suggest that the existence of spatial or social structure may actually increase the effectiveness of individual agent learning, and may increase the robustness and stability of the collectivities' ability to problem solve in the face of change among the constituent members.

18.4 IMPLICATIONS FOR SYSTEMS ENGINEERING AND MANAGEMENT

Implications in two areas are considered: support for collaborative work, and understanding organizational learning. As to the first area, much of the research to date has focused on providing communication tools and databases in support of collaborative or distributed work. However, the work discussed on mental models and social networks can be read as suggesting the need for a different type of support. In particular, teamwork requires transactive memory, that is, knowledge of who knows what and knowledge about how to find things out. Teamwork also requires having some level of shared understanding. Thus tools that support the development of a shared mental model or the construction and maintenance of the informal social network should facilitate collaborative work.

As to organizational learning, the implication is that computational models are providing important insights and future progress will require more detailed computational models. Modeling organizations as collections of intelligent adaptive agents acting more or less concurrently is key to understanding issues of organizational learning and design. Organizational learning focuses on performance improvement and environmentally triggered adaptation. As an example, Elofson and Konsynski (1993) apply AI and machine learning techniques to the analysis of organizational learning for the purpose of monitoring and analyzing decisions relative to the organization's structure and for monitoring change as part of the learning cycle. They demonstrate that knowledge caching can increase flexibility and so provide a means of realizing an explicit organizational memory where information and processing capabilities are distributed among personel. Carley (1991b, 1992) used an approach akin to neural networks to represent hierarchies, and demonstrated that when organizational learning was embedded in the relationships between agents and not just in the agents, the organization was more robust in the face of various problems, such as turnover and information error. Results from work on the coevolution of intelligent adaptive agents suggests that the concurrent interaction among agents when combined with access to different forms of communication media can effect radical changes in the ability of subgroups to acquire novel information and to be socialized (Carley, 1995a). Moreover, work in this area suggests that simple access to different collaborative or communication technologies will not in and of itself be sufficient for guaranteeing access to new ideas, and thus may not lead to quality or performance improvements. Collectively, these and other results from ongoing research in the organizational, social, and psychological sciences suggests that organizations of agents often exhibit complex and high nonlinear behavior. As such, traditional methods for modeling these systems as systems may not suffice. In many engineering disciplines, engineers employ simulations to capture the complexity of higher-order systems. The same is true in systems engineering, where we will need to utilize simulations much more frequently if we are to assess the impact of the nonlinearities present in distributed and colaborative work.

18.5 CONCLUSION

In a way, these diverse approaches are growing together. Carley and Newell (1994) in their discussion of what it takes to create a model social agent point out that socialness and the ability to act like an organizational agent derives both from limitations to an agent's cognitive capabilities and acquisition of multiple types of knowledge as the agents tries to operate within a certain type of environment. Agents that are too capable cognitively, have no need for social interaction or learning. Agents that are not in a complex enough situation and do not have certain knowledge cannot engage in certain actions. Complex social and organizational phenomena emerge from concurrent interactions among even simple agents (Shoham and Tennenholtz, 1994), but the nature of the social dynamics and the speed with which they emerge are determined, at least in part, by the agents' cognitive abilities (deOliveira, 1992; Collins, 1992; Carley 1998b) and their sociocultural–historical position (White, 1992, Carley, 1991a; Kaufer and Carley, 1993). This development is seen both in the new work in social networks, in which there is a growing recognition of the cognitive abilities of the nodes and in multiagent models, in which there is a growing recognition of the need to incorporate more structural constraints on agent communication.

On the network front, researchers are increasingly examining both the individual's social network position and demographic and psychological characteristics. This research suggests that bringing the individual back into the social network affords a better understanding of actual organizational behavior (Krackhardt and Kilduff, 1994). Krackhardt and Kilduff argued that an observer's perception of an individual's performance was influenced by whether or not the observer perceived the individual as having an influential friend. Network theorists often argue that structure influences actions, decisions, attitudes, and so forth (Burt, 1982). By combining these perspectives, researchers in organizational decision making can examine how the structural position of the organizational agents influences what information they attend to and how they use that information, their perception of the social structure in making attributions about others and themselves, and how these attributions then affect their decisions and actions. Such a combination of perspectives leads to the argument that it is not structure per se, but individuals' perception of structure and differences in their perception of structure that influences their decisions, attitudes, and evaluations of self and others.

On the computational organization theory front, multiagent models of organizations, in which the agents have more restricted cognitive capabilities, exhibit a greater variety of social behaviors. By increasing the realism of the agent, either by restricting its cognitive capability, or by increasing the amount or type of knowledge available to the agent or the situation in which it must act, the researcher is able to produce models that are more capable of producing social behavior and a wider range of organizational behavior. For example, the agents in Plural-Soar (Carley et al., 1992) are more restricted than the boundedly rational agent used in AAIS (Masuch and LaPotin, 1989). The agents in the AAIS model, however, effectively had access to more types of social information than did the Plural-Soar agents. Combining the two models led to an agent that was capable of exhibiting a greater range of social behaviors (Verhagen and Masuch, 1994) than either of the parent models.

We began by noting that organizations do not make decisions, people do. The research on organizational decision making indicates that although this point is incontestable, the decisions that individuals made are highly constrained by the task they are doing, their position in the organization, and their sociohistorical–cultural position. The goal now is to present the specific way in which these factors influence the decisions made in teams.

REFERENCES

Ashby W. R. (1956). Principles of self organizing systems. In *Modern Systems Research for the Behavioral Scientist* (W. Buckley, ed.). Chicago: Aldine.

Axelrod, R. M. (1987). The evolution of strategies in the iterated prisoner's dilemma. In *Genetic Algorithms and Simulated Annealing* (W. Davis, ed.), pp. 32–41. London: Pitman.

Axelrod, R. M., and Dion, D. (1988). The further evolution of cooperation. *Science* **242:** 1385–1390.

Baird, I. S., and Thomas, H. (1985). Toward a contingency model of strategic risk taking. *Acad. Manage. Rev.* **10,** 230–243.

Barnard, C. (1938). *The Functions of the Executive.* Cambridge, MA: Harvard University Press.

Bechtel, R. (1997). Developments in computer science with application to text analysis. In *Text Analysis for the Social Sciences: Methods for Drawing Statistical Inferences from Texts and Transcripts* (C. W. Roberts, ed.). Mahwah, NJ: Erlbaum, pp. 239–250.

Becker, G. M., DeGroot, M. H., and Marschak, J. (1964). Measuring utility by a single-response sequential method. *Behav. Sci.* **9,** 226–232.

Becker, M. H. (1970). Sociometric location and innovativeness: Reformulation and extension of the diffusion model. *Am. Sociol. Rev.* **35,** 267–282.

Bell, D.E. (1982). Regret in decision making under uncertainty. *Oper. Res.* **30,** 961–981.

Bem, D. (1965). An experimental analysis of self-persuasion. *J. Exp. Soc. Psychol.* pp. 199–218.

Bernoulli, D. (1738). Exposition of a new theory on the measurement of risk. *Comment. Acad. Sci. Imp. Petropolitan.* **5,** 175–192; translated in *Econometrica* **22,** 23–36 (1954).

Bond, A. H., and Gasser, L. (1988). *Readings in Distributed Intelligence.* San Mateo, CA: Morgan Kaufmann.

Bonini, C. P. (1963). *Simulation of Information and Decision Systems in the Firm.* Englewood Cliffs, NJ: Prentice-Hall.

Boorman, S. A., and White, H. C. (1976). Social structure from multiple networks. II. Role structures. *Am. J. Sociol.* **81,** 1384–1446.

Burt, R. S. (1973). The differential impact of social integration on participation in the diffusion of innovations. *Soc. Sci. Res.* **2,** 125–144.

Burt, R. S. (1976). Positions in networks. *Soc. Forces* **55,** 93–122.

Burt, R. S. (1977). Positions in multiple network systems, Part 1: A general conception of stratification and prestige in a system of actors cast as a social topology. *Soc. Forces* **56,** 106–131.

Burt, R. S. (1980). Innovation as a structural interest: Rethinking the impact of network position innovation adoption. *Soc. Networks* **4,** 337–355.

Burt, R. S. (1982). *Toward a Structural Theory of Action*. New York: Academic Press.

Burt, R. S. (1992). *Structural Holes: The Social Structure of Competition*. Cambridge, MA: Harvard University Press.

Carley, K. (1986a). Measuring efficiency in a garbage can hierarchy. In *Ambiguity and Command* (J. G. March and R. Weissinger-Baylon, eds.), pp. 165–194. New York: Pitman.

Carley, K. (1986b). Efficiency in a garbage can: Implications for crisis management. In *Ambiguity and Command* (J. G. March and R. Weissinger-Baylon, eds.), pp. 195–231. New York: Pitman.

Carley, K. (1986c). An approach for relating social structure to cognitive structure. *J. Math. Sociol.* **12**(2), 137–189.

Carley, K. M. (1990a). Group stability: A socio-cognitive approach. In *Advances in Group Processes* (E. Lawler, B. Markovsky, C. Ridgeway, and H. Walker, eds.), Vol. 7, pp. 1–44). Greenwich, CT: JAI Press.

Carley, K. (1990b). Structural constraints on communication: The diffusion of the homomorphic signal analysis technique through scientific fields. *J. Math. Sociol.* **15**(3–4), 207–246.

Carley, K. M. (1991a). A theory of group stability. *Am. Sociol. Rev.,* **56**(3), 331–354.

Carley, K. M. (1991b). Designing organizational structures to cope with communication breakdowns: A simulation model. *Ind. Crisis Q.* **5**, 19–57.

Carley, K. M. (1992). Organizational learning and personnel turnover. *Organ. Sci.* **3**(1), 2–46.

Carley, K. M. (1994). Extracting culture through textual analysis. *Poetics* **22**, 291–312.

Carley, K. M. (1995a). Communication technologies and their effect on cultural homogeneity, consensus, and the diffusion of new ideas. *Sociol. Perspect.* **38**(4), 547–571.

Carley, K. M. (1995b). Computational and mathematical organization theory: Perspective and directions, *Comput. Math. Organ. Theory* **1**(1), 39–56.

Carley, K. M. (1996) A comparison of artificial and human organizations. *J. Econ. Behav. Organ.* **896**, 1–17.

Carley, K. M. (1996). A comparison of artificial and human organizations. *J. Econ. Behav. and Organ.* **31**(1), 175–191.

Carley, K. M. (1997). Network text analysis: The network position of concepts. In *Text Analysis for the Social Sciences: Methods for Drawing Statistical Inferences from Texts and Transcripts* (C. W. Roberts, ed.). Mahwah, NJ: Erlbaum, pp. 79–100.

Carley, K. M., and Kaufer, D. S. (1993). Semantic connectivity: An approach for analyzing semantic networks. *Commun. Theory* **3**(3), 183–213.

Carley, K. M., and Lin, Z. (1996). A theoretical study of organizational performance under information distortion. *Manage. Sci.* **25**(1), 138–168.

Carley, K., and Newell, A. (1994). The nature of the social agent. *J. Math. Sociol.* **19**(4), 221–262.

Carley, K. M., and Palmquist, M. (1992). Extracting, representing and analyzing mental models. *Soc. Forces* **70**, 601–636.

Carley, K. M., and Prietula, M. J. (eds.). (1994a). *Computational Organization Theory*. Hillsdale, NJ: Erlbaum.

Carley, K. M., and Prietula, M. J. (1994b). ACTS theory: Extending the model of bounded rationality. In *Computational Organization Theory* (K. M. Carley and M. J. Prietula, eds.), pp. 55–88. Hillsdale, NJ: Erlbaum.

Carley, K. M., and Svoboda, D. (1988). Modeling organizational adaptation as a simulated annealing process. *Sociol. Methods Res.*

Carley, K. M., and Wendt, K. (1991). Electronic mail and scientific communication: A study of the soar extended research group. *Knowl. Creation, Diffus. Util.* **12**(4), 406–440.

Carley, K. M., Kjaer-Hansen, J., Prietula, M., and Newell, A. (1992). Plural-soar: A prolegomenon to artificial agents and organizational behavior. In *Artificial Intelligence in Organization and Management Theory* (M. Masuch and M. Warglien, eds.), pp. 87–118. Amsterdam: Elsevier.

Chapman, L. J., and Chapman, J. P. (1967). Genesis of popular but erroneous psychodiagnostic observations. *J. Abnorm. Psychol.,* **72,** 193–204.

Chapman, L. J., and Chapman, J. P. (1969). Illusory correlation as an obstacle to the use of valid psychodiagnostic signs. *J. Abnorm. Psychol.,* **14,** 743–749.

Cohen, M. D. (1986). Artificial intelligence and the dynamic performance of organizational designs. In *Ambiguity and Command: Organizational Perspectives on Military Decision Making* (J. G. March and R. Weissinger-Baylon, eds.), pp. 53–70. Marshfield, MA: Pitman.

Cohen M. D., March, J. G., and Olsen, J. P. (1972). A garbage can model of organizational choice. *Admin. Sci. Q.* **17,** 1–25.

Coleman, J. S., Katz, E., and Menzel, H. (1966). *Medical Innovation: A Diffusion Study.* New York: Bobbs-Merrill.

Collins, R. J. (1992). *Studies in Artificial Evolution* (CSD-920037). Los Angeles: University of California, Computer Science Dept.

Crowston, K. (1994). Evolving novel organizational forms. In *Computational Organization Theory* (K. M. Carley and M. J. Prietula, eds.), pp. 19–38. Hillsdale, NJ: Erlbaum.

Crowston, K. (1998). An approach to evolving novel organizational forms. *Comput. Math. Organ. Theory.*

Cyert, R, and March, J. G. (1963). *A Behavioral Theory of the Firm,* 2nd ed. Cambridge, MA: Blackwell.

Dawes, R. M. (1988). *Rational Choice in an Uncertain World.* New York: Harcourt Brace Jovanovich.

Dawes, R. M. (1989). Statistical criteria for establishing a truly false consensus effect. *J. Exp. Soc. Psychol.* **25,** 1–17.

Dawes, R. M. (1990). The potential non-falsity of the false consensus effect. In *Insights in Decision Making: A Tribute to Hillel J. Einhorn* (R. M. Hogarth, ed.), pp. 179–199. Chicago: Chicago University Press.

Dawes, R. M., and Mulford, M. (1996). The false consensus effect and overconfidence: Flaws in judgment or flaws in how we study judgment? *Organ. Behav. Hum. Decis. Process.* **65**(3), 201–211.

Dennison, D. R. (1996). What is the difference between organizational culture and organizational climate? A native's point of view on a decade of paradigm wars. *Acad. Manage. Rev.* **21**(3), 619–654.

de Oliveira, P. P. B. (1992). *Enact: An Artificial-life World in a Family of Cellular Automata,* Cogn. Sci. Res. Pap., CSRP 248. Brighton (East Sussex), England: University of Sussex, School of Cognitive and Computing Sciences.

DiMaggio, P. J. (1986). Structural analysis of organizational fields: A blockmodel approach. *Res. Organ. Behav.* **8**(3), 35–370.

Duncan, R. B. (1973). Multiple decision-making structures in adapting to environmental uncertainty: The impact on organizational effectiveness. *Hum. Relat.* **26,** 273–291.

Durfee, E. H. (1988). *Coordination of Distributed Problem Solvers.* Boston: Kluwer Academic Publishers.

Durfee, E. H., and Montgomery, T. A. (1991). Coordination as distributed search in a hierarchical behavior space. *IEEE Trans. Syst., Man, Cybernet.* **21**(6), 1363–1378.

Dutton, J. M., and Starbuck, W. H. (1971). *Computer Simulation of Human Behavior.* New York: Wiley.

Eccles, R. G., and Crane. D. B. (1988). *Doing Deals: Investment Banks at Work*. Boston: Harvard Business School Press.

Eden, C., Jones, S., and Sims, D. (1979). *Thinking in Organizations*. London: Macmillan.

Edwards, W. (ed). (1992). *Utility Theories: Measures and Applications*. Boston: Kluwer Academic Publishers.

Elofson, G. S., and Konsynski, B. R. (1993). Performing organizational learning with machine apprentices. *Decis. Support Sys.,* **10**(2), 109–119.

Ennals, J. R. (1991). *Artificial Intelligence and Human Institutions*. London and New York: Springer-Verlag.

Ericsson, K. A., and Simon, H. A. (1984). *Protocol Analysis: Verbal Reports as Data*. Cambridge, MA: MIT Press.

Fararo, T. J., and Skvoretz, J. (1984). Institutions as production systems. *J. Math. Sociol.* **10,** 117–182.

Fauconnier, G. (1985). *Mental Spaces: Aspects of Meaning Construction in Natural Language*. Cambridge, MA: Bradford Books, MIT Press.

Festinger, L. (1954). A theory of social comparison processes. *Hum. Relations* **7,** 114–140.

Fischoff, B., Lichtenstein, S., Slovic, P., Derby, S., and Keeney, R. (1981). *Acceptable Risk*. New York: Cambridge University Press.

Forester, J. W. (1961). *Industrial Dynamics*. Cambridge, MA: MIT Press.

Freeman, L. C. (1984). Impact of computer-based communication on the social structure of an emerging scientific specialty. *Soc. Networks* **6,** 201–221.

Galbraith, J. R. (1973). *Designing Complex Organizations*. Reading, MA: Addison-Wesley.

Galbraith, J. R. (1977). *Organization Design*. Reading, MA: Addison-Wesley.

Gasser L., and Huhns, M. N. (eds.). (1989). *Distributed Artificial Intelligence,* Vol. 2. San Mateo, CA: Morgan Kaufmann.

Gasser, L., and Majchrzak, A. (1992). HITOP-A: Coordination, infrastructure, and enterprise integration. In *Proceedings of the First International Conference on Enterprise Integration,* pp. 373–378. Hilton Head, SC: MIT Press.

Gasser, L., and Majchrzak, A. (1994). ACTION integrates manufacturing strategy, design, and planning. In *Ergonomics of Hybrid Automated Systems IV* (P. Kidd and W. Karwowski, eds.), pp. 133–136. Netherlands: IOS Press.

Gasser, L., and Toru, I. (1991). A dynamic organizational architecture for adaptive problem solving. *Proc. 9th Natl. Conf. Artif. Intell.,* Anaheim, CA, pp. 185–190.

Gazdar, G., and Mellish, C. S. (1986). *Computational Linguistics* (CSRP 058). Brighton, England: University of Sussex, Cognitive Studies Programme, School of Social Sciences.

Glance, N. S., and Huberman, B. A. (1993). The outbreak of cooperation. *J. Math. Sociol.,* **17**(4), 281–302.

Glance, N. S., and Huberman, B. A. (1994). Social dilemmas and fluid organizations. In *Computational Organization Theory* (K. M. Carley and M. J. Prietula, eds.), pp. 217–240. Hillsdale, NJ: Erlbaum.

Golumbic, M. C. (ed.). (1990). *Advances in Artificial Intelligence: Natural Language and Knowledge-Based Systems*. New York: Springer-Verlag.

Granovetter, M. S. (1973). The strength of weak ties. *Am. J. Sociol.* **68,** 1360–1380.

Granovetter, M. S. (1974). *Getting a Job: A Study of Contacts and Careers*. Cambridge, MA: Harvard University Press.

Greenwald, A. G., and Pratkanis, A. R. (1984). The self. In *Handbook of Social Cognition* (R. S. Wyer and T. K. Srul, eds.), Vol. 3, pp. 129–178, Hillsdale, NJ: Erlbaum.

Harrison, J. R., and Carrol, G. R. (1991). Keeping the faith: A model of cultural transmission in formal organizations. *Admin. Sci. Q.* **36,** 552–582.

Heider, F. (1958). *The Psychology of Interpersonal Relations.* New York: Wiley.

Heise, D. R. (1979). *Understanding Events: Affect and the Construction of Social Action.* New York: Cambridge University Press.

Heise, D. R. (1991). Event structure analysis: A qualitative model of quantitative research. In *Using Computers in Qualitative Research* (N. G. Fielding and R. M. Lee, eds.). Newbury Park, CA: Sage.

Higgins, E. T., and Bargh, J. A. (1987). Social cognition and social perception. *Annu. Rev. Psychol.* **38,** 369–425.

Holland, J. H. (1975). *Adaptation in Natural and Artificial Systems.* Ann Arbor: University of Michigan Press.

Holland, J. H. (1992). Genetic algorithms. *Sci. Am.* **267,** 66–72.

Holland, J. H., and Miller, J. (1991). Artificial adaptive agents in economic theory. *Am. Econ. Rev., Pap. Proc.* **81,** 365–370.

Holland, J. H., Holyoak, K., Nisbett R., and Thagard, P. (1986). *Induction: Processes of Inference, Learning, and Discovery.* Cambridge, MA: MIT Press.

Hutchins, E. (1990). The technology of team navigation. In *Intellectual Teamwork* (J. Galegher, R. Kraut, and C. Egido, eds.), pp. 191–220). Hillsdale, N.J: Erlbaum.

Hutchins, E. (1991a). Organizing work by adaptation. *Organ. Sci.* **2,** 14–39.

Hutchins, E. (1991b). The social organization of distributed cognition. In (L. B. Resnick, J. M. Levine, and S. D. Teasley, eds.), *Perspectives on Socially Shared Cognition* pp. 238–307. Washington, DC: American Psychological Association.

Janis, I. (1982). *Groupthink,* 2nd ed. Boston: Houghton Mifflin.

Johnson-Laird, P. N. (1983). *Mental Models: Toward a Cognitive Science of Language, Inference, and Consciousness.* Cambridge, MA: Harvard University Press.

Jones, E. E., and Nisbett, R. E. (1972). The actor and the observer: Divergent perceptions of the causes of behavior. In *Attribution: Perceiving the Causes of Behavior* (E. E. Jones, D. E. Kanouse, H. H. Kelley, R. E. Nisbett, S. Valins, and B. Weiner, eds.). Morristown, NJ: General Learning Press.

Kahneman, D., and Tversky, A. (1979). Prospect theory: An analysis of decision under risk. *Econometrica* **47,** 263–291.

Kahneman, D., Slovic, P., and Tversky, A. (1982). *Judgment Under Uncertainty: Heuristics and Biases.* London: Cambridge University Press.

Kaufer, D., and Carley, K. M. (1993). *Communication at a Distance: The Effect of Print on Socio-Cultural Organization and Change.* Hillsdale, NJ: Erlbaum.

Kirkpatrick, S., Gelatt, C. D., and Vecchi, M. P. (1983). Optimization by simulated annealing. *Science* **220,** 671–680.

Klimoski, R., and Mohammed, S. (1994). Team mental model: Construct or metaphor? *J. Manage.* **20**(2), 403–437.

Knorr-Cetina, K. (1981). *The Manufacture of Knowledge.* Oxford: Pergamon.

Kontopoulos, K. M. (1993). Neural networks as a model of structure. In *The Logics of Social Structure,* pp. 243–267. New York: Cambridge University Press.

Krackhardt, D. (1987). Cognitive social structures. *Soc. Networks,* **9**(2), 109–134.

Krackhardt, D. (1990). Assessing the political landscape: Structure, cognition, and power in organizations. *Admin. Sci. Q.* **35,** 342–369.

Krackhardt, D. (1991). The strength of strong ties: The importance of philos in organizations. In

Organizations and Networks: Theory and Practice (N. Nohira and R. Eccles, eds.), pp. 216–239. Cambridge, MA: Harvard Business School Press.

Krackhardt, D. (1994). Graph theoretical dimensions of informal organizations. In *Computational Organization Theory* (K. M. Carley and M. J. Prietula, eds.), pp. 89–239, Hillsdale, NJ: Erlbaum.

Krackhardt, D., and Brass, D. (1994). Intra-organizational networks: The micro side. In *Advances in the Social and Behavioral Sciences from Social Network Analysis* (S. Wasserman and J. Galaskiewicz, eds.), pp. 209–230. Beverly Hills, CA: Sage.

Krackhardt, D., and Kilduff, M. (1994). Bringing the individual back in: A structural analysis of the internal market for reputation in organizations. *Acad. Manage. J.* **37**(1), 87–108.

Krackhardt, D., and Porter, L. W. (1985). When friends Leave: A structural analysis of the relationship between turnover and stayer's attitudes. *Admin. Sci. Q.* **30**, 242–261.

Krackhardt, D., and Porter, L. W. (1986). The snowball effect: Turnover embedded in communication networks. *J. Appl. Psychol.* **71**, 50–55.

Lant, T. K. (1994). Computer simulations of organizations as experiential learning systems: Implications for organization theory. In *Computational Organization Theory* (K. M. Carley and M. J. Prietula, eds.), pp. 195–216. Hillsdale, NJ: Erlbaum.

Lant, T. K., and Mezias, S. J. (1990). Managing discontinuous change: A simulation study of organizational learning and entrepreneurial strategies. *Strategic Manage. J.* **11**, 147–179.

Lant, T. K., and Mezias, S. J. (1992). An organizational learning model of convergence and reorientation. *Organ. Sci.* **3**(1), 47–71.

Larkey, P. D., and Caulkins, J. (1991). All above average and other pathologies of performance evaluation systems. In *National Conference on Public Management.* Syracuse, NY: Syracuse University, Maxwell School.

Latour, B., and Woolgar, S. (1979). *Laboratory Life.* Beverly Hills, CA: Sage.

Leblebici, H., and Salancik, G. R. (1989). The rules of organizing and the managerial role. *Organ. Stud.* **10**(3), 301–325.

Levinthal, D., and March, J. G. (1981). A model of adaptive organizational search. *J. Econ. Behav. Organ.* **2**, 307–333.

Levitt, R. E., Cohen, G. P., Kunz, J. C., Nass, C. I., Christiansen, T., and Jin, Y. (1994). A theoretical evaluation of measures of organizational design: Interrelationship and performance predictability. In *Computational Organization Theory* (K. M. Carley and M. J. Prietula, eds.). pp. 1–18. Hillsdale, NJ: Erlbaum.

Lin, N., and Burt, R. S. (1975). Differential effects of information channels in the process of innovation diffusion. *Soc. Forces* **54**, 256–274.

Lin, Z. (1994). A theoretical evaluation of measures of organizational design: Interrelationship and performance predictability. In *Computational Organization Theory* (K. M. Carley and M. J. Prietula, eds.), pp. 113–160. Hillsdale, NJ: Erlbaum.

Lin, Z., and Carley, K. M. (1998). Organizational response: The cost performance tradeoff. *Manage. Sci.* **43**(2), 217–234.

Loomes, G., and Sugden, R. (1982). Regret theory: An alternative theory of rational choice under uncertainty. *Econ. J.* **92**, 805–824.

Luce, R. D., and Raiffa, H. (1957). *Games and Decisions: Introduction and Critical Survey.* New York: Wiley.

MacCrimmon, K. R., and Wehrung, D. A. (1986). *Taking Risks: The Management of Uncertainty.* New York: Free Press.

Machina, M. J. (1982). Expected utility analysis without the independence axiom. *Econometrica* **50**, 277–323.

Macy, M. W. (1991a). Learning to cooperate: Stochastic and tacit collusion in social exchange. *Am. J. Sociol.* **97**(3), 808–843.

Macy, M. W. (1991b). Chains of cooperation: Threshold effects in collective action. *Am. Sociol. Rev.* **56**, 730–747.

Majchrzak, A., and Gasser, L. (1991). On using artificial intelligence to integrate the design of organizational and process change in US manufacturing. *Artif. Intell. Soc.* **5**, 321–338.

Majchrzak, A., and Gasser, L. (1992). HITOP-A: A tool to facilitate interdisciplinary manufacturing systems design. *Int. J. Hum. Factors Manuf.* **2**(3), 255–276.

Malone, T. W. (1987). Modeling coordination in organizations and markets. *Manage. Sci.* **33**, 1317–1332.

March, J. G. (1981). Decisions in organizations and theories of choice. In *Perspectives on Organization Design and Behavior* (A. H. Van de Ven and W. F. Joyce, eds.). New York: Wiley.

March, J. G., and Romelaer, P. (1976). Position and presence in the drift of decisions. In *Ambiguity and Choice in Organizations* (J. G. March and J. P. Olsen, eds.). Bergen: Universitetsforlaget.

March, J. G., and Shapira, Z. (1987). Managerial perspectives on risk and risk taking. *Manage. Sci.* **33**, 1404–1418.

March, J. G., and Simon, H. A. (1958). *Organizations*. New York: Wiley.

Markus, H. (1983). Self-knowledge: An expanded view. *J. Personality* **51**, 543–565.

Marschak, J. (1950). Rational behavior, uncertain prospects, and measurable utility. *Econometrica* **18**, 111–141.

Marschak, J. (1955). Elements for a theory of teams. *Manage. Sci.* **1**, 127–137.

Masuch, M., and LaPotin, P. (1989). Beyond garbage cans: An AI model of organizational choice. *Admin. Sci. Q.* **34**, 38–67.

Masuch, M., and Warglien, M. (1992). *Artificial Intelligence in Organization and Management Theory*. Amsterdam: Elsevier.

McClelland, J., and Rumelhart, D. (1986). *Parallel Distributed Processing: Explorations in the Microstructure of Cognition*, Vol. 2. Cambridge, MA: MIT Press.

McGuire C. B., and Radner, R. (1986). *Decision and Organization*. Minneapolis: University of Minnesota Press.

Miller, D., Kets de Vries, M. F. R., and Toulouse, J. M. (1982). Top executive locus of control and its relationship to strategy-making, structure, and environment. *Acad. Manage. J.* **25**, 237–253.

Mobley, W. H. (1982). *Employee Turnover: Causes, Consequences, and Control*. Reading, MA: Addison-Wesley.

Monson, T. C., and Snyder, M. (1977). Actors, observers, and the attribution process: Toward a reconceptualization. *J. Exp. Soc. Psychol.* **13**, 89–111.

Mowday, R. T., Porter, L. W., and Steers, R. M. (1982). *Employee-Organization Linkages: The Psychology of Commitment, Absenteeism, and Turnover*. New York: Academic Press.

Orbell, J. M., and Dawes, R. M. (1993). Social welfare, cooperators' advantage, and the option of not playing the game. *Am. Sociol. Rev.* **58**(6), 787–800.

Orbell, J. M., van de Kragt, A. J. C., and Dawes, R. M. (1988). Explaining discussion-induced cooperation. *J. Personality Soc. Psychol.* **54**(5), 811–819.

Padgett, J. F. (1980). Managing garbage can hierarchies. *Admin. Sci. Q.* **25**(4), 583–604.

Padgett, J. F. (1997). The emergence of simple ecologies of skill. In *The Economy as a Complex Evolving System*, II. (B. Arthur, S. Durlauf, and D. Lane, eds.). Santa Fe Institute Studies in the Science of Complexity. Reading MA: Addison-Wesley.

Plous, S. (1993). *The Psychology of Judgment and Decision Making*. New York: McGraw-Hill.

Polanyi, M. P. (1958). *Personal Knowledge: Towards a Post-Critical Philosophy*. Chicago: The University of Chicago Press.

Prietula, M. J., and Carley, K. M. (1994). Computational organization theory: Autonomous agents and emergent behavior. *J. Organ. Comput.* **41**(1), 41–83.

Pruitt, D. (1971a). Choice shifts in group discussion: An introductory review. *J. Personality Soc. Psychol.* **20**, 339–360.

Pruitt, D. (1971b). Conclusions: Toward an understanding of choice shifts in group discussion. *J. Personality Soc. Psychol.* **20**, 495–510.

Radner, R. (1993). The organization of decentralized information processing *Econometrica* **61**(5), 1109–1146.

Reger, R. K., and Huff, A. S. (1993). Strategic groups: A cognitive perspective. *Strategic Manage. J.* **14**, 103–124.

Rice, R. E., and Aydin, C. (1991). Attitudes toward new organizational technology: Network proximity as a mechanism for social information processing. *Admin. Sci. Q.* **2**, 219–244.

Roberts, C. W. (1989). Other than counting words: A linguistic approach to content analysis. *Soc. Forces* **68**, 147–177.

Roberts, C. W. (ed). (1997). *Text Analysis for the Social Sciences: Methods for Drawing Statistical Inferences from Texts and Transcripts*. Mahwah, NJ: Erlbaum.

Roethlisberger, F.J., and Dickson, W. J. (1939). *Management and the Worker*. Cambridge, MA: Harvard University Press.

Ross, L., Amabile, T.M., and Steinmetz, J. L. (1977). Social roles, social controls, and biases in the social perception process. *J. Personality Soc. Psychol.* **35**, 485–494.

Rumelhart D., and McClelland, J. (1986). *Parallel Distributed Processing: Explorations in the Microstructure of Cognition,* Vol. 1. Cambridge, MA: MIT Press.

Rutenbar, R. A. (1989). Simulated annealing algorithms: An overview. *IEEE Circuits Devices Mag.* **5**, 12–26.

Salancik, G. R., and Leblebici, H. (1988). Variety and form in organizing transactions: A generative grammar of organization. *Res. Sociol. Organ.* **6**, 1–31.

Salancik, G. R., and Pfeffer, J. (1978). A social information professing approach to job attitudes and task design. *Admin. Sci. Q.* **23**, 224–253.

Savage, L. J. (1954). *The Foundations of Statistics*. New York: Wiley.

Schank, R. C., and Abelson, R. P. (1977). *Scripts Plans and Goals and Understanding*. New York: Wiley.

Schelling, T. (1978). *Micromotives and Macrobehavior*. New York: Norton.

Shoham, Y., and Tennenholtz, M. (1994). *Co-learning and the Evolution of Social Activity* (STAN-CS-TR-94-1511). Stanford, CA: Stanford University, Dept. of Computer Science.

Simon, H. A. (1945). *Administrative Behavior: A Study of Decision-Making Processes in Administrative Organization*. New York: Macmillan Company.

Simon, H. A. (1959). Theories of decision-making in economics and behavioral science. *Am. Econ. Rev.* **49**(3), 253–283.

Simon, H. A. (1979). Rational decision making in business organizations. *Am. Econ. Rev.* **69**(4), 493–513.

Simon, H. A. (1981a). *The Sciences of the Artificial,* 2nd ed). Cambridge, MA: MIT Press.

Simon, H. A. (1981b). Studying human intelligence by creating artificial intelligence. *Am. Sci.* **69**(3), 300–309.

Sowa, J. F. (1984). *Conceptual Structures*. Reading, MA: Addison-Wesley.

Staw, B. M. (1982). Counterforces to change. In *Changes in Organizations* (P. S. Goodman, ed.), pp. 87–121. San Francisco: Jossey-Bass.

Staw, B. M., and Ross, J. (1989). Understanding behavior in escalation situations. *Science* **246,** 216–220.

Tambe, M., Johnson, W. L., and Shen, W. (1997). Adaptive agent tracking in real-world multi-agent domains: A preliminary report. *Int. J. Hum. Comput. Stud. (IJHCS)*.

Tversky, A., and Kahneman, D. (1974). Judgment under uncertainty: Heuristics and biases. *Science* **185,** 1124–1131.

Tversky, A., and Kahneman, D. (1981). The framing of decisions and the psychology of choice. *Science* **211,** 453–458.

Verhagen, H., and Masuch, M. (1994). TASCCS: A synthesis of double-AISS and plural-SOAR. In *Computational Organization Theory* (K. M. Carley and M. J. Prietula, eds.) pp. 39–54. Hillsdale, NJ: Erlbaum.

von Neumann, J., and Morgenstern, O. (1947). *Theory of Games and Economic Behavior*. Princeton, NJ: Princeton University Press.

Wasserman, P. D. (1989). *Neural Computing: Theory and Practice*. New York: Van Nostrand Reinhold.

Wasserman, P. D. (1993). *Advanced Methods in Neural Computing*. New York: Van Nostrand.

Weick, K. E., and Roberts, K. A. (1993). Collective mind in organizations: Heedful interrelating on flight decks. *Admin. Sci. Q.* **38,** 357–381.

White, H. (1992). *Identity and Control: A Structural Theory of Action*. Princeton, NJ: Princeton University Press.

White H.C., Boorman, S. A., and Breiger, R. L. (1976). Social structure from multiple networks. I. Blockmodels of roles and positions. *Am. J. Sociol.* **81,** 730–780.

Winghart, O. (1988). Roles, Events and Saying Events in Expository Discourse. (AI88-85). Austin: University of Texas, *Artifi. Intell. Lab.*

Ye, M., and Carley, K. M. (1995). Radar-soar: Towards an artificial organization composed of intelligent agents. *J. Math. Sociol.* **20**(2–3), 219–246.

Zock, M., and Sabah, G. (eds.). (1988). *Advances in Natural Language Generation: An Interdisciplinary Perspective*. London: Pinter.

Zweben, M., and Fox, M. S. (eds.). (1994). *Intelligent Scheduling*. San Mateo, CA: Morgan Kaufmann.

19 Human Error and Its Amelioration

PATRICIA M. JONES

19.1. INTRODUCTION

Human error as a topic in its own right has come about partly because of its prominence as a cause or contributing factor in disasters and accidents such as Three Mile Island, Chernobyl, and Bhopal. Investigations of human error are inseparable from those of human performance; the same mechanisms and processes that presumably give rise to "correct" human performance give rise to "incorrect" human performance. Thus, conceptual distinctions in generative mechanisms of, and remedies for, human error are an integral part of human–machine systems engineering. Reason (1990) quotes Ernst Mach in this regard: "Knowledge and error flow from the same . . . sources; only success can tell the one from the other." Performance that is later tagged "human error" is done so in hindsight and is dependent on the consequences of actions (Woods, 1990).

Many researchers and practitioners have examined human error from a variety of perspectives. Safety engineers and risk assessment experts have focused on quantifying probabilities of different sorts of errors and the consequences associated with those errors. Typically, the analysis of human error is in terms of phenomenological categories such as errors of omission, commission, and substitution. Cognitive and engineering psychologists have focused primarily on the generative cognitive mechanisms for error (e.g., lapses in memory and attention) that give rise to observable error types and at the interactions between system design and human performance as the nexus for error. In looking at group work and interpersonal interaction, human communication researchers have studied misunderstandings in conversational and face-to-face interaction. More recently, sociologists and organizational theorists, as well as a growing number of cognitive engineering researchers, have focused on the historical and cultural contexts in which organizations make decisions and assess risks, and have examined the ways in which practices, norms, and the like have influenced actions that, in retrospect, may have contributed to disaster.

The organization of this chapter is shown in Table 19.1, which illustrates perspectives (phenomenological, cognitive, human–machine, human–human, sociocultural/ organizational) against analytic goals (descriptive taxonomies, generative mechanisms, remedies). Thus, inquiries into human error may focus on describing different kinds of errors, on underlying causes of error, and on how to prevent or minimize the impact of errors. A comprehensive approach to examining human error spans perspectives, from simple "physical" categories to cultural and organizational influences.

In particular, Section 19.2 provides an overview of definitions of terms. Section 19.3 is organized around Table 19.1, where Section 19.3.1 discusses phenomenological, Section

Handbook of Systems Engineering and Management, Edited by A. P. Sage and W. B. Rouse
ISBN 0471-15405-9 ©1999 John Wiley and Sons, Inc.

TABLE 19.1 Organization of this Chapter: Perspectives on Human Error and Goals of Analysis

Perspective	Descriptive Taxonomies	Analytic Goal Generative Mechanisms, Theory, or Techniques	Remedies, Prevention, Amelioration
Phenomenological	Categories of action, such as commission, omission, repetition, substitution, out-of-order	• Signal detection theory • Probabilistic risk assessment (PRA) • Human reliability analysis (HRA) • Work safety analysis (WSA) • Failure modes and effects analysis (FMEA) • Hazard and operability analysis (HAZOP)	"Human factors" design, especially of controls and displays; operating procedures, training, redundant backup systems; fault-tolerant computing, testing procedures
Cognitive Viewpoint and Human–System Interaction	• Heuristics and biases in human reasoning ("cognitive economy") • Slips and mistakes • Skills-rules-knowledge	• Cognitive resources; bounded rationality; • Attention, memory, inference, reasoning • Mismatch between human and environment • Contextual control model	Operating procedures, training, display and control design, representation aiding; joint cognitive systems design; intelligent support systems
Human–Human Interaction	• Violation of Cooperative Maxims (Grice) • Encoding/decoding • Bandwidth/noisy channel • Vocabularies, referents • Coherence	• Communication theory • Alignment talk (Ragan)	• Standard dictionary • Communication and collaboration systems • Redundancy • Documentation • Training
Sociocultural and Organizational	• Organizational metaphor: as information processing systems, brains, machines, cultures, family systems • Organizational structure, patterns of communication, and information flow	• Normalization of deviance (Vaughan) • Negotiation, interpretation	• "Safety" as the prime directive for org leaders • Nurturing a "safety culture" • Decentralized decision making and flexible organizational structures for rapid response • Audits by outsiders • Support and document organizational learning

19.3.2 discusses cognitive, Section 19.3.3 discusses interpersonal, and Section 19.3.4 discusses social, cultural, and organizational aspects of human-error research. Within each section, a suite of approaches, theories, and guidelines is presented. Section 19.4 summarizes practical implications for organizations.

Throughout this chapter, two examples are used to illustrate concepts. One example is operation and maintenance of a car; the other draws upon the Three Mile Island incident in which a variety of complex factors contributed to a near-accident in the Three Mile Island nuclear power plant (see Leveson, 1995).

19.2 BASIC DEFINITIONS

A goal of systems engineering is to design, construct, and maintain safe and reliable systems that work as intended. With respect to human error and its amelioration, we use the following definitions:

Risk. Risk includes the perceived likelihood of disastrous or hazardous event(s), their severity, and the duration of their effects. Typical approaches for quantifying risk are based on estimating the probability of occurrence of a particular disaster or hazard and the extent of damage. Sage (1992, p. 248) provides the following concise definition: "For our purposes, risk is the probability or likelihood of injury, damage, or some loss in some specific environment and over some stated period of time. Thus, risk involves two elements, probability and loss amount." Grabowski and Roberts (1996) separate these concepts: they characterize systems in terms of hazard and risk, where hazard refers to the severity of loss amount and risk refers to the probability of the hazard's occurrence. Vaughan (1996) emphasizes that risk is socially constructed and historically constituted. Leveson (1995) defines risk as a combination of "hazard level" (which includes hazard severity and its likelihood of occurrence), "danger" (likelihood of the hazard leading to an accident), and "latency" (duration or exposure of hazard).

Safe. Safe systems are defined as those for which "overall" risk is zero or as "freedom from accidents or losses" (Leveson, 1995, p. 181).

Reliability. Reliability is the probability that a system performs "some specified end-user function under specified operating conditions for a stated period of time" (Sage, 1992, p. 248).

Definitions of human error abound; indeed much of this chapter is devoted to delineating perspectives on error. Here, we distinguish between error (in actions taken by one or more humans, where actions include verbal, cognitive, and manual actions) vs. faults (in system operation; e.g., mechanical failures). A reasonable synthesis of opinions, as suggested by Senders and Moray (1991), is the following: *Human error* occurs when human action is performed that was either (1) not intended by the actor, (2) not desired according to some specified set of rules or by some external observer, or (3) contributed to the task or system "going outside its acceptable limits." This latter point emphasizes the posthoc nature of classifying human error. In terms of performance, humans may commit unintended or undesirable acts, but then have time to recover the system before a "real problem" occurs (Woods, 1990). What it means to "go outside acceptable limits" or "be a real problem" depends on the particular context of the system or task, but it is important to note that the notions of intentionality and consequences are critical aspects of the definition. Furthermore, as many authors have pointed out, error is necessary for learning.

19.3 PERSPECTIVES ON ERROR TYPES, CAUSES, AND AMELIORATION

19.3.1 Phenomenological

A phenomenological view of error simply seeks to describe the overt phenomenon of error. Here we are concerned with simple descriptive labels for error types, such as errors of omission, commission, repetition, substitution, and sequencing. That is, we can analyze human performance in retrospect as having (1) failed to perform some action, (2) performed some extraneous unnecessary action, (3) repeated an action unnecessarily, (4) substituted one action for another, (5) performed actions in the wrong order, and so on.

The simplest sort of categorization of error may be exogenous vs. endogenous; that is, those errors that originate outside or inside the individual person (Senders and Moray, 1991). If we take the human–machine system as the unit of analysis, however, we may be more inclined to speak of demand–resource mismatches in which the demands of the environment are not met by the resources of the person involved (Woods, 1990). For example, as Senders and Moray (1991) discuss, if an operator misreads a dial because of glare, that seems to be more properly categorized as an exogenous error, or a mismatch between the demands of the situation and the physical capabilities of the human operator.

A simple model of human problem solving, of which there are innumerable variations, is of two interleaved stages of situation assessment and response planning (Roth et al., 1994). Within this framework, we can generate categories of error, such as (1) errors in detection of a problem, (2) errors in diagnosis of a problem, and (3) errors in planning and executing actions (Sage, 1992). Sage (1992, p. 233) notes that "[f]ailure in problem detection and diagnosis may occur [in] many different forms . . . [including] Setting improper thresholds; Failure to generalize; Failure to anticipate; Failure to search for and process potentially available information." A large body of work related to errors in problem detection is found in the literature in signal detection theory (e.g., Swets, 1996), in which errors are classified as misses (a signal occurs and the person has missed it) and false alarms (a signal does not occur and the person believes that it did). For planning and execution, a common distinction is between slips (correct intention, wrong action; as in a typographical error) and mistakes (wrong intention) (Norman, 1988; Reason, 1990).

Hollnagel (1993) proposes a more elaborate taxonomy of phenotypes (overt manifestations of errors) that distinguishes between four error modes, eight simple phenotypes, and seven complex phenotypes. The four error modes are "action in wrong place," "action at wrong time," "action of wrong type," and "action not included in current plans." These four error modes map in systematic ways to an array of simple phenotypes, which are repetition, reversal, omission, delay, premature action, replacement, insertion, and intrusion. These map systematically to the complex phenotypes of restart, jumping, undershoot, side-tracking, capture, branching, and overshoot.

A phenomenological approach to error implies a focus on systems engineering techniques for identifying potential error, inducing situations, rather than an exploration of underlying cognitive or social mechanisms that may lead to human error. A range of well-known techniques for systematically articulating possible fault and error conditions include probabilistic risk assessment (PRA), which uses fault tree and error tree representations and human reliability analysis (HRA) techniques such as the technique for human error rate prediction (THERP) (Reason, 1990; cf. Swain and Guttmann, 1983). Other techniques include work safety analysis, failure modes and effects analysis (FMEA), and hazard and operability analysis (HAZOP) (Kirwan and Ainsworth, 1992; Kirwan, 1994; Leveson, 1995).

Fault trees and event trees are two diagramming techniques that are used to systematically examine relations among system events ("what leads to what"). Both rely on hierarchical "tree" structures in which a top-level node is decomposed or related to other types of nodes at more detailed levels. Fault trees use Boolean logical structures (AND and OR gates) to articulate combinations of events that can lead to hazards. The "top event" of the fault tree is a particular fault condition (e.g. "wrong dosage administered"), and logical relations (AND, OR, NOT) are used to articulate the ways in which more elementary "basic" or "primary events" can lead to that top event. Events below AND gates are causally related to events above; events below OR gates are reexpressions, not causally related, to events above (Leveson, 1995). Event trees are also diagrams in a tree structure format that represent events. Unlike fault trees, event trees start with an initiating event and articulate subsequent possible events as comprehensively as possible. These subsequent events are formatted into two branches; the upper branch for successful performance, and the lower for failures. Both fault trees and event trees can be used as the basis for quantitative assessments of risk; both are more suitable for discrete-event systems with fairly simple temporal constraints, and both assume that the system is already quite well defined.

THERP is a comprehensive methodology for assessing human reliability that includes task analysis, error analysis, and quantification of human error probability (where human error probability is defined as the number of errors that actually occurred divided by the number of opportunities for error). The basic idea is to explicitly represent human operator tasks, systematically describe possible errors of omission and commission for each task, and quantify the probabilities of occurrence for each of these on the basis of expert judgments and/or available quantitative data from simulation exercises or real-life incidents.

Work safety analysis (WSA) is a technique to systematically identify hazards and potential corrective measures, particularly for physical operational kinds of work. The work is decomposed into steps, typically with hierarchical task analysis (Kirwan and Ainsworth, 1992). Then for each step, hazards and associated causative factors are described on the basis of expert judgment and available data. Next, each hazard is classified in terms of its relative probability [on a five-point scale from 0 (hazard eliminated) to 5 (very probable)] and severity of consequences [on a five-point scale from 1 (insignificant; only first aid required) to 5 (very serious; over 300 days of disability)], and its relative risk is calculated by multiplying probability by consequences. Finally, for each hazard, corrective measures are described. Work safety analysis can thus be used to systematically identify the most cost-effective and high-impact improvements in safety.

Failure modes and effects analysis (FMEA) grew out of hardware reliability engineering. Traditionally, the technique tabulates probability of failure in terms of system events or components and modes for each of the specified failures (e.g., an amplifier's failure modes may be open, short, or other; Leveson, 1995), tabulates percentage of failures by mode, and categorizes these failure modes as critical or noncritical (Leveson, 1995). FMEA for human reliability analysis has also been described as very similar to event trees with more elaboration of consequences and corrective measures for possible error types (Kirwan and Ainsworth, 1992).

HAZOP was originally developed in the 1960s by the British chemical industry. It is a "qualitative technique whose purpose is to identify all deviations from the design's expected operation and all hazards associated with these deviations" (Leveson, 1995). It is a very systematic technique, in which particular "guidewords" are used to organize discussion of potential hazards (e.g., MORE refers to "more of any physical property than expected," such as higher temperature, more viscosity; REVERSE refers to a logical opposite, such as backflow instead of frontflow). Data for HAZOP includes conceptual flowcharts and preliminary layout diagrams for preliminary HAZOP and control logic diagrams, piping and instrumentation diagrams, plant layout, draft procedures, manuals, and more for more detailed HAZOP. However, the later that HAZOP is carried out, the less impact its results will have on plant design.

A variety of practical guidelines and principles have been developed to prevent or minimize the impact of human error in systems engineering. Indeed, much of the entire field of human factors and human–machine systems engineering addresses this issue; for example, guidelines for the design of displays, controls, training, and standard operating procedures (e.g., Sanders and McCormick, 1987). Part of the rationale for the engineering design process itself is to provide safe and reliable systems; for example, redundant back-up systems, fault-tolerant computing, and comprehensive integration and testing procedures.

With respect to our example of operation and maintenance of a car, let us consider the case of the car running out of gas while being driven in the city. This might have happened because the driver failed to notice that the gas gauge showed that the gas tank was nearly empty or failed to plan to purchase more gas in time (error of omission, error in detecting a fault, or an error in planning). Or, the gas gauge itself might be broken; or perhaps environmental circumstances led the driver to not purchase gas (e.g., being lost in an unfamiliar neighborhood). With respect to human factors design, perhaps the gas gauge was not readable or clear, or there was no warning system to alert the driver that the gas tank was nearly empty.

The hazards associated with running out of gas include possible damage to the car itself (its engine), possible injury to the driver and passengers, and damage to car due to accidents (e.g., getting rear-ended by other vehicles on the road), and the inconvenience and expense of getting more gas for the car (e.g., getting the car towed to a service station or walking to a service station).

An example of a partial fault tree for this situation can be seen in Figure 19.1.

The Three Mile Island example also illustrates errors of omission (missed signals) and commission (incorrect actions taken on the basis of mistaken intentions).

19.3.2 Cognitive Mechanisms and Human–System Interaction

Hollnagel (1993) distinguishes between phenotypes and genotypes for error analysis. *Phenotypes* refer to overt manifestation in behavior; in other words, the phenomenological categories described in the previous section. *Genotypes* refer to cognitive mechanisms that contribute in part to the appearance of phenotypes. Such cognitive mechanisms are the subject of this subsection.

A variety of frameworks have been articulated for such cognitive error analysis. An implicit or explicit thread running through most of this work is the view of the human actor as having "limited rationality" or "limited cognitive resources." The person has limits of perception, memory, attention, reasoning, and motor response; these limited resources contribute in part to human error. Natural cognitive strategies seek to economize, to

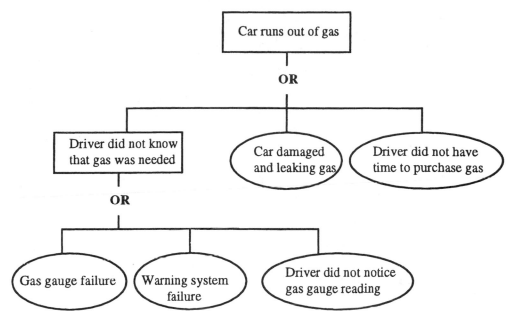

Figure 19.1 Partial fault tree for the event "car runs out of gas." Circles denote basic events that are not decomposed further; boxes represent events that should be decomposed into more basic events and conditions.

minimize resource usage, to take shortcuts and establish routines, to cope with task and environmental demands. Such "cognitive economy" can also lead to systematic errors.

One tradition of work focuses on judgment and decision-making tasks and explicates apparent ways in which human judgment is systematically biased or "errorful" (Kahneman et al., 1982). For example, in estimating the frequency of events, people apparently employ the "availability heuristic," in which the frequency or probability of events is assessed by the ease with which instances are brought to mind (Tversky and Kahneman, 1974). This heuristic often works, but there are other factors that influence availability in addition to frequency and probability. For example, salience ("impact" or noticeability) and recency also contribute to availability. Hence, salient or recent classes of events are perceived as more frequent or probable than they really are.

A second tradition of work that is the focus of this section is generative cognitive mechanisms for human error in work tasks beyond simple "one-shot" judgment tasks. Rasmussen (1986), Reason (1990), and Norman (1988) have proposed cognitive models that systematically organize discussions of human cognitive generative mechanisms for error. As noted in the Introduction to this chapter, an important part of many of these models relates to goals and intentions: mistakes relate to incorrect intentions, and slips or lapses refer to incorrect execution and/or storage of an action sequence (regardless of whether the intention was correct) (Reason, 1990). *Slips* are potentially observable, overt actions; *lapses* refer to covert error forms such as failures of memory.

Sage (1992) provides the following interpretation of Reason's taxonomy that distinguishes between mistakes, lapses and slips:

```
Human action is performed.
Was there a prior cognitive intention?
        IF YES: were action plans implemented properly?
                IF YES: Did actions accomplish intent?
                        IF YES: Success
                        IF NO: Intentional and mistaken act of poor
                               planning
        IF NO: Was unintended act cognitive?
                        IF YES: Mental, memory lapse
                        IF NO: Physical, action slip
        IF NO: Was the action itself intentional?
                IF YES: Spontaneous or subsidiary act
                IF NO: Nonintentional involuntary act
```

Norman (1988) distinguishes between a variety of slips, including capture, mode, description, data-driven, loss-of-activation, and associative activation errors. His distinctions are based on cognitive mechanistic explanations (i.e., particular failures of attention and memory).

The generic error modeling system (GEMS) (Reason, 1990) relies on Rasmussen's framework of skills, rules, and knowledge (Rasmussen, 1983, 1986) to articulate error types. The skills, rules, and knowledge (SRK) framework distinguishes reasoning used in highly routine and practiced situations (skill-based), familiar tasks in which if–then rules are used (rule-based), and novel situations that require more in-depth reasoning (knowledge-based). Reason (1990) summarizes error at these levels in the following way:

> Errors at . . . [the skill-based] level are related to the intrinsic variability of force, space, or time coordination. . . . Errors [at the rule-based level] are typically associated with misclassification of situations leading to the application of the wrong rule or with the incorrect recall of procedures. . . . Errors at . . . [the knowledge-based] level arise from resource limitations ("bounded rationality") and incomplete or incorrect knowledge. (Reason, 1990, p. 43)

With respect to the latter, Woods (1990) further distinguishes between missing or incomplete knowledge, "buggy" knowledge (i.e., incorrect knowledge), and inert knowledge. The latter refers to knowledge that fails to get activated in the appropriate context; the person "has" the knowledge, but cannot make use of it when needed.

Sage (1992, pp. 563–564) provides a summary of four categories of error and associated design guidelines:

1. Errors related to learning and interaction— allow errors that can be learned from
   - Make acceptable performance limits visible
   - Visible feedback
   - Reversible actions
   - Predictive displays/feedback
   - Consistent unique natural mappings between signs, symbols
   - Simulate and evaluate

2. Interference among competing cognitive control structures—strategies
   - Provide overview displays
   - Integrated displays
   - Support formal knowledge-based reasoning
3. Lack of cognitive and physiological resources—so automate
   - Data and information integration
   - Representation design
4. Intrinsic human variability
   - Memory aids and representation design
   - Support for standard procedures

Many human factors and user interface design guidelines have similar themes (e.g., Sanders and McCormick, 1987; Wickens, 1992). Another relevant body of research is on how to design intelligent support systems or "joint cognitive systems" to work with human operators in complex dynamic worlds (e.g., Woods, 1986). Supporting effective human–computer cooperative problem solving can be organized around high-level principles such as "be sure the human is ultimately in charge of the system" and "support mutual intelligibility between human and computer" (by, for example, making the state of the computer's reasoning visible to the human) (see Jones and Mitchell, 1995; Jones and Jasek, 1997).

With respect to our simple example of a car running out of gas, a variety of cognitive mechanisms could be used to account for driver behavior, including failures of attention that led to the gas gauge being ignored (a type of slip) or overconfidence that there was still enough gas in the tank to get to a service station even if the gauge showed "zero" (related to a mistake of poor planning). Related to human factors design guidelines, the lack of a warning system, or the lack of salience of cues that would alert the driver to the situation, could also be contributing factors. In the case of advanced intelligent transportation systems, it is easy to imagine a situation in which many complex displays might be competing for the driver's attention, and again leading to missed cues for an empty gas tank.

In Three Mile Island, the human operators were faced with an incredibly complex cognitive task that required them to do a lot of integration of information and inference to assess the current state of the plant. Furthermore, the layout of the control room and the display formats used did not support operator problem solving. For example, as noted in Leveson (1995), some instruments went off scale during the incident, thus depriving the operators of necessary information in order to make decisions; there was no direct indication of the amount of water turning into steam; over 100 alarms went off in the first 10 seconds of the incident; and relevant controls and displays were not grouped together to support quick and effective action by the operators.

In contrast to the approaches discussed in this section, a "situated" approach to system design for humans is proposed in Greenbaum and Kyng (1991). Such an orientation differs from traditional systems engineering and cognitive psychology approaches in that it is oriented toward actual practice, concrete contexts of work, and relationships. In particular, it emphasizes situations and breakdowns over problems; social relationships over information flow; knowledge over tasks; tacit skills over describable skills; mutual competencies over expert rules; group interaction over individual cognition; and experience-based work over rule-based procedures (Greenbaum and Kyng, 1991, p. 16). Part of what this view implies is the importance of considering human communication and social relationships in the work context. Human communication is the topic of the next section.

19.3.3 Human Communication

The study of human communication has a long and rich history. With respect to systems engineering of complex sociotechnical systems, interpersonal interaction is of interest from several points of view: (1) as a contributing factor in accidents (e.g., miscommunication between pilot and air-traffic control); (2) as a facet of distributed cognitive systems in which group work in fundamental to the task at hand; and (3) as carrying implications of power and control that intersect with issues of organizational sources of error.

An influential model of communication is the "code model" (Sperber and Wilson, 1995): a speaker encodes a message and transmits it over a channel to a receiver, who then decodes and recognizes or interprets the message (Cherry, 1965). In this framework, we can characterize a number of error types with respect to our previously discussed phenomenological types:

- The speaker encodes the wrong message (a mistake).
- The speaker does a poor job of encoding the message (poor choice of words or symbols, missing words, etc.; errors of commission, omission, etc.).
- The channel does not carry the message properly (e.g., insufficient bandwidth, noise, and so on; an exogenous error).
- The receiver decodes the message incorrectly.
- The receiver misunderstands the message.

This superficial analysis points to a number of important issues. A common distinction in communication research is between syntax (the surface form of a message), semantics (the meaning of a message), and pragmatics (the goals to be achieved by saying the message). Most of the preceding "engineering" communication analogy focuses on processes relating to syntax and semantics. Certainly, incorrect syntax can lead to messages that cannot be understood by the receiver. Part of the issue of semantics has to do with common knowledge of what words refer to; establishing the referent and using consistent terminology for referents is another prerequisite for effective communication. Hence, errors in communication may arise because of differences in vocabulary and lack of agreement about the referent.

However, as Sperber and Wilson (1995, p. 9) succinctly state, "there is a gap between the semantic representation of sentences and the thoughts actually communicated by utterances." Sperber and Wilson (1995) use Paul Grice's approach to meaning as a starting point for an alternative model of communication that is oriented around processes of inference. Sperber and Wilson present Strawson's reformulation of Grice's original work (Sperber and Wilson, 1995, p. 21):

To mean something by X, the Speaker must intend

(1) that his or her utterance of X will produce a certain response R in a certain Audience

(2) that the Audience will recognize the Speaker's intention of (1)

(3) that the Audience's recognition of the Speaker's intention (1) will function as at least part of the Audience's reason for its response R

Part of Grice's contribution here was to argue that recognition of the speaker's intention is sufficient to enable communication; human inferential processes make communication possible even without a code. Sperber and Wilson (1995) argue that both the code model and inference model are useful to theories of human communication and provide the notion of relevance as central to their integrative theory.

With respect to implications for human error, the inferential model suggests that errors in inference making will contribute to miscommunication. In particular, the speaker may not have intended response R or may have designed an utterance for a different audience; the audience may not recognize the speaker's intention of an intended response R; and the audience may not use this intention as part of designing its response.

An explanatory model of the inferential model of communication is suggested by Grice's work on the "Cooperative Principle" and associated maxims of cooperation (Sperber and Wilson, 1995; Mura, 1983). The "Cooperative Principle" is a general principle that conversational participants are expected to observe: "Make your conversational contribution such as is required, at the stage at which it occurs, by the accepted purpose or direction of the talk exchange in which you are engaged" (Grice, 1975, cited in Sperber and Wilson, 1995, p. 33). Grice proposed four associated maxims of quantity, quality, relation, and manner. Mura (1983) summarizes these maxims respectively, as "be succinct, yet complete," "be truthful," "be relevant," and "be clear and orderly." Thus, effective communication means that participants orient themselves to each other and rely on this principle and associated maxims as aids in inferencing.

At first glance, we may consider violations of these maxims as categories of errorful or "uncooperative" behavior in human communication: being long-winded or leaving out important information, lying, being irrelevant, and being unclear and disorganized. Participants may mislead others by "quiet" violations of maxims or may blatantly violate one or more; they may "opt out" of the interaction if they do not want to be cooperative; and most interestingly, they may have to make tradeoffs among maxims and sacrifice one for another (Grice, cf. Sperber and Wilson, 1995). In this latter case, participants may engage in "metatalk" to cue hearers of their intention; for example, the Speaker may say "Sorry it's going to take me a while to explain this, but I think it's important" as a way of indicating that quality is more important than quantity. By drawing explicit attention to these violations, the speaker is in effect asking for license to do so, and thus reestablishing a cooperative relationship to the audience (Mura, 1983).

More generally, "alignment talk" refers to conversational strategies to repair misunderstandings and provide appropriate framing to other participants (Ragan, 1983). This includes categories of talk such as accounts, disclaimers, "motive talk," and other remedial processes. Ragan (1983) argues that alignment talk can be viewed more generally as a way to manage conversations and maintain conversational coherence.

Amelioration of human error in this perspective focuses on the explication, documentation, and training of standard terms, definitions, and communicative practices; and providing flexible support for collaboration through technology (e.g., chat boxes, telephone, audio).

Returning to our Three Mile Island example, human communication issues were certainly pertinent. Operators needed to collaborate in problem-solving tasks, and as the incident developed, other engineers and managers were in the control room, discussing and debating the current status and future actions of the situation.

Even more pertinent to this issue are the aviation accidents that are documented by the National Transportation Safety Board. Conversations between the cockpit and air-traffic control, and talk within the cockpit itself, have been cited as contributing factors in accidents, partly due to the ambiguity of referential terms and the ambiguity, for example, of air-traffic control repeating a request from the cockpit—which could be interpreted as simply an acknowledgment or as permission to carry out the request.

19.3.4 Sociocultural and Organizational Perspectives

Another critical aspect of human error analysis and amelioration has to do with social, cultural, and organizational issues. There are a number of ways in which social norms in the workplace, organizational culture, structure and practices, and the like can encourage a "safety culture," including systematic and honest evaluation of hazardous situations, effective response to emergency situations, and effective design and training.

Communication, information flow, and organizational structure are three interrelated aspects of the organization of particular importance (Grabowski and Roberts, 1996). Effective communication is needed for the organization to maintain its own knowledge of safety and human error issues. Information needs to be distributed to relevant people in order to do effective problem solving. The organizational structure may encourage or discourage information sharing and communication.

Leveson (1995) describes three classes of root causes of accidents: flawed safety culture, flawed organizational structures, and ineffective technical activities. In particular, the safety culture of an organization or industry may be defective in several major ways: (1) overconfidence and complacency, leading to practices and thought patterns that discount risk or assess risk unrealistically, place too much reliance on redundancy, ignore high-risk, low-probability events, assume that risk decreases "by itself" over time, and ignore warning signs; (2) low priority assigned to safety (as realized in particular by personnel and financial resources); and (3) poor resolution of conflicting goals, in particular placing safety as a low-priority goal and not working hard to meet both safety and other goals.

Ineffective organizational structures are also a barrier to an effective safety culture. If responsibility and authority for safety are diffused throughout the organization without a central coordinating function, problems can easily arise (Leveson, 1995). If safety personnel are low status and are not independent of the projects they are supposed to be evaluating, that can lead to poor "watchdog" behavior. Lastly, as noted earlier, poor communication and information sharing contribute to a flawed safety culture.

Finally, Leveson describes a range of poor safety engineering practices, including superficial safety efforts, ineffective use of the knowledge of risks, poor design, increased complexity of the system, failure to evaluate safety after changes are made, and poor documentation and use of information.

Thus, obvious guidelines for good safety engineering practices include (1) a central coordinating safety function, (2) high-status and independent safety personnel, (3) taking safety seriously throughout the organization, (4) having safety professionals be partners rather than adversaries, (5) good documentation and information dissemination practices, and (6) continuous monitoring and improvement of health and safety issues. Several techniques for making explicit the organizational impacts on error exist, including the management oversight risk tree (MORT) technique (Kirwan and Ainsworth, 1992).

As a final note, Morgan (1986) discusses organizations from a variety of metaphorical perspectives: organization-as-information-processing system, organization-as-brain, organization-as-machine, organization-as-culture. Each perspective highlights certain features of organizational life and has implications for how we might view organizational error. For example, in information-processing terms we may be concerned with the speed and dissemination of information in an organization; in biological terms we may be concerned with maladaptation to environmental influences; and in cultural terms we may look at power, authority, and other political or family issues that inhibit effective performance (also see Star, 1995; Latour, 1996; Perrow, 1984; Vaughan, 1996). As an example, a sociologist's analysis of the *Challenger* disaster highlights a cultural view of NASA and a social constructivist view of risk (Vaughan, 1996). For example, Vaughan's analysis of the decision-making process about the solid rocket booster joint argues that the work group (Vaughan, 1996, p. 65):

> normalized the deviant performance of the SRB joint . . . the behavior that the work group first identified as technical deviation was subsequently reinterpreted as within the norm for acceptable joint performance, then finally officially labeled an acceptable risk . . . Risk had to be renegotiated. The past—past problem definition, past method of responding to the problem—became part of the social context of decision making

Returning to our driving and Three Mile Island examples, social, organizational, and cultural factors are potentially at work as well. Coping with crying children in the car, hurrying to work or on a business trip, or being distracted by the social situation while driving may all form part of the context to explain why a driver may fail to notice that the gas gauge reads "empty." (On the other hand, the potential social embarrassment of running out of gas could motivate the driver to avoid this situation!) As summarized in Leveson (1995), contributing factors to the Three Mile Island accident included management complacency, inadequate training and procedural specification, lack of a good system for disseminating "lessons learned," and deficiencies in the licensing process.

19.4   CONCLUSIONS

The study of human error and its amelioration is a never-ending social and technical challenge for systems engineering (Nieves and Sage, 1998) and many other areas. This chapter has provided broad outlines of approaches to consider in the analysis of human error: starting with simple labels, progressing to more in-depth analysis of cognitive and communicative competencies, and finally to a broader view of the social, organizational, and cultural context. A comprehensive approach to the study and amelioration of human error is an ongoing multidisciplinary effort among engineers and social scientists that puts together both traditional, formal systems engineering approaches and rich, "thick" ethnographic descriptions of the work context to look at all these perspectives and continuously try to improve these facets of working life.

TABLE 19.2   Summary of Practical Systems Engineering Guidelines

Systems Engineering Phase	Associated Guidance
System Definition	• Establishment of Safety Culture, including: safety considerations are a visible and explicit part of the system definition process; safety, failure, and human error analysts are part of the design and deployment team; establishment of documentation procedures in terms of safety, risk, and human error; safety reporting system clear lines of communication clear and cooperative organizational roles training and personnel selection issues • Function allocation to minimize human error possibilities: automate routine tasks that are likely candidates for slips • Automation philosophy to keep humans in the loop and aware of the current and predicted state of automation:
System Development	• Detailed analyses of proposed human operator activity in terms of risk, reliability, human error, cognitive, and physical resources required • Detailed analysis of planned communication between human operators (managers/workers, between shifts, etc) to develop standard vocabulary, procedures, etc. • Articulation of design criteria with respect to human error, safety, and reliability • Use of human factors guidelines and checklists to critique system design to minimize/eliminate human error • Hazard analysis to assess potential impacts of human error • Plans for data collection and reporting (audit program)
System Deployment	• Revisit hazard analysis, pilot empirical studies of human error in context of activity • Data collection for human error • Verification and validation of system with human input • Audit program • Training • Further development of standard and contingency or emergency operating procedures
System Operation and Maintenance	• Configuration Control • Continued hazard analysis, data collection, training • Iterative human factors analysis of system with new activities • Keeping documentation up to date • Audit program/reporting system • Continued development of emergency and contingency procedures • Articulation of "lessons learned" for future reference

Systems engineering recommendations for human error amelioration, in the context of a traditional systems engineering process, are shown in Table 19.2 (partially based on Leveson, 1995). In a sense, this table reorganizes the guidelines outlined in Table 19.1 in the context of the typical systems engineering processes that are enacted by most organizations.

ACKNOWLEDGMENTS

The author gratefully acknowledges the editorial assistance of Andrew Sage and William Rouse and the comments of anonymous reviewers in the laborious writing of this chapter.

REFERENCES

Cherry, C. (1965). *On Human Communication,* (2nd ed.). Cambridge, MA: MIT Press.

Grabowski, M., and Roberts, K. H. (1996). Human and organizational error in large scale systems. *IEEE Trans. Syst., Man, Cybernet.* **26**(1), 2–16.

Greenbaum, J., and Kyng, M. (eds.). (1991). *Design at Work: Cooperative Design of Computer Systems.* Hillsdale, NJ: Erlbaum.

Grice, H. P. (1975). Logic and Conversation in *The Logic of Grammar* (D. Davidson and G. Harmon, eds.) Encino, CA: Dickenson, pp. 64–75.

Hollnagel, E. (1993). *Human Reliability Analysis: Context and Control.* San Diego, CA: Academic Press.

Jones, P. M., and Jasek, C. A. (1997). Intelligent support for activity management (ISAM): An architecture to support distributed supervisory control. *IEEE Trans. Syst., Man, Cybernet.,* **27**(3, Spec. Issue), 274–288.

Jones, P. M., and Mitchell, C. M. (1995). Human-computer cooperative problem solving: Theory, design, and evaluation of an intelligent associate system. *IEEE Trans. Syst., Man, Cybernet.,* **25**(7), 1039–1053.

Kahneman, D., Slovic, P., and Tversky, A. (eds.). (1982). *Judgment under Uncertainty: Heuristics and Biases.* Cambridge, UK: Cambridge University Press.

Kirwan, B. (1994). *A Practical Guide to Human Reliability Assessment.* London: Taylor & Francis.

Kirwan, B., and Ainsworth, L. (eds.). (1992). *A Guide to Task Analysis.* London: Taylor & Francis.

Latour, B. (1996). *ARAMIS: Or the Love of Technology.* Cambridge, MA: Harvard University Press.

Leveson, N. (1995). *Safeware: System Safety and Computers.* Reading, MA: Addison-Wesley.

Morgan, G. (1986). *Images of Organization.* Beverly Hills, CA: Sage Publ.

Mura, S. S. (1983). Licensing violations: Legitimate violations of Grice's conversational principle. In *Conversational Coherence: Form, Structure, and Strategy* (R. Craig and K. Tracy, eds.), pp. 101–115. Beverly Hills, CA: Sage Publ.

Nieves, J. M. and Sage, A. P. (1998). Human and organizational error as a basis for process reengineering: with applications to systems integration planning and marketing. *IEEE Trans. Syst., Man, Cybernet.* **SMC-28**, 742–762.

Norman, D. (1988). *The Psychology of Everyday Things.* New York: Basic Books.

Perrow, C. (1984). *Normal Accidents: Living with High-risk Technologies.* New York: Basic Books.

Ragan, S. (1983). Alignment and conversational coherence. In *Conversational Coherence: Form, Structure, and Strategy* (R. Craig and K. Tracy, eds.), pp. 157–173. Beverly Hills, CA: Sage Publ.

Rasmussen, J. (1983). Skills, rules, and knowledge: Signals, signs, and symbols and other distinctions in human performance models. *IEEE Trans. Syst., Man, Cybernet.* **SMC-13,** 257–267.

Rasmussen, J. (1986). *Information Processing and Human-machine Interaction: An Approach to Cognitive Engineering.* New York: Wiley.

Reason, J. (1990). *Human Error.* Cambridge, UK: Cambridge University Press.

Roth, E. et al. (1994). *An Empirical Investigation of Operator Performance in Cognitive Demanding Simulated Emergencies.* (NUREG/CR-6208). Washington, DC: Westinghouse Science and Technology Center, Report prepared for U.S. Nuclear Regulatory Commission.

Sage, A. P. (1992). *Systems Engineering.* New York: Wiley.

Sanders, M., and McCormick, E. (1987). *Human Factors in Engineering and Design,* (6th ed.). New York: McGraw-Hill.

Senders, J., and Moray, N. (1991). *Human Error: Cause, Prediction, and Reduction.* Hillsdale, NJ: Erlbaum.

Sperber, D., and Wilson, D. (1995). *Relevance: Communication and Cognition* 2nd ed. Oxford: Blackwell.

Star, S. L. (ed.) (1995). *Ecologies of Knowledge: Work and Politics in Science and Technology.* New York: State University of New York Press.

Swain, A., and Guttman, H. (1983). *Handbook of Human Reliability Analysis: With Emphasis on Nuclear Power Plant Applications.* (NUREG/CR-1278) Washington, DC; Nuclear Regulatory Commission.

Swets, J. (1996). *Signal Detection Theory and ROC Analysis in Psychology and Diagnostics: Collected Papers.* Mahwah, NJ: Erlbaum.

Tversky, A., and Kahneman, D. (1974). Judgment under uncertainty: Heuristics and biases. *Science* **185,** 1124–1131.

Vaughan, D. (1996). *The Challenger Launch Decision: Risky Technology, Culture, and Deviance at NASA.* Chicago: University of Chicago Press.

Wickens, C. D. (1992). *Engineering Psychology and Human Performance,* 2nd ed. New York: HarperCollins.

Woods, D. D. (1986). Cognitive technologies: The design of joint human-machine cognitive systems. *AI Mag.* **6,** 86–92.

Woods, D. D. (1990). Modeling and predicting human error. In *Human Performance Models for Computer-aided Engineering* (J. Elkind, S. Card, J. Hochberg, and B. Huey, eds.), pp. 248–274. San Diego, CA: Academic Press.

20 Culture, Leadership, and Organizational Change

JOYCE SHIELDS, CHARLES S. HARRIS, and BETTY K. HART

20.1 INTRODUCTION

20.1.1 A Behavioral Approach to Change

Change is in; stability is out. From large companies and nonprofit organizations to federal, state, and local governments, there is a broad array of efforts underway to improve the way organizations operate. These innovations range from reengineering to reorganization and from downsizing to strategic planning. Despite the general recognition of the merits of organizational improvement, surveys reveal that the overwhelming majority of organizations that have undergone systematic change efforts have been dissatisfied with the results. One particularly egregious approach has been the tendency to equate "downsizing" with "reengineering."[1] While a changed business process that reduces the number of steps or phases in the process can lead to workforce reduction, initiating change with the goal of downsizing may result in a less effective organization.

Recently many analysts, consultants, and change leaders concluded that failure to make successful change happen results from inadequate attention to the "people side" of reengineering and related organizational improvement efforts (Champy, 1995; Kotter, 1995; Kriegel and Brandt, 1996; Steininger, 1994; Price Waterhouse, 1996). All too often reengineering efforts focus on tangible, visible processes and technologies and do not explicitly consider people's capabilities for working and managing in new ways. This approach to organizational change we term the *structural* model. The *behavioral* model, in contrast, defines the future work environment in terms of the changed behaviors that will be required of participants; those who will need to, make the process, technology, and organizational changes happen. The behavioral approach to change, which is the focus of this chapter, is critical to implementing organizational improvements and realizing performance improvements. As James Champy (1995) observes:

> It's only now, when businesses are recognizing that their dependency is on people—especially as they reengineer their work processes—that they find themselves forced to get serious about their values and behavior of all their employees.

---

[1]William Safire's "On Language" column in the New York Times Magazine examined "downsize" as a euphemism for layoffs and dismissals. Safire concluded: "All these [terms] are more vivid than *downsize*, which in turn has more blood in its veins than the recent *restructure* and *re-engineer* . . ." (Safire, 1996).

Handbook of Systems Engineering and Management, Edited by A. P. Sage and W. B. Rouse
ISBN 0471-15405-9   ©1999 John Wiley and Sons, Inc.

A premise implicit in many reengineering and other business process and management models is the concept of a Cartesian or mechanistic universe. Such an approach presumes a rational model of organizations and, in turn, of human behavior. One could characterize the approach as tinkering with organizations so that they run like clockwork. Such a model is reductionist—individuals become another part of the larger mechanism (the organization). Also absent is any sense of the importance of culture or of the political, social, and economic environment within which both organizations and individuals operate.

20.1.2 Chapter Overview

This chapter presents a model of change and outlines a strategy for implementing organization-wide change at the highest levels of public and private organizations. The approach builds on the premise that when prior efforts at implementing major organizational change have failed, a common cause is too little attention to the *people* side of reengineering. Change implementation is reduced to a mechanistic model, the extent of built-in resistance has been underestimated, and the role of the external environment in sustaining change has been ignored (Barttett and Ghoshal 1994, 1995; Beckhard and Pritchard, 1992; Goldman et al., 1995).

Section 20.2, "Setting the Context: Culture," explores the social landscape in which individuals work. The discussion builds on the premise that work environments shape individual and group behaviors. Such environments can take several forms. The functional or traditional model emphasizes specialization and hierarchical authority. The newer forms, including process, time-based, and network or matrix cultures, stress speed, flexibility, integration, and innovation.

Section 20.3, "The Role of Leadership," highlights the part that individuals play in transforming organizations. Indeed, effective leadership at the top is the most common factor that distinguishes organizations that are successful in implementing desired change from those that are not. Analysis of leadership styles suggests that there are common sets of competencies that are shared by successful leaders. Furthermore, leaders need to be savvy in their manipulation of their organization's levers, including personnel, communication, information management, and measurement systems.

Section 20.4, "Applying the Change Model," details the process change leaders can employ to achieve desired change. It details *how* the change model can be implemented through a three-phase, seven-step process. The process stresses the behavioral side of change and is designed to complement any ongoing reengineering efforts. It outlines the ways leaders communicate, convince, and move their people to new work behaviors, organizational concepts, and practices. Leaders then use the various organization levers to sustain the change. Among the major levers are values and culture, management processes and systems, and reward and recognition.

Section 20.5, "Profiles in Change," presents three case studies of organizations that have achieved a new direction and culture, realigned organizational levers, and created a cadre of motivated employees. A charismatic visionary leader transformed GE into a world-class, highly competitive organization; the Department of the Army overcame bureaucratic inertia to develop a new, comprehensive management system; and the Defense Printing Service implemented a new vision and culture that replaced the traditional typographical-based production process. In selecting these three examples we have purposely included two public-sector ones, because such successful efforts are often overlooked in favor of corporate cases.

In the *Conclusion* we observe that, while the challenges to successful change occur in all three change phases, they are most significant at phase one—design—and phase two—deploy. In the former case, it is top leadership that must initiate major change (Beckhard and Pritchard, 1992; Kotter and Heskett, 1992; Kouzes and Posner, 1995). They must agree on what the vision comprises and then translate that into a message that resonates with the core values of the men and women who will ultimately be responsible for making the vision real. Among the key factors important in the case of deploy are the time and resources required to accomplish an organization-wide transformation. Such a wholesale change necessitates developing new roles and competencies, transforming communication practices, developing training and development programs, and creating new evaluation and compensation procedures.

20.2   SETTING THE CONTEXT: CULTURE

Drawn from anthropology, culture refers to the relatively enduring set of values and norms that underlie a social system (Burke and Litwin, 1992). Passed from one generation to the next, a culture is slow to develop and not readily amenable to change. Culture also operates on at least two levels. At the deeper and less visible level, it is constructed around the values shared by a group that persist even when the group's membership changes. If a group culture emphasizes security and predictability over risk taking and innovation, substituting new values for old may be a major effort.

At a more visible level, culture represents the patterns of behavior that new employees are expected to adopt for doing work. We term this visible level of culture *work culture*. Since participants are conscious of these cultural elements, they are somewhat more malleable than their less visible counterparts (Kotter and Heskett, 1992).

20.2.1   Culture as a Dynamic

Beliefs generate behavior. Environmental constraints and opportunities also shape and constrain the customary ways in which beliefs are demonstrated through behavior. When revolutionary changes occur, business practices (and legal and regulatory requirements) lag in taking advantage of the changes. Initially, the arrival of the "iron horse" in the United Kingdom was accompanied by legal requirements necessitating a man walking down the track in front of the "horse" (steam engine) swinging a lantern. New technologies require a conscious reconsideration of the way business is done, in order to increase organizational capability and effectiveness.

The initial growth of large-scale manufacturing and the revolutionary application of interchangeable parts helped to influence the requirements for the new "industrial" man, who had to work by the clock, under the clock, and in a controlled environment—very different from the hand work and cottage industries predominant prior to the application of steam power to manufacturing.

Information technology represents the second great power revolution impacting on the world. Just as the steam engine generated a new industrial structure and increased productivity manyfold, so will the power of information technology play a key transformative role in corporate and public enterprises. Distributed access to information implies distributed decision making rather than centralized, hierarchical management in stovepiped processes.

The information-technology revolution, particularly in the last decade, facilitates a new strategic approach—a new mind-set-characterized by its constant attention to changing contexts and requirements combined with an ability to prepare and respond immediately. The telecommunications revolution has accelerated not only the amount of information available, but the speed of transmission and the rapidity with which it can be arrayed (Negroponte, 1995; Goldman et al., 1995). Such a revolution has affected the entire environment and introduced increasing turbulence. Wall Street analysts have commented since the 1980s that the financial markets can move as much in a day as previously in a month or even a year. This compression of market behavior has emerged from the telecommunications revolution.

Information networking is the antithesis of hierarchical power. Organizational agility is dependent on distributed information and decision making in order to make the frequent adjustments required by the increasingly turbulent environment. Enterprise integration includes discovering a shared strategic direction carried on throughout and down the organization; integration of internal functional organization and external partners; and development of management of end-to-end processes that cross functional and organizational boundaries. Such integration requires establishment of a cooperative culture and involves complex interaction between technological and behavioral enablers.

20.2.2 Corporate and Government Cultures

The mechanistic approach to the universe and the emerging industrial revolution spanning the 18th and 19th centuries ignored culture. Its underlying philosophic premises supported the growth of large corporations and the professionalization of government through introduction of merit principals in lieu of patronage in the industrializing Western economies.

The emergent model for success, which reached its zenith during the first decades of the 20th century, can be termed the *Functional Work Culture.* This classic 19th century industrial concept derives from a time when there was an emphasis on control, conformity, and continuity. Frederick Taylor and other management theorists in the early 20th century continued this ultimately rational approach to organizations by positing "scientific management" as the solution to business problems.

Government agencies, particularly, have reflected the machine model. Nineteenth and early 20th century reformers' response to government inefficiencies, fraud, and nepotism was to create a system of rules, regulations, and standards. Although "merit principles" replaced nepotism, the top layers of government hierarchies are still staffed by political appointees at both the federal and state level (Holden, 1991; Mintzberg, 1996; Osborne and Gaebler, 1992; Wilson, 1989).

Every organization is made up of "hardware" and "software." The hardware of any enterprise consists of its structure, its assets, and its businesses. The software consists of its people and its operating work cultures. Public as well as private organizations are composed of people who have a usual way of doing business (Wilson, 1989). These people—whether professional, technical, administrative, or craft—are subject to common motivations, concerns, and attitudes toward change. Understanding the particular motivations and attitudes of people making up the work force contributes to developing an approach to leading and managing change that is tailored to the organization's requirements and vision.

20.2.3 Four Culture Models

Classifying and describing work cultures provides heuristic models against which to diagnose current cultural attributes and determine preferred new ones.

Traditional management practices conform to the philosophic underpinnings, technological constraints, and educational background of most workers during the 19th and early 20th centuries. Traditional management practices complemented the Functional Work Culture's emphasis on specialization, role clarity, and control, which enabled the creation of economies of scale and the first billion-dollar corporations. The Functional Work Culture was undergirded by a mechanistic view of the universe that also enabled managers to view workers as interchangeable parts; such a perspective encouraged resistance to labor attempts to upgrade working conditions, benefits, and rights. Traditional management—with its emphasis on clearly delineated functional and hierarchical boundaries, combined with limited access to the information necessary to decision making—fails to motivate and energize today's workers.

The shifting paradigm for organizational success is closely related not only to changes in technology, but the vastly changed expectations and skills brought by many individuals into the workplace. New work cultures all emphasize issues related to flexibility, innovation, and integration of efforts, which in turn presuppose very different roles and expectations for everyone in the workforce.

Functional work cultures are driven by the need to accumulate resources, achieve economies of scale (initially linked to standardization and long production runs), and integrate behavior through deep management hierarchies where decision-making authority is clearly delineated. With the emergence of the total quality management movement (TQM), a second organizational model—the Process Work Culture—has evolved. The process model is driven by customer satisfaction and focus on continuous quality improvement. Organizations conforming to this model design work around processes and team efforts. Getting close to the customer requires linkages between customers, the organization's process team, and suppliers. Many retail organizations, notably firms like McDonald's and Federal Express, conform to the process model. Time-based work cultures emerged coincident with the telecommunications advances in the 1980s where the strategic priorities required flexibility and agility as well as reliability and quality in order to maximize the return on fixed assets. Effective management requires a workforce comprises individuals who have multifunctional expertise and competencies. The approach is particularly focused on acquiring the ability to dominate highly changeable markets in their high-profitability phases; the approach requires the agility to quickly integrate technological change or move to another market sector when returns fall. Companies such as Intel and Microsoft are examples of time-based cultures.

The Network Work Culture focuses on responsiveness and flexibility relative to the changing needs of the customer. Work is formed around temporary alliances that bring together the necessary proficiencies to successfully complete a venture. The production of a motion picture would provide a relatively pure model of the network model. Such a model has been inherent in military task forces created for short-term assignments; disaster relief scenarios where a variety of federal, nonprofit, and private organizations come together for an intense, relatively short lifetime of effort; and in the construction industry where a general contractor links a number of subcontractors to complete a project, whether it is residential remodeling or the building of a multistory commercial enterprise. Once the task at hand is completed, the network organization ceases to exist.

The new work cultures all foster capabilities aimed at:

- Taking the lead in changing environments to gain competitive advantage.
- Harnessing new technologies to effectively rework requirements and how people organize to accomplish tasks.
- Exploiting the increasing ability to rapidly aggregate, transmit, segment, differentiate, and act upon information.
- Leveraging individual competencies by pulling together task forces, teams—particular configurations of competencies—to address specific ventures or situations.

These practices radically change customary ways of doing business. They are reflective of the new realities of the emerging postindustrial economy in that manufacturing is no longer the preeminent driver. The Industrial Revolution aggregated labor and capital, generating economies of scale derived from long manufacturing cycles, standardized products, reaching markets made accessible by the advent of steam power. Rather than emphasizing economies of scale only, today's organization success hinges upon the ability of the entire workforce to respond to an increasing demand for customization. Cultural alignment for an integrated enterprise requires a shared mind-set focusing on horizontal integration, mission, customers, and continuously adjusting to new requirements and demands. The traditional functional work culture, organized in a series of functional boxes and in vertical stovepipes, must decentralize and also flatten the hierarchy in order to push decision making closer to action taking. Successful cultural alignment requires sharing information not only laterally and upward but outward to key influencers, customers, and stakeholders as well as in a cascade down through the organization.

While identifying the desired work culture is critical for improving organizational effectiveness, it should be used as a model not a micrometer. Seldom will an organization fall within the parameters of a single model. Most organizations are variations and combinations of the four work-culture models (e.g., a combination of process and network) and large organizations comprise multiple, overlapping cultures. In effect; each work culture has the select characteristics summarized below (Hay Group, 1996):

Functional Work Culture: features uniformity and control of internal functions.

Process Work Culture: looks outward in response to customers.

Time-based Work Culture: emphasizes rapid response to external environmental challenges.

Network or Matrix Culture: emphasizes bringing together temporary configurations of experts and resources to meet specific needs.

When Jack Welch began the organizational transformation of GE, he used a sports analogy to describe the change he envisaged. He described GE as having been organized like a football model whereby everyone has a carefully defined role and plays with the expectation of gaining only a few yards at a time. However, what Welch envisioned was a time-based culture favoring speed and agility—a hockey game played with a free-flowing style and a blurring of individual roles as the play moves at high speed across the rink. Welch envisaged such players slamming opponents into the wall—fast moving and goal-oriented—yet willing to sit down after the game for a friendly drink (Tichy and

Charan, 1995). Welch has always held out the ideal of managing the "quantum leap" or the ability to achieve a decisive goal.

Thomas Flannery, Dave Hofrichter, and Paul Platten from the HayGroup (1996) use succinct sports analogies: football also resembles the prototypical Functional Work Culture in organizational structure. Not only do individual players have separate and distinct roles, there are separate defensive and offensive teams and special teams. While everyone wants to win, they concentrate on their own assignments. The game is also hierarchical—the coaches (the executive team) call the plays; the quarterback (middle management) directs; the team members (employees) execute by functional area.

Basketball, like hockey, reveals a Time-based Work Culture. While team members start out as specialists, their roles blur during a fast-moving game. Cross-functionality is key and requires multiple competencies. While the coaches may be present on the sidelines, a basketball team is self-directed in its effort to achieve more baskets than its competitors within a limited window of opportunity.

In contrast to basketball and hockey, soccer models a Process Work Culture. Unlike basketball, where players move quickly up and down the court focusing on the hoops, soccer team members maintain their positions or roles, so that the outcome can be highly dependent upon individual competencies. In soccer, the team as a whole continually shifts in response to the everchanging demands determined by the position of the ball in relation to the competition.

Baseball truly represents the Network organization, according to the HayGroup (1996). Although baseball is billed as the competition between two teams, it consists of a series of interactions and relationships between two or usually at most, three players. The pitcher throws the ball. The moment the batter hits the ball, the network shifts. The interaction focuses on the runner, the opposing team member who fields the ball, and the first baseman. The rest of the network is standing by. Like many business networks, downtime can be a reality for many of the members. (Many a Little League wanna-be has been picking dandelions when a ball finally makes it to that player's outfield location.) Managers in baseball resemble producers or strategists who pull a team together and provide the game plan, then expect the individual "expert" talents to "do their thing," rather than calling in each play as typical of the football coach (HayGroup, 1996).

To create change, the change leader needs to assess the current work culture—or patterns of work—and determine the most effective and appropriate patterns to support new processes and practices. Greater organizational effectiveness can result from using work-culture models as a diagnostic tool. An array of different techniques and tools to assess and then to close the gap between existing and desired work cultures is available (Ashkenas et al., 1995; HayGroup, 1996; Kouzes and Posner, 1995).

20.3   THE ROLE OF LEADERSHIP

Leaders focus on the future: "The single most visible factor that distinguishes major cultural changes that succeed from those that fail is competent leadership at the top" (Kotter and Heskett, 1992). Our model poses a top-down approach to leadership; initially change is instituted by upper-level leaders, and subsequently is assured by middle- and lower-level leaders. In this way change is embedded throughout the organization and sustained beyond the tenure of a small grouping of leaders. In effect, leaders give voice to the vision for change and motivate employees to participate in the process of transformation. They also serve as behavioral role models for their followers (Burke and Litwin, 1992). The importance of followers' perception of leaders and their competencies are included in the model for several reasons. Most importantly, the leadership phenomenon is a *relational* process, involving two or more participants (Farkas and Wetlaufer, 1996; Handy, 1995; O'Toole, 1996).

Leaders must motivate, and followers, in turn, must want to act in a manner both different from their conventional behavior and consistent with the leader's expectations. Such a scenario is not easily achieved and is beyond the ability of many managers. Indeed, managers maintain organizations; leaders transform them. Leaders are actors on a stage against a cultural backdrop rather than bit players performing rote tasks.

20.3.1 Skills, Competencies, and Actions

There is a generally agreed-upon cluster of attributes associated with successful change leaders. These individuals challenge existing processes or procedures, inspire a shared vision in their staff, enable others to act toward achieving the vision, model the way through their own behavior, and motivate and encourage their followers (Kouzes and Posner, 1995). Such leaders recognize the complexity of human behavior and are able to move employees to act "outside the box." Viewed from another perspective, successful leaders realize that, ultimately, only people transform organizations.

Leaders are individuals charged with making change happen in any organization. To achieve enterprise integration leaders must:

1. Articulate a vision coupled with a sense of urgency.
2. Form a powerful guiding coalition.
3. Commit resources to crossover of a few key technologies and processes.
4. Communicate values and goals up and down the organization.
5. Inculcate integrity throughout the organization.
6. Energize and empower others to make change happen.
7. Plan and create short-term wins.
8. Consolidate and institutionalize improvements and continuing change.
9. Possess the stamina to sustain changes across five or more years.

These generic characteristics help inform our understanding of leadership in general. Successful leadership also involves aligning individual attributes with an organizational culture (Farkas and Wetlaufer, 1996); some leaders are simply more likely to succeed in transforming a particular organization than are others. Typically, change agents seek to move the organization from a *functional* culture with its focus on control, reliability, and an inward orientation to another form. The leader's goal may be to create a *time-based, process,* or *network* culture. A successful leader must thoroughly understand component elements of the new work culture, make a commitment to change, and develop a strategy for achieving a "new way of doing business." This individual must understand the current organization and its culture. Some observers maintain that some degree of insider knowledge is essential if change is to succeed (Kotter and Heskett, 1992). Successful leadership, from this perspective, can be defined as the process of mobilizing others to achieve a work culture that supports the desired vision and strategy.

Hay/McBer (1995) recently completed research involving over 60 multinational CEOs and key leaders. Based on the research, Hay identified two distinct systems of competencies working in the most successful leaders. These are *international adaptability* and *universal competencies*. International adaptability is the ability to make deliberate choices about how to conduct business successfully in different parts of the world. Hay/McBer found that the most successful CEOs adapt their leadership styles along three key dimensions. For each dimension the research identified a continuum of business styles. The three key dimensions are:

Building Business Relationships. This continuum reflects the manner in which an executive develops business relationships with strategic partners, suppliers, customers, government officials, and others. It varies from the personal relationship end of the continuum where "trust and mutual respect" are essential to contractual relationships where "the deal's the thing."

Basis for Action. The study found two opposing cultural styles for carrying out actions, planning vs. implementation. At the planning end of the continuum leaders place emphasis on "getting the plan right." At the implementation end leaders focus on "just doing it" by choosing the right people to carry out the work.

Exercising Authority. All successful leaders use authority. The difference found in this research focused on opposing cultural styles. The continuum varies from centralized authority where a leader "leads through self" vs. participatory authority where the leader "leads through others."

The most successful international leaders were those who could master and adapt the universal competencies to different requirements in different parts of the world. Leaders fail when they are unable to identify and use the specific sets of competencies required for their organization and the culture in which their organization operates (Hay/McBer, 1995).

The research found that the dimensions of international adaptability combined with three universal competency clusters are necessary for sustained organizational success. The universal competency clusters necessary to CEO and key managers include:

Sharpening the Focus. Successful CEOs demonstrate strength in three areas: (1) broad scanning, or the ability to quickly assess changing conditions to include recognizing threats and opportunities; (2) analytic and conceptual thinking; (3) ability to provide clear, focused direction for the organization.

Building Commitment. This competency cluster centers upon organizational savvy; a sense of timing, or the ability to recognize when to take action; communications skills including significant ability to listen and respond; and team leadership attributes.

Driving for Success. Three critical competencies necessary to long-term organizational success are intuitive self-confidence; a driving need to achieve; and a sense of social responsibility (core values extending beyond the profit motive).

Another key consideration for the organizational change is the need to initiate change quickly. Effective leaders recognize that they have little time to spare. Change agents in both the private and public sectors are expected to "hit the ground running," work toward reorienting their organization, and produce bottom-line results within a year or two. These individuals must quickly communicate and successfully model a clear and compelling vision while working toward a flatter, more permeable organization. They must assure that initiative is rewarded rather than thwarted and springs from all levels in the organization. Leaders have to establish their credibility quickly, avoiding negative comparisons with their predecessors. While previous analysts saw a CEO honeymoon, more current thinking views this initial period as one of high vulnerability.

Current thinking must move beyond merely identifying a certain set of individual characteristics or training employees to lead. Rather, effective leadership revolves around the kind of relationship top leaders have with their many constituencies (Farquar, 1995). When these ties are characterized by effective communication, integrity, mutual respect, attention to customers, and trust, good leadership is present. Central to these characteristics is the recognition that leadership occurs when an authentic **relationship** is present.

An effective relationship between leaders and employees occurs within an organizational context. As Figure 20.1 suggests, organizations vary in their complexity; and change itself varies in its intensity. In cases where complexity and intensity are low, changes can be managed through normal management processes; reasonably competent managers can effectively shepherd such routine transitions or incremental change. As change intensity and organizational complexity increases, however, more fundamental alterations are required. This form of change, which is the focus of this chapter, is termed *discontinuous change*. Discontinuous change involves a "complete break with the past and a major reconstruction of almost every element in the organization" (Nadler et al., 1995). Such major change must be initiated at the top of the organization and, to succeed, draws on the

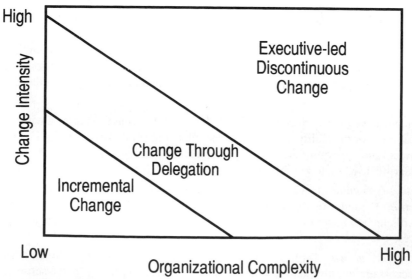

Figure 20.1   Organizational change continuum. (Adapted from Nadler et al., 1995)

complete repertoire of executive leadership skills. Such change requires sustained commitment of time by the change leaders.

Successful change leaders have moved beyond the paradigm of scientific management. Scientific management helped to instill objective measures and procedures into the corporate and public environment, but leached away emotional commitment and a sense of shared purpose. Leaders intuitively understand that authenticity incorporates passion (Drucker Foundation, 1996).

The imperative of contemporary leadership is to decentralize, de-layer, and loosen up organizations, positioning them to survive and thrive in a chaotic environment while creating internal strategic unity. "Only one element has been identified as powerful enough to overcome . . . centripetal forces, and that is trust" (O'Toole, 1996). Trust emanates from the relationships where leaders model and inspire a shared vision and purpose.

Organizations are buffeted by volatile environments. If their leaders are to respond appropriately to environmental forces, they must achieve a coherent and appropriate organizational direction and culture, and revise and realign key organizational levers. These elements are necessary conditions; each must be in place if successful change is to occur. The behavioral approach showcases the importance of organization participants thinking and acting in a manner consistent with achieving an organization that can succeed in an unpredictable environment. Not all change is equal, and organizational changes can take many forms. While incremental changes can occur at many levels, more intense and complex changes must be led from the top

20.3.2 Organizational Levers

Levers are the most visible components of an organization aside from employees. To achieve successful change, leaders must become adept at using levers: work processes and business systems, communications, individual competencies, job design, reward and recognition, and management processes. As the lever metaphor suggests, these components are not immutable. Organizational change levers (Fig. 20.2) can be turned on or off, modified, or totally redesigned. All organizations are constrained by their environment: organizations where there are unions face different considerations in changing personnel practices than do network organizations in the software industry. For government organizations such as the Department of Defense some levers (personnel regulations) may be "moved" through external forces (congressional legislation and Office of Personnel Management regulations). Other changes may be the result of internal decisions.

Organizational levers are captured in a range of sources on organizational charts, in documents, even in floor plans. When faced with describing organizational change, many observers point to altering structural components. *Reorganization* as commonly practiced is essentially the rearrangement of organizational units without necessarily altering substantive work processes, decision making, or communication flows. However, from a behavioral perspective, such change is incomplete; genuine organizational transformation entails much more than realigning levers. Indeed, some organizational change can occur without realigning levers. One savvy leader may manipulate a lever (personnel regulation) to achieve a desired change. Another may work to transform a lever to ensure long-term change. In a real sense levers act both as tools of change and as barriers to change. A charismatic leader may model and energize a new way of business, but upon that person's departure, the previous work culture will "spring back" into being. Under ideal conditions

Figure 20.2 Organizational change levers.

levers permit the change agent to bring major organizational resources to bear to accomplish a goal. Under adverse conditions they create insurmountable barriers; an outmoded major procedural requirement or legal barrier may impede even the most highly motivated change agent from accomplishing a goal.

20.4 APPLYING THE CHANGE MODEL

The change process includes three phases and seven steps (see Fig. 20.3). The discussion in this chapter focuses on the behavioral side of discontinuous change, particularly on how to alter the work environments, attitudes, and activities of large and diversified organizations. The behavioral process is to occur coincident with technically focused change methodologies, such as business process reengineering. The change process can be applied to all of an organization or to components within the organization.

Broadly speaking, organizational change moves through three phases: first, change is designed, then it is directed, and finally it is deployed. While several organizational levers may be manipulated across all phases, some levers are more inherently associated with a phase than are others. By virtue of their pervasive nature, values and culture, for example, must be addressed initially—in the design phase. In contrast, rewards and recognition are key levers most effectively manipulated in the final, or deploy change, phase.

These broad changes are associated with seven action steps (see Fig. 20.4). These steps have been built on a review of the literature on discontinuous change, in both corporate and government organizations, and through an analysis of interview with successful private- and public-sector change agents. Further, each step has been tied to the organization lever that is manipulated (e.g., Step 1.2 in Fig. 20.4, Design Work Processes, is tied to the core work processes lever). Collectively the steps move across time, covering all three change phases, and across the organization, affecting all major levers or components. The following sections detail and illustrate how these steps can be taken to implement successful discontinuous organizational change.

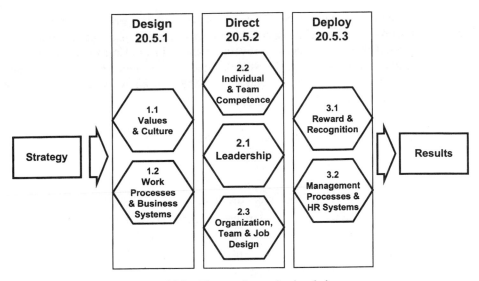

Figure 20.3 Phases of organizational change.

20.4.1 Phase 1: Design

Leadership is critical. The senior management team begins to define the vision for change in terms of concrete decisions and actions. This team starts the process of building support for change both inside and outside the organization.

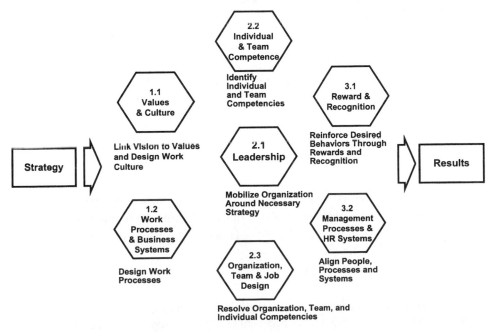

Figure 20.4 Organizational change steps.

Like any successful design effort, thoughtful planning is integral to significant organizational transformation. Furthermore, when people are the objects of change, even more effort is required. This section addresses several organizational elements that need to be incorporated early on if the change effort is to proceed. These elements include the nature and content of organizational cultures and values, the flow of work processes and of information, and the configuration of supporting structures.

Step 1.1 Link Vision to Values and Redesign Work Cultures

Key Questions

- Do you know the values and behaviors that support success?
- What are the core values that characterize the organization now?
- What are the key values we will need to implement change successfully?
- What stays the same and what core values need to be translated to meet the new environment? What other values might require change?
- Does behavior match espoused values?
- Is excellence in leadership valued?
- Does the organization value people?
- Are leaders willing to live the values?

When the desired future state closely corresponds to the current culture and values, this can be emphasized as a benefit in promoting organizational change. Discontinuous change runs a high risk of resistance and failure unless key leaders intensively manage the culture change. James L. Houghton of Corning was among the first American executives to commit his company to Deming's "total quality" program. Corning achieved great success. Houghton devoted approximately one-third of his time to ensuring that total quality management (TQM) was implemented across Corning (O'Toole, 1996).

With discontinuous change, the future is uncertain, and leaders are asking people to take a leap of faith into the unknown. An effective way to help people make this leap is to appeal to the organization's core work values when communicating the vision. Explicitly identifying such values allows leaders to validate the past while focusing on the future. It helps people over the normal resistance they show based on fear of the unknown and provides a bridge for individuals to decide to change. Successful change leaders do this by:

- Explicitly identifying existing core values and selecting those that support the vision;
- Including values (in words and symbols) in the initial messages about change;
- Defining the values underlying work behaviors required to address key challenges and future opportunities;
- Visibly living the values and vision.

Clearly culture affects performance. Along with core values, a common shared culture produces a unity of action and defines accepted employee behaviors. However, a strongly shared work culture can also act as a detriment to success in a radically altered organizational environment:

> (G)ood long-term performance can cause or reinforce a strong culture. But with much success, that strong culture can easily become somewhat arrogant, inwardly focused, politicized, and bureaucratic . . . . It can blind top management to the need for new business strategies. It can also make strategic change, even when attempted, difficult or impossible to implement. (Kotter and Heskett, 1992, p. 24)

To create change, the change leader needs to assess the current work culture—or patterns of work—and determine the most effective and appropriate patterns to support new processes and practices. Greater organizational effectiveness can result from using work-culture models as a diagnostic tool. An array of different techniques and tools to assess and then to close the gap between existing and desired work cultures is available (Hay Management Consultants, 1996; Executive Forum, 1994, 1996).

Step 1.2 Design Work Flows and Structures

Key Questions

- Have you identified primary work processes required to support strategy?
- Are your key work processes efficient and effective?
- Are organizational hand-offs effectively managed?
- Is there clear work process ownership?
- Do process members understand overall process requirements?
- Are business systems and policies aligned with work processes?

Once the change vision, organizational values, and targeted work culture have been linked, leaders must move to create a flow of work and structure to support the new work culture. Here it is important to assess the alignment of the targeted work culture and current work processes and to determine the magnitude, significance, urgency, and potential complexity of needed change in work processes. The major outcomes of this phase include identification of the required changes in how work and tasks are performed, and the major barriers and challenges in making these changes (Rouse, 1993; Shields and Adams, 1991; Price Waterhouse Change Integration Team, 1996).

Other chapters in this handbook focus on the tools and techniques required to "reengineer" an organization's processes. The focus of this chapter highlights the importance of examining the differences, or gaps, between the organization's culture as it is and the organization in its desired future state. Successful reengineering moves from redefining work processes only, to incorporating the changed behaviors that will be required of participants at all organizational levels. Such behavioral change can be accomplished by the application of a variety of levers available to leaders. The behavioral approach recognizes cultural considerations as integral to the successful design and application of work flows and structure. The three-phase change model outlined in this chapter helps to identify related barriers, enablers, and opportunities that will impact the change process or the outcome desired.

Change leaders must ensure that the changed business processes support the desired work cultures by doing the following.

1. Developing a template against which to determine management practices that support altered work flows. This task can be assigned to a design team, which tackles questions such as

    (a) How does the flow of work and tasks that accomplish that flow need to change?

    (b) How can both vertical and lateral barriers to the flow of information be overcome?

    (c) How can the flow of work be reduced or made more permeable?

    (d) What new behaviors are required of managers and operators?

2. Forming cross-functional teams to put systems and structures into place. Comprising line management, personnel lists, and manpower specialists, such teams will reconfigure positions and measures of task performance. Unlike work groups, teams are characterized by having shared goals and shared accountability. Teams become critical for bringing together the right skills, and they can be configured in a number of ways. Whether teams succeed or fail depends on such factors as whether

    (a) Their composition and resources are appropriate to their task.

    (b) Team members have a performance goal and mutual accountability.

    (c) The team is designed to focus on significant issues (and does so).

    (d) Team members have received adequate task-appropriate training.

3. Developing a comprehensive plan to inform and persuade participants to "buy into" the new work processes. The design and implementation of an effective communications strategy is a major challenge to teams directing change. Nadler et al. (1995) note that reengineering efforts often fail because of ineffective change management, particularly because of ineffective communications.

20.4.2 Phase 2: Direct

This phase marks the implementation of discontinuous change for an organization. More and more individuals are actively engaged by the change effort. Steps in this phase focus on how leaders at all levels mobilize individuals and groups to buy in to the new work culture, build work teams, restructure jobs, and clarify new competency requirements.

Step 2.1 Mobilize Organization Around Necessary Strategy

Key Questions

- Is there a compelling vision that inspires the organization?
- Do leaders create clear expectations for others?
- Do leaders communicate vision and expectations?
- Do leaders change themselves?

- Do leaders establish a foundation of trust and confidence in leadership so that people are willing to take the risks associated with change?
- Do leaders create and reward early successes?
- Do leaders clear a path and eliminate obstacles?
- Do leaders build support for and anticipate resistance to change among constituencies external to the organization?

All our research shows that while strong leaders are pivotal to either turning around or sustaining organizations, the companies that do endure are those that have strong systems of leadership—leadership that is really bred into people at different levels. Murray Dalziel (HayGroup Conference, 1995)

Leaders often approach planning for major organizational change the same way that they would approach project implementation planning. As described in Section 20.1, re-engineering efforts typically have focused on tangible, visible processes and technology, often overlooking the significance of the requirements placed on people to work and manage in new ways. In contrast to the "structural" approach, a "behavioral" approach also addresses the new skills, characteristics and behaviors that will be required. The behavioral approach considers the context, the competencies, and the timing required. Change leaders must be just as focused on planning for the behavioral, or "software" of change as they are on processes and technology, or the "hardware" of change.

20.4.2.1 *Senior Management.*
Senior management's role is crucial to the change effort—they must lead. Senior managers are the primary drivers or resisters of change (Ghoshal and Bartlett, 1995; Kotter, 1995; Mintzberg, 1996; Tichy and Charan, 1995). They can accomplish desired change by using organizational levers (personnel, communication, and information management systems) and their own personal attributes (coaching, listening, political influence skills). Given the repertoire of potential resources that they can bring, either in support of or in resistance to, the change effort, getting senior management support is a vital initial step. Since public-sector change leaders cannot make changes to the management team as easily as leaders in the private sector can, alternate methods for constraining resistant senior managers are generally used. These tactics are also available anywhere when dismissal is not an option.

Successful senior change managers have emphasized the importance of devoting large amounts of time to moving senior managers toward the new vision, new processes, and new requirements. These change leaders continuously act to build support while simultaneously applying a number of tactics to defuse both covert and overt resistance on the part of senior players. Some often cited techniques for bringing along top management are summarized in Table 20.1.

20.4.2.2 *Communications.*
In addition to determining a strategy, senior managers must begin communicating about the change to the rest of the organization. It is also critical to craft an initial direct message and frame the issues since the initial message becomes the basis for ongoing communications. Effective change leaders do not appeal to values to justify downsizing the work force. Facts provide the rationale for downsizing. The initial message does not *achieve* buy-in; it only sets the stage. The initial message must be

TABLE 20.1   Tactics for Defusing Resistance to Change

Actions to Build Support	Actions to Manage Resistance
Articulate a case for change; Initiate team building to engender trust and open communication between the members of the senior team. Build support by including the senior management team in refining and further developing the vision. Assess individual perspectives on the change and select only those managers who will effectively support change to be key members of the change team.	Prepare to spend personal time one-on-one with resistant senior managers—listening to their complaints and concerns about the change in order to diffuse their opposition and provide an outlet for them to vent their negative reactions. Give resistant (or incompetent) managers assignments where their views will not be communicated downward. Provide individuals with new performance standards. Help individuals determine whether they will stay or leave the organization. Encourage early retirement. Assign managers to a part of the organization where their presence is less intrusive.

Source: Hay Management Consultants (1996).

abbreviated, direct, and focused on the next 12 to 18 months. The integrity and framing of the initial message establishes the climate for transforming communications to support the desired work culture. Often, the issue of trust hangs in the balance.

To improve the chances of people accepting the change message, individuals must understand that their work lives depend on it. An abbreviated, direct change message would contain the four elements indicated below (Larken and Larken, 1996):

Initial Change Message Elements

- Emphasizes the urgency of the need to change;
- Widely distributes factual information;
- Provides *honest assessment* of how the change will immediately impact people;
- Provides a forward-looking vision of the future.

20.4.2.3   Early Successes.   Getting commitment to a new vision and new behaviors can be almost overwhelming. Change leaders must get people to want to change, or to buy into the vision. Short-term successes provide tangible evidence that change is possible. According to Jeffrey Pfeffer (1992):

> Because of social influence processes and how they operate, momentum is very important . . . . Once a social consensus begins to develop, it is difficult to change. . . . The implication is that affecting how decisions are viewed, very early in the process, is absolutely critical in affecting outcome.

Early success helps to bring the change vision down to a level that is understandable and tangible. Often "small wins" and visible signs of success create a pattern that attracts people. Success undermines the arguments of naysayers in the organization who resist any vision in their determination to do "business as usual." Not only do most people want to be associated with winners and successful work units, the successes also serve to establish the credibility of the project, the team, and the management that supports them.

In addition to creating early wins, creating momentum requires a process of publicizing success and rewarding the people involved. This process reinforces their behavior, sustains commitment, and communicates to the rest of the organization that there are personal benefits from the change.

Early Success Characteristics

- Visibility (changes in work processes that involve some form of organizational change are often the most visible);
- Strong likelihood for success defined in terms of desired outcomes;
- Targeted to organizational units where key managers are strong supporters of the change agenda;
- Involves a number of people involved in the change;
- Provides recognition and rewards for those who achieve the desired outcomes and behaviors.

In an environment where there is a requirement for massive change combined with significant ambiguity, issues about job security, power, control, and influence all become extremely important to the workforce. Breaking the vision into doable pieces provides an opportunity for people to experience short-term success. Success helps to reestablish feelings of control and self-worth; successes increase commitment to the change process.

20.4.2.4 *External Environment.*

Finally, to achieve change, senior management must also address the external environment. The leader must target key influencers in order to build support for and manage resistance to change. Ideally, external constituencies that either influence the environment or can derail change are brought in line with requirements for success. The external constituency includes those individuals or constituencies outside the purview of the change leader (Wilson, 1989).

A lack of internal consensus can also impact the external environment. High levels of discord can generate multiple appeals and complaints to significant stakeholders such as unions, as well as to Congress, regulatory bodies, and other external players. Visible conflict draws media attention, thereby potentially arousing stockholders and attracting added regulatory scrutiny. Clear, consistent communication helps avoid creating a backlash before the change case can be made to external influencers who have the power to derail the change process.

The approach for targeting key influencers includes Identifying key constituencies that will have concerns about the proposed change. Targeting key influencers is equally a requirement for public and private organizations. Critical Elements Include:

- Developing a communications plan to deal with each constituency;
- Using multiple means to influence targeted constituencies, including multiple avenues of communications; and

- Incorporating plans to work affirmatively with the media, since they provide a significant avenue for transmitting information and educating significant opinion makers in various constituencies. Such plans may be contingency only.

As an example, the secretary of defense looks to elements outside the Department of Defense (DoD) (such as key congressional committees), whereas a defense agency may also look to elements in the Office of the Secretary of Defense (such as senior DoD officials, including the secretary) and other players in the DoD structure Table 20.2 provides an example of key constituencies for DoD, it arrays interests in specific DoD components. The scope and level of effort spent on targeting key influencers should mirror the extent of change planned and the size of organization undergoing the transformation. Successful leaders demonstrate the following when instituting change:

- Accept that leaders are a component of culture and must change themselves;
- Maintain personal integrity and honesty;
- Remain personally involved in the effort, particularly by taking decisive action when people run into obstacles;
- Ensure that there is broad participation from all major constituencies in the organization in the planning, design, and implementation; and
- Recognize that planning for discontinuous change needs to be more iterative and exploratory than is typically the case with planning for incremental change.

Step 2.2 Identify Individual and Team Competencies

Key Questions

- Have new roles people will be required to play been identified?
- Is there a clear picture of what excellence looks like?
- What level and variability of performance is acceptable?
- Do you know which individual and team competencies are needed for success?
- Do you have an adequate talent pool?
- Is there a process to identify and learn from superior performance?
- Have outstanding performers been matched to pivotal roles?

Many reengineering efforts rely solely on training and education as the tool for turning the reengineered work processes into reality. While training may be necessary, leaders must first identify the new roles, responsibilities, and competencies needed to perform each new or modified process. The process of identifying roles and competencies must focus on the *work required,* and the characteristics required of the *people* who will be doing the work. The process involves:

- Identification of skills, abilities, and competencies that would be required of all people, leaders, and roles.
- Assessment of the existing talent pool to match people to the new roles and to determine the extent to which the new competencies need to be recruited or developed.
- Alignment of all human resource systems, including recruiting, selecting, training, developing, rewarding, and supporting.

TABLE 20.2 Identification of Key Constituencies

Entity Undergoing Change	Senior Team	Senior Official(s)	Agency Heads	Key Constituencies			Private-sector (Defense Contractors, Trade Groups)	Media
				Congress	Unions			
OSD	SecDef and key subordinates; Chairman,	President and National Security Council	Senior appointees	Members, esp. committee chairs; other political leadership	National president; national conventions		CEOs, boards of directors	National media
JCS	JCS, service chiefs and key subordinates							
Agencies and military services	Senior political appointees, SES, senior military	SecDef, OJCS	Senior appointees and career civilians	Congressional staff, committee staff	President and key subordinates		Key persons responsible for market and policy	Esp. trade publications

CEO = chief executive officer; JCS = Joint Chiefs of Staff; OJCS = office of the JCS; OSD = office of the Secretary of Defense.

Many companies undergoing change start with the tools or measures they have in hand and adapt them to fit their new strategy. The competency approach starts the other way. It begins with the strategy, values, work culture, and works back to discover the critical competencies necessary for achieving the desired business results. A competency assessment begins with an analysis of what leads to superior performance. Research has shown that there are unique competencies that are associated with different organizations. A methodology to identify and tailor the competencies most appropriate to a specific organization's mission and work culture grew out of David McClelland's research at Harvard many years ago (Spencer and Spencer, 1993).

Job competencies are the critical characteristics that cause or predict outstanding job performance in an organization (HayGroup, 1996; Jones, 1995; Lawler, 1994; Spencer and Spencer, 1993). The concept of job competencies arose out of a methodology called Job Competency Assessment, which is formulated around three basic assumptions:

- In every job, some people perform more effectively than others. These people also approach their jobs differently from the average worker.
- These differences in approach relate directly to specific characteristics or competencies of the superior performers that are often absent in the "fully successful"[2] performers.
- The best way to discover the characteristics that relate to effective performance in an organization is to study its top performers.

Often our "gut feelings" tell us when someone is a high performer in a job, but we cannot tell exactly why it is so; the qualities that lead to high performance levels are often difficult to identify. However, traditional job analysis approaches focus entirely on the tasks performed in a job, rather then on top performers. An error in this approach is that it will usually lead to a definition of what is required for average or even minimum performance. Job competency assessment takes a more direct approach by focusing on the superior rather than the average performers, and identifying what they do when they are in critical job situations.

Job competency models can be developed to describe job requirements. These models comprise several individual competencies, which include knowledge, skills, attitudes, values, and motives (see Fig. 20.5). These characteristics exist at various levels in a person, with the inner levels (values and motives) representing characteristics that are more enduring, and often have a wider range of effect on the individual's behavior. Further, the more complex the job, the more important job competencies are relative to technical or task mastery (HayGroup, 1996; Jones, 1995).

Different types of competencies are associated with different aspects of human behavior and with the ability to demonstrate such behavior. For example, an "influence" competency would be associated with specific actions, such as having an impact on others, convincing them to perform certain activities, and inspiring them to work toward organizational objectives. In contrast, a "planning" competency would be associated with specific actions, such as setting goals, assessing risks, and developing a sequence of actions to reach a goal. These two types of competencies involve different aspects of human behavior on the job.

---

[2]"Fully successful" is a performance evaluation term commonly given to average performers. Average performers are not the ones who lead the organization to new successes.

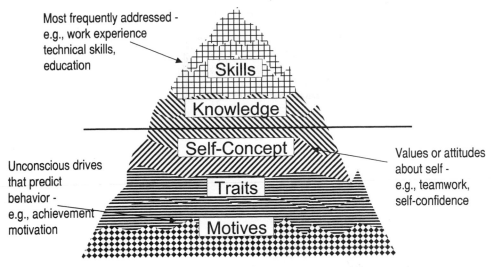

Figure 20.5   Individual competency components.

The importance of competencies is often overlooked in favor of specialized job-related knowledge. This is partly because skill and knowledge have a more observable effect on an individual's performance. Research has shown that although knowledge is important, it is essential to focus on the competencies that allow employees to use the facts and concepts contained in the knowledge. When asked to describe the best people that ever worked for you, managers describe how workers accomplished their work, not their technical knowledge.

A job-competency model can be developed for leadership and team management roles. The job-competency model consists of competencies with associated behavioral indicators. Organizations need to recognize and develop leaders, formal and informal, at every organizational level. New leader competencies relate to the new job compact between employers and employees. Changing workplace demographics, combined with the need to respond rapidly to requirements, demand competencies related to communications, coaching, and performance management.

Identifying competency clusters provides the base from which a corporation can build a new mode of leadership by following a multistep process (see Fig. 20.6). Steps include assessing the levels of leadership competence and using these initial findings to help their executives recognize their own limitations and need for improvement. The organization can conduct a needs assessment determining the current level of leadership competencies and identify any current leadership gaps. It can perform competency surveys for determining whether to train or recruit for leadership competencies. It can design competency-based leadership alternatives to build the requisite competencies. Once competencies are identified, the pay and rewards can be tied to demonstrated performance.

Approaches to determining competencies for future roles include: (1) using expert panels to construct critical incident scenarios in order to identify likely accountabilities and competencies; (2) benchmarking to analogous jobs, or isolating competencies previously identified in job elements of other organizations in order to develop standards for new models. For instance, a U.S. telecommunications firm needed a model for senior repre-

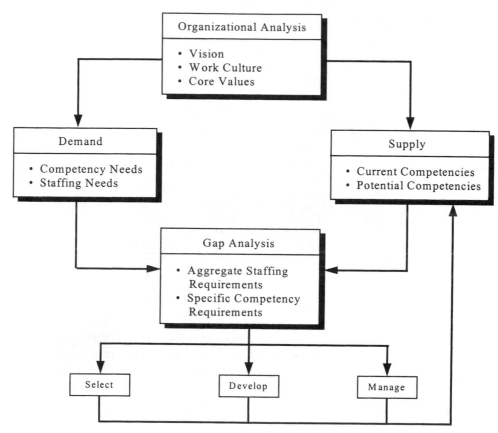

Figure 20.6   Competency-based human resource management system.

sentatives capable of marketing to European government and European community officials. Analysis of the requirement indicated it would require elements of both diplomatic and high-technology sales jobs. Already existing competency models for diplomats and high-tech sales provided the competencies identified below. These competencies fit the vision of the international marketing positions:

Diplomatic Model	High-Tech Sales Model
Cross-cultural interpersonal sensitivity	Achievement orientation
Overseas adjustment	Consultative selling skills
Speed of learning foreign political networks	

The development of job competency models and applications based upon those models are an important part of an organization's human resource planning and operations. The center or core of this approach is the job-competence assessment (JCA) methodology, from which a job-competency model is developed and validated.

Step 2.3 Resolve Organization, Team, and Individual Competencies

Key Questions

- How effectively do people and teams work together?
- Is authority and accountability widely distributed to those closest to the action?
- Is Information open and accessible to those who need it?
- Is work performed where it makes sense?
- Are roles correctly assigned and clearly understood?
- Does the structure efficiently support the required work processes?
- Does the way resources are allocated support the strategy?
- Are the core values honored and followed?

People, groups and whole organizations not only have to learn new ways of thinking, working, and acting, but they also have to 'unlearn' the habits, orientations, assumptions and routines that have been baked into the enterprise over time. (Nadler et al., 1995)

Jack Welch's emphasis on "soft issues" and "empowerment" at GE are actually short-hand for ensuring that resources and information are available to those who have the accountability for performance. With accountability, all the new work cultures demand the necessary authority. A formula could be derived to measure whether "empowerment" is platitude or reality on the work site:

$$2A + I = \text{empowerment}$$

or *authority* and *accountability* plus *information* necessary to make decisions provides *empowerment*. Discontinuous change requires rethinking the old functional work culture and changing to a new way of working. A responsive, flexible workforce is a key component in all discontinuous change, regardless of the particular shape the work and organizational structure takes. In additional to individual competencies, this new workforce will need a broad spectrum of behavioral competencies as well. Communications, coaching, cooperation, and influence abilities are increasingly important for most of the workforce as more and more people are required to make decisions, work in teams, and negotiate across traditional functional boundaries. Organizations undergoing change must systematically develop these abilities in their people—otherwise, the old ways of doing business will persist.

20.4.2.5 Training. Training helps to ensure that organization members are given a common vocabulary, common measures, a common approach to problem solving, and a common vision of the new work culture and enterprise. Training programs are defined around competencies that align with business strategies. Organizations such as GE and Allied-Signal have required thousands of managers and employees to learn how to determine customer needs and how to create solutions.

Three areas that successful managers address when moving to a process, time-based, or network work culture include:

- Facilitator training for all team leaders, supervisors, and managers.
- Team member training, including practice in active listening, and practice in identifying and clarifying discussion issues.
- Diversity training including not only race, gender, and ethnicity, but also training styles, individual learning styles, and communication styles.

Acquiring these behavioral skills requires hands-on training rather than education, which is traditionally more lecture, reading, and discussion-driven. Employees without the necessary skill sets or who have very limited skill need to practice, and to correct, and to practice again. Practice does not make perfect—*perfect practice makes perfect.*[3]

20.4.2.6 *Communications.*

The need to establish trust is a critical success factor in successfully accomplishing discontinuous change. "The First Law of Leadership: If we don't believe in the messenger, we won't believe the message. This is a principle that every leader must acknowledge" (Kouzes and Posner, 1995). People must have trust in the communicator, built by the integrity of the message and by the ways in which communications occur. Without accurate and timely information, employees find it difficult to take empowered actions on their own. In times of rapid change, widely sharing information is even more important than in less stressful times. Management needs communications to establish trust. In one situation, for example, the director of the Defense Printing Service set out to ensure that the rumor mill would not lower morale and increase stress already caused by heavy downsizing and loss of traditional work patterns (Hay Management Consultants, 1996).

Effective communication can only be achieved through ensuring that communications are two way, having channels for both upward and downward communication flow. Listening may well be one of the most effective communication tools used during the change process. This allows individuals to express dissatisfaction with the current state and gives participants sufficient time to let go of past ways of acting.

Additional considerations to achieving effective communications include (Ashkenas et al., 1995):

- Aligning the messages to be transmitted with appropriate channels (see Fig. 20.7).
- Engaging both formal and informal communication channels.
- Making messages both simple and complex.
- Using repetition and diversity in message delivery.
- Sharing both positive and negative information.
- Invoking symbols and values along with content.
- Delivering messages through modeling appropriate behavior.

Chrysler's Lee Ioccaca appointed a person to play "devil's advocate" during every staff meeting (Bennis and Townsend, 1995). GE used "workouts" that exposed and educated participants to the decision-making process. Such forums have resulted in senior managers handing the decision back to teams. The GE example also demonstrates that two-way

---

[3]George Allen, coach of the Washington Redskins, forcefully voiced and trained to this premise.

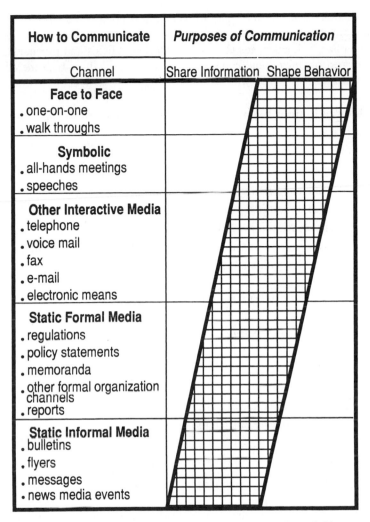

How to Communicate	*Purposes of Communication*	
Channel	Share Information	Shape Behavior
Face to Face . one-on-one . walk throughs		
Symbolic . all-hands meetings . speeches		
Other Interactive Media . telephone . voice mail . fax . e-mail . electronic means		
Static Formal Media . regulations . policy statements . memoranda . other formal organization channels . reports		
Static Informal Media . bulletins . flyers . messages . news media events		

Figure 20.7   Aligning formal communication channels. (Adapted from Ashkenas et al., 1995)

communications vehicles can have a powerful educational and developmental impact. The open process facilitates candor, which in turn maintains trust. In addition to GE, other organizations, including New York Life, Digital Equipment Corporation, Exxon, SmithKline Beecham, and the World Bank, have used town meetings as a means of removing vertical boundaries (Ashkenas et al., 1995).

20.4.3 Phase 3: Deploy

When an organization has experienced significant change, the transformation may still not be complete. Workflow may be different, behaviors may be cross-functional, and decision making may be more rapid. While major in scope, these changes may be short-lived. True transformation must extend into the next generation. This phase addresses the steps leaders take to sustain change. Leaders install new, supportive policies and procedures, align the newly transformed organizational components, and assess and readjust the course of organizational change.

Step 3.1 Reinforce Desired Behaviors Through Rewards and Recognition

Key Questions

- Is your reward structure aligned with strategy and values?
- Are you attracting and retaining the right people?
- Is reward focused on internal equity or on results?
- Is there a clear tie between rewards and measurement of results?
- Do rewards discourage or encourage people to develop new capabilities?
- Are pay grades tied to value-added activities?

As Kouzes and Posner (1995) note, the role of the leader is to create the appropriate work culture wherein employees motivate themselves. The time-based, network, and process work cultures in fact require self-motivation in order to be effective.

Common Characteristics of the New Work Cultures

- Link pay to performance;
- Reject the notion of perfection;
- Encourage creative thinking outside the box;
- Avoid early negative feedback when a new idea is presented;
- Reward initiative; and
- Honor risk takers.

Instead of exercising command and control, leaders must work to mobilize and motivate others. Command and control management involves close monitoring and controlling of subordinates' behavior and sends a message that the leader *lacks trust* in the ability and judgment of people within their purview (Steininger, 1994). Command-and-control style managers often fail to provide subordinates sufficient information for the subordinate to make effective decisions and initiate actions. As a result, such control-oriented managers tend to discourage creativity and initiative, and thereby slow the organization's ability to respond to environmental pressures.

The traditional functional organization pays people primarily through base salaries based on the specific job; to be competitive with salaries paid in the relevant marketplace; and to establish equity. Employees look for pay increases resulting from promotions, cost of living, and merit raises. Point-based pay systems have been used to effectively provide measures such as scope or responsibility and degree and kind of technical qualifications required in order to establish "end-to-end" pay grades within each of the functional areas of an organization.

Base pay closely tied to an individual job is less relevant to a process or time-based culture. The new GE de-layered and moved toward true "boundarylessness" and team effort. In the new culture, movement up a hierarchical, functional ladder is limited, and security is minimal. GE aligned pay with desired cultural attributes by eliminating numerous pay grades in favor of five broad pay bands. Unlike pay grades, bands can overlap; individuals can receive raises within bands rather than having to receive a promotion to increase base salary. Broadbanding (a term adapted from radio) as opposed to "single-frequency" correlates with a boundaryless organization where tasks and processes cross departments and demand team skills. In 1996, the annual sales award trip for GE Information Services (GEIS) was awarded to more technical representatives than sales people (reflecting both team achievements and the boundarylessness or permeability of the organization's functional components). Cost-of-living adjustments have been eliminated, and bonuses and other awards make up a greater part of the compensation plan. Middle managers can present on-the-spot cash awards of up to $2,500; executives can award $5,000 in cash. Up to 20 percent of employees receive stock options. Compensation restructuring occurred after major changes in the culture were well underway. In aligning pay with desired cultural attributes, GE used job competencies to determine the skills and competencies needed to create value for the organization (HayGroup, 1996).

Leaders can create a compensation and reward system that reinforces the new organizational values and work culture. The appropriate actions can be summarized as nine principles of dynamic pay (HayGroup, 1996). They include:

1. *Align Compensation with the Organization's Culture, Values, and Strategic Business Goals.* In the last analysis, the failure of most compensation programs can be traced to a lack of alignment. When leaders rely on older, established programs, they may fail to support the evolving work cultures.

2. *Link Compensation to Other Changes.* While pay and benefits are necessary to motivate and retain employees, they are not a panacea. Compensation must support and reinforce other change initiatives and be integrated with them. It is a tool that must not be presented by itself, but in the context of other organizational changes.

3. *Time the Compensation Program to Support the Change Initiatives.* Leaders must not only know where the organization is and where it is headed, but when to introduce new compensation plans. These plans are powerful motivators, but cannot be expected to lead the change process.

4. *Integrate Pay with Other People Processes.* Leaders must also recognize that while compensation plans are important, they must be developed in the context of more complex human resources systems. Such systems look at how work is designed, how people are selected, and how their performance is managed.

5. *Democratize the Pay Process.* If leaders are going to give people more power in their organization, then they have to be willing to decentralize the authority for giving employees rewards for work done well.

6. *Demystify Compensation.* People cannot be accountable for something they do not understand. The hows and whys of compensation should be known throughout the organization, rather than kept in the heads of a few personnelists.

7. *Measure Results.* If pay is to be tied to performance, performance must be measured. The measures used depend on the organization and its goals. They may be based on financial performance, quality improvement, productivity, or customer service. Employees must understand how they will be evaluated and how their efforts are linked to broader organizational goals.

8. *Refine. Refine Again.* Pay is not a static process. Just as organizations, groups, and individuals evolve, so should compensation programs. Leaders should place more emphasis on what is working, and de-emphasize efforts that are less than successful.

9. *Be Selective.* When it comes to compensation, there is a seemingly endless supply of models and strategies about how people should be paid. Leaders need to carefully examine alternative options and determine the solutions that were appropriate for the organization, culture, and changes that they are going through.

Step 3.2 Align People, Processes, and Systems

Key Questions

- How well are your human resource systems aligned?
- Are there appropriate selection systems?
- How adequate are succession systems?
- Are there appropriate development systems?
- Are there adequate performance management systems?
- How well are your planning systems aligned?
- How well do your management systems align?
- Are there effective links between monitoring and management systems?

Studying and working closely with some of the world's most visionary organizations has made it clear that they concentrate primarily on the process of "alignment", not on crafting the perfect "statement". (Collins and Porras, 1994)

The final step in deploying major organizational change involves aligning the new organizational levers. Alignment serves three functions: it assures that the organization operates in a coordinated way; it increases the chances that changes in place persist over time; and it facilitates the organization becoming a self-correcting entity. In effect, the changes and processes implemented need to be institutionalized. Senior management can spearhead the move to change systems and policies that drive the way people work together and are organized, selected, evaluated, developed, and promoted.

Policies must be created or modified that address such components as:

- *Reporting Relationships, "Span of Control," and Control Processes.* Position management must reinforce the move from hierarchical relationships to team and other horizontal forms.
- *Position Management.* The static nature of position management in the civilian workforce, which focuses on the mechanics of establishing and paying jobs rather than people, must be addressed. The newer nonfunctional work cultures require an increasingly flexible workforce in which assignments and tasks may be broader than those in a traditional position description.
- *Staffing Processes.* These processes encompass recruitment, selection, and promotion, and incorporate the new competency requirements. For example, selection and promotion practices should be altered to ensure that new managers possess many of the required leadership competencies.

- *Appraisal System.* Performance standards should be altered to reflect the extent to which people now display behaviors consistent with new work cultures. For example, employees may now be rated on the extent to which they keep their subordinates informed.
- *Training and Career Development.* For organizations moving into a process work culture, horizontal career moves become more important; employees are encouraged to broaden their skill base. A common way of accomplishing this goal is to require managers to rotate among different organizational components.
- *Performance Management System.* This component involves a day-by-day process for establishing a shared understanding between worker and rater of what is to be achieved, and how. This approach to managing people increases probability for achieving success.

In changing the culture, leaders and managers must model, mentor, and coach employees, rather than direct and determine their actions. Such dynamic, real-time performance management must become a feature throughout the workforce. Performance management is not a paper and pencil exercise. It is rooted in core strategy and comprises good communications, mentoring, coaching, monitoring, and rewarding good performance.

Making the changes just outlined is a daunting task, and most change leaders' efforts fall short of the mark. One senior civilian change leader at the Department of Defense's Defense Logistics Agency (DLA) implemented a major, long-term change. He and his managers formalized personnel changes to ensure that the new work culture would be sustained. In DLA's change vision, headquarters was to be focused on policy and the field on execution. The leadership team took the following actions to sustain the vision over time:

- Increased the span of control of Senior Executive Service (SES) positions. This change permitted an immediate transfer of some SES slots (and the people occupying them) out of headquarters, emphasizing the decentralization of decision making for operations.
- Established a formal rotation process for the headquarters' senior SES position. Under this new arrangement the agency's most senior civilian (the deputy director) would rotate out of that position into another headquarters position every three years. The intended result was for the top SES position to rotate, with no single individual having the opportunity to become ensconced long-term in the top permanent civilian position. Another result was that the key change leader ensured that future "empire building" would be counterproductive. Further, no long-term career civilian would be in a position to have sole influence over the selection or approval of the agency's senior cadre of career civilians, and would be less likely to be considered the de facto leader of the organization. Having a single permanent senior civilian as a deputy director to a military director or to a political appointee can build a bias against change, or at least to a circumscribed view of what the organization needs (Hay Management Consultants, 1996).

By deliberately departing from traditional DoD organization structure, the agency change leaders were able to set up an operating concept that would persist in supporting a cross-functional perspective and openness to change that would otherwise not have been possible.

All organizations undergoing discontinuous change must also establish a means of monitoring their progress toward achieving agreed-upon goals in order to create feedback mechanisms. Rather than being separate activities, these mechanisms are tailored to the specific change objectives and woven into the organizational management fabric. The team responsible for developing and implementing new measures needs to follow a number of guidelines (Osborne and Gaebler, 1992). These recommendations include:

- Recognize that there is a significant difference between measuring process and results. The former is relatively easy to capture; results are more difficult but also more important to document.
- Be aware of the vast difference between documenting efficiency and effectiveness. Efficiency is a measure of the cost of production, while effectiveness captures the quality of output. Both phenomena need to be included in a comprehensive monitoring effort.
- Involve all participants in developing change measures. This guideline assures that employees will "buy into" the change measurement process. Initially senior management and cross-functional teams are involved in the effort; later all affected managers contribute.
- Try to strike a balance between using a few and too many measures. If too few are used, not all change objectives will be measured. By creating too many, the power of all measures may be diluted and managers will be overwhelmed with information.
- Focus on maximizing the use of performance data. Performance measures assist managers in planning and conducting their work. Other measures, in contrast, are typically seen as burdensome and merely as reporting requirements.
- Plan to subject measures to an annual review and modification. The fluid nature of discontinuous organizational change means that measures that appeared ideal when originally developed may not work out as well in practice.

20.5 PROFILES IN CHANGE

A number of organizations, both public-sector and private, have successfully applied the key principles we have outlined. Under competent leadership, these organizations have achieved a new direction and culture, realigned organizational levers, and created a cadre of motivated employees. The following sections illustrate three examples of implementing discontinuous change. A charismatic visionary leader transformed GE into a world-class, highly competitive organization; the Department of the Army overcame bureaucratic inertia to develop a new, comprehensive management system; and the Defense Printing Service implemented a new vision and culture that replaced the traditional typographical-based production process. In selecting these three examples we have purposely included two public-sector cases because such successful efforts are often overlooked in favor of corporate cases. The Department of Defense employs over half the civilian workforce of the federal government. Its nearly three-quarters of a million people represent great diversity and density in technical skills. The DoD also allocates a high percentage of "discretionary dollars" available to congressional direction. This increases the challenge of managing diverse interests under the scrutiny of multiple stakeholders.

20.5.1 GE: Transformation into a World-class Competitor

Jack Welch took over as Chairman and CEO of GE in April 1981. He had the support of his board and a key group of insiders that he could draw upon for support across the next decade. His precepts began with: "face reality"; "change before it's too late"; and "be ready to rewrite your agenda as necessary." His *vision* was to make GE "a company known around the world for unmatched excellence . . . the most profitable, highly diversified company on earth, with world-quality leadership in every one of its product lines" (Slater, 1994). During the early 1980s, Welch focused on the hardware of his company. He continuously redefined his agenda, starting with the determination to be number one or number two in any key business. He then identified GE businesses as those fitting into one of three circles: core, high technology, or service. Welch intended to leverage GE's strengths to include its highly skilled workforce and deep coffers. To this end, he jettisoned "commodity" products. The vision determined where resources would be allocated, where divestitures and acquisitions would occur, and shaped the current GE. In 1982, *Newsweek* labeled Welch "Neutron Jack" for his willingness to downsize the number of people and leave the buildings standing. Between 1982 and 1992, major acquisitions totaled $22 billion and major divestitures represented $14 billion total. During the decade, GE earnings grew at an annual rate of 10 percent, one and one-half times the rate of growth of the GNP; most of its businesses were dominant players in their sector (Tichy and Sherman, 1994).

Welch early determined that he wanted a "cultural revolution"; he explicitly set out to accomplish cultural change in 1989. Welch had always poured money into Crotonville (GE's management training center) and visited the center at least once a month in order to find out what was going on across his company. Welch funded Crotonville and the training required to develop new competencies out of a central fund fueled by the money from his divestitures. This strategy meant that any cost savings attributable to new processes and new ways of doing business dropped straight to the bottom line in subsidiary businesses. Key management could see and show the impact of change in a very measurable way. Welch would take the stage in "the pit"—the stage of the large lecture hall with elevated seats. Welch used the platform as a "listening post . . . debating mechanism . . . pulpit. . . ." to communicate his vision and agenda (Slater, 1994).

Jack Welch routinely went down into "the pit" at Crotonville to field questions from employees. He initiated a concept called the "workout" in an attempt to bring the same honesty to all parts of GE. Workout is the term for the process Jack Welch developed and applied; it is used in common business parlance today. Early workouts focused on getting at the "low hanging fruit." Such early successes laid the foundation for the really difficult issues that concerned later workouts. Welch's vision of a culture presumed that "managing less is managing better." Welch insisted that a key role of any manager and supervisor is making sure everyone in your business has the information to make decisions—understanding that a manager's role is to provide resources, dollars, and ideas, then get out of the way. GE has explicitly targeted changing the work culture to produce fast and effective decisions.

Workouts are similar to town meetings and represent a way to shift authority and also educate people (both managers and associates) into their real degrees of freedom and responsibility. A typical workout involves a group of GE employees (associates) who are in some way connected (report to the same person or work a common business process). Usually at least two levels of management are present. Meetings are off-site and provide the forum for debate and discussion. The workout is preceded by work in small groups for one to two days to generate ideas for removing hindrances and improving business. The senior business leader chairs the session. The leader answers each proposal on the spot: approved, disapproved (with a rationale), or sometimes returned for more work. Someone is assigned to make sure decisions are implemented (Ashkenas et al., 1995).

In essence, the new vision for GE discards the notion of control by a traditional hierarchy as too limiting, and looks to shared values to allow employees to manage themselves. In this, Jack Welch's bedrock belief in the value of ideas and the capabilities of the average person comes through. In 1992 Welch articulated this commitment to the capacity of all the organization's employees when he defined the four types of executives present in a business organization and GE's response to them: The one who delivers on commitments, both financial or otherwise, and shares the values (love and reward); the second leader who does not meet commitments nor share values (less pleasant but also easy call); the third leader who misses commitments but shares the value (usually gets a second chance, often in a different spot). Welch then defined the fourth type, which is the most difficult for many organizations to deal with: the person who delivers on commitments, makes the numbers, but does not share the values. "This person forces performance—the autocrat," Welch stated. "Where we can convince and help these managers to change—recognizing how difficult that can be—or part company with them if they cannot, will be the ultimate test of our commitment to the transformation of this Company and will determine the future of the mutual respect and trust we are building . . . , (General Electric, 1992)

Great candor in a public document from a man who several months earlier had dismissed just such a visible, senior producer—the person who delivered the numbers, but did not share the values. The message to all GE associates was very clear: actions had preceded the words. Demonstrated commitment to the individual who can produce, who can share the values, coupled with opportunity for personal and professional growth (as opposed to continuing promotion) offers the basis for a new "psychological contract" with the company (General Electric, 1995; Tichy and Sherman, 1994). Trust and candor are the glue that hold the workforce together.

20.5.2 MANPRINT: Focus on the Ultimate Consumer

A major cultural change that radically altered traditional functional behaviors characteristic of public agencies resulted from the successful application of an integrated systems approach to the army's acquisition process. The army's traditional, stovepiped functional way of doing business was replaced by a newer, more encompassing approach that focused attention on the ultimate customer: the commander and soldier user in the field.

The vision of the army's key military leaders was to define a system as an organization with people and equipment, superseding the more parochial view that the system was just the equipment. This new comprehensive management system, designated MANPRINT in the army, required continuous integration of six functional areas throughout the development and acquisition process. A major barrier facing the planners was the complexity of effort required, combined with bureaucratic inertia. The effort was able to seize the moral high ground and simplify the concept by focusing on the fact that the army was going to design and build equipment with soldiers in mind. The slogan simply said: "MANPRINT:

Remember the Soldier." This slogan tied the new vision directly to basic army values (Booher, 1990).

The army introduced hundreds of new weapon and equipment systems into the force in the late 1970s and early 1980s. Force modernization was occurring just when the military went to an all-volunteer mode. As the equipment was acquired and put in the field, the army found that the high-tech equipment failed to meet user expectations.

The development of the Bradley fighting vehicle represents one example of the failure of the traditional functional culture. The engineering and acquisitions process operated in stovepipes separate from the acquisition and training processes, and failed to coordinate effectively with the field operating and maintenance commands. The mid-1970s prototype Bradley held one less soldier than the number assigned to the infantry squad. Designers had failed to accommodate the amount of equipment actually carried by each infantryman. The solution was to eliminate one position in the squad. It took the army's combat development theorists and manpower planners several years to reduce the doctrinal (i.e., the "how to man" and "how to fight" policies) requirements determining the size of an infantry squad to fit the "space available" in the vehicle.

Despite extensive documentation of problems and a committed senior leadership, the old culture rewarded program managers for moving a system forward on a tight time schedule and within costs. Many saw focusing on the ultimate customer; safety; integrating human factors; and manpower, personnel, and training (MPT) as increasing the "investment" costs and throwing up potential roadblocks to fielding systems on time. Many different parts of the bureaucracy reacted with concern, particularly where the various activities had been in competition so long they found it difficult to cooperate in an interdisciplinary approach.

In the early 1980s a key figure in changing the army's process of acquiring, designing, and fielding new equipment emerged—General Maxwell Thurman. General Thurman began the massive process of changing the army's multibillion dollar acquisition system to focus on the ultimate customers—the soldiers and leaders that would use and maintain the system. General Thurman changed the army's acquisition system through a multiyear, multipronged approach that engaged key leaders and personnel at all levels in many stovepiped organizations with huge bureaucratic structures, such as the Army Material Command; the Training and Doctrine Command; the army staff; the Congress; and the multibillion dollar defense industry. These changes impacted tens of thousands of people—military and civilian personnel as well as private industry employees (Blackwood and Riviello, 1994).

General Thurman began his push for massive changes by clearly articulating the need and the disastrous consequences of purchasing systems that could not be supported, maintained, or operated because of the shortage of people with the right skills. He created the vision and energized the community to make massive changes in the highly regulated and proceduralized acquisition process that was documented in a multiphased process. Thurman was the major change agent in initiating the program called MANPRINT. This involved six functions (manpower, personnel, training, human factors, system safety, and health hazards).

Through the leadership of General Thurman, the army introduced the MANPRINT concept using all of the change levers. General Thurman articulated the vision in speeches, actions, and decisions. The acquisition program includes many milestones and decision points that involve large meetings of key stakeholders. At these meetings he refocused the decision points on the final customers—the soldiers and commanders. Prior to this time decisions were made on the equipment and hardware capabilities; now, his questions centered on the performance of the system in the hands of the soldier.

As General Thurman articulated his vision, he also initiated activities to examine and change the formal work and decision processes used to write requirements, design systems, award contracts, and test and field the systems. The army initiated major retraining activities that involved the most senior leaders (civilian and military) who set direction and policy, the key program managers, and leaders who managed the many aspects of the acquisition process and the thousands of people engaged in the day-to-day processes.

The army, the Congress and the defense industry were changed. The whole process of awarding major contracts was altered. Contractors now had to meet MANPRINT criteria to win the large, lucrative contracts (Booher, 1990).

In order to expedite learning and minimize the impact of an early failure, a number of pilot projects were selected. These were chosen because they provided experience in each phase of the acquisition process and provided experience in procurement procedures. The Light Helicopter Experimental (now the Commanche) was selected as the key pilot program because of its visibility, need, and industry environment. Six functional areas—manpower, personnel, training, human factors, system safety, and health hazards—of the army were identified and empowered to act as a project team. This process gave broad ownership, facilitated involvement, and *provided many opportunities for success*. Supervision of critical components was retained at a high level.

The high visibility of the light-helicopter project unmistakably demonstrated tremendous senior leadership commitment. The pilot project was the subject of an active information campaign ranging from keeping the senior leadership informed to routine briefings to newly assigned personnel. Such visibility not only sustained support, but allowed it to grow (Blackwood and Riviello, 1994).

The senior military leadership in concert with Thurman identified a critical issue, created a vision, and linked the vision to an absolute core value: Remember the Soldier. Careful planning, visibility, sustained high-level support, early wins in highly visible pilot efforts, involvement of the total army community, and institutionalization of the process helped drive necessary work culture and systems change.

20.5.3 Defense Printing Service: Transition to a High-technology Team

In the Defense Printing Service (DPS), the new director had created a new high-level vision for the organization just consolidated from the previously self-contained air force, army, and navy printing establishments. This consolidation took place at the same time that congressionally mandated downsizing was heavily impacting the Department of Defense. The director built support for a new vision by being factually honest about the downsizing that would go on across several years and by involving his senior team in its further development.

Initially, the director pulled together the area directors in three different off-site, facilitated conferences over a period of six months. This in itself was a new way of doing business, because the (now) 25 people had never been in the same room together. "First, as individuals, we had to work on ourselves," then develop the new vision for the Defense Printing Service tied to a new role as a "corporate board of area directors" (Hay Management Consultants, 1996).

Once the area directors began to coalesce into a team, they developed their vision, which included:

- "Going digital" or radically changing the nature of their business from printing paper documents to offering customers a new, value-added service: automating documents by capturing their information in computer-ready formats.
- Creating a "team concept" all the way down through the organization where employees were no longer "Army," "Navy," or "Air Force." Instead, everyone would become a member of the worldwide DPS team so that, to their customers, geographic location would become immaterial.
- Setting a new goal for communications: replacing the grapevine as the avenue for communicating change. "We wanted the bottom person in the organization as aware of and as enthusiastic about selling to customers as the top" (Hay Management Consultants, 1996).

For Defense Printing, reinventing themselves took senior management's building trust as a team and learning to "think outside a box where we had been thinking in a box".

At a different level, the key leadership at DPS gave printers the opportunity to mourn the loss of their presses combined with the tactile experience of and ability to produce finely crafted, crisp paper products when computer technology replaced the century old offset presses. Leaders transformed the pride in workmanship into continuing to take pride in having technology that would provide the customer the most effective, efficient service possible. And, the new vision for DPS also built on the concept of each print shop opening the door to a "worldwide" team. This new integrated approach enables a leveraging of skills and capabilities to bring added technical resources to bear anywhere in the world. This team could bring technical know-how to bear for the user, by producing documentation that was more easily updated, transported, stored, and accessed for the ultimate user.

While DPS was undergoing downsizing and significant change, the director identified the various constituencies (see Fig. 20.5), and determined a need to target the union, congress, and private contractors.

At DPS unions are very significant. The director met with the national head of the employee union, and key DPS senior management worked with union members to ensure that a consistent message was being delivered internally to employees and to the union. The director and senior managers clearly articulated the changing environment and the need for moving forward with their change vision to internal employees and to union officials. They worked with the union to accomplish change.

Congressional scrutiny of DPS's change effort was intense. The environment was so hostile that one senior DoD official was maligned as a "liar" before a congressional committee. The director conferred with key committee staff with whom he had developed relationships in order to deliver, once again, a consistent message concerning the changing environment and the requirement for discontinuous change.

The printing contractors who had been printing massive amounts of DoD documents were concerned about losing business. Their concerns generated pressures from Congress and negative articles in trade publications. The director took the DPS case to the printing community using multiple channels of communication. He attended national conferences and talked with trade publications, emphasizing the fact that DPS needed to maintain certain core functions. Everything else would be put "on the table" for contracting to the private sector. As one senior DPS official noted, "The private sector printing community thought they could do things better. We said, 'You'll be a partner'" (Hay Management Consultants, 1996).

The report *Progress in Reinventing the Defense Printing Service: Towards Document Automation* summarized initial dramatic outcomes. Traditionally, DoD organizations have had technical manuals printed in bulk, destroying old inventory whenever manuals were updated. One supply center disposed of 11.9 million obsolete technical manuals in a two-year period. Using the DPS Technical Manual Publish On Demand System (TPODS) means fewer errors and less cost. Printing on demand at the local level coupled with the ability to input revisions by using a computer, and distribution by CD-ROM reduced the average charge for a technical manual to $9.09 from $23.21 for traditional printing, storage, and distribution.

The proverbial red tape (generated by the desire of hierarchies for controls) was reduced significantly in the procurement and billing cycles by pushing down authority from head-quarters levels, resulting in reduced internal directives by 35 percent and reporting requirements by 50 percent. Another outcome was the reduction of both headquarters' staff and field office overhead staff by 20 percent by the end of 1995. DPS headquarters made itself accountable by signing a contract with each field office concerning headquarters responsibilities to provide service to the field. Field offices evaluate headquarters' services annually, and those evaluations affect the performance ratings of senior managers.

> Product and service innovations tend to come from customers, clients, vendors, people in the labs and people on the front lines, while process innovations tend to come from people doing the work. (Kouzes and Posner, 1995)

Successful change leaders, whether in the private sector or in public organizations, depend upon the energy, creativity, and commitment of their entire workforce. The director of DPS understood that the focus on empowering employees—predominantly skilled union crafts people—would result in "reinvention in action." Reengineering was imperative to the organization's survival, but focusing on components of the culture was the platform for successful process innovation (Hay Management Consultants, 1996).

Jack Welch's dictum that "managing less is managing better" holds equally true in the public sector, as demonstrated by the kind of success achieved by DPS. The significant role of training as a means of influencing the culture is also a distinctive element in the transformation of GE, the weapons acquisition process in the army, and in the DPS.

20.6 CONCLUSION

> In this and like communities, public sentiment is everything. With public sentiment nothing can fail. Without it nothing can succeed.
> —Abraham Lincoln (First Lincoln-Douglas Debate)

Collectively, the three phases of change and their associated steps represent a model for designing, initiating, and implementing discontinuous change. The realization of such a daunting undertaking requires significant resources and high levels of commitment by participants. Each of the components of change identified earlier in this chapter must be in place if the process of change is to succeed. As Figure 20.8 illustrates, when key elements of change are absent, planned changes are derailed. Adopting a static rather than a dynamic responsive vision, for example, leads to diffuse and directionless actions and decision making by the organization and its leaders. Similarly, the failure to align the

Static Vision	– Many Initiatives – Lack of Direction
Failure to Identify and Change Work Culture and Values	– Confusion – Conflict – Mixed Messages
Out of Date Leadership Behaviors and Competencies Inadequate Rewards	– Inadequate Solutions – Failure – Morale Problems
Non-Aligned Structure, Processes, Technology, Systems, and Measures	– Failure to Sustain Change – Inefficiencies

Figure 20.8   Consequences of omitting change components.

levers of structure, processes, technology, and measures results in a continuation of inefficiencies and an overall failure to sustain planned innovations.

Challenges and barriers to the process of change occur at all points. They are, however, most significant at two watersheds: the design and deploy phases. The initial step of the design phase, Step 1.1 (Link Vision to Values and Redesign Work Cultures) proved the catalyst for initiating the macrolevel changes outlined in this chapter. It is the top leadership at any large organization that must initiate major change. Further, senior leaders are the source of the new vision. They must agree on what the vision includes and translate it into a message that resonates with the core values of the rank-and-file who will ultimately make the vision real.

Within any complex organization there are an array of demographic and occupational groups. If such diverse groups are to be persuaded that change is in their interest, leaders must tailor the message to each constituency. Leaders are faced with the challenge of connecting with core values while presenting an alternative way of doing business. Furthermore, the scale of change required necessitates the early involvement of a senior leader group acting as a single voice, articulating the change message consistently. Leaders must invest a considerable amount of energy and thought early in the change process if the discontinuous change effort is to succeed. The message must be broad enough to appeal to all groups; leaders at all levels of the organization must help group members coalesce around a shared vision of the new work culture.

At the deploy phase a different set of challenges appears. Now the organization has undergone major transformation and is better prepared to meet its changing mission. Recently minted, the changes initiated may be tenuous. If individuals in the organization are to be discouraged from reverting to former, now inappropriate behaviors, renewed effort has to be devoted to maintaining the new structures. Leaders must ensure that the organization continues to respond to changing circumstances, both internal and external. This phase has two primary goals: formalizing the organization's responses to desired behaviors, and achieving alignment among the new organizational levers. Motivated employees are the centerpiece in a successful organization. Step 3.1 (Reinforce Desired Behaviors

Through Rewards and Recognition) focuses on the need to tailor compensation policies to the new work culture and the composition of the organization's workforce. Leaders are at the center of this effort, particularly in the continuing need to model new behaviors themselves, to monitor others' behavior, and to align the compensation and reward system with desired outcomes.

The final step, 3.2 (Align People, Processes, and Systems), is an essential one to accomplish if the organization is to function as a coordinated entity. Simply put, alignment means that components of the organization are not working at cross purposes. Older, inappropriate ways of doing business are gone, replaced with new planning, human resource management, and monitoring systems that collectively move the organization forward. Such a transformed organization is a dynamic one, prepared to effectively confront a continuously changing political, social, and economic landscape.

In summary, all initiatives must be aligned to achieve the desired results. A key to accomplishing alignment is integrating change throughout the process. This process begins by defining the case for change and painting a clear picture of the desired future state in terms of work processes, organization structure, and new behaviors. It is sustained by identifying multiple change leaders to build commitment and reduce resistance to the change and by developing a process to involve all employees in the change effort. And finally, a major, frequently overlooked activity is the development of a detailed plan to sustain the change process, identifying critical steps, clarifying roles and accountabilities, and defining measures of success.

REFERENCES

Ashkenas, R., Urich, D., Jick, T., and Kerr, S. (1995). *The Boundaryless Organization: Breaking the Chain of Organizational Structure.* San Francisco: Jossey-Bass.

Bartlett, C. A., and Ghoshal, S. (1994). Changing the role of top management: Beyond strategy to purpose. *Harv. Bus. Rev.,* November/December, pp. 79–88.

Bartlett, C. A., and Ghoshal, S. (1995). Changing the role of top management: Beyond systems to people. *Harv. Bus. Rev.,* May/June, pp. 132–142.

Beckhard, R., and Pritchard, W. (1992). *Changing the Essence: The Art of Creating and Leading Fundamental Change in Organizations.* San Francisco: Jossey-Bass.

Bennis, W., and Townsend, R. (1995). *Reinventing Leadership: Strategies to Empower the Organization.* New York: Wm. Morrow.

Blackwood, W. O., and Riviello, R. N. (1994). *Organizational Change: Lessons Learned from MANPRINT,* Final Report. Washington, DC: Department of the Army.

Booher, H. R. (ed.), (1990). *MANPRINT: An Approach to Systems Integration.* New York: Van Nostrand-Reinhold.

Burke, W. W., and Litwin, G. (1992). A causal model of organizational performance and change. *J. Manage.* **18**(3), 523–545.

Champy, J. (1995). *Reengineering Management: The Mandate for New Leadership.* New York: Harper Business.

Collins, J. C., and Porras, J. (1994). *Built to Last: Successful Habits of Visionary Companies.* New York: Harper Business.

Director, Defense Printing Service. (n.d.) *Progress in Inventing the Defense Printing Service Towards Document Automation.* Washington, DC: Defense Printing Service.

Drucker Foundation. (1996). *The Leader of the Future: New Visions, Strategies, and Practices for the New Era*. (F. Hesselbern et al.) eds. San Francisco: Jossey-Bass.

Executive Forum. (1994). *Executive Forum on Changing the Enterprise*. Fairfax, VA: George Mason University.

Executive Forum. (1996). *Executive Forum on Enterprise Planning and Management Change Strategies*. Fairfax, VA: George Mason University.

Farkas, C. M., and Wetlaufer, S. (1996). The ways chief executive officers lead. *Harvard Bus. Rev.*, May/June, pp. 110–122.

Farquar, K. W. (1995). Guest editor's note: Leadership transitions—current issues, future directions. *Hum. Resour. Manage.*, Spring, pp. 3–10.

General Electric. (1992). *1991 Annual Report: To Our Share Holders* (J. Welch and E. Hood) Fairfield, CT: GE.

General Electric. (1995). *1994 Annual Report*. Fairfield, CT: GE.

Ghoshal, S., and Bartlett, C. A. (1995). Changing the role of top management: beyond structure to processes. *Harv. Bus. Rev.*, January/February, pp. 86–97.

Goldman, S. L. et al. (1995). *Agile Competitors and Virtual Organizations: Strategies for Enriching the Customer*. New York: Van Nostrand-Reinhold.

Handy, C. (1995). Trust and the virtual organization. *Harv. Bus. Rev.*, May/June, pp. 40–50.

HayGroup. (1996). *People, Performance, and Pay: Dynamic Compensation for Changing Organizations* (T. P. Flannery, D. Hofricher, and P. Platten). New York: Free Press.

HayGroup Conference. (1994). Ensuring leadership for the future. *Developing Global Leadership* (Hay/McBer). Philadelphia: HayGroup.

Hay Management Consultants. (1996). *Beyond Reengineering: A Behavioral Approach to Leading Change in the Department of Defense* (Contract DASW01-95-F1931). Arlington, VA: Hay Management Consultants.

Hay/McBer Innovational and Leadership Center. (1995). Leadership for the 21st century. *Newsbreak Bull.*, August 2 (2).

Hirschhorn, L., and Gilmore, T. (1992). The new boundaries of the "boundaryless" corporation. *Harv. Bus. Rev.*, May/June, pp. 105–115.

Holden, M., Jr. (1991). Why entourage politics is volatile. *Managerial Presidency*, pp. 61–77.

Jones, T. W. (1995). Performance management in a changing context, monsanto pioneers a competency-based developmental approach. *Hum. Resour. Manage.* 34 (3, Fall), 425–442.

Kelly, R., and Caplan, J. (1993). How Bell Labs creates star performers. *Harv. Bus. Rev.*, July/August, pp. 128–136.

Kotter, J. P. (1995). Leading change: Why transformations fail. *Harv. Bus. Rev.*, March/April, pp. 59–67.

Kotter, J. P., and Heskett, J. L. (1992). *Corporate Culture and Performance*. New York: Free Press.

Kouzes, J. M., and Posner, B. Z. (1995). *The Leadership Challenge: How to Get Extraordinary Things Done in Organizations*. San Francisco: Jossey-Bass.

Kriegel, R., and Brandt, D. (1996). *Sacred Cows Make the Best Burgers: Paradigm-busting Strategies for Developing Change-ready People and Organizations*. New York: Warner Books.

Larken, T. J., and Larken, L. (1996). Reaching and changing frontline employees. *Harv. Bus. Rev.*, May/June, pp. 95–104.

Lawler, E. E. (1994). From job based to competency based organizations. *J. Organ. Behav.* 15(1), 3–15.

Mintzberg, H. (1996). Managing government, governing management. *Harv. Bus. Rev.*, May/June, pp. 75–83.

Nadler, D. A., Shaw, R. B., and Walton, A. E. (1995). *Discontinuous Change: Leading Organizational Transformation.* San Francisco: Jossey-Bass.

Negroponte, N. (1995). *Being Digital.* New York: Knopf.

Osborne, D., and Gaebler, T. (1992). *Reinventing Government: How the Entrepreneurial Spirit is Transforming the Public Sector.* Reading, MA: Addison-Wesley.

O'Toole, J. (1996). *Leading Change: The Argument for Values-based Leadership.* New York: Ballantine Books.

Pfeffer, J. (1992). *Managing with Power: Politics and Influence in Organizations.* Boston: Harvard Business School Press.

Price Waterhouse Change Integration Team. (1996). *The Paradox Principles: How High-performance Companies Manage Chaos, Complexity, and Contradiction to Achieve Superior Results.* Chicago: Irwin Professional Publishing.

Rouse, W. B. (1993). *Catalysts for Change: The Strategy of Innovative Leadership.* New York: Abington Press.

Safire, W. (1996). On language: Downsized. *N.Y. Times Mag.,* May 26, pp. 12, 14.

Shields, J. L. (1995). *Civil Service Reform: Changing Times Demand New Approaches.* Testimony before the Subcommittee on Civil Service, Committee on Government Reform and Oversight, House of Representatives, October 26.

Shields, J. L., and Adams, J. (1991). *Competency Requirements of Managerial Jobs in the Public and Private Sectors: Similarities and Differences.* Arlington, VA: Hay Systems.

Slater, R. (1994). *Get Better or Get Beaten! Leadership Secrets from GE's Jack Welch.* New York: Irwin Professional Publishing.

Spencer, L. M., and Spencer, S. M. (1993). *Competence at Work: Models for Superior Performance.* New York: Wiley.

Steininger, D. J. (1994). Why quality initiatives are failing: The need to address the foundation of human motivation. *Hum. Resour. Motivation* **33**(4, Winter), 601–617.

Tichy, N. M., and Charan, R. (1995). The ceo as coach: An interview with Allied Signal's Lawrence A. Bossidy. *Harv. Bus. Rev.,* March/April, pp. 68–79.

Tichy, N. M., and Devanna, M. A. (1996). The transformational leader. *Train. Dev. J.,* July, pp. 27–35.

Tichy, N. M., and Sherman, S. (1994). *Control Your Destiny or Someone Else Will: Lessons in Mastering Change—From the Principles Jack Welch is Using to Revolutionize GE.* New York: Harper Business.

Wilson, J. Q. (1989). *Bureaucracy: What Government Agencies Do and Why They Do It.* New York: Basic Books.

21 Model-based Design of Human Interaction with Complex Systems

CHRISTINE M. MITCHELL

21.1 INTRODUCTION

This chapter proposes a methodology for designers of human interaction with complex systems. It contains a generic characterization of the proposed methodology—model-based design—and a description of a *specific* methodology, which uses the operator function model (OFM) (Mitchell, 1996a). The chapter begins with an overview of the current state of design for human–system interaction in complex systems and presents a set of critical design issues that model-based design can help address.

21.2 HUMAN INTERACTION WITH COMPLEX SYSTEMS: THE SYSTEMS, TASKS, AND USERS

Human–system interaction addresses the set of issues concerning the manner in which a user or operator interacting with a system carries out activities to meet system or user objectives. The particular forms or styles of human–system interaction govern the extent to which both the overall system and users meet objectives in efficient and safe ways and reduce the occurrence of human error.

In many systems, human–*system* interaction includes human–*computer* interaction and its associated design issues. Given the proliferation of computers throughout the home, office, and industrial workplace, humans often interact with systems through a computer medium. Thus, humans "program" a computer that in turn instructs the system. Human–system interaction via a computer medium occurs in the kitchen (we program the microwave for a given power level and time), family room (we program the television/VCR to record a desired show), office (we configure word processing and spreadsheet templates), and workplace (pilots program the flight management system and manufacturing shop-floor personnel "teach" desired operations to a computer-based machining station).

Human–system interaction includes a wide variety of systems and a wide variety of humans who interact with them. One useful way of organizing and sorting out the myriad of associated design issues is to consider types of users and types of systems with which the users interact. Design issues and characteristics are widely acknowledged to be context-dependent, that is, both task- and user-dependent. As most human factors engineering practitioners (Cardosi and Murphy, 1995) and human–computer interaction texts (Kantowitz and Sorkin, 1983; Preece, 1994) point out: identifying and understanding the in-

Handbook of Systems Engineering and Management, Edited by A. P. Sage and W. B. Rouse
ISBN 0471-15405-9 ©1999 John Wiley and Sons, Inc.

tended users and specifying the users' needs are *the* fundamental prerequisites to effective design.

This chapter concerns systems that are sufficiently complex that personnel who manage them must be trained in system operation or maintenance. As such, we characterize both the types of systems and design issues of interest, and circumscribe, to some extent, the generalizability of the discussion.

The term *operator* is used rather than the more general term *user* to indicate the types of systems and human–system interactions under discussion. This chapter concerns work-related systems, as opposed to, for example, home-entertainment systems. The operators of such systems are usually highly trained, professional personnel. They have a set of reasonably well-specified goals with respect to system performance and, likewise, a set of reasonably well-defined activities with respect to system operation and maintenance. Thus, operator refers to a person who is "well-trained and well-motivated" with respect to the operation of a particular system (Baron, 1984; Sheridan and Ferrell, 1974).

Users who are casual, intermittent, or naive do not characterize the types of human–system interactions addressed in this chapter. We hasten to note that such interactions are not unimportant: for some systems, each of us is a naive or casual user and Norman's (1988) *Psychology of Everyday Things* speaks poignantly to the importance of effective design for such systems. Design issues, however, are far different for operators who are trained in system operation than they are for casual users. Systems with operators are typically complex and the cost of error may be high, and occasionally catastrophic. Often such systems are dynamic; that is, the state of the system evolves over time without waiting for operator input. Error-free activity or designs that allow operators to recognize and recover from errors are important, since the costs of unrecovered errors may be unacceptable. Examples of such systems include aerospace, aviation, air-traffic control, process control, telecommunications, and manufacturing systems.

21.3 EMERGING TECHNOLOGY AND DESIGN

New computer-based information and control technologies require careful consideration of human factors and human–system interaction issues. New technologies succeed or fail based on a designer's ability to reduce incompatibilities between the characteristics of the system and the characteristics of the people who operate, maintain, and troubleshoot it (Casey, 1993). Moreover, well-designed human interfaces are essential to ensure reliable human performance and system safety (International Nuclear Safety Advisory Group, 1988; Moray and Huey, 1988; O'Hara, 1994; Woods et al., 1994). O'Hara (1994) suggests that safety depends, in part, on the extent to which a design reduces the chances of human error, enhances the chances of error recovery, or safeguards against unrecovered human errors, e.g., provides an *error-tolerant* system (Billings, 1997; Nagle, 1988; Rasmussen, 1987; Woods et al., 1994).

Experience in a wide variety of systems and applications suggests that the use of sophisticated computer technology in the human interface, such as computer displays, controls, and operator aids, raises serious issues related to the way humans operate, troubleshoot, and maintain these systems (Billings, 1997; Casey, 1993; Sheridan, 1992; Woods et al., 1994). Several recent studies highlight the importance of the "human factor" when incorporating computer technology in safety-critical systems. One study, conducted by a subcommittee convened by the Federal Aviation Administration (FAA), found that interfaces between flight crews and modern flight decks to be critically important in achieving the administration's zero-accident goal (Abbott, 1996). The committee, however, noted a wide range of shortcomings in designs, design processes, and certification processes for current and proposed systems. Other examples in which the human factor is identified as a significant contributor abound. Two recent studies of incidents in nuclear power plants identified human factors as the primary concern. One study, which classified failures in nuclear plants that included digital subsystems, found that human factors issues, including human–machine interface errors, were a "significant" category (Ragheb, 1996). The other study noted that over the 13-year study period, the trend in most error categories was decreasing or flat, whereas events attributable to inappropriate human actions "showed a marked increase" (Lee, 1994). Woods and his colleagues (1994) report that a survey of anesthetic incidents in the operating room attribute between 70 and 75 percent of the incidents surveyed to the human element. Moreover, in a set of diverse applications, including medical, aerospace, process control, and aviation, Woods et al. (1994) identify characteristic and recognizable human–system interaction problems that manifest themselves repeatedly across a broad and varied range of systems. These systems, however, all have in common newly introduced "automation" or human interaction with advanced technology systems.

As a result there is a widely held belief that while overall reliability in complex and/or safety-critical systems is increasing, the number of problems attributed to human error is increasing or, at least, not decreasing. The FAA committee on human factors stated that human-related problems, as a percentage of the whole, are increasing (Abbott, 1996). Billings (1997) notes. . . "that it has long been an article of faith that from 65% to 80% of air transport accidents are attributable in whole or part to human error" (p. 4).

Human factors engineers and researchers are quick to note that many of these problems are *design* problems, not inherent deficiencies of the technology (Abbott, 1996; Sheridan, 1992; Wiener, 1989; Woods et al., 1994). Skillful design that uses emerging technology effectively can make a system safer, more efficient, and easier to operate (Abbott, 1996).

Many industries currently make extensive use of digital technology: Fossil-fuel plants, a wide range of process control industries (e.g., textile, steel, paper), manufacturing, aerospace, aviation, and air-traffic control systems increasingly use digital technology for operator displays and controls, aids, and control systems themselves. Despite many computer glitches, most industries perceive a substantial benefit to overall system safety and effectiveness by incorporating digital technology. The most striking example may be in aviation. Although there are many areas that require improvement, incorporation of digital technology in commercial aircraft is widely believed to have increased overall safety and system efficiency (Abbott, 1996). Reviews of cockpit automation such as those appearing in *Aviation Week and Space Technology* (1995a, b) in early 1995, note problems or glitches in the human interface, but none of the parties involved in the flight-deck dialog (e.g., pilots, airlines, air frame manufacturers, or regulatory bodies) suggests that these problems necessitate a return to conventional analog technology. The belief is that digital technology, despite many serious and persistent problems, is often beneficial, and, with more effective design and evaluation, will continue to improve the effectiveness of the overall system (Abbott, 1996; Parasuraman and Riley, 1997). Thus, the challenge is to *identify* important human–system interaction issues and to develop design methodologies to address these issues so that safe and effective systems are assured.

21.4 HUMAN–SYSTEM INTERACTION ISSUES

The proliferation of inexpensive computers and computer-interface technologies raises a seemingly uncountable set of design opportunities and challenges. Historically, design and implementation of new concepts or technologies were slow and change was incremental. For the past 10 to 20 years, however, there have been tremendous economic and social pressures to use rapidly emerging and relatively inexpensive digital technologies to enhance system performance and safety, while concurrently reducing human error. Human factors specialists suggest, however, that the research base supporting the design of human–system interaction in advanced technology systems, in comparison to knowledge available to previous generations of designers, has not kept pace with technology development (O'Hara, 1994). O'Hara and his colleagues suggest that there are many more *issues* than *answers* for how to design computer-based interactions with complex systems.

Some guidance does exist. In fact, some argue that too many guidelines exist, with the result that designers are often confused and even the best guidelines ignored (Lund, 1997; Woods, 1992). Lund (1997), in a recent review of human–computer interface guidelines, reported on recent guideline development efforts including the compilation of 3,700 guidelines for the Belgian government with an accompanying bibliography of 1,200 entries (Vanderdonckt, 1994). Available guidelines include the widely cited Smith and Mosier (1988) guidelines for "generic" human–computer interaction, Space Station Freedom Human-Computer Interface Guidelines [National Aeronautics and Space Administration (NASA), 1988], Human Factors in the Design and Evaluation of Air Traffic Control Systems (Cardosi and Murphy, 1995), and User Interface Guidelines for NASA Goddard Space Flight Center (NASA, 1996). The guidance contained in such documents is limited, however. Anthologies of guidelines primarily address low-level physical or generic human–computer interaction issues rather than higher level cognitive issues (Wadlow, 1996; Woods, 1992). Cognitive issues, such as mode error, cognitive workload, and situation awareness, are increasingly important in complex systems that incorporate high levels of computer control and automation (Parasuraman and Mouloua, 1996).

Guidance that extends beyond guidelines is often conceptual; Lund (1997) calls such guidelines *design maxims*. Most maxims are unarguably correct, but there is no consensus on how to implement them in an actual design. For example, Billings's (1997) *human-centered automation* is a timely concept that should permeate all human–system interaction. But knowledge or methods for how to create a design that is human centered, or to evaluate a design or prototype to ensure that it is human centered do not exist. Design maxims are sometimes formulated as features to avoid rather than characteristics that a design should possess. Wiener's (1989) notion of *clumsy automation* describes an undesirable design feature: automation that "helps" when workload is low but makes excessive demands when workload is high. No one would disagree with the maxim that a design should *not* be clumsy, but methods to avoid creating clumsy automation do not exist.

Such maxims might be thought of as *conceptual normative characteristics*. They provide concepts that a good design should possess or avoid, but are specified at such high levels that there is a great deal of difficulty or uncertainty in understanding how to apply effectively a particular maxim in a specific application.

Finally, because the science and engineering basis of human factors for computer-based systems is so new, little guidance is generally applicable. Most concepts must be carefully implemented as a prototype, and subsequently evaluated in the context of the application under consideration (Cardosi and Murphy, 1995; O'Hara, 1994).

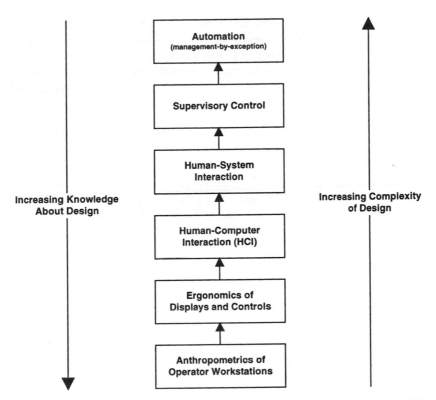

Figure 21.1 Design knowledge for human interaction with complex systems (HICS).

Figure 21.1 depicts a hierarchy of issues related to human factors in the use of advanced technologies in complex systems. The amount of existing knowledge is inversely related to the levels in the hierarchy. Thus, more generally accepted and widely applicable design knowledge is available for lower-level issues. Moving up the hierarchy, design knowledge is less detailed and more conceptual, and design experience is often not applicable across a range of applications (e.g., from office automation to control rooms).

21.4.1 Anthropometrics of Operator Workstations

At the lowest level, anthropometry—the science of establishing the proper size or height of equipment and space—there is a good deal of knowledge. Like guidelines for conventional displays and controls, the hardware associated with computer-based workstations is no longer subject to widespread debate. There are standards and recommendations for computer-based workstations that specify working levels, desk height, foot rests, document holders, task lighting, and viewing distances (Cakir et al., 1980; Cardosi and Murphy, 1995).

21.4.2 Ergonomics of Displays and Controls

At the next level are issues that specify the characteristics of computer-based displays and controls. Issues include font size, use of color, input devices, and types of displays (e.g., visual, audio). Knowledge here blends commonly accepted guidelines with emerging research results, which are often task-specific or user-dependent. For example, despite many disputes when first introduced in the late 1970s, a computer-input device, called a mouse, was empirically shown to produce performance superior to available alternatives for pointing tasks. Today, mice are routinely packaged with computer hardware. The number of buttons on a mouse, however, still varies from one to three. Recommendations for the best design vary depending on task and user requirements, as well as designer and user preferences.

The issue of the "ideal" or "best" number of buttons on a mouse illustrates the state of a great deal of human factors engineering knowledge. Often there is no single, best answer (Wadlow, 1996). In some cases, within some range, some characteristics may not make a discernible difference and users can readily adapt to a range of characteristics. In other situations, an acceptable solution is task- and/or user-dependent. Mock-ups or prototypes and empirical evaluations are needed to ensure that a proposed design meets user needs.

21.4.3 Human–Computer Interaction

Level-three, human–computer interaction (HCI) is the area to which the most study has been devoted. This area receives widespread academic and industry attention. Most current guidelines address this level of consideration. Issues include style of windows, window management, and dialog type. HCI guidelines address application-free, or generic, characteristics of the human–computer interface. From the point of view of design in a complex system, such issues might be thought of as the computer medium's *interface syntax*. Most guidelines for HCI specify attributes that are *likely* to be effective that *may* be desirable, or that *must be evaluated* in the context of the application (see Cardosi and Murphy, 1995). Again, at this level, there is a wide range of acceptable designs and no indication that a single, best design is emerging. Following routine HCI guidelines, this level of human factors issues can be adequately, though not optimally, addressed.

21.4.4 Human–System Interaction

The transition to the fourth level of consideration, human-system interaction, marks the point where many serious issues concerning human capabilities and limitations and human attributes that affect safe and effective system operation emerge. This is also the level where there are many more questions than even marginally satisfactory answers. At this time, the majority of issues arises, and must be addressed, in the context of the application-specific tasks for which a computer interface will be used.

Early conceptual issues, still not adequately resolved, include "getting lost" and the "keyhole effect" (Woods, 1984), gulfs of evaluation and execution (Hutchins et al., 1986; Norman, 1988), and the inability of designers to aggregate and abstract the vast amount of data made available by computers into meaningful information (Mitchell and Miller, 1986; Mitchell and Sundström, 1997; Rasmussen and Goodstein, 1988). Essentially, issues at this level concern the *semantics* of the computer interface: how to design information displays and system controls that enhance human capabilities and compensate for human limitations.

"Getting lost" describes the phenomenon in which a user, or operator, becomes lost in a wide and deep forest of display pages (Woods, 1984). As computer displays proliferate, so too does the potential number of display pages available to operators. One recent description of a new digital nuclear power plant reported that operators have as many as 15,000 display pages from which to choose. Operators are often overwhelmed with choices, and empirical research shows that some operators use information suboptimally in order to reduce the number of transitions among display pages (Mitchell and Miller, 1986; Woods, 1984). When issues of across-display information processing are ignored, the computer screen becomes a serial data presentation medium with which the user has a *keyhole* through which data are observed. The limitations on short-term memory suggest that a keyhole view can severely limit information processing and increase cognitive workload, particularly in comparison to parallel displays used in analog control systems.

The gulfs of evaluation and execution describe attributes of a design that affect cognitive workload. They describe the conceptual distances between decisions and actions that an operator must undertake and features of the interface to support them. The greater either distance, the less desirable the interface (Hutchins, 1983; Hutchins et al., 1985; Norman, 1988). The gulf of evaluation characterizes the difficulty with a particular design as a user moves from perceiving data about the system to making a situation assessment or a forming an intention to make a system change. The gulf of execution characterizes the difficulty with a particular design as a user moves from forming the intention to make a change to executing the action or series of actions required to bring about the change. Display characteristics such as data displayed at too low a level or decisions that require the operator to access several display pages sequentially, mentally or manually extracting and integrating data along the way, are likely to create a large gulf of evaluation. Likewise, control procedures that are sequential, complex, or require a large amount of low-level input from the operator are likely to create a large gulf of execution.

Finally, and particularly true of control rooms in which literally thousands of data items are potentially available, defining information—that is, the useful part of data—is a serious concern. The keyhole effect and getting lost are due in part to the vast number of display pages that result when each sensed datum is presented on one or more display pages in a format close to the manner in which the data are recorded. Rasmussen (1986) characterizes such displays as representative of single-sensor–single-indicator (SSSI) design. Reminiscent of analog displays, and because displays may be used for many different purposes, data are often presented at the lowest level of detail possible, typically the sensor level. There is rarely an effort to analyze the information and control needs for specific operational activities and to use that knowledge to design operator displays and controls that support system control goals given current system state.

A good deal of conventional wisdom characterizing good human–system integration is available with the goal of reducing the cognitive workload associated with information extraction, decision making, and command execution in complex dynamic systems. As noted before, Woods (1984) cautions against designs that create a keyhole effect and proposes the concept of *visual momentum* to mitigate it. Visual momentum proposes that display-page design attempts to limit the number of transitions among pages for a single monitoring or control task and that pages be consistent for the task at hand; operators should not be required to dig around or aggregate low-level data to obtain needed information. Hutchins et al. (1986) propose the concept of *direct engagement* to bridge the gulfs of evaluation and execution, that is, the use of direct manipulation to implement displays and controls. Others propose the use of system or task models to organize, group, and integrate data items and sets of display pages (Kirlik, et al., 1994; Mitchell, 1996b; Rasmussen et al., 1994; Vicente and Rasmussen, 1992; Woods, 1995).

Such concepts are well understood and have broad agreement at high levels. This agreement, however, does not translate into specific or widely accepted design methods or guidelines. For example, everyone agrees that a newly designed interface should not raise the level of required problem-solving behavior as defined by Rasmussen's (1983) SRK (skills–rules–knowledge) problem-solving paradigm, yet agreement for how to design displays that respect this principle do not exist.

Thus, at this time, design of operator workstations is often an art, depending on designer intuition and skills. As such, design prototypes, walk-throughs, and empirical evaluations are all necessary components to ensure effective design (Cardosi and Murphy, 1995). A prototype communicates some things directly from designers to users, such as the set of available displays, display formats, and alternative input options. User preferences, however, do not guarantee effective human–system interaction. Recent data show that user performance does not necessarily improve with designs based on user preferences. When implemented, preferred styles sometimes failed to enhance performance, and in some cases actually degrade it (Andre and Wickens, 1995). Thus, prototypes alone are insufficient. Prototypes and their implementation for carrying out operator control activities must be assessed. Walk-throughs with actual users performing realistic tasks help to ensure effective design for higher-level cognitive and system control issues. Even walk-throughs alone, however, are not sufficient. Karat et al. (1992) compared walk-throughs with experiments. The experiments yielded analyzable data of user and system performance. Subjects in both walk-throughs and experiments were actual users who carried out realistic tasks from the domain in which they were expert. The results showed that experimental evaluations identified new, and sometime very serious, problems that walk-throughs failed to identify. Thus, a proposed design must be rigorously and empirically evaluated to ensure that it enhances significantly operator and overall system effectiveness (Cardosi and Murphy, 1995).

21.4.5 Supervisory Control

Introduced by Sheridan in 1960, the term *supervisory control* characterizes the change in an operator's role from manual controller to monitor, supervising one or more computer-based control systems (Sheridan, 1976). The advent of supervisory control raises even more concerns about human performance. Changing the operator's role to that of a predominantly passive monitor carrying out occasional interventions is likely to tax human capabilities in an area where they are known to be quite weak (Bainbridge, 1983; Wickens, 1991). Specific issues include automation complacency, out-of-the-loop familiarity, and loss of situation awareness.

Concern about automation complacency is widespread in aviation where the ability of pilots to quickly detect and correct problems with computer-based navigation systems is essential for aircraft safety (Abbott, 1996; Billings, 1997; Wiener, 1989). As a result, demands for human-centered flight decks abound (Billings, 1997; Parasuraman and Mouloua, 1996). Yet, to date, there are no proven methods a designer can use to specify displays, aids, or warning systems that enable operators to monitor automation effectively and intervene when necessary. To the contrary, research has shown that it is difficult to design automation that an operator will override even when it is necessary (Roth et al., 1987; Smith et al., 1997).

As with the concepts of visual momentum and direct engagement, the set of conceptual normative characteristics dictate that keeping the operator in the loop and cognizant of computer-based operations are important goals (Bainbridge, 1983; Billings, 1997; Sheridan, 1992). There is, however, no consensus as to *how* to achieve these goals. Most operational designs address the out-of-the-loop issue by periodically requiring the operator to acknowledge the correctness of the computer's proposed solution, despite research data to the contrary (Roth et al., 1987), or have the operator log system state variables at various times (Thurman and Mitchell, 1994). The latter feature is similar in principle to a software-based deadman's switch: it guarantees that the operator is alive, but not necessarily cognizant.

How to achieve a design that supports situation awareness is a recent but pervasive topic (Garland and Endsley, 1995). Pew (1995) notes that "Achieving situation awareness has become a design criterion supplementing more traditional performance measures" (p. 7). He and many other human factors practitioners and cognitive engineers, however, have yet to agree on a definition for situation awareness, measurement methods, or even if the concept of situation awareness is anything more than an abstraction (Garland and Endsley, 1995). If situation awareness is only an abstraction, Billings (1995) concludes that it can be neither measured nor quantified.

Moreover, as supervisory control has become the dominant paradigm, advanced information systems and relatively inexpensive computer technology has made possible a wide variety of aids to help operators, including intelligent displays, computer-based aids/assistants/associates, electronic or intelligent checklists, and error-tolerant systems. These concepts are all appealing. A display or aid that utilizes computer-based knowledge with stable, accessible, and up-to-date information about nominal and off-nominal operations has obvious advantages. Up-to-date information has the potential to enhance greatly operator decision making and problem solving. To date, however, research has not produced prototypes or design guidance that consistently lives up to the expected outcome.

In principle, smart displays can mitigate the SSSI problem and provide display pages that directly support specific operator activities given current system state. Research results are mixed, however. Some designs that sought to enhance performance with direct perception (Kirlik and Kossack, 1993) or direct engagement (Benson et al., 1992; Pawlowski, 1990) found that while such techniques helped during training, the design did not continue to enhance performance once the operator was no longer a novice. Some research has shown that some design methods enhance performance for trained operators (Mitchell and Saisi, 1987; Thurman and Mitchell, 1994), but these methods are in initial stages of testing and are not uniformly known or agreed upon.

As with displays, there are mixed results concerning the effectiveness of specific designs for computer-based aids. When empirically evaluated, some aids failed to enhance performance and in some cases actually degraded it (Kirlik and Kossack, 1993; Knaeuper and Morris, 1984; Resnick et al., 1987; Zinser and Henneman, 1988).

Electronic checklists or procedures are another popular concept. Such checklists or procedures are technically easy to implement and reduce the substantial overhead associated with maintaining up-to-date paper versions of procedures and checklists. The Boeing 777 flight deck includes electronic checklists, and several European nuclear power plants are evaluating them as well (Turinsky et al., 1991). Two empirical studies, however, demonstrate the mixed results that occur when electronic checklists are compared to traditional paper-based versions. In a full-motion flight simulator at NASA Ames Research Center, a study showed that pilots made *more* mistakes with computer-based and smart checklists than with the conventional paper versions (Degani and Wiener, 1993). Another study, in a simulated nuclear power plant control room with certified power plant operators as subjects, also had mixed results. The data showed that during accident scenarios, while computer-based procedures resulted in significantly fewer errors, time to initiate a response was significantly longer than with traditional paper-based procedures (Converse, 1995).

The design of error-tolerant systems is similar. An error-tolerant system is a system in which a computer-based system compensates for human error or helps operators recognize errors and recover before undesirable consequences occur (Morris and Rouse, 1985; Uhrig and Carter, 1993). This notion remains ill-defined. For example, researchers currently do not agree on whether once an error-tolerant system detects a potential error, it notifies the operator, who may take remedial action, or whether the computer-based system assumes control and corrects the problem without human intervention or possibly even notification. Yet error-tolerant systems are necessary to meet new and higher safety goals.

21.4.6 Automation

In the continuum from manual control to full automation, operators are increasingly removed from system control, and in-the-loop familiarity fades. In some systems, control will be fully automatic; anomalies will cause the system to fail safe; and personnel will be notified and eventually repair the automation and mitigate any problems with the controlled system. "Lights out" automation in manufacturing (Shaiken, 1985) and on-going projects in NASA (Brann and Mitchell, 1995) and Navy (Campbell, 1998) are current examples. The design of several recent models of AirBus aircraft is a step in this direction. Some AirBus aircraft such as the AirBus A-320 and A-300, have an electronic envelope that overrides some pilot inputs for given system states (*Aviation Week and Space Technology,* 1998).

There are numerous human performance issues associated with fully automatic systems in which the operator is no longer in the control loop. One issue is whether automation in which the operator is a periodic manager can ever be considered human centered. Another is whether *any* fully automated system can be designed so that an operator can swiftly assume control when the automation fails or engineers and software designers can quickly locate and correct software problems. If so, how should such automation and its interface be designed? What characteristics must the human–system interaction possess to support an operator engaged in fault management rather than control? What characteristics must the human–system interaction possess to enable an operator to assume control? What characteristics must the human–system interaction possess to support engineers and software designers to enable quick and accurate correction of the conditions in the automation software that that caused it to fail?

It is important to note that even with fully automated systems there will always be humans involved with the system. Fault-free software does not exist (Leveson, 1995). Thus, when automation reaches its inevitable limits or system conditions change beyond those for which the software was designed to cope, operators, engineers, and software designers will interact with the system to diagnosis and correct the problems. The design issue is how to specify the human–automation interface(s) so that required personnel can quickly assume system control, carry out fault management activities, and correct the automation software.

21.5 MODEL-BASED DESIGN

This chapter proposes a design methodology to address some of the issues discussed in the previous sections. The methodology, model-based design, proposes that designers *de-*

velop and *use* a comprehensive model of operator activities to specify *all* human-system interactions. That is, all design decisions about operator roles, workstation design, aids, and so on, should be linked to the descriptions and needs specified in the model. The use of an operator model may be one way of realizing a human-centered design concept. The process of creating such a model is almost certain to ensure that designers understand both broadly and in detail, the users and users' needs. Model-based design may decrease the use of "design by intuition" that occurs all too frequently, in both simple and complex systems.

Figure 21.2 depicts some of the issues model-based design can address. This section, as well as Figure 21.2, is generic in form. The sections that follow propose a set of model characteristics for a model intended to guide design, describe a model-based design methodology, and illustrate how to use the methodology to address the design issues in Figure 21.2. After this generic presentation, the next section describes a specific model-based design methodology, one based on the OFM (Mitchell, 1996a).

21.5.1 Background

Design based on a model of operator monitoring and control activities offers one way of addressing many of the unresolved issues in human–system interaction in complex work domains. Displays, for example, can use this type of model to identify and tailor information needed to support various operator activities, given the current system state. Aids can use such a model to structure a knowledge base, which in turn can advise, remind, or assist operators in contextually relevant ways.

New insights, theories, and models are needed, however, if one of the intended uses of a model is to guide design of human-system interaction in contemporary systems (Miller, 1985). As technology evolves and is introduced, systems change, and operator behaviors change from primarily manual actions to predominantly cognitive and perceptual actions. Models that represent operator activities in contemporary systems must include realistic system features and operator behaviors (Baron, 1984). Historically, human–machine sys-

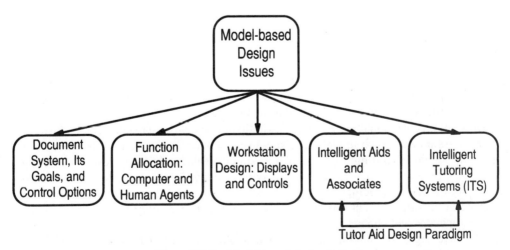

Figure 21.2 Model-based design issues.

tem models often lacked the needed structure to represent fully such systems, their constraints, and the range of operator activities, including discrete manual actions or non-manual actions such as cognitive and perceptual actions (Baron and Levison, 1980; Miller, 1985). Moreover, these models often do not use the increasingly powerful yet inexpensive computing technologies or the knowledge representation techniques developed by the artificial intelligence community over the past 30 years.

Below, a set of characteristics is proposed that an operator model intended to guide design of human–system interaction should possess. They are probably neither sufficient nor comprehensive. The intent is to define a *minimal* set of characteristics. Models for other purposes, for example, cognitive science models whose goal is to represent processes that specify how a human takes input and produces output, are likely to require very different characteristics and structures.

21.5.2 Model Characteristics

Models should reflect the *actual* work domain and make explicit the types of information and range of control activities operators use to interact with the system (Mitchell, 1996b; Morris and Rouse, 1985; Rasmussen, 1988). As Rasmussen cautioned the Human Factors and Ergonomics Society in his 1988 keynote address, context-free models or theories based on research using *simplistic* systems, *unrealistic* tasks, or *naive* subjects are likely to provide little design guidance for real-world applications (Rasmussen, 1988) and may in fact be misleading.

Models of the operator should provide an explicit and detailed representation of operator activities and task constraints. Models should reflect the dynamic nature of the system and environment as perceived by operators given the evolving system state and changing system goals. Representations of operator activities should identify and structure information about important system variables, important collections of states, and define variables that aggregate data or compute measures from low-level data that operators use to assess the current state and to anticipate its evolution. Models should represent *all* operator actions: cognitive and perceptual actions, such as situation assessment and monitoring, as well as the manual actions that traditional task analysis techniques represent (Kirwan and Ainsworth, 1992).

An operator model is likely to be *hierarchic* in form and to represent at least three elements of the overall system: the controlled system, the control system, and the operator or team of operators supervising both. Each level of the hierarchy has important relationships to the levels above and below it. For a given level of abstraction, the current level defines what activities the operator wishes to undertake. The level above indicates *why* the activities need to be performed, while the level below specifies *how* the required activities are to be carried out (Mitchell, 1996a; Rasmussen, 1985).

Operator models should represent the flexibility inherent in human interaction with most systems. As a result, such models are likely to be *heterarchic* at high levels, since most operators juggle several high-level activities simultaneously. At lower levels, flexibility might be represented by a feasible set of activities from which an operator can choose. In this case, the model specifies a set of *acceptable* activities given the current state and higher-level constraints, but does not designate a particular activity from within that set that the operator will choose.

Models of operators in complex systems can take a variety of forms—see Jones and Mitchell (1987) for a more detailed discussion. At one end of the continuum are *conceptual* models; at the other are *analytic* or *computational* models. Concepts from model theory and philosophy of science help to distinguish among various model types (Mitchell and Miller, 1993). From a model-theoretic perspective, a conceptual model is really a *theory* (Sheridan and Ferrell, 1974); whereas, a model is a physical or formal *structure,* for example, a physical scale model, blueprints, or a mathematical or computational formalism, that implements the concepts contained in the theory (Mitchell and Miller, 1993). For example, a conceptual model of troubleshooting, which posits that an operator first uses symptomatic search, followed by topographic search if symptomatic search fails, is really a theory (Rasmussen, 1986; Rouse, 1978); whereas, computational representations of this problems-solving theory implemented in a work domain are models in the sense previously defined (Fath et al., 1990; Hunt and Rouse, 1984; Knaeuper and Rouse, 1985). For design, the complete spectrum of models from conceptual to computational is important.

21.5.3   A Model-based Design Methodology

Model-based design begins with the development of an operator model. Given a model, each design application and design decision within an application is linked to the model. Essentially the model provides a rationale for *all* design decisions.

21.5.3.1   Step I. Develop an Operator Model.   For the system of interest, develop a detailed operator model. Rouse (1980) suggests that one of the primary advantages of modeling is that the modeling process itself is very beneficial. For complex systems, mathematical or computational models help ensure that the complexity and breadth of the system has been taken into account. As Rouse (1980) notes

> (O)nce the model is sufficiently detailed to allow the computer program to produce predictions of performance, then it is not unusual to find predictions to be ridiculous when compared to performance of the real system. . . .Then, the modeling process iterates and testing provides many insights with respect to the system being studied. (p. 2)

The model should posses the characteristics described earlier. It may well include additional features. The following are useful procedures for the construction of an operator model to guide human–system interaction design.

21.5.3.2   Procedures for Model Development

Step 1. Prepare a high-level written description of the system of interest. Specify system goals, exogenous system inputs, major system components, and high-level activities— both those carried out by the automation and those performed by operations personnel. Conduct extensive on-site observations of normal, that is, nominal and off-nominal,[1] operations. In addition, conduct interviews with various personnel and review system

---

[1]Most complex systems define normal operations as nominal—everything goes as planned—or off-nominal— one or more unexpected events occur, but operators are trained to recognize and cope with them. Off-nominal conditions can be thought of as anticipated anomalies. For example, pilots practice procedures for coping with a failed engine: first stabilize the aircraft, then land with one engine.

documentation. Verify the information in the document with various personnel involved in the system, including system designers, management, and, most importantly, operators. Note that for a new system, which has no previous instantiation of any kind, such as a workstation for the International Space Station, the operator model must evolve with the other parts of design. Most systems, however, that involve introducing or upgrading information technology or computer-based control systems have some previous version, even if it significantly different in appearance or operation. The transitions from steam-gauge airplanes and analog nuclear control rooms to glass cockpits and digital nuclear control rooms typify such changes.

Step 2. Prepare an initial, high-level draft of the operator model, that is, identify the high-level activities that operators perform. Again, conduct extensive on-site observations of operations. Ask operators to describe *what* they do to attain system goals defined in Step 1. Interview a range of operators.[2] Review available documentation that describes the system, and most importantly operator functions or procedures. Periodically review the evolving model with operators and other relevant personnel.[3]

Step 3. Define the *heterarchy*. Identify and annotate when and how operators initiate, terminate, or transition among high-level activities. Such annotations relate previous system state and the results of operator actions to the current state or new expectations about operator activities.

Step 4. For each high-level activity, specify an associated hierarchy. For each decomposition and at every level of abstraction, define conditions that initiate or terminate activities. For existing systems, the model may represent operator actions as its lowest level. For a proposed design, the development of an operator model forces specification to the action level—a level often overlooked by system designers. Note that actions can be cognitive, perceptual, manual and so on. Avoid "oversequencing" activities. Many groups of activities do *not* have to be performed sequentially. Assuming a deterministic operator or fixed sequences of activities is a serious flaw in many operator models.

Step 5. Validate the model via direct observation of operators as they carry out control operations under various conditions, nominal and off-nominal. Map operator actions bottom-up into the model. That is, for each observed action, including cognitive or perceptual actions, ensure that the observed action links to the appropriate higher-level activity in the current context. On-line validation while operators carry out normal activities is always preferred. Occasionally, validation may have to be *post hoc*. Then modelers should use, as appropriate, videotapes, extensive notes, display screen snapshots, and event and command logs, to recreate operational contexts. They should then ask one or more operators *what* was done—the actions—and *why*. That is, how the

---

[2]Operators, like all other people, bring various perspectives to a problem or set of problems. It is important that the model synthesizes the perspectives of a variety of operators and, when necessary, resolves seeming contradictions.

[3]In complex systems, many people play important roles, including management, system designers, and operators of similar systems. The goal is to obtain an accurate and balanced model of operator activities. Typically operator verification is the final resolution of differing perspective among various groups involved in the design process. It is not unusual to find that management has vague or even an incorrect understanding about detailed operator activities and procedures (Bradley, 1997).

actions link to currently active activity decompositions or activities that should have been active but were not. The latter case is a model error! Avoid validation that depends on interviews. Interviews provide answers obtained out of context. In complex systems, particularly dynamic systems, context is of the utmost importance. In this process attempt to distinguish between model errors and operator error. An operator activity or action identified in this step, which is *not* represented in the model for the current context, may indicate an error of commission. An activity predicted by the model, which the operator fails to perform or to perform fully, may indicate an error of omission. It is important to note, however, that each of these situations can be due to *model,* as opposed to *operator,* error.

Step 6. Refine the model iteratively. Conduct multiple on-site observations of actual operations to periodically validate and refine the model. Review the model with operators, management, and designers. Continue to review new documentation as the system evolves, particularly new operations or operator training manuals. With each change or addition of design information, update the model accordingly and validate it both by on-site observation of actual operations and by asking domain experts to review the "new" model.

21.5.3.3 *Step II. Use the Model to Guide Design of Specific Applications.* Figure 21.2 depicts various design issues that an operator model can help address. The figure organizes the discussion below, showing how an operator model guides design in various applications.

Issue 1. Document the System, Its Goals, and Control Options. The development of a formal model is one way to ensure and make explicit the most widely touted of all design maxims: "know thy user, and YOU are not thy user" (Lund, 1997). A model of the operator-system interaction helps to ensure that designers have developed a detailed understanding of the overall system, its goals, the *controlled* system, the *control* system, and operator *activities* available to meet system goals. Prepare a short description of the system and its controls, linking goals, limitations, and characteristics to the operator model. This description is typically based on the description prepared in Step I of the model development process. Compare the model-based description with other available system documentation to ensure there are no serious mismatches.

Issue 2. Function Allocation: Human and Computer Agents. The *role* of operators in complex systems should be as well understood and specified as system hardware and software. It is surprising how often system designers consider operators and their roles at the very end of the design process. Hardware is purchased, software is specified, and only then are the roles of operators considered—and that role is often to fill the gaps between computer-based subsystems (Kantowitz and Sorkin, 1983). Designers normally think about the operator's role at a very high level: ensure everything works!

One reason for this situation may be the lack of tools to specify the roles of operators. Hardware and software designers have much better tools than do human–system interaction designers, a situation that well-defined operator models may help to remediate. As noted earlier, currently, design guidance is often very general—maxims or rules-of-thumbs. Guidelines typically address *syntactic* HCI issues, which are far less important than *semantic* human–system interaction issues, for example, the content of displays and the design of high-level operator controls. Furthermore, although there is wide agreement about semantic design principles at high levels, there are essentially no methods to help designers apply them in actual systems.

Increasingly, designers can choose how to allocate functions between human and computer agents. Sophisticated computer technology provides the capability to begin to automate activities previously performed by operators. These are typically activities for which no analytic model currently exists, so expert operators pass on their knowledge to novices by extensive on-the-job training.

Function allocation is not a straightforward decision. Human factors practitioners (Fitts, 1951) suggest that function allocation begin by considering the individual characteristics, strengths, and limitations of both humans and machines/computers, and that then functions should be allocated according to what each does best—a good design heuristic. The oft-repeated acronym MABA/MABA (men are better at/machines are better at) typifies this philosophy. As with most conceptual normative guidance, although this principle sounds good in principle, it is hard to apply and frequently ignored (Chapanis, 1970).

Typically, the limits of technology define operator activities. In many contemporary systems, the prevalent design philosophy dictates that everything that *can be* automated *should be* automated (Kantowitz and Sorkin, 1983). Operators are responsible for performing those activities that remain and, of course, ensuring that *all* activities, human and computer, are carried out effectively in the face of changing or unanticipated system and environmental conditions.

Allocation by default often results in "islands of automation," with operators in such systems acting as bridges between automated subsystems (Rouse, 1985; Shaiken, 1985). Few studies investigate whether operators perform "bridge" functions well. Studies that have been conducted show that operators frequently perform low-level activities such as tending the computers or manually transferring computer-generated output from one computer to another (Brann et al., 1996; Roth et al., 1987). At best, operators performing such activities are underutilized; at worst, they are error-prone (Brann et al., 1996; Roth et al., 1987). Allocation of function by default may result in humans performing tasks for which they are poorly suited or in operator workload that is so low that humans, who are retained in the system to monitor the automation and compensate when it reaches its inevitable limits, are so removed from the system that situation awareness is not maintained and fault compensation is slow or inadequate (Bainbridge, 1983; Sheridan, 1992; Wickens, 1991).

Dynamic or *collaborative* function allocation is an alternative to static allocation or allocation defined by technology limitations (Chambers and Nagel, 1985; Hammer, 1984; Johannsen, 1992; Rouse, 1994). A human-computer collaborative strategy is one in which workload is balanced between operators and computers in response to changing task demands and system state (Jones and Jasek, 1997). Dynamic task allocation and effective human–computer collaboration potentially offer a means of realizing Fitts's principle, with allocation tailored by environmental demands, dynamic control requirements, and an operator's own style (Billings, 1997; Bushman et al., 1993; Jones and Jasek, 1997; Jones and Mitchell, 1995; Rouse, 1981, 1994; Rouse et al., 1990).

Dynamic function allocation, however, requires a detailed, and computationally tractable, representation of operator activities that can guide computer implementation of operator activities and specify the details of how to execute each. Moreover, if an operator model, which the operator understands, guides computer behavior, some of the problems with human–automation communication and understanding may be reduced. An operator model can provide a common frame of reference.

Issue 3. Workstation Design: Displays and Controls. The impact of new technology is most obvious in recent generations of operator workstations, both displays and controls. Historically, control systems were analog (power plant control rooms, cockpits display, process control systems) and sometimes distributed (e.g., manufacturing systems, refineries) (Zuboff, 1991). With the introduction or extensive upgrade of inexpensive computer-based technology into control rooms, operators often work in centralized areas away from the plant and use sophisticated workstations to monitor and control various computer-based subsystems that make up the overall control system.

As discussed briefly earlier, the design of workstations for operators of complex systems must be considered at two levels: syntactic and semantic. *Syntactic* issues concern human–computer interaction. For displays, syntactic issues include formatting, dialog style, windowing, color, and consistency among windows and display pages. For controls, syntactic issues address the choice of input device(s) such as command line, mouse, joystick, or touch screen. *Semantic* issues concern human–system interaction and address issues concerning display content and control functionality. Designers, however, often confuse, or fail to distinguish between, syntactic and semantic issues, or, worse, they fail to realize that *two* sets of issues exist. As a result, semantic issues are often ignored.

In complex systems, however, semantic issues are critical. Research on the design of interfaces for complex systems repeatedly demonstrates that semantic issues, such as display content, level of abstraction, visual momentum across display pages, control functionality, and wide gulfs of execution or evaluation, dominate the effects of primarily syntactic aspects of the human–computer interface (Mitchell, 1996a; Woods, 1992). Most operators, within some envelope, compensate for and adapt to less-than-perfect design at the syntactic level, if they have at their disposal sufficient information and effective controls at the semantic level (Mitchell and Sundström, 1997; Woods et al., 1994).

The design challenge is to understand and represent system goals, structure, and behavior together with required operator activities in such a way that workstation semantics can be specified. Effective displays and controls should support operator activities for which they are used rather than being a potpourri of what is possible to sense or change. Too frequently display content and control commands reflect system hardware and software, sensed data, and hardware-level commands. For example, one system required operators to encode and input commands in hexadecimal representation.

An operator model can help designers bridge the gaps between high-level operator needs and low-level system data and hardware-oriented control commands. Effective displays should provide the correct information, at the correct time, at an appropriate level of detail. Likewise, operator controls should correspond to high-level operator control intentions rather than low-level, often arcane, commands defined by system hardware.

Displayed information should enable operators to develop and maintain a coherent and accurate representation of the current system. Moreover, such displays should allow operators to anticipate accurately how the system will evolve (Bainbridge, 1983) and to be aware of upcoming changes such as indirect (that is, programmed rather than human-activated) mode transitions (Woods et al., 1994). Displays should mitigate problems such as those Wiener (1989) identifies for many pilots with respect to the highly automated flight management systems. Pilots often ask: *"What* is it doing now? *Why* is it doing that? *What* will it do next?" An operator model can help identify needed data, guide the abstraction and aggregation of low-level data, and suggest how to link one or more display pages to support each operator's activities, particularly monitoring, thus, potentially mitigating or reducing the gulf of evaluation (Hutchins et al., 1985).

Controls should permit operators to modify the system quickly and accurately in response to evolving system conditions, failures, or changing system goals. Controls should not require operators to input long sequences of low-level commands. Operator commands should concatenate hardware-level commands into semantically meaningful high-level commands, thus, potentially mitigating or reducing the gulf of execution.

Below is a set of suggested procedures for developing an operator model to help design and validate operator workstations, including specification of the number of monitors that make up the workstation and the sets of operator displays and controls. Model-based design assumes that designers have developed a detailed operator model.

The model construction steps proposed below assume initially that designers have the ability to mock up each proposed display page in software as simple as PowerPoint. Before carrying out the third step (Step c), more sophisticated software is needed to permit dynamic display transitions. Various products such as Visual Studio provide this capability at very little expense. The last step (Step d) requires a simulator, though simulator fidelity need only be sufficiently high to replicate the look and feel and human–computer interaction of the actual operator workstation with the system, thus allowing operators to carry out realistic monitoring and control activities in real time.

21.5.3.4 *Procedures for Developing and Validating a Workstation Prototype*

Step a. Top-down Assessment of Display and Control Prototypes. For each operator activity contained in the model, identify the set of display pages designed to support it. Perform an analysis to ensure that the data are aggregated and presented in a way that meets operator decision-making needs for that activity without imposing excessive workload. Operators should not be forced to perform computer functions such as data gathering, adding, multiplying, and synthesizing low-level data in order to produce information needed for operator control assessments and decisions. Having a computer do the things it does best, that is, quickly and accurately collect and integrate data, is likely to reduce operator error due to workload mandated by real-time computations. This top-down approach ensures that operators are likely to have the right information, at the right time, and at the right level of detail for routinely executed operator activities.

Next, for each operator command—that is, commands designed to reflect high-level operator control intentions, repeat the preceding process: identify the sequence of interface actions required to carry out the command. Even for "conceptual commands" such as situation assessment, analysis should ensure that required display page requests or data synthesis are reasonable and manual operator input is sufficiently simple and meaningful so that operators can execute it in a manner that is likely to be rapid and error free.

Step b. Bottom-up Assessment of Display and Control Prototypes. Identify the collection of displays that designers expect operators to use frequently for routine monitoring and fault detection. This process is bottom-up: for each display page, designers should provide a rationale, with respect to the operator model, for *why* the page was designed, *why* each datum is included on the page, and *why* the format in which the datum is displayed was chosen, for example, level of abstraction and/or aggregation.

Step b is not meant to exclude traditional or hardware-oriented displays. Unanticipated events may require access to low-level system data. Step b is intended to augment the set of displays available to operators and ensure that displays designed to support operator activities in fact do so.

Similarly, identify the collection of operator commands designed to support operator control activities. Again the process is bottom-up: for each command, designers should provide a rationale, with respect to the operator model, for *why* the command was created, *why* the format in which the command is implemented was chosen, for example, checkpoints or required operator input. Commands should be effective and relatively easy to use.

Again note, Step b in not intended to eliminate operator access to low-level commands. The intent is to reduce errors and operator workload by combining into a set of high-level control commands collections of low-level commands that operators frequently use together to achieve specific purposes or system goals.

Step c. Conduct Walk-throughs to Evaluate the Proposed Operator Workstation. This is the stage where the number of monitors composing the workstation should be specified. The number of monitors should be justified by the required activities specified in the operator model: the number of monitors should reflect the tradeoff between display page manipulation and the available space vs. the ability to display all or most of the needed information concurrently for each activity the operators performs.

Jointly, designers and users should evaluate the assignment of displays, or the default assignment of displays, to monitors. A workstation design that uses display real estate — that is, concurrent displays — to support operator activities helps operators maintain situation awareness, and supports visual momentum, thus, reducing the keyhole effect and cognitive workload and enhancing overall system effectiveness. A multimonitored workstation designed along the lines of *system* characteristics, as opposed to day-to-day operator needs, may result if one is not careful to consider workload. With the latter design, operators often have to dig through an increasingly large collection of display pages and synthesize low-level data to obtain required information. Such designs add unnecessary workload.

Next, given a specific workstation prototype, that is, the exact number of monitors and the full sets of displays and controls, conduct an operational walk-through. One or more operators should use the proposed workstation to carry out each activity specified in the operator model. This step begins to integrate the operator workstation with the dynamic system and ensure that the two work together effectively. This exercise, though sometimes labor-intensive, is much more likely to detect problems that increase operator workload and error, or reduce overall system effectiveness, than a traditional workstation design review.

Workstation design reviews are a common technique for obtaining user input. Designers present display pages individually, usually on overhead projectors, and operators are invited to comment or critique the proposed designs. The audiences for such presentations are typically very silent and passive. Thus, designers assume that the proposed workstation is acceptable and, as dictated by good human factors engineering practice, they have obtained user input.

Walk-throughs, on the other hand, actively involve users in the evaluation process.

Designers can observe how operators use the proposed workstation to carry out activities specified in the operator model. Flagrant flaws or omissions will become obvious. Walk-throughs help to ensure that transitions between display pages are few in number and that the overall design supports Woods's (1984) concept of visual momentum, that is, the operators can quickly and effectively obtain needed information and carry out necessary activities. At the conclusion of the walk-through, users and designers, in conjunction with the operator model, should assess the degree to which the overall workstation reduces the gulfs of evaluation and execution.

Demonstrations, walk-throughs, and all other evaluations should include *both* management and users, in addition to system designers. It is important that all the people responsible for the system—designers, managers, operators—share a common understanding of the overall set of operator activities, how each is carried out, and the demands placed on operators during nominal, off-nominal, and unanticipated situations.

Step d. Conduct an empirical evaluation. The design process concludes with a rigorous experiment in which numerous operators use the proposed workstation to carry out a range of routine operator activities, as specified in the operator model. As Karat et al. (1992) found, walk-throughs (Step c) sometimes are not sufficient. Rigorous experiments enable users and designers to detect new, and occasionally very serious, workstation limitations that walk-throughs do not reveal. Experimental data should be evaluated to ensure that the workstation as a whole provides both usability and utility. The data should demonstrate that the proposed design *significantly enhances* operator and system performance (utility).

Issue 4. Intelligent Aids and Associates. In the face of numerous reports of human error in operational settings, the urge to design intelligent systems to guide operator actions is almost irresistible. With hindsight, it is clear that if a current state or consequence of an action was known, the error could have been avoided.

Yet, given decades of research on expert systems, advisory systems, and operator aids, there are few generalizable operational success stories. Empirical studies show that advice-giving systems frequently fail to enhance overall system performance. Sometimes operators do not request proffered advice (Zinser and Henneman, 1988); sometimes operators do not like the "tone" of the advice-giving agent, and thus refuse to follow its advice (Knaeuper and Morris, 1984); and sometimes operators do not think the advice is worth seeking (Resnick et al., 1987). Other aids are effective for novices, but as operators acquire skill, the aid no longer has a discernable effect on performance (Benson et al., 1992; Kirlik et al., 1994).

Despite the lack of demonstrated effectiveness of such systems, advice-giving and warning systems proliferate in operational settings. It is therefore not surprising that operators often question their usefulness. For example, "Bellowing Bertha" on the flight deck commands "Pull-up! Pull-up!" The initial response of pilots may be to silence the alarm first and only subsequently attempt to identify the system state that triggered it. Such behavior was also observed during the accident at Three Mile Island.

Advice-giving and warning systems manifest prototypical limitations across a wide variety of systems. First, such aids rarely integrate multiple limit violations in plausible ways. Warnings, particularly audio warnings, are frequently linked to a single sensor or measurement—an auditory example of SSSI design. In operational settings, a single out-of-limits value for one variable may be acceptable in the context of the values of other variables. Second, control software often lags behind operational requirements.

Factors such as software errors or changes in system dynamics may necessitate a software upgrade that is frequently preceded by long delays. Delayed updates lead to alarms that are obsolete, meaningless, or expected. A visit to almost any control room shows alarms or warnings to which an operator responds "Oh, ignore that. It's not important" or "That value (displayed in red) is now within limits."

Design issues for computer-based aids, associates, and assistants are varied and many. Should the aid be stand-alone, that is, compute and recommend a solution? Or should it have the capability to interact with the operator, who can guide and refine its reasoning? How can operators identify when an aid gives bad advice or has reached the limits of its capabilities? How should knowledge and related advice be structured? Can knowledge be monolithic, that is, compiled, or should the aid help the operator understand the *process* by which it arrived at its recommendation? Should the aid recommend a solution that the operator must accept or reject? Or should the aid help the operator evaluate solutions that the operator rather than the computer proposes?

Billings (1997) suggests that the ability to engage in mutual-intent inferencing by collaborative human and computer agents is a necessary prerequisite of human-centered automation. An aid whose knowledge is structured by an operator model can quickly interpret what the operator is doing and why, and notify the operator if discrepancies between model expectations and operator actions are detected. Moreover, representing the knowledge and rationale of an advisory or aiding systems by means of an operator model is likely to facilitate rapid operator understanding of what the aid is doing and why.

For pilot aiding in particular, and operator aiding in general, intent inferencing provides a first step in defining an intelligent computer assistant: task knowledge and a definition of a common frame of discourse (Callantine et al., 1998; Geddes, 1989; Rouse et al., 1987; Rubin et al., 1988). Models of operator activities, goals, and plans have served as the basis for numerous successful strategies to infer operator intent (Callantine et al., 1998; Geddes and Hammer, 1991; Rubin et al., 1988). Moreover, the model, in computational form, is one way to provide the "intelligence" required by an aiding or associate system.

Extending intent inferencing to aiding or collaboration, an operator model can help designers specify different levels of assistance. The model-based system can structure advice or offer assistance at multiple levels of abstraction, dynamically aggregate and abstract data into useful information, and control its timing using a computational implementation of the operator model that represents the operator's current state as well as system state. Bushman et al. (1993), Funk and Lind (1992), Geddes (1989), Geddes and Hammer (1991), Jones and Mitchell (1995) have all successfully used this paradigm: model-based intent inferencing and an aiding system that uses the model of operator intent as the basis for advising operators or providing assistance.

Issue 5. Training Systems. In contemporary systems, more frequent, formalized, and efficient training to accommodate rapidly changing systems is a necessity. Historically, change in complex systems was slow, and both system and operational knowledge was embedded in organizations by virtue of staff with very low turnover. On-the-job training, with one generation of operators training the next, was common. Today's rapidly changing systems, however, require more frequent training.

The widespread use of digital technology exacerbates this need. Digital systems are almost always characterized by rapid change; and with rapidly changing technologies come rapid changes in how systems are controlled and even configured. Each digital upgrade or added function may result in a changed system.

Further increasing training problems, workforces in modern systems change more rapidly than those of previous generations. Workforce mobility is increasing and a system, which regularly had operators with 5, 10, or 20 years of experience, may now have an operational staff that changes completely on an annual basis. Thus, relying upon on-the-job training is no longer sufficient.

As such, training has become an increasingly important and challenging issue. In lieu of an institutional memory whose repository is personnel with extensive and varied operational experience, an organization must explicitly define and document operational procedures, fault management techniques, and make explicit the implicit knowledge that was traditionally informal, heuristic, and passed on to new generations of operators implicitly. Moreover, such information must be communicated to new employees, or to existing employees being trained on new and possibly much more sophisticated systems, in timely and cost-efficient ways.

Computer-based training systems and intelligent tutors, that is, computer-based training systems that tailor instruction to individual student needs, have great potential to partially fulfill growing training needs. Inexpensive computer technology offers a range of tools to support knowledge consolidation (Mitchell, 1997b; Mitchell et al., 1996; Morris and Mitchell, 1997; Ockerman and Mitchell, 1998) that can also be used as the basis of knowledge to construct an operator model and use it to design and implement training systems that in turn teach this knowledge to operations personnel (Chappell et al., 1997; Chu et al., 1995; Govindaraj et al., 1996; Mitchell, 1997a; Vasandani and Govindaraj, 1994).

An intelligent tutoring system (ITS) typically has three components: an expert model, a student model, and a pedagogical component. For systems that train operators to control a dynamic system, a fourth component is often added: a simulated environment to emulate, from the user's point of view, the actual work environment (Fig. 21.3). Chu

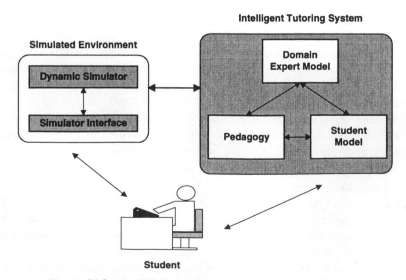

Figure 21.3 An ITS for operators of complex dynamic system.

et al. (1995) suggest that an intelligent tutoring system for operators of a dynamic system consists of three knowledge components: declarative knowledge about the system; procedural knowledge about how to control it; and operational skill that integrates declarative and procedural knowledge into real-time control of the system. The expert model encodes expertise that may include declarative and procedural knowledge as well as contextualized operational skill. The student model represents a student's current level and range of understanding and skill. The pedagogical component structures and controls instruction. When the system includes a simulator, the pedagogical component, in conjunction with expert and student models, use the simulator to construct and control "interesting" scenarios and provide contextually appropriate instruction or coaching. Moreover, for dynamic systems, a simulator, running in conjunction with a computer-based training system, provides a *protected* environment that allows a system to train operators and provides an opportunity for operators to practice operational skills for nonroutine or unsafe events that occur rarely in the actual domain.

The effectiveness of an intelligent tutor depends extensively on the quality of two models: the expert model and the student model. An operator model is a natural representation for the procedural knowledge and operational skill portions of the expert model. A large portion of the expertise embedded in an operator model is precisely the knowledge an intelligent tutoring system attempts to teach the student—the expert model. Student models can take many forms. Overlay models, however, have been widely and successfully used in intelligent tutoring systems (Carr and Goldstein, 1997). An overlay student model has the same structure as the expert model and, over time, the ITS annotates the student model to indicate that the student has acquired another piece of knowledge or skill as represented in the expert model. Comparison of these models indicates the degree to which the student model matches the expert model, that is, the extent to which the student has learned the material encoded in the expert model or areas in which additional instruction is needed. Thus, an ITS using a model of operator activities as the basis of both its expert and student models addresses two important ITS design issues.

Issue 6. The Tutor-aid Paradigm. Figure 21.2 depicts a relationship between training and aiding. Chu et al. (1995) call this concept the tutor-aid paradigm. The expert model in a tutor and the knowledge base embedded in an aid contain essentially the same information. The development of structurally similar models and model-based architectures for both tutors and aids holds out the promise of combining into a single software system the knowledge and resources currently required to build and maintain two separate systems. If successful, such integration may greatly reduce costs while increasing the reliability and sophistication of the resulting training–aiding system. As important from an operator's point of view, a tutor that smoothly evolves to an aid has the potential of becoming a well-understood and trusted assistant.

21.5.4 Conclusions

As the above discussion suggests, operator models for complex systems potentially provide designers with well-defined knowledge for systematically specifying and verifying many components of human interaction with complex systems. Given this background, the first, and probably most important, issue is the selection or construction of a suitable model.

The next section describes such a model—the operator function model (OFM)—and illustrates how OFM-based design is carried out.

21.6 MODEL-BASED DESIGN USING THE OPERATOR FUNCTION MODEL

The operator function model (OFM) was developed at Georgia Tech's Center for Human-Machine Systems Research (Mitchell, 1987, 1996a). The model itself has changed and matured over the years in response to the needs of a range of domains and applications to which it has been applied. It is certainly not the only modeling representation that can be used for model-based design, but it is one that has a demonstrated track record of real-world applications together with empirical data that support the effectiveness of the resulting applications.

The subsections that follow provide a background and description of the OFM and OFMspert, the OFM's computational implementation. The section concludes with a summary of lessons learned using an OFM-based design methodology. Most lessons were encouraging, but some applications were somewhat less than successful; in particular, they illustrated a *gap* between model-based design and implementation.

21.6.1 Background

The operator function model grew out of Miller's discrete control model (1978, 1985). Miller suggests that traditional control-theoretic representations of the human operator in complex systems limit both the types of systems and the types of models that can be constructed. In particular, traditional models can only represent tasks that are little more than simple tracking tasks. As an alternative, he proposes that models of operators in realistic applications consist of cognitive symbol and knowledge-manipulation tasks, with each task characterized by a control context, goal structure, and set of feasible actions.

Miller (1978, 1985) initially proposed the discrete control model as a metatheory, a formal characterization of a system in which the operator fills the roles of adaptive controller, coordinator, and supervisor. This metatheory specifies the types of objects and relations that a theory must contain to extend traditional manual-control models to account for actual operator behaviors in current and future systems. Miller (1985, pp. 180–192) first instantiated his metatheory of discrete control as an abstract system-theoretic structure, specifically a hierarchic/heterarchic network of finite-state systems. Finite-state systems are similar to finite-state automata,[4] which are defined by a collection of nodes and arcs, which cause transitions from one node to another (Bobrow and Arbib, 1974).

The first application of the discrete control model was to analyze data from a simulated

---

[4]Like finite-state automata, finite-state systems (nodes) in a discrete control model may be deterministic, stochastic, nondeterministic, or some combination thereof. In a deterministic system, the state transition function maps the current state into a *single* next state. In a stochastic system, the state transition function maps the current state into a set of *feasible* next states, with the set of current-state–next-state pairs described by a *probability function*. In a nondeterministic system, the state transition function maps the current state into a *feasible set* of next states, but does not specify the element from the feasible set.

antiaircraft artillery (AAA) system (Miller, 1979, 1985). The task was to model the behavior of a three-person team controlling the AAA system. The AAA system involved some continuous tasks, such as manual gun tracking. Due to the level of automation, however, lower-level behaviors consisted primarily of *discrete* actions, such as switch settings, and higher-level behaviors consisted of *cognitive* tasks, such as selection of overall strategy, mode choice, and crew coordination.

It is interesting to note that there were at least two other attempts to model the AAA system. Both used or extended Baron et al. (1970a,b) optimal control model (OCM)—a formulation using traditional control-theoretic mathematics to represent both the system and human operator. Phatak and Kessler (1977) modeled only the gunner and the associated gun-tracking task, essentially a compensatory tracking task. Zacharias et al. (1981) modeled a two-person AAA system consisting of a gunner and commander. They modeled the gunner's continuous tracking task with the OCM and used the procedure-oriented crew model (PROCRU) to model remaining crew activities. PROCRU, proposed by Baron (1984) to extend the OCM, models tasks and systems that include discrete actions and more complex tasks than tracking together with structures to represent multiple operators, procedures, and somewhat unconstrained control and monitoring behaviors. Thus, these two models implicitly or explicitly acknowledge the validity of Miller's concerns: exclusive use of control-theoretic representations limits the types of possible engineering theories and the types of tasks that can be modeled with such representations.

Miller (1985) characterizes discrete control models as an attempt to capture in a mathematical or computational form the ways in which team members and operators might decompose a complex system into simpler parts and how they then manage to coordinate their activities and system configuration so acceptable overall performance is achieved. The model attempts to address questions of knowledge representation, information flow, communication, and decision making in complex systems. Miller suggests that the result might be thought of as an operator's internal model of the system plus a control structure that specifies how the model is used to solve the decision problems that make up the control tasks.

At the highest levels, the discrete control model identifies system goals, exogenous inputs to the system, and possible strategies. The goal in the AAA application is to maximize time on target. Exogenous inputs include type of target and target arrival. The commander specifies crew strategy. At the lowest level, the discrete control model of the AAA system accounts for all system outputs; these are typically switch settings that in turn correspond to individual manual crew actions. Levels in between represent the structural coordination of subsystems by upper level systems. A heterarchical control structure shifts the focus of control to the proper subsystem at the proper time (Miller, 1985).

The discrete control model of the AAA system is *descriptive* in nature. It was used to guide analysis of data obtained from teams controlling the AAA simulation. The AAA discrete control model clearly illustrates the decision-making strategies employed by individual teams. Using the model, situations of high confusion, errors, and misunderstandings are readily detectable. Miller (1985) summarizes the results as follows:

> As research progressed it became apparent that the issues of problem organization and complexity management on the part of the team members were much more interesting than the problems of quantifying the mode-switching behavior. In this light, the discrete control network, rather than details of individual nodes, provided much insight. Through the *network*

which was developed, it became clear that much of the *complexity* in the problem was *superficial*. Once things were *structured,* it became apparent that the system was controlled *by a very small number of key decisions.* (p. 195; emphasis added)

Miller proposes a general strategy for developing discrete control models. He notes that there are at least two types of discrete control models: descriptive and prescriptive. This in turn leads to two slightly different modeling strategies. Both begin by identifying system goals and exogenous inputs. If the model, like Miller's discrete control model of the AAA system, is intended to support data analysis, that is, it is a *descriptive* model, available data guide model development and it typically proceeds bottom up. If the intent is to use the model for design, that is, it is a *prescriptive* model, model development proceeds top down, with specification of the higher-level components or activities first. High-level nodes are decomposed until the lowest level nodes represent observable data. Miller notes that prescriptive models tend to be more operator-centered. Both strategies are likely, however, to require iterative refinement. A descriptive model may include top-down analysis; similarly, a prescriptive model may require bottom-up analysis.

Mitchell and Miller (1983, 1986) extended the discrete control model to a discrete control model of operator function for the control of dynamic systems. The purpose was to explore the use of the discrete control model to identify high-level operator control functions and to specify the design of computer-based displays required to carry out the functions. As a proof-of-concept, the discrete control model of operator function was used to identify operator functions in a simulated test-repair routing system. Since Mitchell and Miller's model was intended for design, the model was prescriptive and constructed primarily in a top-down manner. The hierarchic/heterarchic constraint network was used to identify possible high-level operator activities together with needed information, that is, display pages, to carry out each.

21.6.2   The Operator Function Model

The OFM is an *operator,* as opposed to a *system,* description (Mitchell, 1987). The OFM evolved from the discrete control model of operator function, with changes due to the need to represent systems that were much more complex than Miller's or Mitchell's initial applications.

The first use of the OFM was to represent a NASA satellite ground control system. At the lowest level of this system, there were often thousands of possible operator actions, including displays requests, system commands, and both cognitive and perceptual actions. Furthermore, operator actions did not correspond directly to system settings or outputs. As a result, the discrete control model of operator function was modified so that nodes represent operator activities at various levels of abstraction and the system is represented only to the extent that system events or outputs initiate, terminate, or sequence operator activities.

Like the discrete control model, the operator function model is a hierarchic/heterarchic network of finite-state systems. Figure 21.4 depicts a generic operator function model together with some of its essential properties. Table 21.1 lists and defines many OFM features.

Network *nodes* represent operator *activities.* The OFM hierarchy decomposes each high-level operator activity into lower-level activities. Activities are defined at multiple levels of abstraction: from high-level planning sequences and operator functions to low-

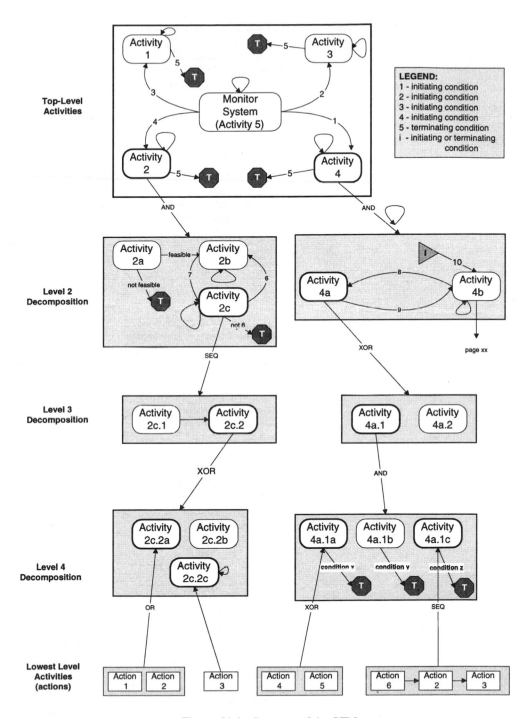

Figure 21.4   Structure of the OFM.

TABLE 21.1 Features that Characterize the Operator Function Model

Feature	Description
Activity/Node	The nodes in the OFM network correspond to activities; as such they are the basic unit of description. An activity occurs at various levels of abstraction; it may be a high-level function or a low-level action
Hierarchy	Activities are decomposed hierarchically. Operators of complex systems organize activities into hierarchies to relate and coordinate them. Operators aggregate low-level activities to high-level activities as a means of reducing cognitive workload. Furthermore, most operator activities are described, documented (manuals and checklists), and taught (training) by means of hierarchies
Heterarchy	The heterarchical structure of the OFM reflects the concurrency of activities for which operators are often responsible. Heterarchical activities are typically defined at the same level of abstraction
Finite-state Systems	In discrete mathematics, a finite-state automation is defined as a finite collection of nodes (states), arcs between nodes, and conditions that cause transitions from one node to another
Initiating and Terminating Conditions	System events, for example, a new air-traffic control clearance, or the results of an operator action, can initiate or terminate activities. They often define the transitions to the finite-state automaton
Nondeterminism	Models may be static (nonchanging), deterministic (for each input there exists exactly one output), stochastic (for each input, there is a set of possible outputs together with a probability distribution that links input–output pairs), or nondeterministic. A nondeterministic model is similar to a stochastic model in that for a given input there is a set of feasible outputs; the difference between a stochastic and nondeterministic model is that the latter does not have a probability distribution associated with input–output pairs
Goals	New Collegiate Dictionary defines a goal as a state to be achieved. All functions, and indeed most activities, are linked either explicitly or implicitly to one or more goals. Pilots may navigate in such a way as to minimize fuel and passenger comfort. Goals are not, however, activities; nor are they represented explicitly in the OFM

Table continued on following page

TABLE 21.1   (*Continued*)

Feature	Description
"Functions, Subfunction, Task" Vocabulary	Operators often use terms such as function, subfunction, task, or procedure to describe their activities. For example, pilot functions are frequently described as aviate, navigate, and communicate. Functions, subfunctions, tasks, etc., however, are closely linked to goals
Actions: Manual, Cognitive, Perceptual	In the OFM hierarchy, actions are typically the lowest level activity. In addition to manual actions, the OFM also represents cognitive actions, for example, situation assessment, and perceptual actions, ensure that no warning indicators are illuminated
Required Information and Actions	Some interfaces require multiple actions, for example, requests for multiple display pages, to obtain desired information. In some OFM applications, information nodes are added to action nodes
Operator Flexibility	The heterarchic structure, together with nondeterminism, allow the OFM to reflect operator flexibility. Typically, heterarchy represents choice, such as a change in the focus of control, for high-level activities. Nondeterminism depicts choice at the same level abstraction and is one type of state transitions in which operators can choose from a set of feasible next actions, given the current state
Operator Error	The OFM models operator error by predicting operator activities starting at the highest level activities down to the lowest level actions. As operators perform actions, the OFM attempts to link new actions to one or more active activity trees for which the new action is included in the feasible set of next actions. A potential error of commission occurs if an action cannot be linked to *any* active activity tree, that is, the action is not contained in any of the feasible set of next actions. A potential error of omission can also be identified when all required lower-level activities in the decomposition are not performed

level actions. In between, collections of nodes can represent procedures, checklists, tasks, or other sets of activities typically carried out together. The *heterarchy* represents activities that the operator may perform concurrently or shifts in the operator's focus of control from one high-level activity to another. Arcs between nodes depict conditions that may initiate, terminate, or sequence activities. At high levels, initiating and terminating conditions add or delete an activity to the current set of activities for which the operator is responsible. For example, in Figure 21.4, arc 1, in the top-level representation, initiates Activity 4, adding it to the set of high-level activities for which the operator is currently responsible. A terminating condition indicates that some activity, or a portion of some activity, is completed and actions to support that activity are no longer expected. In this example, arc

5 terminates Activity 4. In many dynamic systems, operators engage in multiple, concurrent activities, with an overall monitoring and control activity that is always active. For example, in Figure 21.4 Activity 5, at the highest level of specification, is expected to be active at all times. Some arcs, particularly at the lower levels, merely sequence activities. Since OFM nodes are *finite-state systems,* a node can be in one of a finite number of states. State transitions are typically *deterministic* or *nondeterministic.* An OFM used for design rarely has enough data to specify a stochastic transition function.

Both the discrete control model and the OFM *implicitly* acknowledge that operator activities are undertaken to achieve one or more goals. Neither model, however, *explicitly* represents goals. Goals in most control systems are implicit—states to be achieved. Thus the goal of each activity, particularly a high-level activity, is to carry out the activity in a timely and correct manner. Moreover, the OFM deliberately uses terms such as "functions, subfunctions, tasks" and more generally, "activity," rather than "goals, operations, and so on," that the GOMS model and methodology use (Card et al., 1983). The reason is simple: the function–subfunction–task vocabulary is the vocabulary that operators themselves use to describe and document how they view and control the system.

In the OFM, actions can take forms other than manual. In particular, various OFM applications include *cognitive* actions, such as situation assessment, or *perceptual* actions, such as scanning for alarms. Some applications of the OFM have linked information requests to actions. This is sometimes necessary when multiple display pages are needed to carry out an action. More frequently with modern graphical user interfaces, multiple actions are alternative *syntactic* actions for carrying out the same *semantic* action. As a simple example, in Microsoft Word 6, users can save their current file (a semantic action) in one of several ways: (1) select the File menu, then the Save option; (2) press crtl-S on a (PC) keyboard; or (3) click the diskette icon on the tool bar.

Flexibility in the OFM is represent by two features: heterarchy and nondeterminism. Operators are rarely deterministic, though procedure manuals often suggest that they are. Operators can, and do, make choices, selecting which of the set of active high-level activities to pursue next. At higher levels, heterarchy represents the choices operators have, within some time frame, to alternate among control activities. For lower-level activities, nondeterminism represents choices operators have about which activity from a feasible set of activities to perform next. Moreover, high-level activities are not necessarily performed to completion; rather operators interleave actions to meet the requirements of the set of high-level activities for which they are currently responsible. For example, in considering a series of operator actions, the first two actions may support Activity A, the next action Activity B, and the fourth action Activity A, which in turn terminates Activity A. Explicit representation of acceptable actions via heterarchy and nondeterminism allows the OFM to identify potentially erroneous actions. If the operator performs an action that does not support any currently active activity or if an action is not contained in any feasible set of actions, the action may be an error.

Figure 21.5 depicts a number of heterarchic and nondeterministic relationships that the OFM uses. It is not comprehensive, but illustrates major relationships together with the most recent OFM modeling syntax. Heterarchy (*a*) depicts an (AND) relationship in which an activity decomposes into a set of activities, all of which must be carried out concurrently. Heterarchy (*b*) depicts a set of activities that must be carried out concurrently only if the associated initiating condition occurs. If the initiating condition does not occur, the operator is only responsible for Activity A. Moreover, since Activity B also has a terminating condition, part of the time there may be two concurrent activities, A and B, and part of

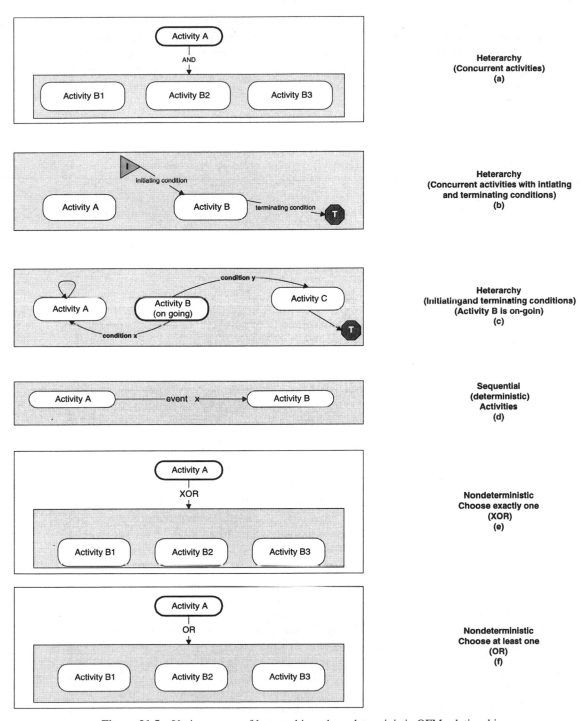

Figure 21.5 Various types of heterarchic and nondeterministic OFM relationships.

the time only one activity, Activity A. Heterarchy (*c*) depicts a high-level heterarchy that is common in control rooms in which operators are responsible for extensive monitoring as well as other activities. Activity B is on-going, that is, the operator is always responsible for Activity B, whereas the operator is responsible for other activities only when the associated initiating condition occurs. As with heterarchy (*b*), when a terminating condition occurs, the associated activity is no longer active. The relationship depicted in Figure 21.5*d* is the standard deterministic state transition function typical of most finite-state automata. The (XOR) nondeterministic relationship depicted in Figure 21.5*e* is one in which the operator must choose exactly one activity from the feasible set of activities: B1, B2, and B3; this activity might be a mode choice. The (OR) nondeterministic relationship depicted in Figure 21.5*f* is one in which the operator may choose one or more activities from the feasible set of activities: B1, B2, and B3. Relationship (*f*) could represent the "save file" example. Typically, a user would select one of the three activities; it would not be an error—or cause harm—if two or more of the actions were performed.

21.6.3 OFMspert: The Operator Function Model Expert System

The operator function model expert system (OFMspert) is a computational implementation of the OFM. The original intent was to use the OFM to predict operator activities, and as an operator performed actions in real time, attempt to interpret those actions by linking them to the upper level activities they were intended to support. As such, it could be thought of as a "living task analysis" (Palmer, 1987).

OFMspert is a form of intent inferencing (Rubin et al., 1988) or activity tracking (Callantine et al., 1998). Originating in the days when great things were expected from artificial intelligence, the theory underlying intent inferencing was that understanding takes place when actions are identified as part of a sequence of actions undertaken to achieve one or more goals (Card et al., 1983; Schank and Abelson, 1977). Callantine et al. (1998) are a bit more restrained. They define activity tracking as a computer capability that seeks answers to the following questions: *What* is the operator doing? *Why* is the operator doing that? *What* will the operator do next? Answers to such questions are possible because an OFM represents an operator acting in a task domain where goals are shared and personnel are trained. Readers might note that the questions just posed are the complement of Wiener's (1989) characterization of questions that pilots often have about automation. Moreover, a human–machine system's ability to respond accurately to both sets of questions is one way to implement Billings' (1997) principle of mutual intent inferencing between the operator and the automation in a human-centered system design.

21.6.3.1 ACTIN: OFMspert's Actions Interpreter. OFMspert uses a blackboard model of problem solving (Nii, 1986b, c, 1994) to construct expectations about current operator activities and to interpret in real-time actions that the operator performs. The blackboard model consists of several components. The blackboard data structure is a global database where a set of partial and full solutions, known as the *solution space,* are kept. The blackboard data structure is both heterarchic and hierarchic. Heterarchies represent alternative or concurrent solutions. The hierarchies are decompositions that correspond to various levels of abstraction in the problem analysis. *Knowledge sources* (KS) are functions that link or transform information on one level of the blackboard into information on another level. As knowledge sources add, delete, and modify blackboard entities, the solution space changes and at any given time it provides the current best hypothesis about the problem.

The original HASP blackboard was designed to interpret sonar signals in real time in order to characterize both the number and type of ships in the area (Nii, 1986a). Thus, the blackboard model of problem solving possesses many of the properties required to implement the OFM in computational form.

Figured 21.6a depicts ACTIN (actions interpreter), OFMspert's blackboard model of problem solving. Implementing a computational form of the operator function model as a blackboard seemed a natural way to represent the OFMspert's most salient properties: representation of dynamically changing state and environment data, hierarchic and heterarchic levels of the OFM, and interpretation of input data (i.e., operator actions) in real time. The heterarchical hypotheses that the HASP blackboard contained matched well the OFM requirement to represent concurrent operator activities. Since the blackboard data structure is hierarchical in form, it allows OFMspert to represent hierarchically decomposed expected activities as well. Operator actions are represented as the lowest level nodes on ACTIN's blackboard. Knowledge sources implement ACTIN's intent inferencing functions, attempting to interpret detected actions as supporting one or more activities posted on the blackboard.

Modeled after the HASP blackboard model of problem solving (Nii, 1986a, b, c), ACTIN contains a blackboard data structure, knowledge sources, and a set of lists to coordinate activities. ACTIN's blackboard data structure is the current best hypothesis about what an operator or team is doing and why. The ACTIN blackboard data structure (Fig. 21.6b) contains activity trees, each a hierarchy of nodes that represent high-level operator activities and their decompositions as specified in the OFM. Operator actions are typically the lowest level(s) of the blackboard, though recent applications depict both *expected* and *detected* actions as the two lowest levels, respectively. Activity trees posted on the ACTIN blackboard are usually *model-derived,* specified by a combination of the operator function model and current system state. Thus, some event, perhaps a change in system state or an operator action, triggers an ACTIN cycle that posts one or more activity trees on the ACTIN blackboard data structure.

Actions that the operator performs are always *data-derived*. Knowledge sources attempt to link each action to all possible upper level activities that the action may support. Rubin et al. (1988) call this property *maximal connectivity*. The assumption is that it is better to overinfer than to guess and be wrong. As shown in Figure 21.6b, sometimes a detected action is posted on the blackboard but has no links. That is, given the current state of OFMspert, knowledge sources are unable to interpret the action at the present time as supporting one or more activities posted on the blackboard. Subsequent processing may allow ACTIN to disambiguate multiple-linked actions or provide links for an action that has none. Occasionally, there are data-derived activity trees. ACTIN may examine a sequence of actions that was not previously understood and infer a purpose. Subsequently this data-derived, or inferred, tree is posted on the blackboard.

Like HASP, ACTIN has three lists (Fig. 21.6a). The future-events list is a time-ordered list where blackboard activities such as periodic assessments or efforts to disambiguate or interpret actions are scheduled for future execution. Processing events on the future-events lists may add events to the events list. The events list contains activities that need to be executed during the current blackboard cycle, that is, before the blackboard processes any new input. The problems list is where anomalies, such as a failed attempt to interpret a detected action, are stored. Like HASP, during every blackboard cycle, ACTIN processes all the lists in the following order: future events, events, and problems.

ACTIN knowledge sources do many things: post activity trees and actions, attempt to

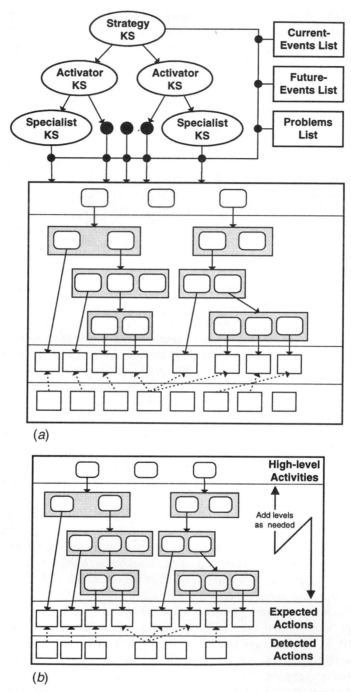

Figure 21.6 (*a*) ACTIN: OFMspert's blackboard. (*b*) ACTIN's blackboard data structure.

link an action to one or more activity trees that it might support, terminate an activity and all of its children, and assess the extent to which an activity has been correctly carried out. Other knowledge sources attempt to disambiguate actions with multiple links. Assessment knowledge sources are often used after a terminating event to assess the degree to which the operator carried out expected activities. Moreover, if OFMspert is used as the basis of an aid or tutor, assessment knowledge sources may check intermittently the progress of the various activities and possibly remind the operator or student that some activities remain incomplete.

The original HASP blackboard had a hierarchy of knowledge sources: strategy, activator, and specialist (Figure 21.6*a*). In the OFMspert application, *strategy* knowledge sources process information from the various lists, that is, future-events, current events, and problems lists. After determining the list from which the next event will come, the corresponding strategy knowledge sources triggers one or more *activator* knowledge sources to manage various types of events or blackboard activities. For example, if a new action arrives, the strategy knowledge source for the events list activates the action-management activator knowledge source. In turn, this knowledge source initiates at least two specialist knowledge sources: post action and interpret action. The first knowledge source posts the new action on the lowest level of the blackboard; the second attempts to link the action to any activity tree currently on the blackboard that the action can support. If the interpret-action knowledge source is unable to link the action, it will place a copy of the action together with a snapshot of the current system context on the problems list and schedule on the future-events list a subsequent attempt to interpret the action. Future operator actions may help other specialist knowledge sources infer a purpose for a set of actions that was not previously understood.

21.6.3.2 OFMspert Architecture.

Figure 21.7 shows the complete OFMspert architecture. In addition to ACTIN, OFMspert has two other major components: the current problem space and the operations model. The current problem space stores information about the system, including state variables, derived variables, and a semantic description of the operator workstation. State variables contain sensor-level information about the system, while derived variables represent operator computations in which they collect and aggregate lower-level data into meaningful information or make a qualitative assessment of one or more state variables. Derived variables are usually identified during the development of the OFM and often have Boolean values. For example, in satellite ground control, a derived value may be data quality: data quality is good if 90 percent or more of the expected data are recorded. In aviation, a derived variable might be "heading out of limits." This Boolean variable is true if the cleared/desired heading differs from the actual heading by more than several degrees. The representation of the operator workstation, that is, the display pages and information that the operator has in view while performing some activity, is also represented in the current problem space. The representation is defined at a semantic level—a description of the information or data currently available to the operator without performing any additional display page or information requests.

The operations model, initially called the enhanced normative model (Rubin et al., 1988), contains static information about the system. Its primary component is the OFM activity model. The operations model may also contain operational heuristics. Operational heuristics are rules of thumb or background knowledge, which are not represented in the operator function model. Satellite controllers, for example, know that a "Wallops pass" requires a certain network configuration. Similarly, pilots have expectations about aircraft

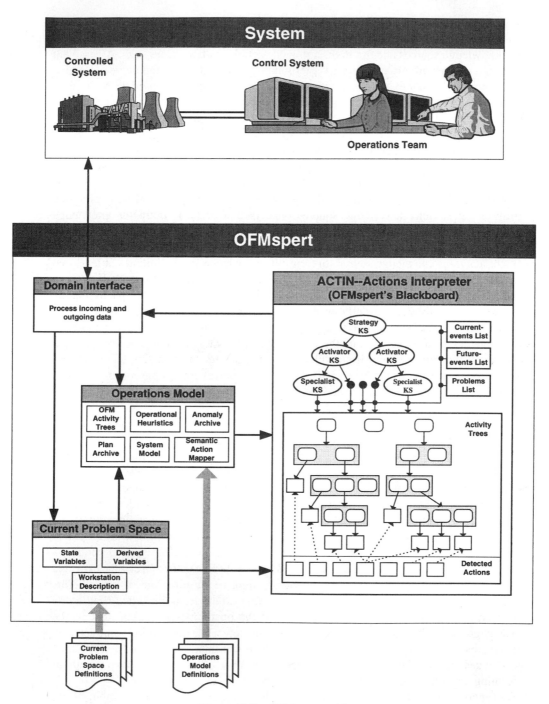

Figure 21.7   OFMspert architecture.

performance based on aircraft type. The semantic action mapper in the operations model is the structure that converts one or more syntactic actions into a single semantic action. Particularly in workstations using modern interface technology, operators often have multiple ways of accomplishing the same objective. The operations model may include an anomaly archive to support on-line troubleshooting and fault management of previously seen and solved problems. Another component of the operations model may be a plan archive. This is a domain-specific database that stores prototypical plans. For example, a plan archive might contain pass-plan templates for satellite control or initial flight plans for a specific company, aircraft, and route. Finally, the operations model may contain a state or system model; such a representation may be important for OFMspert applications such as tutors or aids in which the "intelligent" system explains its recommendations to an operator or contextualizes information for a student. Rubin et al. (1988) provide a detailed description, at the implementation level, of the components that made up the initial OFMspert.

21.6.4 Constructing an Operator Function Model

The process of constructing an OFM, like most modeling and design activities, is characterized as a process of initial conceptualization, iterative evaluation and refinement, and final specification (Jones et al., 1995). Generally, the steps described in the generic methodology for model construction and validation apply equally well to the construction of an OFM.[5] The generic steps are summarized as follows:

1. Prepare a high-level *written* description of the system of interest, specifying system goals, exogenous system inputs, major system components, and high-level activities—both those carried out by the automation and those performed by operations personnel

2. Prepare an initial, high-level draft of the operator model—that is, identify the high-level activities that operators perform

3. Define the *heterarchy*. Identify and annotate when and how operators initiate, terminate, or transition among high-level activities

4. For each high-level activity, specify an associated *hierarchy*. For each decomposition and at every level of abstraction, define *conditions* that initiate or terminate activities

5. *Validate* the model via direct observation. Map operator actions bottom up into the model, including cognitive and perceptual actions, ensure that the observed action links to the appropriate higher-level activity in the current context. Identify each discrepancy between the model and observed activities; resolve whether it is due to a model or operator error

6. *Refine* the model iteratively as the proposed system evolves

Exceptions or additions include the following. In Step 2, make extensive use of the *visual form* of the operator function model (e.g., Fig. 21.4). The pictorial form allows

[5]*Caveat:* All these steps assume extensive on-site observation of operations and periodic interaction with the various groups involved: managers, designers, and operators. Operators are always the final arbiters if there is some disagreement in how operations are carried out.

domain experts to easily and iteratively inspect, correct, and refine the evolving model. In Step 3, construction of an OFM heterarchy, make use of OFM heterarchic properties such as those displayed in Figures 21.4 and 21.5 and described in Table 21.1. Step 4 specifies the hierarchic decomposition for each high-level activity. This involves defining the relationship among a set of activities resulting from a decomposition. It is important to define conditions that initiate (make active) and terminate (make inactive) activities defined at the same level of abstraction. Since mathematically the OFM is a hierarchic/heterarchic network of finite-state systems, it is important to specify the state transition functions. Identify the conditions that *initiate* or *terminate* a node, define a *sequence* between nodes, and *enable* transitions among the nodes that decompose the activity above them. As shown in Figure 21.5, activities may decompose into subactivies where the state transition is nondeterministic, but mutually exclusive (XOR), that is, a feasible set of activities from which exactly one activity can be selected. Others may have a nondeterministic next state that consists of a feasible set of activities where one or more (OR) activities may be chosen. Some sets of activities must all be done (AND), but the activities may be carried out in any order. The number of levels in the OFM hierarchy depends entirely on the application of interest. Although the original OFM application had four levels of activities (e.g., functions, subfunctions, tasks, and actions), subsequent OFMs have had many different numbers of levels, including phase, subphase, subtask, mode, and procedure. Recall that actions can be cognitive, perceptual, and manual. In the current OFMspert, the two lowest levels of decomposition consist of *expected* actions (model generated) and *detected* actions (operator performed). Step 5, model validation, is critical. The visual form, the use of operator language, and level of detail in the model facilitates verification and validation. Direct observation of operations or *post hoc* analysis of videotaped operations is essential. Validation includes bottom-up mapping of observed operator actions to OFM activities that are currently active. This is straightforward, as the OFM decomposition supports this type of validation. The last step, iterative refinement, is as important as initial model definition and validation. In work domains, particularly those that are changing due to the introduction of new computer-based or information technology, systems change quickly during design and implementation. As a result, particularly at lower levels, the model changes. The OFM drawings should be updated to reflect all changes and the visual form used to validate the modeler's new understanding by domain experts. Direct observation, however, remains the most reliable verification and validation method.

The generic model-based design methodology ends with Step 6, interative refinement. An OFM-based design methodology may also include development of OFMspert, the computational form of the operator function model. Some types of intelligent displays, aids, and tutors need a software implementation of the objects and relations specified by the operator function model in order to define the intelligence of the artifact. Implementing OFMspert, given a well-defined operator function model, is conceptually straightforward, though potentially, like all large software development efforts, quite time-consuming.

21.7 OPERATOR-FUNCTION-MODEL-BASED DESIGN: ILLUSTRATIVE APPLICATIONS

As described in the generic section on model-based design, there are numerous applications for which a model-based design methodology proves quite effective. The following sections illustrate a number of such applications. Each was conceived, implemented, and evaluated by members of Georgia Tech's Center for Human-Machine Systems Research.

Although the operator function model and OFMspert have been used in many applications, (e.g., aviation, air-traffic control, manufacturing), the applications discussed in this section are all in the context of NASA Goddard's Satellite Ground Control System. These examples are not necessarily the newest or most sophisticated, but they illustrate OFM-based design without requiring the reader to understand in detail numerous complex domains. References to applications in other domains are also provided.

21.7.1 Operator-function-model-based Design Methodology

Phase I. Develop an Operator Function Model

Prepare a High-level Written Description of the Domain of Interest (Step 1). The main function of NASA Goddard Space Flight Center (GSFC) is to design, launch, and control near-earth satellites. These satellites are unmanned spacecraft that gather science data about the weather, atmosphere, sun, and the Earth itself. The satellites periodically transmit data to an earth ground station that in turn forwards the data in real time to NASA Goddard control rooms. During these short periods of time, called *passes*—as they occur as the satellite passes over a ground station—spacecraft-specific controllers command the spacecraft, initiate transmission of science data to the ground, and ensure that the current state of the spacecraft is within nominal limits. The goal is to collect as much good-quality data as possible.

The Multisatellite Operations Control Center (MSOCC) is a specific NASA Goddard system that coordinates the use and monitors the effectiveness of communication and computer systems shared by a number of satellites. Since contact with a given satellite is brief, satellite-specific controllers share this equipment. The MSOCC system is responsible for scheduling, configuring, and deconfiguring the shared equipment. During real-time contacts, MSOCC operators are responsible for ensuring the continuous functioning of computers and communications networks as well as the integrity of the data flowing through them. Exogenous inputs to the MSOCC system are satellite pass and equipment schedules and the status of MSOCC hardware.

MSOCC is a good candidate for increased use of automation. The current system is labor-intensive—four operators per shift, three shifts a day, seven days a week. Most operator functions are straightforward and many manual activities can be automated by means of simple rules. When this research began in 1984, NASA was in the process of upgrading the system with state-of-the-art automation (Computer Sciences Corporation, 1983).

The most significant information obtained during Step 1 was that the schedule for the new MSOCC system would always be at least 12 hours old. This was due to constraints of other systems with which MSOCC needed to interact and to budget limitations for the upgrade to the MSOCC system itself. All scheduled configurations and deconfigurations are carried out automatically. Operators, however, had to perform manually any activity that was not included in the predefined MSOCC schedule.

Prepare an Initial High-level Draft of the OFM (Step 2). Despite the fact that three operators would staff the system 24 hours a day, seven days a week, management and

786 MODEL-BASED DESIGN OF HUMAN INTERACTION WITH COMPLEX SYSTEMS

system designers initially said that operators would not have *any* functions except monitoring the automation, as the system would be highly automated. Demonstrating the importance of developing a well-defined model that reflects multiple perspectives, it turned out that even with the addition of automation, MSOCC operators would have at least five major functions (top-level activities of Fig. 21.8). MSOCC operators are expected to monitor continuously any current spacecraft-to-ground transmission and, when necessary, engage in fault detection and management whenever the automation reaches its limits, which for this system is any time a change to a 12-hour-old schedule is made.

Define MSOCC's Heterarchy for the Top-level Activities (Step 3). Determine the relationships among the top-level, heterarchic activities, on-going or ones with initiating and terminating conditions. As Figure 21.8 shows, there is one on-going activity:[6] monitoring currently active spacecraft. The other activities all have both initiating and terminating conditions. Recall, at high levels, heterarchy and the combination of initiating and terminating conditions model operator flexibility and refocus of attention. Initiating or terminating conditions typically add or delete high-level activities to the set of activities for which the operator is currently responsible.

Decompose Each High-level Activity and Define Conditions that Initiate, Terminate, or Sequence Activities at the Same Level of Abstraction (Step 4). The remaining portion of Figure 21.8 shows the decomposition of one high-level activity, "configure to meet support requests." The first level of decomposition is a sequence of activities (SEQ). If activity one or three in this sequence is not feasible, then the upper level activity terminates, together with all its subactivities, and the operator reports that the request cannot be met.

The lower portions of Figure 21.8 show the remaining decomposition for activities associated with "configure to meet support requests" down to the actions level. The activity terminates (normally) if the operator is able to configure the requested network; otherwise, the operator informs controllers that the addition of the requested pass violates a system constraint or that there is insufficient hardware at the time requested.

Validate the Model via Direction Observations of Operations and Refine the Model Iteratively over Time (Steps 5 and 6). Extensive on-site observation of operations allowed modelers to link observed actions to upper level activities that they were hypothesized to support, and both validate or update the model. Figure 21.8 and the remaining seven pages that made up the MSOCC OFM were intermittently validated over a period of 12 to 18 months as MSOCC evolved; see Mitchell (1987) for the complete MSOCC OFM. The visual form of the OFM greatly facilitated this process.

Develop an MSOCC OFMspert (Step 7). Although development of a computer implementation of the model is not a required step in the proposed methodology, several of types of applications require a computer implementation of the model. OFMspert is one such computer implementation and it formally guides the transition from the knowledge and structure contained in the OFM to a rigorous software specification. OFMspert for MSOCC served both as a proof-of-concept for the OFMspert architecture and as the

---

[6]MSOCC operators are *always* responsible for monitoring any network, from control room to spacecraft, that is currently transmitting data. Other activities are always performed in addition to the "control of current missions" activity.

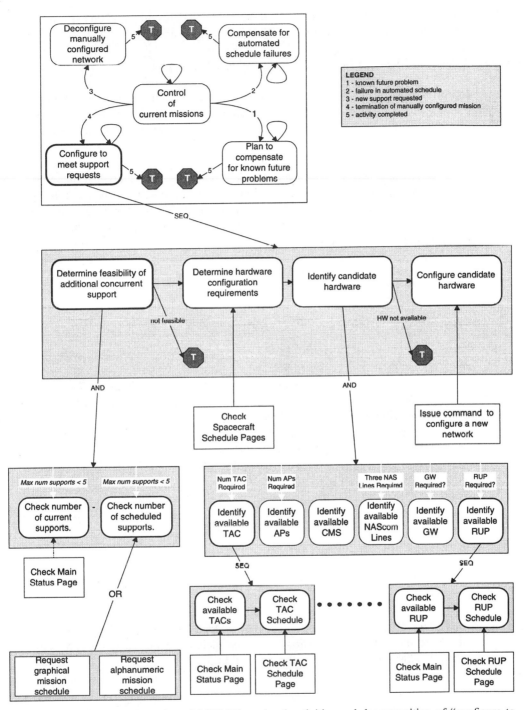

Figure 21.8   Representation of MSOCC top-level activities and decomposition of "configure to meet support requests."

basis of several studies to ensure that it performed adequately and could be used as the intelligence for computer-based artifacts such as assistants and tutors.

Phase II. Use the Model to Guide the Design of Specific Applications

The MSOCC operator function model and its computational implementation—OFMs-pert—have served as the basis of a variety of applications. Examples for each design issue depicted in Figure 21.2 are given. References to more detailed publications for the applications are also provided.

Prepare a High-level Written Description of the Domain of Interest (Step 1). During the initial interaction with MSOCC managers and designers, it became clear that operators were vital to the success of the system, though neither managers nor designers initially believed this. A number of "limits to automation" reduced the extent of the automation that was finally implemented and required operators to weave together various automated subsystems into a coherent whole. MSOCC operators were responsible for ensuring that the system worked and quickly correcting any faults. The goal is to collect as much "good" science data as time limits for a pass allow.

Function Allocation: Human and Computer Agents (Step 2). Although it is a mantra of all human factors engineering that each function should be allocated to the agent (human or computer) that can best carry it out (Chapanis, 1970), function allocation for the new MSOCC system proceeded typically: all activities that could be automated—within budget constraints—were automated, leaving operators to perform all other activities. Moreover, operators were expected to fill the gaps between automated subsystems and compensate when the limits of automation were reached. The evolving MSOCC operator function model clearly showed the high-level activities that operators were likely to perform, in addition to monitoring the system.

MSOCC designers and NASA management could have used the MSOCC operator function model as an engineering tool to evaluate staffing needs, project operator workload, specify workstation hardware, and design workstation software. Instead, and somewhat arbitrarily, NASA planned to staff the new MSOCC system as they had staffed the old system: around the clock with at least three operators on console, that is, each operator sitting at a MSOCC workstation, and a supervisor. A thorough review of system documentation never showed any rationale or analysis to justify the planned staffing levels for the new, highly automated system.

Workstation Design: Displays and Controls (Step 3). This subsection describes two workstations: the baseline workstation proposed by NASA and a workstation designed using OFM-based design methodology. We begin with a description of the base-line workstation.

> *The baseline MSOCC operator workstation.*[7] Most systems specify workstation hardware and software long before details on specific operator activities are considered. This was also true for the new MSOCC system. Given staffing information, designers planned to have four MSOCC workstations—one for each of

---

[7]The baseline workstation used in this study extensively replicated, from the operator's point of view, the proposed MSOCC workstation (Computer Sciences Corporation, 1983). The baseline workstation is very similar to the workstation that MSOCC eventually implemented around 1986, and, with some modification, is still in use today.

the three operators and the supervisor. Each MSOCC workstation would consist of several monitors and a keyboard for command-line input. Since the designers were unaware of any operator functions other than monitoring the automation, they had little rationale for configuring the workstation or its associated display pages, except along physical and functional system characteristics.

There are three obvious MSOCC functional features: (1) information about schedules; (2) information about hardware status and active networks; and (3) data and data-quality counts for each piece of equipment composing the overall system. Given these three functions, designers specified a three-monitor MSOCC workstation. The left monitor contained all schedule-related information. There were hundreds of schedule pages corresponding to the overall system, individual spacecraft, equipment classes, and specific pieces of equipment. The second (middle) monitor consisted of one graphics display that depicted the current hardware status for every piece of MSOCC equipment and the networks that were currently supporting spacecraft contacts. This page was completely filled by the specified information. No other MSOCC equipment, using the proposed design, could be added; no consideration was given to the inevitable addition of hardware or how its status would then be displayed. Finally, as the system's primary responsibility was to ensure the transmission of good-quality data between the spacecraft and the ground system, the third monitor contained dozens of alphanumeric pages displaying the amount and quality of data received, both incoming and outgoing, at every piece of MSOCC equipment, whether the equipment was operating or not.

These displays, organized by system, rather than operator function, required operators to do a tremendous amount of work. For example, it was not unusual for the operator to request ten or more display pages, copy low-level data onto paper, and perform extensive paper-and-pencil calculations to ensure that data were flowing through each active network properly and were acceptable. Thus, monitoring and troubleshooting, if a problem was suspected, or carrying out most of the other major operator functions, were difficult and error-prone tasks.

Essentially, this workstation violated every step of the proposed workstation design construction and validation process. There was no operator model; thus, most model-development steps could not be carried out at all. Traditional design reviews were the only methods of assessment.

An OFM-based MSOCC operator workstation. In the first application of the operator function model for workstation design, Mitchell and Saisi (1987) used the OFM to specify workstation hardware requirements and sets of display pages and information requests as specified by the MSOCC OFM. The descriptions below summarize corresponding features of the baseline vs. the OFM-based workstation. The OFM-based workstation design methods were used to assess the baseline workstation and to specify, and partially validate, the utility of, the OFM.

The MSOCC OFM-based workstation had two very different types of displays. Each type corresponded to one of two major operator functions: monitoring and fault detection vs. system reconfiguration. As Figure 21.8 shows "control of current mission" is an on-going monitoring activity. Its decomposition shows that this activity consists of both routine monitoring of each network and more detailed monitoring or troubleshooting when a fault is suspected. Each currently active network was represented by a window containing two icons: a spigot icon and an hourglass timer. The spigot icon integrated the features of the network important to the MSOCC operator: Are the data flowing? Are they flowing at the correct rate for the given network configuration? Is the error count reasonable? The width of the spigot icon depicted the *expected* data flow at the end point of the system—the spacecraft-specific control room. The width of the fluid flowing from the spigot depicted the *actual* flow rate at the same point. Significant differences in widths indicated a possible problem. The fluid representing data rate flowed from the spigot icon into a bucket. Error accumulation was depicted by red dots at the bottom of the bucket. The second icon, the hourglass timer, depicted the time remaining in the pass. By examining remaining time and the current number of errors, MSOCC operators could estimate whether the error count was reasonable or not. These metaphors defined a high-level qualitative monitoring display: there was one spigot and one hourglass timer for each active network. If a fault was suspected, the operator could request additional information about the suspect network, which was displayed as a network of spigot icons depicting expected and detected flow rates and error counts at each component in the network. The second window was a more detailed, though still a qualitative display, and permitted the operator to troubleshoot quickly and accurately a suspected network. These two windows shared a single monitor, supporting monitoring and fault-detection.

The second type of activity, the conditional activities depicted at the top level of Figure 21.8, all relate to system reconfiguration and require a different type of display than the monitoring and troubleshooting representations. The OFM was used to design a series of help requests that corresponded to each high-level activity except "control of current missions." By requesting help, the operator obtained required information to either carry out the system configuration or decide that it was not possible. This information was a set of alphanumeric windows displayed on the same page. The contents of each window were tailored by the needs specified by the OFM and current system state. For example, for the activity decomposition in Figure 21.8, "configure to meet support requests," the operator would receive a request from a spacecraft, say ERBS, to see if they could have a pass for the next five minutes. The operator input the help command: help configure ERBS 5. This command brought up a set of windows corresponding to the decomposition of that activity in the OFM. A window determined the feasibility of "configure to meet support requests" by checking to ensure that adding another network for the next five minutes would not violate the five-network constraint. If this constraint would be violated, another window provided this information to the operator, who then communicated to the controller that the support was not possible. If the support was possible, the help request automatically displayed a spacecraft-specific network template together with a window for each equipment type. Equipment windows showed individual components that were available for the entire time in question. Given this information, the operator could easily configure an ERBS network with the correct equipment types and individual components that were free for use. This type of interface was easy to use and resulted in few operator errors due to memory constraints or input error.

Thus, the OFM workstation consisted of two monitors: one support monitoring and fault detection with the qualitative graphical icons depicted at two levels of abstraction, and another to provide contextually relevant information, displayed on the same page, to enable the operator to evaluate the feasibility of a manual activity.

Results. The OFM-based workstation was designed to meet the activities specified by the operator function model: in number of monitors, types of displays, display contents, and operator commands. Essentially the OFM guided software developed to provide contextually relevant information in response to help requests and presents the iconic panels to support on-going monitoring and, as needed, troubleshooting.

Many walk-throughs of the OFM-based workstation were conducted. Each walkthrough led to improvements until the designers/researchers felt that the workstation was ready for experimental evaluation. The project concluded with a controlled experiment. A between-subject experiment compared performance using the two workstations on a variety of measures. Not surprisingly, the evaluation demonstrated that the OFM-based workstation, on almost all measures, significantly enhanced operator and system performance. Where performance was not significantly better, performance with the OFM-based workstation was *always* qualitatively better (Mitchell and Saisi, 1987).

A subsequent project in the area of model-based design for operator workstations, extended the OFM to include information flow and intentionally designed human–computer interaction into the workstation (Thurman, 1995, 1997; Thurman and Mitchell, 1994, 1995). The latter feature—an interactive operator workstation—was an attempt to mitigate the inevitable boredom associated with extensive monitoring in increasingly automated systems (Bainbridge, 1983). The enhanced OFM and the proposed design methodology were implemented in proof-of-concept form for another NASA satellite control system, the spacecraft-specific control room. The empirical evaluation improved on the Mitchell and Saisi results. The OFM-based workstation consistently and significantly improved operator performance on *all* measures. Moreover, a number of its features have become part of the standard NASA suite of software available to individual spacecraft control systems.

A similar result was obtained in a manufacturing application in which an operator function model was used to specify information for displays used by a human supervisory controller of a manufacturing system. Again, the OFM-based workstation resulted in significantly better system performance (Dunkler et al., 1988).

Intelligent Aids and Associates (Step 4). Knowledge acquisition and knowledge representation are the most difficult parts of creating an intelligent, advice-giving system (Buchanan and Wilkins, 1993; Forsythe and Buchanan, 1993). Obtaining accurate and complete operational knowledge, by designers often lacking domain and operational knowledge, is very difficult. Knowledge acquisition is typically performed by identifying a single-domain expert, and perhaps supplementing the expert's views with less rigorous interviews with other personnel, such as other operators, management, and designers, and by reviewing system documentation. These methods have obvious limitations. When a knowledge base depends primarily on a single-domain expert, the broad range of experience, alternative techniques, and so on, of the set of operators is not represented. Interviews and other off-line information-gathering techniques are obtained out of context. Context, however, is very important in dynamic systems. It is often necessary in order to understand the conditions under which something occurred: the what, how, and why.

The process of developing an OFM is an alternative to typical knowledge engineering. The extensive on-site observations required to develop and, subsequently, verify and validate the OFM ensures that extensive operational control and system knowledge is gathered and represented formally in the OFM's hierarchic/heterarchic network.

The OFM and OFMspert both facilitate knowledge representation that can be inspected, refined, or repaired. The visual form of the OFM helps operators and other domain experts ensure that the model is complete and correct. As implemented in OFMspert, the ACTIN blackboard provides a means for representing dynamic information visually. Operators and other domain experts can validate OFMspert by observing ACTIN, as the system dynamically posts expectations for operator activities and attempts to interpret detected operator actions. Furthermore, the ACTIN display and corresponding information assessments provide a straightforward mechanism that operators and designers can use to further inspect, correct, or verify the system. Moreover, as both the OFM and OFMspert use a representation that closely emulates the *manner* in which operators carry out required activities, a mutually intelligible framework for discourse is defined. As noted earlier, mutual intent inferencing between operators and automation is one of Billings' (1997) primary principles of human-centered automation.

Intent inferencing. The first activity in creating an artifact with computer-based intelligence is to specify the intelligence itself. For control of a dynamic system, the OFM provides a prescriptive specification of operator activities to meet various system goals over a range of nominal and off-nominal operating conditions. Thus, the OFM, or a similar prescriptive model of operator control, provides an initial definition of intelligence.

The definition of OFMspert as derived from an OFM is intended to specify, in a principled and formal manner, a mechanism to encode the intelligence embedded in an operator function model so that it can be used computationally by a range of applications. As noted earlier, in the Mitchell and Saisi model-based workstation, the MSOCC OFM guided the design and implementation of workstation hardware and software, but it did so in an *informal* manner.

The original OFMspert was implemented in proof-of-concept form for the MSOCC system (Rubin et al., 1988). Its current problem space included state variables to represent spacecraft networks, equipment status, and MSOCC schedules, and data and error block counts at each piece of equipment. The operations model, originally called the *enhanced normative model* (Rubin et al., 1988), contained both activity trees derived from the OFM and triggers that OFMspert used to post trees on the ACTIN blackboard. The operations model also contained triggers that allowed ACTIN to post operator actions on the blackboard and attempt to link them to current activities. Finally, the operations models for MSOCC contained operator heuristics, not included in the OFM, that operators often used to troubleshoot suspected networks.

The initial proof-of-concept implementation showed that OFMspert could indeed represent dynamically changing state data for the MSOCC system, correctly post corresponding activity trees, and typically link actions accurately to posted activities. The primary performance measure in this first step, however, was to show that the OFMspert concept could be implemented, run in real time, predict reasonably correct operator activities, and interpret in a reasonably correct manner actions that supported them (Rubin et al., 1988). The next step was to demonstrate empirically that OFMspert could perform dependable-intent inferencing or operator activity tracking. Jones et al. (1990) undertook an extensive evaluation of OFMspert as implemented for MSOCC. The experiment involved comparing previously collected MSOCC data from the Saisi and Mitchell experiment as well as verbal protocols from two additional subjects performing the complete MSOCC experiment in the control condition. The data were paired: for each action, there was an interpretation from either a domain expert (original MSOCC data) or the subjects and an interpretation from OFMspert. A nonparametric sign test showed significantly good matches, that is, matches that could not have occurred by chance, for all system commands and most display requests. *Post hoc* analysis suggested that when there were poorer matches, that is, not significant matches, they occurred for activities that were less structured, for example, planning or browsing. The latter was an unanticipated result, but one that has consistently reappeared in similar applications (Callantine et al., 1998). Operators like to interact with the system, testing it, confirming their understanding, and so forth, particularly when no other active operational activities require their attention.

Another implementation of OFMspert, tailored to navigation for the Boeing B-757 flight deck, showed similar results (Callantine et al., 1997, 1998). While "flying" a B-757 desktop simulator, ten certified B-757 line pilots working for the same airline performed 2,089 actions. OFMspert correctly interpreted 92 percent of them. The *post hoc* analysis showed that the remaining 8 percent consisted primarily of browsing actions.

One OFMspert implementation demonstrated at least one OFMspert limitation. OFMspert was implemented for navigation of a Boeing B-727 aircraft (Smith, 1991; Smith et al., 1990). The B-727 is a mixed-mode aircraft in that pilots can choose to hand fly it using continuous control actions with throttle, stick, and rudders, or they can use discrete actions on the mode control panel to set a desired heading or altitude. The autopilot then controls the plane to fly to the specified heading or altitude, and hold it until it receives further pilot commands. OFMspert for this application accurately interpreted the discrete actions involving inputs to the mode control panel, but had a great deal of difficulty interpreting continuous actions associated with manually controlled flight.

Aiding and assistance. The ultimate goal of systems such as OFMspert is to use its dynamic understanding of the operator or operations team to either provide advice or act as an electronic assistant. Bushman et al. (1993) first used the MSOCC OFMspert to create Ally. Ally used OFMspert for its intelligence about current system state and to hypothesize operator activities. In addition, Ally augmented OFMspert with control capabilities through which the operator could delegate a control task or portions of a control task at various levels of abstraction and aggregation. The OFM defined the various choices Ally offered to the operator; for each high-level OFM activity, Ally offered advice or assistance with options corresponding to activity decomposition. An experiment compared two teams controlling the MSOCC system: one team consisted of two human operators and the other of a human operator and Ally. The analyses of variance of operator actions were within subject designs and the second operator in the two-operator team was a confederate. The data showed that there were no significant differences between the two teams for any measure of performance. Not surprisingly, however, *post hoc* analysis revealed that operator strategies were quite different depending on whether a team had a human or computer-based assistant.

In another NASA satellite ground control system application, Jones and Mitchell (1995) again used OFMspert as the basis for the design of an operator assistant. The goal was to define principles of human–computer collaboration for control of dynamic systems. Georgia Tech Mission Operations Cooperative Associate (GT-MOCA) blended the proposed principles of collaboration with OFMspert inferences about operator needs and current workload in order to specify an operator-controlled assistant. An experiment, similar to Bushman et al.'s, showed that, with NASA operators serving as subjects, two-person teams, and teams consisting of one NASA operator and GT-MOCA were indistinguishable on all measures.

Intelligent Tutoring Systems (Step 5). Models comprise two of the primary components of an intelligent tutoring system (ITS) (Figure 21.3): the expert model and the student model. OFMspert, with its embedded OFM, offers a potentially productive starting place for development of both models. A well-verified and extensively validated OFMspert can serves as an expert model. For use in an ITS, however, it requires extensions that support novices rather than experts. Moreover, if the student model is an overlay model—that is, a model with the same form as the expert model—OFMspert again provides a robust foundation. As instruction and student–system interaction progress, the ITS annotates the student model and tailors instruction based on comparisons between expert and student models.

Chu et al. (1995) proposed an architecture for an ITS for students learning to control dynamic systems. Both her expert and student models began with OFMspert; extensions were added to accommodate tutoring. Georgia Tech Visual and Inspectable Tutor and Aid (GT-VITA), a general architecture for tutoring operators of complex systems, included declarative knowledge, procedural knowledge, and operational skill. Declarative knowledge was added to entities in the OFMspert current problem space to explain purpose, function, and behavior. Procedural knowledge and operational skill were added to the OFMspert operations model. A separate pedagogical module was also added. Pedagogy defines teaching styles. GT-VITA demonstrates operational skill and coaches students who are practicing operational skill using various levels of assistance. For example, as student skills increase, tutor interventions decrease. The ACTIN blackboard was used to represent expected operator activities, given the current state. Knowledge sources provided information to the tutor to guide students through procedural and operational activities. GT-VITA was implemented in proof-of-concept form for the same NASA satellite control system used by Jones and Mitchell (1995).

Evaluating something that did not exist before, at least in a tractable form, is very difficult. Since current NASA training is almost exclusively informal, on-the-job training, extending over a three to six month period, it was impossible to conduct a controlled experiment to compare the tutor with the actual alternative. Thus, a study was conducted, similar to traditional classroom assessment, to determine the extent to which students learned the material that the tutor was designed to teach. Sixteen employees of NASA Goddard who worked in the area of satellite ground control, but were novice operators—that is, they did not have previous training in satellite control—participated in this study. The overall results, both objective and subjective, showed that participants were able to describe correctly all declarative and most procedural knowledge for satellite ground control. For operational skill, participants performed well for *all* on-line tasks. Participants, however, had problems learning to perform correctly off-line tasks (e.g., bookkeeping), but neither the operator workstation nor the tutor concentrated on these activities. The highest compliment to this work is that the tutor, GT-VITA (Chu et al., 1995), is still in operational use today at NASA Goddard, almost seven years after its initial demonstration and evaluation.

Tutor-aid Paradigm (Step 6). In addition to making a contribution to intelligent tutoring, the two sets of research performed by Jones and Mitchell (1995) and Chu et al. (1995) proceeded well down the path toward demonstrating the viability of a *student-aid paradigm,* as depicted in Figure 21.2. Both used the same domain of application, and both were based on the same OFM/OFMspert models. In GT-MOCA, OFMspert was the expert model for the intent inferencer. In GT-VITA, OFMspert provided the basis for both the expert and student models. As noted before, a tutor that evolves into a well-known and well-understood assistant may alleviate some "automation trust" concerns and certainly reduces the expense and possible errors associated with maintaining as many as three different models for two different software systems.

21.7.2   Lessons Learned

The series of research projects that led to the model-based design methodology described in this chapter taught many lessons over the 20 years that the work spans. These include limitations of using microworlds, especially if your results are surprising, the importance of including software specification in the interface between a research theory and a proof-of-concept implementation, the importance of a correct model, and the possibility that model-based design may compensate in part for individual differences among operators of complex systems.

21.7.2.1 Microworlds. In some respects, failures, which are often not published, teach more than successes. The results of the experiment evaluating Mitchell and Miller's (1983, 1986) model-based displays designed with the discrete control model of operator function were both unexpected and undesirable: neither set of model-based displays enhanced performance over a typical SSSI static display, and one model-based display type significantly decreased performance. Analysis of the data showed that the "unexpected" and "surprising" results could be attributed to any number of factors, and thus both the researchers and the research community were left with few conclusions or explanations for these results. Initially, we speculated that perhaps the model itself was incorrect, but then concluded that if this was true, *both* model-based display conditions should degrade performance, but, the data did not support this hypothesis (Mitchell and Miller, 1986). Another explanation was that the results were due to a "fatal flaw" in the experiment itself. Our experimental domain was a microworld—a contrived system. In specifying the system, it turned out, somewhat serendipitously, that the system required exactly one dedicated display. The microworld had so few state variables that all the required information fit nicely on a single display page. This display page, however, was intended as a surrogate for

analog control rooms containing hundreds or even thousands of displays. It recalls the new nuclear power plant control room with 15,000 display pages that operators may select. For this experiment, this single page, with state-variable values updating, dynamically served as the control condition against which two sets of displays were designed using a discrete control model of operator function. In retrospect, it seems that a poor choice of a surrogate for the dedicated display condition as well as the specification of the contrived system itself might explain the experimental results.

After the unhappy results of this experiment, we never again used a contrived system for research intended to produce operational relevance. Thus, for all future work, research projects always started by identifying a generalizable, but real, system and emulating, from the operator's point of view, the look and feel of the operator–system interaction and dynamics. This typically meant that we were novices with respect to the system, but that the system itself was real and any results applied at least minimally to the system under study; in addition, they might well generalize to a wider class of systems.

The MSOCC research projects were the first to reflect this philosophy. The projects began with construction of a simulator with which to evaluate experimental work. The Georgia Tech Multi-Satellite Operations Control Center (GT-MSOCC) simulator replicated to a high degree the proposed MSOCC system. Each display page, several hundred in all, for the proposed system was exactly replicated in GT-MSOCC. The simulator carefully emulated operator–system interaction and dynamics. For example, the actual control system updates data and error block counts every 20 seconds, as does GT-MSOCC.

GT-MSOCC and its interfaces were used as the baseline experimental condition for most of the studies previously described. From NASA's point of view, these experiments had high face validity and generalized to an entire class of satellite control systems. Moreover, since traditionally analog systems continue to convert to digital technologies, it is reasonable to conjecture that the experimental results, particularly those involving model-based workstation design, generalize more widely, as human–system interaction issues are often quite similar across real-time control systems [Woods et al. (1994)].

21.7.2.2 Structure to Guide Software Development. Prior to research projects using OFMspert, a tutor for MSOCC, ITSSO (intelligent tutoring system for satellite operator), was designed using the MSOCC OFM to guide software design (Mitchell and Govindaraj, 1991). Research has many failures as well as successes, and like the OFMspert developed for navigation on the B-727, this project was not a success. Similar to the Mitchell and Saisi workstation, the MSOCC OFM was used to guide software development for this tutor, but in an unstructured and informal manner. Since this was an MSOCC experiment for which there was a controllable real-time simulator, a control condition was possible and a transfer-of-training experiment conducted. The experiment compared performance of subjects after the normal MSOCC training and system introduction (the control condition) to performance of subjects who were trained with the ITSSO tutor. To everyone's surprise, ITSSO failed to enhance performance on all but one measure. Moreover, on another measure, performance was significantly better in the no ITSSO condition.

After the unexpected results with ITSSO—for the most part the tutor did not help (Mitchell and Govindaraj, 1991)—it seemed clear that exactly *how* the OFM is used to guide software development is also important. Although OFMspert began as an exploration for the design of a dynamic tasks analysis and as the first step in constructing an operator's assistant, over time, it became clear that four facets composed the OFMspert project. First, there is the *theory* of OFMspert. Next, there is the OFMspert *architecture,* that is, a well-defined OFM specification intended for software design. Next, there is a *proof-of-concept implementation,* and fourth, an *evaluation.* From the perspective of the research community and potential operational systems, though it is a prerequisite that the proof-of-concept implementation perform as advertised, the more interesting OFMspert facets are its theory and architecture. The architecture is the tool that enables designers to use OFMspert in other domains.

There are many ways to move from a theory to an architecture that implements the theory. For example, the OFM theory states that a hierarchic/heterarchic network of finite-state systems is a useful way to represent how operators control complex systems. Moving from this theory statement to OFMspert required much thought, experimentation, and evaluation, and resulted in a variety of mistakes.

OFMspert continues to evolve. The OFMspert architecture specifies the manner in which to construct a current best hypothesis about operator activities and to interpret incoming operator actions. Moreover, though at this time only manual actions can be validated, it is important to represent cognitive and perceptual actions as well; in fact, representation of cognitive and perceptual actions is essential for a tutoring system.

The HASP blackboard data structure helped initially to define the current problem space, but other requirements manifested themselves over time and individual applications. For example, OFMspert's current problem space needs information about the current system state (state variables) and how operators interpret various sets of state variables to form intentions or validate anticipated expectations (derived variables). Moreover, to understand whether the operator is accessing the needed information to make a well-informed decision, the current problem space needs a structure to represent the display pages an operator currently has in view or that have been in view recently enough that OFMspert can reasonably assume that the operator remembers the information contained on them. This requirement dictated the inclusion in the current problem space of a semantic description of the operator workstation.

In a similar way, OFMspert's operations model grew. It began with a static representation of the OFM and triggers to enable ACTIN to post activity tress at the appropriate time. It also included information about all possible operator actions—at the semantic, not the keystroke, level—and a table defining how to link each action to all possible upper level activities that the action could support. Other research projects added knowledge structures to the operations model. Weiner and Mitchell (1995) and Weiner et al. (1996) enhanced the operator workstation with a model-based anomaly archive. This component includes a case base of previously seen and resolved spacecraft or ground system problems together with models that organize the cases and help operators recognize an applicable case at the appropriate time—in real time. Moreover, both Jones and Mitchell (1995) and Callantine et al. (1998) realized that their applications, as opposed to the original MSOCC application, which had very little variation in the overall set of operator functions, required inclusion of a plan that guided operator activities, and thus OFMspert, over time. For Jones and Mitchell the plan was the pass plan that NASA operators follow carefully for each pass. For Callantine et al. it was the flight plan, which can change dynamically over the course of a flight. Without planning knowledge, it was impossible to predict or understand what operators were doing at particular times. Thus, for many applications, the operations model includes a plan archive that instantiates a plan for each mission—for example, flight or pass—for which OFMspert is implemented. The need for a semantic action mapper in the operations model grew as technology changed. Even in the original baseline MSOCC operator workstation there are multiple actions that operators can use to accomplish the same thing, typically different means of obtaining the same information. The fact that different *syntactic* actions are the *same semantic* action became even more obvious with the proliferation of graphical user interfaces in control rooms. Command-line input was not eliminated, but it was augmented by point-and-click and drag-and-drop options. This trend is likely to continue. Thus, OFMspert's semantic action mapper will be an increasingly vital component of its operations model.

21.7.2.3 Operator vs. Model Error.

Another application highlights an important lesson in model-based design. With expert operators, and less-than-expert designers with respect to the domain, it is important to be very careful about assuming *operator,* as opposed to *model,* error. For Georgia Tech Crew Activity Tracking System (GT-CATS), Callantine (1996) and Callantine et al. (1998) substituted the ACTIN blackboard and its notion of *maximal connectivity* of operator actions with an operator function model and an associated software implementation that dynamically specified a preferred course of activities. The preferred or expected course of activities extended from high-level activities, such as predicted selection of a flight navigation mode, to low-level operator actions. Callantine et al. included a revision process to interpret valid, alternative choices that were available, but not preferred. Callantine et al.'s data showed that, even with certified Boeing-757 pilots from the same airline, flying under near-perfect conditions, almost 50 percent of pilot actions triggered the revision process, that is, pilots chose some other path than the path GT-CATS predicted as preferred. This is an example of model error, or at least a model limitation. When queried, pilots always had a reasonable explanation for why they chose a particular series of actions, but the explanation was not included in the model. Reasons included personal style, weather, or traffic in the area.

21.7.2.4 Subject-to-Subject Variability and Model-Based Design.

Another interesting result has occurred several times as researchers began to use model-based design methodologies, realistic simulations of actual human–system interaction, and subjects who are experts in the domain of study. The result involves individual differences and the propensity of model-based designs to influence how individual operators make decisions: the data have begun to suggest that model-based design may help *reduce* individual differences.

The first case occurred in Miller's (1985) reexamination and analysis of Mitchell and Miller's data (1986). Although Mitchell and Miller's displays did not produce the expected outcome, Miller concludes: "From the models, it is clear that the (types of) displays are not the same in terms of their propensity to encourage good decision-making strategy (p. 244)."

Although Georgia Tech engineering students were used as subjects, the Mitchell and Saisi (1987) research on OFM-based workstation design analyses of variance showed that subjects were not significant factors affecting the measures of performance. This result occurs rarely in between subject designs.

Two more recent studies confirmed these results with real operators as subjects. Thurman et al.'s *interactive* operator workstation methodology was implemented for another NASA ground control system (Thurman, 1997; Thurman and Mitchell, 1995). The data show that the OFM-based workstation significantly enhanced subject performance on all measure of performance, whereas the subject was not statistically significant for several measures. In another proof-of concept implementation (Weiner et al., 1996), evaluation of the operator's workstation showed significant improvement in all performance measures with subjects not significant for more than half of the measures.

21.7.2.5 Summary. As the OFM and OFMspert are used for new applications and in the context of new domains, there will be additional lessons learned. Probably the most important contribution of these two models, in addition to their successful use as the foundation of a model-based design methodology, is the continuity of this research. Rather than traditional, one-shot laboratory experiments, the methodology, models, and applications have been implemented, evaluated, and refined over a long period of time and in a variety of real-world situations.

21.8 BASIC RESEARCH AND OPERATIONAL RELEVANCE TO REAL-WORLD DESIGN

The research base to guide effective design of computer applications is very weak and, as a result, has little, or no, impact on the design of real-world systems (Campbell, 1998; Carroll and Campbell, 1989; Mackie, 1984; O'Hara, 1994). It is vital, however, to conduct meaningful research and to provide guidance to real-world designers. Computers are ubiquitous and enormous strides are being made in technology. As a result, computer applications proliferate, affecting almost every aspect of modern life. This is true in everyday applications, safety-critical applications, and applications in which system failure is extremely costly, for example, telecommunications, power, and electrical networks. Ineffective design produces applications that are hard to use, expensive to train on, or possibly dangerous. Effective and usable design guidance is mandatory, and a primary source should be the enormous amount of research conducted in academic and research-and-development laboratories.

The failure to transfer laboratory results to real-world applications is often attributed to the methods employed by the research community, particularly academic researchers (Carroll and Campbell, 1989; Mackie, 1984; Wolfe, 1966). Academic research related to human factors is often guided by the paradigm and methods of experimental psychology. Experimental research has traditionally favored the study of narrow, artificial tasks, selected not because they illuminate anything typical or significant, but because they are tractable for study in the laboratory. Mackie's (1984) paper on research relevance and the information glut describes the problem succinctly:

> Experimental psychologists are busily engaged in the production of prodigious amounts of information which, whatever its scientific merits, is having little impact except on small groups of other experimental psychologists. It is an unfortunate fact that very little of this large, high-quality scientific output offers useful direction for solving problems of the kind on which we claim to be working. . . . This is certainly not because the amount of ongoing scientific effort is lacking in either quality or quantify. Rather, it is because so much of our research is the wrong research. . . . In the interests of scientific control, psychologists have been taught to employ simplistic, highly constrained experimental tasks that are often *invented* by the experimenter to suit the convenience of hypothesis testing. . . . Guided by their hypotheses and the particular characteristics of conveniently available experimental apparatuses, they invent perceptual games, decision-making games, learning games, and 'stress' conditions that may admirably serve their experimental objectives but will also ensure that great difficulty will be encountered in operationalizing their results, whether the attempt is made by others or even by the behavioral scientists themselves. (p. 1)

As a result, it is not surprising that laboratory-bound research has produced little if any, impact on the design of computer systems or applications (see Carroll, 1991). Mackie's critique of traditional research methods and why there is so much research and so little information is particularly applicable to the problem of transferring to real-world applications research on the design of human interaction with complex systems conducted in laboratories.

This section proposes a metatheory for conducting basic research, particularly research in university and research-and-development laboratories, that is more likely to produce believable and usable results for real-world designers. The metatheory attempts to address the underlying causes of the widespread belief that laboratory research has little, if any, impact on real-world design. First, some history is provided to explain in part why traditional methods were employed and the subsequent inability to transfer these results to realistic applications. Second, the processes that compose the metatheory are presented, both for field-study research and artifact development. Finally, generalization of results from a particular application to a wider class of systems is discussed.

21.8.1   Historical Origins

One frequently given reason for the lack of relevance of university and laboratory-based research is the research methods employed, including toy tasks and naive subjects (Carroll and Campbell, 1989; Mackie, 1984). These methods, however, were used in part because of technology constraints.

Before the advent of inexpensive and portable workstations, researchers were confined to laboratories that housed immobile computers and restricted to available subject populations, often university students with little or no domain experience. Physically, it was difficult or even impossible to conduct controlled experiments in realistic conditions with subjects who were actual users or operators in the domain of application. Thus, research employed contrived tasks with uncertain relevance to real-life applications. Experimental subjects were students who may have had 30 minutes, or at best 6 to 10 hours, of training — significantly different from real-world training programs that span weeks or months. Given these constraints, it is difficult to convince engineers and designers in the real world that such results are valid or generalize beyond university or laboratory walls. Historically, moreover, hardware and software incompatibilities were such that even if a concept or artifact was highly successful in the laboratory and potentially useful in the real world, transfer to an actual work environment was too expensive, too hard, too slow, or all three.

The emergence of portable, inexpensive computers challenges the need to continue the traditional academic research paradigm. Advanced technologies allow researchers interested in understanding and aiding human interaction with complex systems to exchange laboratories, microworlds, and naive subjects for the challenges and opportunities of actual work domains. Workstation technology, together with standardized hardware and software platforms, permit realistic demonstrations and empirical evaluations to be conducted on-site with actual users. Research based on actual tasks and evaluated by real users has high face validity within the domain of application, and is much more likely to generalize to comparable systems.

In addition to new challenges and opportunities, new technology permits new, and perhaps demands, more effective research methodologies. Rasmussen (1988), in his keynote address to the Human Factors and Ergonomics Society, summarizes the challenge as follows:

> In a way, the effect of computerization is to activate rather suddenly the problems that already existed but were latent. . . . Automation creates higher-level tasks of diagnosis and disturbance control; centralization makes the consequences of errors less acceptable; and the flexibility of information technology requires that designers know how to match the interface to a user's resources and preferences . . . (I)t creates a need for methods to analyze complex work settings, not well-defined laboratory experiments. (p. 2–3)

Thus, the tasks must be realistic, evaluations must reflect both the tasks and the context in which the tasks are performed, and, finally, participants in evaluations must be actual users. For those long-standing practitioners of traditional academic research, field-oriented research will be significantly different and very challenging. The end, however, operational relevance, may be well worth the effort.

The following section proposes a process for conducting academic research that may be more likely to produce results that address real-world problems. It encompasses two major approaches: field studies and artifact construction and evaluation.

21.8.2 A Metatheory for Conducting Academic Research

Figure 21.9 depicts a metatheory for how to conduct research that mediates between the historical problems of field studies and laboratory-based research and the concerns about validity and generalizability of such research in real-world applications. Whether the research goal is generalization of insights from one or more field studies or the construction and evaluation of an artifact, the first phase of the process requires extensive observation of one or more real-world systems.

In Phase 1, researchers first select an *interesting* system. Such a system is one that is available for observation by researchers, and one in which research results are likely to generalize beyond the specific system being studied. The second step in the observation phase dictates extensive field study to identify relevant issues and understand salient characteristics of the work domain (Mackie, 1984; Meister, 1988; Rasmussen and Goodstein, 1988).

The second phase, hypothesize, is one that is often implicit. Research, as opposed to technology development, however, requires formulation of a theory to characterize either insights that describe the work domain and its operations personnel, or system features and limitations that a research artifact can potentially ameliorate. Theory forms the basis of the research hypothesis or thesis. If the research approach is artifact or tool development (i.e., engineering), a second step in the hypothesize phase is required: specification of a domain-independent architecture to address the issues raised in the theory. For example, the widely cited Pilot's Associate project began with the premise that fighter pilots need assistance. The hypothesis was that a computer-based assistant could be built and would be useful. The domain-independent architecture was independent in the sense of being independent of a particular aircraft type or pilot. It is important to note that the architecture is platform-, hardware-, and software-independent. Essentially, the architecture is a concept with more definition than a theory and less definition than a computer implementation.

Figure 21.9 depicts the process of field study research in the lower half of the figure. Much of the insight that guides design at high levels results from this type of research. Essentially, concepts, such as human-centered automation (Billings, 1997), clumsy automation (Wiener, 1989), the keyhole effect and visual momentum (Woods, 1984), auto-

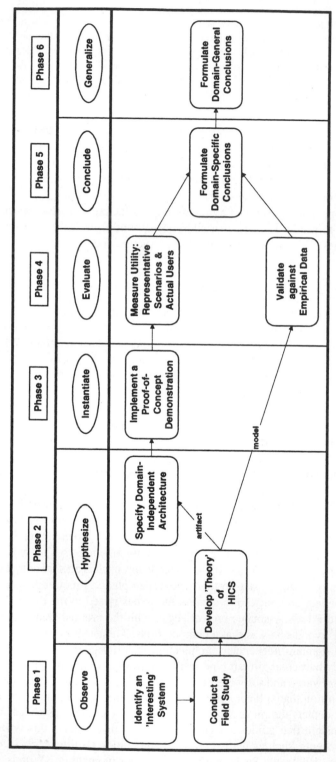

Figure 21.9 A metatheory for conducting research with operational significance in the design of HICS.

800

mation complacency (Sheridan, 1992), and the need for design that supports maintenance of accurate operator models (Sarter and Woods, 1995a, b), are derived from extensive field studies both within a domain and across domains (Woods et al., 1994). Given a theory such as the keyhole effect or visual momentum, researchers typically validate the theory by conducting additional field studies either in the same or a related domain. Next, domain-related conclusions or recommendations are proposed. Finally, since the goal of all research is to provide results that generalize, the last phase of the proposed process requires researchers to carefully specify under what conditions the theory generalizes to other systems.

The top portion of Figure 21.9 depicts a research process that has many points of correspondence with the traditional academic research paradigm. Where it differs, however, the differences are significant. As noted earlier, the first phase is to identify and study an actual system. Contrived tasks or microworlds are not adequate substitutes. This process requires a detailed understanding of a real domain, its goals, limitations, problems, and complexity. For the design of computer-based artifacts, it is imperative to state explicitly a theory that your research will prove or disprove. Specification of a theory provides the basis for subsequent generalization. The next step is specification of an architecture that embodies the characteristics specified in the theory. Phase three is a modeling or implementation phase in which the architecture is instantiated and typically linked to the system under study or a high-fidelity simulation of the system.[8] The link, either the real system or a high-fidelity simulation, allows controlled evaluation. Evaluation should address both usability and utility. Usability and associated testing has become the watchword of human–computer interaction design. Usability studies typically seek to determine the ease with which a system can be learned or used. Designers of technology that affects operators responsible for complex systems must also ensure *utility*. Evaluation of utility seeks to measure the added value derived from use of the proposed artifact. Experimental evaluation allows the determination of whether the artifact significantly enhances performance. Moreover, and perhaps most importantly, portable and inexpensive workstations permit evaluations to be conducted on-site with actual users. Like field studies, the next phase is to draw conclusions, most of which are domain-specific. The last phase discusses how to generalize research results more widely.

21.8.3 Generalizability

Given the proposed research process, two levels of conclusions can be drawn: the first with respect to the reference system that motivated the research; the second to systems that share important features with the reference system. The latter are conjectures intended to generalize to a class of applications. By specifying salient characteristics of the domain of application, researchers can characterize other domains to which the research results are likely to apply. Clancey (1989) calls such conclusions *domain general,* as opposed to *domain independent.* He suggests that for realistic applications there are likely to be few domain-independent results and, thus, researchers should carefully construct research programs and specify concrete characteristics of the system that can produce conclusions that

---

[8]As noted before, if a simulator is used, its fidelity should closely resemble both the look-and-feel and human–system dynamics. A test for simulator adequacy might be whether or not operators of the original system, including experimental participants, believe that results obtained via the simulator apply to the real system.

generalize to a set of related systems. Research has as its goal the evolution of sets of design methods and principles appropriate to large classes of application—not just a single system, for example, the SAMPEX NASA spacecraft, nor a class of related systems, for example, the set of NASA near-earth satellite. The proposed metatheory for conducting research with operational significance attempts to address this issue: results have high face validity in the original domain of application, and, given careful specification of the salient characteristics of the domain from the "human-interaction-with-complex-systems" point of view, may well generalize more broadly to a larger class of systems that possess the specified characteristics held in common with the original application.

21.9 CONCLUSION

This chapter attempts to lay out important design issues for effective human interaction with complex systems in a world in which technology allows, or even encourages, systems to change quickly and often to include automation that replaces activities in complex systems that historically have been carried out by humans. First, design guidance and levels of human–system interaction were discussed, with limitations in design knowledge pointed out. Next, a set of important issues in human interaction with complex systems was proposed. The heart of the chapter is the description of a design methodology—model-based design. A generic formulation of the methodology was initially presented. A specific implementation using the OFM as the specific representation. To illustrate the use of the methodology, one or more applications using OFM-based designs for each issue were presented. A lessons-learned section discussed issues raised by various OFM model-based design applications. The chapter concludes with a challenge to human factors practitioners and others concerned with operational relevance of human–system interaction research, to use methods that better reflect real-world characteristics and actual users. Inexpensive and powerful computer technology creates many opportunities to conduct research with much more operational significance. A metatheory for a process with which to conduct academic research that is an alternative to the traditional experimental research paradigm is proposed. The author's and her colleagues' work, as well as much of the work cited in the chapter, follows successfully some variation of this paradigm and produces academic results that real-world systems find applicable and useful.

ACKNOWLEDGMENTS

Many organizations, colleagues, and friends made the research underpinning this chapter possible. At an organizational level they include Georgia Tech's School of Industrial & Systems Engineering, NASA Goddard, NASA Ames, National Science Foundation, Delta Airlines, Federal Express, NCR, and Chrysler. I wish to thank my technical monitors over the years for their support, both financial and moral: Karen L. Moe (NASA Goddard), Walt Truszkowski (NASA Goddard), Bill Stoffel (NASA Goddard), Everett A. Palmer III (NASA Ames), and the many "operators" who familiarized ivory-tower novices with the real-world of operations in complex systems. The operators taught us the importance of people in ensuring overall system safety, reliability, and efficiency, along with the challenges in the changing roles of operators as, inevitably, new information and control technology are introduced into these systems. We interacted closely with many NASA Goddard satellite controllers, airline pilots, airline dispatchers, airline instructors, and manufacturing shop-floor personnel. As the citations indicate, the majority of the research that allowed this methodology to manifest itself was conducted in collaboration with my colleagues. Individual sections are often based on papers that summarize thesis research, including that of Donna Saisi, Janet Fath, Serena Smith, Julie Chronister, Olaf Dunkler, Jim Bushman, Patty Jones, Rose Chu, Tom Pawlowski, Jim Williams, Andy Weiner, Todd Callantine, Dave Thurman, and Alan Chappell. Although Kenny Rubin was *just* a Georgia Tech undergraduate looking for a "fun" senior design project and Al Miller was my Ph.D. advisor, as the chapter indicates, each played a major role in the research and my own professional development. As always, I gratefully acknowledge the continued support and insight of Dr. Govindaraj, who at various times participated in the research and at other times provided suggestions that facilitated the projects when we seemed to be wondering. Finally, a "thank you" to the editors, Bill Rouse and Andy Sage. When asked, I always say "yes," swear at them and myself for the length of time it takes me to write and repeatedly rewrite the promised material, and conclude by being grateful for the opportunity to integrate related but distributed ideas and projects.

REFERENCES

Abbott, K. (1996). *The Interfaces Between Flightcrews and Modern Flight Deck Systems.* Washington, DC: Federal Aviation Administration.

Andre, A. D., and Wickens, C. D. (1995). When users want what's **not** best for them. *Ergon. Des.,* **12**(1), pp. 10–14.

Aviation Week and Space Technology. (1995a). Automated cockpits: Who's in charge? Part 1. *Aviat. Week Space Technol.,* January 30, pp. 52–53.

Aviation Week and Space Technology. (1995b). Automated cockpits: Who's in charge? Part 2. *Aviat. Week Space Technol.,* February 6, pp. 52–53.

Aviation Week and Space Technology. (1998). NTSB warns of A300 display reset problem. *Aviat. Week Space Technol.,* February 9, pp. 76–77.

Bainbridge, L. (1983). Ironies of automation. *Automatica* **19**(6), 775–779.

Baron, S. (1984). A control theoretic approach to modelling human supervisory control of dynamic systems. In *Advances in Man-Machine Systems Research* (W. B. Rouse, ed.) Vol. 1, pp. 1–47. Greenwich, CT: JAI Press.

Baron, S., and Levison, W. H. (1980). The optimal control model: Status and future directions. *Proc. IEEE Conf. Cybernet. Soc.,* Chicago, pp. 90–100.

Baron, S., Kleinman, D. L., and Levison, W. H. (1970a). An optimal control model of human response. Part I: Theory and validation. *Automatica* **6,** 357–369.

Baron, S., Kleinman, D. L., and Levison, W. H. (1970b). An optimal control model of human response. Part II: Prediction of human performance in a complex task. *Automatica* **6,** 371–383.

Benson, C. R., Govindaraj, T., Mitchell, C. M., and Krosner, S. M. (1992). Effectiveness of direct manipulation interaction in the supervisory control of flexible manufacturing systems. *Inf. Decis. Technol.* **18,** 33–53.

Billings, C. E. (1995). Situation awareness measurement and analysis: A commentary. *Proc. Int. Conf. Exp. Anal. Meas. Situation Awareness,* Daytona Beach, FL, pp. 1–6.

Billings, C. E. (1997). *Aviation Automation: The Search for a Human-centered Approach.* Hillsdale, NJ: Erlbaum.

Bobrow, L. S., and Arbib, M. A. (1974). *Discrete Mathematics.* Philadelphia: Saunders.

Bradley, D. (1997). Operations in control rooms. Personal communications.

Brann, D. M., Thurman, D. A., and Mitchell, C. M. (1995). Case-based reasoning as a methodology for accumulating human expertise for discrete system control. *Proc. IEEE Int. Conf. Syst., Man, Cybernet.,* Vancouver, BC, pp. 4219–4223.

Brann, D. M., Thurman, D. A., and Mitchell, C. M. (1996). Human interaction with lights-out automation: A field study. *Proc. Sympo. Hum. Interact. Complex Syst.,* Dayton, OH, pp. 276–283.

Buchanan, B. G., and Wilkins, D. C. (1993). Overview of knowledge acquisition and learning. In *Readings in Knowledge Acquisition and Learning: Automating the Construction and Improvement of Expert Systems* (B. G. Buchanan and D. C. Wilkins, eds.), pp. 1–4. San Mateo, CA: Morgan Kaufmann.

Bushman, J. B., Mitchell, C. M., Jones, P. M., and Rubin, K. S. (1993). ALLY: An operator's associate for cooperative supervisory control systems. *IEEE Trans. Systs., Man, Cybernet.* **23**(1), 111–128.

Cakir, A., Hart, D. J., and Stewart, T. F. M. (1980). *Visual Display Terminals: A Manual Covering Ergonomics, Workplace Design, Health and Safety, Task Organization* New York: Wiley.

Callantine, T. J., (1996). Tracking operator activities in complex systems. Ph.D. Thesis, Georgia Institute of Technology, School of Industrial and Systems Engineering, Atlanta.

Callantine, T. J., Mitchell, C. M., and Palmer, E. A. (1997). *GT-CATS: Tracking Operator Activities in Complex Systems* (CR-A-1997a). Moffett Field, CA: NASA Ames Research Center.

Callantine, T. J., Mitchell, C. M., and Palmer, E. A. (1998). GT-CATS as a model of pilot behavior in the 'glass cockpit': An application and empirical evaluation. *Submitted for publication.*

Campbell, G. (1998). When opportunity knocks. *Cogn. Times* **3** (1), 2–3.

Card, S., Moran, T., and Newell, A. (1983). *The Psychology of Human-computer Interaction.* Hillsdale, NJ: Erlbaum.

Cardosi, K. M., and Murphy, E. D. (eds.). (1995). *Human Factors in the Design and Evaluation of Air Traffic Control Systems.* Springfield, VA: National Technical Information Service.

Carr, B., and Goldstein, I. (1997). *Overlay's: A Theory of Modeling for Computer-aided Instruction* (AI Lab Memo 406; Logo Memo 40). Cambridge, MA: MIT University Press.

Carroll, J. M. (1991). Introduction: The Kittle House manifesto. In *Designing Interaction* (J. M. Carroll, ed.) pp. 1–16. Cambridge, UK: Cambridge University Press.

Carroll, J. M., and Campbell, R. L. (1989). Artifacts as psychological theories: The case of human-computer interaction. *Behavi. Inf. Technol.* **8**(4), 247–256.

Casey, S. (1993). *Set Phasers on Stun and Other True Tales of Design, Technology, and Human Error.* Santa Barbara, CA: Aegean Publishing.

Chambers, A. B., and Nagel, D. C. (1985). Pilots of the future: Human or computer? *Commun. ACM* **28**(11), 1187–1199.

Chapanis, A. (1970). Human factors in systems engineering. In *Systems Psychology* (K. B. DeGreene, ed.), pp. 28–38. New York: McGraw-Hill.

Chappell, A. R., Crowther, E. G., Mitchell, C. M., and Govindaraj, T. (1997). The VNAV Tutor: Addressing a mode awareness difficulty for pilots of glass cockpit aircraft. *IEEE Trans. Syst., Man, Cybernet.* **27**(3), 327–385.

Chu, R. W., Mitchell, C. M., and Jones, P. M. (1995). Using the operator function model and OFMspert as the basis for an intelligent tutoring system: Towards a tutor/aid paradigm for operators of supervisory control systems. *IEEE Trans. Syst., Man, Cybernet.* **25**(7), 1054–1075.

Clancey, W. J. (1989). Viewing knowledge bases as qualitative models. *IEEE Expert,* pp. 9–23.

Computer Sciences Corporation. (1983). *Multisatellite Operations Control Center (MSOCC) Data Operations Control System (DOCS) Operator Interface Design Review* (Des. Rev. NAS 5-27555). Greenbelt, MD: NASA Goddard Space Flight Center.

Converse, S. A. (1995). *Evaluation of the Computerized Procedures Manual II (COMPMA II)*(NUREG/CR-6398). Washington, DC: U.S. Nuclear Regulatory Commission.

Degani, A., and Wiener, E. L. (1993). Cockpit checklists: Concepts, design, and use. *Hum. Factors* **35**(2), 345–359.

Dunkler, O., Mitchell, C. M., Govindaraj, T., and Ammons, J. C. (1988). The effectiveness of supervisory control strategies in scheduling flexible manufacturing systems. *IEEE Trans. Syst., Man, Cybernet.* **18**(2), 223–237.

Fath, J. L., Mitchell, C. M., and Govindaraj, T. (1990). An ICAI architecture for troubleshooting in complex dynamic systems. *IEEE Trans. Syst., Man, Cybernet.* **20**(3), 537–558.

Fitts, P. M. (1951). Engineering psychology and equipment design. In *Handbook of Experimental Psychology* (S. S. Stevens, ed.), Vol. 65, pp. 423–432. New York: Wiley.

Forsythe, D. E., and Buchanan, B. G. (1993). Knowledge acquisition for expert systems: Some pitfalls and suggestions. In *Readings in Knowledge Acquisition and Learning: Automating the Construction and Improvement of Expert Systems* (B. G. Buchanan and D. C. Wilkins, eds.), pp. 117–125. San Mateo, CA: Morgan Kaufmann.

Funk, I., K. H., and Lind, J. H. (1992). Agent-based pilot-vehicle interfaces: Concept and prototype. *IEEE Trans. Syst., Man, Cybernet.* **22**(6), 1309–1322.

Garland, D. J., and Endsley, M. R. (eds.). (1995). *Proceedings of an International Conference on Experimental Analysis and Measurement of Situation Awareness.* Daytona Beach, FL: Embry-Riddle Aeronautical University Press.

Geddes, N. D. (1989). Understanding human operator intentions in complex systems. Ph.D. Thesis, Georgia Institute of Technology, Center for Human-Machine Systems Research, School of Industrial and Systems Engineering, Atlanta.

Geddes, N. D., and Hammer, J. M. (1991). Automatic display management using dynamic plans and events. *Proc. 6th. Int. Sympo. Aviat. Psycho.,* Columbus, OH, pp. 83–87.

Govindaraj, T., Su, Y. I., Vasandani, V., and Recker, M. M. (1996). Training for diagnostic problem solving in complex engineered systems: Modeling, simulation, and intelligent tutors. In *Human/Technology Interaction in Complex Systems* (W. B. Rouse, ed.), Vol. 8, pp. 1–66. Greenwich, CT: JAI Press.

Hammer, J. M. (1984). An intelligent flight-management aid for procedure execution. *IEEE Trans. Syst., Man, Cybernet.* **14**(6), 885–888.

Hunt, R. M., and Rouse, W. B. (1984). A fuzzy rule-based model of human problem solving. *IEEE Trans. Syst., Man, Cybernet.* **14**(1), 112–121.

Hutchins, E. (1983). Understanding Micronesian navigation. In *Mental Models* (D. Gentner and A. L. Stevens, eds.), pp. 191–225, Hillsdale, NJ: Erlbaum.

Hutchins, E. L., Hollan, J. D., and Norman, D. A. (1985). Direct manipulation interfaces. *Hum. Comput. Interact.* **1,** 311–338.

Hutchins, E. L., Hollan, J. D., and Norman, D. A. (1986). Direct manipulation interfaces. In *User Centered System Design* (D. A. Norman and S. W. Draper, eds.), pp. 87–124. Hillsdale, NJ: Erlbaum.

International Nuclear Safety Advisory Group. (1988). *Basic Safety Principles for Nuclear Power Plants* (Saf. Ser. No. 75-INSAG-3). Vienna: IAEA.

Johannsen, G. (1992). Towards a new quality of automation in complex man-machine systems. *Automatica* **28**(2), 355–373.

Jones, P. M., and Jasek, C. A. (1997). Intelligent support for activity management (ISAM): An architecture to support distributed supervisory control. *IEEE Trans. Syst., Man, Cybernet.* **27**(3), 274–288.

Jones, P. M., and Mitchell, C. M. (1987). Operator modeling: Conceptual and methodological distinctions. *Proc. 31st Annu. Meet. Hum. Factors Soc.,* New York, pp. 31–36.

Jones, P. M., and Mitchell, C. M. (1995). Human-computer cooperative problem solving: Theory, design, and evaluation of an intelligent associate system. *IEEE Trans. Syst., Man, Cybernet.* **25**(7), 1039–1053.

Jones, P. M., Mitchell, C. M., and Rubin, K. S. (1990). Validation of intent inferencing by a model-based operator's associate. *Int. J. Man-Mach. Stud.* **33,** 177–202.

Jones, P. M., Chu, R. W., and Mitchell, C. M. (1995). A methodology for human-machine systems research: Knowledge engineering, modeling, and simulation. *IEEE Trans. Syst., Man, Cybernet.* **25**(7), 1025–1038.

Kantowitz, B. H., and Sorkin, R. D. (1983). *Human Factors: Understanding People-system Relationships.* New York: Wiley.

Karat, C.-M., Campbell, R., and Fiegel, T. (1992). Comparison of empirical testing and walkthrough methods in user interface evaluation. *Proc. CHI '92,* pp. 397–404.

Kirlik, A., and Kossack, M. F. (1993). Perceptual augmentation to support skilled interaction. *Proc. IEEE Int. Conf. Syst., Man, Cybernet.,* Le Touquet, France, pp. 896–901.

Kirlik, A., Kossack, M. F., and Shively, R. J. (1994). *Ecological Task Analysis: Supporting Skill Acquisition in Dynamic Interaction.* Atlanta: Georgia Institute of Technology, Center for Human-Machine Systems Research, School of Industrial and Systems Engineering.

Kirwan, and Ainsworth (eds.). (1992). *A Guide to Task Analysis.* Washington, DC: Taylor & Francis.

Knaeuper, A., and Morris, N. M. (1984). A model-based approach for online aiding and training in process control. *Proc. IEEE Int. Conf. Syst., Man, Cybernet.,* pp. 173–177.

Knaeuper, A., and Rouse, W. B. (1985). A rule-based model of human problem-solving behavior in dynamic environments. *IEEE Trans. Syst., Man, Cybernet.* **15**(6), 708–719.

Lee, E. J. (1994). *Computer-based Digital System Failures* (Tech. Rev. Rep. AEOD/T94-03). Washington, DC: U.S. Nuclear Regulatory Commission.

Leveson, N. G. (1995). *Safeware: System Safety and Computers.* New York: Addison-Wesley.

Lund, A. M. (1997). Expert ratings of usability maxims. *Ergon. Des.* **5**(3), 15–20.

Mackie, R. R. (1984). Research relevance and the information glut. *Hum. Factors Rev.* pp. 1–11.

Meister, D. (1988). References to reality in the practice of human factors. *Hum. Factors Soc. Bull.* **31**(10) 1–3.

Miller, R. A. (1978). *A Formal Model of the Adaptive and Discrete Control Behaviors of Human Operators* (RF 760373/784382). Columbus: Industrial & Systems Engineering, Ohio State University.

Miller, R. A. (1979). *Identification of Finite State Models of Human Operators* (OSU-RP-784556). Columbus: Ohio State University, Industrial & Systems Engineering.

Miller, R. A. (1985). A system approach to modeling discrete control performance. In *Advances in Man-machine Systems* (W. B. Rouse, ed.), Vol. 2, pp. 177–248. Greenwich, CT: JAI Press.

Mitchell, C. M. (1987). GT-MSOCC: A domain for research on human-computer interaction and decision aiding in supervisory control systems. *IEEE Trans. Syst., Man, Cybernet.* **17**(4), 553–572.

Mitchell, C. M. (1996a). GT-MSOCC: Operator models, model-based displays, and intelligent aiding. In *Human/Technology Interaction in Complex Systems* (W. B. Rouse, ed.), Vol. 8, pp. 67–172. Greenwich, CT: JAI Press.

Mitchell, C. M. (1996b). Models for the design of human interaction with complex dynamic systems. *Proc. Cogn. Eng. Process Control Workshop,* Kyoto, Japan, pp. 230–237.

Mitchell, C. M. (1999). Horizons in pilot training: Desktop tutoring systems. In *Cognitive Engineering in the Aviation Domain* (N. Sarter and R. Almerberti, eds.). Mahwah, NJ: Erlbaum (in press).

Mitchell, C. M. (1997b). *Technology and Education: Issues, Applications, and Horizons in the Workplace*. Washington, DC: National Research Council, Board on Assessment and Testing.

Mitchell, C. M., and Govindaraj, T. (1991). Design and effectiveness of intelligent tutors for operators of complex dynamic systems: A tutor implementation for satellite system operators. *Interact. Learn. Environ.* **1**(3), 193–229.

Mitchell, C. M., and Miller, R. A. (1983). Design strategies for computer-based information displays in real-time control systems. *Hum. Factors* **25**(4), 353–369.

Mitchell, C. M., and Miller, R. A. (1986). A discrete control model of operator function: A methodology for information display design. *IEEE Trans. Syst., Man, Cybernet.* **16**(3), 343–357.

Mitchell, C. M., and Miller, R. A. (1993). A framework for model validity with applications to models of human-machine systems. In *Advances in Man-machine Systems Research* (W. B. Rouse, ed.), Vol. 6, pp. 233–293. Greenwich, CT: JAI Press.

Mitchell, C. M., and Saisi, D. L. (1987). Use of model-based qualitative icons and adaptive windows in workstations for supervisory control systems. *IEEE Trans. Syst., Man, Cybernet.* **17**(4), 573–593.

Mitchell, C. M., and Sundström, G. A. (1997). Human interaction with complex systems: Design issues and research approaches. *IEEE Trans. Syst., Man, Cybernet.* **27**(3), 265–273.

Mitchell, C. M., Morris, J. G., Ockerman, J. J., and Potter, W. J. (1996). Recognition primed decision making as a technique to support reuse in software design. In *Naturalistic Decision Making* (C. E. Zsambok and G. A. Klein, eds.), pp. 305–318. New York: Erlbaum.

Moray, N., and Huey, B. (1988). *Human Factors and Nuclear Safety*. Washington, DC: National Research Council, National Academy of Sciences.

Morris, J. G., and Mitchell, C. M. (1997). A case-based support system to facilitate software reuse. *Proc. IEEE Int. Conf. Syst., Man, Cybernet.*, Orlando, FL, pp 232–237.

Morris, N. M., and Rouse, W. B. (1985). The effects of type of knowledge upon human problem solving in a process control task. *IEEE Trans. Syst., Man, Cybernet.* **15**(6),698–707.

Nagle, D. C. (1988). Human error in aviation operations. In *Human Factors in Aviation* (E. L. Wiener and D. C. Nagle, eds.), pp. 263–303. Orlando, FL: Academic Press.

National Aeronautics and Space Administration (NASA) (1988). *Space Station Freedom Human-computer Interface Guidelines* (NASA USE-100). Washington, DC: NASA, Johnson Space Flight Center.

National Aeronautics and Space Administration (NASA). (1996). *User-interface Guidelines* (NASA DSTL-95-033). Greenbelt, MD: NASA Goddard Space Flight Center.

Nii, H. P. (1986a). *Blackboard Systems* (KSL 86-18) Stanford, CA: Stanford University, Knowledge Systems Laboratory.

Nii, H. P. (1986b). Blackboard systems: Part I. *AI Mag.* **7**(2), 38–53.

Nii, H. P. (1986c). Blackboard systems: Part II. *AI Mag.* **7**(3), 82–110.

Nii, H. P. (1994). Blackboard systems at the architecture level. *Expert Syst. Appl.* **7**, 43–54.

Norman, D. A. (1988). *The Psychology of Everyday Things*. New York: Basic Books.

Ockerman, J. J., and Mitchell, C. M. (1999). Case-based design browser to support software reuse. *Intern. J. Hum. Comp. Stud.* (accepted for publication).

O'Hara, J. M. (1994). *Advanced Human-system Interface Design Review Guideline*, Vols. 1 and 2 (NUREG/CR-5908). Washington, DC: U.S. Nuclear Regulatory Commission.

Palmer, E. A. (1987). Using the operator function model as a 'living' task analysis. Personal communications.

Parasuraman, R., and Mouloua, M. (eds.). (1996). *Automation and Human Performance*. Mahwah, NJ: Erlbaum.

Parasuraman, R., and Riley, V. (1997). Humans and automation: Use, misuse, disuse, and abuse. *Hum. Factors* **39**(2), 230–253.

Pawlowski, T. J. (1990). Design of operator interfaces to support effective supervisory control and to facilitate intent inferencing by a computer-based operator's associate. Ph.D. Thesis, Georgia Institute of Technology, Center for Human-Machine Systems Research, School of Industrial and Systems Engineering. Atlanta.

Pew, R. W. (1995). The state of situation awareness measurement: Circa 1995. *Proc. Int. Conf. Exp. Anal. Meas. Situation Awareness,* Daytona Beach, FL, pp. 7–15.

Phatak, A. V., and Kessler, K. M. (1977). Modeling the human gunner in an anti-aircraft artillery (AAA) tracking task. *Hum. Factors* **19**(5), 477–494.

Preece, J. (1994). *Human-computer Interaction.* New York: Addison-Wesley.

Ragheb, H. (1996). Operating and maintenance experience with computer-based systems in nuclear power plants. *Int. Workshop Tech. Support Licens. Comput. Based Syst. Important Saf.,* Munich, Germany, pp. 143–150.

Rasmussen, J. (1983). Skills, rules, and knowledge: Signals, signs, and symbols, and other distinctions in human performance models. *IEEE Trans. Syst., Man, Cybernet.* **SMC-13** (3), 257–267.

Rasmussen, J. (1985). The role of hierarchical knowledge representation in decisionmaking and system management. *IEEE Trans. Syst., Man, Cybernet.* **15,** 234–243.

Rasmussen, J. (1986). *Information Processing and Human-machine Interaction: An Approach to Cognitive Engineering.* New York: North-Holland.

Rasmussen, J. (1987). Reasons, causes, and human error. In *New Technology and Human Error* (J. Rasmussen, K. Duncan, and J. Leplat, eds.), pp. 293–303. New York: Wiley.

Rasmussen, J. (1988). Information technology: A challenge to the Human Factors Society? *Hum. Factors Soc. Bull.* **31**(7), 1–3.

Rasmussen, J., and Goodstein, L. P. (1988). Information technology and work. In *Handbook of Human-computer Interaction* (M. Helander, ed.), pp. 175–202. New York: North-Holland.

Rasmussen, J., Pejterson, A. M., and Goodstein, L. P. (1994). *Cognitive Systems Engineering.* New York: Wiley.

Resnick, D. E., Mitchell, C. M., and Govindaraj, T. (1987). An embedded computer-based training system for rhino robot operators. *IEEE Control Syst. Mag.* **7**(3), 3–8.

Roth, E. M., Bennett, K. B., and Woods, D. D. (1987). Human interaction with an "intelligent" machine. *Int. J. Man-Mach. Stud.* **27,** 479–526.

Rouse, W. B. (1978). A model of human decisionmaking in a fault diagnosis task. *IEEE Trans. Syst., Man, Cybernet.* **8**(5), 357–361.

Rouse, W. B. (1980). *System Engineering Models of Human-machine Interaction.* New York: North-Holland.

Rouse, W. B. (1981). Human-computer interaction in the control of dynamic systems. *Comput. Surv.* **13**(1), 71–99.

Rouse, W. B. (1985). Man-machine systems. In *Productions Handbook* (J. A. White, ed.), pp. 52–66. New York: Wiley.

Rouse, W. B. (1994). Twenty years of adaptive aiding: Origins of the concepts and lessons learned. In *Human Performance in Automated Systems: Current Research and Trends* (M. Mouloula and R. Parasuraman, eds.), pp. 28–33. Hillsdale, NJ: Erlbaum.

Rouse, W. B., Geddes, N. D., and Curry, R. E. (1987). An architecture for intelligent interfaces: Outline for an approach supporting operators of complex systems. *Hum.-Comput. Interact.* **3**(2), 87–122.

Rouse, W. B., Geddes, N. D., and Hammer, J. M. (1990). Computer aided fighter pilots. *IEEE Spectrum,* **27**(3), 38–41.

Rubin, K. S., Jones, P. M., and Mitchell, C. M. (1988). OFMspert: Inference of operator intentions in supervisory control using a blackboard architecture. *IEEE Trans. Syst., Man, Cybernet.* **18**(4), 618–637.

Sarter, N. B., and Woods, D. D. (1995a). How in the world did we ever get in that mode? Mode error and awareness in supervisory control. *Hum. Factors* **37**(1), 5–19.

Sarter, N. B., and Woods, D. D. (1995b). *Strong, Silent, and Out-of-the-Loop': Properties of Advanced (cockpit) Automation and Their Impact on Human-automation Interaction* (CSEL Rep. 95-TR-01). Columbus: Ohio State University, Cognitive Systems Engineering Laboratory.

Schank, R. C., and Abelson, R. P. (1977). *Scripts, Plans, Goals, and Understanding.* Hillsdale, NJ: Erlbaum.

Shaiken, H. (1985). The automated factory: The view from the shop floor. *Technol. Rev.,* January, pp. 17–25.

Sheridan, T. B. (1976). Toward a general model of supervisory control. In *Monitoring Behavior and Supervisory Control* (T. B. Sheridan and G. Johannsen, eds.), pp. 271–281. New York: Plenum.

Sheridan, T. B. (1992). *Telerobotics, Automation, and Human Supervisory Control.* Cambridge, MA: MIT Press.

Sheridan, T. B., and Ferrell, W. R. (1974). *Man-machine Systems: Information, Control, and Decision Models of Human Performance.* Cambridge, MA: MIT Press.

Smith, P. J., Layton, C., and McCoy, C. E. (1997). Brittleness in the design of cooperative problem-solving systems: The effects on user performance. *IEEE Trans. Syst., Man, Cybernet.* **27**(3), pp. 360–372.

Smith, S., and Mosier, J. (1988). *Guidelines for Designing User Interface Software* (ESD-TR-86-278). Washington, DC: U.S. Department of Defense, Office of Management and Budget.

Smith, S. C. (1991). Modeling the pilots and constructing an intent inferencer for a Boeing 727 cockpit. M.S. Thesis, Georgia Institute of Technology, Center for Human-Machine Systems Research, School of Industrial and Systems Engineering, Atlanta.

Smith, S. C., Govindaraj, T., and Mitchell, C. M. (1990). Operator modeling in civil aviation. *Proc. IEEE Int. Conf. Syst., Man, Cybernet.,* Los Angeles, pp. 512–514.

Thurman, D. A. (1995). Improving operator effectiveness in monitoring complex systems: A methodology for the design of interactive monitoring and control interfaces. M.S. Thesis, Georgia Institute of Technology, Center for Human-Machine Systems Research, School of Industrial and Systems Engineering, Atlanta.

Thurman, D. A. (1997). The interactive monitoring and control (IMaC) design methodology: Application and empirical results. *Proc. 41st Annu. Meet. Hum. Factors Ergon. Soc.* Albuquerque, NM, pp. 289–293.

Thurman, D. A., and Mitchell, C. M. (1994). A methodology for the design of interactive monitoring interfaces. *Proc. IEEE Int. Confe. Syst., Man, Cybernet.,* San Antonio, TX, pp. 1738–1744.

Thurman, D. A., and Mitchell, C. M. (1995). A design methodology for operator displays of highly automated supervisory control systems. *Proc. 6th IFAC/IFIP/IFOR/SEA Symp. Anal., Des., and Eval. Man-Mach. Syst.,* Cambridge, MA, pp. 821–825.

Turinsky, P. J., Baron, S., Burch, W. D., Corradini, M. L., Lucas, G. E., Matthews, R. B., and Uhrig, R. E. (1991). *Western European Nuclear Power Generation Research and Development* (FASAC Tech. Assess. Rep.). McLean, VA: Science Applications International Corporation.

Uhrig, R. E., and Carter, R. J. (1993). *Instrumentation, Control, and Safety Systems of Canadian Nuclear Facilities,* JTEC/WTEC Monogr. Baltimore, MD: Loyola College, World Technology Center.

Vanderdonckt, J. (1994). *Tools for Working with Guidelines (Bibliography).* Belgium: University of Namur.

Vasandani, V., and Govindaraj, T. (1994). Integration of interactive interfaces with intelligent tutoring systems: An implementation. *Mach.-Media. Learn.* **4**(4), 295–333.

Vicente, K. J., and Rasmussen, J. (1992). Ecological interface design: Theoretical foundations. *IEEE Trans. Syst., Man, Cybernet.* **22**(4), 589–606.

Wadlow, M. G. (1996). Visual interaction design. *SIGCHI Bull.* **28**(1), 32–33.

Weiner, A. J., and Mitchell, C. M. (1995). FIXIT: A case-based architecture for computationally encoding fault management experience. *Proc. 6th IFAC/IFIP/IFOR/SEA Symp. Anal., Des., Eval. Man Mach. Syst.,* Boston, pp. 196–200.

Weiner, A. J., Thurman, D. A., and Mitchell, C. M. (1996). FIXIT: An architecture to support recognition-primed decision making in complex system fault management. *Proc. 40th Annu. Meet. Hum. Factors Ergon. Soc.,* Philadelphia, pp. 209–213.

Wickens, C. D. (1991). *Engineering Psychology and Human Performance.* New York: Harper-Collins.

Wiener, E. L. (1989). *Human Factors of Advanced Technology ("Glass Cockpit") Transport Aircraft* (Tech. Rep. 117528). Moffett Field, CA: NASA Ames Research Center.

Wolfe, D. (1966). Social problems and social science. *Science* **151,** 1160–1178.

Woods, D. D. (1984). Visual momentum: A concept to improve the cognitive coupling of person and computer. *Int. J. Man-Mach. Stud.* **21,** 229–244.

Woods, D. D. (1992). Are guidelines on human-computer interaction a Faustian bargain? *Comput. Syst. Tech. Group Bull.* **19** (1), 2.

Woods, D. D. (1995). Towards a theoretical base for representation design in the computer medium: Ecological perception and aiding human cognition. In *Global Perspectives on the Ecology of Human-machine Systems* (J. Flach, P. Hancock, J. Caird, and K. Vicente, eds.), Vol. 1, pp. 157–188. Hillsdale, NJ: Erlbaum.

Woods, D. D., Johannesen, L. J., Cook, R. I., and Sarter, N. B. (1994). *Behind Human Error: Cognitive Systems, Computers, and Hindsight.* Wright-Patterson AFB, OH: Crew Systems Ergonomics Information Analysis Center (SOAR/CERIAC).

Zacharias, G. L., Baron, S., and Muralidharan, R. (1981). *A Supervisory Control Model of the AAA Crew* (October 4802). Boston: Bolt, Beranek, & Newman.

Zinser, K., and Henneman, R. L. (1988). Development and evaluation of a model of human performance in a large-scale system. *IEEE Trans. Syst., Man, Cybernet.* **8** (3), 367–375.

Zuboff, S. (1991). *In the Age of the Smart Machine: The Future of Work and Power.* New York: Basic Books.

22 Evaluation of Systems

JAMES M. TIEN

22.1 INTRODUCTION

The purpose of this handbook is to demonstrate the use of the theory and practice of systems engineering and systems management. In this chapter, we present a systems engineering approach to the evaluation of systems. In this regard, four points should be made. First, as considered in other chapters of this handbook, almost any device, product, operation, or program can be considered to be a system or subsystem. While the evaluation of electromechanical devices or products is important and necessary, the more challenging problem is that of evaluating a human-centered system or program. Consequently, our focus herein is on the evaluation of human-centered systems or programs.

Second, evaluation is simply purposeful analysis. According to *Webster's Dictionary,* to evaluate means "to examine and judge"; thus evaluation can be thought of as analysis (i.e., examination) with a purpose (i.e., judgment). In this regard, since every analysis should be purposeful (otherwise it is unclear as to why the analysis is being undertaken), all analyses should be conducted within a purposeful and systematic evaluation framework, such as the one presented in Section 22.2.

Third, it is generally recognized that the reason most evaluations or purposeful analyses fail—or are not valid—is because their research or evaluation designs are lacking (Chelimsky and Shadish, 1997). Although there is no stock evaluation design that can be taken off the shelf and implemented without revision, there should be a systematic approach or process by which valid and comprehensive evaluation designs can be developed. We present such an approach in this chapter; in particular, a systems engineering approach.

Fourth, the systems engineering approach to program evaluation that is detailed herein was first advanced by Tien (1979); it has since been successfully employed in a number of evaluation efforts [see, as examples, Colton et al. (1982), Tien and Cahn (1986), and Tien and Rich (1997)]. Indeed, this chapter summarizes Tien's earlier work (1979, 1988, 1990), and extends that work through its current emphasis on an approach that is not only systems-engineering-oriented but also modeling-based. In particular, a simple but insightful stochastic model for determining valid sample sizes is also developed herein.

The remainder of this chapter comprises five sections that address, respectively, the evaluation field, the evaluation framework, the evaluation components, an example evaluation modeling effort, and some concluding remarks.

22.2 EVALUATION FIELD

In the United States, evaluation has emerged as a formal field of practice over the past three decades, beginning in 1965 when Congress passed the Great Society education leg-

Handbook of Systems Engineering and Management, Edited by A. P. Sage and W. B. Rouse
ISBN 0471-15405-9 ©1999 John Wiley and Sons, Inc.

islation with an evaluation mandate. Since then the evaluation mandate has been attached to other social programs and big-ticket items at all levels of government. Indeed, even some private organizations seek evaluations of their major program investments. Evaluation is growing because it helps governments and organizations to make difficult choices by providing answers to such questions as: Does the program work? Is it worth the cost? Can and should it be implemented elsewhere? The need for conducting evaluations becomes more critical as systems or programs become more complex and more costly and, concomitantly, as the tax base or resources for their funding remain fixed or are decreased.

As the evaluation of devices, products, operations, and programs grows and plays a more central role in societal decision making, it is critical that evaluations be conducted in an objective manner. This is indeed a challenge, as opposing political, financial, and environmental forces tug at the ethical values of evaluators or analysts. House (1993, p. xvi) cautions against "clientism (taking the client's interests as the ultimate consideration), contractualism (adhering inflexibly to the contract), managerialism (placing the interests of the managers above all else), methodologicalism (believing that proper methodology solves all ethical problems), pluralism/elitism (including only the powerful stakeholders' interests in the evaluation), and relativism (taking all viewpoints as having equal merit)."

While evaluation has become a multidisciplinary field, it is not surprising that given its initial social program focus, its roots are in the social sciences, especially in the discipline of psychology. Obviously, the conduct of evaluation has extended beyond social programs and includes, as examples, technology assessments (Porter et al., 1980) and evaluation of computer aids (Sage, 1981).

The social sciences have also provided the field of evaluation with a common understanding and shared terminology. Basically, in regard to a program's life cycle, it is assumed, first, that the program design and its evaluation design are developed concurrently (so that the program is indeed amenable to evaluation), and, second, that the traditional paradigm of evaluation is in effect (i.e., evaluation provides feedback to the program administrator or decision maker, who decides whether the program should be refined, rejected, and/or transferred). In regard to the evaluation process, every unit (i.e., subject, group, site, or time period) can be designated as being either test or control. During the period of evaluation, pretest or pretreatment measurements are first made of both sets of units, followed by the administration of the program intervention on each test unit, and concluding with appropriate posttest or posttreatment measurements. There may, of course, be several test units, control units, program interventions, pretest measurements, and posttest measurements.

22.3 EVALUATION FRAMEWORK

Our evaluation framework is based on a dynamic rollback approach that consists of three steps leading to a valid and comprehensive evaluation design. The "rollback" aspect of the approach is reflected in the ordered sequence of steps. The sequence rolls back in time from (1) a projected look at the range of program characteristics (i.e., from its rationale through its operation and anticipated findings); to (2) a prospective consideration of the threats (i.e., programs and pitfalls) to the validity of the final evaluation; and to (3) a more immediate identification of the evaluation design elements. The logic of the sequence of steps should be noted; that is, the anticipated *program characteristics* identify the possible *threats to validity,* which in turn point to the *evaluation design elements* that are necessary to mitigate, if not to eliminate, these threats. The three-step sequence can also be stated in terms of two sets of links that relate, respectively, an anticipated set of program characteristics *to* an intermediate set of threats to validity *to* a final set of design elements. Although some of the links between program characteristics and threats to validity are obvious (e.g., a concurrent program may cause an extraneous event threat to internal validity), an exhaustive listing of such links—for purposes of a handbook—will require a significant amount of analysis of past and on-going evaluations. Similarly, the second set of links between threats to validity and design elements will also require a significant amount of analysis. Both sets of links are briefly considered herein.

The "dynamic" aspect of the approach refers to its nonstationary character; that is, the components of the framework must constantly be updated throughout the entire development and implementation phases of the evaluation design. In this manner, the design elements can be refined, if necessary, to account for any new threats to validity that may be caused by previously unidentified program characteristics.

In sum, the dynamic rollback approach is systems-oriented; it represents a purposeful and systematic process by which valid and comprehensive evaluation designs can be developed. The first two steps of the design framework are elaborated on in the next two subsections, while the third step is considered in Section 22.4.

22.3.1   Program Characteristics

In general, the characteristics of a program can be determined by seeking responses to the following questions: What is the program rationale? Who has program responsibility? What is the nature of program funding? What is the content of the program plan? What are the program constraints? What is the nature of program implementation? What is the nature of program operation? Are there any other concurrent programs? What are the anticipated evaluation findings?

Again, it should be noted that the purpose of understanding the program characteristics is to identify the resultant problems or pitfalls that can arise to threaten the validity of the final evaluation. The possible links between program characteristics and threats to validity are considered in the next subsection, following a definition of the threats to validity.

22.3.2   Threats to Validity

After almost two decades, the classic monograph by Campbell and Stanley (1966) is still the basis for much of the on-going discussion of threats to validity. However, their original 12 threats have been expanded by Tien (1979) to include eight additional threats. As indicated in Figure 22.1, the 20 threats to validity can be grouped into the following five categories:

1. *Internal validity* refers to the extent that the statistical association of an intervention and measured impact can reasonably be considered a causal relationship.
2. *External validity* refers to the extent that the causal relationship can be generalized to different populations, settings, and times.

Threats to Internal Validity

1. *Extraneous events* (i.e., history) may occur during the period of evaluation, inasmuch as total test or experimental isolation cannot be achieved in real-world experimentation.
2. *Temporal maturation* of subjects or processes (e.g., growing older, growing more tired, becoming wiser, etc.)—including cyclical maturation—may influence observed impacts.
3. *Design instability* (i.e., unreliability of measures, fluctuations in sampling units or subjects, and autonomous instability of repeated or equivalent measures) may introduce biases.
4. *Pretest experience,* gained from a response to a pretest measurement (e.g., questionnaire, test, observation, etc.) may impact the nature and level of response to a subsequent posttest measurement.
5. *Instrumentation changes* (e.g., changes in the calibration of a measurement instrument, changes in the observers or evaluators used, etc.) may produce changes in the obtained measurements.
6. *Regression artifacts* may occur due to the identification of test or control subjects (or periods) whose dependent or outcome measures have extreme values—these extreme values are artificial and will tend to regress toward the mean of the population from which the subjects are selected.
7. *Differential selection*—as opposed to random selection—of subjects for the test and control groups may introduce biases.
8. *Differential loss* (i.e., experimental mortality) of subjects from the test and control groups may introduce biases.
9. *Selection-related interaction* (with extraneous events, temporal maturation, etc.) may be confounded with the impact of the intervention, as, for example, in the case of a self-selected test group or in test and control groups which are maturing at different rates.

Threats to External Validity

10. *Pretest-intervention interaction* (including "halo" effect) may cause a pretest measurement to increase or decrease a subject's sensitivity or responsiveness to the intervention and thus make the results obtained for a pretested population unrepresentative of the impacts of the intervention for the unpretested universe from which the test subjects are selected.
11. *Selection-intervention interaction* may introduce biases which render the test and/or control groups unrepresentative of the universe from which the test subjects are selected.
12. *Test-setting sensitivity* (including "Hawthorne" and "placebo" effects) may preclude generalization about the impact of the intervention upon subjects being exposed to it under non-test or non-experimental settings.
13. *Multiple-intervention interference* may occur whenever multiple interventions are applied to the same subjects, inasmuch as the impacts of prior interventions are usually not erasable.

Threats to Construct Validity

14. *Intervention sensitivity* may preclude generalization of observed impacts to different or related interventions—complex interventions may include other than those components responsible for the observed impacts.
15. *Measures sensitivity* may preclude generalization of observed impacts to different or related impact measures—complex measures may include irrelevant components that may produce apparent impacts.

Threats to Statistical Conclusion Validity

16. *Extraneous sources of error* (including "post hoc" error) may minimize the statistical power of analysis.
17. *Intervention integrity* or lack thereof may invalidate all statistical conclusions.

Threats to Conduct Conclusion Validity

18. *Design complexity* (including technological and methodological constraints) may preclude the complete and successful conduct of the evaluation.
19. *Political infeasibility* (including institutional, environmental and legal constraints) may preclude the complete and successful conduct of the evaluation.
20. *Economic infeasibility* (including hidden and unanticipated costs) may preclude the complete and successful conduct of the evaluation.

Figure 22.1   Threats to validity. (*Source:* Tien, 1979.)

3. *Construct validity* refers to the extent that the causal relationship can be generalized to different interventions, impact measures, and measurements.

4. *Statistical conclusion validity* refers to the extent that an intervention and a measured impact can be statistically associated—error could be either a false association (i.e., Type I error) or a false nonassociation (i.e., Type II error).

5. *Conduct conclusion validity* refers to the extent that an intervention and its associated evaluation can be completely and successfully conducted.

It is clear from Figure 22.1 how program characteristics can engender threats to validity. For example, a program's test setting may cause a "Hawthorne" effect that refers to the tendency of test subjects who are aware of the test to be biased by the test-setting. The effect was identified by Roethlisberger and Dickson (1939), who concluded that their observation of an increased work effort at the Hawthorne Plant of Western Electric in Chicago could not be attributed to changes in light intensity or to the addition of music, since responses to these innovations were being confounded by the subjects' awareness of the study. As another example, a long program evaluation period would tend to exacerbate the various threats to internal validity.

Although the contents of Figure 22.1 are self-explanatory, it is helpful to highlight three aspects of threats to validity. First, the threats to external and construct validities are threats to the *generalizability* of the observed impacts. Generalization involves the science of induction that causes a number of problems that are, according to Campbell and Stanley (1966),

> . . . painful because of a recurrent reluctance to accept Hume's truism that *induction or generalization is never fully justified logically*. Whereas the problems of *internal* validity are solvable within the limits of the logic of probability and statistics, the problems of external validity are not logically solvable in any near, conclusive way. Generalization always turns out to involve extrapolation into a realm not represented in one's sample. Such extrapolation is made by *assuming* one knows the relevant laws, (p. 17)

Although generalization is difficult to undertake, it is a fundamental aspect of social program evaluation. While the classic sciences (i.e., physics, chemistry, biology, etc.) emphasize *repeatability* in their experiments, the social sciences emphasize *representativeness* in their experiments, thus facilitating extrapolations or generalizations.

Second, it can be seen that the threats to validity listed in Figure 22.1 are overlapping in some areas and conflicting in other areas. For example, seasonal effects could either be identified as extraneous events or a result of temporal maturation. Additionally, factors that mitigate threats to conduct conclusion validity would most likely be in conflict with those that mitigate the other threats to validity. It is, however, *essential* that the threats to conduct conclusion validity be borne in mind when developing an evaluation design; the field of evaluation is littered with studies that were not concluded because of the design's complexity or because of the political and economic infeasibilities that were initially overlooked.

Fourth, the threats to validity can be regarded as plausible rival hypotheses or explanations of the observed impacts of a program. That is, the assumed causal relationships (i.e., test hypotheses) may be threatened by these rival explanations. Sometimes the threats may detract from the program's observed impacts. The key objective of an evaluation design is then to minimize the threats to validity, while at the same time to suggest the causal relationships. The specific evaluation design elements are considered next.

22.4 EVALUATION DESIGN ELEMENTS

We have found it systematically convenient to describe a program evaluation design in terms of five components or sets of design elements, including test hypotheses, selection scheme, measures framework, measurement methods, and analytic techniques.

22.4.1 Test Hypotheses

The test hypotheses component is meant to include the range of issues leading up to the establishment of test hypotheses. In practice and as indicated in the dynamic rollback approach, the test hypotheses should be identified only after the program characteristics and threats to validity have been ascertained.

The test hypotheses are related to the rationale or objectives of the program and are defined by statements that hypothesize the causal relationships between dependent and independent measures, and it is a purpose of program evaluation to assess or test the validity of these statements. To be tested, a hypothesis should (1) be expressed in terms of quantifiable measures; (2) reflect a specific relationship that is discernible from all other relations; and (3) be amenable to the application of an available and pertinent analytic technique. Thus, for example, in a regression analysis the test hypothesis takes the form of an equation between a dependent measure and a linear combination of independent measures, while in a before–after analysis with a chi-square test, a simple test hypothesis, usually relating two measures, is used.

In the case of a complex hypothesis, it may be necessary to break it down into a series of simpler hypotheses that could each be adequately tested. In this manner, a measure that is the dependent measure in one test could be the independent measure in another test. In general, input measures tend to be independent measures; process measures tend to be both independent and dependent measures; while impact measures tend to be dependent measures.

Another difficulty arises in the testing process. Analytic techniques exist for testing the *correlation* of measures, but correlation does not necessarily imply *causation*. Inasmuch as causation implies correlation, however, it is possible to use the relatively inexpensive correlational approach to weed out those hypotheses that do not survive the correlational test. Furthermore, in order to establish a causal interpretation of a simple or partial correlation, one must have a plausible *causal* hypothesis (i.e., test hypothesis) and at the same time no plausible *rival* hypotheses (i.e., threats to validity) that could explain the observed correlation. Thus, the fewer the number of plausible rival hypotheses, the greater is the likelihood that the test hypothesis is not *disconfirmed*. Alternatively, if a hypothesis is not disconfirmed or rejected after several independent tests, then a powerful argument can be made for its validity.

Finally, it should be stated that while the test hypotheses themselves cannot mitigate or control for threats to validity, poor definition of the test hypotheses can threaten statistical conclusion validity, since threats to validity represent plausible rival hypotheses.

22.4.2 Selection Scheme

The purpose of this component is to develop a scheme for the selection and identification of test groups and, if applicable, control groups, using appropriate sampling and randomization techniques. The selection process involves several related tasks, including the identification of a general sample of units from a well-designated universe; the assignment of these (perhaps matched) units to at least two groups; the identification of at least one of these groups to be the test group; and the determination of the time(s) that the intervention and, if applicable, the placebo are to be applied to the test and control groups, respectively. A more valid evaluation design can be achieved if random assignment is employed in carrying out each task. Thus, random assignment of units to test and control groups increases the comparability or equivalency of the two groups, at least before the program intervention.

There are a range of selection schemes or research designs, including *experimental* designs (e.g., pretest–posttest equivalent design, Solomon four-group equivalent design, posttest-only equivalent design, factorial designs), *quasi-experimental* designs (e.g., pretest–posttest nonequivalent design, posttest-only nonequivalent design, interrupted time-series nonequivalent design, regression–discontinuity design, ex post facto designs), and *nonexperimental* designs (e.g., case study, survey study, cohort study). In general, it can be stated that nonexperimental designs do not have a control group or time period, while experimental and quasi-experimental designs do have such controls, even if it is just a before–after control. The difference between experimental and quasi-experimental designs is that the former set of designs have comparable or equivalent test and control groups (i.e., through randomization), while the latter set of designs do not.

Although it is always recommended that an experimental design be employed, there are a host of reasons that may prevent or confound the establishment   through random assignment—of equivalent test and control groups. One key reason is that randomization creates a focused inequity because some persons receive the (presumably desirable) program intervention, while others do not. Whatever the reason, the inability to establish equivalent test and control groups should not preclude the conduct of an evaluation. Despite their inherent limitations, some quasi-experimental designs are adequate. In fact, some designs (e.g., regression–discontinuity designs) are explicitly nonrandom in their establishment of test and control groups. On the other hand, other quasi-experimental designs should only be employed if absolutely necessary and if great care is taken in their employment. Ex post facto designs belong in this category. Likewise, nonexperimental designs should only be employed if it is not possible to employ an experimental or quasi-experimental design. The longitudinal or cohort study approach, which is a nonexperimental design, is becoming increasingly popular.

In terms of selection scheme factors that could mitigate or control for the various threats to validity, it can be stated that randomization is the key factor. In particular, most, if not all, of the internal and external threats to validity can be mitigated by the experimental designs which, in turn, can only be achieved through randomization. Thus, *random assignment* of units—especially *matched* units—to test and control groups can control for all the threats to internal validity except, perhaps, extraneous events; *random identification* of a group to be the test group and *random determination* of time(s) that the intervention is to be applied can control for selection-related interaction threats to internal validity; and, *random sampling* can allow for generalization to the universe from which the sample is drawn.

22.4.3 Measures Framework

There are two parts to the measures framework component. First, it is necessary to specify the set of evaluation measures that is to be the focus of the particular evaluation. Second, a model reflecting the linkages among these measures must be constructed.

In terms of evaluation measures, Tien (1979) has identified four sets of measures: input, process, outcome, and systemic measures. The input measures include program rationale (objectives, assumptions, hypotheses), program responsibility (principal participants, participant roles), program funding (funding level, sources, uses), program constraints (technological, political, institutional, environmental, legal, economic, methodological), and program plan (performance specifications, system design, implementation schedule). The process measures include program implementation (design verification, implementation cost), program operation (system performance, system maintenance, system security, system vulnerability, system reliability, operating cost), and concurrent programs (technological, physical, social). The outcome measures include attitudinal, behavioral, and other impact considerations. The systemic measures include organizational (intraorganizational, interorganizational), longitudinal (input, process, outcome), programmatic (derived performance measures, comparability, transferability, generalizability), and policy (implications, alternatives) considerations.

In general, the input and process measures serve to "explain" the resultant outcome measures. Input measures alone are of limited usefulness since they only indicate a program's potential—not actual—performance. On the other hand, the process measures do identify the program's performance, but do not consider the impact of that performance. Finally, the outcome measures are the most meaningful observations, since they reflect the ultimate results of the program. In practice and as might be expected, most of the available evaluations are fairly explicit about the input measures, less explicit about the process measures, and somewhat fragmentary about the outcome measures. Indeed, Nas (1996) argues that a key merit of cost-benefit analysis is that it focusses on outcome measures.

The fourth set of evaluation measures—the systemic measures—can also be regarded as impact measures, but have been overlooked to a large extent in the evaluation literature. The systemic measures allow the program's impact to be viewed from at least four systemic perspectives. First, it is important to view the program in terms of the *organizational* context within which it functions. Thus, the program's impact on the immediate organization and on other organizations must be assessed. Second, the pertinent input, process, and outcome measures must be viewed over time, from a *longitudinal* perspective. That is, the impact of the program on a particular system must be assessed not only in comparison to an immediate "before" period but also in the context of a longer time horizon. Thus, it is important to look at a process measure like, for example, average response time over a five-to-ten-year period to ascertain a trend line, since a perceived impact of the program on the response time may just be a regression artifact. Third, in an overall *programmatic* context, the evaluator should (1) derive second-order systems performance measures (e.g., benefit cost and productivity measures) based on the first-order input, process, and outcome measures; (2) compare the program results with findings of other similar programs; (3) assess the potential of transferring the program to other locales or jurisdictions; and (4) determine the extent to which the program results can be generalized. In terms of generalization, it is important not only to recommend that the program be promulgated, but also to define the limits of such a recommendation. Fourth, the first three systemic contexts can be regarded as *program-oriented* in focus as compared to the fourth context, which assesses the program results from a broader *policy-oriented* perspective. In addition to assessing the policy implications, it is important to address other feasible and beneficial alternatives to the program. The alternatives could range from slight improvements to the existing program to recommendations for new and different programs.

The second part of the measures framework concerns the linkages among the various evaluation measures. A model of these linkages should contain the hypothesized relationships, including cause-and-effect relationships, among the measures. The model should help in identifying plausible test and rival hypotheses, as well as in identifying critical points of measurement and analysis. In practice, the model could simply reflect a systematic thought process undertaken by the evaluator, or it could be explicitly expressed in terms of a table, a block diagram, a flow diagram, or a matrix.

In conclusion, concise and measurable measures can mitigate the measures-related threats to validity. Additionally, the linkage model can help to avert some of the other threats to validity.

22.4.4 Measurement Methods

The list of issues and elements that constitute the measurement methods component include measurement time frame (i.e., evaluation period, measurement points, and measurement durations), measurement scales (i.e., nominal, ordinal, interval, and ratio), measurement instruments (i.e., questionnaires, data-collection forms, data collection algorithms, and electromechanical devices), measurement procedures (i.e., administered questionnaires, implemented data-collection instruments, telephone interviews, face-to-face interviews, and observations), measurement samples (i.e., target population, sample sizes, sampling technique, and sample representativeness), measurement quality (i.e., reliability, validity, accuracy, and precision), and measurement steps (i.e., data collection, data privacy, data codification, and data verification).

Clearly, each of the preceding indicated measurement element has been the subject matter of one or more theses, journal articles, and books. For example, data sampling— a technique for increasing the efficiency of data gathering by the identification of a smaller sample that is *representative* of the larger target database—remains a continuing hot research area in statistics. The dilemma in sampling is that the larger the sample, the greater the likelihood of representativeness, but likewise the greater the cost of data collection.

Measurement methods that could mitigate or control for threats to validity include a multimeasurement focus, a long evaluation period (which, while controlling for regression artifacts, might aggravate the other threats to internal validity), large sample sizes, random sampling, pretest measurement, and, of course, techniques that enhance the reliability, validity, accuracy, and precision of the measurements. Further, judicious measurement methods can control for the test-setting sensitivity threat to external validity, while practical measurement methods that take the political and economic constraints into account can control for the conduct conclusion threats to validity.

22.4.5 Analytic Techniques

Analytic techniques are employed in evaluation or analysis for a number of reasons: to conduct statistical tests of significance; to combine, relate, or derive measures; to assist in the evaluation conduct (e.g., sample size analysis, Bayesian decision models); to provide data adjustments for nonequivalent test and control groups; and to model test and/or control situations.

Next to randomization (which is usually not implementable), perhaps the single most important evaluation design element (i.e., the one that can best mitigate or control for the various threats to validity) is, as previously alluded to, modeling. Unfortunately, most evaluation efforts to date have made minimal use of this simple but powerful tool. Larson (1975), for example, developed some simple structural models to show that the *integrity* of the Kansas City Preventive Patrol Experiment was not upheld during the course of the experiment, thus casting doubt on the validity of the resultant findings. As another example, Tien (1988) employed a linear statistical model to characterize a retrospective "split area" research design or selection scheme, that was then used to evaluate the program's impact. Finally, to demonstrate the usefulness of evaluation modeling, a stochastic model is developed in the next section to assist in the identification of valid sample sizes.

22.5   EVALUATION MODELING

The model that is developed in this section was instrumental in modifying the design requirements of a large, multicity federal demonstration program (Tien and Cahn, 1986). Without being specific about the particular federal program, the problem can be simply stated in the context of business establishments that were experiencing a high level of burglaries. The establishments were being encouraged to buy and install an electronic system (consisting of cameras, silent alarms, etc.) that ostensibly could help prevent burglaries. It can be assumed that an establishment would elect to be "treated" (i.e., buys and installs the system) with probability t (where $0 < t < 1$). The question is how many business establishments should be targeted within an area so that a one-year evaluation effort would yield valid impact results. The federal demonstration program initially called for areas with 40 to 60 establishments each: Was that a reasonable sample size?

In attempting to resolve this issue, first we note that if the area contains N establishments and each has a t probability of being treated, then on average Nt establishments would be treated. Next, we derive an expression for the expected minimum percent decrease (MPD) in total burglaries for all the treated establishments at a level of significance of 0.05. Let,

$b =$ number of successful pretreatment or "before" burglary attempts on each business establishment per year, with an expected value of $E(b)$ and a standard deviation of $S(b)$.

$A =$ total number of successful posttreatment or "after" burglary attempts on all the treated establishments for a one-year period following treatment.

$H_0 =$ null hypothesis where there is no statistically discernible change in the total pretreatment and posttreatment burglary levels for all the treated establishments.

$H_1 =$ alternative hypothesis where, if the treatment is successful, the total posttreatment burglary level for all the treated establishments is less than the total pretreatment figure.

Under H_0 and assuming that burglary attempts are independent, we have

$$A = b_1 + b_2 + \cdots + b_{N(t)} \tag{22.1}$$

where $N(t)$ represents the number of establishments (out of N) selecting the treatment; it is a function of t and is binominally distributed as

$$P_{N(t)}(n) = C_n^N t^n (1 - t)^{N - n}, \qquad n = 0,1,2, \ldots ,N \tag{22.2}$$

with

$$E(N(t)) = Nt \qquad \text{and} \qquad S(N(t)) = Nt(1 - t) \tag{22.3}$$

Note that A is actually a random sum of random numbers, which, through transform techniques and differentiation, can be shown to have an expected value of

$$E(A) = NtE(b) \tag{22.4}$$

and a variance of

$$S^2(A) = Nt\, S^2(b) + (E(b))^2\, Nt(1 - t) \tag{22.5}$$

Since H_1 is located to the left of H_0, the rejection region under H_0 should be to the left; that is, there is an A_{max} value such that

$$P(A \le A_{max}) = 0.05 \tag{22.6}$$

or, assuming $N > 30$ (so that A can be approximated by a normal distribution),

$$(A_{max} - E(A))/S(A) = -1.64 \tag{22.7}$$

The expected MPD can be stated as

$$\text{MPD} = (E(A) - A_{max})/E(A))100\% \tag{22.8}$$

Combining Equations (22.3), (22.4), (22.5), (22.7), and (22.8) and collecting terms results in:

$$\text{MPD} = 164\, [((\text{CoV})^2 + (1 - t))/Nt]^{1/2}\% \tag{22.9}$$

where

$$\text{CoV} = \text{coefficient of variation} = S(b)/E(b) \tag{22.10}$$

It should be noted that Equation (22.9) is a plausible expression, since MPD increases with $S(b)$ and decreases as $E(b)$, N, or t is increased. Figure 22.2 depicts these tradeoffs, assuming $t = 0.5$ (i.e., half the establishments are treated).

As indicated in Figure 22.2, even with $N = 60$, we would have to see an almost 40 percent decrease in burglaries at a $\text{CoV} = 1$ and over 120 percent decrease (which is, of course, an impossibility) at $\text{CoV} = 4$. After reviewing similar results to those shown in Figure 22.2, the federal program did indeed double the sample size requirements (i.e., from 40 to 60 *to* 80 to 120 establishments) before the demonstration program began. A postanalysis justified this change, since t was indeed found to be about 0.5 and the CoV of an individual establishment's burglary rate was determined to be between 3 and 4.

In sum, the simple model illustrated in Figure 22.2 provides critical insight into identifying appropriate sample sizes; it, for example, did indeed ensure the integrity of a large, multicity federal demonstration program, fortunately before the fact in that instance, as opposed to, unfortunately, after the fact in the case of the Kansas City Preventive Patrol

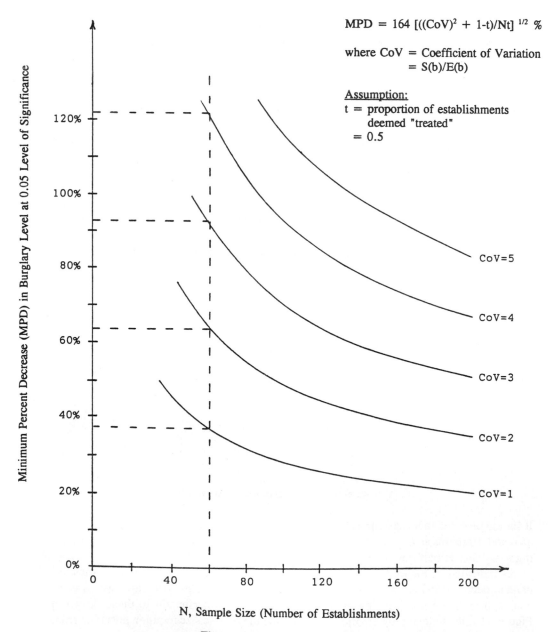

Figure 22.2 Impact of sample size.

Experiment (Larson, 1975). Again, this example effort demonstrates the power of evaluation modeling in mitigating threats to validity.

22.6 CONCLUDING REMARKS

In conclusion, three remarks should be made. First, it should be stated that the evaluation approach presented herein is very much dependent on identifying and understanding the potential threats to validity. It is through these threats that program characteristics are linked to evaluation design elements, which seek to mitigate, if not eliminate, the threats.

Second, while this chapter provides a purposeful and systematic approach or guide to the development of a sound evaluation, it does not constitute a "cookbook" or handbook. An adequate handbook on the evaluation of systems will probably not be forthcoming in the near future; it will require many more years of evaluation experience and a careful analysis of that experience.

Third, as noted in Section 22.2, the need for evaluation is growing, and it will continue to grow in the foreseeable future. Government at every level is increasingly being required to justify the value of its programs. Furthermore, increased federal deregulation, increased domestic and foreign competition, and ever more frequent "future shocks" have combined to bring similar pressures on private industry. Given a growing need for evaluation, it is critical that proper procedures exist for the development of valid evaluations. Certainly, the systems engineering and evaluation modeling approach presented herein attempts to provide such procedures.

REFERENCES

Campbell, D. T., and Stanley, J. C. (1966). *Experimental and Quasi-experimental Designs for Research.* Chicago: Rand McNally.

Chelimsky, E., and Shadish, W. R., Jr. (eds.). (1997). *Evaluation for the 21st Century.* Thousand Oaks, CA: Sage Publ.

Colton, K. W., Brandeau, M. L., and Tien, J. M. (1982). *A National Assessment of Command, Control, and Communications Systems.* Washington, DC: National Institute of Justice.

House, E. R. (1993). *Professional Evaluation: Social Impact and Political Consequences.* Thousand Oaks, CA: Sage Publ.

Larson, R. C. (1975). What happened to patrol operations in Kansas City? A review of the Kansas City preventive patrol experiment. *J. Criminal Justice* **3**, 267–297.

Nas, T. F. (1996). *Cost-benefit Analysis: Theory and Application.* Thousand Oaks, CA: Sage Publ.

Porter, A. L., Rossini, F. A., Carpenter, S. R., and Roper, A. T. (1980). *A Guidebook for Technology Assessment and Impact Analysis.* New York: North-Holland.

Roethlisberger, K., and Dickinson, W. (1939). *Management and the Worker.* Cambridge, MA: Harvard University Press.

Sage, A. P. (1981). A methodological framework for systemic design and evaluation of computer aids for planning and decision support. *Comput. Elect. Eng.* **8**, 87–101.

Tien, J. M. (1979). Toward a systematic approach to program evaluation design. *IEEE Trans. Sys., Man, Cybernet.* **9**(9), 494–515.

Tien, J. M. (1988). Evaluation design: Systems and models approach. In *Systems and Control Encyclopedia* (M. G. Singh, ed.), pp. 1559–1566. New York: Pergamon.

Tien, J. M. (1990). Program evaluation: A systems and model-based approach. In *Concise Encyclopedia of Information Processing in Systems and Organizations* (A. P. Sage, ed.), pp. 382–388. New York: Pergamon.

Tien, J. M., and Cahn, M. F. (1986). Commercial security field test program: A systematic evaluation of the impact of security surveys. In *Preventing Crime in Residential and Commercial Areas* (D. P. Rosenbaum, ed.), pp. 192–206. Beverly Hills, CA: Sage Publ.

Tien, J. M., and Rich, T. F. (1997). *Early Experiences with Criminal History Records Improvement*. Washington, DC: Bureau of Justice Assistance.

23 Systems Reengineering

ANDREW P. SAGE

23.1 INTRODUCTION

Responsiveness is very clearly a critical need today. By this, we mean, of course, organizational responsiveness in providing products and services of demonstrable value to customers, and thereby in the provision of value-added capabilities to organizational stakeholders. This must be accomplished by efficiently and effectively employing leadership and empowered people such that systems engineering and management strategies—including organizational processes, human resources, and appropriate technologies—are brought to bear on the production of high-quality, trustworthy, and sustainable products and services. Virtually all of the discussions in this handbook have indicated the need for continual revitalization in the way in which we do things, such that they are always done better. This is the case, even if the external environment were static and unchanging. However, when we are in a period of high-velocity environments, then continual organizational change and associated change in processes and product must be considered as a fundamental rule of the game for progress.

Change has become a very popular word in management and in technology. It has been said that today we expect change everywhere except perhaps in an automatic vending machine. There are a variety of change models and change theories. Some seek to change in order to survive, others seek to change in order to retain competitive advantage. In this chapter, we examine intentional engineered change, or reengineering. There are a variety of names given to the number of change-related terms now in use: reengineering, restructuring, downsizing, rightsizing, redesign, and many others. Reengineering is probably the most often used word, and it is the one chosen for this chapter. Figure 23.1 indicates some of the approaches that have been taken to reengineering. We briefly examine many of these in this chapter.

Figure 23.2 represents a generic view of reengineering. We can approach a discussion of reengineering from several perspectives. First, we can discuss the structural, functional, and purposeful aspects of reengineering. Alternately, or in addition, we can examine reengineering at the level of systems management, process, or product. We can examine reengineering issues at any, or all, of the three fundamental systems engineering life cycles—research, development, test, and evaluation (RDT&E); systems acquisition, procurement, or production; or systems planning and marketing—that are discussed in our introductory chapter and elsewhere, such as Sage (1995), a reference on which this chapter is based. Within each of these life cycles, we could consider reengineering at any or all of the three generic phases of definition, development, or deployment. At the level of systems management, we examine the enterprise as a whole and consider all organizational processes within the company for improvement through change. At the level of process

Handbook of Systems Engineering and Management, Edited by A. P. Sage and W. B. Rouse
ISBN 0471-15405-9 ©1999 John Wiley and Sons, Inc.

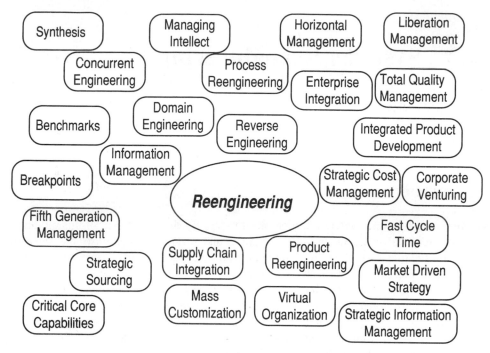

Figure 23.1   Some approaches to reengineering.

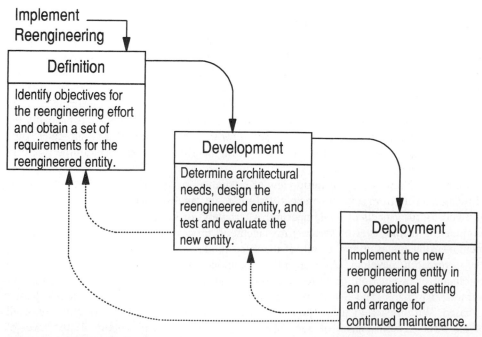

Figure 23.2   Generic implementation of reengineering at the level of product, process, or systems management.

reengineering, as we define it here, only a single process is redesigned, and with no fundamental or radical changes in the structure or purpose of the organization as a whole. Changes, when they occur, may be radical and revolutionary, or incremental and evolutionary, at the level of systems management, process, product, or any combination of these. The scale of improvement efforts may vary from incremental and continuous improvement, such as generally advocated by quality management efforts, to radical change efforts that affect systems management itself.

One fundamental notion of reengineering is, however, the reality that it must be top-down directed if it is to achieve the significant and long-lasting effects that are possible. Thus, there should be a strong purposeful and systems management orientation to reengineering, even though it may have major implications for such lower-level concerns as structural facets of a particular product.

Notions of change are not at all new. A half century ago, Kurt Lewin (1947) was among the first to consider organizational change, and his studies of group dynamics and culture change in organizations are among the early seminal writings in this area. Many of these concepts have played a major role in the development of change and learning organizations.

1. A best way of understanding a system is to try to change it
2. Successful change involves participative management.
3. Participative management suggests that the gatekeepers who control a situation must be involved in planning for change or they will most likely resist change.
4. Successful change involves the three activities of unfreezing, movement, and refreezing.

Some aspects of learning organizations are considered in Chapters 20 and 30 of this handbook.

This chapter is organized as follows. We first provide some definitions of reengineering. Then, we indicate some of the many perspectives that have been taken relative to reengineering. Finally, we indicate a number of approaches, such as illustrated in Figure 23.1, that may be utilized to accomplish reengineering.

23.2 DEFINITION OF AND PERSPECTIVES ON REENGINEERING

In this section, we provide definitions and perspectives on what we consider to be three related but different types of systems reengineering efforts: reengineering at the levels of

- Product
- Process or product line
- Systems management

There have been a number of definitions, formal and informal, of reengineering. The word is occasionally spelled as re-engineering. We choose the former spelling here; both are correct.

23.2.1 Product Reengineering

The term reengineering could be used to mean some sort of reworking or retrofit of an already engineered product. This could well be interpreted as maintenance or refurbishment. Or, reengineering could be interpreted as reverse engineering, in which the characteristics of an already engineered product are identified, such that the original product can be subsequently modified and reused or so that a new product with the same purpose and functionality can be obtained through a forward engineering process. Inherent in these notions are two major facets of reengineering.

1. Reengineering improves the product or system delivered to the user for enhanced reliability, maintainability, or for an evolving user need.
2. Reengineering increases understanding of the system or product itself.

We see that this interpretation of reengineering is almost totally product focused. We will call it product reengineering.

The following definition is offered:

> Product reengineering is the examination, study, capture, and modification of the internal mechanisms or functionality of an existing product in order to reconstitute it in a new form and with new functional and nonfunctional features, often to take advantage of newly emerged technologies, but without major change to the inherent purpose of the product.

This definition indicates that product reengineering is basically structural reengineering with, at most, minor changes in purpose and functionality of the product that is reengineered. This reengineered product could be integrated with other products having rather different functionality than was the case in the initial deployment. Thus, reengineered products could be used, together with this augmentation, to provide new functionality and serve new purposes. A number of synonyms for product reengineering easily come to mind. Among these are: renewal, refurbishing, rework, repair, maintenance, modernization, reuse, redevelopment, and retrofit.

A specific example of a product reengineering effort might be that of taking a legacy system written in COBOL or Fortran, reverse engineering it to determine the system definition, and then reengineering it in C^{++} or Ada. Depending upon whether or not any modified user requirements are to be incorporated into the reengineered product, we would either "forward engineer" the product just after reverse engineering had determined either the initial development (technical) system specifications, or after reverse engineering far enough to determine user requirements and user specifications, and we update these. This reverse engineering concept (Rekoff, 1985), in which salient aspect of user requirements or technological specifications are recovered from examination of characteristics of the product, predates the term product reengineering and occurs before the forward engineering that comprised the latter portions of product reengineering.

Figure 23.3 illustrates product reengineering conceptually. An IEEE Software standards [Institute of Electrical and Electronics Engineers (IEEE) (1991)] states that "reengineering is a complete process that encompasses an analysis of existing applications, restructuring, reverse, and forward engineering." The IEEE Standard for Software Maintenance (1992) suggests that reengineering is a subset of software engineering that is composed of reverse

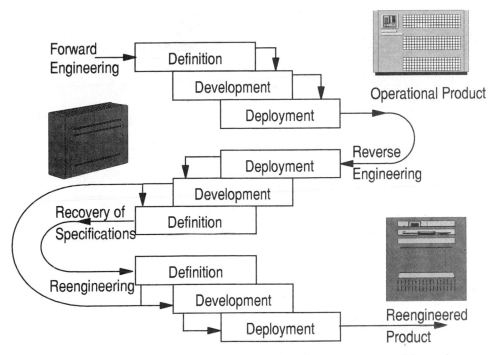

Figure 23.3 Basic notions concerning product reengineering as a sequence of forward, reverse, forward engineering.

engineering and forward engineering. We have no disagreement with the sort of definition at all; but, we prefer to call it product reengineering for the reasons just stated. There are two other very important forms of reengineering, and it is necessary to consider reengineering at the levels of processes and systems management if we are to take full advantage of the major opportunities offered by generic reengineering concepts. Thus, the qualifier "product" appears appropriate and desirable in the context used here.

23.2.2 Process Reengineering

Reengineering can also be considered at the levels of processes and systems management. At the level of processes only, the effort would be almost totally internal. It would consist of modifications to whatever standard life-cycle processes are in use in a given organization in order to better accommodate new and emerging technologies or new customer requirements for a system. For example, an explicit risk management capability might be incorporated at several different phases of a given life cycle and accommodated by a revised configuration management process. This could be implemented into the processes for RDT&E, acquisition, and systems planning and marketing. Basically, reengineering at the level of processes would consist of the determination, or synthesis, of an efficacious process for ultimately fielding a product on the basis of a knowledge of generic customer requirements, and the objectives and critical capabilities of the systems engineering organization. Figure 23.4 illustrates, conceptually, some of the facets of process reengineer-

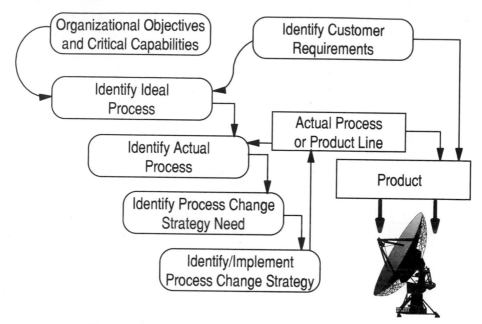

Figure 23.4   Conceptual illustration of process reengineering.

ing. Process reengineering can be accomplished because of the desire to obtain better products, or it can be accomplished as support to efforts to obtain a "better" organization. There are three ways in which we can attempt process improvement:

- New process development
- Process redevelopment, or process reengineering
- Continuous process improvement over time

New process design would be called for because of some strategic-level change, such as might occur when a previously outsourced development efforts was insourced and there is no present process from which to base the new process. Benchmarking, which we discuss here, is one way of accomplishing this new process engineering. Process engineering, or reengineering, should occur because the present process is currently dysfunctional or because the organizations wishes to keep abreast of changing technology or changing customer requirements, or other change. Continuous process improvements are improvements that are less radical in nature and that can be made incrementally over time. Each of these involves leadership, strategy, and a team to accomplish the effort. It seems reasonable to call each of these "reengineering," and this is done here.

In accordance with this discussion and by analogy to our definition of product reengineering, we offer the following definition:

Process reengineering is the examination, study, capture, and modification of the internal mechanisms or functionality of an existing process, or systems engineering lifecycle, in order to reconstitute it in a new form and with new functional and nonfunctional features, often to take advantage of newly emerged or desired organizational and/or technological capabilities, but without changing the inherent purpose of the process itself that is being reengineered.

As a systems engineering effort, we could reengineer either the process for RDT&E, system acquisition or production, or systems planning and marketing. Among the first discussions of this sort of effort at business process reengineering, although the word redesign was used rather than reengineering, is in a contemporary paper by Davenport and Short (1990). This discussion was greatly expanded upon in a recent and seminal text by Davenport (1993), which does make use of the term reengineering. We provide an overview of this major work in a later section.

It should be recognized that redevelopment of processes only, and without attention to reengineering at a higher level than processes, may in many instances represent an incomplete and not fully satisfactory way to improve organizational capabilities if they are otherwise deficient. Thus, all the processes considered as candidates for reengineering should be high level managerial as well as operational. Information technology is considered to be a major enabling catalyst for process reengineering. Whether this should be in the form of new process engineering, existing process reengineering, or continuous improvement of existing processes, depends upon individuating circumstances. There is a plethora of literature relating to process reengineering and redesign (Huber and Glick, 1993; Bowman and Kogut, 1995; Harrington and Harrington, 1995; Keen, 1997).

23.2.3 Reengineering at the Level of Systems Management

At the level of systems management, reengineering is directed at potential change in all business or organizational processes, and thereby the various organizational life-cycle processes as well. Many authors have discussed reengineering the corporation. Arguably, the earliest use of the term *business reengineering* was by Hammer (1990), and it is more fully documented in a more recent work on reengineering the corporation (Hammer and Champy, 1993). There is a small number of related works, as we will soon discuss.

Hammer's definition of reengineering, "Reengineering is the fundamental rethinking and radical redesign of business processes to achieve dramatic improvements in critical, contemporary measures of performance, such as cost, quality, service and speed," is a definition of what we will call reengineering at the level of systems management. There are four major terms in this definition:

- *Fundamental* refers to a large-scale and broad-scope examination of virtually everything about an organization and how it operates. The purpose is to identify potential weakness that are in need of diagnosis and correction.
- *Radical redesign* suggests disregarding existing organizational processes and structures, and inventing totally new ways of accomplishing work.
- *Dramatic improvements,* in Hammer's view, suggests that reengineering is not about making marginal and incremental improvements in the status quo; it is about making "quantum" leaps in organizational performance.
- *Processes* represent the collection of activities that are used to take input materials, including intellectual inputs, and transform them into outputs and services that have value to the customer.

Hammer suggests that reengineering and revolution are almost synonymous terms. He identifies three types of firms that attempt reengineering: those in trouble, those who see trouble coming, and those who are ambitious and seek to avoid impending troubles. Clearly, it is better to be proactive and be in this latter category, rather than to be reactive and seek to emerge from a realized crisis situation.

Hammer indicates that one major catalyst for reengineering is the creative use of information technology. Reengineering is not just automation, however, it is the ambitious and rule-breaking study of everything about the organization to enable more effective and efficient organizational processes to be designed. We essentially share this view of reengineering at the level of systems management. Our definition is similar:

> Systems management reengineering is the examination, study, capture, and modification of the internal mechanisms or functionality of existing system management processes and practices in an organization in order to reconstitute them in a new form and with new features, often to take advantage of newly emerged organizational competitiveness requirements, but without changing the inherent purpose of the organization itself.

We make no representation that this definition, or the other two for that matter, of reengineering is at all the same across the many works that we discuss. Figure 23.5 represents this conception of reengineering at the level of systems management. Life-cycle process reengineering occurs as a natural by-product of reengineering at the level of systems management. This may or may not result in the reengineering of already existing

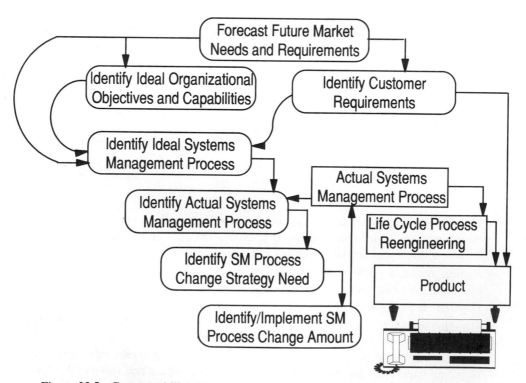

Figure 23.5 Conceptual illustration of reengineering at the level of systems management.

products. Generally it will; new products and new competitive strategies are each a major underlying objective of reengineering at the level of systems management, or organizational reengineering, as it is more commonly called.

23.2.4 Perspectives on Reengineering

This very brief discussion of reengineering suggests that we can consider reengineering at three levels: systems management, life-cycle processes, and product. The major purpose of reengineering, regardless of whether it is at the product level or the process level or the level of systems management, is to enable us to produce a better product for the same cost, or a lower-cost product that performs in a manner comparable to initial product performance. Thus, reengineering improves competitiveness of the organization in coping with changing external situations and environments. We may approach reengineering, at any or all of these levels, from either of three perspectives:

- *Reactive,* because we realize that we are in trouble and perhaps in a crisis situation, and reengineering is one way to bring about needed change.
- *Interactive,* because we wish to stay abreast of current changes as they evolve.
- *Proactive,* because we wish to position our organization now for changes that we believe will occur in the future, and to emerge in the changed situation as a market leader.

We could approach reengineering from an inactive perspective, although this suggests not considering it at all, and this is likely to lead to failure to adapt to changed conditions and requirements. In the next section, we examine some of the many contemporary ideas that have been expressed about the subject of reengineering.

23.3 AN OVERVIEW OF REENGINEERING APPROACHES

As we have noted, it is possible to consider reengineering at the levels of systems management, process or product line, or product. In this section, we provide an expanded overview of each of these forms of reengineering by means of an overview and interpretation of contemporary literature on reengineering.

Without question, more has been written about reengineering at the levels of strategy or systems management than at the other levels. This is not unreasonable since reengineering efforts at the level of organizational strategy have direct implications for management at the levels of management control, and thence at the levels of process to implement management controls, and product.

23.3.1 Business Process Improvement

Our first overview is of a seminal work on business process improvement by Harrington (1991). The major thesis of the work is that it is business and manufacturing processes that are the key to error-free performance. His view—that the process is the problem and not the employees—is essentially that of Deming and others in the total quality management (TQM) areas. Harrington defines a process as a group of activities that takes inputs, adds value to them, and produces an output that is in support of an organization's objectives. Two generic types of processes are identified: production processes are directly concerned with yielding the output product or service; business processes support production processes. These latter include design of production processes, payroll processes, and engineering change processes.

Many works on reengineering recognize a dichotomy between organizational processes and organizational functionality. Most organizations are structured into vertically functioning groups, or hierarchies, and most processes are organized into horizontal phases for workflow. Three desirable attributes of business process improvement (BPI) cited by Harrington are:

- Efficiency, in terms of minimizing the cost of the resources used.
- Effectiveness, in terms of producing desired results.
- Adaptability, in terms of flexibility in accommodating changing customer and organizational needs.

To do this requires processes with the following well-defined characteristics: ownership and accountability, boundaries and scope, interfaces and responsibilities, work tasks and training requirements, measurement and feedback controls, customer-related measurements and targets, cycle times, and formalized change procedures.

Harrington suggests five phases for (BPI), as illustrated in Figure 23.6. These are as follows:

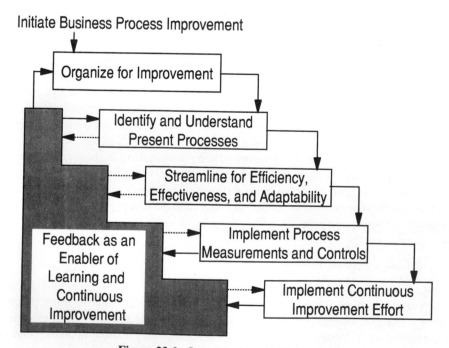

Figure 23.6　Interpretation of BPI process.

1. *Organize for Improvement.* This phase involves several steps. First, it is necessary to establish an executive improvement team (EIT). Then a BPI champion is appointed and executive training is provided. A normative improvement model is identified and BPI objectives are communicated to employees. Next, it is necessary to review business strategy and anticipated customer requirements. This enables the organization to identify and select critical processes for improvement and to appoint process owners. Finally, performance improvement team (PIT) members are selected.

2. *Develop an Understanding of the Various Processes Currently in Use.* This phase includes defining the scope and objectives of current processes and the boundaries within which they are functional. This includes a variety of analysis functions, including structuring current processes, and detection and diagnosis of areas for potential improvement.

3. *Streamline Organizational Processes for Enhanced Efficiency, Effectiveness, and Adaptability.* Current processes are corrected through various streamlining approaches that are responsive to the diagnosis performed in the last phase. This includes eliminating bureaucracy, reducing the opportunity for errors, reducing non-value-added activities, and otherwise simplifying processes such as to reduce cost and increase effectiveness.

4. *Implement a Program of Systematic Measurement and Controls.* In this phase, a program of systematic measurements is used as a quality control and monitoring system in order to both maintain the new process productivity and to keep from regressing into poorer process implementations only because they have been used in the past. An extensive program of measurements, feedback, and action is suggested. This includes audits of what are denoted as poor quality cost (PQC) facets.

5. *Continue the Evolutionary Improvement.* Periodic reviews of the effort are used to enable detection, diagnosis, and correction of difficulties as improvement continues. A formal program of business process qualification is suggested through use of a six-level PBI scale for process maturity status. The levels are unknown, understood, effective, efficient, error-free, and world class.

The experience of the author in efforts of this sort is much in evidence in this high-caliber writing. These efforts have much in common with those associated with strategic quality management. The work also has much relevance to benchmarking, a topic we will soon discuss, and to the systematic measurement subjects considered in Chapter 15.

23.3.2   Intelligent Enterprise

The efforts of James Brian Quinn are focused on knowledge-based services as a necessary compliment to manufacturing efforts. His view of an organization is basically that of a collection of service activities, and that service- and product-oriented organizations alike will each obtain their major competitive advantages not from superior physical facilities and materials alone, but from knowledge and service based capabilities. In three works (Quinn et al., 1990a,b,c) concerning technologies in services, the claim is set forth that it is services and not manufacturing activities that provide the major sources of value to consumers. This does not suggest that services replace manufacturing, but rather that if it were not for the value added by the services that are associated with a manufactured product, there would be far diminished value to the product itself. Several characteristics of new organizations that focus on services are cited in these works:

- "Infinitely flat," or horizontal.
- "Spider's-web"-like, or nonhierarchical and with highly networked interconnections.
- "Hollow corporations," in which outsourcing of both products and services becomes an increasing reality.
- Demolished bureaucracies and vertical integration.
- "Intellectual holding companies," in which intellectual technologies are critical.

It is represented that this will lead to precise and swift strategy execution, the leveraging and retention of key people, and "creative management for profits."

A recent text by Quinn (1992) represents a definitive integration and synthesis of earlier efforts on this subject. It is much concerned with concentrating organizational strategy on core intellectual competencies and core service competencies. He suggests four key rules to follow in order to generate success in this regard.

1. Focus internal organizational resources on those relatively few basic sources of intellectual strength and service strength that will create and sustain a real and meaningful distinctiveness to the customer over the long term.
2. Approach the remaining capabilities as a noncritical set of service activities that may be supplied internally or outsourced from external suppliers who compete well in functional activities related to these capabilities.
3. Sustain success by building entry barriers around those selected critical core capabilities to prevent a competitor from assuming a substantial market position.
4. Plan and control outsourcing so as never to become either dependent upon, or dominated by, external suppliers.

Strategic sourcing, including outsourcing, is a major ingredient in these maxims, and we discuss it later. Appropriate interfaces between production efforts and service efforts are also stressed, as is the management of knowledge-based intellect and professional intellect.

Seven types of innovative organizations are identified in this work:

1. Basic research organizations support large RDT&E units. They select products for development on the basis of careful and conservative tradeoffs among risk and potential profit.
2. Large-system producers develop large-scale systems that generally cost a great deal and that must perform in a reliable manner for a very long time.
3. Dominant market-share-oriented companies are often not the first to introduce an emerging technology into the marketplace. They often support large research units and plan market entry and product evolution to obtain maximum penetration, decreased risk, great product reliability, and lower overall costs.
4. State-of-the-art-technology development companies are preeminent in their knowledge of a specific technology. Often they operate in markets in which technical performance criteria are the major drivers of demand, and are therefore able to self-define the characteristics of next-generation technology.

5. Discrete freestanding product-line companies form strong research divisions and act as entrepreneurial units. They depend more on technology push for their initiatives than demand pull. They will often introduce new products on a small scale and obtain real-time market tests of these, rather than pay for very expensive marketing studies that may not be as effective as small-scale product introduction and subsequent interactive modification of the product to meet consumer needs.

6. Limited-volume or -fashion companies provide small-scale and limited-quantity products to a specialized market niche.

7. Job-shop or custom-design companies provide one-of-a-kind products that meet an individual customer's requirements. These may often involve flexible manufacturing or mass customization approaches to enable highly specialized design to met individual customer requirements.

Clearly, this is not a mutually exclusive listing. Nor is the listing collectively exhaustive. Each of the industry types may be associated with an organizational strategy and configuration that is tuned to providing maximum success opportunities. Each type will have a different propensity for organizational growth, maturity, decline, and rebirth. In his effort, Quinn discusses the typical product life cycle for organizations with each of these seven organizational characteristics. This includes the mix of attributes that describe typical innovations, and recommended organizational structures for RDT&E, acquisition or production, planning and marketing, and interfaces between marketing and customers. Approaches for managing the intelligent enterprise are also suggested. These involve efforts that also encompass core capabilities, strategic sourcing, TQM, and benchmarking, as well as some other reengineering activities identified in Figure 23.1.

23.3.3 Process Innovation

In a text by Davenport (1993), a careful distinction is made between process improvement and process innovation. His fundamental distinctions are that improvement is continuous and incremental in nature, deals with the existing process, can be accomplished in a relatively short time, is a bottom-up activity, and is a narrow-scope effort with relatively moderate attendant risks. On the other hand, innovation is generally a discrete phenomenon that is revolutionary and radical in nature. It starts with a zero base as contrasted with the existing process, can only be accomplished over a long time, is a top-down activity, is a broad-scope effort that cuts across all of the functional areas and processes in the organization, and is usually characterized by high associated risks.

Of course, there may be questions of whether an innovation is a major improvement and whether a set of incremental improvements does not add up to an innovation. Rather than a binary scale to separate these two, perhaps it would be best to consider a continuous scale. This would enable us to consider incremental improvement at one end and radical innovation at the other. Obviously, either can be appropriate or inappropriate in specific situations. Change increments may be so small that centuries would be required to accomplish any change with appreciable value added. On the other hand, radical innovation may be so dramatic, and otherwise so inconsiderate of humans in the organization, that the organization is culturally and otherwise unable to adapt. Culture shock within the organization, and customer revolt as they sense increasing dysfunctionality, is often the result.

Information technology is suggested as a major enabler of process innovation, together with the organizational and human enabler, and an enabler based on measurements associated with process information and management of the information environment. The process innovation process itself comprises five phases, and a number of activity steps within each phase, as represented by Figure 23.7. We note that there are opportunities both for traditional incremental improvement and the more extreme innovative improvement. The overall process is iterative in nature and this leads to continual innovation, as suggested by the feedback from phase V to phase I in the interpretation of Figure 23.7. Not explicitly shown in this figure is the organizational communications that occur at each phase of the effort and the commitment building that is also needed. Davenport examines a number of enablers of specific processes including: planning, research, development, design, production, marketing, and sales.

The role of organizational culture in shaping desirable innovation strategies is considered briefly here. In a related work, Davenport et al., (1992), this is considered in greater detail. Five models of information culture, denoted as information politics, are defined.

1. *Technocratic utopianism* is a formal analytical approach to information management. It stresses highly quantitative approaches, major reliance on emerging technologies,

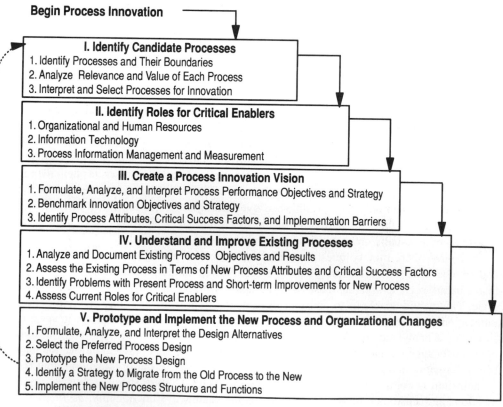

Begin Process Innovation

I. Identify Candidate Processes
1. Identify Processes and Their Boundaries
2. Analyze Relevance and Value of Each Process
3. Interpret and Select Processes for Innovation

II. Identify Roles for Critical Enablers
1. Organizational and Human Resources
2. Information Technology
3. Process Information Management and Measurement

III. Create a Process Innovation Vision
1. Formulate, Analyze, and Interpret Process Performance Objectives and Strategy
2. Benchmark Innovation Objectives and Strategy
3. Identify Process Attributes, Critical Success Factors, and Implementation Barriers

IV. Understand and Improve Existing Processes
1. Analyze and Document Existing Process Objectives and Results
2. Assess the Existing Process in Terms of New Process Attributes and Critical Success Factors
3. Identify Problems with Present Process and Short-term Improvements for New Process
4. Assess Current Roles for Critical Enablers

V. Prototype and Implement the New Process and Organizational Changes
1. Formulate, Analyze, and Interpret the Design Alternatives
2. Select the Preferred Process Design
3. Prototype the New Process Design
4. Identify a Strategy to Migrate from the Old Process to the New
5. Implement the New Process Structure and Functions

Figure 23.7 Interpretation of the process of process innovation of Davenport.

and full information assets. There is an underlying assumption that technology will resolve all problems. Organizational and political issues are generally ignored, perhaps because they are considered unmanageable. In reality, the major focus may be on the technologies used to process data rather than on the information content in the data. The major objective is modeling and use of organizational data. Data engineering, and some aspects of information engineering, are considered king. All that is really needed for an organizational paradise is the selection of the best hardware and software, and use of the most appropriate computer-aided software-engineering (CASE) tools to assist in constructing the most appropriate entity relationship diagrams and data flow models.

2. *Feudalism* was the model of information politics most often encountered in this study. In this model, a number of people in several distinct feudal departments individually control information acquisition, representation, storage, analysis, transmission, and use. The language used for information representation is often different across the various departments. Critical information that affects the organization as a whole is often not collected. When an individual department collects a subset of this, it may not be passed on outside the boundaries of the department. Making organizationally informed decisions for the common good is often very difficult, as a result of the information-poor environment. Sometimes, however, strategic alliances across feudal departments are possible, and innovation within these units may be possible.

3. *Monarchy* is a pragmatic solution to the difficulties inherent in the feudal model. Information management is centralized and there is little autonomy concerning information policies. A benign monarch who is enlightened concerning information technology and organizational needs for information may set up, potentially through a chief information officer (CIO), a very effective and efficient system. A constitutional monarchy may well accomplish this; a despotic monarchy seldom will perform satisfactorily. One problem with a constitutional information monarchy, however, is the problem of succession when the monarch dies, retires, or is overthrown. This can lead to information anarchy.

4. *Anarchy* is the result of complete absence of any information policy. Usually this is not a willfully imposed model, but a result of a breakdown of one of the centralized approaches, such as a monarchy. This results in individuals and units in an organization managing their own information resources and developing information reports that serve their own needs rather than the needs of the organization as a whole. The major shortcoming of an information anarchy is not the wasted effort involved in the redundant information-processing and storage effort across units, but the Tower of Babel effect that results in terms of providing information that is useful for the entire organization. There will seldom exist any interoperability across units, at the level of either data or information. As a consequence of this, the various units may well present entirely different results from using the same data on the generally very different alternative model management systems, including intuitive model management systems, that will be in use.

5. *Federalism* involves the use of negotiations (Kennedy, 1994) to bring competing and noncooperative individuals and units into consensus on information-related issues. Strong central leadership and an organizational culture that encourages learning, cooperation, and consensus are required in order for this model to work.

Understanding by top organizational management of both information techno-logies and the value of information is a need if a shared-information vision is to be created.

It seems quite clear that these models are neither mutually exclusive nor collectively ex-haustive. A given organization may well function with a hybrid model, and the model that is in use at a given time may evolve from one form to the other with changes in organi-zational dynamics.

In a study of the information culture at 25 companies, Davenport and his colleagues found that the feudal model was the most common model in 12 companies. Federalist, monarchist, and technocratic utopianist models were found with approximately equal fre-quency in that 8, 7, and 9 companies were found predominantly following the prescriptions of these models. Only four organizations followed the anarchist model. Davenport et al. also ranked these five alternative models of information cultures according to their scoring on four attributes of information value:

- Commonality of vocabulary
- Access to information
- Information quality
- Information-management efficiency

Each of the four attributes had equal weight in this study. Davenport et al.'s conclusion—that the federalist and monarchist models are most appropriate, in most situations—seems inescapable. The feudalist and anarchist models, which actually seem to have a lot in common, were the poorest performers.

In order to determine a "best" model for a given organization, we need to know the current model that is in use, and the evolutionary or revolutionary model for the infor-mation culture to which the organization should be moving. Four suggestions are given; our interpretation of these is as follows:

1. Match information-management strategies to the extant organizational culture and the organizational culture the organization is desirous of adopting.
2. Practice technological realism, both in terms of information technologies themselves, interoperability across potentially different platforms, and the value of information that is not always easily captured in electronic form.
3. Select appropriate information managers, both from the standpoint of technical skills and broad-scope organizational understanding skills.
4. Avoid building information empires, such as through creation of despotic monar-chies and information czars.

Explicit recognition of information-management cultures and managing them construc-tively is suggested as the bottom line.

Effective use of information leads to major potential changes in systems and organi-zational management practice, through efforts that have been called "information ecology." Davenport (1997) has described several attributes of information ecology in an innovative work.

1. *Integration of Diverse Types of Information.* An information ecology prospers because of information diversity, just as the classic biological ecology flourishes because of species diversity. This requires the integrated management of dissimilar information species and is driven primarily by the need to better provide for and enhance the utility of information acquired in unconventional frameworks. This information integration is unlikely to occur without conscious attention to developing receptive organizational structures.

2. *Recognition of Evolutionary Change.* Information systems must be flexible and adaptive to change in order to accommodate the currently unknown information environments of the future. This may potentially be accomplished through use of such approaches to systems modeling and acquisition as iterative prototyping and modeling by agents that will allow for different forms of organizational rationality and information imperfections.

3. *Emphasis on Observation and Description.* From essentially the very beginning, natural ecologists have attempted to describe their biological world as a complex system such that they could comprehend the way in which particular species adapted into their natural environment and the resulting environmental change dynamics of environmental change. This poses particular challenges for such efforts as identifying the requirements for a system that is to fulfill some useful organizational purpose. Organizational information environments are complex and change in unpredictable ways. Thus describing and understanding the present organizational environment and the external environment surrounding the organization, and the associated provision of architectural and functional capabilities to adapt present capabilities to generally unknown future needs, is the appropriate approach to strategic programming and planning.

4. *Focus on People and Information Behavior.* Attempts to describe a natural ecological system necessarily involve many complex dynamic interactions. Information ecology is similarly complex if it is to enable the wise use of information to support humans, organizations, and technology in the many complex interactions involved in the private-and public-sector efforts needed to ensure economic and social progress and sustainability. The major challenge here is not just to make data available for use through efficiency and effectiveness of data mining, but to facilitate the conversion of data into information and knowledge for use in problem solving.

Davenport proposes a model that indicates many interconnected elements associated with components of an information ecology. There are three fundamental environments contained in this model and a web of interactions among the various elements.

1. The Information Environment. The inside circle of this model comprises the information environment internal to the organization and represents the core elements for ecological information management. The six most critical components of information ecology are interconnected in the internal sheath of this web:

   (a) Information architecture, representing the structure and location of internal organizational information.

   (b) Information behavior and behavior, representing the individual behavioral patterns in acquisition, analysis, interpretation, and utilization of information and the resulting information culture of the organization.

(c) Information politics, representing the organizational style for dealing with information and that can be categorized on a control continuum from monarchist to anarchist, as we have just discussed.

(d) Information processes, representing the activities performed by information workers in acquiring, analyzing, interpreting, and using information.

(e) Information staff, representing the set of people who provide, interpret, and utilize information and knowledge in the organization.

(f) Information strategy, representing the strategic approach of an organization concerning information, in other words, an organization's "information intent."

2. Internal Environment. The internal environment of the organization is the site of most management initiatives of an organization. These initiatives depend upon the strengths and weaknesses of an organization.

3. External Environment. The external environment affects the workings of many organizations. Threats and opportunities arise from the external environment. Many of these, such as changing government regulations, changing customer requirements, the emergence of competitors, and culture changes, are uncontrollable by the organization.

An information ecology web, illustrated in very rudimentary form in Figure 23.8, is the set of complex interactions across these three environments, and a change in one environment will affect the others. We desire to determine how an organization's information strategy affects their business strategy and vice versa. We desire to determine how infor-

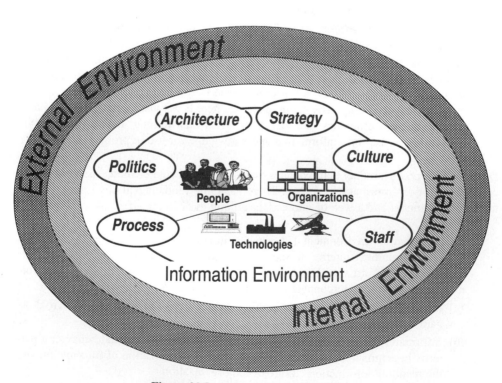

Figure 23.8   Information ecology webs.

mation politics mirrors and influences broader organizational politics, and the critical factors that relate to overall success of the organization. This human-centered model contains a technology component to be sure, but there are many factors in addition to technology that influence an organization's success: Technology, people, and organizations and their environment inextricably related in this model, just as they are in more general systems engineering efforts. Davenport suggests this model as a visionary model and not as a rigid prescription for determination of all of the events and elements that are important to an organization. He sees it as an empirical framework for experimentation to determine the most important organizational information strategy for success.

23.3.4 Corporate Reengineering

In the work by Hammer and Champy (1993), the forces of three Cs are suggested as combining together to require massive, discrete-level, transformations in the way organizations do business.

- Customers, who demand customized products and services that are of high quality and are trustworthy.
- Competition, which has intensified on a global scale in almost all niches.
- Change, which now becomes continuous.

Radical and dramatic reengineering of fundamental organizational strategy and of all organizational processes is suggested as *the* path to change for many organizations.

The authors are very concerned with organizational processes that have a number of common characteristics. Our interpretation of these is as follows:

1. The steps and phases in the process are sequenced in a logical order in terms of earlier phase results being needed for later activities. The phases are not necessarily in a linear order. They are sequenced in a concurrent fashion whenever possible, so as to generally enable obtaining results from the effort in minimum time.
2. The various business processes are integrated throughout the organization and, often, a number of formerly distinct efforts are combined in order to produce savings in costs and increased effectiveness as well.
3. There are multiple versions of many processes such that mass customization is thereby made possible.
4. Work is shifted across organizational boundaries to include potential outsourcing, and is performed in the most appropriate setting.
5. Decision-making efforts become a part of the normal work environment, and work is compressed both horizontally and vertically.
6. Reactive checks, controls, and measurements are reduced in frequency and importance in favor of greater use of interactive and proactive approaches.
7. There is always a point of contact, or "case manager" that is empowered to provide service to each individual customer, and a customer need never go beyond this point of contact.
8. Organizational operations are a hybrid of centralized and decentralized structures, as best suited to the particular task at hand.

Several, generally beneficial changes are claimed to result from this. Work units change from functional departments to multifunctional, process-oriented teams. Performers of simple tasks now are able to accomplish multidimensional work. People become empowered rather than controlled. The major needed job preparation changes; it becomes education rather than training. The focus of measures and performance shifts to results rather than activities. Promotion or transfer to a new organizational assignment is based on ability for the new assignment and not performance in a former assignment. Values change from reactive and protective to productive and proactive. Managers become coaches as well as supervisors, executives become leaders and not just scorekeepers, and organizational structures shift away from the hierarchical and to the flat. And, as we have noted, information technology is represented to be a major enabler of all of this. Chapter 30 of this handbook is devoted to an explicit consideration of information technology within the context of systems-engineering-guided knowledge organizations.

In *Reengineering the Corporation,* Hammer and Champy (1993) described a revolution in the way that American companies and developed nation organizations, in general, accomplish work. This book defined reengineering and its process components. It also suggested how jobs were different in the reengineered organization. Hammer and Stanton (1995) have written *The Reengineering Revolution: A Handbook,* Champy (1995) has authored *Reengineering Management: The Mandate for New Leadership,* and Hammer (1996) has authored *Beyond Reengineering: How the Process Centered Organization Is Changing Our Work and Our Lives.* Each of these works discusses extensions of the original efforts of Hammer and Champy.

All three of these latter works seem motivated by potential difficulties in implementing reengineering. Each acknowledge reengineering failures and present strategies to overcome them. The experiences of many suggest that radical reengineering effort, or process innovation, is not always successful. The major difficulty is a failure to cope with the impact of reengineering on people and their potential resistance to change. Other potential difficulties are inadequate team-building and the failure of senior management to appropriately convey the need for change and to be fully aware of the human element. While Hammer and Stanton's book is a handbook of techniques and practical advice, Champy focuses on management as the single critical influence of reengineering success. His focus is not only on reengineering processes through innovation, but also on reengineering at the level of systems management.

The major thrusts of the Hammer and Stanton work is that only top-level managers have the breadth of perspective, knowledge, and authority required to both oversee the effort from beginning to end and to overcome the resistance that will occur along the way. In their view, top-level managers make decisions to reengineer and then create a supportive environment that results in transformation of organizational culture. The reengineering team is also a major ingredient in success or failure. This team accomplishes the following:

- Develops an understanding of the existing process and customer requirements to enable a *definition* of the reengineering requirements.
- Identifies new process architectures, and undertakes *development* of the new process(es)
- Provides for *deployment* of the process and the new way of doing work.

The environment in which reengineering is done is one of uncertainty, experimentation, and pressure, and, based on these characteristics, the essential success characteristics of a reengineering team are identified. These include a process orientation, creativity, enthusiasm, persistence, communication skills, diplomacy, wholistic perspective, and teamwork.

Dealing with the human element in the organization that may be disoriented by the immense changes brought about by reengineering is important, and several strategies are suggested. Resistance to change is acknowledged to be natural and inevitable, with reasons for this resistance being a function of how people feel about the new situation. It is suggested that the imposition of a new process on people who have become attached to a familiar process will create natural resistance unless a five-step process for implanting new values is adopted:

1. Articulate and communicate the new values in an effective manner.
2. Demonstrate commitment of the organizational leadership to the new values.
3. Hold to these values in a consistent manner.
4. Ensure that the desired values are designed into the process.
5. Measure and reward the values that the organization wants to install.

Thus, the advocacy here is centered on customers and on the end-to-end processes that create value for them. By adhering to these principles, the organization should operates with high quality, tremendous flexibility, low cost, and exceptional speed.

Champy (1995) also examines the successes and failures of contemporary process reengineering innovations. He suggests that the failure of management to change appropriately is the greatest threat to success of reengineering efforts and that managers must change how they work if they hope to realize the full benefits of reengineering. In other words, reengineering of the lower-level work details is the focus of many contemporary efforts. However, reengineering of management itself is at least as significant a need and the representation is made that this has not been explored sufficiently at present. Such exploration, and subsequent action, are the major objectives in reengineering at the level of systems management. The intent here, as in *Reengineering Management* (Champy, 1995), is to identify concepts and methods that organizational administrators, managers, and leaders may use to reengineer their own executive functions for enhanced efficiency and effectiveness.

Champy begins with the impact of reengineering on managers and suggests that the greatest fear of executives is loss of control. The role of executives in a knowledge-based society is not to command or manipulate, but to share information, educate, and empower. They must have faith in human beings and their ability, if led properly, to do a better job for the customer. This is called *existential authority*. To bring this about requires a change in purpose, culture, processes, and attitudes toward people. Champy suggests that managers must focus on the answers to four questions to enable these changes:

1. What is the purpose of the organization?
2. What kind of organizational culture is desired?
3. How does the organization go about its work?
4. What are the appropriate kinds of people for the organization?

He suggests that management processes provide support for management reengineering, and defines new core management processes for the reengineered executive.

As a consequence of reengineering at the level of systems management, steep hierarchies will be flattened. Culture will be more of a determinant of how the organization runs than structure. A major need is for managers to organize high-performance, cross-functional teams around the needs of changing product lines or processes. Profit and principle must be congruent. Five core management processes, each of which potentially needs to be reengineered to be in tune with the core capabilities and mission of the organization, are identified:

1. *Mobilizing* is the process through which an organization, including the humans in the organization, is led to accept the changes that will be brought about by reengineering.

2. *Enabling,* or empowering, involves redesigning work so that humans are able to use their capabilities to the maximum extent possible and must be associated with a culture that motivates people to behave in the way the organization needs to have them behave.

3. *Defining* is the process of leadership through continual experimentation and empirical efforts. This process includes the development of experiential learning from these efforts and learning to act on what is learned from them.

4. *Measuring* is focused on identifying important process results, or metrics, that will accurately evaluate organizational performance.

5. *Communicating* involves continually making the case for changes that will lead to organizational improvement and being concerned not only with the what and how, but also the impacts of actions on employee lives. As suggested by many, managers are now coaches and must provide tools needed to accomplish tasks, remove obstacles hindering team performance, and challenge imaginations through the sharing of information. This relates strongly to empowerment, and to trust building, which is a goal of communications.

This "people focused management" requires "deep generalists," who are broadly focused and can respond to changing work demands, changing market opportunities, evolving and changing products and services, and changing demands of customers. Of course, it is necessary that these generalists also bring deep expertise in some specialty area and well-established skills to the organization.

23.3.5   Business Process Reengineering

Another view of process reengineering is provided by a team of authors from Coopers & Lybrand (Johanson et al., 1993). They view a given organization as being driven by market forces and by production forces. Organizations are viewed as having a balance of process-oriented horizontal operations and bureaucratic hierarchical operations. The more successful contemporary organizations will be market driven and will have a process orientation. A metric called *value,* based on four facets of value, is defined as:

$$Value = \frac{(quality)(service)}{(cost)(cyle\ time)}$$

where these four facets are, in turn, defined in terms of a number of lower-level attributes.

Process reengineering is indicated to be one of a family of three related, and process-oriented, methodologies:

- Total quality management (TQM)
- Just-in-time (JIT) manufacturing
- Breakpoint business process reengineering

There is much in common in these three efforts. According to these authors

> A BreakPoint is the achievement of excellence in one or more value metrics where the marketplace clearly recognizes the advantage, and where the ensuing result is a disproportionate and sustained increase in the supplier's market share.

Thus, a breakpoint is defined as the achievement of excellence in one or more value metrics.

Breakpoints are indicated to be necessary as metrics to support process reengineering. The two notions, breakpoints and business process reengineering, are viewed as virtually inseparable. *Business process reengineering* (BPR) is the means by which an organization achieves radical change in performance on those value-added facets of organizational products and services that are measured by cost, cycle time, service, and quality. BPR results from the application of tools and techniques that focus on an organization as a set of integrated customer-oriented core processes, rather than a set of functions. A breakpoint represents the achievement of excellence in one of the facets of value. The combination of the breakpoint construct and the process reengineering construct is termed *breakpoint business process reengineering*.

There are three essential phases suggested for BPR. These, which appear totally equivalent to the generic life-cycle phases of definition, development, and deployment that we have suggested here and illustrated in Figure 23.2, may be described as follows:

1. *Discovery* is the phase in which the organization identifies and provides definition for a strategic vision and needs to achieve that vision.
2. *Redesign,* or development, involves detailed planning and engineering of the strategic vision in terms of business processes.
3. *Realization,* or deployment, of the developed strategic plan is needed in order to operationalize the strategic vision.

Each of these phases is described in terms of a number of supporting steps. As with the other approaches to process reengineering noted earlier, the major focus is upon redefinition, redevelopment, and redeployment of all major organizational processes for greater organizational responsiveness and productivity as measured by customer satisfaction.

Defining a strategic vision for an organization is a key part of this effort. In a very insightful pair of articles, Schoemaker (1991, 1992) discusses scenario planning and the identification of strategic vision and core capabilities for an organization and their linkages. The framework for defining a strategic vision is as follows.

1. Generate broad-scope scenarios of possible futures for the organization.
2. Conduct a competitive analysis of the fields of interest of the organization.

3. Analyze the core capabilities of the organization and its competitors.

4. Define and develop a strategic vision and identify possible strategic options.

Figure 23.1 can easily be tailored to represent the business process reengineering perspective and strategic visioning described here.

In the process of defining a strategic vision, an organization must necessarily consider a number of related facets, such as organizational leadership and culture, and forecasted future scenarios for these. This leads to selection of strategic options that implement the strategic vision. This approach of Schoemaker appears most useful for completing the effort at BPR suggested here.

Another key characteristic of this approach is the identification of breakpoints, and the identification of core business processes in which appropriate breakpoints may exist. There are four critical processes where appropriate breakpoints may exist:

1. Processes that require very radical process reengineering due to large discrepancies between theoretical and realized process capabilities.

2. Processes associated with major potential improvements in customer receptivity.

3. Processes that could become the source of critical competitive products.

4. Processes that are responsive to either external competitive pressures or regulatory requirements.

Process mapping is suggested as an approach to understanding existing processes well and in suggesting potential breakpoints for exploitation. Essentially, this involves structural modeling and simulation of processes so as to enable determination of a vision for new processes.

23.3.6 Breakpoints

In the last subsection, and in others in this chapter, we have referred to the notion of a breakpoint; a breakpoint represents the achievement of excellence in one of the facets of value. This is a, more or less, internal definition. In a recent work concerning breakpoints, Strebel (1992) provides an external perspective on a breakpoint with the notion that it is a sudden radical change in the rules of the business game, in particular of market conditions, that will shape the future course of an industry or organization. A given organization may, however, create a breakpoint by proactive creation of competitive barriers, in the form of performance gaps, that others must then attempt to overcome. These two interpretations of the concept are very similar, although the inherent discontinuity notion does not seem to be necessarily associated with the first interpretation. In reality, a breakpoint is an internal concept that is intended to drive organizational activities to produce results that represent continuous improvement. It is also externally focused in that external reference points are needed in order to establish a breakpoint in a meaningful way.

We see that an organization may approach breakpoints from an inactive, reactive, interactive, or proactive perspective:

1. An *inactive* approach is one in which an organization simply does not concern itself with the market, and the status of the products and/or services of the organization relative to the market, at all. Thus, there is no need for breakpoints

2. A *reactive* approach to breakpoints is associated with marshaling organizational forces to a crisis situation, relative to the status of the organization's products and the market for these, after a crisis has eventuated. Ultimately, the organization attempts to exploit the breakpoints that are found to exist after diagnosing the damage that has been done and determining an appropriate correction and implementing it.

3. An *interactive,* or anticipatory, approach to breakpoints would be associated with attempting to keeping abreast of changing market situations as they occur, rather than after they have eventuated and caused damage.

4. A *proactive* approach to breakpoints would involve forecasting conditions such that the organization can prepare itself to create change and then position itself to take maximum advantage of the resulting competitive breakpoint.

An inactive approach, which is actually not discussed, is a path to disaster. Similarly, a reactive approach should be avoided as it is generally preferred to avoid difficulties rather than to emerge from them after they have materialized. Sometimes, however, it will not be possible to anticipate crisis situations. Thus, reactive strategies are always needed, even though we should also implement interactive and proactive approaches that minimize the need for reaction. An anticipatory and interactive approach works generally better than a reactive approach. To reach a high degree of excellence, a proactive approach is needed.

We see that there is much in common between the breakpoint concept and that of systematic measurements, which are discussed in Chapter 15. We can view a breakpoint as a metric, or we can view it as a process. Actually there are three processes involved, one each for reactive, interactive, and proactive breakpoint management. We can use the results of the breakpoint management process to affect the products of an organization, the product line of the organization, or we can attempt systems management strategy shifts that will change the essential characteristics of both processes and products. Figure 23.9 illustrates the three generic phases for breakpoint process management, and suggests the use of breakpoints to alter systems management, product-line processes, or products. As we have also noted, we may implement benchmark process management from either a reactive, interactive, or proactive perspective, or ideally a combination of these perspectives. Figure 23.10 illustrates this. It shows breakpoint strategy formulation and deployment in terms of reactive, interactive, and proactive benchmarking efforts.

In his effort, Strebel describes what we would call here the three life cycles for proactive, interactive, and reactive breakpoint management processes. The terminology creating, anticipating, and exploiting breakthrough is used. Our interpretation of these lifecycles is illustrated in Figures 23.11 through 23.13. Each of the phases illustrated, for each life cycle, is described in a number of formulation-, analysis-, and interpretation-like steps. The discussion by Strebel of these life cycles is rich and occupies much of the aforenoted work.

23.3.7 Benchmarking

A benchmark is very much like a breakpoint. The concept of a benchmark predates that of a breakpoint. Considerable discussion of benchmarking is available in Camp (1989), Liebfried and McNair (1992), Watson (1992), and Harrington and Harrington (1996a,b).

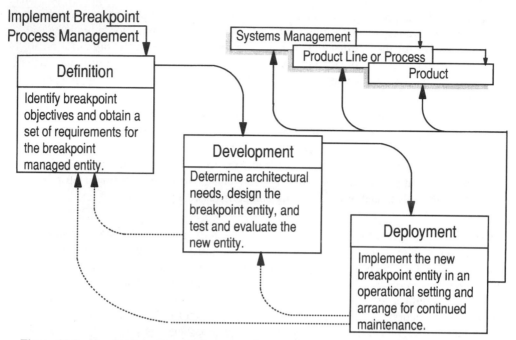

Figure 23.9 Breakpoint process management for systems management, product line, and product enhancement.

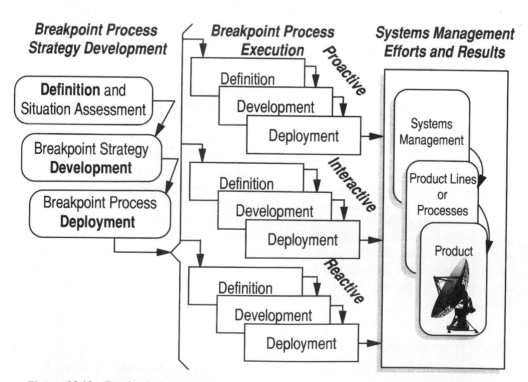

Figure 23.10 Breakpoint strategy development for reactive, interactive, and proactive benchmarking.

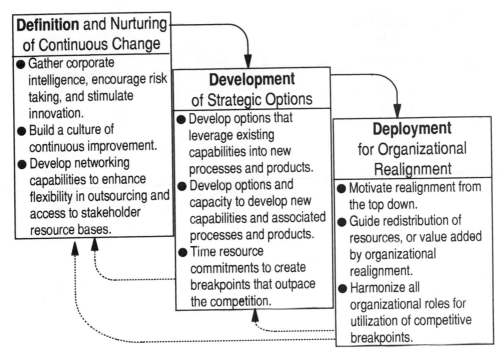

Figure 23.11   Interpretation of proactive breakpoint management process.

The definitive work by Watson (1992) provides an excellent overview of many current efforts concerning benchmarking and identifies four types of benchmarking efforts and a number of organizations that have implemented these benchmarking forms:

1. *Internal benchmarking* involves observations taken entirely within the same organization, generally of the best practices that have resulted in one segment of the organization that are desired to be transferred into anther segment of the organization. Internal benchmarking efforts at Hewlett-Packard are described.

2. *Competitive benchmarking* involves targeted best practices in an external organization. These can be at the level of systems management, processes, or product. Often, much secondary research is used for competitive benchmarking. The competitive benchmarking studies of the Ford Motor Company, and their use of these in developing Ford Taurus, are described.

3. *Functional benchmarking* is concerned with performance investigation within a specific functional area for an industrywide function. The results of functional benchmarking are especially suited to identifying process-improvement-related benchmarks. The efforts of the General Motors Corporation in September 1994, which led to major GM strides to enhance product quality and reliability, are described.

4. *Generic benchmarking* is concerned with studying the best processes in actual use in a given organization such as to enable analogous development of enhanced processes by the organization(s) sponsoring the benchmarking study. A generic bench-

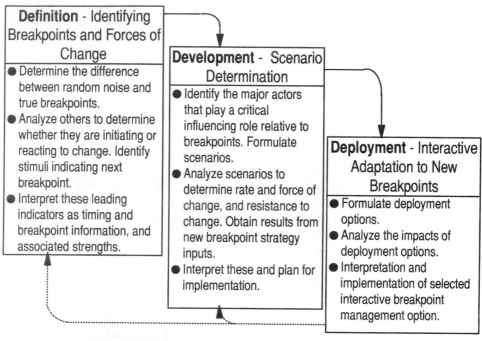

Figure 23.12 Interactive breakpoint management life cycle.

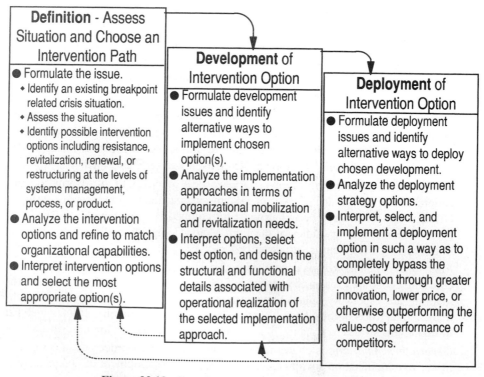

Figure 23.13 Reactive breakpoint management process.

marking study by Xerox, which ultimately led to understanding and analogous modification and adoption of facets of the shipping processes activities of the catalog casual-clothing sales organization L. L. Bean, are described. This effort was so successful that it led to establishment of a *benchmarking effectiveness strategy team* (BEST) network to transfer lessons learned throughout the Xerox Corporation.

The terminology used in this listing is that of Watson. In our terminology, as established in the introductory chapter, we would use the term *functional product benchmarking* to refer to benchmarking of the directly measurable aspects of a product. These aspects serve to describe the operational capabilities that are supplied to the user of the product, or service. The nonfunctional attributes of a product are those constraints, alterables, or limitations that relate to the structural or purposeful properties of the product that are not a part of the functional properties. The nonfunctional attributes of a product relate to such important characteristics of a product as reliability, maintainability, quality, and availability. Thus what was accomplished here would not be described, using our terminology, as functional product benchmarking. In our terminology, it is really nonfunctional product benchmarking. The study conducted by GM was not restricted to automotive products. It was a study of quality processes at 11 cooperating organizations in a variety of product areas. They had established 10 hypotheses for empirical evaluation, and these relate very closely to quality processes. It would therefore be very appropriate to denote this type of benchmarking as functional process benchmarking.

Harrington and Harrington (1996a,b) also describe the benchmarking process. They identify five types of benchmarking efforts:

1. *Internal benchmarking* is suggested as the starting point and that it should be considered before other forms, primarily because of its simplicity and low cost, and because it can serve as an education and training ground for accomplishing external benchmarking. There is a high degree of cooperation in the process, and the information obtained is generally very relevant; however, the potential significance of the findings is low.

2. *External competitive benchmarking* will generally involve reverse engineering of the products, services, and processes of a competitor. This is often accomplished by purchasing a competitor's product, and associated materials, and studying them. Often, the competitor will be unwilling to fully cooperate in an effort this sort. However, the information obtained may be quite useful, and the potential significance of the findings may fully justify the cost of the product familiarization capabilities necessary to accomplish this satisfactorily.

3. *External industry (compatible) benchmarking* is external benchmarking across a set of congruous products and services that are not direct competitors. The degree of cooperation to be expected here is higher than in external competitive benchmarking, as is the potential significance of the findings. It is generally a more time-intensive effort than external competitive benchmarking.

4. *External generic (transindustry) benchmarking* involves organizations that produce disparate products and services. This may require quite a considerable amount of time to accomplish. However, the potential benefits include major improvements possible through fundamental changes in product lines and products.

5. *Combined internal and external (competitive, industry, and generic) benchmarking* is benchmarking that involves all of the basic four forms. Naturally, this is the form of benchmarking with the most potential for improvement. However, it will generally be expensive and time-consuming.

Five phases are suggested for the accomplishment of benchmarking, as illustrated in Figure 23.14. These generally correspond to the generic definition, development, and deployment phases identified in the introductory chapter. Twenty and 144 steps are associated with these five phases; they are discussed in Harrington and Harrington (1996a,b).

Like a breakpoint, a benchmark involves a baseline comparison of existing practices of an organization with those best practices as used by one organization, or perhaps many other organizations. These practices may be at the level of product, process, or systems management. Breakpoints and benchmarks have the same ultimate purpose, that of enabling and enhancing organizational improvements for customer satisfaction. The use of a benchmark can be reactive, in which case we desire to emulate the best practices of others that have already been established, and emulate or perhaps excel in these. It can be interactive or anticipative, in which case, we desire to foresee efforts that are now occurring and adjust organizational practices, more or less in real time, to keep abreast of the competition. We can use benchmarks for forecasting purposes such that we evolve entirely new forms of organizational results in the form of system management strategies, processes, and products that are each world class in quality and trustworthiness. In all cases, benchmarks are based on systematic measurements, both internal and external measurements, and serve as a catalyst for action and results. In this regard, benchmarks are also very much related to critical success factors (CSF) (Rockart, 1979; Rockart and Bullen, 1986). The two terms are, for most purposes, synonymous.

We see that a benchmark is a standard of excellence or achievement, or CSF, that provides a baseline against which to measure or evaluate, or otherwise judge, similar entities. While the benchmarking notion may seem quite simple at first, there are really a

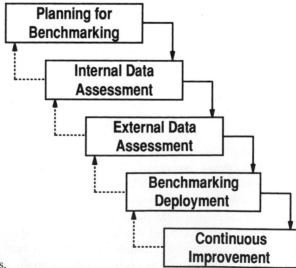

Figure 23.14   Benchmarking phases.

number of benchmarking features that need to be examined. There are at least seven important and interrelated questions that need to be asked before we benchmark:

1. Why do we benchmark?
2. What do we benchmark?
3. Which benchmark measures do we need?
4. Who do we benchmark?
5. Where do we benchmark?
6. When do we benchmark?
7. How do we benchmark?

We can and should ask these questions of our organization and, to the extent possible, we can and should ask these questions of those organizations that represent competitors. We can and should ask these questions in the search for structural, functional, and purposeful answers. A similar set needs to be asked after we benchmark. Benchmarking efforts can be conducted relative to issues at the level of systems management, process, or product. Answers to these questions need to be obtained in terms of implications for each of these levels.

There can be some obvious ethical, moral, and legal concerns relative to benchmarking. Some aspects of benchmarking may seem to amount to spying, or espionage. Clearly, successful benchmarking should not, and cannot, involve behavior that is either immoral, unethical, or illegal. One of the major activities of the International Benchmarking Clearinghouse Committee of the American Productivity and Quality Center (APQC) has resulted in a set of guidelines, known as a Code of Conduct, for benchmarking. These codes have been subscribed to by a considerable number of companies and are generally felt to be above reproach (Watson et al., 1992). There are nine principles in the Benchmarking Code of Conduct. They may be summarized as follows:

1. *Legality.* The acquisition of trade secrets is proscribed, as is doing benchmarking without first requesting approval by the benchmarkee. All actions and intents should be legal ones.
2. *Exchange.* The same type and level of information should be provided both by the benchmarked organization and the organization doing the benchmarking in an openly communicated exchange.
3. *Confidentiality.* Benchmarking communications should be considered as confidential to the organizations involved, and obtained information concerning benchmarking should not be divulged outside the concerned organizations without prior consent of all parties.
4. *Use.* Benchmarking information should only be used for formulation of improvement options for processes or products within organizations participating in the study. Attribution of benchmarking partner names requires prior permission, and benchmarking information must not be used for marketing or sales purposes.
5. *First-party Contact.* Benchmarking information contacts with a partner organization should be made through the point of contact established by the benchmarking partner. Mutual agreement must be reached for delegation of responsibility in this regard to other parties.

6. *Third-party Contact.* Prior permission must be obtained before divulging the name of the benchmarking point of contact to a third party or in an open forum.

7. *Preparation.* A benchmarking contact with the point of contact at another organization should only be made after proper planning and preparation of an interview guide, in order to make the encounter efficient and effective.

8. *Completion.* Each benchmarking study must be completed to the satisfaction of all benchmarking partners in a timely manner, as agreed to prior to the study.

9. *Understanding and Action.* All partners should be treated with mutual respect and understanding, and information should be used as mutually agreed upon.

It seems quite clear that these conduct codes are also applicable as ethical codes of conduct, as well as being very useful standards for benchmarking practice.

The American Productivity and Quality Center has established a comprehensive framework for benchmarking. It comprises 13 steps, each representing an organizational process, and the first two levels of the 13 step framework are as follows:

1. Understand markets and customers
   1.1. Determine customer needs and wants
   1.2. Measure customer satisfaction
   1.3. Monitor changes in market or customer expectations
2. Develop vision and strategy
   2.1. Monitor the external environment
   2.2. Define the business concept and organizational strategy
   2.3. Design organizational structure and relationships between units
   2.4. Develop and set organizational goals
3. Design products and services
   3.1. Develop new product/service concept and plans
   3.2. Design, build, and evaluate prototype products and services
   3.3. Refine existing products/services
   3.4. Test effectiveness of new or revised products or services
   3.5. Prepare for production
   3.6. Manage the product/service development process
4. Market and sell
   4.1. Market products or services to relevant customer segments
   4.2. Process customer orders
5. Produce and deliver for manufacturing
   5.1. Plan for and acquire necessary resources
   5.2. Convert resources or inputs into products
   5.3. Deliver products
   5.4. Manage production and delivery process
6. Produce and deliver for service-oriented organization
   6.1. Plan for and acquire necessary resources
   6.2. Develop human resource skills
   6.3. Deliver service to the customer
   6.4. Ensure quality of service
7. Invoice and service customers
   7.1. Bill the customer

    7.2. Provide after-sales service

    7.3. Respond to customer inquiries

8. Develop and manage human resources

    8.1. Create and manage human resource strategies

    8.2. Cascade strategy to work level

    8.3. Manage deployment of personnel

    8.4. Develop and train employees

    8.5. Manage employee performance, reward, and recognition

    8.6. Ensure employee well-being and satisfaction

    8.7. Ensure employee involvement

    8.8. Manage labor–management relationships

    8.9. Develop human resource information systems (HRIS)

9. Manage information resources

    9.1. Plan for information resource management

    9.2. Develop and deploy enterprise support systems

    9.3. Implement systems security and controls

    9.4. Manage information storage and retrieval

    9.5. Manage facilities and network operations

    9.6. Manage information services

    9.7. Facilitate information sharing and communication

    9.8. Evaluate and audit information quality

10. Manage financial and physical resources

    10.1. Manage financial resources

    10.2. Process finance and accounting transactions

    10.3. Report information

    10.4. Conduct internal audits

    10.5. Manage the tax function

    10.6. Manage physical resources

11. Execute environmental management program

    11.1. Formulate environmental management strategy

    11.2. Ensure compliance with regulations

    11.3. Train and educate employees

    11.4. Implement pollution-prevention program

    11.5. Manage remediation efforts

    11.6. Implement emergency response programs

    11.7. Manage government agency and public relations

    11.8. Manage acquisition/divestiture environmental issues

    11.9. Develop and manage environmental information system

    11.10. Monitor environmental management program

12. Manage external relationships

    12.1. Communicate with sharcholders

    12.2. Manage government relationships

    12.3. Build lender relationships

    12.4. Develop public relations program

    12.5. Interface with board of directors

    12.6. Develop community relations

    12.7. Manage legal and ethical issues

13. Manage improvement and change
    13.1. Measure organizational performance
    13.2. Conduct quality assessments
    13.3. Benchmark performance
    13.4. Improve processes and systems
    13.5. Implement TQM

This *process classification framework* is intended to supply a universal view of the processes that exist in organizations of all types and to assist in understanding organizations from a horizontal and process-oriented viewpoint, as contrasted with a vertical and functional viewpoint. Thus, the process classification framework is intended to represent major processes and subprocesses in an organization, and not functions. Obviously, there may exist processes that are not included here in any specific organization, and there may be processes described here that may not exist in any given organization. It should also be remarked that this framework is an evolving framework and subject to continuous improvement over time.

The APQC has also established three awards for benchmarking: a research prize, a benchmarking study prose award, and an award for excellence in benchmarking. *The Benchmarking Research Prize* is presented to recognize accomplishments in development of the technical aspects of benchmarking. There were four award attributes for the 1997 prizes:

- Creativity and innovation
- Technical sophistication
- Demonstrated or potential applicability
- Practicality and portability across benchmarking efforts

Each of these attributes is associated an equal weight of 250 points.

The *Benchmarking Study Award* is intended to encourage and recognize excellence in the way in which benchmarking studies are undertaken. There are four attributes:

- Planning the study (100 points)
- Collecting information (50 points)
- Analyzing data (50 points)
- Results and improvements obtained (100 points)

The major awards are *Awards for Excellence in Benchmarking*. For the 1997 competition, there were six achievement categories for benchmarking, and these are further disaggregated into a number of subcategories. Figure 23.15 illustrates an attribute tree, and provides attribute weights for the benchmarking excellence awards. The total weight of 1,000 points is distributed such that results and measured improvements from the benchmarking process is the most important attribute with 300 points or a normalized weight of 0.30. In this category, specific benchmarks that have been adopted over the past three years are examined to determine results obtained in terms of customer satisfaction, quality and productivity, cycle time, employee well-being and morale, and environmental impact. Also examined are the measures used to evaluate and improve the benchmarking process within the organization, and how the objectivity and validity of the benchmarking study

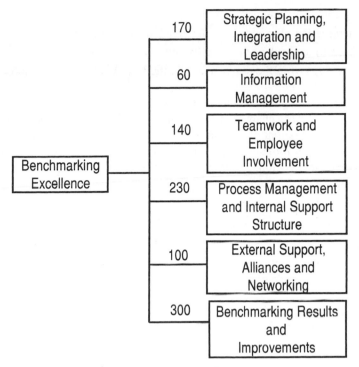

Figure 23.15   Major attribute and attribute weights for benchmarking recognition award criteria.

results are established.[1] Detailed information and application forms for these awards are available from the APQC, which maintains an extensive Web site.

There are several ways in which a benchmarking study can be conducted. In each case, there needs to be formulation, analysis, and interpretation at each of the phases in a benchmarking study. We generally need to define what it is we wish to benchmark. We need to develop an appropriate benchmarking process in terms of determining how to go about benchmarking. Then we need a deployment strategy in terms of an effort that puts the development tactics into operational implementation. The questions we asked earlier in this subsection each need answers in order to develop the actual benchmarking plans and implementation effort. The references cited earlier provide a number of relatively specific implementation discussions and case studies of benchmarking. Watson, for example, illustrates how Hewlett-Packard, Ford, General Motors, and Xerox each performed benchmarking studies. Often, it is suggested that organizations follow the plan-do-check-act cycle of Schewhart and Deming associated with TQM, as discussed in Chapter 7. This would involve:

---

[1]Further information on these awards should be available from: The American Productivity and Quality Center, Benchmarking Award Administrator, 123 Post Oak Lane, Houston TX 77024-7797. Phone: 713-681-4020; fax: 713-681-8578; e-mail: apqcinfo@apqc.org

1. Planning for the benchmarking process.
2. Implementing the plan and obtaining benchmarking information.
3. Checking and analysis of the benchmarking information.
4. Acting by improving the organization on the basis of the analyzed information.

Benchmarking is not altogether dissimilar from the other related improvement approaches we describe in the present chapter. It is also very closely related to the strategic quality management approaches that are discussed in Chapter 7. Nor is it unrelated to generic strategies for successful imitation (Schnaars, 1994). The metrics and systematic measurements associated with benchmarking are, for the most part, the metrics discussed in these chapters and in Chapter 15.

23.3.8 Competitive Advantage and Competitive Strategy

Michael Porter has been much concerned with competitive advantage and competitive strategy and has written four seminal texts (Porter, 1998a–d), which are editions of works first appearing in the decade of the 1980s. He identifies four forces: threats from potential new competitors, bargaining power of suppliers, bargaining power of potential customers, and threats due to substitutable products that combine together to induce and intensify organizational competition. Together with threats from existing competitors, the net result is *rivalry and competition in the marketplace.*

These are the external factors influencing competition. Figure 23.16 represents our interpretation of these five factors and their influence on marketplace rivalry. The extent

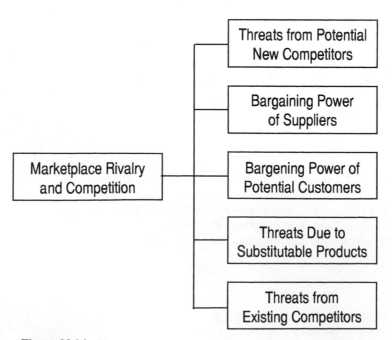

Figure 23.16 Five major factors influencing marketplace competition.

to which each of these factors is a dominant influence in the marketplace is said to depend upon industry structure. The internal structure within an individual organization is also a major factor for that organization. Structural change in a given industry sector may change the way in which these potential factors influence actual competition. Changes in an individual organization may have a constructive or destructive effect upon a given industry. A given organization may have little or much ability to cause structural change in the industry in which it operates.

Seven major barriers may act to prohibit an organization from entering a given market area:

1. Economies of scale may make it very difficult for a firm, especially one without very deep pockets, from entering a new market area in which it cannot obtain a competitive cost advantage until it has captured a significant share of the market.

2. Product differentiation may pose another barrier to entry, especially when the marketing efforts for currently competing products has been very successful and they are perceived as highly differentiated products with superior functional features and performance characteristics.

3. Capital requirements to enter a given market may be very high.

4. Experiential familiarity barriers to entry may exist whenever there is major experiential-familiarity-based expertise associated with competitors, and where this is not available to organizations potentially desirous of entering the market with a new product. This generally leads to cost disadvantages that will exist even for organizations with such a large available financial resource base that they might use to ultimately sustain themselves to obtain economies of scale if they could obtain the needed, but unavailable to them, experiential base needed for success.

5. Access to product distribution channels may be difficult to obtain for an organization with a new product.

6. Government policy and controls may, through licenses and regulations, make it very difficult if not impossible for an organization to enter a product market that is new to them.

7. Fear of retaliation from a larger competitor now in the market, particularly if the smaller firm is also a supplier to the larger competitor, may provide a significant barrier to entry.

There are, of course, other barriers to entry into a given market area. Supports and barriers for the external factors influencing marketplace entry, illustrated in Figure 23.16, are generally provided by Porter. The conditions under which customer and supplier groups are powerful, and the threats of new entrants and substitute products are major influencers on competitive advantage of an organization and combine with market conditions to support or deny competitive advantage for a given organization.

Two basic facets of *competitive advantage* are identified in these works:

1. *Cost* is a major influencer of competitive advantage. If the functional and nonfunctional characteristics of a product, or service, are presumed to be the same, positive competitive advantage will accrue to a lower-cost product, or one that can be produced, delivered to customers, and supported more efficiently. It can be sold at a

lower price than a higher-cost product of comparable value, and this results in greater profits per unit product delivered than would be the case if production costs were higher.

2. *Differentiation* is a second major influencer of competitive advantage. Among products or services that cost the same to produce, there will be a positive competitive advantage to those products that have greater perceived differentiated value, as the weighted sum of functional and nonfunctional features of the product are greater.

We see that lowering production costs, through greater production efficiency, will enable competitive advantage. In a similar way, increasing product effectiveness or differentiated value will increase competitive advantage. Competitive advantage is one major ingredient of competitive strategy. The competitive advantage for an organization's products will be very dependent upon how the organization manages all of the organizational processes (RDT&E, acquisition or production, marketing, and such related subprocesses as warehousing, supplier relations, and distribution) that relate to the final product or service. Porter uses the terms "value chain" and "value system" to refer to the linkages created by these various processes, and the entire system.

Competitive focus, or *competitive scope,* of products and services is a third major generic ingredient of competitive strategy. There are two market foci, or competitive scope targets:

1. A *broad-target* general market exists for some products and services.
2. A *narrow-target* and highly focused, or segmented, or niche, market exists for other products and services.

Together, these make up competitive scope. An organization can target its products broadly or narrowly. It can do this with low-cost products that are undifferentiated in their features, or with higher-cost products that are differentiated in their features from the products of competitors.

At first glance, it appears that there are two other types of competitive advantage strategies that are possible. One could attempt to offer high-cost products that are undifferentiated in their features. Clearly, this would only result in major competitive disadvantage as a special, but rather undesirable, form or competitive advantage. Only some form of sales and marketing camouflage could make this succeed, but the success would generally endure only over the short term. Alternately, an organization could conceivably have a product that is differentiated in features from its competitors, and which also has lower cost. In this highly desirable situation, there would exist a major competitive advantage due both to lower cost and greater differentiation. So, we see that these two types of competitive advantage are not necessarily mutually exclusive.

Porter identifies four generic competitive strategies. These involve a mix of the two identified types of competitive advantage and the two types of competitive focus:

1. Differentiation strategies involve offering a large number of highly competitive and specialized products that appeal to a wide segment of the market, often at premium prices.
2. Focused differentiation strategies involve targeting narrow-market segments with highly specialized products offered at very premium prices.

3. Cost-focused strategies involve offering relatively simple basic and generally un-differentiated products to a narrowly targeted market at low prices.

4. Cost leadership strategies involve offering many types of products to a broadly targeted market that are of good but generally not superior quality, and which are offered at very low prices.

Since there are a variety of ways in which a given organization can focus its product(s) and differentiate among competing products in order to appeal to different market segments, there are a potentially large number of variations of a given strategy. No one of these four strategies can be called universally "best" or most appropriate. Industry and organizational structure will strongly influence the range of competitive strategy options that will be available and that can be successful if adopted.

Figure 23.17 illustrates the three generic competitive strategy dimensions:

- Product cost
- Product differentiation
- Market scope

It also shows our interpretation of the eight possible strategies that may be implemented. Two of these are disaster strategies, that of entering any of the two generic markets with a product with no competitive advantage whatever. While it might appear highly desirable to attempt to implement several of the strategies simultaneously, that is to say entering

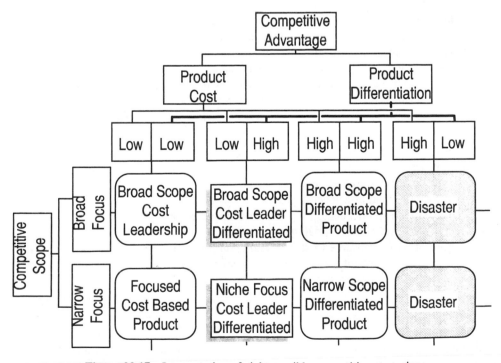

Figure 23.17 Interpretation of eight possible competitive strategies.

both a broad-scope and a narrow-scope market simultaneously, this will generally be a foolish move and a strategic recipe for poor performance. Thus, there are some combination situations or strategies in this figure that may be bad.

Generally, in fact, only one strategy should be adopted for a given specific organizational product. Of course, there can be strategy migration over time. For example, an organization might initially have a product that is a low-cost, highly differentiated product and attempt to enter a broadly targeted market. This is what we have called a broad-scope, cost leadership, highly differentiated product. It may well occur that the initial cost leadership opportunities vanish over time for any of a number of reasons, such as a new technology that makes low-cost production opportunities that are available to many organizations. The organization may then wish to reposition itself as a producer of a broad-scope highly differentiated product and forego the initially held competitive advantage position of cost leadership. This might well lead, however, to image difficulties due to the changed marketing strategy that is needed. In such a situation, it might have been better for the organization to have established its competitive advantage on the basis of product differentiation only. It could have converted the initial cost leadership position into increased profits, while the cost leadership existed, and market the product as a broad-target, highly differentiated product only. It is for reasons like this that Porter does not suggest the two competitive strategies associated with low-cost, high-differentiation competitive advantage, and does encourage use of only one of the four generic strategies we described earlier.

Porter suggests three conditions under which a firm can concurrently market low-cost and high-product-differentiation competitive advantage:

1. The other competitors are "stuck in the middle," in the sense that they have tried to concurrently implement several generic strategies, but are not successful in implementing any one of them. Thus, they are not powerful enough to force a situation where low cost and high product differentiation become inconsistent. Eventually, however, one or more competitors will satisfactorily implement one of the four generic strategies suggested by Porter. They will place enormous pressure on either the low-cost or the product-differentiation advantage held by the organization that once excelled in both and will ultimately cause it to yield on one of these, unless it wishes to also become stuck in the middle.
2. The cost leadership position is more strongly determined by the large market share that is associated with a specific product, rather than by process and technology or service factors. A cost leadership position will often develop when an organization is the first to market a specific product. It can also occur when an organization has sufficiently deep pockets to market a product at a price lower than the development and deployment cost for small quantity production would otherwise justify, and the organization is then able to sustain and endure the loss situation until the product has captured sufficient market share that economies of scale result in lower production cost to the organization and result in a profit for each product sold.
3. An organization has the benefit of a major new innovation and is the only organization that can benefit from this innovation.

This does not say, however, that an organization should not attempt efforts aimed at both low cost and high product differentiation. Clearly, an organization should purse low cost when it does not sacrifice product differentiation. It should also pursue high product differentiation through measures that are not very expensive. In this sense, it needs to strive for Pareto optimality (Sage, 1983) such that there will be no allocation of effort or reallocation of the resources of the organization that will increase differentiation without increasing costs, and no reallocation that will decrease costs that will also not decrease differentiation. Figure 23.18 suggests the notion of Pareto optimality and a desirable Pareto frontier relative to the tradeoff among low cost and product differentiation. An organization should select a single competitive advantage perspective, of either being a high-differentiation or a low-cost producer, and adopt a competitive strategy that is supportive of this.

Five causal factors that may result in innovations and new ways of competing are identified:

- New technologies
- New or changing customer needs
- Emergence of new industry segments
- Changing cost or availability of the input factors to production
- Changes in government regulation

These factors will, of course, influence organizational competition and potential. It is not only necessary to establish competitive advantage, it is necessary to sustain it. Porter indicates three approaches to accomplish this:

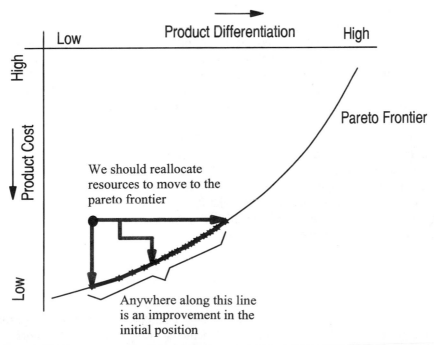

Figure 23.18   Pareto optimality and improvement in cost leadership and differentiation.

1. The source of competitive advantage may be sustained. Some sources of competitive advantage are relatively easy to reproduce or emulate. This is particularly the case for lower-cost advantages, such as those due to lower labor costs and cheap raw materials. Product differentiation advantages generally will be much more difficult to replicate. These advantages are primarily due to innovative RDT&E products or innovative and speaialized marketing, customer relations and service, or product reliability and quality. This reality strongly supports the strategic quality management efforts described in Chapter 6. Often, lower-cost advantages can be offset quite dramatically by the greater product differentiation that results from careful and sustained investments in people, processes, technologies, and organizational facilities.

2. The number of distinct sources of competitive advantage is as important for sustaining advantage as it is for initially establishing it. If an organization obtains competitive advantage due only to one source, this may be easily overcome. A low labor cost producer may be upset because of technological improvements that increases the productivity of highly educated, motivated, and well-paid empowered workers.

3. Constant improvement and upgrading is cited as the most important sustainer of competitive advantage.

An organization can approach competitive advantage and its sustenance in a reactive, interactive, or proactive fashion. If the approach is reactive, then there will likely be a constant challenge of continuing to react to competitors after their new and superior products have emerged. Change is costly, and it is difficult to establish the motivation and momentum required when the existing products and services have initial competitive advantage. Thus, sustaining competitive advantage may be more difficult for a given organization than was the effort in initially establishing it.

There are systems management risks associated with choice of either, or both, of the competitive advantage strategies. These are that either cost leadership is not sustained, or that high product differentiation is not sustained, or both. With no doubt at all, we can state that the risks associated with not sustaining both cost and differentiation leadership is greater than the risks associated with not sustaining either one of these. It is not immediately clear that the risk of loosing either the advantage due to cost leadership or the advantage due to high differentiation standard is greater than either the risk of loosing cost leadership when only that is established or the risk associated with loosing product differentiation when only that is initially established. Porter provides compelling verbal arguments that suggest that this will almost always be the case, however. There are also risks associated with choice of a narrow-target, or niche, market. Essentially, these are the risks that the selected competitive advantage is successfully imitated, or that the targeted niche becomes unattractive because either the structure of that segment of the market changes greatly or the product demand vanishes in that specialized niche. The risk management approaches in Chapter 3 should be of value in suggesting appropriate risk management controls.

Four interacting causes of national competitive advantage in international markets are cited in the works on competitive advantage of nations and developing nations:

1. *Factor* conditions are needed in order to enable an industry to compete satisfactorily in a given market. These conditions include human resources, educated profession-

als, information technology and knowledge resources, raw materials and physical resources, physical infrastructure, capital resources, and other factors of production. Today, the advanced information and knowledge-based factors may be more important than traditional basic factors, such as raw materials. The factor proportions that are needed to be deployed efficiently and effectively in various industries, however, differ widely.

2. *Demand* conditions for a given product, or service, must generally exist in the home nation. Porter found that quality is more important than quantity in this regard, and that it is very advantageous if home-based customers are anticipatory of international demand for a product in terms of their own present demands. It is generally difficult to sustain competitive advantage when the only demand for a product is far away geographically.

3. Related and supporting industries, which are internationally competitive, are generally needed. This provides for *competitiveness* conditions in supplier industries through continuous patters of innovation and process improvement. It is rare that a given nation can maintain competitive advantage if it must import high-quality items, or significant quantities of well-educated labor, necessary to establish a high-quality supplier base.

4. Firm *strategy, structure,* and *rivalry* are important, and often the most important, factors.

Organizational culture and leadership, and the resultant process maturity, are clearly of vital importance as well. Of major importance also are individual objectives and motivations and the influence that national leadership provides to enhance human resources. Rivalry among competitive organizations is needed to motivate the necessary development of appropriate organizational strategy and purpose. It acts as a major catalyst for needed change for enhanced national competitive advantage. National or governmental leadership and chance play roles as influencers, also. A major caution is given that attempts to preserve everything will most likely result in lowering national standards, not raising them. On the other hand, attempts to create a single national industry leader will generally fail as success would significantly impede development of a very needed national rivalry in that industry sector from which the leader could emerge.

Porter makes a major argument that productivity is the single most useful measure of national economic strength. National productivity depends less upon the traditional macroeconomic management of money supply and other macroeconomic controls than it does upon successful government programs that encourage continuous improvement on all fronts, especially human-related fronts. It is argued that balancing the money supply may lead to short-term advantage and not the long-term competitive advantage that is needed. Those who advocate total quality management for high productivity and customer satisfaction make similar arguments.

That each of the four factors of national competitive advantage is mutually supportive and causal to each of the other three factors is very significant. It suggests that one of the four factors might well be absent, and that this will delay and reduce the amount of competitive advantage ultimately possible. But, it will not prevent it. As with the cost and product-differentiation facets that led to competitive advantage for a given organization, we need to associate these national competitive advantage strategies with a suitable market focus or scope, which now becomes international, and therefore establish a very broad and distributed geographic scope in order to obtain an international competitive strategy.

It would not be possible in a few brief pages here to capture other than the highlights of these very seminal and important works. They are true pacesetters and they set forth many prescriptions for reengineering, especially at the level of systems management and processes, that involve government and organizational policies and strategy for enhanced advantage. Much additional information is available in the four cited references (Porter, 1998a–d) and in two reprint works (Porter, 1991a,b). There are, of course, many other works that discuss related approaches concerning managing and forecasting for strategic success (Thomas, 1993), managing to cope with dual strategies for ensuring success today *and* success tomorrow (Abell, 1993), and other important strategy needs for competitive positioning (Rumelt et al., 1994) and markets structure influences on organizations (Shy, 1995).

23.3.9 Time and Reengineering

Time is of much concern in virtually all systems engineering and management, including systems reengineering, efforts. In this subsection, we discuss recent efforts to manage time related facets for enhanced productivity and trustworthiness.

Many researchers have commented on the relationship between time and productivity (Thomas, 1990; Bower and Hout, 1988). Schmenner (1988), for example, has documented the results of several studies of productivity improvement and concluded that the only approaches that were demonstrably successful were those that focus on time as the single most important determinant of overall productivity. This "productivity" included not only delivery of products faster, but also improved quality and process effectiveness.

George Stalk (1988) has noted that "as a strategic weapon, time is the equivalent of money, productivity, quality, even innovation." Time-based competition is therefore a potential source of competitive advantage. In this work, flexible manufacturing, just-in-time manufacturing, and other modes of time-based competition are viewed as generally superior strategies than more traditional ones that rely on low wages, which, it is claimed, reduce responsiveness as well as cost. Time-based management also reduces costs, but it accomplishes this through greater efficiency and effectiveness of organizational processes, and thereby enhanced responsiveness. Stalk and Hout (1990) indicate several tasks that must be satisfactorily performed in order to bring about time-based competition as a, if not the, critical parameter for strategic management:

1. The value-delivery system of the organization must become much faster and much more flexible than that of competitors.
2. The organization must identify the ways their customers relate to value and responsiveness issues, and then focus on those customers with the greatest sensitivity to value and responsiveness.
3. This responsiveness must be used to stay close to customers such that, in effect, customers become dependent upon the organization, and competitors then have to deal with "less attractive" customers.
4. The organization must identify and implement strategies that "surprise" competitors with the resulting innovative and time-based competitive advantage.

Four very beneficial results of implementing time-based competition are identified:

1. Organizational productivity is increased.
2. Product prices may be increased.
3. Organizational risks are reduced.
4. Market share is increased.

We should *not* believe that time-based competition is simply equivalent to having everyone work faster and harder. It does not at all advocate a shortening of time horizons and concentration on short-term profits so often obtained by disabling the organization. It is much more concerned with working smarter, and primarily through process improvements and the resulting life-cycle time compression. In addition, competing on time as a competitive advantage recognizes and supports the basic competitive advantages of low price and high product differentiation and such supports for differentiation as

- *Quality, consistency,* and *conformity* in satisfying customer requirements and in fulfilling customer needs.
- *Adaptability, innovativeness,* and *perceptiveness* in coping with many organizational environments and in being able to respond to changes in organizational environments and customer needs.

These linkages suggest, again, the major role of processes in determining successful products and markets for those products. it provides further evidence of the role of competitive strategy in determining the paths to success.

A recent work by Meyer (1993) is concerned with implementing a fast-cycle-time (FCT) strategy. It is concerned both with a systemic organizational strategy and associated tools and measurements to bring this about. Establishing leadership, multifunctional teams, people empowerment, and reengineering of core organizational processes are the primary activities associated with this effort. Organizational systems and organizational learning are the critical foundations on which FCT results are constructed. The fundamental premise of FCT is that "the competitor who consistently, reliably, and profitably provides the greatest value to the customer first wins." According to Meyer, "fast cycle time is the ongoing ability to identify, satisfy, and be paid for meeting customer needs faster than anyone else."

It is continually emphasized that FCT results are obtained by fundamental change in organizational processes to achieve reduced costs and increased product quality. Again, it is not simply working faster; nor is it only associated with the introduction of new tools. While new tools may be introduced, this introduction follows from needs established by the process reengineering effort. On some occasions, a new tool or method may well be the driver of process change. But the major improvements in organizational responsiveness occur from process change for effectiveness. These changes are generally top-down directed from the level of systems management, rather than bottom-up directed for efficiency improvement from tools and methods for production.

Four basic principles support the FCT strategy:

1. Organizations and their strategies should be driven by the source of value-added revenue: paying customers.

2. FCT organizations should strive for continuous improvements in processes to produce results, and consider results that fail to meet expectations as symptomatic of process errors.

3. FCT organizational units are interdependent systems that are managed using cycle-time measures.

4. FCT organizations deploy their quick learning and rapid change abilities for competitive advantage.

FCT organizations are not structured in the traditional line and stafflike steep-horizontal structure. Nor is the structure that of a matrix, although the structure does resemble a matrix structure more than the traditional functional line organization. The major elements of the FCT organization include cross-functional and multifunctional teams that are networked both to each other and to centers of excellence. The organizational structure is focused on support for processes that result in high-value delivery to the customer. Organizational learning, and unlearning, are major by-products of the FCT organization. We could now suggest a hypothetical structure for an FCT organization. As we would then see from our later discussions, there are major similarities between a FCT organization, and an integrated product and process development (IPPD) organization, a horizontal management organization, and TQM and learning-focused organizations.

Considerable attention should be devoted to identifying an appropriate life-cycle process for FCT reengineering. This might comprise four generic phases, as illustrated in Figure 23.19:

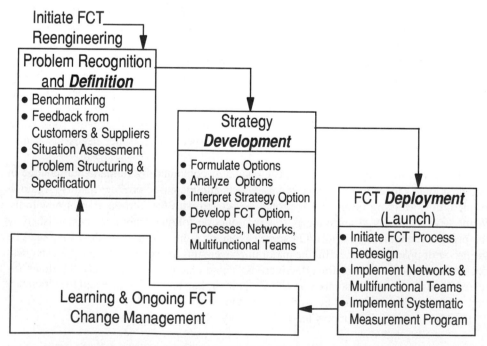

Figure 23.19 Four fundamental phases and learning for continuous improvement in fast-cycle-time reengineering.

- Recognition and definition of the problem.
- Development of FCT strategy.
- Deployment of FCT strategy as a set of organizational tactics.
- Ongoing change management and learning as the process changes are implemented, experience with FCT is obtained, and as the FCT reengineering process is continued for improvement in subsequent efforts.

A number of steps within this life cycle are also identified in the work by Meyer (1993) and are shown in this figure. One of the major issues in any sort of organizational change effort is that of obtaining the needed momentum and commitment of energy for successful implementation. Meyer also makes a number of suggestions to this end.

These studies are concerned with reducing the overall cycle time for a product or service. This is composed of the cycle time for RDT&E, product acquisition or manufacturing, and planning and marketing. Some works have seen specifically concerned, at least in part, with lead-time reduction for RDT&E efforts (Roussel et al., 1991; von Hippel, 1988; Howard and Guile, 1992). There have also been efforts specifically directed at reduction of acquisition, or production, lead time (Smith and Reinertsen, 1991; Millson et al., 1992). Robertson (1993) is specifically concerned with reducing market penetration cycle times. First, concerns relative to product development time, including RDT&E time as needed, are separated out from the overall cycle time. He suggests five guidelines to enable fast market penetration time:

1. Get the organization's product, or service, on the market first.
2. Announce forthcoming availability of the product or service before it actually reaches the market.
3. Innovate continually, with respect to both product and marketing.
4. Saturate the marketplace with multiple products to occupy the entire range of product differentiation and positioning foci, and develop wise scope alliances to establish the equivalent of a de facto standard for the product or service in question.
5. Evaluate and monitor market penetration in accordance with several customer decision stages: awareness, attitude formation toward product, initial product purchase and test, and repeat purchase by the same customer.

Necessary caveats, that must be associated with each of these suggestions, are discussed. It must also be recognized that, to some considerable extent in many cases, there is necessarily time overlap and concurrency in efforts devoted to the RDT&E, production, and marketing life cycles. This strongly suggests approaches that consider the interdependence of each of these life cycles, or alternate awareness of the associated concurrency that may exist. Concurrent engineering, integrated product and process development, and mass customization also address time issues. We will soon examine these approaches to reengineering.

23.3.10 Strategic Cost Management

The purposes of a management accounting system include the provision of timely and useful information to enable management of the affairs of an organization, including management control of the various organizational life-cycle processes. Many volumes have been written on the subject of management accounting systems. No one seriously questions whether finance and accounting systems are a very necessary part of an organization. On the other hand, there is concern that many management accounting systems are neither sufficient nor adequate to cope with contemporary technological, organizational, and societal change.

We can view the major purposes of organizational management and leadership as

- Definition of appropriate strategies, including the formulation, analysis, and interpretation of a variety of strategy options to enable selection and communication throughout the organization of the selected strategies.
- Development of tactics that implement the strategies, including the formulation, analysis, and interpretation of a variety of tactical options and selection and design of appropriate tactics for implementation as management controls.
- Deployment of these controls as operational task controls, obtained through another formulation, analysis, and interpretation cycle to result in implementation of task controls to ensure efficient and effective performance of operational tasks.

There is organizational learning (Argyris and Schön, 1996), hopefully at least, and iteration of the process for continued improvement. We may view this as a sequence of three life-cycle phases for the overall organization, much as we have viewed the other life cycles we examine here. Alternately, we may view it as a three-level hierarchy of strategic planning, management control or systems management, and task control as in the seminal work of Anthony (1998). Figure 23.20 illustrates each of these views of the organizational

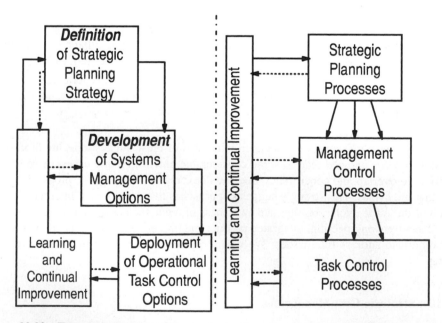

Figure 23.20 Two alternative views of organizational management systematic measurement needs exist at all phases and for all processes.

management life cycle. Obviously, there are many opportunities for systematic measurement, including accounting information, at all of the phases, including the learning phase, and for all processes in this life cycle.

In a definitive work, Shank and Govindarajan (1993) indicate that management accounting is not an end in itself but a means to help achieve organizational success. Furthermore, specific accounting systems must be evaluated in terms of the extent they support this end. They recognize that particular management accounting information constructs and information may not necessarily have, nor are they intended to have, knowledge that is of value for everyone. Nevertheless, each accounting system must serve some purposeful need in supporting humans in fulfilling some organizational role. Accounting systems need to be evaluated with respect to the ends they serve. A system appropriate for an organization attempting competitive advantage through being a low-cost producer may not be at all appropriate for one desirous of obtaining highly differentiated product status. Entirely analogous statements apply to systematic measurements in general.

Shank and Govindarajan pose three questions about accounting systems that we rephrase about systematic measurements in general:

1. Does the systematic measurement program serve organizational objectives and purposes?
2. In terms of the specific purposes each measurement serves, does the relevant systematic measurement process enhance possibilities of obtaining the supported objective?
3. Does the objective whose attainment is facilitated fit strategically within the overall organizational plan.

This suggests the relationship between cost ideas and cost information and the organizational hierarchy illustrated in Figure 23.20.

These authors approach what they term *strategic cost management* (SCM) from three perspectives:

- Value chain analysis
- Strategic positioning analysis
- Cost driver analysis

We discuss each of these related concepts here.

Porter (1998a–d) has identified the value chain of an organization as the linked set of value-creating activities that exist throughout the organization. The links of this value chain begin with material and subsystem acquisition from suppliers, and include such other linkages as RDT&E, production, marketing, and delivery of a product to the customer, and associated maintenance. The value-chain concept is intended to be a broad, externally based concept that also includes the value chain of the other units that interface and interact with the organization. It is not just an internal concept based only on value-added notions. It involves much use of life-cycle costing throughout the various organizational processes, and consideration of the effects of possible changes in these on other processes. The approach suggested by Shank and Govindarajan for construction and use of value chains is interpreted as follows:

1. Formulate the organization's value chain, detect costs and revenues that are associated with the various value-chain activities, and identify potential competitive advantage options.
2. Analyze the value chain such as to diagnose the cost drivers that cause the various costs and revenues associated with value-chain activities.
3. Interpret the results of the analysis such as to select an appropriate and sustainable competitive advantage option for implementation. Accomplish this through implementation and control of cost drivers that are better than those of competitors, or through restructuring of the value chain itself if no appropriate cost drivers result from the initial formulation.

We see that value-chain concept is not at all independent of either the strategic positioning or the cost driver concept.

There are three basic strategic positioning strategies:

- *Building,* for increased market share.
- *Holding,* to retain an organization's present market position.
- *Harvesting,* to maximize short-term earnings often at the expense of market share over the longer term.

The most appropriate strategy in any given market depends upon whether the organization obtains competitive advantage through low cost or high differentiation of the product in question. These three market strategies may be disaggregated into a number of others. In particular, we can split each of them up into broadly focused or selectively (narrowly) focused strategies. This enables us to concurrently adopt more than one strategy. For example, we can simultaneously pursue a selective harvesting in one market area while building or holding in a different market segment.

Generally, strategic planning is more important for a building strategy than it is for a harvesting strategy. On the other hand, more formal financial analysis and decisions are needed for harvesting than for building. The criteria for capital expenditures depend on information that is not strongly financial, such as that relating to market share and efficient use of RDT&E expenditures. For harvesting, there is much more interest in cost–benefit and other strictly financial information. Holding strategies are in the middle relative to these considerations. In a similar way, the compensation system of a building market strategy organization depends upon mostly non-cost-related criteria and subjective judgments associated with success through a building strategy. For a harvesting organization, compensation based on financial formulas is recommended.

Generally, a high-product-differentiation focus is associated with greater risk than low-cost focus due to three factors:

- Technological innovation is needed for differentiation.
- Broader mass-customization-type product lines are often needed for differentiation.
- Higher prices associated with highly differentiated products may be difficult to sustain.

Changing environmental conditions, internal and external, also influence strategy choice or mission. Finally, the market scope or focus of the organization needs to be considered.

Thus, we need to devote attention simultaneously on mission and competitive advantage needs, as suggested by the authors.

We could also consider the interaction of the following factors on strategic costs and competitive advantage for each of these three market positioning strategies:

- Strategic mission
- Competitive advantage
- Competitive scope or focus

It seems apparent that a building strategy is difficult to accomplish for an organization with a low-cost product-focused competitive advantage. This would seem to be even more the case where the product is focused on a broad and potentially large market. In a similar way, a harvesting strategy is difficult and risky to bring about for an organization whose competitive advantage is based upon product differentiation. This becomes even more difficult when the market scope is narrow. Figure 23.21 suggests the 12 cells that exist, many of which need to be considered in determining organizational controls and cost and revenue management systems, for implementation of one of the three possible strategic missions. Each of these cells needs to be analyzed in order to determine cost and other competitive strategy facets, such as risk.

The strategic competitive advantage position of the organization determines the appropriate ways to use strategic cost and management accounting information. As we recall

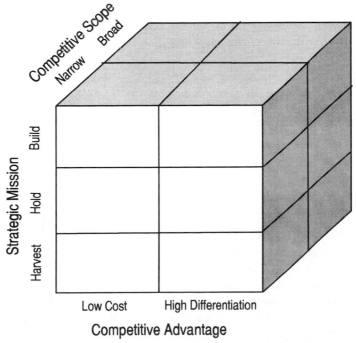

Figure 23.21   Illustration of twelve cells influencing strategic mission determination and associated organizational controls.

from our discussion of Michael Porter's competitive advantage concept, this may range from cost leadership to product differentiation. If, for example, the organization competitive advantage for a given product is associated with high product differentiation, then the role of marketing is crucial since the product must be targeted and represented very correctly. One the other hand, if the organization is attempting competitive advantage through cost leadership, then the most important cost management issues are those that directly affect product cost, such as manufacturing and engineering costs and competitive product pricing. There are, of course, a number of noncost or nonfinancial performance measures that need to be considered as well as financial ones in determining an appropriate strategic mission in any given situation.

Cost drivers are the third important ingredient in strategic cost management. Among the suggested strategic cost drivers or factors are

- Scale of the investments needed in the life-cycle processes for RDT&E, acquisition or production, and planning and marketing.
- Scope of these processes in terms of vertical integration.
- Technological leadership and the extent to which new and emerging technologies are used throughout the organization's value chain.
- Experiential familiarity of the organization with use of these processes for this product.
- Complexity and breadth of the products and services to be offered to customers.
- Capacity utilization needed to produce at a given output level.
- Plant layout efficiency.
- Workforce empowerment and effectiveness.
- Process and product quality.
- Product effectiveness, in terms of function and purpose.

The generally accepted way to determine cost is to assume that it is some multiplicative function of the various factors, as weighted by importance; thus we obtain

$$\text{Cost} = C_0(F_1)^{m_1}(F_2)^{m_2}(F_3)^{m_3}(F_4)^{m_4}(F_5)^{m_5}(F_6)^{m_6}\ldots$$

where C_0 represents the nominal cost of some basic product. This can be written in the product form, where there are N cost factors

$$\text{Cost} = C_0\prod_{i=1}^{N} F_j^{m_i}$$

and the cost driver analysis proceeds in a standard way. Product line and product complexity are each major sources of the factors that determine product cost. The authors of this work suggest activity-based costing (ABC) (Burk and Webster, 1994; Hicks, 1992; Kaplan and Cooper, 1997) as an adjunct to this approach for cost driver analysis. The ABC approach and the similar work breakdown structure, described in Chapter 6, are very useful in estimating costs of systems engineering and management activities. In either approach, such factors as the cost of low quality and the cost of technological innovation can be included in the results of the strategic cost analysis.

23.3.11 Strategic Sourcing: Outsourcing and Insourcing

Several of our commentaries have mentioned the notion of outsourcing to obtain component products and services by the organization responsible for a product or service. Outsourcing can be mandated when the component product or subsystem needed simply cannot be produced by the organization needing it. In many cases, however, whether or not to outsource is a matter of choice. The choice should be based on the myriad issues that are involved in strategic policy and planning determination. The basic issue is that of determining the extent to which a necessary part of a product or service could be obtained more efficiently and effectively from inside or outside the organization itself. An issue of equal, and perhaps greater importance in some cases is whether insourcing or outsourcing support or inhibit the organization relative to related efforts that are also of importance. The strategic sourcing question is, therefore, a very important one. To answer this it is necessary to consider extrapolation of the make or buy decision to include strategic systems management and process-related considerations that do really need to be examined for almost all make or buy decisions.

The principles supporting strategic sourcing decisions are conceptually simple and nicely stated by Venkatesan (1992):

1. The organization should focus on those components and subsystems that are crucial to the product itself and where the organization has critical core capabilities (Prahalad and Hamel, 1990; Stalk et al., 1992; Meyer and Utterback, 1993; Hamel and Prahalad, 1994) that support the efforts required and that the organization desires to sustain. This enables an organization to exercise judgment concerning subsystems that are strategic and those that are nonstrategic. It potentially eliminates difficulties that result from conflicting priorities and sourcing decisions.

2. Components and subsystems should be outsourced where potential suppliers with a distinct competitive advantage at producing these exist. These competitive advantages could be either those of lower-cost producers or higher subsystem differentiation.

3. Outsourcing should always be used in such a manner that it supports continuing employee commitment and empowerment. It is necessary to outsource this in such a manner that there is minimum opportunity for exploitation and hollowing of the organization, including its people, by the external supplier.

These principles lead to a process for strategic sourcing, and Venkeatesan suggests a multistage process for strategic sourcing. Figure 23.22 presents our interpretation of this.

1. The first phase of the strategic sourcing process is actually concerned with subsystems sourcing decisions. The organization distinguishes possible options for strategic sourcing at the subsystem level and distinguishes strategic subsystems from nonstrategic subsystems through a formulation, analysis, and interpretation process. A major architectural issue in doing this is that of determining the proper architectural level of abstraction for the various options that will be subject to analysis. This architecture can comprise individual components or entire large subsystems. In exceptional circumstances, an entire product or system might be outsourced and the parent organization may act as a marketing agent only. The nonstrategic subsystems generally comprise mature technologies for which there exist an external supplier with outstanding competitive advantage relative to their

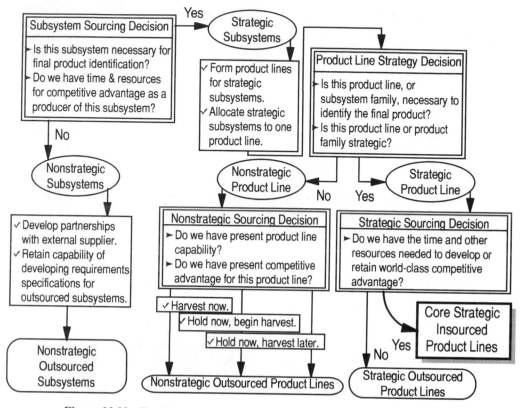

Figure 23.22 Decision-oriented representation of the strategic sourcing process.

production. The strategic subsystems generally comprise those subsystems that the organization has significant reason for retaining or for developing the capabilities needed to produce them.

2. In the second major phase of the strategic sourcing process, the strategic subsystems identified in the first phase are grouped into a number of families, each of which have common product line requirements. Each product line, or process, is able to produce the entire family assigned to it. The major decision made at this phase is whether each of these product lines is strategically important to the organization. If the answer is no, then this product line is an appropriate candidate for either a hold or a harvest decision relative to retaining the existing capability at this product line in the firm.

3. The final phase in our interpretation of this strategic sourcing process results from the need to deploy the strategic product line decision outcomes and the nonstrategic product line outcomes reached by the decision at the previous phase. Thus, this phase can be described in two parts, or subphases, depending upon whether we are dealing with strategic or nonstrategic product lines.

    a. We need a subphase that represents deployment for nonstrategic product lines. If we have present capabilities and competitive advantage, we deploy a holding strategy for the present and identify plans for gradual harvesting, potentially over a sustained time period. If we have present capabilities but are not competitive, we adopt a holding strategy for a very short period and then harvest. On the other hand, if our present product line capabilities are incapable of coping with present needs, we immediately outsource this entire product line.

b. If the product line is of strategic importance to the firm, we should deploy necessary resources and time, if feasible, in order to either hold or build the requisite capabilities. If either the time or the resources are unavailable, we outsource the associated product line for which this is the case. If the resources and time are available, we insource and hold or build the requisite capability to retain or obtain world-class competitive advantage. Thus this subphase is concerned with sourcing decisions for strategically important lines.

These three phases are associated with four significant questions and answers. The first two are obtained relative to

- Strategic subsystems
- Nonstrategic subsystems

The strategic subsystems are aggregated into a generally smaller number of product lines. These product lines are then partitioned into

- Nonstrategic product lines, which are outsourced, at least over time
- Strategic product lines

This leads to two additional questions and answers. The strategic product lines are insourced wherever possible and outsourced only when there is insufficient time and other needed resources to bring about world-class competitive advantage.

As a result of this, we have nonstrategic outsourced subsystems, nonstrategic outsourced product lines, and strategic outsourced product lines. These sourcing strategies and decisions will need to be reexamined over time. In particular, it will often be desirable to convert an initial decision to outsource strategic product lines to an insourcing decision when and if resources and timing becomes supportive of such a decision. This may occur because of technological or other changes. Indeed, we see that sourcing questions are fundamentally reengineering questions as well. There are other approaches that are similar to the one discussed here. A step-by-step approach for sourcing, including the development of performance assessment and decision-making guides, is presented by Maromonte (1998). RDT&E, production, and purchasing services are included, as well as for other enterprise functions.

In works concerned primarily with information outsourcing (Lacity and Hirschheim, 1993a, b), three generic types of outsourcing are identified:

1. *Body-shop outsourcing* is a way to meet short-term demands that cannot be met by people internal to the organization even though the decision would otherwise be favorable to insourcing.
2. *Project management outsourcing* involves the use of external suppliers to furnish a subsystem or service activity, such as training. This would seem closely equivalent to product line or subsystem outsourcing.

3. *Total outsourcing* exists whenever an external supplier is responsible for all, or a very major portion of, a complete turn-key-like information system function.

Each of these outsourcing options needs to be considered in the development of an effective strategic sourcing decision. There are two major issues to be considered:

- *Transactions costs* for outsourcing are composed primarily of direct production costs and indirect coordination, or support, costs. Transactions, as exchanges of resources, may be described in terms of frequency and degree of specificity, or customization.
- *Organizational systems* of influence and culture are composed primarily of ingredients that may support or inhibit information system influences on the organization. We have said much about these issues in our last chapter and in Section 8.2.3 on process innovation.

Lacity and Hirschheim indicate six primary reasons for considering outsourcing possibilities:

1. We desire to improve operational efficiency. This improvement can be obtained by determining whether the information system service can be externally supplied in a more effective manner than it can be internally supplied.
2. We need to acquire resource capabilities, such as hardware or expertise, not available internally to the organization and perhaps to support a new start-up organization or an organization in a crisis situation relative to information system unfilled needs.
3. We desire to replicate the success stories of others who have outsourced their information system functions.
4. We wish to reduce uncertainties and risks associated with unpredictable demand for products and services.
5. We wish to eliminate a potentially troublesome function or process and let others cope with it.
6. We wish to enhance credibility for an organization, unit, or individual.

These reasons were obtained from a survey of senior management in some 13 organizations and specifically concerned information systems outsourcing. This behavior would appear to be much more descriptive than either normative or descriptive. It also may be reflective of the fact that information systems may well be regarded as very necessary but either as nonstrategic subsystems or nonstrategic processes in many organizations.

The major conclusions from these studies are as follows:

1. Commonly available sources may paint an overly optimistic picture of information system outsourcing benefits.
2. Information system outsourcing may be symptomatic of inherent difficulties in demonstrating the value of information system capabilities to non-information-system top managers.
3. Cost-efficiency improvement may well not be a dominant reason for implementation of information system outsourcing.

4. The internal information system department may not be inherently more efficient than an external supplier.

5. An internal information system department may achieve results of comparable or greater efficiency and effectiveness than an external vendor by internal reengineering of the information system structure and function.

6. The contract with an outsourcing information system supplier is the only mechanism, as contrasted with partnership considerations, that may ensure accountability and satisfaction of outsourcing expectations. Outsourcing suppliers will only become partners when and if the profit motive is shared.

7. The belief that an information system is a nonstrategic utility, commodity, or process is erroneous.

The bottom-line summary message of these authors is that one really cannot outsource the management of information systems. There seems to be much agreement with this, especially as concerns information support for the highest level managerial decisions (Benjamin and Blunt, 1992; Boynton et al., 1992; Willcoks and Lacity, 1998). Particularly when there is considerable outsourcing to external suppliers, there will be a major mandate for very careful integration of all aspects of the supply chain and an understanding of the relationships between the supply chain and the value chain.

23.3.12 Mass Customization

Mass customization is also concerned with quick responsiveness. It also calls for flexibility in turbulent and changing environments by continuing to adapt to changing customer needs and in being able to supply these needs with high-quality and trustworthy products and services. The focus is upon highly customized goods and services that fill these needs. Thus, it is an approach for a line of adaptable highly differentiated products.

A text by Pine (1993) and associated companion works (Pine et al., 1993; Gilmore and Pine, 1997) provide much insight into this process. The need for mass customization comes about because of the usual mass-production process and its assumption of long product development life cycles and the resulting low-cost, consistent-quality, low-differentiation product that results from this. This may be an acceptable approach in a homogeneous market with stable demand. In the mass-production system, a producer enters the marketplace with a low-cost consistent-quality nondifferentiated product, and a homogeneous mass market with stable demand is said to result from this. As long as there is no alternative product available, these dynamic influences will continue and will result in the long product development life cycle that is necessary to support the mass production and low product cost.

Pine indicates that this mass-production system ran into trouble in the 1960s. The difficulties exacerbated in the 1970s, but with management becoming very able to detect this only in the 1980s. While the difficulties were detected, the attachment to mass production were so great and so culturally ingrained that the causes of the difficulties were not then diagnosed. Corrective attempts began in the 1990s, and this has resulted in efforts to change to a new set of guiding principles and rules. While it is indicated that this has led to continuous improvement schools of thought and such emphasis areas as total quality management, it is stated that these continuous-improvement approaches do not generally allow for questioning the basic design tenants of the products being built and the hidden assumption that the design specifications are really indicative of what the customer wants. Instead, they tacitly assume markets that are relatively stable and predictable.

It is suggested that mass customization goes one step further by allowing for high-velocity markets and environments, rapidly changing technologies, and associated changing of customer requirements and needs. Continuous improvement, through processes like TQM, are viewed as generally necessary for mass customization. The change is presumed to require augmenting, and perhaps even disbanding, cross-functional teams with teams obtained through the forming of dynamic networks. Whereas the mass-production system encouraged vertical integration of teams, each of which performs an isolated incremental task, continuous improvement generally replaces vertical integration needs with horizontal integration needs. In horizontal integration, individual teams have responsibility and authority over task control areas.

This trend is said to continue with the advent of mass customization, and is associated with the organizational structure becoming more horizontal. Processes, or product lines, become flexible modules that are coordinated and potentially linked together in a dynamic networklike structure. This is said to reverse the causal dynamics of the system such that it becomes possible to respond to turbulence in demand changes and heterogeneous markets. This leads to recognition of the need for high-quality customized products and mass-customized processes that are responsive to this need.

While many, if not most, economic and competitive structure paradigms would associate high price with high-quality, rapidly changing, and highly differentiated products; the mass customization approach is said to accommodate these differentiated product attributes at a low product cost. In mass production, a product is designed first and the processes that will eventuate in the product are developed. In mass customization, this flow is said to be reversed; processes are generally created first and support the rapidly changing characteristics of the products that follow. This enables a low-cost highly differentiated, set of products that appeal to a variety of niche markets.

In Gilmore and Pine (1997), faces of mass customization are identified. The product or service may or may not be customized for individual customers. In a similar way, the representation of the product or service may be changed or not changed for individual customers. This leads to four approaches to customization, and four types of customizers are identified:

1. *Collaborative customizers* work with customers to help them identify needs and the sort of customized products that will fill these needs. This is the usual customer most often associated with mass customization.

2. *Adaptive customizers* provide a standard product that can be adjusted by users to best fulfill their individual needs. Many commercial software products fit into this category.

3. *Cosmetic customizers* present a standard approach to potential users in different ways, depending upon the needs of the user. Ballpoint pens that can be engraved with a customer's company name are cosmetically customized products.

4. *Transparent customizers* are able to provide customized and unique products and services to users without their explicit knowledge that these have been customized for them.

These four approaches to customization are used to suggest a framework to enable the engineering of customized products, or services, or processes.

McCutcheon et al. (1994) note the potential conflict between customization and rapid response. This results because rapid product acquisition is generally dependent upon much standardization of the production processes and products. Customization requires a great deal of innovation and flexibility. Anticipatory use of knowledge perspectives and forecasting of demand can potentially result in the production of highly differentiated and customized products before there is an actual demand for them. While this will reduce the response time, there are obvious risks involved if the anticipated market does not materialize.

These authors note that customization and product differentiation can be infused into a product early or late in the acquisition life cycle. If differentiation occurs early in the production process, it is more difficult for an organization to be responsive than if it is possible to accommodate differentiation at a late stage in the life cycle. If differentiation must occur early in the life cycle, if customers demand quick delivery, and especially if there is a high degree of customization required; then a "build-to-forecast" (BTF) approach is appropriate, though it is necessarily associated with risk. Otherwise, make-to-stock (MTS), make-to-order (MTO), or assemble-to-order (ATO) approaches, which are inherently less risky, are more appropriate.

A three-phase process is suggested:

1. Analyzing customer expectations.
2. Assessing the organization's capabilities.
3. Selecting and implementing appropriate tactics.

This process can lead to one of six possible options to cope with the customization responsiveness squeeze:

1. Change the design of the production process through use of concurrent approaches, cellular manufacturing, and other flexible process technologies.
2. Change the product design through use of such approaches as computer-aided design, and standardized subsystems that may be integrated together to produce a variety of customized products.
3. Demand management may be improved through better forecasting approaches such as reducing the very significant inventory and other difficulties that occur when demand is forecast incorrectly.
4. Supply management can be improved: internally, through development of insource capabilities to manufacture long production time subsystems, and through development of a production management approach that is tuned to customer demand; and externally through altering arrangements with suppliers to provide for rapid delivery of outsourced subsystems.
5. Slack resources, when they occur, can be used in an optimum manner.
6. Build-to-forecast strategies may be based on production orders that are based on periodic demand forecasts; incoming customer orders that can be adapted to nearly finished products that can be finished completely; and retrofitting approaches that may be used to make late life-cycle phase modifications to products in order to accommodate customer demand in a rapid manner.

These options are not at all mutually exclusive, and they may be blended with other existing practices.

The relations between continuous improvement, innovative process reengineering, and mass customization are interesting. Although continuous improvement is necessary for successful mass customization, they are two rather different concepts. Continuous improvement and innovative process reengineering emphasize the redesign of existing products and processes, while mass customization is focused on the formation of new products and services, primarily and whenever possible, from existing processes. In a mass customization effort, organizations become dynamic networks and managers become coordinators across a horizontal structure. This is orchestrated such as to facilitate the recombination of, and incremental improvement in, existing processes to yield products and services that satisfy customer demands. Success in mass customization depends upon incremental improvement and coordination of existing processes. Ideally, this coordination process is seamless, in the sense that organized product teams must work well with one another from the start and be integrated. Information technology plays a key role in enabling this. The major potential achievement of information technology is not at all that of automating mass production, but an ability to empower people and to fully utilize their capabilities. Mass customization is thus enabled by integrated product and process teams, which we discuss later in this chapter, and information technology and knowledge management, which we discuss in Chapter 30.

Fundamentally, the judgment relative to the degree of customization that will be accommodated for any given product line and product is a very strategic one. Thus, we see that the seemingly oxymoronic term *mass customization* is therefore both appropriate and descriptive of a concept well worth detailed exploration for specific applications. Some specific applications of mass customization to agile product manufacturing are considered in Anderson (1997).

23.3.13 Market-driven Strategy

In a work on market strategy, Day (1990) also notes that new specialty niche market opportunities appear as the era of mass marketing reaches its final days. This is said to suggest that conventional understandings of organizations and their strategies for competitive advantage necessarily must be augmented, if not replaced, by a market-driven approach to competitive strategy. The new competitive strategies should result in choice of:

- Markets to be served and customer segments to be targeted.
- Product differentiation for competitive advantage.
- Communication and distribution channels to realize market potential.
- Scale and scope for support activities to convert competitive strategies into competitive success stories.
- Future growth areas and strategies for continued success.

These choices are driven by three contemporary realities:

1. Information technologies enable major changes in the way organizations operate. They may, and generally will, accelerate the pace of decision making, and the results of this may often blur boundaries between organizations and markets.

2. Rapid technological change and innovation reduces the time during which a product may be successfully deployed. This occurs because of customer demand for increasingly higher levels of product functionality and quality. Technological change potentially enables shortening the development life-cycle time as well and the resulting improvement in responsiveness.

3. Former mass markets are becoming fragmented into narrow specialty, or niche, markets that may be global in scope.

Day emphasizes the major role for a shared strategic vision, which is driven by the need to be responsive to customer requirements and market needs, in realizing successful strategy.

According to Day, *strategy is a set of integrated actions implemented for the pursuit of competitive advantage.* He is very concerned with planning processes that yield adaptable and effective strategy. A successful strategy comprises

1. Definition of the organization and strategic strengths that comprise the basis of competitive advantage.

2. Development of an associated set of organizational objectives and investment strategics that specifies both resource sources and uses.

3. Deployment of these strategies in terms of functional programs that support the strategies.

Several measures of effectiveness, stated in the form of four test questions, are suggested to determine soundness of identified strategy options:

1. Are the assumptions used to formulate and analyze the strategy option valid?

2. Is implementation of the option feasible, supportable, and consistent with other organizational strategies?

3. Will the strategy option create and maintain competitive advantage, through either low cost or product differentiation?

4. Do potential returns from implementing the option justify the risks of failure, is the strategy option associated with unacceptable internal or external risks, and have appropriate risk management strategies been identified?

These very useful questions and the associated measures of effectiveness, which are not independent, relate well to the

- Meaningfulness of the inputs used to generate the potential benefits.
- Amount of the competitive advantage results created by implementation of the option and the associated risks.
- Question of whether the option itself is consistent with other aspects of organizational mission.

Day suggests benefits and disbenefits to both bottom-up incremental planning and top-down strategic planning, and indicates that a balanced or adaptive blend of these two is generally superior to either one would be, when taken alone and implemented on an exclusive basis. Mintzberg expresses similar conclusions in recent works on strategic planning (1994a,b) in his indications that strategic planning, as it is often practiced, is not strategic thinking but rather the analysis and programming of already existing strategies. He suggests that strategic thinking and the resulting strategy change require the synthesis of new approaches and not the incremental rearrangement of old ones. He rejects detachment from real problems and overformalization of approaches, and suggests that strategic planners should be strategy finders and not just analysts and forecasters. This "finding" is associated with "learning by walking around" and other bottom-up approaches to understanding organizational realities and customer needs. This sort of adaptive planning approach to strategy (Mintzberg and Quinn, 1998) enhances the ability to learn from experience.

There are four essential phases in Day's adaptive planning strategy:

1. Situation assessment involves identification of current strategies and their strengths and weaknesses in terms of the market and general environment, present and potential competitors, and organizational resources and objectives. It also involves examination of the present and potential future market and identification of potential threats and opportunities.

2. Strategic thinking involves the generation, or definition, of optional courses of action, or strategies. It also includes analysis and choice of options for further development.

3. Negotiation and decision making is associated with the efforts involved in the development of the selected options in terms of resource allocations for specific tactical efforts.

4. Implementation represents the activities undertaken to deploy the developed tactics.

The first two of these phases appear equivalent to the generic definitional phase of a planning life cycle. The other two phases are directly analogous to definition and deployment. Adaptive learning occurs in that the various phases are exercised both in a top-down and in a bottom-up fashion.

Following these phases come efforts to understand the competitive situation and the nature of competitive advantage. An organization may place different degrees of emphasis on customers and competitors. If it emphasizes neither, it is self-centered; if it places a major emphasis on customers and minor emphasis on competitors, it is customer oriented; and if it places major emphasis on competitors and only minor emphasis on customers, it is competitor-centered. The more appropriate realization of an organization's competitive advantage is obtained when it places major emphasis on both customers and competitors. These are denoted as *market-driven* organizations.

Customer-oriented assessments and competitor-oriented assessments are each important. Included in this are customer perceptions of the competitive position of our organization vs. that of competitor organizations. Value chains, including supplier value chains, customer value chains, and organizational value chains, are each important. Market considerations, related to customers and competitors, are the driver and integrator of these value chains and assessments alike. This results in something like the 12 competitive strategy cells we illustrated in Figure 23.21, and resulting suggestions for development of a competitive strategy option. All of this results in the decision concerning how to compete. Deciding where to compete and gaining access to the selected markets are discussed as well. The bottom-line message of this work is that strategies based on understanding and implementing the strategy for developing an appropriate market leadership position, continually seeking new approaches and methods that result in competitive advantage, and measuring progress against achievements made and using this information to guide future efforts will when taken together assure considerable market-driven success.

23.3.14 Horizontal Management

There have been an abundance of recent studies, such as Dertouzos et al. (1989), which identify critical attributes among what are called *best-practice firms:*

- Concurrent improvement in quality, delivery, and cost.
- Closer interaction with customers and suppliers alike.
- Effective use of technology for strategic advantage.
- More flexible organizations that are less hierarchical and compartmentalized in order to give employees greater responsibilities.
- Continuous learning, teamwork, participation, and flexibility through enlightened human resource policies.

These identified critical success factors are of particular interest here, as they relate most strongly to systems reengineering efforts. Other authors reach similar conclusions. In very thought-provoking works, Thurow (1992, 1996) discusses the technological and economic forces that are shaping the world of today and tomorrow. He provides five suggestions concerning a game plan that amounts to reengineering:

1. Develop abilities to cooperate effectively with direct competitors.
2. Focus on process innovation as contrasted with product innovation.
3. Encourage intense development of seven key technologies.[2]
4. Provide a major focus on redeveloping a highly skilled and educated workforce.
5. Emphasize both individual and team achievements.

Most of these suggestions are centered on the rediscovery of quality in process and product, and enhancement of manufacturing processes through effective systems management and through advantageous use of information technologies. These findings indicate that we need to rediscover the importance of manufacturing. They suggest that it is a counterproductive myth that there can be a successful postindustrial economy that is not based to a very large extent on the manufacturing of high-quality technological products and systems (Cohen and Zysman, 1987). While the information, and/or knowledge, technologies can be expected to play a truly major role in augmenting support for advanced and appropriate manufacturing, they will in no way replace manufacturing. Of course, manufacturing efforts must often be reengineered to assure a competitive edge.

Various other contemporary investigations have suggested strategies for reinventing the factory (Harmon and Peterson, 1990) so as to enhance participation in the globalized

---

[2]These seven key technologies are microelectronics, biotechnology, new materials, civilian aviation, telecommunications, robots and machine tools, and computers and software.

markets. Competitiveness through increased advancement of emerging information technologies is a major thrust of many of these works. In Brandin and Harrison (1987), for example, it is argued strongly that information technology studies and developments extend much beyond the neoclassic engineering of data processing to incorporate intellectual property laws, public- and private-sector policy considerations, and economic and systems management considerations. Two of the major strategies suggested in this effort relate to redefining the technology base to include information as an essential ingredient and leading in the development and application, often through technology transfer, of new and emerging technologies. There are many similar discussions of information technology in organizational settings (Bloomfield et al., 1997). The discussions concerning information technology and knowledge management in Chapter 30 are very relevant to reengineering efforts, just as they are for all of systems engineering and management.

There are a number of important suggestions in these efforts, particularly as related to the importance of quality, human resources, and processes. Our emphasis in this subsection is upon the need for flatter organizations and the use of these flatter organizational hierarchies, or structures, to enhance organizational productivity through information technology and knowledge management, the abolition of the steep organizational hierarchy, and the teamwork of ad hoc, cross-functional teams.

In two inspiring works, Savage (1990, 1998) illustrates five recognitional stages, denoted as "days," in the life of many contemporary organizations. These findings may be described as follows:

1. The organization is organized into a set of hierarchically related personnel and applications. These applications carry such titles as R&D, engineering, manufacturing, sales, service, and accounting. They report in a traditional line structures and the various functional units do not interact.

2. In order to cope with the need for interaction, various application groups are set up. This creates a necessary linkage, or network, between one functional unit in the organization and the others that are needed for a particular application, such as product development or product marketing. Since the people in the various functional units cannot communicate well, or even at all, with one another, as they speak different languages than those in other functional units, a "translator" or "expediter" is needed. Since there are many applications in a given organization, the network linkages become numerous, as does the need for expediters and translators.

3. The difficulties of working in parallel across functional units become apparent, and ways are sought to cope with the resulting complexity. Someone suggests having customer expectations as the thematic drivers of considerations that relate to such nonfunctional efforts as process, quality, market, and service.

4. Concerns arise with respect to how the various cross-functional teams are to be managed. Organizational vision is suggested as the monitor and controller of the cross-functional teams through the resultant strategic plans, organizational mission statements and objectives, and realistic management controls. Knowledge is recognized as a valuable resource in this regard in terms of various "knows." This knowledge is responsive to the same sort of questions used in benchmarking, except that it relates to a common knowledge base and capability for describing the various elements needed for each of the applications for which a cross-functional team is responsible. Thus, a knowledge base is needed in terms of

- Know why
- Know what
- Know which
- Know who
- Know where
- Know when
- Know how

This represents the organizational knowledge base. It is what the various cross-functional teams bring to bear on various applications, such as product development. It suggests a role for the original departments as "centers of excellence" or repositories for critical core capabilities, or "virtual resources," but not as actual working line units.

5. In the last stage of development, the potential fragmentation of the organization due to the cross-functional teams is dealt with in terms of strategies for accountability, focus, and coordination. This leads to strategies for integrating the organization through human networking in such a way as to build a continual learning capability. These networks are not just informal networks of humans communicating with one another. This capacity is augmented by networks of information processing systems that enable interrelating various knowledge patterns for enhanced capability and competitiveness.

The strategies for human networking and enterprise integration result from the reality that the traditional resources for production—land, labor, and capital—are now augmented by a most important information and knowledge resource.

It is primarily this that has led to the major need to replace the traditional steep hierarchical structure found in most organizations by cross-functional teams and human networking as to enable people empowerment and enterprise integration for enhanced responsiveness and competitiveness. The need is *not* to computerize steep hierarchies. Figures 23.23 through 23.26 indicate the evolution from first- through fourth-generation management. In second-generation management, steep hierarchies are introduced, and we show just a simplified representation of one in Figure 23.24. Matrix management, or third-generation management, accomplishes the change in organizational structure to enable horizontal communications through adding additional management complexity. In a sense, this is accomplished in fourth-generation management without the additional management complexity through networking the organization. This involves the major use of information technology and the need for integrating information technology products and services into organizational environments such as to enhance knowledge management and intellectual capital.

There has been much contemporary discussion of this augmentation of the traditional physical resource base with the information and knowledge base. The *Coming Post-Industrial Society* (Bell, 1973) was perhaps the first to indicate this shift some two decades ago. Zuboff (1988) uses the term *informate* to describe the effort of humans in simultaneously working at and on multiple levels of abstraction, each suited to a particular purpose. Such efforts will be prototypical and characteristic of the efforts of many in the emerging knowledge-based networked society. There are a number of useful contemporary works that discuss such topics as the emergence of top-management computer use (Rockart

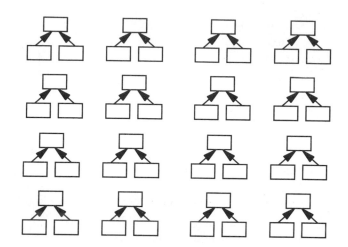

Figure 23.23 Simplified illustration of many individual proprietorships.

First Generation Management - Individual Proprietorships Made Possible by Land and Labor

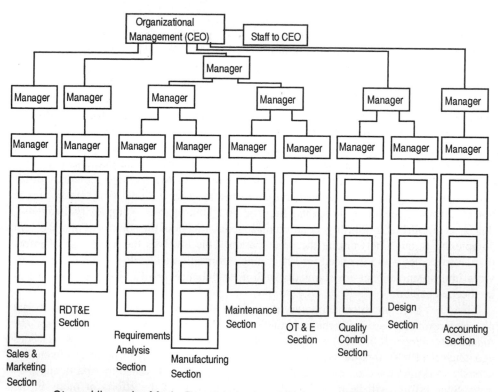

Steep Hierarchy Made Possible by Land, Labor, and Capital

Figure 23.24 Simplified illustration of steep organizational hierarchy.

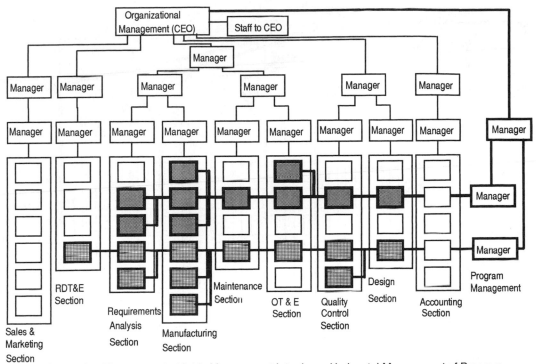

Third Generation Management - Matrix Management Introduces Horizontal Management of Programs

Figure 23.25 Simplified illustration of program management superimposed on steep hierarchy.

and DeLong, 1988), business design through information technology (Keen, 1991), and managing information technologies in the 1990s (Harvard Business Review Editors, 1990). More recent works are concerned with information technology and organizational transformation (Morton, 1991), information technology as an integrating force for such organizational efforts as marketing (Blattberg et al., 1994), strategic planning for information technology through information movement and management (Boar, 1993), and how to increase organizational competitiveness and efficiency through use of information as a strategic tool (McGee and Prusak, 1993). Peters (1992) has denoted the effort to integrate information technology and organizational considerations, and organizational needs for competitive positioning in a rapidly changing environment, as *liberation management.*

In his works, Savage describes five generations of management organizations. The first of these is based on individual proprietorships and resulted from the use of two resources: land and labor. While organizations might take on a hierarchical appearance, they were not steep hierarchies. They did not need to be since the capital available to an individual proprietor would not allow this. In the later portions of the Industrial Revolution, large amounts of capital became available. Without major knowledge resources, steep hierarchies naturally evolved as the large organizational model of choice. This represents the second generation of organizational structure. With recognition of the difficulties brought about by steep hierarchies, basically that of coordination across the various functions in the hierarchy, matrixed organizational structures resulted as third-generation management

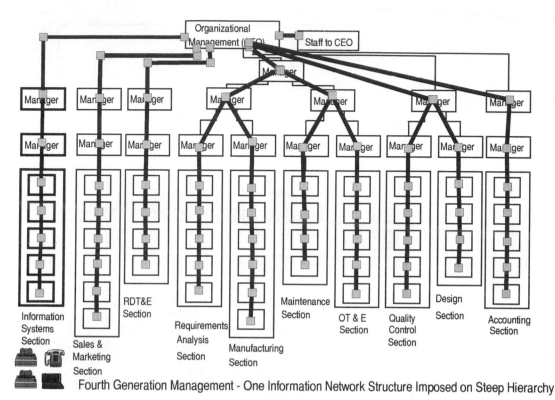

Fourth Generation Management - One Information Network Structure Imposed on Steep Hierarchy

Figure 23.26 Fourth-generation management through information technology and networking.

organization structure. These matrixed structures brought about problems in that the notion of one person but more than one boss of that person creates many difficulties. This has led to the general demise of matrix management structures.

Networking is suggested as a fourth-generation management remedy for the dilemmas brought about by steep hierarchies and vertical management. In fourth-generation management, horizontal and vertical communication linkages are established for what was initially the second-generation model of management structural organization. This is not really enterprise integration in that the organization is not truly integrated, except in a narrow technological sense by the wires and software that one person uses in interfacing with another. It is really just computerizing a steep hierarchy, with perhaps 10 to 20 layers of management. Savage identifies six issues that generally emerge from fourth-generation management:

1. Ownership of information issues, as information becomes "turf" in a steep functional hierarchy.
2. Managed and massaged information system issues, as various functional units present selective information that best supports their unit.
3. Hidden assumptions imbedded in various software representations of levels of abstractions and associated information presentations.
4. Inconsistent terms and definitions across different applications, which are due to lack of organizational standardization of information architectures and dictionaries.

5. Accountability and social value of information.
6. Organizational information politics.

Some of these are not inherent issues in information technology and knowledge networking, but are aided and abetted by retention of the steep hierarchy form of management. It would appear that these are very related to and influenced by organizational cultures and the information cultures examined earlier in the process innovation approach of Section 23.3.3. Figures 23.23 through 23.26 are representations of the first four generations of management structure evolution. Second-, third-, and fourth-generation management each involve steep hierarchies. Third-generation management attempts to improve on horizontal communication needs through a matrix management structure, but this superimposes additional management layers and brings about the difficulty of one worker reporting to more than one supervisor. There are communication difficulties as well. Fourth-generation management attempts the needed horizontal communication through technology, but does not fully ameliorate the other difficulties we have just cited.

Savage presents a number of illustrations in an attempt to show that the computerization of steep hierarchies, or fourth-generation management, really will not work. He presents five needs, in the terms of a set of interrelated conceptual principles, that form the nexus of early fifth-generation management and that will enable the desired transformation:

1. *Peer-to-peer networking* is a major need. This involves three major ingredients: technologies, information, and people. Peer-to-peer networking enables communication from any individual in the organization to any other individual without the necessity of having to go through the conventional steep hierarchical structure. It allows people to work together in a cross-functional manner. Far from eliminating hierarchies, this results in a redefinition of the role and function of the hierarchy and a resulting hierarchy that is much flatter than before networking. While there are major hardware and software difficulties in bringing this about, the human and organizational issues are larger and more complex.

2. *Work as dialogue* is another important need. This involves listening, visioning, remembering, and using knowledge relative to both process and product.

3. The *human time and timing* need is concerned with developing an understanding of past, present, and future patterns such that it becomes possible to see and anticipate future patterns of the basis of experiences and knowledge.

4. An *integrative process* across people, technologies, and the organization allow for continuous change and teamwork in the organization. This should be contrasted with the pattern of unfreezing, change, and refreezing (Fishbein and Azjin, 1975), typically in a reactive fashion, that does not support continued improvement over time except at the discrete time instances where refreezing and change occur.

5. *Virtual task-focusing teams* is the final major need and results only from satisfaction of the first four needs.

On the basis of these principles, Savage suggests ten pragmatic organizational considerations for enabling fifth-generation management:

1. Develop a technical networking infrastructure that is flexible and adaptable to organizational needs and continual change.

2. Develop a data integration strategy.

3. Develop functional centers of excellence.

4. Develop and expand the organizational knowledge base.

5. Develop organizational learning, unlearning, and relearning capabilities that are continuously updated and rejuvenated.

6. Develop visioning capacities so that the context for judgments and decisions is visible to all through knowledge of strategic plans and organizational objectives, mission statements, and values.

7. Develop behavior norms, a sense of values, a reward structure, and measurements that support task-focusing teams.

8. Develop the organizational ability to identify, support, and manage multiple functional-task teams.

9. Develop the organizational capacity and capability to support the teamwork of teams.

10. Develop virtual task-focusing teams that are formed of suppliers, customers, and appropriate people from within the organization.

The first two are primarily technology-based needs, although there is much need for human interaction with the technologies to be implemented. The last needs have very much to do with people. Those in the middle represent organizational needs. There is not a sharp cleavage between technology, organization, and people needs, and these were identified in Figure 23.1 as major ingredients for reengineering.

Notably lacking in these works is a very clear and comprehensive explication of the structure of a fifth-generation organization. It is indicated that the horizontally functional units in the typical organization are wrapped around into a circle, and that there are linkages to customers and suppliers, and organizational cognitive connections to knowledge and vision. But, aside from these connections no explicit structure is shown. Perhaps it is simply too early in the development of fifth-generation management strategies, purposes, and functions to be concerned with specific organizational architectures. Perhaps it is simply best to let an individual organization adapt to a structure that fits best with its evolution to the fifth generation.

Denton (1991) is much concerned with horizontal management, and associated refocusing of leadership, culture, and cognizance within the organization. Horizontal management involves much participatory management and people empowerment, as well as the development of a nearly flat pyramid-like structure for an organization that is well focused on cross-functional activities and results.

Denton suggests a *next organization as customer* (NOAC) strategy that has been used at a number of successful organizations. In this way, every person concerned with an individual process, product line, or operation within an organization, is encouraged to consider those who supply the inputs to that particular process, or part of a process, as if they were suppliers or providers, and those who receive the outputs as if they were customers or users. Thus, the identity or suppliers and providers changes as one moves from one process, or subprocess, to another. It is maintained that considering the next unit to receive the output from any given process phase as if it were a customer allows identification of needed communications paths within the organization.

Major objectives in the NOAC effort include identifying needed communication flows within the organization and process change to enable this and associated improvements.

The effort is particularly focused on requirements and on being sure that there is conformity between the external or final customer requirements and the plethora of "internal customers" who are identified as the process is traced from the external customer through the front end of the life-cycle phases. Process improvement is obtained by selecting the process phase with the greatest potential for improvement and improving it first. In some cases, this will occur through elimination of unnecessary or redundant activities. Measures of effectiveness are determined and improvements implemented in order to enable performance of necessary activities to be right the first time, and such that all activities are in conformance with a meaningful process to ensure final customer satisfaction and internal customer satisfaction as well. It is indicated that Metropolitan Life and Motorola have implemented versions of NOAC, and these implementations are discussed in the referenced text.

Denton defines a horizontal management maturity model that comprises five steps that indicate various horizontal-maturity gradations with the ultimate being described as a level-one organization, or completely flat, hierarchy. Moving up these steps increases individual decision-making responsibility and empowerment from relatively narrow to relatively broad areas. The time span of the decisions at increasing steps is also increased. Figure 23.27 illustrates climbing the steps of the horizontal management ladder until level-one management is reached. These steps are not necessarily easy to climb. The first step allows worker empowerment over very operational issues such as monitoring quality-control charts and determining work breaks. At the second step, schedules are arranged, weekly

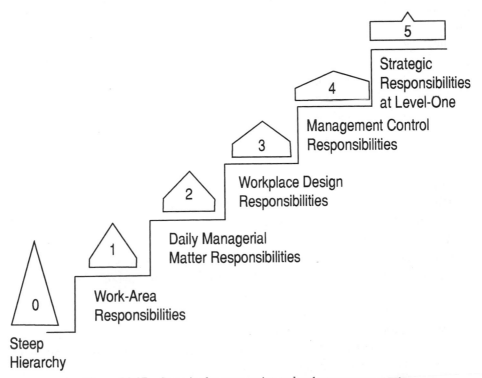

Figure 23.27   Steps in the progression to level-one management.

recordkeeping is accomplished, and short-term forecasting is done at the lowest possible level. At Step 3, selection of vendors and technologies and intermediate-range forecasting are done at the lowest possible level. At Steps 4 and 5, the hierarchy flattens further as management control decisions, long-range forecasting, and strategic issues are all accomplished at the lowest level.

Denton identifies a number of organizational culture and leadership issues that potentially need to be addressed in order to bring about *employee involvement* (EI), which is indicated as the cornerstone of moving to a true horizontal management organization. The efforts of the Ford Motor Company to bring about EI are described in some detail. It is indicated that horizontal management requires streamlining—through elimination, simplification, and combination (ESC)—of many workplace efforts and that this can only be accomplished satisfactorily through much employee involvement. A number of vertical motivators and horizontal motivators are suggested. Among the horizontal motivators are various TQM-related measurement approaches, the NOAC philosophy, employee stock ownership plans (ESOP), decentralization of almost everything, and participative management. The needed transformations to bring this about are a reduction in the knowledge gap and use of knowledge incentives across the organization, creation of a sense of full partnership and lines of communication across the organization, fluid leadership based on competency, and much teamwork across an equitably treated workforce.

Notions of horizontal management as an approach to reengineering, or perhaps more appropriately, as the result of reengineering have been suggested by many. We have attempted here to provide a salient overview of two of the major works on horizontal management. Essentially all of these suggest an approach toward creation of a horizontal organization that involves the sort of definition, development, and deployment phased efforts illustrated in Figure 23.28. As we have discussed throughout much of this chapter, the major precepts in implementing this and other approaches for total quality management and strategic reengineering involve the following precepts:

1. Making customer satisfaction the major driver of the organization, and the major driver of organizational performance.
2. Maximize contact and interaction with customers and suppliers.
3. Educate, train, and informate all organizational personnel with respect to knowledge about the organization's mission-relevant areas, as well as with respect to general problem formulation, analysis, and interpretation abilities.
4. Organize cross-functional, and generally multidisciplinary teams that are self-managing and accountable for their performance responsibilities.
5. Use cross-functional teams to manage virtually everything.
6. Reward team performance.
7. Define the organization's missions and critical objectives.
8. Identify the strategic processes, and those that are nonessential or redundant as candidates for abolition.
9. Organize around processes and not functions or departments.
10. Assign a process owner to each process.
11. Flatten the hierarchy to reduce unnecessary layers of management and administration and to increase worker empowerment.

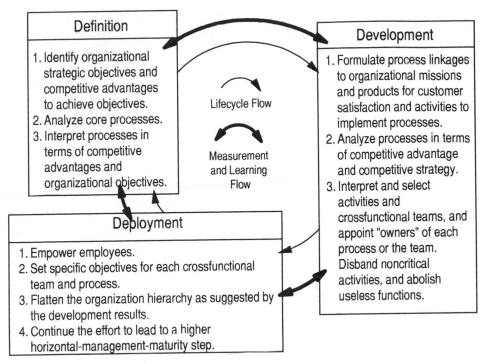

Figure 23.28   Life-cycle phases in implementation of horizontal management.

12. Develop an interactive and proactive measurement system to assist in guiding the development and deployment of the strategy as a results-oriented set of activities.

It is important to note that maximization of short-term, or long-term, for that matter, profits is not one of the fundamental objectives or activities. This is also true for all of the other approaches discussed in this chapter. Perhaps it should be stated strongly, and arguably, that profit is really, and should be regarded as, an objectives measure. If this is not done, ultimately the organization suffers, through the short-range perspective that typically results. In other words, if profit is the only objective function to be maximized, it will not be maximized!

Denton identifies Metropolitan Life, Motorola, Aluminum Company of America (ALCOA), Glaxo, and Ford Motor Company as among the organizations moving in the direction of horizontal management. According to a study in *Business Week* (Editors, 1993), AT&T, the Eastman Chemical unit of Eastman Kodak, General Electric, Lexmark International (a former IBM Division), Motorola, and Xerox are among the American companies also moving toward a horizontal management structure.

There are many challenges associated with implementation of a horizontal management outlook. Among these are the culture and leadership issues we addressed in the introductory chapter. These relate to many of the concerns associated with bringing about total quality management (Grant et al., 1994), such as the implementation of work teams (Manz and Sims, 1993; Katzenbach and Smith, 1993; Katzenbach, 1997), which will generally be self-managing, and implemented for cross-functional purposes to meet organizational performance challenges for competitive advantage through people (Pfeffer, 1994; Mankin et al., 1996; Fitz-Enz, 1997).

23.3.15   Concurrent Engineering

Often, it is desired to produce and field a system in a relatively rapid manner. The life-cycle processes needed to bring this about could indeed be potentially accelerated in time if it were possible to accomplish phases of the relevant life cycles in a more or less concurrent, or simultaneous, manner. *Concurrent engineering* is a systems engineering approach to the integrated coincident design and development of products, systems, and processes (Rosenblatt and Watson, 1991; Kusiak, 1993; Prasad, 1996a,b). Concurrent engineering is intended to cause systems engineering and management efforts to be explicitly planned so as to better integrate user requirements such as to result in high-quality cost-effective systems, and thereby to reduce system development time through better integration of life-cycle activities. Here, we view the subject from a systems reengineering perspective. Chapter 9 of this handbook discusses configuration management in some detail.

The basic tasks in concurrent engineering are much the same as the basic tasks in systems engineering and management. The first step is that of determining what it is that the customer wants. After the customer requirements are determined, they are translated into a set of technical specifications. The next phase involves program planning for development of a product that will satisfy the customer. Often, especially in current engineering, this will involve examining the current process, especially at the systems management or management controls level. This process is usually refined to best deliver a superior quality product as desired by the customer, within cost and schedule constraints.

In concurrent engineering, the very early and very effective configuration of the system's life-cycle process takes on special significance. This is so because the simultaneous development efforts need to be very carefully coordinated and managed or the opportunities abound for significant cost increases, significant product time increases, and significant deterioration in product quality. The use of coordinated product design teams, improved design approaches, and careful and critical use of standards are among the aids that can be brought to bear on concurrent engineering needs. These needs are particularly critical in concurrent engineering because of the group nature and simultaneous nature of the system development effort.

There is much need for a controlled environment for concurrent engineering, and for system integration needs as well. This requires several integration and management undertakings:

1. *Information Integration and Management.* It must be possible to access information of all types easily. It must be possible to share design information across the levels of concurrent design in an effective and controlled manner. It must be possible to track design information, dependencies and alterations in an effective manner. It must be possible to effectively monitor and manage the entire configuration associated with the concurrent life-cycle process.
2. *Data and Tool Integration and Management.* It must be possible to integrate and manage tools and data such that there is interoperability of hardware and software across the several layers of concurrency.

3. *Environment and Framework Integration, or Total Systems Engineering.* It must be possible to ensure that process is directed at evolution of a high-quality product; and that this product be directed at resolution of the needs of the customer or user in a trustworthy manner that is warmly endorsed by the customer. This requires integration of the environment and framework, or the processes, for the systems engineering and management efforts.

Thus, there is also a close relationship between concurrent engineering and systems integration, as discussed in Chapter 14. Andrews and Leventhal (1993), Kronlöf (1993), and Schefstrom and van den Broek (1993) provide a number of details concerning the method, tool, and environment integration needed to bring about concurrent engineering and other systems engineering efforts.

Concurrent engineering clearly requires much up-front planning such that simultaneous development of the various process and product subsystems can occur in a trustworthy manner. Compression of the phases of the life cycle, and at least partial parallel accomplishment of some of them, is somewhat more problematic. The macroenhancement approaches to systems engineering, especially software systems engineering (Sage and Palmer, 1990), would appear particularly useful in this regard. These include prototyping as a means of system development, use of reusable (sub)systems, and expert system and automated program generation approaches. Each of these allows, at least in principle, compression of the overall time to exercise the resulting parallel subsystem life cycles in a manner that is compatible with the engineering of a trustworthy product or service.

Through use of approaches such as these, it is hoped to obtain systems that are characterized by:

- High quality, in terms of system performance, suitability, and reliability in a large variety of operational environments.
- Short deployment time, for new product and service designs, and for delivery and maintenance of existing product designs.
- Low life-cycle costs, for system design, development, maintenance, and retrofit or phaseout.

In an insightful work, Winner et al. (1988) identify three critical activities and a number of technical capabilities that support these:

1. Obtain early, complete, and continuing understanding of customer requirements and priorities. This requires capabilities for obtaining information concerning comparable products, processes, and support. It requires identification of complete and unambiguous information about this new system and product, including support needs. Finally, it requires synthesis or translation of user requirements into design specifications for the system product and the validation of design specifications.
2. Translate the system requirements and specifications into optimal products and manufacturing and support processes that can be performed concurrently and in an integrated fashion. There are many required capabilities. These include managing information and data concerning the system product and development process; and dissemination of product, process, and support data to the concurrent teams.

3. Continuously review and improve the system product, product line, and support characteristics. This includes intelligent oversight to enable impact assessment of changes, and proactive, concurrent availability of current design.

These provide a very useful set of critical success factors for the concurrent engineering process. Several recommendations are made that will enhance success probabilities for concurrent engineering efforts:

- There is much needed leadership for change, and this must come from the very top of the organization through executive-level commitment.
- Pilot projects may be very advantageous in acceleration of the deployment of an appropriate concurrent engineering methodology in a given organization.
- Concurrent engineering should be considered in addition to other approaches for reengineering.
- Education and training for concurrent engineering will be needed across the organization.
- It will generally be necessary to identify and remove a number of barriers to successful implementation of concurrent engineering efforts.

It appears reasonably clear that there is no single unique best way to approach concurrent engineering. It would appear to be more of a philosophy of and approach to strategic management than anything else. Of course, this strategic management needs to be translated into management controls or systems management, and thence to task- and operational-level effort that results in a process for ultimate production of a product or system. How a particular organization approaches concurrent engineering will be very much a function of their organizational traditions, leadership, and culture.

Formally, there is little that is very new in the subject of concurrent engineering. We simply speed things up by doing them concurrently, at least on the surface. But, we should be very careful to not dismiss the much strengthened needs for the strategic planning and systems management, and the major need for much attention to processes that are well deployed, and the resulting integration that is necessary to ensure success in concurrent systems engineering.

Concurrent engineering approaches are also discussed in Nevins and Whitney (1989) and Shina (1991). While these works concentrate on product manufacturing, more recent works also consider concurrency for information and other service systems. Carter and Baker (1992) indicate that success in concurrent engineering depends very much on maintaining a proper balance between four important dimensions of a concurrent engineering environment:

- Organizational culture and leadership, and the necessary roles for product development teams.
- Communications infrastructure for empowered multidisciplinary teams.
- Careful identification of all functional and nonfunctional customer requirements, including those product and process facets that impact on customer satisfaction.
- (Integrated) process and product development.

Carter and Baker identify approaches at the levels of task, project, program, and enterprise to enable realization of the proper environment for concurrent engineering across each of these four dimensions. Each of the four dimensions is associated with a number of critical factors, and these may be approached at any or all of the levels suggested. It is suggested that the general equivalent of the matrix in Figure 23.29 be completed both for the present situation and the desired situation. This will enable identification of the needed development areas to ensure definition, development, and deployment of an appropriate concurrent engineering process environment. We show some very hypothetical development needs in this figure. The cited reference provides a wealth of pragmatic details concerning determination of concurrent engineering (CE) process needs.

It is also noted that there are often five major roadblocks that exist to impede development of a concurrent engineering process environment:

1. The currently available tools are not adequate for the new CE environment.
2. There are a plethora of noninteroperable computers, networks, interfaces, operating systems, and software in the organization.
3. There is a need for appropriate data and information management across the organization.
4. Needed information is not communicated across horizontal levels in the organization.
5. Correct decisions, when they are made, are not made in a timely manner.

Approaches are suggested to remove each of these roadblocks to enable development of a concurrent engineering process. Presumably, this needs to be implemented in a contin-

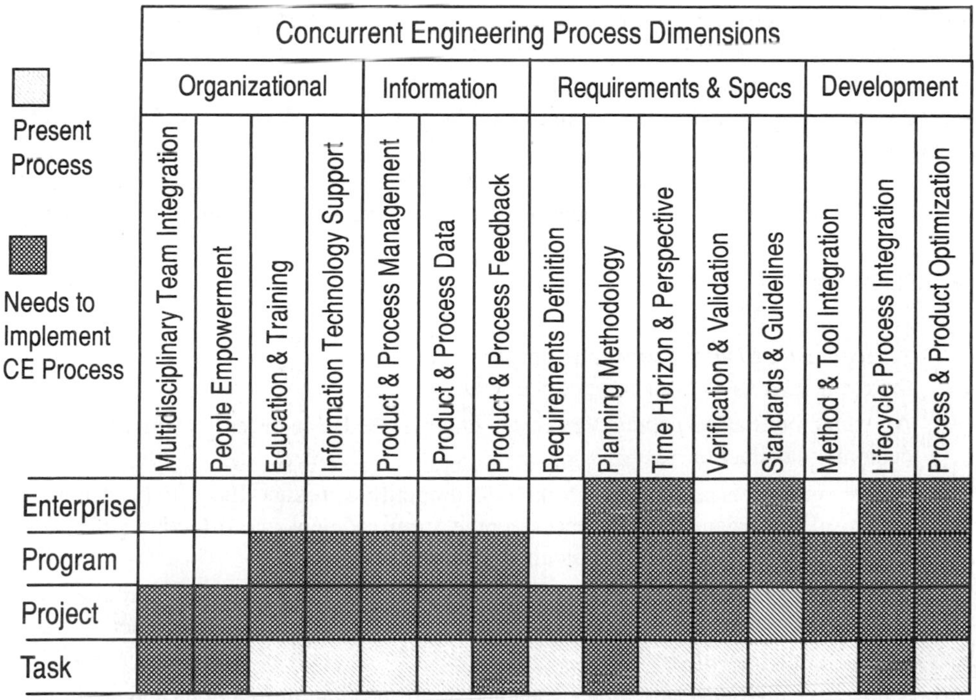

	Concurrent Engineering Process Dimensions														
	Organizational				Information			Requirements & Specs					Development		
	Multidisciplinary Team Integration	People Empowerment	Education & Training	Information Technology Support	Product & Process Management	Product & Process Data	Product & Process Feedback	Requirements Definition	Planning Methodology	Time Horizon & Perspective	Verification & Validation	Standards & Guidelines	Method & Tool Integration	Lifecycle Process Integration	Process & Product Optimization
Enterprise									■			■		■	
Program			■	■	■	■									
Project	■	■	■	■	■	■	■	■	■	■	■	■	■	■	■
Task	■	■	■	■	■	■	■	■	■	■	■	■	■	■	■

□ Present Process

■ Needs to Implement CE Process

Figure 23.29 Hypothetical needs to implement a concurrent engineering process.

uous fashion over time, as appropriate for a given organization, rather than attempting a revolutionary or radical change in organizational behavior. A number of worthwhile suggestions to enable this implementation are provided.

23.3.16 Integrated Product and Process Development

In many ways, integrated product development (IPD) is an extension of concurrent engineering. Fiksel (1993), in a work that focuses on the importance of requirements management, states that concurrent engineering is more accurately known as integrated product development. It is also closely related to the other reengineering approaches we describe here. The notion of integrated product development really cannot be carried out and orchestrated effectively without simultaneous consideration of integrated process development. Thus, at this time, the concept is more commonly called *integrated product and process development (IPPD)* (Usher et. al., 1998). The associated concepts also relate very closely to enterprise integration, which we discuss next.

The following definition of integrated product and process development seems appropriate:

> Integrated product and process development is a systems engineering and management philosophy and approach that uses functional and cross-functional work teams to produce an efficient and effective process for the ultimate deployment of a product or service that satisfies customer needs through concurrent application and integration of all necessary life-cycle processes.

We see in this many of the same terms that we have already used in this handbook. IPPD involves systems management, leadership, systems engineering processes, the products of the process, concurrent engineering and integration of all necessary functions and processes throughout the organization, to result in a cost-effective product or service that provides total quality and customer need satisfaction, generally in a rapid just-in-time fashion.

Thus, IPPD is an organization's product and process development strategy. It is focused on results. It also addresses the organizational need for continual enhancement of efficiency and effectiveness in all of its processes that lead to a product, or service. There are many focal points for IPPD. Twelve are particularly important:

1. A *customer satisfaction focus* is needed as a part of a competitive strategy and is the result of a successful competitive strategy.
2. A *results focus* and a *product or service focus* is needed in order to bring about total customer satisfaction.
3. A *process focus* is needed as high-quality competitive products that satisfy customers and result in organizational success come from efficient and effective processes. This necessarily requires process understanding.
4. A *strategic planning and marketing focus* is needed to ensure that product and process life cycles are fully integrated throughout all organizational functions, external suppliers, and customers.
5. A *concurrent engineering focus* is needed to ensure that all functions and structures

associated with fulfilling customer requirements are applied throughout the life cycle of the product to ensure correct people, correct place, correct product, and correct time deployment.

6. An *integration engineering focus* is needed to ensure that relevant processes and the resulting processes fit together in a seamless manner.

7. A *teamwork and communications focus* is needed to ensure that all functional, and multifunctional, teams function synergistically and for the good of the customer and the organization.

8. A *people empowerment focus* is needed such that all decisions are made by qualified people at the lowest possible level that is consistent with authority and responsibility. Empowerment is a responsibility and not just an entitlement and entails commitment and appropriate resource allocation to support this commitment.

9. A *systems management reengineering focus* is needed, both at the levels of radical and revolutionary change, as well as for evolutionary change, so as to also result in radical, revolutionary, or evolutionary changes in processes and product.

10. An *organizational culture and leadership focus* is needed in order to successfully accommodate changed perspectives relative to customers, total quality, results and products, processes, employees, and organizational structures.

11. A *methods, tools, and techniques focus* is needed, as methods, tools and techniques are needed throughout all aspects of an IPPD effort, even through they alone will not bring about success.

12. A *systematic measurements focus,* primarily on proactive measurements, but also on interactive and reactive measurements, is needed, as we need to know where to go and where we are now, in order to make progress toward getting there.

All of this should bring about high-quality, continual and evolutionary, and perhaps even radical and revolutionary, improvement for customer satisfaction. Each of these could be expanded into a series of questions, or a checklist, and used to evaluate the potential effectiveness of a proposed integrated product development process and team. While our discussion of IPPD may make it seem, as an approach, particularly and perhaps even uniquely suitable for system acquisition, production, or procurement; it is equally applicable to the products of the RDT&E and marketing life cycle.

We see that IPPD is a people-, organization-, and technologies-focused effort that is tightly linked together through a number of life-cycle processes through systems management efforts. These ingredients are major ingredients for all systems engineering and management efforts, as suggested in the information ecology web of Figure 23.8. The major result of IPPD is the ability to make optimum decisions within available resources and to execute them efficiently and effectively in order to achieve three causally linked objectives:

1. To integrate people, organizations, and technology into a set of multifunctional and networked product development teams.

2. To increase the quality and timeliness of decisions through centrally controlled, decentralized, and networked operations.

3. To completely satisfy customers through quality products and services that fulfill their expectations and meet their needs.

The bottom line is clearly customer satisfaction through quality, short product delivery time, reduced cost, and improved performance and functionality. Equally supported by IPPD are organizational objectives for enhanced profit, well-being of management, and a decisive and clear focus on risk and risk management and amelioration.

Figure 23.30 illustrates a suggested sequences of steps and phases to establish an IPPD endeavor. The approach is not entirely different from that suggested for successful product and process development by those who do not use the IPPD framework (Bowen et al., 1994).

1. Understand the core capabilities, and core rigidities, of the organization.
2. Develop a guiding vision in terms of the product or service concept, the project and process vision, and an organization vision that will assure an understanding of the relationships between organization, customers, and process and product.
3. Push the frontiers of the organization, process, and product or service in order to identify and achieve the ultimate performance capabilities for each.
4. Develop leadership and an appropriate structure to manage the resulting process and product or service engineering.
5. Develop commitment at the level of organizational management, the integrated product team (IPT), and the individual team members to assure appropriate ownership of the IPPD effort.

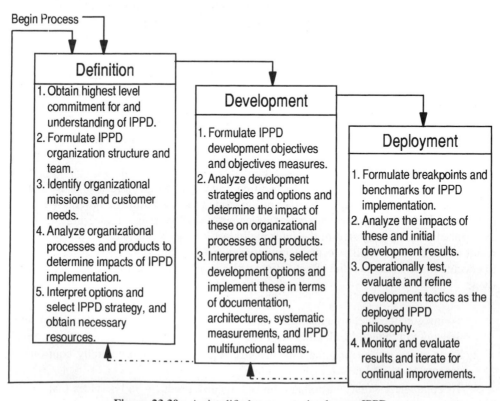

Figure 23.30 A simplified process to implement IPPD.

6. Use prototypes to achieve rapid learning and early evolution and testing of the IPPD concept.

7. Assure integration of people, organizations, and technologies to assure success of the IPPD concept.

As with other efforts, this embodies the definition, development, and deployment triage we have used so often in this handbook.

Appropriate references to IPPD include Hunt (1993) and Usher et al. (1998). At this point, IPPD is a relatively new concept used very often within the U.S. Department of Defense (Air Force Material Command, 1993; Department of Defense, DoD 1996). In this latter document ten key tenets of IPPD are identified:

1. *Customer Focus.* The primary objective of IPPD is to satisfy the needs of the customer in a more efficient and effective manner. Customer needs are the major determining influence of the product or service definition and the associated product lines.

2. *Concurrent Development of Products and Processes.* It is necessary to develop processes concurrently with the products or services that they support.

3. *Early and Continuous Life-Cycle Planning.* Planning for both the product, or service, and process begins early and extends throughout the IPPD life cycle.

4. *Maximize Flexibility for Optimization and Use of Contractor Unique Approaches.* Requests for Proposals should provide flexibility for optimization and use of contractor-unique processes and commercial specifications, standards, and practices.

5. *Encourage Robust Design and Improved Process Capability.* Advanced robust design and manufacturing techniques that promote total quality and continuous process improvement should be emphasized.

6. *Event-driven Scheduling.* The scheduling framework should relate program events to their desired accomplishments and should reduce risk by ensuring product and process maturity prior to undertaking actual development.

7. *Multidisciplinary Teamwork.* Multidisciplinary teamwork is essential to the integrated and concurrent development of product and process.

8. *Empowerment.* Decisions should be taken at the lowest level commensurate with appropriate risk management, and resources should be allocated at levels that are consistent with authority, responsibility, and ability. The team should be given authority, responsibility, and resources. The team members should accept responsibility, manage risk appropriately, and be held accountable for the results obtained.

9. *Seamless Management Tools.* A single management system should be established to relate requirements, planning, resource allocation, execution, and program tracking over the entire life cycle.

10. *Proactive Identification and Management of Risk.* Critical cost, schedule, and technical specifications should be identified from user requirements. Systems management of risk, using appropriate metrics, should be established in order to provide continuing verification of achievements relative to appropriate product and process standards.

The objectives in this DoD document are to obtain reduced time to deliver operationally functional products and services, to reduce the costs and risks associated with obtaining deployed systems, and to improve their quality. IPPD is a central feature of the DoD Standards for a Major Defense Acquisition Program (MDAP) and a Major Automated Information System (MAIS) acquisition program as established in DoD Regulation Directive 5000.2-R.

The subject of IPPD is closely related to corporate information management and enterprise information integration, and we now look at these approaches for systems reengineering.

23.3.17 Enterprise Information Integration and Enterprise Architecture

Corporate information management (CIM) and enterprise integration through information technology, which we denote as enterprise information integration (EII), and development approaches that focus on the role of information technology in supporting development. The Center for Information Management, Defense Information Systems Agency (DISA), has been chartered to support the Director of Defense Information (DDI) by providing information management technical services to the DoD community. These services are an integral part of the CIM program. The CIM program represents an effort to streamline organizational operations and processes to support definition, development, and deployment of high-quality, cost-effective standard information systems.

Five support efforts are considered essential for CIM:

1. CIM provides assistance to functional managers in identifying better ways of doing business, in part by providing standard methods and tools for developing improved business methods and practices that are tied to quantifiable measures of performance.
2. CIM promotes efficiencies and standardization in information technology and software engineering through appropriate tools and methods.,
3. CIM assists in integrating common and standardized information systems within each functional area, and across functional areas, of an organization.
4. CIM promotes the use of open systems standards, as discussed briefly in Chapter 6, to allow use of commercial-off-the-shelf (COTS) products, vendor-neutral commercial products, and to facilitate open competition for services. This should facilitate porting of applications among platforms, and enable the emergence of applications operating in common environments.
5. CIM efforts should assist in planning for and managing development of an efficient and effective information technology infrastructure. Such reengineering technologies as benchmarking are suggested for these efforts.

The CIM effort is intended to be customer oriented and to provide support for development and implementation of improved business practices and information management capabilities. The ultimate aspiration is enhanced mission capability with reduced costs.

The scope of the CIM effort is large. It includes:

• Standard methods and tools for improved business practices.
• Standard methods and tools for information engineering, software engineering, and infrastructure engineering.

- Standardizing information engineering, data administration, reuse, and software engineering practices.
- Integrating common information systems within and across functional areas.
- Identifying and promoting open systems standards.
- Providing wide-scope and common system-acquisition frameworks.
- Planning and engineering of a full-service utility that will provide technical support to customers on a fee-for-service basis.

Thus, it seems that an inferred definition of CIM might be that corporate information management is an activity that connects humans across the organization in order to facilitate access by appropriate people, in a timely and cost-effective manner, to appropriate information. This requires information access, and infrastructure to ensure pertinent information integration, and decision support for enterprise management. The CIM initiative appears to be based on the premise that there are two fundamental processes in modern organizations, one each for converting raw material into products, and converting data into information.

The major objective of the CIM initiative is to provide support for information technologies in support of the production function. While this is appropriate, the major relations between organizational leadership and culture, process innovation needs, and information support concerns also need to be considered, and in a major way. As stated, the major focus in the CIM effort appears directed at the product and single-process level, and improvements in these through information-technology-based supports. While this is fully appropriate, even greater benefits may result from considerations directed at the level of systems management and with respect to organizational leadership and cultural issues as well.

In more recent efforts (Office of the Secretary of Defense/ASD, 1994), the CIM initiative has been broadened to include more of a focus upon strategic and organizational issues. The resulting effort has been named "enterprise integration" and is intended to align the major elements of the CIM model and prepare the way for implementing improved information technology capabilities on a common infrastructure of platforms, software, communications, and applications. An enterprise roadmap is suggested. This is composed of a five-phase life cycle for bringing about enterprise information integration. Figure 23.31 represents the key aspects of this life cycle. There is intended to be much feedback and iteration across the various steps shown for each of the five major phases and an *Integrated Computer-Aided Manufacturing DEFinition* (IDEF) software-based modeling approach is used in the referenced document to display a plethora of possible interconnections among these elements, to yield an activity model for this life cycle.

The activities and critical core processes of the DoD have been described in an enterprise model. This model is composed of four major activities, 15 supportive core processes, and 54 procedures that result in successful products from the processes. The major activities and core processes are:

- Establish Policies and Plans
    Establish Policies
    Determine Requirements

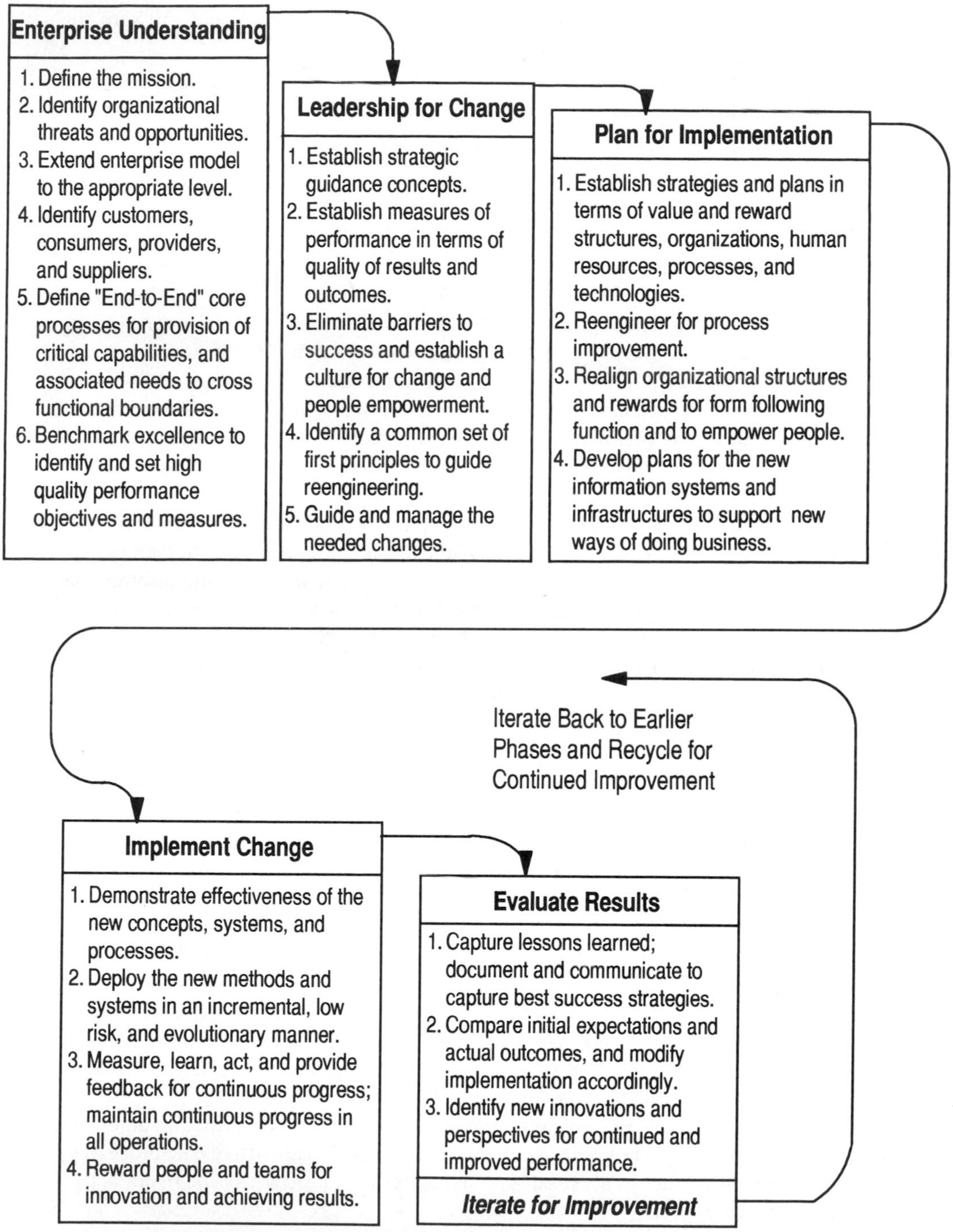

Enterprise Understanding

1. Define the mission.
2. Identify organizational threats and opportunities.
3. Extend enterprise model to the appropriate level.
4. Identify customers, consumers, providers, and suppliers.
5. Define "End-to-End" core processes for provision of critical capabilities, and associated needs to cross functional boundaries.
6. Benchmark excellence to identify and set high quality performance objectives and measures.

Leadership for Change

1. Establish strategic guidance concepts.
2. Establish measures of performance in terms of quality of results and outcomes.
3. Eliminate barriers to success and establish a culture for change and people empowerment.
4. Identify a common set of first principles to guide reengineering.
5. Guide and manage the needed changes.

Plan for Implementation

1. Establish strategies and plans in terms of value and reward structures, organizations, human resources, processes, and technologies.
2. Reengineer for process improvement.
3. Realign organizational structures and rewards for form following function and to empower people.
4. Develop plans for the new information systems and infrastructures to support new ways of doing business.

Iterate Back to Earlier Phases and Recycle for Continued Improvement

Implement Change

1. Demonstrate effectiveness of the new concepts, systems, and processes.
2. Deploy the new methods and systems in an incremental, low risk, and evolutionary manner.
3. Measure, learn, act, and provide feedback for continuous progress; maintain continuous progress in all operations.
4. Reward people and teams for innovation and achieving results.

Evaluate Results

1. Capture lessons learned; document and communicate to capture best success strategies.
2. Compare initial expectations and actual outcomes, and modify implementation accordingly.
3. Identify new innovations and perspectives for continued and improved performance.

Iterate for Improvement

Figure 23.31   The life cycle phases for enterprise integration.

Develop Plans

Allocate Resources

- Acquire Assets

Manage Acquisition Process

Conduct Research, Development, Test, and Evaluation

Produce Assets

- Provide Capabilities

Manage Assets Through Appropriate Allocations

Maintenance and Support of Assets

Provide Administrative Services

Embody, Train and Develop Unit Capabilities, and Assess Readiness

- Employ Forces

Constitute Operational Forces

Provide Operational Intelligence

Conduct Operations

Sustain Operations

Understanding the enterprise is very much enhanced by understanding this enterprise model and expanding it as needed to accommodate the other needs that are associated with this phase and are identified in Figure 23.30.

Notions of enterprise integration and enterprise architecture include much more than functional integration of organizational information systems. They include:

- Integrated planning and direction through development and use of a functionally oriented enterprise model that enables cross-functional management-oriented efforts.
- Internal and external integration of the organization, such that it is able to deal with transactions between partners and suppliers and to bring about appropriate teaming.
- Integrated organizational processes to enable continuing cross-functional improvements to these processes.
- Integrated human resources that will encourage empowerment and adaptation to change all types.
- Integrated financial resources to assure standardization of financial policies, practices, and procedures.
- Integrated information and information systems that assure maximum communication across the organization and transition of legacy systems into an integrated framework.
- Integrated physical assets for flexibility and adaptability through removal of functional, managerial, and technical barriers to IPPD, and to enable establishment of end-to-end functional and managerial processes for enhanced efficiency and effectiveness.

Clearly, the life-cycle process suggested for enterprise integration is very thorough, at least in a conceptual fashion. It includes notions of information systems integration (Spewak,

1992) and embodies many of the other approaches to reengineering that we have discussed in this chapter.

This concludes our discussion of approaches for reengineering at the level of process and systems management. There are a number of methods that can be used to assist in reengineering at these levels. Clausing (1994) is very concerned with the use of quality function deployment and other analysis approaches for process reengineering. Workflow approaches to process reengineering are described in Kobellas (1997), Hannaford (1996), and Fischer (1995). Software for reengineering process life cycles are described in Spurr et al. (1994), Andrews and Leventhal (1993), Ould (1995), Hansen (1994), and Andrews and Stalick (1994).

23.3.18 Product Reengineering

Reengineering at the level of product has received much attention in recent times, especially in information technology and software engineering areas. This is not a subject that is truly independent of reengineering at the levels of either systems management or of a single life-cycle process. In our final subsection concerning contemporary reengineering approaches, we examine some facets of product reengineering. Product reengineering is generally needed whenever development of an entirely new product is too expensive, when there is no suitable and available commercial product, and when the current system is not suitable in the sense of not fulfilling some of the functional or nonfunctional requirements, such as trustworthiness.

As we noted earlier, much of product reengineering is very closely associated with reverse engineering to recover either design specifications or user requirements. This is then followed by refinement of these requirements and/or specifications and the forward engineering to result in an improved product. The term reverse engineering, rather than reengineering, was used in one of the early seminal papers in this area (Chifosky and Cross, 1990) that was concerned with software product reengineering. In this latter paper, as well as in a related chapter on the subject (Cross et al., 1992), the following efforts represent both the taxonomy of and phases for what we denote here as *product reengineering:*

1. *Forward engineering* is the original process of defining, developing, and deploying of a product, or realizing a system concept as a product.
2. *Reverse engineering,* sometimes called inverse engineering, is the process though which a given system or product is examined in order to identify or specify the definition of the product either at the level of technological design specifications or system-or user-level requirements.
   2.1 *Redocumentation* is a subset of reverse engineering in which a representation of the subject system or product is recreated for the purpose of generating functional explanations of original system behavior and, perhaps more importantly, to aid the reverse engineering team in better understanding the system both at a functional and structural level. There are a number of redocumentation tools for software available and some of these are cited in these works. One of the major purposes of redocumentation is producing new documentation for an existing product where the existing documentation is faulty, and perhaps virtually absent.

2.2 *Design Recovery* is a subset of reverse engineering in which the redocumentation knowledge is combined with other efforts, often involving the personal experiences and knowledge of others about the system, that lead to functional abstractions and enhanced product or system understanding at the level of function, structure, and even purpose. We would prefer to call this deployment recovery, development recovery (which would include design recovery), and definition recovery, depending upon the phase in the reverse engineering life cycle at which the recovery knowledge is obtained.

3. *Restructuring* involves transformation of the reverse engineering information concerning the original system structure into another representation form. This generally preserves the initial functionality of the original system, or modifies it slightly in a purposeful manner that is in accord with the user requirements for the reengineered system and the way in which they differ from the requirements for the initial system. For our purposes, the terms *deployment restructuring, development restructuring,* and *definition restructuring* seem to be appropriate disaggregations of the restructuring notion.

4. *Reengineering* is, as defined in these efforts, equivalent to redevelopment engineering, renovation engineering, and reclamation engineering. Thus, it is more related to maintenance and reuse than the other forms of systems management and process reengineering that we have discussed in this chapter. Reengineering is the recreation of essentially the original system in a new form that has improved structure but generally not much altered purpose and function. The nonfunctional aspects of the new system may be considerably different from those of the original system, especially with respect to quality and reliability.

Figure 23.3, which illustrates product reengineering, involves essentially these six activities.

We can recast this by considering a single phase for definition, for development, and for deployment that is exercised three times. We then see that there is a need for recovery, redocumentation, and restructuring as a result of the reverse engineering product obtained at each of the three basis phases.

This leads us to suggest Figure 23.32 as an alternative way to represent Figure 23.3 and as our interpretation of the representations generally used for product reengineering. Many discussions, such as those just referenced, utilize a three-phase generic life cycle of requirements, design and implementation. Implementation would generally contain some of the detailed design and production efforts of our development phase and potentially less of the maintenance efforts that follow initial fielding of the system. The restructuring effort, based on recovery and redocumentation knowledge obtained in reverse engineering, is used to effect deployment restructuring, development restructuring, and definition restructuring. To these restructured products, which might well be considered as reusable products, we augment the knowledge and results obtained by detailed consideration of potentially augmented requirements. These augmented requirements are translated, together with the results of the restructuring efforts, into the outputs of the reengineering effort at the various phases to ultimately result in the reengineered product.

For the most part, this is the perspective taken on reengineering in a definitive reprint book on software reengineering (Arnold, 1993), especially in the lead article by the editor

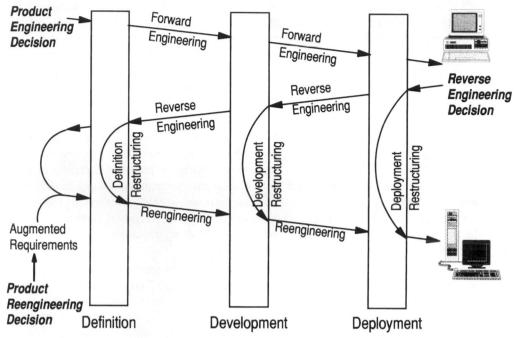

Figure 23.32 Expanded notion of product reengineering.

of this work that takes an inherently transformational view of product reengineering. This reprint book is much concerned with the major ingredients needed for software product reengineering as summarized below:

Reengineering process
Reengineering cost–effectiveness
Reengineering risks
Reengineering to reduce software maintenance
Technology and tools for reengineering
Data reengineering
Source-code analysis for reengineering
Software restructuring and translation
Reverse engineering and design recovery
Reengineering for reuse
Reengineering to object-oriented architectures
Reengineering through knowledge-based program analysis and understanding

There are three basic classes of transformational views of use in product reengineering:

1. *Nonprocedural views* are metalevel views, such as decision tables, event trees, attribute trees, data schemas, user requirements, and system specifications. These

views do not represent views of the actual entity but a view of a view of the entity or, in other words, salient characteristics of the entity. A nonprocedural view is a purposeful view.

2. *Procedural views* contain direct information about procedures or representations, or information intimately associated with this information. Source code and the objects and entities of object-oriented languages are procedural views. A procedural view is a functional view.

3. *Psuedoprocedural views,* or architecturally oriented views, contain perspectives of both procedural and nonprocedural views. Hierarchy charts, structural models, data flow diagrams, entity-relationship diagrams, and petri nets are examples of psuedoprocedural views. A psuedoprocedural view could also be called a structural view.

We can also have views that are derived from analysis of one, of in some other way derived from one, of the three basic view categories. Arnold denotes these as analysis views. For the most part, purposeful views or nonprocedural views are associated with the definitional phases of the life cycle for product acquisition. They concern user requirements and technological specifications. Functional or procedural views tend to be associated with the very end of the development phase of the life cycle and the deployment phase when systems may be thought of in terms of their input–output characteristics. Psuedoprocedural, or architectural or structural, views tend to be associated with the earlier phases of system development. One of the major purposes of both forward and reverse engineering, and tools that support these, is to enable transformation from one view to another such as to ultimately obtain a functionally useful product.

There are a number of objectives in, potential uses for, and characteristics of product reengineering. These include, but are neither mutually exclusive nor collectively exhaustive, the following (Arnold, 1993a,b; Sneed, 1995):

1. Reengineering may help reduce an organization's risk of product evolution through what effectively amounts to reuse of proven subproducts.

2. Reengineering may help an organization recoup its product development expenses through constructing new products that are based on existing products.

3. Reengineering may make products easier to modify for purposes of accommodating evolving customer needs.

4. Reengineering may make it possible to move the product, especially a software product, to a less expensive operational environment, such as from COBOL to an object-oriented language, or from a mainframe to a server.

5. Reengineering may be a catalyst for automating and improving product maintenance, especially through obtaining smaller subsystems with more well-defined interfaces.

6. Reengineering a product may result in a product with much greater reliability.

7. Reengineering may be a catalyst for application of new technologies, such as CASE tools and artificial intelligence.

8. Reengineering may prepare a reengineered product for functional enhancement.

9. Reengineering is big business, especially considering the major investment in legacy systems that need to be updated and maintained and improved in functionality.

In short, product reengineering provides a mechanism that enables us to understand systems better, such that we are capable of extending this knowledge to new and better systems that have enhanced functional and nonfunctional features. Thus, it enhances both product understanding and product improvement abilities.

Planning for product reengineering is essential, just as it is for other forms of reengineering. The planning effort would involve:

1. *Definition Phase.*
   (a) Formulation of the reengineering issue such as to determine the need for and requirements to be satisfied by the reengineered product, and subsequent identification of potential alternative candidates for reengineering.
   (b) Analysis of the alternatives to enable determination of costs and benefits of the various alternatives.
   (c) Interpretation and selection of a preferred plan for reengineering.
2. *Development Phase,* in which the detailed specifications for implementation of the reengineering plan are determined.
3. *Deployment Phase,* in which operational plans, including contracting, are set forth to enable reengineering of the product in a cost-effective and trustworthy manner.

A number of authors have suggested specific life cycles that will lead to determination of a decision to, or not to, reengineer a product and, in support of a positive decision, enable a product reengineering life cycle (Ulrich, 1993; Olsem, 1993). There are a number of needed accomplishments. These include the following:

1. Initially, there exists a need for formulation, assessment, and implementation of definitional issues associated with the technical and organizational environment. These issues include organizational needs relative to the area under consideration, and the extent to which technology and the product or system under reengineering consideration supports these organizational needs.
2. Identification and evaluation of options for continued development and maintenance of the product(s) under consideration, including options for potentially outsourcing this activity, is a need.
3. Formulation and evaluation of options for composition of the reengineering team, including insourcing and outsourcing possibilities, is a need.
4. Identification and selection of a program of systematic measurements that will enable demonstration of cost-efficiency of the identified reengineering options and selection of a chosen set of options is required.
5. The existing legacy systems in the organization need to be examined in order to determine the extent to which these existing systems are functionally useless at present and in need of total replacement, functionally useful but with functional and nonfunctional defects that could potentially be remedied using product reengineering to create renovated systems, or systems that are fully appropriate for the current and intended future uses.
6. A suite of tools and methods to enable reengineering needs to be established. Method and tool analysis and integration provides for multiple perspective views across the various abstraction levels (procedural, psuedoprocedural, and nonprocedural) that will be encountered in reengineering.

7. A reengineering process for product reengineering needs to be created on the basis of the results of these earlier steps that will provide for the reengineering of complete products, or reengineering of systems, and for incremental reengineering efforts that are phased in over time.

8. There must be major provisions for education and training such that it becomes possible, through education and training, to implement whatever reengineering process eventuates.

This is more of a checklist of needed accomplishments for a reengineering process than it is a specification of a life cycle for the process itself. Through perusal of this checklist, we should be able to establish an appropriate process for reengineering in the form of Figures 23.3 or 23.32.

We will not describe the large number of tools available to assist and support in the product reengineering process. These vary considerably across they type of product that is being reengineered. A number of tools for software reengineering are described in Muller (1996) and the extensive bibliography to his paper.

There are several needs that must be considered if a product reengineering process is to yield appropriate and useful results.

1. There is a need to consider long-range organizational and technological issues in developing a product reengineering strategy.

2. There is a need to consider human, leadership, and cultural issues, and how these will be impacted by the development and deployment of a reengineered product as a part of the definition of the specifications for the reengineered product.

3. It must be possible to demonstrate that the reengineering process and product are, or will be, each cost-effective and of high quality, and that they support continued evolution of future capabilities.

4. Reengineered products must be considered within a larger framework that also considers the potential need for reengineering at the levels of systems management and organizational processes, as it will generally be a mistake to assume that technological fixes only will resolve organizational difficulties at these levels.

5. Product reengineering for improved postdeployment maintainability must consider maintainability at the level of process rather than at the level of product only, such as would result in the case of software through rewriting source-code statements. Use of model-based management systems or code generators should yield much greater productivity, in this connection, than rewriting code at the level of source code.

6. Product reengineering must consider the need for reintegration of the reengineered product in with existing legacy systems that have not been reengineered.

7. Product reengineering should be such that increased conformance to standards is a result of the reengineering process.

8. Product reengineering must consider legal issues associated with reverse engineering.

The importance of most of these issues is relatively self-evident. Issues surrounding legality are in a state of flux in product reengineering, in much the same way as they are for benchmarking. They deserve special commentary here.

It is clearly legal for an organization to reverse engineer a product that it owns. Also, there exists little debate at this time on whether inferring purpose from the analysis of the existing functionality of a product and without any attempt to examine the architectural structure or detailed components that make up the existing product, and then recapturing the functionality in terms of a new development effort (the so-called black-box approach), is legal. Doubtless, it is legal. Major questions, however, surround the legality of "white box" reverse engineering, in which the detailed architectural structure and components of a system, including code for software, are examined in order to reverse engineer and reengineer it. The major difficulty appears to surround the fair use provisions in copyright law, and the fact that the fair use provisions are different from those associated with the use of trade secrets for illicit gain. Copyrighted material cannot be a trade secret since the copyright law requires open disclosure of the material that is copyrighted. In particular, software is copyrighted and not patented, so trade-secret restrictions do not apply. There is a pragmatist group that says white-box reengineering is legal, and a constructionist group that says it is illegal (Samuelson, 1990; Sibor, 1990). Those who suggest that it is illegal argue that it is not the obtaining of trade secrets that is illegal, but rather the subsequent use of these for illicit gain. These issues will be the subject of much debate over the near term. Many of the ethical issues in product engineering are similar to those in benchmarking and other approaches to process reengineering.

Some useful guidelines applicable primarily to product reengineering are as follows:

1. Reverse engineering procedures can be performed only on products that are the property of the reengineering organization or that have come into their possession through legal means.

2. There is no patent in existence that would be infringed through a functional clone of the computer program, and we cannot be under a contractual obligation not to reverse engineer the original product.

3. A justifiable procedure for reverse engineering is to apply input signal to the system or product being reengineered, observe the operation of the product in response to these inputs, and characterize the product in a functional manner based upon observed operation. An original product, computer code in the case of software, should then be engineered in order to achieve the functional characteristics that has been observed.

In the case of computer programs, it is permissible to disassemble programs that are available in object-code form, for purposes of understanding the functional characteristics of the programs. The purpose of disassembly is utilized only to discover how the program operates and it may be utilized only for this purpose. The functional operating characteristics of the disassembled computer program may be obtained, but original computer code should be prepared from these characteristics so as to achieve the functional operating characteristics of the original program. This code must serve a functional purpose.

Reengineering is accompanied with a variety of risks that are associated with processes, people, tools, strategies, and the application area for reengineering. These risks can be managed using the methodologies discussed in the handbook chapters concerning risk management and metrics. Arnold (1992) has identified many of these product reengineering needs in the form of risks that must be managed during the reengineering effort. These risks are associated with a variety of factors for product reengineering, as suggested in the following list:

- *Integration risk* is the risk associated with having a reengineered product that cannot be satisfactorily integrated with, or interfaced to, existing legacy systems.
- *Maintenance improvement risk* is the risk that the reengineered product will exacerbate, rather than ameliorate, maintenance difficulties.
- *Systems management risk* is the risk that the reengineered product attempts to impose a technological fix on a situation where the major difficulties are not needed for greater support, but for organizational reengineering at the level of systems management.
- *Process risk* is that associated with having a reengineered product that might well represent an improvement in a situation where the specific organizational process in which the reengineered process is to be used is defective and in need of reengineering.
- *Cost risk* is associated with having major cost overruns in order to obtain a deployed reengineered product that meets specifications.
- *Schedule risk* is associated with having schedule delays in order to obtain a deployed reengineered product that meets specifications.
- *Human acceptance risk* is the risk associated with obtaining a reengineered product that is not suitable for human interaction, or one that is unacceptable to the user organization for other reasons.
- *Application supportability risk* is that risk associated with having a reengineered product that does not really support the application or purpose it was intended to support.
- *Tool and method availability risk* is associated with proceeding with the reengineering of a product based upon promises for a method or tool, needed to complete the effort, that does not become available or that is faulty.
- *Leadership, strategy, and culture risk* is that associated with imposing a technological fix in the form of a reengineered product, in an organizational environment that cannot adapt to the reengineered product.

Clearly, these risks are not mutually exclusive, the risk attributes are not independent, and the listing is incomplete. For example, we could surely include legal and ethical risks. We can use this list as the basis for a multiattribute-type utility assessment that could be a part of the model-based management system of a decision support system (Sage, 1991) design that supports risk assessment and management for product reusability.

There are a number of cost and benefit factors that influence cost–benefit analysis of potential reengineering strategies. Some of these are listed below (the time value of money will need to be considered in determining overall costs and benefits):

- Current system
    - Current operating costs
    - Current maintenance costs
    - Current business value
    - Estimated life of current system
- Reengineering process
    - Reengineering risk factor

       Estimated reengineering costs

       Estimated reengineering time

- Reengineered system

       Estimated operating costs of reengineered system

       Estimated business value after reengineering

       Estimated maintenance costs after reengineering

       Estimated life of reengineered system

There is clearly a very close relationship between product reengineering and product reuse. The reengineering of legacy software and the reuse-based production of new software are closely related concepts. It often occurs that the cost of developing software for one or a few applications is approximately the same as the cost of developing domain reuse components and reengineering approaches to legacy software. Ahrens and Prywes (1995) describe some of these relations in an insightful work.

23.4   SUMMARY

In this chapter, we have considered a number of issues related to systems reengineering. We indicated that reengineering can take place at either, or all, of the levels of

- Product
- Process
- Systems management

Reengineering at any of these levels is related to reengineering at the other two levels. Reengineering can be viewed from the perspective of the organization fielding a product as well as from the perspective of the customer, individual, or organization receiving the product. From the perspective of either of these, it may well turn out to be the case that reengineering at the level of product only may not be fully meaningful if this is not also associated, and generally driven by, reengineering at the levels of process and systems management. For an organization to reengineer a product when it is in need of reengineering at the levels of systems management and/or process is almost a guarantee of a reengineered product that will not be fully trustworthy and cost-efficient. An organization that contracts for product reengineering when it is in need of reengineering at the levels of systems management and/or process is asking for a technological fix and a symptomatic cure for difficulties that are institutionally and value-related. Such solutions are not really solutions at all.

Thus, there are potential needs for reengineering at the levels of product, process, and systems management. While it may well be the case, for example, that product reengineering may occupy many resources, the combined total of resources needed for systems management and process reengineering may be not insubstantial. Resources expended upon product reengineering only, and with no investigation of needs at the systems management and process levels, may well not be wise expenditures, from either the perspective of the organization producing the product or the one consuming it.

In an insightful study (Hall et al., 1993), it is indicated that organizations often squander resources that look very promising, but which fail to produce long-lasting results of value for the organization. Four major ways to fail are identified:

1. Assigning average performers to the reengineering effort, often because the more valuable people are needed for other more important efforts, will guarantee mediocre performance of the product of the reengineering effort.
2. Measuring the reengineering plan and activities only, and not the results, will often produce deceptive measurement results.
3. Allowing new and innovative ideas for reengineering to be squelched through opportunistic politics and extreme risk aversion will preserve the status quo rather than encourage implementation of beneficial activities in the form or results.
4. Failing to communicate wisely and widely during implementation will almost always frustrate success.

To this list, we might add failure to obtain real commitment from the highest levels of the organization for the reengineering effort. It might be argued, of course, that the failure of commitment leads to such things as assignment of average and mediocre performers to the reengineering effort. These authors also offer five factors said to enhance success at reengineering:

1. Set aggressive reengineering performance targets in terms of results.
2. Commit a significant portion of the CEO's time to the reengineering effort, especially during deployment of reengineering operations.
3. Assign a very senior executive to head the reengineering effort, especially during deployment.
4. Perform a comprehensive review and analysis of customer needs, organizational realities, strategic economic issues, and market trends as a prelude to reengineering.
5. Conduct a pilot study and prototype the reengineering effort in order to obtain results useful both to refining the reengineering process and to enhancing communications and building enthusiasm.

This study was based primarily on organizational, or systems management, reengineering efforts. However, there are clear implications in these suggestions for all three types of reengineering efforts.

In an insightful article, Venkatraman (1994) identifies five levels for organizational transformation through information technology. We can expand on this slightly through adoption of the three levels for reengineering we have described here and obtain the representation shown in Figures 23.33 through 23.35. These figures show our representation of these five levels, two for organizational reengineering, two for product reengineering, and one for process reengineering. These can be expanded to a greater number of levels. It would be reasonable to identify integration levels for processes and systems management, as well as for products.

Venkatraman notes technological and organizational enablers and inhibitors that will affect desired transformations at both evolutionary and revolutionary levels of transfor-

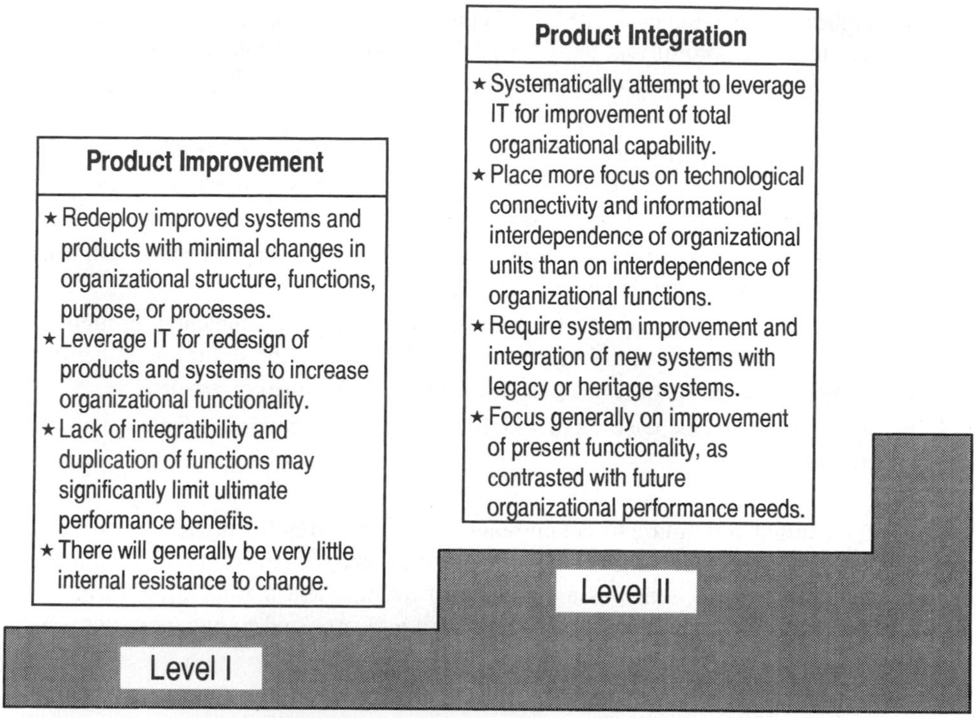

Product Integration

★ Systematically attempt to leverage IT for improvement of total organizational capability.
★ Place more focus on technological connectivity and informational interdependence of organizational units than on interdependence of organizational functions.
★ Require system improvement and integration of new systems with legacy or heritage systems.
★ Focus generally on improvement of present functionality, as contrasted with future organizational performance needs.

Product Improvement

★ Redeploy improved systems and products with minimal changes in organizational structure, functions, purpose, or processes.
★ Leverage IT for redesign of products and systems to increase organizational functionality.
★ Lack of integratibility and duplication of functions may significantly limit ultimate performance benefits.
★ There will generally be very little internal resistance to change.

Level II

Level I

Figure 23.33 Representation of two levels for product reengineering and associated characteristics.

mation. The technological enablers include increasingly favorable cost-effectiveness trends for various information technologies and enhanced connectivity possibilities. Technological inhibitors include the lack of currently established standards that are universally accepted and the rapid obsolescence of current technologies. Organizational enablers include managerial awareness of the need for change and existing leadership. Organizational inhibitors include financial limitations and managerial resistance to change. While both product reengineering and organizational reengineering desires will ultimately lead to change in organizational processes, changes for the purpose of producing a product with greater cost-effectiveness, quality, and (external) customer satisfaction will be generally different and more limited in scope than those made for the purpose of improvement in internal responsiveness to satisfaction of present and future customer expectations.

Top-down directed changes, from Level V to lower levels is often directed at capability and effectiveness enhancement. Efforts directed from Level I up are generally concerned with efficiency enhancement. It is generally at the level of improved processes that enhancement in efficiency and effectiveness may both be realized.

It is generally the case that organizations should develop strategy first, then determine appropriate processes, and then choose appropriate information technology and other products that are most appropriate. There are exceptions, however, and an interesting case study is described in Yetton et al. (1984) of an organization, with a high organizational learning capacity and a mature approach to risk management, in which incremental adoption of information technology was a driver of strategic change.

Process Reengineering

★ Systematically attempt to redesign new innovative processes for the production of more cost effective, higher quality, and customer responsive products.
★ This may be implemented in response to competitor challenges and/or to better satisfy customers.
★ This may be restricted to a single process, implemented in response to operational crises, and without potential need for organizational culture and leadership change if implemented for product improvement reasons only.
★ This may be responsive to planned organizational change issues.

Level III

Figure 23.34 Representation of process reengineering and associated characteristics.

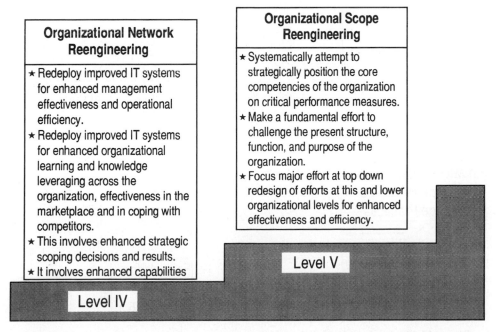

Organizational Network Reengineering

★ Redeploy improved IT systems for enhanced management effectiveness and operational efficiency.
★ Redeploy improved IT systems for enhanced organizational learning and knowledge leveraging across the organization, effectiveness in the marketplace and in coping with competitors.
★ This involves enhanced strategic scoping decisions and results.
★ It involves enhanced capabilities

Level IV

Organizational Scope Reengineering

★ Systematically attempt to strategically position the core competencies of the organization on critical performance measures.
★ Make a fundamental effort to challenge the present structure, function, and purpose of the organization.
★ Focus major effort at top down redesign of efforts at this and lower organizational levels for enhanced effectiveness and efficiency.

Level V

Figure 23.35 Representation of two levels for organizational reengineering and associated characteristics.

So, change can be initiated at any of these levels. Systems management deals with appropriate changes at all of these three levels and with efficiency and effectiveness, and also with the explicability and equity issues necessary to bring this about and to ensure a better tomorrow. It involves a number of perspectives and has roles for a great many professionals. Some of these are illustrated in Figure 23.36. This figure shows some of the many roles and identities associated with systems engineering and systems management, and a number of three-level concerns that form an inherent part of this professional area of effort and that have been discussed here.

We should be careful in distinguishing sincere efforts at reengineering from hype and fad management programs. As we noted earlier in this chapter, a number of approaches to reengineering, particularly process reengineering, have failed, generally because they did not adhere to fundamental principles and strategies that might pay off in the long run — both in a more productive organization. It is only natural to ask "How much of this is hype, and how much of it is meaningful dialog?" We need to be able to separate the two. In an insightful work, Eccles et al. (1992) seek to discover the essence of management and to separate this from much of the speechcraft describing management actions. They suggest seven underlying principles of management-initiated action:

1. Managers often act under conditions of uncertainty. Full, complete, and perfect information will generally never be obtained. Thus, risk management is an important facet of management.

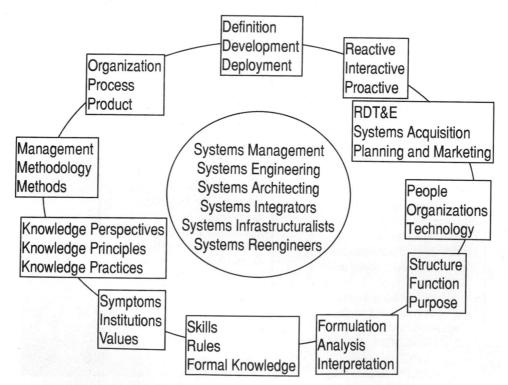

Figure 23.36 Some of the roles for and facets of systems engineering and management.

2. Management constantly allows for and provides flexibility. Often, situations will change and what constitutes appropriate action at one time may well have to be changed in order to adapt to changing conditions.

3. Management is politically astute. It is recognized that organizational and opportunistic politics will generally arise, despite the potential harm that may be done to the organization in the long run. Skillful managers become knowledgeable about the potentially hidden agendas and actions of others and attempt either to thwart these or to reorganize them as appropriate to best achieve organizational objectives.

4. Management has a critical sense of time. They know when to hold the cards, and when to fold the cards. They know when to act, and when to hide or run. They know that what is an appropriate action today may well not be an appropriate action tomorrow.

5. Management is necessarily associated with good abilities at judgment and decision making.

6. Management is able to use rhetoric and oratory effectively.

7. Management must necessarily deal with many agendas in a simultaneous or concurrent fashion.

Many of these observations relate to the observations of Winograd and Flores (1986), who indicate that managers:

- Cannot avoid acting.
- Cannot retreat to contemplate on potential actions.
- Cannot predict the outcome of actions with certainty.
- Do not have a stable assessment or representation of the situation.
- Cannot provide unequivocal truth, but instead can only provide representations of actions and outcomes.
- Must realize that their language must have action interpretations by others.

In short, management is necessarily down-to-earth and pragmatic.

Winograd and Flores suggest four language needs in strategic statements that are intended to define organizational purpose and to serve as guides for organizational action:

1. The core concepts of the language must provide fresh and powerful insights.

2. The language must be such as to allow clear guidelines for action to be defined.

3. The language must enhance the ability to communicate strategy and clarify the implications of actions in terms of accomplishing purposes of the organization.

4. It should be possible to use the language for multiple organizational purposes.

In this chapter, we have examined a considerable number of approaches for improvements in organizational productivity. It would be especially interesting to evaluate the extent to which the "new languages" offered by these approaches comply with these guidelines.

Strategy and strategic plans should lead to action. Any organization will have a number of past and present strengths and weaknesses. At present, and in the future, there will be a number of external threats and opportunities. Strategy is intended to traverse these internal characteristics, generally associated with the past and the present, so as to enable the organization to cope with present and future threats and opportunities. From the perspective of actions to follow, effective strategies are said to have the following desirable qualities:

- They act as a bridge that traverses the past, the present, and the future.
- They are both consciously planned and opportunistic.
- They involve a very wide range of alternative courses of action.

It is necessary that organizational processes and structure be tuned to strategy and that systematic measurements, which are generally at the level of process and product, provide useful guidelines for strategy. These observations lead to a number of useful suggestions to put new ideas into a useful perspective and focus on the action and pragmatism that is management. As well documented by Mintzberg et al. (1998), there has been an evolution in strategic thought and strategic management over the years. Clearly, this evolution continues.

Effective management pays particular attention to the human element. In a recent work (Fitz-Enz, 1997), eight practices of exceptional companies are described:

- Balanced value fixation
- Commitment to a core strategy
- Culture−system linkage
- Massive two-way communication
- Partnering with stakeholders
- Functional collaboration
- Innovation and risk
- Never being satisfied

In addition, guidelines are presented that will enable the enduring human-asset management practices that will make such efforts as reengineering long-term successes, for the *human element* is a major part of reengineering, just as it is in all of systems engineering and management.

REFERENCES

Abell, D. F. (1993). *Managing with Dual Strategies*. New York: Free Press.

Air Force Material Command Guide (1993). *Integrated Product Development*. Washington, DC: U.S. Air Force.

Ahrens, J. D., and Prywes, N. S. (1995). Transition to a legacy and reuse based software life cycle. *IEEE Comput.* **28**(10), 27–36.

Anderson, D. M. (1997). *Agile Product Development for Mass Customization: How to Develop and Deliver Products for Mass Customization, Niche Markets, JIT, Built to Order, and Flexible Manufacturing*. New York: McGraw-Hill.

Andrews, C. C., and Leventhal, N. S. (1993). *FUSION - Integrating IE, CASE, and JAD: A Handbook for Reengineering the Systems Organization*. Englewood Cliffs, NJ: Prentice-Hall.

Andrews, C. C., and Stalick, S. K. (1994). *Business Reengineering: The Survival Guide*. Englewood Cliffs, NJ: Yourdon Press Prentice-Hall.

Anthony, R. N. (1988). *The Management Control Function*. Boston: Harvard Business School Press.

Argyris, C., and Schön D. A. (1996). *Organizational Learning II: Theory and Practice*. Reading, MA: Addison-Wesley.

Arnold, R. S. (1992). Common risks of reengineering. *IEEE Comput. Soc. Reverse Eng. Newsl.*, April, pp. 1–2.

Arnold, R. S. (ed.). (1993). *Software Reengineering*. Los Altos, CA: IEEE Computer Society Press.

Bell, D. (1973). *The Coming Post-Industrial Society: A Venture in Social Forecasting*. New York: Basic Books.

Benjamin, R. J., and Blunt, J. (1992). Critical IT issues: The next ten years. *Sloan Manage. Rev.* **33**(4), 7–19.

Blattberg, R. C., Glazer, R., and Little, J. D. C. (eds.). (1994). *The Marketing Information Revolution*. Boston: Harvard Business School Press.

Bloomfield, B. P., Coombs, R., Knights, D., and Littler, D. (eds.). (1997). *Information Technology and Organizations: Strategy, Networks, and Integration*. New York: Oxford University Press.

Boar, B. H. (1993). *The Art of Strategic Planning for Information Technology*. New York: Wiley.

Bowen, H. K., Clark, K. B., Holloway, C. A., and Wheelwright, S. C. (eds.). (1994). *The Perpetual Enterprise Machine: Seven Keys to Corporate Renewal Through Successful Product and Process Development*. New York: Oxford University Press.

Bower, J. L., and Hout, T. M. (1988). Fast-cycle time capability for competitive power. *Harv. Bus. Rev.* **66**(6), 110–118.

Bowman, E., and Kogut, B. (eds.). (1995). *Redesigning the Firm*. New York: Oxford University Press.

Boynton, A. C., Jacobs, G. C., and Zmud, R. W. (1992). Whose responsibility is IT management. *Sloan Manage. Rev.* **33**(4), 32–38.

Brandin, D. H., and Harrison, M. A. (1987). *The Technology War: A Case for Competitiveness*. New York: Wiley.

Burk, K. B., and Webster, D. W. (1994). *Activity Based Costing*. Fairfax, VA: American Management Systems.

Camp, R. C. (1989). *Benchmarking: The Search for Industry Best Practices That Lead to Superior Performance*. Milwaukee, WI: Quality Press, American Society for Quality Control.

Carter, D. E., and Baker, B. S. (1992). *Concurrent Engineering: The Product Development Environment for the 1990s*. Reading, MA: Addison-Wesley.

Champy, J. S. (1995). *Reengineering Management: The Mandate for New Leadership*. New York: HarperCollins.

Chikofsky, E., and Cross, J. H. (1990). Reverse engineering and design recovery, a taxonomy. *IEEE Software* **7**(1), 13–17.

Clausing, D. (1994). *Total Quality Development: A Step by Step Guide to World-Class Concurrent Engineering*. New York: ASME Press.

Cohen, S. S., and Zysman, J. (1987). *Manufacturing Matters: The Myth of the Post Industrial Economy*. New York: Basic Books.

Cross, J. H., II, Chikofsky, E. J., and May, C. H., Jr. (1992). Reverse engineering. (Yovits, M. C. ed.), *Adv. Comput.* Academic Press **35**, 199–254.

Davenport, T. H. (1993). *Process Innovation: Reengineering Work through Information Technology*. Boston: Harvard Business School Press.

Davenport, T. H. (1997). *Information Ecology: Mastering the Information and Knowledge Environment*. New York: Oxford University Press.

Davenport, T. H., and Short, J. E. (1990). The new industrial engineering: Information technology and business process redesign. *Sloan Manage. Rev.* **31**(4), 11–27.

Davenport, T. H., Eccles, R. G., and Prusak, L. (1992). Information politics. *Sloan Manage. Rev.* **34**(1), 53–65.

Day, G. S. (1990). *Market Driven Strategy: Processes for Creating Value.* New York: Free Press.

Denton, D. K. (1991). *Horizontal Management: Beyond Total Customer Satisfaction.* New York: Lexington Books.

Department of Defense (DoD) (1996). Guide to Integrated Product and Process Development, (Feb 5, 1996), office of the Under Secretary of Defense (Acquisition and Technology), Washington DC.

Dertouzos, M. L., Lester, R. K., and Solow, R. M. (1989). *Made in America: Regaining the Productive Edge.* Cambridge, MA: MIT Press.

Eccles, R. G., Norhia, N., and Berkley, J. D. (1992). *Beyond the Hype: Rediscovering the Essence of Management.* Boston: Harvard Business School Press.

Editors of Business Week. (1993). The horizontal cooperation. *Bus. Week,* December 20, pp. 76–81.

Fiksel, J. (1993). Computer aided requirements management for environmental excellence. *Proc. Annu. Meet. Natl. Counc. Syst. Eng.* Alexandria, VA, pp. 251–258.

Fischer, L. (ed.). (1995). *New Tools for New Times: The Workflow Paradigm: The Impact of Information Technology on Business Process Reengineering:* New York: Future Strategies.

Fishbein, M., and Azjin, I. (1975). *Belief, Attitude, Intention, and Behavior.* Reading, MA: Addison-Wesley.

Fitz-Enz, J. (1997). *The 8 Practices of Exceptional Companies: How Great Organizations Make the Most of Their Human Assets.* New York: AMACOM Publishers.

Gilmore, J. H., and Pine, B. J., II. (1997). The four faces of mass customization. *Harv. Bus. Rev.* **75**(1), 91–101.

Grant, R. M., Shanu, R., and Krishnan, R. (1994). TQM's challenge to management theory and practice. *Sloan Manage. Rev.* **35**(2), 11–24.

Hall, G., Rosenthal, J., and Wade, J. (1993). How to make reengineering really work. *Harv. Bus. Rev.* **71**(6), 119–131.

Hamel, G., and Prahalad, C. K. (1994). *Competing for the Future: Breakthrough Strategies for Seizing Control of your Industry and Creating the Markets of Tomorrow.* Boston: Harvard Business School Press.

Hammer, M. (1990). Reengineering work: Don't automate, obliterate. *Harv. Bus. Rev.* **68**(4), 104–112.

Hammer, M. (1996). *Beyond Reengineering: How the Process-Centered Organization is Changing Our Work and Our Lives.* New York: Harper Business.

Hammer, M., and Champy, J. (1993). *Reengineering the Corporation: A Manifesto for Business Revolution.* New York: Harper Business.

Hammer, M., and Stanton, S. (1995). *The Reengineering Revolution.* New York: Harper Business.

Hannaford, S. (1996). *Workflow Reengineering.* New York: Hayden Publ. Co.

Hansen, G. A. (1994). *Automating Business Process Reengineering: Breaking the TQM Barrier.* Englewood Cliffs, NJ: Prentice-Hall.

Harmon, R. L., and Peterson, L. D. (1990). *Reinventing the Factory: Productivity Breakthroughs in Manufacturing Today.* New York: Free Press.

Harrington, H. J. (1991). *Business Process Improvement: The Breakthrough Strategy for Total Quality, Productivity, and Competitiveness.* New York: McGraw-Hill.

Harrington, H. J., and Harrington, H. J. (1995). *Total Improvement Management: The Next Generation in Improvement Performance*. New York: McGraw-Hill.

Harrington, H. J., and Harrington, J. S. (1996a). *High Performance Benchmarking: 20 Steps to Success*. New York: McGraw-Hill.

Harrington, H. J., and Harrington, J. S. (1996b). *The Complete Benchmarking Implementation Guide: Total Benchmarking Management*. New York: McGraw-Hill.

Harvard Business Review Editors. (1990). *Revolution in Real Time: Managing Information Technologies in the 1990s*. Boston: Harvard Business School Press.

Hicks, D. T. (1992). *Activity Based Costing for Small and Mid-Sized Businesses: An Implementation Guide*. New York: Wiley.

Howard, W. G., Jr., and Guile, B. R. (1992). *Profiting from Innovation: The Report of a Three Year Study from the National Academy of Engineering*. New York: Free Press.

Huber, G. P., and Glick, W. H. (eds.). (1993). *Organizational Change and Redesign: Ideas and Insights for Improving Performance*. New York: Oxford University Press.

Hunt, V. D. (1993). *Reengineering: Leveraging the Power of Integrated Product Development*. Essex Junction, VT. Oliver Wright Publications.

Institute of Electrical and Electronics Engineers (IEEE). (1991). *Software Engineering Glossary, IEEE Software Engineering Standards*. New York: IEEE Press.

Institute of Electrical and Electronics Engineers (IEEE). (1992). *IEEE Standard for Software Maintenance* (IEEE P1219/D14). New York: IEEE Standards Dept.

Johanson, H. J., McHugh, P., Pendlebury, A. J., and Wheeler, W. A., III. (1993). *Business Process Reengineering: Breakthrough Strategies for Market Dominance*. Chichester: Wiley.

Kaplan, R. S., and Cooper, R. (1997). *Cost & Effect: Using Integrated Cost Systems to Drive Profitability and Performance*. New York: McGraw-Hill.

Katzenbach, J. R. (1997). *Teams at the Top: Unleashing the Potential of Both Teams and Individual Leaders*. Boston: Harvard Business School Press.

Katzenbach, J. R., and Smith, D. K. (1993). *The Wisdom of Teams*. Boston: Harvard Business School Press.

Keen, P. G. W. (1991). *Shaping the Future, Business Design through Information Technology*. Boston: Harvard Business School Press.

Keen, P. G. W. (1997). *The Process Edge: Crating Value Where It Counts*. Boston: Harvard Business School Press.

Kennedy, G. (eds.). (1994). *Field Guide to Negotiations*. Boston: Harvard Business School Press.

Kobellas, J. (1997). *Workflow Management Strategies*. Boston: I D G Books Worldwide.

Kronlöf, K. (ed.). (1993). *Method Integration: Concepts and Case Studies*. Chichester: Wiley.

Kusiak, A. (ed.). (1993). *Concurrent Engineering: Automation, Tools, and Techniques*. New York: Wiley.

Lacity, M. C., and Hirschheim, R. (1993a). The information systems outsourcing bandwagon. *Sloan Manage. Rev.* 35(1),73–86.

Lacity, M. C., and Hirschheim, R. (1993b). *Information Systems Outsourcing: Myths, Metaphors, and Realities*. Chichester: Wiley.

Lewin, K. (1947). Group decision and social change. In *Readings in Social Psychology* (T. N. Newcomb and E. L. Hartley, eds.). Troy, MO: Holt, Rinehart & Winston.

Liebfried, K. H. J., and McNair, C. J. (1992). *Benchmarking: A Tool for Continuous Improvement*. New York: HarperCollins.

Mankin, D., Cohen, S. G., and Bikson, R. (1996). *Teams and Technology: Fulfilling the Promise of the New Organization*. Boston: Harvard Business School Press.

Manz, C. C., and Sims, H. P., Jr. (1993). *Business Without Bosses*. New York: Wiley.

Maromonte, K. R. (1998). *Corporate Strategic Business Sourcing.* New York: Greenwood Publishing Group.

McCutcheon, D. M., Raturi, A. S., and Meredith, J. R. (1994). The customization-responsiveness squeeze. *Sloan Manage. Rev.* **35**(2), 89–99.

McGee, J., and Prusak, L. (1993). *Managing Information Strategically: Increase Your Company's Competitiveness and Efficiency by Using Information as a Strategic Tool.* New York: Wiley.

Meyer, C. (1993). *Fast Cycle Time: How to Align Purpose, Strategy, and Structure for Speed.* New York: Free Press.

Meyer, M. H., and Utterback, J. M. (1993). The product family and the dynamics of core capability. *Sloan Manage. Rev.* **34**(3) 29–48.

Millson, M. R., Raj, S. P., and Wilemon, D. (1992). A survey of major approaches for accelerating new product development. *J. Prod. Innovat. Manage.* **9**(1), 53–69.

Mintzberg, H. (1994a). The rise and fall of strategic planning. *Harv. Bus. Rev.* **72**(1), 107–114.

Mintzberg, H. (1994b). *The Rise and Fall of Strategic Planning.* New York: Free Press.

Mintzberg, H., and Quinn, J. B. (1998). *Readings in the Strategy Process.* Upper Saddle River, NJ: Prentice-Hall.

Mintzberg, H., Ahlstrand, B., and Lampel, J. (1998). *Strategy Safari: A Guided Tour through the Wilds of Strategic Management.* New York: Free Press.

Morton, M. S. S. (ed.). (1991). *The Corporation of the 1990s: Information Technology and Organizational Transformation.* New York: Oxford University Press.

Muller, H. A. (1996). Understanding software systems using reverse engineering technologies research and practice. *Proc. Int. Conf. Software Eng.,* Berlin.

Nevins, J. L., and Whitney, D. E. (eds.). (1989). *Concurrent Design of Products and Processes: A Strategy for the Next Generation in Manufacturing.* New York: McGraw-Hill.

Office of the Secretary of Defense ASD. (1994). (C³l), *The DoD Enterprise Model,* Vols. 1 and 2. Washington, DC: U.S. Department of Defense.

Olsem, M. R. (1993). Preparing to reengineer. *IEEE Comput. Soc. Reverse Eng. Newsl.* December, pp. 1–3.

Ould, M. A. (1995). *Business Processes: Modeling and Analysis for Re-engineering and Improvement.* New York: Wiley.

Peters, T. (1992). *Liberation Management: Necessary Disorganization for the Nanosecond Nineties.* New York: Knopf.

Pfeffer, J. (1994). *Competitive Advantage Through People.* Boston: Harvard Business School Press.

Pine, B. J., II. (1993). *Mass Customization: The New Frontier in Business Competition.* Boston: Harvard Business School Press.

Pine, B. J., II Victor, B., and Boynton, A. C. (1993). Making mass customization work. *Harv. Bus. Rev.* **71**(5), 108–119.

Porter, M. E. (ed.) (1991a), *On Competition and Strategy.* Harv. Bus. Rev. Paperback 90079. Boston: Harvard Business School Press.

Porter, M. E. (ed.) (1991b). *The State of Strategy,* Haro. Bus. Rev. Paperback 90082. Boston: Harvard Business School Press.

Porter, M. E. (1998a). *Competitive Advantage: Creating and Sustaining Superior Performance.* New York: Free Press.

Porter, M. E. (1998b). *Competitive Strategy: Techniques for Analyzing Industries and Competitors.* New York: Free Press.

Porter, M. E. (1998c). *The Competitive Advantage of Developing Nations.* New York: Free Press.

Porter, M. E. (1998d). *The Competitive Advantage of Nations.* New York: Free Press.

Prahalad, C. K., and Hamel, G. (1990). The core competence of the corporation. *Harv. Bus. Rev.* **68**(3), 60–74.

Prasad, B. (1996a). *Concurrent Engineering Fundamentals: Integrated Product and Process Organization.* Upper Saddle River, NJ: Prentice Hall.

Prasad, B. (1996b). *Concurrent Engineering Fundamentals: Integrated Product Development.* Upper Saddle River, NJ: Prentice Hall.

Quinn, J. B. (1992). *Intelligent Enterprise: A Knowledge and Service Based Paradigm for Industry.* New York: Free Press.

Quinn, J. B., Paquette, P. C., and Doorley, T. (1990a). Beyond products: Service based strategies. *Harv. Bus. Rev.* **68**(3), 58–68.

Quinn, J. B., Paquette, P. C., and Doorley, T. (1990b). Beyond products: creating organizational revolutions. *Sloan Manage. Rev.,* Winter, pp. 67–78.

Quinn, J. B., Paquette, P. C., and Doorley, T. (1990c). Technology in services: Rethinking strategic focus. *Sloan Manage. Rev.,* Winter, pp. 79–87.

Raiffa, H. (1982). *The Art and Science of Negotiation.* Cambridge, MA: Belknap.

Rekoff, M. G., Jr. (1985). On reverse engineering. *IEEE Transa. Syst., Man, Cybernet.* **15**(2), 244–252.

Robertson, T. S. (1993). How to reduce market penetration time. *Sloan Manage. Rev.* **35**(1), 87–96.

Rockart, J. F. (1979). Chief executives define their own data needs. *Harv. Bus. Rev.* **57**(2), 81–93.

Rockart, J. F., and Bullen, C. V. (eds.). (1986). *The Rise of Managerial Computing.* Homewood, IL: Dow Jones-Irwin.

Rockart, J. F., and DeLong, D. W. (1988). *Executive Support Systems: The Emergence of Top Management Computer Use.* Homewood, IL: Dow Jones-Irwin.

Rosenblatt, A., and Watson, G. F. (1991). Concurrent engineering, *IEEE Spectrum,* July, pp. 22–37.

Roussel, P. A., Saad, K. N., and Erickson, T. J. (1991). *Third Generation R&D: Managing the Link to Corporate Strategy.* Boston: Harvard Business School Press.

Rumelt, R. P., Schendel, D. E., and Teece, D. J. (eds.) (1994). *Fundamental Issues in Strategy: A Research Agenda.* Boston: Harvard Business School Press.

Sage, A. P. (1983). *Economic Systems Analysis: Microeconomics for Systems Engineering, Engineering Management, and Project Selection.* New York: North-Holland/Elsevier.

Sage, A. P. (1991). *Decision Support Systems Engineering.* New York: Wiley.

Sage, A. P. (1995). *Systems Management for Information Technology and Software Engineering.* New York: Wiley.

Sage, A. P., and Palmer, J. D. (1990). *Software Systems Engineering.* New York: Wiley.

Samuelson, P. (1990). Reverse engineering someone else's software: Is it legal? *IEEE Software* **7**(1), 90–96.

Savage, C. M. (1990). *Fifth Generation Management: Integrating Enterprises through Human Networking.* Burlington, MA: Digital Press.

Savage, C. M. (1998). *Fifth Generation Management: Co-Creating Through Virtual Enterprising, Dynamic Teaming, and Knowledge Networking.* London: Butterworth-Heinnemann.

Schefstrom, D., and van den Broek, G. (eds.) (1993). *Tool Integration: Environments and Frameworks.* Chichester, UK: Wiley.

Schmenner, R. W. (1988). The merit of making things fast, *Sloan Manage. Rev.* **30**(1), 11–17.

Schnaars, S. P. (1994). *Managing Imitation Strategies: How Later Entrants Seize Markets from Pioneers.* New York: Free Press.

Schoemaker, P. J. H. (1991). When and how to use scenario planning: A heuristic approach with illustrations. *J. Forecast.* **10**, 549–564.

Schoemaker, P. J. H. (1992). How to link strategic vision to core capabilities. *Sloan Manage. Rev.* **34**(1), 67–81.

Shank, J. K., and Govindarajan, V. (1993). *Strategic Cost Management: The New Tool for Competitive Advantage.* New York: Free Press.

Shina, S. G. (1991). *Concurrent Engineering and Design for Manufacture of Electronics Products.* New York: Van Nostrand-Reinhold.

Shy, O. (1995). *Industrial Organization: Theory and Applications.* Cambridge, MA: MIT Press.

Sibor, V. (1990). Interpreting reverse engineering law. *IEEE Software* **7**(4), 4–10.

Smith, P. G., and Reinertsen, D. G. (1991). *Developing Products in Half the Time.* New York: Van Nostrand-Reinhold.

Sneed, H. M. (1995). Planning the reengineering of legacy systems. *IEEE Software* **12**(1), 24–34.

Spewak, S. H. (1992). *Enterprise Architecture Planning: Developing a Blueprint for Data, Applications, and Technology.* New York: Wiley.

Spurr, K., Layzell, P., Jennison, L., and Richards, N. (1994). *Software Assistance for Business Re-Engineering.* New York: Wiley.

Stalk, G., Jr. (1988). Time - The next source of competitive advantage. *Harv. Bus. Rev.* **66**(4), 41–51.

Stalk, G., Jr., and Hout, T. M. (1990). *Competing Against Time: How Time Based Competition is Reshaping Global Markets.* New York: Free Press.

Stalk, G., Jr., Evans, P., and Shulman, L. E. (1992). Competing on capabilities: The new rules of corporate strategy. *Harv. Bus. Rev.* **70**(2), 57–63.

Strebel, P. (1992). *Breakpoints: How Managers Exploit Radical Business Change.* Boston: Harvard Business School Press.

Thomas, P. R. (1990). *Competitiveness through Total Cycle Time.* New York: McGraw-Hill.

Thomas, R. J. (1993). *New Product Development: Managing and Forecasting for Strategic Success.* New York: Wiley.

Thurow, L. C. (1992). *Head to Head: The Coming Economic Battle Among Japan, Europe, and America.* New York: Morrow.

Thurow, L. C. (1996). *The Future of Capitalism: How Today's Economic Forces Shape Tomorrow's World.* New York: Morrow.

Ulrich, W. M. (1993). Re-engineering: Defining an integrated migration framework. In *Software Reengineering* (R. S. Arnold, ed.), pp. 108–118. Los Altos, CA: IEEE Computer Society Press.

Usher, J. M., Roy, U., and Parsaci, H. R. (1998). *Integrated Product and Process Development: Methods, Tools, and Technologies.* New York: Wiley.

Venkatesan, R. (1992). Strategic sourcing: To make or not to make. *Harv. Bus. Rev.* **70**(6), 98–107.

Venkatraman, N. (1994). IT-enabled business transformation: From automation to business scope redefinition. *Sloan Manage. Rev.* **35**(2), 73–88.

von Hippel, E. (1988). *The Sources of Innovation.* Oxford, UK. Oxford University Press.

Watson, G. H. (1992). *The Benchmarking Workbook: Adapting Best Practices for Performance Improvement.* Cambridge, MA: Productivity Press.

Watson, G. H. (1993). *Strategic Benchmarking: How to Rate Your Company's Performance Against the World's Best.* New York: Wiley.

Watson, G. H., Bookhart, S., et al. (1992). Applying moral and legal considerations to benchmarking protocols, In *Planning, Organizing, and Managing Benchmarking: A User's Guide,* App. 2. Houston, TX: American Productivity and Quality Center.

Willcoks, L. P., and Lacity, M. C. (eds.). (1998). *Strategic Sourcing of Information Systems: Perspectives and Practices.* New York: Wiley.

Winner, R. I., Pennell, J. P., Bertrand, J. P., and Slusarczuk, M. M. G. (1988). *The Role of Concurrent Engineering in Weapons System Acquisition,* Tech. Rep. R-338. Alexandria, VA: Institute for Defense Analyses.

Winograd, T., and Flores, F. (1986). *Understanding Computers and Cognition.* Reading, MA: Addison-Wesley.

Yetton, P. W., Johnson, K. D., and Craig, J. F. (1994). Computer aided architects: A case study of IT and strategic change. *Sloan Manage. Rev.* **35**(4), 53–68.

Zuboff, S. (1988). *In the Age of the Smart Machine: The Future of Work and Power.* New York: Basic Books.

24 Issue Formulation

JAMES E. ARMSTRONG, JR.

24.1 INTRODUCTION: PROBLEM AND ISSUE FORMULATION

Formulation is the most important, as well as one of the most challenging and rewarding, part of any systems engineering effort. It is crucial because the main product of a systems engineering formulation effort is a set of feasible alternatives. Eventually, one of the alternatives generated during formulation will be selected for implementation. No matter how elegantly the alternatives are analyzed and modeled or how thoroughly you plan for implementation, the value of the entire effort depends on having developed good alternatives in the first place.

Unfortunately, many books on systems engineering and problem solving, in general, pay scant attention to this critical step. Often only a superficial treatment of formulation is offered. A major contribution of this chapter is the detailed treatment of the all-important formulation step. This chapter covers formulation methods and techniques with examples that should help to unveil the mystery surrounding formulation. Those who master the art and science of formulation are the most sought after and highly rewarded systems engineers. A problem well begun is a problem half done.

Formulation begins with a definition of the problem or design issue to be resolved and an assessment of the situation surrounding the problem. The term *situation assessment* is important to understand. Situation implies that there are many aspects of the problem and its environment that we need to know about and that many of these factors are dynamic and subject to change with the passage of time. Assessment means that we need to make a careful appraisal of these various factors. Generating good alternatives, ones that truly satisfy the effective needs of our customers and significantly impact our clients' objectives, depends on an in-depth situation assessment.

24.2 SITUATION ASSESSMENT

A complete situation assessment contains three parts. Each part answers one of three important questions, as shown in Figure 24.1. The first part is the definition of the goal or set of objectives. This normative component of a situation assessment answers the question, *What should be?* The second part of a situation assessment examines the current status or position of things relative to the goal. It is a descriptive component and answers the question, *What is?* Understanding all environmental factors influencing the past, present, and future is the third part of a situation assessment. This is the horizon component and it answers the question, *What factors over time matter?* Therefore, we have a complete situation assessment once we know what we want to accomplish, the current state of affairs

Handbook of Systems Engineering and Management, Edited by A. P. Sage and W. B. Rouse
ISBN 0471-15405 9   ©1999 John Wiley and Sons, Inc.

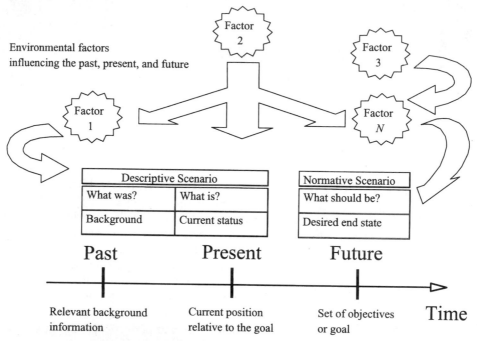

Figure 24.1   The parts of a complete situation assessment.

relative to our goal, and the relevant factors in the environment and their associated time frame.

As you can see, situation assessment is very important, since it leads to the end result of any systems engineering effort, which like any other engineering endeavor, is a plan. The plan answers the analytic question, *What to do and how?* In other words, a situation assessment is an important first step in determining the best alternative from the goal definition, knowledge of the current state, and knowledge of the environment and time frame. By adding the analytic component to situation assessment, we have the four key components of planning (Sage, 1977):

1. Definition of a goal.
2. Knowledge of the current position with respect to the goal.
3. Knowledge of the environmental factors influencing the past, present, and future.
4. Determination of a plan to achieve the goal given knowledge of components 1 and 2.

Systems engineers excel at planning because it is an innovative process that requires both broad comprehension and goal-oriented discipline. Usually, a systems engineering plan consists of the user requirements and system specifications for the recommended alternative. Figure 24.2 shows the various parts of a typical systems engineering design or plan. To be relevant and attractive, plans must be framed within the context of the

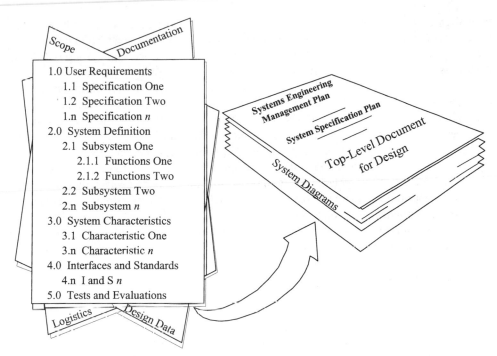

Figure 24.2 Typical systems engineering plan.

situation in which an issue is imbedded or within the context of the environment in which a system is to be deployed.

There is a large variety of environments in which various systems must operate. A number of different perspectives—economic, legal, political, environmental, and social—can be important for understanding a system's environment. When an adequate appreciation of all the important perspectives is achieved, then we say we have preserved the contextual integrity of the system or design issues under consideration. Developing worthwhile alternatives depends on an accurate and timely situation assessment.

Making such a situation assessment depends on several factors. First, the experience that the systems engineer has with the design issues or system under consideration is very important. For example, it is unlikely that someone with very little background knowledge about nuclear power plants would be able to develop, without considerable study, useful alternatives for ensuring safe operation of such a facility. But you need more than just knowledge about the specific problem or design issues at hand.

Second, the familiarity that someone has with the environment into which the system is imbedded is also crucial. For example, the design of a successful information system for airline reservation agents is made much easier by someone with an in-depth knowledge of the various tasks that such agents perform and the environment that agents must cope with as they perform these tasks. You could learn a lot about designing such a system from someone who is an expert at developing human–computer interfaces. But you can also appreciate how helpful it would be to interview some experienced ticket agents and to find out about the difficulties they experience when talking to customers and booking their flights. It is the job of systems engineers to make sure that the relevant knowledge of experienced people is brought into a project. Capturing this knowledge often requires the systems engineer to skillfully conduct both individual interviews and group meetings. A discussion of some useful techniques for this is just a few sections ahead.

On the other hand, extensive familiarity with an existing system may blind someone to truly new and innovative approaches. People with extensive experience with an existing system may find it difficult to think about new solutions, so it is also the job of the systems engineer to overcome the mental stumbling blocks that can be caused by thinking that anchors too strongly on existing approaches. Applying a systems engineering life-cycle process will help the systems engineer avoid these pitfalls. Therefore, even if systems engineers lack specific knowledge about the particular domain of a project, they can make important contributions to the effort by putting their systems methodology and systems management knowledge to good use.

Besides talking to key people, another way of accomplishing a situation assessment is to observe the current system, if one exists. An objective appraisal of the problems with the current system can be very helpful. A simple listing of the observations in a logbook by date and time is often worthwhile. Be sure to note the main ideas of people managing and working with the system. Often, problems will not be evident on the first visit. Subsequent visits will reveal more. People also are more open with you after they get to know you and feel comfortable with you. Information gained by observing the current system and talking to people involved with it, is most valuable for writing a descriptive scenario.

A *descriptive scenario* tells why the current system came into being, how it works, and what problems it has. You should include the main actors or stakeholders involved with the current system in your descriptive scenario. These stakeholders will usually be owners, users, customers, clients, managers, administrators, or regulators of the current system. But, for a complete situation assessment you will also need a normative scenario.

A *normative scenario* describes how the stakeholders want the system to be in the future. This is not as easy as it may at first seem. Often important stakeholders have conflicting opinions of what the future should be. Some have unrealistic expectations of what technology can deliver in the time frame under consideration. Others are reluctant to see any changes because they think their job or relationship to a system may be put in jeopardy. A skilled systems engineer must broker these various interests to create enthusiasm and support for a technically feasible, trustworthy project. One way of gaining support for an improvement to an existing system or for a completely new system, is to involve stakeholders in the process. Helping people visualize the future in meaningful ways will go a long way toward creating a useful normative scenario. But how can you see into the future?

If there is not a current system to observe, then you need to create a way of seeing how an entirely new system might work. A number of useful ways exist for this. A group of people with relevant knowledge can meet and talk through how a new system might work. Some people can be assigned roles to play. For example, a new software system for managing home security could be thought about by people pretending to use the new system as homeowners about to go on vacation. "Now I am using the system to set lights to go on and off at various times in different rooms so it looks like we are at home." says one participant. As ideas emerge, the system engineer can record them.

Many other ways exist for examining possibilities for new systems. Prototypes and simulations can be built and potential users can experiment with them. A *prototype* is a first-cut approximation of what a new system might be. The purpose of a prototype is to quickly gain information from the user about what seems to work and what is needed. For example, a prototype of a new information system might be a set of screens that a user could view and comment as to how useful the screens are in helping provide needed information.

A *simulation* can imitate the operation of a new system. To build a simulation requires a model or some representation of the new system. The purpose of a simulation is to get an idea of how the new system might actually work if it were put into operation. For example, a city wanting to build an airport at a new location might like to know if the land available can accommodate the number of planes that they anticipate will be landing and taking off from their new facility. A simulation of an airport that fits within the city's land constraint can be tested to see if any airport layouts can successfully meet anticipated passenger and freight demands.

To a systems engineer, accomplishing a situation assessment means writing both a descriptive and a normative scenario. It also means making a careful appraisal of the factors that matter as we think about how to get from where we are now to where we want to be tomorrow. Sage (1977) describes some of the factors that matter over time as alterables. *Alterables* are factors that can be changed to help you realize the desired end state or normative scenario. Sage (1977) recommends that you divide the alterable factors that you identify into two categories: *controllable* and *uncontrollable*. Factors that are possible to control are ones that you can change or modify to help you accomplish the project's goals. Therefore, much of your situation assessment effort should be spent on understanding the controllable alterables. Uncontrollable alterables are things like the weather or the current state of technology that may impact your efforts, but which are beyond your abilities to control. Depending on the situation you are analyzing, however, you may need to monitor and understand how the uncontrollable factors can impact your efforts. Once you understand where you are now, where you want to go, and what factors are important to consider, you are well on your way to the main task of systems engineering—building a plan or a set of feasible *alternatives* that can get you from today to a better future.

24.3   PROBLEM OR ISSUE IDENTIFICATION

The toughest part of most systems engineering efforts is identifying the problem correctly. Many engineering efforts, although great feats of engineering skill, fail because they were designed to solve the wrong problem or because the solution created more problems. For example, consider the Aswan High Dam in Egypt built in the 1960s. Hundreds of then Soviet engineers set about and successfully built one of the biggest dams and hydroelectrical facilities in the world. However, as the Egyptians soon found out, the effort was, in many respects, a grand failure because it destroyed much of the precious, arable land in the Nile River valley. The Nile River's natural flooding process had continually replenished the soil downstream and made it possible for many crops to grow in fertile river-bottom land. Once the dam was in place, the natural floods no longer replenished the land and vast amounts of formerly rich agricultural land were lost forever. Egypt, which had been able to feed itself, soon became dependent on food exports. Further, the dam flooded forever many priceless ancient ruins under deep water that were upstream from the massive dam.

Why didn't the engineers consider the upstream and downstream impacts of the project? We can guess that the narrow perspective taken by the engineers led them to conclude that the problem was to block tremendous amounts of water behind a dam for generating electricity and preventing floods. A broader perspective of the problem would have accounted for the upstream ancient ruins and the downstream arable lands. A systems engineer must always consider the potential problems or ramifications caused by any proposed solution to the original problem.

As the Aswan High Dam issue shows, identifying a problem or issue for a large-scale project is not easy, because there is not just one simple question or difficulty to solve. The large-scale projects that systems engineers work on require the identification of multiple questions or difficulties that must be resolved. And there is not a single correct answer. Instead, there are usually a number of alternatives that can be devised, depending on which difficulties the stakeholders believe to be most important. Usually, some success on one important aspect of a large-scale project will come at the expense of success on another part of the project. These tough decisions are called *tradeoffs* by systems engineers. As you can learn in Chapter 28, one of the important skills of a systems engineer is to understand how to analyze and communicate these tradeoffs.

24.3.1 Scoping and Bounding the Problem

How can you successfully identify the relevant questions and difficulties in a large-scale project? Properly scoping and bounding a systems engineering effort is the answer. To *scope* a project means to understand why the project is necessary, what the stakeholders intend to accomplish with the project, and how to measure project success. In systems terms, this means identifying the needs, objectives, and criteria for the project. The *needs* tell why the project is necessary. Related to the needs, the *objectives* describe in detail everything that the project is intended to accomplish. *Objectives measures* or *criteria* measure success in achieving the objectives.

To *bound* a project means to understand the limitations associated with the project, the changes that can be made to achieve desired objectives, and the important quantities that are likely to change as a result of the project. In systems terms, this means identifying the constraints, parameters, and variables for the project. *Constraints* are the limits that must be observed for the project. Constraints include realistic considerations related to things such as money, time, people, organizations, and society. For example, most projects have budget, time deadlines, and environmental impact constraints.

Parameters[1] are factors that define an alternative and determine its behavior. The value to which parameters are set restrict what results are possible to achieve with an alternative. For example, in the design of a mass-transit system for a city, some parameters might be the number of buses and trains. How many people can be moved by any specific mass-transit alternative depends on the value to which these parameters are set—how many buses and trains there are for an alternative. Once a system is in operation, parameters do not change much. Every day that the city operates the mass-transit system, the same number of buses and trains are part of the system. Parameters are factors that the systems

---

[1] Parameter, as used here, means a determining characteristic, feature, or prominent factor. It is closely related to the mathematical definition of a parameter as a special variable because a parameter is a controllable factor that has a main role in determining the basic form and function of an alternative. This is like a parameter in a mathematical function or statistical distribution because the value to which a mathematical or statistical parameter is set determines the specific form of the function or the shape of the distribution.

engineer manipulates to create alternatives and alter their performance. To distinguish parameters from the quantities associated with mathematical models, you can refer to them as alterables.

In contrast to parameters, variables are likely to change once the system is in operation. *Variables* are measurable quantities that you want to monitor as the system operates. For example, the number of passengers that use the buses or trains each day is a variable that changes as the system operates. For large-scale systems the number of variables can be overwhelming. That is why we use the concept of state variables.

State variables are a collection of variables that we choose to monitor to inform us about the status of a system. The specific variables chosen depend on why we want the information. If we are systems engineers investigating the adequacy of a currently existing mass-transit system, then we probably want to know a number of variables at different times of the day and week. For example, two state variables might be the number of passengers per route and the number of operational buses and trains. State variables give the systems engineer important snapshot information about the operation of a system.

24.3.2 System Definition Matrices

A convenient way for the systems engineer to document the scoping and bounding of a problem is to use a *system definition matrix* like the one shown in Figure 24.3. There are many different ways to construct a system definition matrix. The matrix shown here has two main sections. One section defines the scope of the project by listing the needs, objectives, and criteria. The other section lists the parameters, variables, and constraints and defines the bounds of the project. The simple framework of the system definition matrix is useful as a checklist for information gathering, documentation, and for communication.

The simple matrix suggested here can be expanded to include more features and elements of a proposed system or project. For example, the purposes and basic functions of a system, its inputs, outputs, major components, human agents, its environment, super-

SCOPE			BOUND		
Needs	Objectives	Criteria	Parameters	Variables	Constraints

Figure 24.3 System definition matrix.

and lateral systems, and interfaces and controls could all be dimensions of the system that are described in a system definition matrix.

If you want to put more details in the matrix, then you can create a system definition matrix for each subsystem of the larger system. For a multiproject program you might want to complete a system definition matrix for each major project in the program, as depicted in Figure 24.4. And for a multistage project, you may want to complete a matrix for each phase of the project. Additionally, you can use these matrices to help you identify and specify the interactions among the various components.

The system definition matrix, when complete, should give you and those that you need to communicate with, a thorough picture of the basic ingredients of an existing or proposed system. It can help you to discover insights into the feasibility and internal consistency of a proposed design. Also it can help you identify different ways that you can make improvements to an existing system. Before we explore more about how to create different design alternatives, let us focus in on two very important sets of elements of the problem formulation effort: needs and constraints.

24.3.3 Needs and Constraints Identification and Analysis

Since the definition of a successful design depends on meeting the effective needs of the customer, it is no wonder that often there is great importance placed on identifying and analyzing the needs for a project. Such efforts have become so important that they are known by their own name—a needs analysis.

What are needs and how are they analyzed? First, a *need* is a lack of something that is desired or required. As Figure 24.5 shows, Maslow (1954) defined a hierarchy of needs

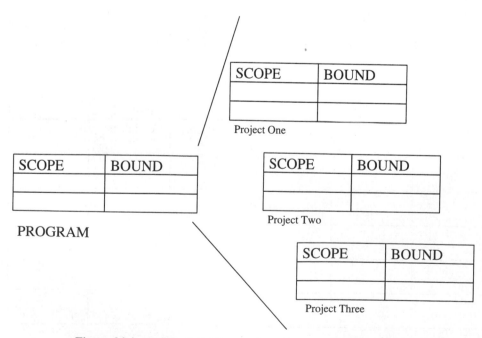

Figure 24.4 System definition matrices for a multiproject program.

Figure 24.5   Maslow's Hierarchy of Needs.

for people that shows how basic lower-level physical needs such as hunger or thirst must be satisfied before higher-level emotional and mental needs such as love and self-esteem become important. As Maslow has shown, needs stem from being human. But humans also desire to engage in many different activities. Many needs come from the desire to improve or enhance human activities. People are always wanting to do things faster, easier, cheaper, and better in some way. Technology itself is often defined as the application of scientific knowledge to enhance some human activity. As we know from the increasing pace of advances in technology, needs are limitless. But notice that we put the word "effective" in front of needs in our definition of successful design.

What is an effective need? This is best answered by what it is not. An effective need is not a primitive need. *Primitive needs* are the unsupported opinions of people involved with a system. Primitive needs are usually narrowly focused assertions or beliefs that people have. These opinions normally reflect their position or role within a system. For example, in an assembly facility that is experiencing problems, primitive needs from assembly-line workers may focus on management expertise as the problem that must be addressed. Management may assert that the workers are the problem.

Where does the truth lie? Like any good detective, the systems engineer, as illustrated in Figure 24.6, cannot begin to say until the investigation has been broadened and evidence gathered and objectively analyzed. This means the systems engineer should look beyond just the immediate, obvious clues at hand. Perhaps there are factors in the environment of the assembly plant itself, beyond the immediate control of either the assembly workers or line management, that are the culprits. A needs analysis will broaden the search beyond primitive need statements and look for evidence to support claims of effective needs. Effective needs are usually broader statements than the primitive needs. For instance, one strong piece of evidence that an effective need exists to bring a new system into being or

Figure 24.6 Systems engineer as detective.

to improve an existing system is that there is someone willing to pay for it! Other evidence can be gathered from historical accounts or logs of system performance.

If a preponderance of the major stakeholders of a system are saying that a need exists to improve some aspect of a system, then it is more likely that an effective need exists. For example, if postal patrons, mail delivery people, and postmasters are all talking and writing about an alarming increase in lost and stolen mail, then evidence is beginning to indicate an effective need. Still, you want to gather better evidence. This means that collecting and analyzing data are often necessary. "Legal authorities report a 25 percent increase of reported lost and stolen mail incidents this year," is a compelling piece of evidence. The bottom line is that effective needs must be supported by organized, convincing evidence. A good systems engineer cites evidence linked with sound rationale when presenting the results of a needs analysis. The following table is a three-column table that can be used to help organize your thinking about the effective needs of a project. You support each need in the table by corresponding entries in both the evidence and rationale columns. The rationale should provide short explanations or underlying reasons that clearly link the evidence cited in column two to the effective need listed in column one.

Needs	Evidence	Rationale
1.0 Need One	1.0 Evidence One to support Need One	1.0 Rationale to link Evidence One to Need One

Effective needs are needs that meet three basic conditions. First, they are needs that can be supported by organized evidence. Analyzing past, present, and future trends related to the tangible need in question can help provide evidence. For example, market research and business trends if properly focused can reveal important insights about the need for new products and services. This organized evidence must in turn be tied to convincing rationale. For example, if market research shows that people are willing to spend some of their disposable income on a new entertainment service, there should be some compelling rationale that explains why people would want to spend their money for this new service instead of some other leisure pursuit or existing entertainment system. This leads to the

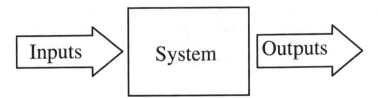

Figure 24.7   Basic input–output model of a system.

next condition, which stipulates that someone must be willing to pay for it! There is not much reason to consider great plans for designing a new system or starting a large improvement project if no one can raise the funds to pay for them.

The last condition simply says that effective needs must be needs that can actually be carried out or satisfied. No matter how much evidence that we compile or how much money we raise, there are some needs that are beyond the human grasp or, at least, are beyond the realm of what is currently possible. Many schemes that play on the hopes and fears of people and attract much attention and money do not meet this last condition. The systems engineer must have the integrity and sound objective judgment to identify futile projects that will only waste scarce resources.

24.3.4   Input–Output Analysis

A very useful device for thinking about the needs and constraints for a proposed system is a basic system input–output model. The familiar system input-output model appears as shown in Figure 24.7. The square box in the input–output model represents the system. The main purpose of the system is to take inputs, shown as an arrow flowing into the box from the left, and transform them into outputs by some process. Outputs are shown as an arrow exiting the box to the right.

Ostrofsky (1977) shows how to create an input–output matrix that examines different types of inputs and outputs over the system's total life cycle. Constructing one of these input–output matrices is fairly simple, as shown in Figure 24.8. You divide the inputs into two thought-provoking categories, intended and environmental. *Intended inputs* are ones that the system's designers and operators can completely determine or control. They

Life Cycle Phases	Inputs		Outputs	
	Intended	Environmental	Desired	Undesired
Design				
Production				
Distribution				
Operation				
Retirement				

Figure 24.8   Input–output matrix.

include things like facilities, resources, procedures, organizational structure, and raw materials. *Environmental inputs* are inputs that may occur but often cannot be controlled. However, it may be possible to partially control, influence or mitigate the effects of environmental inputs. Some examples are weather, customer demand, labor relations, and government regulations.

To stimulate your thinking about the system's outputs, divide the outputs into two further categories—desired and undesired. *Desired outputs* are the results or main products of the system. They are the outputs that justify its existence. As you specify the desired outputs, you should ask yourself if these outputs are truly needed. Also, you should consider if there are other desired outputs that are also needed. Usually you need to maximize the desired outputs. *Undesired outputs* are the by-products of the system's transformation process and often they suggest constraints that the new system must meet. Usually you need to minimize undesired outputs. Examples of undesired outputs are waste products, pollution, accidents, and excess inventory. For example, when you realize that the operation of a system produces pollutant by-products, then that suggests that you should investigate possible constraints related to these by-products that government regulations or industry standards may stipulate.

To make a complete input–output matrix, you should consider the four input–output categories for each phase of the system's life cycle, as shown in Figure 24.8. To begin such a matrix, you will find it easiest to start with the operations phase of the life cycle. Beginning with the operations phase is easier because defining what a new system should accomplish while it is operating is usually more straightforward than thinking about the different inputs and outputs for the other life-cycle phases. Once you define the inputs and outputs for the operations phase, you can use that definition as a reference to help stimulate thinking about the other phases.

While searching for evidence to support effective needs, a systems engineer will come across constraints or limits related to the effort. For example, when investigating the question of whether or not someone is willing to pay for a project, information about how much someone is willing to pay will logically follow. Therefore, a budget constraint has been tentatively identified. In a similar manner, other important constraints will emerge during a needs analysis effort. These constraints deserve careful scrutiny of their own and should be thoroughly explored before too much work is expended to meet what turns out to be an artificial constraint.

Arthur Hall (1962) recommends examining a wide range of boundary conditions for discovering the constraints of a system design. Here is a list and short explanation of his recommended questions that systems engineers should strive to answer concerning needs and constraints:

- *Situation.* What kind of systems design situation are you considering? Are you designing a new system, or is this a project to improve an existing system? If it is an improvement, what aspect of the life cycle of the existing system needs to be improved? Is it the operating, maintaining, or manufacturing of the system that must be improved?

- *Expertise.* What field of design is most critically related to the need implied by the situation? Is it increasing the functionality, performance, and attractiveness of the system, or is it a matter of reducing the cost, or some combination of all of these factors?

- *Risk.* Do the stakeholders of the proposed system want to be bold and search for new possibilities, or are they more comfortable with less risk and content to align closely with the existing environment?

- *Spillover Effects.* What are the expected impacts of the resulting system on other systems, on other areas in the environment, and on the rest of the business sponsoring the venture?
- *Knowledge.* What is the current state of knowledge about the proposed system and its environment, especially relevant available technologies?
- *Viewpoints.* What are the views that different classes of consumers or users have about the proposed system's features and costs?
- *Experience.* How much experience does the systems team have at developing similar systems or projects?
- *Kind of Need.* What sort of need does this system satisfy? Is it an isolated need or does it interact with other needs? Will this need still exist if the related needs disappear or are satisfied in some other way?
- *Frequency.* What is the frequency of this need? Is it something the consumer wants to have over and over again, or is it a finite need?
- *Urgency.* How urgent is the need for the consumer? What are the time limits that the consumer has for making a decision about how to satisfy this need?
- *Limits.* What are the physical limits related to this need such as size or weight constraints?
- *Tolerances.* Are there any tolerances that must be observed when satisfying this need such as speed or capacity constraints?

24.3.5 Partitioning the Problem

Because there are many *elements,* or fundamental parts, in a complex system, you probably have several different lists or sets of elements that all concern the problem and its many aspects. What you want to find out is how all of these issue formulation elements (needs, objectives, criteria, constraints, parameters, alterables, variables, inputs, outputs, and more) relate to each other and how you can organize them into a sensible structure. *Interaction matrices* can help you explore the structure of a set of elements and the linkages between the elements of two or more sets. Each entry in an interaction matrix indicates the existence or nonexistence of a relationship between each possible pair of elements. From the matrix, you can draw conclusions about the structure of a set of elements, and you can portray this structure in a graphical representation. This means you can make conclusions about which elements seem to be most important or central, which elements are relatively separate from others, and which elements form a cluster of related elements. You can also identify the *direct* and *indirect* or *higher-order* relationships between elements. For instance, a second-order interaction is an indirect interaction because the influence between the two elements occurs by means of a third element. A first-order interaction is called a direct interaction because the influence between the two elements does not occur by means of a third element.

There are two kinds of interaction matrices that you will find useful depending on the nature of the elements and the linkages to be explored. A *self-interaction matrix* describes the interactions or linkages between elements of the same set. To construct a self-interation matrix, you create a triangular matrix like the one shown in Figure 24.9 that contains

	S_6	S_5	S_4	S_3	S_2	S_1
S_1	0	1	0	1	0	CN
S_2	0	0	1	0	ME	
S_3	1	1	1	MA		
S_4	1	0	NH			
S_5	0	RI				
S_6	VT					

Figure 24.9   Self-interaction matrix for 6 New England states.

$n(n-1)/2$ entries for n elements. Down the side of the first column, you place a numbered letter to identify each element in the set. As depicted in Figure 24.9, since we are building a self-interaction matrix for the New England states of the United States based on the contextual relation that "element j is adjacent to element k," we use S_1, S_2, S_3, S_4, S_5, and S_6 to represent Connecticut, Maine, Massachusetts, New Hampshire, Rhode Island, and Vermont respectively. Note that across the top of the matrix, the numbered letters are listed in reverse order. For clarity, you should label each row to the right with a short name that describes the element represented by that row.

Once you have the matrix framework outlined, use the contextual relation that you defined to fill in the boxes of the matrix with entries that represent the relationship that exists between each pair of elements. For example, in row S_1, which stands for Connecticut, place a "1" if Connecticut is adjacent to the state represented by that column and put a "0" if it is not. For example, since Connecticut is adjacent to Massachusetts, you see that there is a "1" entered in the box formed by the intersection of row S_1 (CN) and column S_3 (MA).

When you have completed the matrix as shown in Figure 24.9, then you can construct an *interaction graph* as depicted in Figure 24.10. From the matrix and the graph, you can

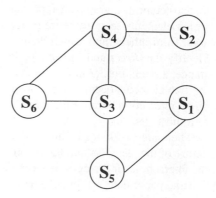

Figure 24.10   Interaction graph constructed from self-interaction matrix.

draw some very useful conclusions. For instance, you can conclude that S_3, Massachusetts, is a central element, while S_2, Maine, is relatively separate or isolated. Also, you can see that S_6, S_4, and S_3 (Vermont, New Hampshire, and Massachusetts) form an interconnected cluster, as do S_1, S_5, and S_3 (Connecticut, Rhode Island, and Massachusetts). From looking at the graph, you can see not just the first-order interactions displayed but also you can clearly see the higher-order interactions. For example, you can see that S_6 has a second-order effect on S_1 by means of S_3 (Vermont has a second-order influence on Connecticut by means of Massachusetts. A possible interpretation of this second-order effect, for a systems study of recreational travel in New England, is that people traveling north on Interstate 91 to ski could experience traffic problems in Connecticut because traffic has backed up in Vermont, which has caused dealys in Massachusetts, which in turn has slowed down on I-91 in Connecticut.)

To define a structure for the simple system of New England states, you can use the interaction graph to create a map of the system as shown in Figure 24.11. Although this is a very simple system to illustrate a technique, you can imagine how helpful it is to have the ability to examine interaction effects in such a systematic way. This technique is especially helpful when considering a complex system made up of many elements where neglected interactions can often cause serious difficulties.

The second kind of interaction matirx is a *cross-interaction matirx*. You use a cross-interaction matrix to explore the linkages between the elements of two or more different sets, such as between needs and objectives or needs and constraints. To set up a cross-interaction matrix, you list the elements of one set across the top of the matrix and the elements of the other set down the right-hand side of the matrix. This format makes it easy to include self-interaction matrices as part of your analysis, as shown in Figure 24.12. In this illustration, you can see how labeling the cross-interaction matrix across the top and right made it easy to include the two self-interaction matrices shown. Note how you put elements across the top in reverse order to accommodate the inclusion of the self-interaction matrices.

Figure 24.12 shows the cross-interaction of the New England states with five outdoor activities. The contextual relation used to fill in the example is "element j interacts or overlaps with element k." A simple binary 1 means that some interaction or overlap exists between the pair of elements, while a 0 indicates no interaction or overlap. But you can enter any number of different codes into the boxes of a cross-interaction matrix. For instance, you can place an X in a box to indicate that a strong or intense interaction exists

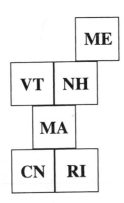

Figure 24.11 Map of the New England states constructed from interaction graph.

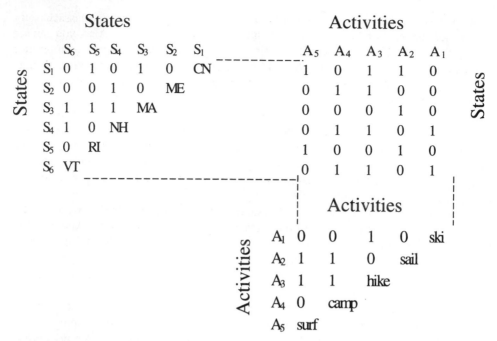

Figure 24.12 Cross- and self-interaction matrices of New England states and associated five outdoor activities.

between a pair of elements, a / to indicate a mild interaction, or just leave the box blank if no interaction exists. Another code is a simple binary where a 1 indicates interaction and a 0 indicates no interaction. Also you can specify and use any contextual relationship that is appropriate for the nature of the sets of elements and the linkages you are exploring. For example, here are some sample contextual relations that may be appropriate to use, depending on the problem you are considering:

- Element j interacts with element k
- Element j is adjacent to element k
- Element j interfaces with element k
- Element j controls element k
- Element j is physically connected to element k
- Element j is electronically coupled to element k
- Element j provides an input to element k
- Element j receives an output from element k
- Element j has a directed interaction from element j to element k
- Element j helps to achieve element k
- Element j is a subsystem of element k
- Element j is a component of element k
- Element j is subordinate to element k
- Element j is necessary or desirable to achieve element k

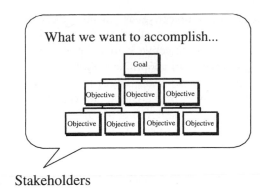

Stakeholders

Figure 24.13 The main reason for an objectives tree.

Whatever contextual relationship that you decide to use, it is important to define it clearly and use it properly, because different contextual relationships result in very different interaction graphs and structures. For example, as you will see in Section 24.4, when you use a contextual relation that has a direction associated with it, like "is subordinate to," you will create a subordination matrix that results in a special oriented or directed graph such as a tree.[2]

24.3.6 Products of Problem Definition

The main result from a problem definition effort is a broader definition of the problem when compared to what was originally conceived. This is because problem definition is an outscoping activity where the stakeholders in the problem, as they meet and discuss it, usually add to the content of the problem. As Sage (1977) points out,

> There is a danger, however, in being too specific initially. There are often advantages in embedding specific problems into more general ones, developing the solution to the more general problem, and then obtaining the solution to the specific problem as a special case of the more general solution. Only in this way can we obtain solutions which are truly optimum. The difficulty is, of course, that the very general problem may also be very difficult to solve. (p. 87)

24.4 VALUE SYSTEM DESIGN

The second logical step of the formulation phase of a systems engineering effort is called *value system design*. It is probably the most controversial and crucial step of the entire process because several very important products result from this step.

First, the system engineer defines *objectives* and structures them in an *objectives tree*. As shown in Figure 24.13, these objectives are an organized picture of what the stake-

---

[2]A tree is a set of elements or nodes joined by directed line segments where the directed line segments branch out from a central point without forming any closed loops. Between each pair of nodes there is only one path so that the entire structure resembles a tree with branches.

holders intend to accomplish with the project. Selecting objectives is crucial because, as Hall (1962, p. 105) points out, . . . it is much more important to choose the 'right' objectives than the 'right' system. To choose the wrong objective is to solve the wrong problem; to choose the wrong system is merely to choose an unoptimized system."

Because stakeholders have different viewpoints about what objectives to pursue and which objectives are most important, developing objectives may create some controversy. Selecting objectives to pursue represents a claim by the group devising the objectives that it is possible to attain value from achieving them. To value something is to appraise the worth of an object, event, or condition. *Value* is a relative term and is used here to refer to various outcomes that result from alternatives in a decision problem.

In large-scale public and private systems projects, decision makers must frequently allocate resources among objectives that compete for the same resources. Hence, a value judgment or decision to adopt a particular alternative is an expression of preference for a particular set of outcomes. Value judgments are often hard to discuss since they can involve both objective and subjective points of view. However, there also is good news here. The systems engineer finds many benefits from resolving these disagreements early in the project's life. If you have a clear, agreed-upon idea of what you are trying to accomplish from the beginning, the chances of accomplishing it are much greater. Since systems engineers are primarily concerned with the relative value of competing alternatives, the term *value system* refers to the set of objectives used for decision-making purposes.

The second important product of value system design is the definition of a set of *objectives measures* or design criteria. These criteria are crucial since you use them to evaluate your alternatives. In other words, these **criteria** measure your success or failure in achieving the project's objectives. Developing useful criteria requires the systems engineer to first help stakeholders identify and define worthwhile objectives.

24.4.1 Defining Objectives

How do you develop and structure objectives? First, you get objectives defined by meeting with groups of stakeholders and recording and refining their concerns. It is best to first capture lists of possible objectives without comparing them or judging them in any way. This encourages the free flow of ideas and helps people reveal their agendas relative to the project under consideration. As ideas begin to take shape, you want to structure each objective in this form:

To (action word) + (object) + (qualifying phrase)

Having people state objectives in this form is helpful for several reasons. It is much easier to understand precisely what they mean. Stated in this form, it is clear what action they want to undertake, what objects will be impacted by the action, and what qualifications they consider to be important. It is also easier to compare different objectives when they are stated the same way. This will help you structure the objectives in a tree, since you can more easily identify objectives stated in the same format as higher- or lower-level objectives. You can also know better how to group objectives together when they are stated in standard form. For example, you can group objectives that have the same (object) together in one branch of an objectives tree.

How can you get started thinking about what objectives to choose? Fortunately, experience has shown that certain kinds of objectives appear frequently in many different types of systems. Here are a number of different types of objectives that Hall (1962, pp. 105–107) has found to appear most often:

- *Profit Objectives.* Profit is typically defined as revenue minus costs and is measured in dollars. Revenue is computed in terms of the number of units of product or service sold multiplied by the unit price of the goods or services. How profit varies over the life of a proposed project is important to consider, since stakeholders may expect some new systems to generate a lot of profit very quickly. On other projects, the stakeholders may be content to generate a modest profit over a longer time.

- *Market Objectives.* Even for nonprofit or public enterprises, it is always important to think about objectives that relate to the amount of product or service delivered per unit of time as a function of time. Sometimes it is measured as a proportion of all competitors or market share. This means that market objectives can sometimes be independent of profit motives.

- *Cost Objectives.* Because budgets are limited, cost is an important factor to consider in almost every system. Minimum cost is often a decision criterion, especially when gross income is fixed. Minimum initial cost may be an appropriate objective if capital budgets are tight. However, be careful not to emphasize cost to the extent that you are "penny wise and pound foolish." Minimizing annual costs, costs of money, depreciation, taxes, and maintenance, can be most important if the stakeholders are responsible for the system during its useful life. Also, remember that most of the time you get what you pay for. Lower cost often means lower quality, which can cost more in the long run. To avoid low initial cost yet expensive projects, you should minimize the life-cycle cost of a system, which considers both initial costs and annual costs.

- *Quality Objectives.* Hall says that quality has both objective and subjective aspects. For example, he says that the objective quality of a television picture is measurable by such things as the number of scanning lines, the size of the screen, and the scan rate. The subjective quality of things is the human response that people have to objective quality in a given environment. For example, the quality of light emitted by a lantern used in a tent may be very pleasing in a camping environment, but it would be very unsatisfactory from a quality standpoint as lighting for your home. Quality objectives are often subjective, especially in the beginning of a project. Understanding what features in a system are important from the user's perception of quality is key to a project's success. The systems engineer will find it a challenge to give quality attributes a physical interpretation and make them quantifiable as the project matures. One way to overcome this challenge is to get end users to experiment with prototypes of the proposed system and to provide feedback on the different features of the design.

- *Performance Objectives.* These objectives depend strongly on the type of system under consideration, since performance relates to the main purpose of a system. Hall uses the example of a communication channel that must include objectives about attenuation and phase characteristics, and signal-to-noise ratios. Normally you want to maximize performance objectives. Such terms as figures of merit, measures of productivity, measures of effectiveness, measures of performance, measures of efficiency, and efficiency factors all relate to a system's performance. Speed, accuracy, response time, throughput, and other concepts are frequently appropriate.

- *Reliability Objectives.* These objectives concern the probability that the system or its components will operate properly for some specified time under normal operating circumstances. Care must be taken to define the intended operating environment correctly, otherwise the system may fail to operate in the users' environment.

- *Competitor Objectives.* Often there are two sides that can determine the success of a new or improved system. When this is the case, as in many military or business systems, objectives must be stated in terms relative to the competition. For example, market share is an objective that considers capturing a certain amount of the total market available.

- *Compatibility Objectives.* Making sure the new system can work well with existing systems in the anticipated environment is an important goal in many situations. Sometimes backward compatibility, making sure the new system can work well even with older "legacy systems" is important.

- *Adaptability Objectives.* These objectives speak to how well a system can adjust to a changing environment. For example, building into new systems the capability to handle growth in demand over time is very important, especially for customer-service-type systems such as telephone and cable communications systems.

- *Flexibility Objectives.* While adaptability means changing to accommodate growth patterns that place more demands on a system's productivity, flexibility objectives are about converting a system to new or multiple uses.

- *Permanence Objectives.* Systems can quickly become old or outdated due to the fast pace of advances in technology. This is especially true today of information technology products and services. For example, software that cannot be modified to incorporate the latest features desired by users will quickly become unwanted. Avoiding technical obsolescence is the goal of permanence objectives.

- *Simplicity Objectives.* Straightforward designs that are easy to explain and implement make for elegant solutions because they are eagerly adopted by users and their managers. Users want systems that are easy to understand since it minimizes their mental overhead or start-up costs of learning the new or improved system. Managers appreciate systems that are easy to install and use, since it minimizes their integration costs, paying for the installation of the systems, and any training on it.

- *Safety Objectives.* These objectives should be a consideration in every system, but how much emphasis to place on them is open to debate. People disagree about safety objectives because they are difficult to measure. For example, how do you measure the cost of an accident, especially if human or environmental damage occurs? But you should resolve these questions with the stakeholders and use objectives that are consistent with ethical, professional practices.

- *Time Objectives.* Objectives related to time are almost always part of any set of objectives. Also time interacts strongly with many other types of objectives. Many objectives are a function of time or require some qualifying phrase about time to make them clear.

By now, you probably have quite a few objectives in mind about some new system or about some improvements to an existing system. But you need to know how to organize and structure these objectives into a tree that you can use to communicate to other engineers and scientists working on the system and to the important stakeholders. Hall (1962, pp. 108–109) offers ten points, summarized in the list that follows, to help guide you on developing and understanding a set of objectives:

1. *List the Objectives.* Create a list of all the possible objectives that are important to the project's stakeholders and others that the systems engineering team believe to be important for the project's success.
2. *Identify Means and Ends.* Place the related chains of means and ends in a tree structure.
3. *Check Relative Importance.* Test to see that objectives are at the correct level of relative importance.
4. *Test Logic at Each Level.* Test to see that the objectives at each level are logically consistent. Identify the inconsistent objectives because they represent tradeoff relations that will have to be carefully understood and specified.
5. *Define Tradeoffs.* Define the terms of trade for related variables. This may be as simple as finding the derivative of one variable with respect to another. Also state the limits of the variables where the trade is valid.
6. *Use Experts.* Use experience from experts (or the best available representatives of stakeholders) who have worked on similar systems to be sure the set of objectives is complete.
7. *Define Criteria.* Define objectives measures or criteria to measure each objective.
8. *Check Needs and Constraints.* Check objectives to be sure they link to the needs (will help to satisfy the effective needs) and are within the project's constraints.
9. *Account for Risks and Uncertainties.* Remember to account for risks and uncertainties in the way you state and measure objectives.
10. *Settle Value Conflicts.* Isolate logical and factual questions from purely value questions. Have all stakeholders' interests represented. Avoid dogmatism, dictatorial methods, and premature voting.

24.4.2   Objectives Hierarchies or Trees

Structuring objectives into a tree takes considerable thought, but as you will soon see, it has enormous payoffs. The first step in constructing the objectives tree is to examine how the objectives relate to each other. A simple *subordination matrix,* like the one shown in Figure 24.14, can be useful for involving a group in the effort. Use the contextual relation

	Objective 1	Objective 2	Objective 3	Objective 4	Objective 5	Objective 6
Objective 1	0	0	0	0	0	0
Objective 2	1	0	0	0	0	1
Objective 3	1	0	0	1	0	0
Objective 4	1	0	0	0	0	0
Objective 5	1	0	0	1	0	0
Objective 6	1	0	0	0	0	0

Figure 24.14   Subordination matrix of objectives.

"attainment of objective j is necessary or desirable in order to attain objective k" to define the term subordinate. A 1 in the matrix means that the objective in that row is subordinate to or "helps to achieve" the objective in that column, or objective j is subordinate to objective k. Since the interactions in the subordination matrix are directed in nature, any time you enter a 1 in the matrix in some location (j, k), you should also enter another mirror-image entry, a 0 in location (k, j). If the row objective is not subordinate to the column objective, then you put a 0 in that (j,k) location, (row j, column k). Since an objective cannot be subordinate to itself, all entires along the main diagonal of the matrix also get a 0. So before you start the dialog with the group, you can place a 0 in each main diagonal box, as shown in Figure 24.14. To check the logic of your work when you finish filling out the matrix, you should be able to pair the 1s together in succession. In other words, if objective j is subordinate to k, and k is subordinate to one or more other objectives, then j should also be subordinate to the same one or more other objectives as k.

Once you have the subordination matrix, it is relatively easy to structure the objectives into a tree, given the information from the matrix. Since all the entries in objective 1's row are 0 in the subordination matrix of Figure 24.14, objective 1 is soubordinate to none of the other objectives; it is also the top objective depicted in the tree of Figure 24.15. Reading the conclusion from the second row of the subordination matrix, objective 2 is subordinate to both objective 1 and objective 6. Looking at the row for objective 6, you can see that objective 6 is only subordinate to 1. Therefore, you can conclude that 2 has a second-order effect on 1. That is, objective 2 only affects objective 1 indirectly through objective 6. Hence, you see that objective 2 is placed beneath objective 6 in the tree. There are more detailed rules and several good examples for the construction of the subordination matrix and other interaction matirices with applicable discussions of graph theory in Sage (1977). Figure 24.15 illustrates the structure of a completed objectives tree to match the subordination matrix in Figure 24.14.

In constructing an objectives tree, note the arrowhead lines that connect the boxes of objectives. These upward pointing arrowhead lines show the contextual relationship between the objectives and, to make your diagram clear, you should provide a key on the diagram that defines the contextual relationship that you used to build the tree. Use something simple and clear like "will assist in" or "contributes to."

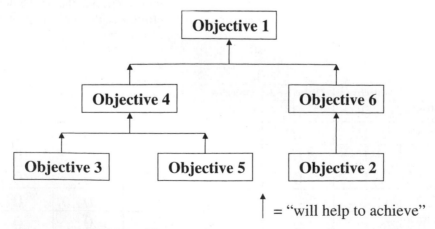

Figure 24.15 Objectives tree.

As the tree nears completion, the top-level objectives usually describe more lofty goal aspirations that reflect strongly held values and effective need statements. These top objectives are normally difficult to measure directly. Lower-level objectives get progressively more specific and are easier to measure. The lowest level objectives suggest activities or functions that stakeholders want to pursue. Lower-level objectives are the means for accomplishing the ends of the higher-level objectives. Figure 24.16 shows the general structure of an objectives tree. As in all system diagrams, there is a key that defines what the lines represent. For an objectives tree, the arrowheaded lines define a relationship between the objectives such as "will assist in." Therefore, a lower-level objective is a means to a higher-level objective if achieving the lower-level objective results in or helps to achieve the higher-level objective. As you examine a completed objectives tree, you should see a causal relation between ends and means.

You can embellish an objectives tree by adding *logic* and *ownership* annotations. There are three kinds of logic connectors that you can add to a tree: AND, inclusive OR, and exclusive OR. The AND means that the higher-level objective can only be accomplished if the preliminary connected subordinate objectives are both accomplished first. The inclusive OR means that the higher-level objective can be accomplished if any of the connected subordinate objectives are accomplished. The exclusive OR means that the higher-level objective can only be accomplished if only one or the other, but not both, subordinate objectives are accomplished. You depict these logic connectors by adding them in small boxes. You can add ownership tags to any objective by adding an ajoining box to any objective to show which stakeholder owns each objective.

As depicted in Figure 24.17, there are five tests of logic you can use to check your objectives tree (Sage, 1992, p. 286). First, looking up any one branch of the tree, higher-level objectives should tell "Why?" for lower-level objectives. Next, looking down any branch of the tree, lower-level objectives should explain "How?" for their higher-level objectives. Third, looking across any one level of any branch, the set of objectives should pass the "Enough" test This means that the set of lower-level objectives in question, if

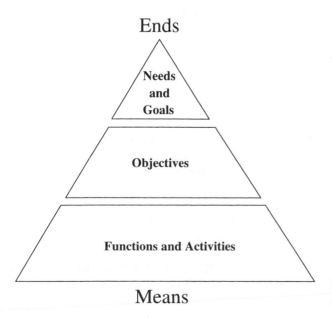

Figure 24.16 The content of an objectives tree.

Figure 24.17   Five tests of logic for an objectives tree.

successfully achieved, should be sufficient to convince the stakeholders that success on the higher-level objective would have to follow.

The fourth test again looks across any one level of any branch and asks the question "Extras?"—are any of the lower-level objectives not needed or extraneous to the higher-level objective? Last, the fifth test of objective tree logic—"Owners?"—requires the systems engineer to identify the owners of each objective. If no stakeholders want to claim an objective, then you must seriously consider why it is included in the tree at all. Be careful in doing this, since some owners may not have a voice in your meeting. Yet, they may have a very strong stake in the outcome, as in many systems where the public has a strongly held interest. In these cases, you may want to have someone represent these interests even if just as a surrogate. Also, it is wise to omit the ownership of objectives during the structuring process itself. Sometimes people will compete to have the objective they own near the top of the tree, despite logic to the contrary, when an ownership tag is attached.

Besides these five tests of objectives tree logic—why, how, enough, extras, and owners—you may find a few more dead branches to prune out of your tree. People often want to include in an objectives tree the various tasks that a systems engineering team needs to accomplish in support of a project. An example of this is an objective "To publish a final technical report," or "To prepare an in-progress review briefing by the end of the quarter." Such objectives are project milestones that belong on a Gantt chart of the systems engineering team, but they do not make any contribution at all to showing your stakeholders everything they intend to accomplish with a project. They want you to improve or create a successful plan for a large-scale system, not to generate paper! So leave the by-products off the objectives tree.

Another problem with objectives is that, after careful inspection, you may find objectives that are duplicates of each other. It is best to define objectives that are significantly different from each other. If you do not eliminate them, they cause you to double count or overemphasize one particular aspect of your design. On the other hand, there may be times when you need to repeat a particular objective in different branches of the same tree.

24.4.3   Objectives Measures or Design Criteria

Now that you have a logical, complete objectives tree, what can you use it for? First, the objectives tree itself is a coherent statement of everything that the project is intended to accomplish. The tree helps the systems engineer clearly communicate with stakeholders. It is easy to see the relative priority of objectives. More important objectives are near the top of the tree. Also, the tree helps the systems engineer identify tradeoffs. By looking across any one level of the tree, it is normally possible to identify objectives that compete with each other in some way. For example, an objective to minimize cost will normally compete with an objective to maximize performance since lower costs usually result in reduced performance.

The objectives tree also helps to uncover hidden agendas of stakeholders. If you ask stakeholders to review and reach agreement on an objectives tree, then they usually make sure that everything they hope to accomplish with a project is shown on the diagram.

One of the most important uses of the objectives tree is that it helps you to define and develop objectives measures or design criteria. These criteria are important since you use them to measure the success of alternatives in achieving the objectives. Figure 24.18 is a picture of how objectives link criteria to the effective needs. Even though higher-level objectives may be difficult to measure, lower-level objectives are usually much easier to measure. Hence, the systems engineer can develop a combination of lower-level criteria to measure the more difficult higher-level objectives. You should strive to develop at least one objectives measure or criterion for each objective.

How do you create criteria? Useful criteria are objectives measures that take a quantifiable form with both a clear definition of the measure and the units associated with the measure. Criteria are often associated with counting such things as money, time, people, and products. Economic, reliability, availability, maintainability, supportability (RAMS), operational, and logistics are useful categories of criteria to consider. For example, the life-cycle cost (LCC) of the system is an economic criterion for the objective of minimizing the cost of a system. The unit of measure for this criterion is the present worth of the alternative in dollars.

Operational criteria relate to the main purposes of a system. For an airport, operational criteria would quantify its main purposes, such as the number of flights per unit time and the number of passengers and amount of freight per unit time that the airport can accommodate. Logistical criteria count the number of people and the amount of supplies that are necessary to keep the system in operation. RAMS criteria quantify the durability of the

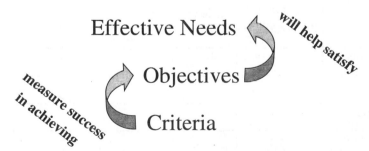

Figure 24.18   How objectives link design criteria to effective needs.

system when it is put into operation and the probability that the system will not fail. The criteria you use depend on the type of system you are designing and the objectives for the system. For information systems, the functionality and usability of the system are important attributes that may require task-oriented criteria measured in user tests or experiments with prototypes. Can the user perform tasks quicker or better with the new system?

One useful technique for developing criteria, as shown in the following table, is to list objectives in the first column of a table and then to brainstorm definitions and associated units of measure in two companion columns.

Objective	Criteria Name	Definition	Units of Measure
1.0 To maximize the capacity of the new airport sufficient to meet future demand	1.0 Annual Service Volume (ASV)	1.0 The number of operations (landings, takeoffs, and touch-and-go's) that the runway configuration can accommodate during one year	1.0 Operations/ Year

The table illustrates an objective to maximize the capacity of a new airport that can be measured by the design criterion of annual service volume in units of operations per year, and it includes a definition that explains what is meant by an operation. Eventually, the systems engineer puts together a scorecard matrix, like the one shown in Figure 24.19, that shows the feasible alternatives for a project down one side and the design criteria across the top. For every column in the matrix, the systems engineer uses a model or simulation, or some other analytical technique to evaluate how well each alternative scores on each criteria. Since the alternatives are significantly different from each other, the

Alternatives	Criteria or Objectives Measures			
	Operational Performance	Life-cycle Cost	Reliability, Availability, and Maintainability	Logistics and Supportability
New, Unique				
Off the Shelf				
Modified or Improved				
Baseline or Existing System				

Figure 24.19 Scorecard matrix.

models will vary somewhat for each alternative, although they will be generally the same for each ciriterion. Chapter 26 has many useful techniques for modeling and analyzing alternatives on various criteria. Often you will see the format of this scorecard matix reversed, with the alternatives listed across the top and the criteria along the right side.

24.5 ITERATION OF THE DESIGN

Going back and repeating the logical steps of systems engineering is very important for achieving results that are successively closer to what is actually desired. As you learn about functional decomposition, business activity models, and quality function deployment, you will see how helpful they are in going back and refining your problem definition and value system design efforts. Also, you will find these formulation techniques helpful for moving forward to the final logical step of formulation, system synthesis: generating alternatives and building descriptions of these alternatives with detailed system requirements and specifications.

Another benefit from an objectives tree is that the lower levels of the tree often suggest functions for the new or improved system. A function is something that the system must do or accomplish to achieve its purposes. Since purposeful activity is a basic characteristic of any system, the systems engineer designs a system to accomplish specific tasks or functions. A function is a definite, discrete action that a system must do to achieve one of the system's objectives. These functions are typically operations that a system must perform to accomplish its purposes, or actions that must be taken to restore the system to operational use. For example, a weapon system normally has a loading function, a firing function, and a set of maintenance actions.

Maintenance actions are part of a functional description of a system. For a car, changing the engine oil and oil filter are two examples of functions related to maintenance activities. The systems engineer can use the lower levels of the objectives tree to help build a functional description or decomposition.

There are several reasons why a systems engineer wants to understand a project from a functional viewpoint. First, a functional description of a system allows the systems engineer to design or plan a system independent of any specific technical solution. It is often very important not to identify any specific technical solution until later in a project. This assures that developers do not adopt a specific technical approach until all the different technical possibilities have been adequately evaluated. Once the best technical approach is known, a functional decomposition provides reasons for the different physical components or equipments selected for the system. When someone asks why certain equipment is needed, the systems engineer can trace the requirement for it to a specific function. For example, this sensor is needed because it performs the detection function, or a certain radar is needed because it performs the tracking function.

Next, thinking of a system in functional terms provides a basis for developing innovative alternatives. Many new ideas have come about from systems engineers experimenting with the reallocation, redistribution, or the duplication and propagation of the functions of a system. One example from naval warfare is the aircraft carrier, which married the functions of an airbase—storage, arming, takeoff, landing, and retrofit of warplanes—with the functions of a ship—seaborne transport. The result was an innovation in warfare that permitted vast advances in the ability to project naval air power both at sea and onshore.

Another example, this one from the computer industry, is the early Apple laser printer. As Apple engineers and scientists were thinking about how to improve the speed and memory capability of their computer, they came to the realization that a lot of the computer's processing power and memory was being taken up by functions associated with the printer. So someone had the brilliant idea of reallocating those functions to the printer itself. The result of this innovation was the first printer with both memory and processing power. In fact, early Apple laser printers were some of the most powerful "computers" on the market. Soon every printer manufacturer followed Apple's lead. Many innovations in the information technology market today came from this kind of thinking. It is easy to see how different combinations of information technologies are leading to new products and services. The telephone, television, video camera and recorder, compact disc stereo, and personal computer are all being put together today in different ways.

Before you can begin to rearrange functions, you need a basic functional decomposition of the system. One way to accomplish this is by constructing a functional-flow block diagram of the system. The next chapter shows you how to do functional decomposition and analysis.

Closely related to functional decomposition is a technique for understanding management systems and business operations that is called *business process reengineering*. See Chapter 23 for a discussion of systems reengineering. Business process reengineering systematically examines a set of interrelated business activities and their transactions for several purposes. First, it helps to identify the most important activities so that they receive proper attention and priority effort. The most important activities are ones that add cost and value to the process. Second, it can identify extraneous or redundant activities that can be eliminated. Also, it is possible to cost activities and find new ways of accomplishing activities to save time, effort, and money. Most importantly, it is possible to construct a coherent, effective data model or information architecture to support your business activities using this approach. Building a business activity model also helps to provide a common language, which makes communication easier about how a business performs its repetitive processes both within the business enterprise and for external uses. Chapter 25 shows you how to build activity models using the integrated definition methodology (IDEF) (pronounced "eye-deaf").

Besides using activity models for a steady-state or static representation of a business process, you can use an activity model to build a simulation of a business process. A simulation helps to illustrate the importance of how activities change with time and can show the sequencing of activities. Most of the elements in an activity model translate easily to a simulation model with the straightforward addition of attributes to activities like activity duration and sequence information. Simulations help to identify the dynamics in business processes that cause bottlenecks or other resource allocation or flow problems that may not be evident in static models. Using these kinds of analytical tools for a formulation effort can be very helpful to you.

24.5.1 Quality Function Deployment

Quality function deployment (QFD) is a frequently used formulation technique of total quality management (TQM) that grew out of the Japanese business practices of Mitsubishi and Toyota in the 1970s (Hauser and Clausing, 1988, p. 63; Clausing, 1994). The purpose of QFD is to help businesses focus on customer requirements across organizational units, especially at the front-end or product development stage of the system life cycle. In other words, QFD is a way to get marketing people, design engineers, and manufacturing staff to work together so that their products reflect customers' desires and tastes (Sullivan, 1986, p. 39). Note that this Japanese concept is different than the U.S. idea of being responsive to customer complaints. It is much more proactive.

QFD concentrates more effort on designing into a product what customers like instead of later fixing their complaints (Sullivan, 1986, p. 40). Sage (1992, p. 283) points out that QFD shifts the emphasis during design from the traditional goal of completely satisfying technical system specifications to a new goal of total satisfaction of customer requirements. This does not mean that the technical or engineering specifications are not important. Instead, it means that what customers want should drive the technical specifications. Any changes to technical specifications should be evaluated in light of their impact on satisfying the customers' requirements.

The center piece of QFD, as shown in Figure 24.20, is the *voice of the customer*. The voice of the customer is the customers' requirements expressed in everyday common language phrases. These phrases are the words customers use to describe products and product features. Hauser and Clausing (1988, p. 65) call these phrases customer attributes, and say that typical applications have somewhere from 30 to 100 of them. Some examples from the auto industry of customer attributes to describe a car door are "easy to close" or "stays open on a hill" (Hauser and Clausing, 1988, p. 69). Customer attributes can be

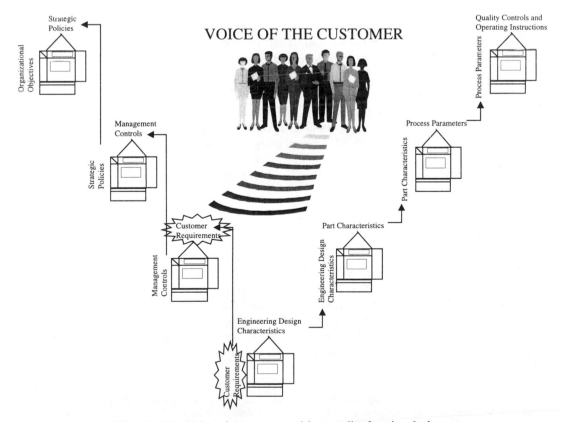

Figure 24.20 Voice of the customer drives quality function deployment.

grouped into bundles or categories to represent an overall customer concern like safety. You should consider customer attributes in the broadest sense of the word. Customers are not just end users. For example, regulators of the automobile industry would insist on customer attributes like "safe in a rear-end collision."

Once the customers' attributes are known, QFD is a systematic way to deploy or translate these customer requirements in two related directions. The first direction, as shown in Figure 24.20, uses the customer requirements to drive the product design and production process. The second direction uses the customer requirements to drive organizational objectives and management policy.

The first direction is called *policy assessment* or sometimes the *house of policy,* while the second is called *product quality deployment* or sometimes the *house of quality,* which is discussed in the next paragraph. In a policy-assessment effort, the objective is to determine how meeting the customers' requirements under consideration is supportive of the long-term objectives of the organization. It also evaluates the competitive position and any advantage to be gained from undertaking the development effort required to meet or exceed the customers' expectations.

The second direction for deploying the voice of the customer referred to in the preceding paragraph is called product or service development. The objective is to translate customer needs into product or service design characteristics. Both efforts use a special type of interaction matrix or quality table planning matrix, which is commonly referred to as a house of quality. Sage (1992, p. 284) highlights three generic steps for constructing a house of quality:

1. Identify customer requirements and needs. Put these in the form of weighted requirements expressing their relative importance.
2. Determine the systems engineering architectural and functional design characteristics or specifications that correspond to these customer requirements.
3. Determine the manner and extent of influence among customer requirements and functional design characteristics and how potential changes in one functional design specification will affect other functional design specifications.

With these three general steps in mind, lets see how to build a house of quality room by room.

24.5.1.1 Customer's Room.

As you might expect, the first room in the left corner of the house, shown in Figure 24.21, contains the customers' attributes and their relative importance. These customer requirements are often called the *Whats* and answer the question, "What do your customers really want?" These attributes are often organized into hierarchical categories or bundles. Most quality function deployment software packages, such as QFD/CAPTURE™ or QFD DESIGNER™, let you organize the attributes in multiple related columns.[3] The rightmost column in the customers' room is the "Importance" column. In it, you put the relative importance of the different customer attributes. You usually express the relative importance weightings in terms of integer numbers on a scale

---

[3]QFD/CAPTURE™ is a software application offered by International TechneGroup Inc., Milford, Ohio, and QFD DESIGNER™ is a software application offered by Qualisoft® Corporation, West Bloomfield, Michigan, and the Amercian Supplier Institute®, Dearborn, Michigan.

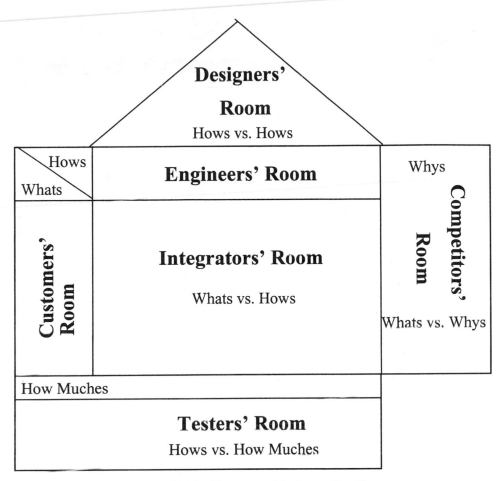

Figure 24.21 The rooms of the house of quality.

from 1 (least important) to 5 (most important). Figure 24.22 shows how you might build a list of customer attributes for designing the ideal lunch from a tutorial in QFD/CAP-TURE.

24.5.1.2 Engineers' Room. The main upper room in the house of quality is the engineers' room. It contains a list of engineering characteristics or product characteristics that can be measured to satisfy the Whats. In this room, you try to translate customer attributes into product engineering characteristics. These engineering characteristics are called the *Hows* and answer the question, "How can we describe the product with measurable engineering design characteristics?" A good way to do this is to try to relate each desired customer attribute to a corresponding engineering characteristic that can be measured. Figure 24.23 shows an engineerings' room with seven design characteristics for an ideal lunch.

These measurable characteristics are important because they are used by designers to assess competitors' products, establish performance objectives, product controls, and de-

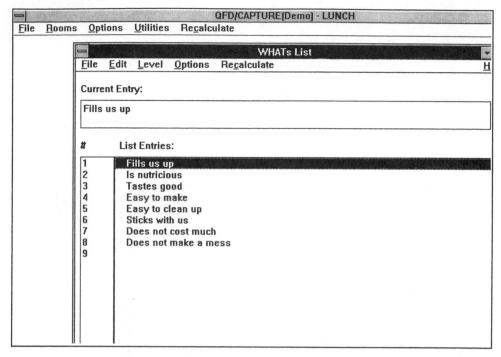

Figure 24.22 Listing customer attributes using QFD/CAPTURE

velop tests and evaluations. You can also group these into a hierarchy of characteristics as depicted in Figure 24.24. Note that each column in the engineers' room for the ideal cellular phone QFD planning matrix corresponds to one lowest level engineering characteristic. You should not include engineering characteristics that are not critical to meeting customer desires or those that are easily achieved. As you look down each column, you should see that each engineering characteristic affects one or more customer attributes. You determine these relationships in the next room.

24.5.1.3 Integrators' Room.   The central room in the middle of the house of quality is the integrators' room. It is a cross-interaction matrix that shows the relationships between customer attributes and engineering characteristics. Figure 24.23 depicts the interactions of customer attributes and design characteristics for an ideal lunch. Figure 24.24 shows the interactions of customer attributes and engineering characteristics for the ideal cellular phone. In each box of this interaction matrix, you place a triangle, open circle, or filled circle to denote a weak, medium, or strong relationship. These relationships are also given numerical values in the legend or key to the house of quality, usually a 1, 3, and 9 for weak, medium, and strong. These values are used to help calculate the absolute and relative importance of each characteristic (see the testers' room). Leaving a box blank means that no relationship exists for that particular customer attribute row and engineering characteristic column pair. These interactions are called the *Whats* vs. *Hows*.

The IDEAL LUNCH

WHATs vs. HOWs Legend		Weight of portion	Percent nutrition requirement	Percent carbohydrate	Time to prepare	Number of dishes used	Cost of ingredients	Number of measured ingredients
Strong	● 9							
Medium	○ 3							
Weak	△ 1							
Fills us up		●	△	○				
Is nutritious		△	●	△			○	
Tastes good			△	●	△		○	
Easy to make				△	●	○	△	○
Easy to clean up					△	●		○
Sticks with us		○		●			△	
Does not cost much		△	△	△			●	

Figure 24.23   Customer attributes vs. design characteristics for an ideal lunch from a QFD/CAP-TURE tutorial.

Figure 24.25 shows a legend of symbols for the house of quality. The roof symbols are for the designers' room and indicate the strength of the interactions among the engineering characteristics. The arrows on the right side of the legend table are used to indicate the direction of improvement for each engineering characteristic. The matrix symbols are for the integrators' room and depict the cross interactions between customer attributes and engineering design characteristics. These and other symbols are used in various rooms of the house of quality.

24.5.1.4   Competitors' Room.   The right-hand-side room of the house of quality is the competitors' room. The top of the room describes the market for the product or service from the customers' point of view. This can be as simple as a graduated customer rating scale from worse to better, as shown in Figure 24.24. It describes *why* you are developing the product or service from a marketing perspective.

The main or lower part of the competitors' room is a cross-interaction matrix that shows the relationship between the customers' attributes and the market descriptors—the *Whats* vs. *Whys*. Typically, you use a different symbol for each competing product or service you are evaluating. For three products you can use a square, a circle, and a triangle. Then you

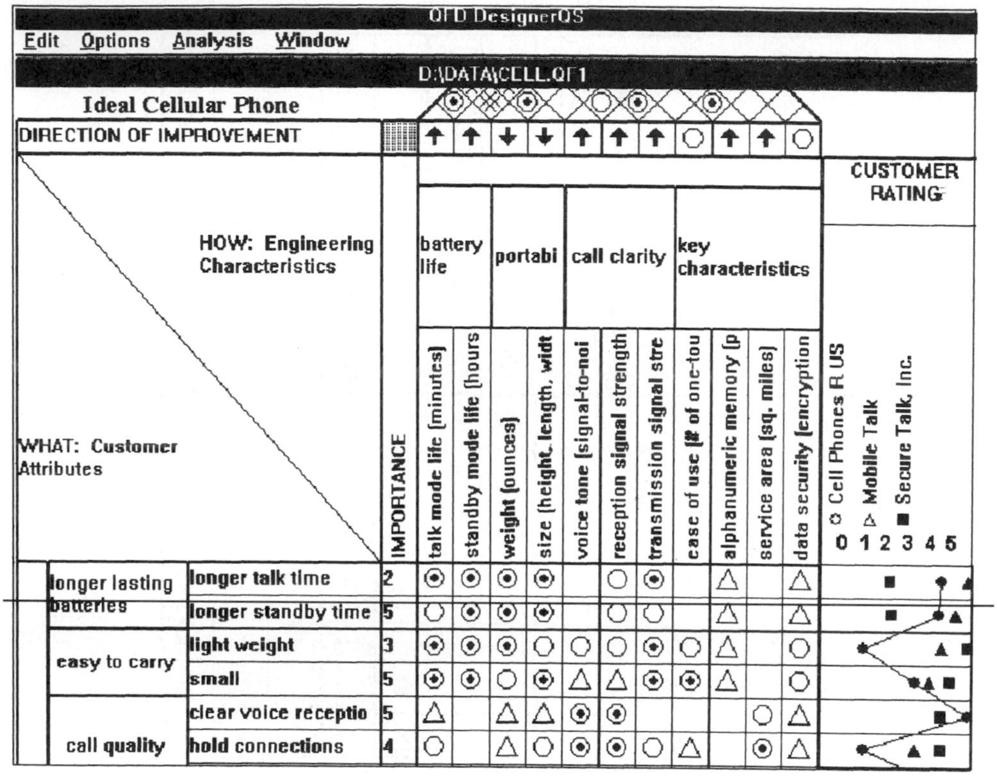

WHAT: Customer Attributes		IMPORTANCE	talk mode life (minutes)	standby mode life (hours)	weight (ounces)	size (height, length, widt)	voice tone (signal-to-noi)	reception signal strength	transmission signal stre	ease of use (# of one-tou)	alphanumeric memory (p	service area (sq. miles)	data security (encryption)
longer lasting batteries	longer talk time	2	⊙	⊙	⊙	⊙		○	⊙	△		△	
	longer standby time	5	○	⊙	⊙	⊙		○	○	△		△	
easy to carry	light weight	3	⊙	⊙	⊙	○	○	○	⊙	○	△		○
	small	5	⊙	⊙	○	⊙	△	△	⊙	⊙	△		○
call quality	clear voice receptio	5	△		△	△	⊙	⊙				○	△
	hold connections	4	○		△	○	⊙	⊙	○	△		⊙	△

Figure 24.24 Whats vs. Hows using QFD DESIGNER.

just place the appropriate company product symbol in the correct column to denote which product is better or worse on each customer attribute. By connecting your company product symbols with a line, you get a very good visual representation of where your product is positioned overall when compared to the competition.

For example, Figure 24.24 shows a customer rating for three different cellular phones, each made by a different company. This example shows a profile of how the "Cell Phones 'R Us" alternative compares with two other cellular phones from competing companies.

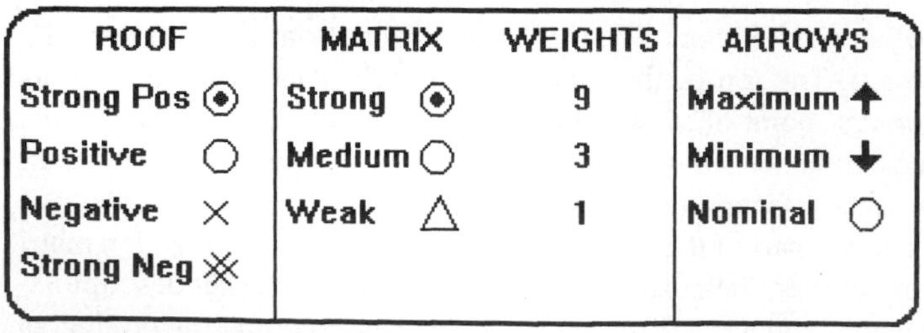

ROOF		MATRIX		WEIGHTS	ARROWS	
Strong Pos	⊙	Strong	⊙	9	Maximum	↑
Positive	○	Medium	○	3	Minimum	↓
Negative	×	Weak	△	1	Nominal	○
Strong Neg	※					

Figure 24.25 Legend for house of quality with QFD DESIGNER.

24.5.1.5 Designers' Room. The designers' room is the top or roof of the house of quality. It is a self-interaction matrix that shows how the engineering design characteristics relate to each other—the *Hows* vs. *Hows*. This room depicts how improvements in one engineering characteristic will affect other engineering characteristics. Usually you have four choices to enter in these boxes. You can put a circle, or a circle with a dot in the center of it, to denote a positive or a strong positive relationship, respectively. This means improvement in one engineering characteristic will cause some improvement or strong improvement in another characteristic. Or you can put an X or a double XX to show a negative or strong negative relationship. This means improvement on one engineering characteristic will degrade or strongly degrade another engineering characteristic. The designers' room is very useful in identifying key tradeoff design decisions that must be made.

For example, Figure 24.26 shows a designers' room or roof that depicts how the design characteristics for an ideal lunch relate to each other. You see, for instance, a strong negative relationship, XX, between percent nutrition and cost of ingredients. This means that when percent nutrition changes in a favorable direction (increases in this case), cost changes in an unfavorable direction (also increases).

24.5.1.6 Testers' Room. The bottom room of the house of quality is the testers' room. As shown in Figure 24.27, the room identifies objective target values, products or services to be benchmarked, and associated technical importance values. These objective target values are called the *How Muches*. Manufacturing, marketing, and engineering must work together to make the decisions on the numerical targets for the engineering or technical characteristics.

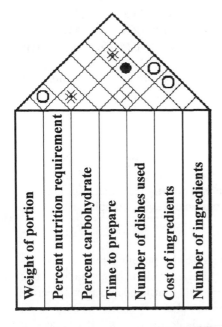

Figure 24.26 Self-interaction of design characteristics for an ideal lunch from a QFD/CAPTURE tutorial.

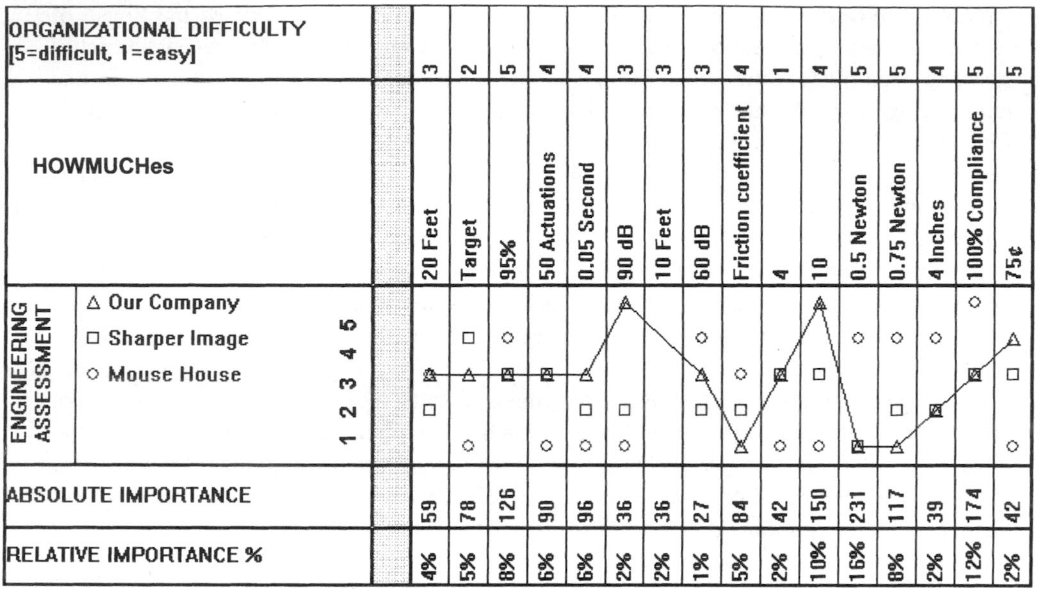

ORGANIZATIONAL DIFFICULTY [5=difficult, 1=easy]		3	2	5	4	4	3	3	3	4	1	4	5	5	4	5	5
HOWMUCHes		20 Feet	Target	95%	50 Actuations	0.05 Second	90 dB	10 Feet	60 dB	Friction coefficient / 4	10	0.5 Newton	0.75 Newton	4 Inches	100% Compliance	75¢
ENGINEERING ASSESSMENT	△ Our Company / □ Sharper Image / ◇ Mouse House																
ABSOLUTE IMPORTANCE		59	78	126	90	96	36	36	27	84	42	150	231	117	39	174	42
RELATIVE IMPORTANCE %		4%	5%	8%	6%	6%	2%	2%	1%	5%	2%	10%	16%	8%	2%	12%	2%

Figure 24.27   Engineering assessment in the testers' room of different mousetraps using tutorial from QFD DESIGNER.

The main part of the testers' room is a cross-interaction matrix that shows the relationships between the target values of the engineering characteristics and the various products or services to be evaluated. It is similar to the competitors' room except, instead of a market assessment, it is an engineering competitive assessment. Figure 24.27 shows an example from QFD DESIGNER™ of an engineering assessment of three different companies mousetraps. The testers' room shows the *Hows* vs. the *How Muches* for different product or service alternatives.

The top of the testers' room is a row that shows the relative difficulty the organization thinks it will experience in trying to meet or exceed the target values. Note that there are two rows along the bottom of the house of quality next to the testers' room. One is labeled "Absolute Importance" and the other "Relative Importance." The numbers in the boxes of these two rows help you analyze a completed house of quality. Higher numbers indicate a greater absolute or relative importance. To calculate these numbers you multiply the customer importance values by the numerical value of each interaction in the integrators' room and then add up all these products down each engineering characteristics column. There are many other things you can determine from analyzing a completed house of quality planning matrix.

24.5.1.7   Analyzing the House of Quality.   To analyze the planning matrix, first examine the relationship matrix or integrators' room carefully. Blank rows may indicate that a customer attribute or requirement has not been successfully translated into an engineering characteristic or technical requirement. Or it may indicate that a customer attribute is not necessary or has been stated incorrectly. Most likely, a blank row means that you need to specify a technical requirement or engineering characteristic to meet a customer attribute. If few strong correlations exist in the relationship matrix, then the correct engineering or technical characteristics have been omitted.

A blank column in the integrators' room relationship matrix means that an engineering or technical characteristic has been specified for which there are no related customer attributes. You should determine if you are missing any customer attributes or if the engineering characteristic is not necessary.

Next you can examine and compare the competitive and engineering data in both the competitors' room and the testers' room. If your company's product or service exceeds the competition in customer attributes, then it should also exceed the competition in the technical characteristics. You should also closely examine the customer attributes and the engineering characteristics with high importance. If your competition is not doing so well on these important requirements, it could be an opportunity for you to make significant design improvements. These key requirements could become important discriminators in the market for your product or service.

Here are some final tips on constructing and analyzing a house of quality planning matrix. When you think your organization or team has completed the effort, get knowledgeable people from outside your group to check the data and verify your conclusions. Be certain that the customer attributes are what the customer really wants and not what some other group of people think the customer wants. This is very important because it is the voice of the customer that is deployed to drive the design and production processes. In this regard, it is important to have an independent market evaluation from outside your organization so that you can have an unbiased assessment of customer needs.

From a policy assessment or business strategy perpsective, it would be necessary to determine the extent to which the specific systems fielding effort under consideration is supportive of the long-term objectives of the organization. For example, one important question is to find out what will happen to the firm's competitive position and market advantage as a result of undertaking the development. As you can see, the QFD approach has uses in strategic management. It can also help you to determine if the management controls and systems management effort, associated with the development of the system under consideration, is supportive of the overall development strategies of the organization.

Sage (1992, p. 285) points out that QFD matrices provide a link between the semantic prose that often represents customer requirements and the functional requirements that would be associated with technological system specifications. In a similar manner, an interaction matrix can be used to portray the transitioning from functional requirements to detailed design requirements. Thus, they provide one mechanism for transitioning between these two phases of the system's life cycle. In this QFD discussion, we have mentioned the words "requirements" and "specifications" numerous times. Therefore, to understand systems engineering, you need to learn what these two terms are all about.

24.5.2 The Systems Engineering Requirements Statement

Requirements and specifications are the product of all of the design decisions about a system. This means that the design or plan of a system is the set of requirements and specifications for that system. But what is the difference between requirements and specifications? When system engineers use the term *requirement,* they are usually referring to an essential attribute or characteristic of a system or one of its components. Figure 24.28 shows that a requirement starts out as a basic or primitive idea.

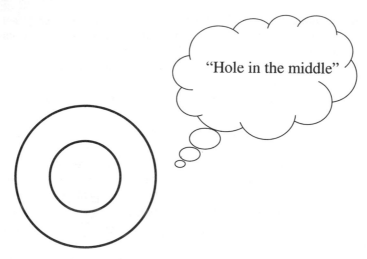

Figure 24.28   A primitive requirements statement starts with a basic idea.

During the formulation effort, systems engineers capture requirements by writing them down as a list of primitive requirement statements (Grady, 1993). Each *primitive requirement statement is numbered from 1 to n* and follows a simple, three-part format: name + relation + value. As shown in Figure 24.29, the name describes the characteristic or attribute to control. The relation details the connection between the attribute and its control value. The value sets a quantifiable number with units or defines a standard. For example, "weight less than 165 lb" is a primitive requirement statement where weight is the name of the attribute, "less than or equal to" is the relation, and "165 lb" is the value. Numerical requirements use one of six possible relations: less than, greater than, equal to, less than

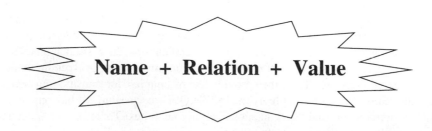

Name = describes the characteristic or attribute to control
Relation = the connection between the attribute and its control value
Value = sets a quantifiable number with units or defines a standard

Example: Weight less than 5 lbs.

Figure 24.29   Primitive systems engineering requirements statement.

or equal to, greater than or equal to, or between a range of values. For nonnumerical requirements, you use words like "is," "be," or "conform to" as the relation.

The list of primitive requirement statements is called a *concept requirements list* (Grady, 1993). This early requirements document is usually made up of multiple pages of primitive requirement statements with accompanying diagrams. As illustrated in Figure 24.30, the engineer preparing the document labels it by system or component title, and numbers and signs it along with the lead systems engineer controlling the project. At some time, systems engineers must translate the primitve requirement statements into more detailed requirements document or specification language.

A *specification* for a system or component, as shown in Figure 24.31, is really a published set of requirements that have been properly refined and formatted into more precise language. Usually each primitive statement will become a short paragraph when converted into specification language. To convert a primitive requirement into specification text, you make several important additions: paragraph number, paragraph title, subject, verb, sentence ending, and any necessary explanatory remarks to avoid misunderstandings. The verb is either "shall," "will" or "should." Most requirement statements use "shall" to express mandatory or binding requirements. Statements of fact or goals in the text use the verbs "will" or "should," respectively. "Should" means that the requirement is not mandatory but desired. Figure 24.32, summarizes the six additions needed to convert a primitive requirement into specification language. Figure 24.33 illustrates an example of converting a primitive requirement for a component of a system, in this case a valve, into more detailed specification language (Grady, 1993, p. 91).

Remember, the reason for developing requirements and specifications is to provide a disciplined framework within which engineers and scientists can creatively work together on many separate but related smaller-scale problems. The challenge is to provide enough, but not too many, requirements to the engineers and scientists so that when their individual

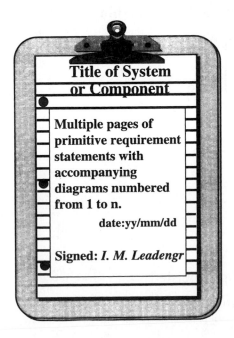

Figure 24.30 Concept requirements list is an early requirements document.

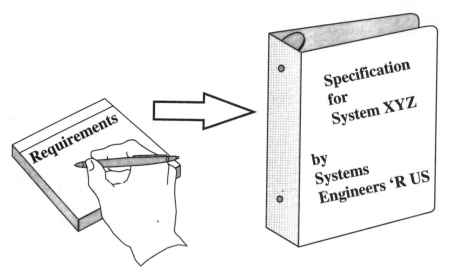

Figure 24.31   A set of requirements statements make up a specification document.

solutions to their smaller-scale problems are put together, the resulting system works well and solves the large-scale systems engineering problem.

Early in the life cycle of a system, it is very useful to bring together all of the information generated during the formulation effort, especially the results of the Needs Analysis, Problem Definition, and Feasibility Studies, into one summarized statement that helps to define

Take the primitive requirements and add six things!

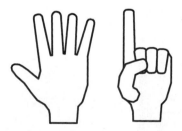

1 - paragraph number
2 - paragraph title
3 - subject
4 - verb
5 - sentence ending
6 - explanatory remarks as needed

Figure 24.32   Converting primitive requirements into specifications.

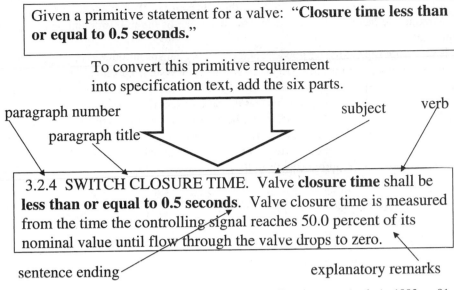

Given a primitive statement for a valve: **"Closure time less than or equal to 0.5 seconds."**

To convert this primitive requirement
into specification text, add the six parts.

paragraph number

paragraph title

subject

verb

3.2.4 SWITCH CLOSURE TIME. Valve **closure time** shall be **less than or equal to 0.5 seconds**. Valve closure time is measured from the time the controlling signal reaches 50.0 percent of its nominal value until flow through the valve drops to zero.

sentence ending

explanatory remarks

Figure 24.33 Example adapted from Grady, *System Requirements Analysis,* 1993, p. 91.

the systems engineering effort. This statement is called the Systems Engineering Problem Statement. It contains the effective needs, objectives, criteria, parameters, variables, and constraints that you have identified so far in your work. It describes the logical relationship of how objectives can, if achieved, satisfy the effective needs, and how the objectives can be measured by the objective measures or design criteria. It also discusses the key performance parameters that define the basic operating characteristics of the system, the important state variables that the system must monitor and control, and the various constraints that must be observed for the system design. You should include the major advantages and disadvantages of the different alternatives and technical approaches identified by the feasibility studies.

The purpose of this statement is to inform the clients and important stakeholders of the overall concept of the system in an easy to understand way. This gives them a chance to make changes in the direction of the design effort early in the system's life, when changes are least expensive, before major commitment of resources. Information about what major functions the system will perform, how the system will accomplish its objectives, when it will be deployed, where the system will operate, and who will use the system is all open to discussion and approval with the clients (Blanchard, 1991, p. 26).

24.5.3 User Requirements

Finding out what users really want or need is difficult for several reasons. Information about user requirements often has a lot of uncertainty. For example, even though an effective strategy is to ask users about their requirements, there are many perceptual and human information processing biases that may limit the value of information obtained from observing or interviewing users. Sometimes users focus on symptoms related to their true problem and ignore the root causes of their difficulties. This means the systems engineer frequently has to help users detect the existence of the true problem and properly diagnose its causes.

It is very important to have a good set of top-level system requirements to guide decision making throughout the design process. This top-level system specification document is called a "Type A" specification. Beam (1990, p. 56) points out, "It is now almost universally accepted in system design circles that the best system designs are carried out in *top-down* [emphasis added] fashion. Designers begin with an overall concept of the system, then work out further details in an orderly and mainly hierarchical manner." These top-level system requirements should orient on the user. As shown in Figure 24.34, the system engineer derives these requirements from these five sources:

1. The Needs Analysis
2. The Problem Definition
3. The Operational Concept
4. The Maintenance Concept
5. The Functional Analysis

You already know about a needs analysis, problem definition efforts, and the functional analysis. So lets describe the operational concept and the maintenance concept.

The *operational concept* depends on a thorough understanding of the missions that the system must perform and the environment that the system will have to operate in. Hence the systems engineer works closely with users or their representatives to write the operational requirements. Together, they describe the major functions and operational characteristics desired for the system. How the system operates, where in the operating environment the system will be distributed, how long the system must operate, and how effective

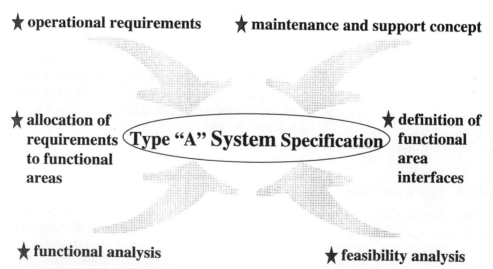

Figure 24.34 The technical, performance, operational, and support characteristics of the system as an entity make up the type A system specification.

the system's performance must be are all part of the operational concept. Care is taken not to specify technical solutions but to describe the performance desired of the new or improved system. Usually the operational concept describes typical mission profiles or operational scenarios of the system.

The *maintenance and support* concept is put together to establish how we envision keeping the system in operation and restoring it to operational use should it break down. These requirements describe the effectiveness of the support elements of the system. Maintenance levels, repair policies, anticipated major maintenance procedures, and logistical requirements related to people and support equipment are part of the maintenance and support concept.

As you can imagine, the requirements and specifications for a system grow considerably as the effort continues. Although the systems engineer is responsible for the system specification, which is prepared early in the design process, the second-level specifications will be developed later by experts in appropriate engineering disciplines such as mechanical or electrical. The systems engineer will often be involved in technical direction and integration efforts to ensure that subordinate specifications comply with the system specification.

Figure 24.35 tabulates the five different types of specification documents. A "Type B" or development specification document contains the technical requirements for any item below the system level where research, design, and development are necessary (Blanchard, 1991, p. 66). A major equipement item or hardware assembly, a computer software program, a facility, or critical support item are examples of the types of items where a Type B specification might be needed. Each specification in a Type B document details the performance and support characteristics that are required.

A "Type C" or product specification contains the technical requirements for any items below the system level that are already in the inventory and can be purchased without requiring any research and development (Blanchard, 1991, p. 67). Examples include many standard or noncomplex fabricated system components such as certain equipment and assemblies, computer hardware and software, cables, spare parts, and tools. The form, fit, and function of the product for its intended use and any interface and interchangeability characteristics are all described by the Type C specification.

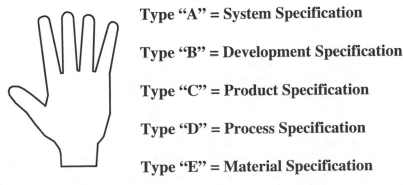

Figure 24.35 Five different types of specification documents (according to Department of Defense Military Standard 490).

As you might expect, a "Type D" or process specification contains details on the services that are required to be performed on a product or material, such as the specifications for a painting, packing, or marking process. A "Type E" or material specification contains the specifications for the raw and semifabricated materials and mixtures used in the fabrication or development of a material (Grady, 1993, p. 131). Sometimes systems engineers find it useful to prepare other types of specification documents such as procurement specifications or interface specification documents (Grady, 1993, p. 131). The purpose of all of these specification documents is to bring discipline to the process of designing and developing a large-scale system so that the resulting system is delivered on-time and within budget and can meet or exceed performance expectations.

Every type of specification (according to IAW MIL-STD-490A) follows the same general six-section format as shown below:

Title of Specification

   1.0 Scope

   2.0 Applicable Documents

   3.0 Requirements

   4.0 Quality Assurance Requirements

   5.0 Preparation for Delivery

   6.0 Notes

The scope or first section of a specification document identifies the name, life-cycle phases (design, production, test), function, and architecture of the system or item being specified. Typically this section includes an architecture block diagram (ABD) of the system or item being specified. In the second section of a specification document, the applicable documents section, you list the other documents that also relate to the requirements for the item or system being specified. The most extensive section of any specification document is the third section, requirements. In the requirements section you address the performance, interface, environmental, specialty engineering (physical characteristics, reliability, maintainability, interchangeability, transportability, human factors, safety, producibility, parts, materials and processes, identification and marking, and other specialty concerns), product, and programmatic (cost, schedule, and manufacturing) requirements (Grady, 1993, pp. 136–175). The fourth section of a specification document, qualification requirements (software) or quality assurance provisions (hardware), lists the inspection procedures, tests, conditions, and data review standards that show that the requirements outlined in the specification document have been met. The fifth secion, preparation for delivery, lists the packaging and markings of packages requirements and is usually only applicable to product specifications. The last section, notes, has any necessary explanatory remarks such as ordering information and does not contain any requirements information.

To help keep track of all these specifications, the systems engineer prepares a *specification or documentation tree*. This tree is based on the system's physical hierarchy or ABD that shows how the system is broken down into segments, prime items or elements, subsystems, major components or assemblies, subassemblies, and individual units and parts. The ABD diagram is a set of simple boxes with text connected by lines that shows how the system is partitioned into its different related elements. Each block contains the name of the element, and there is also an architecture dictionary that contains a more detailcd description of each element. The lines of the ABD diagram show the ordered arrangement of all of the parts that make up the system. The ABD is also used in conjunction with the functional flow block diagram to create a functional requirements allocation matrix (FRAM) to detail how the various functions are mapped onto or allocated to the system's physical components.

The systems engineer builds a specification tree by identifying the different types of specifications that should go with each element shown on the system hierarchy or ABD diagram. Often times the specification tree is created by simply annotating the architecture block diagram with the document specification details (Grady, 1993, p. 309). You annotate each block of the system hierarchy with the number and title of the specification and the name of the organizational department and engineer responsible for that specficiation document. For example, the Type A system specification is annoted to the top block or system block of the architecture block diagram as shown in Figure 24.36. The tree, when completed, shows the hierarchy of specifications from the system specification on down. It shows who is responsible for each specification. The specification tree also helps to resolve conflicts that arise over which specification should take precedence.

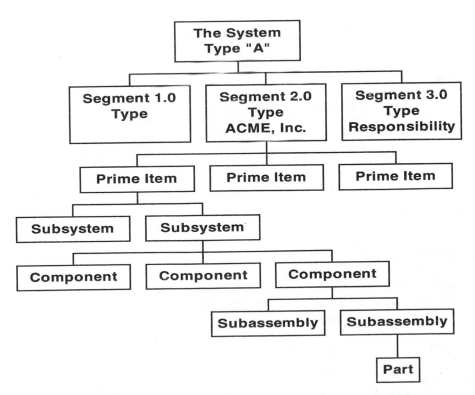

Figure 24.36 Sample structure for a specification tree on the ABD.

24.6 GENERATION OF POTENTIAL ALTERNATIVES OR SYSTEM SYNTHESIS

One of the biggest payoffs from making a good objectives tree and describing a system from a functional perspective, is the way you can use it to help you generate alternatives. How do you do this? It is a relatively easy *down-and-across technique*. Working down first, you create a list of activities for each lower-level objective in the tree or each function in a block diagram. This list of activities is a set of different ways for accomplishing each lower-level objective or function. Once you have a good list of activities for each lower-level objective, then you create alternatives by combining these activities together by working across the lists. You can select one activity from each list or you can select several. Each synthesis or combination of activities represents one draft alternative that possibly merits further consideration and refinement.

Consider a simple example of this approach. In Figure 24.37 you see an objectives tree that shows what a group of stakeholders hopes to accomplish with a new software product or service to help computer users better manage their multiple computer accounts. Underneath each lower-level objective is a list of activities that represent different ways of achieving the lower-level objectives. The circled activities and links between them represent one possible synthesis or alternative. For example, the circles in Figure 24.37 describe an alternative for easing computer account security that includes using a pop-up window that displays a list of the user's accounts and passwords with on-screen buttons that a user can click on to login to an account with an alarm to remind the user before an account expires. There are many more possible alternatives that you can generate from the three lists of activities in the example. You can appreciate the power of this approach. It helps you quickly generate many different concepts for potential alternatives.

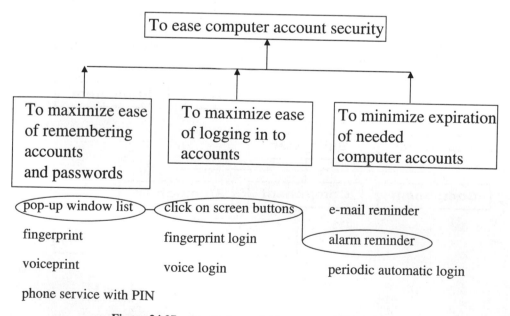

Figure 24.37   Combining activities to create **alternatives**.

Of course, the quality of alternatives generated in this way depends heavily on the quality of the lists of activities that were used to create the alternatives in the first place. For this reason, most large-scale projects require the systems engineer to involve a group of experts to create these lists of activities and to synthesize these activities into viable alternatives. Hence, systems engineers need to know a lot about group dialog techniques such as brainstorming, brainwriting, and Delphi approaches.

24.6.1 Group Dynamics

Often there is no one person with sufficient knowledge about a complex situation to develop a set of elements that describes it. Usually, available information about a complex situation is incomplete, uncertain, or even wrong. The model that represents a complex situation may also be unverified and perhaps incomplete. Developing useful alternatives requires finding a way to overcome these information problems and find as much truth as we can.

We can classify the information about a complex situation into three types: speculation, opinion, and knowledge (Sage, 1977). As speculation becomes knowledge, the probability of truth varies from 0 to 1. We would rather not base decisions on speculations because the probability of truth is so low. However, sometimes the knowledge about a complex issue is such that we have little detailed, reliable information. Therefore, in actual practice, the systems engineer must find ways to gather people's opinions and develop from them better information that can inform and enlighten a complex situation. Hence, systems engineers often work with groups to solicit better information.

The appeal of groups is based on the idea that "two heads are better than one." Committees, juries, boards of directors, and panels are all examples of this. They represent some of the many mechanisms for pooling minds to generate ideas. We hope that their ideas are based on wise collective opinions or expert knowledge, each with a high probability of being correct.

However, the systems engineer must realize that there are both advantages and disadvantages to reliance on group opinions. On the favorable side, many times a group will interact to compensate for the bias of individual members of the group. Or knowledge of one group member can compensate for ignorance or speculation on the part of other members. Also, people who participate in a group tend to develop a stronger stake in the group's enterprise. One way of getting more support for a large project is to include many potential stakeholders in the planning group for such an effort.

Unfortunately, there are drawbacks to having a group involved. When too many diverse interests go unchecked in a group, you may end up with the camel effect: "a horse designed by a committee." Sometimes the opinion of the group can be strongly influenced by the person who talks the most and the loudest. This can be upsetting to a group since there is little necessary correlation between knowledge and loudness or talking a lot. Another problem is that group activity can lapse into a "bull session" unless the group is well organized. Bull sessions create lots of talk about matters of individual and group interest, but avoid discussing the main issues of the group's charter. Often there is strong pressure for group conformity and avoidance of unpopular viewpoints. Martino (1972) provides a good list of group disadvantages that the systems engineer should try to control when working with committees to evolve expert opinion:

- *Misinformation.* There is as much misinformation available to a committee as there is available to any of its members. It is hoped that correct information held by some group members cancels out wrong information held by another group member. There is no guarantee that this will happen.

- *Social Pressure.* A group often exerts strong social pressure on its members and encourages all members to agree with the majority even when several individuals feel that the majority view is wrong.

- *Vocal Majority.* Often the number and volume of comments and arguments for and against a position is more influential in determining results than the validity of these comments. Again, a strong, loud majority may overwhelm the group.

- *Agreement Bias.* A group is often more concerned with reaching an agreement than reaching well-thought-out conclusions. The result is usually a statement of mild philosophical views that can offend no one rather than concrete, specific suggestions that may offend some group participants.

- *Dominant Individual.* The dominant individual often has an undue impact on the final results of the group unless there is strong and impartial group leadership. This dominance may be due to active and loud participation, a persuasive personality, or extreme persistence.

- *Hidden Agendas.* Hidden agendas and vested interests on the part of some group members may lead to a game in which the objective is convincing the group of their view, rather than striving for what might be a better group decision.

- *Premature Solution Focus.* The entire group may possess a common bias for a particular alternative or technology.

24.6.2   Brainstorming

So how does a systems engineer cope with the challenges of group dynamics? Some of the drawbacks of groups may be made better by the process of brainstorming. A classic brainstorming exercise involves a small group, a well-defined problem, prior awareness of the problem by the group, a leader, a secretary, and a blackboard. Classic rules for a brainstorming session, which normally lasts for 60 to 90 minutes, are that the leader reminds the group of the problem at hand and the rules for brainstorming. The leader's role is to ensure that all participants join in the discussion. Leaders should suppress their own ideas as long as the group is generating ideas. If the group has a lull in their efforts, then the leader may inject new ideas to spark the group forward. The leader does not allow criticism of any ideas. The group is encouraged to keep the ideas short and to save full details for later. As participants talk, the leader writes short, two-word descriptions of all ideas on the blackboard. The secretary captures more details. To stimulate new ideas, the leader may review aloud the ideas and elements already generated by the group.

Many variations of the classic brainstorming exercise are possible. For example, participants may supply ideas in writing to the leader before the meeting. The group hears these ideas and discusses them without necessarily knowing who thought of the ideas. The classic brainstorming approach, if led properly, can minimize the negative aspects of misinformation, dominant individuals, and hidden agendas. Anonymous brainstorming, as just described, can minimize undue majority social pressure and overemphasis on consensus as well.

It is possible to modify the brainstorming exercise in other ways. You can do this by not announcing the problem to the group prior to the meeting. This helps to get people to reveal what they really think, since they do not have an opportunity to coordinate their position on issues beforehand. Also, you can hold a series of brainstorming meetings instead of just one session. In the first meeting, the leader presents ideas and elements that the group criticizes in all possible ways. In later meetings, the group finds alternatives to the difficulties generated in the first meeting. Any of these modifications to brainstorming still has advantages and disadvantages.

24.6.3 Brainwriting

One very popular way to modify brainstorming is to replace most of the verbal communication with writing. This is called *brainwriting*. In this approach participants write a number of relevant ideas, often limited to three, on a sheet of paper. The paper is then passed to another group participant, and the ideas originally developed by one group member are further developed by another group member. These new ideas and elements are added, and the augmented pages are passed to another individual.

When the brainwriting cycle is complete and each person receives the sheet of paper they first used, the beginning phase of the session is over. The group leader collects the sheets of paper in preparation for the next phase. In this next phase, the written contribution from the first phase is circulated to the entire group. The object of the second phase is to revise the ideas or elements developed in the first phase. This process can be repeated in similar phases until enough information is collected to serve the meeting's purpose.

Again, several variations to brainwriting are possible. The papers can be put into a pool rather than circulated to the participants, and an individual who puts a sheet in the pool takes another sheet from the pool and further expands upon the ideas and elements on that sheet. Deciding what sort of brainstorming or brainwriting exercises to use depends on the particular group you are working with and the issues you want them to consider.

24.6.4 Groupware

Computer and communications technology may be used to increase the speed and productivity of group dialog and to make it possible to have larger brainstorming or brainwriting groups. In fact, a number of researchers and private companies have made it their business to specialize in providing hardware, software, and expertise for groups. This area of specialization that helps people work together better is called *collaborative computing* (Hsu and Lockwood, 1993). Collaborative applications for groups, usually called *groupware,* help people work in groups by making sharing ideas easier in three ways: common task sharing, environment sharing, and time and place sharing. Common task-sharing systems help many people in a workgroup work on the same task. Environment-sharing systems help people stay abreast of a project and what other participants are doing on the same project. Time-and-space-sharing systems help people to work together in several different modes: same time and place (synchronous), same place at different times (asynchronous), same time at different places (distributed synchronous), or different times and places (distributed asynchronous).

One example of a groupware application is GroupSystems V® from Ventana in Tucson, Arizona. This package provides a group-decision support environment on desktop or laptop computers arranged in a meeting room configuration and linked by an Ethernet or local area network (LAN). Basic tools for group processes in GroupSystems V are a meeting manager, group link, agenda, briefcase, electronic brainstorming, categorizer, vote, topic commenter, group dictionary, alternative evaluation, and policy formulation. Most of these tools are self-explanatory. *Meeting manager* helps the session leader manage face-to-face workgroups by initiating certain planned activities, and recording and printing session reports. *Group Link* is a utility that people can use away from a meeting room, such as at their desk, to collaborate with other people on group projects. *Briefcase* provides several convenient tools that workgroup participants may find handy during a meeting, such as an electronic calculator, calendar, and notepad.

The *electronic brainstormer* in GroupSystems V allows people to work together in parallel by using the computer to capture and broadcast their comments and ideas to others simultaneously and anonymously. *Vote* allows people in a workgroup to express their evaluation of issues in seven quantifiable ways: rank order, multiple choice, agree or disagree, yes or no, true or false, 10-point scale, and allocation of points. Results from a vote session are electronically recorded, tabulated, and displayed graphically or printed in text on the screen or on paper. *Policy formulation* helps a group to jointly write and edit statements by an electronic, interactive process of review and revision.

As you might expect, companies like Ventana are improving their groupware by adding more functions and features. For example, advanced tools for GroupSystems V include group outliner, survey, group writer, idea organization, group matrix, stakeholder identification, and questionnaire. Again, most are self-explanatory, but here is a short description of a few of them.

Group matrix helps a group fill in the boxes of an interaction matrix. Participants choose a relationship between the ideas or items in rows and columns by entering a numeric value or predefined word. The system displays the average group response and indicates by colors which cells have consensus or disagreement so the group can focus its work. *Stakeholder identification* helps a workgroup to carefully identify and consider the impact of stakeholders on a proposed plan. But, Ventana's GroupSystems V is only one example of the many groupware products on the market. While GroupSystems V provides mainly text-based interactions for groups, other groupware applications are using the power of computer graphics.

An example of the kind of groupware graphics power available today is CM/1® developed by Corporate Memory Systems Inc. of Austin, Texas. CM/1 is a Windows-based hypertext tool for creating a *graphical discussion tree* with a workgroup (Gottesman, 1993). The tree helps a group explore complex issues with unknown answers by structuring the discussion using *issue nodes, position nodes,* and *argument nodes.* Discussions start with the creation of an issue node that is labeled with an open-ended question. Then group participants offer their solutions by creating a position node and labeling it with their idea. The link between an issue node and a position node is a "responds to" arc or line. Argument nodes "support," "object to," or "specialize" (refine) a position. Participants can raise new issues that expand or challenge previous issues, positions, or arguments. Eventually, the group can enter a *decision node* to show that an issue has been resolved.

The graphical discussion tree encourages a logical, focused approach to group work and promotes understanding among participants. *Reference nodes* and *note nodes* allow participants to put in links to documents and other evidence to support their arguments. CM/1 and GroupSystems V are just two examples of the many computer-based workgroup applications that you might want to consider. Groupware can help the groups you work with generate ideas, develop action plans, refine work, make decisions, and negotiate compromises.

24.6.5 Delphi Methods

In traditional group discussion methods, effective communication is sometimes inhibited by the disadvantages inherent in groups. These include many psychological factors, problems with dominant individuals, the consequent problem that all points of view are not heard, and the lack of documented, concise statements about what actually happened in the group dialog. Brainstorming and brainwriting exercises try to get the most from group dialog by a systematic and controlled exploration of the elements and factors of a particular issue. Still, even computer-assisted brainstorming and brainwriting sessions suffer from many group drawbacks.

However, the procedure known as Delphi, developed by the Rand Corporation, eliminates many of the disadvantages of group dialog (Dalkey, 1969). The Delphi approach does this because of three features not found in either classic brainstorming or brainwriting. These are anonymity, iteration with controlled feedback, and statistical group response. Anonymous response is made possible by the use of a formal questionnaire to get responses from group members. Iteration and controlled feedback happens because Delphi is a systematic exercise conducted in several iterations. Between each iteration there is carefully controlled feedback. At the conclusion of the final iteration, group opinion is combined from individual opinions to form a statistical group response based on the relative expertise of members with the issue under consideration. These three essential features of the Delphi technique make it a very successful method of soliciting and refining group opinion.

Although a Delphi exercise is really a modification of both brainstorming and brainwriting, there are two primary variations that make it different. First, there are simultaneous individual contributions from each participant at every step, without participants having knowledge of inputs supplied by others for that particular step. Second, the sources of all inputs are anonymous. This anonymity of input is maintained through the entire Delphi dialog.

The Delphi approach tries to minimize the biasing effect of irrelevant dialog between individuals and of group pressure toward conformity. Delphi helps to ensure that group interactions about the issue at hand compensate for the biases of individuals, and that the knowledge of several members of the group will compensate for ignorance on the part of others. Now let us describe the classic Delphi followed by several well-known variations.

1. *To Begin.* A Delphi sequence interrogates a group using a series of questionnaires. Each iterative summation of a questionnaire to a group is referred to as a round in the original Rand terminology. The questionnaire distributed to the group asks questions of the group; it also provides information to group members concerning the degree of consensus that has resulted from previous rounds. It also gives diverse arguments presented anonymously by various group members. A director leads the group or panel through each round. Each round calls for different activities on the part of either the panel, the director, or both. The subject area for the panel to deliberate is clarified before the first round. The director resolves any questions about the rules of the Delphi exercise with the panel members before the first round.

2. *Round One.* The questionnaire distributed during the first round is completely unstructured. It requests the group to discuss elements or make a forecast about the specific problem under consideration. If the group is truly expert and very knowledgeable about the subject area under discussion, then this approach to round one is most useful. It allows

the group to utilize its expertise to maximum advantage. If the questionnaire used during round one is too structured and restricted, the group may well overlook elements and events discussed in the questionnaire.

After the panel responses have been returned to the director, they must be consolidated and placed into a single set. Individual members of the panel may present their discussion of events, elements, and forecasts in the form of a narration. The director will then disaggregate the pieces into a set of discrete events. Other individuals may perhaps have given a list of events and elements arranged in chronological form. The main task of the director at the end of round one is to identify events, elements, and forecasts; consolidate those that are similar; eliminate those that are unimportant for the purpose at hand; and prepare a final list of elements, events, and forecasts in the most clear fashion possible. This final list becomes the input or second questionnaire for round two.

3. *Round Two.* In round two, the panel examines the list of consolidated events prepared from round one and estimates dates when these elements or forecasts may occur. Panel members give reasons why they expect these dates to be correct. After these forecasts and estimated dates from round two are complete, the director prepares a consolidated statistical summary of the panel's opinion. A brief discussion of the reasons for the group's opinion is also prepared. From this information, the director prepares the questionnaire for round three.

4. *Round Three.* The third questionnaire consists of the list of elements, events, forecasts; the group median date and the upper and lower quartile dates for the occurrence of each event, element and forecast; and a summary of the reasons for the choice of dates. This questionnaire starts round three. The panel reviews the arguments and prepares a new estimate of the date on which each event, element, or forecast may occur. If any of the estimated dates falls later than the upper quartile or earlier than the lower quartile, the panel members of the group must give reasons to justify this view as well as to comment on the opposing views of others.

If an individual's estimate is earlier or later than that of three-fourths of the group, the individual must justify this estimated date and indicate why previous group arguments in support of an earlier or later date are wrong. These arguments typically reference outside factors that other individuals in the group neglect and cite facts that others may not have considered. Therefore, individuals in the group are just as free to raise objections and make arguments as they would in any face-to-face confrontation. However, their arguments are now anonymous.

Results from the round-three questionnaire now go to the director. Preparation of a questionnaire for round four is similar to previous rounds. Group estimates must be summarized; new medians and quartiles must be computed; summary arguments for both sides of any dispute must be prepared. All of these are put together for the round-four questionnaire.

5. *Round Four.* After the panel members get the new questionnaire (a new list of events, elements, and forecasts; a statistical description of the estimates of dates on which these will occur; and summary arguments), the group makes a new forecast. In classic Delphi, round four is the final round. Often there is no need for the director to analyze the group's arguments at the end of round four. However, if the panel has not been able to reach a consensus, the director may want to get final arguments from both sides in a dispute. With these final arguments, the director can prepare an effective statement about this lack of consensus as the final result of the Delphi exercise.

6. *To End.* The final Delphi output is a written report from the director with a list of events, elements, and forecasts with the median and quartile dates estimating when they will occur and a summary of relevant arguments (Martino, 1968, pp. 138–144). It is not necessary that any particular event, element, or forecast last through all four rounds of the Delphi exercise unchanged. If the group agrees that a particular event, element, or forecast will never take place, then it can be dropped at the round where this agreement occurs. It does not need to appear in later rounds. Sometimes originally stated events, elements, or forecasts must be restated to make them clearer, or perhaps one will be divided into several events. Some events or elements may need to be combined.

In general, most Delphi exercises observe an outcome with three characteristics:

(a) *Wide Initial Distribution of Responses.* On the first round, there is a wide distribution of individual responses to the questionnaire.

(b) *Convergence of Responses.* The distribution of individual responses converges in later rounds as iteration and feedback take hold.

(c) *Accuracy.* The group response becomes more accurate from round to round. Accuracy is defined here as the median of the final individuals' responses that can never be reduced to zero since we are usually talking about future events.

7. *Variations.* Many variations on the original Delphi approach are possible. You can appreciate that a very useful modification is to use a computer-based approach to help make the job of the director easier and faster. Most variations preserve the three fundamental principles of a Delphi exercise: anonymity, iteration, and statistical response. You can also understand that some panelists who are not familiar with statistics may at first find it difficult to make assessments using this method. Again, the computer with its graphical display capabilities can make it easier for participants to understand statistical terms like quartile and median. Other variations actually modify one or more of the Delphi principles. Here are some variations you may want to consider.

(a) *Begin with a Blank Sheet of Paper.* The group is now completely unstructured and no precise guidelines tell the participants how or where to start. This approach may allow for more creativity. A drawback of this variation is that the elements, events, or forecasts produced by the group may be totally irrelevant to the director's purpose.

(b) *Start with a List of Events, Elements, or Forecasts Generated by External Processes Prior to the Delphi Exercise.* Use this list to begin the Delphi sequence. This is the same as accomplishing round one with another group and then transferring the results from round one with the first group to a second group to start round two. This variation helps to focus the Delphi exercise, but it also may inhibit the creativity of the second group. If you use more than one group in a Delphi effort, there is no reason why some of the same people cannot be members of more than one group.

(c) *Use Ratings of Individual Expertise.* If the panelists vary considerably with respect to their expertise about the issue being considered, then the director may ask each individual to indicate, on a simple scale, their level of expertise on each question they answer. The director combines the individual estimates in a weighted average using the self-rating of expertise as the weights. In this way, answers from panelists with the most knowledge count for more.

(d) *Provide Background Briefings to Panelists.* When highly technical issues influence the results of the Delphi exercise, it may be very helpful to brief all the panelists with necessary background information. For example, when determining futures for oil supply and demand, it might be very helpful to inform the group of likely developments in solar and other alternative energy sources, especially if the group lacked expertise in this area. This provides the entire group with a starting baseline of knowledge and can avoid needless arguments about technical details.

(e) *Attach Names to Responses.* You may want the influence or position of the originator of an idea to sway the judgment of others. This defeats one of the basic principles of Delphi, which holds that the purpose of anonymous responses is to judge responses solely on their merits. However, one panelist may be so truly expert in the area that is under discussion that it may be beneficial to identify that expert's responses.

(f) *Reduce Feedback.* If you eliminate all feedback, then individual opinions on the second and subsequent rounds might just be a repetition of the first round's responses without any reexamination of these initial responses. This could cause the reinforcement of wrong responses by repetition. It is the well-known phenomenon that "if you tell a lie often enough, then people will soon begin to believe it to be true." Sometimes feedback may result in over convergence to a given median. This happens when people want to avoid providing an argument for not shifting to the median. One way to overcome this is to only provide a portion of the feedback, eliminate the previous group median information and give only the upper and lower quartile information.

Which variation is right for your project depends a lot on the problem under consideration and the people you are bringing together to work on it. Here are some good guidelines that should help you successfully conduct a Delphi exercise (Martino, 1972):

- *Get Willing Agreement from the Individuals Who you Want to Serve on the Delphi Panel.* A few willing participants is better than a bigger group with a "bad apple."
- *Explain the Delphi Procedure Completely to the Group.* You may find it helpful to run a practice round or two on an easy problem so that everyone understands the rules.
- *State Events, Elements, and Forecasts in Simplest Form.* Avoid using compound events.
- *Avoid Ambiguous Statements of Events, Elements, and Forecasts.*
- *Make the Questionnaire as Easy as Possible.*
- *Use no more Questions than an Individual Can Adequately Consider.* Fewer is better than too many.
- *Explain Why Contradictory Events, Elements, or Forecasts are Included, so that Panelists will not Think that the Director is Trying to "Trap Them."*
- *The Director Should Never Inject Personal Opinions into Group Feedback.* If the director knows that the group has overlooked significant factors, then the group output should be discarded and the group considered unqualified. Repeat the exercise with a more qualified group. The director must not meddle in the deliberation of the group.
- *Compensation of Group Members for Work Load Involved in a Delphi Exercise Should be Appropriate to the Service Rendered.* It should also be appropriate to the type of organization, profit or nonprofit, for example, requesting the Delphi exercise.

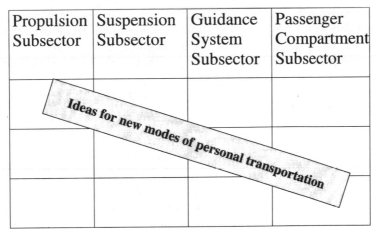

Figure 24.38   Gibson's example of using Zwicky's morphological box.

- *Strive for the Best Qualified Panel that you Can Find.* Remember that the knowledge that the panelists have about the issues under discussion and the clarity of the panelists' thinking processes play the most important part in determining the quality of the Delphi results.

24.6.6   Morphological Box Approach

One very useful technique for stimulating a group's creative thought is called Zwicky's morphological box.[4] Zwicky's morphological box approach stimulates thought by encouraging people to think of new ideas at the subsystem level and then putting these ideas and elements together into new and untried combinations.

1. *To Begin.* Define all of the functional classes that make up the basic subsystems of the problem under consideration. Again, the lower levels of the objectives tree can help to suggest what some of these functional classes should be. Gibson (1991, p. 133) presents an excellent example of Zwicky's approach that focuses on the problem of thinking about new modes of personal transportation. As depicted in Figure 24.38, he divides this problem into four functional classes, or what he calls major subsectors: the propulsions subsector, the suspension subsector, the guidance system, and the passenger compartment. Note that these four sectors are not technological solutions such as a private vehicle of some sort. Instead, these subsectors or functional classes focus on human passengers and the various activities that must be done to successfully transport human passengers from one place to another. In the next step of Zwicky's box, you concentrate on all the different ways, including technological activities, for accomplishing each sector.

2. *Sector Brainstorming.* In this step, the group focuses on one sector at a time and creates a list of activities for accomplishing that one sector. This is important because you do not want to limit your thinking about one sector because of preconceived constraints

---

[4]In his time, Zwicky became very well known for his book, *Discovery, Invention, and Research,* published in 1969, according to Gibson (1991). Zwicky's main idea in the book was that people can be very creative if they use techniques that force them out of their traditional modes of thinking.

about how one sector relates to another. You are looking for new ideas, not the standard conventional solutions in operation today. Brainstorming, brainwriting, and Delphi are all useful techniques that you can use to help a group create these sector lists. In Gibson's example, he presents a list of ideas for the suspension sector. Gibson stresses the importance of concentrating on one sector at a time. He says that you are accustomed to conventional means of transportation, and if you permit yourself to think of complete transportation units, apparently incongruous combinations may be suppressed, simply because nobody wants to appear silly before oneself and one's colleagues. But when you consider only one sector at a time, who is to say what is silly? Thinking of ideas for one sector at a time releases you from the artificial bounds of conventional approaches (Gibson, 1991, p. 134).

Since the ideation of activities for each sector is an out-scoping mental effort, there should be no criticism of ideas for the list while it is being created. Gibson suggests that the group lists all conceivable ways of accomplishing each sector without regard for practical limits or consideration of the other sectors. After creating exhaustive lists for each sector, the group is ready to try to synthesize or forge the ideas from the lists together into a set of workable alternatives.

3. *Combining Elements.* Picking at least one element from each sector list, a combination or *synthesis* of elements is put together to form an alternative. At this stage, it is again important not to criticize these combinations, even though some will seem ridiculous at first glance. All possible combinations should be exhausted so that no new directions or innovations are overlooked. Gibson (1991, p. 133) relates the example of a popular children's toy that is a book with three different sectors of pages running through it. The top section has pages of different cartoon heads while the bottom two sections have pages of funny bodies and legs. By turning the three set of pages in different ways, the book amuses the children by the unexpected figures of weird cartoon characters that they can create by combining different heads, bodies, and legs. Although this seems like a funny example, Gibson says that it makes the important point that creativity often lies in unexpected directions and that is why every combination of the morphological box should be explored.

At this point, you might ask why we bothered with the morphological boxes or sector lists. Why did we not just go directly into creating complete alternatives? Again Gibson has a good answer (1991, p. 126): "Because we find that creating 'complete solutions' initially often results in premature inscoping. 'Complete solutions' offered at this point tend to be old and tired ideas. They seldom, if ever, are imaginative or take full advantage of the situation." Now, with Zwicky's box, you have many alternatives, some of which were never even imagined as the individual sector lists were being created.

4. *Making Combinations Work.* After you assemble all of the possible combinations from the elements in the different sector lists, the group focuses its creative energies on making the combinations work. The group should strive to think about how to make even the most unusual combinations work. Gibson (1991) suggests, "Don't ask, 'will it work?' Don't ask, 'what do you think of the following combination?' Those are killer phrases designed to bring out the critical faculty. It is better to say, 'do a preliminary design of a device with the following attributes . . .'" (p. 134). As work progresses, you can discard combinations that are impractical and build more detail into the descriptions of promising alternatives. These promising alternatives will undergo more scrutiny in feasibility studies.

5. *Variations.* As you can imagine, there are many successful ways to use brainstorming, brainwriting, and Zwicky's morphological box together to work on tough issues. For

example, you may want to use brainstorming to create the different functional classes or sectors in Zwicky's box. Then, you may want to involve the group in a brainwriting exercise to develop the lists of ideas for accomplishing each sector. Putting the elements from the lists into different combinations and elaborating how each combination might work could be done by a Delphi exercise. Here are some important variations on Zwicky's box that you may find useful, especially in corporate settings:

(a) *Match Combinations to Objectives.* One way of determining whether the combinations from a Zwicky's box approach will work is to simply compare them to the objectives and attributes of the corporation. If the combinations seem to support the corporation's objectives and match up well with the firm's attributes, then it is likely that those combinations deserve serious attention.

(b) *Order the Selection.* Another way to change Zwicky's procedure is to impose some ordering discipline on the selection of elements from the various morphological sectors or categories. A benefit of this approach is that it reduces the number of impractical combinations generated. Using the transportation example again, if you select a propulsion idea related to human self-propelled power such as pedaling, it is unlikely that you would select an idea from the passenger compartment sector that is large and heavy and that clearly would not work with pedal power. The drawback to this approach is that it clearly limits the consideration of many ideas, especially unconventional ones.

(c) *Use Profiles on a Field.* This variation of Zwicky's box, developed by John Warfield (1979), is very useful for stimulating creative thought about potential business ventures for a corporation. The first step is to create a morphological field of the various sectors that make up a functional description of the corporation's business. Next, make a list of all the possible attributes for each of the functional categories. The categories, together with their lists of attributes, make up the options field. Then you can create a profile that describes your corporation by selecting attributes from each functional category that best describes your organization. In a similar manner, you can create a profile for each of the different ventures that your organization is considering. By comparing the venture profiles to the corporate profile, you can provide some preliminary recommendations about whether or not the corporation should take on these ventures. Ventures that have profiles that align nicely with the corporate profiles promise a high expectation of success.

24.7 ALTERNATIVES AND FEASIBILITY STUDIES

Many systems engineers use the term feasibility study to encompass the entire formulation effort because the end product of formulation is a set of feasible alternatives. However, since we have discussed most of the formulation effort already, this section uses the term feasibility study to focus on how to screen a set of alternatives to eliminate those that are not viable.

24.7.1 Feasibility Screening

In the previous section, you learned how to combine the ideas of experts and stakeholders to generate many alternatives. Now you need a way to scope down all of these alternatives to a sensible number of viable alternatives. We call this type of effort a feasibility screening because it examines alternatives in sufficient detail to determine if they meet the minimum requirements of the project's stakeholders. A feasibility screen is not an effort to fine-tune the alternatives and make them the best they can be, nor is it an effort to select the best alternative. Such efforts come later in the analysis and interpretation of alternatives. In a feasibility study, you want to identify those alternatives that cannot meet even the minimum requirements of the project's stakeholders and society. Another goal of a feasibility study is to make changes, if possible, to alternatives identified as infeasible so that they are viable. However, the overall result of a feasibility study is usually a tremendous reduction in the number of alternatives.

Why do you want to reduce the number of alternatives by conducting a feasibility study? First, analyzing and interpreting alternatives takes considerable time, money, and effort. It would be a waste of valuable resources to analyze and interpret alternatives that could never be placed into operation anyway. Second, since design is an iterative process, developing high-quality alternatives takes lots of working and reworking of each alternative. Scattering this effort across many alternatives reduces the amount of work that can go into developing each alternative to make it the best it can be. Most importantly, you conduct a feasibility study to make certain that flawed alternatives are not eventually selected for implementation. How sad it would be to create a new systems design only to realize, after lots of time and money, that the design will never work in actual practice because of some limitation that could have been discovered much earlier.

Now that you understand why you want to do a feasibility study, how do you go about it? The first challenge that you must overcome is the sheer number of alternatives that exist at the beginning of a feasibility study. Don't worry; this is something to be happy about. It means that you and your group of stakeholders and experts have been very successful and creative! One way to reduce the number of alternatives to consider at the very beginning of the feasibility study is to categorize or group the alternatives into similar sets. For example, a project that is developing alternatives for a computer system for its sales force might have a very large number of alternatives. But a grouping of the alternatives into several categories, such as desktop, laptop, subnotebook, and handheld, might prove very useful. Now your efforts can focus on picking representative alternatives from these four categories for further study.

If major variations within or across the categories of alternatives exist, you may need to consider more than one representative alternative from each category. For example, installing a multimedia capability on a computer device is a significant variation that will impact these four categories in different ways. Therefore you may want to consider two alternatives from each of the four categories, one with multimedia capability and one without it. The main point to remember is that it is much easier to work with a few representative alternatives from a large category than to try and work with every single possibility. If the alternative you have chosen is truly representative of the entire class, then you often find that if it is not feasible, the entire class is also not feasible. One way of testing whether or not you have divided the alternatives into proper representative categories is to determine if the categories are significantly different from each other. Alternatives should be significantly different from each other, and therefore categories of alternatives should be very significantly different from each other.

24.7.2 Architecture and Standards

Another good way to categorize alternatives and to reduce the number of alternatives quickly at the beginning of a feasibility study is to use architecture and standards considerations. Architecture is the scheme of arrangement of the components of a system, and it describes features that are repeated throughout the design and explains the relationship among the system's parts. Architecture can be very useful for categorizing or eliminating alternatives. In the example of a computer for a sales force, you could categorize alternatives by their hardware or software architecture such as by their central processing units or by their operating systems. Also you may be able to eliminate a large number of alternatives because, for example, corporate headquarters may have placed a constraint on what is possible by imposing a certain architecture as a companywide standard.

A standard is an accepted or approved way of doing things. Government can require and enforce standards such as the safety belts and air bags in automobiles. Industry can recommend standards such as the RS-232 interface cable for computers. Some standards are de facto. They simply evolve into common practice because it makes good business sense to adopt them.

Standards are very useful because they make it possible for many things to work together better and help achieve many economies. For example, you know when you buy an electrical appliance that you can take it to any office or home in the United States and it will easily plug into the electrical system because of the standard plug, receptacle, and power source. You can imagine the frustration of users if they had to worry about whether or not some electrical appliance would work in their house or they had to buy some expensive adapters to make machines work at the office.

But standards can also limit innovation. Perhaps there are easier and safer ways to accomplish hooking up an appliance to a home power source, but the standard may inhibit the use of such new ways. So you have to be careful to balance the benefits and costs of following different standards. Also, standards must be evaluated periodically to see if they are really necessary. If simple guidelines or functional standards will do, then they should be used in lieu of rigid technical standards.

But oftentimes, there are many useful standards that ought to be followed as part of a large-scale design project. Therefore, another way of reducing the number of alternatives you must consider is to screen out those alternatives that cannot be adapted to standards that you know the project must use.

24.7.3   Feasibility Criteria

You can now see that much of the feasibility study involves identifying criteria that can be used to screen the alternatives, eliminating those that cannot meet the minimum criteria. What are these feasibility criteria? Although it depends on the specifics of the system being designed and the system's environment, there are a number of different categories of feasibility criteria that are useful to consider. For example, economic, safety, environmental and social impact are important feasibility considerations in engineering design.

Ostrofsky (1977, pp. 262–264) recommends screening a set of candidate systems by eliminating those that are not physically realizable, economically worthwhile, or financially feasible. He defines *physical realizability* as whether or not the components of each subsystem are physically compatible. One way of doing this, as shown by Ostrofsky, is to make a two-column matrix that lists each component in one column, with a second column that lists the corresponding incompatible components. Then you can use this matrix to screen each alternative to identify compatibility problems and make decisions to either eliminate the alternative from further consideration or modify the alternative so it is physically realizable. For example, an alternative can possibly be made compatible by buying

Feasibility Criteria

Alternatives	Cost (< $)	Performance (> Value)	RAM (> Value)	RECAP
New	G	G	G	G
Modified	G	G	G	G
Baseline	G	NG	G	NG

Figure 24.39 Feasibility screening matrix.

or making a translator or adapter to overcome the incompatibility problem between two components.

Economic worthwhileness examines the expected costs and potential revenues and benefits of the alternatives. This effort requires forecasting the future and understanding how to evaluate cash flows and incorporate risk considerations. You can learn these and other analysis techniques in later chapters. Closely related to economic worthwhileness is financial feasibility. In short, this techniques really examines the budget required to accomplish each alternative and determines whether the stakeholders funding the project are willing to pay for it. This is necessary because even though an alternative may eventually pay off in the long term, the shorter-term outlays may exceed what is financially feasible to do.

Another very important feasibility category is *legal acceptability*. The systems engineer must be careful to consider the various legal ramifications of large-scale projects. Without doubt, there are many local, state, and government agencies and private organizations that can have a significant impact on a project's viability. Again, careful planning in the early stages of an effort can avoid lots of wasted time and money that can be caused by regulatory work stoppages or litigation.

A thorough feasibility screening effort will examine each of these suggested categories and develop from them appropriate specific feasibility criteria. For each criterion, you establish a minimum or maximum level of attainment that you can use to make a "Go" or "No Go" decision. Figure 24.39 shows a feasibility screening matrix that summarizes these kinds of results. Note that there is a recapitulation column that makes a decision about the overall viability of each alternative. A No Go in any column for an alternative results in a No Go overall for that alternative. Only alternatives with an overall Go are carried forward to the next step: modeling and analysis.

24.8 SUMMARY AND CONCLUSIONS

One way to summarize this chapter is to first recount the three steps of issue formulation: problem definition, value system design, and system synthesis. Second, we mention the products of these steps and the various techniques that you learned in this chapter to help you create these products. Finally, it would also be good to remind you of how important iteration is to accomplishing a good formulation effort. However, the very best summary is to offer you some final sage thoughts about formulation from one of the great systems analysts of all time, Jack Gibson.

In 1990 Jack Gibson wrote *The Systems Analyst's Decalog* at the University of Virginia, a summary of his wealth of experience in ten rules to remember. Not only are these rules good guideposts for the practice of systems engineering, but they also help you understand what the systems approach is really all about and why it is different than the traditional approach to engineering. Here are three of Gibson's ten rules for practicing systems analysis that are an excellent conlcusion to the subject of issue formulation:

1. *There Always Is a Client.* Gibson's point here is that systems analysis is all about solving real problems for real people. He emphasizes the importance of not focusing so much on the mathematical techniques of systems analysis that you forget that your real purpose is to use the techniques on the client's real-world problem. Working with a client will ground you in the reality of the problem you are trying to resolve and help you avoid developing abstract solutions to abstract problems. Gibson also points out that there is usually more than one person or stakeholder involved on a large-scale problem and it is important for the analyst to understand the roles each person plays. For example, people may act as a client, a decision maker, a sponsor, or even as another analyst, or some combination of these. The client is typically the person fulfilling the role of a customer with whom the analyst interacts most frequently while solving the problem. However, the client may not be the ultimate decision maker who decides if the work is acceptable or not. The client may be representing an executive decision maker who does not have the time to spend working with the systems analyst on a frequent basis. In some cases, there may be several people or organizations interested enough in resolving a problem that they help pay for it or provide some other kind of help.

2. *The Client Does Not Understand the Problem.* Gibson says that the client has called in a systems analyst to consult on a problem because the client does not understand the problem very well in the first place. If the client did understand the problem, then the client could probably fix the problem without any assistance from the analyst. Therefore, Gibson thinks that the systems analyst should work with the client to help them develop a proper understanding of the correctly defined problem and to find the best solution. How do you do this? According to Gibson, one of the best ways to help the client is by practicing active listening.

Active listening (Gibson, 1990, p. 8) has five levels: surface, logic, agenda, emotional, and feedback. The surface level is simply listening carefully to the words and sentences of the client and encouraging the client to keep talking. At the logic level of active listening, the systems analyst should be trying to determine the client's conceptual organization of the problem. Gibson (1990, p. 9) suggests phrases like, "O.K., good, but I don't see why that point follows from what you said a moment ago."

The third level of active listening is the agenda level. At this level, you are trying to determine whether the client has any hidden agendas. Gibson says, "What is the client trying to say without saying it or trying to conceal [from you]?" The next level closely relates to the hidden agendas level. It is the emotional level where you are trying to discover the client's attitude toward the problem and the problem's elements, such as other people in the organization that are perhaps part of the problem, from their body language, tone of voice, and general emotional demeanor.

The fifth level, feedback, is the most important level in the active listening process according to Gibson. During feedback listening, you describe to the client what you thought you heard and give the client a chance to correct or elaborate on your perceptions. Basically, you just repeat back to the client in your own words what you think you heard and understood from them. Gibson cautions that, during feedback listening, you are not telling the client what you think about what was said. Rather you are telling the client what was said to make sure you listened correctly and completely.

3. *You Must Generalize the Problem to Give it Contextual Integrity.* If you have done your active listening properly, you will begin to develop the client's original problem statement. But be careful, because Gibson's third rule says the client's original problem statement is too specific. This means that unless you understand the context of the client's problem, you will likely miss some of the important aspects of the client's problem or you may even miss the client's problem all togeher! Gibson emphasizes that this is one of the things that makes the systems engineering approach to problem solving different from traditional engineering.

Traditional engineering design tries to abstract or isolate a problem from its context and divide it up into smaller parts for optimization and perhaps further analysis, and then reassembles all the parts of the problem into a final solution. This *bottom-up approach* of working from the specific to the more general is best at problems that are well-defined and have well-known and established solution procedures. For more complex, ill-defined problems with, as yet, unknown solution procedures, Gibson recommends a top-down approach.

In a *top-down approach,* instead of "drilling down to a solution" by subdividing the problem, you "climb up" one or two levels by generalizing the original problem statement so you understand it better in light of its now richer context. Then you proceed to resolve the problem by proceeding from the general to the more specific. You do this by the top-down formulation approach explained in this chapter. That is, you first define the effective need, determine objectives that if successfully achieved will fulfill the client's need in a cost-effective, high-quality way. Gibson relates an excellent real-world example of what he means by generalizing the original problem and using a top-down approach—the Woodward Avenue Subway study.

In the Woodward Avenue Subway study (Gibson, 1990, pp. 20–21), the mayor of Detroit asked Jack Gibson and a group of systems analysts to recommend whether or not to go ahead with the expensive construction of a new 4.5-mile subway line from Cobo Hall on the Detroit River up Woodward Avenue to the New Center area near the GM and Fisher buildings. A bottom-up approach would have immediately subdivided the proposed 4.5-mile subway into smaller sections and examined the difficulties and costs and benefits associated with constructing the new stretch of line. The top-down approach to the proposed Woodward line took a different route.

The top-down approach first extended the 4.5-mile question to look at an 8-mile, then a 22-mile subway all the way out to the city of Pontiac. Eventually, the original Woodward subway question was generalized several levels up to examine Detroit's major arterial transportation needs. As it turned out, the mayor's real problem was growing traffic congestion in downtown Detroit caused by all routes around the Detroit area, even those traveled by long-distance truck haulers just passing through the Detroit area having to travel through the middle of the city. The original problem statement, building the 4.5-mile Woodward line, would have done little to reduce the inner-city traffic congestion.

A bottom-up approach by itself would probably never have discovered the client's real problem.

Although Gibson is a champion of a top-down systems approach to problem solving and design, he does offer a few important cautions. First, he says you have to be careful not to generalize the client's problem too far. Even a simple problem can become too difficult or too expensive to resolve if overgeneralized. For example, trying to examine the entire interstate transportation system of the United States would have been inappropriate for the mayor of Detroit's problem about city traffic congestion. Gibson says that any study that just recommends more study as its only conclusion is a failure. Real success, according to Gibson (1990, p. 20), is answering the client's real question on time and within budget.

Another important caution from Gibson is that you must recognize that there is room for error in systems analysis because the systems engineer extends the analytic or mathematical approach in traditional engineering design to issues that depend on the problem context. And often, there is no set of equations that can completely describe these issues. This has both positive and negative consequences. On the positive side, the systems engineer can treat many problems not amenable to traditonal methods alone. On the minus side, the systems engineer must be cautious because there is more chance of error when problems are ill-defined and have no well-known mathematical solution procedure. Gibson (1990, p. 13) says that the best way to reduce the likelihood of error in a systems analysis is to generalize the problem and use a top-down technique.

In short, if you want to make sure you are solving the right problem and if you want to maximize your chances of properly understanding the problem under consideration, then your formulation efforts should focus on generalizing the client's original problem and using a top-down problem-solving process.

REFERENCES

Beam, W. R. (1990). *Systems Engineering: Architecture and Design.* New York: McGraw-Hill.

Blanchard, B. S. (1991). *Systems Engineering Management.* New York: Wiley.

Clausing, D. (1994). *Total Quality Development: A Step by Step Guide to World-Class Concurrent Engineering.* New York: ASME Press.

Dalkey, N. C. (1969). An experimental study of group opinion—the Delphi method. *Futures,* (September issue), pp. 408–426.

Gibson, J. E. (1990). *The Systems Analyst's Decalog.* Charlottesville: University of Virginia.

Gibson, J. E. (1991). *How to do Systems Analysis.* Charlottesville: University of Virginia.

Gottesman, B. Z. (1993). Best of breed–groupware: Are we ready? *PC Mag.* **12**(11), 276–284.

Grady, J. O. (1993). *System Requirements Analysis.* New York: McGraw-Hill.

Hall, A. D. (1962). *A methodology for systems engineering,* pp. 103–107. Princeton, NJ: Van Nostrand.

Hauser, J. R., and Clausing, D. (1988). The house of quality. *Harv. Bus. Rev.* May-June, pp. 63–73.

Hsu, J., and Lockwood, T. (1993). Collaborative computing: Computer-aided teamwork will change your office culture forever. *Byte Magazine,* March, pp. 112–116. New York: McGraw-Hill.

Martino, J. (1968). An experiment with the delphi procedures for long range forecasting. *IEEE Trans. Eng. Manage.* **EM-15,** 138–144.

Martino, J. (1972). *Technological Forecasting for Decision Making*. New York: American Elsevier.

Maslow, A. H. (1954). *Motivation and Personality*. New York: Harper & Row.

Ostrofsky, B. (1977). *Design, Planning, and Development Methodology*. Englewood Cliffs, NJ: Prentice-Hall.

Sage, A. P. (1977). *Methodology for Large-scale Systems,* pp. 60–75. New York: McGraw-Hill.

Sage, A. P. (1992). *Systems Engineering*. New York: Wiley.

Sullivan, L. P. (1986). Quality function deployment. *Qual. Prog.,* June, pp. 39–50.

Warfield, J. N. (1979). Systems planning for environmental education, *IEEE Trans. Syst., Man., Cybernet.,* **19**(12), 816–823.

Zwicky, F. (1969). *Discovery, Invention, and Research*. New York: Macmillan.

25 Functional Analysis

DENNIS M. BUEDE

25.1 INTRODUCTION

Functional analysis is performed in systems engineering, software systems engineering, and business process reengineering as a portion of the design process. These design processes typically involve the steps of requirements definition and analysis, functional analysis, physical or resource definition, and operational analysis. This last step of operational analysis involves the marriage of functions with resources to determine if the requirements are met.

Functional analysis addresses the activities that the system, software, or organization must perform to achieve its desired outputs; that is, what transformations are necessary to turn the available inputs into the desired outputs. Additional elements include the flow of data or items between functions, the processing instructions that are available to guide the transformations, and the control logic that dictates the activation and termination of functions. Various diagrammatic methods for functional analysis have been developed to capture some or all of these concepts.

This chapter examines the elements of functional analysis, functional decomposition, systems engineering requirements statements and functional analysis, and diagrams and software support for functional analysis.

25.2 ELEMENTS OF FUNCTIONAL ANALYSIS

There are four elements to be addressed by any specific functional analysis approach. First, the functions are represented as a hierarchical decomposition, in which there is a top-level function for the system or organization. This top-level function is partitioned into a set of subfunctions that use the same inputs and produce the same outputs as the top-level function. Each of these subfunctions can then be partitioned further, with the decomposition process continuing as often as it is useful.

Second, functional analysis diagrams can represent the flow of data or items among the functions within any portion of the functional decomposition. As we examine the first and subsequent functional decompositions, it is common for one function to produce outputs that are not useful outside the boundaries of the system or organizations. These outputs are needed by other functions in order to produce the needed and expected external outputs.

Processing instructions are a third element that appears in some functional analysis diagrams. These instructions contain the needed information for the functions to transform the inputs to the outputs. Also included here are the activation and termination conditions associated with each function in the functional hierarchy.

Handbook of Systems Engineering and Management, Edited by A. P. Sage and W. B. Rouse
ISBN 0471-15405-9   ©1999 John Wiley and Sons, Inc.

The fourth element is the control flow that sequences the termination and activation of the functions so that the process is both efficient and effective. Questions addressed here include: (1) Can these functions work serially or must they be processed concurrently? (2) Are these functions activated once or a series of times? and (3) What circumstances dictate that one function be activated rather than another function?

25.3   FUNCTIONAL DECOMPOSITION

First we define the concept of system modes, followed by simple and complete functionalities. Modes and functionalities have long been thought to be crucial to the establishment of an understanding of the logical aspects of a system.

System *modes* are defined here to be distinct operating states of the system during which some or all of the system's functions may be performed to a full or limited degree. Other authors (e.g., Wymore, 1993) define the modes of a system to be functions of the system; that is not the definition presented here. All systems have at least one standard or fully operational mode. Most systems have operating modes during which they are partially operational. For example, an elevator system has a maintenance mode during which one or more of the elevator cars can be stopped for maintenance, while the others continue in operation. Often systems have a start-up and/or shutdown modes. The laptop computer, on which I am writing this paragraph, has several modes of operation that correspond to the power that is being supplied; all of the laptop's functions are available in each of these modes, but not with the same performance characteristics. Finally, systems often have a number of unwanted failure modes; car manufacturers have installed switches to enable the use of an extra gallon of gasoline to try to avoid the failure mode of no gas.

Now we define simple and complete functionalities:

Simple Functionality. Simple functionality is an ordered sequence of functional processes that operates on a single input to produce a specific output. Note there may be many inputs required to produce the output in question, but this simple functionality only addresses one of the inputs. As a result the simple functionality may not include all of the necessary functional processes needed to produce the output. Nor does this simple functionality trace the only possible sequence of these functional processes. Note, each simple functionality has a specific order associated with it; for this reason we cannot say that a simple functionality is an element of the power set of functional processes because there is no order associated with an element of the power set. (The power set of set A is the set of all subsets of A.) Also we cannot say that this simple functionality is a mathematical function, since a given input can be mapped into more than one output.

Complete Functionality. Complete functionality is defined as a complete set of coordinated processes that operate on all of the necessary inputs for producing a specific output. There is usually no specific order associated with the complete set of functional processes; rather only a *partial order* of the functional activities. There is a well-defined set of inputs, which are an element of the power set of inputs associated with each output.

A *functional architecture* can be defined at several levels of detail:

1. A logical architecture that defines what the system must do, a decomposition of the system's top-level function. This definition of the functional architecture is represented as a directed tree.
2. A logical model that defines how the system transforms its inputs into its outputs. There are many graphical methods for accomplishing this.
3. A logical model to which input–output requirements have been traced to specific functions.

It is possible to complete the functional architecture without resorting to any graphical techniques. Text and tables are sufficient to represent all of the information conveyed by any of the graphical techniques. However, Jones and Schkade (1995) provide convincing evidence that most systems and software professionals resort to graphical techniques during the system or software engineering process. The graphical techniques contain much greater information in a format that can be more efficiently manipulated mentally.

25.3.1 Decomposition vs. Composition

Decomposition, often referred to as top-down structuring, begins with the top-level system function and *partitions* it into several subfunctions. This decomposition process must conserve all of the inputs to and outputs from the system's top or zeroth-level function. By "conserve," we mean use/produce all and add no new ones. Next, each of the several first-level functions are decomposed (partitioned) into a second-level set of subfunctions. Note, it is not necessary to decompose every function; only those for which additional insight into the production of outputs is needed.

The success of decomposition is predicated on having a sound definition of the top-level function of the system and the associated inputs and outputs, that is, a compete set of requirements. A major difficulty of decomposition is that the partitioning process is somewhat unguided. The best decomposition is one that will match the partitioning of the system's physical resources, the physical architecture. This way the flow of data and physical items that cross the internal interfaces between components will be clearly identified. But functional analysis is often done prior to or in parallel with the definition of physical resources; in this case an iterative definition of the functional and physical architectures is needed.

The opposite approach, composition, is a bottom-up approach. With composition one starts by identifying the simple functionalities-associated simple scenarios involving only one of the outputs of the system. For complex systems this initial step is a substantial amount of work. After all of the many functionalities have been defined, one begins the process of grouping the functions in all of the functionalities into similar groups. These groups are aggregated into similar groups. This process continues until a hierarchy is formed from bottom to top.

The advantage of this approach is that the composition process can be performed in parallel with the development of the physical architecture so that the functional and physical hierarchies match each other. Second, this approach is so comprehensive that it is less likely to omit major functions. The drawback is that the many functionalities must be easily accessible during the composition process so that all of this work can be successfully used. This is actually the way that functional architectures were created in the 1960s, and 1970s, when systems engineering was in its infancy; many systems engineers continue to prefer this approach. There is no empirical evidence that either approach is better than the other.

Ultimately, it is wisest to use a combination of decomposition and composition. Often, one makes use of simple functionalities associated with specific scenarios defined in the operational concept to establish a "sense" of the system. Then it is common to posit a top-level decomposition that is likely to match the top-level segmentation of the physical architecture and proceed to do decomposition.

Before we proceed, it is important to discuss some valuable properties of the functional hierarchy. Besides the obvious design implications that are embodied in this hierarchy, it is also important as a communication tool. This communication is important for both other engineers and the stakeholders. Substantial psychological research (Miller, 1956) has shown that humans, whether they are engineers or nonengineers, have certain mental processing limitations. For this reason, there should be three to seven subfunctions for each node in the functional tree.

There are a number of keys one can use to partition a function into subfunctions. At the top of the hierarchy we would expect to see functions devoted to the system's operating modes, if there are any. For functions that have multiple outputs, we could partition the function into subfunctions that correspond with the production of each output. Similarly, we could key on the inputs and controls to find a partition of the function. More appropriate than either of these is to decompose on the basis of stimulus–response threads that pass through the function being decomposed. Finally, there is often a natural sequence of subfunctions for a particular function. For example, at the bottom of the functional architecture we would expect to see functions such as receive input, store input, and disseminate input; or retrieve output, format output, and send output.

Hatley and Pirbhai (1988) developed an architectural template for representing the physical architecture of the system (see Fig. 25.1). This template suggests a generic partition of six subfunctions for any function of the functional architecture:

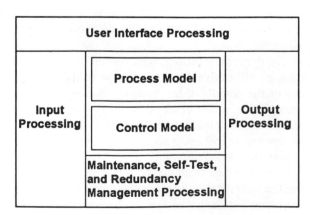

Figure 25.1 Architecture template. (Hatley and Pirbhai, 1988, p. 195).

- *Provide User Interface.* Those functions associated with requesting and obtaining inputs from users, providing feedback that the inputs were received, providing outputs to users, and responding to queries of those users.
- *Format Inputs.* Those functions needed to receive inputs from external interfaces (nonhumans) and other system components, and to process (e.g., analog-to-digital conversion) those inputs to put them into a format needed by the system's processing functions.
- *Transform Inputs into Outputs.* The major functions of the system.
- *Control Processing.* Those functions needed to control the processing resources or the order in which these processing functions should be conducted.
- *Format Outputs.* Those functions needed to convert the system's outputs into the format needed by the external interfaces or other system components and then place those outputs onto the appropriate interface.
- *Enable Maintenance, Conduct Self-test, and Manage Redundancy Processing.* Those functions needed to respond to external diagnostic tests, monitor its own functionality, detect errors, and enable the activation of standby resources.

As the decomposition of system functions proceeds, we would expect to find smaller subsets of these six generic functions being embedded within one of the higher-level functions. Figure 25.2 illustrates the functional decomposition by showing likely decompositions within the top-level functions.

McMenamin and Palmer (1984) describe a system's functions as being composed of essential or fundamental activities and custodial activities. All of the functions implied by the Hatley and Pirbhai architecture template are fundamental activities. Custodial activities maintain the system's memory so the system knows what it needs to know to perform its fundamental activities. This knowledge is called the *essential memory* of the system, the storage of data items between the time they become available and the time they are used by the fundamental activities. McMenamin and Palmer recommend separating the custodial activities and the fundamental activities. This separation is not possible at the top level with the taxonomy suggested by the Hatley–Pirbhai template, nor is it often desirable at this high level. However, it becomes possible to achieve this separation at lower levels of the functional decomposition.

A functional architecture can be evaluated for shortfalls and overlaps. A *shortfall* is the absence of a functionality that is required to produce an output. The most common types of shortfall are responses to unexpected inputs and to failure modes within the system. For example, an ATM system must be able to address a user's input of an incorrect access code and a user's decision to cancel a transaction part way through a transaction. Less obvious unexpected inputs are a loss of power and an attempted break-in to the ATM's cash stock. The systems engineer, however, must always ensure that the functional decomposition can account for every possible input and the production of every output. When shortfalls are found, functionality must be added to correct the situation.

An *overlap* is a redundancy that is not needed to achieve additional performance, for example, reliability. Functional overlaps, unlike physical overlaps for redundancy, are not needed and therefore can only cause problems.

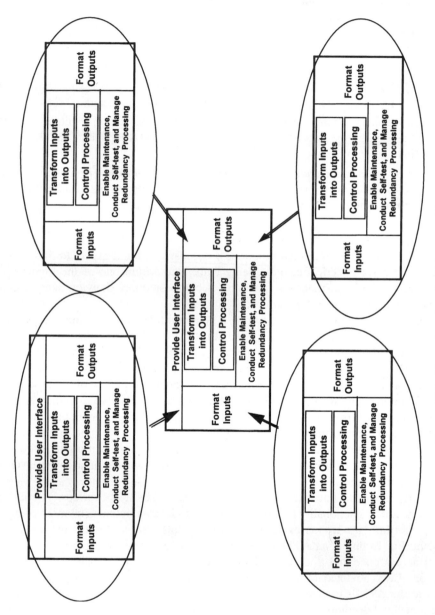

Figure 25.2 Exemplary functional decomposition.

25.4 THE SYSTEMS ENGINEERING REQUIREMENTS STATEMENT AND FUNCTIONAL ANALYSIS

The development of requirements involves creating an operational concept, defining the system's boundaries and external systems, and explicating the system's objectives. The operational concept is prepared from the perspective of the stakeholders of the system and describes how they expect the system to fit into their world, which contains a number of external systems and has a certain context. The objectives of each stakeholder group are suggested here. The operational concept defines the system and external systems in very general terms (often as a block diagram) and establishes a number of scenarios in which the system will be used by the stakeholders. These scenarios suggest functions that the system will be expected to perform. The second step makes the boundaries between the system and external systems crystal clear, leaving no doubt in anyone's mind where the system starts and stops. In addition, we establish all of the inputs to and outputs of the system here, as well as the external system or context with which each input and output is associated. The third step clarifies the objectives of the stakeholder groups and formulates a coherent set of objectives for the system. The creation of the originating requirements, followed by the translation of these requirements into system requirements is the fourth step. The originating requirements are created by an analysis of the operational concept for system functions, an exhaustive examination of the system's inputs and outputs, the specification of interfaces of the external systems with which the system must interact, a thorough examination of the system's context and operational concept for systemwide and technology constraints, a detailed discussion with the stakeholders to understand their willingness to trade off a wide range of nonmandatory but desirable system features, and the complete specification of test requirements needed to verify and validate the system's capabilities from the stakeholders perspectives. Before proceeding, these requirements must be examined to ensure that a feasible design exists that meets the requirements; for example, we could not build a supersonic transport aircraft that had a production cost of $1,000. The sixth step is the development of requirements for the test system needed to verify and validate the resulting system. Finally, the stakeholders must approve the requirements documents.

25.4.1 Requirements Taxonomy

Wymore (1993) identifies six types of system design requirements; it is this structure, slightly modified, that we will adopt. We will incorporate these six types of requirements into each relevant phase of the system's life cycle (development, production, deployment, training, operation and maintenance, refinement, and retirement), starting with the phase we would be in at the beginning, namely development. Hierarchically, we will have each relevant phase, and within each phase the six types of requirements. Table 25.1 provides examples of various types of requirements, which have been collected from a wide variety of sources.

1. *Input–output requirements* include sets of acceptable inputs and outputs, trajectories of inputs to and outputs from the system, interface constraints imposed by the external systems, and eligibility functions that match system inputs with system outputs for the life-cycle phase of interest. Clearly there are a number of requirements in this category during the operations phase of the life cycle. However, the system may

TABLE 25.1   Exemplary Requirements in the System/Programmatic Taxonomy

Requirements Category	Exemplary Requirements
Output performance	Accuracy
	Quantity of an output, e.g.,
	Distance
	Throughput
	Amount
	Survivability against a defined threat, requiring a system output
	Coverage (area or volume covered)
	Timeliness (time to create an output)
	Availability of an output
Input–output interface constraint	Required format/trajectory of an input–output
	Timing constraint associated with an input
	Accuracy associated with an input
Input–output functional constraints	Functions required to transform inputs into outputs
	Functional constraints on transforming inputs into outputs
Functional responses to undesired or unexpected events	Unexpected or undesired inputs and appropriate response
	Bounds on expected inputs and appropriate response
	Appropriate response to internal system malfunctions
Technology/systemwide performance	Usability
	Weight of the system
	Form (volume) and fit (dimensions) of the system
	Survivability of the system
	Availability, reliability, maintainability of the system
	Durability (or operational life) of the system
	Supportability of the system
	Safety of the system
	Trainability of the system
	Testability of the system
	Extensibility (expected changes/growth potential) of the system
	Affordability or operating and maintenance cost of the system
	Production cost (manufacturability) of the system
	Deployment cost of the system
	Decommissioning cost of the system
Performance, cost, and tradeoff	Performance parameters with value curves and weights
	Cost parameters with value curves and weights
	Performance and cost parameters with weights
Test	Acceptable tolerance, based on test data, of a requirement
	Amount and type of test data for a requirement

have inputs and outputs in all portions of the system's life cycle (e.g., training stimulations, standardized internal interfaces for product improvement); if so, the requirements for these activities would be found in this category in the appropriate life-cycle stage. We can partition the input–output requirements into four subsets: (a) inputs, (b) outputs, (c) external interface constraints, and (d) functional requirements.

2. *Technology and systemwide requirements* consist of constraints and performance index thresholds (e.g., the length of the operational life for the system, the cost of the system in various life-cycle phases, and the system's availability) that are placed upon the physical resources of the system. Many of the requirements from each life-cycle phase are found in this category because they specifically relate to the physical manifestation of the system. This category can be partitioned into four subsets: (a) technology, (b) the "ilities," (c) cost, and (d) schedule (e.g., development time period, operational life of the system).

3. *Performance requirement* is an algorithm (computable expression) that defines how the relative performance of any two alternate designs can be compared in terms of the system's performance requirements. These performance requirements are defined within the input–output and the noncost, systemwide requirements. The performance requirement specifically defines how the performance parameters are to be compared to each other. This algorithm could be logical or decision analytic [i.e., value curves and swing weights (see Buede and Bresnick, 1992; Buede and Choisser, 1992)].

4. *Cost requirement* is an algorithm that defines how the relative cost of any two alternate designs can be compared across all cost parameters (life-cycle phases) of interest to the stakeholders. Note, dollars spent at different times may not be comparable as simply as present-value computations when there are different bill payers.

5. *Tradeoff requirement* is an algorithm for comparing any two alternate designs on the aggregation of cost and performance.

6. *System test requirements* have four primary elements: (a) observance: to state how the estimates (test data) for each input–output, systemwide, performance, cost, and tradeoff requirement will be obtained; (b) verification plan: to state how the test data will be used to determine that the real system conforms to the design that was developed; (c) validation plan: to state how the test data will be used to determine that the real system complies with the originating performance, cost, and tradeoff requirements; and (d) acceptability: to state how the test data will be used to determine that the real system is acceptable to the stakeholders. Note, the test requirements associated with the first objective define the basis for the requirements for the suite of test systems (e.g., simulations, instrumented test equipment) needed for the system under development. It is common to have technology and systemwide requirements that limit the flexibility to develop new test equipment.

Figure 25.3 traces the origins of the performance requirements to the objectives hierarchy by showing that the objectives hierarchy defines the performance parameters requiring nonpoint requirements. These performance parameters can fall within the categories of input, output, "ilities," cost, and schedule requirements. The thresholds and goals for these tradable requirements are defined as part of the input, output, "ilities," cost, and schedule requirements. The algorithms that define the tradable space over the performance parameters are documented in the performance, cost, and tradeoff requirements. These algorithms define the isovalue lines in the tradable space; these isovalue lines will be the basis for design tradeoffs.

If every set of requirements contained the information defined by Wymore (1993), there would be far fewer problems in system development efforts. Very few requirements documents contain performance, cost, and tradeoff requirements as defined by Wymore. These

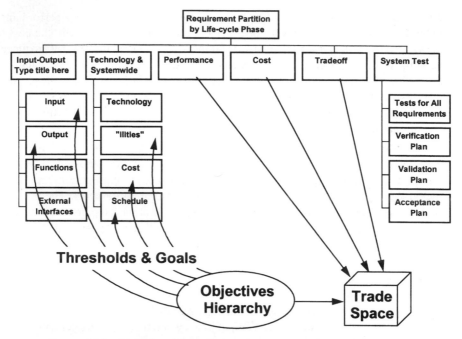

Figure 25.3   Objectives hierarchy, requirements partition, and trade space.

elements should be defined in the originating requirements document from the stakeholders' perspective; otherwise, the systems engineers must guess at the ultimate tradeoffs of the stakeholders; our ability to do a complete and effective job of this is questionable at best.

25.4.2   Requirements Criteria

A number of authors, including Frantz (1993), Davis (1990), and Mar (1984), have developed various numbers of attributes for requirements. The literature is not in total agreement about the meaning of these attributes. Listed below is a consistent examination of the literature. We have divided the characteristics that are related to individual requirements and those relevant to groups of requirements.

Individual Requirement Attributes

1. Unambiguous—every requirement has only one interpretation.
2. Understandable—the interpretation of each requirement is clear.
3. Traced—each requirement can be traced to some document or statement of the stakeholders.
4. Traceable—each derived requirement must be traceable to an originating requirement via some unique name or number.
5. Correct—a requirement the system is in fact required to do.
6. Concise—no unnecessary information is included in the requirement.
7. Verifiable—a finite, cost-effective process has been defined to check that each requirement has been attained.

Attributes of the Set of Requirements

8. Unique—requirement(s) is(are) not overlapping or redundant with other requirements.
9. Complete
   (a) Everything the system is required to do throughout its life cycle is included
   (b) Responses to all possible (realizable) inputs throughout its life cycle are defined
   (c) The document is defined clearly and self-contained
   (d) There are no "to be defined" (TBD) or to be reviewed (TBR) statements completeness is a desired property but is not known at the time of requirements development, or perhaps ever.
10. Consistent
    (a) Internal—no two subsets of requirements conflict
    (b) External—no subset of requirements conflicts with external documents that the requirements are traced from
11. Comparable—the relative necessity of the requirements is included.
12. Modifiable—changes to the requirements can be made easily, consistently (free of redundancy), and completely.
13. Design independent—requirements do not specify a particular solution.
14. Attainable—solutions exist within performance, cost, and schedule constraints.

In any systems engineering effort we must develop as many correct requirements as possible; these correct requirements are defined as being verifiable. In addition, we must try to eliminate as many incorrect requirements as possible. The requirements document should contain a complete, consistent, comparable, design-independent, modifiable, and attainable statement of the design problem.

25.4.3   Operational Concept

An operational concept (Lano, 1990) is a shared vision from the perspective of the system's stakeholders of how the system will be developed, produced, deployed, trained, used and maintained, refined, and retired to achieve the stakeholders' operational needs and objectives. Often a block diagram that characterizes the interaction between the system and important external systems is developed for each life-cycle phase. The operational concept includes a collection of scenarios; one or more for each group of stakeholders for each relevant phase of the system's life cycle. Each scenario addresses one way that a particular stakeholder(s) will want to use, deploy, fix, and so on, the system and how the system will respond to produce a desired end. Included in each scenario are the relevant inputs to and outputs from the system. It is critical that the shared vision be consistent with the collection of scenarios making up the operational concept.

Hunger (1995) uses the phrase "mission analysis" for the development of the operational concept. The collection of scenarios in the operational concept include sortie missions or scenarios and life missions. Sortie missions are scenarios that describe how the system will be used during the operational phase, capturing the reasons the system has for existing. The life missions address the nonoperational life-cycle aspects of the system, resulting in scenarios for each life cycle phase and some that cross life-cycle phases.

The shared vision and scenarios define the system's mission and give us the first hints as to the boundary of the system. The system's inputs and outputs cross this boundary defining the input—output requirements of the system and the external interfaces. The operational concept also suggests the fundamental objectives (objectives hierarchy of the stakeholders). This objectives hierarchy becomes the basis of the system's performance requirements. Finally, the system's major functions can be identified by examining the operational concept. Thus the operational concept also leads to the functional requirements. Hunger has suggested using time lines to better define these system scenarios (or sorties as he calls them).

In order to generate these scenarios, it is natural to start with the key stakeholder, the operator/user, and generate a number of simple scenarios. Then we expand to other stakeholders while staying simple. Finally, we add complexity to all scenarios for each stakeholder, explicitly addressing atypical weather situations, failure modes of external systems that are relevant, and identifying key failure modes, constraints, standards, and external system interfaces that the system should address *in every phase of the life cycle*. In all scenarios the focus should be on *what* the stakeholders and external systems do, not on *how* the system accomplishes its tasks.

25.4.4   System's Boundary and External Systems

The single largest issue in defining a new system is where to draw the system's boundaries. Everything within the boundaries is open to change, subject to the requirements, and nothing outside of the boundaries can be changed, leading to many of the system's constraint requirements. The inputs to the system are those items that cross the system's boundaries from the outside; the system's outputs are those items that the external systems are expecting to receive.

Who is responsible for drawing these boundaries? We argue that all of the stakeholders have a say in drawing these boundaries. There are substantial cost and schedule implications, however, so the procurer of the system typically has a major input. Nonetheless all of the stakeholders should be prepared to discuss the impact upon them of various boundary-drawing options. The systems engineer is responsible for guiding this boundary-drawing process to a conclusion that the stakeholders understand and accept. The systems engineer uses these boundaries to establish and maintain control of the system's interfaces.

The system's boundaries need to be drawn early in the systems engineering process because so much else in the design phase is dependent upon them. As we discuss next, the fundamental objectives or measures of effectiveness of the system need to be focused just beyond the external interfaces of the system. The operational concept relies upon knowing where the boundaries are for each stakeholder. The interface requirements capture the implications of the boundaries on the system design.

Every functional modeling technique (e.g., IDEF0, behavior diagrams, data flow diagrams) can be used to define the system boundary. In addition to the usual syntax and semantics requirements of these diagrammatic methods, an external systems diagram introduces several new constraints for the diagram to be valid. First, all of the outputs of the system's function have to go to one of the external systems' functions on the page and cannot exit the diagram. If the output did exit the page, there would be an external system that was not included in the diagram, invalidating the purpose of the effort. Similarly, all of the external systems must be receiving at least one output of our system; otherwise, the system should be part of the context. In some cases, part of the context could be shown on the external systems diagram to emphasize the importance of a particular input to the system.

25.4.5   System's Objectives

Traditionally systems engineers have used the terms measure of effectiveness (MOE) and measure of performance (MOP) or figure of merit (FOM). A MOE describes how well a system carries out a task or set of tasks within a specific context; a MOE is measured outside the system for a defined environment and state of the context variables. Note, that the further outside the system that the MOE measurement process is established, the more influence the external systems have inside the measurement window, yielding less sensitivity in the measurement process for evaluating the effectiveness of the system. There is typically one or a few MOEs for a given system, often one for each major output of the system.

A MOP (or FOM) describes a specific system property or attribute for a given environment and context; a MOP is measured from within the system. There are many possible and relevant MOPs for a specific system output; examples include accuracy, timeliness, distance, throughput, workload, and time to complete.

Since the systems engineering design process is decision-rich, it is important to introduce some concepts from decision/risk analysis. Value-focused thinking (Keeney, 1992) emphasizes the proper structuring of decisions in terms of a fundamental objective. The *fundamental objective* is the essential set of objectives that summarizes the current decision context, and is yet relevant to the evaluation of the options under consideration. Generally this fundamental objective can be subdivided into value objectives that more meaningfully define it, thereby forming a *fundamental objectives hierarchy* or value structure. Keeney distinguishes this hierarchy from a *means–end objectives network,* which relates means or "how to" variables (the design options) to the fundamental objective.

The process that Keeney describes for defining this situation-based fundamental objectives hierarchy involves working from both ends, by generalizing means–ends objectives and operationalizing strategic objectives. Means–ends objectives are ways to achieve the fundamental objective. Strategic objectives are beyond the time horizon and immediate control of options associated with the current system design decision situation. As an example, one of the set of fundamental objectives for the operation of a new ATM would be "minimize customer time per transaction." The set of fundamental objectives define value tradeoffs among the stakeholders of the ATM. A strategic objective would be to "improve the profitability of the bank"; there are too many other factors beyond the ATM that will determine whether this objective is met for it to be a fundamental objective. A means–ends objective would be to "use a fuzzy logic error checker"; this statement addresses a means for achieving an objective.

The fundamental objectives hierarchy is developed by defining the natural subsets of the fundamental objective. Keeney (1992) gives the following example of a fundamental objectives hierarchy: maximize safety (the fundamental objective) is disaggregated into minimize loss of life, minimize serious injuries, and minimize minor injuries. The tradeoffs among these objectives clearly involve one's values. This subdivision is contrasted with a means–ends breakout of maximize safety that starts with minimize accidents and maximize the use of safety features on vehicles, both of which are means oriented and involve outcomes for which value tradeoffs are difficult.

The objectives hierarchy (a directed tree) usually has two to five levels. The objectives in the hierarchy may include stakeholders explicitly and often include context (environmental) variables (e.g., weather conditions, peak vs. nonpeak loading) from the scenarios in the operational concept. If present, these scenarios are usually at the top of the hierarchy.

The objectives hierarchy is typically used throughout the systems engineering design process as the cornerstone of all the trade studies that compare one design alternative with another. In doing trade studies we must evaluate which of several design alternatives is preferred; each design alternative will commonly have one advantage over the others, such as operational cost, reliability, or accuracy of outputs.

In order to make use of the objectives hierarchy for trade studies, additional information must be added, information that Wymore (1993) calls the *performance requirement*. Wymore suggested using value curves for each objective at the bottom of the objectives hierarchy and value weights for comparing the relative value of swinging from the bottom of each value scale to top. Figure 25.4 illustrates the value curves for a simplified objectives hierarchy for an ATM. See Sailor (1990) for another example.

25.4.6   Requirements Development

Now, the input–output requirements are defined on the basis of the inputs and outputs in the external systems diagram. *The systems engineering team must examine each input, control, and output in detail to discover every requirement associated with it.* For example, the potential users of an ATM expect feedback about specific inputs that they provide. See Table 25.1 for examples of requirements that may be associated with inputs or outputs.

The environment (e.g., weather and elements outside control of the system or "context") is typically defined as part of the scenarios of the operational concept. The questions typically addressed are:

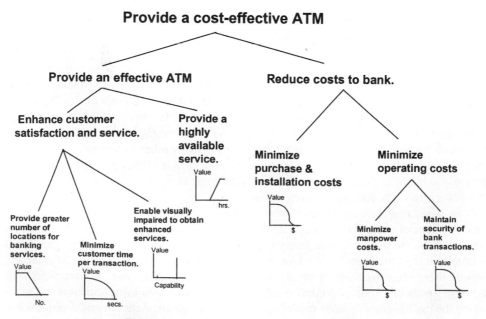

Figure 25.4   Objectives hierarchy for an ATM.

1. What elements of the environment matter?
2. How much variation in the environmental elements must be planned for? At what priority?
3. How well can these variations be forecast (predicted)? Can these forecasts be part of the system?
4. Can the environment be controlled by the system or external system? Must the system protect itself from the environment?
5. How do the answers for these questions impact the *functions* of the system?

Requirements are generally considered the cornerstone of the systems engineering process. Originating requirements are those requirements initially established by the system's stakeholders with the help of the systems engineering team. The systems engineering design process is a mixture of establishing requirements and partitioning the physical resources of the system into components. This partitioning process is decision-rich in that many important decisions are made by the systems engineering team that will ultimately affect the performance of the system and the happiness of the stakeholders.

The three key points that we have made in this chapter concerning the systems engineering design process are as follows: (1) all stakeholders have originating requirements, which taken together, address every stage of the system's life cycle. Capturing the complete set of originating requirements ensures a concurrent engineering process. (2) The set of originating requirements should ensure a decision-rich design process by not overconstraining the design. The following attributes of requirements are meant to ensure the process is not overconstrained: traced, correct, unambiguous, understandable, design independent, attainable, comparable, and consistent. (3) At the same time, the originating requirements should not underconstrain the design, because we want the stakeholders to be happy with the system that we create. Complete, verifiable, and traceable requirements should guarantee this.

Our proposed systems engineering design process involves the development of an operational concept for each stakeholder group, external systems diagram for each life-cycle phase, and an objectives hierarchy for each stakeholder group. These three concepts are then used to develop the originating requirements, organized by life-cycle phase (see Fig. 25.5). We have adopted and modified Wymore's partition of requirements as being particularly relevant for a decision-rich design process: input–output requirements, technology and systemwide requirements, performance requirement, cost requirement, tradeoff requirement, and system test requirements. In particular the tradeoff information defining stakeholder values that is needed to support design decisions is included in the performance requirement, cost requirement, and tradeoff requirement. This initial systems engineering phase is complete when the existence of at least one feasible solution is verified, the requirements for the test system are defined, and the stakeholders have approved the originating requirements document (ORD).

25.4.7   System Requirements and Specifications

Figure 25.6 depicts yet another representation of the systems engineering design process; a sequential decomposition of requirements and the operational architecture (functions mapped to physical resources) by moving from left to right and top to bottom. There is a difference between a requirement and a specification. A *requirement* is one of many state-

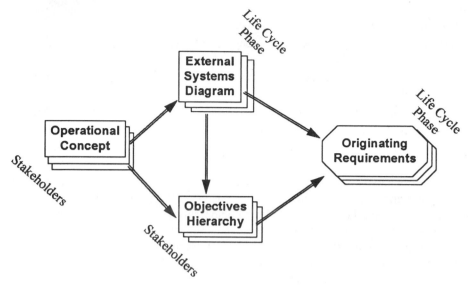

Figure 25.5   Summary of originating requirements development.

ments that constrain or guide the system's development in such a way that it is useful to one or more of its stakeholders. A *specification* is a collection of requirements that completely define the constraints and performance requirements for a specific physical entity that is part of the system. The systems engineering design process involves defining all of the system's requirements and then bundling them by segmenting and refining into a specification for each of the system's segments, elements, and then components.

Operational Need	System Design	Segment Design	Element Design	Component Design
Operational Requirement ⟹	System Operational Architecture ⇊			
	Segment Specs ⟹	Segment Operational Architectures ⇊		
		Element Specs ⟹	Element Operational Architectures ⇊	
			Component Specs ⟹	Component Operational Architectures ⇊
				CI Specs

Figure 25.6   Design decomposition of architectures and specs.

Originating requirements are statements by the stakeholders of the system that define the constraints and performance parameters within which the system is to be designed. Systems engineers take these high-level originating requirements and derive a consistent set of more detailed engineering statements of requirements as the design progresses. We can aggregate the previous requirements taxonomy into constraints and performance indices. Some constraints are simple, for example, the system must be painted a specific shade of green. Other constraints are the minimally acceptable level associated with a performance requirement. A performance requirement defines a desired direction of performance associated with an objective of the stakeholders for the system. For an ATM a performance requirement might be to "minimize customers' time per transaction." For any performance requirement there must also be a minimum acceptable performance constraint or threshold associated with the index; this threshold dictates that no matter how wonderful a design's performance is on other objectives, performance below this threshold on this requirement makes the design unacceptable. This is a very strong statement of needs and so minimal acceptable thresholds must be established very carefully.

Virtually every major organization, governmental or commercial, has established its own guidelines for system or product development. The names and organizations of the requirements document vary somewhat, but cover the same material. Table 25.2 summarizes the common major requirements documents that are produced during the design phase of the systems engineering process. The contents of these documents, as shown in this table, are a generalization of what you may find in practice. The Problem Statement (or Mission Element Need Statement in the military) gets the process rolling and identifies a problem for which a solution in the form of a system (new or improved) is needed. This document supports and documents a decision-making process to start a system development effort. The SEMP then defines the bounds of this effort.

Originating requirements are found in the operational needs or requirements document. This document is produced with or by the stakeholders and is written in their language(s). Systems engineers need to be involved in a substantial way in this activity, although not all systems engineers share this view. Experience has shown that if this document is left to the stakeholders it will be incomplete; the systems engineers can play a major facilitation role among the various groups of stakeholders, as well as bring an assortment of tools to bear on a difficult problem (the creation of this document) that ensure a greater completeness. A major focus of the previous discussion was on the methods available to define these requirements.

The systems engineer then begins restating and "deriving" requirements in engineering terms so that the systems engineering design problem can be solved. It is critical that the requirements in all of these documents address "what" and "how well" the system must perform certain tasks. Requirements do not provide solutions, but rather define the problem to be solved.

The systems requirements validation document serves as the acceptance test or validation of the originating requirements process. It should demonstrate that if the systems engineering process continues, an acceptable solution is possible. Unfortunately, this document is seldom produced in practice, which is a major downfall of those systems engineering efforts that later either have to modify their originating requirements in significant ways or admit failure. This happens all too often.

Systems engineers have always desired to demonstrate the importance of requirements and getting the requirements right, for example, complete, consistent, correct. In the mid-1970s three organizations, GTE (Daly, 1977), IBM (Fagan, 1974), and TRW (Boehm,

TABLE 25.2   Typical Requirements Documents

Document Titles	Document Contents
Problem situation or mission element need statement and systems engineering management plan (SEMP)	• Definition of stakeholders and their relationships • Stakeholders' description of the problem and its context • Description of the current system • Description of major objectives in general terms • Definition of the systems engineering management structure and support tools that will be responsible for developing the system
Operational need or operational requirement (ORD)	• Definition of the problem needing solution by the system (including the context and external systems with which the system must interact) • Definition of the operational concept on which the system will be based • Creation of the structure for defining requirements • Description of the requirements in the stakeholders' language in great breadth but little depth • Trace of every requirement to a recorded statement or opinion of the stakeholders • Description of tradeoffs between performance requirements, including cost and operational effectiveness
System requirements (SRD)	• Restatement of the operational concept on which the system will be based • Definition of the external systems in engineering terms • Restatement of the operational requirements in engineering language • Trace of every requirement to the previous document • Justification of engineering version of the requirements in terms of analyses, expert opinions, stakeholder meetings • Description of test plan for each requirement
System requirements validation	• Document analyses to show that the requirements in the SRD are consistent, complete, and correct, to the degree possible • Demonstrates that there is at least one feasible solution to the design problem as defined in the SRD

1976), conducted independent studies of software projects. These studies addressed the relative cost of fixing a problem based on where in the system cycle the problem originated. Boehm (1981) and Davis (1993, p. 25) compared the results (see Table 25.3). The costs have been normalized so that the relative cost to repair a problem found in the coding phase is 10. Table 25.3 shows the median value, followed by the range of findings in the several studies examined. Getting the requirements right is probably the most difficult task, and therefore a task that is fraught with errors. Unfortunately, many of these errors are not caught until much later in the life cycle, causing the expenditure of significant money. The analyses of Boehm (1981) and Davis (1993, pp. 25–31) document these difficulties.

The format for an ORD, (as follows) should include sections for a brief overview of the system, relevant documents from which the originating requirements have been traced, and the requirements themselves. The requirements themselves should be organized by

TABLE 25.3   Comparison of the Relative Cost to Fix Software in Various Life-Cycle Phases

Phase of Life Cycle	Relative Cost of Repair Given the Fix is Found in this Phase Median [Range]
Requirements	2 [1, 3]
Design	5 [3, 6]
Coding	10
Unit test	20 [15, 50]
Acceptance test	50 [20, 80]
Maintenance	200 [40, 400]

life-cycle phase. Within each phase the taxonomy of six types of requirements discussed should be used. Given this organization, we now need an overall tradeoff requirement (Section 3.8 of the ORD) that addresses comparisons across life-cycle phases.

Originating Requirements Document

1.0 System Overview
2.0 Applicable Documents
3.0 Requirements
    3.1 Development Phase (Programmatic) Requirements
        3.1.1 Input–Output Requirements for Development
        . . .
        3.1.6 Test Requirement for Development
    3.2 Manufacturing Phase Requirements
        . . .
    3.3 Deployment Phase Requirements
        . . .
    3.4 Training Phase (If Present) Requirements
        . . .
    3.5 Operational Phase Requirements
        3.5.1 Input–Output Requirements for Operations
            3.5.1.1 Input Requirements for Operations
            3.5.1.2 Output Requirements for Operations
            3.5.1.3 External Interface Requirements for Operations
            3.5.1.4 Functional Requirements for Operations
        3.5.2 Systemwide/Technology Requirements for Operations
        3.5.3 Performance Requirement for Operations
        3.5.4 Cost Requirement for Operations
        3.5.5 Tradeoff Requirement for Operations
        3.5.6 Test Requirement for Operations
    3.6 System Improvement/Upgrade Phase Requirements
        . . .
    3.7 Retirement Phase Requirements
        . . .
    3.8 Overall Tradeoff Requirement
Appendix A. Operational Concepts by Phase
Appendix B. External System Diagrams by Phase

25.5   DIAGRAMS AND SOFTWARE FOR FUNCTIONAL ANALYSIS

We address the three primary modeling approaches used as part of systems engineering: data modeling, process modeling, and behavior modeling. Within each of these approaches there are a number of methods that are currently being used in systems and software engineering (see Table 25.4). In addition, we discuss two relatively new object-oriented approaches to systems engineering.

25.5.1   Diagrammatic Methods

25.5.1.1   Data Modeling.   Entity-relationship (ER) diagrams model the data structure or relationships between data entities. Entity types are shown in boxes, relationships are shown in diamonds or as labels on the arcs. If diamonds are used, the graph has no directed edges and the relationships are read from left to right or from top to bottom; otherwise the edges are directed (Fig. 25.7). A unique relationship is that of supertype/subtype, which has become known as class/subclass relationship and is shown in Figure 25.8. Many of the entities and relationships associated with systems engineering that we have discussed so far, as well as a few others, are shown in Figure 25.9.

Higraphs were introduced by Harel (1988) as a generalization of Venn diagrams and ER diagrams. Figure 25.10 shows a higraph for a subset of the ER diagram of systems engineering shown in Figure 25.9. An entity is considered to be set with multiple elements, called a *blob*. A blob is represented as an enclosed area (see systemwide requirement in Fig. 25.10). Atomic sets are blobs with no other blobs contained within them; the only nonatomic blobs in Figure 25.10 are requirements, time, and components. (To be correct we should have placed blobs inside the six intersections of originating and derived requirements with input–output, systemwide, and test requirements. However, this would have compromised the readability of the figure.) The is-a relationship from ER diagrams is replaced by representing one entity as a subset of another. Cartesian products (unordered

TABLE 25.4   Modeling Approaches and Methods

Modeling Approaches	Modeling Methods
Data modeling	Entity-relationship diagrams (ERDs)
	Higraphs
	IDEF1
	IDEF1X
Process modeling	Data flow diagrams (DFDs)
	IDEF0
	N^2 Charts
Behavior modeling	Control flow diagrams (CFDs)
	Function flow block diagrams (FFBDs)
	Behavior diagrams (BDs)
	State transition diagrams (STDs)
	Statecharts
Object-oriented modeling	Object modeling technique (OMT)
	Real-time object-oriented modeling (ROOM)

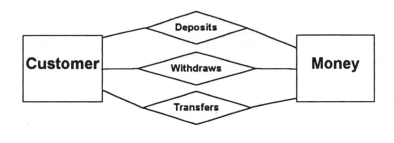

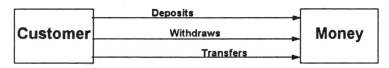

Figure 25.7 Simple entity-relationship diagram.

n-tuples) are shown by placing a dashed line between blobs inside a larger blob representing the *n*-tuple. See the time blob, representing a four-tuple of year–month–day–hour in Figure 25.10 that is not in Figure 25.9.

The relations are shown in diamonds with a line entering them and an arc leaving them to indicate which way the relation is read.

IDEF1 models data using entity classes and relations among entity classes. An entity class has attributes that describe the entity. The relations that are possible between classes come from entity relationship diagrams and address mainly one-to-one, one-to-many, and the like, issues. IDEF1 is an approach for modeling the structure of information as it is maintained in an organization, including the business rules (Griffith, 1994).

IDEF1X also models data using entity classes and relations among the classes. IDEF1X allows for a fuller definition of subtypes and attributes in terms of their aliases, data type, length, definition, primary key, discriminator, alternate keys, and inversion entities. Similarly, the relationships in IDEF1X may be defined on the arcs and include one-to-one, one-to-many, and the like, issues, as well. IDEF1X is used for designing relational data bases (Griffith, 1994).

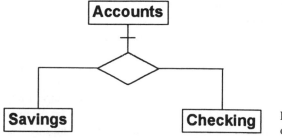

Figure 25.8 Class/subclass relationship diagram.

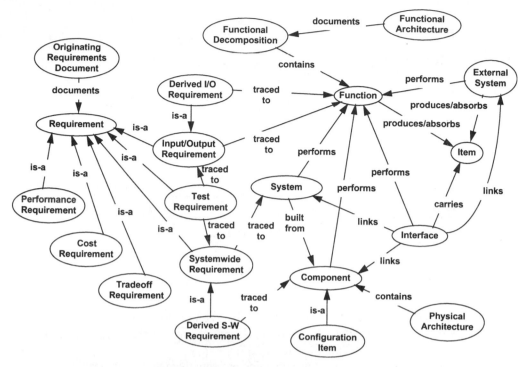

Figure 25.9 A complex ER diagram of systems engineering.

25.5.1.2 Process Modeling

Structured Analysis. The structured analysis and design technique (SADT) was developed between 1969 and 1973. SADT is a graphical modeling language and a comprehensive methodology for developing models. In the 1970s, the U.S. Air Force incorporated SADT into its integrated computer-aided manufacturing (ICAM) program as the definition language for manufacturing systems, yielding the phrase IDEF for ICAM *Def*inition. IDEF0 focuses on functional models of a system. Recently, the Department of Commerce issued a Federal Information Processing Standard (183) that redefines the IDEF0 acronym to be integrated definition.

Within IDEF0, a *function* or activity is represented by a box, described by a verb—noun phrase, and numbered to provide context within the model. Inputs enter from the left of the box, controls enter from the top, mechanisms (or resources) enter from the bottom, and outputs leave from the right. A *flow* of *material* or *data* is represented by an arrow or arc that is labeled by a noun phrase. The label is connected to the arrow by an attached line, unless the association is obvious.

An IDEF0 model has a purpose and viewpoint and comprises two or more diagrams. The A0 page is the context diagram and establishes the boundaries of the system or organization being modeled. The A0 page, shown in Figure 25.11, defines the decomposition of the A0, or top-level, function by two to six functions for display reasons. The decomposition of a parent function (A0 in this case) preserves the inputs, controls, outputs, and mechanisms of the parent. There can be no more, no less, and no differences. Every

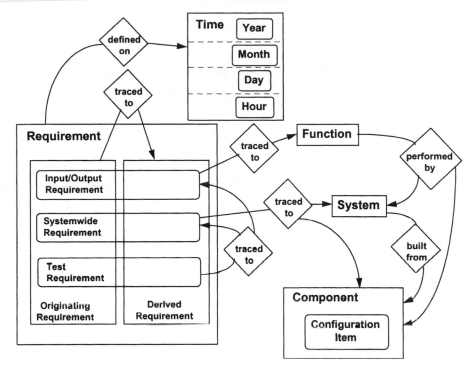

Figure 25.10 Partial higraph representation of the systems engineering ER diagram.

function must have a control. An input is optional. Boxes are usually placed diagonally. Arcs are decomposable, just as functions are. Feedback is modeled by having an output from a higher-numbered function on a page flow upstream as a control, input, or mechanism to a lower numbered function.

IDEF0 models can also address the interaction of the system with other systems. This interaction is modeled on the A1 page, which takes the A0 function and places it in context with other systems or organizations. This representation is often critical to understand the relationship of the system being addressed to its outside world, and establishing the origin of inputs and controls and the destination of outputs.

In order to emphasize the readability and understandability of IDEF0 models, the IDEF0 community has been very strict in establishing criteria for constructing IDEF0 diagrams correctly. In fact, the National Institute of Standards and Technology has released standards for IDEF0 and IDEF1.

Data Flow Diagrams. Data flow diagrams (DFDs), dating from the early 1970s, are one of the original diagramming techniques used in the software and information systems communities. The basic constructs of DFDs are the (1) function or activity, (2) data flow, (3) store, and (4) terminator.

The circle is the most standard representation for a function, defined as a verb phase. Arcs again represent the flow of data or information between functions, or to and from stores. Double-headed arcs are allowed; these represent dialog between two functions, for

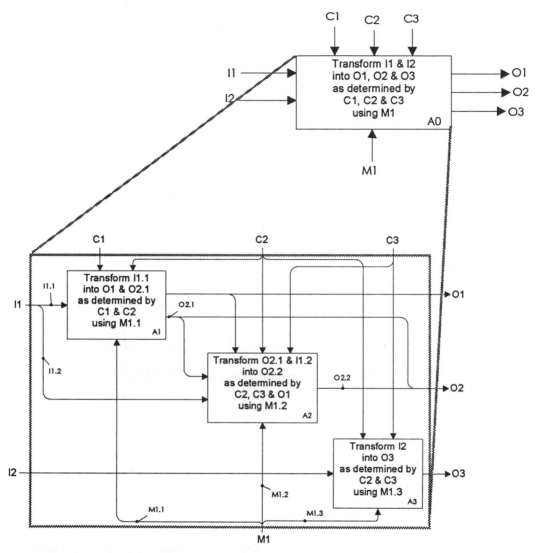

Figure 25.11 Functional decomposition in an IDEF0 model showing the preservation of inputs, controls, outputs, and mechanisms.

example, a query and a response. The labels for an arc are noun phrases and are placed near each arrow. Branches and joins are allowed and are depicted as forks.

A new concept is introduced, the store or buffer, a set of data packets at rest. Again there are several legal representations of a store. A store can be represented as a noun phrase between two diagonal lines that are not connected, connected at one end by a straight line, or connected at both ends by semicircles.

The final syntactical elements of DFDs are terminators, or other systems. In fact, context diagrams that show the interaction between the external systems, or terminators, and the system being designed or analyzed, are standard in DFDs (see Fig. 25.12). Terminators are shown in boxes.

A DFD model is typically a set of "leveled" DFD diagrams showing the flow of data

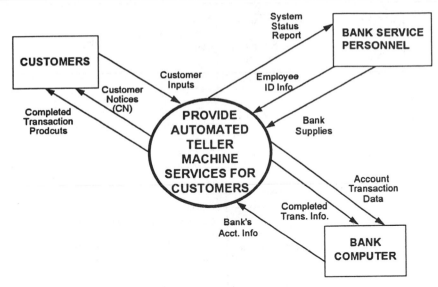

Figure 25.12   Context diagram using a data flow diagram.

between functions as well as the hierarchical decomposition of the functions. No standard has developed within DFDs, as many practitioners have modified the basic DFDs as described previously to suit their needs. For example, continuous data flow is often represented as a double arrow. The flow of events is represented by a broken line, with the continuous flow of events represented by a broken line with a double arrow.

N^2 *Charts.* N^2 charts were created with the behavior method, function flow block diagrams (FFBDs), to depict the data or items that are the inputs and outputs of the functions in the functional architecture. The charts are called N^2 because for a set of N functions the chart contains N^2 boxes to show the flow of items within (or internal to) the N functions. The N functions are placed along the diagonal (see Fig. 25.13). Items flowing from function i to function j are defined in the i,j box. Additional boxes along the top and down the right are needed to add the flow of external items into and out of the set of N functions, respectively. N^2 charts provide the set of information comparable to IDEF0 and DFDs.

25.5.1.3   Behavior Modeling

Control Flow Diagrams. Control flow diagrams (CFDs) are sometimes used in conjunction with DFDs, either as separate but parallel diagrams or superimposed on DFDs. Control flow is information that is transmitted between functions or between a function and the outside to determine how the functional processes must operate under specific changes in the operating modes. These operating modes may dictate that certain functions are present or absent, or change the way in which these functions perform. This control flow is typically shown as broken arcs; therefore, control flow is not used in conjunction with the expanded distinctions of event flows in DFDs.

Function Flow Block Diagrams. This was the original approach to functional decomposition in systems engineering, dating to the late 1960s and early 1970s. FFBDs show the

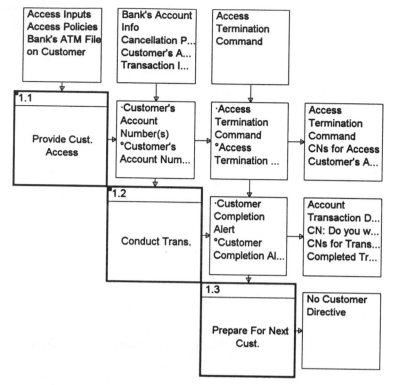

Figure 25.13   An N² chart showing external inputs and outputs plus the internal inputs and outputs.

functions at a given level and a control structure that dictates the order in which the functions can be executed. Functions are typically shown in boxes with their associated number.

In the *original,* or *basic,* FFBD syntax there were four types of control structure that were allowed: series, concurrent, selection, and multiple-exit function. A set of functions defined in a series control structure must all be executed in that order; the second cannot begin until the first is finished, and so on. Control passes from left to right along the arc shown from outside (depicted by a function in a box with broken top and bottom lines) and activates the first function. When the first function has been completed (that is, its exit criterion has been satisfied), control passes out of the right face of the function and into the second function.

The concurrent structure allows multiple functions to be working in parallel; thus, it is sometimes called "parallel." However, the concurrent structure should not be confused with the concepts of parallel in electric circuits or redundant systems. Essentially control is activated on all lines *exiting* the first AND node, and control cannot be closed at the second AND node until all functions on each control line are completed (see Fig. 25.14).

A selection structure and a multiple-exit function achieve essentially the same purpose; the possibility of activating one of several functions. The multiple-exit function (1.2.2.5 in Fig. 25.14) achieves this by having a function placed at the fork to make the selection process explicit. When the selection function has been completed, one of the several emanating control lines is activated. Once all of the functions on the activated line have finished execution, control passes through the closing OR node. The exit criteria for the

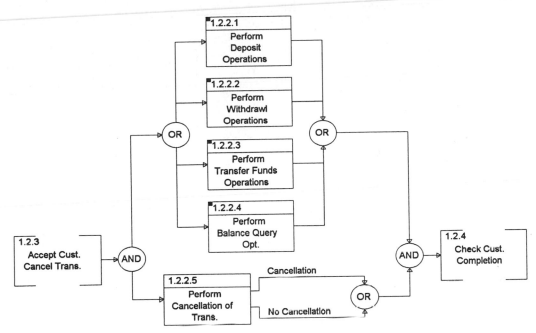

Figure 25.14 Concurrent, selection, and multiple-exit functions in an FFBD.

multiple-exit function are labeled. (Note, there is an exit criterion for every single-exit function, but it does not appear on the diagram.)

For the selection construct (functions 1.2.2.1 through 1.2.2.4 in Fig. 25.14), which is an exclusive OR, the first OR node passes control to one of the exiting control lines in a manner that is unspecified on the diagram. This control line stays active until the set of functions on that control line is completed; control then passes through the second OR node.

Recently additional control structures have been added to FFBDs to form what we might call *enhanced* FFBDs: iteration, looping, and replication. Iteration (see the top half of Fig. 25.15) involves the repetition of a set of functions as often as needed to satisfy some domain set; this domain set must be defined based upon a number or an interval. Looping (see the bottom half of Fig. 25.16) provides a similar control structure, but in this case it is possible to exit the loop if the appropriate criterion has been satisfied. Finally, replication involves repeating the same function concurrently using identical resources.

Behavior Diagrams. Behavior diagrams (Alford, 1977) originated as part of the Distributed Computer Design System of the Department of Defense. System behavior is described through a progressive hierarchical decomposition of a time sequence of functions and their inputs and outputs. Functions are represented as verb phrases inside boxes. There is a control structure represented by lines that flow vertically, from top to bottom, through the boxes. The control structures (see Fig. 25.17) are identical to those just described for FFBDs. The control lines have only one entry path into a function, but may have multiple exit control paths. Input and output items are represented in boxes with rounded corners; their entry to and exit from functions is depicted by arcs that enter and exit the boxes, respectively.

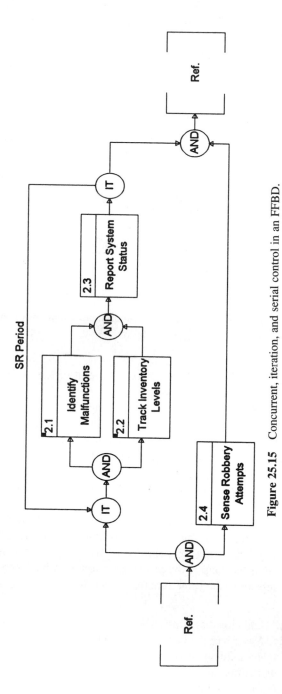

Figure 25.15 Concurrent, iteration, and serial control in an FFBD.

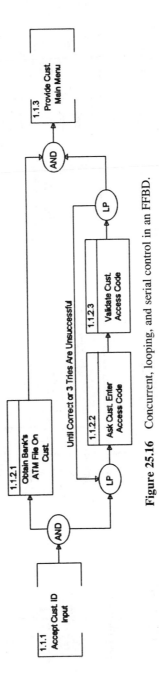

Figure 25.16 Concurrent, looping, and serial control in an FFBD.

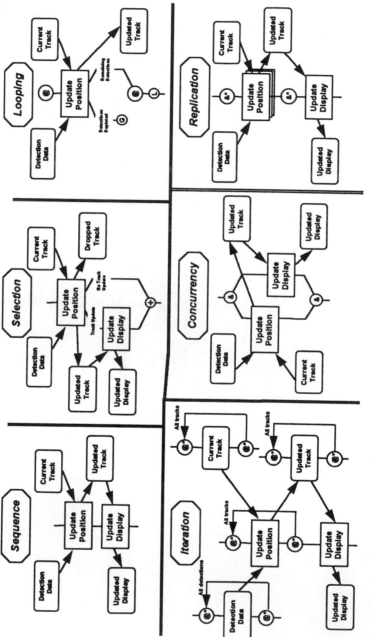

Figure 25.17 Control structures for behavior diagrams.

1026

Specific control structures for sequence, selection, iteration, looping, concurrency, and replication have been defined within behavior diagrams, just as they have been in FFBDs. A sequence of functions is connected to a vertical straight line. A selection function is denoted by a function with two or more control lines emanating from the bottom of the function. The emanating control lines must be labeled to denote the exit criterion associated with each control line. The multiple control lines must also be joined lower in the diagram at a select node, a small circle with a + inside. Figure 25.18 shows such a selection function.

An iterate control structure is set off on a control line by two nodes. Each node is a circle with an @* inside; there is an arc from the bottom iterate node to the top iterate node with a DomainSet label that defines at what frequency or how many times the functions inside the iterate structure are to be exercised.

An exit-loop control structure uses a selection function to determine the point at which the repetition of a function (or set of functions) should be terminated. The exit-loop control structure is set off by two vertically placed nodes (circles with an @ inside) that are connected with an arc going from the bottom node to the top node. The selection function that is responsible for ending the repetition has multiple exit control lines, one of which ends at an G node or circle with G inside. When the exit criterion for the G node is satisfied within the function, control emanates out the control line with the G node and drops below the bottom iterate node to the L node.

The control structure denoting that functions can be executed concurrently (Fig. 25.18) is depicted by two vertically placed nodes designated by circles with an & inside. In this special control structure all of the control lines below the first concurrent node are activated and the control line below the bottom concurrent node cannot become active until all of the functions on the concurrent control lines are finished executing.

Two vertically placed nodes with &* inside denote a replication control structure, which is a special case of a concurrent control structure. In this case, an identical function is executed concurrently, presumably by multiple copies of the same resource. The number of concurrent resources is labeled by a DomainSet on a line that connects the upper and lower replication nodes. The fact that there are multiple resources executing the same function is made visual by a "stack of papers" symbol on the main control line between the upper and lower replication nodes. There may be a Coordination function on the line with the DomainSet label.

Definition of the items within the behavior diagram is equally important. First, it is possible to use the sequence, concurrent, and replication control structures to organize the items (or inputs and outputs) associated with functions. Second there are various categories of items. Items that enter the system from outside or are produced by the system for outside consumption are called *External Items;* all other items are called *Internal Items.* The round-cornered box for external items is larger than that for internal items. All items can be hierarchically decomposed just as functions can. An item that is decomposed is called a *Time Item* and is represented by a clear box with a solid little square in the upper left corner. An item that is at the bottom of a decomposition is called a Discrete Item; it is represented in a shaded, round-cornered box. Discrete Items are classified as either message, state, temporary, or global items. A message item is sent from a function on one control line (or process) to a function on a different control line (or process) and the message item triggers the receiving function to execute as soon as it is enabled by the control structure. Global items do not trigger the receiving function to execute. State items are input to and output from functions on the same control line, and are therefore always Internal Items. A state item is not a trigger. Temporary items are for special purposes.

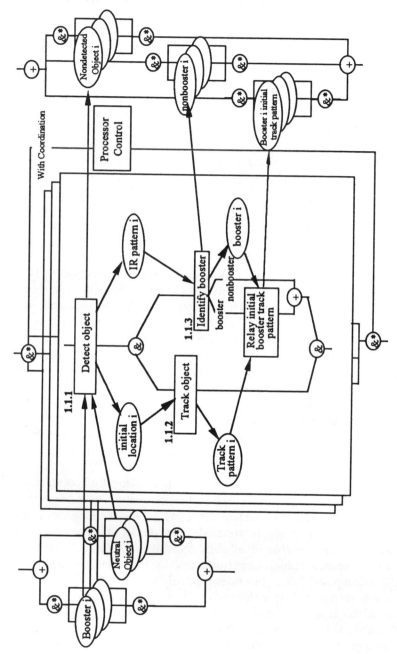

Figure 25.18 Concurrent control structure.

Finite-state Machines. Finite-state machines (FSMs) (Denning et al., 1978) are a subset of machines that have only discrete-valued inputs, outputs, and Internal Items. Continuous machines, the second and last subset in the partition of machines, allow continuous and discrete inputs, outputs, and Internal Items. Continuous machines are sometimes called analog machines. When digital computers became more popular than analog computers, FSMs became the major focus of attention in engineering due to the finite-state nature of digital computers. Even so, continuous and discrete signals are usually handled very differently by a digital computer. The continuous variable (e.g., speed or internal temperature of the elevator car) is represented by a word that typically contains many more bits than the variable has significant digits. On the other hand, a digital variable (e.g., operating mode, such as fully operational or partially operational or not operational, and direction of a specific elevator car, such as up or down) is usually represented by a symbolic word that has a relatively small number of states, say less than 10.

FSMs are usually divided into sequential and combinational; see the machine partition in Figure 25.19. We focus on the sequential FSM, as represented by a state-transition diagram (STD). A combinational FSM is one in which its current outputs are characterized only by its current outputs, a condition of having no memory that is often not met. The sequential FSM allows past inputs to play a role in the determination of the current outputs, thus enabling the FSM to have a memory.

The STD models the event-based, time-dependent behavior of a system. The *state* of a system is defined to be its status, as defined by as many variables as needed to determine its ability to meet its missions. The *mode* of a system is its operating condition, such as off, idling, or moving, for an automobile. It is the mode of a system that should be modeled by an STD. Functionality within the system is often modeled instead, as shown in Figures 25.20 and 25.21.

Boxes (or ovals) and arcs are the syntactical elements of STDs; the boxes represent system modes and the arcs represent the direction of mode change. Typically the arcs are labeled to show both the input stimulus or event that triggers the mode change and the action or output taken by the system in response to the event. The event and output are typically separated by a slash or horizontal line: event/output. Figure 25.20 shows a partially completed STD for an automatic teller machine. This STD is incomplete because the transitions to the four customer choices are not labeled; the transitions from the four customer choices are not depicted by arcs. It is possible that each might be completed successfully or canceled. The withdrawal might be denied. In each case the customer can choose another transaction or not.

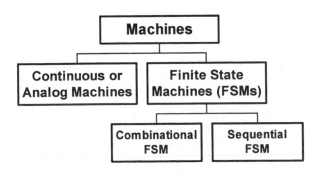

Figure 25.19 Partition of machines.

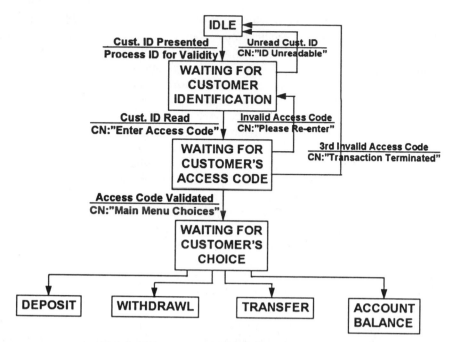

Figure 25.20   State-transition diagram for an ATM.

It is important to note that differences between the view provided by an STD and that provided by one of the process modes (DFD, IDEF0). The STD makes no attempt to provide a functional partition of the top-level system function or any function that is part of its partition. Rather the STD focuses on key triggering events that will cause the system to transition from one operational mode to another and identify any key system outputs produced as a result of that transition. Similarly, process models are not required to capture the system's operating modes. Figure 25.21 shows an STD for an elevator car [this figure is a modification of one found in Gomaa (1993)].

Statecharts. Statecharts are a generalization of higraphs by Harel to extend the notions of STDs. A major criticism of STDs has always been that the entire diagram must be contained on one level, meaning that an STD for a large system quickly becomes unintelligible. Statecharts, by exploiting the subset properties of higraphs, provide a means to develop hierarchical STDs, as shown in Figure 25.22 for a cruise-control system (CCS).

Another drawback of STDs is that an event such as an interrupt can cause a transition from many states to a single state; this results in many arrows to depict the effect of a single event. Since an arrow can go from a state (blob in higraphs) containing several atomic states (blobs), it is easy to see how the number of these arrows can be reduced with statecharts. See the transitions between "not off" and "off" in Figure 25.22. The atomic blobs in a statechart are singleton, or atomic, states.

Arcs in statecharts are labeled, just as they in STDs. The initial state is identified by finding the arc that emanates from a black dot; the state that this arc enters is the initial state of the system.

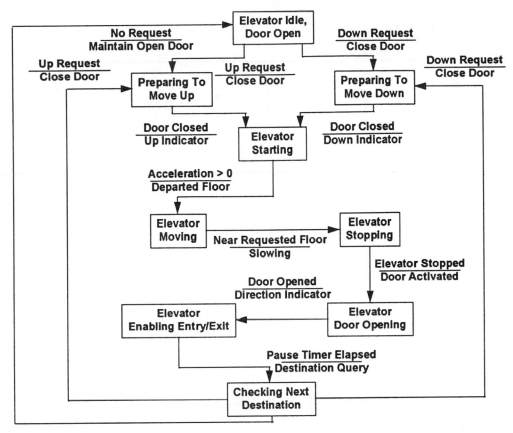

Figure 25.21 State-transition diagram for an elevator car.

The ability to represent unordered n-tuples in higraphs enables statecharts to depict states as being the orthogonal composition of elements from sets of states. When the initial state is an n-tuple, there must be n initiating arcs to define which element of the set of n-tuples is the initial state. Similarly, when there is a transition from (to) a state that is part of an n-tuple to (from) one that is not, the arc must be joined by an arc from (must branch to) $n - 1$ other arcs.

Another extension of statecharts is the ability to nest transitions by using labels such as a/b. This means that transition a will cause another transition b, located elsewhere in the statechart, to occur. Harel calls this broadcasting, because one event can broadcast a trigger that generates a chain reaction of one or more transitions throughout the statechart.

25.5.1.4 Object-oriented Modeling. While structured analysis has its roots in systems engineering, the object-oriented paradigm is a product of computer science and software systems engineering. Indeed the original concepts of objects and classes that encapsulate procedures and data in one entity originated in the 1960s with the development of the Simula programming language. This approach has continued to flourish because it allows software designers to deal with the complexity of modern software systems. Recently, as traditional hardware systems have become dependent on computer software components,

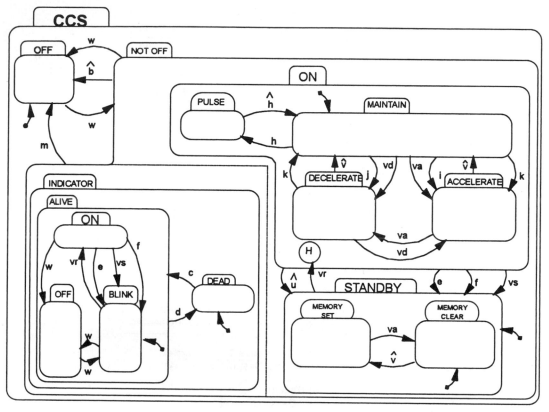

Figure 25.22   Statechart for a cruise-control system. (From Charbonneau, 1996).

system designers have been exploring the notion of using the object-oriented approach, not only for the design of software, but for the total system engineering process.

Object-oriented analysis focuses on the concept of objects that act; that is, essentially on the physical resources of the system. The object-oriented view is that a system is composed of individual agents that have a well-defined behavior. These objects do things, and they are asked to do things by messages that are sent to them. When they accomplish their activity, they may respond with a message if that is the nature of their behavior. To design systems using objects is to organize cooperative collections of objects such that their interactions will cause a higher level of overall behavior.

Object-oriented models of a system have multiple views. Booch (1991) proposed a canonical model of an object-oriented design that comprises an object class view, showing the inheritance hierarchy, and an object model view, showing the composition hierarchy. Classes in the former view are mapped into the latter view via relationships. The object modeling technique (Rumbaugh et al., 1991) uses three views: an object model view, a dynamic view, and a functional view. The object model view builds upon ER diagrams (i.e., data models); the functional view employs DFDs; and the behavior view uses both state transition and nested state diagrams to model the behavior, and event traces to describe the behavior dynamics.

Selic et al. (1994) have developed another popular object-oriented design approach

TABLE 25.5 Software for Functional Analysis

Software Product Name	Platform	Vendor Name	Address/Phone
BACHMAN Analyst	PC	Bachman Information Systems, Inc.	8 New England Exec. Park Burlington, MA 01803 800-bachman
BPWin	PC	Logic Works	1060 Route 206 Princeton, NJ 08540 800-783-7946
CASETS	PC	Rockwell	12214 Lakewood Blvd. Downey, CA 90241 310-922-1791
Composer	PC, Sun	Texas Instruments	6550 Chase Oaks Blvd. Plano, TX 75023 214-575-4404
CORE	PC	Vitech Corp.	2070 Chain Bridge Rd., Suite 105 Vienna, VA 22182 703-883-2270
Cradle	PC	Mesa Systems Guild	60 Quaker Lane Warwick, RI 02886 401-828-8500
DesignIDEF	PC, MAC	Meta Software, Corp.	150 Cambridge Park Dr. Cambridge, MA 02140 617-576-6920
EasyCASE Pro	PC	Evergreen CASE Tools, Inc.	8522 154th Ave. NE Redmond, WA 98052 206-881-5149
Excelerator	PC	Intersolv, Inc.	3200 Tower Oaks Blvd. Rockville, MD 20852 301-230-3200
FORESIGHT	UNIX	NuThena Systems, Inc.	1430 Spring Hill Rd., Suite 220 McLean, VA 22102 703-356-5056
ObjectMaker	PC, MAC	Mark V Systems, Ltd.	16400 Ventura Blvd. Encino, CA 818-995-7671
Objecttime	UNIX	ObjectTime, Ltd.	340 March Rd., Suite 200 Kanata, Ontario Canada K2K 2E4 613-591-3400
Oracle Designer 2000	PC	Oracle Corp.	500 Oracle Parkway Redwood Shores, CA 94065 415-506-7000
RDD-100	SUN, PC	Ascent Logic, Corp.	180 Rose Orchard Way San Jose, CA 95134 408-943-0630

TABLE 25.5 (*Continued*)

Software Product Name	Platform	Vendor Name	Address/Phone
System Architect	PC	Popkin Software and Systems, Ind.	11 Park Place New York, NY 10007 212-571-3434
SLATE	UNIX	TD Technologies, Inc.	2425 N. Central Exprway, Suite 200 Richardson, TX 75080 800-669-4998
Teamwork	PC, VAX	Cadre Technologies, Inc.	222 Richmond St. Providence, RI 401-351-5950

called ROOM for real-time object-oriented modeling. Objects in ROOM are called *actors*. Actors communicate by messages based upon predefined protocols. Actors are organized into hierarchical structures. Process and behavior modeling are accomplished via ROOM-charts, a generalization of statecharts. This generalization addresses less than fully reliable and instantaneous communication of messages and transitions that take longer than "zero time."

25.5.2 Software Support for Functional Analysis

In the past 10 years commercial software to support functional analysis has risen from nothing to become a major industry. Table 25.5 contains many of the current products that support one or more of the data, process, behavior, or object-oriented modeling techniques. These products vary from packages that support only the functional analysis process to packages that support most of the systems engineering process, starting with requirements definition and management through functional analysis and performance modeling. Additional packages are available that only address requirements management. In the next five years there will be many extensions of these packages and many new packages.

REFERENCES

Alford, M. W. (1977). A requirements engineering methodology for real-time processing requirements. *IEEE Trans. Software Eng.* **SE-3**(1), 60–69.

Boehm, B. W. (1976). Software engineering. *IEEE Trans. Comput.* **C-25**(12), 1226–1241.

Boehm, B. W. (1981). *Software Engineering Economics*. Englewood Cliffs, NJ: Prentice-Hall.

Booch, G. (1991). *Object-Oriented Design*. Redwood City, CA: Benjamin/Cummings.

Buede, D. M., and Bresnick, T. A. (1992). Applications of decision analysis to the military systems acquisition process. *Interfaces* **22**(6), 110–125.

Buede, D. M., and Choisser, R. W. (1992). Providing an analytic structure for key system design choices. *J. Multi-Criteria Decis. Anal.* **1**, 17–27.

Charbonneau, S. M. (1996). Generation of originating requirements: Use of functional decomposition and state transition diagrams. M.S. Thesis, George Mason University, Fairtax, VA.

Daly, E. (1977). Management of software development. *IEEE Trans. Software Eng.* **SE-3**(3), 229–242.

Davis, A. M. (1990). A comparison of techniques for the specification of external system behavior. In *System and Software Requirements Engineering* R. H. Thayer and M. Dorfman, (eds.). New York: IEEE Computer Society Press.

Davis, A. M. (1993). *Software Requirements: Objects, Functions, and States.* Englewood Cliffs, NJ: Prentice-Hall.

Denning, P. J., Dennis, J. B., and Qualitz, J. E. (1978). *Machines, Languages, and Computation.* Englewood Cliffs, NJ: Prentice-Hall.

Fagan, M. (1974). *Design and Code Inspections and Process Control in the Development of Programs* (IBM Rep. IBM-SDD-TR-21-572).

Frantz, W. F. (1993). Requirements: A practical, tested approach for break-through systems. In *Systems Engineering in the Workplace* J. E. McAuley and W. H. McCumber, (eds.), Third Annual International Symposium of INCOSE, pp. 801–810.

Gomaa, H. (1993). *Software Design Methods for Concurrent and Real-Time Systems.* Reading, MA: Addison-Wesley.

Griffith, P. B. (1994). Different philosophies/different methods: RDD and IDEF. In *Systems Engineering: A Competitive Edge in a Changing World.* (J. T. Whalen, D. McKinney, and S. Shreve, eds.). Fourth Annual International Symposium of INCOSE, pp. 489–495.

Harel, D. (1988). On visual formalisms. *Commun. ACM* **31**(5), 514–530.

Hatley, D. J., and Pirbhai, I. A. (1988). *Strategies for Real-Time System Specification.* New York: Dorset House.

Hunger, J. W. (1995). *Engineering the System Solution.* Upper Saddle River, NJ: Prentice Hall.

Jones, D. R., and Schkade, D. A. (1995). Choosing and translating between problem representations. *Organ. Behav. Hum. Decis. Proces.* **61**(2), 214–223.

Keeney, R. L. (1992). *Value-Focused Thinking.* Boston: Harvard University Press.

Lano, R. J. (1990). The N^2 chart. In *System and Software Requirements Engineering* (R. H. Thayer and M. Dorfman, eds.) New York: IEEE Computer Society Press, pp. 244–271.

Mar, B. W. (1984). Requirements for development of software requirements. *Proc. 4th Annu. Int. Symp. Nat. Coun. Syst. Eng.,* Vol. 1, pp. 39–44.

McMenamin, S. M., and Palmer, J. F. (1984). *Essential Systems Analysis.* Englewood Cliffs, NJ: Prentice-Hall.

Miller, G. A. (1956). The magical number seven, plus or minus two. *Psychol. Rev.* **63**(2), 90–114.

Rumbaugh, J., Blaha, M., Premerlani, W., Eddy, F., and Lorensen, W. (1991). *Object-Oriented Modeling and Design.* Englewood Cliffs, NJ: Prentice-Hall.

Sailor, J. D. (1990). System engineering: An introduction. In *System and Software Requirements Engineering* (R. H. Thayer and M. Dorfman, eds.). New York: IEEE Computer Society Press, pp. 35–47.

Selic, B., Gullekson, G., and Ward, P. T. (1994). *Real-Time Object-Oriented Modeling.* New York: Wiley.

Wymore, W. (1993). *Model-Based Systems Engineering.* Boca Raton, FL: CRC Press.

26 Methods for the Modeling and Analysis of Alternatives

C. ELS VAN DAALEN, WIL A. H. THISSEN, and
ALEXANDER VERBRAECK

26.1 INTRODUCTION

Mathematical modeling may serve many purposes in a systems engineering project. It may be used for analyzing an existing system and for specifying and analyzing design alternatives. In this chapter, a general cycle for developing mathematical models is presented. A number of phases in modeling are recognized: defining goal and function, conceptualization, model construction, validation, and model use. The discussion focuses on issues that are common to many types of model. Following this, several different modeling methodologies are described according to the model cycle. Aspects specific to these methodologies are discussed, as well as requirements for use of the methodology. Three of these methodologies are aimed at developing causal system models, and allow investigation of alternative system configurations: physical systems modeling, system dynamics, and discrete-event simulation modeling. Physical systems modeling is meant for modeling technical/physical systems, whereas system dynamics is a methodology that has been developed primarily for analyzing and modeling business and socioeconomic systems. Discrete-event simulation models are stochastic models in which separate entities are recognized. These three methodologies result in causal models that are able to explain system behavior. Time-series models, on the other hand, are black-box models that are not aimed at explaining, but at forecasting only. Black-box models may be used when it is not necessary to describe the exact internal workings of the system, or when the underlying causal mechanisms are unknown, but sufficient data are available. The modeling methodologies just cited allow investigation of system behavior over time, but do not include financial aspects. When designing a system, cost aspects should be taken into account. Therefore, cost–benefit analysis is discussed briefly. Although all modeling methodologies are discussed in separate sections, a combination of various approaches is usually necessary for large-scale systems. Finally, some recent advances in modeling are discussed, including group model building and animation possibilities.

26.2 QUANTITATIVE MODELS AND METHODS

26.2.1 Introduction

In the course of a systems engineering project, it may be necessary to investigate the behavior of a system. Investigation of the behavior of a system may be carried out by

Handbook of Systems Engineering and Management, Edited by A. P. Sage and W. B. Rouse
ISBN 0471-15405-9 ©1999 John Wiley and Sons, Inc.

experimenting with the actual system. Many systems do not lend themselves to experimentation, however, as this may be expensive (e.g., wind-tunnel experiments), dangerous (e.g., in nuclear power plants), impractical or impossible (e.g., social systems). In these situations, models may be developed in order to investigate the system.

A model is a simplified representation of an object, system, or idea in some form other than that of the entity itself. The term *model* is very broad and may, for example, refer to a physical model made of tangible components, such as a scale model, a symbolic model such as a mathematical model or diagram, or a mental model that exists only in the mind (Neelamkavil, 1987). A widely used category of symbolic model is the mathematical model. In systems engineering, models are used as tools for description, prediction, and communication. Often, there is an existing system that has to be changed or improved. In that case, both the current situation (the as-is or "ist" situation) and alternative system configurations (to-be or "soll" situations) have to be studied. This means that in a modeling exercise a descriptive model should be made first, after which models are developed that are aimed at exploring the consequences of alternatives, to support design or selection of a "best" alternative.

An important focus of this chapter will be the discussion of different types of mathematical models of system behavior, and methodologies to create and use these models (Sections 26.2 to 26.6). In addition to models of system behavior, cost–benefit models will be addressed in Section 26.7. Section 26.8 deals with the choice and combination of modeling methods, as well as with some recent advances in modeling. This chapter is concerned only with modeling and the model cycle, and will not consider the broader issues of defining system needs, specifying alternatives, and decision analysis. It is recognized, however, that model-based analysis of an existing system may be very helpful in identifying alternatives, and that the analysis of alternatives provides input for decision analysis and/or optimization. Optimization of alternatives and control are discussed in Chapter 27.

26.2.2 A Model Cycle

Numerous different cycles for model development, which usually only differ very slightly, have been described in the literature. A general iterative scheme for model development is shown in Figure 26.1. After the goal and function of the model have been determined and the modeling methodology has been chosen, the rest of the conceptualization phase, model construction, verification, and validation usually proceed according to the modeling methodology that has been selected. Because systems engineering involves large and complex systems, it is usually not sufficient to develop a single model. The development of a number of different models may be required: descriptive models of existing systems, as well as models of alternatives, and models of different aspects of the system or of different subsystems. A separate model cycle is carried out for each model that is to be developed.

Some general considerations regarding the steps in the model cycle are mentioned in this subsection. The parts that are specific to certain modeling methodologies are discussed in the relevant section on a specific methodology.

A. *Defining Goal and Function.* The purpose of models in general is to acquire knowledge about a system. Thus a modeling process should start with defining what specific knowledge is desired. In addition, thought must be given to the function the model fulfills within a systems engineering project. For example, gaining basic understanding of the

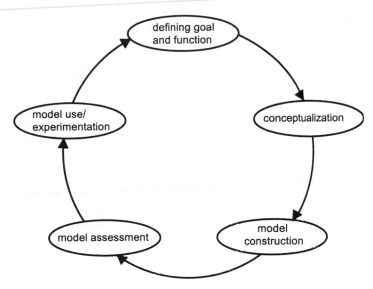

Figure 26.1   A model cycle.

chemistry and physics of some industrial process will require different models from models optimizing the operations of a specific production plant. Goal and function largely determine the extent and detail to which aspects are included in the model (Hoover and Perry, 1989), the degree of abstraction that is still acceptable, and the modeling methodologies that are appropriate.

B. *Conceptualization.* Models must cover only those aspects of the system that are considered essential for their goal and function. At the conceptual stage, both the structure and complexity of a model are largely determined. A reduction of complexity may be brought about by selection (leaving out aspects that are considered irrelevant for the model), aggregation (combining different components into a single entity), and idealization (approximation of structure or behavior). The result is a set of variables and their relationships. Goal and function of the model determine which of the variables will be endogenous (i.e., values computed within and/or output by the model) and exogenous (i.e., values entered into the model).

Choice of modeling methodology. Based on the goal of the model and the characteristics of the system, a modeling methodology is chosen. A *methodology* consists of a set of systematic procedures based upon knowledge accumulated over a number of years, for tackling a class of problems (Neelamkavil, 1987). Most modeling methodologies include procedures for continuing the conceptualization and for model construction and validation.

Various classes of modeling methodologies are discussed in this chapter. These include physical systems modeling, system dynamics, discrete-event simulation modeling, and statistical modeling. Although the choice of a modeling methodology should depend on the problem to be solved, in practice it often also depends on the experience of the analyst in developing certain kinds of models.

Related to the choice of a modeling methodology, is the choice of the modeling software that is used. Most of the available software relates directly to a specific modeling methodology, and often supports conceptualization, model construction, model validation (to a certain extent), and allows model use. In the sections on specific modeling methodologies, some of the available software packages are mentioned. The Appendix includes a table of names and vendors of software packages that are discussed in this chapter.

Conceptual model. The result of the conceptualization phase consists of one or more conceptual models highlighting different aspects of the system. A conceptual model is a simplified representation of a system in terms of elements and their relationships, using an explicit convention and with the aim of enhancing comprehension of the system. A conceptual model is data void. An explicit convention is a set of agreements with respect to syntax and semantics of the representation of elements and relationships.

- *Diagrammatic or Linguistic Conceptual Model.* During a study, the first type of conceptual model that is drawn up is often either diagrammatic or linguistic. Examples of types of diagrammatic conceptual models are a causal-loop diagram and a means–ends diagram. An object definition, consisting of a list of relevant objects and their relevant attributes, is an example of a linguistic conceptual model. In practice, modeling methodologies often include preferred types of conceptual models, modeling language and modeling tools. The types of conceptual models that are used in certain modeling methodologies are discussed in the relevant sections below.

- *Mathematical Conceptual Model.* Following the development of a diagrammatic or linguistic conceptual model, a mathematical conceptual model is developed. This model consists of a description of the behavior of variables (usually over time) using (standard) mathematical notation. Actual numerical values or model parameters are not specified yet, hence the model is still at a conceptual level.

Mathematical models may be characterized along a number of dimensions. Most modeling methodologies are closely related to a specific mathematical modeling technique, and therefore the resulting model will possess certain characteristics. There are various characteristics that may be distinguished:

- *Continuous and Discrete.* In a continuous model the values of the variables may change at anytime and on a continuous scale, and differential equations are commonly used for mathematical representation of the relationships between the variables. Discrete models are models for which the state variables either change at discrete points in time or may only take on discrete values. The choice for a discrete or a continuous model depends on the goal of the model and the key system characteristics considered to be essential. For instance, car traffic can be modeled as a continuous flow, or by identifying individual cars and using a discrete approach. Not all models have to be completely continuous or discrete. Models that contain elements of both are called *hybrid models*.

- *Deterministic and Stochastic.* Deterministic models have predictable relationships between input and output variables. Stochastic models make use of probabilities.

- *White Box and Black Box.* A white-box model is based on an explicatation of causal mechanisms provided in theories on the subject matter, and its internal structure thus consists of relationships between real-world variables. In a black-box model the basis for the relationships is empirical rather than theoretical. The black-box approach is based on statistical techniques.

In Sections 26.3 to 26.6 these characteristics will be related to specific modeling methodologies.

C. *Model Construction.* During the construction phase, a quantitative model is constructed on the basis of the results of the conceptualization phase. In contrast to the conceptual model, which is data void, the quantitative model includes numeric data that will allow the values of the variables to be calculated over time, and thus allow system behavior to be analyzed quantitatively. This means that quantitative information is needed with respect to the relations between variables (i.e., values of model parameters) and the variables that are considered to be exogenous (input values). Depending on the modeling methodology and the characteristics of the system, different ways of estimating the parameters may be used. Data may be available in the form of historical data, or a special experiment may be conducted with the system in order to obtain the data. However, there are a number of difficulties associated with measurements that relate to the reliability and validity of measuring instruments (Flood and Carson, 1990). Measuring instruments may not be totally reliable, that is, the (numerical) assignment process may introduce an error (e.g., inability of a measuring instrument to filter noise). Validity relates to whether the instrument is measuring the attribute of interest. In situations where it is not possible to determine parameter values on the basis of data, it may be necessary to obtain parameter values from a panel of experts. At the end of the model construction phase, the quantitative model will have been represented in the computer (executable model), so that the behavior of the model can be investigated.

D. *Model Assessment.* Model assessment is carried out by way of verification and validation. *Verification* refers to building the model "right," and *validation* refers to building the "right" model.

Verification. Verification requires investigating that the translation from one modeling product to the other has been carried out correctly, and making sure that the model does not contain logical errors. In other words, the model is tested for consistency.

Validation. Model validation can be defined as the process of substantiating that the model within its domain of application provides the required functionality (input and output variables), and that the values it computes are sufficiently accurate for the intended use. Thus, investigations of both functionality and accuracy should be performed. In order to do this, criteria should be identified that will allow the model to be judged. Models are not true or false, but rather they are useful and appropriate for the analysis at hand (Hoover and Perry, 1989). All models have a limited domain of validity. A model should not be used outside the area it has been validated for (Ljung and Glad, 1994).

With respect to validation, black-box and white-box models must be distinguished (Barlas, 1996). In a black-box model the internal structure does not contain causal relations. Therefore, the internal structure is not investigated and the question is whether the model output is accurate enough for the purpose intended. This type of investigation can often be cast as a statistical testing problem. In a white-box model, however, the internal structure of the model should also be validated, since the model must be able to adequately explain how the behavior is generated. Thus, for white-box models the structure must be validated prior to validating behavior.

Investigation of the structure of a model requires, for example, studying whether the relationships and assumptions are based on general accepted theory (theoretical validation) and whether all variables considered relevant have been taken into account. It is also useful to determine whether the model equations are still valid when the variables take on extreme values. In addition to these direct structure tests, structure-oriented behavior tests are carried out (Barlas, 1996). Direct structure tests involve an investigation of the model without running it on the computer. There is some discussion in the literature as to whether these tests are part of the verification of a model. However, there is no discussion about the nature of the investigations that have to be carried out.

Structure-oriented behavior tests assess the validity of the structure indirectly by running the model (Barlas, 1996). One method is to study the behavior of the system at extremes, for instance, by entering extreme input values. Another example of this is sensitivity analysis. This entails an analysis of the model assumptions and of the influence of plausible variations in parameters, structure, and possible exogenous variables. The sensitivity of the outputs in response to changes in parameters and structure can be studied to see whether these are realistic. Modes of behavior, frequencies, mechanisms causing behavior, and other characteristics should be studied to see whether they are as expected (Sterman, 1984).

Replicative validation entails a comparison of the model to the system that has been modeled. Many quantitative techniques are available for model testing. Only data that have not been used for model construction should be used for validation purposes; otherwise the model would be tested against itself instead of against the actual system. The choice of statistical measures that are used during replicative validation depends to a large extent on the modeling methodology that is adopted, and will be dealt with in the sections on specific modeling methodologies.

In the case of a descriptive model, replicative validation will be easier than in the case of a model of a system that does not (yet) exist, because empirical data will not be available. In these cases it is often appropriate to involve domain experts in the judgement of the accuracy of the model. A way of doing this is using an approach based on the Delphi method (Dalkey, 1969). A questionnaire is given to a panel of experts a number of times. The experts are asked to predict the response of the system when confronted with certain data input. A variant of the so-called Turing test [originally described by Turing (1963) as an experiment to investigate machine intelligence] is the reverse of the Delphi method. Reports based on output of the quantitative model and on measurements of the real system are presented to a team of experts. When they are not able to distinguish between the model output and the system output, the model is said to be valid.

Sometimes model results can be compared to models that have been developed in the same or similar fields. There may be other models that tackle part of the problem in detail, where certain variables can be compared.

The usability of the model will also have to be investigated. The model has to be geared to problems and questions that are thought to be important by the relevant actors. In the case of a white-box model, it should be understandable and be instrumental in improving the understanding of the behavior of the system.

In the model cycle (Fig. 26.1), model verification and validation are technically defined to take place after model construction. In practice, however, this takes place in every stage of the methodology (Barlas, 1996). Model validation is not separable from model building. Validation techniques should be used during the model-building process. Thus, confidence in the model is increased step by step (Kheir et al., 1985). Figure 26.2 shows a summary of the verification and validation investigations in relation to the modeling products that were discussed earlier.

E. *Model Use/Experimentation.* When a model is used to study a problem and to learn more about a system, several experiments will be carried out with the model in order to determine the effect of different values of the parameters, or of a change in model structure,

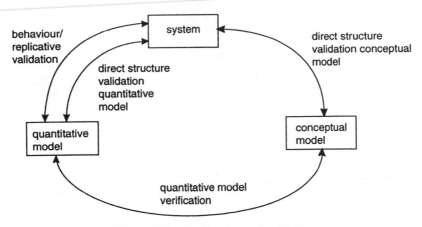

Figure 26.2   Verification and validation.

for instance, through addition, deletion, or change of variables, entity types, attributes, actions, and/or relationships. The kinds of investigations that are carried out will depend on the goal of the study. Specific types of model use are discussed in the sections on specific modeling methodologies.

26.2.3 Modeling Methodologies

Various modeling methodologies are discussed in this chapter. These include physical systems modeling, system dynamics, discrete-event-simulation modeling, and statistical modeling. The goal of this chapter is to give some insight into the potentials, limitations, and applicability of different modeling methodologies, so that an informed choice can be made as to the appropriate modeling methodology. In each section a different modeling methodology is explained, and the techniques within the model cycle that are specific to the methodology are discussed. The references point to more detailed material for those who want to actually apply the methods.

26.3 PHYSICAL SYSTEM MODELS

26.3.1 Introduction

Modeling of physical systems by way of mathematical equations is carried out in many different disciplines of science and engineering. A mathematical model allows for the design of new systems and for experiments that are not possible or not feasible in the real world. Although the physical systems and processes encountered in engineering practice differ widely, the basic approach of deriving the equations describing these systems varies only slightly (Smith et al., 1970). The process of deriving these equations, the similarities between models in very different scientific disciplines, and the ways of analyzing a model once it has been quantified are discussed in this section.

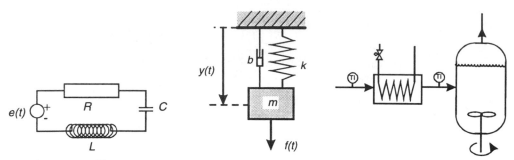

Figure 26.3 Examples of schematic diagrams of different systems.

26.3.2 Outline of the Approach

26.3.2.1 Conceptualization. Many scientific disciplines have their own conventions for schematizing the physical systems that are studied in that field. For instance, in electronics, circuit diagrams are drawn, and in process technology process diagrams are used. Some examples of diagrams are shown in Figure 26.3.

There are also various types of diagrams, such as signal flow graphs (Kheir, 1996), bond graphs (Karnopp et al, 1990; Van den Bosch and Van der Klauw, 1994), and block diagrams, that are applicable to a wide range of systems and are independent of the problem domain. In contrast to the schematic pictures, which are based on the physical construction of the system, these diagrams are based on the logical description (Ljung and Glad, 1994). A block diagram, for example, shows how a system's components are interconnected and also represents cause–effect relationships between variables (see, e.g., Kheir, 1996; Flood and Carson, 1990). Block diagrams may contain a mathematical representation of the system, in order to make the mathematical model more transparent or to represent it in the computer. However, block diagrams may also be drawn at a higher level of abstraction, without entering the exact mathematical relationship in the blocks (Fig. 26.4). The more abstract block diagrams can be used prior to entering mathematical equations. Various examples of block diagrams are shown in Figure 26.5.

The mathematical relationships between physical quantities can usually be divided into two groups: conservation laws and constitutive relationships (Ljung and Glad, 1994). Conservation laws relate quantities of the same kind. A general procedure specifying the conservation laws is suggested by Smith et al. (1970):

1. Define the quantity conserved, that is, the variable for which the equations are to be written, and select the independent and dependent variables. If only one independent variable appears, an ordinary differential equation will result.

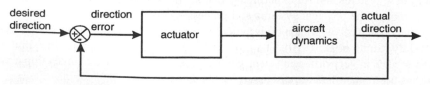

Figure 26.4 High-level block diagram of an autopilot system.

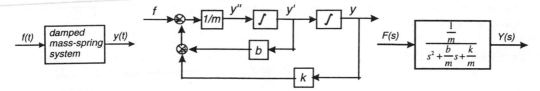

Figure 26.5 Various block diagrams of a damped-mass-spring system ($my'' = -ky - by' + f(t)$).

2. Apply the physical laws that yield the input and output expressions for the quantity conserved.

3. Substitute these expressions into the appropriate terms in the equation:

(rate of) input − (rate of) output = (rate of) accumulation of quantity conserved.

4. Develop the necessary boundary conditions.

5. Solve the resulting differential equation and boundary conditions. The solution technique may be analytic, numerical, or otherwise.

Constitutive relationships relate quantities of different kinds. An example is the relationship between voltage and current for a resistor. These relationships are static in the chosen variables (Ljung and Glad, 1994).

If two systems are described by the same differential equations, these two systems are said to be analogs. Analogous systems with similar solutions exist, for example, for electrical, mechanical, thermal, and fluid systems. The existence of analogous systems and solutions provides the analyst with the ability to extend the solution of one system to all analogous systems with the same mathematical description (Dorf and Bishop, 1995).

In a dynamic system two types of dependent variables can be distinguished. These are level variables (also called *across* variables), which are influenced by a rate of change, and the rate of change variables (also called *through* variables). The relationship between the level and rate variables is expressed in terms of the system parameters. Table 26.1 shows different types of level and rate variables for different areas. System parameters are parameters that are considered given by the system and cannot be chosen by the designer.

TABLE 26.1 Examples of Types of Level and Rate Variables

Physical Area	Level Variable	Rate Variable
Electrodynamics	Voltage	Current
Electrostatics	Electric potential	Flux
Dynamics (mechanical)	Displacement or velocity	Force
Elasticity	Strain	Stress
Fluid mechanics	Pressure	Flow rate
Particle diffusion	Concentration	Mass-transfer rate
Heat transfer	Temperature	Heat flux

Source: Adapted from Karplus, 1958.

Design parameters can be chosen in order to give the system/model the desired properties (Ljung and Glad, 1994).

26.3.2.2 Model Construction. In order to quantitatively investigate the behavior of a model, the parameter values, values of the exogenous variables, and initial values of the level variables have to be determined. Certain parameter values can be determined by direct measurement, other parameters may have to be estimated by calibrating the model to fit historical data. In the case of an existing system, the data can be obtained by carrying out experiments that are especially designed to be able to quantify the model.

26.3.2.3 Modeling Tools. A physical system model is usually represented in the computer using specialized software, such as MATLAB®, VisSim™, MATRIX$_x$®, or 20-SIM™. VisSim is based on block diagrams, as is the environment SIMULINK within MATLAB and the environment SystemBuild within MATRIX$_x$. The 20-SIM software also allows bond-graph modeling.

In order to solve the differential equations, the computer uses a numerical integration method. The user has to choose an appropriate numerical integration method and the time step that is suitable for the model at hand. The solution of the differential equations depends on the choice of integration method and time step. Although it is easy to analyze a model using the present-day software, a lack of theoretical knowledge may lead to incorrect answers. The numerical methods that are most often employed are the Runge-Kutta methods. The Runge-Kutta methods may be termed differently by different authors. Many elementary textbooks on differential equations (e.g., Boyce and DiPrima, 1986) and on modeling in general (e.g., Close and Frederick, 1978) describe a range of numerical methods.

26.3.2.4 Validation. Model validation in general was discussed in Section 26.2. In contrast to models of socioeconomic systems, physical systems models have the advantage that validation may often be carried out more thoroughly, as more theory and quantitative information about the system are often available. On the other hand, the development of a physical system model is often aimed at obtaining accurate knowledge of the workings of the system and at a detailed investigation of the effects of a specific design, for systems such as nuclear reactors, which makes a thorough validation indispensable. In some cases, it may be possible to carry out special experiments in order to validate the model.

26.3.2.5 Model Use. Once represented in the computer, the model can, among other things, be used to analyze the consequences of alternatives, which may comprise alternative system structures, different parameter values or different input values, or to design control systems. Different system designs can be compared to each other by calculating the value of a performance measure. The performance measure should be stated explicitly in the form of a mathematical function. The aim is usually to minimize the performance measure (or maximize it, depending on how the measure is stated). Alternatively, different performance measures, such as costs, accuracy, and reliability, can be computed, leaving tradeoffs to the client.

In an existing system, the goal of modeling may be to design control strategies with the aim of eliminating certain undesirable properties of the system. Possible aims are minimizing the effects of disturbance, following a desired trajectory, or stabilization. The control strategies are designed according to optimal control theory.

To be useful, a system must be stable. Stability at all times is viewed as the most important characteristic of a system. A stable system is defined as one that demonstrates a bounded response if the input is also bounded in magnitude (Kheir, 1996). Several methods have been developed to investigate stability of systems. For linear systems this will entail an investigation of the eigenvalues of the model.

26.3.2.6 *Large-scale Systems.*

The preceding description introduces physical modeling, and mainly concerns small-scale systems, whereas systems engineering is aimed at large-scale systems (Sage, 1977). The key to successful treatment of large-scale systems is to fully exploit their structure interconnection (Jamshidi, 1996). An example of model reduction is separation of time constants (Ljung and Glad, 1994). In the same system there may be time constants of different magnitudes (e.g., several seconds for a certain production system and several years for certain economic effects), whereas the interest and goal of the model may be focused on a certain time scale. In these cases the modeling is concentrated on phenomena whose time constants are of interest when considering the intended use of the model. Subsystems whose dynamics are considerably faster are approximated with static relations and variables whose dynamics are appreciably slower are approximated as constants. By ignoring very fast and very slow dynamics, the order of the model is lowered. Several models can be used for the same system, for example, there may be a sequence of models for different operating conditions. It is also possible to have one model for fast responses and one for slow responses. Another possibility is to work with a hierarchy of models with different levels of accuracy and complexity to solve different problems. Jamshidi (1983) mathematically treats a large number of large-scale systems model reduction methods.

Models of real systems may not only be large, but are often also stochastic and nonlinear. This implies that the theory of deterministic, linear systems does not hold. For further reading there are references concerning stochastic systems (Borrie, 1992) and nonlinear models of physical systems (Gu, 1996; Vidyasagar, 1978).

For complex geometrical problems, instead of using a differential equation approach, the finite-element method is often used. Areas of interest include structural, fluid, and thermal analysis. By using the finite-element method (Brauer, 1988), the difficulty of mathematically solving large, complex geometric problems is transformed from a differential equation approach to an algebraic problem, where the building blocks, or finite elements, have all the complex equations solved for their simple shape (e.g., triangle, rod, beam). The representation of the relationships of the important variables is determined for each element. Once this is done, a matrix of a size equal to the number of unknowns for the element can be produced that represents the element. The element matrices are then assembled, the boundary conditions are introduced, and the system of equations can be solved.

26.3.2.7 *Conditions for Use.*

The approach that has been discussed in this section, and that results in a model in the form of differential equations may be used for dynamic technical/physical systems. This approach can only be used when it is possible to express the essentials of the system in the form of a set of differential equations, and if a method exists for solving the equations.

When considering incorporation of socioeconomic variables in a model (rather than considering a purely technical/physical system), a system dynamics approach may be preferred (see Section 26.4). If the aim is to model separate entities and events or discrete states, rather than aggregate values, a discrete-event approach should be considered. In a continuous model, separate entities are not regarded. For example, when modeling container transport, a continuous model could render information about the average number of containers per unit of time, whereas in a discrete-event model, the containers would be modeled as separate entities (see Section 26.5).

When developing a continuous physical system model, depending on the goal of the model and the key characteristics of the system, it should be determined whether the model can be linearized and whether it is possible to use a deterministic model, or if stochastic differential equations are needed for adequately describing system behavior.

26.3.3 Examples

Mathematical models are increasingly being applied for optimizing design and operation of municipal wastewater treatment plants. Prior to the availability of these models, scale models (pilot plants) were used, which is very costly and time-consuming. Wastewater treatment usually takes place in activated sludge plants, where biomass (bacteria) in suspension (the activated sludge) purifies the wastewater. Many models of wastewater treatment plants use the so-called International Association on Water Quality (IAWQ) Activated Sludge Model No. 1 (Henze et al., 1987). In this model, the wastewater is disaggregated and characterized into several components that are considered relevant for modeling. The model defines the biological processes taking place in activated sludge systems and their stoichiometry (relative molecular quantities of chemical substances) and kinetics. Full-scale plants are typically modeled as a series of continuously stirred-tank reactors (CSTRs). Examples of software packages for modeling activated-sludge systems that implement the IAWQ model include SIMBA (1995) (developed by Ifak, Germany) and GPS-X (1994) (Hydromantis Inc., Canada). On the basis of input variables (e.g., influent flow and concentrations), stoichiometric and kinetic parameters (which are site- and wastewater-specific) and flowsheet parameters, these programs calculate output variables such as effluent concentrations, oxygen consumption, and surplus sludge production.

The problem of climate change is an extremely complex problem. The IMAGE 2.0 model (Alcamo, 1994), for example, is a multidisciplinary model designed to simulate the dynamics of the global society–biosphere–climate system, in order to provide scientific and policy-relevant information concerning climate change. The model can, for instance, be used to asses the possible consequences of different policies aimed at mitigating climate change. It consists of three linked submodels: energy–industry, terrestrial–environment, and atmosphere–ocean. The energy–industry models compute the emissions of greenhouse gases in a number of world regions as a function of energy consumption and industrial production. The terrestrial–environment models simulate the changes in global land cover based on climatic and economic factors, and the flux of carbon dioxide and other greenhouse gases between the biosphere and the atmosphere. The atmosphere–ocean models compute the build-up of greenhouse gases in the atmosphere and the resulting temperature and precipitation patterns. The IMAGE 2.0 model is a very large model. Among, for example, mathematical equations describing economic functions, it also contains a large number of differential equations describing certain physical phenomena, which are modeled in the way described earlier.

26.4 SYSTEM DYNAMICS

26.4.1 Introduction

The system dynamics approach to modeling and analyzing dynamic systems was developed by Forrester (1961) in response to the finding that many of the existing problem-solving methods did not provide enough insight into the strategic problems associated with complex systems. Forrester combined ideas from three areas (Meadows, 1976) that were relatively new at that time:

- Control theory; the concepts of feedback and self-regulation.
- Cybernetics; the nature of information and the role of information in control systems.
- Organization theory; the structure of organizations and the ways of decision making.

According to Wolstenholme (1989) system dynamics is:

> A rigorous method for qualitative description, exploration, and analysis of complex systems in terms of their processes, information, organisational boundaries and strategies; which facilitates quantitative simulation, modelling and analysis for the design of system structure and control.

During the earlier years of development of the method, applications were largely industrial (Forrester, 1961). Gradually the applications became broader, such as urban development (Forrester, 1969) and world dynamics (Meadows et al., 1972). During the end of the 1970s and during the 1980s the field expanded. At present there is a wide range of applications.

The basis of the system dynamics paradigm is the notion that the behavior of a system is largely caused by the structure of the system. The structure of the system contains elements such as feedbacks, gains, and time lags in the same way as may be found in the physical systems that were discussed earlier. The concept of feedback is one of the most important characteristics of system dynamics.

26.4.2 Outline of the Approach

26.4.2.1 Conceptualization. In system dynamics, formulating a diagrammatic conceptual model is seen as a separate and important phase in the development of a model. Developing and using the diagrammatic conceptual model is sometimes called *qualitative system dynamics*. The diagrams are often developed in close cooperation with domain experts and people involved in the problem situation.

A type of diagram used in system dynamics is a causal-loop diagram. A causal-loop diagram shows causal influences of variables on each other. The left-hand side of Figure 26.6 shows an example of a causal-loop diagram for a container of water with a flow into and out of the container. The signs adjacent to the arrows indicate the polarity of the influence. A variable A has a positive influence on B if A adds to B, or if a change in A results in a change in B in the same direction. A variable A has a negative effect on B if A subtracts from B, or if a change in A results in a change in B in the opposite direction (Richardson, 1986). Causal-loop diagrams are often used in the early stages of model conceptualization.

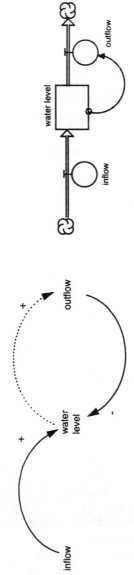

Figure 26.6 Causal-loop diagram and system dynamics flow diagram of the same system.

A special diagramming technique has been developed for system dynamics. These diagrams are called *flow* diagrams, *stock-and-flow* diagrams, or *pipe* diagrams. The most important concepts in a system dynamics flow diagram are levels and rates. Levels represent the parts of a system in which accumulation occurs. Rates are variables that cause the value of a level to change. For instance, CO_2 emissions cause the atmospheric concentration of CO_2 to change. In this example, CO_2 emissions represent a rate and the atmospheric concentration is a level. The right-hand side of Figure 26.6 shows an example of a system dynamics flow diagram.

In the causal-loop diagram in (the left-hand side of) Figure 26.6, a distinction has also been made between rate-to-level links (solid arrows) and proportional links (dotted arrows). This distinction is not often made in the literature and all lines are usually drawn as solid lines. Richardson (1986), however, who discusses a number of problems with causal loop diagrams, argues that making the distinction increases the likelihood that a simple causal-loop diagram can be "read" correctly and that its dynamic behavior to some extent can be inferred.

After a diagrammatic conceptual model has been drawn up, the mathematical relationships between the variables are determined. This may be done by the modeler (after having consulted domain experts) and/or in a group setting (Vennix, 1996). In mathematical terms, system dynamics is based on differential equations, where a level is equal to the integral of rate (or flow) variables over time.

26.4.2.2 Model Construction.
After having determined the form of the mathematical relationships, the parameters, initial values and values of the exogenous variables are determined and experiments to be carried out with the model are designed in cooperation with domain experts and people involved in the problem situation. Parameter and input values are obtained by measurements or by consulting people who have knowledge of the system. The system dynamics method is not meant for providing exact predictions or solutions, even though the variables are quantified. The aim is to analyze the general patterns of behavior of the system. The process of modeling is used as a way to improve understanding of the system. This understanding should change perceptions and increase the capability of those involved to react to future problems (Wolstenholme, 1989).

26.4.2.3 Modeling Tools.
Representing and solving system dynamics models is possible in any computer language. However, there is special system dynamics software available. Examples of packages are DYNAMO, ithink™ (STELLA™), Powersim®, and Vensim®. These allow easy representation of system dynamics models. First, the graphical model (usually a system dynamics flow diagram) is drawn on the screen, using special graphic tools, after which the equations, and parameter and input values are connected to the relevant parts of the diagram. The model is (implicitly) specified in terms of differential equations. To investigate the behavior of the model over time, a numerical integration method and step size has to be chosen by the user.

26.4.2.4 Validation.
Model validation in general was discussed in Section 26.2. As mentioned these, part of the formal validation of a model consists of comparing the behavior of the model to the behavior of the system being modeled using historical or experimental data. In system dynamics the emphasis of testing the behavior of the model is on pattern prediction (periods, frequencies, trends, phase lags, amplitudes) rather than on

point prediction. This is a logical result of the long-term policy orientation of system dynamics models (Barlas, 1996).

The role of "statistical significance" testing for validation of system dynamics models is controversial. The reasons for this are both technical and philosophical and are discussed in Barlas (1996). This, however, does not mean that statistical measures should not be used. Sterman (1984) and Barlas (1996) discuss several statistical measures that may be used in the investigation of system dynamics models.

26.4.2.5 Model Use.   System dynamics models are often used to investigate the behavior of an existing system and then to evaluate alternative solutions. These solutions may consist of alternative system structures or changing design parameters. Experiments with the model are carried out to investigate the impact of the alternatives. The alternatives can be identified by a group of people, or domain experts can be consulted separately by the modeler.

26.4.2.6 Conditions for Use.   Although some of the modeling tools allow discrete representation, system dynamics models are continuous time, and continuous event, models. The system dynamics approach is geared toward systems that are not purely technical/physical, but also contain socioeconomic aspects. The level of aggregation is usually higher than in physical systems modeling. Physical system models are usually aimed at providing detailed quantitative information about behavior of a system, whereas in system dynamics models the patterns of behavior are more important. System dynamics is used more often in a managerial context than in a hard engineering approach. A system dynamics study is usually aimed at providing managers with a deeper knowledge of the workings of a system, or of the ways in which they can improve system behavior, rather than providing numerical answers. A word of caution may be in order. Given the objectives of many system dynamics modeling exercises, and the associated choice of the type of variables, it is often difficult to perform rigorous empirical validation of the model. Care should be taken to always verify model-based insights and conclusions in the real-world system.

26.4.3   Examples

Many examples of managerial applications of system dynamics can be found in Roberts (1978). System dynamics models may also be part of interactive games, or "management flight simulators," as they are called in system dynamics. An example of a commercial management flight simulator is Boom & Bust Enterprises™, which allows users to play the role of management when a new product is launched. With a goal of maximising cumulative profits, users make decisions regarding marketing, pricing, and capacity expansion. The model contains various linked sectors (or submodels): a competitor sector, customer sector, price and marketing sector, distribution sector, financial sector, and capacity sector. The idea is that by taking risks and experimenting with new ideas, managers can learn about their business and industry.

Another example of the use of system dynamics is an investigation of the influence of so-called feebates to encourage the sale of cleaner vehicles in Southern California (Ford, 1995). With a feebate system, fees would be imposed on dirty vehicles to finance rebates for clean vehicles. The aim of the model was to investigate whether the state of California could operate a feebate system in a financially prudent manner. The model consists of three linked sectors (or submodels): a vehicle sector, a fund manager sector, and a utility sector. The utility sector represents the effect of extra electricity loads that would arise from the use of electric vehicles. The vehicle sector represents the decisions by people who purchase a new vehicle and is based on results of a survey. The fund manager sector simulates the operation of a state fund to encourage the sale of cleaner vehicles. The model is meant to be used interactively to help the planners learn about likely effects of incentives, rather than predicting the future.

System dynamics has also been applied to the development of improved resource management policies at the Minnesota Department of Transportation (Wile and Smilonich, 1996). The structure that was investigated consisted of operating units responsible for maintenance of approximately 1,600 miles of highway. The project was intended to broaden the perspective of management for more effective policy making. The model allows insights into the effectiveness of hiring policies, side effects of outsourcing strategies, effectiveness of training, and cross-bargaining of resources. The client team at the Minnesota Department of Transportation had much of the responsibility for testing, validating, and calibrating the model. A management flight simulator was developed to allow other teams to experientially learn some of what the core modeling team learned during development.

26.5 DISCRETE EVENT SIMULATION MODELS

26.5.1 Introduction

Systems can be characterized by defining the variables that describe the system and by determining the relationships between these variables. In physical systems modeling and in system dynamics this is done by using continuous mathematical (differential) equations. For discrete-event simulation models, a slightly different approach is taken: a set of variables is used that describe the system at discrete points in time. The values of these variables define the state of the system. The state of a system at a particular moment can be defined as the set of relevant properties that that system displays at that time (Shannon, 1975). Events can be defined as the points in time where a state change occurs, or defined as an instantaneous occurrence that may change the state of the system (Law and Kelton, 1991). The challenge in making discrete-event models is defining the set of variables that gives an adequate representation of the system, and describing the events, or the activities or processes that lead to events, and their effect on the values of system variables. Because the occurrence of an event and the time of an event can usually not be predicted exactly, stochastic representation plays an important role in most discrete-event (simulation) models. The main difference between several forms of discrete-event simulation models is formed by the way the simulation time advances. The simulation time is part of the state of the system and stored in a variable called the *simulation clock*. The simulation clock can be incremented by fixed time intervals or by variable time intervals. Fixed intervals are, for instance, applied in economic models that are modeled on a monthly or yearly basis. Variable intervals are used in most discrete-event simulation software languages. After changing the state of the system when an event has occurred, the simulation clock is advanced to the time the next event will occur. In a model of a post office, the simulation clock might jump instantaneously from 18:00 to 07:00 the next day if no events within the model occur in between. Event-driven models are set up using the idea that "nothing relevant" happens between successive state transitions (Kreutzer, 1986).

When studying systems using discrete-event simulation models, a discrete worldview has to be used when modeling the system. Therefore, discrete objects from the system are often chosen as the focus of attention. Take a freeway with cars as an example. When looking at this system from a continuous point of view, we model the number of cars on a stretch of road, the properties of the road, and the continuous influence of these variables on the mean speed of the cars, which, in its turn, influences the number of cars. When looking at the system from a discrete point of view, we "see" individual cars with their properties (speed, lane) and their behavior (change speed, change lane) as a result of events (state changes from the environment, state changes of other cars, e.g., brake lights from the previous car). In other words, objects or entities within the system with attributes and behavior are usually the main building blocks of discrete-event simulation models.

26.5.2   Outline of the Approach

26.5.2.1   Conceptualization.   In the conceptualization phase, a consistent set of structure elements for describing the system is chosen as well as a modeling language to formally represent these structure elements. The initially vague knowledge of the system is translated into a written or schematized description, including the tentative system boundaries. When a discrete-event simulation model has to be constructed, the "object−attribute− action" worldview is an example of a fruitful way to conceptualize the system (see Dur, 1992; Sol, 1982). This worldview is close to the object-oriented view, well known from the software engineering field. Object-oriented programming, however, usually includes more aspects like inheritance, encapsulation, and polymorphism, which are not always needed for making a simulation model. Using the object−attribute−action worldview, the structure of the system is modeled by identifying the discrete objects in the system and describing their properties in terms of attributes and their behavior in terms of actions. Another approach can be characterized as the "resource-oriented approach." In that case some of the objects are followed on their path through the system, and everything that happens to these objects is included in the model.

To illustrate the difference between the active (action) and the passive (resource) way of description, a factory can be taken as an example. The active view will study the machines and look at the effects each machine has on the attributes of the materials processed. The passive view will follow the materials through the factory as they are processed by the machines (resources) from raw material into finished goods. Usually, a combination of both views works best, where one view is chosen to write down the conceptual model and the other view is chosen to check the completeness of the model. The choice of the primary view and the description language often depends on the tools that are used to implement the model in the model construction phase. During conceptualization the model should be structured in such a way that the efforts for detailed, low-level data gathering for creating the quantitative model can be focused and minimized. For creating a model of the problem situation, data have to be gathered from the system by using techniques such as observation, measurements, studying reports, conducting interviews, brainstorming, and other types of group meetings.

26.5.2.2   Model Construction.   For investigating the model and thereby describing, diagnosing, or predicting the behavior of the system under different circumstances, the building blocks of the conceptual model have to be detailed and implemented in some model construction (simulation) language. One of the first steps, which usually takes place before or during the conceptualization phase, is to choose the language and the software tools that are used for implementation. The most difficult and time-consuming task is to gather data from the system. Therefore, one of the most important tasks in model construction is reduction. Reduction can have several forms: leaving out less relevant object classes or attributes, combining different object classes into a new class (e.g., combine bus, car, truck into one new object class, vehicle), leaving out actions, or replacing a complex set of measurements by a variable or attribute drawn from a distribution function.

When reducing the model by using a stochastic variable instead of a set of causal relations, it is not necessary to incorporate all (possible) measurements from variables within the system into the model. One could say that instead of using the white-box approach for that part of the system (the causality for the appearance of the value of a variable is modeled), a black-box approach is chosen (the distribution of the values of the variable fits, but the relationship with the rest of the model is lost). When modeling the processing time of a machine in a factory, for example, it is possible, but usually not necessary, to model all factors that cause the production time to be a little longer or a little shorter. Instead, a distribution function (either user-defined or standard, like Normal or Gamma), based on measurements of production times under different circumstances, is usually sufficient to describe that production time in a larger model. Of course, when the aim of the simulation model is to study the machine and the fluctuations of the throughput in detail, a stochastic function is not sufficient. The objective of the simulation study determines where reductions can be applied and where they cannot be applied.

26.5.2.3 *Modeling Tools.* Five components are common to all discrete-event simulation languages and executable simulation models (Kreutzer, 1986):

1. A description of model structure, with objects, relationships, and actions; the product of the conceptualization.
2. A clock for time management.
3. Random processes for describing less essential aspects. Instead of describing each process in detail, some of the processes are described as "in 13 percent of the cases A happens, and in 87 percent of the cases B happens," abstracting from the causes that result in either A or B.
4. Facilities for gathering, reporting, and analyzing data from the simulation. Usually, statistical instruments are necessary to analyze and compare simulation results.
5. Some kind of run-time control system that keeps track of state and time, and enables the other four components to work together.

The main difference between discrete-event simulation languages and tools is formed by the way the model's behavior is arranged. We can distinguish between event-driven simulation, where the events themselves are modeled; process-driven, where the emphasis is on the flow of entities through the model; object-oriented, where the objects encapsulate attributes and methods, and interaction is handled through messages; and activity-scanning, where starting and finishing actions for procedures are defined and the run-time control system scans the conditions of all starting and finishing entry points at each event (Kreutzer, 1986). For the user of the tools, other aspects are important as well, such as the way the model has to be entered, and debugging, data entry, and statistical analysis tools that are available.

Some examples of well-known discrete-event simulation languages and tools are GPSS (Schriber, 1974), Simscript II (Russel, 1976), Arena/Siman (Pegden et al., 1990), Slam II (Pritsker, 1986), AutoMod and MODSIM II®.

26.5.2.4   *Validation.* Model validation in general was discussed in Section 26.2. Before experiments with the model can take place, the model has to be checked to see whether it is a (statistically) valid representation of the system that is being studied. If this check is passed, the model can be used to identify causes and effects. With regard to replicative validation of discrete-event simulation models several statistical techniques are often used, like comparing the means of variables using the *t*-test, and comparison of variances using the *F*-test. With this approach, some problems arise, because in most cases, several actions that occur in the real system, have been deliberately excluded from the model or reduced. Measurements from the real system include the effects from these actions, which the output variables lack. Therefore, a failure of the statistical tests does not automatically mean that the model is invalid.

26.5.2.5   *Model Use.* The simulation models can be experimented with in order to study the (effects of) changed input variables or alternatives in detail. In order to study the effects of the alternatives, a simulation experiment is conducted. In the experimentation phase, the simulation model is exposed to a so-called treatment (Sol, 1982), which consists of the following elements:

- Specification of input data;
- Collection of input data;
- Initialization conditions;
- Run control conditions,
- Specification of output data.

A *run* or *replication* is the combination of a treatment and the collected output data. Usually, several replications with the same treatment are carried out, just with different initialization conditions, that provide statistical independence. An experimental design is a specific set of treatments. An experiment is a set of replications with the same treatment. The results of the experiments are analyzed by conducting statistical tests to compare alternative models with the as-is simulation model or to compare simulation results from different experiments.

26.5.2.6   *Conditions for Use.* Discrete-event simulation models are usually used in situations where it is relevant to take the behavior and properties of discrete objects in the system into account. Examples are simulation studies of factory processes, logistics problems, or even business problems (Dur, 1992; De Vreede, 1995). Many simulation tools allow for visualization (animation) of the results, which increases the usability of discrete-event simulation for many application domains.

One of the conditions for use of discrete-event simulation is to have a team that is able to structure the system and the model, and that has at least one person who is thoroughly trained in statistics. Statistics are an integral part of a discrete-event simulation study and are needed for choosing, defining, and using distributions for reduction of complex variables and for input probabilities, for experimental design (run length, number of replications, random streams), and for the analysis of simulation output data (Law and Kelton, 1991).

26.5.3 Example

A brief example is given here of a discrete-event simulation model for a steel company, worked out in more detail in Verbraeck and De Vreede (1993). The company had some problems with the conveyance of continuously casted steel slabs to train wagons, using two double conveyors and four cranes. Two semiportal cranes transport the red-hot slabs that weigh 10–20 tonnes, from the end of the conveyor belt to 16 stack positions that are used for sorting the slabs. Two full-portal cranes transport entire stacks to the train wagons. The main problems were that the throughput of the system was too low, that cranes could not operate continuously because they moved through each other's working area, and that safety regulations require halting the cranes when trains are being shunted. The aim of the study was to find the optimal sequence of crane movements, without breaking the safety regulations. Furthermore, the steel company was interested in the behavior of the system under heavy load.

The conceptual model identified the following entities: slab, test slab, conveyor belt, semiportal crane, full-portal crane, stack position, wagon, train, controller. For each object, the main attributes and processes were identified. In the specification phase, reduction was applied to the conceptual model. Less relevant attributes of, for example the slab were left out, such as temperature, serial number, and steel quality. Some entity types were left out as well: the train was, for example, not modeled. Interactions, such as decision processes for choosing stack positions or exact train wagons, were simplified by using distribution functions. Some attributes were replaced by stochastic values as well (breakdown duration of cranes, time to pick up a slab). The entity types "slab" and "test slab" were combined into one entity type.

Data were gathered from the system for the model components that remained. The simulation language Siman was chosen for implementation of the model (Pegden et al., 1990), and the model was decomposed into submodels that were implemented in the Siman environment. Animation layouts were added as well, both for debugging of the model and for discussing model behavior and model results with employees of the steel company. Face validity tests were conducted involving experts from the problem domain, like crane operators, controllers, and engine drivers. The replicative validity was tested by comparing model outcomes with real-world data. After the model was considered valid, several experiments were conducted with different scenarios using faster cranes, other safety measures, other working areas for the cranes, and so on. The model outcomes helped the organization to rank the different alternatives and to get a feeling for the effectiveness of the proposed changes in the slab yard.

26.6 TIME-SERIES ANALYSIS

26.6.1 Introduction

A time series is a historical series of data. The series of data can be fitted with a mathematical model, which may then be applied for prediction purposes. Time-series analysis differs from causal modeling in that the predictions are based only on historical data and not on causal relationships between variables (based on accepted or proven theories), as in the models that were discussed earlier. There is a large variety of methods for time-series analysis, such as smoothing methods, decomposition methods, autoregressive integrated moving average (ARIMA) models, and state-space models. Makridakis et al. (1983) discuss a large number of different methods. Smoothing and decomposition methods are relatively easy methods, whereas ARIMA models are more complicated. However, the latter also have a more thorough statistical basis. A comprehensive text on ARIMA models is Box et al. (1994). Other texts on time-series models are Wei (1990), Kendall and Ord (1990), who also discuss recent developments in time-series analysis, and Vandaele (1983) who devotes three chapters to using ARIMA models for real-time series. The general concepts concerning ARIMA models (or Box–Jenkins models) will be explained below. Following this, several other methods will be mentioned briefly. A problem in time-series analysis is the choice of the most appropriate method. Some general indications about when to use certain methods will also be given. The main source used for this section on time-series analysis is Makridakis et al. (1983).

26.6.2   Outline of the Approach

ARIMA models have a wide range of application. They are a combination of two kinds of models autoregressive (AR) models and moving average (MA) models. A pure AR model has the following structure:

$$Y_t = a + b_1 Y_{t-1} + b_2 Y_{t-2} + \cdots + b_p Y_{t-p} + e_t \tag{26.1}$$

where the Y_t is the value of the time series at time t, and e_t is the error term at time t. The order of the AR model is equal to the number of time lags p that are needed to describe the time series.

A pure MA model has the following structure:

$$Y_t = a + b_1 e_{t-1} + b_2 e_{t-2} + \cdots + b_q e_{t-q} + e_t \tag{26.2}$$

Note that the term relates to a moving average of the error terms, whereas in other contexts the term moving average often relates to calculating the average of a number of data points along the data series itself.

A combination of these two models (AR and MA) allows many time series to be appropriately fitted. However, this combination does not take into account time series in which the (local) mean and/or variance do not remain the same, that is, nonstationary models. A series has to be stationary in order to model it with a combination of an AR and MA model. This is where the integrated part of an ARIMA model comes in.

An example of a model that is nonstationary in the mean is the following:

$$Y_t = Y_{t-1} + e_t \tag{26.3}$$

This model has a similar appearance to the AR model previously described; however, the coefficient belonging to Y_{t-1} is equal to 1. Equation 26.3 can be rewritten to show that $Y_t - Y_{t-1}$ forms a random model. Therefore, instead of using the time series Y_t, the series

$Y_t - Y_{t-1} = W_t$ is used in the time-series analysis, provided this series is stationary. If this first difference is not stationary, the series is differenced again until it is stationary. The word integrated as used in an ARIMA model is confusing, and refers to the process of differencing. The number of times the series is differenced is denoted by d.

The standard notation is ARIMA (p, d, q), where the p relates to the order of the AR part, the d to the degree of differencing, and the q to the order of the MA part of the ARIMA model. In practice, a time-series analysis can usually be carried out using an ARIMA model with values of $p, d,$ and q of 0, 1, or 2.

An ARIMA $(1, 0, 1)$ model has the following structure:

$$Y_t = \phi_1 Y_{t-1} + \mu' + e_t - \vartheta_1 e_{t-1} \qquad (26.4)$$

Where μ' is not quite the same as the mean of the Y series, but is equal to $\mu - \phi_1\mu$. The first part of the model is the AR part and the last part is the MA part. The series is assumed stationary in the mean and in the variance.

When a time series is available, and the objective is to use the series for forecasting purposes, an appropriate model structure has to be chosen. The model structure is determined by choosing values for $p, d,$ and q. There are several methodological tools that are used in time-series analysis using ARIMA models: calculating autocorrelations, calculating partial autocorrelations, and constructing a line spectrum of the time series. By combining the results of these calculations, an informed choice of model structure can be made. There are some rules of thumb that will often enable a choice of model structure to be made. The tools will first be explained, after which the steps to be taken in the Box–Jenkins method of time-series analysis will be discussed.

- *Autocorrelation.* The autocorrelation coefficient r_k for $1, 2, 3, \ldots, k$ time lags can be calculated as follows:

$$r_k = \frac{\sum\limits_{t=1}^{n-k} (Y_t - \bar{Y})(Y_{t+k} - \bar{Y})}{\sum\limits_{t=1}^{n} (Y_t - \bar{Y})^2} \qquad (26.5)$$

In the equation, $\bar{Y}$ denotes the mean of the time series. This is the correlation of the time series with itself, lagged by k time steps. In nonseasonal (nonperiodic) time series, the largest autocorrelation will usually be the coefficient corresponding to one time lag. Autocorrelations can be used to determine whether there is any pattern (AR, MA, ARMA, or ARIMA) in a set of data.

- *Partial Autocorrelation.* Partial autocorrelations are used to measure the degree of association between Y_t and Y_{t-k} when the effects of other time lags (i.e., up to $k-1$) are partialed out. The partial autocorrelation at lag k is the correlation between Y_t and Y_{t-k} in excess of that already explained by autocorrelations at lower lags. The partial autocorrelation coefficient of order p is defined as the last autoregressive coefficient of an AR(p) model. In the following equations the partial autocorrelations are given by ϕ':

$$Y_t = \phi_1' Y_{t-1} + e_t$$

$$Y_t = \phi_1 Y_{t-1} + \phi_2' Y_{t-2} + e_t \qquad (26.6)$$

$$\cdots$$

$$Y_t = \phi_1 Y_{t-1} + \phi_2 Y_{t-2} + \cdots + \phi_{p-1} Y_{t-p+1} + \phi_p' Y_{t-p} + e_t$$

It is quite time-consuming to calculate the partial autocorrelations, therefore estimates are usually calculated (using the so-called Yule–Walker equations). The partial autocorrelations are constructed to fit an autoregressive process. When the partial autocorrelations do not exhibit a drop to random values after a number of time lags, but decline exponentially, it can be assumed that the generating process is a moving-average one.

- *Line Spectrum or Fourier Analysis.* The time series can also be decomposed into a set of sine waves of different frequencies. The result will be a diagram in the frequency domain, which shows the amplitudes of the sine waves at the different frequencies. This can, for example, be helpful in identifying randomness and seasonality in a time series. Randomness will show up when the amplitudes are the same for all frequencies, and seasonality may be present when certain frequencies are predominant in the series.

Box and Jenkins (1976) have put together the relevant information required to understand and use univariate ARIMA models. Their approach consists of three phases: identification, estimation and testing, and application. The approach is called the Box–Jenkins method. These phases are discussed under the same headings as the models that were discussed in the previous sections, that is, developing a conceptual model, model construction, validation, and model use.

26.6.2.1 Conceptualization.
When confronted with a temporal series of data with the aim of forecasting, the first step is to plot the data series. The visual display can, for instance, show whether there is a sudden change from one year to another, which may have a definite reason, and it is thus not advisable to use a time-series analysis. A choice as to the most appropriate method of time-series analysis should also be made.

Next, it will have to be determined whether the data are stationary. This may be done by investigating the autocorrelations. In stationary data the autocorrelations will drop to zero after the second or third lag, whereas in nonstationary data they are significantly different from zero for several time periods. Nonstationarity should be removed before developing the actual model. This is done by differencing, as was explained earlier. If differencing once is not sufficient to achieve stationarity, the differenced series is differenced. In practice it is seldom necessary to go beyond second differences.

The presence of seasonality is then investigated. Seasonality can be found in a stationary series by identifying those autocorrelation coefficients of more than two or three time lags that are significantly different from zero. In the line spectrogram, seasonality will manifest itself in large amplitudes at specific frequencies. In exactly the same way that consecutive points might exhibit AR, MA, ARMA, and ARIMA properties, so data separated by a whole season may exhibit the same properties. Thus, once it has been established whether seasonality is present, this will have an influence on the next step, that is, choosing a model. If the seasonality is present every s time lags, in addition to looking at autocorrelations and partial autocorrelations for consecutive time lags, we also look at the autocorrelation and partial autocorrelation pattern only regarding time lags s, $2s$, $3s$, ... The ARIMA notation can be extended to handle seasonal aspects, and the general notation is: ARIMA (p, d, q) $(P, D, Q)^s$, where the small letters indicate the nonseasonal part of the model and the capital letters indicate the seasonal part of the model.

Following this, a model structure is chosen. The process of identifying a Box–Jenkins ARIMA model is difficult and requires experience, but there are some general principles. If the process is an autoregressive one of order one, its autocorrelation coefficients will decline to zero exponentially. If the order of the AR process is two, then the autocorrelation coefficients will show damped sine-wave decay. Furthermore, if the process is autoregressive, the partial autocorrelations may be used to determine the order of the autoregressive process. The order is equal to the number of significant partial autocorrelations.

If the generating process is MA rather than AR, then the partial autocorrelations will not indicate the order of the MA process, since the partial autocorrelations are constructed to fit an AR process. In fact, they introduce a dependence from one lag to the next that makes them decline to zero exponentially when the order of the MA process is one, and the partial autocorrelations will show a damped sine-wave decay when the order of the MA process is two. The autocorrelations can be used to determine the order of the moving average process. The order is equal to the number of significant autocorrelations.

In a combined ARMA(1,1) process, both the autocorrelations and the partial autocorrelations will decay exponentially.

26.6.2.2 Model Construction.
After having chosen a first model structure, the parameters have to be determined. The parameters are usually estimated iteratively. If parameters have to be estimated for nonlinear functions, then ordinary least-squares procedures may not apply. Minimizing the sum of squared residuals is the usual criterion (as in linear estimation), but nonlinear estimation is an iterative procedure and there is no guarantee that the final solution is the global minimum.

26.6.2.3 Validation.
After the parameters of the model have been identified, the performance of the model is investigated:

- Study the residuals to see if any pattern remains unaccounted for. The residual errors should be random noise, which should have no significant autocorrelations, no significant partial autocorrelations and consistently high amplitudes across the range of frequencies in the line spectrum.
- Sampling statistics are studied to investigate whether the model can be simplified. The optimum values of the coefficients should have small standard errors and be almost uncorrelated. If a coefficient is almost zero, model order is reduced and the summary statistics are calculated again; if this does not have an effect on the statistics, the simplified model may be used.

26.6.2.2 Model Use.
When using the model to forecast $Y_{t + 1}$, a term involving $e_{t + 1}$ will be needed. However, the error term one time-step ahead is not known, and is therefore set at zero. As forecasts are made further away, more error terms will have to be set at zero. Furthermore, when forecasting further ahead, the forecast will have to be based on previous forecasts instead of on known data, since the equations will contain future values of Y, that is, forecasts.

26.6.2.3   Other Methods of Time-series Analysis.   Apart from the Box–Jenkins method, there are many methods that may be used for time-series analysis, such as smoothing and decomposition methods, of which many varieties exist. The following description of exponential smoothing and decomposition methods is based on Makridakis et al. (1983).

- *Exponential Smoothing Methods.* In order to predict a future value, exponential smoothing methods use a weighted average of past values of the time series, where the weights decrease exponentially as the observations get older. The easiest way of exponential smoothing is stated below:

$$F_{t+1} = F_t + \alpha e_t \tag{26.7}$$

where e_t is the forecast error (actual Y_t minus forecast F_t) for period t. The reason why this is called exponential smoothing can be seen by recursively substituting for F_t:

$$F_{t+1} = \alpha Y_t + (1 - \alpha)F_t$$
$$F_{t+1} = \alpha Y_t + (1 - \alpha)\alpha Y_{t-1} + (1 - \alpha)^2 F_{t-1} \tag{26.8}$$
$$F_{t+1} = \alpha Y_t + (1 - \alpha)\alpha Y_{t-1} + (1 - \alpha)^2 \alpha Y_{t-2} + (1 - \alpha)^3 F_{t-2}, \ldots$$

In order to calculate the value for F_{t+1} a starting value F_1 is needed. One method of initialization is to use the first observed value Y_t as the first forecast. A choice for parameter α has to be made, based on minimizing a certain criterion. A large α provides little smoothing of the data, and a small α will cause more smoothing. This method, called *single exponential smoothing,* is not suitable for nonstationary data. For nonstationary data a forecast of the trend is also needed. There are various ways of including trend, for example, it may be done by treating trend individually with a new parameter. However, there are also methods that still only involve one parameter, such as Brown's one-parameter linear method, which includes double exponential smoothing (exponential smoothing of smoothed data). Most exponential smoothing methods are not suitable for data including seasonality. However, there are methods, such as Winters' three-parameter trend and seasonality method, which takes seasonality into account using a seasonality index.

- *Decomposition Methods.* The decomposition approach decomposes the time series into cyclical (C), seasonal (I), trend (T), and random (E) subpatterns. The difference between cyclical and seasonal patterns is that seasonal patterns have a constant, relatively short time duration, or length, whereas cyclical patterns are due to longer term (economic) fluctuations, and therefore vary in length. All components are analyzed individually and recombined to obtain predictions. The general notation is $Y_t = f(I_t, T_t, C_t, E_t)$. The exact form of the function depends on the decomposition method that is used. Decomposition is carried out in a number of steps. A multiplicative form may be assumed $Y_t = I_t * T_t * C_t * E_t$ as is, for example, done in the ratio-to-moving-averages method. First, a moving average of the original data set is calculated using the length of the season as the length of the moving average. In this way, seasonality is averaged out of the series. Some of the randomness will be averaged out as well. The averaged series can then be used to isolate trend, by identifying the appropriate form of the trend (linear, exponential, etc.). When the trend has been identified, the value that can be attributed to the trend can be calculated for each of the points in time, using the (linear, exponential, etc.) formula that has been identified. When dividing the averaged data points (actual data points without seasonality) by the calculated trend data points, only the cyclical factor of the data will be left. Estimating a cyclical factor is difficult and requires some knowledge of the level of economic or industry activity during the period to be forecast. When dividing the actual data points by the averaged data points without seasonality, a seasonal factor will be left for each data point. The average seasonal factor for certain points can then be calculated (e.g., average seasonal factor for the month June could be 0.8).

A new forecast may be made by calculating the trend value and multiplying this by the relevant seasonality and cyclical indices for that specific point in time.

- *State Space Models Methods.* Another broad class of models is the class of state-space models, for which optimal forecasts can be computed using a recursive estimation procedure called the Kalman filter. The latter is used widely in control engineering. The state-space form allows a wide range of time-series models to be handled. State-space models and the use of the Kalman filter is explained in detail by Harvey (1989) and Wei (1990).

26.6.2.4 Conditions for Use. The objective of a time-series analysis is to discover a pattern in the historical data series and use that pattern to predict future values. Time-series models are black-box models that are meant to allow future values of variables to be estimated rather than giving insight into the reasons for the behavior of the system, as do the other, so-called white-box modeling methodologies that have been discussed in this chapter. Time-series analysis can be used when sufficient data are available, and it is either not necessary or not possible to model the causal structure. It may not be necessary to develop a white-box model, for example, because it may be too time-consuming in relation to the goal of the model, and it may not possible to develop a white-box model due to a lack of theory. Time-series models can often be used more easily to forecast, whereas causal models can be used with greater success for policy and decision making (Makridakis et al., 1983). Time-series analysis should only be used when there is reason to believe that underlying patterns or relationships remain the same.

The decomposition and ARIMA methods can deal with a wide variety of patterns. Decomposition methods are the strongest for dealing with cyclical components. With regard to the time horizon of forecasting methods, smoothing methods are usually best for the immediate or short term, and decomposition and ARMA methods are usually better for the short to medium term (Makridakis et al., 1983). The type of data is also important; with microdata, exponential smoothing methods are more appropriate than more statistically sophisticated methods. When many analyses have to be carried out, ARIMA methods may become quite time-consuming. According to Harvey (1989), a practical problem with ARIMA models is that unless one has some experience in time-series analysis, it is easy to select an inappropriate model. There is some discussion about the accuracy of ARIMA methods in domains of business and economic applications where the level of randomness is high and where constancy of pattern cannot be assured. Makridakis and Hibon (1995) conclude that an important factor in forecasting accuracy of an ARIMA model is the way the data are made stationary in its mean, and discuss how this can be improved.

Many adaptations to the Box–Jenkins method have been described in the literature (e.g., the *International Journal of Forecasting*) and state-space models methods have become increasingly popular when a thorough analysis of a problem is required.

26.6.3   Examples

Wu et al. (1991) describe a case study in which they investigated whether time-series methods can be appropriate for business planning. They compared Box–Jenkins models and an adaptive smoothing method to the judgmental forecasts of company planners of a computer company concerning product shipments. Various seasonal Box–Jenkins transfer-function models were investigated. The best Box–Jenkins models were better than those obtained from the judgmental forecasts.

Two cases for which the Box–Jenkins method was applied in the way described in this chapter were carried out in the area of hydrology (Van der Kloet and Van Geer, 1983). One of the cases is the forecasting of groundwater levels and the other is forecasting rainfall runoff (into sewers). For the forecasting of groundwater levels data of two locations in The Netherlands were available. For each location, the groundwater levels had been recorded once a month for 17 years, and 150 of the 204 data points were used for the identification of the ARIMA models. Seasonality was described by a cosine function and subtracted from the data before identification of the models, instead of using a seasonal model. On the basis of corellograms, both series were differenced once. For each location several ARIMA models were considered, which were then validated against the unused data, by investigating the residues. Finally, one location was modeled with an ARIMA(0,1,2) model and the other by an ARIMA(2,1,2) model. The rainfall runoff model uses the Box–Jenkins method in another way. Here, the model is a combination of a deterministic model and an ARIMA model. An area of 2 hectares was modeled, for which historical data were available. First, a deterministic physical systems model, consisting of a (Nash-)cascade of vessels was used to calculate the response of the system to rainfall. The error between this model and the actual values (based on a series of historical data) was then represented by an ARIMA model. By adding the forecasted error calculated by the ARIMA model to the physical systems model, more accurate results could be obtained than by using only the linear physical systems model.

26.7   ECONOMIC MODELS OF COSTS AND BENEFITS

26.7.1   Introduction

Economic aspects play an important part in any system design or evaluation effort. When functional performance specifications are fixed, minimum-cost design and cost control during design, realization and use are a prime concern. These issues are discussed in Chapter 6 of this handbook. More often, the choice among alternative designs or systems involves tradeoffs between functional performances and costs. In both public and business environments, the balance of the benefits obtained from system performance and of the costs of system realization and operation plays a central role in decision making about the system. While the modeling approaches described in the preceding sections primarily focus on capturing the essential behavior of a system, they do not address the issue of evaluation of expected system performance where, typically, a variety of costs and benefits occur over time. That is the subject of the present section.

There are a number of problems associated with modeling, estimation, and interpretation of costs and benefits of alternative systems or policies. We discuss a limited selection of issues here. Rather than attempting to cover the broad field of principles and theories for modeling economic systems, we will limit ourselves to a number of practical principles and concerns directly related to evaluating and comparing alternative systems or policies. For more in-depth treatment, the reader is referred to one of several texts in the field (Mishan, 1976; Sugden and Williams, 1978; Bussey, 1978; Sassone and Schaffer, 1978; Sage, 1983).

26.7.2 Time Preference

Costs and benefits related to the design, realization, operation, and phase-out of systems generally occur during the full period of a system's life-cycle. Typically, an important fraction of the costs will be incurred during system design and realization, while the major benefits are expected during system operation. As, in general, availability of goods in the present is preferred over availability of the same goods at some time in the future, this poses the question of comparability of costs and benefits that occur at different points in time. The general approach is to compute equivalent present values of benefits and costs that occur at some future time using the principle of discounting:

$$V_0 = \frac{V_t}{(1 + r)^t} \tag{26.9}$$

where
V_t = the (monetary) value of costs or benefits at year t in the future
V_0 = the equivalent present value
r = the discount rate
t = the number of years
The discount rate r expresses the difference in value attributed to availability of a good a year from now to availability of the same good at the very present moment. If r is, for example, given a value of 0.1, availability of a good a year from now is valued at 10 percent less than availability now.

Using this discounting principle, an overall measure of economic value of a system or project can be computed if cost and benefit estimates over its entire life cycle are available. For example, the net present value (NPV) of a system or project is computed as

$$\text{NPV} = \sum_{t=0}^{n} \frac{B_t - C_t}{(1 + r)^t} \tag{26.10}$$

where
n = the system's or project's lifetime
C_t = estimated system costs in year t
B_t = estimated system benefits in year t

If the NPV is positive, the project is said to be worthwhile to undertake from an economic point of view.

An alternative measure of performance is the benefit–cost ratio, computed as the ratio between total benefits at present value to total costs at present value. If the benefit–cost ratio is larger than one, the project is said to be worthwhile. It is pointed out that, often, net present values or cost–benefit ratio's are used for economic comparison and ranking of alternatives rather than for absolute judgments about their desirability.

The determination of the discount factor is not trivial. Yet, the value used may significantly affect the computed present values and rankings of alternatives, particularly if projects or systems with significantly different time patterns of costs and benefits have to be compared. There is no one agreed-on procedure for determining the value of the discount factor. Time preference of individuals may be different. For a business enterprise able to freely borrow and lend in the capital market, the present and expected cost of obtaining capital (i.e., the risk-free market interest rate) may be an appropriate indicator. If the discount rate is set equal to the market interest rate, projects with an expected economic return on investment less than the market interest rate will have a negative NPV, or a cost–benefit ratio less than one. In such a case, it would be more profitable to lend money at the market rate (or not borrow it) rather than invest it in the project. In practice, additional concerns that are, for example, related to the risk in undertaking a new project, generally lead businesses to require a higher project yield than the financial market rate.

For public systems or projects, factors other than financial come into play. A public agency may not be able to freely borrow and lend money in financial markets. The decision maker's own time preferences, social concerns, and environmental concerns may also be important in public decision making. The use of a (high) discount rate has been criticized from the perspective of long-term environmental sustainability. A discount rate, based on economic rationalities, effectively eliminates potential long-term benefits from the assessment. Environmentalists have therefore proposed to use lower or no discount rates in cases where long-term environmental effects of projects or systems are valued in financial terms.

The systems or policy analyst may wish to select a range of values to explore the sensitivity of the assessment to the discount rate. In practice, to ensure comparability among projects, many governments have set a required test rate of discount for evaluating and comparing public projects.

26.7.3   Costs

Project or system realization requires inputs of different kinds, such as labor and equipment. Market prices provide a good starting point to estimate costs of inputs that must be bought. However, often facilities already available to the firm will be used as well. Part of the costs of these have already been accounted for in the context of other uses. Therefore a different approach is needed to estimate their costs. One approach is to use the marginal costs of using the facilities, for example, extra depreciation and energy/maintenance costs of equipment. Marginal cost figures may not be readily available, as traditional financial accounting has mainly been concerned with average costs. Another possibility is to estimate the so-called opportunity costs of using the inputs. These are defined as the costs of what is foregone if the project is realized. For example, a piece of equipment may either be used in project A, or be used in another project B, yielding a benefit of, say, $10,000. Then the opportunity cost of using the equipment in A would equal $10,000. Note that the opportunity cost approach will generally not lead to the same results as the marginal cost approach.

In more complex cases, it may be necessary for the analyst to derive a cost function or cost model, based on insight in the various cost components of the products or systems under investigation.

For public investments, another issue deserves attention. Part of the market costs of many inputs (goods, labor) are incurred because of government taxes. For example, if a project contemplated by a government agency involves labor costs, part of these expenses will flow back to the government as tax revenues. A correction of the market price may be in place in such cases.

26.7.4 Distribution of Impacts, Market Effects

Intended benefits of investments in many public systems such as defense and transport are other than expected revenues for the government. This provides a number of additional complications for cost–benefit analysis. Both the estimation of economic impacts (positive and negative) and the distribution of benefits over different parties are relevant. This means that models are needed to identify:

- The parties involved in or affected by the system.
- The impacts on markets, particularly on market prices and, through changing market prices, on demand.
- The benefits accruing to different market parties.

In an open and free market, changes in supply of a good, for example, a transport facility, will affect the price of that good, and hence the income of the new supplier as well as that of other suppliers. For consumers, increased supply may lead to price reduction and hence to a benefit. Also the number of consumers may change as a result of price changes. A market equilibrium model, based on information on supply and demand elasticities may be used to estimate the extent of these changes. Generally, a distinction is made between the consumer surplus and the producer surplus. Both are derived from comparison of the market situation and benefits to each of these groups before and after the new system or project has been realized.

26.7.5 Indirect Effects and Input–Output Analysis

The exploitation of a new facility will not only lead to changes in the market for which it produces goods or services, it will also lead to increased demand for its inputs, for example, raw materials, equipment, and labor. As a result, turnover in these other economic sectors will increase, leading to additional changes in demand for their inputs, and so on and so forth. The description of these mutual interdependencies between economic sectors is the core idea of economic input–output models (Leontief, 1966). The basic conceptualization of the economic system is that each sector j, in order to produce one unit of its product j, needs A_{ij} units of product i as input. Then, for the economic system to be in equilibrium, the total production of sector i, X_i, must equal the total use of product i by all sectors plus a quantity used for final consumption, denoted as Y_i:

$$X_i = \sum_{j=0}^{N} A_{ij}X_j + Y_i \tag{26.11}$$

where N is the number of sectors.

Then, as under economic equilibrium conditions, this equality holds for all sectors i, it follows that

$$\mathbf{X} = A\mathbf{X} + \mathbf{Y} \tag{26.12}$$

where $\mathbf{X}$ and $\mathbf{Y}$ are vectors and A is a matrix.

This is the basic equation for economic input–output analysis. The dependencies A_{ij} are called *technological factors*. A next step is to solve Equation (26.12) for $\mathbf{Y}$:

$$\mathbf{X} = (I - A)^{-1}\mathbf{Y} \tag{26.13}$$

Thus, by computation of $(I - A)^{-1}$ (the so-called Leontief-inverse), we can directly estimate the total impacts of changes in final consumption Y on total productivity in each of the production sectors i.

Input–output analysis may, for example, be used to estimate the overall economic impacts (in terms of output, number of jobs, etc.) of additional government spending on military electronics.

Price effects can also be included, but this would go beyond the scope of the present chapter. Problems or limitations with respect to input–output analysis include:

- The assumptions of linearity and market equilibrium may not hold.
- Technological factors will change over time.
- Available data will typically lag several years behind.

Therefore, it is advised to use the method for order-of-magnitude assessments only.

26.7.6   External Effects

The term external effects or externalities is used to indicate relevant impacts on goods or states of nature that are not traded or tradable in a market, and for which therefore the usual methods of economic valuation cannot be applied. Examples include noise, environmental pollution or degradation, and human health impacts. A multicriteria or utility assessment approach may be taken to evaluate the multiple impacts of alternative systems or designs. However, because, in many cases, factors that have not been valued in economic or financial terms have been largely ignored in decision making, considerable attention is being given to approaches for estimation of the economic values of such "invaluable" goods. These include:

- *Costs of Mitigation.* Estimates are based on costs of measures that will effectively mitigate the undesirable effects. This may, for example, be the costs of cleanup in case of waste dumps, costs of water sanitation after pollution, or costs of insulation in the case of excessive noise, for example, near highways or airports.
- *Prevention Costs.* These are costs that would be necessary to prevent the negative impacts to occur in the first place. For example, costs of changing production facilities to prevent water pollution, costs of low-noise engines, costs of increased safety measures in traffic.

- *Compensation Costs.* Compensation costs indicate the value of (financial or other) compensations agents are willing to accept in exchange for their acceptance of negatively valued impacts. For example, residents may be willing to put up with excessive noise or stench if paid a certain amount of money per year. Such payments may also be seen as a compensation for losses incurred, for example, the loss of market value of homes if a new highway passes close by.
- *Curation Costs.* Perhaps the clearest example are the costs of traffic accident and/or alcohol abuse in terms of hospital and health care, lost labor days, psychological costs.

26.6.7   Closing Remarks on Cost–Benefit Models

As previously illustrated, an analyst engaged in performing an economic cost–benefit analysis must make many choices. Most of these choices are not value-free. Therefore, it is strongly recommended, first, to be as explicit as possible about these choices and other uncertainties in the analysis, and second, to communicate extensively about these choices with the decision maker(s) for whom the analysis is being done.

Considering the support of decision making more broadly, one should be aware of the limitations of the approach. There is an understandable tendency to attempt to translate all efforts and revenues in financial terms, thus arriving at a single measure of performance. This tendency is enforced by the economic notion that most decision making, in the end, is based on economic or financial criteria. However, many other issues play a part in most decisions, for example, impacts of a decision on a firm's strategic position; relations with other firms, organizations, or individuals; issues of political gain or loss; long-term environmental sustainability; and concerns related to equity and legitimacy. Many of these cannot easily or not at all be translated into financial terms. Therefore, we suggest to limit cost–benefit studies to those aspects that can be translated into the economic framework without much controversy. Other measurable objectives may be taken into account by complementing cost–benefit analysis with a multiobjective or decision-analysis approach.

Finally, we feel that analytic results of the approaches discussed here should be seen as information to support decision makers in their judgment, but that they can never replace that judgment.

26.8   EVALUATION AND DISCUSSION

For specific applications or types of systems, a variety of modeling approaches other than those described here exist. For example, decision/event-tree models may be very appropriate in situations where a number of distinct decision options may result in a (limited) number of outcomes, and estimates of outcome probabilities given a specific decision may be made (see Chapter 28). Fault-tree models may be used as design aids when system reliability is crucial, and information is available on failure modes of system components and on the relation of component failure to overall system failure. Specialist literature in a variety of domains will reveal a wealth of specific models, for example, traffic-flow models for traffic management and network models for transport planning. The general principles of modeling as outlined in the introduction to this chapter, however, apply equally to these modeling methods.

Rather than attempting to provide a complete catalog, we briefly discuss a number of more general issues at the end of this chapter, namely the choice and selection of modeling approach(es), model complexity, and trends in the modeling field.

26.8.1   Choice and Combination of Methods

For complex systems analysis and design, it is generally advisable to use a combination of modeling approaches rather than to rely on a single method. Different system components may best be modeled using different types of mathematical descriptions, for example, discrete and continuous.

In general, the choice of method or combination will be determined by:

- *The Relevant Characteristics of the Processes to be Modeled.* For example, a queuing process or logistics problem will generally be modeled using the principles of discrete-event models, while changes in key variables of continuous processes in fields such as energy conversion, water control, electricity production, or ecological systems may best be modeled using (continuous) physical systems models or system dynamics. Inventory problems may be modeled either way, depending on whether the discrete characteristics of the inventory are important to system behavior and performance, or not.

- *The Types of Questions to be Answered by the Modeling Analysis.* Discrete-event simulation methods and times-series analysis methods generally aim at quantitative predictions of system performance based on statistical methods, while the system dynamics approach focuses more on behavior patterns and trend analyses rather than precise numerical values.

- *Availability of Data and Knowledge About System Behavior.* Discrete-event and physical system modeling generally assume the existence of accepted theories describing the fundamental mechanisms underlying system behavior, as well as the availability of sufficient data for model estimation and validation. Time-series models are particularly suited in cases where ample data are available, but underlying theories are missing. The system dynamics approach, on the contrary, is often mainly based on descriptions of explanatory mechanisms, and less on detailed matching with data.

These principles are illustrated in Figure 26.7, with respect to four of the modeling approaches discussed in this chapter. The diagram should be considered as indicative rather than prescriptive.

Within one and the same system or policy design project, a combination of modeling methods is often required for appropriate resolution. By way of example, we briefly discuss the design of a flood safety system in an estuary in The Netherlands.

After a severe flood, The Netherlands decided to launch a large-scale flood-protection program. For one of the estuaries in the southwest of the country, the choice at the most general level was between:

- Enforcing and heightening all the existing dikes around the estuary.
- Building a new dike at the narrow entrance to the estuary, effectively closing off the estuary from the sea.

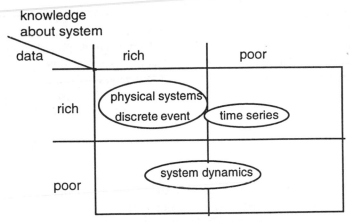

Figure 26.7   Indication of the suitability of modeling approaches.

- Building a storm-surge barrier that would be closed only in cases of severe surges, and would allow tidal flows in the estuary at other times.

While building a new dike closing off the estuary from the sea was the fastest and cheapest solution, it met with strong resistance as it would radically change the estuary's unique ecological tidal system.

Therefore, a systems analysis was set up to support the principal choice, with a focus on safety, costs, and ecological impacts as main criteria. To analyze and evaluate the three options, a combination was used of statistical extrapolations of storm-surge heights and time patterns, a physical model of water levels and wave development in the estuary, other physical models of dike failure, and a probabilistic model to estimate damages from flooding for different flooding situations. Of course, cost-estimation models were included as well (Goeller et al., 1977).

On the basis of study results, policy makers decided to adopt the costly storm-surge barrier solution as a political compromise between the safety and ecological objectives. Next, a number of model-based analyses were made to refine barrier design and to support the choice of control policies for the barrier. Again, combinations of modeling methods were used at different levels and for different aspects of design. In combination with physical storm-surge and hydraulic models for estuary behavior, an event-tree model was constructed to analyze the consequences of various possible events and policies for barrier control and estimate their associated probabilities. For more in-depth analysis of different designs for warning and manning procedures of the barrier after a meteorological storm warning, a separate substudy using discrete-event simulation was performed. Cost estimates were made of different alternatives for setting up and supporting the decision system. More details can be found in Thissen (1989).

26.8.2   Model Complexity

In addition to the combination of modeling approaches, the storm-surge barrier example illustrates the dissection of a design problem in separate subelements, and the use of one or more appropriate modeling methods for each of these subissues. This approach is favored over attempts to describe all system aspects at different levels in one large and complex model. Attempts at such large-scale modeling have displayed serious problems with coordination and management of modeling efforts, associated problems with model estimation and validation, and lack of model transparency. Furthermore, a complex, general system model will often not satisfy all requirements of a specific design situation. Therefore, it is advised to aim for as simple and as transparent models as possible, tuned specifically to the design issue at hand.

When it is important to model a broad range of aspects and elements of a system for overall analysis and evaluation, adoption of a metamodeling approach is suggested. In metamodeling, relatively simple relations between system parameters at a high level of aggregation are identified, based on more detailed analyses or simulations of system components. In the storm-surge-safety case, for example, simple relations between storm conditions and barrier-control operations on the one hand, and loads on the dikes surrounding the estuary on the other, were derived on the basis of numerous detailed, hydraulic simulations of water behavior in the estuary. Similarly, conditional failure probability estimates of the manning procedure were based on a separate simulation study. The aggregate relations were then used as elements in an overall model, linking storm-surge statistics, barrier control options and reliability assessments, and dike statistics to assess overall safety.

26.8.3   Trends and Innovations in Modeling

We conclude by pointing out a number of recent trends in model development and use. First, attention for participative modeling approaches is increasing. Models of complex systems with the involvement of numerous parties or actors can be built more effectively if experts from the parties are more heavily involved in the model-building process. Increased acceptance of and support for modeling results is one of the main advantages of a close cooperation between stakeholders and specialists in modeling. Transparency and comprehensibility of modeling concepts are necessary conditions for participative modeling. Group model building is discussed in detail by Vennix (1996) and by De Vreede (1995).

Second, and related to the interest in participative approaches, development of graphical conceptual modeling approaches deserves a lot of attention. Graphical representations and diagrams are more accessible for communication and discussion than textual, mathematical, or computer-code instructions. Particularly for crucial issues such as the selection of system boundaries and the identification and discussion of fundamental mechanisms underlying system behavior, the use of graphical conceptual models is very helpful. Increasingly, computer simulation software includes a graphical conceptualization language, enabling the user to encode system structure before entering mathematical and quantitative specifications (e.g., ithink (STELLA), Powersim, Vensim, and Arena).

Third, visualization of model behavior has proven to be very helpful in explanation and validation of models. Most simulation packages include animation features that schematically display model operation in a way directly recognizable to users and stakeholders. Pictorial representations of the simulation, for example the start and finishing of tasks and growth of queuing phenomena, enable operators and other stakeholders to "see" how the model works, and thereby suggest model improvements as well as build confidence in model results (see, e.g., De Vreede and Verbraeck, 1996).

Last, but not least, developments in information technology open up opportunities for innovative modeling approaches. Increased computer capacity facilitates access of large databases and coupling of models in so-called decision support systems or environments. Knowledge-based technology and artificial intelligence can be used to enrich traditional, quantitative system modeling approaches. Developments in computer graphics allow for more attractive and interactive presentation of simulation results. An example is the extension of animation into "virtual reality" display of simulation results. High-speed networks enable complex simulation models to take advantage of distributed computer power. Users are enabled to participate in simulation models as one of the "playing" objects, thereby interacting and influencing the model, and understanding the system they are studying or designing.

It can be concluded that modeling is an important building block in systems engineering. However, thorough knowledge of modeling methodologies and the appropriate choice and application of techniques are prerequisites for a successful modeling study. Recent technological developments will strengthen the role of modeling in investigating system behavior and will allow for improved graphical conceptual modeling and visualization possibilities.

ACKNOWLEDGMENTS

We thank Stefan Weijers of Eindhoven University of Technology for contributing the wastewater treatment plant example. Our colleague Pieter Bots is kindly acknowledged for his comments on an earlier version of this chapter.

APPENDIX: MODELING AND SIMULATION SOFTWARE

Table 26.A1 contains a summary of the modeling and simulation software packages that have been mentioned in this chapter. This is only a very small selection of the available packages, and new software is released regularly. For up-to-date and comprehensive in-

TABLE 26.A1   A Selection of Modeling and Simulation Software

Name	Type of Model	Vendor
MATLAB®	Physical system	The MathWorks, Inc., Natick, MA
MATRIX$_x$®		Integrated Systems, Inc., Santa Clara, CA
20-SIM™		Controllab Products, Enschede, The Netherlands
VisSim™		Visual Solutions, Inc., Westford, MA
DYNAMO	System dynamics	Pugh-Roberts Associates, Cambridge, MA
ithink™ & STELLA™		High Performance Systems, Inc., Hanover, NH
Powersim®		Powersim AS. Isdalstø, Norway
Vensim®		Ventana Systems, Inc., Belmont, MA
ARENA	Discrete event	Systems Modeling Corporation, Sewickley, PA
AutoMod		AutoSimulations, Inc., Bountiful, UT
GPSS/H™		Wolverine Software Corp., Annandale, VA
MODSIM II®		CACI Products Company, La Jolla, CA

formation about simulation software, we refer the reader to *The Directory of Simulation Software,* which contains descriptions of the software, information about vendors, platforms and pricing, and is published annually by the Society for Computer Simulation. The magazine *OR/MS Today* includes surveys on modeling tools on a regular basis. An example of such a survey is Swain (1995), which, besides vendor and pricing information, also covers typical applications, primary markets and new features of the software.

REFERENCES

Alcamo, J. (ed.). (1994). *IMAGE 2.0: Integrated Modeling of Global Climate Change.* Dordrecht, The Netherlands: Kluwer Academic Publishers.

Barlas, Y. (1996). Formal aspects of model validity and validation in system dynamics. *Syst. Dyn. Rev.* **12**(3), 183–210.

Borrie, J. A. (1992). *Stochastic Systems for Engineers.* Englewood Cliffs, NJ: Prentice-Hall.

Box, G. E., and Jenkins, G. M. (1976). *Time Series Analysis* (rev. ed.). San Francisco: Holden-Day.

Box, G. E., Jenkins, G. M., and Reisel, G. C. (1994). *Time Series Analysis: Forecasting and Control,* 3rd ed. Englewood Cliffs, NJ: Prentice-Hall.

Boyce, W. E., and DiPrima, R. C. (1986). *Elementary Differential Equations and Boundary Value Problems.* Chichester: Wiley.

Brauer, J. R. (ed.). (1988). *What Every Engineer Should Know About Finite Element Analysis.* New York: Dekker.

Bussey, L. E. (1978). *The Economic Analysis of Industrial Projects.* Englewood Cliffs, NJ: Prentice-Hall.

Close, C. M., and Frederick, D. K. (1978). *Modeling and Analysis of Dynamic Systems.* Boston: Houghton Mifflin.

Dalkey, N. C. (1969). An experimental study of group opinion - The DELPHI method. *Futures,* September pp. 408–426.

De Vreede, G. J. (1995). Facilitating organizational change; The participative application of dynamic modelling. Ph.D. Thesis, Delft University of Technology, The Netherlands.

De Vreede, G. J., and Verbraeck, A. (1996). Animating organizational processes. Insight eases change. *Simul. Prac. Theory* **4**, 245–263.

Dorf, R. C., and Bishop, R. H. (1995). *Modern Control Systems.* Reading, MA: Addison-Wesley.

Dur, R. C. J. (1992). Business reengineering in information intensive organizations. Ph.D. Thesis, Delft University of Technology, The Netherlands.

Flood, R. L., and Carson, E. R. (1990). *Dealing with Complexity; an Introduction to the Theory and Application of Systems Science.* New York: Plenum.

Ford, A. (1995). Simulating the controllability of feebates. *Syst. Dyn. Rev.* **11**(1), 3–29.

Forrester, J. W. (1961). *Industrial Dynamics.* Cambridge, MA: MIT Press.

Forrester, J. W. (1969). *Urban Dynamics.* Cambridge, UK: Productivity Press.

Goeller B. F., A. F. Abrahamse, J. H. Bigelow, J. Bolten, D. M. DeFerranti, J. C. Dehaven, T. F. Kirkwood, and Petrusche II, R. L. (1977). *Protecting an Estuary from Floods - A Policy Analysis of the Oosterschelde* (R-2121/-NETH, Vols. I–VI). Santa Monica, CA: RAND Corporation.

GPS-X User's Guide. (1994). Hydromantis, Inc. Hamilton, Ontario, Canada.

Gu, E. Y. L. (1996). Nonlinear systems analysis and modeling. In *Systems Modeling and Computer Simulation* (N. A. Kheir, ed.), 2nd ed., pp. 89–167. New York: Dekker.

Harvey, A. C. (1989). *Forecasting, Structural Time Series Models and the Kalman Filter*, pp. 100–167. Cambridge, UK: Cambridge University Press

Henze, M., Grady, C. P. L., Jr., Gujer, W., Marais, G. V. R., and Matsuo T. (1987). *IAWPRC Task Group on Mathematical Modelling for Design and Operation of Biological Wastewater Treatment*, Sci. Tech. Rep. No. 1, Activated Sludge Model No. 1. London: IAWPRC.

Hoover, S. V., and Perry, R. F. (1989). *Simulation: A Problem-solving Approach*. Reading, MA: Addison-Wesley.

Jamshidi, M. (1983). *Large-Scale Systems: Modeling and control*. New York: North-Holland.

Jamshidi, M. (1996). Introduction to large-scale systems. In *Systems Modeling and Computer Simulation* (N.A. Kheir, ed.), 2nd ed., pp. 581–624. New York: Dekker.

Karnopp, D. C., Margolis, D. L., and Rosenberg, R. C. (1990). *System Dynamics: A Unified Approach*, 2nd ed. Chichester: Wiley.

Karplus, W. J. (1958). *Analog Simulation: Solution of Field Problems*. New York: McGraw-Hill.

Kendall, M., and Ord, J. K. (1990). *Time Series*, 3rd ed., London: Edward Arnold.

Kheir, N., Damborg, M., Zucker P., and Lalwani C. S. (1985). Credibility of models (panel discussion). *Simulation* **45**, 87–89.

Kheir, N. A. (1996). Continuous-time and discrete-time systems. In *Systems Modeling and Computer simulation* (N. A. Kheir, ed.), 2nd ed., pp. 27–88. New York: Dekker.

Kreutzer, W. (1986). *System Simulation Programming Styles and Examples*. Reading, MA: Addison-Wesley.

Law, A. M., and Kelton, W. D. (1991). Simulation Modeling and Analysis, 2nd ed. New York: McGraw-Hill.

Leontief, W. W. (1966). *Input Output Economics*. London: Oxford University Press.

Ljung, L., and Glad, T. (1994). *Modeling of Dynamic Systems*. Englewood Cliffs, NJ: Prentice-Hall.

Makridakis, S., and Hibon, M. (1995). *ARMA Models and the Box-Jenkins Methodology*, INSEAD Working Paper Ser. No. 95/45/TM. Fontainebleau, France.

Makridakis, S. G., Wheelwright, S. C., and McGee, V. E. (1983). *Forecasting: Methods and Applications*. Chichester: Wiley.

Meadows, D. H. (1976). The unavoidable a priori. In *Elements of the System Dynamics Method* (J. Randers, ed.), pp. 23–57. Cambridge, MA: MIT Press.

Meadows, D. H., Meadows, D. L., Randers, J., and Behrens, W. W. (1972). *The Limits to Growth: A Report for the Club of Rome's Project on the Predicament of Mankind*. New York: Universe Books.

Mishan, E. J. (1976). *Cost-Benefit Analysis*. New York: Praeger.

Neelamkavil, F. (1987). *Computer Simulation and Modelling*. Chichester: Wiley.

Pegden, C. D., Shannon, R. E., and Sadowski, R. P. (1990). *Introduction to Simulation Using SIMAN*. New York: McGraw-Hill.

Pritsker, A. A. B. (1986). *Introduction to Simulation and Slam II*, 3rd ed. New York: Halsted.

Richardson, G. P. (1986). Problems with causal-loop diagrams. *Syst. Dyn. Rev.* **2**(2), 158–170.

Roberts, E. B. (ed.). (1978). *Managerial Applications of System Dynamics*. Cambridge, MA: MIT Press.

Russel, E. C. (1976). *Simulation with Processes and Resources in Simscript II.5*. Los Angeles: CACI Inc.

Sage, A. P. (1977). *Methodology for Large-scale Systems*. New York: McGraw-Hill.

Sage, A. P. (1983). *Economic Systems Analysis, Micro-Economics for Systems Engineering, Engineering Management, and Project Selection*. New York: North-Holland/Elsevier.

Sassone, P. G., and Schaffer W. A. (1978). *Cost-Benefit Analysis*. New York: Academic Press.

SIMBA User's Guide. (1995). Fa. Otterpohl Wasserkonzepte GbR, Luebeck, Germany.

Schriber, T. J. (1974). *Simulation using GPSS*. Chichester: Wiley.

Shannon, E. (1975). *Systems Simulation, the Art and Science*. Englewood Cliffs, NJ: Prentice-Hall.

Smith, C. L., Pike, R. W., and Murril, P. W. (1970). *Formulation and Optimization of Mathematical Models*. Scranton, PA: International Textbook Company.

Sol, H. G. (1982). Simulation in information systems development. Ph.D. Thesis, University of Groningen.

Sterman, J. D. (1984). Appropriate summary statistics for evaluating the historical fit of system dynamics models. *Dynamica* **10** (Part 2), 51–66.

Sugden, R., and Williams, A. (1978). *The Principles of Practical Cost-Benefit Analysis*. London: Oxford University Press.

Swain, J. J. (1995). Simulation survey: Tools for process understanding and improvement. *OR/MS Today,* August, pp. 64–79.

Thissen, W. (1989). Safety impact analysis for control of the Oosterschelde stormsurge barrier. In *Impact Forecasting and Assessment, Methods, Results, Experiences* (P. van der Staal and F. van Vught, eds.). pp. 109–123. Delft: Delft University Press.

Turing, A. M. (1963). Computing machinery and intelligence. In *Computers and Thought* (E. A. Feigenbaum and J. Feldman, eds.), pp. 11–35. New York: McGraw-Hill.

Vandaele, W. (1983). *Applied Time Series and Box-Jenkins Models*. New York: Academic Press.

Van den Bosch, P. P. J., and Van der Klauw, A. C. (1994). *Modeling, Identification and Simulation of Dynamical Systems*. Boca Raton, FL: CRC Press.

Van der Kloet, P., and Van Geer, F. C. (1983). *Application of ARIMA Models to Rainfall Runoff Relations and Groundwater Levels,* Part B (In Dutch). Delft: Delft University of Technology.

Vennix, J. A. M. (1996). *Group Model Building: Facilitating Team Learning Using System Dynamics*. Chichester: Wiley.

Verbraeck, A., and De Vreede, G. J. (1993). Animation as a communication vehicle in simulation studies. In *Modeling and Simulation ESM93*. (A. Pave, ed.), pp. 670–674. San Diego, CA: Society for Computer Simulation.

Vidyasagar, M. (1978). *Nonlinear Systems Analysis*. Englewood Cliffs, NJ: Prentice-Hall Electrical Engineering Systems.

Wei, W. S. (1990). *Time Series Analysis*. Reading, MA: Addison-Wesley.

Wile, K., and Smilonich, D. (1996). Using dynamic simulation for resource management policy design at the Minnesota Department of Transportation. *Syst. Dyn.* **2**, 569–572.

Wolstenholme, E. F. (1989). *System Enquiry: A System Dynamics Approach*. Chichester: Wiley.

Wu, L. S. Y., Ravishanker, N., and Hosking, J. R. M. (1991). Forecasting for business planning: A case study of IBM product sales. *J. Forecast.* **10**, 579–595.

Boom & Bust enterprises is a trademark of MicroWorlds (Cambridge, MA).

27 Operations Research and Refinement of Courses of Action

KEITH W. HIPEL, D. MARC KILGOUR, and SIAMAK RAJABI

27.1 INTRODUCTION

The objective of this chapter is to explain how operations research (OR) can complement systems engineering approaches to decision making by generating and refining alternative solutions to complex engineering problems. Subsequent to a general discussion of OR and its utilization in problem solving, the philosophical underpinnings of OR and systems engineering are compared and evaluated. Their complementary capabilities allow the re ductionist-scientific approach of OR to be coupled with the holistic methodologies of systems engineering to furnish a powerful tool to solve many real-world problems. A range of useful OR techniques are described next, followed by a discussion of how OR can be used to generate and screen actions and alternatives. Because of the great importance of multiple-criteria decision making and multiple-participant decision making in OR, these sets of useful tools are discussed in separate sections. The last section of this chapter is devoted to a brief discussion of heuristic methods in OR.

27.2 OPERATIONS RESEARCH

In this section, the history, development, and applications of OR are put into perspective. First, a brief history of OR is presented, with emphasis on some of its early applications. Subsequently, a general definition of OR is set out and its domains of application specified. Finally, the different phases of an OR study are explained.

27.2.1 History of Operations Research

To understand the nature of OR and its current position in management and engineering, it is useful to review its history. The first OR studies were initiated in Europe just prior to the commencement of World War II to solve some complex military problems. In July 1938, the British High Command ordered that *research* be carried out with respect to the *operational* aspects of radar systems, in order to coordinate defensive actions against potential air attacks. Throughout the war, most branches of the U.K. armed forces employed multidisiplinary OR teams to solve large-scale military problems, especially under conditions of scarce resources. British OR teams, for example, are thought to have contributed significantly to winning the Battle of Britain, during the summer and early autumn of 1940.

Handbook of Systems Engineering and Management, Edited by A. P. Sage and W. B. Rouse
ISBN 0471-15405-9   ©1999 John Wiley and Sons, Inc.

Later in the war, the U.S. armed forces also employed OR teams to address a variety of challenging problems. In the Pacific theater, OR was crucial in the successful campaign to recapture Pacific islands to establish air bases close enough to Japan to carry out saturation bombing of Japanese cities. Other examples of successful applications of OR in military operations include antiaircraft fire control, fleet convoy sizing, and enemy submarine detection (Ravindran et al., 1987). Many secret documents describing the use of OR in World War II have been released, and a host of fascinating articles on this topic are available (see, for instance, Blackett, 1962; Waddington, 1973).

After World War II, the theory and practice of OR was significantly expanded to address a wide range of challenging systems problems arising in systems engineering, management sciences, industrial engineering, water resources, environmental engineering, logistics, and many other fields. Moreover, the development of powerful electronic computers has facilitated the utilization of OR algorithms with high computational time and/or storage requirements. As well, user-friendly decision support systems have made OR methodologies and techniques accessible to practitioners in many fields. Hillier and Lieberman (1990) claim that, except for electronic computers, OR has had the greatest impact on the management of organizations.

Even though OR began as a discipline during World War II, its roots date back much further. Indeed, Quesnery, in 1759, and Walras, in 1874, developed elementary constrained optimization programs for use in economics. Von Neumann (1937), and Kantorovich, in 1939, proposed more sophisticated economic models. Jordan, Minkowski, and Farkas developed linear mathematical models reminiscent of OR techniques, in 1837, 1896, and 1903, respectively. Other pioneering research includes the work of Markov on dynamic models, the model of queuing (waiting-line) processes developed by Erlang, and other studies published during the 1920s in industrial engineering and business journals (Wagner, 1975).

Petroleum companies were among the first industrial firms to use linear programming for production planning. Steel producers and other large industries also employed OR-based production-planning techniques. Traffic-control studies for the Port of New York during the early 1950s were perhaps the first applications of OR in the service sector.

OR societies have now been formed in most industrialized countries, along with the publication of OR periodicals. For example, in 1952 and 1953, the Operations Research Society of America (ORSA) and the Institute of Management Science (TIMS), respectively, started in the United States. Recently, these two societies have joined to form the Institute For Operations Research and Management Sciences (INFORMS).

27.2.2 Definition of Operations Research

ORSA defines OR as (Ravindran et al., 1987):

> A set of procedures and techniques concerned with scientifically deciding how best to design and operate man-machine systems, usually under conditions requiring the allocation of scarce resources.

According to this traditional definition, OR techniques are more appropriate for use in studying tactical or operational decision-making problems, which have well-defined objectives and constraints that can be expressed using quantitative measures. In the early stages of OR's history, OR techniques certainly were used primarily for tackling highly structured problems, which often arise at the tactical or operational level of decision making. More recent advances in OR have tended to address the strategic level, where problems are less well defined, and both qualitative and quantitative information must be taken into account.

Figure 27.1 depicts aspects of engineering decision making, based on the analyses of Hipel (1992). The left side of the figure displays the main feasibility considerations in large-scale engineering solutions, such as alternative ways to satisfy future water demand. In addition to a sound physical design, environmental, economic, financial, political, and social feasibility must be taken into account. These factors are important, not only in designing a new system, but also in planning and improving operations at an existing facility. A successful OR approach to decision making generally includes all of these considerations.

The right side of Figure 27.1 displays four main characteristics of a decision process within a hierarchical framework. As one proceeds from the tactical level to the strategic level, problems change from highly structured systems with quantitative constraints and quantitative information, to unstructured systems having soft qualitative constraints and qualitative information. It is hardly surprising that a wide range of OR and other tools are useful in systematically implementing engineering decisions.

Many quantitative OR methods have been developed for use in tactical decision making. Accordingly, some researchers have suggested that it would be an important contribution to design decision methods for use at the strategic level, where multiple decision makers,

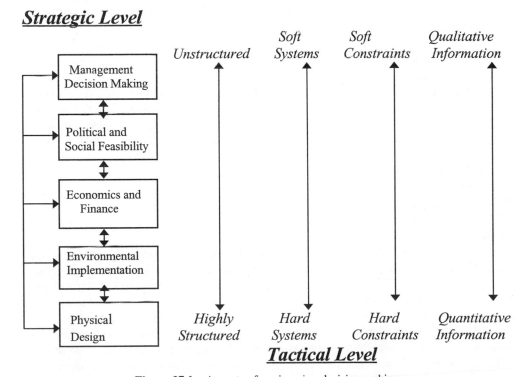

Figure 27.1   Aspects of engineering decision making.

each with multiple objectives, must be taken into account (Ackoff, 1981; Rosenhead, 1989; Hipel, 1990; Hipel and McLeod, 1994, Ch. 1). Another advance relevant to strategic-level OR is the development of techniques that require only ordinal and qualitative data (Cook and Kress, 1992).

27.2.3   The Nature of Operations Research

OR can be considered as both an art and a craft. The art consists of the definition and structuring of problems, while the craft is the application of mathematical techniques to solve them. The art and the craft of OR are best understood in the context of the phases of an OR study, which most researchers agree, include the following stages, more or less (see, for example, Taha, 1971; Hillier and Liberman, 1990; Wagner, 1975; Winston, 1991).

1. Define the problem;
2. Construct a mathematical model;
3. Solve the problem in the context of the model;
4. Test the validity of the model; and
5. Implement the solution.

Note that only the third phase follows a fixed set of rules; the other phases require techniques that depend on the nature of the problem under study and other factors. It is important to keep in mind that the models do not make the decisions—they are only tools to aid decision makers (DMs) in the decision process, and their contributions generally include clarifying the problem, suggesting solutions, and measuring the effectiveness of the solution. Below, each of the phases is described in more detail.

27.2.3.1   Defining the Problem.   A common observation is that most real-world problems are ill-defined and are presented to the analyst in a vague way; often much work is required to comprehend a problem, mold it into a well-defined structure, and interpret possible solutions. In the first stage of an OR study, the problem is explicitly defined and expressed. This step includes identification of the main goal(s) of the study, the parties who have influence on the problem, the objectives, both hard and soft constraints, and the feasible courses of action.

Problem definition is a very important stage of any OR study, since finding the optimal solution to an inappropriate problem does not generally lead to a useful recommendation. Yet, there is surprisingly little research in the OR literature on problem definition. Section 27.5, on generating and screening actions and alternatives, discusses this issue in more detail.

In the problem-definition stage, all objectives that may be important to various parties are elicited. Relationships among objectives are investigated; those that can be aggregated, or given low priority in a hierarchy of objectives, are identified. Traditionally, OR approaches tend to use one overall objective such as profit maximization, and to incorporate any remaining objectives into the set of constraints. However, it is now widely recognized that these approaches often do not provide satisfactory solutions. Even in for-profit organizations, the use of profit maximization as the *only* objective is often not appropriate. Other objectives such as increasing market share, maintaining stable prices, and improving company prestige may be important, and are sometimes in conflict with short-term profit maximization.

Another problem with traditional OR modeling is that usually social, political, and other constraints and objectives, which are often intangible, are ignored. Experience suggests that one reason why success is limited in many OR studies is lack of attention to these crucial factors.

27.2.3.2 *Constructing a Mathematical Model.*

The second stage of an OR study is to construct a mathematical model of the problem in order that a solution can be obtained by a formal procedure. Using a mathematical model also helps the analyst to express the problem more concisely and accurately, to find out what additional information is required, to express relationships among variables and parameters, and to explain the problem more precisely to DMs.

Despite the many advantages of mathematical modeling, there are pitfalls that an analyst should avoid. For instance, real-world problems are often too complex to be modeled exactly. The best compromise is to construct a simplified model that captures the essence of the problem. But the analyst, and the DM, must remember that the optimal solution of the model does not necessarily represent the best course of action for the actual problem.

27.2.3.3 *Solving the Problem Within the Model.*

After formulating the problem as a mathematical model, the next step is to solve the model. Many OR models can be solved using standard algorithms, often implemented in efficient computer software. However, many practical problems cannot be described using standard OR models; special routines, or simulation models, are needed to model and solve them. If the problem can be modeled as a standard OR problem, then the solution step is the easiest part of the study. Hillier and Lieberman (1990) call this phase the "fun part" for most practitioners.

According to the traditional definition, the aim of OR is to find the optimal solution of the problem under study. However, for the following reasons this approach may not be sufficient to solve the problem:

1. A solution may be optimal for the model, but not for the problem. In the best case, the solution is a close approximation to the optimal solution of the problem.

2. In some situations, such as multiple objective programming, there may not exist a solution that optimizes *all* the objectives simultaneously. In this case, one generally searches for an *efficient* solution, a solution that cannot be improved for all objectives.

3. For many organizations, a solution that is as close as possible to a *goal* is more acceptable than an optimal solution. This reflects the fact that usually real-world problems are dynamic, prone to error in measurement, and ill-defined. Simon (1976) introduced *satisficing* (combining the ideas *satisfying* and *optimizing*) to characterize solutions that are good enough but not necessarily optimal. The goals to be satisfied may be based on past performance, the DM's intuition, the level of competition, and so on. Moreover, these goals are not fixed, and can be changed to reflect a circumstance such as the difficulty of reaching previous goals. Eilon (1971) says that "optimizing is the science of the ultimate and satisficing is the art of the feasible." In fact, the idea of using *heuristics* to address a problem is inspired by the idea of

satisficing; an acceptable and usually good solution in hand is better than an optimal solution in the bush.

After obtaining an optimal or satisficing solution, one can investigate how this solution is influenced or changed due to changes in specified model parameters. This step provides additional information on the behavior of the model, and is called *postoptimality* or *sensitivity analysis*. For example, a solution that changes little when the model is solved for different sets of model parameters is robust, and one can have more confidence in its results. When small parameter changes can have a dramatic effect on the solution, the analyst should try to estimate these parameters as accurately as possible.

27.2.3.4   Testing the Validity of the Model. Before implementing a decision, one must ensure that the solution has been obtained from an appropriate model. Different tests can be carried out to examine the validity of the model. Sometimes important interrelationships among decision variables, or relevant constraints, may have been excluded from the model. It is then easy to verify that the model gives plausible results. Alternatively, one can check the recommendations of the model in some prespecifed situations. Comparing the best solution with historical data, checking the optimal solution with a simulation model, using extreme values of some parameters, and reexamining the problem structure and problem definition are among other methods to test the validity of the model.

27.2.3.5   Implementation. The last phase of an OR study is the implementation of a solution that has been tested and approved by the DM. In many situations, translating the optimal solution into a practical policy at the operational level is very complex. Usually, the success of the implementation depends on the cooperation of the OR team and the operating personnel. Most often, adjustments are necessary in both operational procedures and OR solutions in order to make the optimal policies practical.

27.3   OPERATIONS RESEARCH AND SYSTEMS ENGINEERING

The objectives of this section are to discuss the main similarities and differences of OR and systems engineering and to investigate how they can complement one another in practice. To achieve these goals, an appreciation of the distinction between two competing philosophies, reductionism and holism, is essential. *Reductionism* is the doctrine that everything can be reduced, decomposed, or disassembled to individual elements, and the overall system can be best comprehended by understanding these simple individual elements. *Holism,* on the other hand, contends that key properties of a phenomenon cannot be fully explained in terms of properties of its elements. This view is summarized in the well-known statement of Aristotle that "the whole is greater than the sum of its parts." Even though holism has its roots in Greek philosophy, it gained new recognition at the turn of century when scientists found that reductionism is not enough to describe some phenomena in biology and psychology. As described below, OR tends to be more reductionist in its outlook, while systems engineering is more holistic.

It is difficult to draw a firm boundary between OR and systems engineering from either the theoretical or the practical point of view, in part because these methodologies have evolved over time and are now much broader than they were originally. Nevertheless, this section attempts to shed some light on the fundamental principles and procedures of OR and systems engineering.

The previous section described the main objectives of OR and the phases of a typical OR study. Ackoff (1962) defines OR as a collection of scientific tools for use in studying management problems. He argues that OR can address any managerial problem for which a DM seeks to select an alternative to meet a specific objective. Sage (1992) expresses the key objectives of systems engineering as improving the effectiveness and reducing the cost of the system under study. He summarizes the phases of a systems engineering study as follows:

- Identification and specification of requirements
- Preliminary conceptual design
- Logical design and specification of system architecture
- Detailed design and testing
- Operational implementation
- Evaluation and modification
- Operation and deployment

Following these phases, the process of systems engineering begins with the generation and testing of alternatives. Usually, many alternatives are identified initially, unacceptable ones are eliminated, and the rest subsequently evaluated more carefully. Sage (1991, 1992) classifies definitions of systems engineering into *structural, functional,* and *purposeful* categories. According to the functional definition, systems engineering is a set of tools and theories carried out for the resolution of large-scale real-world problems. Arguing that these tools and theories are mainly OR methods, Sage concludes that OR provides the main set of tools used to tackle complex problems in systems engineering.

Even though both OR and systems engineering constitute tools and approaches to deal with complex management problems, their attitudes toward the problems are different. The main distinction between OR and systems engineering is their approach to the behavior of phenomena. OR is based on the scientific method and is therefore grounded on reductionist concepts. It tries to comprehend phenomena by understanding the individual parts and their relationships; because in many situations these relationships are very complex, it cannot capture the whole picture, at least not in practice. OR models are always simplifications of real-world problems. All system behavior is assumed to be determined by cause-and-effect relationships, which are usually modeled by linear or nonlinear equations. OR largely follows Ackoff's (1962) definition: solving a problem means choosing one action among a set of potentially good actions. Accordingly, OR is analytical, and its approaches apply to hard, but often only tame, problems (Keys, 1992). In other words, traditional OR techniques rely on quantitative information and are generally applicable only to well-defined problems at the tactical level (see Fig. 27.1). Therefore, a crucial step in an OR study is translating the real-world situation into a quantitative format. Finally, OR often focuses on aspects of problems related to the consumption of resources (see the ORSA definition of OR given in Section 27.2), and its major aim is to advise on appropriate tradeoffs between effectiveness and efficiency.

Systems engineering, on the other hand, approaches problems according to holistic principles. (Haimes and Schneiter, 1996). It concentrates on the entire system, rather than its components, attempting to understand phenomena by adapting an expansionist procedure. Unlike traditional OR, systems engineering is capable of incorporating technology, institutional perspectives, and value judgments into the study of problems (Sage, 1992, Chap. 2). Systems engineering can assist DMs and analysts in tackling systems containing complex and random behavior without trying to explain that behavior, whereas OR largely deals with modeling and analyzing systems components in an attempt to build a micro- to macro-level understanding (Sage, 1990; Klir, 1991).

Systems engineering is less analytical and more qualitative in nature and, hence, more applicable to unstructured and complex problems. It tries to provide theories and procedures that can help DMs to resolve problems where scientific approaches such as OR may fail. Unlike OR, which often explains how a phenomenon works, systems engineering constructs theories of the behavior of phenomena and uses them to explain why the system works. One can argue that since every phenomenon in the real world acts as a system, any method for understanding and solving it should be grounded on the theory of systems and systems thinking. Accordingly, the concepts of systems theory can be employed with any scientific method, including OR.

In fact, from their origins as distinct disciplines, OR and systems engineering have viewed each other as connected fields of study. Ideas from OR can be used in many systems engineering studies, and vice versa. Systems concepts can effectively describe the overall structure of a system problem as well as relationships among its components, while OR can investigate specific properties of components and their effects on the system. In summary, systems engineering and OR complement each other, so one cannot discard either in favor of the other.

27.4 OPERATIONS RESEARCH METHODS

27.4.1 Classification of Operations Research Techniques

Many practical OR techniques and associated methodologies are available for application purposes. It is important to select the most appropriate method, or set of techniques, to capture the key characteristics of the problem under study. Therefore, it is informative to classify each OR method according to its associated inherent characteristics. As shown in the left column in Table 27.1, an OR technique can typically be described by its objectives, constraints, decision variables, parameters, and number of DMs. The second column from the left provides possible characteristics for each element in the first column. The last row gives different solution techniques to find solutions for a given model. The set of columns labeled as models contains a column for each specified model in which the key characteristics of the model under consideration are indicated using check marks or else described in written form. For example, linear programming is an optimization method with a linear objective function, linear inequality and equality constraints, continuous decision variables, and deterministic parameters that may be given or estimated from the data. The linear programming model usually reflects the viewpoint of one DM and the simplex method or an interior point algorithm can be used to find the optimal solution. Because the unique characteristics of the graph model for conflict resolution are not specifically listed, they are entered in written format. It is noteworthy that many practical problems have a combination of attributes that lend themselves to study using several different methods. For instance, within the engineering decision process depicted in Figure 27.1, optimization

TABLE 27.1 The Main Characteristics of OR Methods

Elements	Characteristics	Models		
		Linear Programming	Integer Programming	Graph Model for Conflict Resolution
Objectives	*Linear*	✓	✓	Ordinal preferences among states for each decision maker
	Nonlinear		✓	
	Single	✓	✓	
	Multiple			
Constraints	*Linear*	✓	✓	Certain moves and countermoves are defined for each decision maker
	Nonlinear		✓	
Decision variables	*Continuous*	✓		Binary option choice for each decision maker
	Discrete		✓	
Parameters	*Deterministic*	✓	✓	Discrete option choices and ordinal preferences among states for each decision maker
	Fuzzy			
	Probabilistic			
Number of decision makers	*One*	✓	✓	
	Several			✓
Solution methodologies		Simplex method Interior point method	Branch and bound Cutting plane Enumeration	A range of solution concepts for defining human behavior under conflict.

methods may be employed for studying problems arising at the tactical level of decision making for a given large-scale engineering project, while the graph model for conflict resolution can be utilized in strategic studies.

27.4.2 Linear Programming

Linear programming (LP) is perhaps the oldest formal OR technique. Both the theory and the practice of LP are now highly developed (Chavatal, 1983), in large part because so many real-world problems can be formulated using LP models. Moreover, developments in LP have formed the basis for other OR methodologies such as integer, nonlinear, and stochastic programming.

In 1947, Dantzig developed the *simplex method* to solve an LP problem (Dantzig, 1963). The introduction of *duality theory* by von Neumann made Dantzig's simplex method extremely important. A typical LP problem can be formulated as follows:

$$\text{Maximize(Minimize)} \; Z = \sum_{i=1}^{n} c_i x_i$$

Subject to:

$$\sum_{i=1}^{n} a_{ij} x_i \leq (\geq) b_j \qquad j = 1, \ldots, m$$

$$x_i \geq 0, \qquad\qquad i = 1, \ldots, n$$

where x_i, $i = 1, 2, \ldots, n$ are the decision variables, and the parameters are a_{ij}, b_j, and c_i, for $i = 1, \ldots, n; j = 1, \ldots, m$. The simplex method depends fundamentally on the convexity of the feasible region, which is the set of points satisfying all constraints. (A set is convex if, for any two points in the set, the straight line joining the points lies entirely within the set.) In any LP model, the feasible region is convex and the objective function is linear, which implies that any extreme (maximum or minimum) value of the objective function must occur at an extreme point of the feasible region. (An *extreme point* of a convex set is a point that is in the set but does not lie in the interior of the straight line joining any two distinct points in the set. For instance, the convex set shown in Figure 27.2 has extreme points, A, B, C, D, E, and F.)

The simplex method finds the optimal solution of a linear program by iterating from one extreme point of the feasible region to an *adjacent* extreme point while increasing (decreasing) the value of the objective function until its maximum (minimum) value is reached. The initial extreme point is usually the origin; if the origin is not feasible, then an initial feasible extreme point is selected by an appropriate algorithm. The computational steps imbedded in the simplex procedure can be explained by the following simple, two-dimensional example:

$$\text{Maximize} \qquad 3x_1 + 5x_2$$

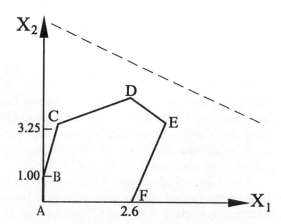

Figure 27.2   Simplex algorithm.

Subject to:

$$-3x_1 + 2x_2 \leq 2$$

$$-x_1 + 2x_2 \leq 5$$

$$4x_1 + 3x_2 \leq 20$$

$$3x_1 - x_2 \leq 8$$

$$x_1 \geq 0, \qquad x_2 \geq 0$$

Figure 27.2 shows the feasible region of the preceding LP maximization problem, consisting of the set of points, (x_1, x_2), satisfying all constraints. The point $A(0, 0)$ is the initial point. From A, the simplex method selects one of the two adjacent extreme points, B or F. To increase the value of the objective function, one moves toward the dashed line in Figure 27.2. Because moving toward B improves the value of the objective function faster than moving toward F, the algorithm pivots to the extreme point B. From B there are two choices, A and C. But A gives a negative improvement to the objective function; hence, C is selected. Then the algorithm moves to adjacent extreme point $D(2.27, 3.636)$. If the procedure reaches an extreme point with no adjacent extreme point that increases the value of the objective function, the current solution must be optimal. In this example, D is the optimal solution since no adjacent extreme point improves the objective function.

For many years, the simplex method and its variants were the only computational procedures that could solve linear programs. But it was eventually shown by Complexity Theory that, in the worst case, the simplex method is an exponential-time algorithm. In other words, the number of computational steps required by the simplex algorithm can grow as an exponential function of the size of the problem.

Khachian's (1979) *ellipsoid method* was the first attempt to develop a polynomial-time algorithm for solving LP problems. The development of Khachian's method proved that LP problems are among the "easy" optimization problems. Khachian's ellipsoid method initiates the solution process by constructing an ellipsoid centered at the origin containing the feasible set. The algorithm then constructs smaller and smaller ellipsoids such that the center of the last ellipsoid lies within the feasible region.

Even though Khachian's algorithm is theoretically superior to the simplex method, in practice its computational performance is very poor. Karmarkar (1984) developed a polynomial-time *interior-point algorithm* for solving large LP problems. In contrast to the simplex method that moves from one extreme point to an adjacent extreme point, an interior-point algorithm moves through the interior of the feasible region until it reaches the optimal solution. Figure 27.3 shows schematically an interior-point search path. For a detailed introduction to interior-point procedures, refer to Ignizio and Cavalier (1994).

In addition to their theoretical advantages, interior-point methods have proved to be very efficient in solving large LP problems. They have been successfully applied in some very large real-world problems such as military airlift logistics and air-crew scheduling (Winston, 1991).

27.4.3 Network Models

An important class of LP problems that has many applications in logistics, distribution, assignment, and scheduling is *network models*. Special-purpose algorithms that have been

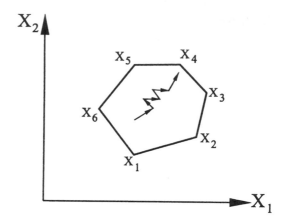

Figure 27.3   Interior-point method.

developed to solve these problems efficiently capitalize on the particular network structure. Some of these algorithms are typically 300 times faster than the simplex method (Ignizio and Cavalier, 1994), although, of course, they only apply to appropriate network problems.

The *transportation problem* is a network problem in which the objective is to satisfy the demands of n customers through transporting a commodity from m suppliers at minimum cost. Let c_{ij} denote the cost of transporting one unit of the commodity from the ith supplier to the jth customer. Denote by S_i and D_j the capacity of supplier i and the demand of customer j, respectively. Then the primary version of the transportation problem is to find the amount of commodity to be transferred from the ith supplier to the jth customer, x_{ij}, as follows:

$$\text{Min} \sum_{j=1}^{n} \sum_{i=1}^{m} c_{ij} x_{ij}$$

Subject to:

$$\sum_{j=1}^{n} x_{ij} = S_i \qquad i = 1, \ldots, m,$$

$$\sum_{i=1}^{m} x_{ij} = D_j \qquad j = 1, \ldots, n,$$

$$x_{ij} \geq 0 \quad \forall i, j.$$

It is noteworthy that some other problems such as production planning can also be modeled as transportation problems.

Sometimes it is advantageous to send a commodity from one supplier to another before transferring it to a customer. Also, it might be more beneficial to transfer a commodity from one customer to another customer. The *transshipment problem* is an extension of the transportation problem that allows these strategies to be included.

One important property of a network model is unimodality. A matrix is *unimodular* if the determinant of any square submatrix is $+1$, -1, or 0. In network models, the coefficient matrix is unimodular. When a problem is unimodular, applying an LP procedure to solve it leads to an integer solution. In other words, all coordinates of every extreme point of a network model are integers. This property permits the avoidance of integer programming algorithms that would be computationally more time-consuming than LP algorithms. Integer programming models are discussed in the next subsection.

The *assignment problem* is a special type of transportation problem in which the right-hand sides of constraints are unity and decision variables are all binary. Equivalently, the primary version of the assignment problem is to assign n persons to n jobs in order to fill the jobs at minimum cost. Suppose c_{ij} is the cost of assigning the ith person to the jth job, and x_{ij} is a binary variable that is 1 if person i is assigned to job j and 0 otherwise. Then the assignment problem can be formulated as follows:

$$\text{Min} \sum_{j=1}^{n} \sum_{i=1}^{n} c_{ij} x_{ij}$$

Subject to:

$$\sum_{j=1}^{n} x_{ij} = 1 \qquad i = 1, \ldots, n$$

$$\sum_{i=1}^{n} x_{ij} = 1 \qquad j = 1, \ldots, n,$$

$$x_{ij} = 0 \qquad \text{or} \qquad 1$$

Note that the optimal values of the decision variables (which solve the assignment problem) must be integers. Due to the unimodality of the assignment problem, however, any LP algorithm can be employed to solve it. However, the most efficient method for solving assignment problems is a specially designed technique called the *Hungarian* method.

27.4.4 Integer Programming

In an integer-programming (IP) model the decision variables can only have integer values. For instance, most models require the number of airplanes to serve a given route, or the number of workers for a job, to take on integer values only. In many decision problems, such as project selection and capital budgeting, the decision variables are restricted to zero and one. This special type of IP, for which all decision variables are binary, is called zero-one programming. If all decision variables are integers, the problem is a *pure* IP problem, whereas if some are continuous and some integer, it is a *mixed* IP problem.

IP problems are inherently much harder to solve than LP problems. Hence, an immediate suggestion to solve an IP problem is to solve the relaxed LP problem and then round off any noninteger variables to the nearest integer. Obviously, this approach is not likely to be used for zero-one problems. Moreover, the rounded value of an LP solution may be infeasible, or in some cases may be quite far from the optimal integer solution. For example, consider the following IP problem:

$$\text{Max } Z = 11x_1 + 4x_2$$

Subject to:

$$3x_1 - 2x_2 \leq 5$$

$$-x_1 + 2x_2 \leq 4$$

$$2x_1 - x_2 \leq 4$$

$$x_1 \geq 0, \qquad x_2 \geq 0$$

x_1, x_2 are integers

The optimal LP solution to the relaxed problem is $x_1 = 2\frac{1}{8}$, $x_2 = 2\frac{3}{16}$, which is far from the optimal integer solution, $x_1 = 2$, $x_2 = 3$. It is quite common that one cannot obtain a good solution by rounding off an LP relaxation solution.

As explained earlier the LP solution of any unimodular problem (such as the assignment problem) has integer values. One can therefore use any conventional LP procedure to solve a unimodular problem. Most IP problems are not unimodular, however, so different approaches had to be developed to solve them.

Cutting plane techniques constitute a set of algorithms in which additional inequalities (cuts) that must be satisfied by integer solutions are added to the constraints, in sequence. The process is terminated when the relaxed LP solution becomes integer. If decision variables are bounded above, one can enumerate, implicitly or explicitly, all feasible solutions to find one that optimizes the objective function. This approach is called the *enumeration method*. Different criteria such as integrality, nonnegativity, and feasibility are used to rule out some of the search directions that cannot improve the current solution. In fact, the efficiency of enumeration methods is highly dependent on these criteria. *Branch-and-bound* algorithms are among the enumeration methods that have been shown to be the most successful for IP problems (Williams, 1990). The *search* algorithm is another enumeration technique that is used for zero-one programming problems.

Although a large amount of effort has been expended on developing efficient techniques to solve IP problems, there is no global algorithm that can solve all types of integer problems efficiently. The theory of complexity suggests that finding a universal IP algorithm is unlikely. Most often, an algorithm is constructed to exploit the special structure of a given problem; of course, it may not be efficient for other IP problems. Moreover, a wide variety of heuristics has been proposed; their purpose is to provide "good" rather than "optimal" solutions (see Section 27.8).

In what follows, some well-known IP solution methods with special structures are described. Each of them has a vast literature that includes both theoretical and computational developments.

The *knapsack problem* is an important IP problem that has been extensively studied and utilized by practitioners and theorists. The simple structure of the knapsack problems allows one to investigate conveniently the combinatorial properties of many other integer problems. Moreover, a number of complex combinatorial optimization problems can be solved using a series of knapsack problems (Martello and Toth, 1990). In practice, many optimization problems, such as capital budgeting, project selection, cutting stock, and cargo loading, can be modeled as knapsack problems.

Suppose that a hiker who has only one knapsack must select some items from a set of items to put in the knapsack in order to maximize its utility. Define a set of binary variables, x_i, $i = 1, \ldots, n$, such that $x_i = 1$ if item i is selected, and $x_i = 0$ otherwise. Let c_i be the utility of, and a_i the size of, item i. Then the knapsack problem can be formulated as follows:

$$\text{Max } Z = \sum_{i=1}^{n} c_i x_i$$

Subject to:

$$\sum_{i=1}^{n} a_i x_i \leq b$$

$$x_i = 0 \text{ or } 1, \, i = 1, \ldots, n$$

where b is the size of knapsack. Some variations on knapsack problems are *multiple-choice knapsack, bounded knapsack, subset-sum, change-making,* and *multidimensional knapsack* problems (Martello and Toth, 1990).

Set covering is another IP problem. Suppose that in a school n courses have to be taught. There are m candidates to teach these courses such that each candidate is qualified to teach some of the courses. The set-covering problem is then to find the least-cost set of candidates to "cover" all courses. Let x_i be a binary variable that takes the value 1 if the ith candidate is selected and is 0 otherwise. Define c_i as the cost associated with the selection of candidate i. Then a basic set-covering problem is as follows:

$$\text{Min } Z = \sum_{i=1}^{n} c_i x_i$$

Subject to:

$$\sum_{i=1}^{n} a_{ij} x_{ij} \geq 1 \qquad j = 1, \ldots, n,$$

$$x_{ij} = 0 \text{ or } 1, \qquad i = 1, \ldots, n, \qquad j = 1, \ldots, n$$

where an element of the coefficient matrix, a_{ij}, is 1 when candidate i is capable of teaching course j, and 0 otherwise. Hence, in a set-covering model the objective function is minimization, all right-hand sides are unity, and all elements of the coefficient matrix are 0 or 1.

If the inequality constraints of the set-covering problem are changed to equality, the problem is transformed to a *set-partitioning problem*. In this case, the objective is to cover all courses using the candidates without any overlap. In a set-partitioning problem, a maximization or minimization objective function may be used.

The *set packing-problem* is similar to the set-partitioning problem, except that the objective function is to be maximized and the constraints are ($\leq$). In set packing, the problem is to gain the maximum utility from assigning candidates to subsets with no overlaps.

One of the most important IP problems is the *traveling salesman problem* (TSP). Generally, the objective of the TSP is to find the shortest possible route visiting each city in a given set. TSP, which exhibits most of the difficulties that can be found in combinatorial optimization problems, has a history going back to the 19th century (Hoffman and Wolfe, 1985). Due to the unique characteristics and many applications of TSP, it is widely used as a benchmark for comparing different IP algorithms and heuristics.

Many practical problems can be formulated in TSP form. For instance, the problem of sequencing jobs on a machine to minimize total setup costs can be modeled as a TSP. An enormous amount of research has been devoted to understanding the characteristics of TSPs, and finding exact algorithms or heuristic procedures to solve them.

The TSP is among the hardest combinatorial problems to be solved for optimality. Some practical TSPs are very large. For example, in very large-scale integrated (VLSI) circuit design, the number of nodes (cities) is more than a million. Moreover, as the number of cities increases, the total number of possible tours grows exponentially. To date, the biggest TSP that has been solved completely is a 4461-city problem (Reinelt, 1994). Recent advances in solving TSPs are due to the rapid improvement in computer hardware, and the development of the theory of *polyhedral combinatorics*.

There are many different ways to formulate a TSP problem. The standard formulation assumes that the cities to be visited are numbered 1, 2, . . . , n. Any feasible solution to the TSP problem is called a *tour*. Let x_{ij} be a binary variable that equals 1 if the tour goes directly from i to j, and 0 otherwise, and let $\mathbf{X}$ be the matrix with elements x_{ij}. If c_{ij} denotes the cost of traveling directly from city i to j, then the following is a TSP model:

$$\text{Min} \sum_{j=1}^{n} \sum_{i=1}^{n} c_{ij} x_{ij}$$

Subject to:

$$\sum_{j=1}^{n} x_{ij} = 1 \qquad i = 1, \ldots, n$$

$$\sum_{i=1}^{n} x_{ij} = 1 \qquad j = 1, \ldots, n$$

$$x_{ij} = 0 \text{ or } 1 \qquad \text{and} \qquad \mathbf{X} \text{ is a tour}$$

Even though the TSP is structurally very similar to the assignment problem (except for the last constraint, the formulations are identical), it cannot be solved by standard LP algorithms. The tour constraint is important, and must be formulated carefully to avoid generating partition-based subtours as solutions. Adding these constraints greatly increases the size of the problem and makes the solution process very difficult. For instance, in a 100-city problem, one must add 9,900 specific constraints to capture the tour requirement. [For more information on integer and combinatorial problems, refer to Nemhauser and Wolsey (1988), and Martello and Toth (1990)].

27.4.5 Nonlinear Programming

When the objective function and/or constraints of an optimization problem are nonlinear, or when there are interactions among some decision variables, the resulting constrained-optimization problem is a nonlinear programming (NLP) problem. NLP problems are much harder than LP problems to solve, because the optimal solutions of NLP problems can be located anywhere in the feasible region, unlike LP models for which the optimal solutions are always extreme points. There is no single technique that solves the many forms of NLP models that can arise in practice. Hence, many procedures have been developed to find the solution(s) of NLP problems. Some of the sources furnishing descriptions of NLP include Fletcher (1981), Dennis and Schnabel (1983), and Gill et al. (1992).

Two main branches of NLP models are the *convex* and *nonconvex* models. A minimization (maximization) problem is convex if the objective function is convex (concave) and each constraint is convex. For a convex problem, the local minimum (maximum) is a global minimum (maximum). Any problem that does not satisfy these convexity conditions is nonconvex, which implies that there can be no assurance that a local optimum is global.

NLP models can also be categorized into *constrained* and *unconstrained* problems. Searches according to the *Fibonacci, golden section,* and *gradient* methods are the most widely used procedures to solve both one-variable and multivariable unconstrained problems.

A *quadratic programming* problem is an NLP problem with linear constraints and a quadratic objective function. When the objective function of a quadratic minimization problem is convex, a modified simplex algorithm can solve the problem efficiently. Note that for a quadratic problem all expressions derived from the *Kuhn-Tucker conditions* for optimality are linear.

Two special cases of nonconvex models are *geometric programming,* which often arises in engineering design, and *fractional programming.* In geometric programming, the objective function is based on a generalization of the arithmetic–geometric mean inequality. In fractional programming, the objective function is in the form of a ratio of two functions. Under certain conditions there are efficient algorithms for solving these two nonconvex models (Ravindran et al., 1987).

Separable programming problems are problems in which the nonlinear objective function can be expressed as the sum of nonlinear functions of single variables. A separable problem can be approximated as a piecewise-linear function. This allows one to use a modification of the simplex method to obtain a global optimum.

27.4.6 Other Operations Research Models

Many other OR methods are described in standard OR textbooks, such as Hillier and Lieberman (1990) and Winston (1991). Some of these are now briefly mentioned here.

Dynamic programming is a general technique for optimization that can be applied fairly widely. Dynamic programming models are not required to have any specific structure; they are applicable to any problem that can be broken into distinct stages. Hence, dynamic programming can be used with any linear, nonlinear, or integer programming problem that can be represented as a multistage problem. The dynamic programming technique was originated by Bellman in the decade following World War II. A basic reference for dynamic programming is Bellman (1957).

The first step in solving a dynamic programming problem is to find the optimal solution for the last (first) stage of the problem. Then move backward (forward), one stage at a time, to find the optimal solution, beginning at the current stage using a *recursive equation* that relates optimality in consecutive stages, keeping the track of the optimal solution from the current stage to the end point. The process continues until the original problem is solved. Inventory control, scheduling, production planning, and integer programming are among the general applications of dynamic programming.

Some real-world problems are too complex to be modeled using a standard mathematical model. The stochastic nature of parameters, for instance, can complicate relationships among decision variables, whereas complex objective functions and constraints in some real-world problems can render the application of standard OR methods very difficult. *Simulation* is a powerful technique for such problems. The main difference between a simulation model and an analytical model is that a simulation model analyzes "what if" questions by running sample problems and generating performance measures for the proposed policy, whereas an analytic model attempts to provide a more formal understanding of a problem.

Queueing theory is the mathematical study of waiting line models. Usually, the inputs of a queueing model are the distribution of an arrival process and characteristics of the system under study. The characteristics of the system include the number of servers, the service order and discipline, and the distribution of service times. The output of a queueing model is a description of the performance attributes of the system under a specific policy. The solution of a queueing model determines, for example, the fraction of time that each server is idle, the expected waiting time of customers, the expected number of customers waiting in the queue, and the number of servers necessary to ensure some level of performance for the system (Gross and Harris, 1985).

The objective of *inventory theory* is to help managers to maintain inventory and meet customer demand optimally. In general, an inventory model specifies the frequency and volume of orders in production or service companies. The objective function of an inventory model is usually the minimization of total cost associated with maintaining and ordering commodities. The cost items include

- Ordering cost (for paperwork, or for setup time at the start or end of the production of a product).
- Holding cost, including storage cost and the opportunity cost of holding inventory.
- Shortage costs, such as the cost of back-ordering and lost sales.

Demand and the lead time are usually probabilistic, so most applications of inventory models are based on probabilistic inventory theory. The difficulty of estimating the costs of losing sales and holding inventory makes inventory models hard to apply. Exchange curves provide a technique for estimating some of their parameters (Peterson and Silver, 1979).

Material requirement planning (MRP) is a branch of inventory theory that deals with controlling inventories and scheduling production of components and subcomponents of a product. For this purpose, MRP establishes the relationship between final product demand and demand for parts and assemblies to schedule production and ordering of components.

Just-in-time (JIT) *Inventory Planning* is an inventory system developed to reduce inventory costs by scheduling production or delivery of items to occur as close as possible to the time that they are needed. Reducing the setup costs is very important in successful implementation of JIT systems, because setup costs are a major factor in ordering large quantities. In contrast to conventional production systems that are based on *push*, JIT works according to *pull*. In push systems, components are pushed into consecutive production steps; in a pull system a workstation does not produce any component until it is requested by the next workstation. The Toyota Kanban system is an example of the successful implementation of JIT.

Multiple-criteria decision-making techniques are designed for decision problems in which multiple, generally conflicting, objectives are to be explicitly modeled. Section 27.6 introduces some basic ideas and techniques of multiple-criteria decision making.

Multiple-participant decision-making methods are mathematical models used to recommend a course of action to a DM (client), when two or more DMs, with conflicting interests, are involved in determining an outcome. A description of some of these methods, including game theory, is presented in Section 27.7.

27.5 GENERATING AND SCREENING ACTIONS

27.5.1 Generating Actions

Generating sets of actions or alternatives is a very important step in the decision-making process, but one to which little study has been devoted (Keeney, 1992; Vincke, 1992). In most decision-making procedures, it is assumed that the DM deals with a well-defined set of predefined actions, and furthermore that these actions are stable during the process of decision making. However, real-world problems deviate from these assumptions; most often before formal decision analysis is carried out some preliminary work is necessary to define, expand, combine, or reduce the set of actions.

There has been relatively little research on procedures for generating actions. Keeney (1992, 1994) introduces the idea of generating actions according to the concept of *value-focused thinking*. He believes that one should think not of decision problems but of decision opportunities. Enthoven (1975) argues that generating one good new action is most likely to be worth more than an evaluation of old actions, no matter how thorough. Miser and Quade (1988) claim that the definition and generation of actions is the most important and least discussed issue in decision making. In fact, the generation of basic actions does not occur at any specific point in a study, and may take place several times during the process of decision making. However, an action that is not generated at an early stage may never be generated. Hence, the best action may not be found, simply because it was not considered (Miser and Quade, 1988).

In general, generating actions upon which alternative solutions can be built requires a good understanding of the problem. Interactions among different parties involved in the problem usually lead to some new ideas for potential actions. Considering the overall goal of the decision problem and related evaluation criteria may help to suggest new actions. One way to expand and generate a diverse set of actions is to consider all constraints as soft conditions, which could possibly change. Keeney (1992) believes that the process of generating actions is constraint-free thinking, and selecting among actions is constrained thinking. He presents several examples and case studies to demonstrate the role of value-focused thinking in generating actions (Keeney et al., 1996).

In some public policy problems, the generation of all actions and alternatives yields a huge number of alternatives. Hence, one feasible approach is to generate a subset of alternatives that is representative of them all. Then evaluation of these alternatives may provide enough insight to generate another subset, this time by concentrating on a specific section of the decision space that is most promising. Brill et al. (1982) develop an iterative mathematical model, called HSJ (hop, skip, and jump), to produce a small number of feasible alternatives in each iteration. By minimizing the sum of the nonzero variables in the previous iteration, HSJ finds a new set of alternatives that is very different from the previous sets. The approach is illustrated in a land-use planning problem. Nakamura and Brill (1979) employ a branch-and-bound procedure to generate alternatives. Their algorithm successively classifies possible actions into mutually exclusive groups, and is illustrated using a regional wastewater planning problem. Stewart and Scott (1995) employ a statistical approach called *response surface fitting* to generate a subset of disparate alternatives. Finally, as Rajabi et al (1998, 1999) show, selecting the best subset of actions can be problematic when actions are interdependent.

27.5.2   Screening Actions

In many practical decision problems the set of potential actions is very large, so it is worthwhile to identify some promising actions to investigate in more detail. This is particularly important when a subset of actions is to be selected, since the number of available alternatives is very large due to the combinatorial nature of the problem. As described in the previous subsection, in the early stages of the decision analysis process, one should attempt to identify actions that seem reasonable. Unfortunately, it is often costly and time-consuming to evaluate actions according to every criterion. In many public policy problems, this evaluation is accomplished through the value judgments of experts in different areas of specialization. Accordingly, it may be quite difficult to collect the required assessments for all actions.

Nevertheless, many actions that have been generated in the early stages can be removed during the screening phase. The main objective of screening is to remove inferior actions from the set of potential actions, so that those remaining can be investigated in more detail, perhaps using more accurate information or additional assessment criteria. Procedures for screening a set of available actions are discussed next.

27.5.2.1   Screening Procedures. This section describes some more general screening procedures that have been applied widely. These methods have been used in the context of certainty as well as uncertainty, in single and multiple-criteria decision problems, and for qualitative as well as quantitative criteria. It is worth mentioning that in most cases a sequence of these methods, rather than a single method, can more reliably eliminate inferior actions. The techniques discussed are the *feasibility test,* the *dominance relation, elementary methods, successive elimination,* and *bounds on performance level.*

> *Feasibility Test.* The feasibility test is a good first stage in the process of screening. An action might be infeasible due to resource constraints, technological constraints, or political issues. One should be careful when removing infeasible actions because in many cases constraints that at first seem hard can be relaxed. In this way, some valuable actions can be considered as feasible and worthy of further investigation.
>
> *Dominance Relation.* Dominance relations have been widely used for screening actions, especially within the multiple-criteria context described in the next main section. An action is dominated when there exists another action that scores at least as well on all criteria and strictly better on at least one criterion. For example, if one action consumes as many resources as another action but its outcome is less desirable, then that action is considered to be dominated.
>
> It is generally agreed to be unreasonable for a DM to select any dominated action. However, one should remove dominated actions only after a complete sensitivity analysis. In other words, one should ensure that an action is indeed worse than some other action under every circumstance before removing it (Hobbs et al., 1990). In most cases, the set of nondominated actions is large. After the screening stage, the DM may still face a complex task of selecting the best action among the many remaining actions. Hence, significant work has been devoted to enriching domination relationships to distinguish and screen out more inferior actions.
>
> In general, including information about the DM's preference structure reduces the decision space, thereby removing more actions from the set of potential actions. For instance, imposing bounds on the importance of criteria, or on the probability

of occurrence of scenarios, can be employed for this purpose (Bana e Costa, 1990; Moskowitz et al., 1992). This procedure is usually called domination in the *reduced decision space*. More complex methods for recognizing domination can be found in the multiple-criteria literature. For example, the techniques of Korhonen et al. (1984) and Koksalan et al. (1984) find dominated actions that fall in the convex cone of inferior actions generated by some nondominated actions. Pairwise comparison of actions is used to construct the cones of inferior actions.

One should keep in mind that in subset selection problems domination relations must be used extremely cautiously to remove individual actions, since it is possible for individually dominated actions to be included in the best subset of actions. Rajabi et al. (1995) describe procedures for determining whether a dominated action might be included in the best subset.

Elementary Methods. There are elementary methods that are quick and do not require much information to rank the set of actions. Actions with low rank are then eliminated. Hobbs et al. (1990) call this procedure the *amalgamation* approach.

Successive Elimination. In the successive elimination approach, which is usually used in the context of uncertainty, one should calculate the best and worst scores of each action under all possible scenarios. Subsequently, an action whose best score is less than the worst score of some other action is dominated and can be removed. In the case of multiple-criteria problems, this can be done by examining different weights for criteria (Sarin, 1977; Hobbs et al., 1990).

Bounds on Performance Level. In this approach, some constraints are placed on the performance of actions according to one or several criteria. Actions that do not meet these constraints are considered to be inferior and unacceptable to the DM. The bounds are most often arbitrary and reflect the subjective values of the DM. The selection of appropriate bounds can be very difficult. If, however, they are chosen properly, then this approach is quite efficient and results in the removal of many actions. At the same time, the best scores for actions on each criterion can give some insights for selection of the bounds. Walker and Veen (1981, 1987) use bounds on criteria to illustrate the screening of water resources actions (see also Miser and Quade, 1988). Bounds can be employed for all criteria, or for some that are considered key.

27.6 MULTIPLE-CRITERIA DECISION MAKING

27.6.1 Background

Today, it is well understood that most decision problems inherently involve choices that ought to be judged according to more than one criterion. In fact, multiple-criteria decision-making (MCDM) problems arise naturally in many situations, both strategic and routine, and MCDM methods have been widely applied in public policy, engineering, and design. For example, in the selection of plans for a road, construction costs and expected rate and severity of accidents are the main criteria. In water-resources planning, criteria such as power-generation capacity, flood-control capability, and environmental impacts may be essential. For selection of a site for waste disposal, considerations such as infrastructure cost, environmental risk, and size of affected population are often taken into account. In designing a gearbox, several criteria, including volume of material, maximal peripheral velocity between gears, width of the gearbox, and distance between axes of input and output shafts, should be minimized simultaneously (Osyczka, 1984). Increasing the output quality level and reducing the overall inspection cost are two conflicting criteria applicable to the design of quality control policies in any production line.

MCDM dates back to the late 19th century, when the concept of equilibrium in consumer economics was introduced by Edgeworth and Pareto (Stadler, 1979). However, MCDM became a useful decision technology in the early 1970s. Specifically, after the first conference on MCDM, held at the University of South Carolina in 1973, the field has been one of the fastest growing in OR, as evidenced by the enormous numbers of books, journal articles, and congresses in both the theory and the application of MCDM (Vincke, 1992; Keeney and Raiffa, 1976; Steuer et al., 1996).

MCDM consists of a set of tools to help a DM or a group of DMs to make a decision by *finding, selecting, sorting,* or *ranking* a set of actions according to two or more criteria, usually conflicting. A possible action may be specified explicitly or implicitly (by identifying the constraints to be satisfied). As pointed out in Section 27.5, the definition and generation of *actions* is an important step in the process of MCDM, but one to which relatively little research effort has been devoted (Vincke, 1992; Keeney, 1992). For most real-world problems there is no preexisting set of well-defined actions. Most often, before any formal decision analysis can be undertaken, some preliminary work to define, combine, expand, or reduce the set of feasible actions is necessary. The set of feasible actions can be reduced by removing some inferior actions, identifying those that do not meet some level of acceptability, or that do not meet key performance standards on criteria (see Section 27.5 for more detailed discussion.).

The criteria by which actions are to be evaluated and compared are usually in conflict with each other, especially if each criterion represents the interest of a specific group of DMs. For example, building a factory may generate job opportunities but, on the other hand, introduce adverse environmental impacts. Increasing the frequency of inspection in a production line decreases the number of defects but, on the other hand, increases the cost of quality control. Thus, it is rare to find an action that is best according to all criteria, and asking for an optimal solution to an MCDM problem does not make sense. Rather, one must search for a compromise solution that appropriately reconciles the different criteria. To find this compromise solution, it is necessary to learn something about the DM's preferences over the criteria. Hence, the role of the DM in MCDM is more explicit, and more crucial, than in single-objective optimization. One should keep in mind as well that due to the behavioral influence of tradeoffs across criteria, it is in many situations impossible to find a solution simply by implementing a mathematical model (Roy, 1990).

Once the actions and the criteria are constructed, one must measure or evaluate each action according to each criterion. Most optimization procedures are based on the assumption that one can assign a real number to represent the consequences of an action according to a criterion. However, in many real-world applications this is often a very difficult task. This issue is more important in MCDM, because in many MCDM applications some criteria are not quantitative. Often the natural way to express the consequences of actions is by using *ordinal* information, whereby the actions are ranked according to each nonquantitative criterion.

MCDM can be classified into two main branches, *multiple-attribute decision making* (MADM) and *multiple-objective mathematical programming* (MOMP). The former applies mainly when there is a small number of actions, the latter when the number of actions is large. Usually MADM applies to decision problems with discrete actions, and MOMP to when the action space is continuous. This section is mainly devoted to MOMP; some concepts of MADM are discussed in Chapter 28.

27.6.2 Overview of Multiple-objective Mathematical Programming

An MOMP problem can be expressed as follows:

$$\text{Max } \{f_1(x), \ldots, f_i(x), \ldots, f_n(x)\}$$

Subject to:

$$x \in S,$$

where $f_i(x)$ is the objective function i, x is the vector of decision variables, and S is the feasible space. In MOMP, a criterion is usually referred to as an *objective*. The main characteristic of an MOMP problem that distinguishes it from single-criterion problems is that there generally does not exist a solution that simultaneously maximizes all of the objectives. A solution which is best according to all objectives is called the *ideal point* and is denoted

$$Z^{**} = (Z_1^*, Z_2^*, \ldots, Z_i^*, \ldots, Z_n^*),$$

where $Z_i^* = \max\{f_i(x)\}$. Some MOMP approaches use this ideal point for assessing other solutions. Most theories of MOMP can be characterized according to the *nondominated* (*efficient* or *Pareto optimal*) solution concept.

An action x in this problem is defined to be *nondominated* if there is no other action x^0 such that:

$$f_i(x^0) \geq f_i(x) \qquad \forall i = 1, \ldots, n$$

and

$$f_k(x^0) > f_k(x) \qquad \text{for at least one } k$$

Figure 27.4 shows a two-criterion problem in which both criteria, f_1 and f_2, are to be maximized. The points on the dotted line are efficient, and other points are dominated. Note that point 2 is efficient, but is convex dominated by points 1 and 3.

The notion of efficiency is especially important in studying deterministic problems, although selected concepts such as stochastic dominance, mean variance dominance, probability, and utility dominance have also been defined in the MCDM literature (Yu, 1985). They have appeared infrequently in practical applications. Note that efficiency is weaker than optimality in the sense that in most cases there exist many efficient points but no ideal point. Hence, after finding the set of efficient solutions the DM must still choose one member of this set.

Most often, the objectives are in conflict with each other. When one of the objectives is improved, one or more other objectives is worsened. What is best for a DM, therefore, depends on the relative importance of the objectives. Furthermore, different DMs may have very different views of the importance of the objectives.

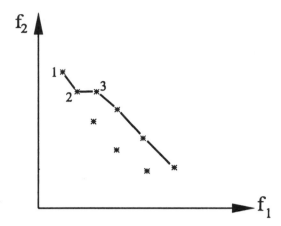

Figure 27.4 Efficient and dominated actions for a two-criterion problem.

One can classify MOMP approaches according to how and when preference information is articulated (Hwong et al., 1979):

1. No preference information,
2. A priori preference information,
3. Progressive articulation of preference information, and
4. Posterior preference information.

27.6.2.1   No Preference Information.   No preference information approaches to MOMP try to give a solution without using any information about the preference structure of the DM. *Maximin* and *elimination by aspects* are among this class of methods. Another approach in this category is the global criterion method, in which some global criterion, usually a relative deviation from an ideal point, is minimized (MacCrimmon, 1973; Hwang and Yoon, 1981).

27.6.2.2   A Priori Preference Information.   A priori preference information methods of MOMP begin with an exploration of the value function of the DM. Once the preference structure of the DM has been assessed, all objectives are aggregated into one, thereby changing the problem to a single-objective optimization problem. In most cases, assessment of the DM's value function is quite difficult and involves a great deal of subjectivity. Many theories and procedures have been developed for determination and characterization of the DM's preference structure for both deterministic and probabilistic cases. The capability of these procedures in conditions of uncertainty, and their usefulness for sensitivity analyses, are among their advantages (Insua, 1990). In addition, their solid theoretical foundations add to their attractiveness. However, they generally require the assumption that a value function exists, and often that it is additive; even if one is prepared to make these assumptions, it may be extremely difficult to construct a value function in practice. It is worth mentioning that these approaches have been mainly used in MADM. In addition to the *additive* form of the value function, others such as *multiplicative, polynomial,* and *partially additive,* have been proposed in the literature. One popular method that is designed primarily for use with a priori preference information is goal programming.

Goal Programming. Goal programming (GP) is perhaps the first formal technique of MCDM. It is recognized as the most popular and most accepted method in MCDM (Rajabi et al., 1996). Different versions of GP have been proposed in the literature. The fundamental idea is that the best solution is as close as possible to some predefined goals. Therefore, one must implement the following two steps before solving the problem:

- Specify the level of goals for all criteria.
- Define a metric to measure the distance of feasible actions from the target.

It is assumed that the DM can specify the desired goal (g_i) for each objective. Therefore, Problem (27.1) can be written as follows:

$$\text{goal } \{Z_1 = f_1(x)\} \qquad (Z_1 \geq g_1),$$
$$\text{goal } \{Z_2 = f_2(x)\} \qquad (Z_2 \geq g_2),$$
$$\vdots \qquad \vdots \qquad \qquad \vdots$$
$$\text{goal } \{Z_n = f_n(x)\} \qquad (Z_n \geq g_n)$$

Subject to:

$$x \in S$$

The information in braces on the left specifies the goals for the different objective functions. In GP the aim is to minimize the deviations (positive or negative) from a target. Define d_i^+ and d_i^- as the overachievement and underachievement, respectively, from a specified target on criterion i. Using different types of distance metrics for measuring the overall distance of the objectives from goals leads to different types of GP techniques. In general, the distance metric is

$$\sum_{i=1}^{n} [(w_i^+ d_i^+)^\alpha + (w_i^- d_i^-)^\alpha]^{1/\alpha}$$

where w_i^+ and w_i^- are the importance of positive and negative deviations, respectively, from the target of ith objective. Note that with $\alpha = 1$, the preceding expression is a simple additive aggregation function; with $\alpha = 2$, it is a Euclidean or L_2-norm; putting $\alpha = \infty$ leads to minimax or Chebyshev GP. Assuming an additive function, the GP formulation of this GP problem is as follows:

$$\text{Min } \sum_{i=1}^{n} w_i^+ d_i^-$$

Subject to:

$$f_1(x) + d_1^- - d_1^+ = g_1,$$
$$f_2(x) + d_2^- - d_2^+ = g_2,$$
$$\vdots \qquad \vdots \qquad \vdots$$
$$f_n(x) + d_n^- - d_n^+ = g_n$$
$$x \in S$$

Note that because the objective function in the original program is maximization, a positive deviation does not correspond to any penalty, and therefore is omitted here.

Another type of GP is lexicographic GP, in which the criteria are ranked rather than weighted (Ignizio, 1985; Romero, 1990; Schniederjans, 1995).

27.6.2.3   *Progressive Articulation of Preference.*

Due to the great difficulty of determining a DM's preference explicitly, many procedures try to elicit them progressively. Methods that alternate between computation and interaction with the DM are called *interactive*. The process starts with little or no preference information. At each iteration a set of solutions (usually, nondominated solutions) is presented to the DM. As each solution is examined, the DM decides upon the updated preference information input into the model. The process terminates when the DM is satisfied with the solution currently proposed by the model. Observe that both the model and the DM learn from each other using an interactive method.

The first interactive method, called STEM, was proposed by Benayoun et al. in 1971. Although originally proposed for solving linear problems, its structure permits it to be applied to integer and nonlinear problems. This procedure is based on reducing the feasible space by adding more constraints, obtained through interaction with the DM. *Augmented weighted Chebyshev* is the method used for assessing the compromise solution at each step.

The GDF method (Geoffrion et al., 1972) is another interactive approach to MOMP. Using an implicit utility function, this method attempts to find the best solution using the Frank–Wolf gradient algorithm.

The method of Zionts and Wallenius (1976) is applicable to linear problems. Relying on the assumption that the DM's utility function is pseudoconcave, this method generates extreme efficient points at each iteration. Adjacent extreme points are compared by the DM and this information is added to the model for the next iteration.

The reference point method (Wierzbicki, 1980, 1982) is based on the DM's aspiration levels (reference point). An achievement scalarizing function is used for projecting the reference point unto the set of nondominated solutions. At each iteration, the DM specifies new aspiration levels; the process terminates as soon as the DM is satisfied by the optimal solution of the achievement scalarizing function.

The method of Steuer and Choo (1983) generates samples of the efficient points by using the augmented weighted Chebyshev norm. Using a filtering algorithm, this method gives a prespecified number of efficient points, that are dispersed throughout the space of efficient solutions, and can be considered representative. For further information about other types of interactive methods, see Steuer (1986) and Vincke (1992).

27.6.2.4 *Posterior Preference Information.*

Posterior preference information methods start by solving the problem without articulating the preference structure. A compromise solution is then obtained by assessing the preference structure. Usually, the first step is carried out by vector optimization, by which the set of efficient solutions, or a subset, is generated. Much research has been devoted to generating efficient solutions for linear, nonlinear, and integer problems (Chankong and Haimes, 1983). There are three main approaches for generating efficient solutions as follows:

The Weighted Approach. This set of approaches constitutes the most common way to generate efficient solutions, especially in the linear case. If a solution x^0 maximizes

$$\sum_{i=1}^{n} \lambda_i f_i(x)$$

where $\lambda_i \in \Lambda$, $\Lambda = \{\lambda \in R^n | \lambda_i > 0, \sum_{i=1}^{n} \lambda_i = 1\}$, and $x \in S$, then x^0 is efficient.

By varying λ within Λ, different efficient solutions will be generated. In multiple-criteria linear problems, the set of efficient solutions obtained using this procedure is the set of nondominated extreme points, and if the DM's true utility function is nonlinear (which is usually the case), then it is possible that the best solution will not occur at an extreme point. To handle this problem, methods have been developed for generating other nonextreme efficient points, usually by taking convex combinations of the extreme efficient points (Steuer, 1986; Ringuest, 1992). It is noteworthy that if all objective functions and the feasible space are convex, then all efficient points can be generated by varying λ_i.

The kth Objective ϵ-Constraint Approach. In this method, one objective (k) is selected for maximization (minimization) while the rest of the objectives are considered as constraints with ϵ or lower (upper) bounds. Hence, the ϵ-constraint formulation of Problem (27.1) is as follows:

$$\text{Max } f_k(x)$$

Subject to:

$$f_i(x) \geq \epsilon_i \qquad i = 1, 2, \ldots, n \qquad i \neq k$$

$$x \in S$$

This technique can generate all efficient solutions for convex and nonconvex cases. However, there is no systematic procedure for selecting the value of ϵ for the objective that is to be maximized.

The kth Objective Lagrangian Approach. Using this approach, Problem (27.1) is converted to the following single-criterion problem:

$$\text{Max } f_k(x) + \sum_{i \neq k} u_i f_i(x)$$

Subject to:

$$x \in S$$

where u_i are Lagrange multipliers.

Other approaches for characterization of efficient solutions that have been presented in the literature are *proper equality constraint, hybrid weighted ϵ-constraint,* and *weighted norms.* In general, posterior preference information approaches first try to obtain maximum information from the objective data, before switching to the task of learning about the DM's preferences.

27.7 MULTIPLE-PARTICIPANT DECISION MAKING

27.7.1 Overview

When interests concerning a decision problem conflict, differences of opinion inevitably arise over its solution. If the real-world outcome reflects the choices of several interested individuals and groups, the conflict is strategic and is amenable to formal study using various analytic techniques. For example, when a firm is considering a new product, there may be conflicting viewpoints within the organization over its design, production, and distribution. Moreover, the firm must take into account how its competitors may react, and how it can expeditiously capture a significant share of the potential market. Obviously, methodologies and techniques from conflict resolution have key roles to play in such systems management and leadership problems.

A variety of formal mathematical models for studying conflict have been developed within the field of *game theory,* which in turn is often considered to be a part of OR. Most multiple-participant decision making (MPDM) techniques are related to game theory in some way, often quite closely. Some MPDM techniques can be considered as optimization methods that determine the best a given participant can achieve within the social constraints of the conflict. Similarly, equilibria can be viewed as stable compromise resolutions from the usually much larger set of possible scenarios. In this way, an MPDM method can act as a screening tool.

The main objective of this section is to provide an overview of MPDM, and point out some methods that may be particularly well suited for certain systems management problems. Standard books furnishing descriptions of a range of game theory and MPDM methods include von Neumann and Morgenstern (1944), Howard (1971), Rapoport et al. (1976), Jones (1980), Owen (1982), Shubik (1982), Fraser and Hipel (1984), Friedman (1988), Brams and Kilgour (1988), Myerson (1991), Binmore (1992), and Fang et al. (1993). Subsequent to a description of the main components of most conflict models, some important MPDM methods are discussed and their contributions explained.

27.7.2 General Structure of a Conflict Model

A *conflict model,* which may be an abstract game model, provides a conceptualization of a real-world conflict within a formal mathematical framework. This conflict model or game constitutes an approximation to the actual conflict that systematically structures its key components. Because the conflict model and related analysis are designed to reflect human perception of conflict, they may be of assistance to DMs in developing and reinforcing their understanding of the conflict. The model and analytical results can also act as a bookkeeping technique to help keep track of what is happening, and to furnish guidance and decision support for participating DMs.

To construct or calibrate a conflict model, the main information required consists of the identification of the DMs, as well as each DM's possible courses of action and preferences. A DM may consist of an individual or a group, such as a governmental, industrial, or military organization. For example, when a buyer and a seller discuss the price of a used car, each DM is a single person. In an international trade conflict, the DMs may represent governments and multinational corporations. DMs are also referred to in the literature as players, actors, stakeholders, or participants.

When each DM chooses a course of action, the result is a *state* or outcome in the game.

Each DM has *preferences* among the possible states that could occur. When states are ranked or ordered, the preferences are said to be *ordinal*. If a DM's preferences are strict ordinal, the states are ranked from most to least preferred, and there are no equally preferred states.

If a DM has cardinal preferences, then the DM's payoff or value for each state is given as a real number. For instance, when the only dimension to consider is cost, the value of each state might be estimated using its dollar value. In *utility theory* (von Neumann and Morgenstern, 1953), there are extensive procedures for assigning a utility value to each state in such a way that expected utilities accurately describe the values of states and the DM's risk attitude. The determination of cardinal preferences is generally quite difficult.

A variety of conflict models have been developed in the MPDM, game theory, and conflict resolution literature. These are briefly described below, followed by an overview of how a given model is used for analysis purposes.

27.7.2.1 Normal Form.

27.7.2.1 Normal Form. The normal form for the abstract game model was originally put forward by von Neumann and Morgenstern (1944) and is commonly used for describing a conflict in which each DM must select a course of action once and for all, in ignorance of the choices of all other DMs. The basic building blocks for the normal form are strategies for each DM, which combine to form the states, as well as each DM's preferences among the states. Note that a strategy is a complete plan of action. When all players have selected strategies, the state or outcome is determined once and for all. A normal form is usually written as a matrix if there are two DMs; if there are many DMs, a multidimensional matrix is used. Each strategy for the first DM is represented by a specific row in the matrix, while each strategy for the second DM is depicted by a column. Therefore, each cell in the matrix represents a possible state in the game. The normal form is commonly referred to as the *matrix form*.

A normal form game with two DMs, one with m strategies and the other with n strategies, is called an $m \times n$ game. In fact, 2×2 games often reflect basic human conflicts in their simplest forms. Well-known 2×2 games include Prisoners' Dilemma and Chicken (Rapoport et al., 1976). Figure 27.5 shows the normal form of Prisoners' Dilemma. In this figure, the rows of matrix represent the available strategies for player 1, and the columns the strategies of player 2. For this game, each player has two possible strategies: cooperating (option C) or defecting (D) and thereby acting alone. The game is illustrated by the decision problems of two bank robbers who face conviction on a lesser charge. If one defects and testifies against the other, the defector will go free and the other

PLAYER 2

	Cooperate	Defect
Cooperate	(3,3) (C,C)	(1,4) (C,D)
PLAYER 1		
Defect	(4,1) (D,C)	(2,2) (D,D)

Figure 27.5 Prisoners' Dilemma in normal form.

will be severely punished. If both defect, their sentences will be longer than if both co-operate.

As shown in Figure 27.5, the four possible states in Prisoners' Dilemma are cells denoted (C,C), (C,D), (D,C), and (D,D). The dual numerical entries in each cell represent the ordinal payoffs in which the first number is the rank associated to that state by player 1, and the second number is player 2's ranking (higher numbers are preferred.)

Prisoners' Dilemma is often used as an explanation of the difficulty of achieving a cooperative outcome in an individualistic world. When both players cooperate, they do quite well and jointly attain state (C,C), but cooperation is unstable because each player would do better, individually, by defecting. In fact, each player would improve its own payoff by defecting, whatever the opponent's choice.

27.7.2.2 Extensive Form. The extensive form was originally presented by von Neumann and Morgenstern (1944) and later refined by Kuhn (1953). A tree structure is used to describe the order of play, the choices a DM may make, the timing and possible outcomes of chance moves, and the information available to each DM at each move. A given node in the tree corresponds either to a chance move or to one of the DMs, and the time and state of information at which the DM makes a choice. Each branch from the node represents a possible choice by the DM. Although the extensive form can portray the play of the game in detail, it is not generally very compact, and becomes extremely difficult to use with even a modestly complex conflict. However, the extensive form can be applied to some practical problems such as compliance to environmental laws and regulations (Kilgour et al., 1992; Hipel and Fang, 1994).

27.7.2.3 Option Form. Howard (1971) developed the option or binary form of the conflict model to represent efficiently a conflict with many DMs and options. A course of action is a selection by the DMs of none, some, or all of their options, and a state is formed when each DM has chosen a course of action or strategy. The option form of Prisoners' Dilemma is shown in Figure 27.6, where a Y opposite an option means the option is selected by the player or DM controlling it, and an N means that the option is not chosen. When players do not choose to cooperate, they decide "by default" to defect. Each column in Figure 27.6 stands for a state. As shown at the bottom of the figure, there is a one-to-one correspondence between each state in the option form here and the normal from shown

States

1. Player 1				
(1) Cooperate	Y	Y	N	N
2. Player 2				
(2) Cooperate	Y	N	Y	N
Option Form States	(C,C)	(C,D)	(D,C)	(D,D)

Figure 27.6 Prisoners' Dilemma in option form.

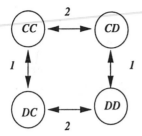

Figure 27.7 Prisoners' Dilemma as an integrated graph model.

in Figure 27.5. To display preference in option form, the states can be ordered from most preferred on the left to least preferred on the right for each DM.

A solution concept mathematically describes how a DM may behave strategically in a conflict. When the solution concepts of Nash stability, general metarationality, and symmetric metarationality are employed with the option form of the game, the approach is referred to as *metagame analysis* (Howard, 1971).

27.7.2.4   *Conflict Analysis.* Conflict analysis (Fraser and Hipel, 1979, 1984) constitutes a reformulation and extension of metagame analysis (Howard, 1971). It employs the option form for representing the conflict, and uses the sequential stability solution concept, as well as the metagame stability concepts, for finding equilibria.

27.7.2.5   *Graph Model for Conflict Resolution.* Both the theory and practice of conflict analysis were significantly improved by the development of the graph model for conflict resolution (GMCR) (Kilgour et al., 1987; Fang et al., 1993). In the conflict model, called a graph model, each DM has a directed graph that records each unilateral move available to the DM to move the conflict from one state to another. Usually, ordinal preferences are used, although a graph model for preferences is also possible (Fang et al., 1993, Chap. 8). The graph model can be employed with the solution concepts developed in metagame and conflict analysis, as well as limited move, nonmyopic, Stackelberg, and other kinds of stability.

The decision technologies contained within the graph model approach have been implemented as the decision support system GMCR (Fang et al., 1993; Hipel et al., 1996). The graph model has been successfully applied to a wide range of real-world conflicts including international trade, resource allocation, environmental, and military conflicts. As pointed out by Kilgour et al. (1995), the decision support system GMCR can be employed as a negotiation preparation system, either prior to, or between, bargaining sessions. Figure 27.7 shows the integrated graph model of Prisoners' Dilemma. Here 1 and 2 stand for the moves under player 1 and 2's control, respectively, among the states defined in Figures 27.5 and 27.6.

27.7.3   Analysis and Results of the Conflict Model

After constructing a conflict model, one can use it as a basic framework within which strategic interactions among the DMs can be modeled and analyzed in detail. In several of the model types, systematic examination of the permissible moves and countermoves by the DMs during possible evolutions of the conflict and the calculation of the most likely

TABLE 27.2 The Main Options of the Softwood Lumber Conflict

Decision Makers	Options
1. Canada	(1) Accept import duty (Duty)
	(2) Take legal action and attempt other sanctions (Legal)
	(3) Impose export tax in lieu of import duty (Tax)
2. Commerce	(4) Retain import duty (Retain)
Department	(5) Drop import duty/accept export tax (Drop)
	(6) Reject the petition (Reject)
3. U.S. Industry	(7) Retain petition (Retain)

resolutions, are carried out at the *stability analysis stage*. Stability analysis is executed using a solution concept, or stability definition, that constitutes a mathematical description model of a DM's behavior in a strategic conflict. Different solution concepts model different patterns of behavior, including different levels of foresight and different attitudes to strategic risk [see Fang et al. (1993) for detailed discussions of solution concepts and their interpretations]. The results of a stability analysis can be used, for instance, to guide and support decisions by DMs taking part in the conflict. States that are stable for all DMs, called equilibria, are interpreted as the most likely resolutions of the conflict. DMs can obtain guidance on how to achieve desirable equilibria.

27.7.4 A Real-world Application of Graph Model for Conflict Resolution

Canada exports large quantities of softwood lumber, worth about $2.3 billion (U.S.), to the United States each year. In the early 1980s, the U.S. domestic lumber industry was in a difficult economic situation. Industrial groups and politicians in the United States blamed Canadian subsidies for the decline of the U.S. industry. The U.S. International Trade Commission ruled on June 1986 that softwood lumber imports from Canada were harming U.S. industry. Following this decision, the U.S. Commerce Department imposed a preliminary 15 percent duty on Canadian softwood imports.

Initially, the Federal Government of Canada vowed to "fight this all the way." Later, however, at a federal–provincial conference, Canada announced its intention to pursue a negotiated settlement. This historical dispute was modeled and analyzed by Fang et al. (1993, Chap.7) in two phases: up to June 1986 and afterward. Here, the second phase of the study is summarized.

The three main DMs in this dispute are the (1) Government of Canada, (2) U.S. Commerce Department, and (3) U.S. Industry. The main options for each of these DMs are given in the Table 27.2.

Table 27.3 shows the 13 feasible states of the conflict model, based on the distinct strategies available to each DM. A dash (—) means a yes or a no, and hence the state numbered 13 in Table 27.3 represents the "combined" state where the U.S. rejects the petition—no matter what other options are selected, the state is exactly the same.

Canada most prefers that the U.S. Industry withdraw its petition. If the petition were withdrawn, Canada would not like to accept the duty or to pursue legal action. On the other hand, whether the U.S. Industry retains or withdraws its petition, the Commerce

TABLE 27.3 States for Phase 2 of the Softwood Lumber Conflict

	1	2	3	4	5	6	7	8	9	10	11	12	13
1. *Canada*													
(1) Duty	Y	N	N	Y	N	N	Y	N	N	Y	N	N	—
(2) Legal	N	Y	N	N	Y	N	N	Y	N	N	Y	N	—
(3) Tax	N	N	Y	N	N	Y	N	N	Y	N	N	Y	—
2. *U.S. Commerce*													
(4) Retain	Y	Y	Y	N	N	N	Y	Y	Y	N	N	N	—
(5) Drop	N	N	N	Y	Y	Y	N	N	N	Y	Y	Y	—
(6) Reject	N	N	N	N	N	N	N	N	N	N	N	N	Y
3. *U.S. Industry*													
(7) Retain	Y	Y	Y	Y	Y	Y	N	N	N	N	N	N	—
State Number	1	2	3	4	5	6	7	8	9	10	11	12	13

Department prefers to do likewise. Figure 27.8 depicts the integrated state-transition graph for the three DMs given as CA for Canada, CD for the Commerce Department, and I for U.S. Industry.

Fang et al. (1993) analyzed this dispute and showed that the parties were able to reach a compromise state in which each side achieved some of its goals. In fact, the predicted equilibria, and the actual resolution, was state number 12.

Table 27.4 shows the evolution of the conflict, in which arrows indicate the option selection changes. As shown in this table, Canada moves from status quo (state 1) to state 3 by proposing an export tax and not accepting an import duty. Subsequently, the Commerce Department moves from state 3 to state 6 by dropping the import duty and thereby

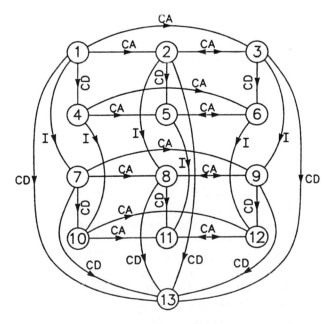

Figure 27.8 State-transition graph for phase 2 of the softwood lumber conflict.

TABLE 27.4   Progression from Status Quo to the Equilibrium State

1. *Canada*						
(1) Duty	Y	$\longrightarrow$	N		N	N
(2) Legal	N		N		N	N
(3) Tax	N	$\longrightarrow$	Y		Y	Y
2. *U.S. Commerce*						
(4) Retain	Y		Y	$\longrightarrow$	N	N
(5) Drop	N		N	$\longrightarrow$	Y	Y
(6) Reject	N		N		N	N
3. *U.S. Industry*						
(7) Retain	Y		Y		Y	$\longrightarrow$ N
State number	1		3		6	12

accepting the export tax. From state 6, the U.S. Industry reaches 12 by withdrawing the petition. Because state 12 is an equilibrium, no DM is motivated to move away from it, and hence the conflict is resolved.

27.7.5   Cooperative Models

The aforementioned abstract game models were mainly developed to analyze the choice of independent, individualistic DMs—the field started as noncooperative game theory. The characteristic function form is another abstract game model; it was originally developed by von Neumann and Morgenstern (1944) for use in cooperative game theory. Cooperative game theory focuses on the development of cooperation, and fair division of the rewards for cooperation, instead of strategic interaction. This game format can be utilized, for example, for equitably allocating costs among users of a resource or facility (see, for instance, Kilgour et al., 1988). Recently, Brams and Taylor (1996) assessed and proposed methods related to game theory for fairly dividing assets.

27.8   HEURISTIC PROGRAMMING

27.8.1   Overview

As discussed previously, especially in Sections 27.4 to 27.6, many real-world problems are modeled using combinatorial structures. A practical combinatorial model is typically very large, and usually no efficient algorithm is known; that is, no algorithm is available that guarantees convergence to an optimal solution in a reasonable time. The cost in time and resources of an extensive search for an optimal solution may not be repaid by the benefits of obtaining one. It may therefore be preferable to have an acceptable (close to optimal) solution that can be obtained quickly, at a lower cost in computational resources.

Heuristics are techniques based on intuition, common sense, or simplification of an exact algorithm. They do not assuredly lead to an optimal solution. In general, however, one can think of the design and use of good heuristic as being part of the art of problem solving. Heuristics are used extensively in computer science and OR.

In general, heuristics are problem-dependent. In other words, a heuristic that finds a solution very close to optimal for one problem may give a very poor result for another problem with a similar structure. For example, the greedy heuristic leads to an optimal solution in some knapsack problems that have special structures, but for other knapsack problems, its result is far from optimal.

For problems with known optimality conditions, such as linear and convex programming, heuristic procedures are usually not very useful. Knowledge of optimality conditions generally provides guidance to the design of efficient algorithms to find optimal solutions.

As described previously, many combinatorial problems do not have efficient algorithms to furnish optimal solutions. Moreover, in many real-world problems the accuracy of the data is questionable; insisting on an optimal solution for a problem with imprecise input may not be wise. Many practical problems are far too complex to be modeled exactly, so simplified models must be constructed. For such models, finding an optimal solution at a high cost in time and other resources is dubious. Therefore, in many situations heuristics are more appropriate than *exact* algorithms.

27.8.2   Conventional Heuristic Methods

In this section we discuss some simple heuristic methods that have been used in many practical combinatorial problems. Usually, these heuristics are utilized as foundations for designing more complex heuristics. In fact, it is common to use a sequence of these fundamental heuristics to solve a problem.

27.8.2.1   Greedy Heuristics. Greedy heuristics are among the most widely used heuristics in decision making and optimization. To apply a greedy heuristic, the problem is modeled so that a solution can be represented either as a subset of some set, or as a sequence of elements of a set. A greedy heuristic emphasizes immediate gain with little if any attention to future consequences. Hence, each element selected for inclusion is the "best" element at that specific stage. Carefully defining the "best" element at each stage is a very important factor in the efficiency of this heuristic.

A greedy heuristic is a single-pass procedure. Therefore, once an element is selected for inclusion in the best set, it is never replaced by another element. For instance, in a knapsack problem, one starts with an empty knapsack and at each stage packs it with the feasible element that is most valuable, smallest, or has the highest ratio of value to size. The packing process continues until there are not enough resources for the inclusion of any further elements.

In some problems, it is more convenient to use the *dual* of a greedy algorithm. In a dual greedy heuristic, one puts all the elements into the solution set and then, starting with this infeasible solution, one removes elements one by one. At each stage, the element that makes the least contribution to the objective function (and/or uses the most resources) is removed. The process continues until a feasible solution is obtained. The method of Toyoda (Senju and Toyoda, 1968) for solving an integer programming problem is an example of this type of dual greedy heuristic.

27.8.2.2   Interchange Heuristics. These heuristics are usually initiated using a randomly selected feasible solution. Then according to an interchange rule, the elements in the current solution are exchanged with other elements not in the solution, or the positions of two elements in the current sequence are reversed. If this change improves the current solution, it is accepted; otherwise, it is not. The process continues until a predefined termination criterion, such as a complete cycle of exchange, is met.

Indeed, interchange heuristics search for a local optimum by comparing the current solution with its neighbors. Hence, they belong to the family of *local search heuristics*. Because, a neighbor will be accepted as a new solution only if it improves the objective function, interchange heuristics can be called *descent* or *ascent search* heuristics for minimization or maximization, respectively.

27.8.2.3 Partitioning Heuristics. A partitioning heuristic is usually used to reduce big problems to manageable smaller problems through decomposition of the whole set of solutions. This heuristic is usually used along with other appropriate heuristics to solve the smaller problems.

27.8.3 Modern Heuristics

Two important recent heuristic techniques that have been widely used in combinatorial and nonconvex programming are outlined in this subsection.

27.8.3.1 Simulated Annealing. The name of this heuristic comes from a simulation model of the annealing process by which a solid approaches a low-energy state. Most simulated annealing heuristics search for an optimal solution in a way that avoids being trapped at a local minimum.

Physical annealing is a process that occurs when a material is melted, then cooled slowly and held for a long time at a temperature close to the fusion (freezing) point. Metropolis et al. (1953) simulate the process of solid annealing by a simple Monte Carlo algorithm. At each iteration of the algorithm, a small displacement of temperature is introduced into the system and the change (δ) in the energy of the system is calculated. If $\delta < 0$, the resulting change is accepted; if $\delta > 0$ the change is accepted with probability $\exp(-\delta/T)$, where T is a constant multiple of the current temperature. The process continues until the lowest energy state is obtained.

Almost 30 years later, Kirkpatrick et al. (1983) used a similar approach to solve a combinatorial problem. Simulated annealing avoids being trapped at a local minimum by allowing controlled uphill moves during the search for a better solution in minimization problems. To apply a simulated annealing heuristic to an optimization problem, the following steps are essential:

- Define a neighborhood for every solution and a mechanism for selecting a specific neighbor. For instance, if the problem is one of finding an optimal sequence, a neighbor is any sequence that can be obtained by pairwise interchange of two elements of the sequence. The efficiency of a simulated annealing heuristic is highly dependent upon this neighborhood definition.
- Set an initial solution, x^0, an initial temperature, T^0, and the number of iterations to be run at each temperature N_t.
- Define stopping criteria.
- Design a cooling schedule, a decreasing sequence of temperatures $(T^0, T^1, T^2, \ldots)$.

Assume that the problem is minimization. At stage i, when the temperature is T^i and the current solution is x^i, a neighbor of x^i, say y, is selected, randomly. If $Z(x^i) > Z(y)$, where Z is the value of objective function, y becomes the new current solution. If $Z(x^i) < Z(y)$, y becomes the new current solution with probability $\exp(-\delta / T^i)$, where $\delta = Z(y) - Z(x^i)$.

This process continues for N^i iterations with the same temperature. Then the temperature is changed to T^{i+1} according to the cooling schedule and the process is repeated with the new temperature. The process of cooling continues until the termination criteria are met.

27.8.3.2 Genetic Algorithms.

This heuristic is inspired by biological systems in which successful organisms evolve over many generations. It is based on operations such as mating, reproduction, cloning, crossover, and mutation. This heuristic was introduced by Holland in 1975 as a useful technique to solve large combinatorial problems. To apply a genetic algorithm (GA) to an optimization problem, the following steps are essential:

- Represent every solution by a string with fixed length.
- Construct an evaluation function to evaluate each solution.
- Develop an appropriate crossover operator for generating the new solutions.

The first step in the GA process is to structure the problem in such a way that every solution can be represented in the form of a string. In a GA, every solution is called a *chromosome,* every variable in the chromosome is a *gene,* and the value of a gene in a chromosome is called an *allele.*

In some optimization problems, expressing the solution by a string is straightforward. For example, in the Traveling Salesman Problem, a feasible route is a fixed-length string. However, formulation of strings is not obvious for some optimization problems such as general integer programming problems.

The GA is applicable only to unconstrained optimization problems. A problem can be made unconstrained by including the constraints in the objective function and imposing suitable penalties for violation. In a GA, the objective function is called the *fitness measure* or *evaluation function.* The form of the evaluation function has great impact on the efficiency of the GA, because the value of this function is used to eliminate or accept the new chromosomes.

The initial population of the solution is usually selected randomly. The size of this population remains fixed during the solution process. From each population a new population is generated iteratively. Each successive population is called a *generation.*

Usually, a certain percentage of the chromosomes in a population is directly transferred to the new generation without any changes. This process is accomplished through either reproduction (probabilistic operation) or cloning (deterministic operation). The rest of the chromosomes for the new population are generated by mating and mutation.

In mating, two parents from the current population are selected to mate. The children become a new string similar to their parents. The process of mating is usually called *crossover.* There are many different crossover procedures, including one-point, two-point, random, and partially matched crossover methods.

Consider two chromosomes with n elements, $a = (a_1, a_2, \ldots, a_n)$ and $b = (b_1, b_2, \ldots, b_n)$. In one-point crossover, first a gene r is selected such that $1 \leq r < n$. Then two children of a and b are generated by exchanging the blocks of alleles following r. Hence, the chromosomes of the two children of a and b are

$$\hat{a} = (a_1, \ldots, a_r, b_{r+1}, \ldots, b_n) \qquad \hat{b} = (b_1, \ldots, b_r, a_{r+1}, \ldots, a_n)$$

In two-point crossover, two genes r are s, such that $1 \le r < s \le n$, are randomly selected and then blocks of alleles between genes r and s are swapped. Hence, the new children are

$$\hat{a} = (a_1, \ldots, a_{r-1}, b_r, \ldots, b_{s-1}, a_s, a_{s+1}, \ldots, a_n)$$

$$\hat{b} = (b_1, \ldots, b_{r-1}, a_r, \ldots, a_{s-1}, b_s, b_{s+1}, \ldots, b_n).$$

The process of mating continues until enough chromosomes for a new population are generated. The process of mutation is random in that a chromosome is randomly selected and one or more alleles changed, or two randomly chosen alleles are swapped.

When there is no improvement in the best solution of the population of several generations, or when a predetermined number of generations has been simulated, the process stops. There have been some successful implementations of GA in combinatorial optimization. Hadj-Alouane et al. (1993) used a GA on a task-allocation problem with 20 tasks and 7 processors. After 110 iterations, they reached a very satisfactory solution. The whole process took a few minutes of CPU time; the same problem had not terminated after three days with a branch-and-bound approach. Some other applications of GA include VLSI circuit layout, image processing, communication link size optimization, job-shop scheduling, multiple-objective optimization, and clustering (Goldberg, 1989).

REFERENCES

Ackoff, R. L. (1962). *Scientific Method*. New York: Wiley.

Ackoff, R. L. (1981). The art and science of mess management. *Interface* **2**, 20–26.

Bana e Costa, C. (ed.). (1990). *Reading in Multiple Criteria Decision Aid*. Berlin: Springer-Verlag.

Bellman, R. (1957). *Dynamic Programming*. Princeton, NJ: Princeton University Press.

Benayoun, R., Demontgolfier, J., Tergny, J., and Laritchev, O. (1971). Linear programming with multi-objective function: Step Method (STEM). *Math. Program.* **1**, 366–375.

Binmore, K. (1992). *Fun and Games*. Lexington, MA: D. C. Heath.

Blackett, P. M. S. (1962). *Studies of War*. Edinburgh. Oliver & Boyd.

Brams, S. J., and Kilgour, D. M. (1988). *Game Theory and National Security*. New York: Basil Blackwell.

Brams, S. J., and Taylor, A. (1996). *Fair Division*. Cambridge UK: Cambridge, University Press.

Brill, E. D., Jr., Chang, S.-Y., and Hopkins, L. D. (1982). Modeling to generate alternatives: The HSJ approach and an illustration using a problem in land use planning. *Manage. Sci.* **28**, 221–235.

Chankong, V., and Haimes, Y. Y. (1983). *Multiobjective Decision Making*. New York: North-Holland.

Chavatal, V. (1983). *Linear Programming*. New York and San Francisco: Freeman.

Cook, W. D., and Kress, M. (1992). *Ordinal Information and Preference Structures*. Englewood Cliffs, NJ: Prentice-Hall.

Dantzig, G. B. (1963). *Linear Programming and Extensions*. Princeton, NJ: Princeton University Press.

Dennis, J. E., and Schnabel, R. B. (1983). *Numerical Methods for Unconstrained Optimization and Nonlinear Equations*. Englewood Cliffs, NJ: Prentice-Hall.

Eilon, S. (1971). Goals and constraints in decision making. *Oper. Res. Q.* **23**(1), 3–16.

Enthoven, A. C. (1975). Ten practical principles for policy and program analysis. In *Benefit-Cost and Policy Analysis* (Zeckhauser et al., eds.), pp. 456–465. Chicag: Aldine.

Fang, L., Hipel, K. W., and Kilgour, D. M. (1993). *Interactive Decision Making: The Graph Model for Conflict Resolution*. New York: Wiley.

Fletcher, R. (1981). *Practical Methods of Optimization*. New York: Wiley.

Fraser, N. M., and Hipel, K. W. (1979). Solving complex conflicts. *IEEE Trans. Syst., Man, Cybernet.* **SMC9**(12), 805–815.

Fraser, N. M., and Hipel, K. W. (1984). *Conflict Analysis: Models and Resolutions*. New York: North-Holland.

Friedman, J. W. (1988). *Game Theory with Applications to Economics*. New York: Oxford University Press.

Geoffrion, A. M., Dyer, J. S., and Feinberg, A. (1972). An interactive approach for multicriteria optimization, with an application to the operation of an academic department. *Manage. Sci.* **19**(4), 357–368.

Gill, P. E., Murray, W., and Wright, M. H. (1992). *Practical Optimization*. San Diego, CA: Academic Press.

Goldberg, D. E. (1989). *Genetic Algorithm in Search, Optimization and Machine Learning*. Reading MA: Addison-Wesley.

Gross, D., and Harris, C. M. (1985). *Fundamentals of Queuing Theory*, 2nd ed. New York: Wiley.

Hadj-Alouane, A. B., Bean, J.C., and Murty, K. G., (1993). *A Hybrid Genetic/Optimization Algorithm for a Task Allocation Problem*. Ann Arbor: University of Michigan, IOE Dept.

Haimes, Y. Y., and Schneiter, C. (1996). Covey's seven habits and the systems approach: A comparative analysis. *IEEE Trans. Syst., Man. Cybernet.* **26**(4), 483–487.

Hillier, F. S., and Lieberman, G. J. (1990). *Introduction to Operations Research*, 5th ed. New York: McGraw-Hill.

Hipel, K. W. (1990). Decision technologies for conflict analysis. *Inf. Decis. Technol.* **16**(3), 185–214.

Hipel, K. W. (ed.). (1992). *Multiple Objective Decision Making in Water Resources*, Monogr. Ser. No. 18. Bethesda, MD: American Water Resources Association.

Hipel, K. W., and Fang, L. (eds.). (1994). *Effective Environmental Management for Sustainable Development*. Dordrecht, The Netherlands: Kluwer, Academic Publishers.

Hipel, K. W., and McLeod, A. I. (1994). *Time Series Modelling of Water Resources and Environmental Systems*. Amsterdam: Elsevier.

Hipel, K. W., Kilgour, D. M., Fang, L., and Peng X. J. (1997). The decision support system GMCR in environmental conflict management. *App. Math. Comput.* 83 (2 and 3), 117–152.

Hobbs, B. F., Chankong, V., and Hamadeh, W. (1990). *Screening Water Resources Plans under Risk and Multiple Objectives: A Comparison of Methods*. Washington, DC: U.S. Army Corps of Engineers.

Hoffman, A. J., and Wolf, P. (1985). History. In *The Traveling Salesman Problem*, E. L. Lawler, J. K. Lenstra, A. H. G. Rinnoy, Kan, and D. B. Shmoys, eds., pp. 1–15. Chichester: Wiley.

Holland, J. (1975). *Adaptation in Natural and Artificial Systems*. Ann Arbor: University of Michigan Press.

Howard, N. (1971). *Paradoxes of Rationality*. Cambridge, MA: MIT Press.

Hwang C. L., Masud, A. S. M., Paidy, S. R., and Yoon, K. (1979). *Multiple Objective Decision*

Making, Methods and Application, Lect. Notes Econ. Math. Syst., Vol. 164. Berlin: Springer-Verlag.

Hwang C. L., and Yoon, K. (1981). *Multiple Attribute Decision Making- Methods and Applications: A State-of-the-Art Survey,* Lect. Notes Econ. Math. Syst., Vol. 186. Berlin: Springer-Verlag.

Ignizio, J. P. (1985). *Introduction to Linear Goal Programming.* Beverly Hills, CA: Sage.

Ignizio, J. P., and Cavalier, T. M. (1994). *Linear Programming.* Englewood Cliffs, NJ: Prentice-Hall.

Insua, D. R. (1990). *Sensitivity Analysis in Multiple Objective Decision Making.* Berlin: Springer-Verlag.

Jones, A. J. (1980). *Game Theory, Mathematical Models of Conflict.* Chichester: Ellis Horwood.

Karmarkar, N. (1984). A new polynomial-time algorithm for linear programming. *Combinatorica* **4,** 373–395.

Keeney, R. L. (1992). *Value Focused Thinking, A Path to Creative Decision Making.* Cambridge, MA: Harvard University Press.

Keeney, R. L. (1994). Using values in operations research. *Oper. Res.* **42**(5), 793–813.

Keeney, R. L., and Raiffa, H. (1976). *Decisions with Multiple Objectives.* New York: Wiley.

Keeney, R. L., McDaniels, T. M., and Ridge-Cooney, V. L. (1996). Using values in planning wastewater facilities for metropolitan Seattle. *Water Resour. Bull.* 32(2), 293–303.

Keys, P. (1992). *Operations Research and Systems: The Systemic Nature of Operational Research.* New York: Plenum.

Khachian, L. G. (1979). A polynomial algorithm in linear programming. *Sov. Math. (Engl. Transl.)* **20,** 191–194.

Kilgour, D. M., Hipel, K. W., and Fang, L. (1987). The graph model for conflicts. *Automatica* **23**(1), 41–55.

Kilgour, D. M., Okada, N., and Nishikori, A. (1988). Load control regulation of water pollution: An analysis using game theory. *J. Environ. Manag.* **27**(2), 179–194.

Kilgour, D. M., Fang, L., and Hipel, K. W. (1992). Game-theoretic analysis of enforcement of environmental laws and regulations. *Water Resour. Bull.* **28**(1), 141–153.

Kilgour, D. M., Fang, L., and Hipel, K. W. (1995). GMCR in negotiations. *Negotiation J.* **11**(2), 151–156.

Kirkpatrick, S., Gelatt, C. D., and Vecchi, M. P. (1983). Optimization by simulated annealing. *Science* **220,** 671–680.

Klir, G. J. (1991). *Facets of Systems Science.* New York: Plenum.

Koksalan, M., M., Karwan, M. H., and Zionts, S. (1984). An improved method for solving multiple criteria problem involving discrete alternatives. *IEEE Trans. Syst., Man, Cybernet.* **SMC-14,** 24–34.

Korhonen, P., Wallenius, J., and Zionts, S. (1984). Solving the discrete multiple criteria problem using convex cones. *Manage. Sci.* **30**(11), 1336–1345.

Kuhn, H. W. (1953). Extensive games and the problem of formulation. *Ann. Math. Stud.* **28,** 193–216.

MacCrimmon, K. R. (1973). An overview of multiple objective decision making. In *Multiple Criteria Decision Making* (J. L. Cochrane and M. Zeleny, eds.), pp. 18–44. Columbia: University of South Carolina Press.

Martello, S., and Toth, P. (1990). *Knapsack Problems: Algorithms and Computer Implementations.* New York: Wiley.

Metroplolis, N., Rosenbluth, A., Teller, M. A., and Teller, E. (1953). Equations of state calculations by fast computing machines. *J. Chem. Phys.* **21,** 1087–1092.

Miser, H. J., and Quade, E. S. (eds.). (1988). *Handbook of Systems Analysis*. New York: North-Holland.

Moskowitz, H., Preckel, P., and Yang, A. (1992). Multiple criteria Robust Interactive Decision Analysis (MCRID), for optimizing public policies. *Eur. J. Oper. Res.* **56**, 219–236.

Myerson, R. B. (1991). *Game Theory, Analysis of Conflict*. Cambridge, MA: Harvard University Press.

Nakamura, M., and Brill, E. D. (1979). Generation and evaluation of alternative plans for regional waste water systems: An imputed value method. *Water Resourc. Res.* **15**(4), 750–756.

Nemhauser, G. L., and Wolsey, L. A. (1988). *Integer and Combinatorial Optimization*. New York: Wiley.

Osyczka, A. (1984). *Multicriteria Optimization in Engineering*, Ser. Eng. Sci. Chichester: Ellis Horwood.

Owen, G. (1982). *Game Theory*. London: Academic Press.

Peterson, R., and Silver, E. A. (1979). *Decision Systems for Inventory Management and Production Planning*. New York: Wiley.

Rajabi, S., Hipel, K. W., and Kilgour, D. M. (1995). Multiple criteria decision making under interdependence of actions. *Proc. IEEE Int. Conf. Syst., Man, Cybernet.*, Vancouver, Canada.

Rajabi, S., Hipel, K. W., and Kilgour, D. M. (1999). Water supply planning under interdependence of actions: theory and applications. Water Res. Res. 35.

Rajabi, S., Kilgour, D. M, and Hipel, K. W. (1996). Multiple criteria zero-one programming under interdependence of actions. *Proc. 12th Int. Conf. on Multiple Criteria Decis. Making*, Hagen, Germany, 1995, pp. 383–392. Berlin: Springer Verlag.

Rajabi, S., Kilgour, D. M., and Hipel, K. W. (1998). Modelling action-interdepence in multiple criteria decision making. Euro. J. Oper. Res. **110**(3), 490–508.

Rapoport, A., Guyer, M. J., and Gordon, D. G. (1976). *The 2 × 2 Game*. Ann Arbor: University of Michigan Press.

Ravindran, A., Phillips, D. T., and Solbreg, J. (1987). *Operations Research, Principles and Practice*, 2nd ed. New York: Wilcy.

Reinelt, G. (1994). *The Traveling Salesman, Computational Solutions for TSP Applications*. Berlin: Springer-Verlag.

Ringuest, J. L. (1992). *Multiobjective Optimization: Behavioral and Computational Considerations*. Boston: Kluwer Academic Publishers.

Romero, C. (1990). *Handbook of Critical Issues in Goal Programming*. New York: Pergamon.

Rosenhead, J. (ed.). (1989). *Analysis for a Problematic World*. Chichester: Wiley.

Roy, B. (1990). Decision-aid and decision making. In *Reading in Multiple Criteria Decision Aid* (C. Bana e Costa, ed.). Berlin: Springer-Verlag, 17–35.

Sage, A. P. (ed.). (1990). *Concise Encyclopedia of Information Processing in Systems and Organizations*. London: Pergamon.

Sage, A. P. (1991). *Decision Support Systems Engineering*. New York: Wiley.

Sage, A. P. (1992). *Systems Engineering*. New York: Wiley.

Sarin, R. K. (1977). Interactive evaluation and bound procedure for selecting multi-attributed alternatives. In *Multiple Criteria Decision Making—TIMS Studies in Management Sciences* (M. K. Starr and M. Zeleny, eds.). Amsterdam: North-Holland.

Schniederjans, M. J. (1995). *Goal Programming, Methodology and Applications*, Boston: Kluwer Academic Publishers.

Senju, S., and Toyoda, Y. (1968). An approach to linear programming with 0-1 variables. *Manage. Sci.* **15**, B196–B207.

Shubik, M. (1982). *Game Theory in the Social Sciences, Concepts and Solutions*. Cambridge, MA: MIT Press.

Simon, H. A. (1976). *Administrative Behavior,* 3rd ed. New York: Macmillan.

Stadler, W. (1979). A survey of multicriteria optimization or the vector maximum problem. Part 1: 1776–1960. *J. Optim. Theory Appl.* **29**(1), 1–52.

Steuer, R. E. (1986). *Multiple Criteria Optimization: Theory, Computation and Application.* New York: Wiley.

Steuer, R. E., and Choo, E. U. (1983). An interactive weighted Tchebycheff procedure for multiple objective programming. *Math. Program.* **26**(1), 326–344.

Steuer, R. E., Gardiner, L. R., and Gray J. (1996). A bibliographic survey of the activities and international nature of multiple criteria decision making. *J. Multi-Criteria Decis. Ana.* **5,** 195–217.

Stewart, T. J., and Scott, L. (1995). A scenario-based framework for multicriteria decision analysis in water resources planning. *Water Resour. Res.* **31**(11), 2835–2843.

Taha, H. A. (1971). *Operations Research: An Introduction.* New York: Macmillan.

Vincke, P. (1992). *Multicriteria Decision-Aid.* New York: Wiley.

von Neumann, J. (1937). Uber ein okonomisches Gleichungssystem und eine Verallgemeinerung des Brouwerschen Fixpunktstazes, Ergebnisse eines Mathematik. *Kolloquium* **8,** 73–83.

von Neumann, J., and Morgenstern, O. (1944). *Theory of Games and Economic Behavior,* 1st ed. Princeton, NJ: Princeton University Press.

von Neumann, J., and Morgenstern, (1953). *Theory of Games and Economic Behavior,* 3rd ed. Princeton, NJ: Princeton University Press.

Waddington, C. H. (1973). *OR in World War 2.* London: Elek Science.

Wagner, H. M. (1975). *Principles of Operations Research.* Englewood Cliffs, NJ: Prentice-Hall.

Walker, W. E., and Veen, M. A. (1981). *Policy Analysis of Water Management for the Netherlands,* Vol. 2, N-1500/2-NETH. Santa Monica, CA: Rand Corporation.

Walker, W. E., and Veen, M. A. (1987). Screening tactics in a water-management policy Analysis for the Netherlands. *Water Resour. Res.* **23**(7), 1145–1151.

Wierzbicki, A. P. (1980). The use of reference objectives in multiobjective optimization. In *Multiple Criteria Decision Making, Theory and Applications* (G. Fandel and T. Gal, eds.). Heidelberg: Springer-Verlag.

Wierzbicki, A. P. (1982). A mathematical basis for satisficing decision making. *Math. Modeling.* **3**(5), 391–405.

Williams, H. P. (1990). *Model Building in Mathematical Programming,* 2nd ed. Chichester: Wiley.

Winston, W. L. (1991). *Operations Research, Applications and Algorithms,* 2nd ed. Boston: Pws-Kent Publ. Co.

Yu, P. L. (1985). *Multiple Criteria Decision Making.* New York: Plenum.

Zionts, S., and Wallenius, J. (1976). An interactive programming method for solving the multiple criteria problem. *Manage. Sci.* **22**(6), 652–663.

28 Decision Analysis

CRAIG W. KIRKWOOD

28.1 INTRODUCTION

Over the last several decades, a philosophy and a body of techniques have been developed to assist in analyzing decisions, and this approach has been used successfully in a wide variety of systems engineering and management situations. The underlying assumption governing the approach is that many significant decisions in complex systems are made through a process involving technical staff and management as well as interested outsiders. Therefore, a key to good decision making is to provide structured methods for incorporating the information, opinions, and preferences of the various relevant people into the decision-making process.

A systematic approach to quantitative decision analysis includes the following steps.

1. Specify objectives and scales for measuring achievement with respect to these objectives.
2. Develop alternatives that potentially might achieve the objectives.
3. Determine how well each alternative achieves each objective.
4. Consider tradeoffs among the objectives.
5. Select the alternative that, on balance, best achieves the objectives, taking into account uncertainties.

In the remainder of this chapter we consider these steps in more detail.

28.2 STRUCTURING OBJECTIVES

This section reviews procedures for developing and organizing objectives to use in analyzing decisions. Methods are first presented that can be used to develop a *qualitative* structure for objectives, and then ways are presented to develop *quantitative* scales that measure the degree of attainment of these objectives. To address these issues, it is useful to define some terminology:

- *Evaluation Consideration.* Evaluation considerations are significant enough to be taken into account while evaluating alternatives. Other terms sometimes used for evaluation considerations are *evaluation concerns* or *areas of concern*.
- *Objective.* An objective is the preferred direction of movement with respect to an evaluation consideration.

Handbook of Systems Engineering and Management, Edited by A. P. Sage and W. B. Rouse
ISBN 0471-15405-9 ©1999 John Wiley and Sons, Inc.

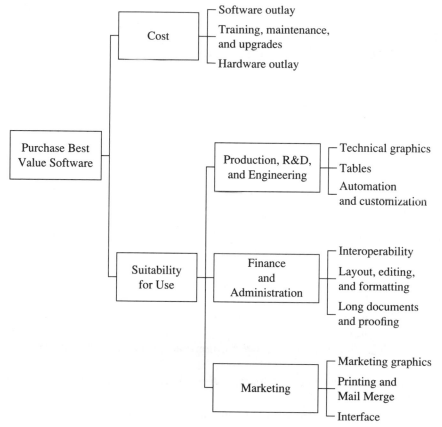

Figure 28.1   Value hierarchy for word processor standardization decision. (From *Strategic Decision Making: Multiobjective Decision Analysis with Spreadsheets* by C. W. Kirkwood. Copyright © 1997 Brooks/Cole Publishing Company, Pacific Grove, CA, a division of International Thomson Publishing, Inc. By permission of the publisher.)

- *Evaluation Measure.* An evaluation measure is a measuring scale for the degree of attainment of an objective. Thus, "annual salary in dollars" might be the evaluation measure for a job seeker's objective of finding a high salary. Other terms that are sometimes used for an evaluation measure are *attribute, measure of effectiveness, performance measure,* or *metric.*
- *Level or Score.* The level is the specific numerical rating for a particular alternative with respect to a specified evaluation measure.
- *Value Hierarchy or Value Tree.* The value hierarchy comprises the entire set of evaluation considerations, objectives, and evaluation measures for a particular decision analysis arranged in a hierarchical, or "treelike" structure. The remainder of this section focuses on methods to develop value hierarchies.
- *Layer or Tier.* The layer refers to the evaluation considerations at the same distance from the top of a value hierarchy.

An example of a value hierarchy is shown in Figure 28.1, which is adapted from Kirkwood (1997). This hierarchy was developed to assist a manufacturing firm that had decided to standardize on the word processing software used by all employees. This figure makes clearer the source of the term *value tree* to refer to such a structure, since the figure looks somewhat like a tree turned on its side with its root at the left side of the diagram. This value structure is *hierarchical;* that is, an evaluation consideration closer to the root of the tree *consists of* the considerations "below" it (that is, in the next further layer from the root of the tree). In the Figure 28.1 value hierarchy, the topmost consideration of "purchase best value software" consists of the considerations "cost" and "suitability for use." Similarly, cost consists of "software outlay," "training, maintenance, and upgrades," and "hardware outlays," while suitability for use consists of "production, R&D, and engineering," "finance and administration," and "marketing."

This example also illustrates that the number of layers may not be uniform across a value hierarchy. There is an additional layer of considerations under the three suitability-for-use considerations, but there is no additional layer under the three cost considerations.

28.2.1   Desirable Properties for Value Hierarchies

The following are desirable properties for value hierarchies (Keeney and Raiffa, 1976; Kirkwood, 1997):

- *Completeness.* For a value hierarchy to be *complete,* the evaluation considerations in each layer of the hierarchy taken together as a group must adequately cover all concerns necessary to evaluate the overall objective of the decision. Thus, if the value hierarchy shown in Figure 28.1 is complete, it is only necessary to know how well an alternative performs with respect to the lowest-layer evaluation considerations to know how well it performs with respect to the overall evaluation consideration of "purchase best value software." (Note that there are twelve lowest-layer considerations.)

- *Nonredundancy.* For a *nonredundant* value hierarchy, no two evaluation considerations in the same layer of the hierarchy overlap. That is, each layer in the hierarchy "divides up" the layer above it into more detailed pieces. For example, in Figure 28.1, the cost evaluation consideration is divided into software outlays; training, maintenance, and upgrades; and hardware outlay. If this is a nonredundant division, then every cost can be assigned to one of the three categories of software; training, maintenance, and upgrades; or hardware outlays. That is, no cost fits into more than one of these categories.

- *Decomposability.* For a *decomposable* hierarchy, the evaluation considerations in a particular layer of the hierarchy are not only nonredundant, but also the value attached to variations in the level (score) of the evaluation measure for any of the lowest-layer evaluation considerations does not depend on the levels of the other lowest-layer considerations. That is, values can be added up across the evaluation considerations.

- *Operability.* An *operable* value hierarchy is one that is understandable for the persons who must use it. The issue of operability can come up in complex systems management situations where technical specialists interact with the public. For example, during the Three Mile Island nuclear power plant incident, technical specialists had difficulty presenting an assessment of risks in a manner that was understandable to journalists or the general public. The evaluation measures that the specialists were using were not operable for the intended audience.

- *Small Size.* Other things being equal, it is desirable to have a *smaller-size* value hierarchy. A smaller hierarchy can be communicated more easily to interested parties, and it requires fewer resources to estimate the performance of alternatives with respect to the various evaluation measures.

28.2.2 Developing Evaluation Measure Scales

Evaluation measures are used to show the degree of achievement of decision alternatives with respect to each of the lowest-layer evaluation considerations in a value hierarchy. Evaluation measure scales can be classified as either *natural* or *constructed,* and also as either *direct* or *proxy.* A natural scale is one that is in general use with a common accepted interpretation. A constructed scale is one that is developed for a particular decision problem to measure the degree of attainment of an objective. A direct scale is one that directly measures the degree of attainment of an objective, while a proxy scale is one that reflects the degree of attainment of its associated objective, but does not directly measure this.

It is possible to have direct scales that are either natural or constructed, and similarly, it is possible to have proxy scales that are either natural or constructed. Furthermore, the sharp distinctions "natural vs. constructed" and "direct vs. proxy" actually represent the extremes of a range of possibilities. For example, is the scale "net present value" natural or constructed? To some extent, this may depend on who you talk to. When this scale was originally developed to represent the time-value of money, it probably met the definition of a constructed scale. However, it has now been in use long enough that for financially literate people it has become a natural scale. On the other hand, there are many people with an interest in money who are not especially financially literate. Some of them probably view net present value as an exotic constructed scale.

There are several questions that often arise while developing evaluation measures:

1. *Natural–Proxy vs. Constructed–Direct Scales.* Natural scales have some nice properties. You do not have to spend time to develop the scale definition. Their use may be less controversial because they are in general use. The difficulty is that natural scales are often not easy to find for a particular decision problem, and therefore you may have to use a proxy scale in order to find a natural scale for some evaluation considerations.

2. *How Much Should Evaluation Considerations Be Subdivided?* This is illustrated by the constructed evaluation measure scale shown in Figure 28.2 to measure ecological impacts of construction of a proposed power plant. Note that this scale considers several different aspects of ecological impact. For example, grassland, shrubland, pinyon-juniper, riparian, and wetland habitats are all considered. This ecological impact evaluation consideration could have been subdivided into lower-layer considerations addressing each of these aspects, but this would have required more analysis and data collection than was warranted for the decision being analyzed. This is because an initial screening had eliminated potential power plant sites with serious ecological impact difficulties.

3. *Should a Precise Natural Scale Be Used that is Technical or a More Operable Constructed Scale that is Less Precise?* This can be a particular issue in situations involving regulation and public stakeholders. For example, suppose that the smell and visual haze associated with sulfur dioxide is a concern for some industrial process. This can be

0: Removal of 6 square miles having >25 percent of cultivated agricultural use.

1: Removal of 6 square miles having <25 percent of cultivated agricultural use with remaining area of grass and shrubland or pinyon-juniper.

2: Removal of 6 square miles of grassland habitat, shrubland habitat, or pinyon-juniper habitat with no important or unique species or habitats present.

3: Removal of 6 square miles of pinyon-juniper/ponderosa pine habitat with no important or unique species or habitats present.

4: Removal of 6 square miles of grassland, shrubland, or pinyon-juniper habitat that includes <10 percent riparian or wetland habitat.

5: Removal of 6 square miles of grassland, shrubland, or pinyon-juniper habitat that is within 1 mile of significant actual or potential raptor habitat.

6: Removal of 6 square miles of grassland, shrubland, or pinyon-juniper habitat that includes <10 percent riparian or wetland habitat and is within 1 mile of significant actual or potential raptor habitat.

7: Removal of 6 square miles of grassland, shrubland, or pinyon-juniper habitat of which <25 percent is actual or potential habitat of threatened, endangered, or otherwise unique species.

8: Removal of 6 square miles of grassland, shrubland, or pinyon-juniper habitat within 1 mile of significant actual or potential raptor habitat, and of which <25 percent is actual or potential habitat for threatened, endangered, or otherwise unique species.

9: Removal of 6 square miles of grassland, shrubland, or pinyon-juniper habitat within 1 mile of significant actual or potential raptor habitat, and of which <25 percent is actual or potential habitat for threatened, endangered, or otherwise unique species.

10: Removal of 6 square miles of grassland, shrubland, or pinyon-juniper habitat including >25 percent actual or potential habitat for threatened, endangered, or otherwise unique species.

11: Removal of 6 square miles of grassland, shrubland, or pinyon-juniper habitat within 1 mile of significant actual or potential raptor habitat, and of which >25 percent is actual or potential habitat for threatened, endangered, or otherwise unique species.

12: Removal of 6 square miles of grassland, shrubland, or pinyon-juniper habitat within 1 mile of significant actual or potential raptor habitat and including <10 percent riparian or wetland habitat and >25 percent actual or potential habitat for threatened, endangered, or otherwise unique species.

Figure 28.2 Evaluation measure scale for ecological impacts at power plant site.

measured precisely by using the natural scale "sulfur dioxide concentration," but this may not be meaningful for some of the decision stakeholders (for example, a city council that has to approve a permit for the facility). A constructed scale that refers to rotten egg smell or uses pictures of visual haze in its definition may be more operable for the decision makers, even though it may be less precise than using sulfur dioxide concentration.

4. *How Carefully Should Scale Levels Be Defined?* Ambiguous scales impede communications. The ideal is to have scales that pass the *clairvoyance test*. That is, if a clairvoyant were available who could foresee the future with no uncertainty, would this clairvoyant be able to unambiguously assign a score to the outcome for each alternative in a decision problem? This test is often difficult to pass, particularly for constructed scales. For example, the scale in Figure 28.2 is fine as long as an alternative meets the specific conditions of one of the defined scale levels. However, for a situation that does not exactly meet one of the definitions, some judgment is required to determine a score. While it is desirable to have precisely defined scales, it is also necessary to develop evaluation mea-

sure scales within a realistic time frame. This can require a tradeoff between the precision of the scale definition and the amount of time available to develop the scale.

See Buede (1986), Keeney (1981, 1992), Keeney and Raiffa (1976, Chapter 2), and Kirkwood (1997, Chapter 2) for further details about structuring objectives.

28.3   DEVELOPING ALTERNATIVES

There is no substitute for a good alternative. The most complete analysis of decision alternatives can only show the best of the identified alternatives. If none of the alternatives is very good, then the best alternative will only be the best of a poor lot. Many people do not cast their nets wide enough when they consider alternatives for a perplexing decision. Failure to consider all the alternatives is encouraged by the technical complexity of many system engineering and management decisions. Important information must be provided by engineers, financial analysts, and other experts. However, these specialists sometimes analyze the options they know how to solve, rather than the preferable ones. In this section we discuss procedures to identify better alternatives.

The root of our difficulty in identifying good alternatives seems to be in basic human reasoning. Since John Locke (1632–1704) there has been increasing support for the view that thought is primarily an *associative* process (Dawes, 1988, p. 68). That is, we think about a new situation by making mental associations with previous situations that seem relevant, and, furthermore, these associations occur with relatively little conscious control on our part. Something "pops into our mind" that seems relevant, and we use this as a basis for structuring our consideration of the new situation.

Often this happens quickly with no conscious effort or control. Sometimes associative reasoning processes can generate amazingly creative thoughts, but often they quickly lock us into a premature conclusion that we have identified the best alternatives for our decision problem. In the remainder of this section we review ways to improve our reasoning about alternatives.

28.3.1   Too Many Alternatives

In situations where there are many potential alternatives, we need a way other than unaided intuition to either analyze all the alternatives or select a smaller set of alternatives that will be analyzed in further detail. Without some systematic procedure, the associative reasoning process will quickly select a smaller set, and, for the reasons discussed in the preceding section, the set that is selected may not be the best one. Hence, the task is to either use a method that can automatically analyze all of the alternatives, or to reduce the set of potential alternatives to a smaller set to be analyzed in more detail in a manner that retains the more desirable alternatives.

28.3.1.1   Automatic Analysis.   In some situations with a large number of alternatives, the primary difficulty is organizing the information about the alternatives. For example, Golabi et al. (1981) developed a computer-based decision support system to aid in selecting proposals for funding from those submitted to the U.S. Department of Energy in response to a request for proposals to design solar energy demonstration projects. It was anticipated that about seventy-five proposals would be submitted, of which about six would be funded.

It can be shown that there are over 200 million different possible combinations of six proposals that can be formed from the seventy-five proposals. It is feasible to assess the seventy-five proposals, but the difficulty comes in considering the 200 million ways that these can be combined. For such situations, the tools of *mathematical programming* (also called *optimization*) can be helpful for considering all the alternatives (Winston, 1994). The various mathematical-programming techniques are able to determine the best alternative without having to consider all possibilities (which would probably take too much time to be feasible). While the theory underlying this approach is complex, you do not need to understand this theory in order to use mathematical programming. Specialized software to do the required calculations is available for personal computers, and spreadsheet programs like Excel now also include the necessary capabilities.

28.3.1.2 *Reducing the Number of Alternatives Considered.*

In some situations, it is not feasible to collect or analyze the data for all the possible alternatives. For example, Kirkwood (1982) presents a decision analysis of where to locate a proposed nuclear power plant. The possible sites included an entire state in the western United States. There were literally an infinite number of possible sites. It was not feasible to collect enough data about the entire state to evaluate the desirability of every spot in the state. Thus, some method was needed to reduce the number of sites under consideration until this number was small enough so that data could be collected for the sites.

In another decision where data cannot be obtained for all possible alternatives, McNamee and Celona (1990) consider a case where a brokerage firm owns a small share of the office complex where its offices are located. The majority owner of the complex offers current shareholders the opportunity to purchase an additional interest in the complex. The brokerage firm must consider this offer, as well as the possibility of moving out of the complex. The decision is made more complex by a variety of uncertainties, as well as the necessity of considering a range of possible investment levels. The decision must be made rapidly, and there is no time to collect and analyze data on all possible relevant alternatives.

For the nuclear power plant site selection, *screening criteria* were established to reduce the number of sites that needed to be considered. For example, a criterion was established that only sites greater than 5 miles from a fault would be considered. Only sites meeting all of the screening criteria were retained. The use of screening criteria involves an approximation. For example, a site that is 4.9 miles from a fault is not infinitely worse than a site that is 5.1 miles from a fault, but the first one will be removed from consideration, while the second will be retained. Thus, it is appropriate to select screening criteria that are relatively loose. That is, we anticipate that the most preferred alternative, once it is identified, will meet all the criteria with ease. For example, we anticipate that the best site will be much further than 5 miles from a fault. If loose screening criteria are not used, then an alternative might be eliminated that is desirable enough with respect to other criteria to overcome its poor rating on the screening criterion that eliminated it.

In the office complex investment decision, the *strategy-generation table* shown in Figure 28.3 was helpful for identifying desirable alternatives to analyze in detail. In this figure, all of the columns except the first one (which is to the left of the dashed vertical line) represent different aspects of the decision alternatives. The name of the office complex was Modena Place, and Figure 28.3 shows columns to the right of the dashed vertical line

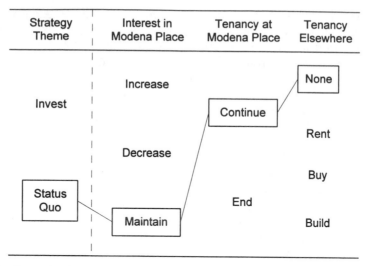

Figure 28.3   Example of a strategy-generation table. (Adopted from McNamee and Celona, 1990.)

for the level of investment interest in Modena Place, whether the brokerage firm should continue its tenancy at Modena Place, and what type of tenancy, if any, the firm should take at another location.

An alternative is constructed by selecting one entry from each of the columns to the right of the dashed vertical line. The column to the left of the dashed vertical line is used to give shorthand names to each strategic alternative that is selected for further analysis. For example, an alternative labeled "Status Quo" is marked on the figure. This alternative includes the entry in each column that is boxed, and the combination of entries is connected with lines. Different alternatives are indicated by different combinations of "interest in Modena Place," "tenancy in Modena Place," and "tenancy elsewhere."

Note that in this strategy-generation table some of the columns show only a few of the total possible variations under the column. For example, the column "interest in Modena Place" only shows the three possibilities "increase," "decrease," and "maintain." In reality, there are many possible levels of investment for either the "increase" or "decrease" cases. A strategy-generation table helps to sort out what types of alternatives make sense to analyze in more detail, but often does not specify exact alternatives. This is the reason that the leftmost column is labeled "strategy theme" rather than "alternative." Once the most promising strategy themes have been identified, then specific alternatives can be defined for each theme.

28.3.2   Too Few Alternatives

When there are too few alternatives, and especially if none of these appears to be very good, then the task is to develop additional alternatives that are better. Associative reasoning processes can be both a help and a hindrance in this process. These reasoning processes can help us because they might generate some ideas that do not seem at first to be relevant, but that turn out to be useful. However, associative reasoning processes can also be a hindrance, because they quickly build a "good story" about why we already have all the alternatives that are possible. Hence, we often have a tendency to "rush to judgment" and select an alternative before we have given careful consideration to other possibilities.

Thus, the basic issue is how to be creative and come up with additional good alternatives. There is a large literature on creativity, and Clemen (1996, Chap. 6), summarizes some of the creativity-enhancing methods that are most relevant for decision analysis. Keller and Ho (1988) also examine methods for developing alternatives. Several of their methods start either with existing alternatives or with a list of objectives for the decision, and attempt to develop new alternatives that are more attractive. Keeney (1992, Chaps. 7, 8), focuses on developing alternatives from decision problem objectives. Methods for developing additional alternatives include:

1. Modify existing alternatives by changing their characteristics. A strategy-generation table can be useful for this purpose. If you develop the columns for a strategy-generation table that represents your existing alternatives, then the table may suggest other entries within some of the columns that result in better alternatives. Or perhaps it will suggest other columns that ought to be in the table, and hence additional desirable characteristics for alternatives.

2. Focus on a single objective. Consider each evaluation consideration in the lowest layer of the value hierarchy for your decision one at a time. Try to develop alternatives that do as well as possible with respect to each evaluation consideration without taking the other evaluation considerations into account. While each of the alternatives may be too "one-dimensional" to be attractive by itself, there may be feasible alternatives that combine the strong points of the (possibly) extreme alternatives that you have developed by this procedure.

3. Maximize a particular objective at a higher layer in the hierarchy. This approach is likely to result in alternatives that are more balanced than if you focus on only one lowest-layer objective.

4. Directly concentrate on developing alternatives that provide balance across all the evaluation considerations.

28.3.3 Developing Alternatives When There Is Uncertainty

The methods just presented for developing alternatives are relevant whether or not there is uncertainty about the outcome of selecting a particular alternative. However, there are some special issues that arise in developing alternatives for decisions under uncertainty. These are illustrated by the decision problem in Crawford et al. (1978). This decision addressed the selection of a transmission-line size and tower configuration for a high-voltage electrical transmission line. The basic tradeoff was between the capital cost of building the transmission line and the operating cost of the line. By spending more on construction (higher capital cost), the resulting transmission line could be larger and hence have lower operating costs because it would have lower electrical losses due to heating.

If there were no uncertainty, then determining the best alternative would be a simple calculation: Select a discount rate and find the alternative with the lowest net present value. When there is uncertainty, however, the situation is not so simple. In the transmission-line decision, the projected operating cost was a function of the cost of the oil that would be used to generate the electricity being transported over the line. At the time the analysis was done, there was substantial uncertainty about the cost of oil over the projected 50-year lifetime of the transmission line.

In this type of situation, some alternatives that are not relevant under certainty can make sense. You might want to *hedge* against the uncertainty by building a medium-size transmission line. This might not be the cheapest alternative with either low or high future oil prices, but it would not do too badly in either case. By selecting this hedging alternative, you assure that you will not do very poorly regardless of what happens. Of course, you also give up the opportunity of doing very well if you happen to guess right about the future.

It may also make sense to select an alternative that allows you to *sequence* your decisions. In the conductor selection decision, an alternative was investigated where large towers would be built but a small transmission line would be installed. Thus, if the price of oil was later high, it would be possible to install larger transmission lines without having to build new towers. In this case, as in many other sequential decisions, you would have to spend some additional money up front (to build larger towers) to preserve the option of responding when the uncertainty is later resolved.

Another possible approach is to *risk share* by taking a partner. When you do this, you lose the opportunity to make all the profit if the uncertainty turns out well, but you also share the loss if things do not turn out so well. Finally, there may be the opportunity to take *insurance* against the risks posed by the uncertainties

28.4 VALUE ANALYSIS

This section presents a multiobjective value analysis procedure to rank alternatives and select the most preferred alternative. This procedure is appropriate when there are multiple conflicting objectives and no uncertainty about the outcome of each alternative. The presentation assumes that evaluation measures and alternatives have been specified as discussed in Sections 28.2 and 28.3. While the methods presented in this section assume that there is no uncertainty about the outcome of each alternative, these methods readily generalize to situations with uncertainty, and such situation are considered starting in Section 28.5.

Suppose that $x_1, x_2, \ldots, x_n$ are the evaluation measures for a decision. Then to select the preferred alternative it is necessary to combine these evaluation measures into a single measure of the overall desirability of an alternative. This is done by developing a *value function* $v(x_1, x_2, \ldots, x_n)$. The most commonly used functional form is a weighted sum of functions over each individual evaluation measure:

$$v(x_1, x_2, \ldots, x_n) = w_1 v_1(x_1) + w_2 v_2(x_2) + \cdots + w_n v_n(x_n)$$

From a practical standpoint, this functional form has the advantage that it is relatively simple to specify for any particular decision. Once the evaluation measures are determined, it is only necessary to specify the *single-dimensional value functions* $v_i(x_i)$ and the *weights* w_i to determine the value function. Procedures to determine these are presented in this section.

The single-dimensional value functions are often scaled so that the least preferred level of x_i has a single-dimensional value of zero and the most preferred level has a single-dimensional value of one. Similarly, the weights are often scaled so that each weight is between zero and one, and the sum of the weights is one. If the value function is scaled in this manner, then the least preferred combination of evaluation measure levels will have an overall value of zero, and the most preferred combination of evaluation measure levels will have an overall value of one.

Note that the weighted-sum functional form implicitly assumes that the increment in overall value received by improving any single evaluation measure by a specified amount is the same regardless of the levels of the other evaluation measures. This will be a reasonable assumption if the evaluation considerations obey the desirable properties described in Section 28.2. Keeney and Raiffa (1976, Chap. 3) and Kirkwood (1997, Chap. 9) review the theory for the weighted-sum value function in more detail.

The remainder of this section reviews a procedure to specify the weighted-sum value function. The procedure is based on the simple multiattribute rating technique with swing weights (SMARTS). Further discussion relevant to the SMARTS procedure is presented in Edwards (1977), von Winterfeldt and Edwards (1986, Chap. 8), and Edwards and Barron (1994). This method of analysis is based on measurable-value theory, which is presented in Dyer and Sarin (1979) and Kirkwood (1997, Chap. 9).

28.4.1 Determining Single-dimensional Value Functions

A single-dimensional value function specifies, in quantitative form, the relative desirability of different levels of a particular evaluation measure. For example, if larger levels are more desirable, then the single-dimensional value function will be *monotonically increasing* with greater levels of the evaluation measure, while if larger levels are less desirable, then the single-dimensional value function will be *monotonically decreasing* with greater levels of the evaluation measure.

Since a single-dimensional value function specifies in quantitative form the relative desirability of different levels of an evaluation measure, different stakeholders for a decision may have different single-dimensional value functions. That is, there is no objectively right or wrong value function since this value function depends on personal preferences. An advantage of specifying single-dimensional value functions in quantitative form is that it then becomes possible to quantitatively investigate the implications of differing views about what are the appropriate value functions.

Two specific forms are often used for the single-dimensional value functions. For each of these forms, it is necessary to ask a decision maker some questions about the relative desirability of different levels of an evaluation measure. The answers to these questions are then used to specify a mathematical form for the single-dimensional value function over the evaluation measure. With both forms, it is assumed that a *range* of possible levels has already been specified for the evaluation measure. More detailed information about the procedures for assessing single-dimensional value functions is presented in Kirkwood (1997, Chap. 4).

Form 1: Piecewise-linear Single-dimensional Value Function. This procedure can be demonstrated with an example. Suppose that a particular decision involves possible process improvements for a manufacturing process, and that there is an evaluation measure scale for productivity enhancement where possible levels of this evaluation measure range between -1 and 2. Larger levels on this evaluation measure scale are more preferred. To determine a single-dimensional value function, the decision maker is asked to consider the value increments obtained from moving between each successive pair of evaluation measure numbers (-1 to 0, 0 to 1, and 1 to 2). Suppose that the value increment between 0 and 1 is the smallest value increment between any of the two neighboring scores on the productivity enhancement scale, and that this value increment is the same as that between 1 and 2. Furthermore, the value increment between -1 and 0 is greater than that between 0 and 1. Specifically, this value increment is twice as great as that between 0 and 1.

To determine a piecewise-linear single-dimensional value function over the productivity enhancement evaluation measure, it is necessary to find values for each of the different levels of that evaluation measure that result in the value increments given in the preceding paragraph. Then the remainder of the value function is approximated by drawing straight lines between each of the determined points. (Hence the name *piecewise linear*.)

This process can be completed using the information in the preceding paragraph by remembering that the total value increment between the lowest possible level of productivity enhancement (which is -1) and the highest possible level (which is 2) is one. The fact that this value increment is the sum of the value increments going from -1 to 0, 0 to 1, and 1 to 2 can be used to determine the necessary values. To do this, let x represent the smallest value increment (that is, the increment going from 0 to 1). Then the increment going from 1 to 2 is also x, and the increment going from -1 to 2 is $2x$. Thus the sum of all the value increments is $2x + x + x$. Hence, $2x + x + x = 1$, and thus $x = 1/4 = 0.25$.

It follows from this result that the increment in value going from 0 to 1 is 0.25, since this increment is equal to x. The increment in value going from 1 to 2 is also 0.25, since this increment is also equal to x. Finally, the increment in value going from -1 to 0 is 0.50, since this increment is equal to $2x$. To obtain the value for each of the defined levels of the productivity enhancement evaluation measure, add up the value increments between the lowest possible level and the level of interest. Thus, the value for -1 is 0 (since this is the least preferred level), the value for 0 is $= 0 + 2x = 0 + 2 \times 0.25 = 0.50$, the value for 1 is $0 + 2x + x = 0 + 2 \times 0.25 + 0.25 = 0.75$, and the value for 2 is $0 + 2x + x = 0 + 2 \times 0.25 + 0.25 + 0.25 = 1$. A graph for the piecewise-linear single-dimensional value function over productivity enhancement is shown in Figure 28.4.

Form 2: Assume an Exponential Functional Form for the Single-dimensional Value Function. With this approach, an exponential functional form is assumed for the single-dimensional value function that has a constant ρ that can be adjusted to fit the particular situation in a specific decision problem. Kirkwood and Sarin (1980) establish conditions under which this form is appropriate, and these conditions are often approximately met in practice. If preferences are *monotonically increasing* over an evaluation measure x_i, then the exponential single-dimensional value function is $v_i(x_i) = \{1 - \exp[-(x_i - \text{low})/\rho]\} / \{1 - \exp[-(\text{high} - \text{low})/\rho]\}$, if $\rho \neq$ infinity, or $v_i(x_i) = (x_i - \text{low}) / (\text{high} - \text{low})$, if $\rho = $ Infinity. If preferences are *monotonically decreasing*, then $v_i(x_i) = \{1 - \exp[-(\text{high} - x_i) / \rho]\} / \{1 - \exp[-(\text{high} - \text{low}) / \rho]\}$, if $\rho \neq$ infinity, or $v_i(x_i) = (\text{high} - x_i)/(\text{high} - \text{low})$, if $\rho = $ infinity.

In these equations, "high" represents the highest level of the evaluation measure that

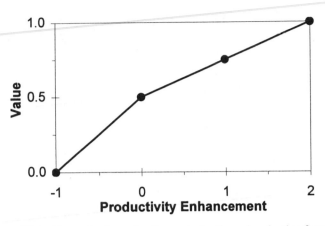

Figure 28.4 Example piecewise linear single dimensional value function.

is being considered, and "low" represents the lowest level that is being considered. Also, "exp" represents the exponential function. The constant ρ is call the *exponential constant,* and its value is adjusted to create different shapes for the single-dimensional value function. Figure 28.5 shows example exponential single-dimensional value functions. In this figure, low = 0, high = 10, and the legend shows different values of the exponential constant ρ. The figure shows that as the magnitude of ρ decreases, the function deviates more from linear. When ρ = infinity, the single-dimensional value function is linear.

The value of ρ is usually assessed by determining the *midvalue* for the interval between "low" and "high." The midvalue of a range is defined to be the score such that the difference in value between the lowest score in the range and the midvalue is the same as the difference in value between the midvalue and the highest score. From the preceding discussion, it follows that the single-dimensional value for the midvalue is 0.5. That is, v_i (midvalue) = 0.5. Since low and high are also known, the only unknown in the exponential single-dimensional value-function equation is ρ. Hence, the equation can be solved for ρ. This must be done numerically, and Kirkwood (1997, Chap. 4) presents the procedure. Realistic values for ρ depend on the range from low to high. If the magnitude of ρ is less than 0.1 $\times$ (high $-$ low), then the single-dimensional value function will be so curved as to indicate that there is some highly unusual situation. If the magnitude of ρ is greater than 10 $\times$ (high $-$ low), then the single-dimensional value function will be essentially a straight line.

28.4.2 Determining the Weights

Once the single-dimensional value functions have been determined for all the evaluation measures, it is only necessary to determine the weights w_i to complete the specification of the value function. To understand the procedure for determining the weights, it is useful to remember some of the properties of a value function. First, the single-dimensional value

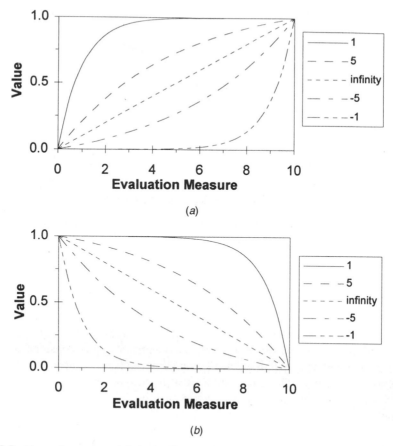

Figure 28.5 Example exponential single-dimensional value functions for different values of the exponential constant ρ. (a) Monotonically increasing preferences. (b) Monotonically decreasing preferences.

functions are specified so that each of them is equal to zero for the least preferred level that is being considered for the corresponding evaluation measure. Similarly, each of the single-dimensional value functions is specified so that it is equal to one for the most preferred level that is being considered for the corresponding evaluation measure.

From these properties of the single-dimensional value functions, it follows that the weight for an evaluation measure is equal to the increment in value that is received from moving the score on that evaluation measure from its least preferred level to its most preferred level. This property provides a basis for a procedure to determine the weights. This procedure is analogous to the procedure used to find piecewise-linear single-dimensional value functions. Specifically, the steps are:

1. Consider the increments in value that would occur by moving (or "swinging") each of the evaluation measures from the least preferred end of its range to the most preferred end, and place these increments in order of successively increasing value increments.

2. Quantitatively scale each of these value increments as a multiple of the smallest value increment received from swinging any evaluation measure over its entire range.

3. Set the smallest value increment for swinging any evaluation measure over its entire range so that the total of all the increments is 1.

4. Use the results of Step 3 to determine the weights for all the evaluation measures.

To illustrate this procedure, consider a decision problem involving upgrading a local area network that has three evaluation measures: productivity enhancement, cost increase, and security. Further, suppose that the range for productivity enhancement is from -1 to 2, the range for cost increase is from 0 to 150 in thousands of dollars, and the range for security is from -2 to 1. Suppose that the value increment from increasing ("swinging") productivity enhancement over its total range from -1 to 2 has the smallest increment of value, the value increment for swinging cost increase over its total range from 150 to 0 is larger, and the largest value increment comes from swinging security over its total range from -2 to 1. This means that productivity enhancement has a smaller weight than cost increase, which in turn has a smaller weight than security.

Suppose further that the swing over the cost increase from 150 to 0 has 1.5 times as great a value increment as the swing over productivity enhancement from -1 to 2, and the swing over security from -2 to 1 has 1.25 times the value increment of the swing over cost increase from 150 to 0. Since the increment in value from swinging an evaluation measure over its total range is equal to the weight from the evaluation measure, it follows from this information that $w_c = 1.5 \times w_p$, and $w_s = 1.25 \times w_c$, where w_c is the weight for cost increase, w_p is the weight for productivity enhancement, and w_s is the weight for security. Since the weights must sum to one, it follows that

$$1 = w_p + w_c + w_s = w_p + 1.5w_p + 1.25 \times 1.5w_p = w_p (1 + 1.5 + 1.25 \times 1.5)$$

Solving this set of equations yields $w_p = 0.23$, $w_c = 0.35$, and $w_s = 0.43$. (Note that these weights do not sum to exactly one due to roundoff error.) More details, as well as alternate approaches for determining both single-dimensional value functions and weights, are presented in Keeney and Raiffa (1976, Chap. 3) and Kirkwood (1997, Chap. 4).

28.4.3   Determining and Interpreting the Overall Value for an Alternative

Determining the overall value for an alternative once the value function is found is just arithmetic. The single-dimensional value for each evaluation measure is determined by applying the appropriate single-dimensional value function to the evaluation measure level (score) for the alternative. Once all the single-dimensional values are determined, then each of these is multiplied by the appropriate weight, and the results are summed to determine the overall value for the alternative. This calculation can be tedious, and in practical decision problems the calculations are typically done using either an electronic spreadsheet or a specialized program. Kirkwood (1997) presents spreadsheet calculation procedures.

As discussed previously, the calculated value numbers will always be between zero and one, but what does a specific value number mean? The value determination procedure has been specified so that an alternative that has the least preferred score on all of the evaluation measures will have an overall value of zero. Similarly, an alternative that has the most preferred score on all of the evaluation measures will have an overall value of one. In many decisions there will not be any actual alternatives that are either this good or bad, but the value numbers for the actual alternatives can be interpreted by comparing these actual alternatives to (possibly hypothetical) alternatives with overall values of zero and one. The value number for a particular alternative gives the proportion of the way, in a value sense, that the alternative is from the (possibly hypothetical) alternative with an overall value of zero to the (also possibly hypothetical) alternative with an overall value of one.

This discussion of the meaning of value numbers shows that no specific meaning can be given to value numbers without knowing the ranges of the evaluation measures that are being used. We showed earlier while presenting the evaluation procedure for weights that the ranges of the evaluation measures must be considered when determining weights for a value function. In a similar way, the ranges of the evaluation measures must be considered when interpreting the overall value numbers for alternatives.

28.5 DECISIONS WITH UNCERTAINTY

This section expands the decision analysis methods presented in earlier sections to situations where there is uncertainty about the outcome of selecting a particular alternative. Research about the mistakes that people make when they think about uncertainty is reviewed, and procedures are presented for thinking more clearly about uncertainty with particular emphasis on probability analysis methods. In this section we focus on situations where there is only a single evaluation measure, and the method is extended to handle multiple evaluation measures in Section 28.6.

28.5.1 Working with Probability

Philosophers and statisticians have debated the meaning of probability for centuries, and a widely held view today is that probabilities are statements about the judgments of individuals. Specifically, a probability is a numerical specification of an individual's *degree of belief* that an uncertain event will occur. Thus, probabilities are ultimately subjective in nature, and it is possible that two different people might have different probabilities for the same event.

For example, if you have not yet looked out the window today, and you hear that the weather forecast on the radio is for an 80 percent chance of rain, then you might assign a probability of 80 percent to the likelihood of rain. In the meantime, your spouse, who has already looked out the window and discovered that it is, in fact, raining, has probably assigned a probability of 100 percent to it raining today. Similarly, two marketing specialists might have different views on the probability that a proposed new product will be a success.

Since probabilities are subjective, they must ultimately be determined by asking someone for them. If the person whose probability we need to determine is familiar with probability concepts, then we may be able to directly ask for the required probabilities. If not, then we need to use some sort of measuring instrument. The *probability wheel* is one instrument for measuring probabilities. Figure 28.6 shows a sketch of a probability wheel. This wheel is shown on its stand, and the wheel is free to spin around its center. The wheel

Figure 28.6 Probability wheel (probability = 0.25). (From *Strategic Decision Making: Multiobjective Decision Analysis with Spreadsheets* by C. W. Kirkwood. Copyright © 1997 Brooks/Cole Publishing Company, Pacific Grove, CA, a division of International Thomson Publishing, Inc. By permission of the publisher.)

has two sectors, one lighter and one darker, and the proportion of the wheel that is of each shade can be adjusted. There is a fixed pointer above the wheel, and after the wheel is spun the pointer will be pointing to either the dark or light portion of the disk.

Suppose the wheel is perfectly balanced so that when it is spun the chances are equally likely that the pointer will end up pointing to any spot along the edge of the disk. Using a probability wheel as a measuring instrument, the probability of an event (for example, that it will rain tomorrow) can be found as follows: consider two different situations. The first is out in the real world where it either rains or it does not rain. Suppose that you will receive an attractive prize if it rains (for example, $100.00), and you will not receive this prize if it does not rain. The second situation involves a single spin of the probability wheel. If, after the spin, the pointer points to the dark sector of the wheel you will receive the same attractive prize ($100.00 for our example), and if the pointer points to the light sector of the wheel you will not receive this prize.

To determine the probability of rain tomorrow, adjust the portion of the probability wheel that is dark until you are indifferent between selecting a spin of the wheel (and hence receiving $100.00 only if the pointer points to the dark sector) or seeing whether it rains (in which case you receive $100.00) or does not rain (in which case you receive nothing). When you have made this adjustment, and are indifferent between participating in these two uncertain situations (one involving the probability wheel and the other involving the weather), then your probability of rain is equal to the portion of the wheel that is dark. Thus, if you are indifferent between betting on the wheel in Figure 28.6 or betting on it raining tomorrow, then your probability that it will rain tomorrow is one-quarter (25 percent or 0.25).

A variety of experiments have been conducted to determine how accurately people can assess their uncertainty. Kahneman et al. (1982) summarize much of this research (see also Tversky and Kahneman, 1974; Dawes, 1988) and have concluded that people who have not received training in assessing their uncertainty underestimate the uncertainty in a situation, often by a substantial amount. That is, people are *overconfident* about their ability to predict the outcomes of uncertain situations. This is true even for people who are trained specialists in a field. Thus, care should be taken in assessing probability information to ensure that this information accurately represents the actual degree of uncertainty in the situation. Merkhofer (1987) and Spetzler and Staël von Holstein (1975) review probability assessment procedures. Shephard and Kirkwood (1994) present a transcript of a probability assessment session.

28.5.2 Expected Value and Decision Making Under Uncertainty

When there is no uncertainty about the outcome of a decision alternative, the complexity in evaluating alternatives comes from the need to consider tradeoffs among the evaluation considerations. When there is only one evaluation measure in such a situation, then there is generally little difficulty in evaluating alternatives: if more of the evaluation measure is preferred to less, then pick the alternative that gives the largest amount of the evaluation measure, but if less of the evaluation measure is preferred to more, then select the alternative that gives the smallest amount of the evaluation measure.

However, when there is uncertainty about what outcome will result from selecting an alternative, then even with a single evaluation measure, the evaluation of alternatives can be difficult. To illustrate the difficulty, consider the following example: suppose that someone offers to sell you for $7,000 a lottery ticket that gives you the right to flip a *fair* coin (that is, one that has equal chances of coming up heads or tails) and receive $10,000 if a head comes up or $5,000 if a tail comes up. Should you buy the lottery ticket?

The answer to this is not immediately clear. If a head comes up, then you will end up with $10,000 − $7,000 = $3,000, which is desirable, but if a tail comes up, then you will end up with $5,000 − $7,000 = − $2,000, which is undesirable. What should you do?

To begin to answer this question, it is useful to consider a quantity called the *expected value,* which is calculated by multiplying each possible outcome by its associated probability and adding up the results. Thus, for the coin-flip lottery, the expected value is $0.5 \times (\$10,000 − \$7,000) + 0.5 \times (\$5,000 − \$7,000) = \$500$. The importance of the concept of the expected value comes from a mathematical result called the Weak Law of Large Numbers, which is presented in probability or statistics textbooks (Olkin et al., 1994; Ghahramani, 1996). This law proves that under very general conditions the probability is very high that the average outcome for a large number of independent decisions will be very close to the average of the expected values for the decisions. Thus, using the expected-value calculation just given, if you pay $7,000 for the coin-flip lottery and repeat this process many times (paying $7,000 each time), then you are very likely to end up making an average of about $500 for each time you play.

This result seems to argue for placing a value on an alternative that is equal to its expected value. This is because if you pay more than the expected value for alternatives, then from the Weak Law of Large Numbers, you are very likely to lose money over many decisions because the probability is high that on average the alternatives will only return about their expected values. Similarly, if you are not willing to pay as much as the expected value for alternatives, then you will lose the opportunity to make additional money that you could have made over the long run because the probability is very high that alternatives will on average return more than you are willing to pay for them.

It is often appropriate to use expected value as a criterion for choosing among alternatives in practical systems engineering and management decisions. However, there are also situations where this is not appropriate. Note that the Weak Law of Large Numbers applies to "a large number of independent decisions." There is nothing in the Law that guarantees that you will obtain the expected value for any particular decision. In fact, for the coin-flip example it is impossible to obtain the expected value for any single flip since you will either make $\$10,000 - \$7,000 = \$3,000$ or lose $\$7,000 - \$5,000 = \$2,000$, and neither of these amounts is equal to the expected value of $\$500$. The Weak Law of Large Numbers guarantees that over many repetitions of this lottery the probability is high that you will average out to a net of $\$500$ for each time you play. However, it may take a while for this to happen. In the meantime, you may go broke and not be able to continue to play the lottery. (A senior executive for a large corporation once commented to me that most of the decisions they analyzed were for a few million dollars, and therefore it was appropriate to use expected value as a decision criterion. Obviously, whether this is true or not depends on your asset position.)

28.5.3   Risk Aversion and Certainty Equivalents

The preceding discussion implies that in situations where the uncertainties are large it may be appropriate to be more conservative in making decisions than would be indicated by using expected value as a criterion to make the decisions. Thus, for the coin-flip example given earlier, it might be appropriate to value the lottery at less than its expected value. The certain amount of an evaluation measure quantity that is equally preferred to an uncertain alternative is called the alternative's *certainty equivalent*. If a decision maker's certainty equivalent for an uncertain alternative is less than the expected value for the alternative and preferences are monotonically increasing over the evaluation measure, then the decision maker is said to be *risk averse*. Similarly, if preferences are monotonically decreasing over the evaluation measure, then the decision maker is said to be risk averse if the certainty equivalent is larger than the expected value.

Analysis methods for decisions when a decision maker is risk averse are presented in detail in decision analysis textbooks (Clemen, 1996; Kirkwood, 1997; McNamee and Celona, 1990). The basic approach is to develop a *utility function*, which quantifies a decision maker's attitude toward taking risks, and then to use this utility function to calculate the certainty equivalents for uncertain alternatives. Once these certainty equivalents are determined, it is straightforward to rank alternatives: if preferences are monotonically increasing over the evaluation measure, then alternatives with larger certainty equivalents are more preferred; if preferences are monotonically decreasing over the evaluation measure, then alternatives with smaller certainty equivalents are more preferred.

Experience indicates that considering risk aversion in a decision analysis often does not change the results. In fact, Howard (1988) notes that he has found it to be of practical concern in only 5 to 10 percent of business decision. Thus, it is natural to ask whether there is a way to quickly investigate the potential impact of risk aversion on a decision without going through a detailed utility analysis.

Theoretical and practical considerations support the use of a particular utility function form called the *exponential* (Kirkwood, 1992). This utility function has a single parameter, ρ, called the *risk tolerance*, which specifies the degree of risk aversion of the decision maker. When this utility function is used, the certainty equivalent (CE) for an uncertain alternative is given by CE $= -\rho \times \ln E[\exp(-x/\rho)]$ when preferences are monotonically increasing, and by CE $= \rho \times \ln E[\exp(x/\rho)]$ when preferences are monotonically decreasing. In these equations, ρ represents the risk tolerance, ln is the natural logarithm, exp is the exponential, and E represents the operation of taking an expected value. In the limit as the risk tolerance gets larger and larger, the certainty equivalent becomes closer and closer to the expected value. Thus, smaller values of ρ represent more risk-averse decision makers.

To illustrate the use of these expressions to calculate a certainty equivalent, suppose that for the coin-flip lottery discussed earlier the risk tolerance is $1,000. With an exponential utility function, the certainty equivalent for the alternative of paying $7,000 for this lottery is CE $= -\$1,000 \times \ln\{0.5 \times \exp[-(\$10,000 - \$7,000) / \$1,000] + 0.5 \times \exp[-(\$5,000 - \$7,000) / \$1,000]\} = -\$1,314$. This contrasts with the expected value of $500 that was determined earlier for this uncertain alternative. Howard (1988) and McNamee and Celona (1990) present procedures for approximately determining the risk tolerance in a particular decision problem, and Kirkwood (1997, Secs. 6.6 and 6.7) discusses these in more detail.

28.5.4   Decision Trees

The expected value and certainty equivalent calculation methods presented earlier allow a straightforward evaluation of alternatives once the probability distributions are known for the evaluation measure. In many practical situations there are several uncertain quantities that must be considered together to determine the probability distribution over the evaluation measure. The discussion below reviews analysis procedures for such situations using an example adapted from Kirkwood (1997, App. D).

An aircraft engine manufacturing company is considering changing its production process. The profitability of the proposed change will depend on the number of engines that are sold during the next year. Discussions with the marketing department make it clear that the sales will differ depending on the state of the economy. This is because the engine is used in corporate aircraft, and companies tend to purchase fewer airplanes when the economy is poor.

Suppose that if the economy is good then there is a 30 percent chance that 40 engines will be sold and a 70 percent change that 30 engines will be sold. On the other hand, if the economy is poor, there is a 25 percent chance that 35 engines will be sold and a 75 percent chance that 25 engines will be sold. The corporate planning department says the chance is 60 percent that the economy will be good and the chance is 40 percent that it will be poor.

The company currently uses a production process that involves mostly purchasing components from subcontractors. As a result, there is no fixed cost of production. The marginal production cost per engine depends on the state of the economy. When the economy is good, the marginal production cost per engine is $125,000, while when the economy is poor the marginal production cost per engine is $110,000.

The proposed modified production process involves more in-house manufacturing, and there would be a fixed production cost of $1,000,000 per year regardless of the number of engines produced. Marginal production costs would still depend on the state of the economy, and would be $90,000 per engine when the economy is good and $85,000 per engine when the economy is poor.

The sale price for the engines does not depend on which production process is used, but does depend on the state of the economy. Specifically, it is $150,000 per engine when the economy is good and $120,000 when the economy is poor. Using expected net profit as the criterion for making a decision, should the company stick with the current production

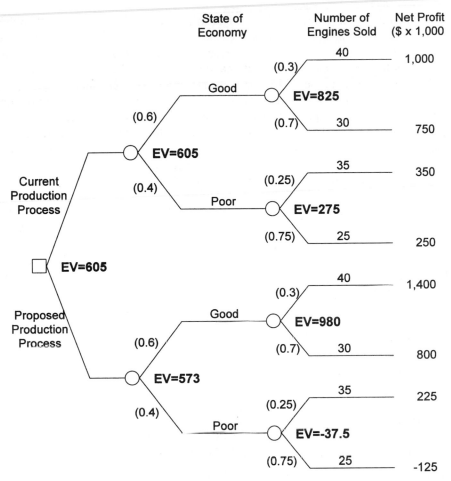

Figure 28.7 Decision tree for production process alternatives. (From *Strategic Decision Making: Multiobjective Decision Analysis with Spreadsheets* by C. W. Kirkwood. Copyright © 1997 Brooks/ Cole Publishing Company, Pacific Grove, CA, a division of International Thomson Publishing, Inc. By permission of the publisher.)

process or switch to the proposed modified process? (*Net profit* is used here to mean the difference between sales revenue and production costs.)

A *decision tree* can be used to organize the information for this decision; such a decision tree is shown in Figure 28.7. In this figure, the square box at the left side of the decision tree is called a *decision node,* and it represents the decision to be made between the current and proposed production processes. The lines emanating from a decision node represent the various alternatives that are available. The circles are called *chance nodes,* and they each represent an uncertainty quantity. The lines emanating from a chance node represent the possible results when the uncertainty represented by the node is resolved. The tree is read from left to right, and it starts on the left from a single decision node (called the *root node*). Each line emanating from a chance node has a number in parentheses that is the probability that the event represented by that line will occur. Furthermore, the probabilities assigned to the lines emanating from each chance node take into account the particular path that leads from the root node to that chance node. That is, the probabilities are *conditional on* the events on branches between the root node and that chance node.

At the right-hand side of the decision tree, the net profits for each path through the tree are given on the end points of the tree. Determining each of these end-point profits requires some calculation. For example, the topmost number (1,000) is calculated in two steps. First, determine the net profit for each engine sold when the current production process is used and the state of the economy is good, which is $150 − $125 = $25 (in thousands of dollars). Then multiply this by the number of engines sold: 40 × $25 = $1,000. A similar calculation is done to determine the net profit for each of the other possible end points.

Once the net profit figures are determined for all of the end points, a procedure called *rolling back the tree* is used to calculate the expected value for each alternative. Clemen (1996, Chap. 4) discusses the rationale for this procedure is further detail. The procedure is as follows: start by calculating the expected value for each of the rightmost chance nodes in the tree. As an example of the calculation, the expected net profit of $825 shown for the upper rightmost chance node is calculated by 0.3 × $1,000 + 0.7 × $750 = $825. After the expected values are calculated for all the rightmost chance nodes in the tree, use these to calculate the expected values for the next set of nodes to the left in the tree. For example, the expected value of $605 shown for the upper leftmost chance node is calculated by 0.6 × $825 + 0.4 × $275 = $605. Figure 28.7 shows that the expected value for the current production process is $605 and for the proposed production process is $573. Thus, the current production process has a greater expected profit.

This procedure can be extended to incorporate attitude toward risk taking. The method can also be extended to consider sequential decisions where an initial decision is made, some uncertainties are resolved, and then additional decisions must be made. See Clemen (1996) or McNamee and Celona (1990) for further details.

28.5.5   Influence Diagrams

The decision tree shown in Figure 28.7 clarifies the production decision; however, a decision tree can quickly become very large if the decision is more complex than the one shown in Figure 28.7. More complex decisions can be addressed with *influence diagrams.* These diagrams specify qualitative information about the interdependencies in a decision model in a more compact way than a decision tree. An influence diagram for the production decision is shown in Figure 28.8. In this diagram, the rectangular box represents the production process decision, and thus corresponds to the decision node in the Figure 28.7 decision tree. Circles or ovals represent uncertainties, and thus correspond to chance nodes in a decision tree. A double-lined circle or oval (for example, Total Cost in Figure 28.8) is called a *deterministic node,* and it represents a quantity whose value is known for certain once the inputs to the node are specified. Finally, a double-lined rounded-corner rectangle (for example, Net Profit in Figure 28.8) represents the quantity whose value is of concern for making the decision. Thus, this node represent all of the end points in a decision tree taken together. This is called the *value node* for the influence diagram.

The arrows pointing into a chance node, deterministic node, or value node represent the influences that determine the value of the quantity represented by the node. Thus,

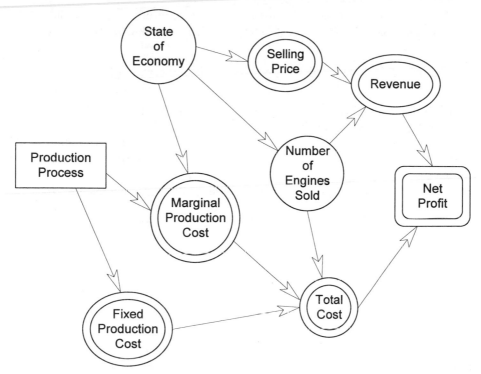

Figure 28.8   Influence diagram for production decision. (From *Strategic Decision Making: Multiobjective Decision Analysis with Spreadsheets* by C. W. Kirkwood. Copyright © 1997 Brooks/ Cole Publishing Company, Pacific Grove, CA, a division of International Thomson Publishing, Inc. By permission of the publisher.)

Figure 28.8 shows that the number of engines sold depends on the state of the economy, while the total cost depends on the number of engines sold, the marginal production cost, and the fixed production cost.

Influence diagrams can compactly represent the structure of a large decision model. For example, Kirkwood (1993) shows an influence diagram on one page that represents a decision model for which a decision tree would have 25,272 end points. See Clemen (1996) or McNamee and Celona (1990) for further details on influence diagram analysis methods.

28.6   MULTIPLE OBJECTIVES AND UNCERTAINTY

This section extends the approaches presented in earlier sections to decisions where there are both multiple evaluation concerns and uncertainty. The weighted-sum value function can still be used to evaluate outcomes of decisions when there is uncertainty, and probabilities are used to analyze uncertainties. The expected value and utility analysis methods presented in the preceding section generalize to situations with multiple evaluation concerns. In particular, for situations with relatively small risks, the expected value of the value function can be used to evaluate alternatives. The well-know result that the expected value of a sum of uncertain quantities is equal to the sum of the expected values of the uncertain quantities can then be applied to the weighted-sum value function to show that

$$EV = w_1 \times E[v_1(x_1)] + w_2 \times E[v_2(x_2)] + \cdots + w_n \times E[v_n(x_n)]$$

where EV is the expected value for the alternative, w_i is the weight for evaluation measure x_i, and $E[v_i(x_i)]$ is the expected value of $v_i(x_i)$.

When the uncertainties are great enough that it is necessary to take risk attitude into account, then theoretical arguments (Keeney and Raiffa, 1976; Kirkwood, 1997) show that it is often reasonable to apply an exponential utility function to the weighted-sum value function in order to determine the certainty equivalent value for an alternative. When this is done, the certainty equivalent value v_{ce} is given by

$$v_{ce} = -\rho_m \times \ln E\{\exp[-v(x_1, x_2, \ldots, x_n)/\rho_m]\}$$

where

$$v(x_1, x_2, \ldots, x_n) = w_1 v_1(x_1) + w_2 v_2(x_2) + \cdots + w_n v_n(x_n)$$

In this equation, ln is the natural logarithm, exp is the exponential, E is the expected value, and ρ_m is a constant called the *multiattribute risk tolerance*. In applications, this last constant is almost always a positive constant greater than 0.2. When ρ_m is greater than 10, the certainty equivalent value is very close to the expected value, and hence expected value can be used as a decision criterion. Often the specific value of ρ_m does not change the ranking of alternatives significantly in applications, and so it may be useful to first conduct a sensitivity analysis over ρ_m before attempting to determine a specific value for this constant. In the event that the ranking of alternatives is not changed when ρ_m is varied over the range from 0.2 to infinity, then it is not necessary to determine a specific value for this constant. Kirkwood (1997, Sec. 7.3) presents procedures for determining ρ_m.

The *certainty equivalent value* v_{ce} for an uncertain alternative has exactly the same interpretation as the value has for an alternative with no uncertainty. That is, v_{ce} is a number between zero and one, and it gives the portion of the way that the alternative is, in a value sense, from the least preferred outcome in the range of evaluation measures being considered to the most preferred outcome in the range of evaluation measures being considered.

In many cases of practical interest, the evaluation measures for each alternative are *probabilistically independent*. That is, the probability distribution for any evaluation measure does not change for different levels of the other evaluation measures once the alternative is specified. When this is true, the expression just given for the certainty equivalent reduces to $v_{ce} = CE_1 + CE_2 + \ldots + CE_n$, where $CE_i = -\rho_m \times \ln E\{\exp[-w_i \times v_i(x_i)/\rho_m]\}$. Thus, when the evaluation measures for a particular alternative are probabilistically independent, the intuitively appealing result holds that the overall certainty equivalent value for the alternative is equal to the sum of the certainty equivalent values for that alternative with respect to each evaluation measure. In this case, spreadsheet procedures to do the required calculations are particularly simple. Kirkwood (1997, Sec. 7.5) presents such procedures.

28.7 DECISION ANALYSIS SOFTWARE

The calculations needed to implement the methods presented in this chapter can be done using standard spreadsheet programs or using specialized decision analysis software. Kirkwood (1997) provides detailed instructions for using a spreadsheet to implement multiobjective value and utility analysis methods. (See Sections 4.7 and 7.5 of Kirkwood, 1997.) He also presents spreadsheet methods for implementing decision tree calculation procedures (Kirkwood, 1997, Section 7.6), but a spreadsheet approach to decision tree calculations becomes unwieldy if the size of the decision tree is large. This is because the size of a decision tree grows *exponentially* with the number of variables in the decision, and hence the construction of the tree in a spreadsheet becomes tedious when the number of variables is large.

Logical Decisions is a Microsoft Windows software package for multiobjective value and utility analysis. In addition to automating the calculations necessary for a multiobjective value or utility analysis, this package provides support for eliciting the information needed for the analysis. Logical Decisions is available from Logical Decisions, 1014 Wood Lily Drive, Golden CO 80401; 303-526-7537, 800-35-LOGIC, gary@logicaldecisions.com, www.logicaldecisions.com/.

Several software packages support analysis of decision trees and/or influence diagrams. The package *DATA* aids in building decision models using influence diagrams and decision trees, and it also supports modeling of Markov processes, Monte Carlo simulation, cost-effectiveness analysis, and multiattribute modeling. It is available for both Microsoft Windows and Macintosh computers. DATA is available from TreeAge Software, Inc., 1075 Main Street, Williamstown, MA 01267; 413-458-0104, 888-TREEAGE, info@treeage.com, www.treeage.com/.

DPL is a Microsoft Windows package that combines influence diagrams, decision trees, and spreadsheets. A decision programming language is included in some versions of DPL that allows the construction of decision models in a non-graphical manner. DPL is available from Applied Decision Analysis, 2710 Sand Hill Road, Menlo Park, CA 84025, 415-926-9251, dpldept@adainc.com.

Sensitivity/Supertree is available for MS-DOS, Microsoft Windows and Macintosh computers, and it supports analysis of decision trees, including links to spreadsheet models. It is available from Strategic Decisions Group, 2440 Sand Hill Road, Menlo Park, CA 94025, 415-854-9000, lhunter@sdgnet.com.

Buede (1996a,b) surveys software for decision analysis and other decision support methods, and these articles provide further information about the packages discussed above and other related software.

28.8 CONCLUDING COMMENTS

The decision analysis procedures discussed in this chapter provide a straightforward, logically defensible approach to decision making in complex systems engineering and management situations. The approaches have been successfully applied to a wide variety of decisions in such areas as bidding, product and project selection, regulation, facility site selection, technology choice, budget allocation, product planning, strategy, and standard setting (Corner and Kirkwood, 1991). Clemen (1996) and McNamee and Celona (1990) present further details on such probability modeling methods as decision trees, influence trees, and Monte Carlo simulation. Kirkwood (1993, 1994) presents methods for modeling and solving very large-scale decision problems.

Kirkwood (1997) presents additional details on multiobjective decision analysis methods, including use of spreadsheets to do the required calculations. He also discusses optimization approaches for using decision analysis to solve complex resource allocation problems. Keeney and Raiffa (1976) and Kirkwood (1997) review theory that is relevant for situations with multiple conflicting objectives.

REFERENCES

Buede, D. M. (1986). Structuring value attributes. *Interfaces* **16**(2), 52–62.

Buede, D. (1996a). Aiding insight III: Decision analysis software survey. *ORMS Today* **23**(4), 73–79.

Buede, D. (1996b). Second overview of the MCDA software market. *J. Multicriteria Decis. Anal.* **5**, 312–316.

Clemen, R. T. (1996). *Making Hard Decisions: An Introduction to Decision Analysis,* 2nd ed. Belmont, CA: Duxbury Press.

Corner, J. L., and Kirkwood, C. W. (1991). Decision analysis applications in the operations research literature, 1970–1989. *Oper. Res.* **39**, 206–219.

Crawford, C. M., Huntzinger, B. C., and Kirkwood, C. W. (1978). Multiobjective decision analysis for transmission conductor selection. *Manage. Sci.* **24**, 1700–1709.

Dawes, R. M. (1988). *Rational Choice in an Uncertain World.* San Diego, CA: Harcourt Brace Jovanovich.

Dyer, J. S., and Sarin, R. K. (1979). Measurable multiattribute value functions. *Oper. Res.* **27**, 810–822.

Edwards, W. (1977). How to use multiattribute utility measurement for social decision-making. *IEEE Trans. Syst., Man, Cybernet.* **SMC-7**, 326–340.

Edwards, W., and Barron, F. H. (1994). SMARTS and SMARTER: Improved simple methods for multiattribute utility measurement. *Organ. Behav. Hum. Decis. Proces.* **60**, 306–325.

Ghahramani, S. (1996). *Fundamentals of Probability.* Upper Saddle River, NJ: Prentice Hall.

Golabi, K., Kirkwood, C. W., and Sicherman, A. (1981). Selecting a portfolio of solar energy projects using multiattribute preference theory. *Manage. Sci.* **27**, 174–189.

Howard, R. A.(1988). Decision analysis: Practice and promise. *Manage. Sci.* **34**, 679–695.

Kahneman, D., Slovic, P., and Tversky, T. (eds.) (1982). *Judgment under Uncertainty: Heuristics and Biases.* Cambridge, UK: Cambridge University Press.

Keeney, R. L. (1981). Measurement scales for quantifying attributes. *Behav. Sci.* **26**, 29–36.

Keeney, R. L. (1992). *Value-Focused Thinking: A Path to Creative Decisionmaking.* Cambridge, MA: Harvard University Press.

Keeney, R. L., and Raiffa, H. (1976). *Decisions with Multiple Objectives: Preferences and Value Tradeoffs.* New York: Wiley.

Keller, L. R., and Ho, J. L. (1988). Decision problem structuring: Generating options. *IEEE Trans. Syst., Man, Cybernet.* **18**, 715–728.

Kirkwood, C. W. (1982). A case history of nuclear power plant site selection. *J. Oper. Res. Soc.* **33**, 353–363.

Kirkwood, C. W. (1992). An overview of methods for applied decision analysis. *Interfaces* **22**(6), 28–39.

Kirkwood, C. W. (1993). An algebraic approach to formulating and solving large models for sequential decisions under uncertainty. *Manage. Sci.* **39,** 900–913.

Kirkwood, C. W. (1994). Implementing an algorithm to solve large sequential decision analysis models. *IEEE Trans. Syst., Man, Cybernet.* **24,** 1425–1432.

Kirkwood, C. W. (1997). *Strategic Decision Making: Multiobjective Decision Analysis with Spreadsheets.* Belmont, CA: Duxbury Press.

Kirkwood, C. W., and Sarin, R. K. (1980). Preference conditions for multiattribute value functions. *Oper. Res.* **28,** 225–232.

McNamee, P., and Celona, J. (1990). *Decision Analysis with Supertree,* 2nd ed. South San Francisco, CA: Scientific Press.

Merkhofer, M. W. (1987). Quantifying judgmental uncertainty: Methodology, experiences, and insights. *IEEE Trans. Syst., Man, Cybernet.* **17,** 741–752.

Olkin, I., Gleser, L. J., and Derman, C. (1994). *Probability Models and Applications.* New York: Macmillan.

Shephard, G. G., and Kirkwood, C. W. (1994). Managing the judgmental probability elicitation process: A case study of analyst/manager interaction. *IEEE Trans. Eng. Manage.* **41,** 414–425.

Spetzler, C. S., and Staël von Holstein, C.-A. (1975). Probability encoding in decision analysis. *Manage. Sci.* **22,** 340–358.

Tversky, A., and Kahneman, D. (1974). Judgment under uncertainty: Heuristics and biases. *Science* **185,** 1124–1131.

von Winterfeldt, D., and Edwards, E. (1986). *Decision Analysis and Behavioral Research.* Cambridge, UK: Cambridge University Press.

Winston, W. L. (1994). *Operations Research: Applications and Algorithms,* 3rd ed. Boston: Pws-Kent Publ. Co.

29 Project Planning: Planning for Action

RUTH BUYS

29.1 INTRODUCTION

The adage that any road will do if you do not know where you are going has many applications. It also has a corollary: Any road to a destination will do if it does not matter how long it takes to get there or how much it costs to make the trip. Few of us can afford to take off in the general direction of some goal without concern for the time, effort, and funds it will take to arrive safely at our destination. On the contrary, most of us have limited resources for such endeavors. Budgets are shrinking while the cost of living grows, requiring that we manage our funds carefully. In the process, we commit funds to known future expenses, allocate others to the routine cost of daily business, and jealously guard any discretionary funds that may become available or accrue through our wise money management techniques. We are careful about how we spend money and are at least somewhat aware of the impact of alternate ways to reach our destinations. And certainly, we are concerned with how long it takes to get there. This approach reflects concern over resources, funds, and the method of achieving our objectives. As rudimentary as these steps are, they contain the basic elements of a plan. And planning before taking action is a crucial component of any systems development activity.

The core elements of planning are familiar to most systems professionals. It is not long before the idea of identifying tasks to be accomplished, finding staff to execute those tasks, and the resources to support them are part of every systems professional's experience and set of skills. For small projects as well as for individual efforts these plans remain largely ad hoc and unstructured. The execution of plans developed in this manner is entirely an individual effort. Each effort is different based on a uniquely individual set of preferences, constraints, and tolerances. These may be, and usually are, expressed in a unique vocabulary with unique assumptions and knowledge about the implementation alternatives.

However, when a project is undertaken by, or on behalf of, more than a few individuals, when the objectives are complex and there are many variables, planning for action becomes a team activity. It must become a more formal process. There must be a common vocabulary, common approach for identifying the work to be done and products to be produced, and a common way to communicate the status and results of the effort. This requires standard, formal methods that can be understood by a wide range of system engineering professionals and shared by colleagues using standard methods of data exchange.

Standard, formal planning techniques all share common goals in that they must allow the definition and tracking of such items as activities, milestones, and deliverables. If the project is large enough, this may require the ability to divide the project into phases.

Handbook of Systems Engineering and Management, Edited by A. P. Sage and W. B. Rouse
ISBN 0471-15405-9 ©1999 John Wiley and Sons, Inc.

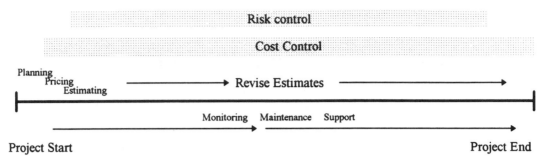

Figure 29.1   Planning for action.

Activities or tasks must be organized into time frames or schedules for their accomplishment. There must also be a mechanism to support the transition of a project between phases.

There are many techniques and methodologies available to systems engineers for planning a project's activities. These include familiar techniques such as work breakdown structures as well as more recent methodologies made possible by technological advances such as network-based planning and project management. Some of the methods focus on tracking the progress of projects. These include techniques such as critical path method (CPM) and program evaluation and review technique (PERT). There are also techniques to support the definition of roles and selection of personnel to staff projects and to assist in the organization of teams to complete the work. Figure 29.1 assembles these elements into a graphic view illustrating their relationship to the entire life cycle of the project.

The following sections of this chapter expand on major topics of interest to systems engineers involved in project planning. The next section, "Network-based Systems Planning and Project Management," describes techniques made possible by recent innovations in network hardware and software technology. The third, "Pricing and Estimating," and the fourth, "Risk and Cost Control," investigate topics tightly linked to establishing and protecting the health of the project. "Maintenance and Support" topics include those associated with ensuring the plan stays on track and monitoring the progress of the project. "Software for Planning Support" addresses automated tools for project planning and related tasks, and the last section, "Presentation and Communication of Results of Systems Planning," covers a variety of techniques for reporting project planning and status information.

29.2   NETWORK-BASED SYSTEMS PLANNING AND PROJECT MANAGEMENT

In any systems development effort that involves more than two or three staff and as many tasks, it is highly probable that one or more tasks will be executed at the same time. When that occurs, the most effective method of representing the tasks and their relationships is through a network diagram. A network diagram is a graphic representation of the plan showing the relationships among the tasks (Cori, 1985a). Establishing these networks requires that certain information be provided. Each task must be uniquely identified and last a certain, defined amount of time. Tasks placed to the left on a graphic representation occur before those to the right, and usually there are no loops in the network. This graphically displays the various tasks, their dependencies, and sequence of execution, as shown in Figure 29.2.

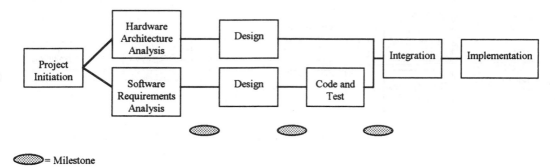

= Milestone

Figure 29.2   Project task network diagram.

Effective systems planning and project management rely on a systematic description of tasks, an understanding of the relationships and dependencies among those tasks, and the scope of the project in terms of schedule, cost, and level of risk. To manage systems development projects successfully, these elements must be assessed at the start of the project and reassessed continually throughout the life of the project so that management decisions are based on current data that are as accurate as possible.

Network-based systems planning and project management is based on the establishment of work or task relationships and the ability to visually represent those relationships along with interdependencies among tasks for monitoring and project management purposes. It relies on the use of automated project management techniques that support the definition and tracking of tasks so that managers can work with an overall picture of the plan regardless of the enormity of the project or number of diverse geographic locations that may be involved.

29.2.1   Task Definition

Tasks must be defined in order to partition responsibilities, determine milestones, and estimate resource requirements. Two separate types of tasks are usually defined. The first is a set of product-related tasks or those technical activities needed to analyze, create, and prepare a system for implementation. The second are those management tasks that support and enable the development of that system. A popular technique is the creation of a work breakdown structure (Thayer and Fairley, 1997). Using this methodology, a series of tasks for managing the project as well as creating the technical product are defined. They are associated through a series of notations that connect the individual management tasks with the product tasks and vice versa. Numbers are often associated with the activities, and alphabetic symbols are associated with the product. When the two are combined, it is possible to determine which activity is to be applied to which element of the product, such as system engineering (2.0) for the algorithm's component (Subsystem B or SB), which results in the task identifier of SB2. Several good examples are provided in Boehm (1981).

Once the workflow has been defined, the next step is to assign the resources. When assigning resources, it is important that the activities be examined thoroughly so that the correct resources are assigned. Also, close attention should be paid to the number of hours or months each resource is being tasked to avoid overtasking a resource. If a resource is overtasked, either the workflow should be changed or the number of resources available increased.

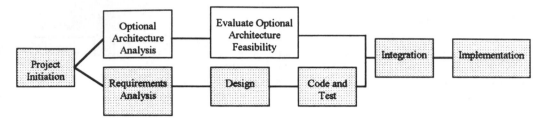

Figure 29.3 Project task network diagram. Shaded boxes indicate critical path.

29.2.2 Schedule Estimates

The order in which the activities will need to be completed in order to meet the project's goal is known as the workflow. The workflow is defined by connecting the activities in a way that defines the sequence in which the activities are to be completed. These connections allow the scheduling software, if any is used, to calculate the critical path in the project. Even if the schedule is not maintained with an automated tool, it is important that the proper sequence of activities required for the successful completion of the project's goals be remembered when performing the schedule estimate.

Milestones are one way managers determine if the project is on schedule. They are often associated with products that are produced at various points on the schedule. These provide a good indicator of how far along the project is and whether or not there are any problems that need attention.

CPM and PERT both provide important information for the tracking and monitoring of project progress (Cori, 1985b). These methods are based on task networks of the type illustrated in Figure 29.2. In such a network, there is a subset of tasks that must be accomplished in order for the project to be completed on time. These tasks represent the "critical path" for the project. There are also estimates of the time required for each task and the calculation of a time frame within which the task must be started and completed for the project to be delivered on time. These time limits are also known as *boundary times* for a task. These parameters include the earliest time a task can begin, the latest time it can start without delaying the project, the earliest time it can be completed, and the latest time it can be completed without delaying the project. There is also the additional time between the earliest start and latest finish, sometimes known as the *float time,* within which the task can be completed and the project remain on time. These are all factors used by scheduling software to determine critical paths and provide other essential information to the manager. An example of a visual indication of a critical path is shown in Figure 29.3. Software support for these scheduling tasks is described in more detail in Section 29.6.

29.3 PRICING AND ESTIMATING

Pricing and estimating are two interrelated activities that must support each other so that the company or business unit can be economically viable. Determining what labor effort, material, and overhead is required for an operation, a project, a system, or to recover costs on a product is a critical task in project management. Estimating system development costs is one of the most difficult and important tasks in any project.

Several different types of estimates can be defined for a systems development project (Ostwald, 1992). The operations cost estimate includes individuals and their tools, such as analysts, engineers, and software and hardware used in the construction of the system.

Product estimates are created for units of items that are intended to be sold. The result of this activity is a single price for the product. Pricing a product is the result of several analyses including determining what it costs to tool up a factory, provide indirect materials, labor, marketing and administrative costs, number to be produced, and a profit margin. All of these elements play a part in determining the price because no matter what price is assigned, it must recover the costs associated with producing the item and begin generating a profit in a reasonable amount of time. One of the analyses needed to price a product is the product estimate or an assessment of what it will take to produce and sell the item. Another analysis that may be required is the cash-flow statement. This may be unnecessary for products that do not incur large costs with respect to the cash the company generates. It takes into account such items as depreciation, the tax rate, and total source of funds for the organization, as well as the expected income from the sale of the product and the costs to generate it. The rate of return is determined for the product to show what it will bring into the corporation over its expected lifetime. A profit-and-loss statement for the product is prepared for some products. This statement outlines the amount that the major organizational units, such as manufacturing and sales, must make available for the product.

A wide variety of product estimating methods exist that are used in manufacturing and other industries. These include the productive-hour method, learning methods that assume costs will decrease as experience with the manufacturing process is gained, and the break-even method. Each of these merit extensive treatment outside the space provided in this chapter and can be found in a number of texts, including Ostwald (1992).

Project bid estimates define the amount a team thinks it will take to complete a project such as an accounts payable system or a new factory addition. System estimates include operations, product, and project estimates in a number of configurations. System estimates are often associated with large programs such as transit systems or other public goals in which profit does not play an explicit part. For all of these the engineer must provide an estimate of the economic value of the item(s) that reflects what the buyer is willing to pay.

The figures generated for these entities are estimates since they have not been spent. When they have been spent, they fall into the realm of cost accounting. Since these estimates are for items or events that have not occurred, it is inevitable that all information pertaining to cost will not be available. But no matter what kind of estimate is required, the engineer needs information that is as accurate and current as possible so that the estimates are as credible and reliable as possible. This information can come from a variety of sources, including past cost and pricing history, standards, marketing activities, personnel department for wages and related items, and other organizations that keep data on economic data and trends. While the motive for preparing each of these estimates is different and there are different techniques for developing them, there are fundamental elements that the approaches for each have in common. For purposes of illustration, the following sections focus on project estimating. The term *system,* used in this context, refers to the configuration of hardware and software typically developed by computer systems engineering projects.

Projects are the type of estimate most often developed by systems. They may require product and operations estimates as input, and, in turn, may become input to larger systems estimation efforts. Sources of information for developing these estimates are similar to those previously described. Since each project is unique, however, estimates developed for other projects can only provide a starting point for a current project's estimated cost. Actual costs can provide even more reliable data for cost estimates, depending on how similar the projects are considered to be. When the final estimate has been reviewed and accepted by management and the customer, it becomes the budget for that project. It is against this budget that the project is compared and tracked to measure progress. Estimates, being nonexact and imperfect, need to be refined as the project evolves to provide an accurate picture of a project's status. A fully developed project estimate can include the following steps:

- It begins with an estimate of the level of work to be accomplished. Section 29.3.1 provides more information on this topic.
- Next an estimate of the time units in staff hours, months, or years required to complete the project is made. That figure is multiplied times the wage, to provide the estimated labor cost.
- An estimate of material, including subcontractor-supplied material, is also provided. This estimate typically refers to raw materials and any standard commercial items required for the project. In most computer-based systems, this category includes supplies such as tape cartridges used for backup or transfer of data.
- The estimate of facilities and equipment cost usually refers to capital equipment and physical plant needed for the project.
- Finally, an estimate of engineering services costs, which includes those incurred for specifications, drawings, or engineering design, is included.

In addition, overhead, interest, contingency, and profit, as well as other components appropriate to a specific company or industry may be added to the estimate.

Pricing the components and estimating the cost of implementing computer-based systems is an exercise that is done several times during a system's life cycle. Hardware components are the tangible elements of a computer and related peripherals that are needed for the system. Computer software has been defined as "computer programs, procedures, rules and possibly associated documentation and data pertaining to the operation of a computer system" [Institute of Electrical and Electronics Engineers (IEEE), 1983]. This definition encompasses both the lines of code (LOC) that are written by a programmer, as well as the infrastructure (analysis, design, configuration management, quality assurance, testing, training, etc.) that is necessary to create a system. As part of software creation, requirements that are gathered from the user are transformed into a language that is understood by the machine. Whether a programmer is using assembler, high-order languages (COBOL, Fortran, etc.), or very-high-order languages (for example, word processors, databases, spreadsheets), LOC are written. Unfortunately, there is no standard for the number of LOC that are created in a given language to transform a given requirement. The LOC are not exactly known until the coding is complete. Also not known is the exact amount of resources needed for the infrastructure support for software. Even with this uncertainty, an estimate of cost must be made very early in order to bid on given software development tasks. The cost estimate primitive—whether it be LOC, function points (counted from the requirements specification), or others—is necessary input to the cost models that help with the process of cost estimation.

The issue of system cost estimation is an important one. A method currently used is to run parametric cost models that are used throughout the systems and software development industry (Stutzke, 1997). The cost estimate usually covers the entire life cycle of development from system definition to operations. In some cases, this must extend to system retirement as well. If the cost estimate does not include all relevant facets of system definition, development and deployment, nasty surprises will generally result. Estimates can be made at the following times during a software development project:

- At proposal time
- After the requirements analysis is complete
- After the preliminary design is complete
- After the detailed design is complete
- At least once, and then as needed, during implementation
- As needed during integration and testing

The reason that estimation should be an iterative process is that project knowledge is increased as the project evolves. As knowledge increases, estimate accuracy improves. This is illustrated in the following paragraphs. The waterfall development model is used to describe the system life cycle for purposes of illustration.

A project begins with the preparation of a proposal to develop a system in response to a stated need, such as the request for proposal (RFP). Usually this is when the least is known about a project and estimates are the most inaccurate. At this time knowledge is based on what was in the RFP, statement of work (SOW), and any referenced documents. Contact with the customer at this time is usually limited. Past project histories play an important part since they provide a means of comparing what has to be done to what has been done before. Further complicating matters, there is usually only a short amount of time to prepare proposals and do the estimates. This also is the time when the estimate needs to be as accurate as possible for two reasons: (1) the final estimate may ultimately decide who receives the contract award, and (2) the final estimate represents the time-and-cost budget within which the system has to be developed. One has to be careful to make sure that the estimate allows for enough money and time to properly do the job without jeopardizing the chances of winning the job.

When a contract is awarded, the first phase of the development cycle is often requirements analysis. The requirements analysis stage builds on the knowledge of what is going to be required in a project gained during the proposal stage. During this stage, additional requirements are derived from negotiations materials, and input from the customer and end users. After the requirements analysis has been completed, a more accurate estimate can be made. The next phase is preliminary design. At the end of the preliminary design stage, the first attempt to design the system has been completed using the requirements identified in the requirements analysis phase. Usually additional requirements are identified at this time. As a result, the total project knowledge increases. The preliminary design is further refined during detailed design along with the requirements, increasing knowledge of the project even more. Estimates done at the end of this stage become even more accurate and reliable. During the next phase, when the system is implemented (in code for software), a few requirements may have been added, deleted, or modified. When this occurs, indeed, whenever requirements change, new estimates should always be done to try to determine the impact of these changes. As a result the total knowledge of the project at the end of this stage allows any estimates done to a high degree of accuracy. During integration and testing, estimates are very accurate and sometimes do not need to be done during the testing stage. But, during testing, if any errors are found that will require significant changes to the system, estimates should be redone to determine the impact of these changes.

29.3.1   Cost Estimation Methods

There are several documented methods for cost estimation. Among these are estimation by analogy, expert judgment, bottom-up, and parametric models (Stutzke, 1997; Roetzheim and Beasley, 1998; Gaffney, 1997). Each of these will be discussed in turn.

29.3.1.1   Estimation by Analogy.   These are the models most intuitive to users. By using the actuals from a previously developed system, the cost estimate for the new system is generated. This estimate is used most often by organizations that have a corporate database with historical data of its projects. For example, if a new development effort includes a medical-forms processing system for a client, the project manager would try to access any information on other systems that may involve forms processing. In accessing the corporate database, they find that a forms processing system was developed in years past to processed forms for employees claiming worker's compensation. The project manager would review the posted actuals and the completed lines of code, if lines of code is the primitive being used for estimating the effort required to implement the system. If the corporate database is robust, it could be very helpful to the project manager. The disadvantage of this method, however, is that often seemingly like projects may be very different depending upon time, budget, and technical constraints.

29.3.1.2   Expert Judgment Models.   More readily used is the expert judgment model that involves using one or more experts to estimate the cost for a new project based on past experience. The problem with this method is that often experts are unable to express completely all of the tasks they perform in a given job, and may inadvertently leave out important aspects. In addition, as in the analogy methods, the system being developed is often new and there are no precedents for the expert to follow. Expert judgment, however, is very useful in estimating the primitives that may be used by other models to estimate cost.

29.3.1.3   Bottom-up Models.   This method involves estimating the cost of individual subsystems and infrastructure costs (quality assurance, configuration management, etc.) and aggregating these costs from the "bottom-up" to determine the overall cost of a system. The people who are responsible for the subsystem or infrastructure area are asked to contribute to this activity. The disadvantage of this method is that it can be time-consuming, and is dependent upon many people to provide the estimates. This method is also undermined if the people involved in the estimation are not aware of all of the requirements of the system. Another issue with this method is that someone is responsible for aggregating the costs from all of the groups, which is tedious.

29.3.1.4   Parametric Models.   Parametric cost models use one or more project variables to estimate cost. The costs are estimated by phase and include infrastructure tasks. They usually include criteria that can be tailored by the project manager to better aid the estimation process (such as the estimated skill level of the programming team). Even though they are popular, parametric models also are not completely accurate. There is much research, however, on ways to better calibrate them so as to create more accurate estimates. The section on software for planning support describes several parametric cost models that have been implemented in software.

29.4 RISK AND COST CONTROL

The most successful project managers have some idea of what risks their projects face and how to deal with them. They also are aware of costs incurred by their projects and how to keep them under control. Although these topics are often discussed separately, in practice they are interdependent. As risk increases, frequently the cost rises as well. On the other hand, if costs are not controlled, the risk increases that the budget will expand over acceptable levels and jeopardize the successful completion of the project. This section is devoted to both a discussion of each topic as well as to a description of how they impact each other.

29.4.1 Risk Control

Risk control is a critical project management activity. Boehm (1991) identifies it as one of two major risk management activities, the other being risk assessment. While the term risk has myriad definitions, a particularly useful one for our purposes is that risk refers to those factors (technical and managerial) that are threats to success or are major sources of problems on systems engineering projects (Boehm, 1991). Boehm translates this into a definition of risk exposure that is the probability of an unsatisfactory outcome times the loss to those affected if the outcome is unsatisfactory, as expressed by this relationship:

$$RE = P(UO)*L(UO)$$

The definition of unsatisfactory outcomes, he adds, is dependent on the perspective of those affected, which includes customers, users, and maintainers of the product. Effective planning requires that those threats and sources of problems be addressed at the beginning of the project so as to reduce the negative impact they may have on a project.

Managing risk is accomplished in several steps. First, risks must be identified, analyzed, and barring unlimited resources and time, prioritized. These steps are devoted to uncovering the issues that can keep the project from staying on schedule or under budget and determining how important they are to that project. Risk control includes planning what to do about risks if they occur, how to eliminate the risks, and tracking the status of the identified risks and any mitigation techniques that may have been employed.

Discovering those issues that pose the greatest risk to a systems development is an exercise unique to each project. Once those issues are identified, controlling the impact on a project is an ongoing task that is revisited all during the project's life.

In his book on systems management, Sage (1995) provides an eight-step process for risk management:

1. Provide infrastructure for risk management strategy.
2. Analyze external factors that influence risk.
3. Identify the organization's strengths and weaknesses with respect to risk.
4. Prepare a strategy at three levels (proactive, interactive, reactive).
5. Prepare budgets and allocate resources to the three levels.
6. Tell all concerned staff about the strategies and plans.
7. Plan and prepare for all contingencies using a variety of risk management analyses, plans, and tactics.
8. Operationalize the risk management plans.

Risk management is about making decisions under uncertainty, uncertainty about the number and variety of inputs, the importance of the different inputs, the number and variety of outcomes, and the impact of outcomes. These elements, while less unknown to those who have more experience, are nonetheless different for each project and must be discovered anew. All attempt to help answer two questions: What can happen and what is the impact of each outcome? and, How likely is it that each outcome will occur?

Controlling risk is a multifaceted process. The objective is to continually update data about the various risks and couple the data with risk mitigation steps that are also continually updated. An essential component is establishing an awareness of the risks that have been identified for the project and being sure that those responsible continue to be alert for their occurrence. This requires constant effective communication.

A risk control or risk management plan outlines the actions that will be taken to deal with risks, should any occur. Moreover, it assigns responsibilities, prioritizes risk mitigation steps, and establishes a budget. Typically the risks in the plan have been judged the most likely to occur and the ones most likely to contribute to a high percentage of the overall project risk. The plan describes sources of risk, estimates of the probability of its occurrence and the consequences, methods to evaluate the level of risk and its impact, and risk management steps.

For instance, if potential delay in the shipment of a computer to be used for software development has been identified as a source of risk serious enough to warrant being included in the plan, the following risk management steps might be proposed:

- Initiate an analysis of alternative hardware environments.
- Initiate a cost–benefit analysis of the alternative development hardware.
- Meet with the technical staff to assess the results of the analysis of alternatives and the cost–benefits of each alternative.
- Report to the project manager on a regular basis on the status of the alternative analysis.
- Monitor the shipping information from the vendor.
- Meet with software engineers to assess the impact of alternative hardware environments on the design and development.
- Meet with the testing and implementation teams to assess the impact of the alternative development environment on the testing phase and integration and transition activities.

- Throughout the early stages of the project reassess this risk and the factors contributing to it, such as shipping announcements from the vendor, and update the risk analysis data.
- Update the risk management steps just listed to reflect new information and adjust the priority assigned to the risk accordingly.

These steps require additional planning and monitoring effort on the part of the project team. Without it, the time spent recovering from the effects of unmanaged risks is potentially much greater. Tracking risk in this way requires that the project team determine that a risk has, in fact, occurred and the steps that have been documented in the plan to manage the risk are taken.

A third element of this is to gather information that will help in future risk management activities. If a database can be set up to function as a repository of this information, it can be accessed by other projects for the following information:

- Descriptive data about the project, including objectives, environment, application, and size characteristics.
- The risks identified for the project with all related data, such as probability of occurrence and assessed impact.
- Analysis and evaluation methods used.
- Management steps identified for the project.
- Risks that were encountered.
- Actual impact.
- Management steps taken and the effects.

A repository of historical risk data will help to refine and increase the effectiveness of future risk control strategies for the organization. This can lead to more productive system development efforts that devote less time to risk control activities that are, at the same time, more efficient and effective.

29.4.2 Cost Control

Cost control addresses the issue of keeping a project or system's development expenses in line with the budget originally estimated as required to accomplish the task. As actual expenses are incurred, the estimates are revised and become more accurate. Similarly, the ability to control costs becomes more objective if not easier to achieve. Increased knowledge and accuracy concerning anticipated expenses as well as a growing database of actual expenses increases the probability that the project will remain under budget and costs will be controlled. Costs must be monitored and reported on a regular basis if they are to be controlled. The following paragraphs illustrate one method for accomplishing this objective.

When a project is first baselined, the original estimate of required expenses becomes the project budget. It includes all of the resources identified in the tasks, which may have been expressed through the work breakdown structure or similar methodology.

For larger projects, it may be important to segment the budget for purposes of tracking and monitoring. This segmentation typically aligns with the project schedule. Due dates, deliverables, and milestones offer good opportunities to review budget items and track progress against anticipated expenses. The amount of work accomplished by a specific time period increases over time. At the end of the project this is the value of the total project budget amount at completion. It is a summation of all costs planned for the project up to that time.

A related data point is the value of work that has actually been completed up to a certain time. This figure accumulates as work is completed. This is critical to the assessment of the budgetary health of the project. If it is done on a routine basis, once a month, for example, then it can be compared to the estimates of the work expected to be completed at the same point in time. This is a good reality check to be sure that overoptimistic projections of work to be completed do not remain unchallenged for long. If estimates are not compared to actuals on a routine basis, management would not be able to easily determine whether the project is truly on time or behind schedule until critical deadlines are missed. Then it would be difficult to determine how far behind schedule the project is as a whole, and certainly, it would be difficult to determine, with any precision, where it began to fall behind and why.

There are several ways to determine how much of the project work has been completed. Among them are level of effort, a time-based method that can be presented in terms of percentages, and completion based on milestones. Each of these is discussed below along with an example to show how a completion estimate is calculated.

The level-of-effort method is the simplest to use and is best applied to tasks such as management, technical consulting, or other tasks that have no real criteria for measuring amount of work accomplished. If you spend what was budgeted, you are on track. Often there is really no way to determine what was actually done. The problem arises when the method is applied to technical tasks such as testing or integration tasks. In this case, if you spend 50 percent of the funds allocated to the task, it may not necessarily mean that 50 percent of the testing or integration activities have been completed. This approach is frequently applied to management tasks, as was just indicated. If, for instance, a task leader's responsibilities are budgeted at a level of $2,000 per month, this means the leader is scheduled to provide $2,000 worth of management tasks on a monthly basis. When the time reports are prepared, charges will be allocated to management activities at a level equivalent to $2,000 per month. Unless there are specific deliverables or other measurable indicators of management work performed, it will be difficult to determine how much and exactly what management tasks were performed. The lack of precision in this method has led to more attempts to define measures for this method, not to eliminate it entirely. That is because it is useful for activities where there is no uniform set of criteria that can be applied to that type of task for all projects. It is a technique that can be useful if its shortcomings are understood and if criteria are applied where possible.

When the amount of work on a task varies according to the time when it is done, it is possible to measure it by percentage of work completed. Often more time is spent on staffing activities, for example, at the beginning of the project than later on, although there will continue to be time spent on them throughout the project. Others will start at a lower level of effort and require more time as the project progresses. The method is typically applied at the task level. When tasks can be subdivided, they should be handled as separate tasks. If that is not possible, this method may be appropriate.

Software-intensive system development provides a good example of this type of task. For instance, a software design may require research before it can be completed. This can be due to the need to assess compatibility of commercial-off-the-shelf components (COTS) or recent upgrades to legacy systems. If it is scheduled to last five months and cost $5,000 to complete, it may be possible to spread the cost out based on knowledge about when the COTS components or upgrades will arrive from the vendor. At the end of the task, when the effort is expected to primarily be preparing reports or getting vendors in to provide demonstrations, the amount allocated will be much less. It may be possible to begin at a low level, increase it during the third and fourth month, and decrease it during the fifth. The issues with this approach are similar to the ones encountered when using the level-of-effort approach. If the analysis is ahead of schedule, there may be no way to know or prove this. The same is true if it is behind schedule.

Completion based on milestones requires that tasks be subdivided into units that describe a logical progression toward completion. Milestones are associated with the completion of each unit. These milestones can reflect an unequal level of effort if appropriate. When a unit is completed, the amount of effort associated with that milestone, for instance, 30 percent or 75 percent, is applied. There are many advantages to this method since accomplishments are recorded based on evidence of their completion. If, for example, several COTS products are required for an implementation effort, the task can be subdivided into ordering, delivery and checkout, integration and test, and operation. When the products are ordered a certain percentage is considered complete, when the product arrives and is checked that it works, another percentage is considered complete, and so on.

The amount of work accomplished should be assessed on a regular and frequent basis. The method used will vary based on the project, task, and contract. Where possible, using milestones to determine the amount of the task or project completed is more objective and informative. If it is possible to develop a task description such that subtasks always have milestones associated with them, then the process is straightforward.

As the project progresses and work is performed, actual costs are accumulated. This is the point where a project's budget is checked against actual costs and the corporate accounting system becomes a major interface for the project. Expenses are collected and reports on the level of expenditure for a project are disseminated at regular intervals. This value provides an objective statement about the amount of funds expended on the project that is not open for discussion unless reallocation of funds is an issue. The problem with many systems, however, is that the collection of expense data and distribution of reports is not as timely as might be desired. If reports are late, it is difficult for project managers to have precise knowledge of the level of expenses incurred by their projects. This has a significant effect on planning for future tasks on the project, since the level of actual expenditures is a major input to the calculation of the new estimates for completion of the project.

The new estimate is the latest projected cost of the project based on both the actuals to date and an estimate of the remaining work. Often the revised estimates, over time, increase such that they represent larger amounts than the anticipated budget at the end of the project. Because the original budget often cannot be changed until there is a modification to the contract, this may occur at several points in the project. It is an indicator that the project is in danger of going over budget by the completion date. Even if it is expected that time or funds will be made up later in the project, this increase should be a warning to managers to review estimated expenses closely. One element of this may be the need to calculate the additional amount required to complete the work. At the end of the project, all actuals are recorded and the final cost to complete the project is available for review.

29.4.3   Project Performance

When a project is underway, its financial health should be assessed at different points. Most program managers are interested in cost and schedule, therefore this section discusses several states of a project based on these two elements. First, a project can be under cost and ahead of schedule. Those projects fortunate enough to be in this state beyond the initial tasks have a responsibility to track their progress and report the methods and techniques that enabled them to be in this condition for the benefit of others.

A project may be on cost and ahead of schedule or under cost and on schedule. This means that the estimated budget is highly accurate, to this point. The project is in no danger, but vigilance should be maintained in the event that schedules begin to slip or costs to increase beyond the estimates.

When a project is on cost and on schedule it is tracking exactly with the original estimates and any revised estimates that have been made since the project began. It bears close monitoring, since any deviations may quickly lead to slippages in schedule or overages in budget. In fact, the budget may require the closest monitoring, since it may reflect actuals that are several weeks or even months old. A project that appears to be on budget may, in fact, be over budget, depending on expenses that have not yet been reported.

A project that is over the estimate for either the schedule or the budget may be in more serious trouble than appears from the reported actuals. This is due to the same issue identified earlier. When actuals reports are delayed, they usually are underreported rather than overreported. Therefore, this state should be investigated and monitored closely to determine the cause of the overruns or schedule slippages and guide the calculation of revised estimates.

A project that is in this last state, over cost and behind on schedule, may be in very serious trouble. A lot depends on how far behind schedule and how much over budget it is. If developments in a higher-level program means there is room for this particular project to slip, or that negotiations are underway to modify the contract to allow for acceptable increases in costs or schedule, then this information is important more for keeping the project within the bounds of any allowable increases. If, however, the original budgets and schedule represent true limits, then the sooner this information is available to management, the better.

Controlling costs is all about monitoring relevant indicators and comparing them to estimates that were created before the project began or were revised as the project progressed. The indicators just cited work for individual tasks or for entire programs. They differ only in the level of detail reported. In a large program, it is even possible to have a large number of tasks over budget and still be on budget, if the tasks that are underbudget have large enough surpluses.

Monitoring costs and schedule is important only if the final cost and deliverable date are important. This statement may seem obvious, but it implies that other data must be available to make a final determination of the health of the project, that is, the final cost of the project, or an estimate of it at any point in the project's life. Without this number, current project status and any revised estimates have little meaning.

29.5   MAINTENANCE AND SUPPORT

Maintenance and ongoing support of the project plan is vital to the success of any system development effort. Knowledge of current project performance supports better satisfaction of customers' needs, the ability to add resources as required, and to improve the systems engineering processes. Also, knowledge of prior project performance supports better estimation and the identification of where our software engineering processes need improvement. However, the plan needs to be current in terms of due dates, task descriptions, task dependencies, and critical tasks if it is to provide management the information it needs. Project managers also need to be able to perform "what if" analyses to determine the impact of changes in due dates, resource levels, or schedule on the interim and final phases of the project. There are a number of techniques that can be used to identify these elements. They include Gantt charts, CPM, and PERT. Each of these methods requires certain information from the ongoing activities of the project.

Each project generates information that supports maintenance of the plan. Each of the items may be analyzed and contribute to management activities to refocus a project, identify problems, anticipate risks, and predict the outcome of the project's implementation strategy. Use of risk and cost data for these purposes is described in Section 29.4. However, maintaining the project plan and supporting its implementation through transitions to subsequent phases of the systems development life cycle requires monitoring of more than the schedule, risks, and costs.

29.5.1 Gantt Charts

Gantt charts were originally called bar charts, and these are still part of the information presented when these charts are used. Gantt charts are created from task definitions that include start and end dates for each task. The duration of the task across a time scale is represented by a bar, hence the name bar charts. The collection of bars shows the start and end time for each task, indicates dependencies, and can also show milestones and deliverables.

29.5.2 Critical Path Method

The CPM originated in manufacturing at the DuPont Company and Remington Rand. It was initially applied to facilities maintenance. The company wanted to determine how much time was needed to conduct repairs to various facilities and to build new ones. The CPM method was developed to identify the tasks that had to be accomplished (the critical path) to complete the project and the time those tasks required. The critical path is the sequence of related tasks that collectively determine the earliest completion time for the project.

Descriptions of the tasks in a CPM diagram often include, among other items, the unique identifier, the portion of the task completed, who is responsible for it, when the earliest start and finish dates are as well as how long it is expected to take, and a description. It may also include an organization code or other pertinent information. The tasks are then linked together based on their recorded start and end dates and dependencies. The critical path is defined by uniquely identifying the tasks that need to be completed to accomplish the project. This can be done by shading the relevant tasks or otherwise highlighting those tasks.

29.5.3 Program Evaluation and Review Technique

The program evaluation and review technique (PERT) came from the U.S. defense community. It was developed for planning and controlling the Polaris missile program that encompassed 250 prime contractors and 9,000 subcontractors. In a program this large, it is inevitable that cost and schedule information would be reported in myriad ways. PERT was developed to address the uncertainties that abound in such situations. It is based on the assumption that estimates for task durations are better described by a probability distribution than by a single data point. It proved to be highly successful, purportedly achieving success over other techniques that had resulted in large cost and schedule overruns.

Three durations are usually provided for each task in a PERT diagram. These are an estimate of the minimum time required to do the task under the most favorable conditions. It is known as the optimistic time. The second is the normal time required, or the one the experienced staff really expect the task to take. The third is the pessimistic time, which is an estimate of how long it could take if nothing goes right. An expected time is derived from these three estimates. The results are presented in terms of a probability distribution for the project that includes the standard deviation. The graphic representation of a PERT chart shows tasks with timelines, dependencies, early and late start and finish dates, and a float factor. They are often identified as network diagrams.

PERT and CPM provide a unique opportunity to visualize the entire project. As such, the graphics are both informative and potentially overwhelming. When sponsors are presented with this information, they usually appreciate the effort and the data included in the charts during the first stages of the project. As the project progresses, however, they may lose interest and find the enormous amount of detail difficult to sort out. In this case, it may be necessary to find a way to segment the information so that it can be viewed in small increments. These techniques are critical to maintaining the viability of the project plan over its life.

29.6 SOFTWARE FOR PLANNING SUPPORT

Project planning tools can assist managers in defining tasks, establishing start and finish dates, and assigning resources. Automated tools increase the ability of managers to control schedules and costs and to reallocate tasks and resources based on progress to date and milestones achieved. Software tools allow more data to be collected and tracked, which supports tighter control and more precise management of the project. Project planning and monitoring software are described in section 29.6.1, and cost modeling tools are presented in 29.6.2.

29.6.1 Project Planning and Monitoring

Some examples of project planning software include such products as Microsoft® Project, Easytrak (Planning Control International, Newport Beach, CA), Primavera Project Planner® from Primavera, and MacProject II® from Claris.

Project planning and monitoring software offers managers highly sophisticated methods of defining and tracking activities, costs, and progress on their projects. They include a variety of monitoring techniques that can frequently be tailored to the specific needs of a project. Many are developed for a specific industry, while others support projects ranging from construction work to software development to research and development. They offer a wide range of features including CPM and PERT, data organization, cost tracking, performance measurement, graphics, and reporting. Once task data are entered, many of the software packages generate time-line charts and resource allocation tables automatically.

Since most project planning software have similar features, Microsoft Project has been selected as representative of the general capabilities that should be available in most project planning packages for the PC environment.

29.6.1.1   *Microsoft Project.*

Microsoft Project includes CPM, PERT, and Gantt charts in its group of support tools for project management. Project planning is approached in three steps: defining the tasks and creating a schedule, maintaining the plan, and creating reports.

The first step is to define tasks and assign start and end dates to each one. Microsoft Project uses the Gantt chart format to define project tasks and schedules. When a new project is initiated, the user enters tasks that need to be done and how long they should take. The user also determines which tasks need to start or end before others can begin. The software automatically calculates dates for each activity.

The user can include resource allocation at the time a new project is defined or this information can be entered at a later time. All of the information entered can be revised as needed at any time during the project. The tools includes customizing features that allow users to create Gantt charts with specialized filters that highlight milestones or other important information. Only specific tasks can be viewed as well as all tasks. The software allows actuals to be entered and compared with the original or revised plan. It supports "what if" analyses by calculating new dates based on schedule changes. It also will trigger a message or visual code to let a user know that a task is slipping behind schedule and needs to be monitored more closely.

Another very important feature of many of these tools is the ability to import and export data to and from other software packages. For instance, Microsoft Project can import and export data from Microsoft Excel files, Lotus 1-2-3 (certain versions), dBASE (certain versions), and FoxPro, among others. The flexibility and depth of capability available in these packages make robust project planning and maintenance techniques available to projects of all sizes and complexities. Some examples of the presentation techniques of these and other packages are described in 29.7.

29.6.2   Cost Models

Although these models may offer task definition and scheduling capabilities that are similar to tools such as Microsoft Project, they were primarily developed to help managers estimate costs. There are several parametric cost models that can be used to develop a cost estimate. The models are similar in many ways. Constructive Cost Model (COCOMO), REVised Intermediate COCOMO (REVIC), PRICE Software (PRICE-S), Galorath Associates System Evaluation and Estimation of Resources (SEER), and QSM Software Life Cycle Model (SLIM) are a few of those available. They are briefly described below.

29.6.2.1   *Constructive Cost Model.*

This nonproprietary model was developed by Dr. Barry Boehm of the TRW Corporation. It is described in his book, *Software Engineering Economics* (Boehm, 1981). It is a regression-based model that originally considered more than 60 programs in three categories: embedded, semidetached, and organic. These modes describe overall software development in terms of size, number of interfaces, and complexity. Separate equations were created relating lines of code (LOC) to effort for each of these categories. More recently, a fourth category, Ada, was added with accompanying equations. A brief description of each mode follows:

ORGANIC: Stand-alone program with few interfaces, a stable development environment, no new algorithms, and few constraints. Usually very small programs.

SEMIDETACHED: A combination of organic and embedded features.

EMBEDDED: Programs with considerable interfaces, new algorithms, or extremely tight constraints. Usually very large or complicated programs.

ADA: Programs developed using an object-oriented analysis methodology or using the Ada language, with emphasis on the separately compilable specifications and body parts of the code.

There are three levels of COCOMO: basic, intermediate, and detailed. The basic model relates staff-months (MM) to LOC (also known as Thousands of Delivered Source Instructions (KDSI)). Intermediate COCOMO contains nominal equations that are similar to the basic equations for each category, but also contains fifteen multipliers that adjust the result of the nominal equations to reflect the unique attributes of a specific program. The detailed level adjusts the multipliers for each phase of the software development cycle.

The primary intermediate COCOMO input is the program size in LOC. However, other factors grouped in four categories are also assessed. These include:

Product attributes: These attributes describe the environment in which the program operates. Includes reliability requirements, requirements volatility, database size, and program complexity.

Computer attributes: These attributes describe the relationship between a program and its host or developmental computer. Includes execution time constraints, main storage constraints, virtual machine volatility, and turnaround time.

Personnel attributes: These attributes describe the capability and experience of the personnel assigned to the system. Includes analysts capability, applications experience, programmer capability, language experience, and experience with the virtual machine.

Project attributes: These attributes describe selected project management facets of a program. Includes modern design practices, use of development tools, and schedule constraints.

Each of the attributes may be described in ratings from very low to extremely high.

29.6.2.2 *REVIC.*

REVIC (1994) predicts the development life-cycle costs for software development from requirements analysis through completion of the software acceptance testing and operations life cycle for 15 years. It is similar to the intermediate form of COCOMO just described. Intermediate COCOMO provides a set of basic equations calculating the effort (staffpower in staff-months and hours) and schedule (elapsed time in calendar months) to perform typical software development projects based on an estimate of the LOC to be developed and a description of the development environment.

The estimate is the output from two equations. The first predicts the staffpower in staff-months based on the estimated lines of code to be implemented and the product of a group of environmental factors. The coefficients, exponents and the factors used in the equations are determined by statistical analysis from a database of completed projects. These variables attempt to account for the variations in the total development environment (such as the programmer's capabilities or experience with the hardware or software) that tend to increase or decrease the total effort and schedule. The results from the first equation provide input for the second equation to determine the time in months needed to perform the complete development. COCOMO provides a set of tables distributing the effort and schedule to the phases of development (system engineering, preliminary design, critical design, etc.) and activities (system analysis, coding, test planning, etc.) as a percentage of the total.

29.6.2.3   *PRICE Software.*

PRICE-S was developed by GE PRICE Systems as one of a family of models for hardware and software cost estimation (General Electric, 1989). It is a proprietary model that can only be accessed through a time-sharing service. The basis for this model is also LOC. In addition, it is a calibratable model that allows for several factors, including productivity, complexity, and platform. Some of the algorithms associated with PRICE-S are proprietary and are not available. PRICE-S computes an estimate in staff-months that can then be converted to dollars. The effort is allocated among three stages of software development: design, code, and test.

29.6.2.4   *Galorath Associates System Evaluation and Estimation of Resources.*

Created by Dr. Randall Jensen, the SEER model uses the Rayleigh–Norden curve to allocate resources during a software project (Galorath Associates, 1988). The model is applicable to all of the phases of the waterfall model. The model is proprietary and must be leased yearly. This model uses LOC and labor rates. It also allows for over 40 inputs grouped into several categories: complexity, personnel capabilities and experience, development and support environment, product development requirements, reusability requirements, and integration. SEER uses PERT; therefore, the user must input a minimum, most likely, and maximum value for all input parameters.

SEER has the ability to allow the user to compare two projects, examine several risk analysis graphs, and see what effect a changed input parameter will have on the overall development and cost and schedule.

29.6.2.5   *QSM Software Life Cycle Model.*

SLIM, developed by Larry Putnam, Sr., CEO of Quantitative Software Management (QSM) Corporation, is proprietary, but may be leased with consulting time from QSM. It computes costs for all waterfall model software development phases except requirements analysis. The primary input is LOC. The model uses PERT to compute expected size from the three LOC estimates: minimum, most likely, and maximum. Other inputs include language, system type, percentage of real-time code, percentage of hardware memory used, experience, modern practices, and productivity factors. The productivity factor for SLIM can be calibrated using historical data.

In his article on industrial-strength project management, Norm Brown (1996) describes nine best practices for management. Of those, the following are directly related to planning and control of the project plan:

- Include formal risk management.
- Use metric-based scheduling and management.
- Provide for program-wide visibility of progress vs. the plan.
- Provide for people-aware management accountability; this means minimizing burnout by responsible scheduling that does not require greater than 40-hour work weeks on a routine basis.

A corollary to these principles is the need for automation to support the planning process, because the amount of information managers need to review is overwhelming without it. Software for planning support, including cost estimation support, provides this.

29.7 PRESENTATION AND COMMUNICATION OF RESULTS OF SYSTEMS PLANNING

All the effort managers devote to planning and maintaining project plans may be of little use if they are not able to communicate the results effectively to staff, higher-level management, and their customers. There is a wide variety of ways the results can be presented, but they fall into two basic categories: text-based reports and visual descriptions using numerous graphic techniques including bar and pie charts, time lines, and animation. The most sophisticated of these supports a full display, including voice, video, and color briefing charts, all tied together in an electronic presentation that proceeds according to a predetermined timeframe and at which there need be no human present. Tufte (1983, 1990, 1997) has written a comprehensive series of books on the visual display of information.

Typically, however, most presentations made by managers employ a fraction of these capabilities and are focused on presentation of data for a specific deliverable or that cover a specific period of time. For these purposes, the effort should be to identify the most effective presentation technique for the information to be conveyed. This means the manager should be aware of the audience, the purpose of the presentation, and the information to be presented so that the best combination of graphic and text is chosen. Dry runs of important presentations are critical and often result in valuable suggestions for improvement. As an introduction to this activity, this section presents an overview of the possible display techniques that can be used.

29.7.1 Tables

Tables can be useful for showing narrative data that include numerical data but that do not lend themselves to other, more graphic forms of presentation. The challenge in these presentations is to provide sufficient explanation that the purpose of the table is clear without including so much information that it is difficult to get anything meaningful from the table. Figure 29.4 shows a portion of a type of responsibility matrix for a software development project. Through this table, project managers, task leaders, and staff can determine their responsibilities for tracking specific elements of project data, what that data are composed of, and what process can be used to locate them. The visual representation provides a description without any more narrative than is necessary to convey the information.

Responsibility Matrix						
Tracking Category	Tracking Item	Responsible Party			Data	Process
		Project Manager	Task Lead	Staff		
Systems engineering	Size estimates		X	X	Lines of code	Source code; size metrics database
	Actual size		X	X	Counted lines of code	System utilities
	Changes			X	Change requests	CM process

Figure 29.4 Example of a responsibility matrix in table form.

29.7.2 Bar Charts

Bar charts are useful for a variety of presentations. They are presented horizontally in Gantt charts, as shown in Figure 29.5, as well as vertically, as shown in Figure 29.6. They are particularly effective when time is a major parameter. In this example, for instance, individual tasks are identified and their start and end dates provided in the rows. To the right of the text are dates from January to March representing individual months during which the tasks can be performed. For each of the tasks, the time period is represented by shading in the row, providing a visual description of the time frame that is more informative than the text alone.

When the bars are vertical, as is shown in Figure 29.6, other information can be presented. Here the bars are divided so that categories of data within the major components can be shown. Each of the major phases has been divided into two subphases. The number of errors found in each subphase is identified. In addition, the types of errors found for

Task ID	Task Description	Early Start	Early Finish	Jan	Feb	Mar
SIF 2.1	Requirements	1 Jan	31 Mar			
SIF 2.2	Feasibility Study	1 Jan	15 Feb			
SIF 2.3.1	Cost/Benefit	15 Jan	15 Feb			
SIF 2.3.1	Vendor Demos	15 Jan	31 Jan			
SIF 2.3.2	Technical Risk	30 Jan	31 Mar			

Progress Bar

Critical Activity

Figure 29.5 Gantt chart with horizontal bars.

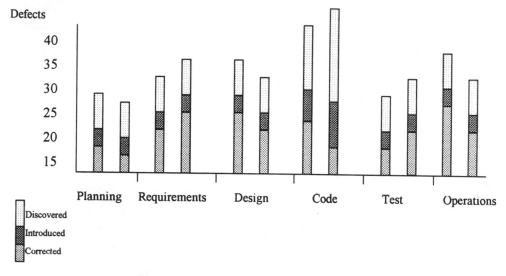

Figure 29.6   Vertical treatment of a bar chart.

each subphase is identified. In this way, it is possible to give a visual description of the magnitude of the type of error found with respect to other types as well as to other phases. This can be difficult to convey without a visual representation. As with other graphic techniques, the amount of information available to viewers from this type of chart is much more than would be available from text or a table alone.

29.7.3   Network Diagrams

Network diagrams, such as the simplified one in Figure 29.7, can be annotated with a variety of information, as indicated in Figure 29.8. If each of the task descriptions, such as Project Initiation, were represented as shown in Figure 29.8, then a great deal more data would be presented. The task name is provided, as in the previous figure, but other information is also shown, including start and end dates, duration, percentage complete, and any other information the users of the system deem appropriate. In addition, these figures show which tasks are dependent on others, which must be started first, and in what order the tasks must be completed. These charts show critical path, if desired, and can show

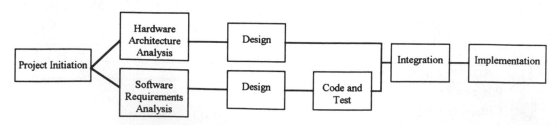

Figure 29.7   Network diagram.

Task ID	Duration	End Date	Percent Complete
Description			Category
Early Start		Early Finish	
Late Start		Late Finish	

Figure 29.8 Task description to be used in a network diagram.

alternative implementations if desired. As such, they supplement narrative descriptions of the tasks as well as provide information on dependencies and precedence.

29.8   PROJECT PLANNING PITFALLS

Even with good intentions and up-to-date skills, the best project plans can go awry. We all have a favorite set of problems that have occurred on projects with which we have been associated. We may even have seen the same one more than once by itself or in combination with one or more others. The outcomes range from projects that are over budget and/or overrun their schedules to systems that do not meet users requirements, or worse, have a negative impact on the productivity of users. Hindsight may help us see them coming earlier the next time, but knowing what to do to prevent their occurrence in the first place is not so easy. Even if we have been in this business for a while, we may be struggling to find reliable methods to prevent these outcomes from occurring in the future.

The impact of project planning on these outcomes has received more and more attention in recent years. As we have understood more about what factors contribute to the failure of systems projects, more research has been conducted on project planning activities. Various organizations, such as Software Productivity Research, Inc., headed by Capers Jones (1996), have accumulated large bodies of data on the impact of decisions made during the project planning phase on the outcome. From these efforts, several failure and success factors emerge as being more common than others. These include technical as well as other factors. Capers Jones, for example, acknowledges the existence of social factors as well as technical factors as causes of project failures. Therefore, while a systematic, proven approach to prevent some of the more common problems is still the subject of discussion and research, good ideas and suggestions are being generated. In this section we describe some of the more common problems that can occur and provide suggestions on what preventative measures to take.

29.8.1   Project Problems

When projects go awry it can be weeks or months before the project team knows it, due to problems such as these:

- The plan looked fine, but it is six weeks into the project and there is no way to tell if the project is still on schedule.

- There was enough money for the requirements phase, but the prototyping software cost more than was estimated and management cannot be sure the costs can be spread across the design and coding phase.
- The requirements team leader and the design team leader are advocating different development methodologies; their arguments sound equally valid. The manager cannot decide which to adopt, and the discussions are adding unplanned time to the schedule.
- The plan included two months of testing, but the design and coding schedule is slipping and the customer says the system must be delivered on time anyway.
- The corporate quality control team insisted on reviewing the quality assurance plan before it is published, but that was six weeks ago and no comments have been returned.

The topics and time frames for each of these can be changed to encompass myriad scenarios. However, each reflects a fundamentally different issue. Many of these can be traced back to the initial project planning stage when the following kinds of questions are answered:

- What are the tasks and to what level of granularity should they be defined?
- What is a reasonable time frame for their accomplishment?
- Who should do what? What are the skills needed for the project?
- What is the best way to use the money available for the project?
- What approach or methodology should be used for the development of the system?
- Who is accountable for deliverables to the project team?

29.8.2   Compensating Strategies

While many project planning techniques and methodologies assist with the answers to these questions, they, like all methodologies, are subject to misinterpretation, reinterpretation, and misuse, with the result that failures are often blamed on the methodology adopted, and the true cause of the problem is never identified. While this aspect cannot be eliminated entirely, there are strategies that can be followed to help reduce the occurrence of these problems and their root causes. Several of these focus on the initial project planning stage. While handling the early project planning in an optimal way will not eliminate all problems, it may provide more options for handling the problems that do occur later on.

29.8.2.1   Automate the Project Plan and Keep It Up-to-date.   The key question this issue raises is whether and when to automate the project plan and what level of resources to commit to maintaining it as the project proceeds. It can be argued that even the smallest of projects requires some type of plan. A single sheet stating the intentions of the team has the advantage of letting any interested party know what to expect of the time and effort that is to be devoted to the project. At the very least, the person or organization funding the effort should be interested in knowing what their money is buying. They may also be interested in knowing what the time frame is and whether others, such as the user group, agree with the objectives and time frame. Once the project size grows beyond a few weeks effort, and engages the interest and participation of more than three or four people, the single sheet will expand to several and may include milestones, interim deliverables, and a series of subtasks.

At the other extreme, projects involving many subsystems, hundreds of people, and spanning several years of effort are clearly unmanageable without some sort of plan. In either case, creation, generation, and maintenance of the project plan becomes increasingly important as the size and complexity of the project grows.

Even if you have a plan, however, it does not guarantee that it will be useful for the life of the project. Many projects spend a great deal of time identifying tasks and subtasks and find, in a very short space of time, that all the paper and charts are worthless when they want to know whether or not they are on track. This can occur for a variety of reasons. First, the project planners may not have really understood what tasks were needed and the plan was developed at too high a level to be useful for tracking purposes. It is possible that tracking data being collected are not useful because the tasks cannot be tracked in meaningful increments. If a client makes major changes in the objectives and the planners do not understand how it affects the structure of the tasks, they may not change the plan and find later that it is useless. Or it may be that the plan was accurate, but changes in personnel, milestones, or task durations that were agreed to in client meetings were not incorporated into the plan, which then evolved, over time, into a useless document. Choosing a good project planning tool and keeping it up to date is crucial to the usefulness of the plan over the life of the project.

29.8.2.2 *Automate the Financial Tracking and Keep It Up-to-date.* The question of whether to automate tracking of the actual expenditures as well as the budget component depends largely on the relationship of the project to the corporate financial system. If the project's budget system is tightly integrated with the corporate system and actuals can be returned to the project in a timely manner, then it may not be necessary to maintain a separate, if informal, tracking system for project actuals. In some cases, however, actuals may not be reported back to the project for weeks or months. If the project manager depends on those reports to determine whether or not it is on track financially, it may be in trouble for a long time before anyone realizes it. As a result, project managers may develop a system of keeping track of their own actuals, which are then reconciled with the corporate reports when those are available. Whether these are major tasks or high-level estimates depends on the needs of the individual projects.

29.8.2.3 *Have Access to Technical Expertise.* If the project involves new technology or application domain, an outside expert should be available. Ideally, this is someone who has similar objectives and goals as the project team and can act as a mentor on technical issues. This is especially important on projects where there is the opportunity for different development methodologies to be adopted, there are strong personalities with different goals, or there is a corporate culture to use one method that is not in the best interests of the project. In these situations much time can be lost as the technical experts debate the pros and cons of each approach. In the extreme, clients can become involved, staff are alienated, and turnover increases. The project can stagnate while the issues become the focus of attention and the team loses sight of goals and priorities.

It may seem obvious that intervention is required in situations such as these. However, it is difficult for those outside the project to know how serious the problems are. The project team will try to solve the problems internally, at least until it is so far behind or over budget that discussions with the client or higher-level managers cannot be avoided. In addition to trying to resolve the issues, those on the project may not realize the extent of the damage until it is too late. Even if management is well grounded in the technical area under debate, it may take an outside opinion to get beyond the personal agendas that can cloud the issues in the best of projects. Moreover, a third party will bring a different set of experiences to the table, in addition to expertise in the technical topic. This contribution can be invaluable in providing insight and validation for any advice or recommendations that may be adopted for the project.

29.8.2.4   *Stick to the Quality Assurance and Quality Control Objectives.*   Often the first thing to be sacrificed when schedules get tight is one or more of the quality assurance tasks. Testing is shortened to weeks or days, walk-throughs are abandoned, test criteria are redefined, or reliability goals for releases are changed. Deadlines are met "by definition," and numerous fixes are left to the maintenance phase as modifications to the original requirements.

While some adjustments are inevitable, quality assurance too often is shortchanged when budgets and time get short. The danger, borne out by reports on how many software-intensive systems are never used, is that whatever is delivered is so short of the users' expectations that the maintenance phase never really starts and all those modifications that were supposed to provide the original functionality are never made.

Sometimes this is because, in the original planning, quality assurance tasks were never adequately defined or staffed. Also, the team may not be supportive of quality assurance tasks, and this translates to lack of cooperation, which undermines the effort and justifies attempts to cut back the quality assurance effort. The result is a loss of standards to guide the work of the development team and an inability to determine whether the product(s) meet users' needs.

One of the most crucial tasks at the beginning of the project is to define the quality control and quality assurance goals for the project and resist attempts to short circuit those goals later on. This means that the staff responsible for quality assurance must be well integrated into the project team and that goals and objectives, as well as tasks, for quality assurance must be clearly stated and articulated.

29.8.2.5   *Understand the Unknowns that Are Specific to the Project.*   The project manager will go a long way toward lowering the risk associated with a project by understanding what the team and management do not know. However, project unknowns are not confined to the team. They can exist among the client staff, users, and maintenance team, as well as among the designers and developers.

Unknowns represent those technical and management areas that are unknown to the individuals who will be working on the project. Sources of unknown areas or topics are many: the methodology chosen for the project, the implementation language, technical implementation requirements, the application area, use of an automated tool, management technique, or the creation of teams from staff who have never worked together, or perhaps in the same company. Any one of these can present major obstacles to the success of the project. Two or more can introduce a high level of uncertainty, since the amount of time and effort required to overcome them can be extremely difficult to determine.

The first challenge for the project is to understand that one or more of these obstacles exist and to appreciate how they can impact progress. This does not mean they can resolve or eliminate the unknown(s) immediately. It simply means that adjustments can be made and plans put into place that will eliminate them over time. Some of these obstacles may be identified during risk assessment activities. However, the focus for risk assessment should include issues over and above the feasibility and achievability of the system product. The ability of the project members to work efficiently and effectively as a team to reach its goal, for example, should be addressed in addition to a formal risk assessment for the project. Determining the impact of unknowns and what to do about them is handled as part of the general risk management techniques.

29.8.2.6 Clearly Define the Obligations to the Project Expected from External Groups This issue is most critical to projects that depend upon feedback and require input from other projects or groups in order to accomplish its goals on time and under budget. In these situations it is crucial to establish clear reporting mechanisms and to identify persons from those external groups who are accountable to the project. In high risk situations it may be necessary to identify workarounds for products or feedback in the event the required input is not forthcoming. Because these situations require an appreciation of the pressures other groups may be facing, it is important to assign staff with strong interpersonal skills to these liaison activities.

29.9 CONCLUSION

These strategies can arm a project team with additional mechanisms to help ensure that they avoid many project development pitfalls.Understanding the team weaknesses as well as strengths can keep fate and other entities from exploiting them and catching a project completely unaware and unprepared. It cannot guarantee that project teams will have no problems. But including this assessment in project planning activities will ensure that the plan is more robust and can mean the difference between success and failure.

Planning for action is complex and difficult, but essential if managers are to increase their chances of completing systems development efforts on time and within budget. This chapter has provided a review of some of the planning components that are critical to the achievement of this goal.

REFERENCES

Boehm, B. W. (1981). *Software Engineering Economics*. Englewood Cliffs, NJ: Prentice-Hall.

Boehm, B. W. (1991). Software risk management: Principles and practices. *IEEE Software* **8** (1) pp. 32–41.

Brown, N. (1996). Industrial strength management strategies. *IEEE Software,* July pp. 94–103.

Cori, K. A. (1985a). "Fundamentals of master scheduling for the project manager," in *Software Engineering Project Management,* 2nd ed (R. Thayer, ed.) Los Alamitos, CA: IEEE Computer Society, p. 173.

Cori, K. A. (1985b). "Fundamentals of master scheduling for the project manager," in *Software Engineering Project Management,* 2nd ed (R. Thayer, ed.) Los Alamitos, CA: IEEE Computer Society, p. 173.

Gaffney, J. E., Jr. (1997). "How to Estimate System Size," in *Software Engineering Project Management,* 2nd ed, p. 246–256, (Thayer, R., ed) Los Alamitos, CA: IEEE Computer Society Press.

Galorath Associates. (1988). *SEER User's Manual*. Marina del Rey, CA: Galorath Associates.

General Electric. (1989). *GE PRICE systems. PRICE-S User's Manual*. Moorestown, NJ: PRICE Systems.

Institute of Electrical and Electronics Engineers (IEEE) (1983). *IEEE Standard Glossary of Software Engineering Terms,* (IEEE-Std 729). New York Dept.

Jones, C. (1996). *Patterns of Software Systems Failure and Success*. Boston: International Thomson Computer Press.

Ostwald, P. F. (1992). *Engineering Cost Estimating,* 3rd ed. Englewood Cliffs, NJ: Prentice-Hall.

REVIC. (1994). *REVIC Software Cost Estimating Model User's Manual,* Versions 9.0–9.2. Arlington, VA: Air Force Analysis Agency.

Roetzheim, W. H. and Beasley, R. A., (1998). *Software Project Cost Schedule Estimating—Best Practices,* pp. 5–25. Upper Saddle River, NJ: Prentice-Hall.

Sage, A. P. (1995). *Systems Management for Information Technology and Engineering*. New York: Wiley.

Stutzke, R. D., (1997). "Software Estimating Technology: A Survey", in *Software Engineering Project Management,* 2nd ed., (Thayer, R. ed.) pp. 221–226. Los Alamitos, CA: IEEE Computer Society Press.

Thayer, R. H. and Fairley, R. E. (1997) "Work breakdown structures (WBS)" in *Software Engineering Project Management,* 2nd ed: (Thayer, R. ed.) pp. 183–194. Los Alamitos, CA: IEEE Computer Society Press.

Tufte, E. R. (1983). *The Visual Display of Quantitative Information*. Cheshire, CT: Graphics Press.

Tufte, E. R. (1990). *Envisioning Information*. Cheshire, CT: Graphics Press.

Tufte, E. R. (1997). *Visual Explanations*. Cheshire, CT: Graphics Press.

BIBLIOGRAPHY

Connell, J., and Shafer, L. (1995). *Object-Oriented Rapid Prototyping*. Englewood Cliffs, NJ: Yourdon Press.

DeMarco, T., and Miller, A. (1996). Managing large software projects. *IEEE Software,* July p. 24–27.

Haimes, Y. Y. (1998). *Risk Modeling, Assessment and Management,* New York: Wiley.

Pfleeger, S. L. (1991). *Software Engineering: The Production of Quality Software*. New York: Macmillan.

Pressman, R. S. (1992). *Software Engineering: A Practitioner's Approach*. New York: McGraw-Hill.

Reifer, D. J. (ed). (1993). *Software Management, 4th ed.,* Los Alamitos, CA: IEEE Computer Society Press.

30 Information Technology and Knowledge Management

WILLIAM B. ROUSE and ANDREW P. SAGE

30.1 INTRODUCTION

Humans have used such natural tools as sticks and stones in order to develop simple products, as well as to break bones, for many thousands of years. In the Old Stone Age—from the beginning of civilization until about 15,000 years ago—human development was primarily dependent upon hunting and fishing using simple stick- and stone-based tools. The New Stone Age was made possible by the development of primitive practices involving animal husbandry and agriculture, and the evolution of building technologies that led to such constructions as the Egyptian Pyramids and Stonehenge. The beginning of manufacturing, using baked clay and soft metals, enabled the development of trade and commerce. The New Stone Age led to the Metal Age and to the development of wind power as a substitute for human muscle power. The printing press, the steam engine, the telescope, metallurgical and mining advances, and continuing agricultural innovations led to the Industrial Revolution. The Industrial Revolution was an advance in power and control technology. Toward the latter portion of the Industrial Revolution, advances in electrical and electronics engineering led to the discovery of the computing machine and the beginning of the Information and Knowledge Revolution.

In the early days of human civilization, development was made possible primarily through the use of human effort, or labor. Human ability to use natural resources led to the ability to develop based not only on labor, but also on the availability of land, which was the classic economic term that implied natural physical resources. At that time, most organizations comprised small proprietorships. The availability of financial capital during the Industrial Revolution led to this being a third fundamental economic resource, and also to the development of large, hierarchical corporations. This period is generally associated with centralization, mass production, and standardization.

Major availability of technologies for information capture, storage, and processing has led to information, as well as its product knowledge, as a fourth fundamental economic resource for development. This is the era of total quality management, mass customization of products and services, reengineering at the level of product and process, and decentralization and horizontalization of organizations, and systems management. While information technology has enabled these changes, much more than just information technology is needed to bring them about satisfactorily.

Commentators such as Bush (1945), Licklider (1965), Bell (1973), Toffler (1980, 1991), and Zuboff (1988) have long predicted the coming of the Information Age. Alvin Toffler writes of three waves: the agriculture, industrial, and information or knowledge ages.

Handbook of Systems Engineering and Management, Edited by A. P. Sage and W. B. Rouse
ISBN 0471-15405-9 ©1999 John Wiley and Sons, Inc.

Within these are numerous subdivisions. For example, the Information Age could be partitioned into the era of vertically integrated and stand-alone systems, process reengineering, total quality management, and knowledge and enterprise integration. Information and knowledge are now fundamental resources that augment the traditional economic resources: land and natural resources, human labor, and financial capital. Critical success factors for success in the third wave, or information age, have been identified (Hope and Hope, 1997) and include: strategy, customer value, knowledge management, business organization, market focus, management accounting, measurement and control, shareholder value, productivity, and transformation to the third-wave model for success.

Major growth in power of computing and communicating, and associated networking is quite fundamental and has changed relationships among people, organizations, and technology. These capabilities allow us to study much more complex issues than was formerly possible. They provide a foundation for dramatic increases in learning and both individual and organizational effectiveness. In large part, this is due to the networking capability that enables enhanced coordination and communications among humans in organizations. It is also due to the vastly increased potential availability of knowledge to support individuals and organizations in their efforts. However, information technologies need to be appropriately integrated within organizational frameworks if they are to be broadly useful. This poses a transdisciplinary challenge of unprecedented magnitude if we are to move from high-performance information technologies to high-performance organizations.

In years past, broadly available capabilities never seemed to match the visions proffered, especially in terms of the time frame of their availability. Consequently, despite these compelling predictions, traditional methods of information access and utilization continued their dominance. As a result of this, comments that go something like "computers are appearing everywhere except in productivity statistics" have often been made (Brynjolfsson and Yang, 1996; Kling, 1996). In just the past few years, the pace has quickened quite substantially and the need for integration of information technology issues with organizational issues has led to the creation of a field of study sometimes called *organizational informatics,* or *knowledge management,* the objectives of which generally include:

- Capturing human information and knowledge needs in the form of system requirements and specifications.
- Developing and deploying systems that satisfy these requirements.
- Supporting the role of cross-functional teams in work.
- Overcoming behavioral and social impediments to the introduction of information technology systems in organizations.
- Enhancing human communication and coordination for effective and efficient workflow through knowledge management.

The Internet, World Wide Web, and networks in general (National Resource Council, 1997; Dertouzos, 1997) have become ubiquitous in supporting these endeavors. Almost everyone is growing up digital (Tapscott, 1998), searching for designs for living in the Information Age (Dyson, 1997), and going on-line to communicate with one another and use digital libraries. It also seems that hardware and software become obsolete as soon as they are unboxed. Indeed, with current trends, boxes themselves may become quaint collectibles. However, organizational productivity is not necessarily enhanced unless attention is paid to the human side of developing and managing technological innovation (Katz, 1997) to assure that systems are designed for human interaction.

Because of the importance of information and knowledge to an organization, two related areas of study have arisen. The first of these is concerned with technologies associated with the effective and efficient acquisition, transmission, and use of information, or information technology. When associated with organizational use, this is sometimes called organizational intelligence, or organizational informatics. The second area, known as knowledge management, refers to an organization's capacity to gather information, generate knowledge, and act effectively and in an innovative manner on the basis of that knowledge. This provides the capacity for success in the rapidly changing or highly competitive environments of knowledge organizations. Developing and leveraging organizational knowledge is a key competency and, as noted, it requires information technology as well as many other supporting capabilities.

The human side of knowledge management is very important. Knowledge capital is sometimes used to describe the intellectual wealth of employees and is a real, demonstrable asset. Sage (1998) has used the term systems ecology to suggest managing organizational change to create a knowledge organization, and enhance and support the resulting intellectual property for the production of sustainable products and services. Managing information and knowledge effectively to facilitate a smooth transition into the Information Age calls for this systems ecology, a body of methods for systems engineering and management (Sage, 1995) that is based on analogous models of natural ecologies. Such a systems ecology would enable the modeling, simulation, and management of truly large systems of information and knowledge, technology, humans, organizations, and the environments that surround them.

It would be very difficult to capture the state of information technology and knowledge management in a handbook chapter without the chapter being hopelessly out of date even before the hardcopy handbook emerged from the publisher. Consequently, our goal in this chapter is to consider several broad trends, which we feel will persist regardless of the specific technologies that enable them. In particular, we focus on effects of information technology on knowledge management, and on the organizational implications of these effects.

These trends pose challenges for systems engineering and management. This chapter elaborates these challenges and suggests the likely impacts on systems engineering and management. Overall, the need for systemic views of the inherently loosely structured systems enabled by information technology provides both a strong challenge and immense opportunity for the field.

30.2 TRENDS

Several trends transcend debates about technology alternatives—currently alternatives such as PCs vs. Net PCs or integrated-services digital network (ISDN) vs. cable. These overriding trends concern directions of computer and communications technologies, and the impacts of these directions on knowledge management and organizations.

30.2.1 Information Technology Trends

The information revolution is driven by technology and market considerations and by market demand and pull for tools to support transaction processing, information warehousing, and knowledge formation. Market pull has been shown to exert a much stronger effect on the success of an emerging technology than technology push. There is hardly any conclusion that can be drawn other than that society shapes technology (Pool, 1997), or perhaps more accurately stated, technology and the modern world shape each other in that only those technologies that are appropriate for society will ultimately survive.

The potential result of this mutual shaping of information technology (IT) and society is knowledge capital, and this creates needs for knowledge management. The costs of the IT needed to provide a given level of functionality have declined dramatically over the past decade—especially within the last very few years—due to the use of such technologies as broadband fiber optics, spectrum management, and data compression. A transatlantic communication link today costs one-tenth of the price that it did a decade ago, and may well decline by another order of magnitude within the next three or four years. The power of computers continues to increase and the cost of computing has declined by a factor of 10,000 over the past 25 years. Large central mainframe computers have been augmented, and in many cases, replaced by smaller, more powerful, and much more user-friendly personal computers. There has, in effect, been a merger of the computer and telecommunications industries into the IT industry, and it now is possible to store, manipulate, process, and transmit voice, digitized data, and images at very little cost.

Current industrial and management efforts are strongly dependent on access to information. The world economy is in a process of globalization and it is possible to detect several important changes. The contemporary and evolving world is much more service-oriented, especially in the more developed nations. The service economy is much more information- and knowledge-dependent and much more competitive. Further, the necessary mix of job skills for high-level employment is changing. The geographic distance between manufacturers and consumers, and between buyers and sellers, is often of little concern today. Consequently, organizations from diverse locations compete in efforts to provide products and services. Consumers potentially benefit as economies become more transnational.

The IT revolution is associated with an explosive increase of data and information, with the potential for equally explosive growth of knowledge. IT and communication technology have the capacity to radically change production and distribution of products and services, and thereby bring about fundamental socioeconomic changes. In part, this potential for change is due to progressively lowered costs of computer hardware. This is associated with reduction in the size of the hardware, which therefore leads to dematerialization of systems. This results in the ability to use these systems in locations and under conditions that would have been impossible just a few years ago. Software developments are similarly astonishing. The capabilities of software increase steadily, the costs of production decrease, reliability increases, functional capabilities can be established and changed rapidly, and the resulting systems are ideally and often user-friendly through systems integration and design for user interaction. The potential for change is also brought about due to the use of IT systems as virtual machines, and the almost unlimited potential for decentralization and global networking due to simultaneous progress in optical fiber and communication satellite technology (Dertouzos, 1997).

The life cycle of IT development is quite short and the technology transfer time in the new "postindustrial," or knowledge-based, society brought about by the information revolution is usually much less than in the Industrial Revolution. IT is used to aid problem-solving endeavors by using technologically based systems and processes and effective systems management. Ideally, this is accomplished through:

- Critical attention to the information needs of humans in problem-solving and decision-making tasks.

- Provision of technological aids, including computer-based systems of hardware and software and associated processes, to assist in these tasks.

Success in IT and engineering-based efforts depends on a broad understanding of the interactions and interrelations that occur among the components of large systems of humans and machines. Moreover, a successful IT strategy also seeks to meaningfully evolve the overall architecture of systems, the systems' interfaces with humans and organizations, and their relations with external environments.

As just discussed, the most dominant recent trend in IT has been more and more computer power in less and less space. Gordon Moore, a founder of Intel, noted that since the 1950s the density of transistors on processing chips has doubled every 18 to 24 months. This observation is often called Moore's law. Moore projected that doubling would continue at this rate. Put differently, Moore's law projects a doubling of computer performance every 18 months within the same physical volume. Schaller (1997) discusses the basis for Moore's law and suggests that this relationship should hold through at least 2010. The implication is that computers will provide increasingly impressive processing power. The key question, of course, is what we will be able to accomplish with this power.

Advances in computer technology have been paralleled by trends in communications technology. The ARPAnet emerged in the 1960s, led to the Internet Protocol in the 1970s, and the Internet in the 1980s. Connectivity is now on most desktops, e-mail has become a "must have" business capability, and the World Wide Web is on the verge of becoming a thriving business channel. The result is an emerging networking market. Business publications are investing heavily to attract readers—or browsers—of their on-line publications (Grover and Himelstein, 1997). Telecommunications companies are trying to both avoid the obsolescence that this technology portends and figure out how to generate revenues and profits from this channel. The result has been a flurry of mergers and acquisitions in this industry.

These strong trends present mixed blessings for end users. The dramatic increases of processing power on desktops is quickly consumed by operating systems, browsers, multimedia demos, and so on. The escalating availability of data and information often does not provide the knowledge that users of this information need to answer questions, solve problems, and make decisions. The notion of "data smog" has become very real to many users (Shenk, 1997). Not surprisingly, this possibility provides opportunities for technologies for coping with data smog. For example, "push" technology enables information to find users (Kelly and Wolf, 1997; Andrews, 1997). Another possibility is "intelligent agent" technology, whereby users can instruct autonomous programs to search for relevant data and information (Maes, 1994). While these types of capabilities are far from mature, they appear quite promising. The specifics of these trends will surely change—probably by the time this chapter appears in print! However, the overall phenomenon of greater and greater processing power, connectivity, and information will be an underlying constant.

That the price of computing has dropped in half approximately every two years over the last two decades or so is nothing short of astounding. Had the rest of the economy matched this decline in prices, the price of an automobile would be in the vicinity of $10. Organizational investments in ITs have increased dramatically and now account for approximately 10 percent of new capital equipment investments by U.S. organizations. Roughly half of the labor force is employed in information-related activities. On the other hand, productivity *growth* seems to have continually declined since the early 1970s, especially in the service sector that makes up about 80 percent of IT investments (Brynjolfsson and Yang, 1996). This situation implies a need to effectively measure IT contributions to productivity, identify optimal investment strategies in IT, and enhance IT effectiveness through knowledge management for enhanced productivity.

On the basis of a definitive study of IT and productivity, Brynjolfsson and Yang (1996) draw a number of useful conclusions. They suggest that the simple relationship between decreases of productivity growth in the U.S. economy and the rapid growth of computer capital is too general to draw meaningful conclusions. In particular, poor input and output data quality are responsible for many difficulties. Many of the studies they review suggest that the U.S. economy would not be enjoying the boom that it is enjoying in the latter half of the 1990s without IT contributions. Their study suggests improvements in accounting and statistical recordkeeping to enable better determination of costs and benefits due to IT. They support the often expressed notion that IT helps us perform familiar tasks better and more productively, but that the greatest improvement is in enabling us to perform entirely new activities. One of these, for example, is the ability to make major new product introductions into the world economy with no macrolevel inventory changes being needed. They indicate some of the new economic and accounting approaches that are needed to provide improved measures for performance evaluation (Kaplan, 1990; Cooper and Kaplan, 1991; Kaplan and Norton, 1996; Kaplan and Cooper, 1997, 1998). For a discussion of activity-based costing and management in systems engineering, see Chapter 6 of this handbook and Sage (1995).

Although IT does indeed potentially support improvement of the designs of existing organizations and systems, it also enables fundamentally new ones, such as virtual corporations and major expansions of organizational intelligence and knowledge. It does so not only by allowing for interactivity in working with clients to satisfy present needs, but also through proactivity in planning and plan execution. An ideal organizational knowledge strategy accounts for future technological, organizational, and human concerns, to support the graceful evolution of products and services that aid clients. Today, we realize that human and organizational considerations are vital to the success of IT.

The challenge for systems engineering and management is to assure that people gain maximal benefit from these capabilities. This is why IT must be strongly associated with information ecology, knowledge management, and other efforts that we discuss here and that will ultimately lead to an effective systems ecology.

30.2.2   Organizational Trends

There was a time when the size of your office and prestige value of your company car—perhaps measured in square feet and horsepower—were the perks of executive life. The technology trends just outlined have resulted in new measures of success—gigahertz and gigabytes. The accouterments of success are now laptops, modems, cellular phones, and the like. Tools for operating in virtual organizational worlds are now the tools of success (Rouse, 1997a).

Associated with this trend is a wide range of new organizational models. Distributed collaboration across organizations and time zones is becoming increasingly common. The motivation for such collaboration is the desire to access sources of knowledge and skills not usually available in one place. The result of such new developments in IT as network computing, open systems architectures, and major new software advances has been a paradigm shift that has prompted the reengineering of organizations; the development of high-performance business teams, integrated organizations, and extended virtual enterprises (Tapscott and Caston, 1993); and the emergence of loosely structured organizations (Rouse, 1997b).

This poses substantial challenges in terms of managing both people and knowledge. The command and control model of leadership is a poor fit for managing organizations where the participants are not bound to the organization by traditional incentive and reward systems (Drucker, 1997). A collaborative effort has to continue to make sense and to provide value to participants for it to be sustained. Otherwise, knowledge and skills are quite portable and the loss of knowledge workers is a major potential downside risk for organizations today.

There are three keys to organizations prospering in this type of environment—speed, flexibility, and discretion (Rouse, 1997b). Speed means understanding a situation (e.g., a market opportunity), formulating a plan for pursuing this opportunity (e.g., a joint venture for a new product), and executing this plan (e.g., product available in stores), in this case all within a few weeks at most. The need for flexibility is obvious; otherwise, there would not be speed. However, flexibility is also crucial for reconfiguring and redesigning organizations, and consequently reallocating resources. Functional walls must be quite portable, if they exist at all.

Discretion is what transforms flexibility into speed. Distributed organizations must be free to act. While they may have to play by the rules of the game—at least the rules of the moment—they need to be able to avoid waiting for permission to proceed. Similarly, they need to be able to pull the plug when things are not working. In this way, resources are deployed quickly and results are monitored just as quickly. Resource investments that are not paying off in the expected time frame, are quickly redeployed elsewhere.

A major determinant of these organizational abilities is the extent to which an organization possesses intellectual capital, or knowledge capital, such that it can create and use innovative ideas to produce productive results. The concept of intellectual capital has been defined in various ways (Brooking, 1996; Edvinsson and Malone, 1997; Stewart, 1997; Klein, 1998; Roos et al., 1998). We would add communications to the formulation of Ulrich (1998) representing intellectual capital to yield

$$\text{Intellectual capital} = \text{competence} \times \text{commitment} \times \text{communications}$$

Other important terms, such as *collaboration* and *courage* could be added to this equation.

Loosely structured organizations and the speed, flexibility, and discretion they engender in managing intellectual capital fundamentally affect knowledge management (Myers, 1997; Ruggles, 1997; Prusak, 1997; Albert and Bradley, 1997; Liebowitz and Wilcox, 1997). Knowledge workers are no longer captive, and hence know-how is not "owned" by the organization. Patents are of much less value, for example, as evidenced by the substantial decline in use of this mechanism (Bean et al., 1998). Instead, what matters is the ability to make sense of market and technology trends, quickly decide how to take advantage of these trends, and act faster than other players. Sustaining competitive advantage requires redefining market-driven value propositions and quickly leading in providing value in appropriate new ways. Accomplishing this in an increasingly information-rich environment is a major challenge, both for organizations experiencing these environments and for those who devise and provide systems engineering and management methods and tools for supporting these new ways of doing business (Allee, 1977; Stacey 1996). There is a major interaction involving knowledge work and intellectual capital, and the communications-driven information and knowledge revolution that suggests many and profound complex adaptive system like changes in the economy for the Twenty-First Century (Shapiro and Varian, 1998; Kelly, 1998).

30.3   SCENARIOS

The effects of the technology and organizational trends just discussed are pervasive. They are most quickly affecting those businesses and other organizations where competitive forces are felt most immediately. Government is also being affected, especially the military where the traditional balances of power and monolithic conflict scenarios have been replaced by a much wider range of smaller magnitude threats and conflict in the Information Age (Arquilla and Ronfeldt, 1997). Academia, always very slow to change organizationally, nevertheless faces the need to rethink its traditional model of narrow disciplinary specialists that many (Sage, 1992; Anderson, 1996; Wulf, 1998) suggest accomplishes little more than to provide students with an unconnected series of myopic views of the world.

In this section, we discuss IT and knowledge management in the context of scenarios within these three sectors of our socioeconomic system. These scenarios set the stage for discussing the ways in which systems engineering and management can support addressing ten central challenges and achieving success.

30.3.1   Impact on Business

The context for this scenario is a company that manufactures consumer products for sale in retail stores. In the "good old days," the sales force called upon a wide range of retail outlets, including a few chains and many individual stores. Orders for toasters, can openers, and so on, were booked and shipments were made from inventory on hand. Product quality was acceptable and profits were reasonable. Then, everything changed. Foreign manufacturers became able to produce high-quality consumer products with very attractive prices, due to both improved processes and lower labor costs. Consumers could now buy higher-quality products for lower prices than traditional manufacturers could match. Manufacturers scrambled to move manufacturing off shore and, somewhat later, to also improve quality.

But, this was not enough. Consolidation in the retail industry resulted in several large retailers dominating the industry. You had to get your products on the shelves at Wal-Mart, K-Mart, Sears, and so on, or you were out of business. To get these "placements," you were required to interact with sophisticated information systems - providing, for example, daily access to point-of-sale data.

The stakes in this industry continue to rise. You need total quality management, reengineered and lean manufacturing, and enterprise resource planning. You also need to be able to create and manage alliances, network with suppliers, and mass customize products to match demographic/lifestyle characteristics of consumers in the market niches you address. You may also need virtual organizations to accomplish some of these ends.

While not inexpensive, all this is possible because of the information and knowledge you can now access and utilize. Networking internally and externally provides information about your operations, your suppliers, and perhaps your competitors. A wide range of databases provide information on consumer characteristics, behaviors, and preferences. All things, it seems, are possible within the new economics of information and its strategic management (Evans and Wurster, 1997; Martin, 1996; Mazarr, 1997; McGee and Prusak, 1993; Brooking, 1996; Sveiby, 1997; Quinn et al., 1997).

However, what information from this available plethora should you access and how should you use it? How can you mine the wealth of data available, convert it to relevant and useful information, and transform this information to knowledge that provides sustainable competitive advantages? Conventional wisdom portrays the key to be enterprise integration, facilitated by data warehouse and data mining tools.

However, what specifically should you do with such leading-edge infrastructure? Certainly you can operate more efficiently if you understand your processes and reengineer them to eliminate unnecessary costs and to improve product effectiveness and quality. But this does not keep you from getting better and better at something that you should not be doing at all. It can only help bottom-line profits to the extent that top-line revenues are there, which requires understanding and creating compelling market propositions. This requires potential reengineering at the level of the organization itself, and it requires knowledge management.

Abilities to create such strategies and to implement them effectively depends on gaining and exploiting market understanding and insights. This requires a knowledge of context and experiential familiarity with the market. Modeling and simulation methods (see Chapter 26) provide powerful means of representing this understanding and gleaning desired insights. Decision support system concepts (see Chapters 18 and 28) can enable making this information and knowledge much more useful and usable. Thus, systems engineering methods and tools can provide the means for taking advantage of the IT infrastructure that is currently receiving so much attention.

30.3.2   Impact on Government

The technology and organizational trends discussed here have affected many aspects of government. However, no other government function has experienced the extent of changes affecting the military. The context of this scenario is the role of the military and the types of missions it must increasingly address.

For almost 50 years following World War II, the United States had a clear military adversary in the U.S.S.R. Massive investments were made to prepare for the potential of large-scale conflicts with the Soviets. The dominant scenario involved Eastern Bloc troops pouring through the Fulda Gap. The United States and NATO allies focused substantial investment and training resources on this dominant scenario. The fall of the Berlin Wall and the disintegration of the U.S.S.R. changed the game completely. No longer is there a single dominant scenario. Conflicts like the Gulf War are more likely, but probably not as likely as engagements such as Bosnia, Ethiopia, and Rwanda. The U.S. military now must organize, train, and equip itself to deal globally with numerous simultaneous small engagements (Fogelman and Widnall, 1996; Fuchs et al., 1997; Larson, 1997).

These engagements range from traditional bombs on targets requirements to showing the flag and humanitarian relief. Most of these engagements will require very rapid response (e.g., 24–48 hours) and will be likely of such a nature that they cannot be planned in advance. Thus, the six months required to stage the Gulf War may not be acceptable in future efforts—and, in fact, had Iraq not paused in Kuwait this length of time would not have been acceptable in that situation.

IT, including telecommunications, is viewed as the means for addressing these challenges (Alberts and Papp, 1997). High bandwidth connectivity will provide the infrastructure for military units to respond rapidly and plan en route. Lean logistics, embedded training, and smart weapons are but a few of the concepts proposed for making such responses possible.

However, the game has changed in terms of more than just rapid response and en route planning. Missions are becoming increasingly information- and knowledge-oriented, with the potential use of physically explosive weapons very limited. Rather than physical fire-power, information is the currency of these missions. Information warfare has become of central interest (Arquilla and Ronfeldt, 1997). Disruption of an adversary's command and control system—or perhaps their banking system—is a more likely goal than bombing bridges and factories. Requirements to organize, train, and equip military forces for such missions dictate major changes in how the military does business. Investments have to shift to the means for flexibility and away from the massive infrastructure the defense establishment has traditionally maintained. Such shifts are economically and politically disruptive and face strong opposition.

Without fundamental reengineering (see Chapter 23) at the level of product, process, and systems management, however, impressive investments in information technology will not yield the desired results. Investments also must be made in the layers that rely on this infrastructure. These investments must focus on manpower, personnel, and training needs (see Chapter 20), as well as decision support capabilities needed to take advantage of the information infrastructure (see Chapters 18 and 28).

30.3.3   Impact on Academia

Life in academia used to be pretty straightforward. You earned your Ph.D., secured a faculty position, gave lectures on your specialty several times per week, and spent the rest of your time doing research to advance the state of knowledge in this specialty. Once every year or so, you would graduate a new specialist with Ph.D. in hand who would pursue the same career path as you.

Of course, there was also undergraduate instruction. You might teach massive-sized introductory courses, at least until you received tenure, or until you secured enough grant money to buy yourself out of such teaching. Fortunately, there were usually many teaching assistants to correct homework and exams, as well as conduct tutoring sessions. And, there were adjunct faculty who could be hired for very small sums of money to teach the sections that you wanted to avoid and for which you billed a very small amount to sponsored research. This system worked pretty well—or at least we thought it did—for many centuries. Since World War II it blossomed, especially the heavy sponsored-research component through heavy nourishment from the U.S. government. We produced masses of bachelor degrees and impressive numbers of graduate degrees. Few questioned the value of higher education and it was perceived to be one of our crown jewels.

Recently, however, questions have arisen about constantly escalating costs. In many institutions, the quality of education has also been questioned. Is it reasonable, for instance, to pay thousands of dollars to be instructed by graduate students with little if any training as a teachers and often with very poor communications skills in the English language? Certainly the diploma—the union card—appears to be worth it, but are students' learning experiences really cost-effective? Is academia becoming a "Temple of Imposters" (Anderson, 1996)? Are there significant academic duties that are being neglected (Kennedy, 1997).

IT provides new means for learning. For example, why can all physics students not receive at least some of their education from the best physics professor in the country? Similarly, why should any student have to endure a mediocre lecture? Telecommunications, multimedia, and simulation technologies, to name but a few technologies, provide opportunities for revolutionizing education. It is now possible in principle to dramatically increase the quality of education while also substantially decreasing the costs.

There are two very sizable obstacles in the way of this happening. First and perhaps foremost, the time students spend in academia is not solely for the purpose of formal education. Academia provides students opportunities for experiences, making choices (and mistakes), and maturing in general. Thus, providing technology-enabled means for everyone to earn a high-quality Ph.D. in physics, for instance, while still living at home with their parents may not be popular with any of the key stakeholders.

The second large obstacle is the fundamental nature of academia itself. Discipline-oriented organizational structures and faculty incentive and rewards systems have been refined and reinforced since their inception in Europe in the 12th century. Trying to change these fundamental attributes of academia is akin to attempting to change a religion or to move an old cemetary—while change does happen, many "heretics" suffer in the process. Of course, this has been denoted as the age of heretics (Kleiner, 1998) in industrial and organizational settings, although few seem to have surfaced or survived in academia.

Despite these impediments, IT has already changed the world via dramatic increases of speed and connectivity (Davis and Meyer, 1998). These trends will eventually radically alter delivery of instruction and evaluation of learning. These changes will profoundly affect the roles of faculty members, especially how they are motivated and rewarded. Very significant organizational challenges (see Chapter 20) are undoubtedly inevitable.

These changes are also likely to dramatically affect perceptions of the costs/benefits of academia's role as a place for young people to mature. This need will certainly persist. However, parents and students themselves are unlikely to be willing to accept constantly increasing prices—and the consequent sizable debts—when the formal education part of the equation becomes much more cost-effective. There would certainly appear to be major opportunities for systems engineering and management, and IT, as catalysts for quality and innovation in higher education (Sage, 1992).

30.3.4   Summary

In this section, we have discussed three scenarios that illustrate the pervasive and substantial impacts of IT. Business faces the challenge of making sense of a wealth of alternatives—including dramatic increases of available data, as well as an evergrowing set of potentially useful methods and tools; the major need to convert data into information and thence into knowledge; and to manage knowledge successfully in the networked economy. The military faces a clearer future, but must accomplish change within an economically and politically complex web of relationships and vested interests. Academia faces the very fundamental challenge of making substantial paradigm shifts in institutions that have refined their current paradigms over centuries and that seem unwilling to change despite the change occurring all around them.

From the perspective of this handbook, the key issue is how systems engineering and management can support business, government, and academia to address these issues successfully. In the remainder of this chapter we discuss ten challenges that this integrative discipline must pursue and resolve if we are to support our key stakeholders.

30.4   TEN CHALLENGES

The technology and organizational trends outlined in this chapter pose substantial challenges for systems engineering and management. Addressing these challenges will require many of the concepts, principles, methods, and tools discussed in earlier chapters. It will also require new paradigms not discussed in these chapters.

30.4.1 Systems Modeling

Our methods and tools for modeling, optimization, and control depend heavily on exploiting problem structure. Understanding the relationships and constraints underlying problem structure enables predicting system behaviors, as well as potentially controlling these behaviors. Decomposing problem structure, associating first principles with the elements resulting from this decomposition, and then recomposing these principles into an overall mathematical or computational model are typical steps of systems modeling (see Chapters 26 and 27).

For the loosely structured systems emerging due to IT, however, behavior does not emerge from fixed structures. Instead, structure emerges from collective behaviors. The distributed, collaborative, and virtual organizations that IT enables are such that the system elements are quite fluid. Distinctions between what is inside and outside the system depend on time-varying behaviors and consequences.

With such systems, modeling must be done on at least two levels. One level concerns the first principles for inclusion of elements in the system. The other level concerns the behaviors of elements if they are included in the system, which depend on what other elements are also included. Thus, we need methods for modeling the inclinations of elements to become parts of systems, the behaviors that tend to result when included, and the ways in which these behaviors affect inclinations to be included. Satisfying this need will significantly challenge typical modeling methods and tools.

30.4.2 Emergent and Complex Phenomena

Meeting the modeling challenge is complicated by the fact that not all critical phenomena can be fully understood, or even anticipated, based on analysis of the decomposed elements of the overall system. Complexity not only arises from there being many elements of the system, but also from the possibility of collective behaviors that even the participants in the system could not have anticipated (Casti, 1997).

An excellent example is the process of technological innovation. Despite the focused intentions and immense efforts of the plethora of inventors and investors attracted by new technologies, the ultimate market success of these technologies almost always is other than what these people expect (Burke, 1996). It was not envisioned, for example, that the primary use of telephones or computer networks would be personal communication. Also, new technologies often cause great firms to fail (Christensen, 1997).

Thus, many critical phenomena can only be studied once they emerge. In other words, the only way to identify such phenomena is to let them happen. The challenge is to create ways to recognize the emergence of unanticipated phenomena and be able to manage their consequences, especially in situations where likely consequences are highly undesirable.

Complexity theory (Kaufman, 1995; Axelrod, 1997; Holland, 1998) is an emerging field of study that has evolved from five major knowledge areas: mathematics, physics, biology, organizational science, and computational intelligence and engineering. Fundamentally, a system is complex when we cannot understand it through simple cause-and-effect relationships or other standard methods of systems analysis. In a complex system, we cannot reduce the interplay of individual elements to the study of individual elements considered in isolation. Often, several different models of the complete system, each at a different level of abstraction, are needed.

There are several sciences of complexity, and they generally deal with approaches to understanding the dynamic behavior of units that range from individual organisms to the largest technical, economic, social, and political organizations. Often, such studies involve complex adaptive systems and hierarchical systems, are multidisciplinary in nature, and involve or are at the limits of scientific knowledge (Arthur, 1994; Coveney and Highfield, 1995; Casti and Karlqvist, 1996; Arthur et al., 1997; Epstein, 1997).

Complexity studies attempt to pursue knowledge and discover features shared by systems described as complex. These include studies such as complex adaptive systems, complex systems theory, complexity theory, dynamic systems theory, complex nonlinear systems, and computational intelligence. Many scientific studies, prior to the development of simulation models and complexity theory, involved the use of linear models. When a study resulted in anomalous behavior, the failure was often incorrectly blamed on noise or experimental error. It is now recognized that such "errors" may reflect inherent inappropriateness of linear models—and linear thinking.

One measure of system complexity is the complexity of the simulation model necessary to effectively predict system behavior (Casti, 1997). The more the simulation model must embody the actual system to yield the same behavior, the more complex the system. In other words, outputs of complex systems cannot be predicted accurately based on models with typical types of simplifying assumptions. Consequently, creating models that will accurately predict the outcomes of complex systems is very difficult. We can, however, create a model that will accurately simulate the processes the system will use to create a given output.

This awareness has profound impacts for organizational efforts. For example, it raises concerns related to the real value of creating organizational mission statements and plans with expectations that these plans will be inexorably executed and missions thereby realized. It may be more valuable to create a model of an organization's planning processes themselves, subject this model to various input scenarios, and use the results to generate alternative output scenarios. The question then becomes one of how to manage an organization where this range of outputs is possible.

Interestingly, most studies of complex systems often run completely counter to the trend toward increasing fragmentation and specialization in most disciplines. The current trend in complexity studies is to reintegrate the fragmented interests of most disciplines into a common pathway. This transdisciplinarity provides the basis for creating a cohesive systems ecology to guide the use of IT for managing complex systems. Whether they be human-made systems, human systems, or organizational systems, the use of systems ecology could more quickly lead to organizing for complexity (McMaster, 1996), and associated knowledge and enterprise integration.

An important aspect of complex systems is path dependence (Arthur, 1994). The essence of this phenomenon begins with a supposedly minor advantage or inconsequential head start in the marketplace for some technology, product, or standard. This minor advantage can have important and irreversible influences on the ultimate market allocation of resources even if market participants make voluntary decisions and attempt to maximize their individual benefits. Such a result is not plausible with classic economic models that assume that the maximization of individual gain leads to market optimization unless the market is imperfect due to the existence of such effects as monopolies. Path dependence is a failure of traditional market mechanisms and suggests that users are "locked" into a suboptimal product, even though they are aware of the situation and may know that there is a superior alternative.

This type of path lock-in is generally attributed to two underlying drivers: (1) network effects, and (2) increasing returns of scale. Both of these drivers produce the same result, namely that the value of a product increases with the number of users. Network effects, or "network externalities," occur because the value of a product for an individual consumer may increase with increased adoption of that product by other consumers. This, in turn, raises the potential value for additional users. An example is the telephone, which is only useful if at least one other person has one as well, and becomes increasingly beneficial as the number of potential users of the telephone increases.

Increasing returns of scale imply that the average cost of a product decreases as higher volumes are manufactured. This effect is a feature of many knowledge-based products where high initial development costs dominate low marginal production and distribution costs. Thus, the average cost per unit decreases as the sales volume increases and the producing company is able to continuously reduce the price of the product. The increasing returns to scale, associated with high initial development costs and the low sales price create barriers against market entry by new potential competitors, even though they may have a superior product.

The controversy in the late 1990s over the integration of the Microsoft Internet Explorer with the Windows Operating System may be regarded as a potential example of path dependence, and appropriate models of this phenomenon can potentially be developed using complexity theory. These would allow exploration of whether network effects and increasing returns of scale can potentially reinforce the market dominance of an established but inferior product in the face of other superior products, or whether a given product is successful because its engineers have carefully and foresightedly integrated it with associated products such as to provide a seamless interface between several applications.

30.4.3   Uncertainties and Control

IT enables systems where the interactions of many loosely structured elements can produce unpredictable and uncertain responses that may be difficult to control. If critical phenomena are unpredictable—and elements can be inside or outside of the system—it can be very difficult to know what variables to control and where to exert control. Prediction and control can be akin to herding cats when you are not sure which cats are part of your herd.

To illustrate, consider a virtual enterprise configured to rapidly exploit an opportunity to create a new consumer product and quickly distribute it nationally. The speed needed to gain competitive advantage may be such that it is not clear who has joined the venture, who is still being wooed, and who has declined. Consequently, the availability of resources (i.e., capabilities for marketing, distribution, etc.) is uncertain, as is the sense of how all the pieces will play together, or potentially conflict. How can you predict and control the behaviors of this system?

The challenge is to understand such systems at a higher level. Control is likely to involve design and manipulation of incentives to participate and rewards for collaborative behaviors. It may be impossible, and probably undesirable, to control behaviors directly. The needed type of control is similar to policy formulation. Success depends on efficient experimentation much more than possibilities for mathematical optimization due to the inherent complexities that are involved. Again, insights from complexity theory may be brought to bear on these situations (Merry, 1995).

30.4.4 Access and Utilization of Information and Knowledge

Information access and utilization, as well as management of the knowledge resulting from this process, are complicated in a world with high levels of connectivity and a wealth of data, information, and knowledge. Ubiquitous networks and data warehouses are touted as the means to taking advantage of this situation. However, providers of such "solutions" seldom address the basic issue of what information users really need, how this information should be processed and presented, and how it should be subsumed into knowledge that reflects context and experiential awareness of related isues.

The underlying problem is the usually tacit assumption that more information is inherently good to have. What users should do with this information and how value is provided by this usage are seldom clear. The result can be large investments in IT with negligible improvements of productivity (Harris, 1994). One of the major needs in this regard is for organizations to develop the capacity to become learning organizations (Senge, 1990; Chawla and Renesch, 1995; Argyris and Schön, 1996) and to support bilateral transformations between tacit and explicit knowledge (Nonaka, 1994; Nonaka and Takeuchi, 1995).

Addressing these dilemmas should begin with the recognition that information is only a means to gaining knowledge and that information must be associated with the contingency-task structure to become knowledge. This knowledge is the source of desired advantages in the marketplace or vs. adversaries. Thus, understanding and supporting the transformations from information to knowledge to advantage are central challenges to enhancing information access and utilization in organizations (Elliott, 1997; Rouse et al., 1998).

For a period of time, organizations addressed the access and utilization challenge by investing in comprehensive and very expensive IT solutions. In recent years, however, investments in solutions have become highly driven by business issues. An organization will consider a particular business issue, such as demand forecasting, and determine the business value of resolving this issue. For example, how would revenues increase or costs decrease if forecasts were more accurate? They next look at the knowledge needed to provide this improvement, and then consider the information that must be accessed and utilized to generate this knowledge. Finally, they determine the life-cycle costs of this information. If these costs do not compare favorably with projected benefits, they do not make this investment. In this way, the challenge has become one of understanding information and knowledge needs in particular rather than in general.

30.4.5 Information and Knowledge Requirements

Beyond adopting a knowledge management perspective, one must deal with the tremendous challenge of specifying information requirements in a highly information-rich environment. Users can have access to virtually any information they want, regardless of whether they know what to do with it or how to utilize the information as knowledge. Users' natural tendencies to hedge and overspecify information needs present problems when these inflated requirements can easily be met.

The difficulty is users' limited abilities to consume information and transform it to the knowledge that will provide them desired advantages. While IT has evolved quite rapidly in recent decades, human information-processing abilities have not evolved significantly (see Chapters 17–19). These limited abilities become bottlenecks as users attempt to digest the wealth of information they have requested. The result is sluggish and hesitant decision making, which in a sense is due to being overinformed.

The challenge is to develop methods for requirements analysis (see Chapter 4) that are able to take into account the consequences of having met requirements. Tradeoffs between meeting one requirement vs. another are often central aspects of requirements analysis. In an information-rich world, all requirements can be met and the key tradeoffs are between meeting and perhaps exceeding these requirements and potentially over-whelming users. As an analogy, if energy were to become free, we then would be faced with tradeoffs between energy usage and the negative consequences of usage, namely, pollution.

A key to dealing with these tradeoffs is recognition that the traditional, crisp view of requirements is much too rigid. The old "procurement model" involved buying products and services from the lowest-price provider that meets requirements. When requirements can commonly be exceeded, the buyer has to decide how to attach value to these product or service attributes. "Nailing down" requirements is replaced by stakeholder-driven creation of innovative value propositions (Rouse, 1991). This can involve many of the complications of multiattribute models (see Chapter 28). Nevertheless, this approach provides many more opportunities for gaining market advantage.

30.4.6   Information and Knowledge Support Systems

Information support systems, including decision support (see Chapters 18 and 28), can provide the means for helping users cope with information-rich environments. Such support systems are difficult to design and develop for loosely structured organizations due to inherently less well-defined tasks and decisions. It is also complicated by continually evolving information sources and organizational needs to respond to new opportunities and threats.

These difficulties can be overcome, at least in part, by adopting a human-centered approach to information and knowledge support system design (Rouse, 1991, 1994). This approach begins with understanding the goals, needs, and preferences of system users and other stakeholders, for example, system maintainers. In particular, stakeholders' perceptions of the validity, acceptability, and viability of potential support functions drive trade-offs and decisions.

This approach focuses on stakeholders' abilities, limitations, and preferences, and attempts to synthesize solutions that enhance abilities, overcome limitations, and foster acceptance. From this perspective, IT and alternative sources of information and knowledge are enablers rather than ends in themselves. Consequently, design and development of support-system concepts can be premised on accessing and utilizing a wider range of resources. Nevertheless, design and development of such systems is challenging for loosely structured environments (see Chapters 16 and 21).

30.4.7   Inductive Reasoning

Prior to the development of simulation models and complexity theory, most studies involved use of linear models and assumed time-invariant processes (i.e., ergodicity). Most studies also assumed that humans use deductive reasoning and technoeconomic rationality to reach conclusions. But information imperfections, and limits on available time, often suggest that rationality must be bounded. Other forms of rationality and inductive reasoning are necessary.

There are a number of descriptive models of human problem solving and decision making. Generally, the appropriate model depends upon the contingency task structure, characteristics of the environment, and the experiential familiarity of humans with tasks and environment. Thus, the context surrounding information and the experiential familiarity of users of the information is most important. In fact, it is the use of information within the context of contingency task structures and the environment that results in the transformation from information to knowledge.

We interpret knowledge in terms of context and experience by sensing situations and recognizing patterns. We recognize features similar to previously recognized situations. We simplify the problem by using these to construct internal models, hypotheses, or schemata to use on a temporary basis. We attempt simplified deductions based on these hypotheses and act accordingly. Feedback of results from these interactions enables us to learn more about the environment and the nature of the task at hand. We revise our hypotheses, reinforcing appropriate ones and discarding poor ones. This use of simplified models is a central part of inductive behavior (Holland et al., 1989).

Models of inductive processes can be constructed in the following way. We first set up a collection of, generally heterogeneous, agents. We assume that they are able to form hypotheses based on mental models or subjective beliefs. Each agent is assumed to monitor performance relative to a personal set of belief models. These models are based on the results of actions, as well as prior beliefs and hypotheses. Through this iterative procedure, learning takes place as agents learn which hypotheses are most appropriate. Hypotheses, or models, are retained not because that are "correct" but because they have worked in the past. Agents differ in their approach to problems and the way in which they subjectively converge to a set of "useful" hypotheses. The question of where these mental models come from is interesting and a subject of much study in the learning literature.

This process may be modeled as a complex adaptive system. As noted, we cannot create models that will accurately predict the outcomes of many complex systems. We can, however, often create a model that will accurately simulate the processes the system uses to create outputs. The major constructs associated with such models are the interactions and feedback relations between the various agents whose choices depend upon the decisions of others, and linearity and return to scale considerations. There are many implications associated with these models. Among them are questions of steady state vs. continued evolutionary behavior, the nature and possibility of time-invariant processes (ergodicity), and questions of path dependence discussed earlier.

30.4.8 Learning Organizations

Realizing the full value of information and knowledge is strongly related to organizations' abilities to learn and become learning organizations. Learning involves the use of observations of the relationships between activities and outcomes, often obtained in an experiential manner, to improve behavior through the incorporation of appropriate changes in processes and products. Thus, learning represents acquired wisdom in the form of abilities for skill-based, rule-based, and knowledge-based—or formal—reasoning (Rasmussen et al., 1994). It may involve know-how, in the form of skills or rules, or know-why, in terms of formal knowledge.

Learning generally involves several processes:

- Situation assessment
- Detection of a problem
- Synthesis of a potential solution
- Implementation of the solution
- Evaluation of the outcome
- Discovery of patterns among the preceding processes

This is a formal description of the learning process. It describes typical problem-solving processes and involves the basic steps of systems engineering and management in an inductive fashion.

While learning appears highly desirable, much of individual and organizational learning that occurs in practice is not necessarily beneficial or appropriate. For example, there is much literature that shows that individuals and organizations use improperly simplified and often distorted models of causal and diagnostic inferences. Similarly, they often employ improperly simplified and distorted models of the contingency task structure and the environment.

Organizational learning results when members of the organization react to changes in the internal or external environment of the organization by detection and correction of errors. Argyris and Schön have developed a theory of reasoning, learning, and action for individual and organizational learning (Argyris and Schön, 1978, 1996). In this model, learning is fundamentally associated with detection, diagnosis, and correction of errors.

The notion of error is singularly important in this theory. Errors are features of behavior that make actions ineffective. Detection and correction of errors produce learning. Individuals in an organization are agents of organizational action and organizational learning. Argyris and Schön have found two information-related factors that inhibit organizational learning: (1) information distortion such that its value in influencing quality decisions is lessened, and (2) lack of receptivity to corrective feedback.

Two types of organizational learning are defined by Argyris and Schön. Single-loop learning is learning that does not question the fundamental objectives or actions of an organization. This type of learning is essential to acting quickly, which is often needed. Such actions are usually based on rule-based or skill-based reasoning. When errors occur, members of the organization may discover sources of these errors and identify new strategic activities that may correct the errors. These activities may be identified either through use of a different rule- or experientially based skill, or through the application of formal reasoning to the situation at hand. The activities are then analyzed and evaluated, and one or more is selected for implementation. Single-loop learning enables the use of present policies to achieve present objectives. The organization may well improve, but this will be with respect to the current way of doing things. Organizational purpose, and perhaps even process, are seldom questioned.

In many cases, this approach is quite valid. Occasionally, however, this approach is not appropriate. For example, environmental control and self-protection through control over others, primarily by imposition of power, are typical managerial strategies. The consequences of this approach may include defensive group dynamics and low production of valid information. This lack of information does not result in disturbances to prevailing values. However, the resulting inefficiencies in decision making encourage frustration and an increase in secrecy and loyalty demands from decision makers. All of this is mutually self-reinforcing. It results in a stable autocratic organization and a self-fulfilling prophecy with respect to the need for organizational control. So, while there are many desirable features associated with single-loop learning, there are a number of potentially debilitating aspects as well. These are quite closely related to notions of organizational culture (see Chapter 20).

Double-loop learning involves identification of potential changes in organizational goals and approaches to inquiry that allow confrontation with and resolution of conflicts, rather than continued pursuit of incompatible objectives, which usually leads to increased conflict. Double-loop learning is the result of organizational inquiry that resolves incompatible organizational objectives through the setting of new priorities and objectives. New understanding is developed that results in updated cognitive maps and scripts of organizational behavior. Studies show that poorly performing organizations learn primarily on the basis of single-loop learning and rarely engage in double-loop learning in which the underlying organizational purposes and objectives are questioned.

Double-loop learning is particularly useful in the case when people's espoused theories of action—the "official" theories that people claim as a basis for action—conflict with their theories in use, which are the theories of action underlying actual behaviors. While people are often adept at identifying discrepancies between other people's espoused theories of action and theories in use, they are not equally capable of self-diagnosis. Further, the dictates of tactfulness normally prevent us from calling to the attention of others the observed inconsistencies between their espoused and actual theories of action. The result of this failure is inhibition of double-loop learning.

Two major inhibitions to learning, distancing and disconnectedness, are noted by Argyris and Schön. *Distancing* is the art of not accepting responsibility for either problems or solutions; *disconnectedness* occurs when individuals are not fully aware of the theories in use and the relationships between these theories and associated actions. Several other factors can interact to result in conflicting and intolerable pressures, including:

- Incongruity between espoused theory and theory in use that is recognized but not corrected.
- Inconsistency between theories in use of different members of the organization.
- Ineffectiveness, as objectives associated with theories in use become less and less achievable over time.
- Disutility, as theories in use become less valued over time.
- Unobservability, as theories in use result in suppression of information by others such that evaluation of effectiveness becomes impossible.

Detection and correction of conflicts between espoused theories of action and theories in use can lead to reductions of inhibitions of double-loop learning.

The result of double-loop learning is a new set of goals and operating policies that become part of the organization's knowledge base. It is when the environment and the contingency task structure changes that double-loop learning is called for. Learning organizations have abilities to accommodate double-loop learning and successfully integrate and utilize the appropriate blend of single- and double-loop learning.

Peter Senge (1990) has extensively discussed the nature of learning organizations. He describes learning organizations as "organizations where people continually expand their capacity to create the results they truly desire, where new and expansive patterns of thinking are nurtured, where collective aspiration is set free, and where people are continually learning how to learn together." Five component technologies, or disciplines, enable this type of learning:

- Systems thinking.
- Personal mastery through proficiency and commitment to lifelong learning.
- Shared mental models of the organization markets, and competitors.
- Shared vision for the future of the organization.
- Team learning.

Systems thinking is denoted as the "fifth discipline." It is the catalyst and cornerstone of the learning organization that enables success through the other four dimensions.

Lack of organizational capacity in any of these disciplines is called a *learning disability*. One of the major disabilities is associated with implicit mental models that result in people having deeply rooted mental models without being aware of the cause–effect consequences that result from use of these models. Another is the tendency of people to envision themselves in terms of their position in an organization rather than in terms of their aptitudes and abilities. This often results in people becoming dislocated when organizational changes are necessary.

Each of the five learning disciplines can exist at three levels:

- *Principles:* the guiding ideas and insights that guide practices.
- *Practices:* the existing theories of action in practice.
- *Essences:* the wholistic and future-oriented understandings associated with each particular discipline.

These correspond very closely with the principles, practices, and perspectives we have used to describe approaches to systems engineering and management in the introduction of this handbook.

Based primarily on works in system dynamics, an approach to the study and modeling of systems of large scale and scope, and on efforts by Argyris and others, eleven laws of the fifth discipline are stated:

- Contemporary and future problems often come about because of what were presumed to be past solutions.
- For every action, there is a reaction.
- Short-term improvements often lead to long-term difficulties.
- The easy solution may be no solution at all.
- The solution may be worse than the problem.
- Quick solutions, especially at the level of symptoms, often lead to more problems than existed initially and, hence, may be counterproductive.
- Cause and effect are not necessarily related closely, either in time or in space.
- The actions that will produce the most effective results are not necessarily obvious at first glance.

- Low cost and high effectiveness do not have to be subject to compensatory tradeoffs over all time.
- The entirety of an issue is often more than the simple aggregation of the components of the issue.
- The entire system, which comprises the organization and its environment, must be considered together.

Neglect of these laws can lead to any number of problems, most of which are relatively evident from their description. For example, failure to understand the last law leads to the fundamental attribution error in which we credit ourselves for successes and blame others for our failures.

On the basis of these laws, several leadership facets emerge. Leaders in learning organizations become designers, stewards, and teachers. Each of these leadership characteristics enables everyone in the organization to improve their understanding and use of the five important dimensions of organizational learning. This results in creative tension throughout the organization. This tension can be addressed in planning, which is one of the major activities of learning organizations, and it is through planning that much learning occurs. Organizational learning is one of the major contemporary thrusts in systems management today (Cohen and Sproull, 1996).

It is important to emphasize that the extended discussion of learning organizations in this section is central to understanding how to create organizations that can gain full benefits of information technology and knowledge management. This is crucial if we are to transform data to information to insights to programs of action.

30.4.9   Planning and Design

Dealing successfully with the challenges just discussed requires that approaches to planning and design be reconsidered. Traditionally, planning and design are activities that occur before solutions are placed into operation (see Chapters 13 and 29). In contrast, for loosely structured systems, planning and design must be transformed to something done in parallel with system operation rather than beforehand.

In the context of traditional systems engineering methods and tools, loosely structured systems pose needs for on-line identification, estimation, and control of time-varying, distributed knowledge-based systems. This is a tall order. Further, as noted earlier, control for such systems may mean manipulating the cost functional rather than directly affecting state variables. In other words, in the absence of being able to control behaviors, the key may be to control the incentives and rewards that determine behaviors.

From this perspective, planning and design become less a problem of specifying tasks, milestones, and schedules, and more an issue of influencing formulation and structuring of goals, priorities, and context. Defining and varying the agenda in light of current trends and events may be the essence of controlling loosely structured systems. This implies the need for assuring influence in the absence of possibilities for control.

An example related to the military scenario discussed earlier provides an excellent illustration of this concept. Traditionally, military operations have focused on threatening the use of, and perhaps employing, decisive and sometimes massive forces to deter, thwart, or defeat adversaries. This tactic is premised on having an adversary that is well defined, and, hence, can be located, identified, and engaged. Loosely structured and potentially virtual adversaries do not satisfy these criteria.

How can you combat virtual adversaries? One answer is to attack the decision to form covert coalitions. One can also attack the decisions of such coalitions to employ military forces. In other words, rather than trying to outwit an elusive adversary on the playing field, you create strong incentives for them to not play the game. Better yet, you can create compelling incentives to not even entertain entering the game. This requires that you understand their motivations and value systems much more then their abilities to execute military operations. Planning and design, in this case, must focus on abstractions somewhat removed from traditional command and control considerations and must give due consideration to the possibilities for cooperation and associated complexities (Axelrod, 1997).

30.4.10 Measurement and Evaluation

Successfully addressing and resolving the many issues associated with the challenges described in this chapter requires that a variety of measurement challenges be understood and resolved. Approaches to modeling loosely structured systems, representing emergent and complex phenomena, and dealing with uncertainties and control all involve measurement (see Chapter 15).

Systems associated with access and utilization of information and knowledge management present particular measurement difficulties because the ways in which information and knowledge affect behaviors are often rather indirect. These inputs to humans often do not produce outputs in terms of observable behaviors. In fact, an important value of information and knowledge is their use in deciding not to act.

For this and a variety of related reasons, it can be quite difficult to evaluate the impact of information technology and knowledge management. Numerous studies have failed to identify measurable productivity improvements as the result of investments in these technologies [see Harris (1994) for a review of many of these studies]. The difficulty is that the impact of information and knowledge is not usually directly related to numbers of products sold, manufactured, or shipped. Successful measurement requires understanding the often extended causal chain from information to actions and results.

Transformations from information to knowledge also present measurement problems. Information about the physical properties of a phenomenon are usually constant across applications. In contrast, knowledge about the context-specific implications of these properties depends on human intentions relative to these implications. Consequently, the ways in which information is best transformed to knowledge depends on the intentions of the humans involved. The overall measurement problem involves inferring, or otherwise determining, the intentions of users of information systems.

Both measuring the impact of information and inferring the intentions that underlie the transformations of information to knowledge present difficulties for evaluation as well as creation of information and knowledge support systems. Many of the concepts discussed in this chapter are only implementable in the context of valid measurement systems. Otherwise, the promises of information technology and knowledge management mainly amount to hand waving.

30.4.11   Summary

In this section, we have posed ten major systems engineering and management challenges. These challenges need to be addressed successfully if the promises of information technology and knowledge management are to be realized. These challenges concern our abilities to deal with:

- Systems modeling
- Emergent and complex phenomena
- Uncertainties and control
- Access and utilization of information and knowledge
- Information and knowledge requirements
- Information and knowledge support systems
- Inductive reasoning
- Learning organizations
- Planning and design
- Measurement and evaluation

As numerous references throughout this chapter have indicated, understanding and addressing these challenges requires making use of a substantial portion of this handbook.

30.5   ECOLOGICAL APPROACHES TO THE CHALLENGES

The remainder of this chapter is devoted to considering how we can make sense of and address these ten challenges as a whole. This requires a deep understanding of the context in which these types of challenges typically emerge. In this section, we discuss ecologically based approaches to gaining this understanding.

Ecology is a science that is concerned with ways in which plants and animals depend upon one another and upon the physical settings in which they live. In some classifications, ecology is considered to be a part of biology. Modern ecology also depends upon a number of other subjects, such as economics and environment. Basically, ecology is the investigation of interactions of organisms in different environments. As a by-product of an ecological study, we learn how nature establishes orderly patterns among living things.

Ecology is basically concerned with interdependence and emphasizes the interdependence of each and every form of life on other living things as well as on the basic natural resources in the environment: air, land, water, and biota. A fundamental thesis of ecology is that people cannot view nature and the natural environment as separated and independent things. Instead, we should accept the view that any action, human or otherwise, that changes the environment affects all the organisms that are in it.

Ecology is concerned with natural balances. As a result of ecological adaptation, every kind of life is suited to the physical conditions of its habitat. Further, the continued existence of this form of life involves considerations of natural balance, or dynamic equilibrium. An equilibrium point is disturbed when changes occur requiring control to achieve a new equilibrium. Affecting this control requires information from nature, and this may or may not be available.

To obtain the information necessary to achieve control, one must understand the biological sciences and the earth sciences. Also, the economic and social sciences are important in a study of ecology, since they relate to the nature and motivations for human interaction with the environment.

There are a few general principles of ecology that can be gleaned from even elementary writings on the subject (Roughgarden, 1998):

- *Principle of Special Environmental Needs of Living Things.* Life patterns reflect the patterns of the physical environment.
- *Principle of Biotic Communities.* Member species (plants and animals) in a given region group themselves into loosely organized units known as communities.
- *Principle of Competition in Communities.* Competition is characteristic of all communities and may be modified through behavioral adjustments, including cooperation, among members of a community.
- *Principle of Succession in Communities.* Orderly and predictable development, or ecological succession, takes place in any area.
- *Principle of Ecosystem Equilibrium.* A community and its living and nonliving environment constitute an ecological system, or ecosystem. The community takes materials from its surroundings and transfers materials to it such that materials are exchanged continuously, ideally such that basic resources are sustained and never exhausted

There are many types of ecosystems and they may exist on land, water, or in the air. The combined total ecosystems of the Earth constitute the biosphere. Inhabitants of an ecosystem are generally classified as producers, consumers, and decomposers. Preservation of a biotic community may require preservation of key member species of that community.

Human-made communities often replace natural communities. Even so, the principles that govern the life of natural communities must be followed if these human-made communities are to survive over the long term and maintain a natural balance. Environmental preservation, conservation of natural communities, and waste abatement may in principle each be achieved through proper ecology. Wise use of technology may support this and also provide for human progress.

A central function in ecology is to study human interactions with the natural environment in order to modify them favorably. Van der Ryn and Cowan (1996) suggest five principles for ecological design:

- Solutions grow from place, in the sense that one needs detailed intimate experiential familiarity, context, and information—in other words knowledge—about a particular place in order to implement appropriate solutions and to inhabit without destroying.
- Ecological accounting must inform design so as to enable definition, development, and deployment of the most appropriate design alternatives.
- Design with nature is needed in order that we respect the needs of all species, and regenerate resources rather than deplete them, while at the same time satisfying our own needs.
- Everyone is a designer in the sense that we must listen to and communicate with all stakeholders in the design process to cultivate a design intelligence that empowers all

and uses the very special knowledge that is brought by all stakeholders to important issues.

- Nature is made visible in the sense that all relevant aspects of the design environment are considered in a continuous learning effort that makes natural cycles and processes visible.

The term "design" is used here in a manner much the same as we use the term "engineer" and not as a term that relates to a portion of the development phase of systems engineering.

The terms information ecology, knowledge ecology, industrial ecology, and systems ecology suggest that models of biological systems and their interactions in nature provide useful analogies for the definition, development, and deployment of systems and processes that relate to information, knowledge management, industry, and systems in general. The ecological model is compelling because of the way in which evolution has enabled biotic and abiotic elements to live off the by-products and wastes of one another. Ecosystem resilience is often achieved, and the analogous ecosystem resilience is a need in these other areas as well.

Not all properties of ecosystems are necessarily and always appropriate; for example, ecosystem efficiency is often quite low. The phrase ecology is used because there is inherent insight in the concept of a natural ecosystem that can be beneficially used in the definition, development, and deployment of systems and processes of all types. The primary objective here is to interpret and adapt understanding of natural ecological systems and to apply the most beneficial of these concepts to the acquisition of human-made systems such that they become efficient, effective, and sustainable.

30.5.1 Information Ecology

Information ecology involves considering the contexts and impacts of information technology on people and organizations (Sage, 1995, 1998). In particular, information access and utilization, as well as knowledge management, occur in organizational contexts whose ecology must be understood if investments in information technology are to yield expected returns.

Davenport (1997) contends that an effective information ecology is associated with four principal characteristics (see Chapter 23):

- Integrates diverse types of information
- Recognizes evolutionary change
- Emphasizes observation and description
- Focuses on people and their information behavior

Fostering these characteristics could lead to major improvements in the ability to accomplish systems engineering and management and to enhance participation by all relevant stakeholders in decision making. This informed participation in decision making is a positive trend; however, there are associated risks. Information and knowledge can make an already powerful elite even more powerful if those who do not now have access are not able to gain this access. There are a number of very complex issues that require attention, such as the impact of information technology on social, ethical, cultural, and family values

(Mason et al. 1995). Nevertheless, information ecology provides a perspective on knowledge management principles that can assist organizations in effectively managing what they know (Davenport and Prusak, 1998).

Incentive and rewards systems provide an important illustration of the effects of ecological forces on information access and utilization. These systems have tremendous impacts on the behaviors of members of an organization (see Chapter 20). Motivations to learn about and take advantage of information systems depends on their value added in the context of overall incentive and reward systems. For instance, if detailed analyses and plans are not valued in organizations, people will not take advantage of information systems that support such analysis and planning.

Therefore, integration of a solution within a particular organizational context requires understanding the roles that information plays in that context. If decision making is driven by advocacy rather than by relevant information and knowledge, people will not rely on databases and knowledge bases, regardless of their relevance and ease of use. In contrast, organizations where analytical lines of reasoning are valued will tend to embrace information systems, warts and all.

30.5.2   Knowledge Ecology

Knowledge can be viewed as transformations of information in the context of a contingency task structure and experiential familiarity. This allows information to have value in such activities as planning and decision making. Knowledge management refers not to direct and explicit management of knowledge, but rather to the environment associated with the transformation of data to information to knowledge such that these transformations are effective and efficient. This requires that particular attention be paid to the context and environmental facets of the contingency task structure associated with knowledge acquisition and use.

By analogy to ecological principles, five principles of knowledge management can be stated:

- *Integration of Diverse Types of Knowledge.* Knowledge ecologies flourish because of knowledge diversity, which requires understanding the relevant facets of organizational activities and the broader context and environment in which the organization operates.
- *Knowledge Capital Must Inform Organizational Change.* Definition, development, and deployment efforts involve identifying the most appropriate alternatives for organizational advancement and then cultivating them for implementation through use of a systems engineering process.
- *Knowledge Management Must Transcend the Particulars of the Organization.* Knowledge workers should be empowered through enhancements associated with competence, commitment, communications, collaboration, and courage.
- *Knowledge Management Becomes Ubiquitous.* Listen to and communicate with all knowledge workers to cultivate an intelligence that empowers all and that encourages bilateral transitions between explicit and tacit knowledge.
- *Focus on People and Knowledge Behavior.* All relevant aspects of the knowledge environment are considered in a continuous learning effort that makes knowledge-acquisition cycles and processes visible throughout the organization.

Organizations are beginning to realize that knowledge is the most valuable asset of employees and the organization. This recognition must be converted into pragmatic action guidelines, plans, and specific approaches. Effective management of knowledge, which is assumed to be equivalent to effective management of the environmental factors that lead to enhanced learning and transfer of information into knowledge, also requires organizational investments in terms of financial capital for technology and human labor to ensure appropriate knowledge work processes. It also requires knowledge managers to facilitate identification, distribution, storage, use, and sharing of knowledge. Other issues include incentive systems and appropriate rewards for active knowledge creators, as well as the legalities and ethics of knowledge management.

Davenport and Prusak (1998) note that when organizations interact with environments, they absorb information and turn it into knowledge. Then they make decisions and take actions. They suggest five modes of knowledge generation:

- Acquisition of knowledge that is new to the organization, and perhaps represents newly created knowledge. Knowledgecentric organizations need to have appropriate knowledge available when it is needed. They may buy this knowledge, potentially through acquisition of another company, or they may generate it themselves. Knowledge can be leased or rented through renting a knowledge source, such as by hiring a consultant. Generally, knowledge leases or rentals are associated with knowledge transfer.

- Dedicated knowledge resource groups may be established. Since time is required in order for the financial returns on research to be realized, the focus of many organizations on short-term profit may create pressures to reduce costs by reducing such expenditures. Matheson and Matheson (1998) describe a number of approaches that knowledge organizations use to create value through strategic research and development.

- Knowledge fusion is an alternate approach to knowledge generation that brings together people with different perspectives to resolve an issue and determine a joint response. Nonaka and Takeuchi (1995) describe efforts of this sort. The result of knowledge fusion efforts may be creative chaos and a rethinking of old presumptions and methods of working. Significant time and effort is often required to enable group members sufficient shared knowledge and to work effectively together and to avoid confrontational behavior. Davenport and Prusak propose five knowledge management principles:

  (i) Promoting appreciation of the value of knowledge and the disposition to support the process of devising it.

  (ii) Identifying appropriate knowledge workers who will be effective in knowledge fusion efforts.

  (iii) Stressing the creative promise associated with the complex fusion tasks and viewing perceptual differences as potential attributes rather than as sources of conflict.

  (iv) Encouraging and directing knowledge-generation efforts toward common goals, and with rewards for achieving these goals.

  (v) Introducing success measures that reflect the value of the knowledge obtained.

- Adaptation through provision of internal resources and capabilities that can be utilized in new ways, and are open to change to the established ways of doing business. Knowledge workers who can acquire new knowledge and skills easily are most suitable to this approach. Knowledge workers with broad knowledge are often the most appropriate for adaptation assignments.
- Knowledge networks are the fifth form of knowledge generation, and may act as critical conduits for innovative reasoning. Informal networks can generate knowledge provided by a diversity of participants. This requires appropriate allocation of time and space for knowledge acquisition and creation.

In each of these efforts, it is critical to regard technology as a potential enabler of human effort, not as a substitute for it.

30.5.3 Industrial Ecology

Information technology and knowledge management are generally associated with economic development and growth. Continuing to meet increasing demands for products will very likely result in a highly polluted environment and a shortage of natural resources. On the other hand, if technology growth and the economic basis for this growth decline, the result is quite likely to be increased pollution and economic stagnation as well. These economic problems may also lead to military or political conflicts. If we can change our attitudes and embrace more functional, less nonrenewable resource-intensive product-oriented consumption, we can direct the wealth of nations toward technological innovations that enable sustainable world growth and development. Industrial ecology (Graedel and Allenby, 1995; Lowe et al., 1997) plays an important role here. Thus, we view this as an essential ingredient in appropriate knowledge-management efforts.

Simply defined, industrial ecology concerns the effective systems engineering and management of industrial processes for the evolution of sustainable products and services. It seeks to integrate simultaneous consideration of product functionality and competitiveness, natural-resource conservation, and environmental preservation to produce sustainable development. Such development rests upon three major pillars:

- Technoeconomic progress.
- Nonconsumptive use of natural resources and environmental preservation.
- Human, social, and cultural progress.

Thus, a successful industrial ecology requires:

- Developing industrial systems in which the wastes of one production process become input sources for others.
- Balancing industrial inputs and outputs with natural system constraints.
- Dematerializing industrial outputs in the sense of reducing the quantity of materials needed in a product to ensure a given functionality.
- Making full use of information technology and knowledge management through information ecology.
- Improving the efficiency and effectiveness of industrial processes or product lines.

- Developing and using renewable natural resources as substitutes for nonrenewable resources.
- Integrating economic and ecological full-cost accounting in policy options.
- Sublimating a product-oriented economy to a functional economy.

These are the principal elements of industrial ecology as a process-focused systems engineering and management endeavor.

Note that there are multiple levels of sophistication of industrial ecology. In particular, three types of industrial ecosystems can be defined. A traditional Type I system is one in which the input natural resources and output wastes are not fundamentally considered except in an economic fashion, and where there is no flow of materials from one life-cycle production process to the other. This is essentially an inactive approach to sustainability, although it can become reactive through imposition of command and control such as pollution restraints. A Type II industrial ecological system is one in which there is recycling of wastes and reuse of by-products of one process, whether in the same process or in another one, in order to reduce the resource input requirements and the output waste by-products of the several processes. This is an interactive approach to industrial ecology. Process redesign and reengineering can potentially enable the Type II system to become a Type III system, in which the overall industrial ecological system is closed and there are no nonrenewable resource requirements and no waste products. This is a proactive approach to industrial ecology and one that ultimately leads to simultaneous human socio-economic development and sustainability.

30.5.4 Toward a Systems Ecology

Systems engineering, as we have so often stressed in this handbook, is the process of engineering high-quality, trustworthy systems embodied in products and services that fulfill the useful purposes of a client group. This involves planning for and definition of a system's requirements and its human interfaces and interactions. It involves realizing the system through engineering development efforts that range from conceptual architectures through production. Deploying the resulting system in an operational setting is also part of systems engineering. Deployment includes integrating the system with legacy systems and assessing system effectiveness. It also often encompasses planning for and implementing maintenance and reengineering efforts to maintain functionality and user satisfaction over an extended life cycle.

Direct technological concerns that affect individual subsystems and specific physical science areas are often the focus of many conventional engineering efforts. Most studies, however, generally show that the major problems associated with the engineering of trustworthy systems have more do with the organization and management of complexity. Typical of these is a National Research Council study (1991) that identified five major categories of national need:

- Systems management for technology development
- Management of complex large-scale processes
- Using technology for competitive advantage
- Technology–organization interactions
- Social impacts of technology

Thus, systems engineering should place major emphasis on process engineering and management. Direct attention to the product or service only, without appropriate attention to the associated processes or product lines, leads to the fielding of a low-quality, expensive product or service that is unsustainable.

As a catalyst for innovation, quality, and productivity, system engineering is associated with several important perspectives:

- Much contemporary thought concerning innovation, productivity, quality, and sustainability can be cast into a systems engineering framework.
- Industrial ecology, information ecology, and knowledge-management efforts, as well as organizational and infrastructure reengineering efforts, are a natural complement to systems engineering and management perspectives.
- The Information Revolution and developments in IT provide a necessary tool base that, when combined with tools of industrial ecology and systems engineering and management, can produce process-level improvements that lead to the development of successful systems.
- Systems engineering constructs are useful not just for managing large systems engineering projects, but also for the creative management of the organization itself. Taking all of these together leads to a systems ecology.

A system's ecology comprises information, knowledge, and industrial ecologies, as well as systems engineering and management concepts, principles, methods, and tools. This ecology enables realization of continued progress and sustainable development that supports the four major players in the modern knowledge organization: professionals, managers, leaders, and support staff (Svelby, 1997; Moingeon and Edmondson, 1996).

A systems ecology should also provide major support to the learning organization (Senge, 1990; Chawla and Renesch, 1995; Argyris and Schön, 1996; Cohen and Sproull, 1996). For a learning organization, organizational intelligence is greater than the sum of the knowledge of each individual in that organization. Organizational intelligence includes historical knowledge inherent in the organization and generative intelligence that results from collaboration between organizational members. Organizational intelligence is the major competitive advantage of a knowledge organization.

30.5.5 Summary

In this section, we discussed ecologically based approaches to making sense of and addressing the ten challenges presented in this chapter. These approaches can enable gaining a deep understanding of the context in which these types of challenges typically emerge. The concepts and principles underlying ecological approaches are reasonably well developed. However, methods and tools for supporting the use of these approaches are in the early stages. We expect that systems engineering and management will play a major role in maturing these methods and tools.

30.6 CONCLUSIONS

Ongoing trends in information technology and knowledge management pose substantial challenges for systems engineering and management. Addressing the ten key challenges elaborated here requires utilizing many of the concepts, principles, methods, and tools presented in this handbook. In addition, it requires a new, broader perspective on the nature of information access and utilization, as well as knowledge management.

Fortunately, systems engineering and management is an inherently dynamic field of study and applications. Thus, the very nature of this discipline makes the ten key challenges and the four ecological "solutions" more tractable. However, this will require the discipline to move beyond structure-bound views of the world and the natural tendency to nail down requirements and constraints before proceeding. The current dynamics of information technology and knowledge management makes such "givens" obsolete almost as quickly as they are envisioned.

REFERENCES

Albert, S., and Bradley, K. (1997). *Managing Knowledge: Experts, Agencies, and Organizations.* Cambridge, UK: Cambridge University Press.

Alberts, D. S., and Papp, D. S. (eds.). (1997). *The Information Age: An Anthology of Its Impacts and Consequences.* Washington, DC: National Defense University Press.

Allee, V. (1997). *The Knowledge Evolution: Expanding Organizational Intelligence.* Boston: Butterworth-Heinemann.

Anderson, M. (1996). *Imposters in the Temple.* Stanford, CA: Stanford University, Hoover Institution Press.

Andrews, W. (1997). Agent makers expand into push in effort to deliver increasingly relevant information to users. *WebWeek* **3**(14), 1.

Argyris, C., and Schön, A. (1978), *Organizational Learning: A Theory of Action Perspective,* Reading, MA: Addison Wesley.

Argyris, C., and Schön, A. (1996). *Organizational Learning II: Theory, Method, and Practice.* Reading, MA: Addison-Wesley.

Arquilla, J., and Ronfeldt, D. (1997). *In Athena's Camp: Preparing for Conflict in the Information Age.* Santa Monica, CA: RAND Corporation.

Arthur, W. B. (1994). *Increasing Returns and Path Dependence in the Economy.* Ann Arbor: University of Michigan Press.

Arthur, W. B., Durlauf, S. N., and Lane, D. A. (eds.). (1997). *The Economy as an Evolving Complex System II.* Reading, MA: Addison-Wesley.

Axelrod, R. (1997). *The Complexity of Cooperation: Agent Based Models of Competition and Collaboration.* Princeton, NJ: Princeton University Press.

Bean, A. S., Russo, M. J., and Whiteley, W. B. R. (1998). Benchmarking your R&D: Results from IRI/CIMS Annual R&D Survey for FY '96. *Res. Technol. Manage.* **41**(1), 21–35.

Bell, D. (1973). *The Coming of Post Industrial Society.* New York: Basic Books.

Brooking, A. (1996). *Intellectual Capital: Core Asset for the Third Millennium Enterprise.* London: International Business Press.

Brynjolfsson, E., and Yang, S. (1996). Information technology and productivity: A review of the literature. *Adv. Comput.* **43**, 179–214.

Burke, J. (1996). *The Pinball Effect: How Renaissance Water Gardens Made the Carburetor Possible and Other Journeys through Knowledge.* Boston: Little, Brown.

Bush, V. (1995). As we may think. *Atlantic,* July, pp. 101–108.

Casti, J. L. (1997). *Would-be Worlds: How Simulation is Changing the Frontiers of Science.* New York: Wiley.

Casti, J. L., and Karkquist, A. K. (eds.). (1996). *Boundaries and Barriers: On the Limits to Scientific Knowledge*. Reading, MA: Addison-Wesley.

Chawla, S., and Renesch, J. (eds.). (1995). *Learning Organizations: Developing Cultures for Tomorrow's Workplace*. Portland, OR: Productivity Press.

Christensen, C. M. (1997). *The Innovator's Dilemma: When New Technologies Cause Great Firms to Fail*. Boston: Harvard Business School Press.

Cohen, M. D., and Sproull, L. S. (eds.). (1996). *Organizational Learning*. Beverly Hills, CA: Sage Publ.

Cooper, R., and Kaplan, R. S. (1991). *The Design of Cost Management Systems*. Englewood Cliffs, NJ: Prentice-Hall.

Coveney, P., and Highfield, R. (1995). *Frontiers of Complexity: The Search for Order in a Chaotic World*. Columbine, NY: Fawcett.

Davenport, T. H. (1997). *Information Ecology*. New York: Oxford University Press.

Davenport, T. H., and Prusak, L. (1998). *Working Knowledge: How Organizations Manage What They Know*. Boston: Harvard Business School Press.

Davis, S., and Meyer, C. (1998). *Blur: The Speed of Change in the Connected Economy*. Reading, MA: Addison-Wesley.

Dertouzos, M. (1997). *What Will Be: How the New World of Information Will Change our Lives*. New York: HarperCollins.

Drucker, P. F. (1997). Toward the new organization. *Leader to Leader,* **3,** 6–8.

Dyson, E. (1997). *Release 2.0: A Design for Living in the Digital Age:* New York: Broadway Books.

Edvinsson, L., and Malone, M. S. (1997). *Intellectual Capital: Realizing your Company's True Value by Finding Its Hidden Brainpower*. New York: HarperCollins.

Elliott, S. (ed.). (1997). *Using Information Technology to Support Knowledge Management*. Houston, TX: American Productivity & Quality Center.

Epstein, J. M. (1997). *Nonlinear Dynamics, Mathematical Biology, and Social Science*. Reading, MA: Addison-Wesley.

Evans, P. B., and Wurster, T. S. (1997). Strategy and the new economics of information. *Harv. Bus. Rev.,* September-October, pp. 71–82.

Fogelman, R. R., and Widnall, S. E. (1996). *Global Engagement: A Vision for the 21st Century Air Force*. Washington, DC: Department of the Air Force.

Fuchs, R., McCarthy, J., Corder, J., Rankine, R., Miller, W., and Gawron, V. (1997). *United States Air Force Expeditionary Forces*. Washington, DC: U.S. Air Force Scientific Advisory Board.

Graedel, T. E., and Allenby, B. R. (1995). *Industrial Ecology*. Upper Saddle River, NJ: Prentice Hall.

Grover, R., and Himelstein, L. (1997). All the news that's fit to browse. *Bus. Week,* June 16, pp. 133–134.

Harris, D. H. (ed.). (1994). *Organizational Linkages: Understanding the Productivity Paradox*. Washington, DC: National Academy Press.

Holland, J. H. (1998). *Emergence: From Chaos to Order*. Reading, MA: Addison-Wesley.

Holland, J. L., Holyoak, K. J., Nisbett, R. E., and Thagard, P. R. (Eds.) (1989) *Induction: Process of Inference, Learning, and Discovery*. Cambridge: MIT Press.

Hope, J., and Hope, T. (1997). *Competing in the Third Wave: The Ten Key Management Issues of the Information Age*. Boston: Harvard Business School Press.

Kaplan, R. S. (ed.). (1990). *Measures for Manufacturing Excellence*. Boston: Harvard Business School Press.

Kaplan, R. S., and Cooper, R. (1997). *Cost and Effect: Using Integrated Cost Systems to Drive Profitability and Performance*. Boston: Harvard Business School Press.

Kaplan, R. S. and Cooper, R. (1998) Cost and Effect: *Using Integrated Cost Systems to Drive Profitability and Performance*. Boston: Harvard Business School Press.

Kaplan, R. S., and Norton, D. P. (1996). *The Balanced Scorecard: Translating Strategy into Action*. Boston: Harvard Business School Press.

Katz, R. (ed.). (1997). *The Human Side of Managing Technological Innovation*. New York: Oxford University Press.

Kaufman, S. (1995). *At Home in The Universe: The Search for the Laws of Self-Organization and Complexity*. New York: Oxford University Press.

Kelly, K., and Wolf, G. (1997). PUSH! Kiss Your browser goodbye: The radical future of media beyond the web. *Wired*. **3**(3) March.

Kelly, K. (1998). *New Rules for the New Economy*. New York: Viking.

Kennedy, D. (1997). *Academic Duty*. Cambridge, MA: Harvard University Press.

Klein, D. A. (1998). *The Strategic Management of Intellectual Capital*. Boston: Butterworth-Heineman.

Kleiner, A. (1996). *The Age of Heretics: Heroes, Outlaws, and the Forerunners of Corporate Change*. New York: Doubleday.

Kling, R. (ed.). (1996). *Computerization and Controversy: Value Conflicts and Social Choices*. San Diego, CA: Academic Press.

Larson, F. (1997). *The Current and Future Operating Environment and the Range of AEF Missions/Applications,* Presentation to Operational Context Panel. Washington, DC: U.S. Air Force Scientific Advisory Board.

Licklider, J. C. R. (1965). *Libraries of the Future*. Cambridge, MA: MIT Press.

Liebowitz, J., and Wilcox, L. C. (eds.). (1997). *Knowledge Management and Its Integrative Elements*. Boca Raton, FL: CRC Press.

Lowe, E. A., Warren, J. L., and Moran, S. R. (1997). *Discovering Industrial Ecology: An Executive Briefing and Sourcebook*. Columbus, OH: Battelle Press.

Maes, P. (1994). Agents that reduce work and information overload. *Commun. ACM*. **37**(3).

Martin, J. (1996). *Cybercorp: The New Business Revolution*. New York: AMACOM Publishers.

Mason, R. O., Mason, F. M., and Culnan, M. J. (1995). *Ethics of Information Management*. Thousand Oaks, CA: Sage.

Matheson, D., and Matheson, J. (1998). *The Smart Organization: Creating Value Through Strategic R&D*. Boston: Harvard Business School Press.

Mazarr, M. (1997). *The Five Paradoxes: Business Competition on the Knowledge Era*. Washington, DC: Center for Strategic and International Studies.

McGee, J., and Prusak, L. (1993). *Managing Information Strategically*. New York: Wiley.

McMaster, M. D. (1996). *The Intelligence Advantage: Organizing for Complexity*. Boston: Butterworth-Heinnemann.

Merry, U. (1995). *Coping with Uncertainty: Insights from the New Sciences of Chaos, Self-Organization, and Complexity*. Westport, CT: Praeger.

Moingeon, B. and Edmondson, A (Eds.) (1996) *Organizational Learning and Competitive Advantage,* Beverly Hills CA: Sage.

Myers, P. S. (1997). *Knowledge Management and Organizational Design*. Boston: Butterworth-Heinnemann.

National Research Council. (1991). *Research on the Management of Technology: Unleashing the Hidden Competitive Advantage*. Washington, DC: National Academy Press.

National Research Council. (1997). *More than Screen Deep: Toward Every-Citizen Interfaces to the Nation's Information Infrastructure*. Washington, DC: National Academy Press.

Nonaka, I. (1994). The dynamical theory of organizational knowledge creation. *Organi. Sci.* **5**(1), 14–37.

Nonaka, I., and Takeuchi, H. (1995). *The Knowledge Creating Company.* Oxford and New York: Oxford University Press.

Pool, R. (1997). *Beyond Engineering: How Society Shapes Technology.* Oxford: Oxford University Press.

Prusak, L. (ed.). (1995). *Knowledge in Organizations: Resources for the Knowledge-Based Economy.* Boston: Butterworth-Heinnemann.

Quinn, J. B., Baruch, J. J., and Zien, K. A. (1997). *Innovation Explosion: Using Intellect and Software to Revolutionize Growth Strategies.* New York: Free Press.

Rasmussen, J., Pejtersen, A., and Goodstein, L. G. (1994). *Cognitive Systems Engineering.* New York: Wiley.

Roos, J., Roos, G., Edvinsson, L., and Dragonetti, N. C. (1998). *Intellectual Capital: Navigating in the New Business Landscape.* New York: New York University Press.

Roughgarden, J. (1998). *Primer of Ecological Theory.* Upper Saddle River, NJ: Prentice Hall.

Rouse, W. B. (1991). *Design for Success: A Human-centered Approach to Designing Successful Products and Systems.* New York: Wiley.

Rouse, W. B. (1994). Human-centered design of information systems. In (J. Wesley-Tanaskovic, J. Tocatlian, and K. H. Roberts, eds.), *Expanding Access to Science and Technology: The Role of Information Technology* pp. 214–223. Tokyo: United Nations University Press.

Rouse, W. B. (1997a). Real estate in a virtual world. *Competitive Edge!,* March/April, p. 72.

Rouse, W. B. (1997b). Connectivity, creativity, and chaos: Challenges of loosely-structured organizations. *Proc. Inter. Confer. Syst., Man, Cybernet.,* Orlando, FL.

Rouse, W. B., Thomas, B. S., and Boff, K. R. (1998). Knowledge maps for knowledge mining: Application to R&D/technology management. *IEEE Trans. Syst., Man, Cybernet.* **28**(3).

Ruggles, R. L. (ed.). (1997). *Knowledge Management Tools.* Boston: Butterworth-Heinneman.

Sage, A. P. (1992). Systems engineering and information technology: Catalysts for total quality in industry and education. *IEEE Trans. Syst., Man, Cybernet.* **22**(5), 833–864.

Sage, A. P. (1995). *Systems Management: For Information Technology and Software Engineering.* New York: Wiley.

Sage, A. P. (1998). Towards a systems ecology. *IEEE Comput.* **31**(2), 107–110.

Schaller, R. R. (1997). Moore's Law: Past, present, and future. *IEEE Spectrum* **34**(6), 52–59.

Senge, P. M. (1990). *The Fifth Discipline: The Art and Practice of the Learning Organization.* New York: Doubleday.

Shapiro, C. and Varian, H. R. (1998) *Information Rules: A Strategic Guide to the Network Economy,* Boston: Harvard Business School Press.

Shenk, D. (1997). Data smog: Surviving the info glut. *Technol. Rev.,* May/June, pp. 18–16.

Stacey, R. D. (1996). *Complexity and Creativity in Organizations.* San Francisco: Berrett-Koehler Publishers.

Stewart, T. A. (1997). *Intellectual Capital: The New Wealth of Organizations.* New York: Currency Doubleday.

Sveiby, K. E. (1997). *The New Organizational Wealth: Managing and Measuring Knowledge Based Assets.* San Francisco: Berrett-Koehler Publishers.

Tapscott, D. (1998). *Growing up Digital: The Rise of the Net Generation.* New York: McGraw-Hill.

Tapscott, D., and Caston, A. (1993). *Paradigm Shift: The New Promise of Information Technology.* New York: McGraw-Hill.

Toffler, A. (1980). *The Third Wave.* New York: Morrow.

Toffler, A. (1991). *The Third Wave*. 2nd ed New York: Bantam Books.

Ulrich, D. (1998). Intellectual capital = competence $\times$ commitment. *Sloan Manage. Rev.* **39**(2), 15–26.

Van der Ryn, S., and Cowan, S. (1996). *Ecological Design*. Washington, DC: Island Press.

Wulf, W. A. (1998). The urgency of engineering education reform. *Bridge* **28**(1), 4–8.

Zuboff, S. (1988). *In the Age of the Smart Machine: The Future of Work and Power*. New York: Basic Books.

INDEX